Neil Jespersen	Department of Chemistry St. John's University
James R. Kincaid	Department of Chemistry Marquette University
Charles H. Lochmüller	Department of Chemistry Duke University
J. West Loveland	Applied Physics Laboratory Sun Oil Company
Ronald E. Majors	Instrument Division Varian
Harry B. Mark, Jr.	Department of Chemistry University of Cincinnati
Horacio A. Mottola	Department of Chemistry Oklahoma State University
Peter Pella	Analytical Chemistry Division National Bureau of Standards
S. Sternhell	Department of Organic Chemistry University of Sydney
Isiah M. Warner	Department of Chemistry Emory University
John R. Wasson	Syntheco, Inc. Gastonia, North Carolina
Lo I Yin	Solar Physics Branch Laboratory for Astronomy and Solar Physics NASA Goddard Space Flight Center
Vaneica Y. Young	Department of Chemistry University of Florida

Instrumental Analysis

Instrumental Analysis

Second Edition

Editors

GARY D. CHRISTIAN
University of Washington

JAMES E. O'REILLY
University of Kentucky

Allyn and Bacon, Inc.
Boston   London   Sydney   Toronto

Copyright © 1986, 1978 by Allyn and Bacon, Inc., 7 Wells Avenue, Newton, Massachusetts 02159. All rights reserved. No part of the material protected by this copyright notice may be reproduced or utilized in any form or by any means, electronic or mechanical, including photocopying, recording, or by any information storage and retrieval system, without written permission from the copyright owner.

Library of Congress Cataloging in Publication Data
Main entry under title:

Instrumental analysis.

   Includes bibliographies and index.
   1. Instrumental analysis.   I. Christian, Gary D.
  II. O'Reilly, James E., 1945–   .   III. Series.
QD79.I5I5 1986   543'.08   85-20032
ISBN 0-205-08640-3
ISBN-0-205-08685-3 (*International*)

Printed in the United States of America

10 9 8 7 6 5 4 3 2 1   90 89 88 87 86

Contents

Preface to the First Edition   xiii

Preface to the Second Edition   xvii

CHAPTER 1   **Introduction to Electrochemical Methods**   1
Henry H. Bauer; James E. O'Reilly

   1.1   Generalities of Electrochemical Methods   1
   1.2   Electrochemical Definitions and Terminology   4
          Selected Bibliography   11
          References   11
          Problems   11

CHAPTER 2   **Potentiometry**   13
James A. Cox; James E. O'Reilly

   2.1   Electrochemical Cells   14
   2.2   The Nernst Equation   15
   2.3   Reference Electrodes   21
   2.4   pH: Definition and Measurement   24
   2.5   Ion-Selective Electrodes   29
   2.6   Potentiometric Titrations   44
          Selected Bibliography   48
          References   48
          Problems   49

Chapter 3 Polarography and Voltammetry 52
James A. Cox; James E. O'Reilly

- 3.1 Introduction and Theoretical Basis   52
- 3.2 Instrumentation and Apparatus   64
- 3.3 Applications   67
- 3.4 Variations of the Conventional Polarographic Method   77
- 3.5 Amperometric Titrations   88
  - Selected Bibliography   93
  - References   93
  - Problems   93

Chapter 4 Coulometry and Electrogravimetry 96
David J. Curran; Kenneth S. Fletcher III

- 4.1 Coulometric Analysis   97
- 4.2 Constant-Current Coulometric Titrations   99
- 4.3 Controlled-Potential Coulometry   105
- 4.4 Electrogravimetry   116
  - Selected Bibliography   119
  - References   119
  - Problems   120

Chapter 5 Conductance and Oscillometry 122
J. West Loveland

- 5.1 Definitions and Units   123
- 5.2 Theory   125
- 5.3 Conductometric Instrumentation   128
- 5.4 Conductance Titrations and Other Applications   132
- 5.5 Oscillometry   139
  - Selected Bibliography   141
  - References   142
  - Problems   142

Chapter 6 Introduction to Spectroscopic Methods 144
Eugene B. Bradley

- 6.1 Theory   144
- 6.2 Application of Quantum Theory to Spectroscopy   149
- 6.3 Instrumentation   150
- 6.4 Applications   158
  - References   160
  - Problems   160

Chapter 7 Ultraviolet and Visible Absorption Spectroscopy 161
K. L. Cheng; V. Y. Young

- 7.1 Molecular Absorption of Radiation: Electronic Spectra   161

7.2 Effect of Structure on Absorption 166
7.3 Magnitude of Absorption of Radiation 173
7.4 Quantitative Absorption Spectroscopy 181
7.5 Spectrophotometric Applications 184
7.6 Apparatus and Instruments 192
Selected Bibliography 207
References 207
Problems 209

CHAPTER 8 Infrared and Raman Spectroscopy   212
James R. Kincaid

8.1 Molecular Vibrations 213
8.2 Instrumentation for Raman Spectroscopy 220
8.3 Instrumentation for Infrared Spectroscopy 227
8.4 Sampling Systems for Infrared Spectroscopy 234
8.5 Qualitative Analysis 240
8.6 Quantitative Analysis 240
Selected Bibliography 245
References 245
Problems 245

CHAPTER 9 Molecular Fluorescence and Phosphorescence   247
Isiah M. Warner

9.1 Principles of Photoluminescence 248
9.2 Fluorescence and Phosphorescence Instrumentation 256
9.3 Applications of Fluorescence and Phosphorescence 263
Selected Bibliography 274
References 274
Problems 276

CHAPTER 10 Flame Emission, Atomic Absorption, and Atomic Fluorescence Spectrometry   278
Gary Horlick

10.1 The Flame as a Source of Atomic Vapor 280
10.2 Flame Emission Spectrometry 290
10.3 Atomic Absorption Spectrometry 295
10.4 Atomic Absorption Measurements 300
10.5 Electrothermal Atomization 304
10.6 Applications 311
10.7 Atomic Fluorescence Spectrometry 315
Selected Bibliography 319
References 319
Problems 320

Chapter 11 Emission Spectroscopy 322
Ramon M. Barnes

11.1 Principles and Theory 323
11.2 Instrumentation 325
11.3 Qualitative and Quantitative Analyses 343
11.4 Applications 349
Selected Bibliography 353
References 353
Problems 355

Chapter 12 Nuclear Magnetic Resonance Spectroscopy 357
S. Sternhell

12.1 Theory and Instrumentation 358
12.2 The Chemical Shift 365
12.3 Time Dependence of NMR Phenomena 371
12.4 Spin-Spin Coupling 373
12.5 NMR Spectroscopy of Nuclei Other Than Protons 382
12.6 Special Topics 383
12.7 Analytical Applications 386
Selected Bibliography 389
Problems 390

Chapter 13 Electron Spin Resonance Spectroscopy 395
John R. Wasson

13.1 The Resonance Condition 396
13.2 ESR Instrumentation 399
13.3 Thermal Equilibrium and Spin Relaxation 401
13.4 ESR Spectra of Free Radicals in Solution 402
13.5 Analytical Applications 407
Selected Bibliography 409
References 410
Problems 410

Chapter 14 X-Ray Spectrometry 412
Peter A. Pella

14.1 Introduction 412
14.2 Instrumentation 414
14.3 Preparation of Samples for X-Ray Fluorescence Analysis 431
14.4 Quantitative X-Ray Fluorescence Analysis 433
14.5 Special Topics and Other X-Ray Methods 440
Selected Bibliography 448
References 448
Problems 449

Chapter 15   Electron Spectroscopy   451
Lo I Yin; Isidore Adler

    15.1   Principles of Electron Spectroscopy   451
    15.2   Instrumentation   460
    15.3   Applications   465
              Selected Bibliography   473
              References   473
              Problems   474

Chapter 16   Mass Spectrometry   476
Michael L. Gross

    16.1   Instrumentation in Mass Spectrometry   477
    16.2   Interpretation of a Mass Spectrum   494
    16.3   Analytical Applications of Electron-Impact Mass Spectrometry   505
    16.4   Other Methods of Vaporization and Ionization   509
    16.5   Other Instrumentation Developments   515
              Selected Bibliography   518
              References   519
              Problems   519

Chapter 17   Thermal and Calorimetric Methods of Analysis   523
Neil Jespersen

    17.1   General Characteristics of Thermal Methods   523
    17.2   Thermogravimetry   525
    17.3   Differential Thermal Analysis   532
    17.4   Differential Scanning Calorimetry   542
    17.5   Thermometric Titration and Direct-Injection Enthalpimetry   546
              Selected Bibliography   557
              References   557
              Problems   557

Chapter 18   Kinetic Methods   560
Horacio A. Mottola; Harry B. Mark, Jr.

    18.1   Types of Kinetic Methods   561
    18.2   Measurement of Reaction Rates   563
    18.3   Mathematical Basis of Kinetic Methods   567
    18.4   Rate Considerations for the Determination of Single Species   568
    18.5   Rate Considerations for Differential Reaction-Rate Methods   574
    18.6   Instrumentation   580
    18.7   Analytical Applications of Kinetic-Based Methods   585
              Selected Bibliography   590
              References   591
              Problems   591

CHAPTER 19 Radiochemical Methods of Analysis 594
William D. Ehmann; Morteza Janghorbani

    19.1   Fundamentals of Radioactivity   594
    19.2   Nuclear Reactions and Types of Radioactive Decay   600
    19.3   Detection of Nuclear Radiation   604
    19.4   Activation Analysis   614
    19.5   Methods Involving Addition of Radionuclide   627
    19.6   Statistical Considerations in Radiochemical Analysis   632
           Selected Bibliography   635
           Problems   636

CHAPTER 20 Fractionation Processes: Solvent Extraction 639
Henry Freiser

    20.1   Phase Processes   639
    20.2   General Principles and Terminology of Solvent Extraction   640
    20.3   Experimental Techniques   646
    20.4   Important Experimental Variables   649
    20.5   Extraction Systems and Examples   652
           Selected Bibliography   656
           References   656
           Problems   656

CHAPTER 21 Solid- and Liquid-Phase Chromatography 658
Ronald E. Majors

    21.1   Introduction   658
    21.2   Basic Principles of Liquid Chromatography   659
    21.3   Theory Related to Practice   665
    21.4   Paper and Thin-Layer Chromatography   675
    21.5   Column Liquid Chromatography   681
    21.6   Uses and Applications of Adsorption Chromatography   697
    21.7   Uses and Applications of Partition and Bonded-Phase Chromatography   706
    21.8   Ion-Exchange Chromatography   712
    21.9   Size-Exclusion Chromatography   718
    21.10 Techniques Related to Liquid Chromatography   722
           Selected Bibliography   725
           References   725
           Problems   726

CHAPTER 22 Gas Chromatography 728
Charles H. Lochmüller

    22.1   The Thermodynamics of Gas Chromatography   729
    22.2   The Dynamics of Gas Chromatography   734

	22.3	Gas-Chromatographic Instrumentation	739
	22.4	Qualitative and Quantitative Analysis	751
	22.5	Applications of Gas Chromatography	754
		Selected Bibliography 762	
		References 763	
		Problems 763	

CHAPTER 23 Introduction to Analog Circuits and Devices   766
F. James Holler

23.1	Data Domains	767
23.2	Electrical Quantities and Basic Circuits	772
23.3	AC Quantities and Measurements	780
23.4	Semiconductor Devices and Power Control	784
23.5	Operational Amplifiers: Principles and Applications	794
23.6	Characteristics of Real Op-Amps	803
	Selected Bibliography 805	
	References 806	
	Problems 806	

CHAPTER 24 Digital Electronics, Data-Domain Conversions, and Microcomputers   809
F. James Holler

24.1	Digital Logic and Gates	809
24.2	Switching	821
24.3	*RC* Circuits and Multivibrators	825
24.4	Counting and Time-Domain Measurements	828
24.5	Data-Domain Conversions	833
24.6	Microcomputers	840
24.7	Data Processing and Enhancement of Signal-to-Noise Ratio	846
	Selected Bibliography 851	
	References 852	
	Problems 852	

CHAPTER 25 Automation in Analytical Chemistry   854
Kenneth S. Fletcher III; Nelson L. Alpert

25.1	Instrumental Parameters for Automated Instruments	855
25.2	Sample Conditioning	858
25.3	Automated Process Control	860
25.4	Automated Instruments in Process-Control Systems	870
25.5	Automation in Clinical Chemistry	879
	Selected Bibliography 894	
	References 895	
	Problems 895	

Appendix A: Units, Symbols, and Prefixes   **897**

Appendix B: Selected Fundamental Physical Constants   **900**

Appendix C: Some Standard Electrode Potentials   **901**

Appendix D: Answers to Selected Problems   **904**

Index   **911**

Preface
to the First Edition

The editors embarked on the venture of editing a textbook dealing with instrumental methods in chemical analysis for several reasons. None of the available texts seemed to us to be as well suited to the types of courses generally given in this area as we would like: coverage of the various types of techniques was uneven in depth, emphasis, and modernity; and in particular there seemed to be insufficient attention to applications of the techniques in practice. We felt that these shortcomings might be minimized if we could have each method discussed by people active in that particular field.

The individual authors were asked to make the theoretical background of each method as brief and qualitative as possible consistent with clarity and accuracy, to limit discussion of instrumentation to general principles as far as possible (i.e., no details of the operation of commercially available apparatus), and to emphasize the utility and actual applications of each method.

Some cynics, particularly in academia, maintain that the last thing ever successfully accomplished by committee was the King James version of the Bible. In a sense, this text has also been composed by a committee. As editors, we have tried to keep the depth of presentation more or less even, consistent with our feelings as to the importance of a particular method for quantitative application; and to provide the continuity of thought and mode of expression so important in a text. We happily allot to the authors whatever credit may accrue for the quality of the individual parts and, as editors, assume responsibility for the shortcomings of the whole.

We thank the authors for their efforts, for the quality of their presentations, for their patience with the various changes and requests made during several drafts of the manuscript, and in particular for their benign attitude to the liberties we have taken with their style and mode of expression.

It is appropriate at this point to elaborate a few of the things we have chosen to do and not do. We have chosen to concentrate particularly on those applications and methods that are, in our opinion, most useful for quantitative analytical measurements — not because these are intrinsically more important than measurements that are more physically oriented, but simply because of the limitations of space. Thus, for example, we have no discussion of optical rotatory dispersion, certainly an important instrumental technique, because it presently has no practical usage as a quantitative method.

The text does not have a description of basic and advanced electronics, except for some discussion of digital electronics in the chapter on computers. There are perhaps two distinct approaches to instrumental analysis: one is in terms of *instrumentation* and instrument design, the second in terms of *instrumental methods* of analysis. We have chosen the latter approach because, first, we believe this to be the more profitable approach for the majority of students in a course of this nature, and second, because there is simply too much material in modern scientific electronics and instrumental methods to cover both comfortably in one semester and do justice to either. There are several excellent texts devoted to scientific electronics and several packages of electronics experiments on the market today; and many universities and colleges now offer separate courses in this area. We have chosen to leave the subject of electronics to these, and to the discretion of the individual instructor.

We have also not included an accompanying set of laboratory experiments. There are several excellent compendia of instrumental analysis experiments available, and a number of quantitative analytical experiments, particularly those involving analysis of "real" samples, appear in the *Journal of Chemical Education* monthly. Moreover, because of the cost and complexity of modern instrumentation, many instructors face a very limited array of instruments, and are forced to drastically redesign experiments for their particular instrument or model anyway. We have chosen to leave the instrumental laboratory to the ingenuity of the individual instructor.

A number of people have helped us in various ways, and we wish to thank them all, while mentioning specifically only a few — Petr Zuman for sharing with us his unmatched experience in the practice of polarographic techniques; Stan Smith for a complete set of NMR problems; Regina Palomo and Ellen Swank for patiently typing, retyping, and reretyping the entire manuscript; and our colleagues for putting up with us through the entire production stage. One of us (JEO) makes Acknowledgment to the Donors of The Petroleum Research Fund, administered by the American Chemical Society, for partial support of this research.

We genuinely hope that instructors will find this book a useful pedagogical aid, and we would appreciate comments concerning shortcomings and errors that we could attempt to rectify should there be the occasion for a later edition.

It is our fondest hope that students will read this text, and feel good about it. We have tried to express things as though we were talking to students, and not with an eye toward impressing an instructor with the depth of coverage or sophistication of the discussion. We hope that, at the end of a semester, the student can emerge with an idea of what the various instrumental methods are capable of doing in order to solve a problem or make a measurement. We hope to impart the overview of the true analytical chemist —

selecting the right tool for the job at hand. Perhaps this can be expressed with a quotation from the Bible:

Thou shalt not have in thine bag divers weights, a great and a small. Thou shalt not have in thine house divers measures, a great and a small. But thou shalt have a perfect and just weight, a perfect and just measure shalt thou have . . .

—DEUTERONOMY 25: 13–15

Preface
to the Second Edition

The second edition of this textbook is overdue. Much has happened in analytical chemistry, the science of chemical measurements, in the eight or more years since the chapters in the first edition were written. Because of the success of the first edition, and the repeated requests from the users of the text for a revision, we undertook the task of preparing this second edition.

    Some of our previous authors chose not to join us in this version. We thank them for the foundation they laid for us. Sadly, Professor Donald Davis of the University of New Orleans has passed away in the meantime. Our discipline is lessened by his passing.

    A number of new major authors and co-authors, however, have joined us for the second edition. We feel the new blood, fresh perspective, and talents of these experts have provided not only a needed updating but also major changes and overall improvements in the text as a whole. Moreover, in this second round, our authors were not writing "blind," so to speak. Our first set of authors had only minimal directions as to the slant and goals of the text. We feel that our first edition was a quality product and that our second edition is even better. The major credit for this is due our authors, and we sincerely thank them for their efforts. As before, we the editors will allot to our authors whatever credit is due for the quality of the parts and will accept the responsibility for the shortcomings of the whole.

    The basic approach of this edition has not changed markedly from that of the first: Our emphasis is still on those methods and applications most useful for quantitative analytical measurements, and our pedagogical style is still *instrumental methods* rather than *instrumentation*. The scope, order of chapters, level of treatment, and so forth, are largely the same. The response to, and success of, the first edition argued against a major change of direction.

One of the major changes in the text is two long chapters (Chapters 23 and 24) on analog and digital electronics and on microcomputers and interfacing, which greatly expand our coverage of electronics. If there was a single most common complaint about the first edition, it was about the paucity of electronics coverage. We bowed to this opinion. In these two chapters, Professor F. J. Holler has presented an entire course in electronics, albeit extremely compressed and very selective.

In addition to overall updating, some of the specific improvements and additions are worthy of note. The addition of Professors James Cox and Gary Horlick to take over the writing of the chapters on potentiometry, polarography, and flame spectroscopy has undoubtedly improved these treatments immensely. Professor David Curran and Dr. Kenneth Fletcher have done an excellent job in taking over the electrogravimetry and coulometry chapter from the late Donald Davis. Dr. J. West Loveland has added pulse methods to the chapter on conductance.

In spectroscopy, Professors K. L. Cheng and Vaneica Young have added a discussion of photoacoustic spectroscopy to the UV/VIS chapter. New authors Professors James Kincaid (IR and Raman) and Isiah Warner (fluorescence and phosphorescence) and Dr. Peter Pella (X-ray) have brought fresh perspectives and presentation in these areas. Professor Sidney Sternhell has greatly increased the coverage of ^{13}C and FT NMR, reflecting the growing importance of these in the chemist's arsenal of techniques.

Professor Horacio Mottola's joining as co-author with Professor Harry Mark has brought increased discussion of the practical applications of kinetic methods to complement the excellent theoretical treatment. The coverage of separation methods by Professor Henry Freiser, Dr. Ronald Majors, and Professor Charles Lochmüller has been increased. The other authors from the first edition who stayed with us have also prepared excellent revisions.

As before, our desire was to put together a text that students will read, will understand, and will even enjoy. It is not uncommon to have a dedication in a book of this nature. If our text needs one, we ask that students who will use and, we hope, profit from this text, to split the dedication with the many, many people who have helped us in the preparation of this work.

Solutions Manual. An addition is the availability of a comprehensive solutions manual with detailed solutions in each chapter. We hope it will be useful.

G. D. C.
J. E. O.

Instrumental Analysis

CHAPTER 1

Introduction to Electrochemical Methods

Henry H. Bauer
James E. O'Reilly

Electrochemistry is a scientific discipline with a well-developed system of theories and quantitative relationships. It has many applications and uses in both fundamental and applied areas of chemistry—in the study of corrosion phenomena, for example, for the study of the mechanisms and kinetics of electrochemical reactions, as a tool for the electrosynthesis of organic and inorganic compounds, and in the solution of quantitative analytical problems. This last area will be emphasized in the next four chapters.

1.1 GENERALITIES OF ELECTROCHEMICAL METHODS

It can be said with some degree of accuracy that, with the exception of the nearly universal use of the potentiometric pH meter, electrochemical methods in general are not as widely used as are spectrochemical or chromatographic methods for quantitative analytical applications. A recent informal readers' survey conducted by *Industrial Research & Development* [1] indicated that about 63% of the laboratories responding used pH meters, ranking them approximately fifth in usage behind analytical balances, fume hoods, laboratory ovens, and strip-chart recorders. Yet where about half of the respondents used visible, ultraviolet, and infrared spectrophotometers and about 30% used atomic absorption spectrophotometers, only about 8% used what were termed polarographic analyzers and 35% used ion-selective electrodes. In general, there is a more widespread usage of electrochemical methods in Europe and Japan than in the United States.

There are probably several reasons why electrochemical methods are not as "popular" as chromatographic or optical methods. One is that electrochemistry and electrochemical methods are not emphasized in typical college curricula. One can cite the nearly universal disappearance of fundamental electrochemistry from beginning general and physical chemistry courses, whereas the interaction of electromagnetic radiation with matter and the energy levels concerned is covered in many first-year courses. Electrochemical theory is really no more complex or abstruse, but probably not so well unified at present, as spectrochemical theory.

A second reason is that spectrochemical methods appear somewhat more amenable to automation or mechanization than electrochemical methods. An extreme example of this can be seen in the clinical analysis laboratory. In 1971, for instance, one particular hospital performed nearly a half-million chemical tests, 91% of which were done with spectrochemical methods and instruments [2]. This, of course, was due to its use of automated clinical analyzers, which are primarily optical in approach. Considerable progress has been made in automating electrochemical instrumentation in recent years. Commercial clinical instruments are available that use potentiometry to measure pH and electrolyte ions in biological fluids such as plasma, or "gas-sensing" electrodes for such gases as CO_2 and O_2.

There are many times when electrochemical methods can provide essentially the same information as other methods, thus offering an alternative approach, and other times when only electrochemistry will provide the answer or will provide the best answer to the problem at hand. A particularly interesting recent development has been the application of electrochemically based detectors in chromatography, such as voltammetric detectors in liquid chromatography, and conductivity detectors in both gas and liquid chromatography.

Advantages of Electrochemical Methods

Although it is difficult to consider electrochemical methods *in general* versus other methods *in general*, electrochemical methods do have certain advantages. First of all, electrochemical instrumentation is comparatively inexpensive. The most costly piece of routine electrochemical instrumentation is priced at about $15,000, with most commercial instrumentation under about $4000. By contrast, some sophisticated nonelectrochemical equipment, such as nuclear magnetic resonance or mass spectrometers, may run well over a quarter of a million dollars.

Second, elemental electrochemical analysis is generally *specific for a particular chemical form* of an element. For example, with a mixture of Fe^{2+} and Fe^{3+}, electrochemical analysis can reveal the amount of each form present, whereas most elemental spectrochemical or radiochemical methods give simply the total amount of iron present almost regardless of its chemical form. Depending on the analytical problem at hand or the question to be answered, one particular method may be "better than others." For example, mercury is a serious environmental pollutant. Elemental or inorganic forms of mercury (Hg^0, Hg^{2+}, Hg_2Cl_2, . . .) are bad, but organic mercury (CH_3Hg^+, $(CH_3)_2Hg$, . . .) is much worse. Perhaps, in a given situation, it is important to know both the total mercury level and the forms the mercury takes.

Another advantage (or disadvantage, depending on the problem at hand) of many electrochemical methods is that they respond to the *activity* of a chemical species *rather than to the concentration*. An example where this may be of importance is the calcium level in serum. Ion-selective electrodes respond to free, aquated Ca^{2+} ions, whereas the usual clinical method for serum calcium is flame photometry, which measures the total calcium present including a large amount tied up as protein-bound calcium. The more important physiological parameter, the measure of the *effective* level of calcium actually available for participation in various enzymatic reactions, may be the free Ca^{2+} level. For another example, lead is a cumulatively toxic substance; plants grown in lead-laden soils can accumulate high levels of lead. If these plants are eaten by humans, toxic levels of lead may be reached. Lead in heavy clay soils, however, is much less available for absorption by plants than is lead in more sandy soils. Perhaps the more useful measure of the arability of land contaminated by trace metals is the metal-ion activity, rather than the total metal concentration.

It can be safely said that in recent years there has been a renaissance of interest in quantitative electrochemical methods. This has been brought about primarily by two factors: the development of ion-selective potentiometric electrodes, which can quantitatively monitor most of the common ionic species in solution (Chap. 2); and the introduction of a new generation of inexpensive commercial voltammetric instrumentation based on pulse methods (Chap. 3), which has increased the sensitivity of electrochemical methods by several orders of magnitude.

Classification of Electrochemical Methods

For the present purpose, an electrochemical method can be defined as one in which the electrical response of a chemical system or sample is measured. The experimental system can be divided as follows: the electrolyte, a chemical system capable of conducting current; the measuring or external circuit, used to apply and to measure electrical signals (currents, voltages); and the electrodes, conductors that serve as contacts between the measuring system and the electrolyte.

Electrodes are classed as *anodes* and *cathodes*. At the anode, *oxidation* occurs—electrons are abstracted from the electrolyte and pass into the measuring circuit; at the cathode, *reduction* occurs—electrons flow from the cathode into the electrolyte. Furthermore, one speaks of *working* or *indicator electrodes*, those at which a reaction being studied is taking place; of *reference electrodes*, which maintain a constant potential irrespective of changes in current; and of *counter electrodes*, which allow current to flow through the electrolyte but whose characteristics do not influence the measured behavior (the latter depends on what happens at the working electrode).

The current flowing in an electrochemical system is determined by the total resistance of the whole circuit. Good experimental design ensures that the magnitude of the current is not influenced by the measuring circuit. That done, one can distinguish two types of methods: those in which the resistance of the electrodes is made negligible, so that one measures the conductance of the electrolyte (see Chap. 5); and those in which the resistance of the electrolyte is made negligible, in which case one studies phenomena occurring at the electrodes (Chaps. 3 and 4).

A multitude of electrochemical techniques based on electrode processes exist; however, only a comparatively small number are of real importance to the analytical chemist. Some of the chief features of these techniques are shown in Table 1.1.

TABLE 1.1. *Analytically Useful Electrochemical Techniques*

Technique	Controlled Electrical Variable	Response Measured	Relative Time Scale for Analysis
Potentiometry	$i (= 0)$	E	Short
Potentiometric titration	$i (= 0)$	E vs. volume of reagent	Long
Voltammetry	E	i vs. E	
Polarography	↓	↓	Medium
Linear-scan or cyclic voltammetry			Short
Pulse methods			Medium
Stripping analysis	E	i vs. E	Medium
Electrogravimetry	i or E	Weight of deposit	Long
Coulometry	i or E	Charge consumed (integrated current)	Long
Coulometric titration	i	Time	Medium
Conductivity	V (AC)	i (AC)	Short

Electrochemical methods can be divided into two classes: those involving no net current flow ("potentiometric"), and all others. In potentiometry, one measures the equilibrium thermodynamic potential of a system essentially without causing electrolysis or current drain on the system—because this would affect the existing equilibrium. In all other methods, a voltage or current is applied to an electrode, and the resultant current flow through, or voltage change of, the system is monitored. The applied waveform is often quite complex. Although this approach may be more complicated than is the case in potentiometry, there are advantages in that one is not forced to deal with the particular equilibrium characteristics of the system. By making the system respond electrochemically to a stimulus, one can gain a good deal of analytical control over it.

1.2 ELECTROCHEMICAL DEFINITIONS AND TERMINOLOGY

As in all other disciplines, electrochemistry has its own terminology with which one needs to be familiar before studying specific electrochemical methods.

Faradaic and Nonfaradaic Processes

Two types of processes occur at electrodes. One type includes processes in which charge (e.g., electrons) is transferred across the electrode-solution interface. In these processes, actual oxidation or reduction occurs. Such processes are governed by Faraday's laws and are called *faradaic* processes.

Under some conditions, an electrode may be in a potential region where charge-transfer reactions do not occur because they are either thermodynamically or kinetically unfavorable. However, such processes as adsorption can occur, and the structure of the electrode-solution interface can change, causing transitory changes in current or potential. These processes are called *nonfaradaic* processes.

Charging Current. An important example of a nonfaradaic process is the *charging* of an electrode (see Fig. 1.1). At some potential E_A (Fig. 1.1A), there is a certain charge per unit area in the metal electrode, with an equal amount of charge of opposite sign present in the solution immediately adjacent to the electrode (forming what is called the *electrical double layer*). If the potential is then changed to E_B, where the charge per unit area is greater, current must flow to bring these extra charges to the interface. This is the *charging current*, and it is a transient current, flowing only until the new charge equilibrium is attained. Then the current stops, because there is no mechanism to cause current to flow *across* the interface, in the absence of redox reactions.

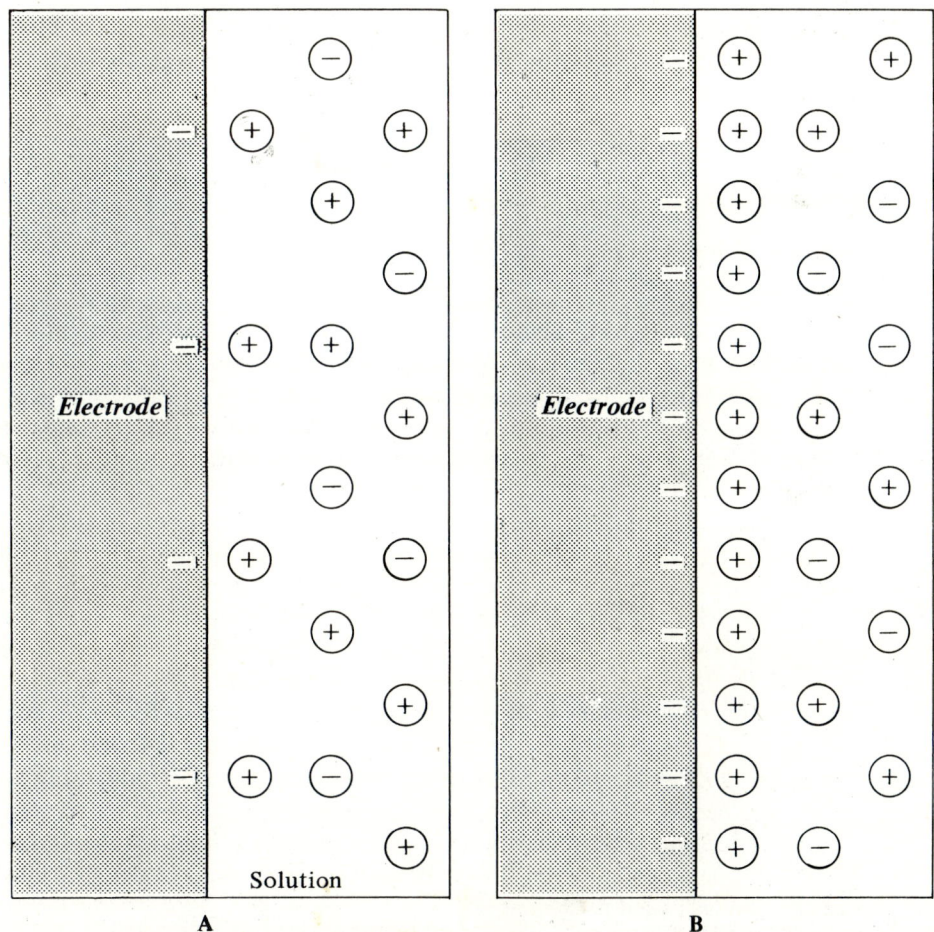

FIGURE 1.1. *Arrangement of charge at the electrode-solution interface. In case B, the electrode is at a more negative potential than in A; hence the greater amount of negative charge at the electrode surface in B.*

SEC. 1.2 Electrochemical Definitions and Terminology

So the charging process is nonfaradaic, and the charging current is a nonfaradaic current.

An electrode at which no charge transfer occurs across the electrode-solution interface, regardless of the potential imposed from an outside source of voltage, is called an *ideally polarized electrode*. No real electrode, of course, can behave in this manner at all potentials; but certain systems approach this behavior over a limited range of potentials. For example, a mercury electrode in contact with a (deoxygenated) NaCl solution acts as an ideally polarized electrode over a range of nearly 2 V. The two *faradaic* processes that can occur—the reduction of Na^+ to sodium amalgam, and the oxidation of the electrode to Hg_2Cl_2—occur at potentials that differ by about 2 V.

Only nonfaradaic processes occur at an ideally polarized electrode.

Capacitance of an Electrode. Since charge cannot cross the interface at an ideally polarized electrode when the potential is changed, the behavior of this interface is similar to that of a capacitor (Fig. 1.1). When a potential is applied across a capacitor, it will charge until it satisfies the relation

$$C = \frac{q}{V} \tag{1.1}$$

where C = capacitance in farads
            q = charge in coulombs
            V = voltage across the capacitor in volts

The time during which the charging or condenser current flows is directly proportional to the capacity of the electrode and the resistance of the solution; for electrodes of constant area immersed in solutions of fairly low resistance, the time during which the charging current is appreciable is very short—small fractions of a second. With electrodes whose area expands with time, for example, at the dropping mercury electrode, the charging current dies down more slowly—it is essentially proportional to the rate of exposure of fresh surface. In terms of analytical applications, the charging current is many times a distinct liability, since it often is the limiting factor in the sensitivity of an electrochemical method. One must use special techniques to distinguish between the current flow due to charging of the double layer and the current flow due to the faradaic reactions of the substance of interest. In general, this is done by using the fact that charging current decreases rapidly with time, whereas faradaic current changes with time much more slowly in typical experiments.

Faradaic Processes. Consider an ideally polarized electrode; only nonfaradaic processes occur, no charges cross the interface, and no continuous current can flow. Upon addition of a substance that can be oxidized or reduced at the particular potential difference, current then flows; the electrode is *depolarized*, and the substance responsible is called a *depolarizer*.

The faradaic process may proceed at various rates, within a wide range of possible rates. If the process is so rapid that the oxidized and reduced species are in equilibrium, the reaction is termed *reversible* and the Nernst equation (1.3) applies.

Reversibility, so defined, depends on the relative rates of the electrode process and on the rapidity of the electrochemical measurement: A particular system may behave reversibly when measurements are made slowly, but irreversibly if the measurement involves short times (pulses of current or voltage, or high frequencies of an alternating electrical signal). Consequently, in modern usage one prefers to talk about electrode processes as *being* fast or slow, and as *behaving* reversibly or irreversibly (rather than the classical usage in which systems were talked of as *being* reversible or irreversible).

Once faradaic current flows, the equilibrium between oxidized and reduced species is disturbed, and can be continually reestablished only if all the steps involved in the electrode process are rapid enough. (These steps include charge transfer, movement of depolarizer to the electrode and of product away from it [mass transport], and possibly adsorption or chemical reactions.) If there is a lag, then the electrode potential changes from its equilibrium value, the magnitude of the change being the *overpotential* or *overvoltage*.

Most systems can show overvoltages; that is, they can become *polarized*. At a polarized electrode, current flows, but the magnitude of the current is less than if the system were behaving reversibly. The current is limited by the rate of one (or more) of the steps in the electrode process. If charge transfer is the slow (limiting) step, the effect is called *activation polarization*; if slow movement of depolarizer or product is responsible, one speaks of *concentration polarization*.

If the electrode process were infinitely fast, then current could be drawn without producing an overvoltage; this would be a *nonpolarizable electrode*. In practice, some electrode systems permit appreciable currents to flow with negligible overpotentials, and such systems are used in reference electrodes.

Sign Conventions and the Nernst Equation

The sign conventions of electrochemistry have caused students and researchers a great deal of difficulty and misunderstanding over the years. All electrochemical cells are considered to be a combination of two half-cells—one for the reduction reaction, one for the oxidation reaction. To have current flow in *any* electrochemical system, both an oxidation and a reduction reaction must occur. Electrons must have someplace to go; they do not simply appear and disappear.

Any half-cell reaction can be written as either an oxidation or a reduction; by convention, they are written as reductions:

$$\text{Ox} + ne^- \rightleftarrows \text{Red} \tag{1.2}$$

where Ox = general symbol for the *ox*idized form of the balanced half-reaction
Red = general symbol for the *red*uced form of the balanced half-reaction
n = number of electrons involved in the half-reaction

By using a table of electromotive forces or standard reduction potentials (E^0 values) for half-reactions, the potential of each half-cell can be calculated by means of the Nernst equation

$$E = E^0 - \frac{RT}{nF} \ln \frac{(\text{Red})}{(\text{Ox})} \tag{1.3}$$

where R = molar gas constant (8.314 J/mole-K)
    T = absolute temperature in kelvins
    F = faraday constant (96,487 coulombs/mole)
   (Red) = activity of the reduced chemical species
   (Ox) = activity of the oxidized chemical species

If the ln term is converted to $\log_{10}$ basis, the value of the constant term $2.303RT/nF$ becomes $0.05916/n$ V at 25°C. The logarithmic term (Red)/(Ox) is simply the thermodynamic equilibrium expression for the electrochemical reaction, *written as a reduction*, and will be affected by changes in concentration of the various chemical species in the same manner as any other equilibrium expression. Further details on the Nernst equation and its use are contained in Chapter 2.

Modes of Electrochemical Mass Transport

In general, chemical species are transported in solution by one or more conceptually distinct processes: *migration, convection,* and *diffusion*.

Migration. Electrical migration is the movement of charged substances in an *electrical gradient*, a result of the force exerted on charged particles by an electric field; this can be viewed as a result of simple coulombic attraction of, for example, a positively charged ion to a negatively charged electrode surface or, alternatively, repulsion from a positively charged electrode. In almost all electrochemical methods of analysis, migration effects serve no useful purpose; they are usually swamped out by the addition of a relatively large amount (perhaps 0.1 or 1 M) of "background" (or "inert" or "indifferent") electrolyte such as KCl or HNO_3. Current can then flow as a result of migration of, for instance, K^+ or Cl^- ions, with negligible migration of the electroactive species, which then moves as a result of concentration differences only (diffusion; see below).

Convection. Convection means, essentially, the mass transport of electroactive material to the electrode by *gross physical movement*—fluid or hydrodynamic flow—of the solution. Generally, fluid flow occurs because of natural convection (caused by density gradients) or forced convection (usually caused by *stirring* of some sort).

Diffusion. Mass transfer by diffusion is the natural transport or movement of a substance under the influence of a *gradient of chemical potential*, that is, due to the *concentration gradient*; substances move from regions of high concentration to regions of low concentration in order to minimize or eliminate concentration differences. Diffusion is perhaps the most widely studied means of mass transport.

The rate of diffusion is given by

$$\text{Rate} = D\frac{dc}{dx} \qquad (1.4)$$

where D = *diffusion coefficient* (in cm^2/sec) of the substance
   dc/dx = concentration gradient

This expression is often approximated by

$$D\frac{\Delta c}{\delta} \tag{1.5}$$

where Δc = difference in concentration across the region where diffusion occurs (the *diffusion layer*)
δ = thickness of the diffusion layer (Fig. 1.2)

The diffusion coefficient, D, is a constant for a given substance under a specified set of solution conditions (temperature, electrolyte nature, and concentration). Since a concentration gradient is established as soon as any electrolysis is begun, diffusion is a part of every practical electrode reaction.

Faradaic current reflects the rate of the electrode process. If the latter is a multistep reaction, then each step has its own inherent rate and the faradaic current reflects the rate of the slowest process in the sequence of steps. That might be an adsorption process, or some chemical reaction in solution involving the oxidant or the reductant, or the charge-transfer process itself, or the rate at which the electroactive species diffuses to the electrode from the bulk of the solution. When a supporting electrolyte is present, in the simplest cases, the movement of the electroactive species is limited by diffusion; therefore, the solution of the equations governing diffusion is relevant to many electrochemical techniques.

Consider an electrode of fixed area immersed in a solution containing an electroactive species and a supporting electrolyte. Initially, the composition of the solution is uniform throughout. When a potential large enough to cause a faradaic reaction to occur is applied, the particles of the electroactive species in the immediate vicinity of the electrode undergo reaction. If the applied potential is sufficiently large, the concentration of the electroactive substance at the surface of the electrode is reduced to zero, as illustrated in Figure 1.2. Immediately after the application of the initial potential, the rate of the reaction, and consequently the magnitude of the current, depends on the rate at which the electroactive species diffuses to the electrode surface. The concentration gradient is steep at first, and the layer of depleted solution (the diffusion layer δ) is thin (see Fig. 1.2); as time goes by, the thickness of the diffusion layer increases, the concentration gradient becomes less steep, and the rate of diffusion decreases. As a result, a large current flows when the potential is first applied, and then the magnitude of the current decreases with time. Solution of the equations for diffusion leads to the relation known as the Cottrell equation:

$$i_t = \frac{nFAD^{1/2}c}{\pi^{1/2}t^{1/2}} \tag{1.6}$$

where i_t = current at time t
n = number of electrons involved in the electrochemical process
A = area of the electrode
c = concentration of the electroactive species

That is to say, the current decreases in proportion to the square root of time from the instant at which electrolysis starts.

We can now proceed, in the following chapters, to consider details of the analytical applications of four classes of electrochemical techniques: in Chapter 2,

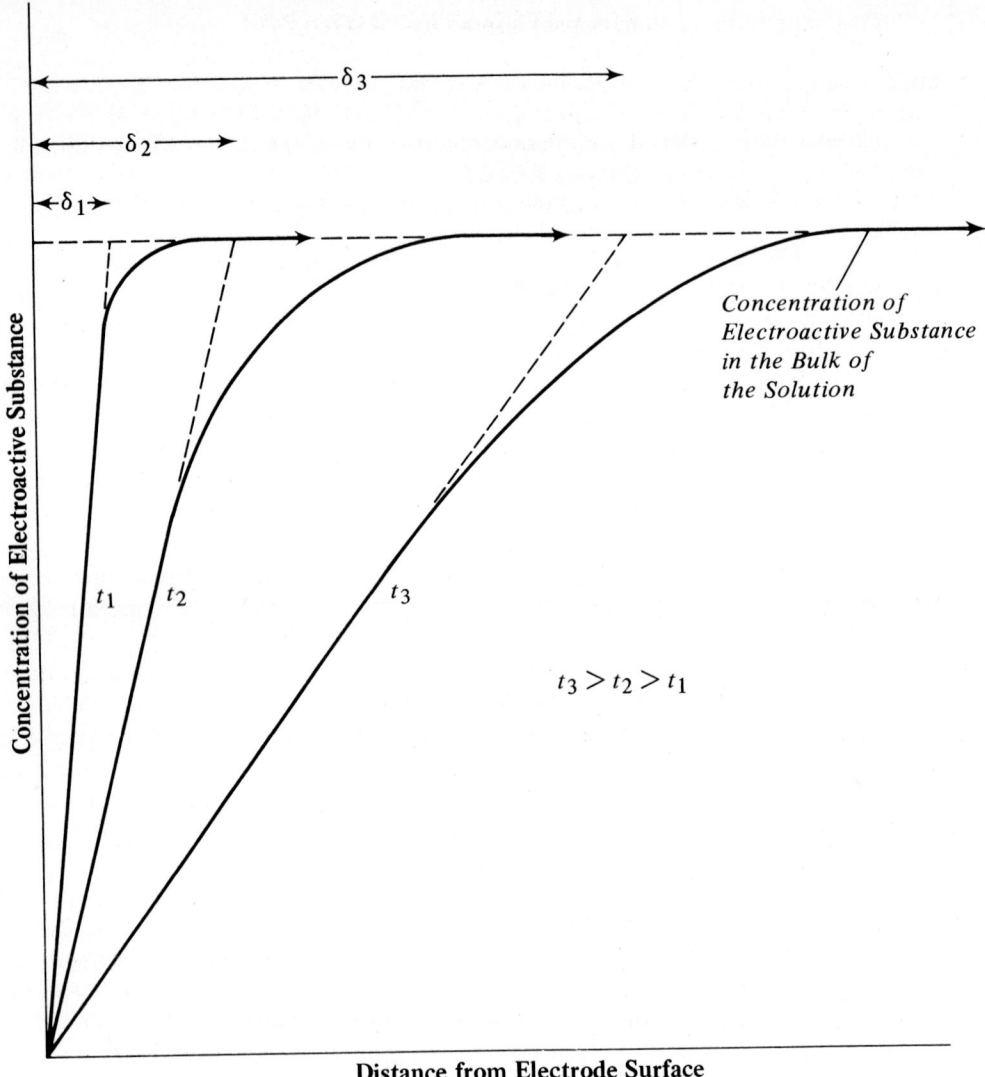

FIGURE 1.2. *Concentration-distance profiles for diffusion of an electroactive substance to an electrode surface at different times. At time zero, a voltage large enough to cause the electrode reaction to occur is suddenly applied to the electrode. Note that with increasing time ($t_3 > t_2 > t_1$), the concentration gradient— the slope of the concentration-distance curve—becomes less steep, and the thickness of the diffusion layer, δ, becomes larger.*

measurements of electrode potentials in the absence of current flow; in Chapter 3, measurements of current flow as potential is varied; in Chapter 4, measurements of the amount of charge required for complete electrolysis (or deposition) of the substance concerned; and in Chapter 5, measurements of the conductances of solutions under conditions where processes at the electrodes themselves are not of concern.

SELECTED BIBLIOGRAPHY

BARD, A. J., and FAULKNER, L. R. *Electrochemical Methods: Fundamentals and Applications.* New York: John Wiley, 1980. *An excellent introduction to and overview of electrochemical methods, particularly the logical development and clear presentations of the underlying equations and theory.*

BAUER, H. H. *Electrodics: Modern Ideas Concerning Electrode Reactions.* Stuttgart: Thieme, 1972.

BOCKRIS, J. O'M., and REDDY, A. K. N. *Modern Electrochemistry*, Vols. 1 and 2. New York: Plenum Press, 1970. *A thorough treatment of electrochemical fundamentals.*

MURRAY, R. W., and REILLEY, C. N. In *Treatise on Analytical Chemistry*, I. M. Kolthoff and P. J. Elving, eds., Part I, Vol. 4. New York: John Wiley, 1963, pp. 2109–2232. *Fundamentals of electrode processes and short introduction to electrochemical techniques.*

SAWYER, D. T., and ROBERTS, J. L., JR. *Experimental Electrochemistry for Chemists.* New York: John Wiley, 1974. *A good introduction to the more popular electrochemical methods; details of cell constructions, instrumentation, purification of solvents and electrolytes.*

REFERENCES

1. C. J. MOSBACHER, *Ind. Res. & Dev.*, 24(2), 176 (1982).
2. G. N. BOWERS, JR., in *Analytical Chemistry: Key to Progress on National Problems*, W. W. Meinke and J. K. Taylor, eds., NBS Special Publication 351, U.S. Government Printing Office, Washington, D.C., 1972, pp. 77–157.

PROBLEMS

1. If the usual units used in electrochemistry are square centimeters for electrode area, seconds for time, and microamperes for current, in what units must the concentration in the Cottrell equation, c, be expressed?

2. As was calculated, the value of the constant term $2.303RT/nF$ in the $\log_{10}$ form of the Nernst equation is $0.05916/n$ V at 25°C. Calculate the value of the term at 0°C, at normal body temperature (37°C), and at 100°C.

3. You are interested in calculating the total charge q, in coulombs, passed as a function of time during what is known as a potential-step experiment, for which the Cottrell equation is valid. Derive the appropriate equation.

4. Derive an equation that will express the charging current, i, to be expected at an ideally polarized electrode if the potential of that electrode is being changed linearly with time at a rate of v V/sec, and assuming the capacitance of the electrode is constant.

5. A typical value for the capacitance of an electrode is about $20\ \mu F/cm^2$. How many coulombs (of surface charge) are there on an electrode of $0.50\ cm^2$ area that is charged up to -1.0 V? Assuming that a typical metal atom occupies about $15\ \text{Å}^2$ of a metal surface, how many "excess" electrons would there be per metal atom on the surface of this electrode?

6. Assume you have a reversible 1-electron redox couple in solution, that is, a couple that obeys the Nernst equation. What is the ratio of the reduced to oxidized forms of the couple *at the electrode surface* if the potential of the working electrode is set exactly equal to the E^0 of that couple? If the bulk solution initially were to contain only the oxidized form of the couple, what would be the ratio of Red to Ox forms *at the electrode surface* if the potential of the electrode were again set at the E^0? Calculate the ratio of Red to Ox forms at the electrode surface if the potential of the electrode were set 10, 100, 200, 500, and 1000 mV negative of the E^0. Assume the temperature is 25°C.

7. When applying a voltage step (a sudden change in potential) to an electrode, the charging current at the electrode decays in an exponential manner with time:

$$i = i_0 e^{-t/RC}$$

where R = total resistance of the solution
C = capacitance of the electrode
RC = time constant in seconds
i_0 = initial current

The initial current is given by Ohm's law:

$$i_0 = \frac{\Delta E_{applied}}{R}$$

where $\Delta E_{applied}$ = magnitude of the potential step in volts

Calculate the RC time constant for cells with solution resistances of 1, 10, and 100 Ω, which are fairly typical values for many cases, and an electrode capacitance of 5 μF. For an applied voltage pulse of 100 mV, calculate the initial current for each of the three cells and the time required for the charging current to decay to 1 nA—a relatively small current for many faradaic methods, which would generally not present a serious interference.

CHAPTER 2

Potentiometry

James A. Cox
James E. O'Reilly

Potentiometric methods in analytical chemistry are based on the relationship between the potentials of electrochemical cells and the concentrations or activities of the chemical species in the cells. Potentiometry is one of the oldest analytical methods still in wide use. The early, essentially qualitative, work of Luigi Galvani (1737–1798) and Count Alessandro Volta (1745–1827) had its first fruit in the work of J. Willard Gibbs (1839–1903) and Walther Nernst (1864–1941), who laid the foundations for the treatment of electrochemical equilibria and electrode potentials.

Applications of potentiometry generally involve use of an electrochemical cell composed of a reference electrode, which maintains a constant potential, and an indicator electrode, which responds to the sample composition. A salt bridge often is used to prevent mixing of the sample and the reference solutions and to minimize the liquid-junction potential. Until very recently, redox indicator electrodes were the main type used in analytical potentiometry, except for pH measurement. These electrodes develop a potential related to the sample composition either by being directly involved in the half-cell reaction or by acting as an "inert" probe of the position of equilibrium of a redox couple in solution. In these cases, the Nernst equation (1.3) is used to relate activity or concentration to the potential.

In recent years, ion-selective electrodes have become more popular in analytical potentiometry than redox electrodes. They are devices that use the potential established across a membrane to make an analytical measurement. The composition of the membrane is designed to yield a potential that is primarily due to the ion of interest. The membrane types include glass (for H^+, Na^+, and certain other monovalent cations), crystals (for F^-, Ag^+, sulfide, etc.), and supported films of liquid ion exchangers that are immiscible with aqueous solutions (for Ca^{2+}, NO_3^-, BF_4^-, etc.). The analytical relationship is similar to the Nernst equation. Ion-selective electrodes will be discussed at the end of this chapter. The initial sections deal with redox electrodes.

2.1 ELECTROCHEMICAL CELLS

An electrochemical cell can be defined as two conductors (electrodes), usually metallic, immersed in the same electrolyte solution, or in two different electrolyte solutions that are connected by a salt bridge. Electrochemical cells are classed into two groups. In a *galvanic* (voltaic) cell, electrochemical reactions occur spontaneously when a series circuit is completed by placing an external connection between the two electrodes. These cells are often used to convert chemical energy into electrical energy. Many types are of commercial importance, such as various batteries and fuel cells. In the case of an *electrolytic* cell, chemical reactions are caused by the imposition of an external voltage greater than the reversible (galvanic) voltage of the cell. Essentially, these cells are used to carry out chemical reactions at the expense of electrical energy. Some important commercial uses of electrolytic cells involve synthetic processes, such as the preparation of chlorine gas and caustic soda from brines, and electroplating procedures.

Electrochemical cells are comprised of a pair of half-cells. An example is shown in Figure 2.1. The porous membrane between the half-cells provides electrical contact by ion migration but prevents gross mixing of the solutions. The half-cell reactions are as follows:

$$Zn^{2+} + 2e^- \rightleftarrows Zn \qquad E^0 = -0.76 \text{ V} \qquad (2.1)$$

$$Cu^{2+} + 2e^- \rightleftarrows Cu \qquad E^0 = 0.15 \text{ V} \qquad (2.2)$$

According to convention, both half-cell reactions are written as reductions [1]; however, because of the more positive standard reduction potential of Reaction 2.2, the net electrolysis that occurs when an external connection is made between the

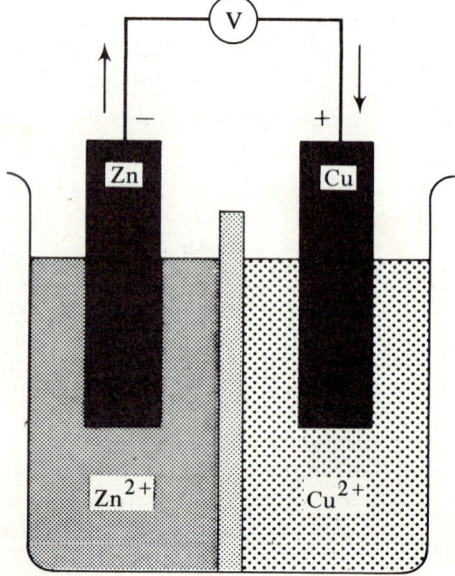

FIGURE 2.1. *Schematic diagram of a simple galvanic electrochemical cell. V is a voltmeter or other voltage-measuring device. The arrows indicate the direction of the spontaneous flow of electrons. The + and − designate the polarity of the cell as measured by a voltmeter.*

copper and zinc electrodes involves the reduction of Cu^{2+} and the oxidation of Zn, that is,

$$Cu^{2+} + Zn \rightarrow Cu + Zn^{2+} \tag{2.3}$$

as long as the concentration of Cu^{2+} is not markedly lower than that of Zn^{2+}.

2.2 THE NERNST EQUATION

A more accurate prediction of the direction of the spontaneous reaction of galvanic cells requires that the effect of concentration on the half-cell potential be considered. Such calculations are made with the Nernst equation. In fact, the Nernst equation is used for all calculations of half-cell potential where the half-cell reaction is at equilibrium (i.e., in reactions such as 2.1 where the rates of the forward reaction [the reduction of Zn^{2+}] and the back reaction are so rapid that during the time scale of the potential measurement, local equilibrium exists at the electrode surface). Such half-cell reactions are termed *electrochemically reversible*. For the generalized half-cell reaction, which is written as a reduction by convention, that is,

$$Ox + ne^- \rightleftarrows Red \tag{2.4}$$

the potential is given by the generalized form of the Nernst equation

$$E = E^0 - \frac{RT}{nF} \ln \frac{a_{Red}}{a_{Ox}} = E^0 - \frac{RT}{nF} \ln \frac{(Red)}{(Ox)} = E^0 - \frac{2.303RT}{nF} \log \frac{(Red)}{(Ox)} \tag{2.5}$$

where
- E^0 = standard electrode potential in volts
- R = molar gas constant (8.314 J/K-mole)
- T = absolute temperature in kelvins
- (Red) or a_{Red} = activity of the reduced form
- (Ox) or a_{Ox} = activity of the oxidized form

If numerical values are inserted for the constants and the temperature is 25°C, the Nernst equation becomes

$$E = E^0 - \frac{0.05916}{n} \log \frac{a_{Red}}{a_{Ox}} \tag{2.6}$$

Pure phases, such as metal electrodes and salt crystals, have unit activity; solvents containing even very large quantities of dissolved solutes are generally assumed also to have unit activity (i.e., they are assumed to approximate pure phases). The partial pressures of gases and the molarities of uncharged solutes are used for the activities in Equation 2.6, when they appear in half-reactions. The molarities of charged solutes differ too much from the activities to be rigorously used in Equation 2.6 directly. The activity and the concentration are related by

$$a_i = f_i c_i \tag{2.7}$$

where f_i = activity coefficient of ion "i"

The Nernst equation (2.5) can be rewritten as follows when the components of the half-cell reaction are solutes:

$$E = E^0 - \frac{RT}{nF} \ln \frac{f_{Red}[Red]}{f_{Ox}[Ox]} = E^0 - \frac{RT}{nF} \ln \frac{f_{Red}}{f_{Ox}} - \frac{RT}{nF} \ln \frac{[Red]}{[Ox]}$$

$$= E^{0\prime} - \frac{RT}{nF} \ln \frac{[Red]}{[Ox]} \tag{2.8}$$

where $E^{0\prime}$ = formal electrode potential

The formal potential, which is somewhat like a standard potential under a given set of experimental conditions, lacks the fundamental thermodynamic significance of the standard potential; but it is often experimentally useful. It can be measured directly as long as the half-cell reaction is reversible.

The departure of activity from concentration for charged species is a function of the ionic strength, I, of the solution:

$$I = \frac{1}{2} \sum c_i z_i^2 \tag{2.9}$$

where c_i = concentration of ionic species "i"
z_i = charge on that ion

The activity coefficient, f_i, can be estimated from the ionic strength using equations developed with Debye-Hückel theory. Various relationships have been derived depending on whether ions are considered point charges in solution or are assigned certain radii for their hydrated spheres [2]. With the point-charge approximation, an assumption that ions do not interact in solution (a reasonable approximation for $I < 0.01$), and a solvent of water at 25°C, the Debye-Hückel limiting law can be used to estimate f_i:

$$-\log f_i = 0.5 z_i^2 I^{1/2} \tag{2.10}$$

For solutions of higher ionic strength, the form

$$-\log f_i = \frac{0.5 z_i^2 I^{1/2}}{1 + I^{1/2}} \tag{2.11}$$

provides a better estimate. It must be noted that only as I approaches zero (the experimentally unreachable "infinite dilution" point) is Debye-Hückel theory rigorous; the estimates worsen as the ionic strength increases.

An example of the application of Expressions 2.6 through 2.11 is the estimation of the formal potential of the couple

$$A^{2+} + e^- \rightleftarrows A^+ \qquad E^0 = 0.400 \text{ V} \tag{2.12}$$

in the following solutions: 0.005 M NaCl and 0.02 M CaCl$_2$, assuming A^{2+} and A^+ do not contribute appreciably to the total ionic strength. In the former, $I = (1/2)(0.005 + 0.005) = 0.005$; at this ionic strength, Equation 2.10 can be applied. For A^{2+}, $z = 2$, and therefore $-\log f_{A^{2+}} = (0.5)(2)^2(0.005)^{1/2}$; so $f_{A^{2+}} = 0.72$. For A^+ ($z = 1$), the result is $f_{A^+} = 0.92$. From Equations 2.6 and 2.7, it is apparent that

$$E^{0\prime} = E^0 - \frac{0.059}{n} \log \frac{f_{Red}}{f_{Ox}} \tag{2.13}$$

and so $E^{0\prime} = 0.400 - (0.059/1)\log(0.92/0.72) = 0.394$ V in 0.005 M NaCl. In the

0.02 M $CaCl_2$ solution, $I = 0.06$. In this case, Equation 2.11 provides a better estimate of the activity coefficients. Here, $f_{A^{2+}} = 0.40$ and $f_{A^+} = 0.80$; so $E^{0\prime} = 0.382$ V. These results demonstrate that higher charge on an ion leads to lower activity coefficients, and higher ionic strengths cause a greater departure of the formal potential from the standard potential. In fact, one method of estimating standard potentials is to measure $E^{0\prime}$ as a function of $I^{1/2}$ and extrapolate to $I = 0$.

Effect of Complexation on Electrode Potentials

Half-cell potentials are established in accord with the equilibrium concentrations of the species involved in the half-cell reaction. This factor becomes especially important when complexing agents are present in solution. If some of the metal ion in solution is tied up in the form of a metal complex, the equilibrium concentration of the free, uncomplexed metal ion is decreased and the half-cell potential is therefore altered. The simplest case is that of a certain ionic species in the presence of various concentrations of a complexing agent. For example, let us again consider the copper electrode of the half-cell in Figure 2.1, where EDTA (ethylenediaminetetraacetic acid) has also been added to the Cu^{2+} solution. The formation of the copper-EDTA complex can be represented by the equilibrium

$$Cu^{2+} + EDTA^{4-} \rightleftharpoons CuEDTA^{2-} \tag{2.14}$$

(where $EDTA^{4-}$ is the basic form of the tetra-acid), for which the formation constant is written as

$$K_f = \frac{(CuEDTA^{2-})}{(Cu^{2+})(EDTA^{4-})} = 6.17 \times 10^{18} \tag{2.15}$$

For the half-reaction involving copper ions and copper (Equation 2.2), the Nernst equation is expressed by

$$E = E^0_{Cu^{2+},Cu} - \frac{RT}{nF} \ln \frac{1}{(Cu^{2+})} \tag{2.16}$$

Combining Equation 2.15 with 2.16 yields the potential of a copper electrode in aqueous $EDTA^{4-}$ systems:

$$E_{Cu} = E^0_{Cu^{2+},Cu} + \frac{RT}{nF} \ln \frac{1}{K_f(EDTA^{4-})} + \frac{RT}{nF} \ln (CuEDTA^{2-}) \tag{2.17}$$

At 25°C, for 0.001 M Cu^{2+} and 0.10 M $EDTA^{4-}$, the potential would be -0.278 V (neglecting activity coefficients), since for a strong complexing agent like EDTA it can be assumed that nearly 100% of the copper is present as $CuEDTA^{2-}$; that is, $CuEDTA^{2-} \approx 0.001$ M and $EDTA^{4-} \approx 0.10$ M.

The shift in electrode potential caused by the complexing agent is contained in the second term on the right side of Equation 2.17. In the present case, it amounts to a shift of -0.526 V. The important practical consequences of chelation and complexation will be discussed in more detail later. For example, one can determine a copper ion by direct potentiometry using a copper ion–selective electrode, or via a potentiometric titration with EDTA using the electrode as an endpoint detector.

Schematic Representation of Cells

To simplify the description of cells, a type of electrochemical shorthand has evolved, which allows an easier depiction of cells and cell components. For example, for the cell in Figure 2.1, one would write

$$Zn/Zn^{2+}//Cu^{2+}/Cu \qquad (2.18)$$

The concentrations are included in parentheses or set off by commas. The slant lines indicate phase boundaries across which are developed potential differences that are included in the measured potential of the entire cell. By convention, double slant lines signify a liquid junction—the zone of contact between two electrolyte solutions. Physically, this may be a porous membrane, as in Figure 2.1, or a salt bridge of some sort.

Also by convention, in a galvanic system the half-cell with the more positive potential is written on the right. This electrode is the cathode (the electrode at which the reduction occurs); the anode is on the left. If there are several components in one electrolyte, they are separated by a comma. For example, for a cell (without liquid junction) comprised of a silver/silver chloride half-cell and a hydrogen-gas electrode, one could write for one set of conditions

$$Pt/H_2 \ (0.5 \ atm), \ HCl \ (0.1 \ M), \ AgCl/Ag \qquad (2.19)$$

Liquid-Junction Potentials

At the boundary between two dissimilar solutions, a junction potential is always set up. The mobilities of positive and negative ions diffusing across the boundary generally will not be equal. Thus, a slight charge separation arises, which results in the junction potential. The solvents, the nature of the electrolytes, and the concentrations of the electrolytes affect the value of the junction potential. Junction potentials can become rather large (50 mV or more), particularly when one of the electrolyte ions, such as H^+ or OH^-, has a very high mobility.

Almost all electrochemical cells contain at least a small liquid-junction potential, generally of unknown magnitude. Only in a few special cases can it be calculated or measured. Experimentally, the most common approach is to minimize the junction potential by use of a concentrated salt bridge between dissimilar solutions. Because the mobilities of potassium and chloride ions are nearly equal, the usual choice for the electrolyte in a salt bridge is a concentrated KCl solution. When potassium or chloride ions cannot be used for chemical reasons, KNO_3, K_2SO_4, or lithium trichloroacetate can be used.

The junction between the bridge and the sample solution is designed to minimize cross-contamination of the solutions. Common styles are a wick of asbestos sealed into glass, a salt-containing agar-gel plug, a porous glass or ceramic plug, or a fine capillary tip. The same precaution must often be taken at the reference electrode/salt bridge junction.

Galvanic-Cell Reactions

An absolute measurement of the potential of a single electrode cannot be made. This fact has led to the establishment of a set of internationally recognized conventions regarding electrode potentials and descriptions of electrochemical cells. For example, the standard hydrogen electrode (SHE), described in the next section, has been assigned a potential of exactly 0 V. All standard half-cell potentials are reported relative to the SHE. Furthermore, all half-cell reactions are written as reductions.

To calculate the potential of an electrochemical cell, the Nernst equation is first used to calculate the half-cell potentials. Unless sufficient information is available to take into account the activity coefficients, Equation 2.8 is used with concentrations; if the formal potential is unknown, the standard potential is substituted into this expression (i.e., all activity coefficients are approximated as unity).

The half-cell reaction with the more positive potential (calculated from Equation 2.8) is written on the right side of the galvanic cell. The overall cell potential is

$$E_{cell} = E_{right} - E_{left} + E_{lj} \tag{2.20}$$

The liquid-junction potential, E_{lj}, is generally assumed to be negligible. If an external connection between the half-cells is made, reduction occurs in the more positive half-cell (right side), and oxidation occurs in the more negative half-cell. The spontaneous reaction proceeds until the concomitant changes in concentration in the half-cells cause E_{right} to be equal to E_{left}. At that point, the overall cell is at equilibrium. From Equation 2.20 and assuming a negligible liquid-junction potential, $E_{cell} = 0$ at equilibrium.

It is important to note that even though the spontaneous reaction involves a net oxidation in the left half-cell, the potential of that half-cell is calculated from the half-cell reaction written as a reduction. This apparent contradiction is explained by the fact that in Equation 2.20, E_{left} is subtracted from E_{right}. This amounts to a change of sign of E_{left} and therefore recognition that the true half-cell reaction is an oxidation rather than a reduction.

As an example, consider the following galvanic cell:

$$Pt/0.10\ M\ Fe^{3+}, 0.20\ M\ Fe^{2+}//0.020\ M\ MnO_4^-, 0.030\ M\ Mn^{2+}, pH\ 2.0/Pt$$

where the balanced half-cell reactions are

$$Fe^{3+} + e^- \rightleftarrows Fe^{2+} \qquad E^{0\prime} = 0.77\ V$$

$$MnO_4^- + 8H^+ + 5e^- \rightleftarrows Mn^{2+} + 4H_2O \qquad E^{0\prime} = 1.51\ V$$

The initial cell potential is calculated by

$$E_{right} = 1.51 - \frac{0.059}{5} \log \frac{[Mn^{2+}]}{[MnO_4^-][H^+]^8} = 1.51 - \frac{0.059}{5} \log \frac{(0.030)}{(0.020)(0.010)^8}$$

$$= 1.32\ V$$

$$E_{left} = 0.77 - 0.059 \log \frac{[Fe^{2+}]}{[Fe^{3+}]} = 0.77 - 0.059 \log \frac{0.20}{0.10} = 0.75\ V$$

$$E_{cell} = E_{right} - E_{left} = 1.32 - 0.75 = 0.57\ V$$

If an external current-carrying path is introduced, the net spontaneous reaction, $MnO_4^- + 5Fe^{2+} + 8H^+ \rightarrow Mn^{2+} + 5Fe^{3+} + 4H_2O$, occurs. Eventually, the depletion of reactants and buildup of products cause the reaction to reach equilibrium, at which $E_{cell} = 0$, and so $E_{right} = E_{left}$.

Analytical Applications of Potentiometry at Redox Electrodes

When potentiometry is used for analytical purposes, the initial cell potential is the measured parameter. The measurement requires an external circuit. Either this potential-measuring circuit has such a high ohmic resistance that the amount of external current flow is so small that significant electrolysis does not occur, or the measurement technique used minimizes any electrolysis. The latter is exemplified by a simple potentiometer where a tap key is used in conjunction with manual variation of a reference potential (Fig. 2.2). A galvanometer detects the external current that flows during the brief period the tap key is depressed. At the balance point, pushing the tap key does not cause detectable current, and the reading of the potentiometer matches the actual cell potential. The widespread availability of voltage-measurement devices with very high ohmic resistances (e.g., pH and pIon meters) has caused a dramatic decrease in the use of the manual potentiometer for such measurements.

A second aspect of analytical potentiometry is that a reference electrode is used in place of one of the half-cells. This is necessary for quantification of the measurement. In the cell

$$Pb/Pb^{2+}//Reference\ Electrode$$

the determination of $[Pb^{2+}]$ can be made only from E_{cell} if the reference potential is known. For example, if $E^{0'}_{Pb^{2+},Pb} = -0.14\ V$ and $E_{ref} = 0.24\ V$, a measured cell potential of 0.42 V corresponds to $E_{left} = -0.18\ V$, and from the Nernst equation the Pb^{2+} concentration, 0.044 M, is calculated.

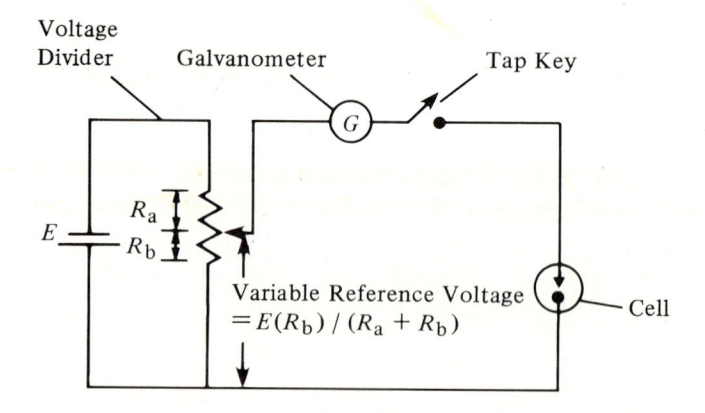

FIGURE 2.2. *Manual potentiometer. When the variable reference voltage is equal to the cell potential, closing the tap key will not cause a current flow through the galvanometer, and hence no deflection.*

2.3 REFERENCE ELECTRODES

The reference electrodes used in analytical potentiometry must have a potential that is unchanged by the passage of the small amount of current (10^{-9} A or less) required to "drive" the measuring instrument—an electrometer, pH meter, or high-impedance voltmeter. Furthermore, one strives to make all liquid-junction potentials either constant or negligible so that the reference electrode/salt bridge combination has a potential that is invariant with respect to the sample-solution composition. Ideally, therefore, the reference electrode is of known and constant potential, with negligible variation in the liquid-junction potential from one test or standard solution to another. In this case, the cell potential of the overall system, Equation 2.20, can be expressed as

$$E_{cell} = E_{constant} + E_{ind} \qquad (2.21)$$

where $\quad E_{ind}$ = potential (varying with the solution composition) of the *ind*icator or sensing electrode

Under these conditions, the indicator electrode can provide unambiguous information about ionic activities in the cell solution. In most analytical work, it is not necessary to know the actual value of the reference-electrode potential because $E_{constant}$ is determined using known standard solutions.

Hydrogen-Gas Electrode

The hydrogen electrode is the ultimate standard electrode, not only for the determination of (relative) potentials, but also for the determination of pH values. Because of experimental difficulties, it is seldom used except for the characterization of secondary reference electrodes and glass pH electrodes. The hydrogen electrode consists of a piece of platinum foil, electroplated with a thin layer of finely divided platinum ("platinized"). This provides a catalytic surface on which the half-cell reaction

$$2H^+ + 2e^- \rightleftarrows H_2 \qquad (2.22)$$

can proceed reversibly. The electrode is immersed in the test solution, and high-purity hydrogen gas is bubbled through the solution and over the electrode surface so that both will be saturated with the gas.

The primary disadvantages of the hydrogen-gas electrode are that it is rather difficult to prepare properly and that it is inconvenient to use. Another disadvantage is that its potential is sensitive to oxidants and reductants in solution—to anything that will oxidize H_2 or reduce H^+. Also, the catalytic Pt surface is poisoned by a variety of substances including As, CN^-, H_2S, Hg, and surface-active compounds such as proteins.

Calomel Electrode

The calomel electrode is the most widely used reference for general electrochemical measurements. This electrode consists of mercury, mercurous chloride (calomel), and a chloride-ion solution:

$$Hg/Hg_2Cl_2 \text{ (satd.), } Cl^- \text{ (x } M) \tag{2.23}$$

The half-cell reaction is

$$Hg_2Cl_2 + 2e^- \rightleftarrows 2Hg + 2Cl^- \qquad E^0 = 0.268 \text{ V} \tag{2.24}$$

Because the activities of solid Hg_2Cl_2 and Hg are unity, the potential is governed by the activity of the chloride ion (x). The most common type of this electrode is the saturated calomel electrode (SCE), in which the solution is saturated with KCl. It is easily made and maintained, and its potential is highly reproducible. A great variety of commercial calomel electrodes are available; two of these are illustrated in Figure 2.3. If chloride ion must be avoided, the mercury/mercurous sulfate couple can be used as the basis of a reference electrode.

The major disadvantage of the SCE is that its potential varies strongly with temperature, owing to the change in solubility of KCl. There is a perceptible hysteresis effect following temperature changes because of the time required for solubility equilibrium to be established. The SCE can be used only at temperatures less than about 80°C, probably due to the disproportionation of mercurous ions to form mercury and mercuric ion.

For accurate work, 0.1 M or 1 M KCl calomel electrodes should be used rather than the SCE, because they reach their equilibrium potential more rapidly and have less temperature dependence. Table 2.1 gives the potentials of several common reference electrodes at selected temperatures.

Silver/Silver Chloride Electrodes

A silver/silver chloride reference electrode is prepared by plating a layer of silver chloride onto a metallic silver wire or sheet. The electrode is immersed in a chloride solution (usually KCl) of known concentration, which is also saturated with AgCl. Since AgCl is appreciably soluble in concentrated chloride media, solid AgCl is

TABLE 2.1. *Potentials of Some Reference Electrodes in Volts versus the Standard Hydrogen Electrode as a Function of Temperature*

Temperature, °C	Calomel[a] (0.1 M KCl)	Calomel[a] (Satd. KCl)	Ag/AgCl[a] (Satd. KCl)
10	0.3362	0.2543	0.2138
20	0.3359	0.2479	0.2040
25	0.3356	0.2444	0.1989
30	0.3351	0.2411	0.1939
40	0.3336	0.2340	0.1835

a. Liquid-junction potential included
Source: Reprinted from R. G. Bates, *Determination of pH*, 2nd ed., pp. 325–35, by permission of the author and John Wiley and Sons. Copyright © 1973 by John Wiley and Sons.

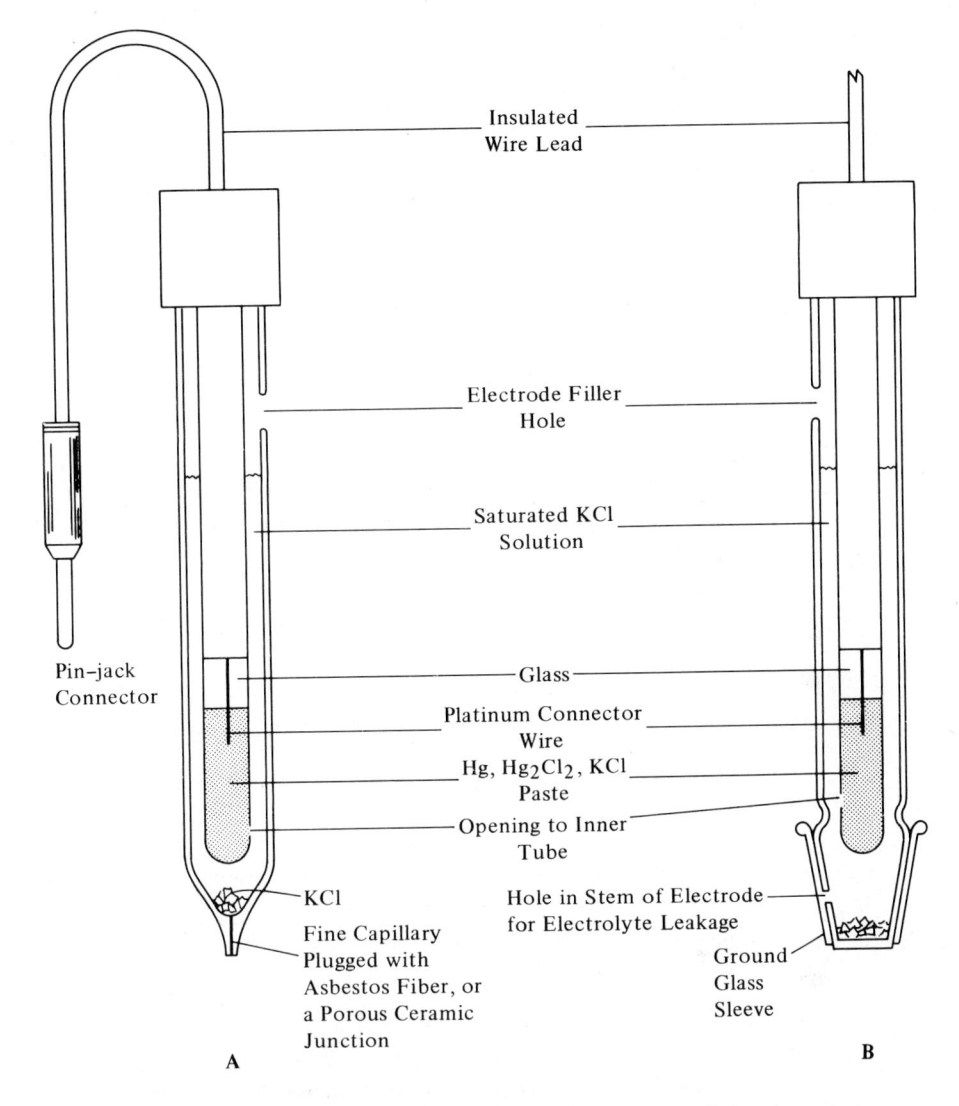

FIGURE 2.3. *Schematic cross-sectional representation of the construction of some typical commercial calomel electrodes. A: Fiber or porous-ceramic junction type. B: Ground glass–sleeve type. The electrolyte leak rate of the fiber or porous-ceramic junction type is quite low, typically 1 to 10 µL/hr; that for the sleeve type around 100 µL/hr.*

usually added to the solution to ensure saturation and prevent dissolution of AgCl from the electrode surface. The half-cell thus constructed may be represented as

$$\text{Ag/AgCl (satd.), Cl}^- \text{ (x } M\text{)} \tag{2.25}$$

for which the half-reaction is

$$\text{AgCl} + e^- \rightleftharpoons \text{Ag} + \text{Cl}^- \qquad E^0 = +0.2223 \text{ V} \tag{2.26}$$

As in the calomel electrode, the potential is governed by the chloride-ion activity. Useful silver chloride electrodes can be prepared by simply anodizing a silver wire in chloride media, but the apparent equilibrium potential of these electrodes may differ by several millivolts from one electrode to another. More care is necessary for the preparation of highly stable and reproducible electrodes [3].

Commercial silver/silver chloride reference electrodes are available in a variety of styles and sizes. They are often used as the internal reference electrodes in glass pH and other ion-selective electrodes. Silver/silver chloride microelectrodes formed from very thin silver wire have found extensive use, for example, in biomedical applications such as *in vivo* studies of biological fluids and intracellular measurements, because of the miniaturization possible with these electrodes.

The Ag/AgCl electrode is also sufficiently stable for use at temperatures up to about 275°C, making it an appropriate alternative to calomel electrodes at elevated temperatures.

Thallium Amalgam/Thallous Chloride Electrodes

The Tl(Hg)/Tl$^+$ reference electrode (Thalamid®) is said to be superior to either calomel or silver chloride electrodes when measurements are made over a range of temperatures, because it attains its equilibrium potential very rapidly after changes in temperature. The half-cell can be written as

$$\text{Tl (40\% amalgam)/TlCl (satd.), KCl (satd.)}$$

Some commercial glass pH electrodes use Thalamid® electrodes as internal reference electrodes.

2.4 pH: DEFINITION AND MEASUREMENT

The earliest definition of pH was given by Sørensen [4], who defined it as the negative logarithm of hydrogen-ion concentration:

$$\text{pH} \equiv \text{p}c\text{H} = -\log\,[\text{H}^+] \tag{2.27}$$

Because of deficiencies in the theoretical assumptions made, this early definition was a measure of neither concentration nor activity. When the concept of thermodynamic activity became established, primarily through the efforts of Lewis and Randall, Sørensen and Linderstrøm-Lang defined pH as the negative logarithm of the hydrogen-ion activity:

$$\text{pH} \equiv \text{p}a\text{H} = -\log\,a_{\text{H}^+} = -\log\,[\text{H}^+]f_{\text{H}^+} \tag{2.28}$$

Unfortunately, since individual ionic activity coefficients cannot be evaluated without extrathermodynamic assumptions, the theoretical thermodynamic elegance and desirability of this pH definition cannot be rigorously related to experimental quantities. For this reason, the modern *operational* NBS (National Bureau of Standards) scale of acidity has been developed.

Operational Definition of pH

For the analytical chemist, an experimentally useful scale of acidity should allow the interpretation of the most important and common measurements as, for example, those with a glass electrode and saturated calomel reference electrode. The operational definition of pH of an aqueous solution is

$$\text{pH} = \text{pH}_s + \frac{(E - E_s)F}{RT \ln 10} \tag{2.29}$$

where E = electromotive force of a cell containing the unknown solution
E_s = electromotive force of a cell containing a standard reference buffer solution of known or defined pH, that is, pH_s

This definition has been endorsed by standardizing groups in many countries and has been recommended by the International Union of Pure and Applied Chemistry. In actual practice, the NBS pH standards were assigned pH_s values from measurements of a hydrogen gas–silver/silver chloride cell without liquid junction,

$$\text{Pt}/\text{H}_2, \text{ Buffer Solution, Cl}^-, \text{AgCl}/\text{Ag} \tag{2.30}$$

while making reasonable assumptions about activity coefficients in such a way as to make pH_s represent as nearly as possible $-\log a_{\text{H}^+}$.

Primary Standards. Since every practical pH electrode can be regarded only as a somewhat imperfect tool that functions more or less unevenly over the whole pH range, and every practical pH reading involves a (possibly variable) liquid-junction potential, the NBS has adopted a series of seven primary standard pH buffer solutions (Table 2.2). The pH of the standards is temperature dependent, primarily because of the variation of the K_a of the buffer system with temperature.

TABLE 2.2. pH_s of NBS Primary Standards

Temp., °C	KH Tartrate[a]	KH$_2$ Citrate[b]	KH Phthalate[c]	Phosphate[d] (Equimolal)	Phosphate[e] (3.5:1)	Borax[f]	Carbonate[g]
0	—	3.863	4.003	6.982	7.534	9.460	10.321
10	—	3.820	3.996	6.921	7.472	9.331	10.181
20	—	3.788	3.999	6.878	7.430	9.227	10.064
25	3.557	3.776	4.004	6.863	7.415	9.183	10.014
30	3.552	3.766	4.011	6.851	7.403	9.143	9.968
40	3.547	3.753	4.030	6.836	7.388	9.074	9.891
50	3.549	3.749	4.055	6.831	7.384	9.017	9.831

a. Saturated at 25°C; NBS Certificate 188.
b. 0.05 m KH$_2$C$_6$H$_5$O$_7$; m = molality (moles/kg).
c. 0.05 m KHC$_8$H$_4$O$_4$; NBS Certificate 185e.
d. 0.025 m KH$_2$PO$_4$, 0.025 m Na$_2$HPO$_4$; NBS Certificates 186-I-c and 186-II-c.
e. 0.008695 m KH$_2$PO$_4$, 0.03043 m Na$_2$HPO$_4$; NBS Certificates 186-I-c and 186-II-c.
f. 0.01 m Na$_2$B$_4$O$_7 \cdot$10H$_2$O; NBS Certificate 187b.
g. 0.025 m NaHCO$_3$, 0.025 m Na$_2$CO$_3$; NBS Certificates 191 and 192.

The primary standards cover the pH range from about 3.5 to 10.5, and were chosen for their reproducibility, stability, buffer capacity, and ease of preparation. From studies of the internal consistency of the seven primary standards, it appears that the total uncertainty of the pH_s values is within ± 0.006 pH. This means that the pH determined by a practical cell such as

$$\text{Pt/H}_2, \text{Solution/KCl (satd.)/Reference Electrode} \qquad (2.31)$$

with properly designed liquid junction will be the same regardless of which of the seven buffers is chosen as a standard. This is also true if a glass electrode with perfect pH response is substituted for the hydrogen electrode.

With regard to the significance of pH values, it can be said that they are at best an *estimate* of $-\log a_{H^+}$, depending on how accurately the liquid-junction potential remains constant for the measurements of the standard and the unknown. For many dilute solutions (less than 0.1 M) between pH 2 and 12, the pH may be considered to correspond to the true hydrogen-ion activity to within about ± 0.02 pH. This is equivalent to ± 1.2 mV in the potential reading, and about a $\pm 5\%$ uncertainty in a_{H^+}.

Secondary Standards. In addition to the seven primary standards, the NBS has designated two secondary standards, one on the acidic end and one on the basic end of the intermediate pH region, in order to affirm the proper functioning of the glass electrode (Table 2.3). Furthermore, a great number and variety of "secondary" pH standard solutions have been collected in various treatises and compendia [3, 5, 6]. A secondary standard may often prove convenient because it may be more easily prepared or be more stable than the primary standards, or it may more closely match the composition and pH of a group of unknowns on which many measurements are to be made over a period of time. Particularly in process-control applications with, for example, very concentrated electrolyte solutions, reproducibility may be the most important concern.

TABLE 2.3. pH_s of NBS Secondary Standards

Temp., °C	K Tetroxalate[a]	Ca(OH)$_2$[b]
0	1.666	13.423
10	1.670	13.003
20	1.675	12.627
25	1.679	12.454
30	1.683	12.289
40	1.694	11.984

a. 0.05 m KH$_3$(C$_2$O$_4$)$_2$·2H$_2$O; NBS Certificate 189.
b. Saturated at 25°C.
Source: V. E. Bower, R. G. Bates, and E. R. Smith, *J. Res. Natl. Bur. Stand.*, *51*, 189 (1953); R. G. Bates, V. E. Bower, and E. R. Smith, *J. Res. Natl. Bur. Stand.*, *56*, 305 (1956).

pH Electrodes

The premier pH electrode, from both a historical and a fundamental point of view, is the hydrogen-gas electrode. As was mentioned in Section 2.3, however, this electrode is difficult to prepare and use, and has been replaced almost universally by the modern glass electrode except for specialized applications.

Glass Electrode. The glass pH electrode is composed of: (a) a thin hydrogen-ion–responsive glass membrane sealed to a stem of high-resistance, nonresponsive glass, and (b) an internal reference electrode with a constant internal hydrogen-ion concentration. The internal electrode may be either Ag/AgCl in HCl or Hg/Hg_2Cl_2 in HCl. The entire cell requires an external reference electrode for operation. The complete cell may be diagrammed schematically as

$$\text{Internal Reference Electrode} \mid \text{Internal Electrolyte} \mid H^+\text{-Responsive Glass Membrane} \mid \text{External Solution} \parallel \text{External Reference Electrode} \quad (2.32)$$

Glass is an irregular three-dimensional arrangement of silicate tetrahedra in which each oxygen atom is shared by two silicate groups. With one type of pH-responsive glass, Na^+ and Ca^{2+} cations are located in this array. Modern glass pH electrodes contain Li^+ and Ba^{2+} to varying degrees, in place of Na^+ and Ca^{2+}, to make the electrode more selective to H^+. When an electrode is immersed in aqueous solution, cations from the surface of the glass are leached out and replaced by protons to form a hydrated silica-rich layer about 500 Å thick, depending on the hygroscopic nature of the glass. The external part of this hydrated gel layer can act as a cation-exchange membrane which has a particularly high degree of selectivity among the various cations. By variation of the glass composition, glass electrodes can be made responsive to a variety of monovalent cations (Sec. 2.5).

When a thin membrane of glass is interposed between two solutions, an electrical potential difference is developed across the membrane that depends on the activities and nature of the ions present in the two solutions, the composition of the glass, and other factors. Although the mechanism by which cations affect the potential developed across the glass membrane is not understood completely, the most generally accepted theory is based on an ion-exchange equilibrium occurring at the solution-glass boundary [7]. Glass electrodes give a Nernstian response to hydrogen-ion activity, at least over a large portion of the pH range.

Because of their convenience and wide use, a great variety of glass pH electrodes are commercially available in various sizes, shapes, temperature ranges, and so forth. Figure 2.4 illustrates the construction of a common type of glass electrode.

A major advantage of glass electrodes is that there is no formal *electron* exchange involved in their functioning; thus, they are uninfluenced by oxidizing and reducing agents in solution, unlike all other pH electrodes. They have a very high impedance (1 to 500 MΩ, typically), necessitating the use of a voltmeter with very high input impedance, or an electrometer. Glass electrodes must also be calibrated fairly often, preferably with buffers within about a pH unit of the pH to be measured. In addition, they cannot be used in acidic fluoride media, and exhibit "acid" and "alkaline" errors. In highly acidic media, glass electrodes exhibit a negative "acid"

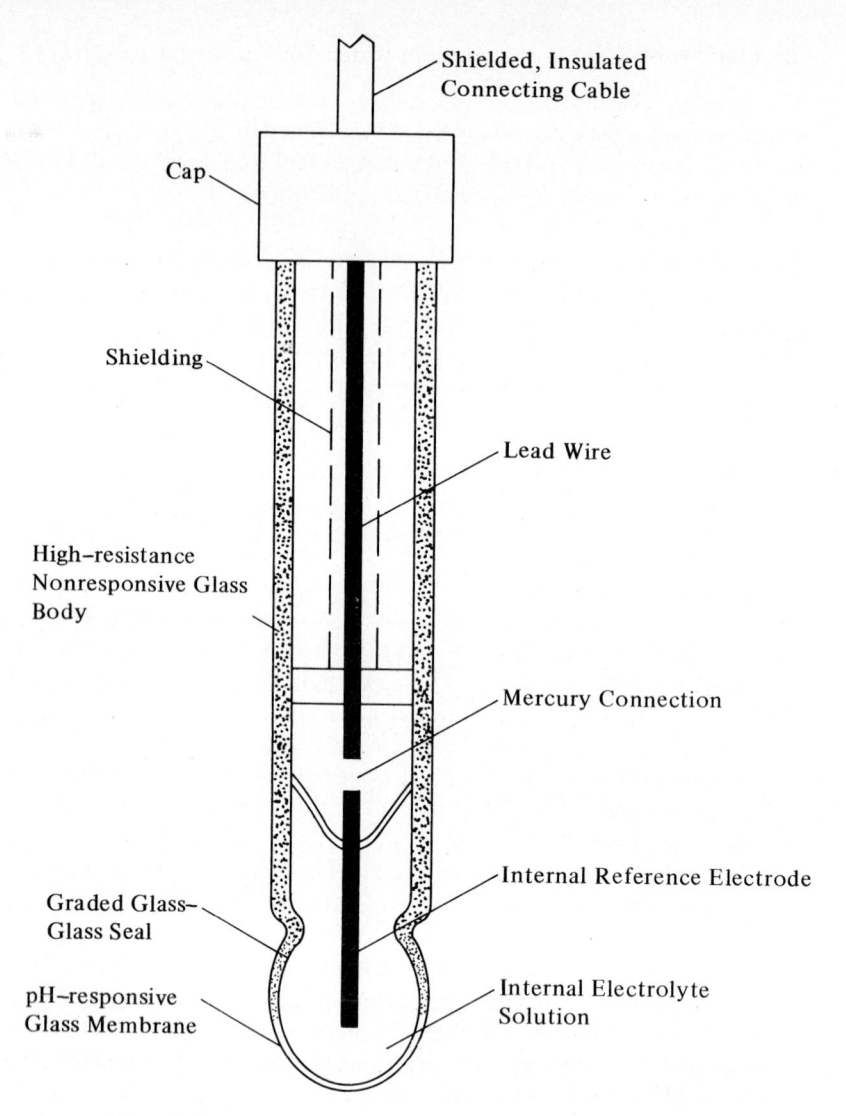

FIGURE 2.4. *Schematic representation of the construction of a typical glass pH electrode.*

error, which is thought to be due to the migration of the anions of the acid into the gel layer or a change in activity of water in the gel layer, thereby affecting the hydrogen-ion activity. In highly basic media, glass electrodes exhibit a positive "alkaline" error owing to the partial exchange of cations other than H^+ (notably Na^+) between the pH-sensitive surface layer and the basic solution.

Quinhydrone Electrode. The quinhydrone electrode is typical of a whole class of electrodes that function as pH sensors due to a reversible organic oxidation-reduction pair involving protons. Quinhydrone (an equimolar compound of quinone and hydroquinone) is only slightly soluble in water. The reversible oxidation-reduction couple

$$\text{Quinone, Q} + 2e^- + 2H^+ \rightleftharpoons \text{Hydroquinone, H}_2\text{Q} \qquad (2.33)$$

involving H^+ will fix the potential of a redox electrode, usually gold or platinum, immersed in a solution, as given by

$$E = E^0_{Q,H_2Q} + \frac{RT}{2F} \ln \frac{(Q)(H^+)^2}{(H_2Q)} \qquad (2.34)$$

The quinhydrone electrode, therefore, responds directly to hydrogen-ion activity in a Nernstian manner as long as the ratio of activities of Q and H_2Q remains constant. The electrode is simple to construct, reaches equilibrium fairly rapidly, and is less disturbed by poisons and by oxidizing and reducing agents than is the hydrogen-gas electrode. It functions well in many nonaqueous and partially aqueous media, and has a relatively low impedance. Its chief disadvantage is that it cannot be used in solutions of pH greater than about 8 because of the air oxidation of hydroquinone to quinone and the acidic dissociation of hydroquinone, both of which cause the ratio $(Q)/(H_2Q)$ to increase and the apparent pH to be too low.

Antimony Electrode. The antimony electrode is perhaps the best representative of a whole class of metal/metal oxide redox electrodes that respond to pH. The potential is probably developed as a result of an oxidation-reduction reaction involving antimony and a skin of Sb(III) oxide that forms on the surface of the metal:

$$Sb_2O_3 + 6H^+ + 6e^- \rightleftharpoons 2Sb + 3H_2O \qquad (2.35)$$

Both antimony and its oxide are solid and therefore have unit activity; so the potential of the electrode can be expressed as

$$E = E^0_{Sb_2O_3,Sb} + \frac{RT}{F} \ln (H^+) \qquad (2.36)$$

if the activity of liquid water is unity, as would be expected for dilute solutions.

In actual practice, the antimony electrode does not give highly accurate, reproducible results. Nevertheless, the ruggedness, simplicity, very low resistance, and low cost of the antimony electrode have made it useful, for example, in continuous industrial-process monitoring when high precision and accuracy are not required.

2.5 ION-SELECTIVE ELECTRODES

Except for the glass pH electrode, the preceding sections of this chapter have dealt with redox electrodes. Redox electrodes are limited in applicability to species that participate in an electrochemically reversible redox half-cell reaction. The time to reach equilibrium depends on concentration. In addition, at low concentrations the potential established may be influenced by another redox couple. Hence, redox electrodes are not useful for trace analysis. Ion-selective electrodes (ISEs) respond through the development of a potential across a membrane rather than a redox reaction. They are therefore useful for a wide range of ions. That the membranes can be made highly selective contributes to the popularity of ISEs as an analytical tool.

Properties of Ion-Selective Electrodes

The response equation for ISEs is similar to the Nernst equation for redox electrodes. For example, in a cell where the reference-electrode potential and the liquid-junction potential are constant and only ion "i" influences the ISE potential, the response is given by

$$E_{cell} = E_{constant} + E_{ISE} = E'_{constant} + \frac{RT}{nF} \ln a_i \qquad (2.37)$$

where n = charge of ion "i"

The electrodes therefore respond logarithmically to activity or concentration. One interesting result of this property is that the relative concentration error for direct potentiometric measurements is theoretically independent of the actual concentration. Unfortunately, the error is rather large—approximately $\pm 4n\%$ per millivolt uncertainty in measurement, which is a serious limitation of ISEs. Since potential measurements are seldom better than ± 0.1 mV total uncertainty, the best measurements for monovalent ions under near-ideal conditions are limited to about $\pm 0.4\%$ relative concentration error; for divalent ions, this error would be doubled. In particularly bad cases, for example where liquid-junction potentials may vary by 5 to 10 mV (as in solutions of high or variable ionic strength), the relative concentration error may be as high as 50%. This limitation may be overcome by using ISEs as endpoint indicators in potentiometric titrations (Sec. 2.6). At the cost of some extra time, accuracies and precisions of 0.1% or better are possible.

Another intrinsic property of ISEs is that they measure activities rather than concentrations; however, they can be made to determine concentrations by appropriate calibration procedures. Activity measurements may be more valuable in certain cases because activity rather than concentration determines rates of reactions and positions of equilibria. Therefore, ISEs can be used to study various physical phenomena.

Interferences

If an ISE membrane, such as the glass sensor in a pH electrode, were permeable only to the ion of interest, Equation 2.37 would always apply, and the electrode would be "ion specific" rather than "ion selective." (This would be true even if a rather small layer of the membrane had this permeability characteristic, which is probably the case with glass electrodes.) Even if the membrane were permeable to a wide range of ions, the electrode would be ion specific if there were no mechanism by which species other than the ion of interest could transfer from the sample solution onto the membrane. In reality, Equation 2.37 must be modified to

$$E_{cell} = E'_{constant} + \frac{RF}{nF} \ln(a_i + k_{ij}a_j^{n/z}) \tag{2.38}$$

where a_j = activity of the interferant ion of charge z
k_{ij} = selectivity coefficient

A separate $k_{ij}a^{n/z}$ term must be included for each interfering ion. The term k_{ij} is a composite of the relative mobilities of ions "i" and "j" in the membrane and the relative constants that describe the distribution of the ions between the solution and the membrane phases. Hence, it is sometimes possible to predict the ions that will interfere with a given ISE.

Selectivity coefficients can vary from about zero for no interference to about 10^3 for different electrodes and different interfering ions. For example, the sodium-ion glass electrode responds about 350 times more strongly to Ag^+ ($k_{ij} = 350$) and about 10^4 times less strongly to Cs^+ ($k_{ij} = 0.0001$) than to Na^+. As the selectivity coefficient is written in Equation 2.38, *small* values of k_{ij} mean a more selective electrode, one less affected by interferences. Selectivity coefficients can be regarded only as approximations, perhaps accurate to within an order of magnitude depending on experimental conditions and solution composition. Most selectivity coefficients, for example, show a dependence on total concentration as well as on the ratio of the interfering ion to the ion of interest.

Figure 2.5 shows a typical calibration curve that might be obtained with a calcium ion–selective electrode. Note that the electrode response is proportional to the logarithm of the calcium activity and has a Nernstian slope. In pure Ca^{2+} solutions, there is departure from the theoretical plot at about $6 \times 10^{-6} M$, as the electrode nears its detection limit—the concentration below which the electrode output is constant, regardless of the Ca^{2+} activity. The calcium electrode, however, also responds somewhat to the Na^+ ion. The curve obtained with Ca^{2+} solutions of varying activity, but also containing 1 M NaCl, exhibits a departure from the theoretical linear portion at about $3 \times 10^{-4} M$ Ca^{2+} activity—a much higher concentration than with the pure Ca^{2+} solutions. In this region, the calcium electrode is responding appreciably to the Na^+ ion present.

Types of Ion-Selective Electrodes

There are various ways of classifying ISEs. The approach taken here will be to subdivide the description of electrodes according to the composition of the membrane sensor.

Glass Electrodes. By varying the chemical composition of the thin, ion-sensitive glass membrane, glass electrodes can be prepared that are differentially responsive to (primarily monovalent) cations. The pH glass electrode already discussed is one member of this class; its general construction and mechanism of operation also hold for other glass electrodes. Table 2.4 shows the typical properties of some commercial glass electrodes. These electrodes show very little response to divalent cations. Note

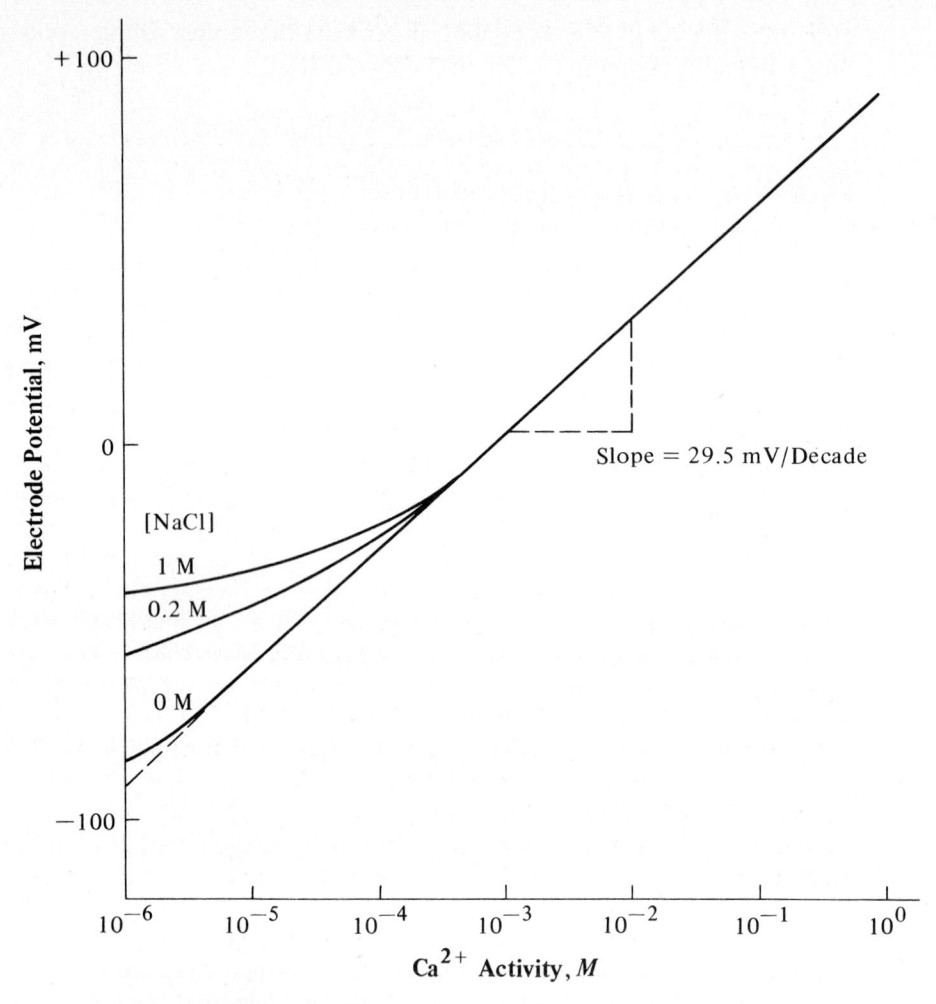

FIGURE 2.5. *Response of a calcium ion–selective electrode to pure $CaCl_2$ solutions, and to $CaCl_2$ solutions with 0.2 or 1 M NaCl also present. Redrawn from J. W. Ross, Jr., in* Ion-Selective Electrodes, *R. A. Durst, ed., NBS Special Publication 314, Chap. 2, Washington, D.C.: U.S. Government Printing Office, 1969, by permission of the author and the National Bureau of Standards.*

that in order to measure Na^+ with a sodium-ion glass electrode, the H^+ concentration must be adjusted to a sufficiently low value to minimize interference.

Solid-State Crystalline and Pressed-Pellet Electrodes. There are two basic types of crystalline-based electrodes. The first is exemplified by the fluoride electrode, with a single europium-doped LaF_3 crystal as the sensor. The LaF_3 crystal is sealed into the end of a rigid cylindrical electrode body made of plastic; an internal electrolyte solution, typically NaF and NaCl, and an internal reference electrode complete the

TABLE 2.4. *Typical Properties of Commercial Glass Ion-Selective Electrodes*

Type of Electrode	Concentration Range, M	Glass Composition	Relative Electrode Response
H^+	10^0–10^{-13} (with corrections)	Li_2O–BaO–La_2O_3–SiO_2 or Na_2O–CaO–SiO_2 (21%) (6%) (72%)	$H^+ \gg Li^+, Na^+ > K^+$
Na^+	10^0–10^{-6}	Li_2O–Al_2O_3–SiO_2 or Na_2O–Al_2O_3–SiO_2 (11%) (18%) (71%)	Ag^+ (350) > H^+ (100) > Na^+ (1) $\gg Li^+, K^+$, Cs^+ (0.001) > NR_4^+, Tl^+ (0.0003) > Rb^+, NH_4^+ (0.00003)
General cation (monovalent)	10^0–10^{-5}	Na_2O–Al_2O_3–SiO_2 (27%) (4%) (69%)	K^+ (33), Rb^+ (17), NH_4^+ (11), Na^+ (4), H^+ (3), Li^+ (2), Cs^+ (1), Tl^+, Cu^+, R_4N^+

construction (Fig. 2.6). At room temperature, LaF_3 is a pure fluoride-ion conductor, thus remarkably free from interferences—an almost "specific" electrode. Virtually the only interferences encountered with the electrode are at low pH, where the fluoride ion forms HF ($pK_a \approx 3$), and at pH values greater than about 8, where OH^- interferes. The fluoride electrode was the first to capture the minds of chemists, because

FIGURE 2.6. *Schematic representation of the construction of a crystal-sensor ion-selective electrode.*

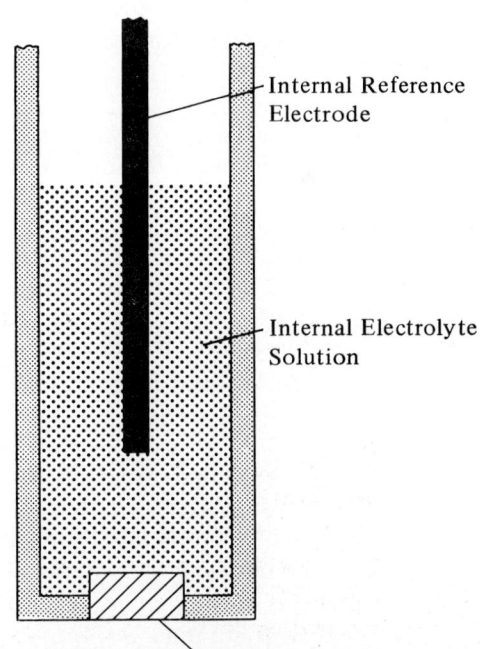

of its reliability and the difficulty of measuring fluoride otherwise. There are probably several hundred publications dealing with the analytical determination of fluoride in such samples as municipal drinking and waste waters, seawater, air particulates, bone, minerals, organic materials, plant tissues, biological fluids, soils, toothpastes, and so forth.

As mentioned previously, membrane response is linked to mobility. Mobility in a crystal is favored by low charge and small ion size. The LaF_3 crystal therefore cannot be used directly to detect La^{3+}. An indirect measurement can be made by equilibrating LaF_3 powder in a sample containing La^{3+}. The resulting F^- activity is related to the La^{3+} activity in the sample and the solubility product of LaF_3, $(F^-) = [K_{sp}/(La^{3+})]^{1/3}$. A measurement of the F^- activity with the fluoride electrode and a simple calculation yields the La^{3+} activity of the sample. Similar procedures work for other crystal-membrane ISEs.

The second class of crystalline sensors is the easily fabricated, low-resistance, selectively permeable cast-disk and pressed-pellet membranes based on Ag_2S. Silver sulfide is an ionic conductor in which silver ions are the mobile species. By itself, it can be used to detect silver ions. The potential-determining mechanism in an Ag_2S electrode is due to the very low solubility product of Ag_2S ($K_{sp} = 10^{-51}$). The silver-ion activity of the test solution on one side of the membrane and the (constant) silver-ion activity of the inner filling solution (Fig. 2.6) establish (assuming no interfering ions in solution) a cell that responds to the Ag^+ activity of the test solution by Equation 2.37.

By making mixed pellets containing AgX–Ag_2S, where X = Cl, Br, I, or SCN, one has an electrode responsive to one of these particular anions. The silver-ion activity at the surface of the electrode is controlled by the activity of X^- in solution via its solubility equilibrium:

$$AgX \rightleftharpoons Ag^+ + X^- \tag{2.39}$$

This in turn controls the electrode potential by being coupled with the Ag_2S solubility equilibrium. By substituting into Equation 2.37,

$$E = E_{constant} + \frac{RT}{F} \ln (Ag^+) = E_{constant} + \frac{RT}{F} \ln \frac{K_{sp}}{(X^-)} = E'_{constant} - \frac{RT}{F} \ln (X^-)$$

$$\tag{2.40}$$

These electrodes, of course, are also responsive to Ag^+ or S^{2-} ions.

In the same manner, electrodes responsive to Cu^{2+}, Cd^{2+}, and Pb^{2+} cations (M^{2+}), which form insoluble metal sulfides, can be made by mixing the appropriate metal sulfide with Ag_2S. In these cases, the M^{2+} activity controls the sulfide-ion activity in solution, which in turn controls the Ag^+ activity and the electrode response.

The selectivity and properties of these various electrodes are basically a function of the solubility products involved. Anything with a lower solubility product than the ion being determined will interfere. For this reason, AgCl membranes are subject to greater interference than the less soluble AgI, AgBr, etc., membranes. Iodide and bromide, respectively, must be reduced to levels less than 5×10^{-7} and 3×10^{-3} times the lowest chloride activity anticipated when using an AgCl membrane. Cast pellets of silver halides alone can also serve as membranes for the respective halide-selective electrodes, but they function less well than the mixed-crystalline sensors. Typical properties of some commercial crystalline-sensor ISEs are listed in Table 2.5.

Another variation of the polycrystalline type of electrode is the Pungor design [8], which involves suspending the active membrane material in a matrix such as silicone rubber. The proportion of material embedded in the matrix must be high enough to produce physical contact between particles.

Liquid-Membrane Ion-Exchange Electrodes. One older design of a liquid-membrane ion-exchange electrode consists of two concentric cylindrical tubes constructed of inert plastic (Fig. 2.7). The inner tube holds the internal reference electrode and an aqueous electrolyte solution containing the ion of interest. The outer compartment contains a charged or neutral organic ion exchanger dissolved in an organic solvent that wicks into a thin hydrophobic membrane; this then forms a thin organic-membrane phase separating two aqueous solutions. The membrane may be, for example, a cellulose-acetate Millipore® filter. An ion-exchange equilibrium is set up at both the inner and outer surfaces of the membrane, and the difference in activity of the ion of interest in the inner electrolyte and outer test solutions gives rise to the potential response of the electrode.

A newer series of commercial ion-exchange electrodes has the entire ion-exchanger assembly contained within a replaceable module that screws into an electrode body. Although each sensor module has only a finite usable lifetime, the need for periodic disassembly and refilling of electrodes is eliminated.

The selectivity and sensitivity of these electrodes are determined primarily by

TABLE 2.5. *Typical Properties of Commercial Crystalline Solid-State Electrodes*

Electrode	Concentration Range, M	Interferences[a]
F^-	sat'd–10^{-6}	$OH^- < 0.1\ F^-$
Ag^+ or S^{2-}	10^0–10^{-7}	$Hg^{2+} < 10^{-7}\ M$
Cl^-	10^0–5×10^{-5}	$2 \times 10^{-7}\ CN^-;\ 5 \times 10^{-7}\ I^-;\ 3 \times 10^{-3}\ Br^-;\ 0.01\ S_2O_3^{2-};$ 0.12 NH_3; 80 OH^-
Br^-	10^0–5×10^{-6}	$8 \times 10^{-5}\ CN^-;\ 2 \times 10^{-4}\ I^-;\ 400\ Cl^-;\ 2\ NH_3;$ $3 \times 10^4\ OH^-$
I^-	10^0–5×10^{-8}	0.4 CN^-; 500 Br^-; $10^6\ Cl^-$
CN^-	10^{-2}–8×10^{-6}	0.1 I^-; 5000 Br^-; $10^6\ Cl^-$
SCN^-	10^0–5×10^{-6}	$10^{-6}\ I^-$; 0.003 Br^-; 0.007 CN^-; 20 Cl^-; 100 OH^-
Cd^{2+}	10^{-1}–10^{-7}	Ag^+, Hg^{2+}, Cu^{2+} absent
Cu^{2+}	10^{-1}–10^{-8}	Ag^+, Hg^{2+} absent
Pb^{2+}	10^{-1}–10^{-6}	Ag^+, Hg^{2+}, Cu^{2+} absent

a. If M is not specified, the value represents the maximum ratio (interferant ion/sensed ion) for no interference.
Source: Selected values have been taken from data supplied through the courtesy of Orion Research, Incorporated: *Guide to Ion Analysis* © 1983.

FIGURE 2.7. *Schematic representation of the construction of a liquid-membrane ion-exchange electrode.*

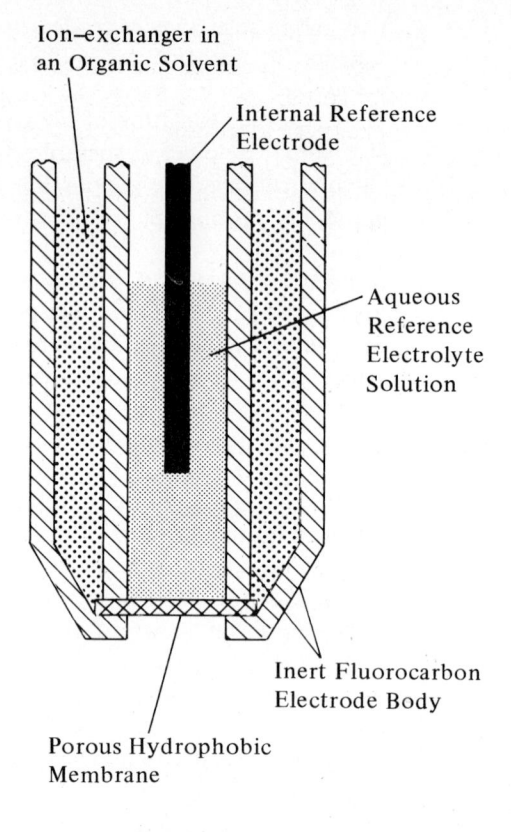

the selectivity of the particular organic ion exchanger for the ion of interest and, secondarily, by the organic solvent used to dissolve the exchanger. Most of the interference is related to the relative stability of the complex formed between the interfering ion and the organic ion exchanger in the membrane because the mobilities of the complexes within the liquid membrane are generally similar.

Phosphate diesters ($(RO)_2PO_2^-$, with R-groups in the C_8–C_{16} range) dissolved in a relatively polar solvent such as dioctylphenylphosphonate show good selectivity for Ca^{2+} in the presence of Na^+, as well as for Ca^{2+} in the presence of other alkaline-earth ions. Less polar solvents such as decanol produce electrodes that give virtually identical response to all the alkaline-earth ions; electrodes of this type are useful for the determination of water hardness.

Ion exchangers of the form $R-S-CH_2COO^-$ (in which the sulfur and carboxylate groups can readily form a 5-member chelate ring with a heavy-metal ion) show good selectivity for Cu^{2+} and Pb^{2+}.

Certain positively charged ion exchangers can be used for anion-selective electrodes. Charged metal salts of appropriately substituted orthophenanthrolines—$M(o\text{-phen})_3^{2+}$—result in good electrodes for nitrate, fluoroborate, or perchlorate by forming ion-association complexes with these anions. The ClO_4^- electrode, in particular, has few interferences; and perchlorate is a difficult ion to measure by almost any other method. A dimethyl-distearyl-ammonium ion, R_4N^+, can be used as an ion exchanger in an electrode that has fair selectivity for chloride.

The organic ion exchanger used need not be charged. Neutral organic ligands, typically dissolved in a low-dielectric liquid such as decane, can show good selectivity for certain cations. These lipid-soluble molecules usually contain a ring arrangement of oxygen atoms that can replace the aqueous hydration shell around cations and thus extract them into organic solvents. This provides a mechanism for transport of the cations across the membrane and, thereby, an electrode response to the cation. The antibiotic valinomycin and the macrotetrolides such as nonactin and monactin are highly selective natural products that can be used for the measurement of K^+ or NH_4^+. The valinomycin electrode, in particular, shows excellent selectivity for K^+ over Na^+ (4000:1), H^+ (20,000:1), and divalent metal ions (approx. 5000:1), which is much better than the best potassium-sensitive glass electrode. The actin-based membranes are about four times more responsive to NH_4^+ than to K^+. Synthetic cyclic polyethers ("crown" compounds) can be used as ion exchangers for univalent and for some alkaline-earth cations, although they generally show much less specificity than the above natural compounds.

The lower limit of detection for liquid ion-exchange electrodes is determined primarily by the solubility of the ion exchanger in aqueous media. As with crystalline solid-state electrodes, Nernstian response is obtained until the activity of the solution is within a factor of about 100 of the solubility of the membrane salt. Then the response deviates and levels off at a constant potential reflecting this solubility.

Typical characteristics of some commercially available liquid ion-exchange electrodes are presented in Table 2.6.

TABLE 2.6. *Typical Properties of Selected Commercial Liquid-Liquid Ion–Exchange Electrodes*

Electrode	Concentration Range, M	Interferences[a]
Ca^{2+}	10^0–5×10^{-7}	10^{-5} Pb^{2+}; 0.004 Hg^{2+}, H^+; 0.006 Sr^{2+}; 0.02 Fe^{2+}; 0.04 Cu^{2+}; 0.05 Ni^{2+}; 0.2 NH_4^+, Na^+; 0.3 Li^+; 0.4 K^+; 0.7 Ba^{2+}; 1.0 Zn^{2+}, Mg^{2+}
Cl^-	10^0–5×10^{-6}	10^{-6} ClO_4^-; 5×10^{-5} I^-, NO_3^-; 10^{-4} OH^-; 2×10^{-4} SO_4^{2-}; 0.001 Br^-, HCO_3^-; 0.003 OAc^-; 0.007 F^-
BF_4^-	10^0–7×10^{-6}	5×10^{-7} ClO_4^-; 5×10^{-6} I^-; 5×10^{-5} ClO_3^-; 5×10^{-4} CN^-; 0.001 Br^-, NO_2^-; 0.005 NO_3^-; 0.003 HCO_3^-; 0.05 Cl^-; 0.2 OAc^-; 0.6 F^-; 1.0 SO_4^{2-}
NO_3^-	10^0–7×10^{-6}	10^{-7} ClO_4^-; 5×10^{-6} I^-; 10^{-4} CN^-; 7×10^{-4} Br^-; 0.002 I^-; 0.02 ClO_3^-; 0.04 CN^-, Br^-; 0.05 NO_2^-, NO_3^-; 0.6 F^-; 1.0 SO_4^{2-}
ClO_4^-	10^0–7×10^{-6}	0.002 I^-; 0.02 ClO_3^-; 0.04 CN^-, Br^-; 0.05 NO_2^-, NO_3^-; 2.0 HCO_3^-, CO_3^{2-}, Cl^-, OAc^-, F^-, SO_4^{2-}
K^+	10^0–10^{-6}	3×10^{-4} Cs^+; 0.006 NH_4^+, Tl^+; 0.01 H^+; 1.0 Ag^+; 2.0 Li^+, Na^+
Water Hardness	10^0–6×10^{-6}	3×10^{-5} Cu^{2+}, Zn^{2+}; 6×10^{-5} Fe^{2+}; 10^{-4} Ni^{2+}; 4×10^{-4} Sr^{2+}; 6×10^{-4} Ba^{2+}; 0.03 Na^+; 0.1 K^+

a. At 10^{-3} M of the analyte ion, level of interferant ion (M) for a 10% error.
Source: Selected values taken from data supplied through the courtesy of Orion Research, Incorporated: *Guide to Ion Analysis* © 1983.

Liquid-membrane electrodes are more sensitive to the solution environment than are solid-state electrodes. Their usable temperature range, generally 0 to 50°C, is more restricted than that of solid-state electrodes. Above this temperature, water may permeate the membrane, or the membrane liquids may bleed excessively into the aqueous solution. Normally, it is best to restrict use of the electrodes to purely aqueous media so that the membrane or ion exchanger will not dissolve in the test solution. The electrodes must be recharged every few weeks with new ion exchanger and internal electrolyte. The membranes cannot be stored for long times. Otherwise they are handled much as glass or solid-state electrodes are.

Enzyme-Substrate Electrodes. Electrodes that can respond to a variety of organic and biological compounds are constructed by coating the surface of an appropriate ISE with an enzyme immobilized in some matrix. Perhaps the most well known of these is the urea electrode [9], which makes use of the enzyme urease to hydrolyze urea (the "substrate"):

$$\text{Urea} + H_2O \xrightarrow{\text{urease}} HCO_3^- + NH_4^+ \qquad (2.41)$$

In this case, the urease is physically entrapped in a polyacrylamide matrix polymerized on the surface of an ammonium-ion glass electrode. The enzyme-gel matrix is supported on the electrode by a sheer Dacron or nylon gauze, about the thickness of a nylon stocking, or it is held by a thin semipermeable cellophane sheet. The urea diffuses to the urease-gel membrane, where it is hydrolyzed to produce ammonium ion. Some of the ammonium ion diffuses through the thin membrane to the electrode surface, where it is monitored by the ammonium-sensitive electrode. The urea electrode is fairly stable, sensitive, specific for urea, has a usable lifetime of 2 to 3 weeks before a new gel layer must be prepared, and has a fairly fast response time (<120 sec). Linear calibration curves can be obtained from about 10^{-4} to 10^{-2} M urea.

There are literally thousands of enzyme-substrate combinations that yield products which could theoretically be measured with ISEs. The high sensitivity of the electrodes and the specificity of enzymes can thus be coupled to produce sensors of value in many biomedical applications. Electrodes responsive to the following substrates are among those that have been described in the literature: urea, L-amino acids, D-amino acids, asparagine, glutamine, amygdalin, creatinine, penicillin, and cholesterol.

With some modifications, an electrode system can be made responsive to *enzyme* levels in test solution by surrounding the ion-sensitive membrane with *substrate* molecules. For example, an electrode that will measure the enzyme activity of cholinesterases in blood fractions has been described. Generally, such electrodes must be designed to replenish the substrate since it is consumed in the reaction.

Gas-Sensing Electrodes. There are several gas-sensing electrodes available, which function by interposing a thin, highly gas-permeable membrane between the test solution and an appropriate sensing element. The dissolved gas passes through the membrane into a small volume of internal filling solution, where a chemical equilibrium is established between the gas dissolved in the test solution and the internal electrolyte. The internal sensor monitors the changes in this equilibrium, and thus produces an output proportional to this concentration of dissolved gas. Potentiometric electrodes for ammonia (ammonium), sulfur dioxide (sulfite), nitrogen oxide (nitrite), and carbon dioxide (carbonate) are available.

The operation of gas-sensing electrodes can be illustrated by considering the sulfur dioxide electrode, which responds directly to dissolved SO_2. Sulfite (SO_3^{2-}) and bisulfite (HSO_3^-) are measured by acidifying the sample to convert these species to SO_2. Dissolved SO_2 diffuses through the gas-permeable membrane until an equilibrium is established in the internal filling solution by the reaction of SO_2 with water:

$$SO_2 + H_2O \rightleftharpoons HSO_3^- + H^+ \qquad (2.42)$$

The hydrogen-ion level is then sensed by the internal sensing element (a conventional glass pH electrode), and is directly proportional to the level of SO_2 in the sample. The electrode has a usable range of 10^{-6} to 10^{-2} M SO_2, and has very few interferences, essentially only volatile weak acids such as acetic acid and HF.

The SO_2 level of stack gases can be measured by drawing a known volume of gas through an absorbing solution, acidifying an aliquot, and measuring SO_2 directly. The sulfite level of pulping liquors can be determined directly after acidification.

Ion-Selective Microelectrodes. Ion-selective electrodes can be fabricated in microassemblies for use in such applications as measuring ion activities in microsamples (10^{-3} mL) and as direct probes in tissues, tubules, capillaries, or even within individual cells. Microelectrodes can be fabricated with sensors made of glass, liquid ion exchanger, or solid-state crystals [10–12].

Closed-tip glass microelectrodes are fabricated by fusing a very small active tip onto a body of insulating glass or by coating the electrode stem with insulation, leaving only the tip exposed. Tip diameters of less than 1.5 μm exposed to a length of 3 to 5 μm can be achieved. Electrodes of this type have been used successfully to measure alkali-metal ion activities in frog skeletal muscle. On a somewhat larger scale, miniaturized glass electrodes have been built within a small hypodermic syringe needle, intended for continuous monitoring of fetal blood pH during birth.

Open-tip liquid ion-exchange microelectrodes can be fabricated by pulling borosilicate glass capillary tubing to an appropriate tip diameter (0.5 to 1 μm), rendering the interior of the tip hydrophobic by coating the surface with an organic silicone compound, and filling the tip with an ion exchanger appropriate for the ion of interest (K^+, Cl^-, Ca^{2+}, etc.). The internal circuit is completed with an electrolyte solution and an Ag/AgCl reference electrode. Such electrodes can measure ion activities in single *aplasia* neurons (0.5 μL total volume), when combined with a suitable micro reference electrode.

The desire to develop miniature ISEs has led to the development of sensors that do not respond by the general mechanism of membrane electrodes but nevertheless provide potential-versus-ion activity data that follow Equation 2.38. Coated-wire electrodes are representative of this class [13]. They are prepared by covering a thin wire, such as platinum, with a polymer into which a crystalline powder has been suspended or a liquid ion exchanger has been dissolved. In addition to being easy to prepare on a miniature scale, such electrodes are inexpensive. They also provide a good student laboratory experiment on potentiometry [14]. Unfortunately, the response mechanism has not yet been established.

pH Meters

Other than the appropriate electrodes, the only major piece of instrumentation needed to perform pH and ISE measurements is a pH meter. This is a very high-impedance electronic voltmeter that draws negligible current from the reference-indicator electrode pair; thus, no error arises from the voltage drop across the inherent (usually high) resistance of the electrochemical cell. A great variety of pH meters are commercially available, with an even greater variety of features. In general, pH meters can be divided into four classes based on price and "accuracy," although there is considerable overlap between classes: *utility* (portable), *general-purpose*, *expanded-scale*, and *research* grades.

Utility-grade pH meters typically cost about $300. Most are battery operated, and thus portable; they generally offer enough sensitivity to be used in many quality-control applications and out in the field. Their relative accuracy is about ± 0.1 pH unit, and they have taut-band meter movements. General-purpose pH meters are more often line operated, and cost about $400–$900. For the extra cost, they usually offer better stability and accuracy (± 0.05 pH or ± 3 mV), larger taut-band scales, and extra features such as a recorder output, millivolt scales, and a constant-current jack for performing polarized electrode measurements or dead-stop titrations such as the Karl Fischer titration for water determination.

For increased accuracy, expanded-scale pH meters (retailing for about $600–$1100, depending on features) generally offer accuracy of about ± 0.01 pH unit. Usually, any 1.0-pH unit or 100-mV range is expandable to full scale; many types that fall into this class have digital (4-digit) displays. For the most demanding applications, research-grade pH meters offer a relative accuracy of about 0.002 pH or ± 0.1 mV (readability to 0.001 pH), for about $1000–$1800. Most of these have digital (5-digit) displays, full-range expanded-scale operation, recorder outputs, and highly adaptable slope and calibration controls. The more sophisticated ones have microprocessors to aid in collection and processing of data.

Because there is such a wide variety of specialized features available on specific pH meters, manufacturers' literature must be consulted to select the best possible unit for the intended use.

Applications of Ion-Selective Electrodes

Ion-selective electrodes are used extensively as quantitative analytical probes in such diverse areas as air and water pollution, fundamental biomedical research, oceanography, geology, agriculture, and clinical chemistry. The ease of making the measurements can mask certain problems that must be considered in order to obtain good data with ISEs.

Practical Considerations

In analytical chemistry, concentrations rather than activities are generally desired. Because ISEs respond to activity, the standard solutions and the samples must be matched in terms of the activity coefficient of the analyte ion. This requires that the ionic strengths of these solutions be essentially the same. Normalization of the ionic strengths of solutions is accomplished by the addition of a concentrated electrolyte to all samples and standards. This electrolyte must consist of ions that do not interfere with the measurement. The ionic-strength adjustor is often a buffer since pH control is also required for many ISEs.

An example is the use of a total ionic-strength adjustment buffer (TISAB) for the fluoride electrode. The fluoride TISAB contains a 1 M acetic acid/sodium acetate buffer at pH 5, 10^{-3} M citrate, and 1 M NaCl. These components have specific roles. The buffer decreases the OH^- concentration to a level at which its interference with the fluoride ion–selective electrode is negligible; it also prevents significant formation of HF (which would not cause a potential response at the ISE). The citrate complexes iron and aluminum and therefore prevents these cations from tying up fluoride. The NaCl serves to increase the ionic strength. The solutions are generally diluted 1:1 with the TISAB prior to measurement. This general approach is common with ISEs, but the makeup of the ionic-strength adjustor varies.

A related consideration is that the ISE responds to the "free"-ion activity rather than to the total concentration of the particular species. For example, only free F^- is determined instead of the total fluorine in the sample, which could include complexed F^- or covalently bound F. Before modifying the sample in any way, the analyst must decide whether the equilibrium activity concentration of the ion in the "raw" sample is desired or whether an attempt should be made to determine the total concentration. If the former is the goal, the sample would be measured without adjustment, and the standard solutions should be prepared to simulate the sample background in some appropriate manner (which may be impossible). If the total concentration is the objective, releasing agents, pH adjustment, ionic-strength normalization, standard-addition calibration, and possibly chemical digestion, are appropriate. Titration (perhaps after preliminary treatment of the sample) with ISE detection is another route to the determination of total concentration.

The temperature, stirring rate, and choice of the reference electrode are other considerations in the use of ISEs. The temperature directly influences the electrode response (Eqn. 2.38). Usually, if all the solutions are at room temperature, the measurements can be made with little concern. An exception would be if a physical constant such as a complexation constant were being measured where a high degree of precision and accuracy is desired; in such cases, a thermostatted cell should be used. If the samples are not at room temperature, care must be taken to match the standards and samples; for example, the samples could be immersed in a water bath for a short time. Finally, certain ISEs do not perform well or have limited lifetimes at elevated temperature. This is particularly true with liquid-membrane electrodes. The manufacturers' literature should be consulted in this regard.

The response of ISEs depends somewhat on stirring rate, especially when they are used near their detection limits. Vigorous stirring, but without vortex formation, should be used. It is important that bubbles not be trapped on the sensor membrane.

Reference-electrode junctions can leak significant quantities of their filler solutions into the sample. Either a reference should be selected that does not contain interfering ions, or a double-junction system should be used (i.e., an extra salt bridge between sample and reference).

Quantification Methods

Concentrations are determined with ISEs either by matching the measured potential in the sample to a *calibration curve* or by using the *standard addition method*. Figure 2.8 illustrates a typical calibration curve for the fluoride ion–selective electrode. Note that the *x*-axis is logarithmic, as required by Equation 2.38; in addition, because fluoride is an anion the slope is negative. In using the calibration-curve method, it is important to control ionic strength, temperature, and stirring rate. In addition, the presence of interfering ions in the sample will cause a positive error, and if a portion of the analyte is held as a metal complex, weak acid, or the like, the measured concentration (the free ion) will be less than the total concentration of that element.

The standard addition method is an alternative to the use of a calibration curve. Often, a single addition of a prescribed volume of a standard solution is added to a known volume of the sample. Alternatively (and preferably), multiple additions can be made. The potential is measured before and after the addition(s).

A reading of the initial potential, E_1, is taken on a sample of concentration c:

$$E_1 = E_{\text{constant}} + \frac{RT}{nF} \ln(fc) \qquad (2.43)$$

where f = activity coefficient

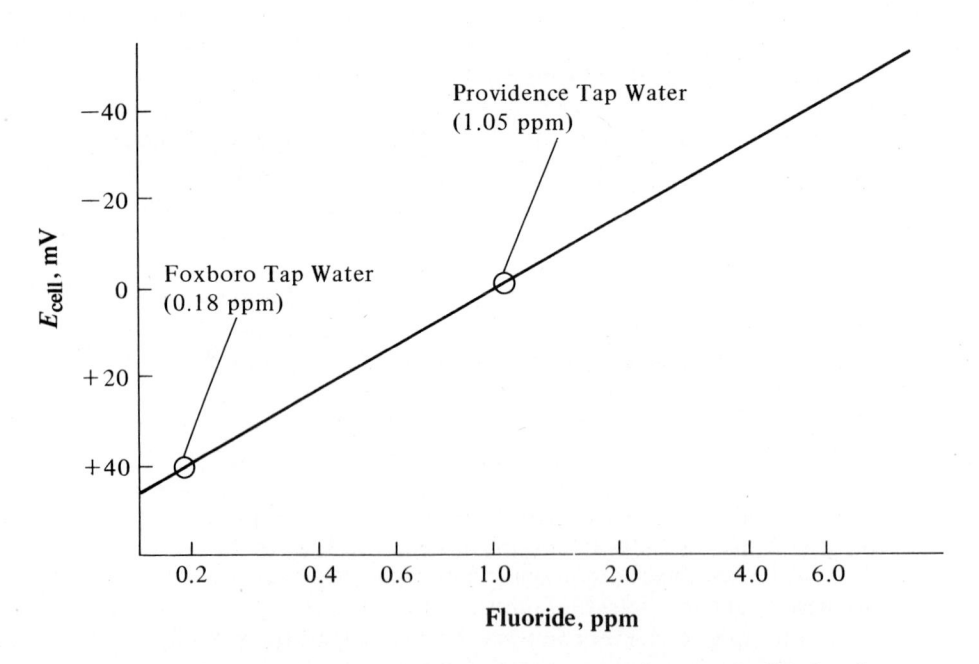

FIGURE 2.8. *Determination of fluoride in municipal water supplies by the calibration-curve method using ionic-strength buffering. Redrawn from T. S. Light, in* Ion-Selective Electrodes, *R. A. Durst, ed., NBS Special Publication 314, Chap. 10, Washington, D.C.: U.S. Government Printing Office, 1969, by permission of the author and the National Bureau of Standards.*

Next, a known amount of the ion of interest is added to the test solution so that the concentration is changed by a known amount, Δc:

$$E_2 = E_{constant} + \frac{RT}{nF} \ln f(c + \Delta c) \tag{2.44}$$

Then the potential is read again. Usually, the standard addition is a small volume of a concentrated solution; so the total solution volume and ionic strength do not change appreciably. Equations 2.43 and 2.44 can be combined to give

$$\Delta E = E_2 - E_1 = \frac{RT}{nF} \ln \frac{(c + \Delta c)}{c} \tag{2.45}$$

Equation 2.45 is solved for c by rearranging the elements and then taking the antilogarithm.

The most accurate determinations are made when Δc is such that the total concentration is approximately doubled. The only requirement is that the electrode be in a linear portion of its calibration curve over the concentration range of interest. The slope of the calibration curve need not be precisely equal to the Nernst factor, RT/nF; if it is not, the empirically determined slope of the calibration curve, S, can be substituted. In solving Equation 2.45, the slope is negative if the analyte is an anion.

When sequential standard additions are made, a graphical method is used to determine the concentration of the unknown. The technique is based upon Gran's method [15] of linearizing Equation 2.37. Rearrangement of Equation 2.43 yields

$$\frac{E}{S} = \frac{E_{constant}}{S} + \log(fc) \tag{2.46}$$

where $S = 2.303 RT/nF$ (or the empirically determined slope)

By taking the antilogarithms, one obtains

$$10^{E/S} = (10^{E_{constant}/S})(fc) \tag{2.47}$$

The constants can be combined to yield

$$10^{E/S} = kc \tag{2.48}$$

where $k = (10^{E_{constant}/S}) f$

A plot of $10^{E/S}$ versus c is linear as long as f and S remain constant. Therefore, with sequential standard additions the antilogarithm of E/S is plotted on the y-axis, versus Δc on the x-axis. The intercept on the x-axis is $-\Delta c'$ (the intercept occurs negative of the origin). The value $-\Delta c'$ is the amount of material that would have had to be removed from the original sample to lower c to zero; hence, $-\Delta c' = c$, the original concentration. Neither standard addition approach will eliminate errors due to interfering ions.

An example of the use of the standard addition method is the determination of ammonia in aquaria and seawater [16] using an ammonia-gas electrode. To a 100-mL sample is added a sufficient number of NaOH pellets to raise the pH above 11 (to convert all ammonium ion to ammonia). The equilibrium potential is read. Then a sufficient volume of standard NH_4Cl solution to approximately double the concentration is added, after which the new equilibrium potential is read. The effective detection limit for this method is approximately 1 µg NH_3 per liter (ppb in dilute aqueous solution); but below about 10 µg/L, considerable time is required for the electrode to stabilize, making the method somewhat impractical in the range 1 to 10 µg/L. Results compare favorably with the phenolhypochlorite spectrophotometric method for ammonia.

The copper-ion content of tap water can be determined with a copper ion–selective electrode using a multiple standard addition procedure [17]. Tap water is mixed 1:1 with a complexing antioxidant buffer (sodium acetate, acetic acid, sodium fluoride, and formaldehyde) to buffer the pH at 4.8, to complex the Cu^{2+} uniformly with acetate, and to complex the Fe^{3+} interferant with fluoride. Copper in tap water can be determined down to about 9 mg/L with a standard deviation of about ±8%. The recovery of Cu^{2+} added to natural waters—an indication of the accuracy of the method—averaged 103% for samples in the range of 3 to 60 mg/L.

A summary of applications of ISEs is provided in Orion Research's *Guide to Ion Analysis* [18].

Advantages and Disadvantages of Ion-Selective Electrodes

After this discussion of the types and applications of ISEs, it is well to consider and review the general advantages and disadvantages of these electrodes as analytical tools. The linear working range of many electrodes is quite large, generally from four to six orders of magnitude. Electrodes will function well in colored or turbid samples, whereas spectral methods generally will not.

In most cases, electrode measurements are reasonably rapid, with equilibrium being reached in less than a minute; but in some cases, usually in very dilute solutions, slow electrode response may require 15 min to an hour for equilibrium. The normally rapid response of ISEs makes them suitable in kinetic studies and for monitoring changes in flowing process-streams. The equipment used is simple, inexpensive, and can be made portable for field operations. The method is virtually nondestructive of the sample (once the sample is in a suitable liquid state), and can be used with samples that are very small (<1 mL).

The greatest disadvantage of ISEs is that they are subject to a rather large number of interferences. The electrodes respond more or less strongly to several ions; and various chemical interferences are possible, including chelation, complexation, and ionic-strength effects. Generally, frequent calibration is necessary. Ion-selective electrodes are not ultra-trace level sensors; some electrodes are good down to only about $10^{-4}\,M$, and most are not usable below around $10^{-6}\,M$. An advantage or disadvantage, depending on the situation, is that ISEs are responsive to only the free-ion form of an element.

2.6 POTENTIOMETRIC TITRATIONS

Many different types of ISEs can be used as endpoint indicators in potentiometric titrations. For example, an acid-base titration can be performed with a glass pH electrode as an endpoint detector, rather than with a phenolphthalein indicator; or calcium can be titrated with EDTA using a calcium ion–selective electrode. During such a titration, the *change* in potential of a suitable indicator electrode is observed as a function of the volume added of a titrant of precisely known concentration. Because it is the change in potential rather than the absolute potential that is of interest, liquid-junction potentials and activity coefficients have little or no effect. Perhaps the primary advantage of potentiometric titrations is that (with the addition of classical titration procedures) they generally offer a large increase in accuracy and precision; $\pm 0.1\%$ levels are not uncommon. (There are, however, certain disadvantages: the increase in analysis time and operator attention required, and the difficulties associated with the preparation, standardization, and storage of standard titrant solutions.)

Another advantage of potentiometric titrations is that substances to which the electrode does not respond can be determined, if the electrode responds to the titrant or to some low level of an indicator substance that has been added to the solution. For example, low levels of Al^{3+} can be determined by titration with standard fluoride solution, using a fluoride electrode [19]. EDTA and other chelators can be determined by titration with standard calcium or copper solution.

The experimental apparatus for a potentiometric titration can be simple: Only a pH or millivolt meter, a beaker and magnetic stirrer, reference and indicator electrodes, and a buret for titrant delivery are needed for manual titrations and point-by-point plotting. Automatic titrators are available that can deliver the titrant at a constant rate or in small incremental steps and stop delivery at a preset endpoint. The instrument delivers titrant until the potential difference between the reference and indicator electrodes reaches a value predetermined by the analyst to be at, or very near, the equivalence point of the reaction. Alternatively, titrant can be delivered beyond the endpoint and the entire titration curve traced. Another approach to automatic potentiometric titration is to measure the amount of titrant required to maintain the indicator electrode at a constant potential. The titration curve is then a plot of volume of standard titrant added versus time, and is very useful, for example, for kinetic studies. The most extensive use of this approach has been in the biochemical area with so-called *pH stats*—a combination of pH meter, electrodes, and automatic titrating equipment designed to maintain a constant pH. Many enzymes consume or release protons during an enzymatic reaction; therefore, a plot of the volume of standard base (or acid) required to maintain a constant pH is a measure of the *enzyme activity*, the amount of enzyme present.

Because potentiometric titrations are an old and well-known technique, particularly in regard to acid-base and oxidation-reduction titrations, only a few selected examples will be presented here. For more detailed treatments, the reader is urged to consult the bibliography at the end of the chapter. It will be assumed that the reader is already familiar with titration curves and their calculation from ionic equilibria and other pertinent data.

Acid-Base Titrations

Acid-base titrations can be performed with a glass indicator electrode and a suitable reference electrode. Titrations of strong acids and bases with suitable strong titrants are relatively simple, since there is a relatively large and abrupt change in pH at the endpoint. Calculated titration curves for the titration of various acids (0.1 M concentration) with 0.1 M sodium hydroxide are shown in Figure 2.9. Note that as the acid titrated becomes weaker (pK_a increases), the sharpness and magnitude of the endpoint break decrease. A somewhat similar situation occurs when the concentration of the substance titrated (and the titrant) is decreased. As a rough rule of thumb, for the titration of a strong acid (or base) with strong titrant, the concentration of the substance titrated should be greater than about 3×10^{-4} M for a 0.1% accuracy. If a 1% accuracy is sufficient, the concentration can be decreased an order of magnitude. For titration of a weak acid with strong base, the product of the acid concentration and its dissociation constant, K_a, should be greater than about 10^{-7} for a 0.1% accuracy, 10^{-9} for a 1% accuracy.

The acetic acid content of household vinegar can be determined by potentiometric titration with sodium hydroxide. Mixtures of carbonate and bicarbonate can be analyzed by titration with HCl.

Acid-Base Titrations in Nonaqueous Solvents. The apparent acidity or basicity of a compound is strongly dependent on the acid-base properties of the solvent. For example, very strong acids such as HCl and HNO_3 cannot be individually titrated in water because water is sufficiently basic that these acids appear to be totally ionized. Very weak bases, such as amines, cannot be successfully titrated with strong acid in water. Many acids or bases that are too weak for titration in an aqueous medium, however, become amenable to titration in appropriate nonaqueous solvents. As a consequence, there are now many neutralization methods that call for solvents other than water.

The earliest advantages recognized for nonaqueous solvents arose from the use of *amphiprotic* solvents, which have both acidic and basic properties. The prototype is water. Significant differences in acid-base properties are seen in both protogenic

FIGURE 2.9. *Theoretical potentiometric titration curves for the neutralization titration of 0.1 M solutions of various acids with 0.1 M NaOH. The number beside each curve is the pK_a value for that acid.*

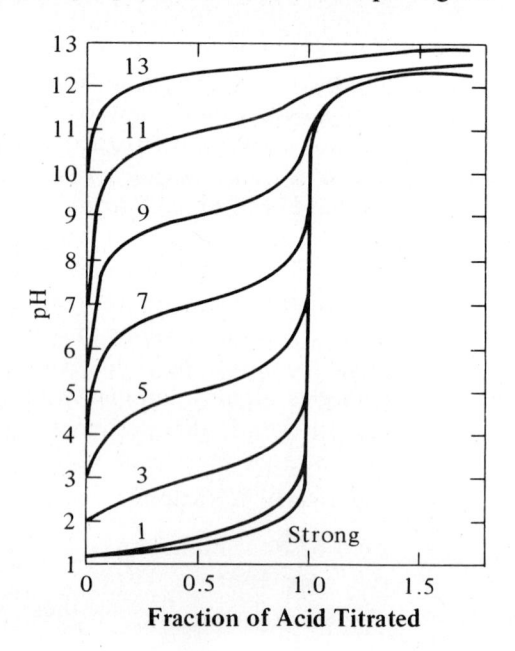

solvents (solvents more acidic than water), for example, acetic acid, and protophilic solvents (solvents more basic than water), for example, ethylenediamine. In the protogenic cases, it was found that bases too weak to be titrated in water could be titrated successfully with a strong acid dissolved in the same solvent. For example, primary, secondary, and tertiary amines can be titrated in acetic acid with perchloric acid in acetic acid as titrant. Medicinal sulfonamides, which have a primary amino group, can be titrated successfully in this manner, as can most of the common alkaloids, and purine compounds—including caffeine and theobromine.

On the other hand, acids too weak to be titrated with strong base in water appear much stronger in a protophilic solvent and can be titrated with a strong base such as sodium methoxide dissolved in a basic solvent or a compatible solvent. 2-Naphthol can be titrated in this manner. Many enols and imides can be titrated in either dimethylformamide or ethylenediamine.

A second type of useful solvent is the *aprotic* (sometimes called *inert*) solvent, which usually exhibits very weak acid properties. Examples are dimethylformamide, dimethylsulfoxide, dioxane, ether, various nitriles, methyl isobutyl ketone, hydrocarbons, and carbon tetrachloride. These solvents often permit differentiation (or stepwise titration) of a series of acidic or basic species that, in water, titrate either together or not at all. For example, perchloric, hydrochloric, salicylic, and acetic acids, and phenol, can be titrated stepwise in methyl isobutyl ketone solvent to obtain discernible endpoints for each compound, using tetrabutyl ammonium hydroxide in isopropyl alcohol as titrant.

In most nonaqueous solvents, precise explanation or prediction of the shape of a potentiometric titration curve, or the possible utility of a specific titration, is usually not possible because of the lack of complete thermodynamic equilibrium constants for the numerous possible processes. In general, the shapes of curves must be determined experimentally, and the behavior of substances in diverse solvents must be considered empirically.

Oxidation-Reduction Reactions

Potentiometric titrations of oxidizing or reducing agents can be performed by making the titrated sample one-half of an electrochemical cell. Typically, the indicator electrode is an "inert" electrode such as a platinum foil or wire that is used to monitor the solution potential; the cell is completed with the addition of a suitable reference electrode such as a saturated calomel electrode.

Thus, in titrating a reducing substance such as Fe^{2+} with a standard solution of an oxidizing substance such as MnO_4^- or Ce^{4+}, the solution potential at equilibrium is given *either* by the formal potential of the titrant couple and the ratio of activities of its oxidized and reduced forms, *or* by the formal potential of the substance titrated and the ratio of its oxidized and reduced forms. For an analytically useful titration, the system of titrant and substance titrated reacts rapidly, and at least one of the electrode couples is reversible at the indicator electrode.

Typical examples of applications include the titration of ferrous ion with permanganate; the titration of As(III) with bromate; the determination of ascorbic acid with iodine; and the determination of organic compounds such as azo, nitro, and nitroso compounds and quinones with chromous ion.

Precipitation and Complexation Reactions

Given the wide variety of ISEs commercially available and the many more specialized ones that can be fabricated, titrations involving the precipitation or complexation of ions are widely used. Halides, cyanide, thiocyanate, sulfide, chromate, and thiols can be titrated with silver nitrate, using the appropriate silver sulfide–based electrode; silver ion can be titrated with sodium iodide. Many metal ions can be titrated with standard EDTA, using the appropriate electrode, and possibly with the addition of an indicator reagent. Molybdate, selenide, sulfate, telluride, and tungstate can be titrated with lead perchlorate and a lead electrode. Aluminum, lithium, phosphate, various rare earths, and zirconate can be titrated with fluoride. Generally, the lower limit for determination is between 10^{-3} and 10^{-4} M.

SELECTED BIBLIOGRAPHY

BAILEY, P. L. *Analysis with Ion-Selective Electrodes*, 2nd ed. London: Heyden, 1980. *Clearly written guide to the theory and application of these sensors.*

BATES, R. G. *Determination of pH: Theory and Practice*, 2nd ed. New York: John Wiley, 1973. *Probably the definitive monograph on the fundamentals of pH.*

DURST, R. A. "Ion-Selective Electrodes in Science, Medicine, and Technology," *Amer. Sci.*, 59, 353 (1971). *A short, clear article on the theory, functioning, and applications of ion-selective electrodes.*

IVES, D. J. G., and JANZ, G. J. *Reference Electrodes*. New York: Academic Press, 1961.

LINGANE, J. J. *Electroanalytical Chemistry*, 2nd ed. New York: Interscience, 1958, Chaps. 2 through 8. *The fundamentals of potentiometry and many applications of potentiometric titrations.*

WEISSBERGER, A., and ROSSITER, B. W., eds. *Physical Methods of Chemistry*, Vol. 1, Part IIA. New York: Interscience, 1971.

REFERENCES

1. T. S. LICHT and A. J. DEBETHUNE, *J. Chem. Educ.*, 34, 433 (1957).
2. H. A. LAITINEN and W. E. HARRIS, *Chemical Analysis*, 2nd ed., New York: McGraw-Hill, 1975, pp. 5–17.
3. R. G. BATES, *Determination of pH*, 2nd ed., New York: John Wiley, 1973, pp. 328–35.
4. S. P. L. SØRENSEN, *Biochem. Z.*, 21, 131, 201 (1909).
5. R. A. ROBINSON and R. H. STOKES, *Electrolyte Solutions*, 2nd ed., New York: Academic Press, 1959.
6. D. D. PERRIN and B. DEMPSEY, *Buffers for pH and Metal Ion Control*, New York: Halsted Press, 1974.
7. R. A. DURST, *J. Chem. Educ.*, 44, 175 (1967).
8. E. PUNGOR, *Anal. Chem.*, 39(13), 28A (1967).
9. G. G. GUILBAULT, *Pure Appl. Chem.*, 25, 727 (1971).
10. D. AMMANN, F. LANTER, R. A. STEINER, P. SCHULTHESS, Y. SHIJO, and W. SIMON, *Anal. Chem.*, 53, 2267 (1981).
11. J. L. WALKER, JR., *Anal. Chem.*, 43(3), 89A (1971).
12. L. R. PUCACCO and N. W. CARTER, *Anal. Biochem.*, 89, 151 (1978).
13. R. W. CATTRALL and H. FREISER, *Anal. Chem.*, 43, 1905 (1971).
14. C. R. MARTIN and H. FREISER, *J. Chem. Educ.*, 57, 512 (1980).
15. G. GRAN, *Analyst*, 77, 661 (1952); F. J. C. ROSSOTTI and H. ROSSOTTI, *J. Chem. Educ.*,

42, 375 (1965).
16. T. R. GILBERT and A. M. CLAY, *Anal. Chem.*, 45, 1757 (1973); R. F. THOMAS and R. L. BOOTH, *Environ. Sci. Technol.*, 7, 523 (1973).
17. M. J. SMITH and S. E. MANAHAN, *Anal. Chem.*, 45, 836 (1973).
18. *Guide to Ion Analysis*, Orion Research, Inc., Cambridge, Mass., 1983.
19. B. JASELKIS and M. K. BANDEMER, *Anal. Chem.*, 41, 855 (1969); E. W. BAUMANN, *Anal. Chem.*, 42, 110 (1970).

PROBLEMS

1. Beginning with the following equation for the response of an electrode that is specific for cation "i," and differentiating, show that the *relative* concentration error incurred in the measurement of a_i by *direct potentiometry* ($\Delta a_i/a_i$) is about $\pm 4n\%$ per millivolt uncertainty in the measurement of E:

$$E = E_{constant} + \frac{RT}{nF} \ln a_i$$

2. Calculate the ionic strength of the following electrolyte solutions, assuming complete dissociation into ions: (a) 0.01 M $CaCl_2$; (b) 0.05 M KCl; (c) 0.1 M Na_2SO_4; (d) 0.001 M $AlCl_3$.

3. Using the simple form of the Debye-Hückel limiting law, calculate the individual activity coefficients for the ions in the following electrolyte solutions: (a) 0.002 M $MgCl_2$; (b) 0.002 M $MgCl_2$ + 0.01 M KCl; (c) 0.002 M $MgCl_2$ + 0.1 M KCl.

4. Using the simple form of the Debye-Hückel law, calculate the pH of a solution that is 0.03 M in the acid, 0.02 M in its conjugate base, and 0.1 M in KCl for the systems: (a) CH_3COOH + CH_3COONa ($K_a = 1.75 \times 10^{-5}$); (b) $CH_3NH_3^+Cl^-$ + CH_3NH_2 ($K_a = 2.0 \times 10^{-11}$).

5. Calculate the theoretical potential of a half-cell composed of a platinum-wire electrode dipped into a solution containing 0.2 M Fe^{3+} and 0.1 M Fe^{2+} versus: (a) the standard hydrogen electrode (SHE); (b) the saturated calomel electrode (SCE). Assume that activity coefficients are unity and that the temperature is 25°C. E_{SCE} is 0.245 V vs. SHE.

6. Calculate the theoretical potential of the following cell at 25°C. Assume activity coefficients are unity; $K_{sp}(AgI) = 8.3 \times 10^{-17}$.

Ag/AgNO₃ (1 M)//KI (1 M), AgI (satd.)/Ag

7. Consider the following cell:

SCE//x M AgNO₃/Ag

(a) Assuming unit activity coefficients, calculate the Ag^+ concentration if the initial cell potential is 0.323 V. (b) If the right half-cell also contained 0.01 M NaNO₃, calculate the initial cell potential assuming the simple form of the Debye-Hückel law holds. (c) If during the measurement of the potential, the instrument allowed the passage of an average of 0.1 mA for about 20 sec, calculate the change in $[Ag^+]$. Assume a cell volume of 5 mL and unit activity coefficients. (d) If the cell were allowed to undergo its spontaneous reaction, calculate the $[Ag^+]$ at equilibrium. Write the balanced spontaneous reaction. Assume that the SCE potential is constant and that the potential in the right half-cell is established only by Ag^+/Ag.

8. Calculate the relative error incurred in the direct potentiometric determination of the calcium concentration of seawater due to magnesium-ion interference. A typical magnesium level is 1300 ppm, and the calcium level is 400 ppm. The selectivity coefficient of the calcium electrode for magnesium is 0.014.

9. What is the maximum concentration of interfering ions that can be tolerated for a 1% interference level when measuring 10^{-4} M Ca^{2+} with a calcium-sensitive liquid ion-exchange electrode? For a 10% interference level? The interfering ions and their selectivity coefficients are: Zn^{2+}, 3.2; Fe^{2+}, 0.80; Pb^{2+}, 0.63; Mg^{2+}, 0.014; Na^+, 0.003.

10. A 0.200-g sample of toothpaste was suspended in 50 mL of fluoride ionic-strength buffering medium (TISAB), and boiled briefly to extract the fluoride. The mixture was cooled, transferred quantitatively to a 100-mL volumetric flask, and diluted to volume with deionized water. A

25.00-mL aliquot was transferred to a beaker, a fluoride ion–selective electrode and reference electrode were inserted, and a potential of -155.3 mV was obtained after equilibration. A 0.10-mL spike of 0.5-mg/mL fluoride stock solution was added, after which the potential was -176.2 mV. Calculate the percentage by weight of F^- in the original toothpaste sample.

11. A sample of skim milk is to be analyzed for its iodide content by the method of multiple standard additions. A 50.0-mL aliquot of milk is pipetted into a 100-mL beaker, 1.0 mL of $5\ M$ $NaNO_3$ is added to increase the ionic strength, and the resultant solution is allowed to equilibrate to room temperature (25°C). A double-junction reference electrode and an iodide ion–selective electrode are inserted, and the solution is magnetically stirred. After equilibrium is reached, the voltage reading, E, is -50.3 mV. Then 100-, 100-, 200-, and 200-μL aliquots of 2.00 mM KI are added sequentially. After each addition, the equilibrium potential is measured: Values are -64.9, -74.1, -86.0, and -94.2 mV respectively. Given that the slope, S, of the iodide calibration curve in this concentration range is 59.2 mV/decade, calculate the original I^- concentration of the milk (in μg/mL).

12. An ISE is used to determine calcium in an aqueous sample. An ionic-strength adjustor and pH buffer are added. The sample contains a complexing agent that ties up 25% of the total calcium at the pH of the buffer. An independent method (e.g., flame spectrometry) yields a value of $4 \times 10^{-5}\ M$ Ca (total) per liter. (a) What answer would you expect by ISE potentiometry with the calibration-curve method? (b) If the complexing agent were in large excess, what value would be obtained by the ISE using the standard addition method?

13. Why are crystal-membrane ISEs for multicharged ions fundamentally less sensitive and more prone to interference than electrodes for singly charged ions?

14. One way to estimate selectivity coefficients for ISEs is to first equilibrate the electrode in a pure solution of the test ion and measure the potential:

$$E_1 = E_{constant} + \frac{RT}{nF} \ln a_i$$

(for a cation-selective electrode). Then, small aliquots of an interferant ion are added, and the potential is measured each time:

$$E_2 = E_{constant} + \frac{RT}{nF} \ln(a_i + k_{ij}a_j)$$

Solve and rearrange these two equations to obtain a single linear equation of the form $y = mx + b$, which can then be used to determine k_{ij}. (Hint: Review the derivation of Equation 2.45, and the suggestion for taking antilogs.)

15. The selectivity coefficient of an iodide ion–selective electrode for bromide ion is to be determined using the equation derived in Problem 14. An iodide ion–selective electrode and an appropriate reference electrode are equilibrated in 50.0 mL of $10^{-4}\ M$ KI at 25°C. The equilibrium potential is -130.2 mV. Then, 0.50-, 1.00-, 2.00-, and 2.00-mL aliquots of $1.00\ M$ KBr are added sequentially. After each addition, the equilibrium potential is measured: Values are -131.5, -133.9, -137.5, and -140.8 mV, respectively. Determine the selectivity coefficient for bromide ion if the slope of the calibration curve for iodide is 59.2 mV/decade. What assumptions have been made in this approach?

16. More precise location of the endpoint (i.e., inflection point) of a potentiometric titration curve can frequently be obtained from a first ($\Delta E/\Delta mL$ vs. mL)- or second ($\Delta^2 E/\Delta mL^2$ vs. mL)-derivative plot. The following data were collected near the endpoint of a titration. Plot the first and second derivatives of the titration curve near the endpoint and compare the endpoint values.

mL	Potential, mV
47.60	372
47.70	384
47.80	401
47.90	512
48.00	732
48.10	748
48.20	756

17. One way of experimentally estimating selectivity coefficients for ISEs is to record the equilibrium potential of the electrode for a series of

solutions of interferant ions of (constant) known concentration. The equilibrium potentials at 25°C for a calcium ion–selective electrode in 10^{-2} M Ca^{2+} solution and in 10^{-2} M solutions of various ions were measured as follows. Neglecting activity coefficients, calculate selectivity coefficients for the interferant ions, assuming that Equation 2.38 is valid, that the $E_{constant}$ for the electrode does not change, and that liquid-junction potentials are constant.

Ion	E	Ion	E
Ca^{2+}	+ 63.3 mV	H^+	+92.9 mV
Zn^{2+}	+113.6	Na^+	−70.4
Pb^{2+}	+101.8	K^+	−84.6
Mg^{2+}	+ 4.2		

CHAPTER 3

Polarography and Voltammetry

James A. Cox
James E. O'Reilly

In Chapter 2, the one electrochemical method involving no net current flow—potentiometry—was discussed. In Chapter 3, two methods will be studied—voltammetry and polarography—in which a voltage is applied to an electrode and the resulting current flow is measured.

For various reasons, classical polarographic techniques became less widely used for routine analytical purposes for some years. Particularly with the advent of flame spectroscopic methods (Chap. 10) for the determination of metals and metalloids, polarography became primarily a tool for more fundamental studies: corrosion processes, electrode mechanisms, and kinetics. However, with the improvement of commercial instrumentation and the introduction of some modern variants of the polarographic method, the use of voltammetric methods for quantitative analytical measurements is again increasing. The newer variations of the method permit selective, parts-per-billion, analyses of a variety of organic and inorganic species. The areas of application include environmental and toxicological studies, biochemistry and pharmacy, geology, and routine industrial quality control.

3.1 INTRODUCTION AND THEORETICAL BASIS

Principles

In *voltammetry*, current-versus-voltage curves are recorded when a gradually changing voltage is applied to a cell containing (a) the solution of interest, (b) a stable reference electrode, and (c) a small-area working or indicator electrode (Fig. 3.1). Usually, the voltage is increased linearly with time. Such curves are generically called *voltammograms*. In the special case where the indicator electrode is the *dropping mercury electrode* (DME), introduced by J. Heyrovský in 1922, the technique is known as

FIGURE 3.1. *Schematic diagram of a simple 2-electrode voltammetric apparatus.*

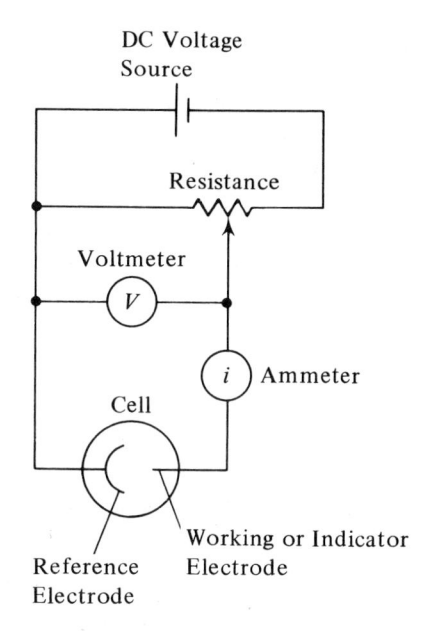

polarography, and the current-versus-voltage curves are called *polarograms*.

The DME consists of a glass capillary attached to a mercury reservoir. Drops of mercury fall from the orifice of this capillary at a constant rate, usually between 10 and 60 drops/min (Fig. 3.2). Each drop serves as the indicator electrode while attached to the column of mercury in the capillary. At the slow rate of voltage scanning generally used in polarography—about 50 to 200 mV/min—the change of potential of the DME during the life of a single drop can be neglected; thus, the current measured on each drop can be considered to be obtained under practically potentiostatic conditions (i.e., at constant potential). To distinguish this method from modern variants, it is sometimes called *conventional* or *DC* (direct-current) *polarography*.

The current-versus-voltage curves obtained with the DME are highly reproducible, because the surface of every new mercury drop is fresh, clean, and practically unaffected by electrolysis at earlier drops. The total amount of electrolysis is very small because of the small area of the electrode and the small currents involved; for example, with 20 mL of a typical solution, 100 polarograms can be recorded without noticeable change in the curve. The small size of the DME permits the analysis of small volumes of solutions; if necessary, less than 1 mL can be used.

Mercury is chemically inert in most aqueous solutions, and hydrogen is evolved on it only at rather negative potentials; consequently, the reduction of many chemical species can be studied at mercury electrodes, but not at electrodes made of most other materials. However, the oxidation of mercury makes it impossible to study reactions at potentials more positive than about 0.4 V versus the saturated calomel electrode (SCE). The positive-potential limit is even less if the experiment is performed in a medium that contains species that can either complex mercury ions or form a precipitate with these ions. For example, in 1 M HCl, the useful potential range of a mercury electrode is about -1.1 to 0 V versus SCE.

Supporting electrolytes are required in voltammetry to decrease the resistance

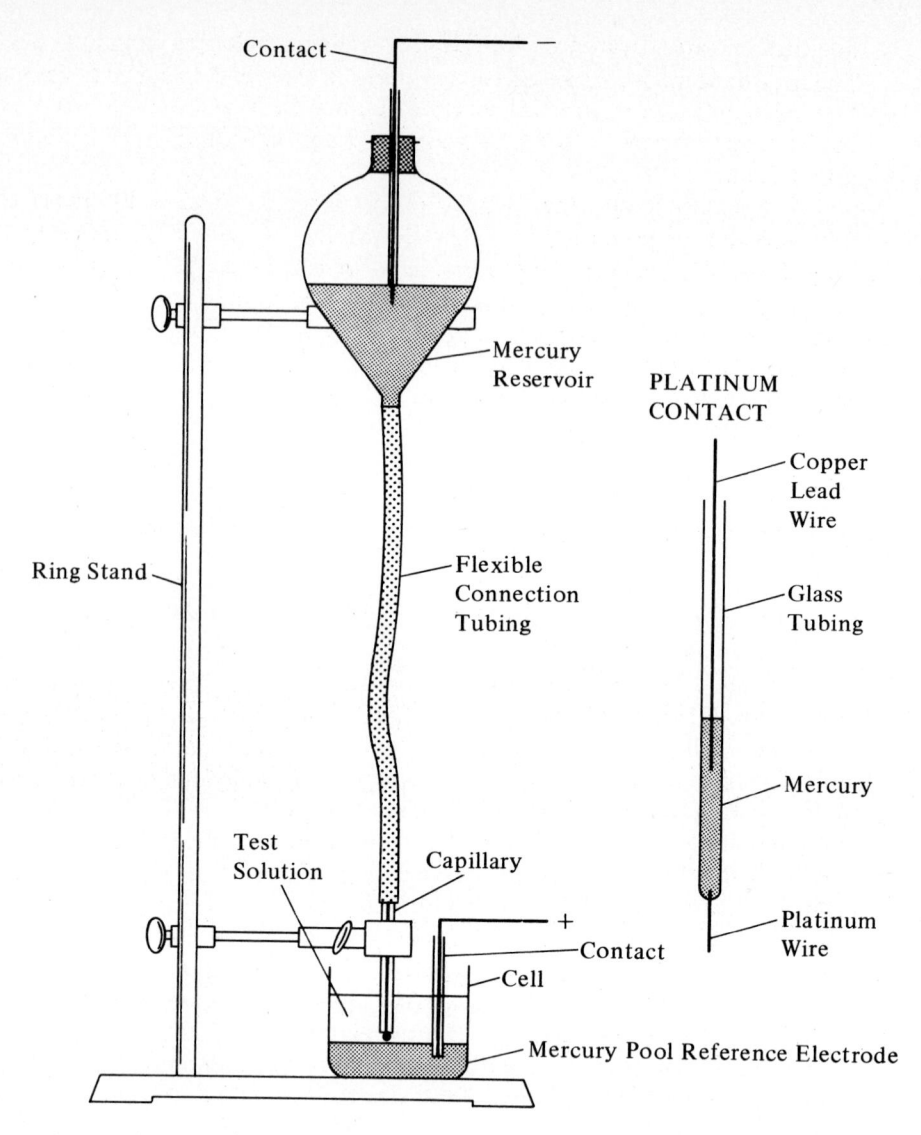

FIGURE 3.2. *Illustration of a dropping mercury electrode and a simple polarographic cell for reductions.*

of the solution and to ensure that the electroactive species move by diffusion and not by electrical migration in the voltage field across the cell. The supporting electrolyte is often chosen also to provide optimum conditions for the particular analysis: for example, buffering at a preferred pH value and elimination of interferences by selective complexation of some species. Solutions of strong acids (e.g., hydrochloric and sulfuric), strong bases (sodium or lithium hydroxide), or neutral salts (e.g., chlorides, perchlorates, or sulfates of either alkali metals or tetraalkylammonium ions) are frequently used, as are buffer solutions or solutions of complexing agents (e.g., tartrates, citrates, cyanides, fluorides, or amines, including ammonia and EDTA). The total concentration of electrolyte is usually between 0.1 and 1 M. To make migration negligible, the supporting electrolyte must be in at least 1000-fold excess over analyte ions; therefore, the practical upper concentration limit of voltammetry is about 5 mM.

A typical polarogram is illustrated in Figure 3.3. The background curve shows the potential range limited by the reactions $2H^+ + 2e^- \rightarrow H_2$ and $2Hg + 2Cl^- \rightarrow Hg_2Cl_2 + 2e^-$. Between the limits, a small residual current flows due to charging of the surface. For example, a net negative charge applied to the Hg electrode causes migration of the cation of the supporting electrolyte up to the electrode surface; this motion produces the *residual current* (commonly termed the *charging* or *capacitive current*). Trace electroactive impurities in solution or in the Hg used as the electrode also contribute to the residual current.

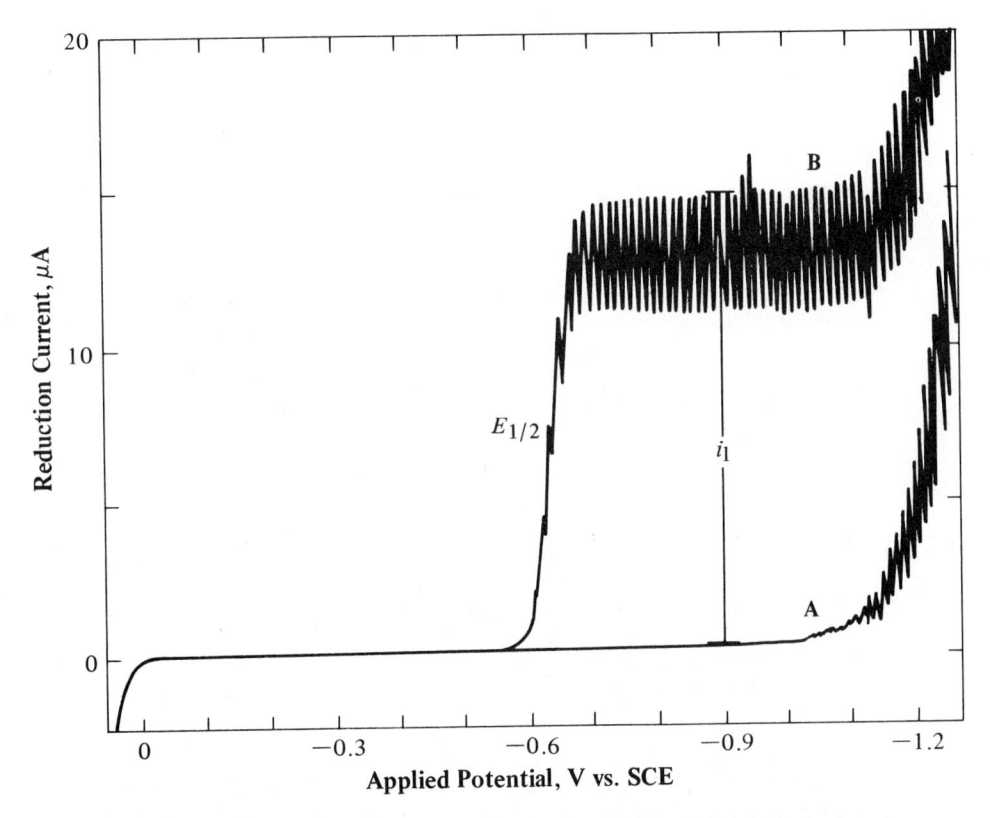

FIGURE 3.3 *Typical polarogram. Curve A: Background, residual-current, or supporting-electrolyte curve (1 M HCl). Curve B: Polarogram of 0.5 mM Cd^{2+} in 1 M HCl. $E_{1/2}$ is the half-wave potential, and i_l is the limiting current of the polarogram. Adapted from D. T. Sawyer and J. L. Roberts, Jr., Experimental Electrochemistry for Electrochemists, New York: Wiley-Interscience, 1974, by permission of John Wiley and Sons. Copyright © 1974 by John Wiley and Sons.*

SEC. 3.1 Introduction and Theoretical Basis

The reduction (or oxidation) of species in solution produces the *polarographic wave*. The resulting current is termed the *faradaic current*. The onset of the wave is near the standard potential of the redox couple for a reversible couple (see the section entitled "Polarographic Wave-Shapes"). The potential at which the current is one-half of the limiting value is called the *half-wave potential*, $E_{1/2}$, and is characteristic for the analyte in a given medium. The plateau current, or *limiting current*, is established for a reduction when the potential is sufficiently negative of the standard potential to cause all of the analyte that reaches the electrode surface to be reduced. Quantitative applications of polarography are generally made under conditions where diffusion is the only means of mass transport of the analyte from the bulk solution to the electrode surface (i.e., where the limiting current is *diffusion controlled*).

The limiting current is measured as the difference between the plateau current and the residual current. If more than one electroactive species is present, a series of polarographic waves will occur. Unless there is interaction between the various species, the polarographic waves are additive. Hence, each wave can be measured from the extrapolated plateau of the previous wave. This extrapolation method can also be used on a one-component system to account for the residual current if a true background polarogram cannot be obtained.

The current oscillations shown in Figure 3.3 are caused by the growth of the drops during the measurement; current (either charging or faradaic) is proportional to electrode area. The trace does not fall to zero current when the drop dislodges because of the limited response time of recorders. The analytical current is usually measured at the maximum point of the drop area (i.e., the current just before the drop dislodges). Most reports and drawings disregard these oscillations and show only the overall polarographic "envelope."

Removal of Oxygen. One difficulty encountered in voltammetric or polarographic analyses is that dissolved oxygen interferes severely and must be removed from test solutions before analysis. One reason for the interference is that oxygen is electrochemically reducible, producing large reduction currents. Furthermore, oxygen or its reduction products react chemically with many analytes. Oxygen is reduced in two steps: a 2-electron reduction to H_2O_2, and then a second 2-electron reduction, to H_2O. The $E_{1/2}$'s are at about -0.1 and -0.9 V versus SCE at pH 7. Air-saturated aqueous solutions have about a 0.25 mM O_2 level and produce a total of about 5 μA diffusion current at the DME.

The usual procedure for removing O_2 from solutions is to bubble the solutions for 5 to 20 min with high-purity nitrogen or argon. Compressed nitrogen in tanks often contains enough residual oxygen to be detected polarographically; it is usually further purified before use by first bubbling it through acidic V^{2+} or Cr^{2+} solution, or passing it through a commercial purification cartridge. Finally, the stream is passed through the solvent used to prepare the sample. This step saturates the gas stream with solvent and eliminates the partial evaporation of the sample solution that would otherwise occur. The deaeration time may be greatly reduced by using a medium- or coarse-porosity fritted-glass dispersion tube in the sample compartment. After deaeration, the gas stream is passed over the top of the sample to prevent oxygen from reentering the sample during the experiment.

The Ilkovic Equation. The current that flows through the polarographic cell depends on the rate of the electrode reaction and on the rate of transport of the electroactive species to the electrode surface. At sufficiently negative potentials (i.e., where the limiting current is observed), the rate of the electrode process is so fast that the rate of transport of the species to the surface becomes the limiting factor.

In the absence of migration (eliminated by addition of supporting electrolyte) and convection (prevented by keeping the electrolyzed solution unstirred), diffusion is the only mode of transport involved. Therefore, the limiting current is proportional to the rate of diffusion, and

$$i_l = nFAD \left(\frac{\partial c}{\partial x}\right)_{x=0} \tag{3.1}$$

where
A = area of the electrode
D = diffusion coefficient of the electroactive species
$(\partial c/\partial x)_{x=0}$ = concentration gradient of the electroactive species at the electrode surface
x = distance from the electrode surface

The DME is virtually spherical; its volume can be calculated from the rate of flow of mercury, m (in mg/sec), the time t (in sec) measured from the beginning of drop growth, and the density of mercury. This gives the radius of the drop, from which the surface area, A (in mm^2), is

$$A = 0.852(mt)^{2/3} \tag{3.2}$$

Solution of Equation 3.1 and the appropriate form of Fick's second law of diffusion yield the Ilkovic equation

$$i_d = 708nD^{1/2}m^{2/3}t^{1/6}c \tag{3.3}$$

where
i_d = diffusion current in microamperes
D = diffusion coefficient in square centimeters per second
c = *analyte* concentration in millimoles per liter (mM)

The subscript "d" signifies that a current limited by the rate of diffusion is considered.

Equation 3.3 is the expression for the maximum current observed during the life of each drop, that is, the current at the end of the life of each drop (see the drop oscillations shown in Fig. 3.3). Modern strip-chart and *x-y* recorders have a sufficiently fast response to ensure that the recorded maximum current can be equated to the theoretical one. In the earlier literature, when fast-response recorders were not common, average currents rather than maximum currents were frequently measured; the theoretical average current is 6/7 of the maximum current; that is, the value of the constant in Equation 3.3 equals 607.

It is evident from Equation 3.3 that the measured maximum current on the plateau of the wave can serve for quantitative analysis, since it is directly proportional to the concentration of the substance being reduced. Moreover, when c is known and when the electrode reaction (i.e., the value of n) is known, measurements of i_d can be used to obtain the diffusion coefficient. In other cases, where the electrochemical reaction is not yet known, one can postulate a plausible value for D and use i_d to calculate n; since the latter must be an integer, it is not necessary that D be known with accuracy.

Currents Controlled by Factors other than Diffusion

The electrochemical reaction may involve, in addition to diffusion and charge transfer, chemical reactions in which the oxidant or reductant is involved and/or adsorption of the electroactive species. Sometimes the magnitude of the current is limited by the rate of a chemical reaction or an adsorption process.

Kinetic Currents. Polarographic currents whose magnitudes are controlled by the rates of chemical reactions are called *kinetic* currents. For example, one may have a system where the analyte, A, is not electroactive in a certain potential range but is in equilibrium with a species that can be reduced:

$$A + X \underset{k_{-1}}{\overset{k_1}{\rightleftarrows}} Ox \tag{3.4}$$

where the equilibrium constant for this reaction, K_{eq}, is equal to k_1/k_{-1}. The reduction of Ox will cause a net reaction between A and X as the system attempts to reestablish equilibrium. The net reaction under conditions where Ox is being electrolytically consumed is

$$A + X \xrightarrow{k_1} Ox + ne^- \rightleftarrows R \tag{3.5}$$

When the rate of the *chemical* reaction is much slower than the rate of diffusion, the current is limited by the rate constant k_1. Many chemical systems that are controlled by this mechanism yield a current that is directly proportional to the analyte concentration. The pH and temperature generally need to be carefully controlled for analytical work based on this mechanism.

An example of a kinetic system is the polarographic reduction of formaldehyde. In aqueous solution, formaldehyde exists in equilibrium with its hydrate:

$$CH_2O(H_2O) \rightleftarrows CH_2O + H_2O \tag{3.6}$$

Only the anhydrous form is polarographically reduced. The polarographic limiting current is therefore controlled in part by the rate of dehydration of $CH_2O(H_2O)$.

Catalytic Currents. Another type of polarographic current is governed by catalytic processes. Such *catalytic currents* are of two types: Either the substance undergoing electrolysis is regenerated in the vicinity of the electrode by a chemical reaction, or the electroreduction of a species is shifted to more positive potentials than would occur in the absence of the catalyst. An example of the former case is the reduction of Fe(III) to Fe(II)—the electrogenerated ferrous ion can be reoxidized back to ferric ion if hydrogen peroxide is present in solution. An example of the second case is the catalytic reduction of hydrogen ions—many substances, proteins for example, catalyze this reduction and shift the corresponding wave to a more positive potential. In both cases, the current is a nonlinear function of concentration (or a linear function only over a certain concentration range).

FIGURE 3.4. *Polarographic curve of methylene blue (0.4 mM), showing the adsorption prewave.*

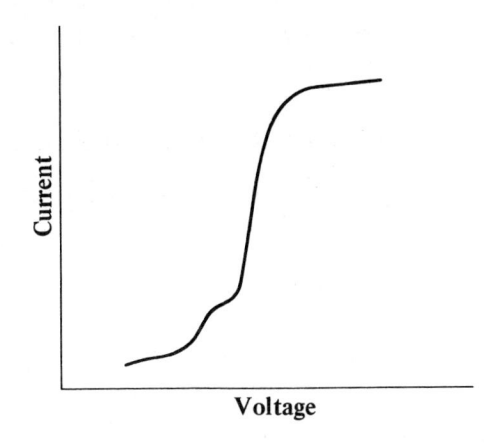

Adsorption Currents. Adsorption of either the oxidized or reduced form of a couple will alter the polarographic behavior. If the oxidized form is adsorbed, its reduction will take place at a more negative potential than in the diffusion-controlled case. The shift in potential is directly related to the free energy of adsorption.

The more common case occurs when the product of the electrolysis is adsorbed:

$$\text{Ox} + ne^- \rightleftarrows R(\text{ads}) \tag{3.7}$$

The adsorption allows the electrolysis by Reaction 3.7 to take place prior to the normal polarographic wave. A prewave such as that shown in Figure 3.4 results. The height of the prewave is proportional to the Ox concentration until complete electrode surface coverage by $R(\text{ads})$ occurs. At that point, the current becomes limited by the availability of fresh surface and not concentration. Eventually, the potential scan reaches a sufficiently negative potential to allow the "normal" reduction (to R in solution) to occur. The total wave height, measured from the background current rather than to the plateau of the prewave, can be used for quantification.

Polarographic Maxima

Sometimes polarograms show currents that are, over certain ranges of potential, considerably higher than diffusion currents—higher by as much as two orders of magnitude. These polarographic maxima may be sharp or rounded and may cover only small regions of potential or quite wide ones. The most common type produces a peak between the $E_{1/2}$ and the onset of the plateau.

Polarographic maxima occur because of spontaneous stirring of the layer of the solution in contact with the electrode surface. They obscure the normal polarogram; and if they are not recognized properly, polarographic maxima can cause the worker to set the instrument sensitivity incorrectly. They do not provide a reliable basis for quantification, and are normally suppressed or eliminated in any analytical work.

Several mechanisms can lead to movement of the solution and the appearance of maxima. The subject is too complicated to be discussed satisfactorily here; the origins of maxima are discussed fully in Reference [1].

Polarographic maxima can be suppressed by adding a surfactant (a material that adsorbs strongly on the electrode surface) to the supporting electrolyte. Triton X-100 in the range 0.001 to 0.005% in the sample is commonly used. Gelatin in the range 0.001 to 0.01% is another common suppressor, but it must be prepared freshly every day or two. Because surfactants affect diffusion coefficients, it is important that all samples and standards receive the same amount. The minimum amount needed should be added because surfactants cause foaming during deaeration and can shorten the capillary lifetime unless meticulous rinsing is performed. In fact, if the plateau is well developed, it may be better to tolerate a maximum than to use suppressors.

Tests for Current-Limiting Processes

Criteria used to distinguish among diffusion, kinetic, adsorption, and catalytic currents include changes in the wave height (limiting current) with (a) concentration of the electroactive species, (b) mercury column height, (c) pH, (d) buffer concentration, and (e) temperature.

Direct proportionality between the limiting current and concentration, as illustrated in Figure 3.5A, is observed for diffusion currents, for the majority of kinetic currents, and for some catalytic currents (e.g., those involving regeneration of a reducible metal ion). Limiting dependences (Fig. 3.5B and C) are observed for adsorption currents and for some catalytic currents.

Varying the height of the mercury column (h) above the orifice of the capillary provides a useful criterion for distinguishing among the various possible current-limiting processes. Diffusion currents are directly proportional to $h^{1/2}$, kinetic currents are independent of h, and adsorption currents are directly proportional to h (Fig. 3.6). The dependence of adsorption currents (i_a) on h should be measured at concentrations where the current is concentration independent; kinetic currents (i_k) are independent of h provided that measurements are carried out under conditions (e.g., pH) where i_k is less than $0.2 i_d$.

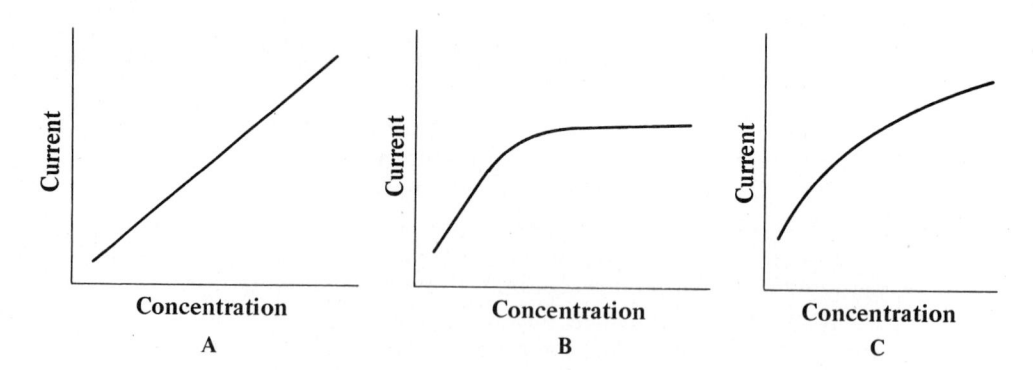

FIGURE 3.5. *Relation of limiting current to concentration. Curve A: Linear dependence observed for diffusion currents and for some kinetic and catalytic currents. Curves B and C: Limiting dependences observed for adsorption and some catalytic currents.*

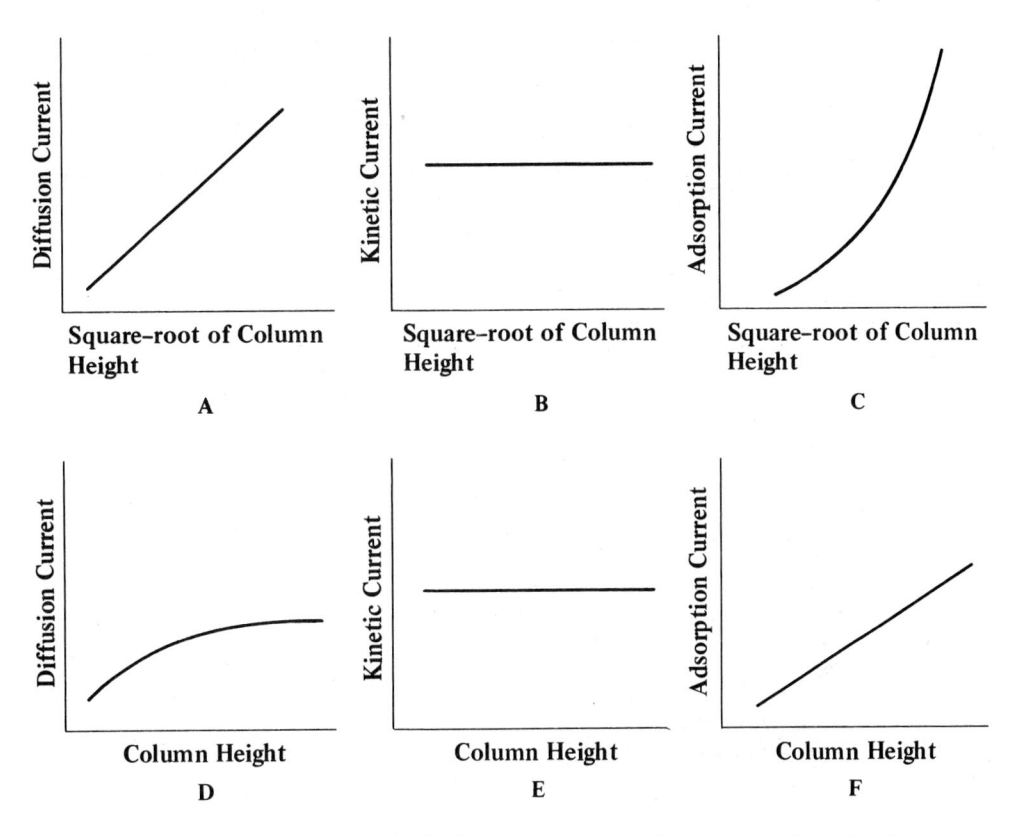

FIGURE 3.6. *Variation of the limiting current with mercury-column height for various types of polarographic currents. Curves A and D: Variation of diffusion current, i_d. Curves B and E: Variation of kinetic current, i_k. Curves C and F: Variation of adsorption current, i_a.*

The variation of current with time during the life of a single drop can be a valuable criterion for determining the current-limiting process. With fast pen recorders, one can obtain a moderately accurate measure of current-versus-time behavior by expansion of the time axis, but oscilloscopic observations are much preferred.

Equation 3.3 showed that diffusion-limited currents vary with time as $t^{1/6}$. This dependence is understandable from the combined effects of the increase in thickness of the diffusion layer with time, and the increase in area of the mercury drop. The first factor leads to a current proportional to $t^{-1/2}$ (Eqn. 1.6), the second to a current proportional to $t^{2/3}$ (Eqn. 3.2), and the combined effect ($t^{-1/2} \times t^{2/3}$) is $t^{1/6}$.

The rates of the homogeneous reactions responsible for kinetic and catalytic currents depend on the volume of solution in which the reactions occur. Because the reactions of interest occur near the electrode, the relevant volume is proportional to the area of the electrode, that is, to $(mt)^{2/3}$, and thus kinetic and catalytic currents usually are proportional to $t^{2/3}$.

Adsorption-limited currents are proportional to the rate at which fresh mer-

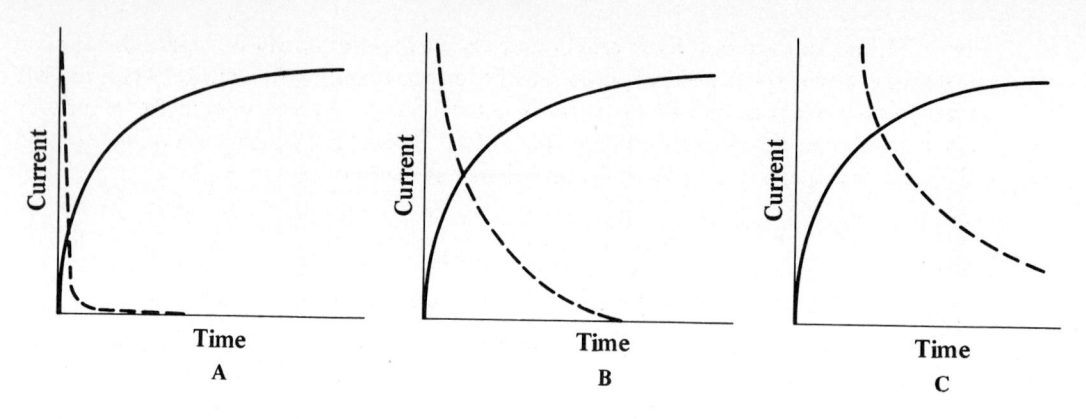

FIGURE 3.7. *Illustration of the relative importance of faradaic and charging current as the concentration of the electroactive species is reduced during the life of a single mercury drop. Solid lines—faradaic current; dashed lines—charging current. Curve A: For a 1 mM solution, the contribution of the charging current is small and can be neglected. Curve B: For a 0.2 mM solution, the contribution of the charging current is still small but not negligible and should be corrected for. Curve C: For 0.03 mM solution, the charging current becomes larger than the faradaic current, and the separation of the two currents becomes necessary. The y-axes on the three curves have been normalized so that the faradaic-current curves (solid lines) are the same size.*

cury surface appears, that is, to the rate of change of the area of the electrode:

$$i_a \propto \frac{d}{dt}(mt)^{2/3} \propto t^{-1/3} \tag{3.8}$$

The residual or charging current is also limited by the rate at which fresh electrode surface is formed, and hence is proportional to $t^{-1/3}$. This fact is of considerable significance in understanding variations of polarographic methods aimed at increasing sensitivity. Since diffusion currents increase with time (as $t^{1/6}$), and charging current decreases (as $t^{-1/3}$), measurement of the current late in the life of the drop gives a better sensitivity ("signal/noise") than measurements of average currents, or of currents early in the drop-life (see Fig. 3.7).

Polarographic Wave-Shapes

In the applications of potentiometry in Chapter 2, it was noted that the Nernst equation could be used only to describe redox couples that were at equilibrium. The time required for a general redox reaction

$$A + ne^- \rightleftarrows B \tag{3.9}$$

to reach equilibrium depends on the rates of the "forward" ($A + ne^- \rightarrow B$) and "back" ($B \rightarrow A + ne^-$) reactions. The rate constants for these reactions, k_f and k_b, depend on the electrode potential. For example, as the potential is made more negative in polarography, the rate constant of the forward reaction, k_f, increases, and k_b decreases (and conversely for positive-potential excursions).

If the forward and back rate constants are high in the region where the applied potential is near the standard potential of the redox couple, Reaction 3.9 can rapidly reach equilibrium at the DME surface. Such couples are considered to be *electrochemically reversible*. A consequence of reversibility is that the Nernst equation (Eqn. 2.8) can be applied to Equation 3.9 and modified as follows. At $T = 25°C$,

$$E_{\text{applied}} = E^{0'} - \frac{0.059}{n} \log \frac{[B]_0}{[A]_0} \tag{3.10}$$

where $[B]_0$ and $[A]_0$ = concentrations at the DME surface

If species B is absent from the bulk (prepared) solution and can be formed only by the reduction of A, then

$$[B]_0 \propto i = ki \tag{3.11}$$

where i = current at the DME

It follows that the concentration of A at the electrode surface is

$$[A]_0 \propto (i_d - i) = k'(i_d - i) \tag{3.12}$$

since $[A]_0$ is the difference between the amount of A that was initially at the electrode (an amount that would produce the limiting current, i_d, if entirely reduced) and the amount remaining after the formation of $[B]_0$. By analogy to the constants in the Ilkovic equation, the proportionality constants k and k' are identical except for the diffusion coefficients of A and B, and so Equation 3.10 becomes

$$E_{\text{applied}} = E^{0'} - \frac{0.059}{n} \log \left(\frac{D_A}{D_B}\right)^{1/2} - \frac{0.059}{n} \log \frac{i}{i_d - i} \tag{3.13}$$

Since the half-wave potential, $E_{1/2}$, is defined as the potential at which $i = 0.5 i_d$,

$$E_{1/2} = E^{0'} - \frac{0.059}{n} \log \left(\frac{D_A}{D_B}\right)^{1/2} \tag{3.14}$$

or

$$E_{\text{applied}} = E_{1/2} - \frac{0.059}{n} \log \frac{i}{i_d - i} \tag{3.15}$$

Equation 3.15 holds for *reversible, diffusion-controlled* electrochemical reactions where the electrolysis product is initially absent in the bulk solution, and is soluble in the solution or in the electrode itself (as an amalgam, which is the case for reduction of many metal ions). A plot of E_{applied} versus $\log[i/(i_d - i)]$ can be used as a test for these conditions (a straight line would be obtained). It is also a means of determining n (from the slope) and $E_{1/2}$. The interpretation requires that in addition to a straight line, a reasonable, integral n-value be obtained before reversibility can be claimed. The $E_{1/2}$-value is useful because it provides an estimate of $E^{0'}$; the term $\log(D_A/D_B)^{1/2}$ is generally small.

If the electrode reaction is not reversible, the rising portion of the polarographic wave is drawn out. This occurs when the rate constants near $E^{0'}$ are too small to allow equilibrium to be reached on the time scale of the experiment. Since k_f increases as the potential is made more negative, even nonreversible couples eventually can become diffusion controlled (i.e., k_f becomes sufficiently large beyond a certain potential that A is reduced to B as soon as it arrives at the electrode surface). A limiting current will then be reached, and the Ilkovic equation can be applied.

Equation 3.15 is not applicable if either species A or B is adsorbed on the electrode or is involved in a chemical reaction other than simple electron transfer. A common example of the latter is the case where A is a metal ion that is in equilibrium with p molecules of a ligand, L, and a metal complex, AL_p:

$$A + pL \rightleftarrows AL_p \tag{3.16}$$

$$K_f = \frac{(AL_p)}{(A)(L)^p} \cong \frac{[AL_p]}{[A][L]^p} \tag{3.17}$$

where   the terms in parentheses = activities
the terms in brackets = molarities
K_f = formational constant of the complex

In this case, the term $[A]_0$ in Equation 3.10 can be replaced by $[AL_p]_0/K_f[L]_0^p$, and, if the concentration of the ligand is in large excess over that of A, the following expression can be derived:

$$(E_{1/2})_{cplx} = (E_{1/2})_{free\ ion} - \frac{0.059}{n} \log K_f - \frac{0.059p}{n} \log [L] \tag{3.18}$$

where $(E_{1/2})_{cplx}$ = half-wave potential in the presence of excess L
$(E_{1/2})_{free\ ion}$ = half-wave potential for the reduction of the metal ion in the absence of L

The treatment assumes that the reduction of the complex is reversible and that the diffusion coefficients of the metal ion and the metal complex are the same. The slope of a plot of $(E_{1/2})_{cplx}$ versus $\log [L]$ yields "p" (if n is known), and the intercept, where $[L] = 1\ M$, can be used to calculate K_f if $(E_{1/2})_{free\ ion}$ is known.

3.2 INSTRUMENTATION AND APPARATUS

The apparatus for voltammetric analysis consists of a suitable cell, electrodes, a potentiostat or polarograph, and a system for removing oxygen from the solution.

Cells

Frequently, a stoppered lipless beaker is quite suitable as a polarographic cell; the stopper should have holes to accommodate the DME, reference electrode, counter electrode (when used), and inlet and outlet tubes for bubbling gas (usually purified nitrogen) through the solution to remove dissolved oxygen. The bubbler can be raised after deaeration to allow a blanket of nitrogen to remain over the solution during the experiment (alternatively, a three-way stopcock can be used to divert the flow over the solution). Only a single exit port should exist to maximize the effectiveness of the nitrogen blanket.

A "remote" or "isolated" reference electrode, separated from the cell by salt bridges, is usually used to prevent contamination of the sample solution by ions from the reference electrode. Commercial SCEs that are used in pH measurements can serve as references in 3-electrode cells. If accurate potentials are not required, a *quasi-reference* can be used. An example is a Pt wire that has been oxidized by momentary electrolysis at 1.2 V versus SCE. Quasi-references are suitable only with the 3-electrode configuration, since they will not be stable if they pass current. A relatively large mercury pool can serve as the reference electrode (Fig. 3.2) in the presence of an electrolyte anion such as chloride, which establishes a fixed half-cell potential.

The cell may be immersed in a constant-temperature bath or may have a jacket through which water from such a bath is circulated (diffusion-controlled currents increase by about 1.3% for each °C rise in temperature). Many voltammetric cells are described in the literature. The range of cells includes designs for accommodating volumes less than 1 mL, monitoring effluents from flow systems such as liquid chromatographs, attaching to vacuum lines, and permitting simultaneous electrochemical and spectrochemical measurements, as well as convenient assemblies for routine measurements. The latter are typified by commercially available units from Metrohm and Princeton Applied Research Corporation.

Electrodes

The DME can be made most economically by using suitable lengths (8 to 30 cm) of capillary tubing (0.05 to 0.10 mm i.d.). The capillary is joined by plastic tubing to a mercury reservoir whose level can be adjusted to produce a suitable drop-time (2 to 6 sec). (The drop-time is inversely proportional to the height of the mercury column, and directly proportional to the length of the capillary.) Electromechanical devices used to obtain a constant drop-life (drop-time) have been described in the literature and are commercially available.

Instrumentation

A simple 2-electrode polarograph can be constructed from a potentiometer and a sensitive current-measuring instrument according to the scheme in Figure 3.1. Manual recording of polarographic curves with such an apparatus is, however, time consuming and therefore cannot be recommended for practical analyses. Furthermore, 3-electrode instruments are generally used since they provide correction for the effect of solution resistance. A number of reliable, low-cost, recording DC polarographs are commercially available.

A wide variety of recording polarographs or potentiostats capable of simple DC polarography and linear-scan voltammetry can be built from modern operational amplifiers for as little as about $200 in parts, depending on the number of extra features and quality of the amplifiers desired. All that is needed in addition is a suitable x-y or strip-chart recorder. Details on the instrumentation are provided in Reference [2].

Potentiostats. Because 3-electrode instruments compensate for a major fraction of the solution and reference-electrode resistances, which draw out and distort the

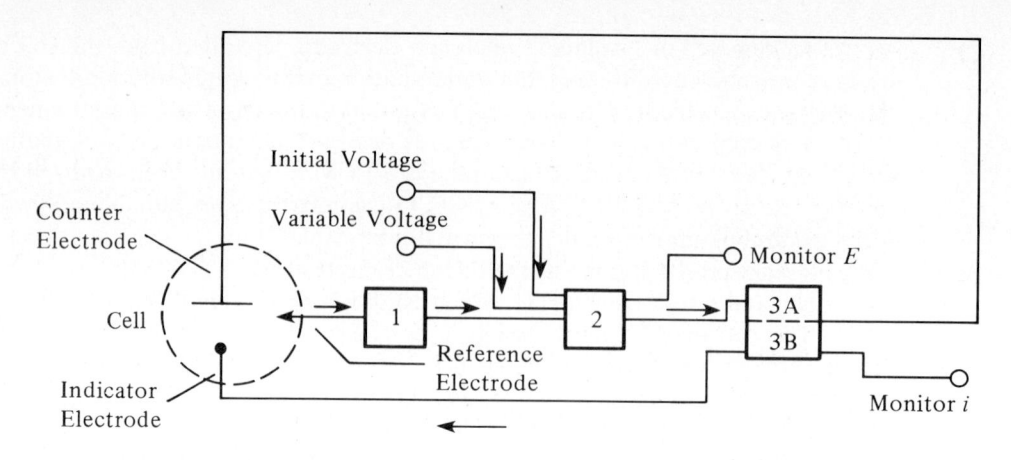

FIGURE 3.8. *Block diagram of a 3-electrode polarograph or potentiostat. Component 1: Unity gain component with a high input resistance and low output resistance. Component 2: Adder (sums the reference, initial, and variable voltages). Component 3: Controller (maintains the input voltages at 3A and 3B at equal values; 3B allows the indicator-electrode current to be measured without affecting the potential). The arrows show a hypothetical direction of voltage propagation. The counter electrode serves only to provide a complete current path through the cell (current cannot flow through the reference electrode in this configuration).*

polarographic wave, they are primarily used in voltammetry. A block diagram of such a unit is shown in Figure 3.8.

The potential of the reference electrode is transmitted across a high-input resistance component (e.g., an operational amplifier in the voltage-follower mode or a field-effect transistor) without significant change in magnitude; the purpose of this component is to block current flow through the reference electrode. The reference potential, the initial potential, and the variable potential (e.g., the voltage scan in DC polarography) are summed and applied to input A of the controller. Input B of the controller is connected to the indicator electrode. The controller maintains its inputs at the same potential; therefore, the indicator-electrode potential can be varied.

The current flow at the indicator must go through a complete series circuit path that includes the electrochemical cell. Because the flow cannot occur through the reference electrode, a third electrode (termed the counter or auxiliary electrode) must be placed in solution and connected to the output of the controller to complete the current path. The B portion of the controller contains additional circuitry to permit current flow from the indicator electrode to occur and be measured without upsetting the condition of $E_A = E_B$ at the controller inputs. In practice, the block components are high-impedance operational-amplifier circuits.

3.3 APPLICATIONS

Measurement of Voltammetric and Polarographic Curves

Although polarography may be used occasionally for qualitative analysis, the typical application is in quantitative analysis. To determine concentrations, the polarographic limiting current is measured.

Numerous methods of measuring the wave height are described in the literature, and a choice among these might seem to be a difficult task. However, most of the methods used give essentially equivalent results, provided that the measurements are carried out with sufficient care and accuracy and that the wave height is measured in the same way for all waves to be compared—that is, for the samples themselves and for the standards used in calibration.

In the measurement of wave heights, corrections for the residual current must be made. This can be done either by recording the residual current separately in a solution containing all the components with the exception of the electroactive species under study, or by extrapolation. The graphical subtraction of residual current is generally considered more reliable, particularly for inorganic species and at lower concentrations. For larger organic compounds, the basic assumption involved—that is, identity of the residual current in the sample solution and in the blank—is not generally fulfilled: Adsorption of the organic compound results in a change in the charging current. In such cases, the extrapolation of portions of the polarographic curves, as shown in Figure 3.9, seems to be the most accurate approach.

Determination of Concentration

Once the height of the wave has been obtained, it needs to be related to the concentration of the solution studied. As with the majority of physical methods used in analytical chemistry, such evaluation is usually based on comparison with a standard.

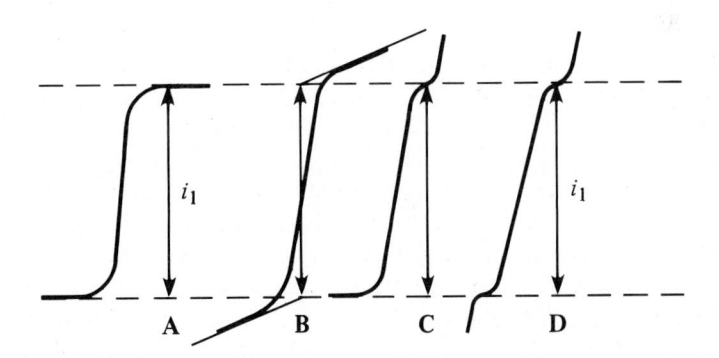

FIGURE 3.9. *Illustration of several methods for measuring the wave height of variously shaped polarographic waves. Curves A and B illustrate the technique of extrapolating the linear portions of the waves before and after the current rise. Curves C and D illustrate the technique of estimating current magnitudes for ill-shaped waves.*

The two methods of comparison most frequently used are that of employing a calibration curve and that of standard addition.

Calibration Curves. The calibration curve is constructed by successively adding increasing amounts of the substance to be studied to a solution of supporting electrolyte. All the components present in the sample solution, other than the analytes, are also included if they are known. The polarographic curves are recorded and measured, and the wave height is plotted as a function of concentration.

Next, curves are recorded for solutions containing samples to be analyzed in the same supporting electrolyte as was used for the construction of the calibration curve. It is essential that the curves for the sample analysis be recorded under *exactly* the same conditions as those used in the construction of the calibration curve. In particular, one uses the same capillary, the same pressure of mercury, and the same temperature. The wave height obtained with the sample is then measured and compared with the calibration curve, and the concentration read off.

The pressure of mercury is kept constant by maintaining the mercury in the reservoir at a constant level. Somewhat more difficult to guarantee is the use of the same capillary. This implies that when a capillary is broken, or behaves erratically (commonly, as a result of penetration of impurities into the bore), a new calibration curve must be constructed. If the highest accuracy is needed, the temperature of the electrolytic cell must also be controlled by using a water jacket or by immersing the cell in a thermostatted bath.

The most difficult condition to meet is making the solutions used for constructing calibration curves identical to those used for sample analysis. Frequently the calibration curves are recorded in solutions containing only the studied compound and supporting electrolyte; however, the preparation of the sample solution often introduces other substances. It can usually be assumed that these substances will have a negligible effect on the waves of the compound to be determined, but sometimes such electroinactive components of the sample can affect the height of the measured wave. If such an effect of electroinactive components cannot be neglected, an attempt can be made to prepare synthetic sample solutions that would contain all components with the exception of the studied substance. Such synthetic samples are then added to all solutions used in the preparation of the calibration curve.

Standard Addition Method. The basic method of standard addition was described in Chapter 2; however, the data analysis is simpler with the polarographic experiment because the signal (the limiting current) is directly proportional to concentration, in contrast to the logarithmic relationship in potentiometry.

In the standard addition method, the limiting current i_1, of the sample solution is first determined as follows:

$$i_1 = kc \tag{3.19}$$

An addition of a known volume V_s of a standard stock solution of the analyte (of concentration c_s) is then made to a known volume of the sample, V_u. After a brief additional deaeration, the limiting current i_2 is obtained under the same conditions as the original measurement (mercury height, capillary length, and potential):

$$i_2 = k'(c + \Delta c) \tag{3.20}$$

where $\quad \Delta c = V_s c_s/(V_u + V_s)$

If the bulk-sample characteristics are unchanged by the standard addition, $k = k'$; hence, taking the ratio of Equations 3.19 and 3.20 yields

$$\frac{i_1}{i_2} = \frac{c}{c + \Delta c} \tag{3.21}$$

Rearranging the latter expression gives

$$c = \frac{i_1}{i_2 - i_1}(\Delta c) \tag{3.22}$$

The condition that $k = k'$ can be assured by making $V_s \ll V_u$ and by following the requirement that the polarographic conditions, especially the potential of the current measurement, be the same.

As in the case of the potentiometric method (Chap. 2), a sequential standard addition method can be used. The graphical data analysis is the same except i is plotted versus Δc.

Pilot-Ion Method. The pilot-ion method is the same as the standard addition technique except that the added species is different from the analyte (e.g., Pb^{2+} may serve as the pilot ion for the determination of Zn^{2+}). More generally, this approach is termed an internal-standard method.

The data analysis for the pilot-ion method is the same as in the standard addition method (Eqns. 3.19 through 3.22) except that the constants k and k' are not generally equal. Comparing Equations 3.3 and 3.19 illustrates that k is the proportionality constant in the Ilkovic equation, $k = 708nD^{1/2}m^{2/3}t^{1/6}$. The capillary factor, $m^{2/3}t^{1/6}$, is not highly potential dependent; thus,

$$\begin{aligned} i_p &= 708n_p D_p^{1/2} m^{2/3} t^{1/6} c_p \\ i_a &= 708n_a D_a^{1/2} m^{2/3} t^{1/6} c_a \end{aligned} \tag{3.23}$$

where $\quad$ subscripts "p" and "a" = designations of pilot ion and analyte, respectively

From the ratio i_p/i_a and a rearrangement of terms,

$$c_a = \frac{i_a n_p D_p^{1/2} c_p}{i_p n_a D_a^{1/2}} \tag{3.24}$$

This treatment assumes that the addition of the pilot ion does not change the analyte concentration; otherwise, a dilution factor must be included. Often, $(D_p/D_a)^{1/2}$ is near unity, and so a reasonable estimate of c_a can be made without considering the diffusion coefficients.

The pilot ion must be either absent from the original sample or present at a known concentration. The reductions of both the analyte and pilot ion must be diffusion controlled for this treatment to be valid.

The pilot-ion method is convenient for the study of multicomponent samples because only two polarograms are needed regardless of the number of components. It is also useful when preparation of a standard solution of the analyte is not practical.

A variation of the pilot-ion method calls for the preparation of a calibration curve from a series of standard solutions of the analyte that also contain a constant concentration of the pilot ion. The curve is plotted as i_a/i_p versus c_a. The samples are made up to contain the controlled concentration of the pilot ion. A single curve of this type can be used for samples of varying temperatures, viscosities, and so forth, and with differing capillaries since it can be assumed that such factors would alter i_a and i_p in the same manner; the ratio i_a/i_p would then be unaffected.

Intercomparison of Methods. The speed of analysis where only one curve of the sample is recorded at the time of the analysis makes the calibration-curve method preferable in serial analysis of a large number of samples of similar composition. When precise results are required, temperature control is necessary, and when the capillary is blocked or broken, additional work for construction of a new calibration curve is necessary. The calibration-curve method is always used when the relation between the measured current and concentration is nonlinear (e.g., when catalytic or adsorption currents are dealt with).

The pilot-ion calibration-curve method can be applied in cases similar to those in which the simple calibration curve is used. Temperature control is unnecessary, but construction of several calibration curves for different concentrations of the "pilot" substance is usually required.

The method of standard addition is useful when analysis of only a limited number of samples is involved. In such instances, the construction of a calibration curve would be too time consuming. The apparatus is simplified by the fact that temperature control is unnecessary. On the other hand, because two curves (with and without addition of the standard) have to be recorded for each sample, the time spent on a single determination is somewhat longer.

The accuracy of the standard addition method is somewhat lower than the procedure based on the use of calibration curves, because, in calculating concentrations, a direct proportionality is assumed between the measured current and concentration. This corresponds to a linear calibration curve passing through the origin, where current and concentration equal zero. This assumption may be approximately valid at concentrations larger than about 10^{-4} M, but is rarely fulfilled in trace analysis. Nevertheless, the method of standard addition remains the first choice in cases where sample components might affect the wave height of the component analyzed and yet it is difficult to obtain a sample with zero content of this component (e.g., biological material).

The relative methods mentioned above, based on comparison with standards, are to be preferred over so-called absolute methods where concentrations are calculated using predetermined and tabulated "diffusion-current constants" or diffusion coefficients, both known to be dependent on experimental conditions.

Scope of Applications

Electroactive Species. Inorganic cations, anions, and molecules can be determined polarographically. Among cations, the transition metals are most profitably determined polarographically, but some alkaline-earth and rare-earth ions also offer useful analytic curves. Strongly hydrolyzed metals (e.g., aluminum, thorium, zirconium) present difficulties (which can sometimes be circumvented by using nonaqueous solvents); so do some elements that form predominantly covalent bonds (e.g., silicon; however, some germanium complexes are reducible). Typical ions frequently determined are those of Cu(II), Cu(I), Tl(I), Pb(II), Cd(II), Zn(II), Fe(II), Fe(III), Ni(II), Co(II), Bi(III), Sb(III), Sb(V), Sn(II), Sn(IV), and Eu(III); and Mo, W, V, Mn, Cr, Ti, N, and Pt in a number of oxidation states. Even when it is possible to determine alkali metals polarographically, flame photometry or some other spectral technique is usually superior.

Anions of the halides, as well as of sulfides, selenides, and tellurides, can be determined by means of anodic waves due to mercury-salt formation. Among the oxygen-containing anions—in addition to those of the metals mentioned above—cathodic reduction waves can be used for determination of bromates, iodates, periodates, sulfites, polythionates, and so on.

Finally, among inorganic molecules, polarography can be used to determine oxygen, hydrogen peroxide, elemental sulfur, some sulfur oxides, and oxides of nitrogen, as well as some undissociated acids.

A great number and variety of organic compounds can be quantitatively determined by reduction at the DME. Only highly polarizable single bonds between carbon and heteroatoms are reducible in the available potential range, for example C–Cl, C–Br, or C–I bonds. Other single bonds, for example, C–O, C–S, or C–N, require the presence of an adjacent activating group such as a carbonyl group or a pyridine ring.

Some single bonds between heteroatoms, such as those in peroxides and disulfides, as well as N–N, N–O, S–O, and similar bonds, are also easily reducible.

Double and triple bonds are frequently reducible, in particular when the reduced bond is part of a conjugated system; for example, unsaturated conjugated or aromatic hydrocarbons, carbonyl compounds and their nitrogen analogs, and nitro, nitroso, and azo compounds.

Because of the relatively easy oxidation of mercury, anodic waves are observed with the DME only for the strongest reducing agents such as hydroquinones, enediols (e.g., ascorbic acid), phenylhydroxylamine derivatives, and certain aldehydes. Numerous organic substances nevertheless yield anodic waves corresponding to mercury-salt formation: for example, thiols and other derivatives of bivalent sulfur, amines, and some heterocycles. The organic compounds, in these cases, are not oxidized, but only make the oxidation of mercury easier (the amount being directly related to the concentration of the compound). The more stable the mercury salt, the more easily the oxidation occurs, that is, the more negative the wave appears on the potential axis.

Nature of the Sample. Since electrolysis is almost invariably carried out in solution, it is necessary first to convert any sample into a solution. The sample itself can be a solid, liquid, or gas. In the last two cases, the dissolution is usually straightforward; for solid samples, procedures used in other wet analytical procedures are followed, with the exception that the use of nitric acid is usually avoided because of the possibility of generating electroactive nitrogen oxides.

Typical examples of samples range from metals, alloys, slags, ores, minerals, and fertilizers to a variety of organic materials such as polymers; petroleum and its products; fibers and textile materials; pesticides; insecticides; herbicides; food and food products, including beverages (beer and wine); biological materials; pharmaceuticals; plants; and soils. Examples of liquid samples subjected to polarographic analysis are body fluids such as blood and urine, natural and industrial water, and seawater. Polluted atmosphere and industrial gases represent samples of gaseous nature. In particular, the ability to determine oxygen in the presence of practically all other gases is often utilized.

Detection Limits, Accuracy and Precision, and Selectivity. The detection limit in DC polarography is usually about 5×10^{-6} M. Below this concentration, the capacitive-current uncertainty is greater than the faradaic current for typical samples. Other polarographic techniques (see Sec. 3.4) offer lower detection limits: 10^{-8} M with differential-pulse polarography, and even lower when preconcentration is used, as in stripping voltammetry. In the latter case, metals that form amalgams and anions that form insoluble mercury salts can be determined down to 10^{-9} and even 10^{-10} M. The typical range for stripping analysis, however, is 10^{-5} to 10^{-9} M, with an absolute detection limit of about 0.1 ng of the element determined. With care taken, the precision is about ± 10–20% at the 10^{-9} M level, and about ± 2–5% for greater concentrations.

Considerably lower detection limits (lower than 10^{-6} M) in DC polarography are observed for some catalytic currents. For example, cobalamine in buffers can be detected even with DC polarography down to about 10^{-8} M.

The final volume of the solution for polarographic analysis is usually of the order of 5 to 20 mL; decrease of the volume down to 0.1 mL does not present any difficulties apart from the need to use special cells. If necessary, polarographic electrolysis can be carried out in as little as 0.01 mL. However, the handling of small volumes requires considerable skill and is much more time consuming than operation on the milliliter level (as is usually the case with microchemical operations, regardless of the particular analytical method used).

The accuracy and reproducibility of the results depend considerably on the shape of the wave under study. For well-developed waves or well-separated peaks, the limiting current (wave height) can be measured with an error of ± 1–2%. Considering the required constancy of the various experimental factors controlling the limiting current and inaccuracies in the preparation of the sample, an overall error of about $\pm 3\%$ can be said to be typical for polarographic determinations. In high-precision serial determinations, the error can be improved to 1% when all factors are strictly controlled.

The accuracy and precision decrease for ill-shaped waves. If, moreover, the composition of the sample with respect to electroinactive materials that affect the shape of polarographic curves varies from sample to sample (e.g., in samples of biological materials), the error may increase to $\pm 5\%$ and perhaps to $\pm 20\%$. However, even this level is often sufficient for trace analysis.

For the entire concentration range from 10^{-3} M to about one order of magnitude above the detection limit of the particular polarographic technique, the relative error of polarographic determinations usually remains almost unchanged. This means that, from an absolute point of view, the accuracy is good or sufficient at low concentration levels but poorer at higher concentrations. This in turn means that polarography is relatively well suited for trace determinations but is suitable for determination of samples containing 30% or more of the active component only when high precision is not required; it is unsuitable for accurate determination of the main component when the sample consists of 90% or more of this component.

In addition to sensitivity, an advantage of polarographic methods is their selectivity. Current-versus-voltage curves often reveal the presence of interfering substances—not the case in optical methods, especially when measurements are carried out at only one wavelength. The selectivity of polarographic methods is particularly great when the determination is carried out in more than one supporting electrolyte.

All the characteristics of polarographic methods as analytical tools indicate that the methods are particularly useful for trace analysis. Components can be determined in the mg/L to μg/L (ppm and ppb, respectively, in dilute aqueous solution) range with accuracy sufficient for most practical studies. The amount of sample needed for analysis varies between a few micrograms and about 50 mg.

Interferences. Two types of interference may be encountered in polarography. The first is identical to problems encountered with other analytical methods in the presence of two or more species that give signals too similar to be distinguished. When two species give waves whose potentials differ by less than about 100 mV in DC polarography (and by less than about 50 mV in linear-scan voltammetry and differential-pulse polarography, Sec. 3.4), such waves or peaks overlap and prevent the determination of the individual components.

In addition, DC polarography has a more specific limitation. Measurement of a more positive wave in the presence of an excess of material reduced at more negative potentials (wave A in Fig. 3.10) can be carried out with maximum accuracy; but when the trace material to be determined is reduced at more negative potentials (wave B′ in Fig. 3.10)—that is, when a small wave follows a large one—measurement of the small, more negative, wave can be carried out only when the concentration

FIGURE 3.10. *Polarograms of mixtures of electroactive substances in different ratios. Curve 1: Reduction of a small amount of a more easily reduced substance in the presence of a larger amount of a substance reduced at a more negative potential. Curve 2: Reduction of a small amount of a substance in the presence of a large amount of a substance reduced at more positive potential.*

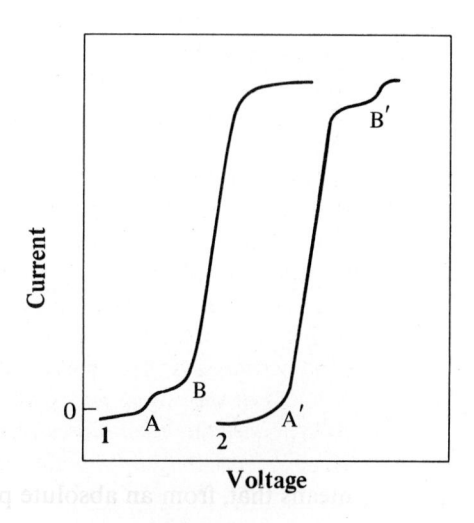

ratio between the excess component and the analyte is less than about 10:1. At larger excess, the accuracy of the determination of the trace component decreases considerably.

Separation of overlapping waves or peaks is frequently made possible by changing the composition of the supporting electrolyte. Differences in complexing properties are used primarily in inorganic analysis, whereas differences in acid-base properties are used in organic analysis. Alternatively, a change in solvent may result in separation of overlapping curves, owing to differences in solvation of the electroactive species.

An example of separation of overlapping waves is the analysis of mixtures of Pb(II) and Tl(I). A mixture of these two ions in neutral media gives a wave that cannot be resolved into the individual waves of Pb^{2+} and Tl^+; the half-wave potentials differ by only about 60 mV. When excess sodium hydroxide is added, the wave for Tl(I) remains at virtually the same potential as in neutral media (since thallium does not form hydroxo complexes), whereas the wave of Pb(II) is shifted to more negative potential by about 300 mV, due to formation of plumbate (Fig. 3.11). Generally, when a metal cation forms a complex, its reduction potential is made more negative—that is, it is more difficult to reduce.

Selectivity depends a great deal on the system studied, and may be either better or worse than in ultraviolet spectrophotometry. This can be illustrated for the simple alkaline cleavage of α,β-unsaturated carbonyl compounds. When cleavage of chalcone ($C_6H_5COCH=CHC_6H_5$) is followed, polarography permits determination of the parent compound, of benzaldehyde, and of the sum of acetophenone and the intermediate ketol $C_6H_5COCH_2CH(OH)C_6H_5$ in the mixture. Because the concentrations of benzaldehyde and acetophenone as final products must be equal, the concentrations of all four components in the mixture can be determined polarographic-

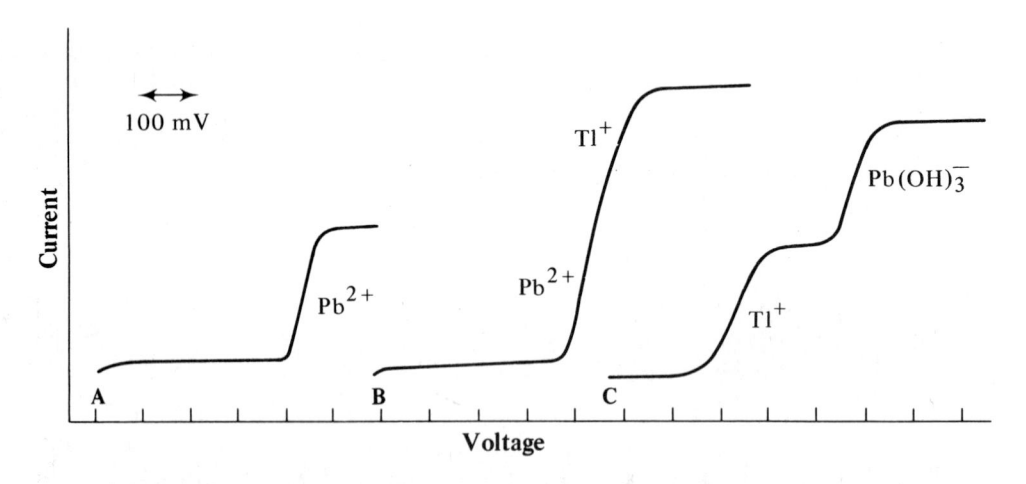

FIGURE 3.11. *Polarographic waves of Pb^{2+} and Tl^+, illustrating the shift of waves with change in background electrolyte. Curve A: Lead reduction in neutral media. Curve B: Merged waves for the reduction of lead and thallium in neutral media. Curve C: Separate waves for the reduction of lead and thallium in excess hydroxide.*

ally. Ultraviolet spectra of the intermediates and products overlap, however, and only the concentration of the starting material can be determined spectrophotometrically without interference.

On the other hand, when the products of the alkaline cleavage of cinnamaldehyde ($C_6H_5CH{=}CHCHO$) are investigated, the polarographic waves of the aldol intermediate and of benzaldehyde overlap, and the acetaldehyde waves are ill developed. Ultraviolet spectra, however, allow the determination of cinnamaldehyde and of the aldol [$C_6H_5CH(OH)CH_2CHO$] in the presence of benzaldehyde. Polarographic and optical methods are thus frequently complementary rather than competitive.

Examples of Practical Applications

Since the invention of polarography in 1922, more than 30,000 papers dealing with this technique have been published. Since over 90% of those papers deal with practical applications, any choice of examples cannot be more than an indication of the possibilities the method offers. The following selection, which is necessarily subjective, was made with the aim of showing applications in a variety of fields.

Manganese and Iron in Ores. In alkaline triethanolamine solution, Mn(III) gives a reduction wave at -0.3 V whereas the Fe(III) wave is at -1.0 V. Since copper, lead, and nickel interfere, the ore is dissolved in hydrochloric acid, and the resulting solution is first reduced with powdered zinc. After addition of triethanolamine and concentrated sodium hydroxide solutions, the mixture is shaken vigorously for about half a minute to ensure oxidation of the manganese and iron complexes to the trivalent state by atmospheric oxygen. The current-versus-voltage curves are recorded after removal of oxygen.

Copper and Other Impurities in Lead. Lead (e.g., that used in storage batteries) is dissolved in nitric acid and the greater part of the lead precipitated by adding sulfuric acid. The supernatant is treated with citric acid and the pH adjusted to about 6 with ammonia. The polarograms show waves of copper at -0.2 V, bismuth at -0.5 V, and lead (the unprecipitated remainder) at -0.4 V. The wave of Fe(III), if present, coincides with that of Cu(II).

To another portion of the supernatant, evaporated to a small volume, is added ammonia/ammonium chloride buffer. The curve recorded in this solution shows waves of copper at -0.25 and -0.5 V, nickel at -1.1 V, zinc at -1.3 V, and manganese at -1.6 V. Cobalt, if present, gives a wave that overlaps that of zinc.

If antimony is present, it can be detected by recording the current-versus-voltage curve in the original acidic solution, where its wave follows a combined wave of copper, iron, and bismuth.

Lead in Tinned Food. The sample, digested with sulfuric and nitric acids, is treated with hydrogen peroxide to remove the oxides of nitrogen. Treating the sample with sodium thiosulfate and nitric acid precipitates stannic acid. Any iron present is reduced by metallic magnesium, and the polarogram of lead is recorded after adding tartaric acid and adjusting the pH to about 5 with base.

Morphine. Morphine can be determined in pharmaceutical preparations after reacting it with nitrite to form a nitro compound. The procedure can also be used to analyze blood or other biological fluids for morphine after separating the fluid into its constituents by paper or thin-layer chromatography. The sample is dissolved in hydrochloric acid, potassium nitrite is added, and the sample is allowed to stand for 5 min at 20°C. After adding an excess of potassium hydroxide and removing oxygen, the polarogram is recorded.

When nitration is carried out under the conditions described, the presence of narcotine, papaverine, or codeine does not interfere. Heroin can be determined after acid hydrolysis of the acetyl group.

DDT. The insecticide DDT (*p,p'*-dichlorodiphenyltrichloroethane) gives a well-developed wave at −0.9 V in 96% ethanol containing lithium and tetraalkylammonium salts.

A mixture of the biologically active *p,p'*-dichlorodiphenyl derivative and the inactive *o,p'*-isomer can be analyzed. Polarography is useful for residue analysis of a number of insecticides, both chlorinated (e.g., hexachlorocyclohexane) and non-chlorinated (e.g., dithiocarbamates, pyrethrins, rotenone).

Ascorbic Acid in Fruit and Vegetables. Anodic waves corresponding to the oxidation of ascorbic acid (vitamin C) can be used for analysis of fruit and vegetables. Soft and juicy fruits (e.g., citrus fruit, currants, melons, gooseberries) and vegetables (e.g., tomatoes) can simply be squeezed, the collected juice mixed with deoxygenated pH 4.7 acetate buffer (to prevent oxidation of ascorbic acid), and the anodic waves recorded. For hard fruit and vegetables, prior homogenization is necessary.

Biological thiols such as glutathione, which interfere with or complicate titrimetric methods, do not interfere with the polarographic determination.

Carbon Disulfide in the Atmosphere. The pulp and paper industry often causes pollution of the atmosphere with carbon disulfide. To determine CS_2 in the atmosphere, a gaseous sample from a chimney, for instance, is drawn into a vessel containing diethylamine solution in 96% ethanol, which converts the carbon disulfide into diethyldithiocarbamate while still at the site. The solution does not deteriorate with time, and so the sample vessel can be left in place for several days to enrich the sample in CS_2. In the laboratory, the sample is diluted with a lithium chloride solution, and a polarogram is recorded.

The method is unaffected by a fivefold excess of hydrogen sulfide and of most mercaptans (phenylmercaptan, however, interferes). If carbon oxysulfide is present, the method must be modified.

Dissolved Oxygen: The "Oxygen Electrode." Compact portable units are available for the determination of dissolved O_2 gas in aqueous solutions. The "oxygen electrode" or sensor probe is really an electrochemical cell composed of a gold or platinum cathode and a suitable reference anode—usually some type of silver electrode. A constant potential of about 0.8 V is applied between the two electrodes. The cell is separated from the test solution by a gas-permeable membrane, typically cellophane, polyethylene, or Teflon. Oxygen diffuses from the test solution through the membrane and is reduced at the cathode; the resulting current is proportional to the oxygen level of the test solution. The entire system must be calibrated with one or more solutions of known oxygen content.

Complete benchtop units are used routinely for monitoring dissolved oxygen in biological fluids and systems. Portable pressure- and temperature-compensated oxygen-electrode systems are available for monitoring oxygen levels in oceans, lakes, and rivers and in effluents and waste waters at depths down to about 30 m.

3.4 VARIATIONS OF THE CONVENTIONAL POLAROGRAPHIC METHOD

Many modifications of polarography have been proposed for a number of specific purposes: to investigate mechanisms of electrode reactions, to decrease the time needed for examination of a given sample, to increase sensitivity, and so on. Here, those modifications that have found significant applications in analytical work will be described briefly.

Pulse Polarography

A major limitation on the application of DC polarography to trace determinations is the detection limit. Concentrations below the $10^{-5}\,M$ level typically cannot be determined because of the magnitude of the capacitive (residual) current relative to the faradaic (electrolysis) current (see Fig. 3.7). Pulse polarography permits an increase in the faradaic-to-capacitive current ratio by taking advantage of the difference in the time dependences of the processes and, in part, by making the current measurements during the latter portion of the drop-life, when the area of the mercury electrode is not changing rapidly with time.

The area of a mercury drop is proportional to $t^{2/3}$. Hence, the change in area with time is very rapid in the early portion of the drop-life, but the area becomes relatively constant in the later stage. In pulse polarography, the potential is not applied to a given drop until the area-time curve has flattened out.

The charging current has a time dependence at an electrode analogous to that of the response of a simple capacitor to a potential pulse: The current flow is initially very high when a voltage is applied but decays exponentially as the voltage across the capacitor approaches its maximum value (i.e., approaches the value of the applied voltage). For the circuit in Figure 3.12,

$$i_c = \frac{E}{R} e^{-t/RC} \qquad (3.25)$$

This circuit is comparable to an electrode-electrolyte interface where R is the solution resistance and C is the double-layer capacitance, typically about $20\,\mu F/cm^2$. The important point in Equation 3.25 is that the capacitive current decays exponentially in an experiment in which a potential pulse is applied to an electrode assumed to have constant area. (Review Sec. 1.2 and Prob. 7 in Chap. 1.)

The faradaic (electrolysis) current also decays in such an experiment. As

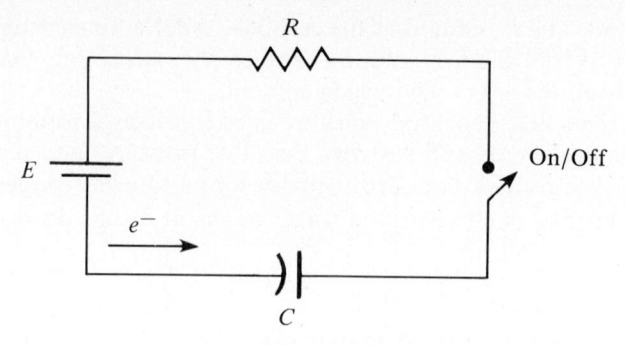

FIGURE 3.12. *Circuit for charging a capacitor in series with the applied voltage. When the on-off switch is closed, the current (electron flow) charges the capacitor C until its voltage equals that of the battery.*

shown in Equation 3.1, this current depends on the slope of the concentration gradient at the electrode surface of the species being electrolyzed, $(\partial c/\partial x)_{x=0}$. After application of the voltage pulse, consumption of the electroactive species in the vicinity of the electrode causes this term to diminish with time; hence, the faradaic current decreases with time. For a diffusion-controlled electrolysis in unstirred solution, at a planar electrode that is assumed to have a constant area A,

$$i_d = \frac{nFAD^{1/2}c}{(\pi t)^{1/2}} \tag{3.26}$$

if the potential pulse is sufficiently negative of E^0 to cause the species to be reduced immediately upon reaching the electrode surface.

The decay of the faradaic current with time is much slower than the decay of the capacitive current ($i_d \propto t^{-1/2}$ vs. $i_c \propto e^{-t}$). Therefore, if the polarographic

FIGURE 3.13. *Waveforms for pulse and differential-pulse polarography. Curves A and D: Excitation signal applied to the working electrode. Curves B and E: Instantaneous current observed at a single drop as a function of time. Curves C and F: The resulting current-versus-voltage curves. In pulse polarography, square-wave voltage pulses of 40-msec duration are applied to the mercury drop, of drop-life mechanically controlled at 2.5 sec (A); t_d, t'_d, t''_d, ... represent successive drops. The overall rate of increase of the amplitude of the voltage pulses is about 0.1 V/min. The instantaneous current at a single drop (B) shows the decay of current, primarily capacitive, during the first 20 msec after the application of the pulse, and the amount of the current flowing during the latter 20 msec of the pulse duration. The response (C) of the system has the familiar sigmoidal shape of an ordinary polarogram. In differential-pulse polarography, constant-amplitude pulses between about 5 and 100 mV are superimposed on a linearly increasing DC voltage ramp (D). The instantaneous current response at a single drop is similar to that for pulse polarography, except that the current is sampled at two places (E). The response is now peak shaped (F).*

CHAP. 3 Polarography and Voltammetry

POLAROGRAPHY

Pulse / Differential Pulse

Excitation Signal Applied

Current Response at a Single Drop

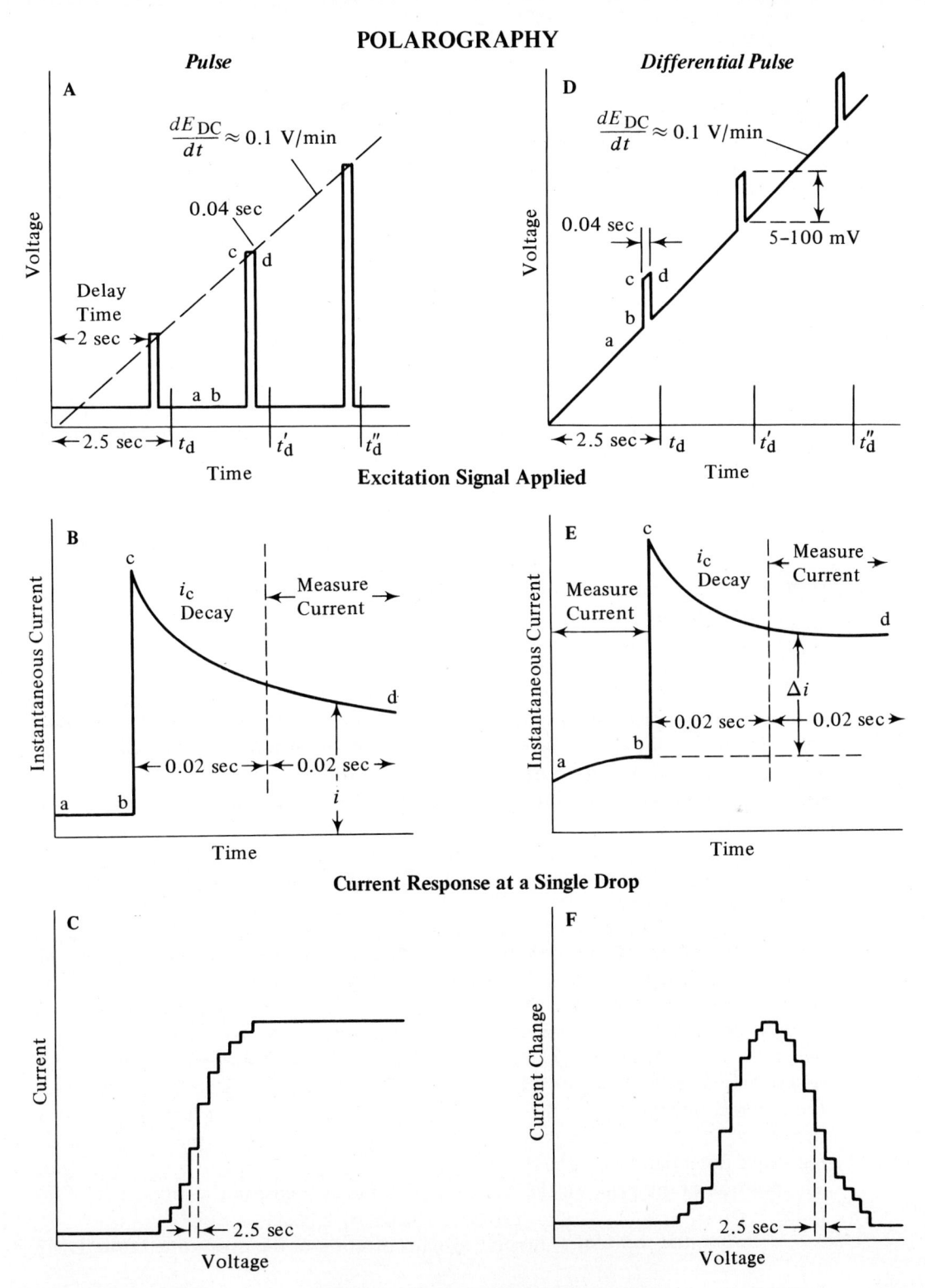

Resultant Current-versus-Voltage Curve

experiment is performed by applying a pulse to a mercury drop during the later stage of its life (to have a relatively slow change of area with time) and then delaying the current measurement for an additional time (so that the capacitive current can decay), the faradaic-to-capacitive current ratio can be increased, and thus the detection limit lowered, relative to DC polarography.

In pulse polarography, two main variants can be distinguished: techniques using gradually increasing amplitude of the voltage pulse (pulse polarography) and those in which the voltage pulse used has a constant amplitude, superimposed on a slowly increasing voltage (differential-pulse polarography).

In *pulse polarography*, a square-wave voltage pulse of about 40 msec duration is applied to the electrode during the last quarter of the drop-life (Fig. 3.13A). At the instant the voltage pulse is applied, the capacitive current is very large, but it decays rather rapidly (exponentially). The current is then measured during the 20 msec of the second half of the pulse (Fig. 3.13B), when the capacitive current is quite small. The amplitude of the applied voltage pulses is systematically increased with time. When the current response is recorded as a function of voltage, the shape of the resulting curve resembles waves in DC polarography (Fig. 3.13C), except that it has a "staircase" appearance because the current is sampled once during each drop-life and stored until the next sample period.

In *differential-pulse polarography*, the duration of pulses is similar to that used in pulse polarography (i.e., 40 to 60 msec); the pulses are also applied during the last quarter of the drop-life (when the surface area of the dropping electrode changes little with the time), but the pulses used have a constant amplitude (usually 5 to 100 mV) and are superimposed on a slowly increasing linear voltage ramp (Fig. 3.13D). Two measuring periods are used, one immediately preceding the pulse, the other very near the end of the pulse. The overall response plotted is the difference in the two currents sampled: one at point "b" in Figure 3.13E, corresponding to the polarographic current that would be measured at the given potential in the absence of a voltage pulse, and one at point "d" in Figure 3.13E, which is the sum of the current at "b" and the current resulting from the application of the additional voltage. The plot of this difference (Δi) as a function of potential is peak shaped (Fig. 3.13F).

By measuring the difference in current before the application of the voltage pulse and toward the end of the pulse, the capacitive-current contribution is further reduced in magnitude.

The peak shape of differential-pulse polarograms results from the relation between the potentials applied during the two sampling periods. When potentials of both sampling periods are either at more positive or at more negative potentials than the rising portion of the DC polarographic wave, the faradaic currents flowing at both potentials are practically the same, and hence Δi approaches zero (or is generally small). However, when at least one of the sampling periods corresponds to a potential on the rising portion of the polarographic wave, the difference between the currents flowing during the two sampling periods is different from zero, and Δi increases. The difference between the two currents is largest in the vicinity of the half-wave potential, where the slope of the DC polarographic wave is usually largest. The position of the peak on the Δi-versus-E curve depends on the amplitude of the pulse.

Both pulse techniques involve synchronization of the drop-time of the DME with the frequency of the applied pulses. This is achieved by mechanical drop-control using a magnetically controlled "hammer" knocking the capillary and causing detachment of drops from the capillary. The frequency of knocking, and hence the drop-time, can be electronically controlled and synchronized with the application of voltage pulses. It has proven useful to synchronize the detachment of the mercury drop with the power-line frequency to minimize electrical noise picked up by the electrodes and metal supporting stands from the surroundings.

Both techniques produce signals that are a linear function of concentration provided that the characteristics of the capillary electrode and the pattern of pulses remain constant. The use of differential pulse assumes constant amplitude of the pulse as well as constant frequency, pulse duration, and location of sampling periods.

The detection limit for pulse polarography is typically about $10^{-7} M$, and about $10^{-8} M$ for the differential-pulse method—although, of course, detection limits do depend on the electrochemical properties of the substance determined, interferences, and other experimental variables. The detection limit for As(III) by differential-pulse polarography, for example, has been reported to be $4 \times 10^{-9} M$ (0.3 ppb), with a linear calibration curve up to $8 \times 10^{-4} M$.

Linear Scan Voltammetry

Principles. Instead of working under essentially potentiostatic conditions with a DME as in DC and pulse polarography, it is possible to carry out the whole potential scan (e.g., from 0 V to -2.0 V or over any other similar potential range) on a constant electrode surface. The experimental setup is the same as in polarography except for the type of indicator electrode. The instrumentation is also the same, with the exception that a means of applying a variety of potential scan rates must be incorporated.

Most experiments involve scan rates in the range 0.05 to 50 V/sec, but occasionally mechanistic studies require a wider range. For quantitative-analysis studies, the slow end of the scan-rate range is usually used so that the data can be obtained on a strip-chart or *x-y* recorder rather than an oscilloscope or with a digital data-acquisition system.

The most common indicator electrode is a stationary mercury drop. Generally, such electrodes are micropipets that can be filled with mercury modified to allow electrical contact with the Hg therein. A fresh Hg drop can be extruded from the capillary for each experiment. These electrodes are available from suppliers of electrochemical equipment such as Princeton Applied Research and Metrohm.

Platinum electrodes are often used because they allow studies in the range $+1.1$ to 0 V versus SCE that is not accessible at Hg electrodes. The erratic background currents associated with Pt make them less suitable than Hg for trace-level studies. Carbon electrodes, such as graphite and pyrolytic graphite, are also used as indicator electrodes in linear-scan voltammetry. They have a useful potential range that overlaps that of Pt, and which, depending on the supporting electrolyte, extends well into the negative-potential region. Because the carbon surface is less easily renewed than Hg, these electrodes are primarily used at positive potentials inaccessible to Hg.

Typical current-versus-voltage curves are shown in Figure 3.14. For qualitative interpretation of the shape of such curves, it is useful to consider two potential

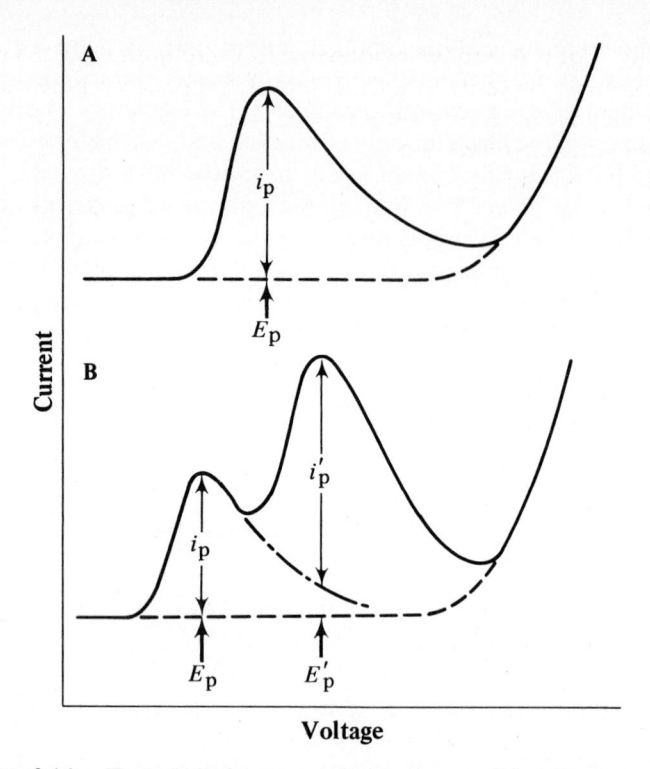

FIGURE 3.14. *Typical linear-scan voltammograms for reductions. Curve A: Reduction of a single species. The peak potential (E_p), and the magnitude of the peak current (i_p) are marked. The dashed line represents the "background" current obtained in the absence of the electroactive species. Curve B: Reduction of two species, with some overlap of peaks. Note that the peak current for the second wave, i'_p, is measured from the (extrapolated) current due to the first peak.*

regions. At potentials more positive than that of the peak (E_p), the increase in current with increasingly negative voltage is caused by the same factors as the rise of the wave in DC polarography, namely, the increased rate of the electrolytic process with increasing potential. On the other hand, at potentials more negative than that of the maximum current, the depletion of the electroactive species in the vicinity of the electrode surface becomes important. At the voltage-scan rates used, diffusion of the electroactive species from the bulk of the solution is not fast enough to replenish that removed at the electrode surface; hence, the concentration at the surface, and thus the current, decreases with time in the same manner as predicted by Equation 3.26, as $t^{-1/2}$. The appearnace that the current is decreasing with potential is coincidental; this misleading display results because the potential scan is generally continued so that the voltage becomes more negative with time. If the scan were stopped and the potential held just beyond the peak, the current would still decrease in the same manner.

Theory. For a reversible redox system where both oxidized and reduced forms are soluble, it can be shown that the current depends on potential in such a way as to pass through a maximum. For the potential of a peak obtained with a planar electrode, it can be demonstrated that

$$(E_p)_{\text{planar}} = E_{1/2} - 1.1\left(\frac{RT}{nF}\right) \tag{3.27}$$

where $E_{1/2}$ = DC polarographic half-wave potential

The potential of the peak corresponding to a reduction process is thus more negative than the half-wave potential by $28/n$ mV at 25°C.

The current (with both spherical and planar electrodes) is directly proportional to concentration and can be used for analytical purposes. For reversible systems, the peak current (i_p) in microamperes obtained with a linear scan at a planar electrode of area A in square centimeters is given by

$$i_p = kn^{3/2}AD^{1/2}cv^{1/2} \tag{3.28}$$

where k = Randles-Sevcik constant (2.69×10^5 at 25°C)
    c = concentration in millimoles per liter (mM)
    v = scan rate in volts per second

The current thus depends on the area of the electrode and on the concentration and diffusion coefficient of the electroactive species. Apart from the difference in the value of the proportionality constant (k), the peak current (i_p) shows a dependence on the number of electrons transferred (n) different from that observed in polarography: Instead of being directly proportional to n, in linear-scan voltammetry the peak current is proportional to $n^{3/2}$. Finally, the essential difference between the currents obtained by the two techniques is in the dependence of the peak currents on the rate of scanning, v, which becomes an important variable.

For irreversible electrode processes, the peak current is often lower than that for a reversible one. Peaks for irreversible processes are also less sharp, and the whole curve is more drawn out; but the peak current is still directly proportional to concentration.

It is interesting to note that the peak current increases with scan rate. Nevertheless, the detection limit cannot be increased by using faster scans. The problem is that the capacitive (residual) current is proportional to v, whereas the faradaic peak current is proportional to $v^{1/2}$. Thus, faster scans decrease the i_p-to-i_c ratio. The detection limit of linear-scan voltammetry is about the same as for DC polarography.

Measurement of Peak Current. Measurement of the peak current is usually carried out by extrapolating the current before the peak and measuring the difference between this extrapolated baseline and the peak (Fig. 3.14A). This presents no problems so long as there is only one peak on the current-versus-voltage curve or, if there is more than one peak, the individual peaks are separated by several hundred millivolts.

In the presence of several peaks that differ by 100 to 200 mV or less, measuring the most positive peak (in reductions) is still simple. Measuring peak currents corresponding to successive processes is more difficult, since extrapolating the decreasing current of the more positive peak is always somewhat arbitrary (see Fig. 3.14B).

Stripping Voltammetry

For cations of amalgam-forming metals, the sensitivity of polarographic techniques can be increased by accumulating the material within the electrode (as an amalgam). This preelectrolysis is generally performed at least 0.2 V negative of the E^0. This electrolysis is essentially an electrochemical preconcentration step wherein the electroactive material is concentrated from the relatively large solution volume, perhaps 5 to 20 cm^3, into the much smaller electrode "volume," perhaps 10^{-3} to 10^{-4} cm^3. Typical increases in concentration are of the order of 10- to 1000-fold. This is followed by a voltage scan from negative to positive potentials. The resulting curves (Fig. 3.15) correspond to the anodic dissolution of the amalgam.

The peak current for the anodic stripping step is quite analogous to the peak current in linear-scan voltammetry, discussed in the previous section. However, the concentration term of Equation 3.28 is the concentration of the metal in the amalgam rather than the concentration of the ion in solution. Because of this preconcentration step, it follows that the anodic stripping peak current will be 10- to 1000-fold higher than the cathodic linear-scan voltammetry peak on a given metal-ion solution.

The increase in sensitivity depends on the concentration of the metal amalgam formed during the preelectrolysis relative to the concentration of the analyte ion in the original solution. This value would correspond exactly to the volume of the sample solution divided by the volume of the mercury electrode (usually a hanging mercury drop) if all of the test ion were electrodeposited into the mercury. In practice, the

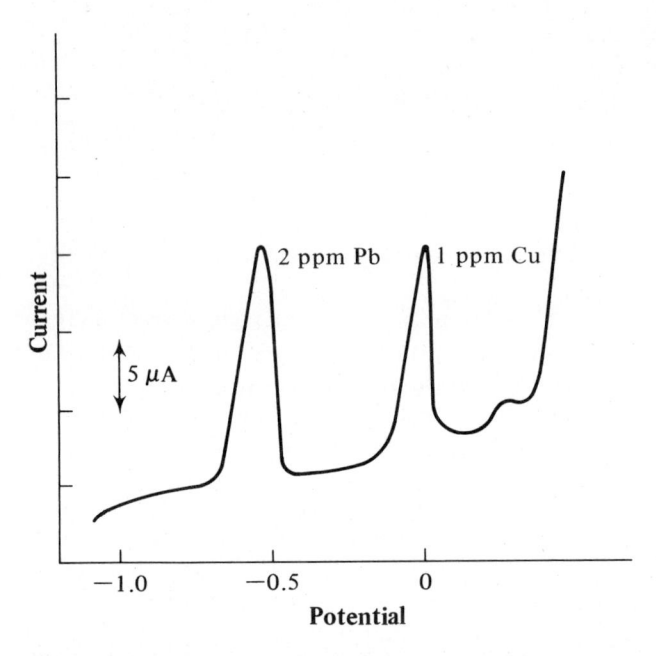

FIGURE 3.15. *Linear-sweep anodic stripping volatammogram, 2 ppm Pb and 1 ppm Cu in 0.1 M HNO_3. Conditions: 5-min plating time at -1.10 V; 15-sec rest time; a thin mercury-film electrode on glassy carbon; voltage-scan rate of 1 V/min for the stripping step.*

experiment is not performed in this manner. Instead, a small, reproducible fraction of the test species is electrodeposited from a series of standard solutions and the unknown. Reproducibility is assured by carefully controlling the volume of the electrode, the rate of stirring during the preelectrolysis (usually with a magnetic stirrer), the flux pattern to the electrode (by careful placement of the stirrer relative to the electrodes, control of the depth of the electrode in solution, and use of the same sample volume), the preelectrolysis potential, and the preelectrolysis time. The latter is usually 5 min at the $10^{-5} M$ level, 30 min at the $10^{-8} M$ level, and of the order of hours at lower concentrations. In addition, a 30-sec rest period after stopping the stirring is generally incorporated between the preelectrolysis and stripping steps. This permits the latter step to be performed under unstirred conditions, which lowers the background current.

The sensitivity of stripping voltammetry can be further increased in two ways. First, the use of differential-pulse rather than linear-scan voltammetry during the stripping step allows study at the $10^{-9} M$ level with only a 5-min preelectrolysis. Second, the sensitivity can be increased by using an indicator electrode that consists of a thin film of mercury deposited on a substrate such as glassy carbon. This approach minimizes the Hg-electrode volume. The differential-pulse method is generally preferred if the instrumentation is available, since mercury-film electrodes are relatively difficult to prepare in a reproducible manner. A further advantage of pulse techniques is that the electrolyte can be much lower, millimolar or even less, which is a critical factor in reducing the overall level of impurities. Very often, the supporting electrolyte solution must be purified by controlled-potential electrolysis at a mercury-pool cathode prior to use in stripping voltammetry.

Stripping analysis in the manner described above is limited to the study of metals that can be reduced to amalgams. Examples include Cu, Pb, Cd, In, Bi, and Zn. Variations of the technique allow the determination of any species that can be collected on an electrode surface and subsequently stripped electrochemically. For example, the halides can be accumulated by an oxidative preelectrolysis of mercury to the corresponding mercurous salts. These salts are only slightly soluble and remain as films on the Hg electrode until they are electrochemically reduced by application of a negative-going potential scan—*cathodic* stripping voltammetry. Arsenite can be reduced to a film on gold electrodes and subsequently determined by anodic stripping. Silver electrodes can oxidatively accumulate Ag^+ salts of ions such as iodide, sulfide, and sulfinate. Lead (as PbO_2) and iron (as Fe_2O_3) can be collected on various solid electrodes by oxidation of Pb^{2+} and Fe^{2+}, respectively. Phosphate and arsenate can be accumulated by the oxidation of Fe^{2+} to Fe^{3+} at a graphite electrode; the Fe^{3+} salts of these anions are slightly soluble.

A recent advance in electrochemistry is the development of electrode surfaces that are irreversibly coated with a layer of a polymer or a film of an organic or inorganic compound. The surface modification can be either by adsorption or by covalent bonding. Generally, such electrodes have been used for catalysis or fundamental studies of redox chemistry, but analytical applications have also been made. For example, a Pt surface modified by adsorption of adenosine monophosphate can accumulate Fe^{3+}. A subsequent negative-going potential scan produces a peak for the reduction of the surface-bound Fe^{3+} to "free" Fe^{2+}. The method is analogous to stripping analysis, but the accumulation is by ion exchange rather than electrolysis [3].

Instrumentation

Instrumentation for pulse polarography is complex. For differential-pulse polarography, the instrumentation requires various timing and sampling circuits, low-drift analog memories to allow storage and subtraction of two sampled currents, and good differential amplifiers to permit the amplification of Δi. Early instruments used tube circuits of limited reliability. Recent instruments involving stable integrated-circuit amplifiers and high-impedance field-effect transistors are reliable and not too expensive. Suppliers include Princeton Applied Research Corporation (EGG PARC) and IBM Instruments.

For stripping analysis, only a special electrode (hanging mercury-drop electrode, mercury-coated carbon electrode, or solid electrode) is needed. Any instrument generating a slow linear voltage scan and recording the resulting current-versus-voltage curves can be used for linear-scan stripping voltammetry. Preferably, the instrument allows curve recording from positive-to-negative potentials and vice versa. For differential-pulse stripping voltammetry, any pulse polarograph can be used.

The measurement of peaks in linear-scan voltammetry and pulse polarography is usually done by measuring the current at a chosen potential (usually that of the peak) and comparing it with the current at the same potential obtained with a blank. Such measurements are sufficiently accurate when the measured peak is not preceded by any other peak. Measuring a second peak at more negative potentials presents all the problems encountered with overlapping spectrophotometric absorption bands or chromatographic elution peaks. An empirical extrapolation of the tailing of the first peak is often used; computer-based analysis of such curves is possible.

Separation of adjacent peaks in differential-pulse polarography, where the symmetry of the peak can also prove to be useful, is usually easier than the separation of consecutive peaks in linear-scan voltammetry. As is the case in conventional polarography, more positive peaks that interfere can sometimes be shifted to more negative potentials and the sequence of peaks inverted by a change in supporting electrolyte.

In linear-scan voltammetry and differential-pulse polarography, the problem of an excess of a species reduced at more positive potentials is of considerably smaller consequence than in conventional polarography. When the current peaks of the species present in excess and that of the components to be determined are separated by more than about 0.3 V in the former and about 0.2 V in the latter technique, the presence of the more positive peak has almost no influence.

Applications

Once again, only a few examples are presented from a very large number of actual reported analyses.

Differential-Pulse Stripping Analysis of Water. The water to be analyzed is deaerated and a voltage of -1.2 V is applied to a hanging mercury-drop electrode for 60 sec while the solution is gently stirred by a magnetic stirrer. The stirring must be constant and reproducible. After 60 sec the stirrer is turned off, and after another 15 sec the stripping curve is recorded using a differential-pulse technique. The peak for zinc appears at -1.0 V, that for lead at -0.5 V and that for copper at -0.1 V. Typical concentrations in tap water might be 2 µg/L for Pb and 10 to 600 µg/L for Zn and Cu [4]. This method was devised for tap-water analysis. When lower metal-ion concentrations are to be determined (e.g., in distilled or deionized water), the preelectrolysis period can be prolonged.

Stripping analysis has been used to determine 10^{-9} M Ag levels in rain and snow samples from clouds seeded with AgI. Precisions are about $\pm 20\%$ at the 0.2 nM level, and $\pm 4\%$ at concentrations above 1 nM.

Differential-Pulse Stripping Determination of Lead in Blood. The blood sample (typically 50 µL) is digested with a mixture of sulfuric and perchloric acids, transferred into the electrolytic cell, and, after removal of oxygen, preelectrolyzed at -0.7 V for about 5 min using a hanging mercury-drop electrode in a stirred solution (the period of deposition chosen depends on the electrode used). The differential-pulse stripping curve shows a peak for lead at -0.4 V. A typical "normal" level of 200 to 300 µg/L of lead in blood will give a large signal, well above background, unless the acids are contaminated [5].

Barbital, Phenobarbital, Pentothal. Barbital can be determined in a borate buffer of pH 9.3 by means of an anodic wave that corresponds to mercury-salt formation. Since the wave height is governed by adsorption at higher concentrations, it is necessary to keep the concentration of barbital below 1×10^{-4} M.

When DC polarography is applied to phenobarbital, the wave is indistinct. However, when differential-pulse polarography is used, easily measurable peaks corresponding to mercury-salt formation are obtained, the total height of which is a linear function of concentration. This procedure has been successfully applied to the determination of phenobarbital in the presence of a number of other drugs in studies of drug metabolism.

Pentothal—ethyl(1-methylbutyl)thiobarbiturate—can be easily determined by simply dissolving the sample in 0.1 M sodium hydroxide and recording the well-developed anodic wave.

Linear-Scan Voltammetry of Tocopherols and Antioxidants in Oils and Fats. Phenolic antioxidants are added to many food products to enhance their stability. In addition, certain foods, particularly vegetable oils, contain significant quantities of natural phenolic materials, the most prominent of which are the various tocopherols (vitamin E group). Linear-scan voltammetric oxidation of vegetable oils dissolved in 2:1 ethanol/benzene solvent (0.12 M sulfuric acid) is a rapid method for estimating the tocopherol content of oils and fats. (Almost all other methods involve considerable sample preparation, such as saponification and extraction, before chromatographic separation and measurement.) The reproducibility of the voltammetric method is good; for a typical α-tocopherol content of 0.3 mg/g of oil, the standard deviation is ± 0.02 mg/g. Quantitation is achieved by the method of standard additions, and a glassy-carbon electrode is used [6].

Differential-Pulse Polarography of Arsenic. For many metals, pulse polarography and flame atomic absorption spectrometry are the most useful methods for trace-level determinations. The generally lower detection limits of the former are offset in many instances by the greater ease of performing atomic absorption experiments. In the case of arsenic, however, pulse polarography shows far greater detectability. In acidic media, As(III) is reduced in two well-developed steps; the first (-0.4 V) is due to the 3-electron reduction to As, and the second (-0.8 V) is due to further reduction to arsine, AsH_3. In 1 M HCl the detection limit by differential-pulse polarography is about 0.3 μg/L (4×10^{-9} M). Linear calibration curves over the range 20 μg/L to 50 mg/L are obtained [7]. The relative standard deviation is about $\pm 16\%$ at 2 μg As per liter and $\pm 2\%$ at 20 μg per liter [8]. The main interferences are lead and tin. One interesting aspect of the pulse-polarographic method is that As(V) is polarographically inactive, so that this method can be used to study the oxidation state of arsenic in various samples. Total arsenic can be determined by prior chemical reduction of As(V) with suitable reducing agents such as hydrazine salts or acidic KI. With the possible exception of neutron-activation analysis, differential-pulse polarography is probably the most sensitive method presently available for arsenic assay.

3.5 AMPEROMETRIC TITRATIONS

A titration in which measurement of the current flowing at a voltammetric indicator electrode is used for detection of the equivalence point is termed an *amperometric titration*. The current measured is almost always a limiting current, which is proportional to concentration, and can be due to the substance titrated, to the titrant itself, to a product of the reaction, or to any two of these—depending on the potential of the electrode and the electrochemical characteristics of the chemical substances involved. The titration curve is a plot of the limiting current, corrected for dilution by the reagent and, if necessary, for any residual current, as a function of the volume of titrant. Ideally, the titration curve consists of two linear segments that intersect at the equivalence point.

Amperometric titrations can be classified into two groups: those using one polarized (indicator) electrode plus a reference electrode, and those using two polarized or indicator electrodes.

One Polarized or Indicator Electrode

Many of the principles of amperometric titrations can be understood by considering an example: the titration of Pb^{2+} with standard sodium sulfate solution. Figure 3.16 illustrates the current-versus-voltage curves for lead ion that could be obtained during the course of an amperometric titration, and the resulting amperometric titration. Under the experimental conditions used, lead ion is reducible, with an $E_{1/2}$ at about -0.4 V, and the sulfate ion is nonreducible. A constant voltage, which may have any value on the diffusion-current plateau, is applied to the indicator electrode; in this case, -1.0 V is applied to a dropping mercury electrode.

At the start of the titration, a polarogram of the test solution would have the appearance of curve "a" in Figure 3.16A (after sufficient deaeration to remove the

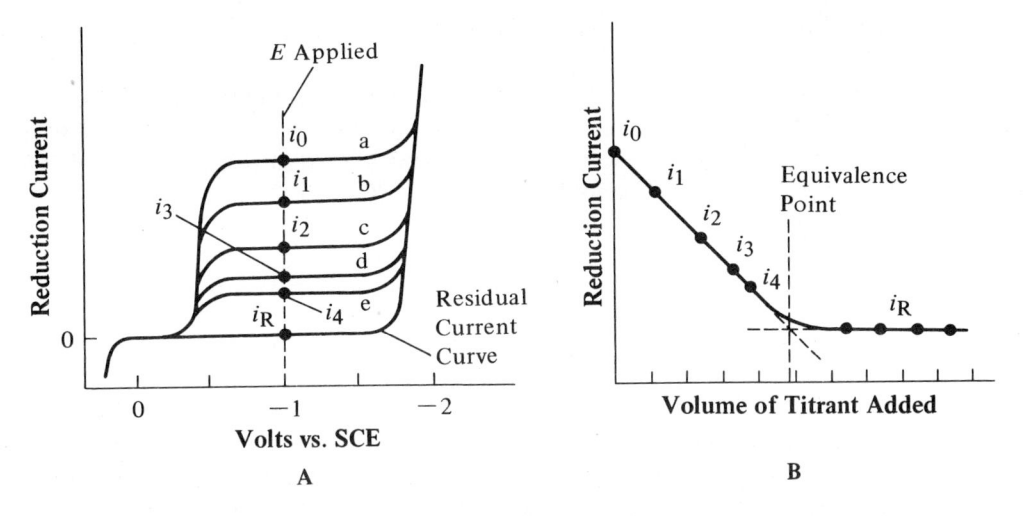

FIGURE 3.16. *Current-versus-voltage curves and amperometric titration curve for the titration of Pb^{2+} with Na_2SO_4 solution. A: Successive current-versus-voltage curves for the reduction of Pb^{2+} ion at a mercury electrode, made after increments of SO_4^{2-} were added. B: Resulting amperometric titration curve for currents (i_0, i_1, i_2, ...) measured at an applied potential of -1 V versus SCE.*

dissolved oxygen). Therefore, the current measured at -1.0 V would have the value i_0. Increments of titrant precipitate $PbSO_4$ and remove some of the Pb^{2+} from solution; since the titrant does not produce a reduction current at the applied voltage, the current decreases with successive additions to i_1, i_2, and so on. When the lead ions have been removed completely from solution, the only current flowing is the residual current, i_R, caused by the supporting electrolyte. A plot of current as a function of titrant volume will have the L-shaped appearance shown in Figure 3.16B.

Normally, there will be some rounding in the vicinity of the equivalence point because of equilibrium effects; the more dilute the solutions used and the more the position of equilibrium favors the reactants, the more pronounced the rounding. In the present example, the finite solubility of $PbSO_4$ will result in some Pb^{2+} ions still being in solution at the equivalence point, and an excess of titrant is necessary to drive the lead-ion concentration to a sufficiently low level that the lead-ion diffusion current is insignificant compared to the residual current.

One advantage of amperometric titrations is that the substance titrated does not have to be electroactive if an appropriate titrant with electrolytic properties is used. For example, sulfate ion can be determined by titration with Pb^{2+}. In this case, an essentially constant residual current flows until there is excess titrant in the test solution. After the endpoint, a linearly increasing current appears that is proportional to the concentration of the excess titrant. The amperometric titration curve will have a shape the reverse of that shown in Figure 3.16B: _/ shaped, or "reverse L shaped."

When both the substance titrated and the titrant undergo electrochemical reactions at the voltage selected, the current will decrease (linearly) up to the equivalence point, then increase again with addition of excess titrant, resulting in a V-shaped titration curve. An example of this is the titration of Pb^{2+} with potassium dichromate in a weakly acidic supporting electrolyte. Dichromate ion is reduced to Cr^{3+} at the DME with $E_{1/2} \approx 0$ V versus SCE. If -1.0 V is applied to the indicator electrode, both Pb^{2+} and $Cr_2O_7^{2-}$ are reducible, and a V-shaped titration curve will result. If, on the other hand, the applied voltage is -0.2 V, only dichromate ion is reducible, and a reverse L-shaped titration curve results.

In general, the best way to predict the shape of amperometric titration curves is to look at or construct the current-versus-voltage curves of the test solution during the course of the electrolysis.

Two Polarized or Indicator Electrodes

The apparatus used for titrations with one polarized electrode, described above, includes a reference electrode whose potential remains fixed during the course of the titration. A second approach involves applying a small, fixed, potential difference (20 to 250 mV) across two identical indicator electrodes; this is often called a *bi-amperometric titration*.

Again, the principles underlying this type of titration can best be understood by considering an example, in this case the titration of ferrous ion (Fe^{2+}) in acidic medium with standard cerate (Ce^{4+}) solution—two reversible redox couples. Figure 3.17 illustrates the current-versus-voltage curves expected during the titration. At

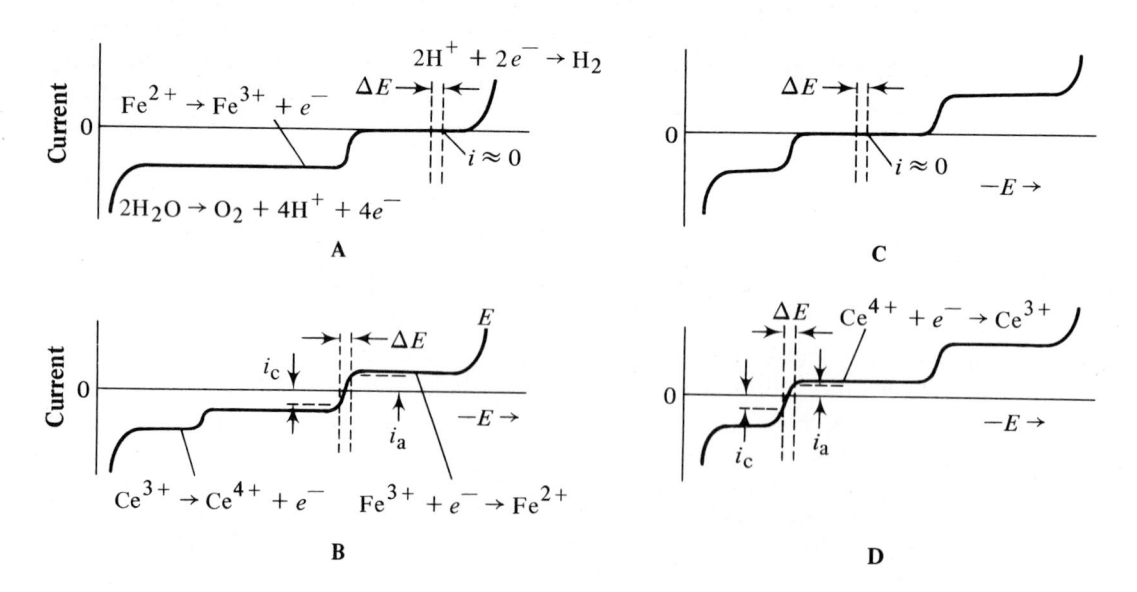

FIGURE 3.17. *Theoretical current-versus-voltage curves at a platinum electrode during an amperometric titration of Fe^{2+} with Ce^{4+} with two polarized or indicator electrodes. Here, ΔE is the constant voltage applied to the two indicator electrodes. A: At the start of the titration. B: At the midpoint of the titration. C: At the equivalence point. D: After the equivalence point.*

the start (Fig. 3.17A), the only electrochemical processes that occur are the oxidation of Fe^{2+} to Fe^{3+} at about $+0.5$ V and the two background processes—reduction of protons to hydrogen gas and oxidation of water to oxygen. The small, fixed, potential difference (ΔE) applied to the indicator electrodes shifts along the potential axis until it stops at the place on the current-versus-voltage curve where the current due to the reduction taking place at the cathode is equal to the current due to the oxidation taking place at the anode. This is at the voltage where the residual current curve crosses the $i = 0$-axis; and the actual current flowing is very close to zero.

Once some Ce^{4+} is added, Fe^{3+} and Ce^{3+} are generated by the chemical redox reaction, and the current-versus-voltage curves for the test solution then have components reflecting the reduction of Fe^{3+} and oxidation of Ce^{3+}, as shown in Figure 3.17B. The ΔE shifts along the current-versus-voltage curve to the point where the anodic and cathodic indicator currents, i_a and i_c, are equal; they are due to the reversible Fe^{2+}/Fe^{3+} oxidation-reduction couple. Prior to the equivalence point, the indicator current, $i = i_a = i_c$, increases until about halfway to the equivalence point and then decreases back to zero at the equivalence point—Figure 3.17C—where the current is once again due only to the small residual current. The voltage applied is insufficient to cause appreciable oxidation of Ce^{3+} and reduction of Fe^{3+}. After the equivalence point, there is some excess Ce^{4+} in solution, ΔE shifts to the potential of the reversible Ce^{3+}/Ce^{4+} couple at about $+1.4$ V, and the indicator current again begins to increase.

The shape of the amperometric titration curve in this case, where both the titrant and the substance titrated undergo reversible redox reactions, is illustrated in Figure 3.18A. When the substance titrated does not have a reversible voltammetric wave, the titration curve will have the shape illustrated in Figure 3.18B. Prior to the equivalence point, the applied voltage is too small to cause both oxidation and reduction of the redox couple of the substance titrated. If the titrant has an irreversible wave, the titration curve will look like that in Figure 3.18C. This type of titration is

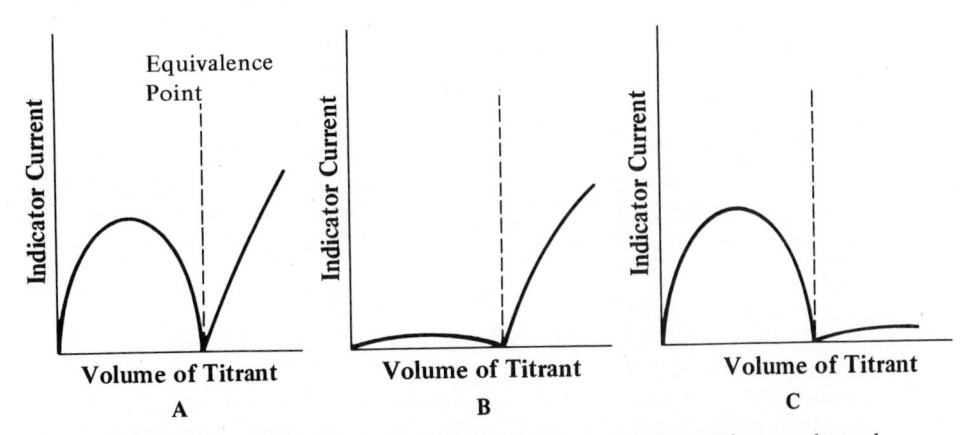

FIGURE 3.18. *Titration curves for amperometric titrations with two polarized or indicator electrodes. A: Both the titrant and the substance titrated have reversible voltammetric curves. B: The substance titrated displays irreversibility, and the titrant reversibility. C: The substance titrated displays reversibility, and the titrant irreversibility.*

commonly called a "dead-stop" titration, because the indicator current falls to zero at the equivalence point.

Applications of Amperometric Titrations

If the stoichiometry of the titration reaction is known and reproducible, amperometric titrations are intrinsically more accurate and precise than direct voltammetric analyses. Precision and accuracy of a few tenths of a percent are commonly attainable with sufficiently concentrated solutions, about $10^{-4} M$ or greater. Precision and accuracy are limited primarily by the errors involved in standardizing the titrant and measuring the volume delivered, and by the abruptness of change in indicator current at the equivalence point. High accuracies require, of course, minimization of or correction for any volume change during the titration. The apparatus for amperometric titrations is quite simple, and requires no prior calibration. Their primary disadvantages, as with potentiometric titrations, are the time required to perform a titration as opposed to a single measurement, and the effort involved in the preparation and storage of standard solutions.

There are numerous examples in the literature [9, 10] of the application of amperometric titrations. One popular titrant is the silver ion, in the form of a silver nitrate solution, coupled with a rotating platinum indicator electrode. Ions such as cyanide; tetraphenylborate; various sulfides; and (singly or mixed) chloride, bromide, and iodide can be titrated. For example, anywhere from 8 μM to 0.1 M cyanide in 0.1 M sodium hydroxide can be titrated amperometrically with silver nitrate solution with good accuracy and precision. Again, the total chlorine in insecticides decomposed with sodium and xylene has been titrated at silver electrodes using biamperometric endpoint detection. An important application is the determination of sulfhydryl groups, especially in proteins and other natural materials. The method is based on the reaction of silver ion—and some other heavy-metal ions such as Hg^{2+}—with sulfhydryl compounds to form highly undissociated mercaptides. For example, SH groups in cysteine, glutathione, certain proteins, and dialyzed human sera can be determined. Under appropriate conditions, as little as about 10 nanomoles of protein can be titrated reproducibly.

Another important amperometric titrant is bromine solution, which undergoes stoichiometric oxidation-reduction reactions with many substances such as As(III), Sb(III), ammonium salts, and others. Often the titration involves adding an excess of KBr to an acidified solution of the substance to be oxidized and then titrating it with potassium bromate solution. Bromine is thereby generated *in situ*:

$$BrO_3^- + 5Br^- + 6H^+ \rightleftarrows 3H_2O + 3Br_2 \qquad (3.29)$$

This avoids the problems involved in storing unstable Br_2 solutions. Bromine can be used to titrate a wide variety of oxidizable organic compounds such as phenols, hydrazines, and anilines. The "bromine numbers" of olefinic hydrocarbons—a measure of the total unsaturation present—are often determined by titrating the hydrocarbon with acidified potassium bromate, or by generating Br_2 electrolytically from an acetic acid/methanol/water solvent containing KBr. Olefinic hydrocarbons generally display no electrochemical properties under the experimental conditions used; however, at the equivalence point the presence of a small excess of bromine increases the current through the indicator electrode pair. Chapter 4 discusses a number of other reagents that can be electrolytically generated *in situ* in order to perform an amperometric titration.

SELECTED BIBLIOGRAPHY

BARD, A. J., and FAULKNER, L. R. *Electrochemical Methods: Fundamentals and Applications.* New York: John Wiley, 1980. *Modern reference on the theory and practice of electrochemical methods.*

BOND, A. M. *Modern Polarographic Methods in Analytical Chemistry.* New York: Marcel Dekker, 1980. *A thorough coverage of the analytical application of electrochemical methods, especially AC polarography, pulse polarography, and stripping analysis.*

GALUS, Z. *Fundamentals of Electrochemical Analysis.* New York: Halsted Press, 1976. *A description of the methods for elucidation of mechanisms of electrode reaction.*

MEITES, L. *Polarographic Techniques.* New York: John Wiley, 1965. *Practical guide to polarographic experimentation.*

SAWYER, D. T., and ROBERTS, J. L. *Experimental Electrochemistry for Chemists.* New York: Wiley-Interscience, 1974. *A description of the practical aspects of various electrochemical methods.*

VYDRA, F., STULIK, K., and JULAKOVA, E. *Electrochemical Stripping Analysis.* New York: Halsted Press, 1976. *Review of the theory and practice of stripping methods.*

ZUMAN, P. *Organic Polarographic Analysis.* Oxford: Pergamon Press, 1964. *Discussion of mechanisms of organic electrode reactions and applications of electrochemistry.*

REFERENCES

1. H. H. BAUER, in *Electroanalytical Chemistry*, A. J. Bard, ed., Vol. 8, New York: Marcel Dekker, 1965, pp. 169–279.
2. A. J. BARD and L. R. FAULKNER, *Electrochemical Methods: Fundamentals and Applications*, New York: John Wiley, 1980, pp. 553–73.
3. J. A. COX and M. MAJDA, *Anal. Chim. Acta*, 118, 271 (1980).
4. H. SIEGERMAN and G. O'DOM, *Amer. Lab.*, 4(6), 59 (1972).
5. Application Note AN-16, Princeton Applied Research Corporation, Princeton, N.J., 1972.
6. H. D. MCBRIDE and D. H. EVANS, *Anal. Chem.*, 45, 446 (1973).
7. Application Brief A-6, Princeton Applied Research Corporation, Princeton, N.J., 1976.
8. D. J. MYERS and J. OSTERYOUNG, *Anal. Chem.*, 45, 267 (1973).
9. J. T. STOCK, *Amperometric Titrations*, New York: Interscience, 1965.
10. J. T. STOCK, *Anal. Chem.*, 54, 1R (1982).

PROBLEMS

1. Diffusion coefficients for divalent metal ions, M^{2+}, are typically about 6×10^{-6} cm^2/sec. Capillaries used in DMEs commonly have m-values of about 2 mg/sec at mercury heights that yield t-values of 5 sec. Calculate a typical limiting current for the reduction of 1×10^{-3} M (1 mM) M^{2+} solution.

2. Beginning with the fact that for a diffusion-controlled process the limiting current, i_d, is given by the Ilkovic equation, demonstrate that

i is proportional to $h^{1/2}$. (Hint: Of the terms in the equation, only m and t are directly influenced by h, the height of the mercury column.)

3. A DC polarogram of an organic compound had a limiting current of 5.00 µA. The following data were obtained from the rising portion of the wave:

E, V	i, µA
−0.475	0.62
−0.490	1.57
−0.510	3.43
−0.525	4.38

Is the reduction electrochemically reversible? If it is reversible, determine n and $E_{1/2}$.

4. You have to determine: (a) formaldehyde and acetaldehyde in 50 samples of white wine per day; (b) 2,4,6-trinitrotoluene in white-powder samples that might be potential material in making bombs—3 to 5 samples per month; (c) copper content in a rare Etruscan vase; (d) a toxic keto compound in an antibiotic, the analysis being done in a production-line quality-control laboratory of a pharmaceutical company. In which cases and why would you use a standard addition method, and when would you use a calibration curve for evaluating current-versus-voltage curves?

5. Deionized water contains ppb levels of heavy metals, mostly zinc, copper, and lead. Solutions of reagent-grade chemicals used as supporting electrolytes contain metal ions at the same concentration or greater. Suggest a supporting electrolyte and describe how the impurity problem could be solved for a study on the determination of metal impurities in deionized water.

6. The *diffusion current constant* I_d is used to correct polarographic diffusion currents for differences in capillary characteristics. For average currents,

$$I_d = \frac{i_d}{cm^{2/3}t^{1/6}} = 607nD^{1/2}$$

For a given electroactive substance under a given set of experimental conditions (temperature, supporting electrolyte, potential of the DME, etc.), I_d should be a constant according to the Ilkovic equation; it should be independent of the capillary characteristics and reproducible in different laboratories or in the same laboratory with different capillaries. Cadmium ion exhibits a reversible 2-electron reduction wave at −0.64 V in 1 M HCl. A 0.50 mM Cd^{2+} solution gave a wave with average limiting current of 3.96 µA; the capillary characteristics were $m = 2.50$ mg/sec, $t = 3.02$ sec. (a) Calculate I_d for Cd^{2+}. (b) Calculate the diffusion coefficient for Cd^{2+} in 1 M HCl.

7. A typical value for the mercury flow rate m for a DME is 2.5 mg/sec, and a typical drop-time is 3.0 sec. What is the maximum area of the mercury drop under these conditions?

8. The oxygen content of aqueous solutions can be estimated by simply measuring the height of its polarographic reduction wave and inserting a known value of the diffusion coefficient D (2.12 × 10^{-5} cm^2/sec) into the Ilkovic equation. A sample of tap water was taken, sufficient solid KCl was added to make a 0.10 M solution, and a polarogram was obtained. The limiting current (maximum) for the first 2-electron oxygen-reduction wave was 2.11 µA. If the capillary used had $m = 2.00$ mg/sec and $t = 5.00$ sec at −0.05 V, what was the oxygen level of the tap water in millimoles per liter (mM)? In ppm?

9. In studying the mechanisms of reduction of organic compounds, one vital parameter is the number of electrons transferred per molecule. An estimate of this parameter can be obtained by assuming a value for the diffusion coefficient. For a particular ketone, we wish to decide whether n is 1 or 2. A millimolar solution yields a (maximum) diffusion current of 6.8 µA at a DME with $m = 2$ mg/sec at $t = 5$ sec. A reasonable value for the diffusion coefficient is 5×10^{-6} cm^2/sec. What is the value of n?

10. The half-wave potential for the 2-electron reduction of a given metal ion to an amalgam in 0.1 M NaClO$_4$ (a noncomplexing electrolyte) was −0.740 V. In the presence of 0.2 M ligand, L, the half-wave potential was shifted to −0.930 V. Assuming that both polarograms are reversible and that the metal-to-ligand ratio of the complex is unity, calculate K_f for the complex.

11. The detectabilities of the various voltammetric methods are determined by the residual-to-electrolysis current ratio and factors that cause uncertainty in this ratio. (a) Why is the presence

of surface-active agents in samples detrimental to detectability? (b) Can detectability be increased in linear-scan voltammetry, anodic stripping voltammetry, or polarography (DC and pulse) by increasing the electrode area? (c) Can the detectability of anodic stripping voltammetry be improved by using longer pre-electrolysis times? (d) What is the effect of the irreversibility of a reduction on the detectabilities of DC polarography and differential-pulse polarography? (e) Sensitivity is defined as the slope of a calibration curve. If a variation of a pulse-polarographic method yielded an improved sensitivity, would it also improve the detectability? (Why)

12. It is often said that many polarograms can be run on a single solution without causing significant depletion of the analyte. Consider an experiment in which the limiting current is monitored for 30 min in 50 mL of a 0.50 mM Cd^{2+} sample. What fraction of the Cd^{2+} is electrolyzed if the average current during that period is 4.0 μA?

13. The following voltammograms were recorded in a suitable supporting electrolyte at a silver electrode versus the SCE: (a) solution of silver ion; (b) solution of chloride ion.

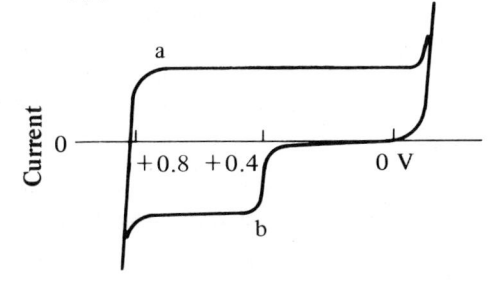

A titration of 10 mL of 0.001 M silver nitrate with 0.001 M sodium chloride was performed, and the endpoint was detected biamperometrically with two silver-wire electrodes. Sketch the biamperometric titration curve for the titration with an applied potential difference of (a) 100 mV, and (b) 600 mV. Assume no Ag^+ or Cl^- can be detected in a saturated AgCl solution. (c) Specify the anodic and the cathodic reactions occurring in each segment of the titration curves.

14. If the density of mercury at 25°C is 13.534 g/cm³, and a polarographic drop can be considered to be a sphere, derive Equation 3.2 for the surface area of a mercury drop as a function of m and t.

15. For the first 2-electron reduction wave of O_2, as described in Problem 8, calculate the peak current, i_p, expected for a linear-scan voltammogram at a planar electrode of the same area as the mercury droplets in that problem, and for scan rates, v, of 0.100 and 1.00 V/sec. Compare the peak currents obtained with the maximum limiting current at the DME.

16. Assume the electrode described in Problem 15 has a fairly typical and constant double-layer capacitance of 20 $\mu F/cm^2$. (a) Calculate the capacitive (charging) current expected at scan rates of 0.100 and 1.00 V/sec. (b) Calculate the expected ratio of faradaic to capacitive currents, i_p/i_c, for the reduction of O_2 under the same conditions as described in Problems 8 and 15.

CHAPTER 4

Coulometry and Electrogravimetry

David J. Curran
Kenneth S. Fletcher III

The electroanalytical methods discussed in this chapter involve exhaustive electrolysis, in which the reaction of interest goes to completion. Such methods are sometimes called stoichiometric or quantitative. Faraday's law of electrolysis states that the quantity of electricity passed through the cell is directly proportional to the quantity of chemical change that occurs at the cell electrodes. Mathematically, this is given as

$$Q = F \text{ (equivalents, eq)} \quad (4.1)$$

where the proportionality constant F, the faraday = 96484.56 ± 0.27 coulomb/equivalent
Q = charge passed in coulombs

The electric current at any time t, i_t, is the rate of change of charge with time, or

$$i_t = \frac{dQ}{dt} \quad (4.2)$$

In chemical terms, the current passing at an electrode may be written as

$$i_t = \pm nF \frac{dN}{dt} \quad (4.3)$$

where n = number of equivalents per mole of electroactive species
dN/dt = rate at which the number of moles, N, changes with time

The plus sign indicates a gain in moles, and the minus a loss. The total charge passed is obtained by direct integration of Equation 4.2:

$$Q = \int_0^t i_t \, dt \quad (4.4)$$

Substituting Equation 4.4 into 4.1 and rearranging, we have several equivalent expressions for the quantity of electroactive material determined:

$$\text{Equivalents} = \frac{1}{F} \int_0^t i_t \, dt \tag{4.5a}$$

$$\text{Moles} = \frac{1}{nF} \int_0^t i_t \, dt \tag{4.5b}$$

$$\text{Grams} = \frac{\text{M.W.}}{nF} \int_0^t i_t \, dt \tag{4.5c}$$

where   M.W. = molecular weight of the electroactive species (A.W. for elements)

Equations 4.5 show that the amount of substance can be calculated from knowledge of the charge passed, or obtained directly from the gain or loss in the weight of an electrode. The latter methods are classified as *electrogravimetric*, and the former as *coulometric*. Because the measured variables in these methods are electricity or weight, very high accuracy and precision are often obtained. Frequently, 0.1% precision can be achieved, and precisions of 0.005% have been reported [1].

4.1 COULOMETRIC ANALYSIS

Although the analytical utility of Equations 4.5 seems obvious, there is a problem that must be considered. We have to be able to distinguish the coulombs passed due to the reaction of interest from any other coulombs that may pass. Unfortunately, it is generally true that the total coulombs exceed those due to the reaction of interest. There is usually current present that arises from nonfaradaic processes, and more than one faradaic reaction may take place at once. The former situation is not addressed by Faraday's law, and the latter does not violate the law. Both of these matters are concerned with the question of current efficiency. An electrode reaction is said to proceed with 100% current efficiency if all of the coulombs passed through the cell are consumed in the reaction of interest. In practice, current efficiency is a relative thing. If an answer to 1% is satisfactory, then only 99% of the coulombs must go to the reaction of interest. We must go beyond Faraday's law to establish the conditions needed to achieve 100% current efficiency. The experimental requirements are contained in the sections that follow, and an introduction is provided here.

As indicated in earlier chapters, for the half-reaction $Ox + ne^- \rightleftharpoons Red$, the Nernst equation at 25°C can be written as

$$E = E^{0\prime} - \frac{0.05916}{n} \log \frac{[\text{Red}]}{[\text{Ox}]} \tag{4.6}$$

Table 4.1 shows values of $(E - E^{0\prime})$ for various concentration ratios and values of n. For $n = 1$, a change in the concentration ratio, [Red]/[Ox], from 0.01 to 100 shifts the potential from 0.1183 V positive of the formal potential of the system, $E^{0\prime}$, to 0.1183 V negative of $E^{0\prime}$, or a total potential change of 0.2366 V. Smaller potential changes are needed for larger values of n. Note that the concentration ratio of Ox and Red changes

TABLE 4.1. *The Nernst Relationship*

	$(E - E^{0\prime})$, V		
[Red]/[Ox]	$n = 1$	$n = 2$	$n = 3$
0.01	0.1183	0.0592	0.0394
0.1	0.0592	0.0296	0.0197
1	0	0	0
10	−0.0592	−0.0296	−0.0197
100	−0.1183	−0.0592	−0.0394

$T = 25°C$

by four orders of magnitude. Although the Nernst equation applies only to equilibrium situations, the idea suggests itself that control of the potential of an electrode at which the reaction $Ox + ne^- \rightleftharpoons Red$ occurs would control the conversion of Ox to Red or of Red to Ox. Methods based on this idea are known under the general term *controlled potential electrolysis*. When the potential is constant, the current is a function of time.

Another feature of the half-reaction is that the electron can be regarded as a reagent that is added to convert Ox to Red, or removed to convert Red to Ox. A mole of electrons contains 6.02205×10^{23} of them, each carrying a charge of 1.60219×10^{-19} coulomb. Thus, the rate at which reagent is added (or removed) can be controlled by external control of the current. This is the basis of *constant-current coulometric titrations*. Note that *both* the current and the potential cannot be controlled simultaneously by means external to the cell.

4.2 CONSTANT-CURRENT COULOMETRIC TITRATIONS

If a substance to be determined reacts with a reagent that can be generated electrolytically, the basis for a coulometric titration exists. When a constant current is used, the analytical result is obtained using Equation 4.5a: The total number of equivalents of reagent is the product of current and time divided by Faraday's number. Constant-current coulometric titration is analogous to conventional volumetric titration: The reagent electrons are metered using a switch rather than a stopcock, and time rather than volume is measured. Both methods require rapid, quantitative, and stoichiometric titration reactions, and some means for endpoint detection. Coulometric reagent addition offers several advantages. There is no need for time-consuming preparation of standard solutions and, hence, no need to store unstable titrants. Furthermore, the ease of metering the reagent by using electric current rather than stopcocks and pumps facilitates automation and remote titrations (e.g., the titration of hazardous radioactive materials).

The apparatus for coulometric titrations is shown in Figure 4.1. The counter electrode is isolated in a tube terminating with a sintered glass disk to prevent products from this electrode from reacting with components of the sample solution. Provision

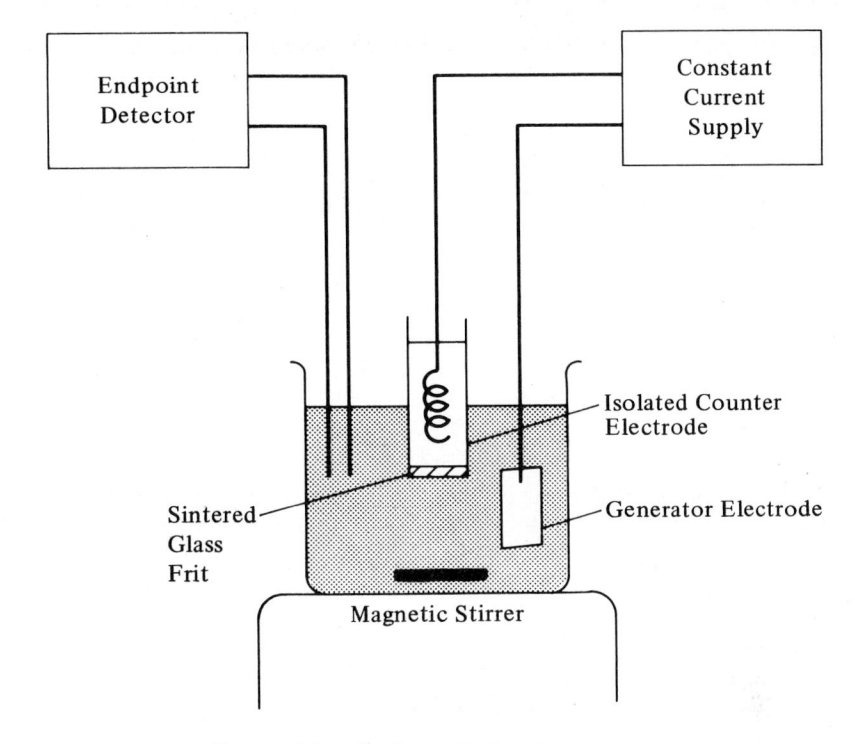

FIGURE 4.1. *Coulometric titration apparatus.*

is often made for removing dissolved air from the solution with a stream of nitrogen, if oxygen-sensitive reagents are to be generated.

The constant-current supply could be high-voltage batteries connected through a large resistance or, more likely, an electronic device. Various types of the latter are available from a number of manufacturers. Typical generating currents are in the range 1 to 200 mA. The current supply includes an on/off switch or push-button which may be manipulated by the operator in a way analogous to a buret stopcock. Current supplies also have counters or timers that run only when the current is turned on. Currents are usually known to 0.1% or better, and time is measured to the nearest 0.1 sec. Therefore, four-significant-figure accuracy is achieved if titrations exceed 100 sec; this is typical for microequivalent-level samples.

Commercially available instruments usually read directly in microequivalents. This is accomplished by setting the current in some multiple of the Faraday constant so that the microequivalents are simply equal to some decimal fraction or multiple of the seconds of generation.

Primary Constant-Current Titrations

The important requirement in coulometric procedures is that the equivalency between electric charge and the substance being titrated must be maintained. Put another way, unwanted electrolytic side reactions must be avoided. To illustrate the experimental parameters affecting this requirement, it is necessary to consider the cell processes in some detail. Two classes of constant-current coulometric titrations, termed *primary* and *secondary*, will be discussed. In the former, the titrant is derived *directly* from the electrode. In the latter, the reagent is generated from a precursor dissolved in the supporting electrolyte.

The use of a silver anode for generation of silver ion for titrations of halides, sulfides, mercaptans, and sulfhydryl compounds is an important example of a *primary coulometric titration*. In the titration of a halide, X^-, AgX deposits directly onto the electrode as Ag^+ is generated according to

$$Ag^0 + X^- \rightarrow AgX + e^- \tag{4.7}$$

As X^- is consumed, its concentration in the solution decreases until the rate at which it can be supplied to the electrode becomes smaller than the rate at which Ag^+ is formed. The precipitation reaction then proceeds into the bulk of solution as the generated Ag^+ diffuses (via stirring) into the solution. Amperometric, potentiometric, and turbidimetric techniques can be used to signal endpoints.

The Hg^{2+} generated from Hg anodes, another primary coulometric titrant, is useful for complexometric titrations. With the exception of Ag^+ and Hg^{2+}, virtually no other metal ion can be generated with high current efficiency from the elemental material. Potentially interesting titrants, such as Cr(II), Fe(II), Fe(III), and Ti(III), are not available using primary coulometry because of the formation of mixed oxidation products. Additionally, the high localized excess of these transition metal ions directly at the electrode surfaces leads to their precipitation as insoluble hydroxides which, unlike silver halide, are electrically insulating and retard flow of current. Fortunately, high and reproducible current efficiencies can be obtained using a different experimental approach, termed *secondary coulometric titration*.

Secondary Constant-Current Titrations

In secondary titrations, it is necessary to choose the appropriate titration precursor, and to adjust its concentration and the applied current density to ensure that every metered electron produces the desired effect. To demonstrate this, it is best to consider the current-versus-voltage curves for the specific chemical system being used. One such system, represented in Figure 4.2, illustrates the coulometric determination of Fe^{2+} by oxidation to Fe^{3+} at a platinum electrode in H_2SO_4 supporting electrolyte. Note in this diagram that, for a given applied current, $i_{applied}$, the anode voltage, V_a, is fixed by the intersection of $i_{applied}$ with the current-versus-voltage curve of the system at hand. As the titration proceeds, Fe^{2+} is oxidized to Fe^{3+}; the current-versus-voltage curve, and hence V_a, shifts toward more positive potential; and the limiting current decreases (curves 1 through 5 in Fig. 4.2).

At some stage of the titration, there is insufficient concentration of Fe^{2+} to support the applied current, and some other process must occur. Curve 4 in Figure 4.2, having the *limiting-current plateau*, i_ℓ, represents the current-versus-potential behavior of the system at a stage during which direct oxidation of Fe^{2+} is sufficient to support the total applied current. As the electrolysis proceeds (i.e., the Fe^{2+} is consumed), the limiting-current plateau will be reduced to a value equal to $i_{applied}$. As i_ℓ is further reduced to less than $i_{applied}$ (e.g., curve 5 in Fig. 4.2), the electrode potential will shift toward positive values sufficient to oxidize water, in order to partially support the

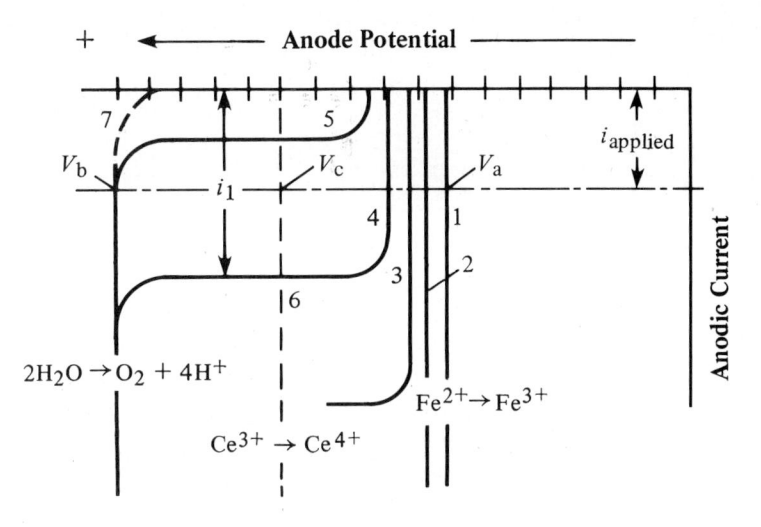

FIGURE 4.2. *Current-versus-voltage curves for the systems* Ce^{3+}/Ce^{4+} *and* Fe^{2+}/Fe^{3+}.

current flow. As the electrolysis continues, an increasingly smaller fraction of the applied current goes to direct oxidation of Fe^{2+}; an increasing fraction goes to oxidize water, at V_b; and the current efficiency for the titration is lost.

To avoid this loss of efficiency, a secondary titrating agent is added to the cell that is oxidized at a potential intermediate between those for Fe^{2+} and H_2O. In the present example, Ce^{3+} is chosen. The current-versus-potential curve for Ce^{3+} oxidation shown as the dotted line in Figure 4.2 (curve 6) will fix the anode potential at V_c after the concentration of Fe^{2+} becomes sufficiently small. Because the electrogenerated Ce^{4+} reacts rapidly with any Fe^{2+} remaining in solution by the reaction

$$Ce^{4+} + Fe^{2+} \rightarrow Ce^{3+} + Fe^{3+} \tag{4.8}$$

the titration efficiency for oxidation of Fe^{2+} is maintained, even though the current efficiency for its direct oxidation is not 100% during the latter stages of the titration. The overall process is equivalent to a conventional volumetric titration of Fe^{2+} with Ce^{4+}. Note, however, that this secondary constant-current titration will require an independent method of endpoint detection, because neither the current nor the potential of the platinum generating electrode provides a good indication of the progress of the reaction.

For some coulometric procedures, the substance being determined takes no direct part in the electrode process at any stage of the titration. In the previous example, the titrant precursor (Ce^{3+}) is recovered in the titration reaction (Eqn. 4.8). An example where this is not the case is the titration of unsaturated olefins with electrogenerated Br_2. Here, a bromide-containing electrolyte provides the precursor, and the Br^- is not regenerated after the chemical reaction. As in the case described above for Ce^{3+} precursor, there will be a limiting current density that must not be exceeded for 100% current efficiency. Although it may vary from case to case (and should be checked empirically), the current density normally should not exceed about 0.025 mA/cm^2-mN. For example, with an electrode of 2 cm^2 area and a 0.1 N concentration of precursor (100 mN), the current should not exceed 5 mA. Obviously, high concentrations of precursor and greater electrode areas increase permissible current range and decrease electrolysis time.

Coulometric titrations can be used for most types of titration reactions. In aqueous solution containing inert electrolyte, the processes

$$2H_2O + 2e^- \rightarrow H_2 + 2OH^- \quad \text{(Cathode)} \quad (4.9)$$

$$2H_2O \rightarrow O_2 + 4H^+ + 4e^- \quad \text{(Anode)} \quad (4.10)$$

at a platinum electrode, are used for generation of hydroxide ion for titration of acids and of hydrogen ion for titration of bases. In this case, very large current densities can be used because a virtually unlimited supply of titrant-precursor, water, is available. Often, the H$^+$ or OH$^-$ must be generated externally and then added to the sample because many samples contain constituents that may electrolyze more readily than water (see below and Fig. 4.3).

Many oxidants [Ce(IV), I$_2$, Mn(III), etc.] and reductants [Fe(II), Ti(III), Cu(I), etc.] are available for redox titrations. Note, however, that oxidants such as MnO$_4^-$ and Cr$_2$O$_7^{2-}$, with bound oxygen, are typically not generated with reproducible current efficiency by electrolysis methods. Some other examples of redox titrations, and of precipitation and complexation titrations, are shown in Table 4.2.

Constant-current coulometry using dual precursors enables slow titration reactions to be performed. Here, an excess of titrant is generated and, after allowing the reaction to go to completion, the excess is back-titrated by reversal of the current. An important dual precursor system is CuBr$_2$, which has been used for determination of "bromine numbers" (total unsaturation) of olefins. The olefin is added to the CuBr$_2$ electrolyte and bromine is generated anodically. After a suitable reaction time, the current is reversed and the excess bromine is titrated with cathodically generated Cu$^+$ [2].

An additional unique feature of coulometry is the simple procedure for dealing with impurities in reagents that cause current and titration inefficiency. The method, termed *pretitration*, simply involves generation of titrant until the preselected endpoint is reached, followed by addition of the sample. If titration progress is monitored potentiometrically, titrant alone is generated in the supporting electrolyte until the

FIGURE 4.3. *Apparatus for the external generation of titrant for coulometric titrations.*

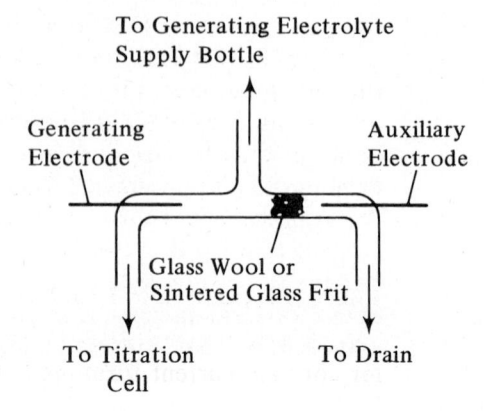

TABLE 4.2. *Methods for Coulometric Generation of Titrant*

Substance Generated	Typical Solution Conditions	Working Electrode	Typical Substances Titrated
Br$_2$	0.1 M H$_2$SO$_4$ 0.2 M NaBr	Pt	Sb(III), I$^-$, Tl(I), U(IV), various organic compounds
I$_2$	0.1 M KI 0.1 M phosphate buffer pH = 8	Pt	As(III), Sb(III), S$_2$O$_3^{2-}$, S^{2-}
Cl$_2$	2 M HCl	Pt	I$^-$, As(III), fatty acids
Ce(IV)	0.1 M cerous sulfate 1.5 M H$_2$SO$_4$	Pt	Fe(II), Fe(CN)$_6^{4-}$
Mn(III)	0.45 M MnSO$_4$ 1.8 M H$_2$SO$_4$	Pt	Oxalic acid, Fe(II), As(III)
Ag(II)	0.1 M AgNO$_3$ 5 M HNO$_3$	Au	As(III), V(IV), Ce(III), oxalic acid
Fe(CN)$_6^{4-}$	0.2 M potassium ferricyanide pH = 2	Pt	Zn(II)
Cu(I)	0.02 M CuSO$_4$	Pt	Cr(VI), V(V), IO$_3^-$
Fe(II)	0.6 M ferric ammonium sulfate 2 M H$_2$SO$_4$	Pt	Cr(VI), V(V), MnO$_4^-$
Ti(III)	0.6 M titanic sulfate 6 M H$_2$SO$_4$	Pt (Hg also used)	Fe(III), V, U(VI), Ce(IV)
Ag(I)	0.5 M HClO$_4$	Ag anode	Cl$^-$, Br$^-$, I$^-$
EDTA (Y^{4-})	0.02 M HgNH$_3$Y^{2-} 0.1 M NH$_4$NO$_3$ pH = 8.3; O$_2$ removed	Hg	Ca(II), Zn(II), Pb(II), etc.
H$^+$ or OH$^-$	Various electrolytes	Pt	OH$^-$ or H$^+$, organic acids or bases

predetermined endpoint potential is reached. In cases where the initial potential is already beyond the desired endpoint, a small amount of sample is added, and titrant is generated to the endpoint. In practice, several successive additions of samples followed by titration can be performed using a single batch of supporting electrolyte. Each titration provides the pretitration for the succeeding analysis.

In cases where the electroactive impurity is in the sample itself, the titrant can be generated *externally* and then added to the sample either continuously or in increments. A suitable arrangement is shown in Figure 4.3. Placing the generating electrode close to the output port of the generator cell ensures a short time delay for delivery of titrant. Alternatively, the output port can be fitted with a stopcock, allowing the titrant to be generated and flushed into the cell in increments.

Instrumentation and Endpoint Detection

Any of the endpoint-detection methods applied to volumetric titrations can be used for constant-current titrations. These include visual methods using dissolved indicators, provided the indicators are not electroactive; and instrumental methods such as potentiometry, amperometry, and photometry. However, the lower limit of sample size that can be titrated using constant-current coulometry is ultimately determined by the sensitivity of the method used to detect the endpoint. This can be demonstrated as follows. Assume that we wish to determine the titer of a 10^{-4} N solution of strong acid. We take 100 mL of Na_2SO_4 supporting electrolyte and carefully pretitrate it to pH 7 using a glass electrode to monitor pH. If we then add a 1-mL sample of the acid to the titration cell, the pH would change to about 6. The total charge required for titration would be 100 millicoulombs, an easily measurable quantity, but the pH change for the titration would be only one pH unit. Because pH measurement is expected to be accurate and reproducible to only about ± 0.05–0.1 pH unit under normal experimental conditions, this titration would give results accurate to only about ± 5–10%. Currents in the microampere range are easily measured and controlled; their use, however, is restricted to titrations where a particularly sensitive method for endpoint detection is available.

A few highly sensitive endpoint-detection systems have been described. A zero-current potentiometric method using an extremely sensitive galvanometer allows the titration of 3 ng of manganese as permanganate (2.5×10^{-10} eq) with electrogenerated iron(II) in a volume of 7 mL [3]; actually, the measurement is done in an amperometric mode by setting the endpoint potential and generating titrant until the off-null galvanometer returns to zero current. The error is only about 9%. Potentiometric endpoint detection to a preset potential has been used to titrate as little as 26 ng of Mn with a precision and accuracy of 2% [4]. An indirect procedure for amperometric [5] and biamperometric [6] titration using a very sensitive current recorder has been used to titrate as little as 9 ng of As(III) (0.24 neq) in a volume of 35 mL with generated bromine. A sufficient excess of bromine is electrogenerated, the sample is then added, and the decrease in the detector current is measured. The titration time is calculated to the nearest 0.01 sec from the slope of the plot of detector current versus time. An accuracy of better than 4% relative is obtained.

The ability to meter reagent by control of electric current, coupled with the ability of dedicated or integrated microprocessors to perform computation, to control the instrumentation, and to manipulate data, has led to the development of many automated titrators, both for laboratory use and for industrial process control. Traditionally, in constant-current coulometric titration, a rigorously controlled constant current is applied to the cell and the analytical result is obtained by accurately measuring the time to reach the endpoint. Because the analytical result is the current-time integral, however, it is not necessary that current be maintained absolutely constant; it can be allowed to vary under control of the computer, using an appropriate algorithm to accumulate the actual charge passed [7]. Feedback control is the important consequence of this. If the progress of the titration reaction can be monitored (e.g., if a pH electrode is used to monitor the titration of an acid with electrogenerated base), the magnitude of the difference between the pH at any point in the titration and the final endpoint pH can be used to set the current. The current can be made large during early stages of the titration and decreased as the endpoint is approached, to avoid overshooting the endpoint. Speed and accuracy are therefore improved.

Since computers operate in the digital domain, coulometric charge, which has traditionally been metered using analog control of current, can alternatively be metered using current or charge *pulses*. This provides an advantage because the coulombic content of the pulses, which is fixed by the current amplitude and pulse width, is controlled by the computer, and the pulses are simply counted.

Automated coulometric titrators generally operate in either of two modes: *fixed level* or *time derivative*. In the fixed-level mode, titration is performed to a known or calculated endpoint level, which can be indicated amperometrically, photometrically, or potentiometrically. In the time-derivative mode, the derivative of the titration curve is calculated as the reaction progresses, and the endpoint is located by detecting the "peak" in the titration curve. In either mode, the processor selects titration currents based on the slope of the titration curve, keeps track of the coulombs applied to the cell, performs real-time computations as the titration reaction proceeds, makes pretitration adjustments, and often controls sampling, provides diagnostic messages, and displays results in the desired units.

4.3 CONTROLLED-POTENTIAL COULOMETRY

In controlled-potential coulometry, total electrolysis of the analyte is accomplished at a fixed potential and the total charge required (coulombs) is measured and used to calculate the equivalents of the analyte. As presented in the beginning of this chapter, Equation 4.3 refers to the state of affairs as it exists at the electrode surface. Because we are interested in electrolysis of the bulk of solution, it is necessary to relate dN/dt at the electrode surface to the rate of change of moles in the bulk of solution. The key to this can be found upon examination of the current-versus-voltage curves in Figure 4.2. In the limiting-current region, the current is independent of the applied potential. The concentration of electroactive species at the surface is said to be zero because all of the species is electrolyzed the instant it reaches the surface. The rate of reaction is limited by the rate of supply of the electroactive species to the electrode surface from the bulk of solution—called *diffusion-controlled electrolysis*. The relationship between the limiting current, i_ℓ, and the concentration in the bulk solution, c_b, at any time t can be written

$$(i_\ell)_t = nF\, Am(c_b)_t \tag{4.11}$$

where A = electrode area in square centimeters
m = mass-transfer coefficient in centimeters per second

Under the conditions described above, the system has reached steady state and the rate of electrolysis at the electrode surface must be equaled by the rate at which the amount of electroactive species is changing in the bulk of solution. This means, then, that both i_ℓ and c_b are changing with time. If V is the volume of solution, we can write $N = c_b V$, and taking the derivative gives

$$-\frac{dN}{dt} = -V\frac{dc_b}{dt} \tag{4.12}$$

Substituting Equation 4.12 into Equation 4.3, and equating the currents in Equations

4.3 and 4.11, we have

$$\frac{dc_b}{dt} = -\frac{Am}{V}(c_b)_t \tag{4.13}$$

which upon integration between the limits 0 to t, and between $(c_b)_0$ (the initial concentration) to $(c_b)_t$, gives

$$(c_b)_t = (c_b)_0 e^{(-Am/V)t} \tag{4.14}$$

Taking Equation 4.11, and letting the decay constant Am/V be equal to p (in sec^{-1}), we can write an equation for the limiting current at any time t as

$$i_t = i_0 e^{-pt} \tag{4.15}$$

where i_0 = initial current

The form of Equations 4.14 and 4.15 is characteristic of any first-order rate process (decay). An important feature is that the fraction of material electrolyzed at any time is independent of the initial concentration. We see that $(c_b)_t/(c_b)_0$ is the fraction left, and therefore the fraction electrolyzed is $[1 - (c_b)_t/(c_b)_0] = [1 - \exp(-pt)]$. If $t_{1/2}$ is the time required to reduce the concentration by one-half, then 10 times $t_{1/2}$ seconds will reduce the concentration to $0.001(C_b)_0$. The time it actually takes to do this depends on the magnitude of p, which, in the interest of speed, should be as large as possible. As Equation 4.14 shows, m and A should be large, and V small, to accomplish this. The manner in which this is handled experimentally provides one way to subdivide the subject of controlled-potential coulometry. Accordingly, we have (a) stirred solutions, (b) flowing streams, and (c) thin-layer cells, to which we add (d) spectrocoulometric titrations.

Stirred Solutions

The basic apparatus for performing coulometry in stirred solutions is shown in Figure 4.4. The potential between the working and reference electrodes is monitored and used to control the variable voltage source, which supplies current between the counter and working electrodes. Instruments that do this automatically and maintain the working-electrode potential at a predetermined value are called potentiostats (see Chap. 3). The cell current is usually measured with an operational-amplifier circuit (see Chap. 23), and the current-time integral is obtained by an integrator. The working electrode (Fig. 4.4) could be platinum. Mercury pools are also widely used. The cell depicted in Figure 4.4 is typical of classical approaches to controlled-potential coulometry. The electrode area is large (about 160 cm^2 for a typical platinum-gauze electrode and perhaps 30 cm^2 for a mercury-pool electrode in a 250-mL beaker). Solution volumes of 25 to 75 mL can be used. Efficient stirring is essential for rapid electrolysis. In the case of the platinum-gauze electrode, if we assume a mass-transfer coefficient of 0.006 cm/sec, we calculate p as 0.0192 sec^{-1}, $t_{1/2}$ as 36.1 sec, and the time for complete electrolysis as 361 sec, or about 6 min. A similar calculation for the mercury electrode suggests 32 min for complete electrolysis. These calculations indicate titration times within the range of about 5 to 50 min, which is typical for controlled-potential coulometry in stirred solutions.

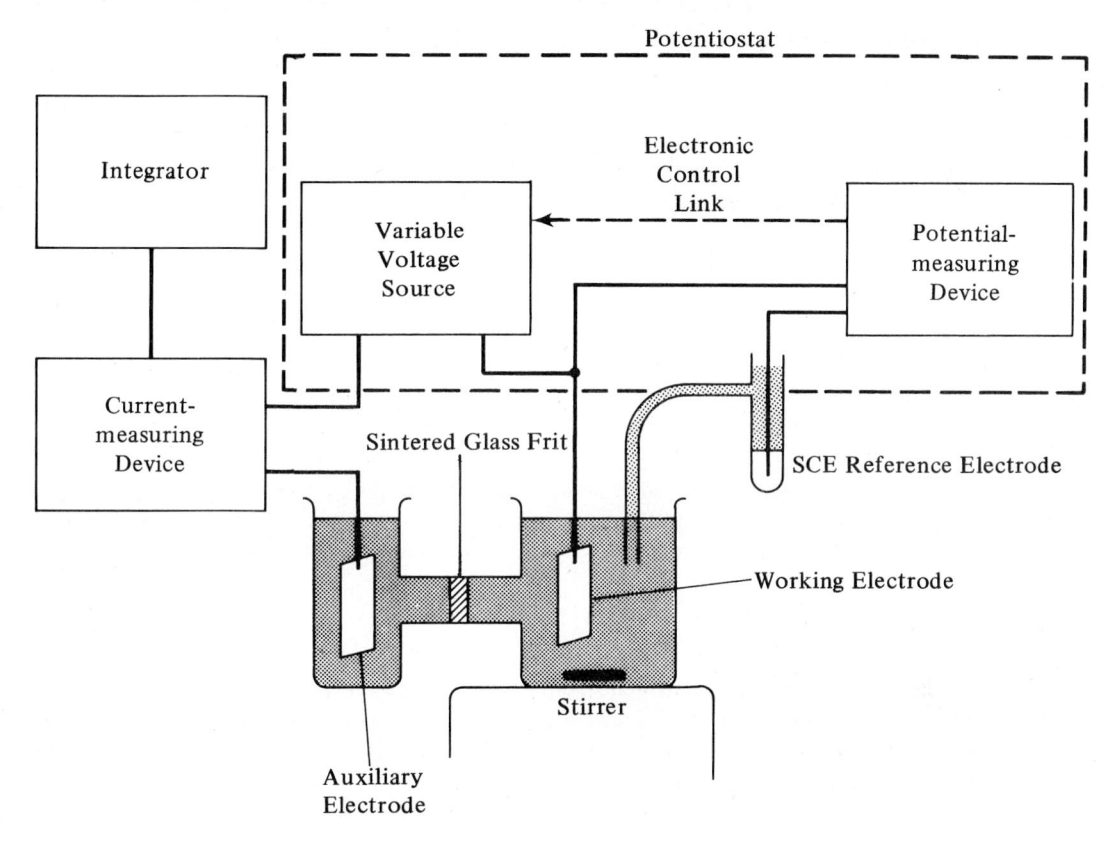

FIGURE 4.4. *Apparatus for controlled-potential coulometry.*

In a comprehensive review [8], Harrar tabulates references to controlled-potential coulometric procedures for 49 of the elements, with representatives from all of the groups of the periodic table except the noble gases. In many instances, the procedures are applicable to fairly complex matrices without requiring prior separation of the analyte from the sample matrix. Six of the actinide elements are included, which illustrates the ease of automation and remote operation of controlled-potential coulometry. Most often, the quantities determined range from about 10 meq to 1 μeq. The upper end of the range corresponds to quantities usually encountered in volumetric titrimetry. The lower limit depends primarily on the precision with which the coulombs from the background or charging current can be measured, because they must be subtracted from the total coulombs passed during the analysis run. The highest accuracy and precision ($\pm 0.05\%$) are gained for samples of about 1 to 10 mg.

Controlled-potential coulometry has also found some use in the study of basic electrochemistry. By most electrochemical methods, polarography for example, it is not always obvious how many electrons are involved in a newly studied electrochemical reaction. Thus, coulometry at controlled potential, in which a known quantity of the substance is electrolyzed and Q is measured, is often used to determine values for n and thereby help elucidate electrode mechanisms. Very slow chemical reactions coupled to the electrochemical reaction can also be studied by the method [9]; other electrochemical techniques usually are suitable only for much faster chemical reactions, with time scales of microseconds to seconds.

Electrolyses do not always follow Equation 4.15. A slow coupled chemical reaction, as mentioned above, is one such example. Another example occurs when the initial concentration of the electroactive substance is too high and the resulting current demand exceeds the maximum output current of the potentiostat. Under these conditions, the potential will shift to regions below the limiting current and the equations cannot apply until the concentration is decreased sufficiently that limiting-current conditions are obtained. In the case when a metal is deposited onto an electrode, high current may lead to deposits that do not adhere well to the electrode. It may be desirable in this situation to begin the electrolysis at a potential less negative than that required to reach the limiting current, and later shift the potential to a more negative value.

A significant effort has been made to reduce the time of electrolysis in controlled-potential coulometry. One direction that has been taken is called *predictive coulometry* and is based on Equation 4.15. The integral of the current is

$$Q = \int_0^t i_t \, dt = \int_0^t i_0 e^{-pt} \, dt = \frac{i_0}{p}(1 - e^{-pt}) \tag{4.16}$$

As t becomes large, Q approaches Q_∞, which is equal to i_0/p. The objective is to predict Q_∞ without actually completing the electrolysis. Harrar [8] describes an approach to this using an on-line digital computer. If Q_t is the quantity of electricity accumulated at any time t in the electrolysis, then

$$Q_\infty = Q_t + Q_R \tag{4.17}$$

where $\quad Q_R =$ coulombs remaining to be accumulated to reach Q_∞

By choosing times t_1, t_2, and t_3 such that $t_3 - t_2 = t_2 - t_1$, and measuring the corresponding coulombs Q_1, Q_2, and Q_3 passed at these three times, it can be shown that [10]

$$Q_R = \frac{(Q_2 - Q_3)^2}{2Q_2} - Q_1 - Q_3 \tag{4.18}$$

and therefore

$$Q_\infty = Q_3 + \frac{(Q_2 - Q_3)^2}{2Q_2} - Q_1 - Q_3 \tag{4.19}$$

The computer is programmed to calculate Q_∞ many times, based on an estimate of Q_R for different values of t_3, say at every 10 sec. As an internal check, the computer is also programmed to calculate values of p to ascertain that Equation 4.15 is actually being followed, and that the automatic background correction being made is not leading to difficulties. The operator ends the experiment when the predicted values become sufficiently constant. For systems that obey Equation 4.15, the results are precise and accurate to 0.1 and 0.2% relative, and require only one-third to one-half of the total electrolysis time normally needed. An additional advantage of the digital computer is the relative ease of data smoothing to improve results.

Flowing Streams

Instead of stirring the solution to produce convective mass transport, the solution can be passed through a porous working electrode. If we let the concentration of electroactive species in the stream at the inlet to the electrode be c_{in}, that at the outlet be c_{out}, and the fraction that is converted be R, then

$$R = \frac{c_{in} - c_{out}}{c_{in}} = 1 - \frac{c_{out}}{c_{in}} \tag{4.20}$$

If R is 1, then electrolysis is complete within the electrode, and $c_{out} = 0$. With the potential of the electrode controlled somewhere on the limiting-current plateau, the current is at the limiting steady state due to the continual supply of material at the electrode inlet. From Equation 4.5b, we have for a constant (steady-state) current

$$i = \frac{nF}{t} \text{(moles)} \tag{4.21}$$

Since moles $= cV$, where c is the concentration in moles per liter and V is the volume in liters, and letting v be the volume flow rate $= V/t$, then

$$i = mFvc \quad \text{and} \quad Q = nFVct = nFN \tag{4.22}$$

Equation 4.22 is merely a statement of Faraday's law for a flowing stream. We see that the measured current is directly proportional to the concentration, which in this case is the inlet concentration. If the current is measured, the inlet concentration is monitored. If the current is integrated, we are performing controlled-potential coulometry in a flowing stream.

It would be helpful if this experiment could be expressed in terms of the flow velocity and physical parameters of the porous electrode, particularly because the flow velocity can be made high enough that R would become less than 1, and a quantitative conversion would no longer take place. Such treatments have appeared in the literature and are effectively presented by Bard and Faulkner [11]. The model for this type of electrode assumes a cylindrical bundle of hollow linear tubes serving as the pores. The treatment ignores iR drop effects, kinetic complications involving the electrode reaction, and current efficiencies less than 1. What is considered is a reaction occurring with 100% current efficiency under limiting-current conditions, which is precisely the case for which analysis by controlled-potential coulometry is applicable. Basically, the concentration of electroactive species that enters the porous electrode decreases exponentially with the distance down the electrode. The objective is, of course, to achieve very close to 100% conversion of the electroactive substance in a reasonable length of time, and with acceptable precision and accuracy.

The theory for this type of system has been developed to enable calculation of predicted conversion efficiencies and the performance of a particular electrode, given such experimental variables as inlet concentration; the length, cross section, and specific surface area of the porous electrode material; the flow velocity of the stream; and so forth.

The apparatus needed for controlled-potential coulometry in flowing streams is very similar to that shown in Figure 4.4 for stirred solutions. Some changes in the electrolysis cell must be made and a means for moving the solution through the electrode, such as a pump, must be provided.

Reticulated vitreous carbon (RVC) is a particularly useful electrode material for flowing-stream coulometry [12, 13]. Its honeycomblike structure is shown in Figure 4.5. Cylinders of this material for use as electrodes are easily cut from blocks of commercially available RVC with a cork borer. Although the counter electrode can simply be located downstream of the working electrode, it is preferable to have a cylindrical counter electrode concentric to the RVC electrode, with a porous separator placed between the two. The reference electrode is most often placed downstream of the working electrode. Reticulated vitreous carbon has a porosity (fraction of open internal volume) of 0.95, and typical electrode dimensions for material of 100 pores per linear inch are 0.3-cm diameter and 2.5-cm length. This will provide for 100% conversion at flow rates up to about 1 mL/min. A typical cell is shown in Figure 4.6. In addition to the instrumentation shown in Figure 4.4, a constant-current source is needed to null out the background current. Unlike the case with stirred solutions, where the background current is often a function of time because of electrolysis of impurities, a flowing stream produces a steady-state background signal which can be effectively nulled. Depending on the size of the electrode and the cell geometry, the background current can be as large as 1 to 10 μA.

FIGURE 4.5. *Photograph of reticulated vitreous carbon showing the porous structure.*

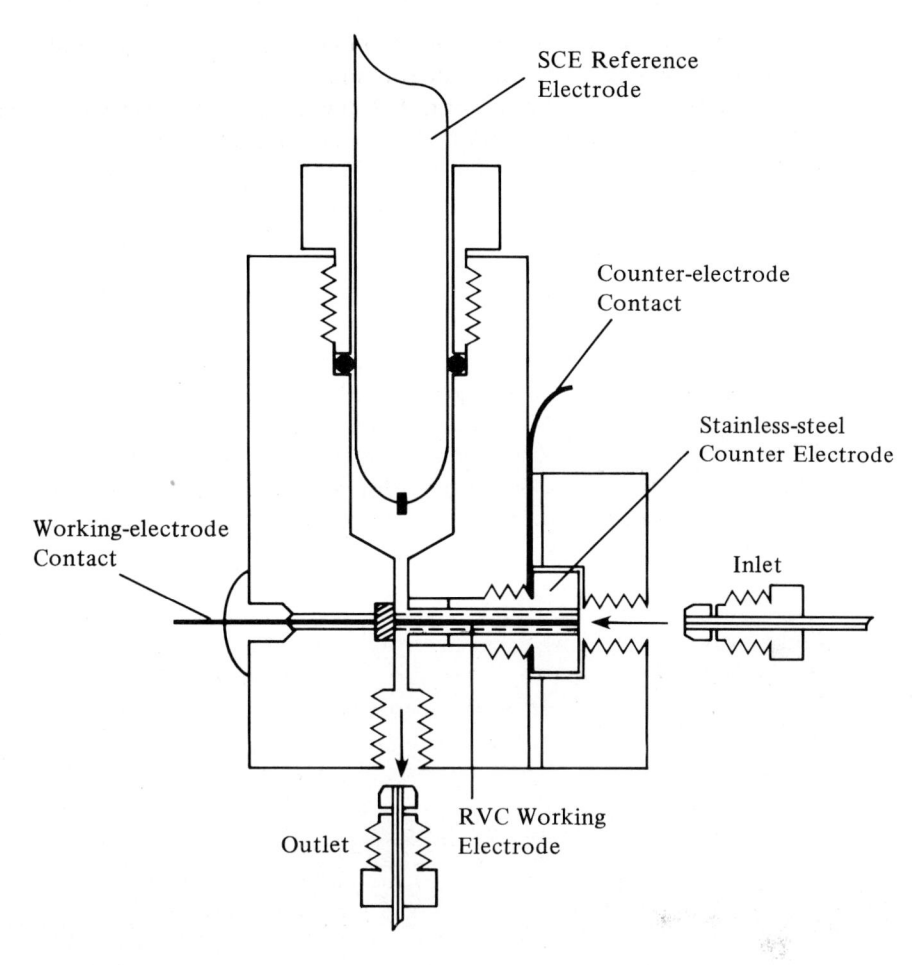

FIGURE 4.6. *Cross section of a flow-through cell using a reticulated vitreous carbon (RVC) working electrode. The body of the cell is made of nylon, and the RVC electrode is wrapped with Millipore® filter material to provide a porous barrier between the working electrode and the concentric stainless-steel counter electrode. Adapted from reference [16] with permission of the copyright holder.*

First-order decay constants for porous-electrode flow-through cells are considerably larger than those for classical stirred-solution cells. Fuginaga and Kirhara [14], using an electrode consisting of a bed of carbon particles, have obtained decay constants greater than $1\ \text{sec}^{-1}$. Decay constants of this magnitude imply electrolysis times of about 7 sec. Generally, however, 100% conversion efficiency can be maintained only with flow rates not exceeding a few milliliters per minute. Consider, however, a 50-mL volume of sample flowing at a rate of 1 mL/min. Although the rate of electrolysis is fast, it would still take 50 min to pump the solution through the electrode.

The obvious way to decrease the time it takes for a determination is to use smaller sample volumes. With the aid of a chromatographic loop injector, 20-μL volumes of sample can be injected into a stream of supporting electrolyte carried in tubing of 0.3 to 0.8 mm in diameter. Such sample-handling techniques fall in the realm of flow-injection analysis (FIA) [15]. Characteristically, the injection process and the flow result in dilution of the sample (a process called dispersion), and the sample might occupy a volume of 200 μL by the time it reaches the cell. At a flow rate of 1 mL/min, it will take 12 sec for the entire sample to pass the front face of the porous electrode. Complete conversion will occur, provided the residence time in the electrode exceeds the time required for complete electrolysis. For the typical RVC electrode cited earlier, the residence time is about 10 sec at a flow rate of 1 mL/min with complete conversion [16]. As an example, the detection limit for the determination of hydroquinone was 55×10^{-12} g using injection volumes of 62.8 μL. Samples were analyzed at the rate of 250/hr. These figures illustrate the important advantages of small sample volumes, low detection limits, and high sample through-put that can be attained using controlled-potential electrolysis in flowing streams.

Thin-Layer Cells

Another way to accomplish rapid electrolyses under conditions of controlled potential is to confine the solution to a very thin layer in front of the electrode. The idea is illustrated in Figure 4.7. It is useful to speak of the thin-layer cavity itself, which contains the working electrode and the solution to be electrolyzed, as distinct from the entire thin-layer cell, which consists of the thin-layer cavity (with working elec-

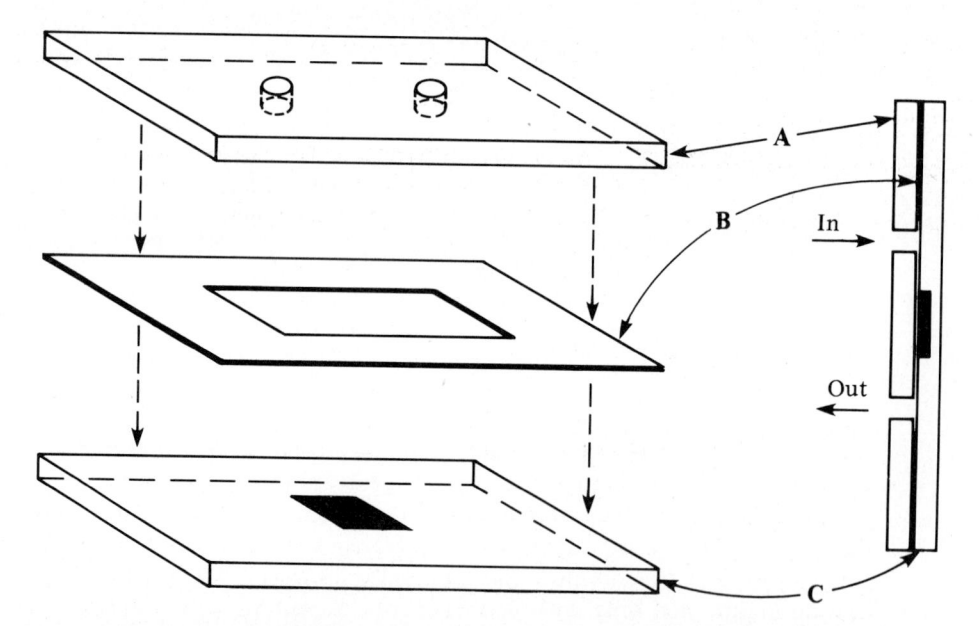

FIGURE 4.7. *Sandwich-type thin-layer cavity assembly. A: Face plate with inlet and outlet ports. B: Teflon spacer with a hole cut in it. C: Back plate with the working electrode mounted in the center.*

trode) and an arrangement of counter and reference electrodes. The reason for this distinction is that the cavity usually cannot be isolated from the world around it. Solution contact to the other electrodes must be provided, and some means to fill the cavity is needed.

The critical dimension of the thin-layer cavity is its thickness. If the wall opposite the working electrode is close enough, the diffusion-layer thickness can grow no further once it reaches the wall, which happens very quickly. Electrolysis then proceeds within a layer of solution thinner than the diffusion layer that would have formed had such a nearby barrier not been there. If the cavity thickness is much thinner than the diffusion-layer thickness that should have prevailed, then the solution within the cavity can be regarded as homogeneous and the mathematics of the diffusion-layer-boundary problem modeling this experimental situation are greatly simplified [17]. Under these conditions, corresponding to later times in the electrolysis, the current-time relationship is

$$i_t = i_0 e^{-pt} \qquad (4.23)$$

where $\quad p = \pi^2 D/4\ell^2$

D = diffusion coefficient of the electroactive species in square centimeters per second

ℓ = thickness of the cavity in centimeters

Again, as in Equation 4.15, there is an exponential decay of the current.

For a diffusion coefficient of 10^{-5} cm^2/sec, and a cavity thickness of 50 μm, the calculated rate constant is 0.99 sec^{-1}, which indicates about 7 sec are required for complete electrolysis. In practice, electrolysis usually takes longer than this because resistance effects produce a nonuniform potential across the face of the working electrode in most thin-layer cells. Nevertheless, the process is rapid. Like the classical controlled-potential experiment, thin-layer experiments require correction for background coulombs. In addition, careful attention to cell design is necessary to minimize contributions to the measured coulombs from electroactive material in solution located outside the cavity. This material, of course, can diffuse into the cavity at the edges (so-called edge effects) and cause error in the quantitative determination.

Spectrocoulometric Titrations

Imagine that the walls and working electrodes of an electrolysis cell are optically transparent. Not only could an electrochemical experiment be monitored but any changes in light transmitted through the cell could also be observed with a suitable detector (see Chaps. 6 and 7 for a discussion of ultraviolet and visible absorption spectrophotometry). A number of combinations of electrochemical and spectrochemical techniques have been developed in recent years [18]. The discussion here will be confined to transmission experiments carried out under conditions of controlled-potential electrolysis. Both ordinary and thin-layer cells have been used. The variable measured in the spectrochemical part of the experiment is the absorbance, A, which is related in the simplest case to the concentration, c, of a single species in solution by the relationship known as Beer's law:

$$A = \varepsilon bc \tag{4.24}$$

where ε = molar absorptivity (in L/mole-cm), characteristic of the absorbing species and a function of the wavelength of the light used
  b = optical path length (in cm)
  c = concentration (in mole/L)

Consider now that we perform a controlled-potential coulometric experiment on the system Ox + ne^- → Red, and that the wavelength is chosen to monitor the absorbance of the reduced species. Provided the solution is chemically homogeneous, the absorbance as a function of time, A_t, is given by

$$A_t = \varepsilon b(c_{Ox})_0(1 - e^{-pt}) \tag{4.25}$$

Like the charge, the absorbance increases exponentially with time. At the end of electrolysis, $\exp(-pt)$ is essentially zero, and the final absorbance, $A_{t=\infty}$, is given by

$$A_{t=\infty} = \varepsilon b(c_{Ox})_0 \tag{4.26}$$

The analytical utility of Equation 4.26 is clear, but it may seem that little has been accomplished because the same information could be obtained from the electrochemical experiment alone. This conclusion is correct, but it arises because the chemical situation chosen as an example was trivial. Nevertheless, the example does illustrate features of the spectrocoulometric experiment and conveys the idea that added information can be obtained compared to the electrochemical experiment alone. We will now consider some more complex chemistry, which will justify the spectrocoulometric experiment.

Knowledge of electron exchange between molecules in biological systems is of great interest and importance in understanding how living organisms function. Biological redox reactions do not differ, in principle, from other types of redox reactions, but the molecules involved are often of high molecular weight, such as proteins and enzymes. As it turns out, many such molecules react slowly or not at all at electrodes. The answer to this problem is to add small molecules to the solution that can undergo rapid electron exchange with both the electrode and the biomolecule under study [19]. These small molecules are called mediators, and a listing of them appears in a publication by Fultz and Durst [20]. The overall scheme can be written as follows:

$$(Med)_{Ox} + ne^- \rightleftharpoons (Med)_{Red} \qquad \textit{Electrode reaction} \tag{4.27}$$

$$(Bio)_{Ox} + (Med)_{Red} \rightleftharpoons (Bio)_{Red} + (Med)_{Ox} \qquad \textit{Chemical reaction} \tag{4.28}$$

where Med = mediator
  Bio = biomolecule

This overall reaction scheme is identical to that of a constant-current coulometric titration, but here the electrode reaction is carried out at controlled potential. Often, the number of electrons transferred at redox sites on biomolecules is not known and the formal potential is of great interest in the study of energy transfer. The spectrocoulometric experiment can provide this information. Known increments of charge, ΔQ, are injected by controlled-potential electrolysis, and the corresponding change in absorbance, ΔA, is measured. From Equations 4.22 and 4.25, we can write

$$\Delta Q = nFV\Delta(c_b)_{Ox} \quad (4.29)$$

$$\Delta A = \varepsilon b \Delta(c_b)_{Red} \quad (4.30)$$

Because the *change* in concentration of oxidant and reductant is the same, we can solve Equation 4.30 for $\Delta(c_b)_{Red}$ and substitute in Equation 4.29. The result is

$$\Delta A = \frac{\varepsilon b}{nFV} \Delta Q \quad (4.31)$$

Thus, a plot of ΔA versus ΔQ has a slope of $(\varepsilon b/nFV)$, and n can be evaluated, provided ε, b, and V are known. Figure 4.8 shows such a plot for the generation and removal of methylviologen (MV^{++}), a widely used mediator [21]. The electrochemical reaction is

$$CH_3-N^+\!\!\!\bigcirc\!\!-\!\!\bigcirc\!N^+-CH_3 + e^- \longrightarrow CH_3-N^+\!\!\!\bigcirc\!\!-\!\!\bigcirc\!\dot N CH_3 \quad (4.32)$$

where the product is the radical cation of methylviologen, $\cdot MV^+$. The slopes of the six lines in Figure 4.8 yielded an average n value of 1.03 ± 0.04. The regeneration of MV^{++} was accomplished by reaction with ferricyanide that was electrochemically generated from ferrocyanide in solution.

The need for optically transparent electrodes (OTEs) is more apparent when

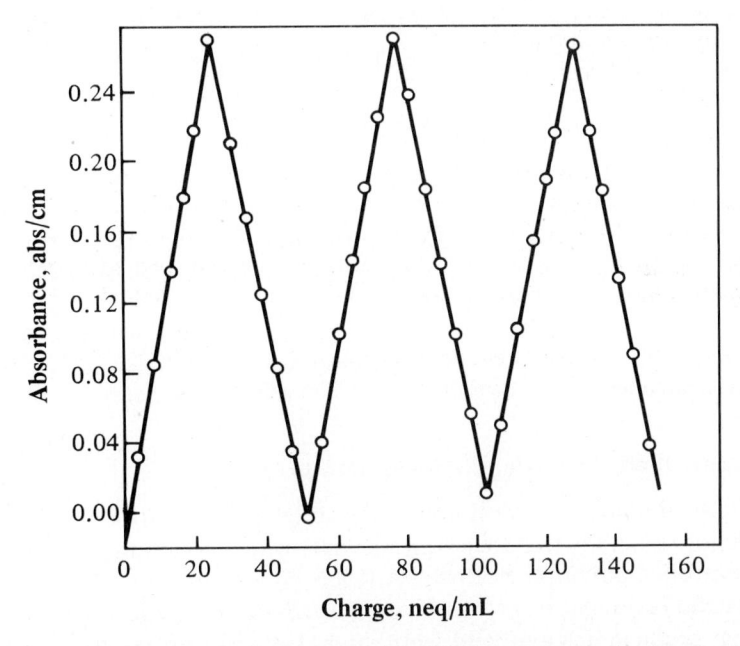

FIGURE 4.8. Plot of absorbance versus charge (corrected) during the generation and removal of methylviologen radical cation. Monitoring wavelength = 605 nm. Solution conditions: Methylviologen dication 0.5 mM, ferrocyanide 1.0 mM, sodium chloride 0.1 M, phosphate buffer pH 7.0. Reprinted with permission from F. M. Hawkridge and T. Kuwana, Anal. Chem., **45**, 1021 (1973). Copyright 1973 American Chemical Society.

thin-layer cells are considered. Space and geometry requirements dictate that the electrodes themselves be optically transparent. Two different types of OTEs have been developed: (a) thin films of conducting material such as Pt, Au, or SnO_2 deposited on transparent substrates such as glass, quartz, and germanium; and (b) minigrid electrodes of Au, Ni, Ag, Cu, or reticulated vitreous carbon, which look something like window-screening materials, but they are not woven. The latter are transparent because of the holes in the grids, which are available with from 100 to 2000 lines per inch (80 to 20% transmittance, respectively). Film thicknesses for deposited electrodes range from 10 to 500 nm (85 to 20% transmittance, respectively).

In thin-layer configurations, the conducting film is one face of the thin-layer cavity, or the minigrid is sandwiched between the two transparent walls of the cavity. These optically transparent thin-layer electrodes (OTTLEs) are especially suited for spectro-electrochemical studies of biomolecules for reasons of economy: They require only small amounts of expensive biochemicals. A gold minigrid OTTLE constructed by Condit [22] with cavity thickness of 200 μm produced a decay constant of about 0.06 sec^{-1}; so the time for 99.9% electrolysis was about 116 sec. Using controlled-potential electrolysis and this thin-layer cell, a direct determination of n is possible in about 2 min. The formal potential of the electrochemically active substance can be determined by scanning the potential and recording the current-versus-voltage curve. In thin-layer cells, the linear-scan experiment is quantitative and the shape of the curve for a reversible couple is gaussian, unlike the normal shape for linear-scan curves (Chap. 3). The formal potential ($E^{0\prime}$) corresponds to the potential at the maximum current.

4.4 ELECTROGRAVIMETRY

Electrogravimetric methods, whether controlled current or controlled potential, involve determination of the weight of a deposit formed electrolytically on an electrode. It is not generally necessary that the material of interest be deposited with ideal current efficiency, but rather that it be deposited exhaustively as a pure adherent substance for subsequent determination of its weight. Selection of controlled-potential or controlled-current methodology is made on the basis of this requirement.

Controlled-Potential Electrogravimetry

Determination of the amount of the sought-for constituent by weighing the material deposited on the electrode is an alternative method to integration of the current during controlled-potential electrolysis. It has an advantage in that simultaneous electrode reactions can occur, provided only the product of interest deposits. It also eliminates the problem encountered with high background currents. However, the major application of controlled-potential electrogravimetry is in performing separations.

Controlled-potential electrolysis can be useful at times for separating large amounts of easily reduced metals from small amounts of less easily reduced materials. This method has the advantage over precipitations in that, with adequate potential control, no coprecipitation occurs and no extraneous reagents need be added to the

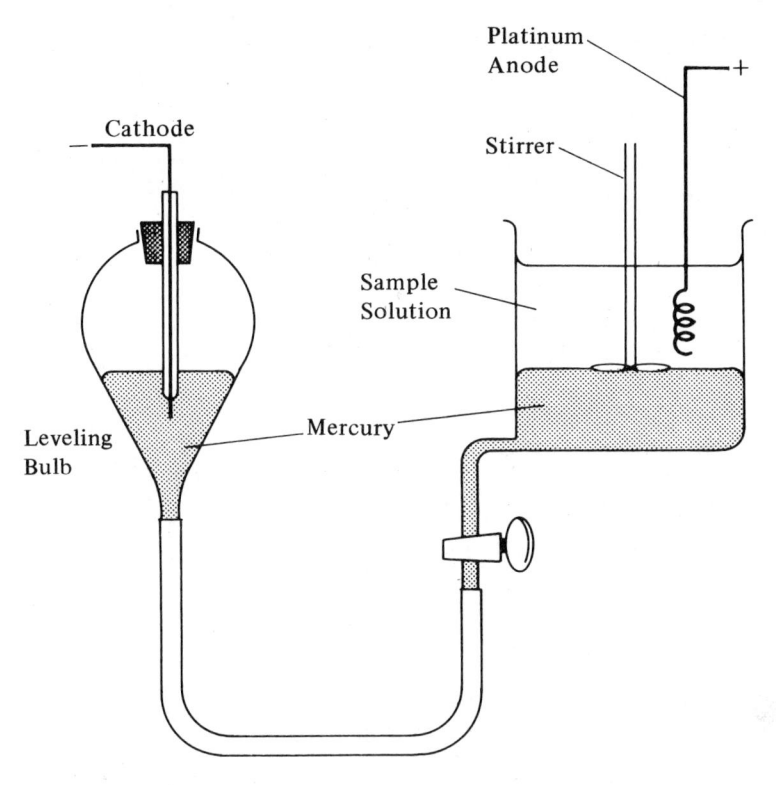

FIGURE 4.9. *Apparatus for electrolysis at a mercury-pool cathode.*

solution. For example, it is possible to remove copper from solutions of copper alloys using a platinum cathode, leaving behind tin, lead, nickel, and zinc. Bismuth and antimony will be removed with the copper at a controlled potential of -0.35 V versus SCE from hydrochloric acid solution. The minor elements can then be subjected to polarographic analysis, which would have failed before because of the large current from copper reduction preceding the smaller currents from the metals of interest. Mercury cathodes can also be successfully applied to a variety of separations. In one application, copper, lead, and cadmium have been concentrated from uranium solutions into a mercury electrode. The mercury is subsequently distilled, leaving behind the concentrated metals.

Controlled-potential electrolysis at a mercury-pool cathode to remove traces of metallic impurities is useful in preparing very pure electrolytes for use in polarography, or for such applications as the "total" removal of heavy-metal ions from solutions to be used in enzyme work. (Even traces of certain metal ions will deactivate some enzymes.) The apparatus shown in Figure 4.9 is suitable for accomplishing this. Furthermore, there are several commercially available units designed specifically for the purification of electrolyte solutions.

Controlled-Current Electrogravimetry

As described earlier for secondary coulometric titration, an understanding of the course of constant-current electrolysis requires analysis of the current-potential behavior of the substances involved in the electrode reactions. For example, as a cathodic electrolysis proceeds, the concentration of the substance being deposited decreases, and at some point in the electrolysis the potential of the electrode shifts to a more negative region where a new electrode process can occur. If this new process is deposition of a second material, then the selectivity of the analysis is lost, because the weight of the deposit does not represent a pure material. If no other material that can be deposited is present, then hydrogen gas is evolved by reduction of water. This may sometimes be helpful, since the gas evolution may improve mass transfer of the material being deposited, thereby shortening analysis time. In other cases, however, gas evolution may lead to poor adherence of deposits, causing their loss during subsequent rinsing and weighing.

A useful technique that avoids gas evolution and additionally imparts selectivity to the constant-current electrogravimetric format uses *cathodic* or *anodic depolarizers*, sometimes called *potential buffers*. These buffers are used to maintain or limit electrode potentials, thereby preventing unwanted oxidation or reduction processes. For example, in the electrodeposition of Cu from a mixture of Cu^{2+} and Pb^{2+}, nitrate ion can be added to the solution to prevent deposition of Pb along with Cu. Reduction of NO_3^- to NH_4^+ by the electrode process

$$NO_3^- + 10H^+ + 8e^- \rightarrow NH_4^+ + 3H_2O \qquad (4.33)$$

occurs at potentials intermediate between those for reduction of Cu^{2+} and Pb^{2+}. As Cu^{2+} is exhausted from the solution, deposition of Pb is prevented because reduction of NO_3^- depolarizes, or fixes, the potential of the electrode at a value more positive than one at which deposition of Pb can occur. During the final stages of deposition of Cu, the simultaneous reduction of NO_3^- exerts no effect on the purity of the deposit. In this particular analysis, Pb can be deposited simultaneously at the anode by the process

$$Pb^{2+} + 2H_2O \rightarrow PbO_2(s) + 4H^+ + 2e^- \qquad (4.34)$$

and determination of both metals can be accomplished by weighing the pure deposit on each electrode.

Hydrazine is an example of an anodic depolarizer, which can be used when chloride is present in the sample. Chloride causes difficulty because chlorine forms at the anode. Chlorine can dissolve some platinum from the electrode, which can then be deposited onto the cathode, or it may reoxidize the deposited material. In either case, this leads to an error in the weight of the deposit. Hydrazine, which is more easily oxidized than Cl^- at the platinum anode, reacts by the process

$$N_2H_4 \rightarrow N_2 + 4H^+ + 4e^- \qquad (4.35)$$

and effectively eliminates all interference by chloride.

Electrogravimetric procedures have been devised for a large number of elements. Relatively noble metals such as copper and silver are frequently determined this way because there are few interferences. The more electronegative metals like cadmium, cobalt, iron, nickel, tin, and zinc can be electrodeposited from alkaline solutions. Under these conditions, the potential for hydrogen evolution is more negative because of the decreased hydrogen-ion concentration. Often, complexing agents such as ammonia or cyanide are added to prevent the metal hydroxide from precipitating and to improve the nature of the deposit. In addition to PbO_2, Tl_2O_3 can also be deposited at the anode, allowing separation of lead and thallium from virtually all other metallic ions. Procedures exist for most common metals and a number of nonmetals [23]. Mixtures of metals can sometimes be analyzed by changing solution conditions or, as noted, by use of depolarizers, but most mixtures are better handled by controlled-potential electrolysis.

SELECTED BIBLIOGRAPHY

STOCK, J. T. *Anal. Chem.*, *54*, 1R (1982); *56*, 1R (1984). *Biennial reviews of the subject.*

LINGANE, J. J. *Electroanalytical Chemistry*, 2nd. ed. New York: Wiley-Interscience, 1958. *Now out-of-date, but a classic reference that is very useful.*

HARRAR, J. E. "Techniques, Apparatus, and Analytical Applications of Controlled Potential Coulometry," in *Electroanalytical Chemistry*, Vol. 8, pp. 1–167, A. J. Bard, ed. New York: Marcel Dekker, 1975. *A thorough review of controlled-potential electrolysis.*

MILNER, G. W. C., and PHILLIPS, G. *Coulometry in Analytical Chemistry*. Oxford: Pergamon Press, 1963. *Treats both controlled-potential coulometry and coulometric titrations.*

KISSINGER, P. T., and HEINEMAN, W. R., eds. *Laboratory Techniques in Electroanalytical Chemistry*. New York: Marcel Dekker, 1984. *A laboratory-oriented text.*

BARD, A. J., and FAULKNER, L. B. *Electrochemical Methods*. New York: Wiley, 1980, Chap. 10. *A short but excellent presentation of bulk electrolysis methods.*

BISHOP, E. *Coulometric Analysis*, Vol. IID of *Comprehensive Analytical Chemistry*, Wilson, C. L., and Wilson, D. W., eds. Amsterdam: Elsevier, 1975. *A comprehensive, in-depth presentation with 2444 references from the time of Faraday to the end of 1969.*

REFERENCES

1. J. K. TAYLOR and S. W. SMITH, *J. Res. Natl. Bur. Stand.*, *63A*, 153 (1959).
2. R. P. BUCK and E. H. SWIFT, *Anal. Chem.*, *24*, 499 (1952).
3. W. D. COOK, C. N. REILLEY, and N. H. FURMAN, *Anal. Chem.*, *24*, 205 (1952).
4. L. B. JAYCOX and D. J. CURRAN, *Anal. Chem.*, *48*, 1061 (1976).
5. G. D. CHRISTIAN and F. D. FELDMAN, *Anal. Chim. Acta*, *34*, 115 (1966).
6. G. D. CHRISTIAN, *Microchem. J.*, *9*, 484 (1964).
7. W. E. EARLE and K. S. FLETCHER, *Chem. Instr.*, *7*, 101 (1976).
8. J. E. HARRAR, in *Electroanalytical Chemistry*, A. J. Bard, ed., Vol. 8, New York: Marcel Dekker, 1975, pp. 1–167.
9. A. J. BARD and K. S. V. SANTHANAM, in *Electroanalytical Chemistry*, A. J. Bard, ed., Vol. 4, New York: Marcel Dekker, 1970, pp. 215–315.
10. F. B. STEPHENS, F. JAKOB, L. P. RIGDON, and J. E. HARRAR, *Anal. Chem.*, *42*, 764 (1970).
11. A. J. BARD and L. B. FAULKNER, *Electrochemical Methods*, New York: John Wiley, 1980, Chap. 10, pp. 398–404.
12. A. N. STROHL and D. J. CURRAN, *Anal. Chem.*, *51*, 353 (1979).
13. A. N. STROHL and D. J. CURRAN, *Anal. Chem.*, *51*, 1050 (1979).
14. T. FUGINAGA and S. KIHARA, *Crit. Rev. Anal. Chem.*, *7*, 223 (1977).
15. J. RUZICKA and E. H. HANSEN, *Flow Injection Analysis*, New York: John Wiley, 1981.
16. T. P. TOUGAS, Ph.D. dissertation, University of

17. C. N. REILLEY, *Rev. Pure Appl. Chem.*, *18*, 137 (1968).
18. W. R. HEINEMAN, *Anal. Chem.*, *50*, 390A (1978).
19. T. KUWANA and W. R. HEINEMAN, *Acc. Chem. Res.*, *9*, 241 (1976).
20. M. L. FULTZ and R. A. DURST, *Anal. Chim. Acta*, *140*, 1 (1982).
   Massachusetts, Amherst, Mass., 1983.
21. F. M. HAWKRIDGE and T. KUWANA, *Anal. Chem.*, *45*, 1021 (1973).
22. D. A. CONDIT, Ph.D. dissertation, University of Massachusetts, Amherst, Mass., 1983.
23. J. A. PAGE, in *Handbook of Analytical Chemistry*, L. Meites, ed., New York: McGraw-Hill, 1963, pp. 5-170–5-186.

PROBLEMS

1. In the coulometric titration of $K_2Cr_2O_7$ with Fe(II), how many micrograms of Cr correspond to 1 μA-sec (1 microcoulomb)?

2. A sample of vegetable oil is being analyzed for olefinic unsaturation. A 1.000-g sample is dissolved in chloroform and made up to 100 mL. A 1-mL aliquot of this solution is placed in the coulometry cell containing $CuBr_2$ electrolyte, and Br_2 is generated at 50.0 mA for 300.0 sec. After waiting a few minutes to allow the reaction to go to completion, the current is reversed and excess Br_2 is determined with electrogenerated Cu(I). This requires 100.0 sec at 50.0 mA. What is the *bromine number* of the oil?

3. A coulometry cell contains a Pt cathode and a Ag anode immersed in a solution of KBr at a pH of 6.00. A volume of 20.0 mL of a strong acid is added to the cell. A current of 15.0 mA is impressed, and after 320 sec the pH is again 6.00. (a) What is the normality of the acid? (b) What is the weight of the product formed on the Ag anode? (c) What are the weight and volume (corrected to standard conditions) of the gas formed at the cathode?

4. In constant-current coulometry, what current would be required so that the time in seconds would be equal to the microequivalents?

5. You want to add electrolytically generated reagent to a flowing stream at such a rate that the normality is equal to the flow rate in milliliters per second. Derive an expression relating current and flow rate.

6. The titer of a flowing stream of strong base is to be determined by adding electrogenerated acid to the stream just before it reaches a pH sensor. The current is adjusted by a feedback loop to hold the pH at 7.00. The flow rate of the stream is v milliliters per second. Write an expression for the normality of the stream in terms of i, the coulometric current in amperes.

7. The thickness of a pure-silver plate on a base metal is to be determined by controlled-potential coulometry. The metal sheet is masked except for a circular area 0.50 cm in diameter; electrical connection is made to the metal, the sheet is clamped in a cell so that the unmasked area is covered with electrolyte, and the silver plate is anodically stripped. Calculate the average thickness of the silver plating in micrometers, if the stripping required 0.600 coulomb and if the density of silver is 10.50 g/cm^3.

8. Prepare a table showing the fraction deposited $(1 - i_t/i_0)$, versus the number of half-times $(t/t_{1/2})$, for fractions deposited of 0.5, 0.90, 0.99, 0.999, and 0.9999.

9. Calculate the decay constant, p, the half-life, $t_{1/2}$, and the time for 99.9% electrolysis for each of the following controlled-potential coulometric experiments: (a) A stirred solution with an electrode of 160 cm^2 area, a solution volume of 75 mL, and a mass-transfer coefficient of 0.008 cm/sec. (b) A thin-layer cell with a cavity thickness of 50 μm and an electroactive species having a diffusion coefficient of 1.00×10^{-5}.

10. The iron in a 0.1000-g sample was converted to Fe^{3+} and titrated coulometrically with electrogenerated titanous ion (Ti^{3+}). A current of 1.567 mA was used, and the time to reach the endpoint was found to be 123.0 sec. Calculate the percentage of iron in the sample.

11. An air sample, polluted with SO_2, is passed through a continuous coulometric cell that automatically maintains a small concentration

of I_2 by electrogenerating it from acidic potassium iodide. The SO_2 is oxidized to SO_3 by the iodine. If the air sample flow rate is 5.0 L/min, and the coulometer averaged an output of 1.40 mA to maintain the I_2 concentration for 10.0 min, what is the concentration of SO_2 in ppm? The density of air may be taken as 1.2 g/L.

12. A protein sample is analyzed by digesting it with sulfuric acid to convert protein nitrogen to ammonium sulfate (Kjeldahl digestion). The digested sample is diluted to 100.0 mL, a 1.00-mL aliquot is adjusted to pH 8.6, and the ammonia produced is titrated coulometrically with electrogenerated hypobromite:

$Br^- + 2OH^- \rightarrow OBr^- + H_2O + 2e^-$

$2NH_3 + 3OBr^- \rightarrow N_2 + 3Br^- + 3H_2O$

The titration is performed at 10.00 mA current, and the endpoint occurs at 159.2 sec. How many milligrams of nitrogen was present in the sample?

CHAPTER 5

Conductance and Oscillometry

J. West Loveland

Electrical conductance occurs in many different materials, either by the flow of electrons (as in metals) or by the movement of other charged species (as in electrolytes or semiconductors). Electrolytic conductance involves the transport of anions to the anode and cations to the cathode while electrons are transferred to and from the ions at the electrode surfaces to complete the current path. Increasing the temperature of electrolytic solutions improves the mobility of the charged species and hence increases the conductance; on the other hand, metallic conduction decreases with increasing temperature because of the increased vibrational-energy barriers created and a consequent loss of mobility or free-energy bands of the electrons. A specialized form of conductance is that observed in the gaseous state (often called a "plasma"), where both ions and electrons can conduct electricity when a potential is applied between two electrodes.

For electrolytic solutions of ions, the magnitude of the electric current depends on the number and types of ions present, their mobility, the type of solvent, and the voltage applied. The number of ions depends on the concentration, but for weak electrolytes it also depends on the degree of ionization, as well as on the temperature.

Ohm's law applies to both metallic conductors and electrolyte solutions. However, anomalies occur under special conditions such as high voltages or very high frequencies. The emphasis here will be to explore electrolytic conductance for analytical uses under the more ideal conditions of low voltage (1 to 100 V) and low frequencies (0 to 5000 Hz). We will also introduce the technique of oscillometry, which is an electrodeless method (using high frequencies) that gives results similar to conductance but is influenced to a large degree by the capacitive and dielectric properties of the system.

5.1 DEFINITIONS AND UNITS

Ohm's law states that the current, i (in A), flowing in a conductor is directly proportional to the applied voltage, E (in V), and inversely proportional to the resistance, R (in Ω), of the conductor. The familiar equation results:

$$i = \frac{E}{R} \quad \text{or} \quad R = \frac{E}{i} \tag{5.1}$$

For a conductor of uniform composition and cross section, the resistance is proportional to the length, l, and inversely proportional to the area, A. The standard unit of resistance for both metallic and electrolytic conductors is called the *specific resistance*, ρ (in Ω-cm), which is the resistance of a 1-cm cube of the material. The resistance expressed in ohms is

$$R = \rho \times \frac{l}{A} \tag{5.2}$$

The reciprocal of Equation 5.2 is the *conductance*, and $1/\rho$ is generally called the specific conductance, κ, with units of $\Omega^{-1}\,\text{cm}^{-1}$ or mho/cm. Conductance G then can be written

$$\frac{1}{R} = G = \kappa \times \frac{A}{l} \quad (\text{mho or } \Omega^{-1}) \tag{5.3}$$

The specific conductances of several different types of materials are given in Table 5.1.

Equivalent Conductance

The specific conductance of electrolytic solutions depends on the concentration of the ionic species present. It becomes useful, therefore, to define the conductance of electrolytes on a basis that takes into account the concentration. This is chosen as the conductance of a hypothetical solution containing one gram-equivalent of solute between two parallel electrodes 1 cm apart. The gram-equivalent weight is equal to the gram-formula (or atomic) weight divided by the charge on the ion. A 1 N solution requires 1000 cm³, and by reference to Equation 5.3, the *equivalent conductance*, Λ, becomes

$$\Lambda = \kappa \left(\frac{1000}{N}\right) \text{cm}^2/(\text{eq-}\Omega) \tag{5.4}$$

As a hypothetical example, a 0.1 N solution requires 10^4 cm³ of solution for one gram-equivalent, or by Equation 5.3, 10^4 cm² of area for each of two electrodes spaced 1 cm apart.

Cell Constant

Obviously, the use of very large platinum electrodes to make conductance measurements is both awkward and expensive. In practice, it is not necessary to fabricate a cell where two platinum electrodes are spaced exactly 1 cm apart to obtain either the specific conductance or the equivalent conductance. Moreover, the potential field

TABLE 5.1. *Specific Conductance of Various Materials*

Material	Temperature, °C	Specific Conductance[a] κ, mho/cm
Silver	20	(6.18×10^5)
Copper	20	(5.81×10^5)
Aluminum	20	(3.55×10^5)
Iron	20	(1.03×10^5)
Mercury	0	(1.06×10^4)
Fused NaCl	850	3.5
1 N HCl	25	3.33×10^{-1}
0.1 N NaCl	25	1.07×10^{-2}
Concentrated H_2SO_4	25	1×10^{-2}
1 N Acetic acid	18	1.32×10^{-3}
0.001 N HCl	25	4.21×10^{-4}
0.001 N Acetic acid	18	4.10×10^{-5}
Bunsen flame	1725	(2.5×10^{-6})
Water[b]	18	0.8×10^{-6}
Acetone	25	6×10^{-8}
Acetic acid	25	1.12×10^{-8}
Ethyl alcohol	25	1.35×10^{-9}
Hexane	18	$(\sim 1 \times 10^{-18})$

a. Values in parentheses calculated from ρ when κ not available from critical tables or handbooks.
b. "Equilibrium water" resulting from dissolution of the CO_2 present in air.

between such large electrodes so far apart usually arches outward between them, and errors will occur in measuring the specific conductance. It is more feasible, therefore, to approximate the cell configuration with smaller electrodes and to determine a *cell factor* using solutions of known specific conductance. Potassium chloride solutions are generally used, since their specific conductances have been determined with high precision. Table 5.2 gives values of κ for several solutions of KCl. More generally, the specific conductance of any aqueous solution of KCl at 25°C can be calculated from the data of Lind, Zwolenik, and Fuoss [1].

The cell factor or cell constant, K, is related to the measured resistance, R, and κ of the solution by the relationship

$$K = \kappa R \quad (cm^{-1}) \tag{5.5}$$

Thus, if $K = 1$, the observed resistance is numerically equal to the reciprocal of the specific conductance of the solution used. Once K has been determined for a cell, the measurement of the resistance of any other solution will provide values of Λ or κ using Equations 5.4 and 5.5, respectively.

Example 5.1. A conductance cell was filled with a KCl solution that has a specific conductance of 0.01288 mho/cm. The measured resistance at 25°C was 48.3 Ω.

TABLE 5.2. *Specific Conductances of Potassium Chloride Solutions for the Determination of Cell Constants*

	Specific Conductance (κ), mho/cm		
Normality	18°C	20°C	25°C
1.000[a]	0.09822	0.1021	0.1118
0.1000	0.01119	0.01167	0.01288
0.01000	0.001225	0.001278	0.001413
0.001000	0.0001271	0.0001326	0.0001469

a. Dissolve 74.555 g KCl (weighed in air) and dilute to 1 L.

(a) What is the cell factor, K? When the same cell was filled with 0.100 N CdCl$_2$, a resistance of 123.7 Ω was obtained. (b) What is the equivalent conductance of the CdCl$_2$ solution?

Solution: (a) $K = \kappa R = (0.01288 \, \Omega^{-1} \, \text{cm}^{-1} \times 48.3 \, \Omega) = 0.622 \, \text{cm}^{-1}$

(b) $\Lambda = \dfrac{1000}{N}\left(\dfrac{K}{R}\right) = \dfrac{1000}{0.100} \times \dfrac{0.622}{123.7} = 50.3 \, \text{cm}^2/(\text{eq-}\Omega)$

5.2 THEORY

The conductivity of electrolyte solutions is equal to the sum of the conductivities of each type of ion present. For a single dissolved salt, the equivalent conductance can be expressed as

$$\Lambda = \lambda_+ + \lambda_- \tag{5.6}$$

where λ_+ = equivalent conductance of the cation
λ_- = equivalent conductance of the anion

For mixtures, Λ would be equal to the sum of all the individual ionic λ_+'s and λ_-'s.

The equivalent conductance of salts or ions increases as the concentration decreases. This phenomenon is directly related to the interionic forces present in solution; a given cation, for example, will have more anions in its vicinity than expected from a purely random distribution. This "ionic atmosphere" has two effects, *electrophoretic* and *time of relaxation*, both of which tend to decrease the ion's mobility. In the former effect, the solvent molecules associated with the ionic atmosphere are moving in a direction opposite to that of the central ion. In the latter, the ionic atmosphere moves slower than the central ion, causing a charge separation (electrostatic retarding force) on the central ion.

As solutions become more dilute, the ionic atmosphere becomes weaker, with the result that both the electrophoretic and time-of-relaxation influences decrease approximately with the square root of the ionic strength of the solution. At infinite dilution, there are no disturbing effects on the mobilities of the ions other than variations in solvent and temperature, and the equivalent conductance reaches its maximum value. Equation 5.6 may be written

$$\Lambda^0 = \lambda_+^0 + \lambda_-^0 \tag{5.7}$$

where Λ^0 = equivalent conductance of the electrolyte at infinite dilution
λ_+^0 = limiting ionic equivalent conductance of the cation at infinite dilution
λ_-^0 = limiting ionic equivalent conductance of the anion at infinite dilution

Onsager [2] has shown that Λ (at finite concentrations) and Λ^0 can be related by the equation

$$\Lambda = \Lambda^0 - (A + B\Lambda^0)\sqrt{c} \tag{5.8}$$

where A = a factor accounting for the electrophoretic effect
B = a factor accounting for the time-of-relaxation effect
c = concentration in gram-equivalents per liter (N)

Table 5.3 gives the limiting equivalent conductances for several ions. Table 5.4 shows the effect of concentration of salts on the equivalent conductance. Two predominant factors should be noted: (a) As concentration decreases, the equivalent conductance increases as predicted by Equation 5.8, and (b) the equivalent conductance decreases more rapidly with concentration as the charge on the ions increases. This effect is seen in comparing Λ for $CaCl_2$ and $NaCl$, and is very evident

TABLE 5.3. *Limiting Equivalent Conductances of Ions in Water at 25°C*

Cations	λ_+^0	Anions	λ_-^0
H^+	349.8	OH^-	198.6
Li^+	38.6	F^-	55.4
Na^+	50.1	Cl^-	76.4
K^+	73.5	Br^-	78.1
Rb^+	77.8	I^-	76.8
Ag^+	61.9	NO_3^-	71.5
NH_4^+	73.3	ClO_3^-	64.6
$(CH_3)_2NH_2^+$	51.8	ClO_4^-	67.4
Hg^{2+}	53.0	IO_4^-	54.5
Mg^{2+}	53.1	—	—
Ca^{2+}	59.5	Formate	54.6
Ba^{2+}	63.6	Acetate	40.9
Cu^{2+}	53.6	Benzoate	32.4
Zn^{2+}	52.8	SO_4^{2-}	80.0
La^{3+}	69.7	CO_3^{2-}	69.3
Ce^{3+}	69.8	$Fe(CN)_6^{4-}$	111.0

TABLE 5.4. *Equivalent Conductance as a Function of Concentration at 25°C*

Concentration, N	HCl	NaCl	AgNO$_3$	1/2 CaCl$_2$	1/2 CuSO$_4$
0.0005	422.74	124.50	131.36	131.93	121.6
0.001	421.36	123.74	130.51	130.36	115.26
0.005	415.80	120.65	127.20	124.25	94.07
0.010	412.00	118.51	124.76	120.36	83.12
0.020	407.24	115.76	121.41	115.65	72.20
0.050	399.09	111.06	115.24	108.47	59.05
0.100	391.32	106.74	109.14	102.46	50.58

Source: Table 5-12, *Handbook of Analytical Chemistry*, Meites, L., McGraw-Hill Book Company, Inc., 1963.

when two divalent ions are together as in the case of CuSO$_4$. One can use the data of Table 5.4 to extrapolate concentrations to zero concentration in order to obtain Λ^0 values.

Conductance Ratio: Weak Electrolytes

One of the early uses of limiting conductances was to determine the degree of dissociation of weak electrolytes. Arrhenius suggested that, at any given concentration, the measured equivalent conductance (when compared to the limiting equivalent conductance where all ions are dissociated) should be a measure of the degree of dissociation, α. This can be expressed as

$$\alpha = \frac{\Lambda}{\Lambda^0} \quad (5.9)$$

To a first approximation, this equation gives values that vary only slightly from the true values. Any variation is due mainly to the fact that activity coefficients and the effect of concentration on the ionic conductances have been neglected. Acetic acid, HOAc, dissociates according to the reaction

$$\text{HOAc} \rightleftharpoons \text{H}^+ + \text{OAc}^-$$

The ionization constant, K_i, is expressed as

$$K_i = \frac{[\text{H}^+][\text{OAc}^-]}{[\text{HOAc}]} = \frac{\alpha[\text{HOAc}] \times \alpha[\text{HOAc}]}{[\text{HOAc}](1-\alpha)} = \frac{\alpha^2[\text{HOAc}]}{(1-\alpha)} \quad (5.10)$$

Using the data from Table 5.3, the limiting equivalent conductance of acetic acid is $\Lambda^0_{\text{HOAc}} = \lambda^0_{\text{H}^+} + \lambda^0_{\text{OAc}^-} = 349.8 + 40.9 = 390.7 \text{ cm}^2/\text{eq-}\Omega)$.

> **Example 5.2.** The equivalent conductance of a 0.0125 N acetic acid solution was determined at 25°C to be 14.4. Calculate both the degree of dissociation and the ionization constant.

$$\text{Solution:} \quad \alpha = \frac{\Lambda}{\Lambda^0} = \frac{14.4}{390.7} = 0.0369$$

$$K_i = \frac{\alpha^2 c}{(1-\alpha)} = \frac{(0.0369)^2 \times 0.0125}{0.9631} = 1.77 \times 10^{-5}$$

5.3 CONDUCTOMETRIC INSTRUMENTATION

The apparatus required for making conductance measurements and performing conductance titrations is generally inexpensive and basically simple in detail. For these reasons, the measurement of conductance finds wide acceptance in industry as an analytical tool, both in the laboratory and in process control.

Conductivity Cells

Various types of conductance cells are used depending upon the application. The most popular type uses platinum electrodes about 1 cm^2 in area, preferably oriented in a vertical position so that solids do not collect on the surface. The electrodes, welded to heavy platinum wire, must be sealed rigidly in Pyrex or some other rigid nonconducting medium so that no movement of the electrodes takes place during stirring. Figure 5.1 depicts three cells suitable for routine use with solutions: (A) a cell used for exact conductance measurement, (B) one used for conductometric titration, and (C) a concentration dip cell for process or laboratory application. Careful cleaning and platinizing of the electrode surfaces are important for accurate and reproducible conductance measurements. More detailed discussion and exact procedures of platinization are available in an article by Jones and Bollinger [3].

A special type of conductivity cell used for detection of gas-chromatographic effluents of carbon-containing compounds was first described by Pringer and Pascalau [4]. The detection system was modified in 1965 by Coulson [5] for the selective

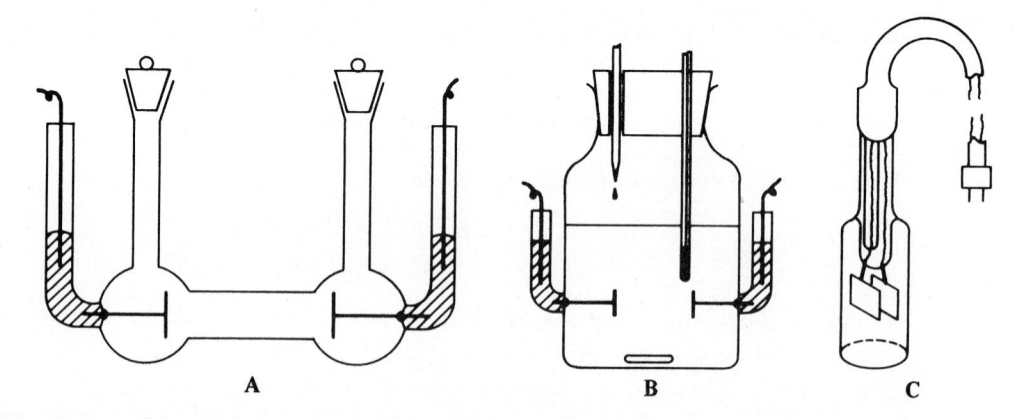

FIGURE 5.1. *Three types of conductance cells. A: Precision conductance cell. B: Conductometric titration cell. C: Concentration dip cell.*

detection of compounds containing halogens or nitrogen. In 1973, Anderson and Hall [6] described a microelectrolytic conductivity cell with a very useful gas-liquid contactor which permits rapid and efficient mixing of the gas stream and liquid solvent. After a series of improvements, the Hall electrolytic conductivity detector (HECD) became commercially established as a unique and sensitive detector for

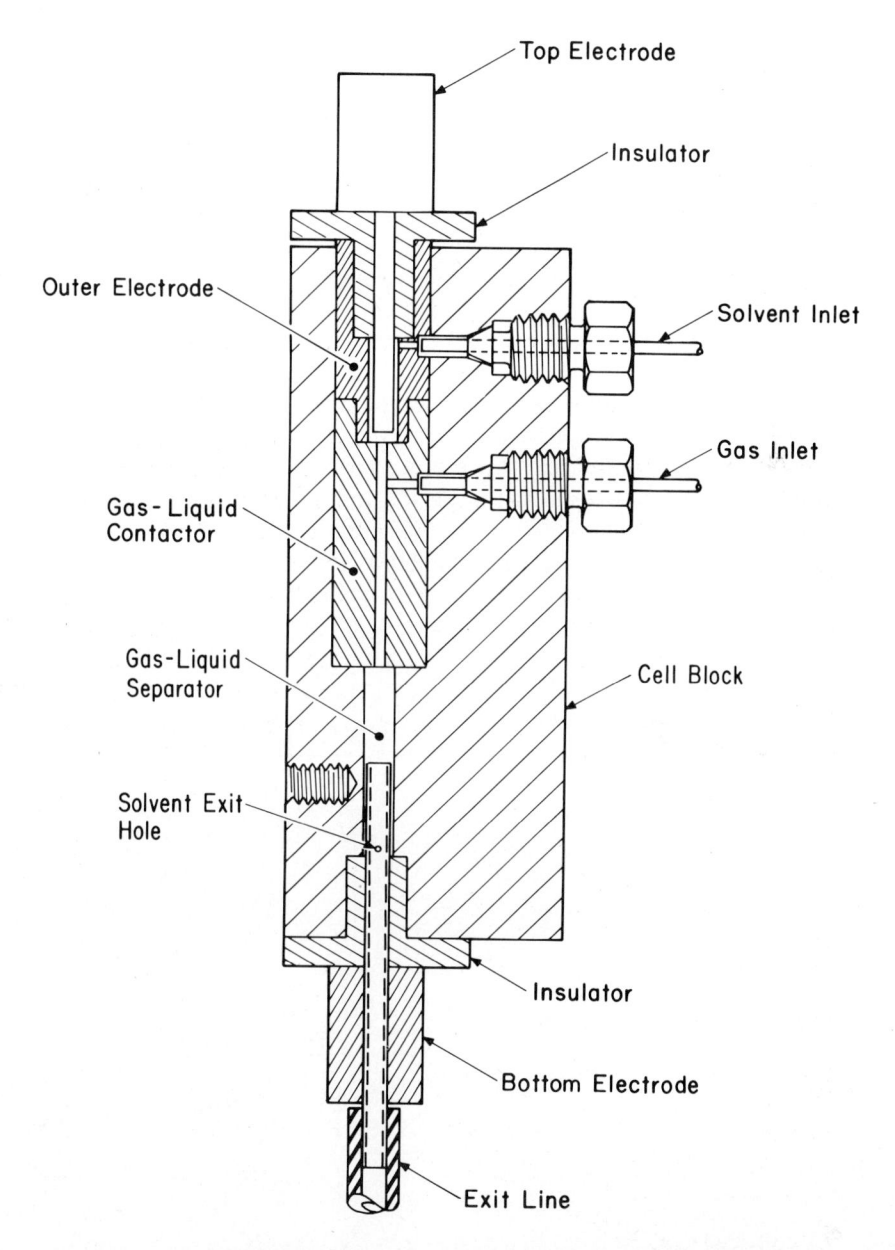

FIGURE 5.2. *Schematic diagram of a Hall electrolytic conductivity detector. Courtesy of Tracor, Inc.*

SEC. 5.3 Conductometric Instrumentation 129

chlorine-, nitrogen-, and sulfur-containing effluents of gas chromatographs. Figure 5.2 shows a basic differential conductivity cell.

The gas-chromatographic effluent is pyrolyzed at temperatures of 700 to 1000°C with an appropriate reaction gas in a microreactor, and the species formed are passed into the differential conductivity cell. The conducting solvent passes through the upper reference cell, mixes with and dissolves the pyrolyzed gases that are soluble. The reaction product/solvent mix then passes through the bottom indicating cell and exits back to the solvent reservoir. The conductivity of the solvent containing the dissolved reaction gases is compared to that of the solvent alone to produce a differential signal for the component of interest. Various scrubbers can be used to make the system respond to one specific element.

Other types of microconductivity cells are used to differentiate ionic components in the effluents of ion-exchange chromatographic systems down to sub-ppm levels of anions and cations. Cells having cell constants from 0.1 to 100 cm^{-1} are available to analyze dilute and concentrated solutions.

Measuring Circuitry

The Wheatstone bridge is used most often for determining the resistance or conductivity of an electrolyte solution. In Figure 5.3, R_c is the resistance of the electrolyte solution and R_1 is the resistance of an adjustable resistance box containing 3 or 4 decades of resistances. Resistances R_2 and R_3 may be fixed known resistors, two halves of a resistance slide wire, or two separate decade boxes of resistors.

The signal generator may be a 60-Hz transformer, a 1000-Hz oscillator, or a variable-frequency oscillator. If earphones are used as a null, the 1000-Hz oscillator

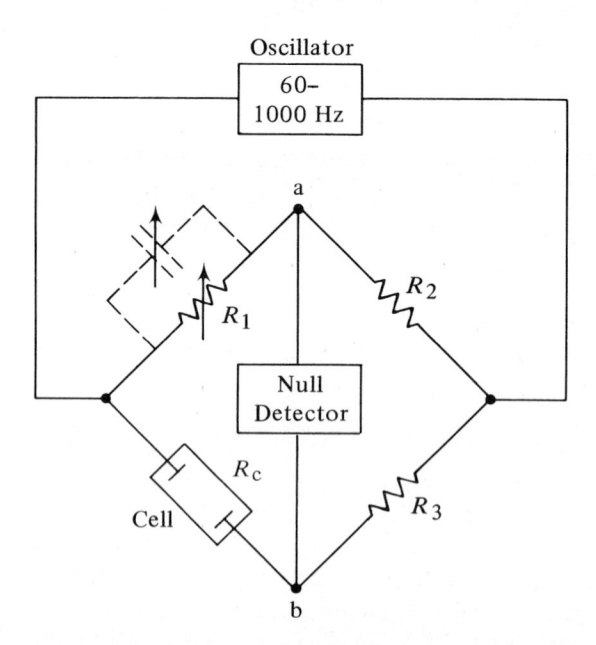

FIGURE 5.3. *Basic Wheatstone AC bridge circuit for measuring conductance.*

is preferred. The null indicator can also be a sensitive microammeter or more elaborate null-point indicators.

Alternating currents are preferred to direct current because little or no polarization of platinized electrodes takes place. During electrolysis, the platinum black adsorbs gases and catalyzes their electrochemical reaction. The alternating current prevents any buildup of material on a given electrode.

In practice, one should attempt to keep R_2 and R_3 in the same range of values. If this is done, then R_1 will have to be adjusted to a range near that of the cell resistance, R_c. This requires that the size and spacing of the electrode be considered in selecting a cell to use in making the conductance measurement. For concentrated electrolyte solutions, small electrodes and long path lengths (high cell constant) are used, whereas for dilute or weak electrolytes, cells with large electrodes and short spacing (low cell constant) should be employed.

When making the measurement, the resistance R_1 is adjusted until a null is observed. Under this condition, there is no potential difference between points "a" and "b" of Figure 5.3, and therefore $E_c = E_1$ and $E_2 = E_3$, where the various Es are the voltage drops across the appropriate resistors; from Ohm's law,

$$i_c R_c = i_1 R_1 \quad \text{and} \quad i_2 R_2 = i_3 R_3 \tag{5.11}$$

and

$$\frac{R_c i_c}{R_3 i_3} = \frac{R_1 i_1}{R_2 i_2} \tag{5.12}$$

Since the current passing through resistances R_1 and R_2 is the same and the current passing through the cell and R_3 is the same (there is no net current flow through the detector at null), the currents cancel and Equation 5.12 can be solved for R_c:

$$R_c = \frac{R_1 R_3}{R_2} \tag{5.13}$$

This is the basic relationship for Wheatstone bridges at balance.

Note that in Figure 5.3 a variable capacitor shunts R_1. This is to balance out any phase shifts in the alternating signal caused by the capacity effects present at the electrode surfaces. It is adjusted to give the sharpest minimum in the null signal. For conductometric titrations, it is generally not needed.

More recent advances in electronics and circuitry design are used for the microcell chromatographic detectors. Classical AC and DC conductance methods have limitations due to electrode polarization, capacitance effects, and cell heating. The latest technology uses *bipolar square-wave pulse conductance*, in which the waveform applied to the electrodes is a train of square-wave pulses of equal but opposite voltages [7]. One type of commercial Hall conductivity detector operates at 1.56 kHz and the peak-to-peak voltage (V_{pp}) is selectable from 14 to 0.014 V. Only one in 10 of the square waves is of full amplitude; the others are about 3% of full amplitude to reduce heating of the cell solution. The capacitance effect, and hence, the associated errors due to capacitance, occurs largely during the first 30% of the square wave at full amplitude; these errors are essentially eliminated by measurement during the last 60% of the full-amplitude positive pulses.

The cell-excitation waveform is applied to both a reference cell and the measurement cell, and the difference in their outputs is measured by a differential amplifier. Any conductances in the measuring cell that are not compensated for by the reference cell are further minimized by additional electronic circuitry. The resultant output of the differential amplifier is due only to the changes in conductivity of the eluant stream produced by the gaseous effluent from the microreactor. Anderson and Hall [6] provide more detailed information on the bipolar-pulse differential-conductivity detector.

Other commercially available conductivity detectors used in ion chromatography and high-pressure liquid chromatography use a 10-kHz waveform and either a 20- or 2-V peak-to-peak voltage depending on the cell used and the particular application, and generally provide for electronic suppression of background conductivity. Sensitivities as high as one part in 30,000 change in conductance are possible.

5.4 CONDUCTANCE TITRATIONS AND OTHER APPLICATIONS

One of the most frequent uses of conductance is in quantitative titrations of systems in which the conductance of the solution varies in such a manner (prior to and after the endpoint) that two intersecting lines can be drawn to indicate the endpoint. The actual shape of the curve depends on the sample, the titrant, and the reactions occurring. To maximize accuracy in all titration work, corrections to the measured resistance may have to be made for dilution by the titrant. To minimize this correction, titrants should be at least 10, and preferably 100, times stronger than the solute. Although the term *conductance* implies that titrations require conductance to be measured, it should be pointed out that the reciprocal of resistance can be plotted and values need only be relative and not absolute.

Volume corrections for the added titrant are made according to the equation

$$R_s = \frac{V}{V + v} R_0 \tag{5.14}$$

where R_0 = measured resistance
 R_s = corrected solution resistance
 V = original volume of solution
 v = amount of titrant added at the time of reading R_0

In general, four to six points are taken prior to the endpoint, and a similar number of points after the endpoint.

Acid-Base Titrations

Many applications of conductance titrations involve acid-base titrations.

Strong Acids and Bases. Because of the high mobilities of H^+ and OH^-, the sharpest and most accurate endpoints are obtained when strong acids are titrated with strong bases and vice versa. Referring to Table 5.3, we see that the equivalent conductance (or mobility) of H^+ is about five times that of the other cations, and that of OH^- is about three times greater than that of other anions.

A typical example is the titration of 100 mL of 0.001 N HCl with 0.1 N NaOH:

$$(H^+ + Cl^-) + (Na^+ + OH^-) \rightarrow (Na^+ + Cl^-) + H_2O \tag{5.15}$$

The relative or even exact resistance values during the titration can be calculated and the shape of the curve predetermined. We will make a calculation for this simple titration as an example to use in approaching the more complicated titrations discussed later.

For the reaction of Equation 5.15, the total conductance is the sum of the conductances due to each of the four types of ions present:

$$G = \frac{1}{R} = \frac{1}{R_{H^+}} + \frac{1}{R_{Cl^-}} + \frac{1}{R_{Na^+}} + \frac{1}{R_{OH^-}} = \frac{\lambda^0_{H^+} c_{H^+}}{1000 K} + \cdots \tag{5.16}$$

To simplify the calculations, it will be assumed that the cell constant, K, is unity and that the λ^0's of the four ions are sufficiently close to those of the 0.001 N solutions to be used without affecting the result. Under these conditions, the specific conductances of each ion and the total specific conductance of the titrated solution are given in Table 5.5.

When these data are plotted, as shown in Figure 5.4, we see that the conductance at first decreases rapidly owing to neutralization of H^+ and then, after the endpoint, increases rapidly as excess OH^- ions are added. The dashed line represents the conductance contribution of the salt (NaCl) formed during the neutralization. This dashed line is significant in the titration of weak and very weak acids or bases—up to the endpoint the conductance generally follows this line closely. If salts are already present in solution, the curve of Figure 5.4 is pushed upward and the relative change (as measured by the conductance bridge) decreases. When salt concentrations are very high, the relatively small change in conductance produces inaccuracy, and better results would be obtained by potentiometric titration.

TABLE 5.5 *Specific Conductances of Ions during the Titration of 100 mL of 0.001 N HCl with 0.1 N NaOH*

mL Titrant	H^+	Cl^-	Na^+	OH^-	NaCl	Total
0.00	3.50	0.76	0.00	0.00	0.00	4.26
0.20	2.80	0.76	0.10	0.00	0.25	3.66
0.50	1.75	0.76	0.25	0.00	0.63	2.76
0.75	0.87	0.76	0.38	0.00	0.95	2.01
1.00	0.00	0.76	0.50	0.00	1.26	1.26
1.25	0.00	0.76	0.63	0.50	1.26	1.89
1.50	0.00	0.76	0.75	0.99	1.26	2.50
1.75	0.00	0.76	0.88	1.49	1.26	3.13
2.00	0.00	0.76	1.00	1.99	1.26	3.75

Specific Conductance $\times 10^4$, mho/cm (25°C)

Note: Specific conductances have not been corrected for volume dilution.

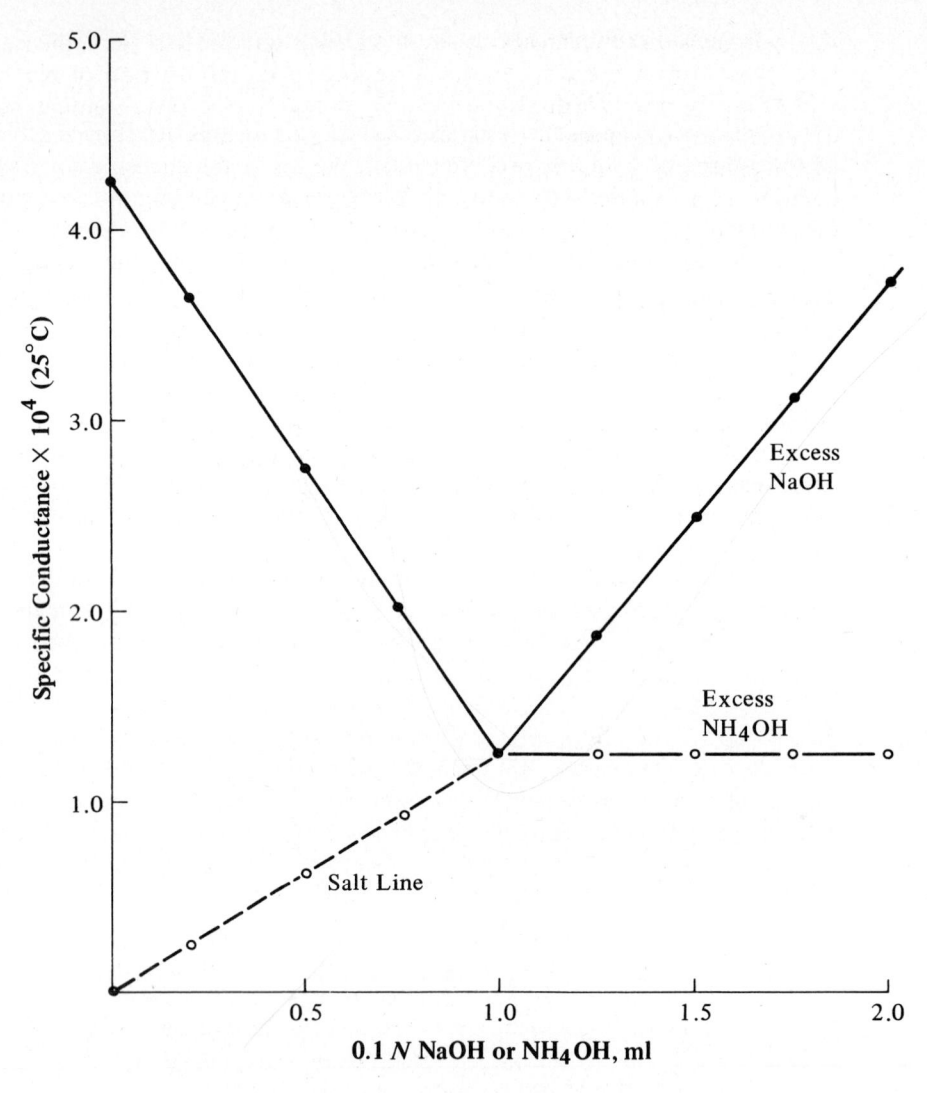

FIGURE 5.4. *Titration curve for the neutralization of* 0.001 *N HCl with* (a) 0.1 *N NaOH and* (b) 0.1 *N NH₄OH* (*the actual salt line and excess NH₄OH line have slightly higher values than shown*).

In the absence of excess salts, accurate measurements can be made equally well on both very dilute and very concentrated solutions. Very dilute solutions, of course, must be protected from contamination by acidic or basic gases in the atmosphere.

Weak Acids and Bases. The titration of weak acids and bases does not result in as sharp an endpoint as is obtained with strong acids and bases.

During the titration of weak acids, the law of mass action applies and, in the case of acetic acid titrated with NaOH, the common ion OAc^- causes a decrease in the hydrogen-ion concentration over and above that due to stoichiometric neutralization. Since the increase in conductance due to the production of Na⁺ and OAc⁻ is less than the decrease due to the loss of hydrogen ions, the net conductance decreases during the early stages of the titration. At some point, depending on the concentration of the weak acid being titrated and its pK_a value, the concentration of H^+ becomes negligible and the conductance of Na^+ and the weak-acid anions follows the salt line indicated in Figure 5.4. However, due to hydrolysis of the sodium acetate, a very slight rounding will take place near the endpoint. Figure 5.5A shows the titration curves of different concentrations of acetic acid with NaOH and also with the weak base NH_4OH at about 0.001 N. Figure 5.5B shows the effect of the ionization constant of acids at concentrations of 0.1 N on the shape of the titration curves.

Curve "c" of Figure 5.5A indicates that the very dilute solution of 0.0001 N acetic acid is dissociated to such an extent that no straight-line portion can be obtained—as in curve "a"—to provide a useful endpoint calculation. Even for the 0.001 N solution (curve "b"), only about the last 20% of the neutralization is linear, and some care must be taken if one is to obtain accurate endpoints.

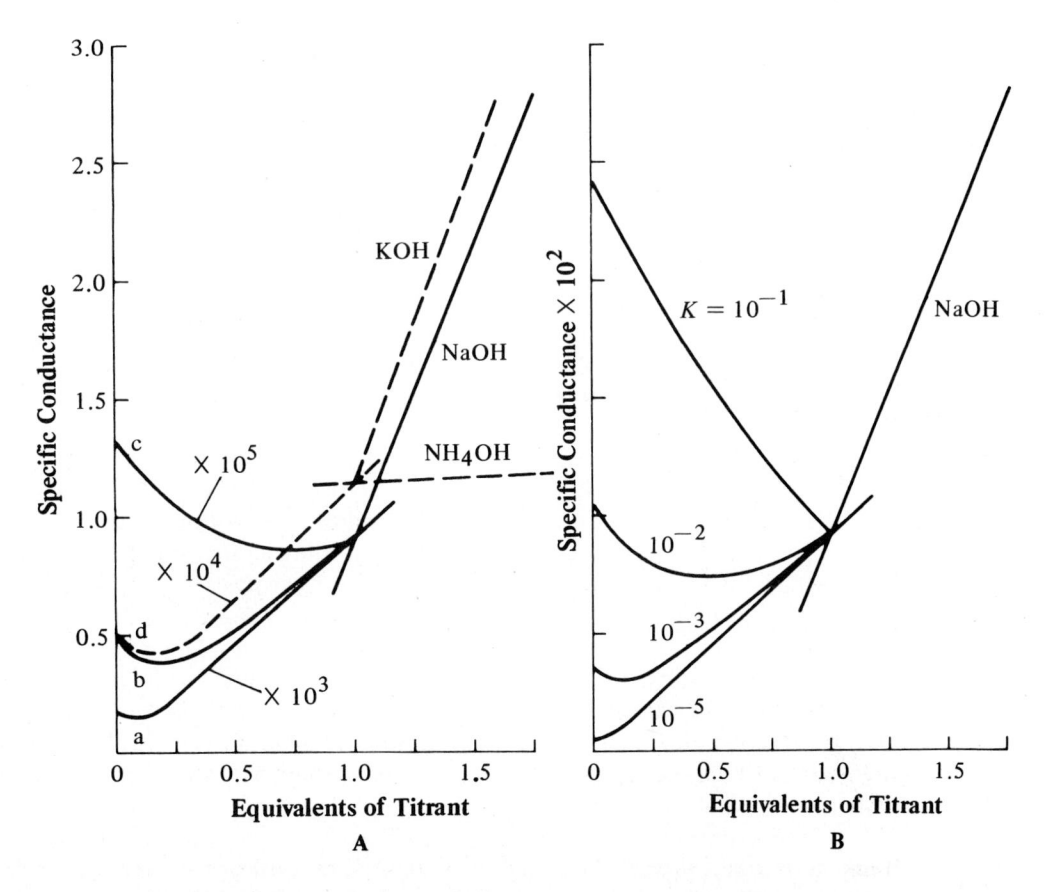

FIGURE 5.5. *A: Titration of various strengths of acetic acid—(a) 0.01 N, (b) 0.001 N, and (c) 0.0001 N titrated with NaOH, and (d) 0.001 N titrated with NH_4OH or KOH. B: Various moderately weak acids with NaOH.*

SEC. 5.4 Conductance Titrations and Other Applications

When NH$_4$OH is the base (curve "d"), a slightly sharper endpoint is obtained. In this particular case, the titration up to the endpoint will proceed as if a strong base were being used. However, after the endpoint, owing to the common-ion effect of NH$_4^+$ on the ionization of NH$_4$OH, the conductance remains essentially constant since NH$_4$OH remains mostly in the un-ionized state. Also, since the conductance of NH$_4^+$ ($\lambda^0 = 73$) is greater than that of Na$^+$ ($\lambda^0 = 50$), the conductance prior to the endpoint increases slightly faster, thereby enhancing the angle at the endpoint.

Figure 5.5B indicates that a moderately strong acid ($K_a = 10^{-1}$) at a concentration of 0.1 N can be titrated if several points are taken between the 50 and 100% neutralization points; a nearly V-shaped curve is obtained, although there is a slight curvature. The situation deteriorates rapidly, however, as slightly weaker acids are titrated; for example, for an acid with a pK_a of 2, no linear portion is available during the neutralization to obtain an endpoint.

Dilution of moderately strong acids will often provide better titration curves. For example, the 0.1 N solution of an acid with a pK_a of 2 will ionize to a much greater extent ($\sim 62\%$ vs. 27%) when diluted 10-fold, resulting in a titration curve similar to that of an acid with a pK_a of 1 at 0.1 N. Further dilution will provide still better V-shaped curves, at least up to a point.

A general rule for obtaining useful curves is that moderately strong acids (and bases) can be titrated as strong acids if their concentration is about 100 times smaller than their ionization constant. Alternatively, they may be titrated as weak acids when their concentration is at least 150 times larger than the ionization constant. For example, the last 25% of the salt line will be followed for an acid with a pK_a of 3 if its concentration is 0.15 N or greater.

When no linear region prior to the endpoint can be easily obtained, one can add alcohol or some other water-soluble organic compound to reduce the dissociation of the acid so that it behaves more like a very weak acid. This often may be the easiest approach for titrating weak and moderately weak acids. If these techniques fail, then potentiometric methods should be tried.

Very Weak Acids and Bases. These compounds may be considered to have pK_a's or pK_b's in the range of 7 to 10. Since they are very weakly ionized, the initial conductance is very low. With a strong base as titrant, the curve follows the salt line from the start of the titration. Rounding of the curve in the vicinity of the endpoint takes place because of the release of OH$^-$ ions by the hydrolysis of the anion formed. The weaker the acid, the more pronounced this effect. After the endpoint, the conductance increases rapidly because of the excess OH$^-$ ions. The endpoint is determined by extrapolating the first straight portion of the neutralization curve and the latter portion of the excess hydroxide curve.

A general guideline is that, if we want to have 50% of the neutralization follow the initial salt line, the K_a value should be greater than $5 \times 10^{-12}/c$, where c is the molar concentration of the weak acid.

A marked improvement with very weak acids can be accomplished by adding an excess amount of weak base such as NH$_4$OH, pyridine, or ethylamine. Addition of such bases causes an increase in the dissociation of the very weak acids according to the equation

$$HA + B \rightarrow HB^+ + A^- \tag{5.17}$$

Therefore, the following reaction occurs during titration:

$$HB^+ + Na^+ + OH^- \rightarrow Na^+ + B + H_2O \qquad (5.18)$$

In effect, the titration becomes one of replacing the HB^+ cation by the Na^+ of the titrant up to the endpoint. After the endpoint, the excess OH^- causes the conductance to rise rapidly. An additional advantage of the use of a weak base is its ability to solubilize many slightly soluble weak acids that otherwise require nonaqueous solvents, in which generally poorer endpoints are obtained.

Salts of Weak Acids or Bases. Salts of weak acids can be titrated with a strong acid, since they are themselves Bronsted bases—that is, proton acceptors. A typical example is the titration of sodium acetate with hydrochloric acid:

$$H^+ + Cl^- + Na^+ + OAc^- \rightarrow Na^+ + Cl^- + HOAc \qquad (5.19)$$

As sodium acetate is titrated, the acetate ion is replaced by the chloride ion, which, owing to its slightly higher ionic-equivalent conductance, causes a slight increase in conductivity up to the endpoint. Beyond the endpoint, excess hydrochloric acid causes large increases. Such titrations are useful when the ionization constant of the liberated weak acid or base, when divided by the salt's concentration, does not exceed 5×10^{-3}.

Precipitation and Complexation Reactions

Whenever a reaction between two compounds or salts produces a change in the conductivities of the ions present before and after the endpoint, conductometry can be considered as a possible analytical method. Precipitation reactions, for instance, involve replacing one ion with another. Silver in $AgNO_3$ solution may be determined by titrating it with the chlorides of sodium, potassium, or lithium. From the ionic-equivalent conductances of the cations involved ($\lambda^0_{Li^+} = 39$, $\lambda^0_{Na^+} = 50$, $\lambda^0_{Ag^+} = 62$, and $\lambda^0_{K^+} = 74$), lithium will give the sharpest endpoint. Since Li^+ is replacing Ag^+, the conductance decreases during the precipitation of silver chloride, but since the amount of NO_3^- remains constant, the conductivity will increase again after the endpoint because of the excess Cl^-.

The concentrations of the salts and the solubility product of the precipitate play an important role in determining whether satisfactory linear portions are obtained prior to and after the endpoint. High solubility of the precipitate causes rounding of the conductance curve at the endpoint; good curves will be obtained if no more than 1% of the precipitate exists in the ionized form. A general rule is that the solubility product, when divided by the concentration OH of the titrant, should not be greater than about 5×10^{-6}. For example, the concentration of $AgNO_3$ with Cl^- titrant should be at least 0.3×10^{-4} N, since AgCl has a K_{sp} of 1.7×10^{-10}.

Complexometric reactions require that stable complexes be formed. A typical example is the formation of the cyanide complex of Hg^{2+} according to the reaction

$$Hg^{2+} + 2NO_3^- + 2K^+ + 2CN^- \rightarrow Hg(CN)_2 + 2K^+ + 2NO_3^- \qquad (5.20)$$

Titrations in Nonaqueous Solutions

In nonaqueous media such as alcohols, not only Arrhenius acids and bases but also Lewis acids and bases can be titrated. (Lewis acids and bases cannot be titrated in aqueous solutions.) Interpreting the curves obtained, however, is more complex than in aqueous solutions. One factor that needs to be considered is the suppression or enhancement of the ionization of the acids or bases by the solvent; another is the viscosity of the solvent (as viscosity increases, the ionic mobility decreases). In Lewis acid-base reactions, factors such as ion-pair formation, hydrogen bonding, and solute-solvent and solute-solute interactions must also be taken into account.

A few examples showing the versatility of nonaqueous titrations are the following:

1. In glacial acetic acid, sulfuric acid gives two endpoints when titrated with a strong base or sodium acetate, since the ionization of the second hydrogen is sufficiently reduced that HSO_4^- acts as a weak acid. When hydrochloric acid is present, three endpoints are observed and both acids may be determined.
2. Phenols in low-dielectric solvents often give sharper endpoints than in aqueous solutions, for instance when 2,3,5-trimethylphenol is titrated with tetrabutylammonium hydroxide in toluene solvent as opposed to water. Similar results are observed in pyridine solvent when sodium isopropoxide is the titrant.

In general, nonaqueous conductance titration is superior to other electrochemical methods for analyzing difficult systems. The major problem is the developmental work needed to establish suitable endpoints for the particular solvents and titrants required.

Single (Batch) and Continuous Measurements of Conductance

Determination of Solubilities. The use of single measurements of conductance in determining the ionization constants of weak acids and bases has been mentioned. In addition, the solubilities of many weakly soluble salts and oxides have been determined by conductance measurements. When the solubility product is less than 10^{-8}, a correction must be made for the specific conductance of water.

Water Purity: Salt Content and Moisture Content. Pure water has a specific conductance of slightly less than 1×10^{-6} mho/cm. Water purity is, of course, important in the laboratory, where a few parts per million of dissolved salts may mask the component being looked for. It is also important in industry, particularly in power plants (where dissolved salts left behind in flash-boiler pipes may clog them) and in plants where water rinsing is employed. In most cases, it is the total salt content that is of importance. Compact and inexpensive instruments are available with special cells in which direct readout in specific conductance, or parts per million, or grams per gallon is provided. Sensors are of the insertion, flow, and submersion varieties. Automated continuous devices are used to control the flow of raw and treated water in demineralizers (ion-exchange beds), to reroute off-test condensate, to control blowdown of boiler water, and for similar operations.

Extensive use of conductometry is now made for controlling pollution in streams, rivers, and lakes, and for detecting sources of contaminants. Work in oceanography uses portable batch or continuous conductivity analyzers which quite often are scaled to read directly in percent salinity.

In the metal industry, acid strength of pickling, caustic degreasing, anodizing, and rinse baths are monitored by conductance measurements.

Some automatic analyzers use conductance to indicate the concentration of some specific component. For example, ambient concentrations of SO_2 in air as low as 0.01 ppm can be recorded continuously. Sulfur dioxide is oxidized to sulfuric acid, after which the increase in conductance due to the hydrogen and sulfate ions is directly proportional to the SO_2 concentration. The moisture content of wood and soil has been measured with special electrodes.

Conductance-Based Chromatographic Detectors

In the sections on conductivity instrumentation, mention was made of the use of microconductivity cells and new solid-state pulse-wave detection systems. The use of these sensitive detectors in both gas- and ion-chromatographic systems provides the analytical chemist with a very selective and important technique for analyzing ppm and even ppb concentrations of chlorine-, sulfur-, and nitrogen-containing compounds important to pharmaceutical, petroleum, and air- and water-pollution studies. The use of the HECD to detect chlorinated pesticides separated on a gas chromatograph is shown in the chromatogram of Figure 5.6A. The minimum detectable quantity of lindane for this system is about 1×10^{-11} g, a very sensitive analysis. Moreover, the detector responds linearly to lindane over almost six orders of magnitude.

The use of the microconductivity cell for the detection of ppm concentrations of anions and cations in tap water as they are eluted from separate ion-exchange columns is shown in Figure 5.6.B.

5.5 OSCILLOMETRY

Oscillometry differs from conductometry in several respects.

1. Electrodes are not in direct contact with the solution but are usually separated from it by the glass walls of the container. This can be a decided advantage in certain situations. For example, very corrosive solutions or solutions that would foul the electrode surfaces can be analyzed.
2. Frequencies used are of the order of 10^6 to 10^7 Hz compared with 10^3 Hz for conductance.
3. Instrument response, often fairly complex, is generally due to a combination of resistance and capacitance of both the solution itself as well as the high-frequency cell.

In conductance, the ions absorb energy, which is translated into heat and motion. In oscillometry, we have not only this aspect, but also the capacitance effect in which molecules absorb and return energy each frequency cycle owing to the induced polarization and alignment of electrically unsymmetrical molecules.

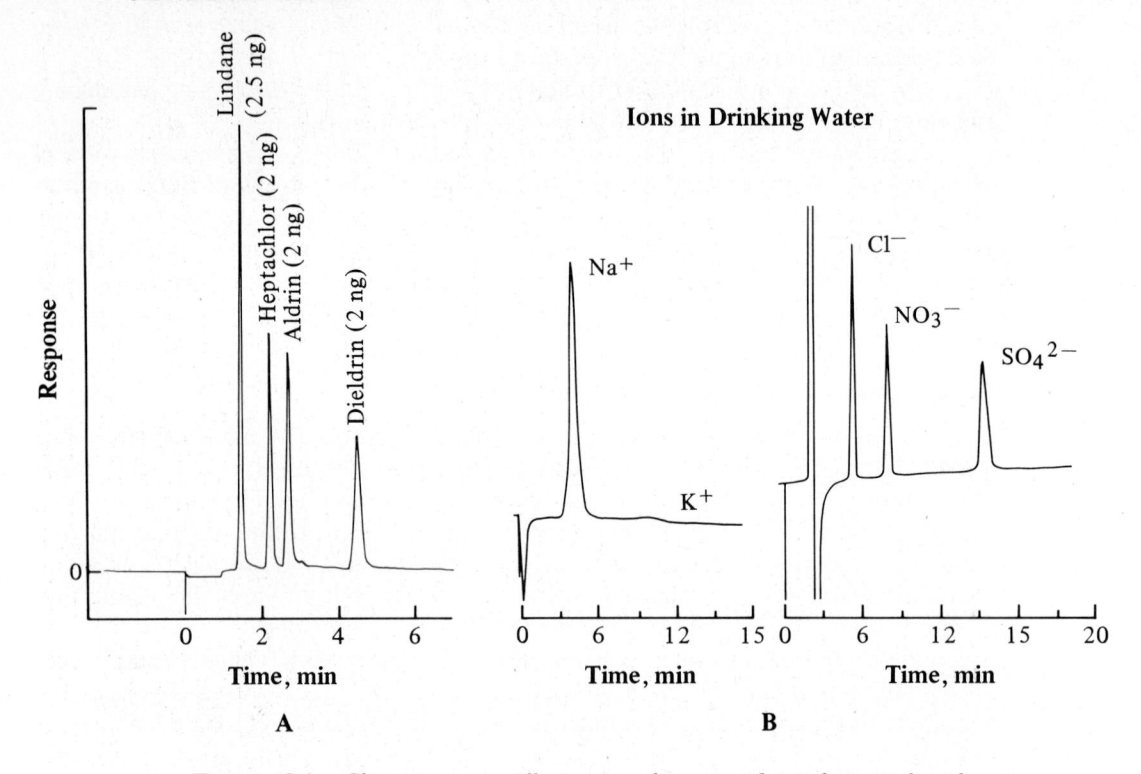

FIGURE 5.6. *Chromatograms illustrating the use of conductance-based detectors in chromatography; A: Gas chromatogram of several important chlorinated pesticides using the Hall electrolytic conductivity detector. (Chromatogram courtesy of Tracor, Inc.) B: Ion-exchange chromatogram of several of the major ions in a sample of ordinary tap water. The two chromatograms were obtained simultaneously on a single sample by using a dual-channel instrument with separate cation- and anion-exchange columns. Each channel has its own pump, eluting solvent (in this case, 2.5 mM HNO_3 for the cations and 4 mM potassium acid phthalate solution, pH 4.5, for anions), and conductivity detector. (Chromatograms courtesy of Wescan Instruments, Inc.)*

Generally, if one is working with solutions of high dielectric constant but low conductivity, the response will be primarily capacitive in nature; where the solution has a low dielectric constant and contains salts, the response will be mainly due to the conductance of the ions present.

The end result is that high-frequency titrations give a variety of responses, including the usual V-shaped curves, nonlinear intersecting curves, and inverted V-shaped curves. The shape of a curve may vary with frequency, or at a given frequency a change in dielectric constant or salt content may change the response.

Reilley and McCurdy [8] provide a comprehensive theoretical discussion on the variables that affect the response of oscillometers. Also, the references at the end of this chapter contain much more detailed information on the types of cells, the instrumentation, and the theory of high-frequency conductance. We will confine ourselves here to a few of the applications of oscillometry.

Dielectric Constant: Measurement of Binary Mixtures. When the low-frequency conductance of a solution becomes very small, the high-frequency response of the oscillator depends primarily on the capacitances of the cell walls and of the solution. If the cell-wall capacitance (C_g) is relatively large compared to that of the solution (C_s), then the overall response will depend mainly on C_s. The value of C_g can be increased by decreasing wall thickness, and C_s decreased by increasing the distance between the walls of the cell. Since the dielectric constant of the sample will alter C_s, the instrument response will follow the dielectric constant.

Table 5.6 gives the dielectric constants for several of the more common liquids. The very large value for water compared to benzene, for example, makes it possible to detect trace amounts of water or some other highly polar material in benzene. Standard curves for each binary mixture have to be prepared at a given temperature. Some examples of binary systems that can be analyzed are hexane/benzene and acetone/water, as well as various lower-molecular-weight alcohols in water.

Moisture in solids such as wood, foods, and textiles can be measured. However, such determinations are more often made by instruments designed to directly measure the dielectric changes between two parallel plates of a condenser.

Reaction Rates. A fascinating application of high-frequency conductance is the determination of rates of reactions. The response must be related directly to the change in the composition of the solution. Thus, for example, the rapid rates of hydrolysis of esters have been determined, as well as the rates of some polymerization reactions.

TABLE 5.6. *Dielectric Constants of Several Common Liquids*

Formamide	109 (20°C)	Phenol	9.78 (60°C)
Water	78.5	Acetic acid	6.15 (20°C)
Formic acid	58.5 (16°C)	Ethyl acetate	6.02
Methanol	32.6	Chloroform	4.81 (20°C)
Ethanol	24.3	Benzene	2.27
Acetone	20.7	Carbon tetrachloride	2.23
Isopropanol	18.3	n-Octane	1.95 (20°C)

Note: Values at 25°C unless otherwise specified.

SELECTED BIBLIOGRAPHY

Conductance

Fuoss, R. M., and Accascina, F. *Electrolytic Conductance.* New York: Interscience, 1959.

Glasstone, S. *Introduction to Electrochemistry.* New York: Van Nostrand, 1942.

Kolthoff, I. M., and Elving, P. J. *Treatise on Analytical Chemistry*, Part I, Vol. 4. New York: Interscience, 1963, Chap. 51.

Robinson, R. A., and Stokes, R. H. *Electrolyte Solutions.* New York: Academic Press, 1959.

WEISSBERGER, A. *Physical Methods of Organic Chemistry*, 3rd ed., Vol. 1, Part IV. New York: Interscience, 1960, Chap. 45.

Oscillometry

BLAEDEL, W. J., and PETITJEAN, D. L. High-Frequency Method of Chemical Analysis. In *Physical Methods in Chemical Analysis*, W. C. Berl, ed., Vol. 3. New York: Academic Press, 1956, pp. 107–34.

PUNGOR, E. *Oscillometry and Conductometry*. Oxford: Pergamon Press, 1965.

REILLEY, C. N. High-Frequency Methods. In *New Instrumental Methods in Electrochemistry*, P. Delahay, ed. New York: Interscience, 1954, pp. 319–45.

REFERENCES

1. J. E. LIND, JR., J. J. ZWOLENIK, and R. M. FUOSS, *J. Amer. Chem. Soc.*, **81**, 1557 (1959).
2. L. ONSAGER, *Phys. Z.*, **28**, 277 (1927).
3. G. JONES and D. M. BOLLINGER, *J. Amer. Chem. Soc.*, **57**, 280 (1935).
4. O. PRINGER and M. PASCALAU, *J. Chromatogr.*, **8**, 410 (1962).
5. D. M. COULSON, *J. Gas Chromatogr.*, **73**, 19 (1972).
6. R. J. ANDERSON and R. C. HALL, *Amer. Lab.*, **12**(12), 108 (1980).
7. D. E. JOHNSON and C. G. ENKE, *Anal. Chem.*, **42**, 329 (1970).
8. C. N. REILLEY and W. H. MCCURDY, JR., *Anal. Chem.*, **25**, 86 (1953).

PROBLEMS

1. The concentrations of three dilute solutions of sodium acetate were measured in a conductance cell in which the parallel electrodes were 1 cm² in area and 0.25 cm apart. Resistances of 274,700, 91,000 and 18,320 Ω were determined for the three solutions. Calculate the normality of each solution. $T = 25°C$.

2. A conductance measurement was made of brackish water containing equimolar concentrations of $MgCl_2$ and NaCl, with traces of other salts, which can be ignored. What is the concentration of chloride ion in ppm when a cell with a cell constant of 5.0 gives a resistance of 1549 Ω? Use the limiting equivalent conductances for the calculation.

3. In a chemical process, an aqueous solution of sodium hydroxide is to be maintained in the range 9 to 14% by weight. This corresponds to roughly 2.5 and 4.0 N solutions with equivalent conductances of 117 and 85, respectively. The commercial conductivity bridge available covers conductance ranges of 0 to 100, 0 to 10,000, and 0 to 100,000 micromhos. The midpoint of each range on the logarithmic scale is a factor of 10 lower in micromhos. (a) Should the cell constant of the conductivity cell be 0.01, 1.0, or 25 cm^{-1}? (b) What range is most suitable for monitoring the solution?

4. A special conductance bridge reads directly over the range 1 to 12% H_2SO_4. The recommended cell constant is 50. Extrapolating from data in handbooks and critical tables, calculate the approximate resistance and micromho range involved at 18°C.

5. Assume 2.4425 g of benzoic acid is dissolved in 1 L of pure water at 25°C. When the solution is placed in a conductance cell having a constant of 0.150, a resistance value of 1114 Ω is obtained. Calculate the equivalent conductance of the solution, the degree of ionization, and the ionization constant.

6. The solubility product of calcium fluoride at 25°C is 3.9×10^{-11}. What was the resistance reading when the cell constant was 0.100 cm^{-1}? Ignore the conductance contribution of the water.

7. In the titration of 100 mL of acetic acid with 1.0 N NaOH, the following relative conduc-

tance readings were observed for the corresponding buret readings. What is the concentration of the acid?

0.00 mL = 0.22	1.60 mL = 1.47
0.10 mL = 0.19	1.80 mL = 1.73
0.20 mL = 0.23	2.00 mL = 2.21
0.40 mL = 0.39	2.20 mL = 2.71
0.60 mL = 0.56	2.40 mL = 3.21
0.80 mL = 0.74	2.60 mL = 3.70
1.00 mL = 0.92	3.00 mL = 4.70
1.20 mL = 1.10	3.40 mL = 5.69
1.40 mL = 1.28	

8. Using the conductance values of Table 5.3, draw the shape of the relative conductance curves for the following titrations: (a) sodium benzoate with hydrochloric acid; (b) silver acetate with lithium chloride; (c) sulfuric acid in glacial acetic acid with sodium hydroxide; (d) mercuric nitrate with potassium chloride; (e) mixture of hydrochloric acid and acetic acid with ammonium hydroxide and with sodium hydroxide; (f) ammonium chloride with potassium hydroxide; and (g) sodium carbonate with calcium nitrate (check the effect of intermediate product formation).

9. In an experiment to determine the solubility of silver chloride, the specific conductance of the demineralized water used was 0.81×10^{-6} mho/cm at 25°C. When solid silver chloride was equilibrated in the same water at 25°C, the specific conductance was 2.62×10^{-6} mho/cm. Determine the solubility product, assuming that the limiting equivalent conductance of silver chloride is 138.3 mho/cm.

10. An oscillometer was used to determine the amount of ethylene glycol in a hydrocarbon layer and of hydrocarbon in the glycol layer. The following calibration curves were established:

0% Glycol 115	95% Glycol 1298
1% Glycol 300	96% Glycol 1324
2% Glycol 436	97% Glycol 1348
3% Glycol 551	98% Glycol 1369
4% Glycol 652	99% Glycol 1386
5% Glycol 743	100% Glycol 1399

The reading observed for the hydrocarbon layer was 371, and for the glycol layer, 1361. Draw the curves and determine the percentage of hydrocarbon in glycol and the percentage of glycol in hydrocarbon.

11. The following ions are to be titrated by conductance measurements. Name the preferred titrant, the concentration range usually covered, and any special solvents or reagents that help endpoint detection or extend the concentration range.

Ag^+, Ba^{2+}, Br^-, CN^-, CO_3^{2-}, Cd^{2+}, Cl^-, SCN^-, SO_4^{2-}, Zn^{2+}

12. For the following analytical problems, select whether conductometry, oscillometry, gas chromatography (HECD), or ion chromatography with a microconductivity detector should be used. (a) Parts-per-million levels of chlorobenzene in a hydrocarbon mix. (b) Percent levels of propionic acid in isopropanol. (c) Parts-per-million levels of Cl^-, Br^-, and F^- (combined with some cation) in an air sample. (d) Percent levels of ethanol in water. (e) A mixture of HCl and chlorinated hydrocarbons in which the percent Cl due to HCl and the percent Cl due to chlorinated hydrocarbon are needed. (f) A sewage sample in which it is suspected that Na_2SO_4, NH_4Cl, K_3PO_4, and slightly soluble organo-sulfur and organo-nitrogen compounds are present. Analysis for all five ingredients is needed. (Assume there are no interferences.)

13. A conductivity analyzer measures trace levels of HF in liquid propane by bubbling 1000 mL/min of gaseous propane into a stream of deionized water flowing at 10 mL/min. The deionized water passes over a pair of reference electrodes, while the water with the extracted HF passes over a pair of indicating electrodes. The difference in current between the two pairs of electrodes is a measure of the concentration of HF. For calibration purposes, a blend of CO_2 in N_2 is bubbled through the same system to give a reading equivalent to 5 ppm (by weight) HF. Flows of propane and N_2/CO_2 blend are identical. Temperature of all components is 25°C, and pressure is 1 atmosphere. Solubility of gaseous CO_2 in water is 0.958 mL per milliliter of water. Equivalent conductances of H^+, F^-, and HCO_3^- are given in Table 5.3. The value of pK_1 of H_2CO_3 is 4.4×10^{-7}. What volume percentage of CO_2 in N_2 should be used as a standard to give a reading equivalent to 5 ppm HF?

CHAPTER **6**

Introduction to Spectroscopic Methods

Eugene B. Bradley

The word *spectroscopy* is widely used to mean the separation, detection, and recording of energy changes (resonance peaks) involving nuclei, atoms, ions, or molecules. These changes are due to the emission, absorption, or scattering of electromagnetic radiation or particles. Spectrometry is that branch of physical science that treats the measurement of spectra.

The experimental applications of spectroscopic methods in chemical problems are diverse, but all have in common the interaction of electromagnetic radiation with the quantized energy states of matter. A group of chemists may wish to determine a molecular structure or the value of an electric dipole moment. They may wish to make an elemental analysis or to verify the presence of a chemical bond. To solve such problems, the chemists choose a particular spectroscopic method, using their knowledge of the possible energy states of matter in specific configurations and the particular wavelengths of electromagnetic radiation that interact with these states.

This chapter is meant to serve as a very general overview of spectroscopic methods; many of the topics will be covered later in greater detail, particularly in Chapters 7 through 9.

6.1 THEORY

The theoretical basis for the interaction between radiation and the energy states of matter is the quantized nature of energy transfer from the radiation field to matter and vice versa. Matter, composed of "particles" such as protons, neutrons, and electrons, sometimes behaves like a wave; radiation, a self-propagating "wave" of crossed electric and magnetic fields, sometimes behaves like a particle. This seeming paradox is reconciled in quantum theory, which is used to calculate quantized energy states.

The wavelike properties of matter are illustrated in the double-slit experiment [1], in which the wave property of *diffraction* is exhibited by an electron beam passing through a double slit. The quantized, particlelike nature of electromagnetic radiation is shown by the photoelectric effect [2], in which the number of electrons emitted by an electrode is shown to be dependent on the number of incoming "packets" of radiation at a certain minimum energy, or frequency—one electron per packet. (This particular effect is useful for some types of detectors discussed later.)

The wavelike character of radiation can be described by its *wavelength*, λ; by the *wavenumber*, $\bar{v}$, which represents the number of waves per unit of distance (the reciprocal of the wavelength); by the speed at which the wavefront advances, the *velocity*, V; and by the number of waves passing a given point in unit time, the *frequency*, v. The relationship among these properties is given by

$$\bar{v} = \frac{1}{\lambda} = \frac{v}{V} \tag{6.1}$$

The velocity of electromagnetic waves in a vacuum is c (the speed of light), which is about 3×10^{10} cm/sec; the velocity in any other medium is lower.

The following units are in common use for measurement of electromagnetic spectra:

$$\mu\text{m (micrometer)} = 10^{-6} \text{ meter} = 10^{-4} \text{ centimeter}$$
$$\text{nm (nanometer)} = 10^{-9} \text{ meter} = 10^{-7} \text{ centimeter}$$
$$\text{Å (angstrom)} = 10^{-10} \text{ meter} = 10^{-8} \text{ centimeter}$$

In recent years, the nanometer (nm) unit has replaced the older unit, the millimicron (mμ), and the micrometer (μm) has replaced the micron (μ).

The terms for wavelength that are customarily used depend on the spectral region being described. The angstrom is commonly used to describe x-ray radiation, the nanometer ultraviolet and visible wavelength, and the micrometer infrared wavelengths.

Electromagnetic radiation is an alternating electrical and magnetic field in space. Its wave properties can be explained in terms of mutually perpendicular electric and magnetic vectors, both perpendicular to the direction of wave propagation and each maintaining the other. A continuously propagating wave motion does not appear subdivisible into discrete units having an independent existence, and could be considered a continuous stream of energy; but when radiation interacts with matter, its properties are those of particles, not waves. A quantitative description of many interactions between radiation and matter is possible by considering radiation as discrete quanta of energy called *photons*. The energy of a photon is proportional to the frequency of radiation. These dual views of radiation as waves and particles are not mutually exclusive. Indeed, this duality is useful for the quantitative description of other phenomena, such as the behavior of electrons or other elementary particles.

The wave nature of radiation is familiarly illustrated by refraction effects in material media, diffraction, and interference phenomena. Line and band spectra are evidences of quantized energy states in matter and of quantized energy transfer between radiation and matter. The amount of energy transferred per photon is given by the Einstein-Planck relation

$$E = h\nu = \frac{hc}{\lambda} = hc\bar{\nu} \tag{6.2}$$

where E = energy in joules
h = Planck's constant (6.62 × 10^{-34} J-sec)
ν = frequency of the radiation in hertz (Hz, or sec^{-1})

Although it may seem strange at first, wavenumbers are easier for most spectroscopists to use. (Frequencies are a factor of 10^{10} greater.) Note that 1 μm = 10^{-4} cm = 10^{-6} m. Thus,

$$\lambda \, (\mu m) = \frac{10^4}{\nu \, (cm^{-1})} \tag{6.3}$$

To convert wavenumbers to electron-volts,

$$E \, (eV) = \frac{12{,}399}{\lambda \, (\text{Å})} \tag{6.4}$$

Table 6.1 presents some conversion factors useful in spectroscopy.

The Einstein-Planck relation indicates that the energy of a photon of *monochromatic* (single-frequency) radiation depends only on its wavelength or frequency. A beam of radiation is more or less intense depending on the quantity of photons per unit time and per unit area, but the quantum energy (E) per photon is always the same for a given frequency of the radiation.

Planck explained correctly the energy *distribution with frequency* of a black body by assuming the atomic oscillators in the body to be quantized according to Equation 6.2. Bohr, in 1914, laid the foundation for the correct interpretation of spectra of atoms and molecules with the following postulates:

1. Atomic systems exist in stable states without radiating electromagnetic energy.
2. Absorption or emission of electromagnetic energy occurs when an atomic system changes from one energy state to another.
3. The absorption or emission process corresponds to a photon of radiant energy $h\nu = E' - E''$, where $E' - E''$ is the difference in energy between two states of an atomic system.

The absorption or emission of radiant energy by matter is one of the most

TABLE 6.1. *Conversion Factors Useful in Spectroscopy*

Unit	ergs/molecule	cm^{-1}	cal/mole	eV/molecule
1 eV/molecule =	1.602 × 10^{-12}	8065.5	23,060	1
1 cal/mole =	6.948 × 10^{-17}	0.34975	1	4.336 × 10^{-5}
1 cm^{-1} =	1.986 × 10^{-16}	1	2.8591	1.240 × 10^{-4}
1 erg/molecule =	1	5.034 × 10^{15}	1.439 × 10^{16}	6.241 × 10^{11}

Source: Adapted from C. E. Meloan, *Elementary Infrared Spectroscopy*, New York: Macmillan, 1963, p. 5.

important "fingerprints" furnished by nature. When a beam of radiation is passed through an absorbing substance, the intensity of the incident radiation (I_0) will be greater than that of the emergent radiation (I). (Quantities I_0 and I are sometimes symbolized by P_0 and P, respectively, since the intensity to which we refer has units of energy per unit time, or power.) Part of the radiation that passes into a substance, instead of being absorbed, may be scattered or reflected when emerging from the substance, or reemitted at the same wavelength or at a different wavelength. In other cases, the radiation may undergo changes in orientation or polarization. The absorption of radiation at various wavelengths is summarized in Table 6.2.

The portions of the electromagnetic spectrum that are useful to chemists are shown in Figure 6.1. The spectrum is divided according to frequency (energy), and corresponding spectroscopic methods are shown in the appropriate frequency ranges. For clarity, these methods and the corresponding energy states of matter are listed in Table 6.3. Notice that the various energy states and basic phenomena are diverse, and so there are different methods and techniques. These methods and techniques are not difficult to learn, however, and one need not be an expert to obtain many extremely important and useful results.

TABLE 6.2. *Interaction of Radiation with Matter*

Radiation Absorbed	Energy Changes Involved
Visible, ultraviolet, or x-ray	Electronic transitions, vibrational or rotational changes
Infrared	Molecular vibrational changes with superimposed rotational changes
Far-infrared or microwave	Rotational changes
Radio-frequency	Too weak to be observed except under an intense magnetic field (see Chaps. 12 and 13)

TABLE 6.3. *Spectroscopic Methods and Corresponding Energy States of Matter or Basis of Phenomenon*

Nuclear magnetic resonance	Nuclear spin coupling with an applied magnetic field
Microwave spectroscopy	Rotation of molecules
Electron spin resonance	Spin coupling of unpaired electrons with an applied magnetic field
Infrared and Raman spectroscopy	Rotation of molecules
	Vibration of molecules
	Rotation/vibration of molecules
	Electronic transitions (some large molecules only)
Ultraviolet-visible spectroscopy	Electronic energy changes
	Impinging monoenergetic electrons causing valence-electron excitations
x-Ray spectroscopy	Inner-shell electronic transitions
	Diffraction and reflection of x-ray radiation from atomic layers

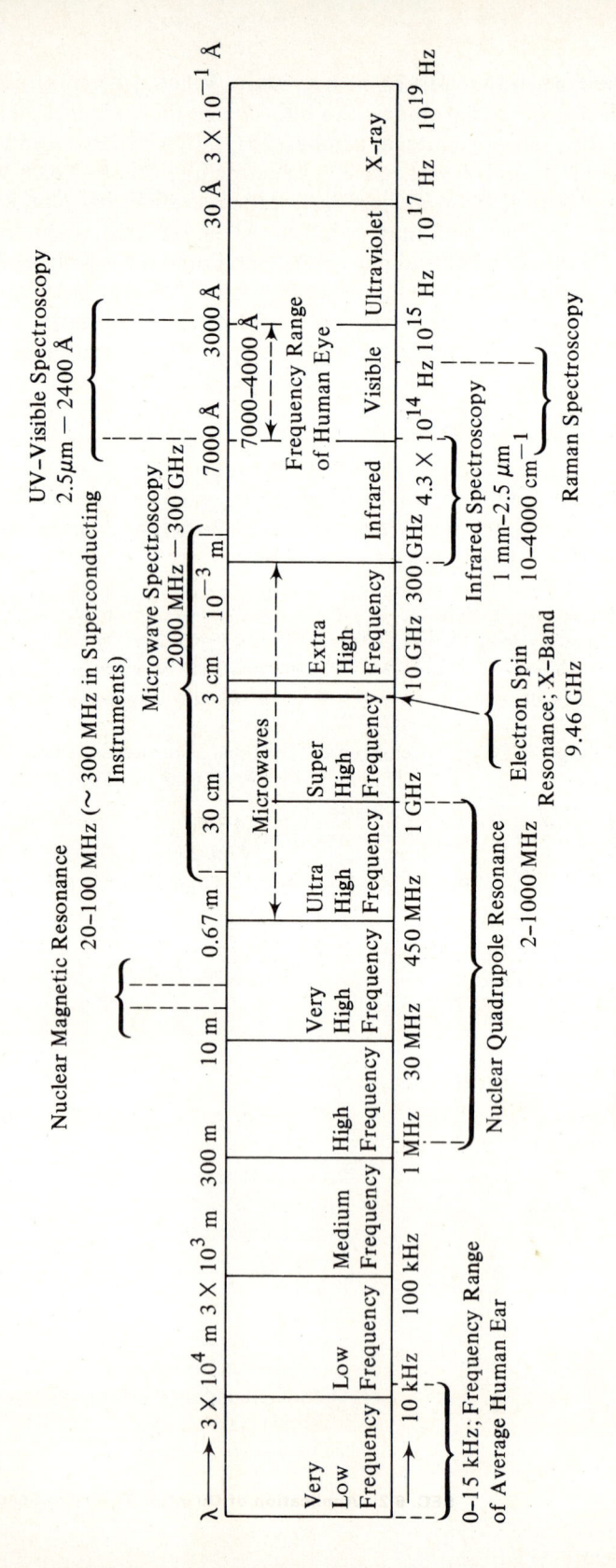

FIGURE 6.1. *Electromagnetic spectrum from DC to x-ray; frequency ranges are shown for different spectroscopic methods.*

6.2 APPLICATION OF QUANTUM THEORY TO SPECTROSCOPY

The laws of classical mechanics, which apply to the energies of objects of ordinary size, such as Ping-Pong balls, cannot be used to understand the behavior of microscopic bodies such as atoms, electrons, and molecules. For example, a Ping-Pong ball can spin (rotate) and bounce with any speed depending on how it is hit; that is, the rotational energy and the bouncing velocity of a Ping-Pong ball can assume any value on a continuous scale. However, a molecule cannot rotate or vibrate freely with any particular energy; it is subject to what are called quantum restrictions, and is limited to only certain discrete values of velocities and energies. The significance of quantum restrictions on a particular motion of a microscopic body depends on the space available for such a motion: If there is a large space for a motion, that motion is less subject to quantum restrictions on its energy.

We learn from quantum mechanics that allowed energy states exist in which a molecule or an atom may spend long or short periods of time. A molecule or atom can exist in intermediate energy states for only a transient time when it is ascending or descending from one level (or state) to another.

Monoatomic substances normally exist in the gaseous state and absorb radiation only through an increase in their electronic energy. It should be remembered that electrons in a given atom occupy discrete energy levels and are thus quantized. These quantized levels take the form of the various subshells illustrated in Figure 6.2. Therefore, electronic absorption of radiation can occur only if the impinging photon has an energy that is equal to the energy difference, ΔE, between two quantized energy levels.

For a polyelectron atom, a multiplicity of absorptions is permissible. The energy required to produce a $3d \rightarrow 4p$ transition (ΔE_1) corresponds to visible radiation; $2s \rightarrow 2p$ (ΔE_2) requires far-ultraviolet radiation; and $1s \rightarrow 2s$ (ΔE_3) requires x-ray radiation.

For polyatomic molecules, electronic transitions involve molecular orbitals; such transitions require energy in the ultraviolet region and are of vital importance in ultraviolet spectroscopy.

Effect of Structure on Absorption

The spectrum is a function of the whole structure of a molecule rather than of specific bonds. Photons of low energy (far-infrared, microwave) can produce changes of

FIGURE 6.2. *Energy levels for the electrons in a polyelectron atom.*

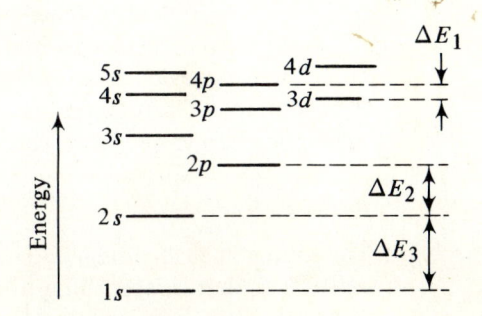

rotational energy. More energetic photons change the energy of molecular vibration as well as rotation (medium-infrared). With visible and ultraviolet light, valence-shell electrons are excited, and these electronic transitions are usually accompanied by changes in vibration and rotation. In the far-ultraviolet, the energies of the photons may even break bonds.

6.3 INSTRUMENTATION

A *spectrograph* is an "instrument with an entrance slit and a dispersing device that uses photography to obtain a record of spectral range. The radiant power passing through the optical system is integrated over time, and the quantity recorded is a function of radiant energy." An *optical spectrometer* has "an entrance slit, a dispersing device, and one or more slit exits, with which measurements are made at selected wavelengths within the spectral range, or by scanning over the range. The quantity detected is a function of radiant power." A *spectrophotometer* "furnishes the ratio, or a function of the ratio, of the radiant power of two beams as a function of spectral wavelength. These two beams may be separated in time, space, or both." [3]

Figure 6.3 shows the block diagram of a basic spectrometer that may be used for the study of various energy states of matter; typical components are listed in each block.

Commercial spectroscopic instruments are readily available; some of these are very sophisticated, for use in exacting research studies that require high precision. Many studies do not require such precision; for these, simple, less expensive models are also available. In some cases, part or all of such instruments may be built by the investigator at a considerable saving.

Usually, the building blocks are the same for any spectrometer, that is, a source of electromagnetic energy, a sample to be investigated, an analyzer to sort out energies that are modified by the sample in some manner, a detector of these energies, and a recorder (which may include an electronic amplifier to boost the power level of the detected energy).

One of the simplest "spectroscopic" (more properly, "optical") methods is termed *colorimetric*. In this technique, white light is passed through a sample and the percentage of energy absorbed is recorded and related to sample properties. If some type of dispersing device such as a prism or grating is used to restrict the white light to a narrow band of frequencies, the method resembles the monochromatic method in which only a single frequency, or very narrow band of frequencies, is viewed at any instant of time.

However, monochromatic methods are used more extensively than simple colorimetric ones because of the frequent need to extract more detailed information from a sample. The method may involve emission, absorption, or scattering of electromagnetic energy, or *fluorescence* (radiation is absorbed and reemitted).

Source

The source of electromagnetic radiation is chosen according to the spectral range to be studied, that is, according to energy requirements (source intensity is usually less

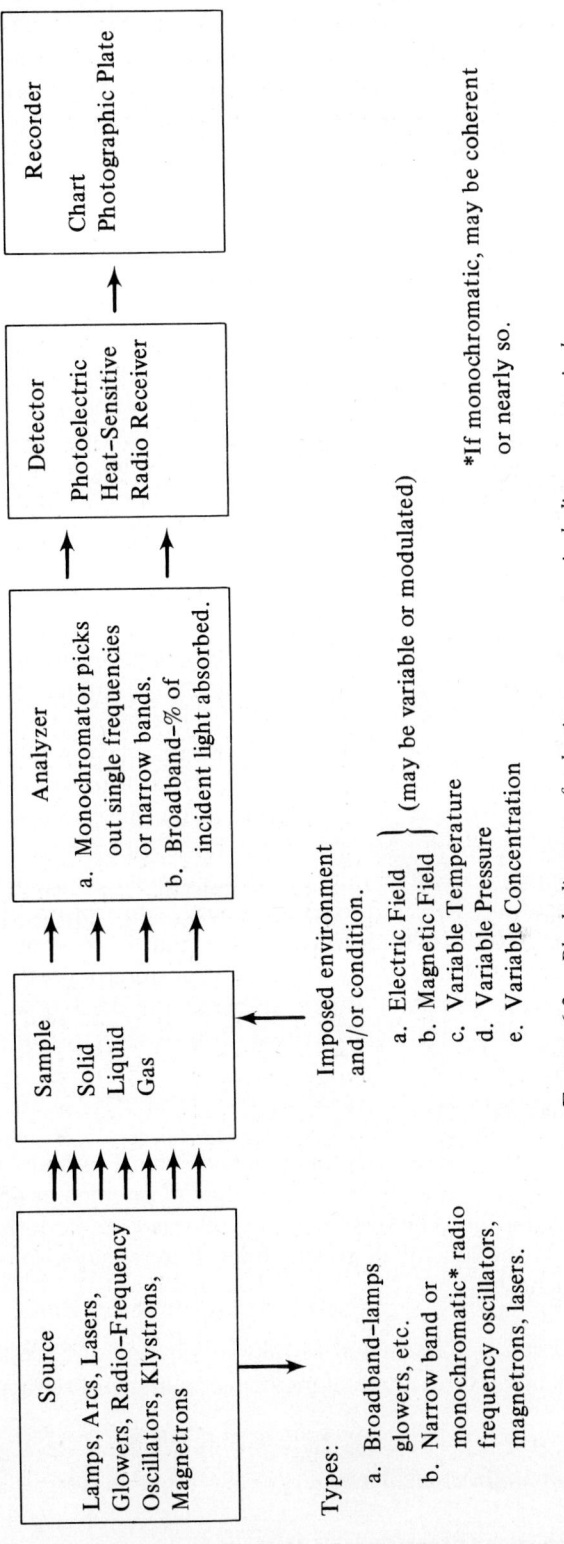

```
┌─────────────┐      ┌─────────┐      ┌─────────────┐      ┌──────────────┐
│   Source    │ ───▶ │ Sample  │ ───▶ │  Analyzer   │ ───▶ │   Detector   │ ───▶ ┌────────────────┐
│             │      │         │      │             │      │              │      │    Recorder    │
│ Lamps, Arcs,│      │ Solid   │      │ a. Monochro-│      │ Photoelectric│      │                │
│ Lasers,     │      │ Liquid  │      │    mator    │      │ Heat-Sensi-  │      │ Chart          │
│ Glowers,    │      │ Gas     │      │    picks    │      │ tive         │      │ Photographic   │
│ Radio-Freq- │      │         │      │    out      │      │ Radio        │      │ Plate          │
│ uency       │      │         │      │    single   │      │ Receiver     │      │                │
│ Oscillators,│      │         │      │    frequen- │      │              │      └────────────────┘
│ Klystrons,  │      │         │      │    cies or  │      │              │
│ Magnetrons  │      │         │      │    narrow   │      │              │
│             │      │         │      │    bands.   │      │              │
│             │      │         │      │ b. Broad-   │      │              │
│             │      │         │      │    band—%   │      │              │
│             │      │         │      │    of       │      │              │
│             │      │         │      │    incident │      │              │
│             │      │         │      │    light    │      │              │
│             │      │         │      │    absorbed.│      │              │
└─────────────┘      └─────────┘      └─────────────┘      └──────────────┘
                          ▲
                          │
              Imposed environment
              and/or condition.
              a. Electric Field  ⎫
              b. Magnetic Field  ⎬ (may be variable or modulated)
              c. Variable Temperature
              d. Variable Pressure
              e. Variable Concentration

Types:
  a. Broadband—lamps glowers, etc.
  b. Narrow band or monochromatic* radio frequency oscillators, magnetrons, lasers.

*If monochromatic, may be coherent or nearly so.
```

FIGURE 6.3. *Block diagram of a basic spectrometer, including some typical devices and conditions.*

important than source energy). The typical sources listed in Figure 6.3 span the electromagnetic spectrum from radio frequencies to x-rays. Some sources such as lasers and klystrons emit nearly monochromatic, nearly phase-coherent radiation, whereas other sources such as lamps and glowers emit a broad spectrum of phase-incoherent frequencies.

Sample

The sample may be a solid, a liquid, or a gas. A certain physical arrangement or enclosure of a sample is usually necessary for successful spectroscopic results; often, environmental conditions are imposed upon the sample to create either a necessary set of conditions for a recordable effect, or a perturbation of some type that produces additional energy states for study. Standard sampling techniques and apparatus suffice for the majority of compounds, but occasionally the chemical nature of the sample is troublesome and the usual techniques fail. A troublesome sample may be a highly corrosive gas, an easily decomposed compound, or a hygroscopic compound; it may be explosive, highly toxic, deeply colored, radioactive, or viscous. In cases like these, the chemist must develop a new sampling technique for the particular problem, and the integrity of the data may depend upon the ingenuity and finesse used to accomplish this.

Experiments require that energy be absorbed or transmitted by a sample, while at the same time gases, liquids, hygroscopic materials, and other sensitive or dangerous compounds must be contained. Therefore, it is often necessary to enclose the sample in a cell of some sort. Each cell must have windows that transmit a particular band of wavelengths and resist particular forms of chemical degradation. As an example, suppose one wished to obtain the pure-rotational energies of the HF molecule (see Sec. 6.4 for some reasons for wanting such information). Hydrogen fluoride is highly corrosive, and the sample must be above room temperature to avoid dimerization. The pure-rotation energy absorption of HF occurs in the far-infrared region (50 to 1000 μm wavelength), and so one must choose a window for a low-pressure gas cell that will withstand atmospheric pressure from the outside, be highly resistant to chemical attack, transmit the desired wavelengths, and not decompose when the sample cell is heated. Polyethylene satisfies these requirements.

Monochromators

Monochromators are frequency (energy) analyzers. For many methods, the analysis is accomplished by varying the incidence angle of prism(s) or grating(s) with respect to the incident radiation. The spectral range scanned is determined by the apex angle of the prism or by the spacing of grating lines. Filters may also be used to pass or reject specific frequencies.

Prisms. A prism is constructed to take advantage of Snell's law. Recall that for a light ray, this law relates angle of incidence to angle of refraction by

$$n \sin \phi = n' \sin \phi' \tag{6.5}$$

where ϕ = angle of incidence

ϕ' = angle of refraction
n = refractive index of the medium of incidence
n' = refractive index of the medium in which refraction occurs

In a prism, two plane surfaces are inclined at some apex angle, α, such that the deviation produced by the first surface is increased by the second surface (see Fig. 6.4). The refraction obeys Snell's law, so that

$$\frac{\sin \phi_1}{\sin \phi_1'} = \frac{n'}{n} = \frac{\sin \phi_2}{\sin \phi_2'} \tag{6.6}$$

The refractive index of the prism material depends on the wavelength of the incident light; so the angle of deviation, ϕ, becomes a function of wavelength:

$$\frac{d\phi}{d\lambda} = \frac{d\phi}{dn}\frac{dn}{d\lambda} \tag{6.7}$$

The factor $d\phi/dn$ is determined by geometrical considerations, such as the apex angle, α, but the factor $dn/d\lambda$ is characteristic of the prism material and it is called the *dispersion*. The value of $dn/d\lambda$ is not constant for a given material, but depends on wavelength. Typical prism materials are quartz, glass, NaCl, KBr, and CsI. Quartz is used in the ultraviolet region, glass is used in the visible region, and the other three materials are used in the medium-infrared region.

If many wavelengths strike the prism simultaneously, each wavelength emerges from the prism at a different angle. In practice, the prism is rotated about an axis perpendicular to the triangular cross section. A radiation detector is placed at a fixed distance from the prism, so that as the prism rotates, successive wavelengths fall upon the detector.

Diffraction Gratings. The diffraction grating was invented by Joseph von Fraunhofer (1787–1826). The word *diffraction* implies effects produced by cutting off portions of wavefronts. A diffraction grating may be used in either transmission or reflection, but the dispersion of incident wavelengths depends upon the geometry of the grating.

A grating is a parallel array of equidistant grooves, closely spaced. The spacing between these grooves is designated as d. A transmission grating is made by ruling parallel grooves on glass with a diamond edge. Reflection gratings are made by ruling parallel grooves on a metal mirror. The great majority of spectroscopic applications of gratings use reflecting optics. In these applications, the reflection gratings are replicated from a master grating.

The reflection of incident light by a grating is shown in Figure 6.5. A portion

FIGURE 6.4. *Refraction of a light ray by a prism.*

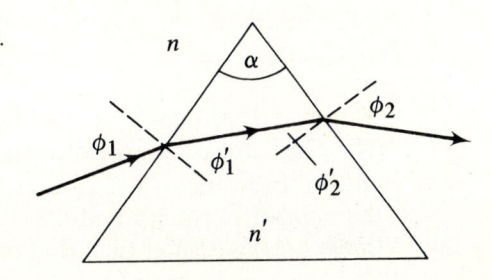

FIGURE 6.5. *Diffraction of incident light by a reflection grating having a groove spacing of d.*

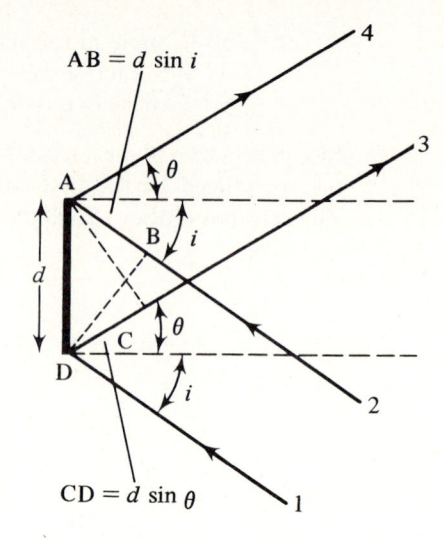

of the collimated light beam incident at angle i is reflected at angle θ. The angle i does not have to equal the angle θ as in reflection. The path difference for incoming rays 1 and 2 is $AB = d \sin i$, and the path difference for outgoing rays 3 and 4 is $CD = d \sin \theta$. The total path difference for an incident and reflected ray is $d \sin i - d \sin \theta$. When this difference is equal to one or more wavelengths, no interference occurs and a bright image is seen. In general,

$$n\lambda = d(\sin i \pm \sin \theta) \quad (6.8)$$

where n (an integer) = *order* of the grating

The path differences are added when both light rays are on the same side of the normal to the grating surface.

Assume that angles i and θ remain constant, and that values of $n > 1$ and smaller values of λ are chosen so that the product, $n\lambda$, remains constant. It is seen that shorter wavelengths may be reflected at the same angle, θ, as were the longer wavelengths corresponding to $n = 1$. These shorter wavelengths are called *higher orders*. To disperse incident light of many wavelengths, the grating is rotated so that the angle i changes. Assuming $n = 1$, different wavelengths will be reflected at the same angle, θ. Higher-order wavelengths will also be reflected at this same angle and these wavelengths will have to be filtered out before they reach the detector.

The dispersion of the grating, $d\theta/d\lambda$, can be calculated by assuming the angle i is a constant. Then,

$$\frac{d\theta}{d\lambda} = \frac{n}{d \cos \theta} \quad (6.9)$$

The dispersion equals the order divided by the product of the grating spacing and the cosine of the angle of reflection. The *resolving power* of a grating is the product of the number of rulings and the order. Hence, the resolving power of a large grating (more area) is greater than that of one with smaller area.

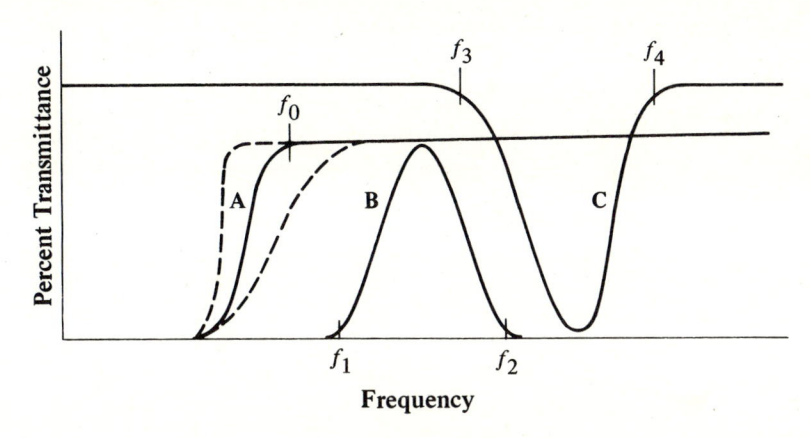

FIGURE 6.6. *Frequency versus percentage transmission for three types of optical filters. A: Cut in (or cut off). B: Bandpass. C: Band rejection.*

Filters. Optical filters are used extensively in spectroscopy to pass or reject certain frequencies or bands of frequencies. These filters are of three main types: (a) cut-in (or cut-off), (b) bandpass, and (c) band rejection. The nomenclature is illustrated by the graph in Figure 6.6. Filter A is a cut-in filter if one reads from left to right along the frequency axis. The cut-in frequency is usually defined as f_0, the frequency at which the transmission begins to be approximately constant. Notice that the rate of cut-in (%T vs. f) may vary depending on the construction of the filter. Filter B (a bandpass filter) rejects all frequencies except those between f_1 and f_2. Practically speaking, the useful frequency range of the filter is between those frequencies where the transmission is greater than 30%. Filter C (a band-rejection filter) passes all frequencies except those in the range f_3 to f_4. This filter is most useful in the region of high attenuation (small-percentage transmission). The shapes and widths of the transmission regions of these various filters may be varied considerably, depending upon the application.

A fourth type of filter, not shown in Figure 6.6, is called a *restrahlen* filter. This type of filter is used in reflection only, and it reflects a relatively narrow range of frequencies. All other frequencies are absorbed. Restrahlen filters are useful, for example, in suppressing unwanted orders from a grating.

It is also possible to use a prism or a grating as a selective filter. The selective transmission or reflection of electromagnetic energy is a function of the incidence angle. For example, a prism may be used in transmission to restrict the number of frequencies incident upon a grating, making it unnecessary to use additional filters to eliminate higher orders of the grating. Such a prism is called a *fore-prism*.

Fourier Transform Spectroscopy

The type of spectroscopy instrumentation discussed thus far is classified as *dispersive*: The electromagnetic spectrum in the region of interest, let us say infrared, is dispersed or spread out and only a small portion or element of the frequencies or wavelengths available is viewed by the detector at one time. The resulting record of intensity versus frequency or wavenumber, what we conventionally understand to be a "spectrum," is said to be a *frequency-domain* spectrum.

A very different way of achieving the same end is to measure the intensity or radiant power of many wavelengths simultaneously as a function of time. This would be *time-domain* spectroscopy, most commonly called *Fourier transform* spectroscopy, because the time-domain spectrum is then converted by a Fourier transform using a digital computer into the conventional frequency-domain spectrum. Perhaps the single most critical distinction in Fourier transform spectroscopy is that the sample sees all the wavelengths in the region of interest at all times, instead of only a small portion at a time. This can result in significant improvement in speed, resolution, or signal-to-noise ratio over conventional dispersive spectroscopy.

Two examples of important modern Fourier transform methods are in infrared and nuclear magnetic resonance (NMR) spectroscopy. In infrared, this is accomplished by recording the intensity at a detector as a function of the *optical path difference* traversed by two light beams in a Michelson interferometer. This is termed the *interferogram*. A mirror in one leg of the interferometer is moved, usually at a constant linear velocity, to provide change in the optical path difference of the two beams of light. The source, the sample, and the detector are positioned in other legs of the interferometer. The particular portion of the spectrum to be scanned is determined by the light source, the thickness and refractive index of the beam splitter in the interferometer, and the choice of bandpass filters.

In Fourier transform or *pulse NMR*, the sample is irradiated periodically with a brief (about 1–10-μsec-wide), intense, square-wave pulse of radio-frequency radiation. The signal induced in the resonating nuclei of the sample by this pulse, the *free induction decay signal*, is followed as a function of time by the detector. Because a square-wave pulse of radio-frequency power is really a summation of many sinusoidal-shaped electromagnetic waves of differing frequencies, a whole range of frequencies is incident on the sample at one time.

A critical advantage of pulse NMR is that an "experiment" is over in so short a period of time: The free induction decay signal drops off very rapidly, and another pulse can be reapplied in, typically, about a second. Thus, although one decay signal is very weak and very noisy, the same digital computer that will later carry out the Fourier transform can be used to store, add together, and average many decay signals. Because "noise" is generally random, addition of many signals will tend to cancel out the noise, while adding up the signal; the signal-to-noise ratio generally improves as the square root of the number of pulses. If one experiment takes only a second, 1000 signals can be obtained and averaged in about 17 min, 10^4 signals in about 2.8 hr, with tremendous improvement in the signal-to-noise ratio.

Resolution

The resolution (ability to separate frequencies) of a dispersive spectrometer is determined by various factors such as the dimensions of the dispersing element, the arrangement of the associated optics, and the spectral limits imposed by a mechanical slit of some sort.

The mechanical slit is placed before the detector to limit further the number of frequencies that impinge simultaneously upon it. Transfer optics may differ from one type of monochromator to another and yield different spatial spreads of the frequencies that arrive at the mechanical slit. An adjustment of the slit width effects a mechanical control of the bandwidth of radiation seen by the detector. The real measure of the radiation seen by the detector is called the *spectral slit width* (in Å or cm^{-1}), which is a measure of the frequency spread seen by the detector. The spectral slit width is a function of the optical geometry, the instrument's dispersion, and the mechanical width of the slit.

If the source is monochromatic or nearly so, or if it is monochromatic and tunable, then very high resolution is possible, provided the frequency stability of the source is good. Tunable sources have been available for some time in the radio-frequency and microwave regions of the spectrum, but it is only recently that tunable lasers have become available in the infrared and visible regions.

An important criterion for a monochromator is the amount of stray radiation reflected and transmitted through the system to the detector. Such extraneous radiation produces signal errors at the detector. Roughly speaking, stray light decreases as the resolution improves. Two monochromators may be coupled optically into a double monochromator to decrease stray light markedly. Also, a monochromator may be a double-pass or a single-pass type, depending on the number of times the radiation is dispersed.

The resolution of a dispersive spectrometer is limited by the grating area and the slit width, but the resolution of an infrared Fourier spectrometer is determined primarily by the reciprocal of the optical path difference. For example, to double the resolution in a dispersive spectrometer, both the entrance and exit slits must be halved, thereby decreasing the energy by a factor of four. In order to maintain the same signal-to-noise ratio, the scan speed must be 16 times slower. To double the resolution of a Fourier spectrometer, one must double the original optical path difference, which only doubles the time required to obtain the interferogram. To achieve the same signal-to-noise ratio, twice as much time must be spent at each data point; so the total time required is only four times longer. Thus, Fourier transform spectroscopy requires *one-fourth* of the time required to obtain the same resolution with a dispersive spectrometer.

Detectors

Research has continued over the years to develop sensitive, noise-free detectors. There are inherent quantum-imposed limits on detectors, which we do not expect to exceed, but in recent years many significant breakthroughs in detector technology have occurred. Detectors are classified into two general groups, selective and nonselective. The response of a selective detector varies with the frequency of the incident radiation, whereas the response of a nonselective detector does not. A listing of typical detectors is shown in Figure 6.3, but each spectroscopic method imposes its special requirements for a detector. Photoelectric detectors, photographic plates, and photoconductive cells are selective detectors, whereas thermocouples, bolometers, pneumatic cells, and square-law crystals have responses (at infrared and microwave frequencies) that are *relatively* insensitive to wavelength and thus can be classified as nonselective.

Photoelectric detectors are useful in the ultraviolet-visible region of the spectrum. When coupled with the appropriate electronic circuitry, these detectors are capable of counting as little as one photon per second. Specific types of photoelectric detectors have frequency (energy) responses that peak in different subregions of the ultraviolet-visible region. Photoelectric detectors are often cooled to reduce random tube noise ("shot" noise).

Photographic plates are extremely sensitive photon detectors in this same region. The speed of data acquisition is, of course, much lower due to the processing time required for the plate. Most photographic plates are insensitive in the infrared region, but infrared phosphors extend their range somewhat into the infrared.

A photoconductive cell is an important selective detector. Such cells show an increase in conductivity when illuminated with infrared light, and they have high sensitivity with fast response. These cells are used extensively in the spectral region 0.5 to 3.5 μm (5000 to 35,000 Å). The range may be extended slightly by cooling the cell with liquid hydrogen.

In a thermocouple—used for measuring infrared radiation—a junction of two dissimilar metals is blackened to increase absorption of incident radiation. The temperature rise at the junction relative to a "cold junction" on which infrared radiation does not fall increases the potential across the junction; this potential is amplified to a usable voltage. Thermocouples have a relatively slow response (thermal lag) and, if the infrared radiation is time varying, it must not vary too rapidly.

Bolometers have a faster response than do thermocouples, and they are useful in the infrared and microwave regions. Metal bolometers have a small thermal capacity that permits a quicker response than thermistor bolometers. Incident radiation falls upon an element that forms one arm of a Wheatstone bridge. The resistance of the arm changes with temperature and the bridge goes unbalanced, producing an error signal, which is amplified. Some types of bolometers are useful in the microwave region at frequencies up to about 60 GHz (1 GHz = 1 gigahertz = 10^9 hertz). Commercial thermistor bolometers are matched pairs of "flakes" that have similar electrical and thermal properties. One of the flakes serves as a compensatory element to eliminate the effect of background radiation.

Pneumatic cells are very sensitive devices, useful from the near-infrared region to the 300 to 150-GHz microwave range. In a pneumatic cell, incident radiation heats a confined gas, which expands, moving a curved diaphragm with a mirror surface. A light beam reflected from this surface to a photocell varies in intensity with the movement of the diaphragm.

6.4 APPLICATIONS

The applications of modern spectroscopic methods are legion. They are used in such widely diverse fields as controlling pollution and detecting art forgeries. A few general applications will be discussed here for each spectroscopic method listed in Table 6.3.

Nuclear magnetic resonance (NMR) is the spectroscopic method used to study proton resonances. Electromagnetic radiation causes the magnetic moment of protons to "flip" in the presence of an external applied magnetic field. Proton chemical shifts are usually measured with respect to an arbitrary reference compound such as tetramethysilane. Proton resonances of different functional groupings occur at characteristic values called τ-values, which depend on the chemical and magnetic environment in which the group occurs. Nuclear magnetic resonance of fluorine, phosphorus, boron, and some other elements with a nuclear magnetic moment is also used extensively. Undoubtedly, the single most important application of NMR has been in the qualitative identification of organic compounds and the elucidation of their structure. Nuclear magnetic resonance, gas-liquid chromatography, and infrared spectroscopy are probably the three most important tools available today for the organic chemist.

Microwave spectroscopy is used to measure dipole moments and moments of inertia of simple molecules in the gas phase. The chemical composition of the molecule and the masses of its atoms are usually known, and so one uses moments of inertia to help determine the structure of a molecule, perhaps the most important application of microwave spectroscopy. Sometimes it is necessary to use x-ray data to supplement the microwave data. From the pattern of the rotational spectrum, it is possible to determine the symmetry of the molecular configuration, that is, whether planar, linear, or polar. The technique is very sensitive to sample concentration and pressure, and it is used in some important air-pollution measurements.

Electron spin resonance is used to map unpaired-electron distributions in molecules and molecular fragments. The method is versatile and may be used to detect free radicals in cancer tissue, for example, or for routine monitoring of vanadium in crude petroleum. Because the effect depends simply on the presence of unpaired electrons, there are many other analytical and practical applications.

Infrared and Raman spectroscopy are complementary methods used, for instance, to study molecular structure, identify compounds and functional groups, determine interatomic forces and bond-stretching distances, perform quantitative and qualitative analyses, and determine thermodynamic properties. The three states of matter may be studied by these methods over wide ranges of temperature and pressure. Selection rules for different molecular structures determine which spectral lines are allowed, and these rules differ for the infrared and Raman methods. This difference is used to advantage in studies of molecular structure, because two types of information are brought to bear on the same problem.

The ultraviolet-visible method is useful for the study of electronic transitions in molecules and atoms. Although various forms of ultraviolet-visible spectroscopy can be used to study a myriad important chemical and physical properties, we will be most concerned with its use in quantitative analysis. It is probably the single most frequently used analytical method, with the possible exception of the analytical balance. For example, a single clinical analysis laboratory in a major hospital may perform a million chemical analyses a year, primarily on serum and urine, and about 70% of these tests are done by ultraviolet-visible absorption spectroscopy. Atomic absorption and emission spectroscopy (Chaps. 10 and 11) are used primarily to analyze for metallic elements in a variety of matrices—serum, natural waters, tissues, and so forth.

X-ray spectroscopy rivals visible spectroscopy as a tool for elemental analysis. Because the energies of x-rays are much higher than those of visible radiation, however, x-rays usually cause transitions of inner-shell electrons rather than of valence-shell electrons. There are many advantages of this method in spectrochemical analysis. A quantitative analysis of a mixture of rare-earth oxides may be performed, or a crystal structure may be determined. A specimen that contains two elements widely separated in atomic number may be studied, or the thickness of a very thin layer of tin plating may be measured. The most widespread use of x-rays has been in the field of metallurgy, but x-rays may also be used to analyze metals, minerals, liquids, glasses, ceramics, or plastics.

REFERENCES

1. R. B. LEIGHTON, *Principles of Modern Physics*, New York: McGraw-Hill, 1959, p. 81.
2. R. B. LEIGHTON, *Principles of Modern Physics*, New York: McGraw-Hill, 1959, p. 67.
3. "Spectrometry Nomenclature," *Anal. Chem.*, 55, 173 (1983).

PROBLEMS

1. Express 2000 cm^{-1} in (a) μm; (b) eV/molecule; (c) cal/mole; (d) ergs/molecule.
2. Convert the wavelength $4.50 \, \mu$m to (a) cm^{-1}; (b) Hz; (c) Å; (d) ergs/molecule; (e) eV/molecule.
3. A far-infrared transition of a particular molecule occurs at a wavelength of $500 \, \mu$m. Calculate (a) the frequency in Hz; (b) the wavenumber in cm^{-1}; (c) the energy in eV/molecule.
4. An argon-ion laser has a transition at 5145 Å (green). Express this wavelength in (a) Hz; (b) cm^{-1}; (c) μm; (d) eV.
5. A Fourier transform spectrometer is adjusted to scan through an optical path difference of 5 cm. (a) Calculate the resolution of the spectrum (in cm^{-1}). (b) Calculate the optical path difference through which this instrument must scan if the resolution is to be doubled, that is, 0.1 cm^{-1} resolution is required.
6. A reflection grating is ruled for 20 lines/mm spacing. Assume angle θ = angle i and the grating is used in first order. (a) Calculate angle θ for $\lambda = 40 \, \mu$m. (b) What spectral region corresponds to $40 \, \mu$m? (c) In second order, what wavelength, λ, will be reflected at angle θ calculated in part (a)?
7. A glass prism, $n' = 1.60$, has a ray of light incident upon one side (one of two sides forming the apex angle) at an angle of 60° measured from the normal to the side. (a) Assume the refractive index of air is 1, and calculate the angle of deviation, ϕ'_1, of the ray. (b) Assume $\phi'_2 = 20.5°$ and calculate the angle ϕ_2.

CHAPTER 7

Ultraviolet and Visible Absorption Spectroscopy

K. L. Cheng
V. Y. Young

Photometric methods are perhaps the most frequently used of all spectroscopic methods, and are important in quantitative analysis. The amount of visible light or other radiant energy absorbed by a solution is measured; since it depends on the concentration of the absorbing substance, it is possible to determine quantitatively the amount present.

Colorimetry involves the determination of a substance from its ability to absorb visible light. Visual colorimetric methods are based on the comparison of a colored solution of unknown concentration with one or more colored solutions of known concentration. In spectrophotometric methods, the ratio of the intensities of the incident and the transmitted beams of light is measured at a specific wavelength by means of a detector such as a photocell.

The absorption spectrum also provides a "fingerprint" for qualitatively identifying the absorbing substance.

7.1 MOLECULAR ABSORPTION OF RADIATION: ELECTRONIC SPECTRA

Molecular absorption in the ultraviolet-visible region depends on the electronic structure of the molecule. In molecules, the valence electrons occupy molecular orbitals which are delocalized over several atomic centers. As an example, we will consider formaldehyde, which absorbs in the ultraviolet region. This molecule (Fig. 7.1) has 12 valence electrons. In Figure 7.1A, the ground-state electronic configuration of this molecule is shown. The orbitals have been classified in terms of both their symmetry designation and the more common σ-π designation. For now,

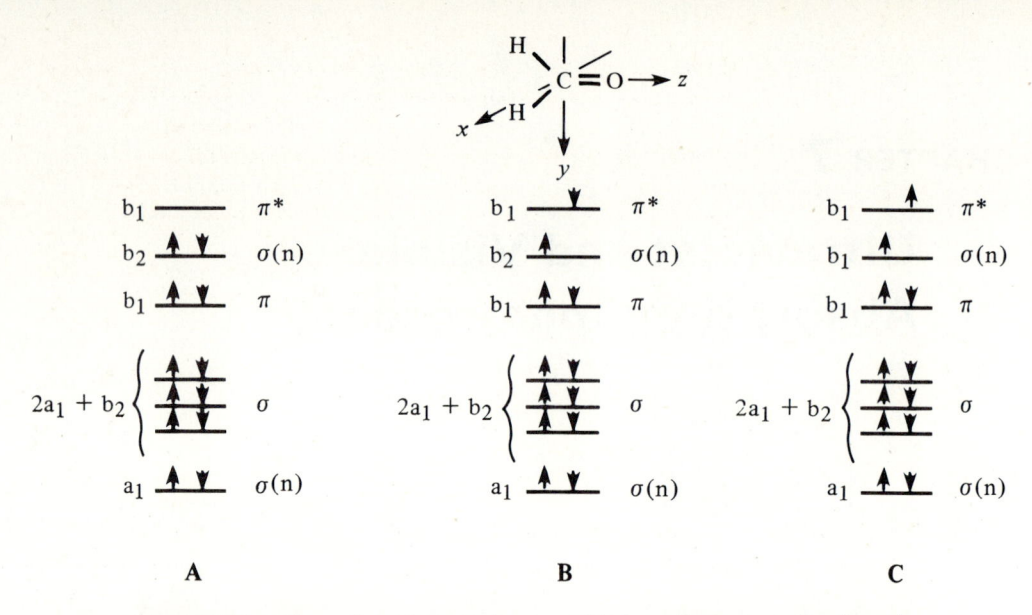

FIGURE 7.1. *Ground- and excited-state orbital configurations for formaldehyde. A: Ground state. B: Singlet electronic configuration after absorption of a photon promoting a σ(n) electron to the π* orbital. C: Triplet electronic configuration after absorption of a photon promoting a σ(n) electron to a π* orbital.*

consider only the σ-π designation. The highest occupied molecular orbital is of σ type, and the σ(n) designation indicates that it is essentially a nonbonding orbital with the electron density concentrated on the oxygen atom. Thus, it is one of the oxygen lone-pair orbitals indicated in the Lewis dot structure. The lowest unoccupied orbital is of π type and is antibonding between the carbon and oxygen atoms. We see that for an electron to be promoted from the highest occupied orbital to the lowest unoccupied orbital requires the input of a precise amount of energy—the energy difference between σ(n) and π*. Hence, energy must be absorbed in quanta, and the final state for the process indicated above is shown in Figure 7.1B. This electronic configuration represents an excited state of formaldehyde. Because the molecular orbitals are discrete, it is evident that only certain states are possible in any molecule. Since the energy difference between any ground state and excited state must equal the energy added by the quantum, only certain frequencies can be absorbed. In many electronic structures, absorption does not occur in the readily accessible part of the ultraviolet region; thus, in practice, ultraviolet spectrophotometry is mostly confined to conjugated systems.

When a beam of radiation is passed through an absorbing substance, the intensity of the incident radiation (I_0) will be greater than that of the emergent radiation (I). The absorption of visible (Table 7.1), ultraviolet, and x-ray radiation usually results in electronic transitions in matter, with accompanying vibrational and rotational changes in the case of molecular substances. In general, the excited atoms and molecules resulting from absorption of radiation return to the ground state very

TABLE 7.1. *Absorption of Visible Light and Color*

Wavelength, nm	Color (Absorbed)	Color Observed (Transmitted) or Complementary Hue
< 380	Ultraviolet	
380–435	Violet	Yellowish green
435–480	Blue	Yellow
480–490	Greenish blue	Orange
490–560	Bluish green	Red
500–560	Green	Purple
560–580	Yellowish green	Violet
580–595	Yellow	Blue
595–650	Orange	Greenish blue
650–780	Red	Bluish green
> 780	Near-infrared	

rapidly, either by losing energy in the form of heat to the surroundings or by re-emitting electromagnetic radiation (luminescence or fluorescence).

The absorption of electromagnetic radiation by molecules is far more complex than absorption by individual atoms, which have no vibrational or rotational energy levels. The total energy may be considered as a sum of contributions from electronic, rotational, and vibrational energies:

$$E_{total} = E_{el} + E_{rot} + E_{vib} \tag{7.1}$$

where E_{el} = electronic energy of the molecule
E_{rot} = energy associated with the rotation of the molecule around its center of gravity
E_{vib} = energy of the molecule due to interatomic vibrations

For each electronic energy state of the molecule, there normally are several possible vibrational states and for each of these, in turn, numerous rotational states (Fig. 7.2). Consequently, the number of possible energy levels for a molecule is much larger than for an atomic particle.

The electronic energy is generally larger than the other two (E_{rot} and E_{vib}), and electronic transitions ordinarily involve energies corresponding to ultraviolet or visible radiation. Pure vibrational transitions are caused by the less energetic infrared radiation (1 to 15 μm); rotational transitions require even less energy (10 to 10,000 μm). Furthermore, changes in vibrational and rotational levels invariably accompany electronic excitation of a molecule. A molecule may jump from any of the vibrational and rotational levels in the ground state to any of a large number of possible vibrational and rotational levels in a given excited state. Because a photon of slightly different energy corresponds to each of the many possible jumps, visible and ultraviolet molecular absorption spectra consist of hundreds or thousands of lines so closely spaced that they appear as continuous absorption bands, in contrast

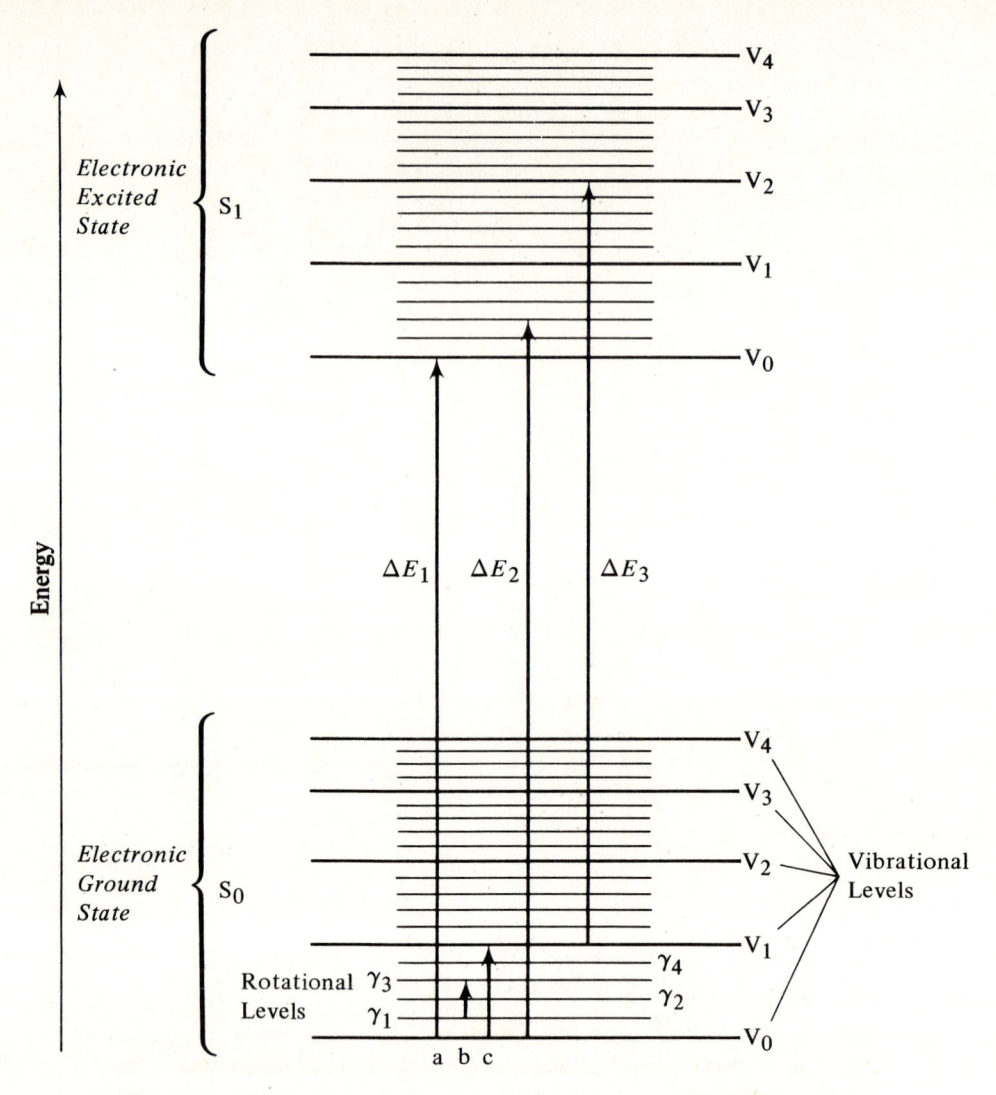

FIGURE 7.2. Molecular energy levels and (a) electronic, (b) rotational, and (c) vibrational transitions.

to the sharp lines that characterize atomic spectra or rotational spectra in the far-infrared region.

Collisions between neighboring molecules in solution cause slight modifications of the various energy levels and lead to further broadening and merging of absorption bands. A dramatic illustration of the effects due to changing from a gaseous state to a liquid state is given in Figure 7.3. The many sharp absorption peaks are an example of vibrational fine structure superimposed on the electronic absorption band; the much broader bands show that a substantial portion of the fine structure is lost because of molecular interaction and collision.

CHAP. 7 Ultraviolet and Visible Absorption Spectroscopy

FIGURE 7.3. *Ultraviolet absorption spectra of benzene. Upper: Benzene solution in ethanol solvent. Lower: Benzene vapor. From R. E. Dodd*, Chemical Spectroscopy, *Amsterdam: Elsevier, 1962, p. 227, by permission of the publisher.*

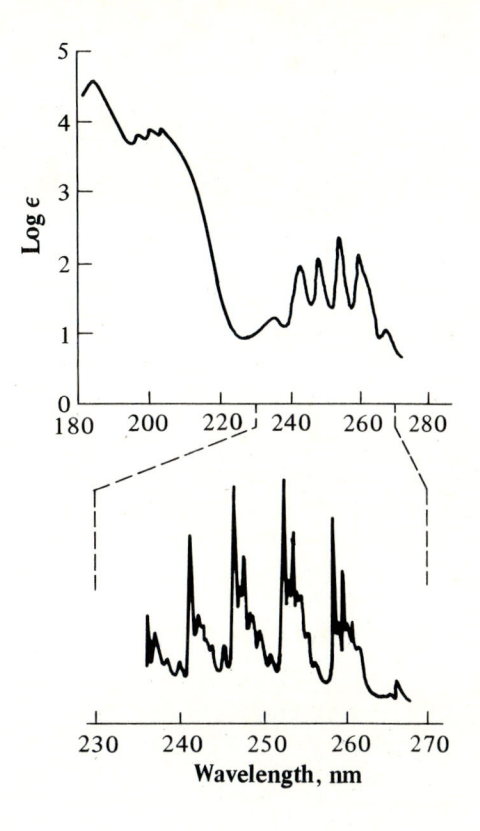

Selection Rules

It is important to realize that even when the energy of a photon matches the energy difference between two levels, the absorption of radiation may not be observed or may be observed with low intensity. The reason for this is that certain requirements, summarized in quantum-mechanical *selection rules*, must be satisfied for a transition to occur with high probability. The most important requirement is that the promoted electron be promoted without a change in its spin orientation. This requirement is stated as $\Delta S = 0$; that is, the system must undergo no change in spin. Returning to Figure 7.1, note that the transition indicated by configuration B occurs with no change in spin and hence satisfies the $\Delta S = 0$ rule. We could also promote an electron from $\sigma(n)$ to π^* to give electronic configuration C. This configuration is more stable than configuration B; however, a transition from configuration A to configuration C violates the $\Delta S = 0$ rule (A is a singlet, whereas C is a triplet). Hence, this transition would occur with very low probability and is said to be *forbidden*.

The second requirement is based on the symmetry of the initial and final states. In Figure 7.1, we have shown the designation of the various orbitals in terms of the symmetry of the molecule. The overall arrangement of the electrons in the electronic state can also be described in terms of its symmetry. For example, the electron configuration shown in Figure 7.1A can be described as an A_1 state, and that in Figure 7.1B can be described as an A_2 state. Thus, in order for the transition indicated in Figure 7.2B to be symmetry allowed, a transition from an A_1 state to an A_2 state must be possible. A detailed treatment of this aspect of absorption is beyond the scope of this chapter. The more advanced reader is referred to two excellent texts [1, 2].

The third requirement concerns orbital overlap. Clearly, we expect transitions to occur with finite probability when there is overlap between the initial and final states, much as we expect bonding or antibonding orbitals to be formed by atomic-orbital overlap. However, forbidden transitions are still observed in many molecules, because intramolecular or intermolecular perturbations cause the rules to relax considerably. Singlet-to-triplet transitions, for instance, occur with increased intensity in the presence of paramagnetic substances such as O_2 or NO or in solvents such as C_2H_5I that contain heavy atoms.

Franck and Condon have suggested an important rule for understanding the nature of electronic transitions. Their principle states that movement of the nuclei is negligible during the time taken by an electronic transition because these transitions are so fast (about 10^{-15} sec) that the positions and velocities of nuclei have no time to change. The idea is obviously closely related to the Born-Oppenheimer approximation in which the various motions of a molecule are considered to be separable. The Franck-Condon principle indicates that electronic transitions will occur only when the internuclear distances are not significantly different in the two states and when the nuclei have little or no velocity.

Nomenclature

Unfortunately, the terms used in spectrophotometry and spectroscopy are confusing. The recommendation of the American Society for Testing and Materials, endorsed by *Analytical Chemistry* [3], is now widely accepted. The recommended terms, symbols, and definitions used in this chapter are given in Table 7.2. The Commission on Nomenclature, Division of Analytical Chemistry of the International Union of Pure and Applied Chemistry (IUPAC), has made recommendations to standardize the terms used in spectrometry. The following guidelines are suggested:

1. Single words are preferred, for example, *wavelength* instead of *wave length* and *absorbance* instead of *optical density*.
2. The commonly accepted metric system is preferred. The IUPAC has recommended the SI system and the U.S. National Bureau of Standards has strongly recommended the SI system.
3. The expression of absorption spectra by plotting the molar absorptivity as a function of wavelength, instead of simply plotting absorbance versus wavelength, is preferred. In addition to showing the maximum absorptions, it gives information about sensitivity for analysis.

7.2 EFFECT OF STRUCTURE ON ABSORPTION

Spectroscopic characteristics are held in common by molecules with some of the same chemical features. It may then be reasonable to expect that the correlation can be extended, and that the presence of a certain chemical feature may be implied by the presence of a certain spectral characteristic.

TABLE 7.2. *Spectrophotometry Nomenclature*

Name	Symbol	Definition	Name Not Recommended
Absorbance	A	$-\log T$	Optical density (O.D.), extinction, absorbancy
Absorptivity	a	$= A/bc$ [a]	Absorbancy index, absorbing index, extinction coefficient
Path length	b	Internal cell or sample length, in cm	l or d
Molar absorptivity	ϵ	$= A/bc$ [b]	Molar absorbancy index, molar extinction coefficient, molar absorption coefficient
Transmittance	T	I/I_0 [c]	Transmittancy, transmission
Wavelength unit	nm	10^{-9} m	mμ (millimicron)
	μm	10^{-6} m	μ (micron)
Absorption maximum	$\lambda_{\max}$	Wavelength at which a maximum absorption occurs	—

a. The concentration is in grams per liter.
b. The concentration is in moles per liter.
c. The ratio of radiant power transmitted to radiant power incident.

Determination of molecular structure and the identification of specific functional groups are extremely important to modern chemistry. Since the interaction of ultraviolet and visible radiation with molecules is governed by the electronic structure of the molecule, these regions of the spectrum are of particular interest to the chemist. At present, most of the reported work has been on the absorption spectra (from 200 to 1000 nm) of organic molecules in dilute solutions [4–6]. In the future, the trend may be to employ greater dispersion and to extend the range of wavelengths investigated.

Electronic Transitions

Electronic transitions in organic molecules are characterized by the promotion of electrons in ground-state bonding or nonbonding molecular orbitals to excited-state antibonding molecular orbitals. If molecular structure is the dominant factor in determining the electronic energies of the ground and excited states, then the photon energy required for n → π^*, π → π^*, and n → σ^* transitions will vary from molecule to molecule, depending on structural and environmental variations. When radiation of a frequency corresponding to one of the fundamental frequencies of a molecule interacts with that molecule, the radiant energy is absorbed to increase the energy content of the molecule by an amount equal to the energy of the quantum absorbed, in accordance with the relation

$$\Delta E = h\nu = \frac{hc}{\lambda} \qquad (7.2)$$

where ν = frequency in hertz
      h = Planck's constant (6.62 × 10^{-27} erg/sec)
      λ = wavelength in centimeters
      c = speed of light in a vacuum (2.998 × 10^{10} cm/sec)

The most common types of electronic transitions are illustrated in Table 7.3.

The $\sigma \rightarrow \sigma^*$ transitions are very energetic and are found only below 200 nm, in the far-ultraviolet region. This is often termed the *vacuum ultraviolet* region because the normal constituents of air, N_2 and O_2, also absorb strongly below about 160 and 200 nm, respectively; and spectra of other substances must be obtained in a "vacuum." The n $\rightarrow \sigma^*$ transitions are also high-energy transitions and generally appear at the shorter ultraviolet wavelengths; for example, absorption by alkyl halides (where the nonbonding electrons are supplied by the halogen) shows a λ_{max} that increases in the order Cl < Br < I, as the electrons are successively easier to excite. The most common examples of $\pi \rightarrow \pi^*$ transitions are found in conjugated polyenes, in which the energy required for the transition decreases with increasing length of the conjugated systems and, correspondingly, λ_{max} increases. The $\pi \rightarrow \pi^*$ transitions are usually the least energetic, which results in their appearance at longer wavelengths.

The molar absorptivity (ϵ) of a compound is a function of the cross-sectional area (θ) of the absorbing species and of the transition probability (P):

$$\epsilon = (9 \times 10^{19})P\theta \qquad (7.3)$$

Using this relation, a molar absorptivity of the order of 10^5 has been calculated for the average organic molecule with an assumed cross section of about 10^{-15} cm^2 and a unit transition probability. The highest values known for ϵ are a few hundred thousand; any value above 10,000 is considered high, and one under 1000 low.

By examining the locations, distribution patterns, and intensities of absorption spectra, one can gain information helpful in the identification of compounds. Unfortunately, interpretation of electronic (ultraviolet-visible) spectra is usually less certain than that of vibrational (infrared) spectra because of the broad overlapping

TABLE 7.3. *Electronic Energy Levels and Transitions*

Transition	Region of Electronic Spectra	Example
$\sigma \rightarrow \sigma^*$	Vacuum ultraviolet	CH_4 at 125 nm
n $\rightarrow \sigma^*$	Far-ultraviolet, sometimes near-ultraviolet	Acetone at 190 nm; methylamine at 213 nm
$\pi \rightarrow \pi^*$	Ultraviolet	Saturated aldehydes at 180 nm
n $\rightarrow \pi^*$	Near-ultraviolet and visible	Acetone at 277 nm; nitroso-*t*-butane at 665 nm

bands that characterize electronic absorption. Even so, a great deal of research effort has been expended in hopes that the structural changes of a molecule and the shifts observed in their electronic absorption spectra can be correlated. On the other hand, absorptivities in the infrared are much lower than at shorter wavelengths, rarely exceeding 1000. As a consequence, electronic (ultraviolet-visible) spectrophotometry is sensitive to a much smaller amount of sample and is quite useful for dilute solutions.

Chromophores

It is a long-recognized fact that colored substances owe their color to absorption of light by one or more unsaturated linkages. Such linkages or groups were named *chromophores* by Witt in 1876. Certain groups that by themselves do not confer color to a substance but that increase the coloring power of a chromophore were called *auxochromes*.

Ultraviolet radiation is usually absorbed by a chromophore rather than by the molecule as a whole. Chromophores are, in most cases, covalent unsaturated groups such as those given in Table 7.4; they are functional groups that usually absorb in the near-ultraviolet or visible region when they are bonded to a nonabsorbing, saturated residue that possesses no unshared or nonbonded electrons (e.g., a hydrocarbon chain). Auxochromes contain functional groups that have nonbonded valence electrons and exhibit no absorption at wavelengths above 220 nm. They do, however, absorb strongly in the far-ultraviolet region (n → σ*). If an auxo-

TABLE 7.4. *Representative Chromophores and Their Approximate λ_{max} and ϵ_{max} Values*

Chromophore	λ_{max}, nm	ϵ_{max}
C=C	185	8000
—C≡C—	175	6000
C=O	188	900
—NH$_2$	195	2500
—CHO	210	20
—COOR	205	50
—COOH	205	60
—N=N—	252	8000
	371	14
—N=O	300	100
	665	20
—NO$_2$	270	14
—Br	205	400

SEC. 7.2 Effect of Structure on Absorption

chrome and a chromophore are combined in the same molecule, the chromophore absorption will typically shift to a longer wavelength and show an increase in intensity. Shifts to longer wavelengths are called *bathochromic* shifts; changes to shorter wavelengths, *hypsochromic* shifts. Increases in intensity of an absorption band are called *hyperchromic* effects, whereas a decrease in intensity is termed a *hypochromic* effect.

In general, molecules containing two or more chromophores show absorption that is the sum of all the chromophores present, provided they are separated by two or more single bonds. If two chromophores are conjugated, they exhibit a much enhanced absorption with an increase in both λ_{max} and ϵ_{max}; three conjugated chromophores result in a further increase in λ_{max} and ϵ_{max}. Such bathochromic shifts are attributed to the formation of a new chromophore from the conjugated systems; the π electrons associated with each chromophore of the conjugated system are able to move with increased freedom throughout the new structure.

Single Bonds and Saturated Compounds

Saturated hydrocarbons contain only single bonds with σ electrons; thus, the only transitions available to these compounds are transitions of the $\sigma \to \sigma^*$ type, which occur at the very short wavelengths of the vacuum ultraviolet. For example, methane and ethane are saturated hydrocarbons with all electrons involved in σ bonds. Electronic transitions are accordingly of the same type as those in the hydrogen molecule, and the separation of the levels is of the same order. Therefore, the first electronic absorption bands for methane and ethane are at 125 and 135 nm, respectively; this band continues to move to longer wavelengths in the larger hydrocarbons, suggesting that the C–C bond is involved.

Because of the excitation of electrons in nonbonded orbitals, saturated molecules that contain atoms with lone pairs of electrons exhibit electronic transitions at longer wavelengths than the corresponding saturated hydrocarbons. Thus, alkyl iodides and monosulfides containing the C–S–C linkage give $n \to \sigma^*$ transitions near 260 and 215 nm, respectively.

The $n \to \pi^*$ transitions associated with carbonyl groups are observed in the 270–290-nm region and are quite useful in the identification of aldehydes and ketones. For example, acetone exhibits three bands—a weak band at 280 nm ($n \to \pi^*$), a more intense band near 190 nm ($n \to \sigma^*$), and a still more intense band near 150 nm ($\pi \to \pi^*$). For these compounds, the $n \to \pi^*$ transition of the carbonyl group varies with the substituents R_1 and R_2 in the molecule:

$$\begin{matrix} R_1 \\ \searrow \\   C{=}O \\ \nearrow \\ R_2 \end{matrix}$$

Substituting a hydroxyl, amino, or halogen group (auxochromes) for hydrogen shifts the transition to higher energy, because these groups donate electron density by a resonance interaction and raise the energy of the excited state with respect to the ground state. In addition, these groups give rise to an inductive effect that withdraws electron density from the carbonyl group, thus lowering the ground state relative to the excited state.

Conjugated Chromophores

As stated above, a molecule that contains more than one chromophore has an absorption band that may be the sum of the separate chromophores, or it may be the result of an interaction between the chromophores. If the two chromophores are separated by a single bond, however, conjugation occurs and the electronic absorption spectra show dramatic changes from the bands due only to the isolated chromophores. One of the simplest examples is 1,3-butadiene, $CH_2=CH-CH=CH_2$, where the two carbon-carbon double bonds separated by a single bond give rise to an absorption spectrum that is shifted to lower energy by conjugation. In conjugated systems, the π electrons are delocalized over a minimum of four atoms; this causes a decrease in the $\pi \rightarrow \pi^*$ transition energy, and the molar absorptivity increases as the result of a higher probability for the transition. The effect of conjugation on $\pi \rightarrow \pi^*$ transitions is considerable. Thus, for the series ethylene (193 nm), 1,3-butadiene (217 nm), hexatriene (258 nm), octatetraene (300 nm), a bathochromic shift accompanied by an increase in molar absorptivity is observed as an additional carbon-carbon double bond is added to each compound in progressing along the series.

Electronic absorption bands for conjugated alkynes are also shifted to lower energy; however, the molar absorptivity is much lower than for the conjugated alkenes. As an example, vinylacetylene, $CH_2=CH-C\equiv CH$, exhibits an absorption band near that of 1,3-butadiene ($\lambda_{max} = 219$ nm); however, its molar absorptivity is only 6500 compared to 21,000 for 1,3-butadiene.

An important feature of ultraviolet absorption spectra is that chromophores that are not conjugated give a summation of $n \rightarrow \pi^*$ and/or $\pi \rightarrow \pi^*$ bands. A $-CH_2-$ group is sufficient to isolate two chromophores, but $-O-$, $-S-$, or $-NH-$ is not. An example of this effect is seen for hexacene, a green compound, and 6,15-dihydrohexacene, a colorless compound whose absorption spectrum is essentially the sum of the spectra of anthracene and naphthalene:

Hexacene   6,15-Dihydrohexacene

Aromatic Hydrocarbons

Benzene, a cyclic conjugated polyene, absorbs at 260, 200, and 180 nm. All of these bands are associated with the π-electron system of benzene. The intense bands at 200 and 180 nm are assigned to transitions to dipolar excited states, and the weak band at 260 nm is ascribed to a forbidden transition to a homopolar excited state.

Electronic transitions in "linear" polycyclic aromatics, such as benzene, naphthalene, and anthracene, exhibit a regular shift toward lower energy with increasing size of the molecule. Figure 7.4 shows that the larger compounds absorb in the same region as benzene, but the bands are more intense. Other compounds in this class, such as phenanthrene, benzanthracene, and pyrene, show absorption

FIGURE 7.4. *Ultraviolet absorption spectra of benzene, naphthalene, and anthracene in ethanol. From K. L. Cheng, "Absorptiometry," in J. D. Winefordner, ed.,* Spectrochemical Methods of Analysis, *New York: Wiley-Interscience, 1971, Chap. 6, by permission of the editor and John Wiley and Sons.*

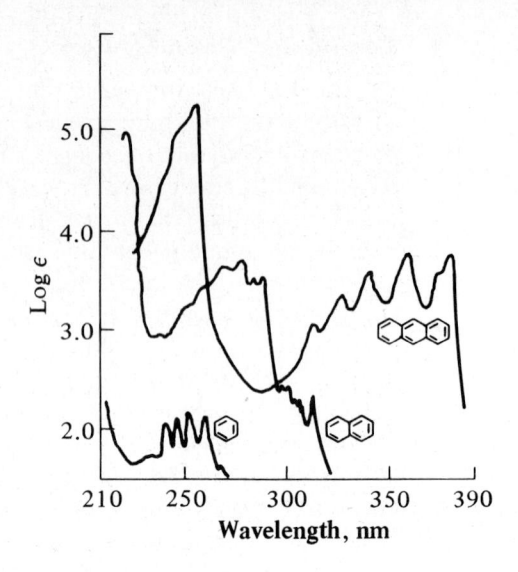

spectra similar to those of the linear ring system but with a more complex pattern.

Resolution of the fine structure of the bands in the spectrum of benzene is highly dependent on two parameters: solvent polarity and ring substitution. Polar solvents tend to merge the bands into a broad hump whereas nonpolar solvents give very good resolution into narrow, separate, peaks. Electronic spectra of benzene in the vapor state exhibit excellent resolution (Fig. 7.3). Upon substitution on the benzene ring, fine structure is diminished considerably and all three bands in benzene are affected markedly.

The effects of substitution on aromatic nuclei have been studied and detailed in the literature [7–9]. Usually, but not always, the absorption maxima shift to longer wavelengths and the intensity of the absorption changes.

Azo Compounds

Straight-chain compounds that contain the —N=N— linkage give rise to low-intensity bands in the near-ultraviolet and visible regions. The long-wavelength bands are thought to arise from n → π^* transitions. For aliphatic azides, the low-energy band at 285 nm is assigned to a $\pi \to \pi^*$ electronic transition, whereas the 215-nm band is considered to arise from a s-p → π^* transition.

For aromatic azides, the —N=N— linkage may be conjugated with the ring π system. In azobenzene, the azo linkage is conjugated with two benzene rings and the π orbitals extend over the whole molecule. The levels are brought closer together and the $\pi \to \pi^*$ transition occurs at 445 nm. This absorption is responsible for the orange red color of azobenzene.

Solvent Effects

Since there is electrostatic interaction between polar solvents and polar chromophores, such as the carbonyl group, these solvents tend to stabilize both the nonbonding electronic ground states and the π^* excited states. This interaction causes the n → π^* transitions, which usually occur at lower energy than the π → π^* transitions, to move to higher energy and π → π^* transitions to move to lower energy. Thus, the π → π^* and n → π^* absorptions of polar chromophores move closer to each other with increasing polarity of the solvent. An example of this phenomenon is the solvent shift of the n → π^* transition to lower energy in the ultraviolet spectrum of N-nitroso-dimethylamine. For various solvents, the order for decreasing n → π^* energy is given by cyclohexane > dioxane > ethanol > water. For a series of hydrocarbon solvents, the effect on λ_{max} and ϵ_{max} is slight and can usually be neglected.

Steric Effects

Electronic interactions may be increased or decreased by steric effects and, in certain cases, totally new interactions may result. Extended conjugation of π orbitals requires coplanarity of the atoms involved in the π-cloud delocalization for maximal resonance interaction. If large, bulky groups are in positions that cause perturbation of the coplanarity of the π system, λ_{max} is usually shifted to shorter wavelengths and ϵ_{max} also decreases. For example, diphenyl (λ_{max} = 246 nm, ϵ_{max} = 20,000) has coplanar rings and shows higher molar absorptivity than its derivative, o,o'-dialkyldiphenyl, which has nonplanar rings (λ_{max} = 250 nm, ϵ_{max} = 2000).

Molar absorptivity increases with conjugation as a result of increased transition-moment length, and reaches its maximum for a displacement of 0.1 to 0.3 nm, corresponding to $\epsilon_{max} = 10^5$. This length is very sensitive to structural changes; in most cases, this effect is more noticeable in the *trans*- rather than the *cis*-isomers. If conjugation is in an open-chain system instead of a constrained-ring system, the effect is also greater. The isomeric absorption difference is clearly demonstrated by comparing the ultraviolet absorption spectra of *cis*- and *trans*-azobenzene [4].

Qualitative Identification

In principle, any organic molecule that contains a chromophore will probably give rise to a characteristic electronic spectrum. This provides a method for identifying structural components in such molecules. In addition to characteristic λ_{max} values, the molar absorptivities are also important in both qualitative and structural applications because this information can sometimes differentiate two chromophores that absorb at the same wavelength. Great care must be taken in suggesting relations from an observed electronic absorption spectrum without fully exploring all possibilities. It can usually be assumed that absorption at a particular wavelength is indicative of a given group, and intensity measurements may lend support to this assumption; however, a small amount of impurity from a substance with high molar absorptivity may result in misleading conclusions.

7.3 MAGNITUDE OF ABSORPTION OF RADIATION

Radiant power (P) is defined as the radiant energy impinging on unit area in unit time. Since the color of a solution is due to the partial absorption of visible light, the power of a beam of light will be reduced as the light passes through a colored solution. The changes in radiant power that occur as monochromatic radiation passes through an absorption cell are illustrated in Figure 7.5. Quantity P_1 is the radiant power of incident radiation, P_0 is the radiant power after passing through one cell wall, P is the radiant power after passing through the absorbing solution or medium, and P_2 is the radiant power after the beam has traversed the last cell wall. An important quantity (see Table 7.2) is the *transmittance*, T, defined as

$$T = \frac{I}{I_0} = \frac{P_2}{P_1} \tag{7.4}$$

which is the quantity that is usually measured in spectrophotometers. Value T_i is the *internal* transmittance of the system:

$$T_i = \frac{P}{P_0} \tag{7.5}$$

Usually, the quantities T and T_i are nearly the same because cells are made of materials that will not appreciably absorb or scatter the radiation used. Any slight difference can be minimized by using matched cells, one containing the sample and the other the reagent blank (a solution containing all the components except the compound of interest). If T is set at 100% for the blank, a measurement of T for the sample gives T_i.

Beer's Law

Bouguer, and later Lambert, observed that the fraction of the energy, or intensity, of radiation absorbed in a thin layer of material depends on the absorbing substance and on the frequency of the incident radiation, and is proportional to the thickness of the layer. At a given concentration of the absorbing substance, summation over a series of thin layers, or integration over a finite thickness, leads to an exponential relationship between transmitted intensity and thickness. This is generally called

FIGURE 7.5. Radiation impinging on an absorption cell whose optical path length is b. From K. L. Cheng, "Absorptiometry," in J. D. Winefordner, ed., Spectrochemical Methods of Analysis, New York: Wiley-Interscience, 1971, Chap. 6, by permission of the editor and John Wiley and Sons.

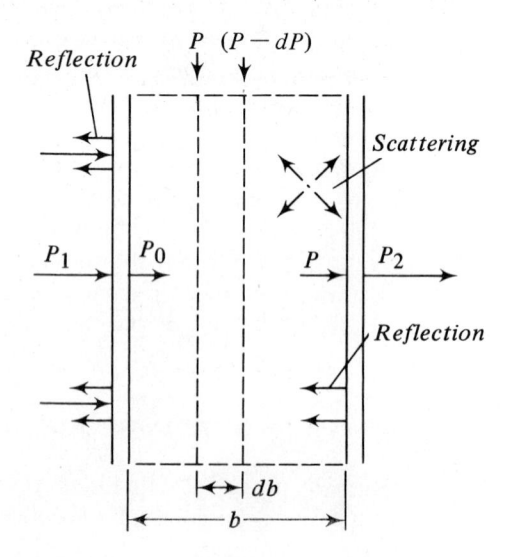

Lambert's law. Beer showed that, at a given thickness, the absorption coefficient introduced by Lambert was directly proportional to the concentration of the absorbing substance in a solution. Combination of these results gives the relationship now commonly known as **Beer's law**. This law states that the amount of radiation absorbed or transmitted by a solution or medium is an exponential function of the concentration of absorbing substance present and of the length of the path of the radiation through the sample.

Beer's law can be derived as follows (see Fig. 7.5). For a layer of infinitesimal thickness, db, the decrease in radiant power $(-dP)$ is given by

$$-\frac{dP}{P} = kc\,db \tag{7.6}$$

where k = a proportionality constant

Integration over the entire absorbing cell length, b—that is,

$$\int_{P_0}^{P} \frac{dP}{P} = -k \int_0^b c\,db \tag{7.7}$$

results in

$$\ln\left(\frac{P}{P_0}\right) = -kbc = 2.303 \log\left(\frac{P}{P_0}\right) \tag{7.8}$$

This gives (Table 7.2)

$$-\log\left(\frac{P}{P_0}\right) = -\log T = A = \epsilon bc \tag{7.9}$$

where $\epsilon = k/2.303$

The constant ϵ is called the *molar absorptivity* when the concentration c is in moles per liter and b is in centimeters. The value of ϵ is characteristic of the absorbing substance at a particular wavelength in a particular solvent and is independent of the concentration and of the path length, b. Equation 7.9 is a fundamental law on which colorimetric and spectrophotometric methods are based. It is known variously as the Bouguer-Beer, Lambert-Beer, or more simply, Beer's law. When other concentration units are used, such as grams per liter, the symbol a (for *absorptivity*, Table 7.3) instead of ϵ is used.

Example 7.1. Palladium reacts with Thio-Michler's ketone, forming a colored 1:4 complex. A 0.20-ppm Pd sample gave an absorbance of 0.390 at 520 nm using a 1.00-cm cell. Calculate the molar absorptivity (ϵ) for the palladium Thio-Michler's ketone complex.

Solution: $0.20\text{-ppm Pd} = \dfrac{2.0 \times 10^{-4} \text{ g/L}}{106.4} = 1.9 \times 10^{-6} \, M$

In a 1.00-cm cell for $A = 0.390$, the molar absorptivity is calculated from Beer's law:

$$A = \epsilon bc$$
$$0.390 = (\epsilon)(1.00)(1.9 \times 10^{-6})$$
$$\epsilon = \frac{0.390}{1.9 \times 10^{-6}} = 2.1 \times 10^5 \text{ L/mole-cm}$$

Beer's law assumes that (a) the incident radiation is monochromatic, (b) the absorption occurs in a volume of uniform cross section, and (c) the absorbing substances behave independently of each other in the absorption process. Thus, when Beer's law applies to a multicomponent system in which there is no interaction among the various species, the total absorbance may be expressed as

$$A_{\text{total}} = \epsilon_1 bc_1 + \epsilon_2 bc_2 + \cdots + \epsilon_i bc_i \tag{7.10}$$

where ϵ_i = molar absorptivity for the ith absorbing species
        c_i = its molar concentration

This equation is the basis of quantitative methods for determining mixtures of absorbing substances.

Deviation from Beer's Law

Beer's law states that a plot of absorbance versus concentration should give a straight line passing through the origin with a slope equal to ϵb. However, deviations from direct proportionality between absorbance and concentration are sometimes encountered. In these cases, a nonlinear working curve may be prepared with solutions of known concentration, and the concentration of the unknown solution found from the absorbance obtained under the same experimental conditions.

The causes of deviations from Beer's law can be categorized into real factors, instrumental factors, and chemical factors. These deviations may result in an upward curvature (positive deviation) or in a downward curvature (negative deviation), as shown in Figure 7.6. A check on instrumental factors can be made by plotting absorbance versus cell length at a constant concentration; this plot will be linear if the instrument is performing satisfactorily. Deviations arising from chemical factors are observed only when concentrations are changed.

Real Factors. In the derivation of Beer's law (see Eqn. 7.6), it has been assumed that the attenuation of P_0 over the entire length of the cell depends only on the

FIGURE 7.6. *Deviations from Beer's law.*
A: Positive deviation. B: Negative deviation.
C: No deviation.

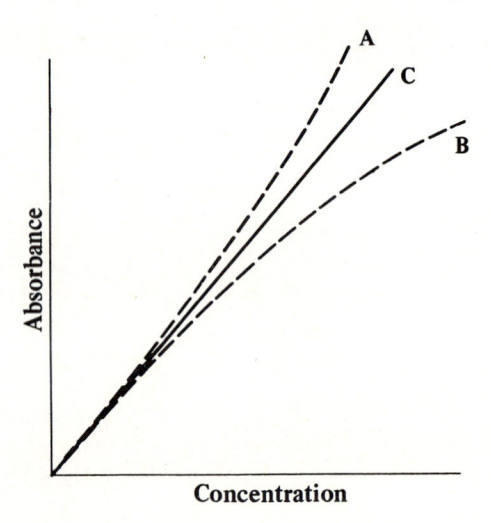

attenuation index of the medium. However, a solution containing a homogeneously distributed absorber can be regarded as an absorbing medium with a complex refractive index $\hat{n} = n - ik$. Both the real component—the refractive index, n—and the complex component—the attenuation index of the medium, k—will contribute to reducing P_0. Hence, the true proportionality constant is not ϵ but the factor $\epsilon n/(n^2 + 2)^2$ [10]. Therefore, deviations occur because of neglect of the changes in refractive index of the solution. The refractive index of a solution increases as concentration increases; this means that $n/(n^2 + 2)^2$ decreases. Thus, an apparent negative deviation from Beer's law will occur at sufficiently high concentrations of absorbers in the situation where absorber-absorber interactions are negligible.

Instrumental Factors. Unsatisfactory performance of an instrument may be caused by fluctuations in the power supply voltage, an unstable light source, or a nonlinear response of the detector-amplifier system. A double-beam system helps to minimize deviations due to these factors. In addition, the following instrumental sources of possible deviations should be understood.

1. *Polychromatic radiation.* Strict conformity of an absorbing system to Beer's law requires that the radiation be monochromatic. However, one always works with a band of wavelengths and not with a single sharp line (laser light sources are available, but there is as yet no tunable laser incorporated into a spectrophotometer). Both the power of the radiation source and the absorptivity will vary with the wavelength.

Let us consider the effect of polychromatic radiation on the relationship between concentration and absorbance. When the radiation consists of two wavelengths, λ and λ', and assuming that Beer's law applies at each of these individually, the absorbance at λ is given by

$$\log\left(\frac{P_0}{P}\right) = A = \epsilon bc \tag{7.11}$$

or

$$\frac{P_0}{P} = 10^{\epsilon bc} \tag{7.12}$$

Similarly, at λ',

$$\frac{P_0'}{P'} = 10^{\epsilon' bc} \tag{7.13}$$

The radiant power of two wavelengths passing through the solvent is given by $P_0 + P_0'$, and that passing through the solution containing absorbing species by $P + P'$. The combined absorbance is

$$A_c = \log\left(\frac{P_0 + P_0'}{P + P'}\right) \tag{7.14}$$

Substituting for P and P', we obtain

$$A_c = \log\left(\frac{P_0 + P_0'}{P_0 10^{-\epsilon bc} + P_0' 10^{-\epsilon' bc}}\right) \tag{7.15}$$

In the very special case where $\epsilon = \epsilon'$, Equation 7.15 simply reduces to Beer's law. However, in the general case where $\epsilon \neq \epsilon'$, the relationship between A_c and c will be nonlinear; therefore, departures from linearity will be greater as the difference between ϵ and ϵ' becomes greater. Furthermore, when $\epsilon > \epsilon'$, the measured absorbance A_c is lower than the true "monochromatic" absorbance at wavelength λ, resulting in negative deviation, and when $\epsilon < \epsilon'$, the measured absorbance A_c is higher, resulting in a positive deviation (Fig. 7.6).

When a broader bandwidth (see below) is used, the lower absorbances toward the edges of the finite band contribute greater total intensities of transmitted light than the higher absorbances at the center of the band, and the summed, "average," absorbance includes those over the bandwidth. It is further noted that the steeper the absorption curve included within the bandwidth, the greater the error. From the same principle, as the concentration is increased the absorption peak becomes narrower, so the error is greater.

2. *Slit width.* The ability of a spectrophotometer to distinguish between two frequencies differing only slightly from each other depends on the widths of the images produced (relative to the separation of two images). The width of the image produced is thus an important measure of the quality of performance of a spectrophotometer. The spread of the image along the frequency, wavenumber, or wavelength scale is defined as the *spectral slit width* or *spectral bandwidth*. It is very closely proportional to the actual width of the slit (the mechanical slit width). The effect of slit width on absorbance is illustrated in Figure 7.7.

The spectral bandwidth of an ultraviolet spectrophotometer is typically of the order of 1 nm. In general, molecular absorption bands are smooth and much broader than 1 nm; so the effect of spectral bandwidth is practically negligible, especially when the absorbance is measured at the maximum absorption. If the absorption band is sharp, or if measurements are made on a steep slope of the spectral band, the absorptivity may be different over the spectral bandwidth, and deviations from Beer's law will be noticed. Figure 7.8 shows the effect of spectral bandwidth: With increasing slit width (also increasing spectral bandwidth), the recorded bands gradually merge together.

FIGURE 7.7. *Effect of slit width on absorbance. From K. L. Cheng, "Absorptiometry," in J. D. Winefordner, ed.,* Spectrochemical Methods of Analysis, *New York: Wiley-Interscience, 1971, Chap. 6, by permission of the editor and John Wiley and Sons.*

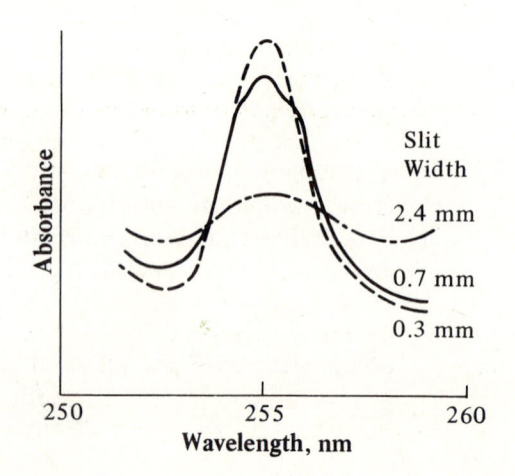

178   CHAP. 7   Ultraviolet and Visible Absorption Spectroscopy

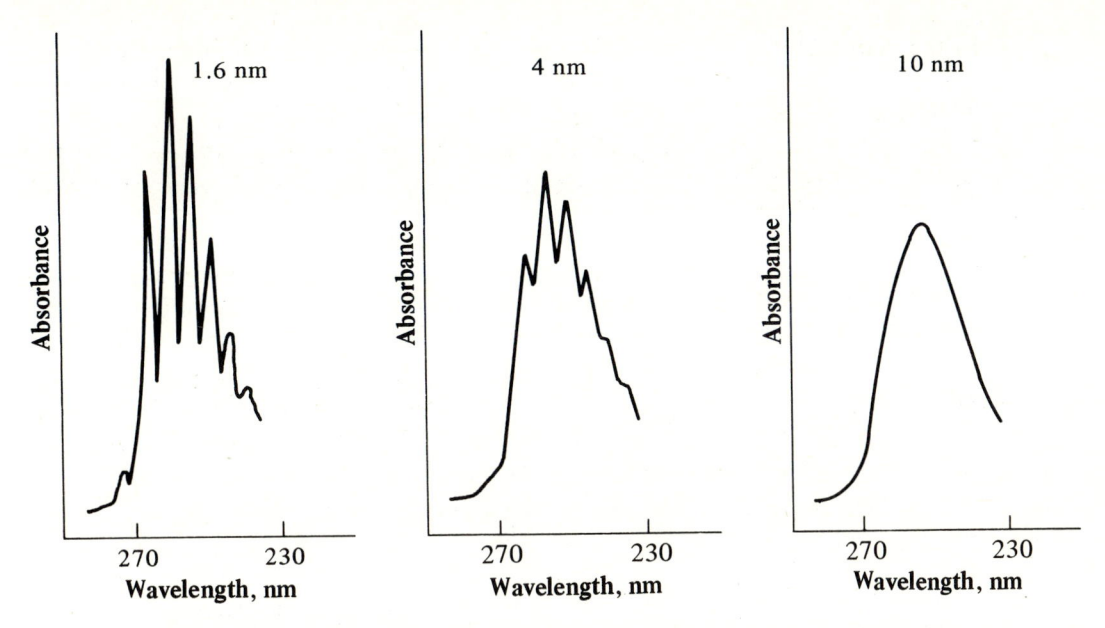

FIGURE 7.8. *Effect of spectral bandwidth on the absorption spectrum of benzene in cyclohexane.*

3. *Stray light.* Stray light that strikes the detector is a potential source of error; the apparent absorbance is decreased as a result:

$$A_m = \log\left(\frac{P_0 + P_s}{P + P_s}\right) \quad (7.16)$$

where P_s = radiant power of the stray light
A_m = measured absorbance

When P diminishes due to increasing concentration and becomes small when compared to P_s, $P + P_s \approx P_s$; and Equation 7.16 becomes

$$A_m = \log\left(\frac{P_0 + P_s}{P_s}\right) \quad (7.17)$$

Thus, there is a negative deviation from Beer's law. Errors due to stray light are more commonly found near the wavelength limits of the instrument components. Many reports of spectra in the ultraviolet region below 220 nm should be checked carefully, since false peaks have been reported. Visible radiation usually presents the most serious stray-light problem for ultraviolet-visible spectrophotometers, because both the spectral radiance of most visible sources and the spectral response of most detectors to visible radiation are high.

Chemical Factors. Apparent deviations from Beer's law are often due to chemical effects such as dissociation, association, complex formation, polymerization, or solvolysis.

Association and polymerization are examples of the process of self-interaction,

SEC. 7.3 **Magnitude of Absorption of Radiation**

and their effects are important in both ultraviolet and visible spectroscopy. Benzoic acid exists as a mixture of the ionized and un-ionized forms, and in dilute aqueous solution it dissociates as follows:

$$C_6H_5COOH + H_2O \rightleftarrows C_6H_5COO^- + H_3O^+ \qquad (7.18)$$
$$(\lambda_{max} = 273 \text{ nm}, \epsilon = 970) \quad (\lambda_{max} = 268 \text{ nm}, \epsilon = 560)$$

The effective molar absorptivity at 273 nm will thus decrease with increased dilution or at high pH.

Another example is observed with unbuffered $K_2Cr_2O_7$ solutions. In pure water, the dichromate and chromate ions are in equilibrium:

$$Cr_2O_7^{2-} + H_2O \rightleftarrows 2CrO_4^{2-} + 2H^+ \qquad (7.19)$$
$$(\lambda_{max} = 350, 450 \text{ nm}) \quad (\lambda_{max} = 372 \text{ nm})$$

The equilibrium constant may be expressed as

$$\frac{[CrO_4^{2-}]^2[H^+]^2}{[Cr_2O_7^{2-}]} = K \qquad (7.20)$$

Obviously, there are deviations from Beer's law when aqueous solutions of chromate or dichromate are diluted with water, and the pH will affect the concentrations of $Cr_2O_7^{2-}$ and CrO_4^{2-}. The effect can be controlled by buffering dichromate with a strong acid or chromate with a strong base.

Occasionally, the absorbance is measured at an *isosbestic point* (or isoabsorptive wavelength)—that is, a wavelength at which the two absorbing species in equilibrium have a common value of ϵ; then, Beer's law holds even though there is a shift of equilibrium. Isosbestic points are often taken as criteria for the existence of two interconvertible absorbing species of a compound, the total quantity of which is constant, though points of common ϵ-value occur also in some irreversible decomposition reactions giving two products.

1. *Solvent.* Dissolution may shift the spectrum of an absorbing substance to longer wavelength (with respect to the spectrum of the gas). This so-called *red shift* or bathochromic effect is greater in solvents of high dielectric constant because the charge displacement for the upper energy state requires less energy in a dielectric solvent than in a vacuum. A *blue shift* (to shorter wavelengths) is generally believed to be associated with a $n \rightarrow \pi^*$ transition involving a nonbonding orbital in the ground state. Dissolution generally causes larger effects on infrared spectra than on ultraviolet spectra, but may cause significant errors even in ultraviolet quantitative work. The greatest effect occurs after mixing dipolar solvated molecules (dissolved in nonpolar solvents) with polar solvents or additives. Many carbonyl compounds are sensitive to changes in solvent media.

2. *Temperature.* Changes in temperature may shift ionic equilibria. In addition, an increase in temperature exerts a bathochromic effect on ions in solution; for instance, the color of a hydrochloric acid solution of ferric chloride changes from yellow to reddish brown on heating. However, temperature is ordinarily not considered an important factor in simple systems, within limits of say $\pm 5°$.

3. *Photo effects.* Fluorescence resulting from frequencies of ultraviolet radia-

tion in a certain range may cause an apparent increase in transmittance with fluorescing substances. Light scattering is found in colloidal systems, the extent depending upon the particle size and shape and the wavelength region used. Photo effects in many organic compounds or indicator solutions may cause *dichroism;* that is, different colors are produced by thick and thin layers. For a polymer or crystal, *pleochroism* may be observed; unpolarized radiation becomes partially polarized on passing through an ordered absorbing substance. Photochemical reactions or photodecomposition, of course, cause a deviation from Beer's law. The effect is usually of little significance unless high-intensity radiation is used or the sample is highly photosensitive, such as the silver thio-Michler's ketone complex.

7.4 QUANTITATIVE ABSORPTION SPECTROSCOPY

Methods based on the absorption of radiation are powerful and useful tools for the analytical chemist. The ultraviolet region is particularly important for the qualitative and quantitative determination of many organic compounds. In the visible region, spectrophotometric methods are widely used for the quantitative determination of many trace substances, especially inorganic elements.

The basic principle of quantitative absorption spectroscopy lies in comparing the extent of absorption of a sample solution with that of a set of standards under radiation at a selected wavelength.

Visual Colorimetric Methods

In its simplest form, colorimetry consists of visual matching of the color of the sample with that of a series of standards. A colored compound is first formed by suitably reacting the constituent to be determined, then the colored solutions are racked side-by-side in Nessler tubes* for viewing from the top. The approximate concentration of the unknown is estimated by finding which standard most closely matches the unknown in color. Visual colorimetry suffers from poor precision since the eye is not as sensitive to small differences in absorbance as is a photoelectric device. The Duboscq colorimeter provides a more refined method of analysis for color comparison. This is equipped with an eyepiece with a split field that permits the ready comparison of beams passing through sample and standard.

Photometric Methods

Photometers equipped with filters are suitable for many routine methods that do not involve complex spectra. Spectrophotometers can provide narrow bandwidths of radiation for accurate work and can handle absorption spectra in the ultraviolet region.

---

*Nessler tubes are essentially large, uniform, flat-bottomed test tubes, about 30 cm in length and perhaps 2.5 cm in diameter.

Choice of Wavelength. When filter photometers are used, a suitable filter is selected in preparing an analytical curve for the unknown substance. With a spectrophotometer, the spectrum of the absorbing substance is determined, and an appropriate wavelength is chosen. Generally, a wavelength close to that of maximum absorption is chosen, for maximum sensitivity; but the wavelength chosen should also fall in a region where the absorbance does not change rapidly with change in wavelength.

Unfortunately, use of the wavelength of maximum absorption is not always feasible because the color-forming reagents often also absorb significantly at the wavelength of maximum absorption of the species being measured. The spectra of 3,3'-diaminobenzidine (DAB) and its monoselenium compound, shown in Figure 7.9, both have absorption maxima at 340 and 420 nm. At 340 nm, the reagent also absorbs strongly. Although it is possible to select 340 nm and subtract the absorbance contributed by the excess reagent, it is difficult to know the amount of excess reagent precisely, and errors increase with increasing absorption by the reagent itself. A better approach is to use a wavelength at which the absorbing substance absorbs rather strongly but at which the absorbance contribution by the excess reagent is minimal—420 nm, in this case.

For systems that are sensitive to pH, and for which an isosbestic point can be located, measurements at the wavelength of the latter are preferred if the pH cannot be readily controlled.

Separation and Formation of Absorbing Compounds. In general, more than one method is available for the spectrophotometric determination of a given substance, and selecting a suitable method plays an important role in successfully analyzing the sample. It is often necessary to separate the absorbing substance before the absorbance measurement; for instance, chromatographic separation of vitamins in natural products is made before the actual spectrophotometric determination. In many cases, the sample compound does not absorb radiation appreciably in the wave-

FIGURE 7.9. *Absorbance curves of toluene solution of 3,3'-diaminobenzidine and its monoselenium compound with λ_{max} at 340 and 420 nm. A: 25 mg Se in 10 mL of toluene. B: 5 mg Se in 10 mL of toluene. C: Diaminobenzidine in toluene. Toluene as blank. From K. L. Cheng, Anal. Chem., 28, 1738 (1956), by permission of the publishers. Copyright © 1956 by the American Chemical Society.*

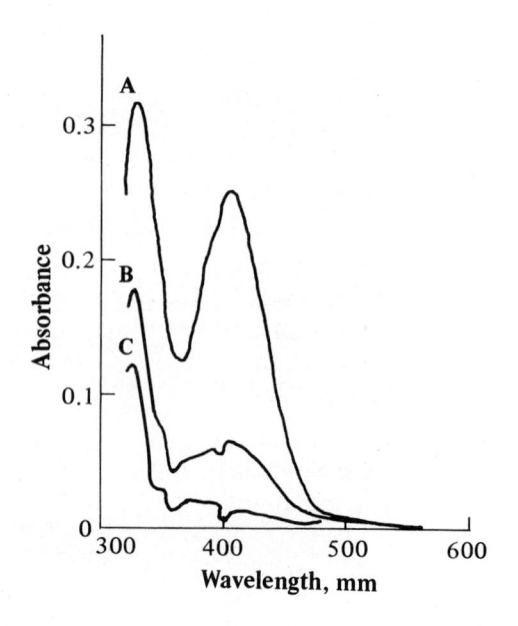

length regions provided; it is then necessary to form an absorbing substance by reacting the compound in question with other reagents. The reagents should be selective in their reactions and should not form interfering absorbing species with foreign substances likely to be present.

The following are some common and important factors involved in the formation of absorbing compounds.

1. *pH.* Since pH plays an important role in complex formation, proper adjustment of pH or the use of a buffer often eliminates certain interfering reactions. For instance, methylthymol blue and xylenol orange (analogs of EDTA) react with many metal ions. Their selectivity for certain metals is much improved in highly acidic media. For example, zirconium can be determined in the presence of hafnium in 1 N perchloric acid [11].

2. *Reagent concentration.* The amount of reagent required is dictated by the composition of the absorbing complex formed. An optimum concentration of reagents should be determined, since either not enough reagent or too much reagent can cause deviation from Beer's law.

3. *Time.* Formation of the absorbing complex may be slow, in some cases requiring several minutes or a few hours. For example, the phosphomolybdate blue method—a common analytical method for phosphate determinations—requires about 15 min standing time for full color development after addition of the reagents.

4. *Temperature.* The optimum temperature should be established in the procedure. Certain reactions require elevated temperature to decrease the time necessary for complete color development.

5. *Order of mixing reagents.* Frequently, it is important to add the reagents in a specified sequence; otherwise, full color development will not be possible or interfering reactions can occur. For instance, the highly selective color reaction of cobaltic nitrilotriacetate (NTA) in the presence of hydrogen peroxide must be preceded by the formation of the cobaltous NTA complex [12].

6. *Stability.* If the absorbing complex formed is not very stable, the absorbance measurement should be made as soon as possible. If the absorbing complex is photosensitive, precautions should be taken in order to avoid its photodecomposition. Certain reagents may sometimes be added to help stabilize the absorbing complex.

7. *Masking.* Very few reactions are truly specific. However, highly selective reactions may be developed through the sophisticated use of *masking.* Masking refers to the addition of a complexing agent to form a metal complex of such stability that, in this case, color-forming reactions with another reagent do not occur to any appreciable extent. For example, in the presence of excess EDTA, ferric ion does not form the colored $FeSCN^{2+}$ complex with thiocyanate ion.

8. *Organic solvent.* Many organic reagents or complexes are only slightly soluble in water. In such cases, it is necessary to add a water-miscible organic solvent to avoid precipitation or to aid color development. In other cases, solvent extraction might be used, for example to separate the colored compound from excess reagent or from interfering substances.

FIGURE 7.10. *Relative concentration error as a function of transmittance.* From K. L. Cheng, "*Absorptiometry,*" in J. D. Winefordner, ed., Spectrochemical Methods of Analysis, *New York: Wiley-Interscience, 1971, Chap. 6, by permission of the editor and John Wiley and Sons.*

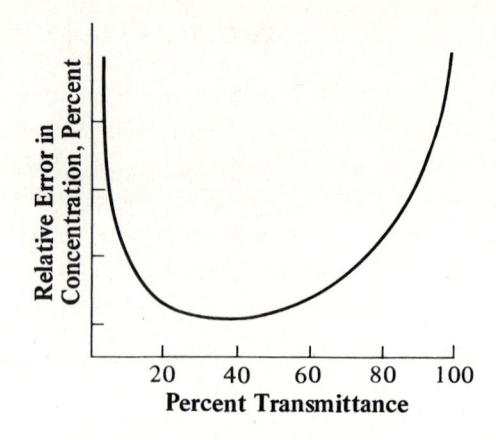

9. *Salt concentration.* High concentrations of electrolyte often influence the absorption spectrum of a compound. This effect may be due to the formation of ion-association complexes that cause a shift in the maximum absorption. This is a type of masking effect and usually causes a decrease in the absorption.

Photometric Errors. Because of the logarithmic relationship between transmittance and concentration, small errors in measuring the transmittance cause large relative errors in the calculated concentration at low and high transmittances. The concentration of a sample solution or the path length or both should be adjusted so that the absorbance will be within the range of approximately 0.2 to 0.7 (i.e., transmittance in the range of 20 to 60%). As shown in Figure 7.10, an absorbance of 0.434 (36.8% transmittance) is considered optimum, but in practice there is little difference in relative error between 0.2 and 0.7 absorbance. The relative-error curve shown in Figure 7.10 is approximately correct, in practice, only for relatively simple instruments with phototube detectors. With photomultiplier detectors, and in most of the more sophisticated commercial spectrophotometers, the usable transmittance range of "minimum" error is extended because of improved electronic signal-to-noise ratio.

7.5 SPECTROPHOTOMETRIC APPLICATIONS

Analysis of Mixtures

Beer's law states that absorbance is an additive property of all the absorbing molecules present in a mixture (see Eqn. 7.10). In principle, n absorbance measurements at n different wavelengths are needed to determine the concentrations of n components in a mixture; this procedure gives n independent simultaneous equations in n unknowns. The molar absorptivities must be known or determined for each individual absorbing species, 1, 2, etc., at each wavelength. If, in a two-component mixture, the values are ϵ_1, ϵ_2 at wavelength λ, and ϵ'_1, ϵ'_2 at a second wavelength, λ', and the absorbance of the mixture is A at λ and A' at λ', for a path length b and unknown concentrations c_1, c_2, then Equation 7.10 becomes

$$A = A_1 + A_2 = \epsilon_1 b c_1 + \epsilon_2 b c_2 \tag{7.21}$$

$$A' = A'_1 + A'_2 = \epsilon'_1 b c_1 + \epsilon'_2 b c_2 \tag{7.22}$$

Thus, the two unknown concentrations are calculated by solving these two simultaneous equations, which are obtained by measuring the absorbance of the mixture at two different wavelengths. Since these equations depend on the use of correct molar absorptivities, large errors may occur in systems where there are deviations from the laws of absorption.

The success of the above method depends on choosing suitable wavelengths for the analysis. For example, we would ideally like a case where at λ_1, ϵ_1 is large and ϵ_2 is small, while at λ_2, ϵ'_1 is small and ϵ'_2 is large. Unfortunately, not all two-component systems fall into this category. It may be that spectral overlap occurs for all λ-values and for every λ, $\epsilon_1 > \epsilon_2$. An alternative procedure for analyzing such mixtures is the graphical method developed by Connors and Eboka [13]. Equation 7.21 can be arranged to give

$$\frac{A}{\epsilon_1 b} = c_1 + \frac{\epsilon_2}{\epsilon_1} c_2 \tag{7.23}$$

Instead of measuring A, ϵ_1, and ϵ_2 at two different wavelengths, these quantities are measured at many different wavelengths. Experimentally, this requires that we obtain spectral scans of the pure compounds and of the unknown. For a discrete number of λ-values, $A/\epsilon_1 b$ and ϵ_2/ϵ_1 are then evaluated. A plot of $A/\epsilon_1 b$ versus ϵ_2/ϵ_1 gives a straight line. The slope of the line gives c_2, and extrapolation to $\epsilon_2/\epsilon_1 = 0$ will give c_1. The method is easily extended to three-component systems. The basic advantage of this method is that use of multiple points across the full spectrum, instead of just two, permits a much higher degree of precision and accuracy.

Isosbestic Point. For a two-component system where the two components are in equilibrium with each other and contribute all the absorption, it can be shown that there is at least one point in the spectrum at which the absorbance is independent of the ratio of the concentrations of the two components. If the bands overlap, there is a wavelength at which the two absorbing species in equilibrium have the same ϵ-value. This wavelength (at which the absorbance depends only on the total number of "equivalents" of the two absorbing species) is called the isosbestic point, or isobistic point, or isoabsorptive wavelength. All curves intersect at this point. The existence of such a point is not proof of the presence of only two components; there may be a third component with $\epsilon = 0$ at this particular wavelength. The absence of an isosbestic point, however, is definite proof of the presence of a third component, provided the possibility of a deviation from Beer's law in the two-component system can be discounted. In one respect, then, the isosbestic point is a unique wavelength for quantitative determination of the total amount of two absorbing substances in mutual equilibrium.

Determination of Stoichiometry

The spectrophotometric method is particularly valuable for studying complexes of low stability. Consider the formation of a complex M_nL_p, where M is a metal ion and L is a ligand:

$$nM + pL \rightleftharpoons M_nL_p \tag{7.24}$$

The molar ratio of the two components of a complex is important. In a quantitative determination, an excess of ligand should be added to force the equilibrium toward completion.

Molar-Ratio Method. In this method, the concentration of one component is kept fixed and that of the other varied to give a series of [L]/[M] ratios. The absorbances of these solutions, measured at an absorption maximum for the complex M_nL_p, increase linearly up to the molar ratio of the complex, at which virtually the whole amount of both components is complexed (assuming little dissociation). Further addition of component L cannot increase the absorbance, and the line becomes horizontal, or shows a break if component L absorbs at the same wavelength (Fig. 7.11). In rare cases, an excess component L may cause a decrease in absorbance owing to the stepwise formation of higher-order complexes that have smaller ε-values at this wavelength. The composition of molybdogermanic acid has been studied by the molar-ratio method [14] showing a ratio of 36 molybdate:1 germanate.

Continuous-Variation Method. The molar ratios may also be varied by changing the concentrations of both components while the total number of moles of both components are kept constant; this is termed the method of continuous variation, or Job's method [15, 16]. The mole fraction of one of the components is plotted on the abscissa scale; the ordinate scale is usually a difference in absorbance, ΔA, representing the difference between the observed absorbance and the summed absorbances of the independent (noncomplexed) components. When the curvature is pronounced and the maximum is not apparent, the apex may be obtained by drawing tangents.

FIGURE 7.11. *Molar-ratio method, showing different curves.* (a) Component L does not absorb at the wavelength of maximum absorption for the complex, for example, Fe(III)-Tiron. (b) Component L absorbs slightly at the wavelength of maximum absorption for the complex, for example, Zn-Pan. (c) An excess of component L causes a decrease in absorbance of the complex, for example, Bi-xylenol orange. From K. L. Cheng, "Absorptiometry," in J. D. Winefordner, ed., Spectrochemical Methods of Analysis, New York: Wiley-Interscience, 1971, Chap. 6, by permission of the editor and John Wiley and Sons.

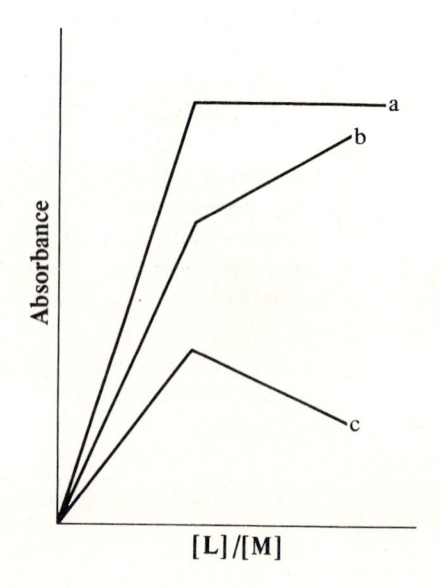

The results may be verified by repeating the process at other wavelengths or total concentrations, since the position of the maximum is independent of wavelength and concentration.

These methods are also applicable at the absorption wavelengths of one of the components—that is, when breaks occur at minimum instead of maximum values of ΔA. The results in Figure 7.12 show the predominant 1 Pd:4 TMK (thio-Michler's ketone) complex and the 1 Hg:3 TMK complex maxima, but they also indicate the formation of 1:1 complex for both Pd and Hg when TMK is not in large excess (the minima coming at 0.5 on the x-axis). The molar-ratio curves for the same complexes give no indication of the existence of a 1:1 complex; in general, Job's method of continuous variation is somewhat more accurate and may provide more information about complex formation. Deviations from Beer's law will result in errors in the direction of larger dissociation constants with either of these methods. The effects may be isolated by varying the path length and concentrations independently or by varying the total concentration of reactants for a given ratio of concentrations. The selection will depend in part on the nature of the deviation from Beer's law. It should be mentioned that these methods are, however, not reliable if the complexes are weak or when several complexes are simultaneously formed in solution.

Vosburgh and his associates [15] have extended Job's method, particularly in dealing with the formation of more than one complex. They investigated the o-phenanthroline (o-phen) complexes of Ni(II) in a range of wavelengths between 500 and 650 nm. The absorption by $[Ni(o\text{-phen})]^{2+}$ at 620 nm, $[Ni(o\text{-phen})_2]^{2+}$ at 580 nm, and $[Ni(o\text{-phen})_3]^{2+}$ at 528 nm were shown with three linear plots.

Bjerrum Method. In the method of Bjerrum, one plots $\epsilon/[M]_t$ versus $[L]_t$ for various constant values of $[M]_t$, where "t" denotes *total* concentration of the designated form. A line drawn horizontally on the graph intersects the experimental curves at points whose coordinates show the composition of the so-called corresponding solutions, which have a given value of $\epsilon/[M]_t$. For complexes of the type ML_n ($n = 1, 2, 3, \ldots$),

FIGURE 7.12. *Job curves of thio-Michler's ketone (TMK) complexes of mercury (○) and of palladium (●). From K. L. Cheng, "Absorptiometry," in J. D. Winefordner, ed., Spectrochemical Methods of Analysis, New York: Wiley-Interscience, 1971, Chap. 6, by permission of the editor and John Wiley and Sons.*

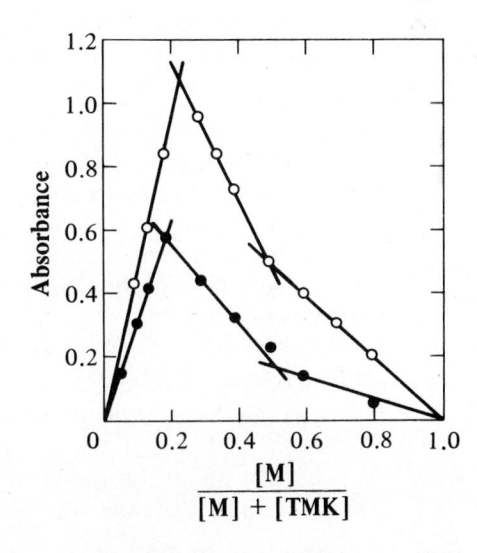

the solutions also have the same value of $\bar{n}$. The value of $\bar{n}$ (an average number of ligands bonded to the central group) may be obtained at various concentrations of L when the total concentrations of such solutions are known. For detailed description, see References [17] and [18].

Studies of Chemical Equilibria

Spectrophotometry can be used to assess chemical equilibria, provided the participating species absorb at markedly different wavelengths.

Determination of Acid-Base Equilibria. Since the absorption spectra of organic molecules with acidic or basic functional groups depend on the pH of the medium, the absorption maxima and intensities vary with the hydrogen-ion concentration. The dissociation constant of an acid or a base may be determined spectrophotometrically as a result of such changes. For a weak acid in water,

$$HA + H_2O \rightleftarrows H_3O^+ + A^- \tag{7.25}$$

$$K_a = \frac{[H_3O^+][A^-]}{[HA]}$$

where K_a = dissociation constant of acid HA

Equation 7.25 may be expressed as

$$-\log K_a = -\log[H_3O^+] - \log\frac{[A^-]}{[HA]} \tag{7.26}$$

$$pK_a = pH + \log\frac{[HA]}{[A^-]}$$

If the pH and the concentrations of HA (acid form) and A^- (basic form) are known, pK_a can be calculated easily. The ratio of [HA] to $[A^-]$ may be found spectrophotometrically if ϵ_{HA} and ϵ_{A^-} are known. These latter values can be determined after converting completely to A^- or HA by adding excess acid or base. As an example, the dissociation constants of several weak acids and bases have been determined photometrically [19]; the base strengths of pyridine derivatives have been determined using a similar procedure [20].

According to Equation 7.26, when $[HA] = [A^-]$, $pK_a = pH$. The pH at this point may be called $pH_{1/2}$, since it occurs at the midpoint of a photometric titration curve, namely, 50% of the titration of acid HA.

This point can be used to calculate K_a: One plots the absorbance at a particular wavelength, say λ_{max}, against the pH of the solution and obtains the midpoint graphically to find $pH_{1/2}$.

King and Hirt [21] have described an instrument called a spectrotitrimeter, which offers a rapid and accurate determination of dissociation constants. A titration flask combined with a pH meter and a spectrometer with an automatic pump as described by Rehm et al. [22] will serve the same purpose.

The pK_a of bromophenol blue has been determined spectrophotometrically [23]. The color change may be followed spectrophotometrically as in Figure 7.13,

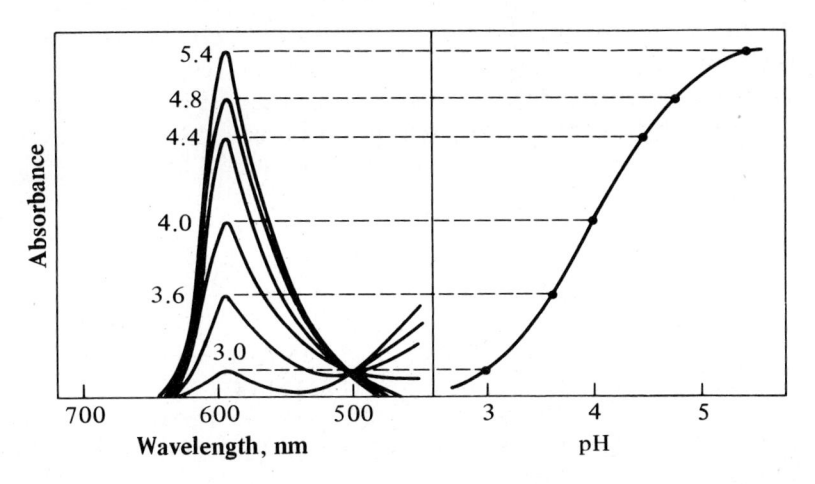

FIGURE 7.13. *Left*: Absorption spectrum of bromophenol blue at various pH values. *Right*: Variation of A_{max} with pH.

which shows the absorption spectrum of bromophenol blue in solutions of pH from 3.0 to 5.4. Usually, photometric titration gives a sigmoid curve; if the curve fails to flatten out at the ends, the midpoint is determined by a graphic method commonly used in polarography for locating $E_{1/2}$ (see Chap. 3). The spectra in Figure 7.13 suggest that the peak at 590 nm is due solely to the conjugate base, A^-, of bromophenol blue (since the peak intensity is reduced by decreasing the pH) and the HA absorption evidently occurs somewhere below 450 nm. There is no problem of overlap in this example. Hence, we may write the equation

$$[A^-] = \frac{A}{\epsilon b} \tag{7.27}$$

where A = absorbance at the maximum
ϵ = absorptivity at the maximum

For a total concentration of bromophenol blue of c, $[HA] = c - [A^-]$ and

$$[H_3O^+] = K_a \left(\frac{c}{[A^-]} - 1 \right) = K_a \left(\frac{c \cdot \epsilon b}{A} - 1 \right) \tag{7.28}$$

$$pH = pK_a - \log \left(\frac{c \cdot \epsilon b}{A} - 1 \right) \tag{7.29}$$

This gives the sigmoid curve shown in Figure 7.13. Such a curve can be obtained experimentally without prior knowledge of K_a or ϵ by measuring the absorbance of HA in various buffer solutions.

Conversely, this experiment offers information about the nature of the indicator conjugate pair itself. Equation 7.28 can be fitted to the plot of A against pH to obtain ϵ and K_a. In that equation $A \neq \epsilon bc$, because c refers to the total concentration, $[HA] + [A^-]$. If the pH is made sufficiently large that practically all the indicator exists as A^-, then $c = [A^-]$ and $A = \epsilon bc$, and hence ϵ is obtained. At the value of A corresponding to $(\epsilon bc)/2$, pH = pK_a.

Kambe et al. [24] determined the dissociation constants of some furfurylidene-p-nitrophenylhydrazones by plotting appropriate absorbance values versus pH. The pK_1 and pK_2 of p-hydroxybenzoic acid have been found to be 4.61 and 9.31 by a spectrophotometric method [25]. Since the equations involve concentration instead of activity, the experimental results for pK values are approximate unless corrected for activity.

Equilibrium Constants. Job [16] has pointed out that, when the formula is known for a complex in solution, the equilibrium constant can be calculated through a relation between concentration and absorptivity. As a part of his continuous-variation studies, Job determined spectrophotometrically the equilibrium constants of many complexes.

Cheng [26] applied Job's method to the determination of the apparent formation constant of the Hf-xylenol orange (XO) complex using the method of mixtures of equimolar solutions. This is a rapid, though probably not too accurate, method of estimating the formation constant of a colored complex with the mole ratio 1:1. Application of this method gave a value of $K = 1.6 \times 10^4$ for the Hf-XO complex formational constant in $0.8\ N$ $HClO_4$. The formation constants of the cerium, titanium, cadmium, and UO_2^{2+} complexes have also been reported [27].

The determination of formation constants may involve the photometric measurement of the complex formed in the presence of a large excess of one of the reagents, and so the formation of the complex may be considered to be essentially complete: This is known as the method of mixtures of nonequimolar solutions and is based on Job's general equation [28, 29] for systems involving mixtures. The method has been applied to the determination of the dissociation constant of Fe(III)-sulfosalicyclic acid mixtures in a pH 5.3 buffer, using sulfosalicyclic acid solutions three, five, and eight times as concentrated as the ferric perchlorate. The best results were obtained by assuming that a 1:1 complex is formed, and K_d was calculated to be 2×10^{-5}. Spectrophotometric methods of determining stability constants are generally unreliable when the complexes are rather weak, or when several complexes are formed in solution. These methods are most suitable for a situation where only one or two complexes are involved in the equilibrium and when absorption by the free ligand is negligible at the wavelength used. Reviews of spectrophotometric methods for determining equilibrium constants are available [18, 30].

Recently, a simplified method for determining equilibrium constants has been published by Brown [31]. It is based on the calculation of the equilibrium concentration of the ligand from the total concentration and the absorbances before and after reaction. A recent study by Sheinker and coworkers [32] has cast doubts on the reliability of stability constants determined by the most commonly used methods. These investigators prepared model systems with preset values for the stepwise formation of 1:1 and 1:2 complexes. They then determined the stability constants using Bjerrum's method, the Landauer-McConnel method, and Yatzimirskii's method. None of the methods gave results in good agreement with the preset values. The generality of this result remains to be verified.

Molecular-Weight Determinations

If an unknown compound can be treated to form a derivative in which a chromophore of known ϵ-value is incorporated, the molar concentration of the chromophore may be obtained spectrophotometrically. This provides a simple method for determining molecular weights. Although the molar absorptivity of the absorption band remains constant in all the derivatives, the absorbance (A) will depend on the molar concentration and hence on the molecular weight of the molecule of interest. The molecular weight (M) may be determined spectrophotometrically from the relation

$$M = \frac{\epsilon w b}{A} \qquad (7.30)$$

where w = weight of the compound in grams per liter of solution
b = thickness of the medium

It is assumed in this method that ε is not affected by intra- or intermolecular forces, and that no interfering bands exist.

Picric acid and the picrate salts of amines absorb at 380 nm with a molar absorptivity of 13,400. An accuracy of $\pm 2\%$ was obtained for the spectrophotometric determination of molecular weights of amines [33]. Molecular-weight determinations have been reported for sugars from the absorption spectra of their osazones [34], for aldehydes and ketones from the absorption spectra of their 2,4-dinitrophenylhydrazones [22, 35, 36], and for saturated alcohols from the absorption of their β-2,4-dinitrophenylpropionyl esters [37].

Reaction-Rate Determinations

The concentration dependence of absorbance has an obvious analytical application in the study of reaction rates. If the absorption spectra of the reactants and products are quite different, we may follow spectrophotometrically changes in concentration of either the reactants or the products during the reaction. For slow reactions, samples can be withdrawn and analyzed at leisure. Absorption spectrometry may play its part in such analysis, but no new features are involved. For fast reactions, spectrophotometry offers advantages, particularly in following the concentration changes of the reactants *in situ*. Chapter 18 discusses the determination of reaction rates in detail.

Purification and Trace Analysis

Trace impurities in a "pure" organic compound may be easily detected or estimated if they have fairly intense absorption bands. As when carrying out a crystallization of a solid compound to a constant (maximum) melting point, the purification should be continued until the molar absorptivity reaches a constant (minimum) value. For example, commercial absolute ethanol commonly contains benzene as an impurity, and the latter is easily detected by spectrophotometric means. The presence of CS_2 in CCl_4 can be detected spectrophotometrically at 318 nm. The absorption data can be taken as truly characteristic of a compound only when its purity has been verified by attainment of constant minimum absorption intensity after repeated fractional purifications. Absorption data have been commonly cited for the purity specifications of some therapeutic solutions of vitamins A, C, and D. Absorption spectra are often used to indicate the purity of unstable biological compounds such as nucleotides or enzymes, because this is often the most convenient, or perhaps the only, way to do so.

7.6 APPARATUS AND INSTRUMENTS

The instruments used in the ultraviolet-visible region of the electromagnetic spectrum fall into three categories, distinguished by complexity of design. Colorimeters generally are the simple visual and photoelectric devices used in the visible region. Photometers include colorimeters, but are more flexible in design so as to include ultraviolet and infrared as well as the visible region. Spectrophotometers are more complex and versatile than either of the others in that they include a monochromator, which provides a narrow band of continuously variable wavelength. A wide variety of spectrophotometers are commercially available.

The choice of source, optical materials, monochromator, and detector depends on the spectral region of interest. This usually imposes a limit on the range of a given instrument. The four ranges for which instruments are presently available are: (a) the visible region (400 to 700 nm); (b) the near-ultraviolet, visible, and very-near-infrared (190 to 1000 nm), using quartz optics; (c) the vacuum ultraviolet (below 190 nm), requiring an evacuated instrument; and (d) the nitrogen ultraviolet region (200 to 160 nm), requiring instruments purged with N_2.

Components of a Spectrophotometer

There are several light sources available for use in the ultraviolet-visible region. *Mercury-vapor* lamps have been used but, owing to the heat evolved by these lamps, thermal insulation or cooling is required. More commonly used for the visible and near-infrared regions are *tungsten-filament* "incandescent" lamps. These are thermal or "blackbody" sources in which the radiation is the result of high temperature of the solid filament material, with only a small dependence on its actual chemical nature. These sources provide continuous radiation from about 320 to 3000 nm—most of it, unfortunately, in the near-infrared. At the usual operating temperature of about 3000 K, only approximately 15% of the total radiant energy falls in the visible region, and at 2000 K, only 1%. Increasing the operating temperature above 3000 K greatly increases the total energy output and shifts the wavelength of maximum intensity to shorter wavelengths, but the lifetime of the lamp is drastically shortened. Inconveniently high temperatures are required for the production of much radiation in the ultraviolet region. The lifetime of a tungsten-filament lamp can be greatly increased by the presence of a low pressure of iodine or bromine vapor within the lamp; with the addition of a fused-silica lamp envelope, these are now called *quartz-halogen* lamps— a popular source at present. Most work in the ultraviolet region is done with *hydrogen* or *deuterium* electrical-discharge lamps typically operated under low-pressure DC conditions (about 40 V with 5 mm gas pressure). These lamps provide a continuum emission down to about 160 nm, but the window material generally limits the transmission at short wavelengths (about 200 nm with quartz and 185 nm with fused silica). Above about 360 nm, hydrogen emission lines are superimposed on the continuum, so incandescent sources are generally used for measurements at longer wavelengths. Deuterium lamps are more expensive but have about two to five times greater spectral intensity and lifetime than a hydrogen lamp of comparable design and wattage.

The continuous radiation from the sources listed above is dispersed by means of monochromators (see Chap. 6).

There are three main types of detectors presently in use. The *barrier-layer* or *photovoltaic type* is illustrated in Figure 7.14A. This device measures the intensity of photons by means of the voltage developed across the semiconductor layer. Electrons, ejected by photons from the semiconductor, are collected by the silver layer. The potential depends on the number of photons hitting the detector. A second type is the photodetector or *phototube* shown in Figure 7.14B. This detector is a vacuum tube with a cesium-coated photocathode. Photons of sufficiently high energy hitting the cathode can dislodge electrons, which are collected at the anode. Photon flux is measured by the current flow in the system. The vacuum phototube type of detector needs further (external) amplification to function properly. The last type of commonly used detector is schematically illustrated in Figure 7.14C. This detector consists of a

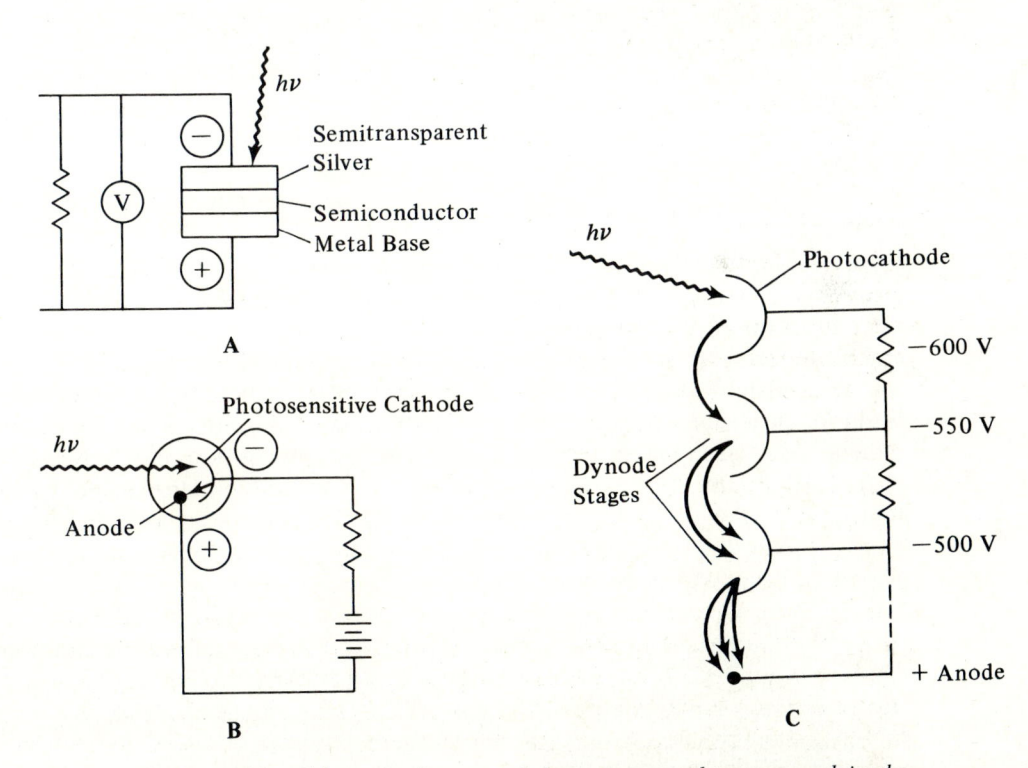

FIGURE 7.14. *Schematic diagram of three common detectors used in the ultraviolet-visible region. A: Barrier-layer or photovoltaic cell. B: Vacuum phototube. C: Vacuum photomultiplier.*

SEC. 7.6 Apparatus and Instruments

photoemissive cathode coupled with a series of electron-multiplying dynode stages, and is usually called a *photomultiplier*. The primary electrons ejected from the photocathode are accelerated by an electric field so as to strike a small area on the first dynode. The impinging electrons strike with enough energy to eject two to five *secondary electrons*, which are accelerated to the second dynode to eject still more electrons. This cascading effect takes place until the electrons are collected at the anode. Typically, a photomultiplier may have 9 to 16 stages, and an overall gain of 10^6 to perhaps 10^9 electrons per incident photon. Reviews of photodetectors have appeared recently [38–40].

Colorimeters

Although colorimetry is much older than many of the spectrophotometric techniques, new instrumentation often appears in the area. In particular, we wish to discuss the Brinkman dipping-probe colorimeters, which use a fiber-optics probe tip which is inserted directly into the sample solution. They can operate in three modes: transmittance, absorbance, or concentration. Models are available with either a single slide-in filter or a multiple-filter wheel, and either analog or digital readout. These instruments are particularly suitable for colorimetric titrations and in-line monitoring.

Single- and Double-Beam Spectrometers

The measurement of absorption of ultraviolet-visible radiation is of a relative nature. One must continually compare the absorption of the sample with that of an analytical reference or blank to ensure the reliability of the measurement. The rate at which the sample and reference are compared depends on the design of the instrument. In *single-beam* instruments, there is only one light beam or optical path from the source through to the detector. This usually means that one must remove the sample from the light beam and replace it with the reference after each reading. Thus, there is usually an interval of several seconds between measurements.

Alternatively, the sample and reference may be compared many times a second, as in *double-beam* instruments. The light from the source, after passing through the monochromator, is split into two separate beams—one for the sample and the other for the reference. Figure 7.15 shows two types of double-beam spectrophotometers. The measurement of sample and reference absorption may be separated in space, as in Figure 7.15A; this, however, requires two detectors that must be perfectly matched. Or, the sample and reference measurement may be separated in time, as in Figure 7.15B; this technique makes use of a rapidly rotating mirror or "chopper" to switch the beam that comes from sample and reference very rapidly. The latter method requires only one detector and is probably the better of the two methods.

There are two main advantages of double-beam operation over single-beam operation. Very rapid monitoring of sample and reference helps to eliminate errors due to drift in source intensity, electronic instability, and any changes in the optical system. Also, double-beam operation lends itself to automation: The spectra can be recorded by a strip-chart recorder.

The above double-beam instruments are single-wavelength spectrophotometers. A modification of these gives rise to the *dual-wavelength spectrophotometer*.

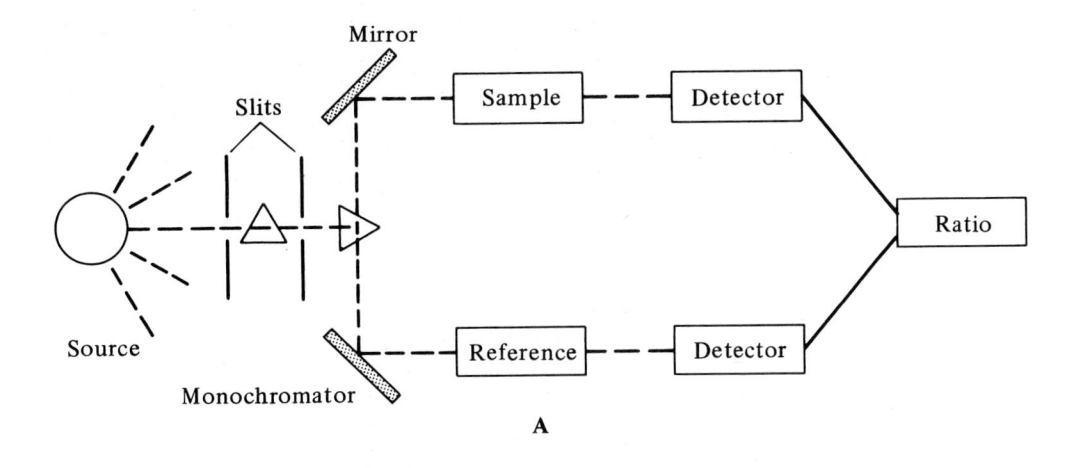

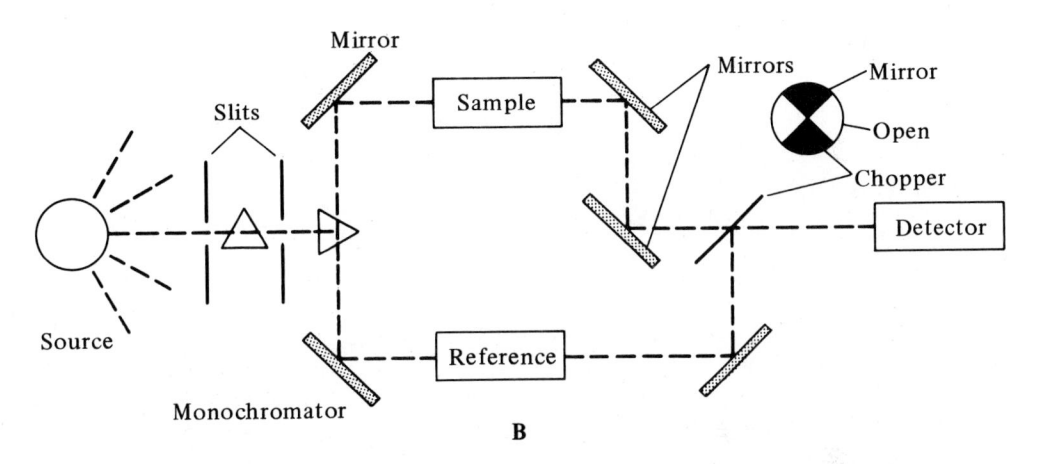

FIGURE 7.15. *Schematic diagram of two types of double-beam spectrophotometers. A: Double-beam-in-space configuration. B: Double-beam-in-time configuration.*

In this arrangement, two beams of differing wavelength are passed through a single sample cell. Figure 7.16 presents a block diagram for an arrangement incorporating two monochromators [41]. The input beam to the sample consists alternately (in time) of λ_1 and λ_2 of equal intensity (beam attenuators are used to reduce the intensity of the more intense λ to that of the less intense λ). An alternative monochromator arrangement [42] uses a single monochromator, but light from the source is chopped and the monochromator is shifted between λ_1 and λ_2 by means of a wavelength control operated by a minicomputer.

Using dual-wavelength spectrometry, the absorbance of an analyte can be determined in the presence of several spectral interferences. Furthermore, the effects of sample settling, scattering, and instrumental stray light are canceled out because a single cuvette is used.

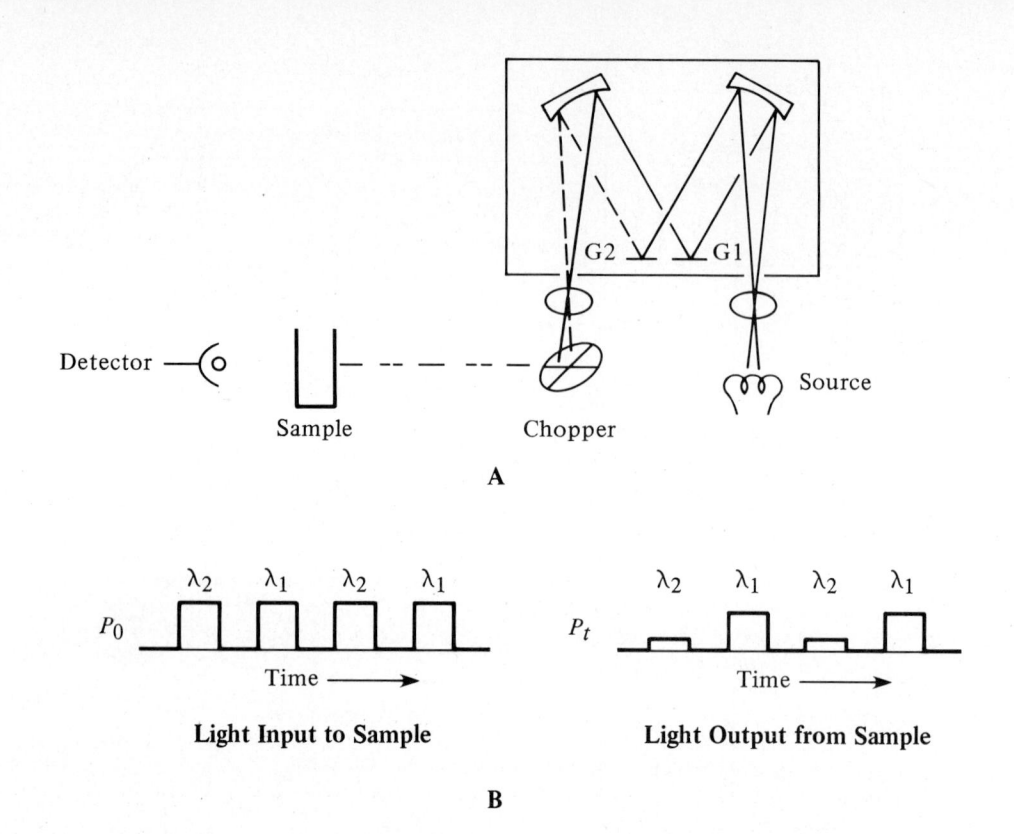

FIGURE 7.16. *Dual-wavelength spectrophotometer using two monochromators.*

Derivative Spectrophotometers

Derivative spectroscopy was introduced by Griese and French [43] in 1955. These authors achieved better resolution by electronically obtaining first and second derivatives of absorption spectra. Subsequently, Hager [44] reported a means of obtaining derivative spectra optically rather than electronically. This is important, since derivatives taken electronically sense changes in intensity with time as well as wavelength, whereas those obtained optically do not. Changes in intensity with respect to time are considered noise; thus, by taking derivatives of these fluctuations, the spectrum actually becomes "noisier." If the technique senses only changes of intensity with wavelength, time fluctuations will not be sensed and noise is minimized.

One important feature of derivative spectra is that peak heights are usually directly proportional to concentration. This is more desirable than the logarithmic relationship in direct absorption spectroscopy. Another important feature is that the sensitivity to concentration depends on the rate of change in molar absorptivity at a particular wavelength, $d\epsilon/d\lambda$, rather than on the absolute magnitude of ϵ itself. Thus, very sensitive analyses are possible for compounds that have sharp absorption peaks. Since absorption spectra are broadened in condensed phases, derivative spectroscopy finds particular application in gas analysis where absorption peaks are much sharper.

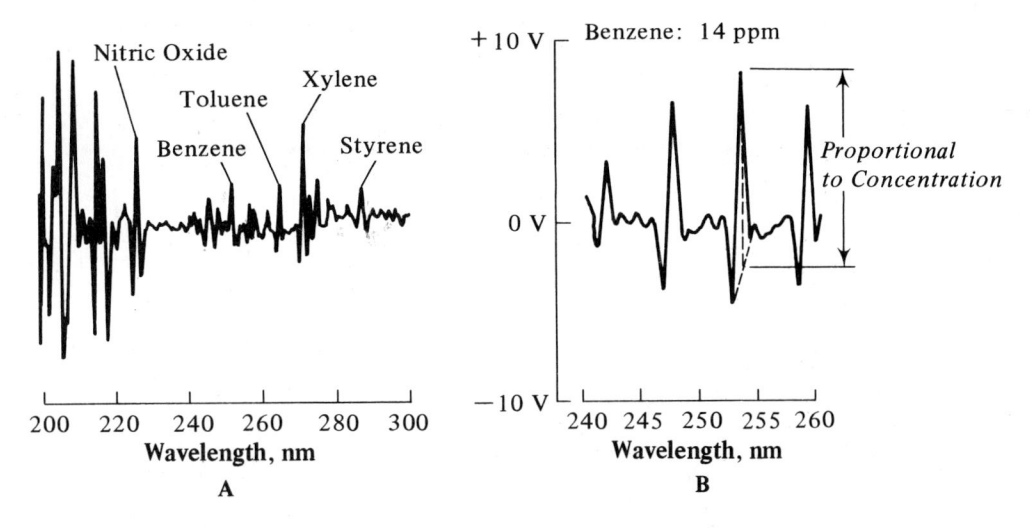

FIGURE 7.17. *Typical second-derivative absorption spectra of gaseous samples. A: Spectrum of an automobile exhaust. B: Spectrum of 14 ppm benzene. Spectra courtesy of Lear Siegler, Inc., Environmental Technology Division, Englewood, Colorado.*

Typical second-derivative spectra are shown in Figure 7.17. The derivative spectra obtained for a given sample can be analyzed for both composition and concentration. Component gases may be identified from the location of their second-derivative peaks, which occur at wavelengths characteristic of the compound. The concentration of each species is determined directly from the peak height. Table 7.5 shows the sensitivity of the technique to various gases.

Using the dual-wavelength spectrophotometer, first-derivative spectra can be obtained optically for solutions. In this mode, $\Delta\lambda$ is fixed and the monochromator is scanned over some desired wavelength range. In the two-monochromator arrangement, the monochromators are mechanically locked together at $\Delta\lambda$, and the locked monochromators are scanned over the desired wavelength range. When $\Delta\lambda$ is sufficiently small (2 to 3 nm), the curves obtained closely approximate the theoretical first derivative.

Rapid-Scan Spectrophotometers

Rapid-scanning spectroscopy (RSS) is a method in which a selected portion of the ultraviolet, visible, or near-infrared spectrum is scanned on a time scale ranging from several seconds to a few microseconds. The applications of this technique to systems in which short-lived transient species exist or large reaction rates are encountered are numerous [45]. Examples include studies of enzyme-substrate complexes [46], mixed complexes in ligand-exchange reactions [47], and flash photolysis [48] or electrochemical reactions [49].

The instrumentation for RSS is divided into two groups: dispersion and multiplex. The more common is the *dispersion* method, in which the light emerging

TABLE 7.5. *Detection Limits of Some Compounds by Use of a Second-Derivative Gas Analyzer*

Compound	Concentration, ppb
Ammonia	1
Nitric oxide	5
Nitrogen dioxide	40
Ozone	40
Sulfur dioxide	1
Mercury vapor	0.5
Benzene	25
Toluene	50
Xylene	100
Styrene	100
Formaldehyde	200
Benzaldehyde	100
Acetaldehyde	400

Note: Signal-to-noise ratio of two used for detection limit.
Source: From R. N. Hager, Jr., *Anal. Chem.*, **45**, 1131A (1973), by permission of the author and publisher. Copyright ©1973 by the American Chemical Society.

from the sample is dispersed by a prism or grating into narrow bands of wavelength which are monitored independently. *Multiplex* methods use mathematical techniques—Fourier or Hadamard transforms—to resolve the spectral bands. Dispersion methods are the faster of the two groups but have lower signal-to-noise ratios. Multiplex methods have their best application in the infrared region.

Several detectors are used in RSS. At present, the one that most closely approaches the ideal is the Vidicon camera tube, shown in Figure 7.18. This detector consists of an array of photodiodes spaced about 15 μm apart. The diodes are biased in sequence by an electron beam that repetitively scans the array in the Vidicon tube. Once the diodes are biased, they are nonconducting and no signal is recovered from the array. The diodes are discharged by photon-generated electron-hole pairs or by leakage. Once discharged, the diodes are conducting, and current flows through the diode. This current is the Vidicon signal and is directly proportional to the number of photons hitting the array. Other high-speed detectors are being refined. Two that show promise are the electrooptic types, which have the fastest scan times (5 μsec), and the charged coupled devices (CCDs), which give the best resolution.

Figure 7.19 shows time-resolved spectra of the reaction of CuCyDTA^{2-} (copper cyclohexanediaminetetraacetate) with ethylenediamine (en). This reaction involves a mixed complex [47] as represented in the following equations:

$$\text{Cu(CyDTA)}^{2-} + \text{en} \rightarrow \text{Cu(CyDTA)en}^{2-} \tag{7.31}$$

$$\text{Cu(CyDTA)en}^{2-} + \text{en} \rightarrow \text{Cuen}_2^{2+} + \text{CyDTA}^{4-} \tag{7.32}$$

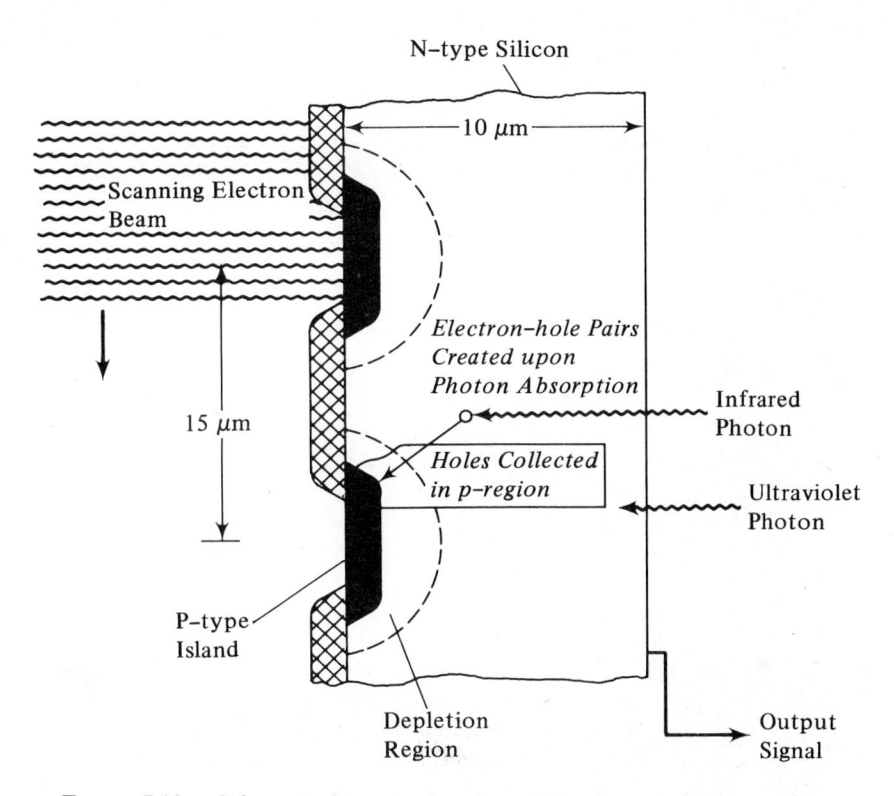

FIGURE 7.18. *Schematic diagram of a silicon Vidicon camera tube. An array of photosensitive diodes are grown on a silicon wafer about* 15 μm *apart. From P. Burke,* Research/Development, *24*(4), 24 (1973), *by permission of the publisher. Copyright © 1973 by Technical Publishing Company.*

This example shows the value of RSS for the study of intermediates in relatively slow reactions. Faster reactions may be studied by using stopped-flow techniques.

For a detailed review of various multichannel detectors, see Reference [50].

Tuned Lasers in Spectrophotometry

Laser light sources have both a high degree of monochromaticity and very high intensity. For molecular absorption measurements, it is desirable to be able to scan the spectrum, and an ideal source for such absorption measurements would be a tunable laser, one whose wavelength could be varied continuously over the spectral range of interest. See Chapter 8 for a discussion of the principles and operation of lasers.

A digital scanning, tunable, dye laser has been constructed for use in the range 358 to 641 nm [51]. The range can be extended to both longer and shorter wavelengths by selecting other dyes, and the tunable range of laser radiation now available is from about 265 to 800 nm. Spectrophotometers using laser radiation as the source routinely exhibit resolution of about 1 nm.

The tuning action of the laser is accomplished by exciting various organic

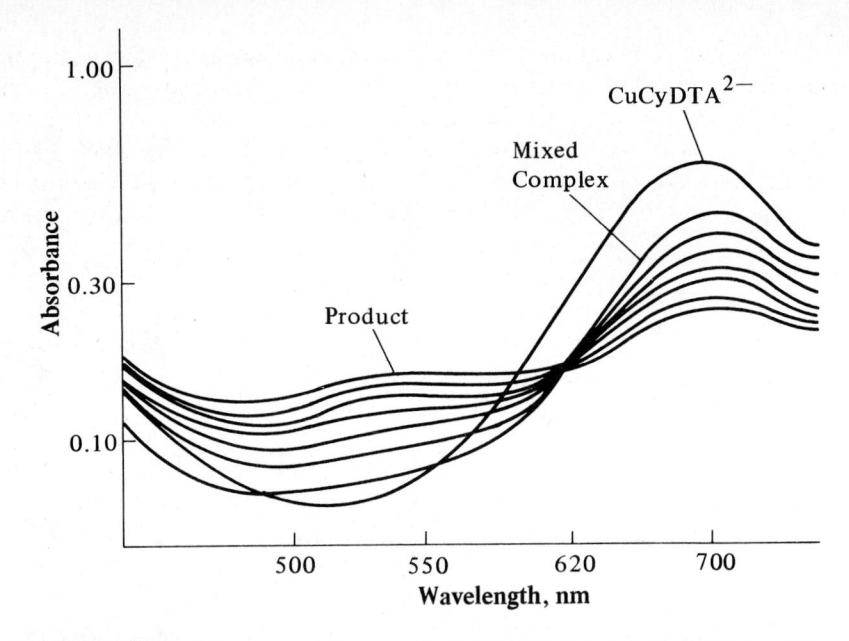

FIGURE 7.19. *Time-resolved spectra of the reaction of CuCyDTA with ethylenediamine. The scan time for one spectrum was 20 msec; spectra were taken every 2 sec. (CyDTA = cyclohexanediaminetetraacetate.) Reprinted with permission from R. E. Santini, M. J. Milano, and H. L. Pardue,* Anal. Chem., **45**, 915A (1973), *by permission of the authors and publisher. Copyright © 1973 by the American Chemical Society.*

dyes with a pulsed nitrogen laser. The dyes presently available are tunable only over a 60–70-nm range (essentially, the width of their absorption bands), and thus several dyes must be used to cover a wide wavelength range. The major limitation at present is the 15% deviation in quantitative studies, owing mainly to instabilities in the laser. A further limitation is that the calibration of a particular laser is dependent upon operating conditions.

Reflectance Spectrometers

In reflectance spectroscopy, one measures the amount of radiant energy reflected from a sample surface. These data are generally reported as percent reflectance:

$$\% R = \frac{I}{I_0} \times 100 \tag{7.33}$$

where I = intensity of reflected radiation
I_0 = intensity of radiation reflected from some "standard" reflecting surface

For a discussion of reflectance spectroscopy, two types of reflectance—specular and diffuse—must be defined. *Specular* reflectance is simply mirrorlike reflectance from a surface and is sometimes called regular reflectance; it has a well-defined reflectance angle. *Diffuse* reflectance is defined as reflected radiant energy that has been partially absorbed and partially scattered by a surface with no defined angle of reflectance. The diffuse reflectance technique is widely used today for industrial applications involving textiles, plastics, paints, dyestuffs, inks, paper, food, and building materials. In the area of basic research, diffuse-reflectance spectroscopy has been used in studies of solid-solid reactions, of species absorbed on metal surfaces, of radiation transfer, and of slightly soluble species.

A common design feature of all commercial diffuse-reflectance instruments is the integrating sphere, which permits the collection of reflected light. Many of the commercial ultraviolet-visible spectrophotometers offer this mode of operation as an accessory. The integrating sphere is usually coated with barium sulfate, which is a highly diffuse reflecting material and serves to "homogenize" the energy being reflected from the sample surface. The intensity of reflected light at any given point in the sphere should be independent of spatial distribution and therefore directly proportional to the diffuse reflectance of the sample. A design feature of some instruments, for example the Cary 1711, allows for exclusion of specular reflectance or, alternatively, inclusion of both types of reflectance to give total reflectance.

An example of an industrial application of reflectance is presented in Figure 7.20. This figure shows spectra taken from various colored papers. In the paper industry, diffuse reflectance is often used to monitor color, whiteness, brightness,

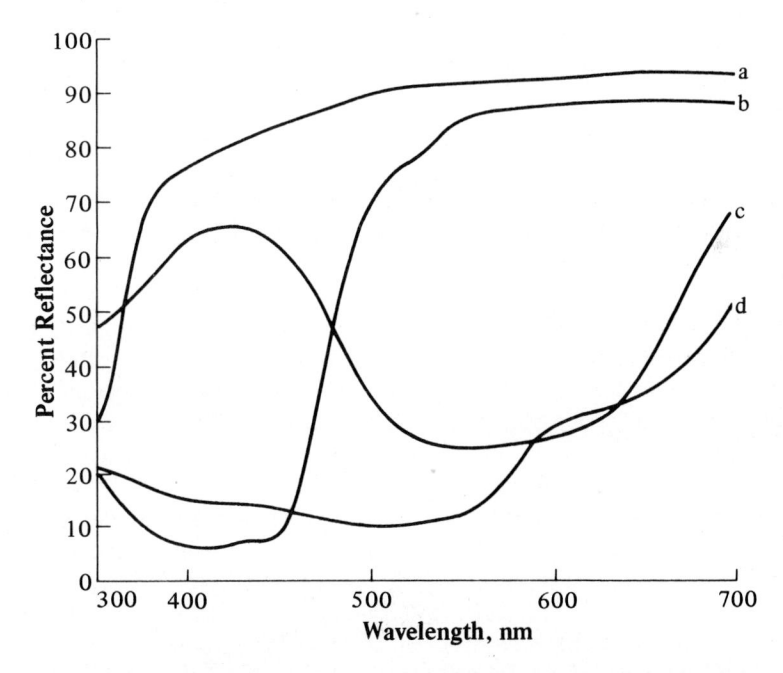

FIGURE 7.20. *Diffuse-reflectance spectra of colored papers:* (a) *off-white,* (b) *yellow,* (c) *purple, and* (d) *maroon. The reference material is* $MgCO_3$. From T. Surles, J. O. Erickson, and D. Priesner, Amer. Lab., 7(3), 55 (1975), by permission of International Scientific Communications, Inc., copyright holder.

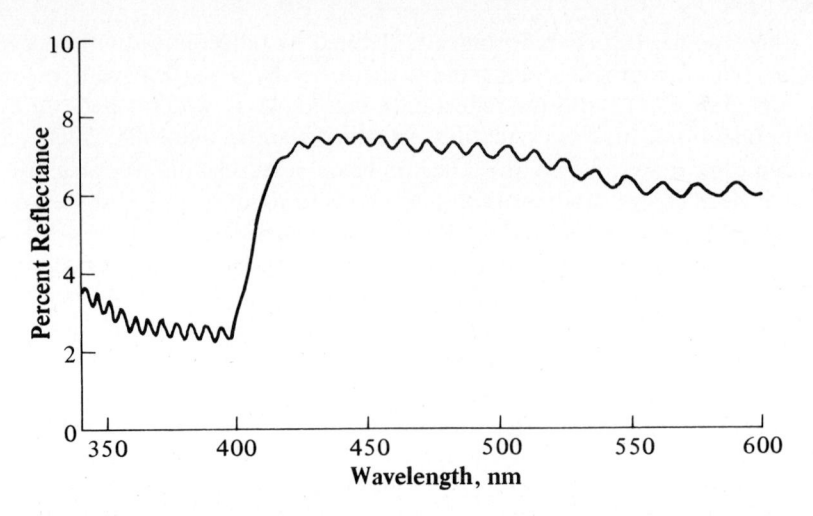

FIGURE 7.21. *Total-reflectance spectrum of clear plastic. The reference material is $MgCO_3$. From T. Surles, J. O. Erickson, and D. Priesner,* Amer. Lab., *7*(3), 55 (1975), *by permission of International Scientific Communications, Inc., copyright holder.*

and gloss of papers. The degree of whiteness is an important parameter that requires a system capable of detecting small variations in reflectance. As little as 0.05% R in the paper's reflectance has a perceptible effect on the whiteness observed by the human eye.

In the area of basic research, the surface properties of plastics are of continuing interest. Figure 7.21 shows the total reflectance of a clear piece of plastic on an expanded scale. An interference pattern is observed (the ripples on the curve), and the thickness of the surface film can be calculated from these data.

Photoacoustic Spectroscopy

Photoacoustic spectroscopy is based on a phenomenon, the optoacoustic effect, discovered by Bell in 1880 [52]. As with many techniques, significant development of photoacoustic spectroscopy has occurred only within the last decade. Its resurgence is due to the development of laser sources, because typical ultraviolet-visible incoherent sources are effective only for the acquisition of low-resolution spectra. The technique has been used to study both gaseous and solid samples. A block diagram of the basic components of a photoacoustic spectrometer is shown in Figure 7.22. A number of different types of cells for gaseous samples have been described [53]. The basic requirements are that the cell be enclosed and have window materials capable of transmitting light. A sample cell suitable for the study of solids is shown in Figure 7.23. Consider first a gaseous sample.

The light source is modulated at an audiofrequency. The sample, absorbing this radiation, becomes a periodic heat source which produces alternating regions of compression and rarefaction in the enclosed gas—that is, an acoustic or sound wave. The acoustic wave is converted to an electrical signal by a microphone. A solid sample

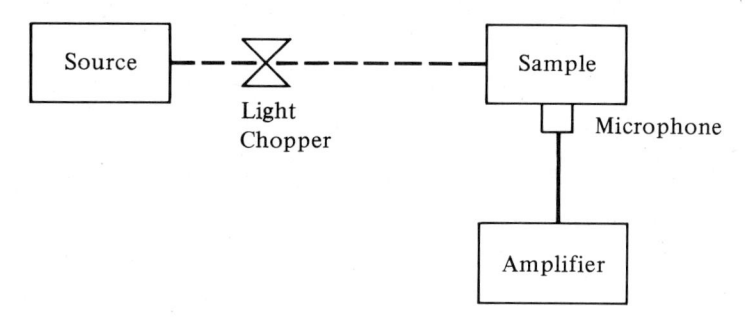

FIGURE 7.22. *Block diagram of a photoacoustic spectrometer.*

is directly irradiated with the modulated source, resulting in production of acoustic waves in the surrounding gas.

The analytical usefulness of ultraviolet-visible photoacoustic spectroscopy in the study of gas-phase systems is virtually unexplored; by contrast, there have been many applications to inorganic, organic, and biological solid samples [54]. The technique has been shown to be appropriate for bulk studies, surface studies, and deexcitation studies. Furthermore, any type of solid or semisolid can be studied, even those that are completely opaque to transmitted light. In Figure 7.24, the photoacoustic spectrum of Cr_2O_3 powder is compared with the corresponding diffuse-reflectance spectrum of crystalline Cr_2O_3. Note that the powder photoacoustic spectrum is equal in quality to the optical-transmission spectrum obtained on a crystal. The powder diffuse-reflectance spectrum is clearly much inferior.

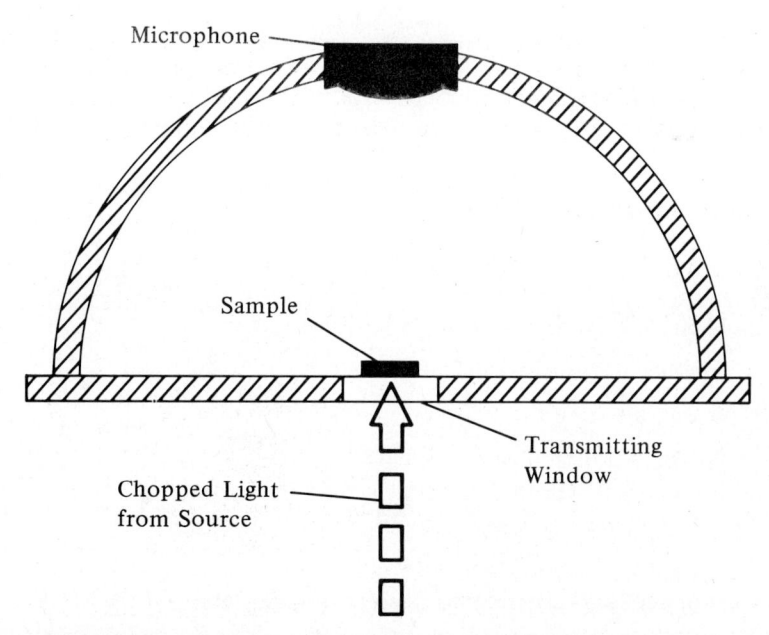

FIGURE 7.23. *Schematic diagram of a photoacoustic sample cell/chamber for solid samples.*

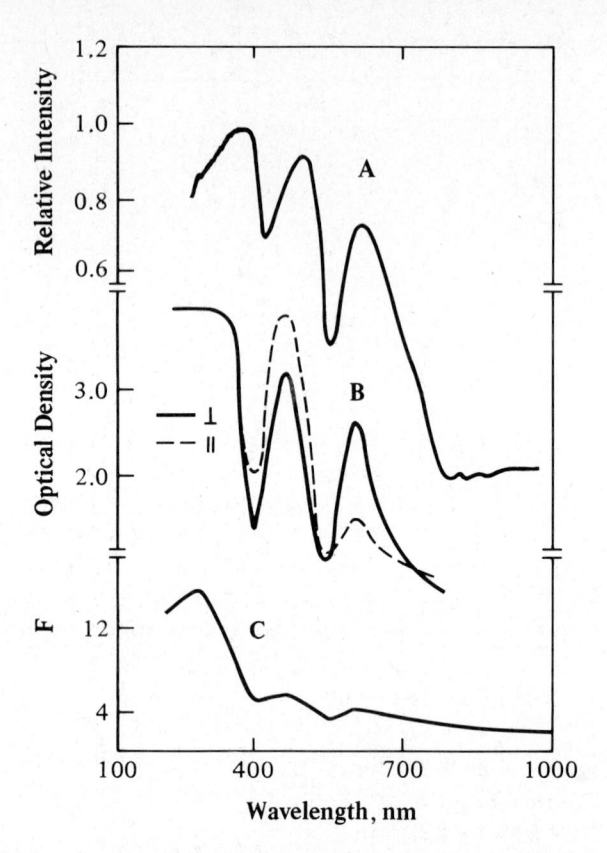

FIGURE 7.24. *Comparison of a photoacoustic spectrum of Cr_2O_3 with other optical spectra. A: Normalized photoacoustic spectrum of Cr_2O_3 powder. B: Optical transmission spectrum of a 4.4-μm-thick crystal of Cr_2O_3. C: Diffuse-reflectance spectrum of Cr_2O_3 powder. All spectra were taken at 300 K. Reprinted from A. Rosencwaig, in Y-H. Pao, ed.,* Optoacoustic Spectroscopy and Detection, *New York: Academic Press, 1977, pp. 193–239. Used with permission.*

Photoacoustic spectroscopy shows promise of being a powerful tool for surface analysis. Adsorbed and chemisorbed species on metals, semiconductors, and insulators can be studied. It has been stated that the greatest potential for photoacoustic spectroscopy lies in the areas of biology and medicine. It is possible to obtain optical data on intact biological matter. This is illustrated in Figure 7.25, where photoacoustic spectra of smears of whole blood, red blood cells, and hemoglobin are shown. Photoacoustic spectroscopy can be used to monitor bacteria growth, and it has been hypothesized that it may be possible to identify a bacterium on the basis of its photoacoustic spectrum.

Microprocessor-Controlled Spectrophotometry

Another recent development in spectrophotometry is the commercial manufacture of microprocessor-controlled instrumentation. Almost every manufacturer of ultra-

FIGURE 7.25. *Photoacoustic spectra of smears of whole blood, red blood cells, and hemoglobin. Reprinted from A. Rosencwaig, in Y-H. Pao, ed.,* Optoacoustic Spectroscopy and Detection, *New York: Academic Press, 1977, pp. 193–239. Used with permission.*

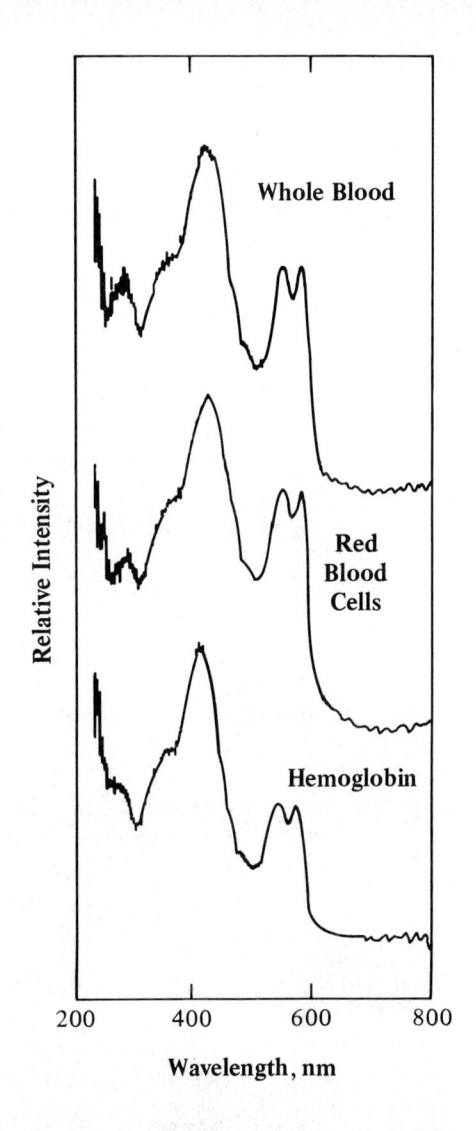

violet-visible instrumentation has at least one model that is microprocessor controlled. Some of these instruments can be purchased for less than $10,000, for example the Bausch and Lomb Spectronic 2000. This is a double-beam, scanning spectrophotometer which can be operated in five different modes: transmittance, absorbance, concentration, first derivative, and second derivative. The microprocessor provides for automated baseline storage and compensation, and even for diagnostics. The 39-key keyboard is marked and color coded by function. Information concerning the tungsten and deuterium lamps can be digitally displayed and also permanently recorded on a built-in printer. Because the exact conditions under which a sample is run can be duplicated a day, a month, or even a year later, the time reproducibility should show much improvement over manually operated instruments. In addition, the results should also show greater operator independence. The reproducibility of sample preparation then becomes the limiting factor.

Precision Spectrophotometry

Precision spectrophotometry (differential spectrophotometry) is a technique that involves comparing an unknown solution with a reference. The reference scale is set at zero using a solution of a highly colored (radiation-adsorbing) species in place of a reagent blank. Concentrations of the unknown higher than the reference are then measured against this zero in the usual way.

Reilley and Crawford [55] described a precision spectrophotometric method that involves the use of two standard solutions to set the 0% and 100% T readings of the photometer. By this means, the full scale can be used for a concentration range much narrower than usual, and precision is thus increased. This method generally requires two standard solutions and several other standard solutions of intermediate concentration to construct a calibration curve, since the measured absorbance is often not a linear function of concentration (deviation from Beer's law at higher concentrations).

A method described by Ramaley and Enke [56] replaces the two-standard and calibration-curve procedures with one standard and isomation. This method involves a titration in which the absorbance of the unknown determines the endpoint. A known amount of solvent is placed in an absorption cell. A standard solution of the sample substance is then added to the cell until the absorbance, and hence the concentration, is identical to that of the unknown solution. To obtain the maximum accuracy, the cell lengths are calibrated as follows. At the endpoint,

$$\frac{c_u}{c_s} = \frac{b_s}{b_u} \qquad (7.34)$$

where c_s = concentration of the standard
c_u = concentration of the sample
b_s = standard-cell length
b_u = sample-cell length

The ratio of the cell lengths may be obtained by using standard solution in both cells and adjusting their concentration until the absorbance is the same for both. The ratio is calculated from Equation 7.34. With knowledge of the cell-length ratio and the concentration of the standard solution, the unknown concentration can be readily determined. With this method, species with molar absorptivity of 10^4 can be determined in the 10^{-6} M concentration range with $\pm 0.2\%$ accuracy.

Photon Counting in Spectrophotometry

Light is a source of discrete photons. When light is measured with a detector such as a photomultiplier, the photons are converted to current pulses which may not be completely resolved in time from each other. Usually these pulses are smoothed to a continuous or continuously varying signal and recorded by a readout device such as a meter. However, the number of photons per unit time reaching the detector can be decreased to a point where the individual current pulses from the photomultiplier tube become resolvable. This can be done by decreasing the light intensity, by isolating a particular wavelength region with a monochromator, and by stopping down the optical aperture of the light beam. If the rate of photoelectron ejection and the frequency response of the measurement system are such that individual current pulses can be resolved, then the number of pulses per unit time can be counted. Because the count rate is a measure of the rate at which photons are striking the photocathode, the measurement technique is appropriately called *photon counting*. In both single- and double-beam instruments, a number proportional to the radiant power of the sample beam is obtained by counting the photoelectric pulses of the sample beam during a precisely controlled time interval [57].

Recently, photon counting has become an important technique in spectrophotometric methods where the radiation is so low in intensity that it is difficult to obtain measurements by conventional means. The ability to deal with low radiation levels with a satisfactory signal-to-noise ratio is, of course, one of the important factors in the photon-counting method. Improvements in precision, resolution, signal-to-noise ratio, and readout are obtainable by photon counting, and this method should be applicable to all spectrophotometric procedures (absorption, emission, reflection, fluorescence, and light scattering) in which a photomultiplier is used.

SELECTED BIBLIOGRAPHY

BAUMAN, R. P. *Absorption Spectroscopy*. New York: John Wiley, 1962.

BOLTZ, D. F., and SCHENK, G. H. "Visible and Ultraviolet Spectroscopy." In *Handbook of Analytical Chemistry*, L. Meites, ed., New York: McGraw-Hill, 1963.

DODD, R. E. *Chemical Spectroscopy*. Amsterdam: Elsevier, 1962.

DONHROW, M. *Instrumental Methods in Analytical Chemistry: Their Principles and Practices*, Vol. 2, *Optical Methods*. New York: Pitman, 1967.

ELVING, P. J., MEEHAN, E. J., and KOLTHOFF, I. M., eds. *Treatise on Analytical Chemistry*, Part I, Vol. 7. New York: Wiley-Interscience, 1981.

OLSEN, E. D. *Modern Optical Methods of Analysis*. New York: McGraw-Hill, 1975.

WALKER, S., and STRAW, H. *Spectroscopy*, Vol. 2. London: Chapman and Hall, 1967.

REFERENCES

1. F. A. COTTON, *Chemical Applications of Group Theory*, New York: Wiley-Interscience, 1971.
2. D. S. SCHONLAND, *Molecular Symmetry*, Princeton, N.J.: Van Nostrand, 1965.
3. H. K. HUGHES, *Anal. Chem.*, 24, 1349 (1952); 40, 2271 (1968); 56, 125 (1984).
4. H. H. JAFFE and M. ORCHIN, *Theory and Applications of Ultraviolet Spectroscopy*, New York: John Wiley, 1962.
5. J. R. DYER, *Applications of Absorption Spectroscopy of Organic Compounds*, Englewood Cliffs, N.J.: Prentice-Hall, 1965.
6. R. M. SILVERSTEIN and G. C. BASSLER, *Spectrometric Identification of Organic Compounds*, 2nd ed., New York: John Wiley, 1967.
7. E. A. BRAUDE, *Determination of Organic Structures by Physical Methods*, New York: Academic Press, 1955.
8. C. N. R. RAO, *Ultraviolet and Visible Spectroscopy*, 2nd ed., London: Plenum Press, 1967.
9. S. F. MASON, *Quart. Rev.* (London), 15, 287 (1961).
10. G. KORTUM and M. SEILER, *Angew. Chem.*, 52, 687 (1939).
11. K. L. CHENG, *Anal. Chim. Acta*, 28, 41 (1963).
12. K. L. CHENG, *Anal. Chem.*, 30, 1035 (1958).
13. K. A. CONNORS and C. J. EBOKA, *Anal. Chem.*, 51, 1262 (1979).
14. R. JAKUBIEC and D. F. BOLTZ, *Anal. Chem.*, 41, 78 (1969).

15. W. C. Vosburgh and G. R. Cooper, *J. Amer. Chem. Soc.*, **53**, 435 (1941).
16. P. Job, *Anal. Chim.*, **9**, 113 (1928).
17. J. Bjerrum, *Kgl. Danske Videnskab. Selskab, Mat-Fys. Medd.*, **21**(4) (1944); H. Olerup, Thesis, "Jarn Kloridernas Komplexitet," Lund, 1944.
18. F. J. C. Rossotti and H. Rossotti, *The Determination of Stability Constants*, New York: McGraw-Hill, 1961; S. D. Christian, *J. Chem. Ed.*, **45**, 713 (1968).
19. L. A. Flexser, L. P. Hammet, and A. Dingwall, *J. Amer. Chem. Soc.*, **57**, 2103 (1935).
20. H. C. Brown and D. H. McDaniel, *J. Amer. Chem. Soc.*, **77**, 3752 (1955).
21. F. T. King and R. C. Hirt, *Appl. Spectrosc.*, **7**, 164 (1953).
22. C. Rehm, J. I. Bodin, K. A. Connors, and T. Higuchi, *Anal. Chem.*, **31**, 483 (1959).
23. W. R. Brode, *J. Amer. Chem. Soc.*, **46**, 581 (1924).
24. M. Kambe, E. Shindo, and M. Marito, *Japan Analyst*, **16**, 1017 (1967).
25. B. N. Mattoo, *Trans. Faraday Soc.*, **52**, 1462 (1956).
26. K. L. Cheng, *Talanta*, **2**, 266 (1959); **5**, 254 (1960).
27. M. Otomo, *Bull. Chem. Soc. Jap.*, **36**, 146 (1962).
28. B. Ricca and G. Fraone, *Gazz. Chim. Ital.*, **79**, 340 (1949); *Chem. Abstr.*, **43**, 8935b (1949).
29. R. T. Foley and R. C. Anderson, *J. Amer. Chem. Soc.*, **72**, 5609 (1950).
30. R. W. Ramette, *J. Chem. Ed.*, **44**, 647 (1967).
31. K. L. Brown, *Inorg. Chim. Acta*, **37**, L513 (1979).
32. V. N. Sheinker, I. V. Petrova, and O. A. Osipov, *Zh. Obshch. Khim.*, **51**, 1716 (1981).
33. K. G. Cunningham, W. Dawson, and F. S. Spring, *J. Chem. Soc.*, 2305 (1954).
34. V. C. Barry, J. E. McCormick, and P. W. D. Mitchell, *J. Chem. Soc.*, 222 (1955).
35. E. A. Braude and E. R. H. Jones, *J. Chem. Soc.*, 498 (1945).
36. C. Djerassi and E. Ryan, *J. Amer. Chem. Soc.*, **71**, 1000 (1949).
37. J. P. Riley, *J. Chem. Soc.*, 2108 (1952).
38. T. P. Lee and T. Li, *Opt. Fiber Commun.*, 593 (1979).
39. A. T. Young, *Methods Exp. Phys.*, **12**, 1 (1974).
40. H. R. Zwicker, *Top. Appl. Phys.*, **19** (2nd ed.), 149 (1980).
41. R. L. Sellers, G. W. Lowy, and R. W. Kane, in *Laboratory Instrumentation: Spectroscopy*, Series II, Volume 11, Fairfield, Conn.: International Scientific Communications, 1977.
42. K. L. Ratzlaff, F. S. Chuang, D. F. S. Natusch, and K. R. O'Keefe, *Anal. Chem.*, **50**, 1799 (1978).
43. A. Griese and C. French, *Appl. Spectrosc.*, **9**, 78 (1955).
44. R. N. Hager, Jr., *Anal. Chem.*, **45**, 1131A (1973).
45. R. E. Santini, M. J. Milano, and H. L. Pardue, *Anal. Chem.*, **45**, 915A (1973).
46. V. Massey and G. H. Gibson, *Fed. Proc.*, **23**, 18 (1964).
47. J. D. Carr, R. A. Libby, and D. W. Margerum, *Inorg. Chem.*, **6**, 1083 (1967).
48. J. I. H. Patterson and S. P. Perone, *Anal. Chem.*, **44**, 1978 (1972).
49. J. W. Strojek, G. A. Gruver, and T. Kuwana, *Anal. Chem.*, **41**, 481 (1969).
50. G. D. Christian, J. B. Callis, and E. R. Davidson, in *Modern Fluorescence Spectroscopy*, E. L. Wehry, ed. Vol. 4, New York: Plenum Press, 1981, pp. 111–66.
51. D. Harrington and H. V. Malmstadt, *Amer. Lab.*, **6**(3), 33 (1974).
52. A. G. Bell, *Proc. Am. Assoc. Adv. Sci.*, **29**, 115 (1880).
53. C. F. Dewey, Jr., in *Optoacoustic Spectroscopy and Detection*, Y-H. Pao, ed., New York: Academic Press, 1977.
54. A. Rosencwaig, *Photoacoustics and Photoacoustic Spectroscopy*, New York: Wiley-Interscience, 1980.
55. C. N. Reilley and C. M. Crawford, *Anal. Chem.*, **27**, 716 (1955).
56. L. Ramaley and C. G. Enke, *Anal. Chem.*, **37**, 1073 (1965).
57. E. H. Piepmeier, D. E. Braun, and R. R. Rhodes, *Anal. Chem.*, **40**, 1667 (1968); M. L. Franklin, G. Horlick, and H. V. Malmstadt, *Anal. Chem.*, **41**, 2 (1969); K. C. Ash and E. H. Piepmeier, *Anal. Chem.*, **43**, 26 (1971).

PROBLEMS

1. A sodium-vapor lamp emits radiation with a wavelength of 5889.97 Å. Express the wavelength in nanometers and calculate its frequency. The speed of light in vacuum is 2.99776×10^{10} cm/sec.

2. The energy of the electronic transition for D-A $\xrightarrow{h\nu}$ D$^+$-A$^-$ may be estimated by the equation $h\nu = I_D$ (ionization potential of donor) $- E_A$ (electron affinity of acceptor) $- C$ (mutual electrostatic energy of D$^+$ and A$^-$). Gaseous NaCl will absorb in the ultraviolet due to a charge-transfer transition. Estimate the energy and wavelength at which absorption may be expected. What does it mean when a negative value is obtained for the energy? $I_{Na} = 5.14$ eV, $E_{Cl} = 3.62$ eV, $C \approx 6.2$ eV.

3. Assuming that C in Problem 2 is directly proportional to the product of the charges of D$^+$ and A$^-$ and inversely proportional to their separation, estimate the energy and wavelength for the charge-transfer transition of gaseous KCl. Assume D$^+$ and A$^-$ just touch. $I_K = 4.34$ eV, $r_{Na^+} = 0.97$ Å, $r_{K^+} = 1.33$ Å, $r_{Cl^-} = 1.81$ Å.

4. The Pd 4,4'-bis(dimethylamino)thiobenzophenone complexation has been reported to be one of the most sensitive color reactions, with a molar absorptivity of 2.12×10^5. Assuming that the minimum measurable absorbance is 0.001 and that a cell with a 10-cm light path is available, what is the lowest possible molar concentration of Pd that can be determined spectrophotometrically? If the volume of the cell is 10 mL, what is the smallest quantity of Pd that can be determined?

5. In 25.0 mL of 0.8 N HClO$_4$, 5.00×10^{-7} mole of Zr forms a 1:1 complex with xylenol orange (XO), giving an absorbance of 0.484 at 535 nm in a 1.00-cm cell. Calculate the molar absorptivity of the Zr-XO complex.

6. As a senior research problem, a chemistry student has been asked to investigate the yield of the products formed in the nitration of aniline as a function of temperature used in the final step.

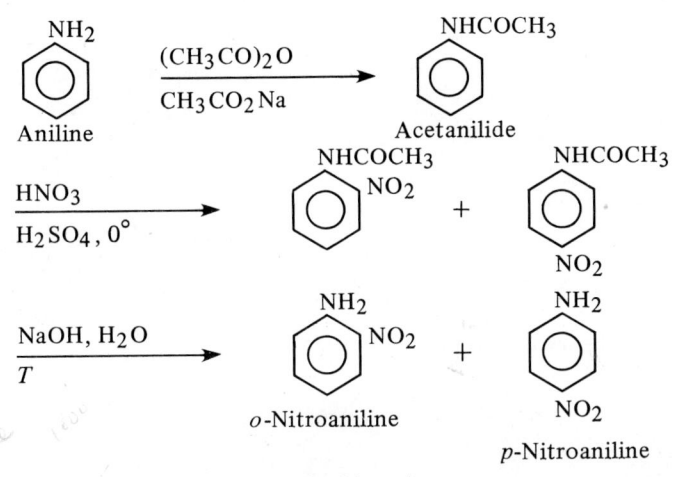

Problem 6

In the third step, any unreacted acetanilide will be converted to aniline. Since aniline is soluble in water while the products are not, the products fall out of solution as crystals. These crystals can be dissolved in alcohol. The student has chosen to use spectrophotometry to determine the percent yields. The following are data obtained for one run with $T = 100°C$. Determine the percent yields of o- and p-nitroaniline.

Aniline: 10.000 g
Product: 12.000 g

Product dissolved in 100 mL of alcohol: 1.75 mg
Cell thickness: 1.00 cm

	Absorbance		Molar Absorptivity	
	285 nm	347 nm	285 nm	347 nm
o-Nitroaniline	—	—	5260	1280
p-Nitroaniline	—	—	1400	9200
Mixture	0.321	0.866	—	—

7. A conjugate base of a weak acid, HA, has an absorption maximum at 520 nm. The following data were obtained by measuring the absorbance of solutions of the weak acid having the same concentration but different pH buffers:

pH	Absorbance
2.0	0.00
4.0	0.00
5.0	0.030
6.0	0.180
7.0	0.475
8.0	0.565
9.0	0.590
10.0	0.590
11.0	0.590
12.0	0.590

What is the approximate pK_a of this weak acid?

8. A colored substance X has an absorption maximum at 400 nm. A solution containing 2.00 mg X per liter has an absorbance of 0.840 using a 2.00-cm cell. The formula weight of X is 150. (a) Calculate the absorptivity of X at 400 nm. (b) Calculate the molar absorptivity of X at 400 nm. (c) How many milligrams of X is contained in 25.0 mL of a solution giving an absorbance of 0.250 at 400 nm when measured with a 1.00-cm cell? (d) How many ppm of X are in the solution of part (c)?

9. A 0.200-g sample containing Cu is dissolved, and a diethyldithiocarbamate colored complex is formed in the presence of EDTA. The solution is then diluted to 50.0 mL and the absorbance measured as 0.260. A 0.500-g sample containing 0.240% Cu is treated in the same manner, and the resulting solution has an absorbance of 0.600. Calculate the percentage of Cu in the sample.

10. A 0.5000-g steel sample is dissolved, and the Mn in the sample is oxidized to permanganate by periodate using Ag^+ as a catalyst. After the sample is diluted to 250.0 mL, the absorbance is found to be 0.393 at 540 nm in a 1.00-cm cell. Calculate the percentage of Mn in the steel. The molar absorptivity for permanganate at 540 nm is 2025.

11. ERIO X forms a 1:1 colored complex with Mg^{2+} at pH 10.00. The ERIO X solution was titrated with Mg^{2+} photometrically. This titration showed that, at the equivalence point, [Mg-ERIO X] = [ERIO X] = 5×10^{-7} M. Calculate the formation constant of the Mg complex.

12. A solution containing 0.0150 g of pure weak acid, HA, was titrated with 0.0500 M NaOH. Only the anion, A^-, in the solution absorbed at 350 nm. The following titration data were obtained at 350 nm. What is the molecular weight of this acid?

NaOH, mL	Absorbance
0.0	0.000
0.50	0.185
1.00	0.370
1.50	0.555
2.00	0.680
2.50	0.750
3.00	0.800
3.50	0.842
4.00	0.870
4.50	0.890
5.00	0.900
5.50	0.910
6.00	0.910
7.00	0.910

13. Plot two Job's curves for the Bi-xylenol orange (XO) complex from the following data obtained in 0.1 N H_2SO_4 at 545 nm. (a) What is the molar ratio of Bi to XO? (b) Estimate the formation constant of the Bi-XO complex. I: [Bi] + [XO] = 2.4×10^{-5} M. II: [Bi] + [XO] = 3.2×10^{-5} M.

$\dfrac{[Bi^{3+}]}{[Bi^{3+}] + [XO]}$	Corrected Absorbance	
	I	II
0.0	0.00	0.00
0.1	0.049	0.070
0.2	0.095	0.145
0.3	0.145	0.208
0.4	0.180	0.260

[Bi³⁺]	Corrected Absorbance	
[Bi³⁺] + [XO]	I	II
0.5	0.198	0.278
0.6	0.190	0.270
0.7	0.160	0.226
0.8	0.108	0.158
0.9	0.058	0.075
1.0	0.00	0.00

14. The following absorbance (A) values were obtained in preparing a spectrophotometric calibration curve: blank, 0.03; 1.00 mM standard, 0.11; 2.00 mM standard, 0.19; 4.00 mM standard, 0.35. Plot the calibration curve. What is the most probable reason for the noncompliance of the data to Beer's law?

15. Literature values for the molar absorptivities of nucleotides are often used to determine the concentration of nucleotide solutions, since even highly purified nucleotides may contain variable waters of hydration or a variable salt content. A sample of the disodium salt of cytidine 5'-monophosphate (5'-CMP, $C_9H_{12}N_3O_8PNa_2$) was weighed out (0.0814 g), dissolved, and diluted to 25.00 mL in a volumetric flask with distilled water. The ϵ-value of 5'-CMP is 13.0×10^3 at $\lambda_{max} = 280$ nm in 0.01 M HCl. An aliquot of the CMP stock solution (0.100 mL) was diluted to 10.0 mL with 0.01 M HCl. The absorbance of this solution was 0.831 in a calibrated cuvette of 0.992 cm path length. (a) Calculate the "nominal" or expected concentration of 5'-CMP, assuming it to be a pure anhydrous salt. (b) Calculate the concentration based on spectrophotometric data. (c) Assuming the difference in concentration is due only to absorbed and bound water molecules, what is the average number of water molecules per 5'-CMP molecule?

16. Suggest a method by which spectrophotometry can be used to follow the course of the following reactions, assuming that the kinetics are sufficiently slow.

(a) Hydrogenation of benzene

$$\bigcirc + 3H_2 \longrightarrow$$

(b)

$$CH_3CH=CHCH_3 \xrightarrow[2. \text{ Zn}, H_2O]{1. O_3} 2CH_3-C\overset{\displaystyle O}{\underset{\displaystyle H}{\diagdown}}$$

(c)

$$\begin{array}{c} H_2C \diagdown H \\ C \\ | \\ C \\ H_2C \diagup H \end{array} \longrightarrow CH_2=CH_2 \quad \bigcirc$$

(d)

$$2\left(\begin{array}{c} H \diagdown H \\ C \\ \| \\ O \end{array}\right)_{aq} \xrightarrow{h\nu} CH_3OH + HCOOH$$

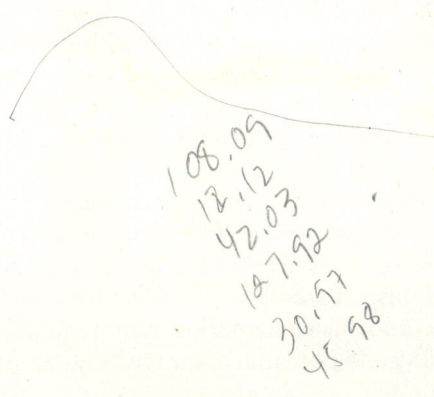

Problems 211

CHAPTER 8

Infrared and Raman Spectroscopy

James R. Kincaid

Infrared and Raman spectroscopy are methods used to investigate molecules in the solid, liquid, and gaseous states. Although both methods provide information about molecular vibrations and rotations (gases), and about molecular structure, there is a fundamental difference between the two techniques with respect to the mechanism of interaction of radiation with the molecular species. Infrared spectroscopy is a particular type of *absorption* spectroscopy, whereas the Raman spectrum arises via the inelastic scattering of photons by molecules of the sample. Both the structure and the electronic distribution of the molecule determine the intensity of a vibrational transition for each technique. In this sense, the methods may be considered to be complementary, and in some cases a combination of both may prove to be especially useful.

There is a high degree of selectivity inherent in these methods because the vibrational spectrum of a molecule is, in general, unique. On the other hand, both methods, but especially Raman spectroscopy, have suffered the disadvantage of relatively low sensitivity. However, developments in instrumentation and sampling methods have greatly improved the situation, resulting in a substantial increase in quantitative analytical applications.

Traditionally, infrared and Raman spectroscopy have been used by physical chemists to obtain detailed structural and bonding information for many types of molecules; several excellent treatments of this aspect are listed in the Selected Bibliography. The purpose of this chapter is to provide a sufficiently detailed description of the theory, principles, and instrumentation requirements to enable the reader to appreciate fully the utility and potential of the techniques for qualitative and quantitative analytical applications.

8.1 MOLECULAR VIBRATIONS

Atoms are never fixed in space but move about continuously. Each atom may be said to possess three degrees of freedom of movement, and thus in an *N*-atom molecule there will be 3*N* degrees of freedom. However, in molecules, movements of the atoms are constrained by interactions through chemical bonds, and the atomic motions may be classified to take into account these restrictions.

The first type of motion—*translation*—corresponds to the movement of the entire molecule through space while the positions of the atoms relative to each other remain fixed. In this sense, the molecule may be considered as a single particle with the mass of the molecule located at its center of gravity and, as a whole, possesses three degrees of *translational freedom*. The translational energy is independent of the energy associated with intramolecular interactions.

Molecules may also undergo *rotational* transitions, in which the interatomic distances remain constant but the entire molecule rotates with respect to three mutually perpendicular axes that pass through its center of mass. Thus, polyatomic molecules generally have three degrees of *rotational freedom*. In the special case of a linear molecule, all the atoms lie on a straight line and only two rotations can be defined.

All other movements of the atoms in a molecule are classified as *vibrations*: That is, the relative positions of the atoms change while the average position and orientation of the molecule remain fixed. Thus, an *N*-atom polyatomic molecule has $3N - 6$ vibrations, or $3N - 5$ if it is a linear molecule. These vibrations can be divided into two main categories: *stretching* and *bending*. The bending vibrations for commonly encountered organic molecules are often further subdivided into *scissoring*, *rocking*, *wagging*, and *twisting* as shown in Figure 8.1.

It is instructive to calculate the energy of a molecular vibration, at least for the simple case of a diatomic molecule, which has only one vibration: stretching of the bond connecting the two atoms. Using the harmonic oscillator model, the energy (in cm^{-1}) (see Chap. 6) associated with the stretching vibration of a diatomic molecule whose atoms have masses m_1 and m_2 can be shown to be

$$\bar{v} = \frac{1}{2\pi c}\sqrt{\frac{k}{\mu}} \qquad (8.1)$$

where $\bar{v}$ = wavenumber of the vibrational band
c = velocity of light
k = force constant of the bond
μ = reduced mass of the atoms involved

The reduced mass is defined by

$$\mu = \frac{m_1 m_2}{m_1 + m_2} \qquad (8.2)$$

where m_1 = mass of the first atom
m_2 = mass of the second atom

For polyatomic molecules, certain groupings may be considered to provide reasonably reliable *group frequencies*. For example, the frequencies associated with

Scissoring   Rocking

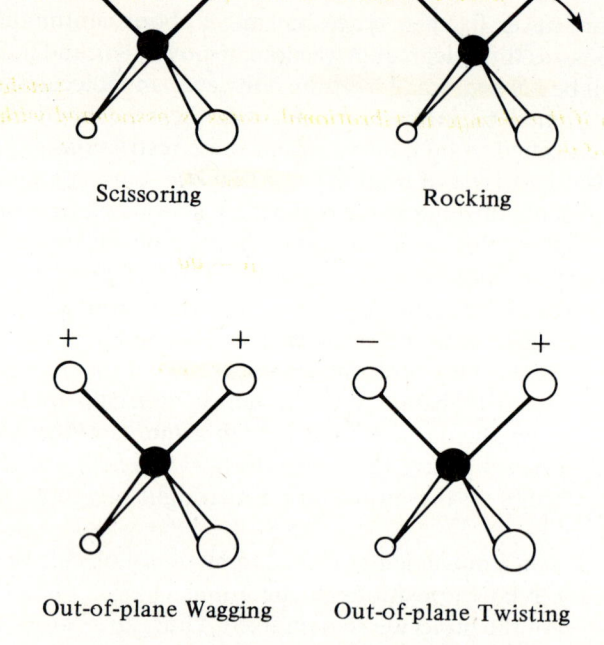

Out-of-plane Wagging   Out-of-plane Twisting

FIGURE 8.1. *Some bending vibrations of organic compounds. The + and − symbols indicate movement of the atom out of and into the plane of the page.*

C=O stretching vibrations of ketones are observed between 1700 and 1800 cm^{-1}, whereas that of the C≡C stretching mode of an alkyne occurs around 2200 cm^{-1}. Actually, these group frequencies are not invariant and may change as a result of interactions of the group in question with atoms associated with it. The result of this *vibrational coupling* is to spread out the group frequency into a range of frequencies. The frequency of the vibration associated with this group generally occurs within this range. In fact, although these interactions weaken the desirable simplicity of the group-frequency concept, they are ultimately responsible for producing the unique features in the spectra of individual compounds. It is this "fingerprint" concept that makes infrared and Raman techniques so useful for unambiguous identification of molecular substances. These group-frequency ranges have been compiled for a large number of molecules, and relatively reliable *correlation charts* have been devised. We will return to this matter later in discussing qualitative applications of infrared and Raman spectroscopy.

Selection Rules

Although both infrared and Raman spectroscopy provide information on vibrational frequency, the distribution of electronic charge and the directional behavior of atomic displacements will determine the extent to which a vibration is *active* in the infrared or Raman spectrum.

Infrared Absorption. Infrared spectroscopy involves absorption of electromagnetic radiation in the infrared region of the spectrum, normally 4000 to 200 cm^{-1}. The energy associated with a quantum of light may be transferred to the molecule if work can be performed on the molecule in the form of displacement of charge. This requirement gives rise to the selection rule for infrared activity: *A molecule will absorb infrared radiation if the change in vibrational states is associated with a change in the dipole moment of the molecule.*

For any given molecule, certain vibrations may be infrared active, while others may not. The electric dipole moment, $\bar{\mu}$, is a vector quantity:

$$\bar{\mu} = q\bar{d} \tag{8.3}$$

where q = electric charge
  $\bar{d}$ = directed distance of that charge from some defined origin of coordinates for the molecule

As the molecule vibrates, the charge distribution may or may not change with respect to that origin, depending on the relative displacements of atoms. Only the vibrations that cause the electric dipole moment to change will be associated with an absorption of infrared radiation.

Raman Scattering. The Raman effect represents an entirely different approach to the study of vibrational and rotational transitions of molecules. If an intense beam of monochromatic light, such as a laser beam, is focused on a transparent sample, most of the light is transmitted. A small fraction will be scattered in all directions by molecules of the substance (*Rayleigh* or "*elastic*" *scattering*) or by small particles such as dust (*Tyndall scattering*). In these cases, the scattered light has the same frequency (energy) as the incident beam. In addition to these scattering processes, a third type of scattering produces light that is shifted in frequency by an increment of energy corresponding to a natural transition of the molecular species present. This effect was first predicted on the basis of theoretical considerations by Smekal in 1923. It was observed in 1928 by C. V. Raman, who was awarded the Nobel Prize in 1931 for the discovery of the effect, which was later named after him.

Although a thorough description of the details of the Raman effect requires a full quantum-mechanical treatment, some insight into the scattering process may be gained from consideration of a simplified but reasonable model of the phenomenon. Referring to Figure 8.2, consider the interaction of a photon with an isolated molecule. The photon, of energy ($h\nu_L$), may be considered to perturb the electronic configuration of the molecule. This is reasonable because electrons can respond to the incoming photon. Such a perturbation may also be described as the creation of some state in which the wavefunctions of the perturbed molecule become some linear combination of all possible wavefunctions of the isolated molecule. For some infinitesimally small period of time, the photon and molecule may be considered to constitute a new state, sometimes called a *virtual state*. The molecule has reached a new, nonstable, state via interaction with the photon; this state is higher in energy by the perturbation energy ($h\nu_L$) associated with the photon. The new state represents an unstable situation and the photon is immediately emitted, while the molecule returns to one of the states associated with the unperturbed molecule. There is an extremely high probability

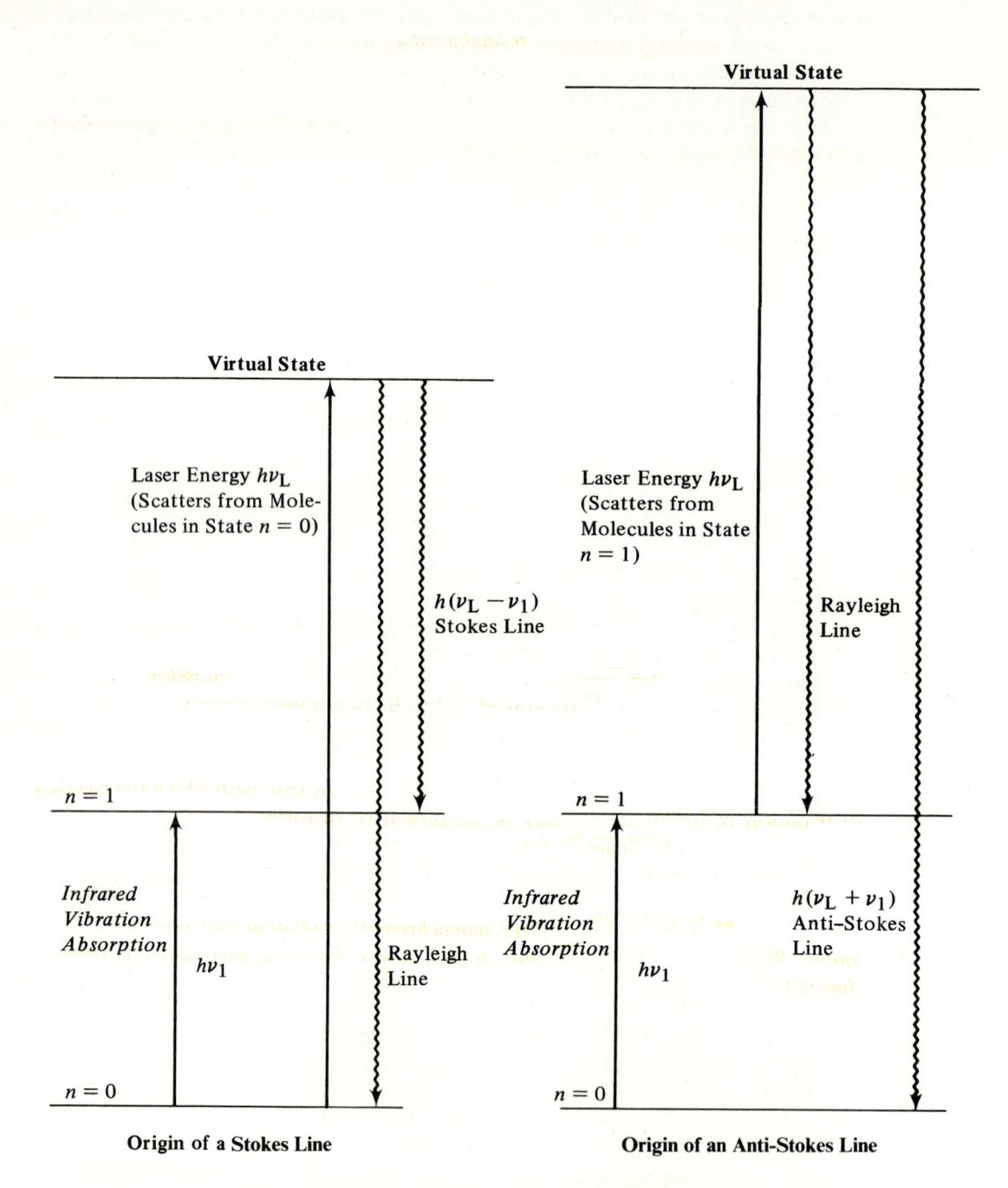

FIGURE 8.2. *Origins of Stokes and anti-Stokes Raman lines.*

that the molecule will revert to its ground state, in which case the photon regains its initial energy, $h\nu_L$. This process is Rayleigh scattering.

With a much lower probability, the molecule can relax to one of the unperturbed states other than the ground state, for example to one of the excited vibrational or rotational states. In this case, the photon loses a fraction of its energy to the molecule, with the loss corresponding to the energy associated with a vibrational or rotational transition of the sample, $h\nu_1$. When the photon is scattered, it is then of lower energy, $h\nu_L - h\nu_1$. Much less frequently, an incident photon encounters a molecule residing in an excited vibrational state. In this case, the photon can gain energy from the molecule and be scattered with an energy higher than that it had originally by an amount equal to the vibrational energy. The scattered photon has energy ($h\nu_L + h\nu_1$). Because the energy of the incident photon is changed in the Raman process, it is sometimes called "*inelastic*" *scattering*.

For historical reasons, the Raman bands that occur at lower frequencies (i.e., lower energy) are called *Stokes lines*, whereas those that occur at higher frequencies are called *anti-Stokes lines*. In both cases, the shifts from the excitation frequency are equal, and are equal to the frequency of a particular vibrational mode of the molecule, ν_1. Because the population of the molecular energy levels follows a Boltzmann distribution, the Stokes lines are stronger than the anti-Stokes lines, the latter diminishing rapidly in intensity for higher frequency modes.

The scattering phenomenon should not be considered as an absorption-emission process in the usual sense. In a normal absorption process, the transition terminates in one of the excited states of the unperturbed molecule which has an associated finite lifetime. When considering the scattering process, it is important to emphasize that the "absorption" of the incident photon and "emission" of the inelastically scattered photon are considered to be simultaneous events.

Let us now consider the creation of this so-called virtual state which has been described as a perturbation of the molecular electronic configuration by the incident photon. The efficiency of this interaction is related to the ease with which the electron distribution of the molecule may be perturbed or distorted. This property of the molecule is called the *polarizability*, α. If the polarizability of the molecule is changed during a particular molecular vibration, then that vibration can be effective in generating a Raman-shifted photon upon interaction with an incident photon of energy $h\nu_L$. A molecule will scatter monochromatic radiation and produce Raman-shifted lines if the molecule vibrates in such a way that its polarizability is changed during the vibration.

Summary of Selection Rules. The charge distribution of a molecule may be such that the dipole moment, $\bar{\mu}$, changes during a vibration (giving rise to infrared absorption) while the polarizability, α, does not change. Conversely, α may change (giving rise to a Raman-shifted line) while $\bar{\mu}$ does not. Thus, the quantum-mechanical selection rules differ for infrared and Raman spectroscopy, and the techniques may be considered to be complementary.

As an example, consider the symmetric and antisymmetric stretching modes of carbon dioxide, CO_2, depicted as follows:

$$\overset{\leftarrow}{O}=C=\vec{O} \qquad \vec{O}=\overset{\leftarrow}{C}=\vec{O}$$
symmetric   *antisymmetric*

The arrows represent the direction of motion of the atoms during one half-cycle of vibrations. At the point of maximum displacement, the directions of motion of the atoms reverse. The center of mass of the molecule remains fixed in space.

During one half-cycle of the antisymmetric stretch, the dipole moment is displaced in one direction. During the next half-cycle, it is displaced in the opposite direction. The result is that the dipole moment oscillates with the frequency of the vibration. This gives rise to an infrared absorption band, and this vibration is said to be *infrared active*. For the symmetric stretching mode, the two oxygen atoms move in opposite directions during each half-cycle of the vibration. This system produces no net change in dipole moment, and the vibration is said to be *infrared inactive*.

The alterations in the internuclear distance will obviously affect the ability of the bond to be polarized; that is, there will be a change in polarizability. Thus, the symmetric stretching mode, wherein both bonds are being stretched simultaneously, will give rise to a Raman line.

The electric-field vector of a Raman line may correspond to partial or total polarization of the electromagnetic radiation in that line, depending upon the degree of symmetry of the molecular motion(s) that spawn the line. Thus, the symmetry of a particular vibrational mode may be inferred from the polarization (or depolarization) of a Raman band.

The polarization of the light in a Raman line is obtained experimentally by measuring a quantity called the *depolarization ratio*, ρ. It is defined by a ratio of the band areas observed when the light is passed through a polarizer turned first perpendicular ($\perp$), then parallel ($\parallel$), to the scattered light. The expression that results is

$$\rho = \frac{3\bar{\beta}^2}{45\bar{\alpha}^2 + 4\bar{\beta}^2} = \frac{I_\perp}{I_\parallel} \tag{8.4}$$

where $\bar{\alpha}$ = isotropic part of the polarizability
$\bar{\beta}$ = anisotropic part of the polarizability

The quantity ρ has no infrared counterpart and is usually reported along with the Raman shift. If $\rho = 0$ ($\bar{\beta} = 0$), then polarized radiation is scattered by a totally symmetric mode of vibration. If $\rho = 3/4$ ($\bar{\alpha} = 0$), then depolarized radiation is scattered by the vibration. When the value of the depolarization ratio is less than 0.2, the $\bar{\alpha}$ term is dominating, but if ρ is greater than 0.2, the $\bar{\beta}$ term is becoming more important and there is some degree of asymmetry in the vibrational mode.

The depolarization ratio is measured by first placing a polarization analyzer between the sample and the entrance slit of the monochromator; ρ is obtained by finding the ratio of the band peak heights when the polarizer is rotated perpendicular and parallel to the scattered radiation coming from the sample. The Raman spectrometer should be standardized first by measuring the polarized 459-cm^{-1} bands of carbon tetrachloride ($\rho = 0.005 \pm 0.002$) and the depolarized band of carbon tetrachloride at 218 cm^{-1} ($\rho = 0.75$).

Depolarization ratios are not obtainable directly on finely ground crystals or powders because refractions and reflections scramble the polarization; however, one may measure depolarization ratios of polymer films, liquids, and gases.

It is often the case for any particular molecule that certain bands will be strong in the infrared while weak in the Raman spectrum and vice versa. The selection rules provide information about which modes of vibration are infrared or Raman active. Using methods based on quantum mechanics and group theory, spectroscopists have used infrared and Raman spectroscopy to effectively investigate molecular structure by correlating an assumed geometric structure and the associated selection rules with the observed vibrational spectral patterns. Several sources dealing with these aspects of the methods are listed in the Selected Bibliography.

Resonance Raman Spectroscopy

Up to this point, the discussion has focused on what may be called the normal Raman effect; that is, we have assumed that the incident frequency is far removed from any electronic transition of the molecular species. Such scattering has been shown to be an extremely inefficient process, and the observed Raman bands are very weak. In fact, it is usually necessary to study "neat" substances (powders or liquids) or relatively concentrated solutions (approx. 0.1 M). This low sensitivity is an obvious disadvantage of Raman spectroscopy from the viewpoint of an analytical chemist. However, under certain conditions it is possible to achieve a remarkable increase in the efficiency of the scattering process and to observe dramatic enhancements of Raman bands.

If the frequency of the incident laser beam is chosen to coincide with or to approach the frequency of an electronic absorption band of the substance, the perturbation process whereby a photon interacts with the molecule may be greatly increased in efficiency. Enhancements of up to six orders of magnitude are not uncommon. This permits the study of molecular species at concentrations of 10^{-6} M and lower.

This resonance effect is clearly demonstrated by the spectra shown in Figure 8.3. This figure shows the Raman spectra obtained for a 0.5-mM solution of the

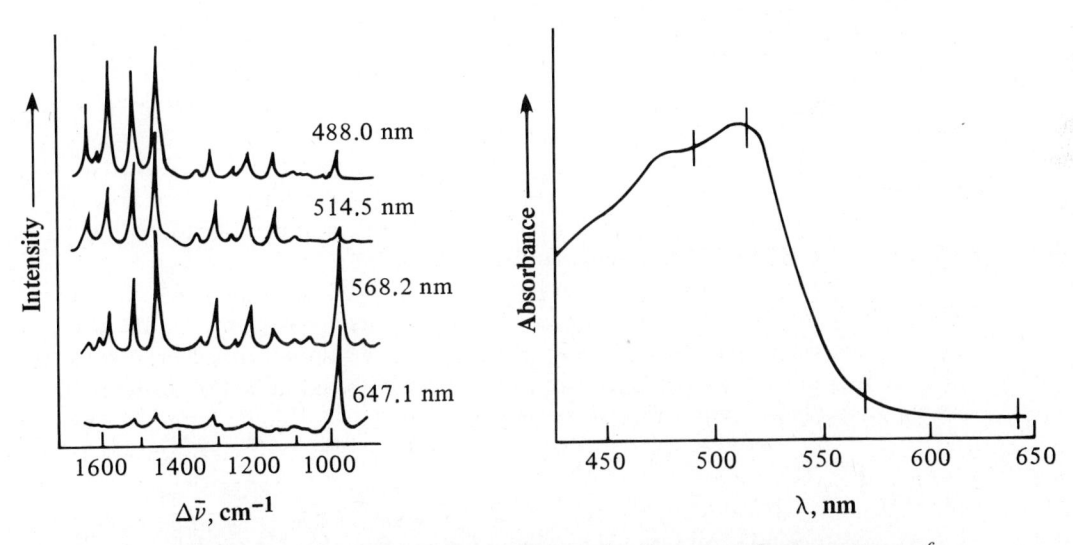

FIGURE 8.3. *Resonance Raman effect. Left: Resonance Raman spectra of 0.5 mM ferrous ortho-phenanthroline using laser lines of different wavelengths. Right: Conventional absorption spectrum of the complex; the positions of the wavelengths used for the resonance Raman spectra are indicated.*

SEC. 8.1 Molecular Vibrations

familiar complex of 1,10-phenanthroline with ferrous ion. The solution also contains 0.5 M sulfate ion as an internal standard which does not exhibit resonance enhancement because the incident light is not near a sulfate absorption band. The complex possesses a strong metal-ligand charge-transfer absorption band centered at about 510 nm, having a molar absorptivity of around 11,000 L/mole-cm. The Raman spectra were obtained using several lines from argon-ion and krypton-ion lasers. The bottom spectrum was obtained with the 647.1-nm line from a Kr^+ laser, which is relatively far removed from the absorption band of the complex. This exhibits a strong band at 983 cm^{-1}, which is associated with the sulfate ion. On the other hand, peaks that may be ascribed to Raman bands of the complex are relatively very weak. The spectrum obtained with the 568.2-nm line from a Kr^+ laser, which lies within the electronic absorption band of the complex, exhibits relatively intense lines which are associated with the complex, as well as with the intense sulfate band. With 514.5-nm excitation from an Ar^+ laser, which is nearly coincident with the maximum in the electronic absorption band of the complex, the resonance-enhanced bands of the complex are much stronger than the band of the sulfate ion, even though the sulfate-ion concentration is nearly 1000 times greater. In this case, an enhancement factor of greater than 10^4 has been observed with change in the incident wavelength. The enhancement at 488.0 nm decreases slightly relative to that at 514.5 nm because this frequency is not on the absorption maximum but instead lies on a high-frequency shoulder.

In addition to this valuable sensitivity enhancement, the resonance Raman effect also provides useful selectivity advantages. In agreement with theory, only those vibrational modes associated with the chromophoric unit are enhanced. All other vibrational modes retain the inherent inefficiency of the normal Raman scattering process. Thus, in systems such as biologically important enzymes and proteins, which may contain a chromophoric group at the active site, the resonance Raman effect may be exploited to obtain vibrational spectral information about the chromophore without interference from the large number of vibrations associated with the protein backbone and side chains. Such selectivity has been most effectively demonstrated in the case of heme proteins, where the observed Raman bands are associated only with the vibrational modes of the tetrapyrrole macrocycle.

8.2 INSTRUMENTATION FOR RAMAN SPECTROSCOPY

The instrumentation for Raman spectroscopy is conceptually very simple since all that is required is to analyze the light scattered by the sample. However, because the observed signals are so weak relative to the incident beam (a factor of about 10^{-7}), very high quality components are needed.

Lasers

Lasers now provide the intense sources of monochromatic radiation essential for obtaining good-quality Raman spectra. In fact, the development of lasers in the early sixties was the major factor responsible for the renewed interest in Raman spectroscopy. Furthermore, because lasers have made such an impact on so many areas of spectroscopy, it is important to have at least an elementary understanding of the basic principles of laser operation and some knowledge of the most common lasers in use at this time.

The acronym *laser* is derived from the concept of *l*ight *a*mplification by *s*timulated *e*mission of *r*adiation. To understand the principle of laser action, it is necessary to consider the process of stimulated emission and the manner in which this process can be exploited to attain light amplification.

The three basic processes involving absorption or emission of electromagnetic radiation are depicted in Figure 8.4. Absorption, depicted in Figure 8.4A, can occur if the energy of the radiation exactly matches the energy difference between two quantized (electronic, vibrational, or rotational) states (E_1 and E_2) of the particles that serve as the medium. The resonant photon is absorbed and an excited state is generated. The extent of absorption by a medium is also controlled by both the probability of interaction of the photon and atoms or molecules of the medium, and the probability that such interactions will produce an excited state. In other words, the rate of photon absorption is proportional to the product of photon density

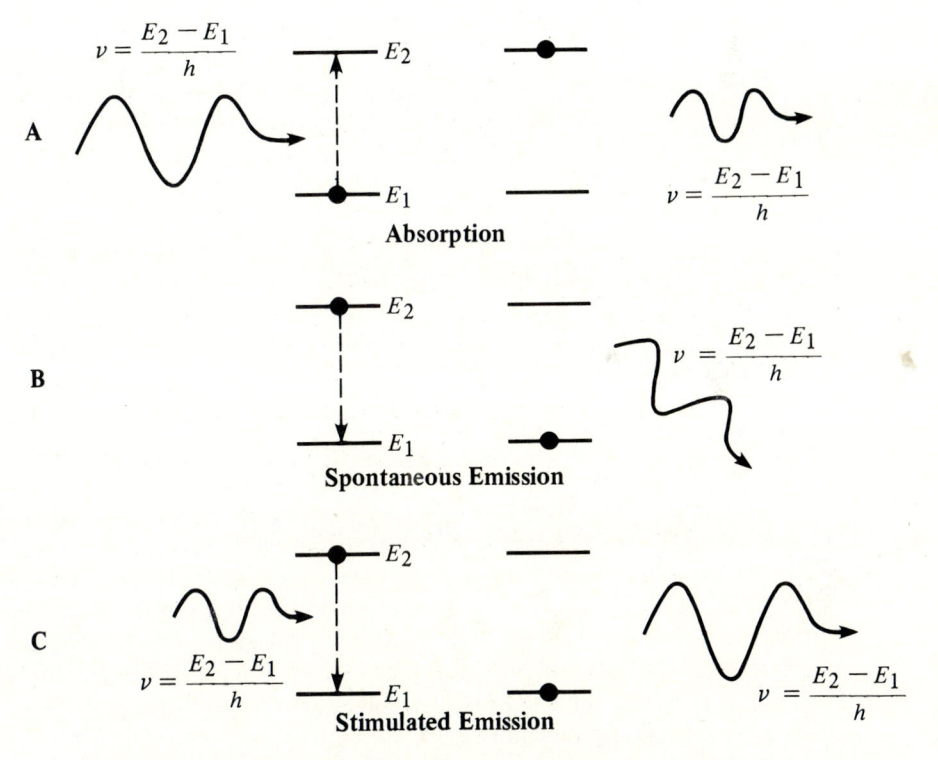

FIGURE 8.4. *Interactions of electromagnetic radiation with matter. Absorption of electromagnetic radiation of energy hν (A) is just equal to the energy difference between the two electronic states. Note that, after absorption, the intensity (vertical amplitude) of the radiation is diminished, but the frequency is unchanged. In stimulated emission (C), the amplitude of the radiation increases after interaction.*

SEC. 8.2 Instrumentation for Raman Spectroscopy

$[P(v)]$ and the number of species in state E_1 (N_1). The proportionality constant is the Einstein coefficient of absorption, B_{12}, which is proportional to the quantum-mechanical probability for the transition:

$$\text{Rate of Absorption} = B_{12}P(v)N_1 \qquad (8.5)$$

Neglecting nonradiative processes, the excited-state species may relax spontaneously by emitting a resonant photon $[v = (E_2 - E_1)/h]$. This process of *spontaneous emission*, which is shown in Figure 8.4B, is a random process, and the fluorescent photon produced by one particle of the medium will vary in the direction of propagation and time of emission from that of another particle. Radiation arising by spontaneous emission is therefore said to be *incoherent*. The rate of spontaneous emission is proportional to the number of species in state E_2 (N_2). The proportionality constant is now the Einstein coefficient for spontaneous emission, A_{21}, which is proportional to the quantum-mechanical probability associated with the transition $E_1 \leftarrow E_2$:

$$\text{Rate of Spontaneous Emission} = A_{21}N_2 \qquad (8.6)$$

Einstein recognized that, just as a resonant photon can interact with a ground-state species to produce an excited state, so too can a resonant photon interact with the excited-state species to produce a ground state. Thus, as shown in Figure 8.4C, in the process of *stimulated emission* the incoming photon stimulates the system to relax with the *simultaneous* emission of a second resonant photon. This second photon is emitted precisely in phase (temporal coherence) and is codirectional (spatial coherence) with the incoming photon. Thus, laser radiation is said to be *coherent*. In a development similar to that used to describe the rate of absorption, the rate of photon production by stimulated emission is proportional to the product of the radiation density $[P(v)]$ and the number of species in state E_2 (N_2). In this case, the proportionality constant is the Einstein coefficient for stimulated emission, B_{21}, which is proportional to the quantum-mechanical probability for the $E_1 \leftarrow E_2$ transition (in the presence of a radiation field) and is *equal in magnitude to B_{12}*:

$$\text{Rate of Stimulated Emission} = B_{12}P(v)N_2 \qquad (8.7)$$

With these facts in mind, it is now possible to examine the consequences of these processes for laser operation. In order to achieve light amplification, it is obvious that the number of photons generated by stimulated emission must exceed the number lost by absorption. Because the two relevant Einstein coefficients (B_{12} and B_{21}) are equal in magnitude, it is clear that achieving a net gain in photon density requires maintaining a larger number of components in state E_2 than are present in state E_1; that is, $N_2 > N_1$. In other words, it is necessary to produce a *population inversion* from that which would normally exist.

Although, for the sake of simplicity, the discussion thus far has considered absorption and emission processes for a simple two-level system, it is important to realize that such a simple system cannot give rise to laser action. This is so because, starting from a normal population where $N_1 \gg N_2$, as soon as the system reaches the condition in which $N_1 = N_2$, the number of photons generated by stimulated emission is equal to the number lost by absorption and the system is transparent or saturated. Thus, two-level systems cannot be used, and various multilevel schemes must be exploited.

Several such "pumping schemes" effective in lasing action are illustrated in Figure 8.5. In one three-level system (Fig. 8.5A), the system is electronically or optically pumped from state E_1 to state E_2, followed by a rapid nonradiative decay to state E_3. This is the crucial process in a three-level scheme because this decay establishes the population inversion between E_3 and E_1 and helps deplete E_2. Since N_3 must be greater than N_1 for the lasing process, it is necessary that E_3 have a fairly long lifetime or that an extremely powerful pumping mechanism be available.

The second three-level scheme shown (Fig. 8.5B) represents a more efficient alternative. In this scheme, the transition responsible for lasing action is between levels E_2 and E_3. It is much easier to attain population inversion in this scheme because the population in E_3 is essentially zero before pumping action starts. Laser action can be maintained as long as the nonradiative transition from E_3 to E_1 is not so slow as to build up level E_3 and thus destroy the population inversion.

The four-level scheme (Fig. 8.5C) represents a very effective method. Because the terminal level of the lasing transition (level E_4) is not the ground state, it is necessary only to pump sufficiently to generate a larger number of particles in state E_3 than are in state E_4. The most efficient systems of this type have rapid radiationless decay from E_4 to E_1. Thus, any practical value of N_3 represents a population inversion. In other words, a population inversion in a four-level laser can be achieved at a lower pumping power than in three-level lasers.

There are two common methods now in use for pumping the active medium. Liquid lasers and solid-state lasers can normally be optically pumped with suitable lamps. Only the part of the light that falls within the absorption band of the *pump transition* can be used. Thus, pumping efficiency improves with increasing bandwidth of the pump transition. Because absorption bandwidths of gases are very narrow at low pressures, optical pumping would be inefficient and gas lasers are electrically pumped.

Thus far, we have established that in order for laser action to be achieved, a population inversion must be maintained by some pumping scheme so that gains

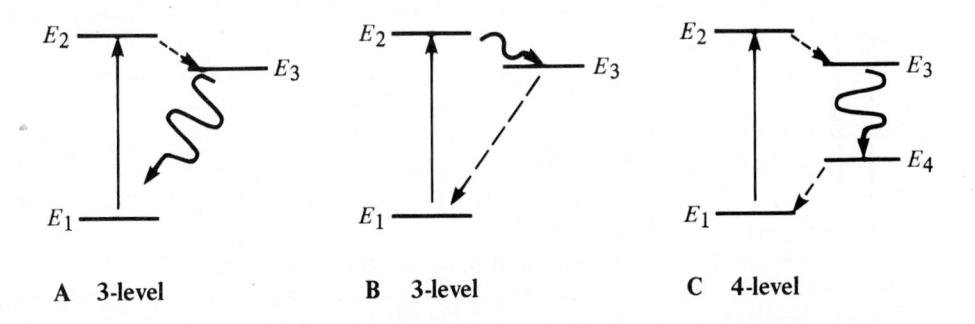

FIGURE 8.5. *Multilevel pumping schemes. Solid lines represent pump transitions, dotted lines represent rapid radiationless decay, and wavy lines represent lasing transitions.*

from stimulated emission can overcome losses by absorption and spontaneous emission. In practice, a third concept is normally utilized—that of the *optical resonator*. The components of a typical laser are shown schematically in Figure 8.6. The optical resonator component is composed of a pair of mirrors that cause the radiation produced by the laser process to pass back and forth through the medium a number of times. Because additional photons are produced with each passage, high amplification is possible. One of the mirrors is only partially reflecting. Thus, a small fraction of the intracavity power is allowed to escape the cavity to form a usable laser beam. Such a design gives rise to a highly directional laser beam because light that is emitted in a nonparallel direction is discriminated against by the stimulated emission process in that greatest intracavity light intensity is parallel to the axis of the resonator.

A large number of molecular and atomic species have been shown to possess the capability for laser action. However, an instructive survey of practical lasers can be given by considering several types most commonly used at this time.

The ionic-crystal lasers, which use absorbing ions in a host crystal lattice, are optically pumped. There are two common examples. The ruby laser, which was the first laser developed, has Cr^{3+} ions substituted (approx. 0.05% by weight) for Al^{3+} sites in a crystal of Al_2O_3. This is a typical three-level device and is usually operated in a *pulsed regime* by surrounding a machined rod (several millimeters in diameter by several centimeters in length) with a low-pressure Xe flash lamp. The output characteristics are given in Table 8.1. The most commonly used solid-state lasers are based on the neodymium ion in different host crystals: Nd:YAG (yttrium aluminum garnet host crystal) or Nd:glass. These are examples of typical four-level pumping schemes. The Nd:YAG is usually operated in the pulsed regime using Xe flash lamps surrounding the crystalline rod. The output characteristics are given in Table 8.1.

Gas lasers (Table 8.1), which are electrically pumped, can be classified into four types:

1. Neutral atom lasers such as He-Ne or He-Cd
2. Ion lasers such as Ar^+ or Kr^+

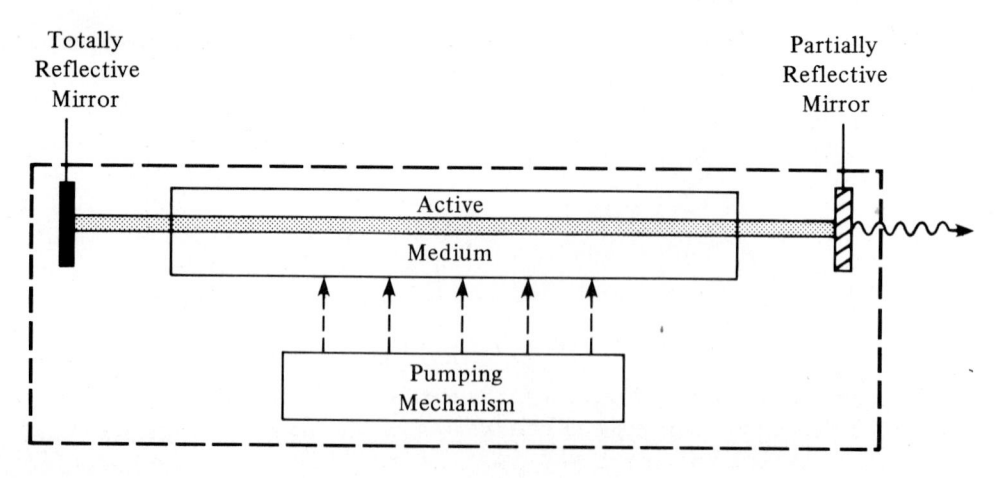

FIGURE 8.6. *Components of a typical laser.*

TABLE 8.1. *Characteristics of Common Lasers*

Laser	Wavelength, nm	Power, W
Ionic crystal		
Ruby[a]	694.3	1–10 MW
Nd:YAG[a]	1064.0	25 MW (8–9 ns)
Gas		
He-Ne	632.8	0.001–0.05
He-Cd	441.6	0.05
	325.0	0.01
Ar$^+$	514.5	7.5
	496.6	2.5
	488.0	6.0
	476.5	2.5
	465.8	7.0
	457.9	1.3
	333.6–363.8 (4 lines)	3.0
Kr$^+$	752.5	1.2
	647.1	3.5
	530.9	1.5
	482.5	0.4
	468.0	0.5
	413.1	1.8
	406.7	0.9
	337.5–356.4 (3 lines)	2.0
Nitrogen[a]	337.1	200 kW (300 ps)

a. Operated in pulsed mode; values given are peak power (pulse width).

3. Molecular lasers such as N$_2$ or CO$_2$
4. Eximer lasers

The He-Ne laser is the laser most commonly used; it is a continuous-output or continuous-wave (CW) device. The laser action occurs on Ne transitions. The He is added to increase pumping efficiency, since two metastable states of He can be electrically pumped, and can populate resonant excited states of Ne via *energy transfer*. Because the lifetimes of the upper states of the lasing transitions are longer than those of the lower states, all three of the lasing transitions can produce continuous laser action by a four-level scheme. A particular transition can be selected by using cavity mirrors that have maximum reflectivity for the wavelength region of interest. Most commercial systems operate on the red (632.8-nm) transition.

The most common ion lasers are the Ar$^+$ and Kr$^+$ lasers. For an ion laser, the pumping process requires two successive electron collisions (one for ionization and one for excitation). Thus, higher current densities are required than for neutral gas lasers. These high ion and current densities produce high temperatures. Therefore, effective cooling (water flow rates of about 3 L/min) and durable tube materials (graphite, beryllium oxide, or tungsten) are essential. Each type of ion laser will produce a number of usable lines. Selection is accomplished by including a prism within the intracavity space, thereby optimizing lasing on a particular transition, depending on the positioning of the prism.

Molecular lasers such as CO_2 and N_2 lase from excited vibrational or electronic states. The CO_2 laser, with $\lambda = 10.6$ μm, is the most powerful continuous-wave laser in operation today, finding many industrial applications. The N_2 laser can be operated only in the pulsed regime and, having an output at 337.1 nm, is very useful for pumping dye lasers.

Dye lasers use solutions of organic compounds that are capable of fluorescing in the ultraviolet, visible, or near-infrared regions. In this case, however, the lower level of the lasing transition is a superposition of many small vibrational energy states upon the lower electronic state. Thus, dye lasers can be tuned over a range of wavelengths of 20 to 50 nm. Tuning to chosen wavelengths with widths of 0.02 nm or less can be accomplished by including various dispersing devices (based on prisms, gratings, or filters) within the laser cavity. Dye lasers are pumped optically by flash lamps or, more commonly, by the pulsed or continuous-wave lasers mentioned above.

The Ar^+ and He-Ne are, by far, the most commonly used lasers for Raman spectroscopy. For resonance Raman studies, the argon-ion and dye lasers (pumped by high-powered argon-ion lasers) are preferred.

The output power of a particular laser can be expressed in a number of ways. For continuous-wave lasers, the power is normally expressed directly in watts. On the other hand, there are several different ways to express the output of a pulsed laser: peak power, average power, or energy per pulse. The energy per pulse is usually given in millijoules. To convert this to peak power, we simply divide by pulse width:

$$\text{Peak Power} = \frac{\text{Energy/Pulse}}{\text{Pulse Width}} \quad (8.8)$$

The average power is obtained by multiplying the energy per pulse by the repetition rate:

$$\text{Average Power} = (\text{Energy/Pulse})(\text{Pulses/Second}) \quad (8.9)$$

Example 8.1. A laser generates 10-ns, 200-mJ pulses at a rate of 100 pulses/sec. Using Equation 8.8, peak power is

$$\frac{0.20 \text{ J}}{10 \times 10^{-9} \text{ sec}} = 20 \times 10^6 \text{ W} = 20 \text{ MW}$$

and the average power is calculated using Equation 8.9:

$$(0.200 \text{ J/pulse})(100 \text{ pulses/sec}) = 20 \text{ W}$$

Sampling Devices for Raman Spectroscopy

Although several types of sampling systems have been devised for Raman spectroscopy, probably one of the most attractive features of this technique is that it may sometimes be possible to avoid special sample cells altogether. Thus, Raman spectra can be obtained directly using the glass containers in which samples are stored. In other cases, one can position a somewhat bulky or otherwise cumbersome solid

Capillary Tube
A

Pellet Spinner
B

Spinning Cell
C

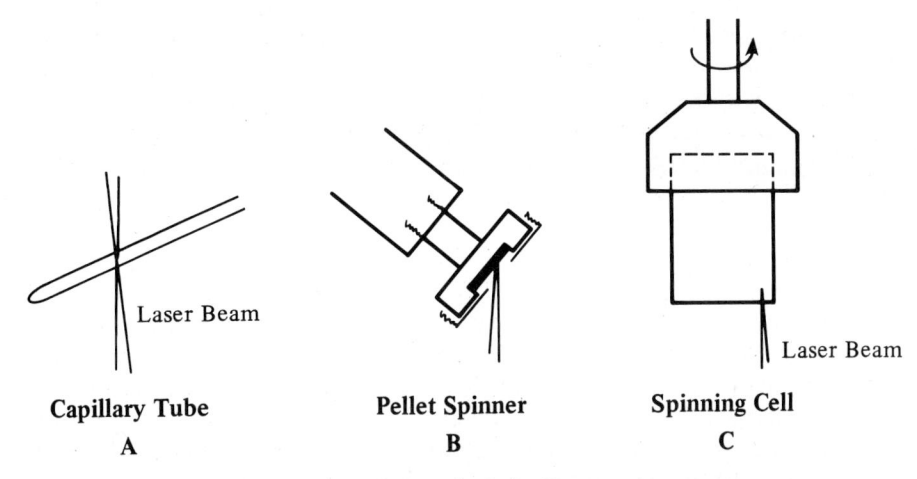

FIGURE 8.7. *Sampling methods for Raman spectroscopy.*

sample directly in the laser beam. A number of conventional and special-purpose sampling devices are shown in Figure 8.7. The most common method is to use an ordinary "melting-point" capillary tube that has been sealed at one end (Fig. 8.7A). Colorless liquids and powders can be analyzed in this way. For samples that can decompose in the laser beam by absorption of the incident light, for example in resonance Raman studies, it is necessary to prevent localized heating. Solids can be pressed into KBr pellets and rotated at high speeds (Fig. 8.7B). For liquids and solutions, one can use the spinning-cell technique (Fig. 8.7C) or circulate the solution through a capillary tube using a peristaltic or some other type of pump. By focusing the laser beam on individual spots, samples on thin-layer chromatography (TLC) plates have been analyzed directly.

Raman Spectrometers

The light scattered from the sample is focused by lenses or mirrors on the entrance slit of a high-quality monochromator. The principles and operation of grating monochromators were discussed in Chapter 6. Because most of the scattered light is at the frequency of the incident beam and only a very small portion is shifted in frequency, it is extremely important that monochromators used for Raman spectroscopy be capable of high stray-light rejection. For this reason, double monochromators, in which the light is passed sequentially through two high-resolution monochromators, are used. Although this design entails some loss in optical through-put, high-sensitivity detectors and *photon-counting* signal processing (Chap. 7) improve detection limits in modern instruments.

8.3 INSTRUMENTATION FOR INFRARED SPECTROSCOPY

Infrared spectroscopy is a type of absorption spectroscopy. Therefore, a dispersive-type infrared spectrophotometer will have the same basic components as the instruments used for the study of absorption of visible and ultraviolet radiation, although the sources, detectors, and materials used for the fabrication of optical elements will be different. Although high-quality dispersive instruments are now in use and will continue to be produced, the most important development in instrumentation for infrared spectroscopy has been increased accessibility of dedicated high-speed computers, which has led to the proliferation of Fourier transform infrared spectrometers.

Sources

Infrared sources are inert solids that are electrically heated to approximately 2000 K and produce intensity-versus-wavelength curves characteristic of blackbody radiation sources (Chap. 6). Intensity is highest at 5000 cm^{-1} and decreases gradually to only 1% of the maximum at 500 cm^{-1}.

The *Nernst glower* is probably the most widely used infrared source. It is usually a cylinder composed of rare-earth oxides that has a diameter of about 2 mm and length of approximately 20 mm. The operating temperature can be as high as 1800 K. It has a negative temperature coefficient of resistance, and thus it is necessary to initially heat this source externally in order to pass sufficient current to maintain a desired temperature. This same property requires that the electrical current be limited in some manner so that the source will not become so hot that it is ruined.

Another source, the *globar*, is a silicon carbide rod about 5 mm in diameter and 50 mm in length. It is operated at lower temperatures (around 1600 K) than the Nernst glower in order to avoid air oxidation. The globar provides a greater output than the Nernst glower in the region below 1500 cm^{-1}.

A third infrared source is a tightly wound coil of *nichrome* wire which is electrically heated to incandescence. This source is of lower intensity in the infrared region than the previous two, but has a longer life.

The generally low intensity of all of these sources is one of the major problems of infrared spectroscopy. In this sense, infrared spectroscopy is *energy limited*. The transform methods discussed later are better suited to this type of situation.

Detectors

Because the available infrared sources are generally of low intensity and the energy of an infrared photon is relatively low, the detection of infrared radiation is more difficult than is the case in the ultraviolet and visible regions. The common phototubes discussed in Chapter 6 are not useful in the infrared region because the photons are not sufficiently energetic to cause photoemission of electrons.

The two general classes of infrared detectors now in use are: (a) *photon detectors*, which are based on the photoconductive effect that occurs in certain semiconductor materials; and (b) *thermal detectors*, in which absorption of infrared radiation produces a heating effect, which in turn alters a physical property of the detector, such as its resistance. In general, quantum detectors have a much faster response and a greater sensitivity to infrared radiation than do thermal detectors; but the former can operate only over a very restricted range of wavelengths because there is a limited range of photon energies that will excite electrons in bound states to the conduction band of a semiconductor. Thermal detectors, on the other hand, are usable over a very wide wavelength range: Essentially, all that is necessary is that the detector absorb a photon; no specific electronic transitions have to occur.

Photon detectors consist of a thin film of semiconductor material, such as lead sulfide, lead telluride, indium antimonide, or germanium doped with copper or mercury, deposited onto a nonconducting glass and sealed into an evacuated envelope to protect the semiconductor from reaction with the atmosphere. Absorption of photons of sufficient energy by the semiconductor material promotes some of the electrons in the bound nonconducting state to the conducting state, resulting in a decrease in the resistance of the material. The excitation of these bound electrons requires a photon having an energy above a certain minimum value. These detectors therefore have definite cutoff points toward the far-infrared. Lead sulfide detectors are sensitive to radiation between 1 and 3 μm in wavelength (10,000 to 2000 cm^{-1}) and have a response time of about 10 μsec. Detectors based on other materials, when cooled to liquid nitrogen or liquid helium temperatures, extend sensitivity to considerably longer wavelengths and have response times as fast as 20 ns.

Thermal detectors may be classified into four types based on the properties of the material that are altered upon exposure of the material to infrared radiation. The *thermocouple*, the most widely used infrared detector, is composed of a small piece of blackened gold foil (the surface for absorbing the incident radiation) welded to two fine wires made of dissimilar metals. A small voltage develops between the two thermocouple junctions, the magnitude of which is dependent upon the difference in temperature of the junctions. One of the junctions (the reference junction) is bonded to a heat sink and carefully shielded from the incident radiation, and thus remains at a relatively stable temperature. Because the incident radiation is chopped, only the temperature change of the thermocouple is important. To minimize conductive heat loss, the entire assembly is sealed in an evacuated housing having an infrared-transmitting window. A *thermopile* is the name given to a detector comprising several thermocouples connected in series so that their outputs are additive. These detectors have response times of about 100 msec and a relatively flat frequency response.

A second type of detector, the *thermistor* or *bolometer*, exhibits a change in resistance when illuminated by infrared radiation. Two matched sensing elements are used as two arms of a Wheatstone bridge (Chap. 5), one of which is shielded from the infrared radiation, the other directly exposed and the surface coated to increase absorption. A temperature difference between the two elements produces a proportional voltage difference. A thermistor is constructed with oxides of metals such as cobalt or nickel that have high temperature coefficients of resistance (4 to 7% per °C). A bolometer uses a fine platinum wire with a temperature coefficient of resistance of about 0.4% per °C.

The *Golay* or "*pneumatic*" *detector* is based on the increase in pressure with temperature of a confined inert gas. Infrared radiation is absorbed by a rigid blackened metal plate sealed to one end of a small metallic cylinder. The heat is transmitted to the gas, which expands and causes a flexible silvered diaphragm affixed to the other end of the tube to bulge outward. The distortion of the thin diaphragm can be measured either by making it part of an optical system in which a light beam reflects from it to a phototube, or by making it one plate of a dynamic parallel-plate capacitor: The distortion of the flexible diaphragm relative to a fixed plate changes the average plate separation and thus the capacitance. The Golay detector has a sensitivity approximately equal to that of a thermocouple for near- and middle-infrared radiation and is not often used for this region. It is superior for the region below 200 cm^{-1} and is useful for instruments designed for the far-infrared region.

The most recently developed infrared detector is the *pyroelectric detector*. Certain crystals such as triglycine sulfate (TGS), deuterated triglycine sulfate (DTGS), lithium tantalate, and some others, possess an internal electric polarization along an axis resulting from alignment of electric dipole moments. Thermal alteration of the lattice spacing caused by absorption of infrared radiation results in a change of the electric polarization. If placed between electrodes consisting of metal plates connected through an external circuit, current will flow in the circuit to balance this charge redistribution. The pyroelectric effect depends on the rate of change of temperature and not on the absolute value. It therefore responds only to modulated radiation and not to slowly varying background radiation. Thus, the pyroelectric detector also operates with a much faster response time and is widely used for Fourier transform infrared spectroscopy.

Dispersive Infrared Spectrometers

By far the most common use of infrared spectroscopy is for the qualitative identification of compounds. Thus, because of the complexity of infrared spectra, most commercial instruments are of the double-beam recording type which cancels background absorption caused by atmospheric gases such as CO_2 and H_2O. A schematic diagram of such an instrument is shown in Figure 8.8.

The radiation from the source is split into two beams, one passing through the sample compartment and the other through the reference compartment. The rotating sector mirror M7 alternately reflects the sample beam or transmits the reference beam onto mirror M9 and, via mirror M10, through the entrance slit S1 into the monochromator. Thus, the sample beam and the reference beam travel through the monochromator in alternate pulses. Each pulse is dispersed by grating G1 (or G2), transmitted through exit slit S2 and focused on the detector (a thermo-

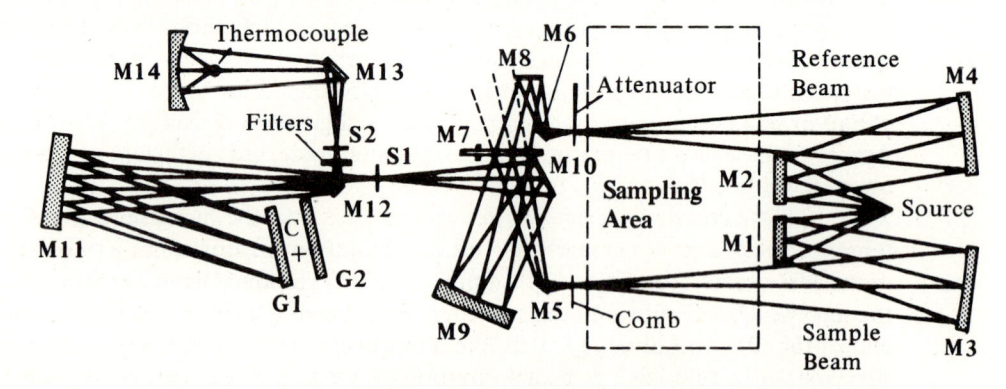

FIGURE 8.8. *Schematic diagram of a typical double-beam infrared spectrometer. The symbols M1, M2, ... indicate mirrors; S1 and S2 indicate slits; and G1 and G2 indicate gratings. Courtesy of the Perkin-Elmer Corporation.*

couple in the figure). No alternating signal is generated by the detector if the intensities of the two beams are equal at the frequency emerging from S2. If they differ, an alternating voltage is developed and amplified and then used to move the attenuator in or out of the reference beam to restore balance. The recording pen is coupled to the attenuator. What the instrument actually records is the position of the attenuator, which is related to the absorption spectrum of the compound.

One fundamental difference between infrared and ultraviolet-visible spectrophotometers is the position of the sample with respect to the monochromator. In ultraviolet-visible spectrophotometers, the sample is placed after the monochromator to minimize the exposure of the sample to the high-energy radiation. In infrared instruments, the sample is placed before the monochromator to minimize the amount of stray radiation (emanating from the sample and the cell) reaching the detector. Stray light is a particularly serious problem in most of the infrared region, more so than in the ultraviolet-visible region. Chopping the light from the source and measuring the alternating signal from the detector helps to alleviate this problem.

Single-beam systems require a background spectrum and a spectrum of the sample plus background. The ratio of the two spectra is found by dividing the two ordinates (i.e., the two intensities) at small frequency increments over the entire range scanned. A plot of these ratios against the frequencies at which each ratio was obtained is the spectrum of the sample. Almost all quantitative and qualitative analysis today is done on double-beam instruments.

Fourier Transform Infrared Spectrometers

The Fourier transform method provides an alternative to the use of monochromators based on dispersion. A brief description of Fourier transform spectroscopy was given in Chapter 6. In conventional dispersive spectroscopy, frequencies are separated and only a small portion is detected at any particular instant, while the remainder is discarded. The immediate result, however, is a *frequency-domain* spectrum $[I(\bar{v})$ vs. $\bar{v}]$ meaningful to the spectroscopist. In contrast, Fourier transform infrared (FT-IR) spectroscopy generates *time-domain* spectra as the immediately available data, in which the intensity is obtained as a function of time.

To illustrate, in Figure 8.9A a frequency-domain spectrum is shown for two component frequencies. The corresponding time-domain spectra are given in Figure 8.9B and C. It is clear that absorption of either frequency component will alter both the spectra shown in parts A and C. Direct observation of a time-domain spectrum is not immediately useful because it is not possible to deduce, by inspection, frequency-domain spectra from the corresponding time-domain waveform.

Thus far, no attention has been given to the collection of time-domain waveforms. In fact, detectors that are sensitive to infrared radiation are not capable of directly producing such a waveform because their response times are much too slow. Thus, it is necessary to *modulate* the incoming beam in such a way as to convert the frequency of the incoming radiation to a frequency that can be followed by appropriate detectors. This optical transformation is accomplished with a *Michelson interferometer*, shown in Figure 8.10.

Incoming radiation of wavelength λ_1 is split by the beam splitter into two beams: A (which travels a fixed distance) and B (whose distance of travel can be

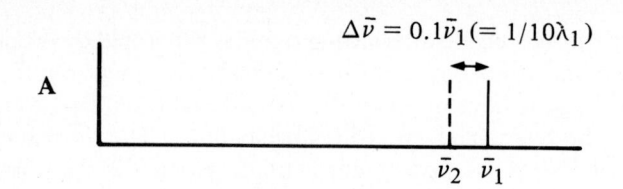

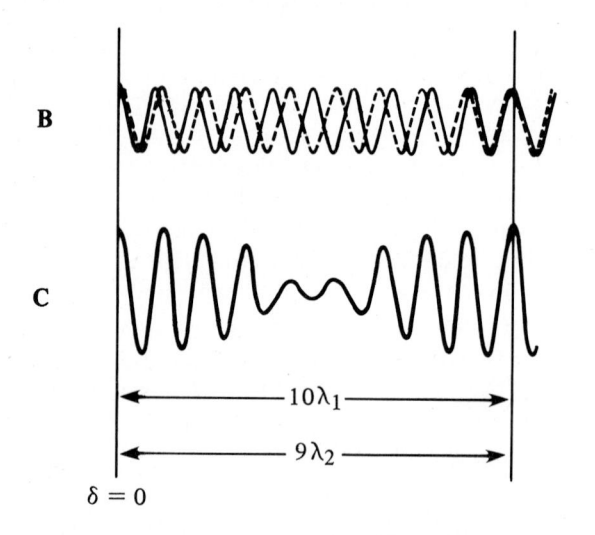

FIGURE 8.9. *Diagram demonstrating that the resolution is determined by the retardation of the interferometer. A: Emission spectrum with a doublet at frequencies $\bar{v}_1$ and $\bar{v}_2$, such that the separation of the lines, $\Delta\bar{v}$, is equal to $0.1\bar{v}_1$. B: Interferograms from the individual lines $\bar{v}_1$ (solid line) and $\bar{v}_2$ (broken line). C: Resultant of interferograms A and B, showing that they become in phase at a retardation $(0.1\bar{v}_1)^{-1}$ cm. Taken from P. R. Griffiths,* Chemical Infrared Fourier Transform Spectroscopy, *Vol. 43 in Chemical Analysis series, New York: Wiley, 1975, p. 16. Copyright 1975 by John Wiley & Sons; used with permission.*

varied by changing the position of mirror B). When beams A and B are recombined, interference can occur. If the path distances of beams A and B are initially identical, then *constructive interference* takes place and the detector has a maximum output. Then, if mirror B is moved by a distance $\lambda_1/4$, then beam B will travel a distance that is different (by a distance $\lambda_1/2$) from that traveled by beam A and *destructive interference* occurs, producing a minimum detector output. If mirror B is displaced further by a distance $\lambda_1/4$ to yield a path difference of λ_1, constructive interference again occurs and the detector will give a maximum response. In practice, mirror B is uniformly and continuously translated to produce a smoothly varying response signal such as those shown in Figure 8.9B. The frequency, f, of the detector signal depends on the translation velocity, v, of mirror B, and the wavelength of the radiation:

$$f = \frac{v}{\lambda/2} = \frac{2v \text{ cm/sec}}{\lambda \text{ cm}} = 2v\bar{v} \text{ sec}^{-1} \tag{8.9}$$

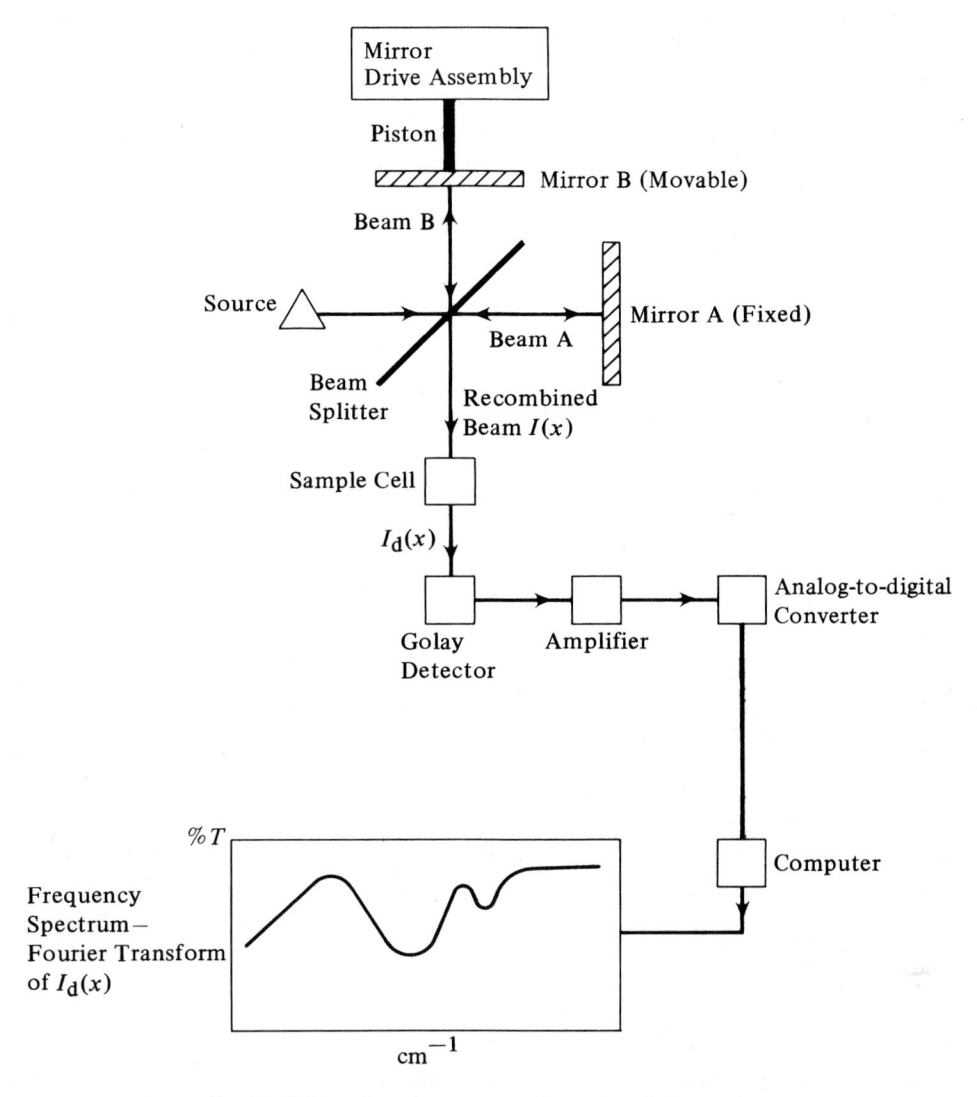

FIGURE 8.10. *Interferometer and associated electronics.*

In this way, the interferometer generates a low-frequency response that is proportional to the frequency of radiation but that can be followed by the detector. From Equation 8.9, it is also apparent that if polychromatic radiation is used, the detector produces a wave of unique frequency for each component wavelength. The detector output as a function of path difference—the *interferogram* $[I(x)]$—is the sum of the waves for each frequency component. If a sample absorbs some component of the radiation, the interferogram will also be affected. Thus, although the interferogram does not directly yield such information, it contains all the information necessary to generate the conventional (frequency-domain) spectrum. The mathematical transformation to a frequency-domain spectrum is complex, and high-speed digital computers are required.

SEC. 8.3 Instrumentation for Infrared Spectroscopy

There are a couple of advantages to using Fourier transform methods. The reduction in measurement time that results from simultaneously measuring all of the radiation, rather than losing most of it with the use of slits, is known as *Fellgett's advantage*, in honor of the astronomer who first realized this potential. In addition, interferometers, especially those equipped with lasers for referencing the path-length difference, are capable of higher resolution than dispersive instruments.

8.4 SAMPLING SYSTEMS FOR INFRARED SPECTROSCOPY

Infrared spectroscopy can be used to examine solids, liquids, or gases. However, a number of problems exist with regard to adequate sample handling. The main problem is that almost all substances absorb infrared radiation. This seriously restricts the choice of materials that may be used for construction of sample cells. The windows of the cell must transmit a range of infrared wavelengths in order to be generally useful. The alkali halides are commonly used. However, these materials are hygroscopic and must be stored in a desiccator; they tend to cloud with use and must be polished periodically. Although it may darken upon exposure to light and is easily deformed, silver chloride is the most widely used material for aqueous solutions or moist samples. For frequencies below 600 cm^{-1}, polyethylene cells and other plastics are appropriate. The infrared transmission characteristics of a number of commonly used materials are summarized in Table 8.2.

Gases

Gas cells for infrared spectroscopy are often constructed with a cylindrical glass body (including ground-glass joints and a vacuum stopcock) and are usually about 10 cm in length. To the ends of the cell are attached disks of appropriate window material with wax, epoxy cement, or pressure plates. Depending on sensitivity and concentration, gas pressures of a few millimeters (for strong absorbers) to several atmospheres (for weak absorbers and low concentration) are encountered. For trace analysis (at the ppm level), such as in air-pollution studies and breath analysis, very long path lengths (up to 30 m) can be obtained using multiple-reflection cells. These are usually

TABLE 8.2. *Infrared Transmission of Cell Materials*

Material	Wavenumber Range, cm^{-1}	Remarks
AgCl	25,000– 425	Useful for aqueous solutions
CaF$_2$	6,700–1,100	Useful for aqueous solutions
NaCl	40,000– 600	Common window material; hygroscopic
KBr	40,000– 400	Common window material; hygroscopic
CsI	10,000– 200	Used mainly in far-infrared
TlBr-TlI(KRS-5)	20,000– 300	Toxic; only slightly water soluble
Polyethylene	600– 33	Useful for far-infrared

designed to provide a multiple-reflection chamber at 90° from the normal beam direction. Gold-surfaced mirrors at each end repeatedly reflect the beam before eventually diverting it to its normal direction.

Liquids and Solutions

A variety of cells are available for liquid samples. A liquid cell may be as simple as a drop of liquid sandwiched between two windows to form a capillary film or as refined as a variable-path cell. The optical path length required for a liquid is usually less than 1 mm and spacers of different thicknesses are used with "demountable" cells to establish a suitable path length. Liquid cells are thus available as variable, fixed, or demountable, as shown in Figure 8.11.

In studying solutions of liquids, solids, or dissolved gases, it is important to select appropriate solvents that transmit over a wide range of wavelengths. The infrared transmission characteristics for several common solvents are shown in Figure 8.12. Consideration must be given to possible solvent interactions with solutes which may affect the infrared spectrum. This is especially true for solutes that are susceptible to hydrogen-bonding effects through such groups as –NH or –OH. Hydrogen bonding lowers the fundamental frequency of these groups. Such hydrogen-bonding effects may be intramolecular, however. It is possible to distinguish inter-molecular from intramolecular effects by carrying out systematic dilution studies. In more dilute solutions, the band due to hydrogen-bonded species (intermolecular) will decrease whereas the band associated with the nonbonded species will increase. Intramolecular hydrogen bonding is not affected by dilution.

Solids

There are two common sampling techniques for powdered solids: mulls and pellets. Thin films or smooth surfaces are examined by reflectance techniques.

For mulls, 2 to 10 mg of a finely ground sample (particle size must be less than the infrared wavelength to prevent excessive loss of signal due to scattering) is further ground with a drop or two of mulling agent. These agents are substances that transmit a wide range of infrared frequencies and help minimize scattering by surrounding the analyte with a medium whose refractive index more closely matches that of the sample than does air. Nujol, refined mineral oil, is commonly used although it is not appropriate for examination of aliphatic CH and CC vibrations. Several other materials are commonly used as well. The resulting mull should have a consistency resembling that of toothpaste. It is spread on a single plate of alkali halide or pressed between two such plates to adjust the thickness of the sample.

By far, the most commonly used sampling method is the KBr-pellet technique. About 1 mg or less of sample is ground with approximately 300 mg of "infrared-quality" KBr in a mortar and pestle or, preferably, with a small ball-mill apparatus such as those used in the dental profession (e.g., "Wig-L-Bug"). The pulverized mixture is transferred to an evacuable die, most commonly 13 mm in diameter, and pressed for several minutes at 8000 to 20,000 psi. High-quality, transparent pellets are readily obtained within a few minutes. An evacuable, heated (approx. 40°C), metal desiccator

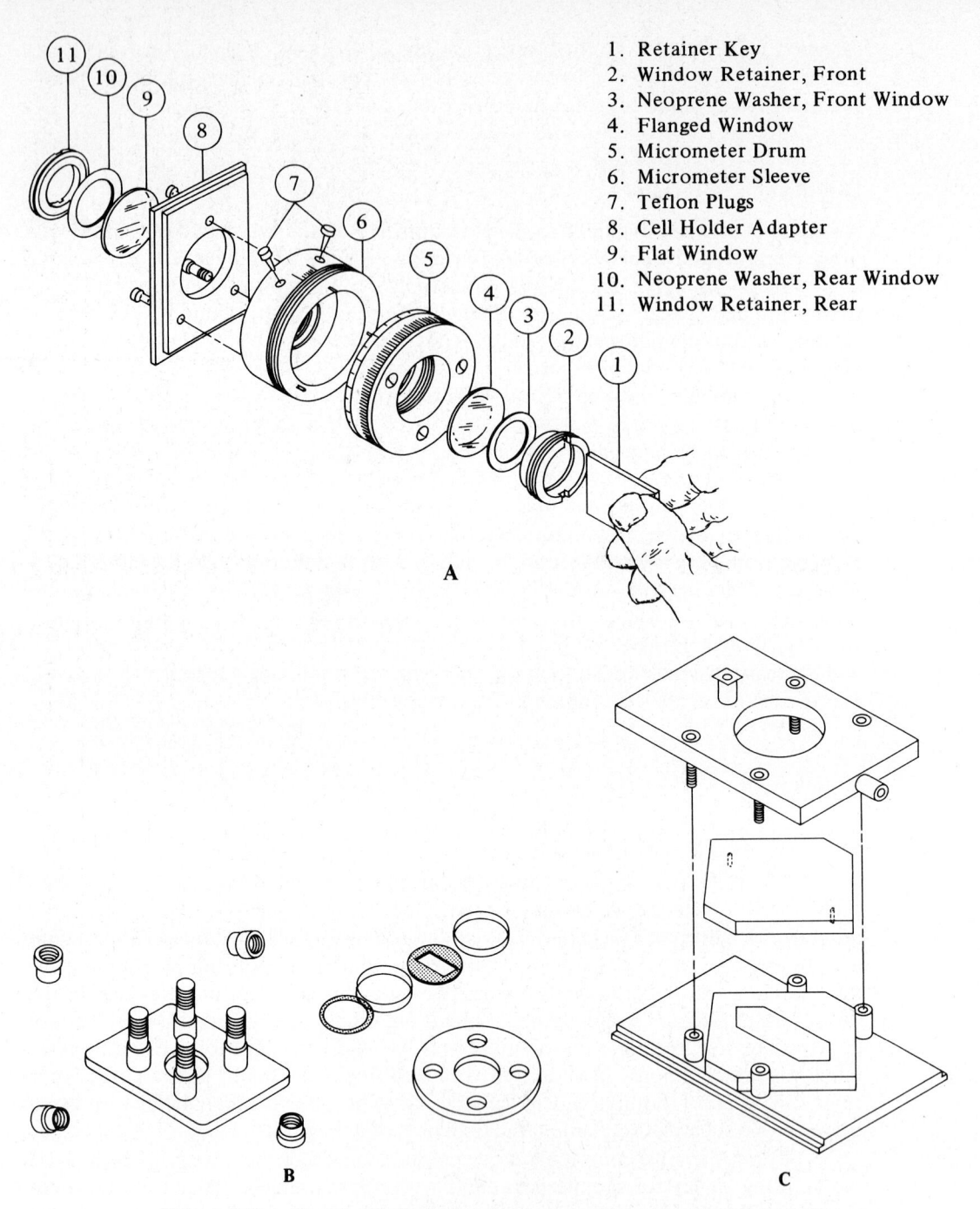

1. Retainer Key
2. Window Retainer, Front
3. Neoprene Washer, Front Window
4. Flanged Window
5. Micrometer Drum
6. Micrometer Sleeve
7. Teflon Plugs
8. Cell Holder Adapter
9. Flat Window
10. Neoprene Washer, Rear Window
11. Window Retainer, Rear

FIGURE 8.11. *Liquid cells for infrared spectroscopy. A: Variable path-length cell. Courtesy of Beckman Instruments, Inc. B: Demountable cell. C: Fixed path-length or sealed cell. Adapted from N. B. Colthrup, L. H. Daly, and S. E. Wiberly*, Introduction to Infrared and Raman Spectroscopy, *New York: Academic Press*, 1964, *by permission of the senior author and publisher.*

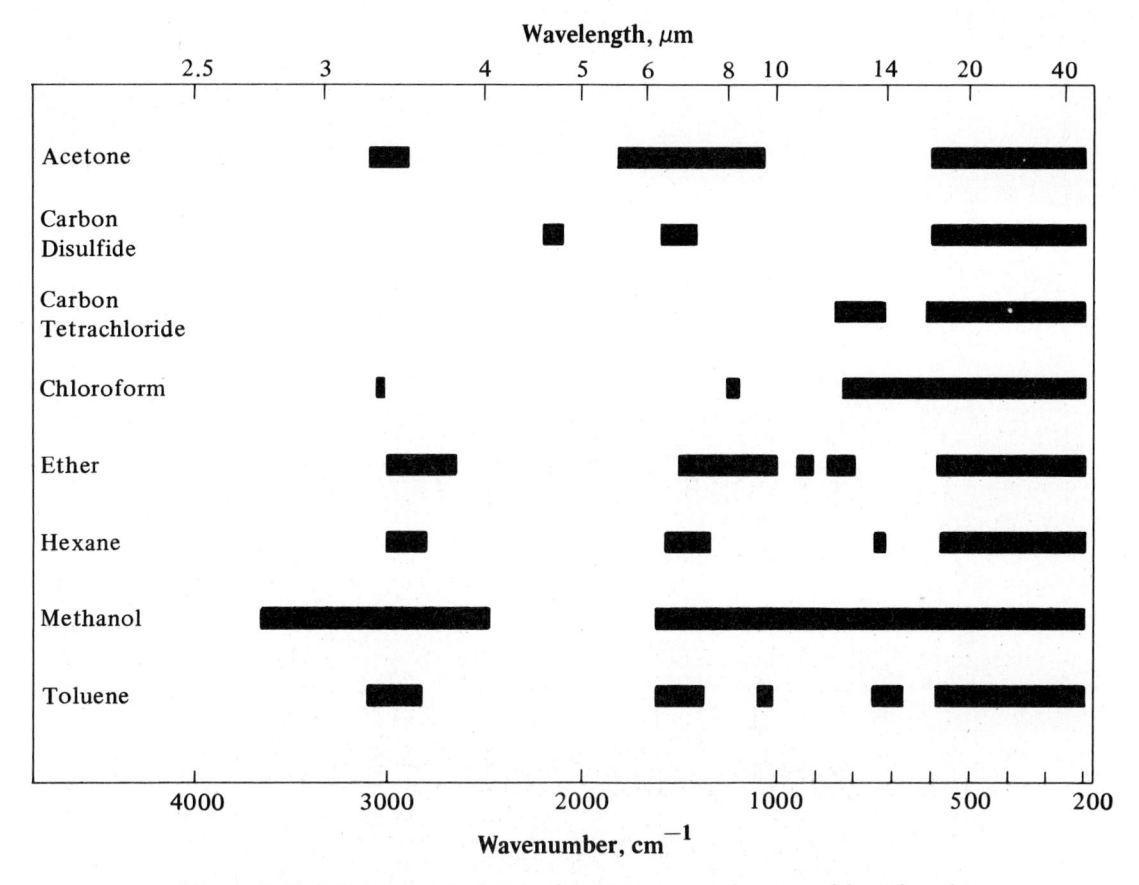

FIGURE 8.12. *Transparent regions of some common solvents used for infrared spectroscopy. The darkened regions are areas in which a 0.1-mm thickness of solvent (in an NaCl cell) transmits 30% or less of the incident radiation.*

is useful for storing die components and KBr. Surface adsorption of water is practically unavoidable but can be minimized by overnight evacuation of the formed pellet at slightly elevated temperature. Other alkali halides such as CsI and CsBr can be used for low-frequency studies.

It is also possible to obtain infrared spectra by reflection of the beam from a surface to be analyzed. Thus, commercial devices for *attenuated total reflectance (ATR)* have become available for obtaining spectra of fabrics, adhesives, foams, plastics, and other materials. The sample is placed on the sides of a material of high refractive index which transmits infrared radiation; thallium bromide/thallium iodide crystal is often used. The beam enters the crystal at a chosen angle (30°, 45°, or 60°) such that the radiation is subject to multiple internal reflections. At each interface, the radiation penetrates a few micrometers into the sample and some absorption can occur. It is important that the material to be analyzed make intimate contact with the surface of the reflecting crystal. Following these multiple internal reflections, the emerging beam is directed to the detector.

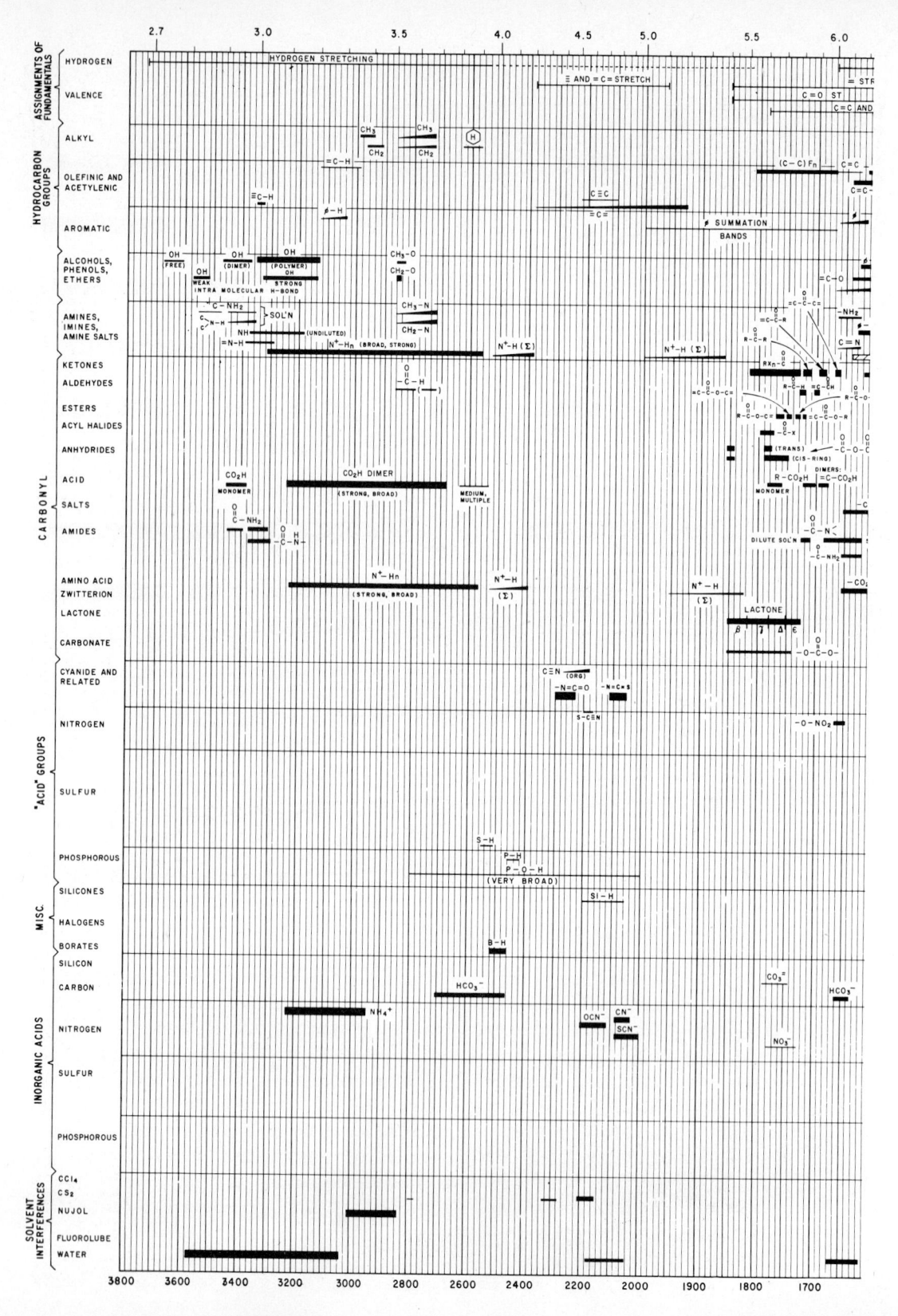

FIGURE 8.13. Correlation chart of group frequencies. Courtesy of Dow Chemical Company.

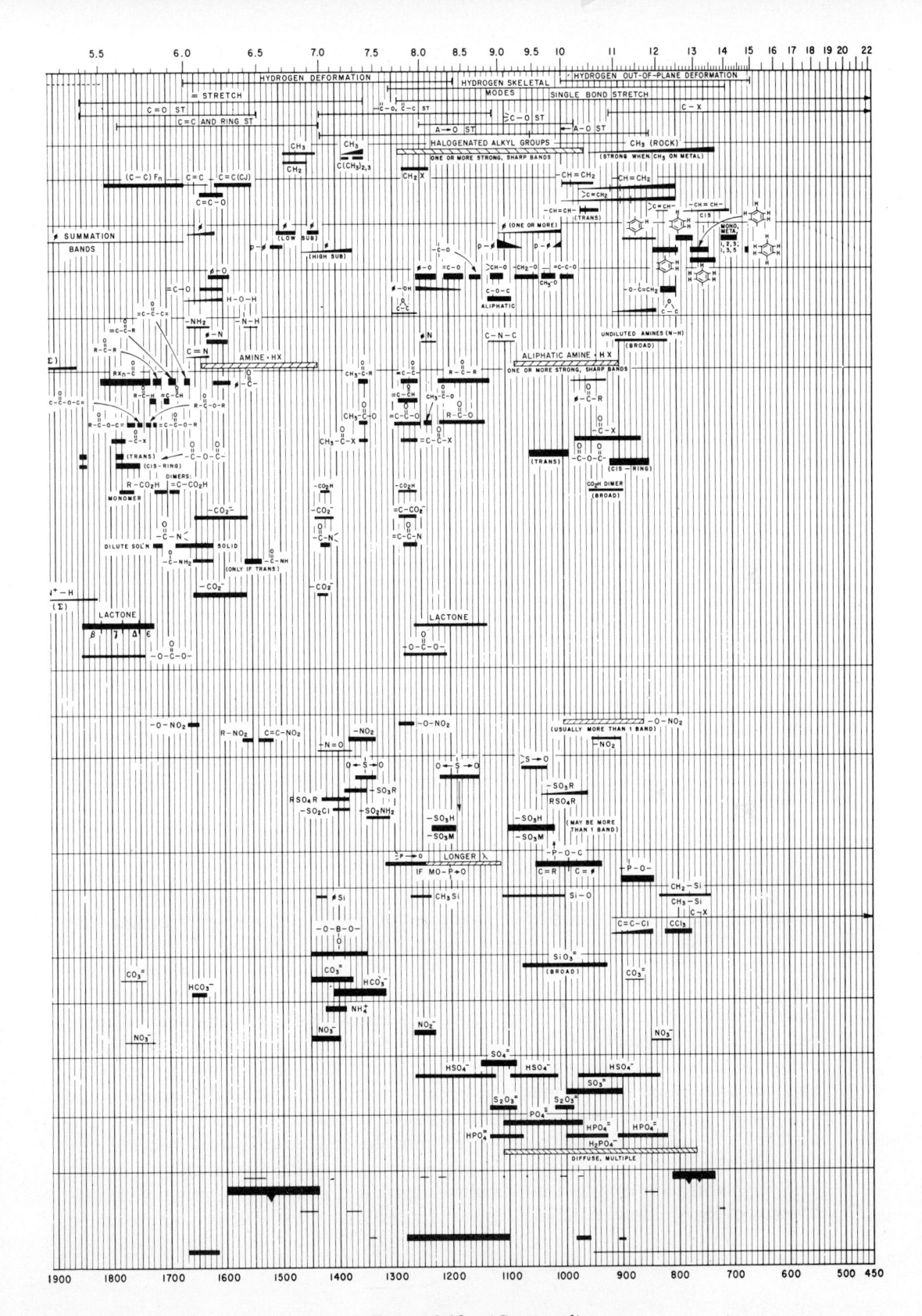

FIGURE 8.13 *(Continued)*

8.5 QUALITATIVE ANALYSIS

The most common use of infrared and Raman spectroscopies is for qualitative identification and structure determination of organic compounds. As was pointed out earlier, the group-frequency concept makes it possible to identify particular functional groups within a molecule, and reliable estimates of group-frequency ranges have evolved over the years. This information is conveniently summarized in *correlation charts* such as that shown in Figure 8.13.

Reliable interpretation of infrared spectra requires practice and access to more detailed information about the dependence of group frequencies on molecular interactions. Organic chemists have devised useful procedures for structure elucidation based on the group-frequency concept, and excellent summaries of these practices are available, some of which are listed in the Selected Bibliography.

Obviously, the group-frequency concept also applies to Raman spectroscopy (the frequency of a particular group is the same), but relative intensities may change from those in the infrared spectrum. Although the number of compounds studied by Raman spectroscopy is not as large as that studied by infrared spectroscopy, excellent summaries of characteristic Raman frequencies have recently become available (see the Selected Bibliography).

8.6 QUANTITATIVE ANALYSIS

All molecular compounds will exhibit infrared or Raman spectral bands. Thus, the scope of these methods is virtually unlimited. In addition, the sharpness of the observed bands and the uniqueness of infrared and Raman spectra for a particular compound lead to the great specificity associated with these techniques, and it is therefore often possible to apply them to the analysis of mixtures of closely related compounds.

Infrared Spectroscopy

Because infrared spectroscopy is an absorption technique, one may expect that analyses can be carried out as are typical analyses in ultraviolet-visible spectrophotometry. However, several difficulties are encountered in the case of quantitative infrared procedures. First, the optical-null method used with dispersive, double-beam instruments introduces some error in setting the 0% transmittance reading because almost no energy is reaching the detector. However, with modern instruments this error is relatively small (approx. 1%), and its effects are not serious if the analyses are limited to transmittances greater than 10%. A second problem is the uncertainty in determining the 100% transmittance points. The use of solvent in a matched cell in the reference beam is not practical for infrared work because it is generally not possible to obtain cells having identical transmission characteristics. The very short path lengths are difficult to reproduce, and the transmission characteristics of the window materials change gradually with use. The best approach to overcoming this difficulty is the *baseline method*, wherein I_0 is approximated by drawing a straight line between points of maximum transmittance on each side of the absorption band, as shown in Figure 8.14. The absorbance, A, is determined from measurement of I and I_0. Finally, the small amount of energy available in the infrared region requires relatively wide slits. This, coupled with the narrow bandwidths of infrared absorption peaks, results in deviations from Beer's law (see Chap. 7). Also, as mentioned earlier, stray light or scattered radiation is often present. Such problems lead to nonlinear relationships between absorption and concentration, and necessitate the construction of calibration curves.

Thus, errors incurred in quantitative analysis by infrared absorption include uncertainty in the measurement of the 0 and 100% transmittance levels and commonly encountered deviations from Beer's law. Therefore, it is seldom possible to obtain the accuracy and precision normally associated with ultraviolet and visible methods.

In order to overcome some of the disadvantages associated with double-beam instruments, a number of sturdy single-beam instruments have been developed for quantitative work. These usually involve nichrome-wire sources and pyroelectric detectors. Wavelength selection is based on filters or variable-thickness filter wedges. One type of instrument uses a variety of interchangeable interference filters that transmit in the range between 3000 and 800 cm^{-1}. Usually, each filter is designed for the analysis of a particular atmospheric pollutant because these instruments find most use as portable monitoring devices for gases at sub-ppm levels. One of the most common instruments of this type is based on the use of three interference wedges mounted on a circular wheel (Fig. 8.15). These wedge-shaped filters increase in thickness as the wheel rotates, and therefore each sector is suitable for isolation of a particular wavelength range depending upon the exact position of the wheel. Microprocessor-controlled positioning of the sector wheel permits rapid wavelength selection in the range between 4000 and 700 cm^{-1} with an accuracy of 0.4 cm^{-1}. Since the analytical wavelength is determined by the instrumental selection of wheel

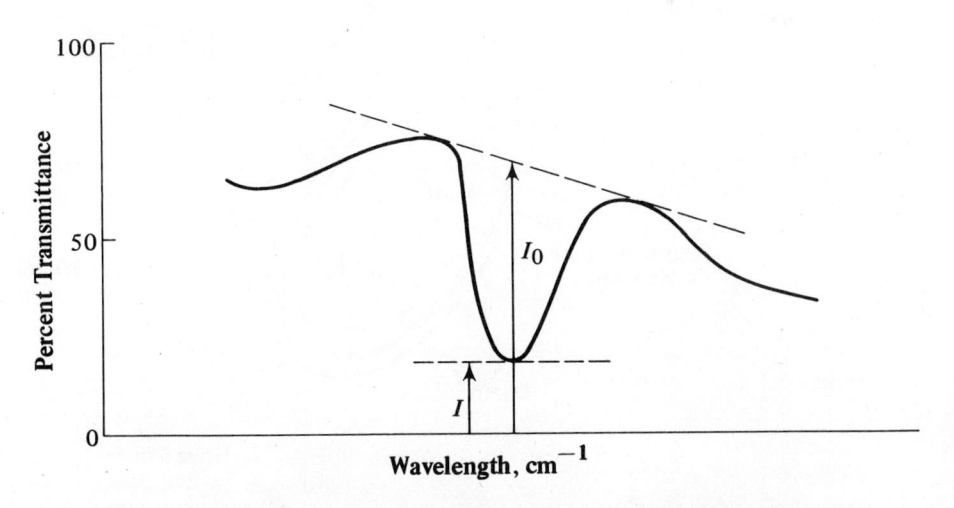

FIGURE 8.14. *Measurement of I_0 and I for an infrared absorption band with a sloping background. The resultant absorbance for the peak is $A = -\log(I/I_0)$.*

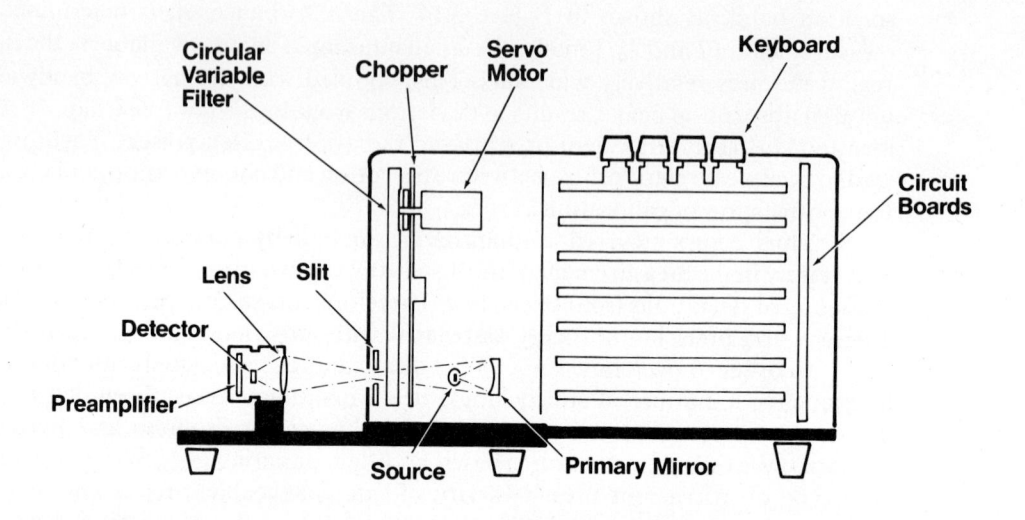

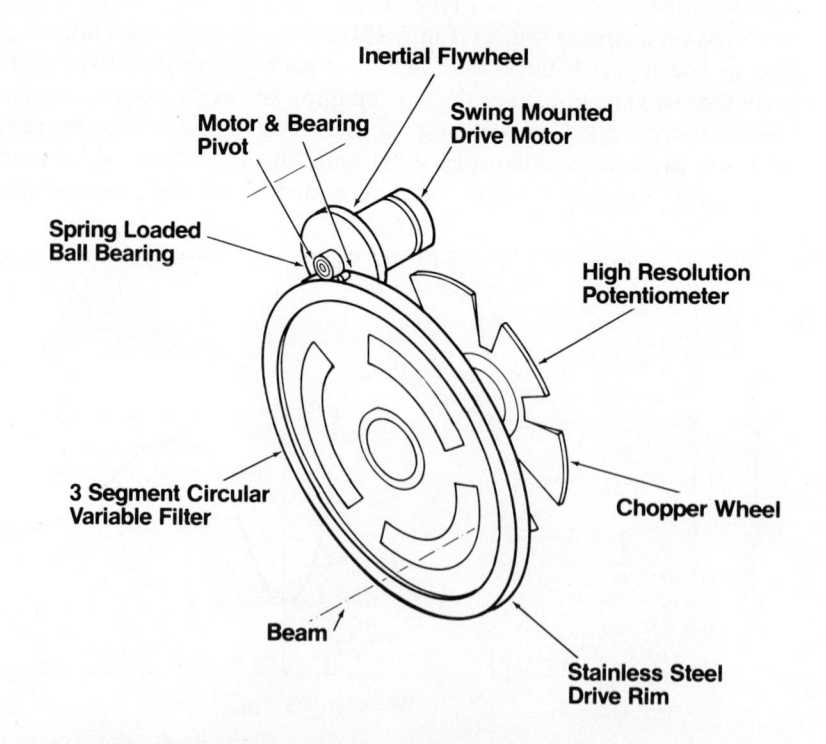

FIGURE 8.15. *Single-beam filter instrument for quantitative infrared analysis. Courtesy of The Foxboro Company, Foxboro, Mass. Used with permission.*

position, the instrument can be programmed to obtain the absorbance at several wavelengths and calculate the concentration of each component in a multicomponent sample.

Applications

Government regulations regarding atmospheric contaminants demand the development of convenient, specific, and sensitive methods for a variety of compounds. Of the several hundred chemicals for which limits of exposure have been set by the Occupational Safety and Health Administration (OSHA), infrared absorption spectroscopy, using instruments such as that shown in Figure 8.15, meets this need more than any other technique, being applicable to over half of the substances. Table 8.3 illustrates the utility of this technique for several atmospheric contaminants.

One particular advantage of infrared analysis is its ability to determine the total amount of a specific functional group present in even a very complex mixture. For example, the total ketone content of a mixture can be determined because almost all ketone carbonyl bands occur at about 1720 cm^{-1} and the intensity of absorption does not vary a great deal from one compound to another. Therefore, one can determine an "average" absorptivity for a particular functional group from known mixtures, and use this value in the analysis of real samples. The total content of a particular functional group is often an important consideration, particularly in industrial situations.

An example of total functional-group analysis is the determination of the carboxyl content of carboxyl-terminated polybutadienes (CTPBs), which are polymers (M.W. approx. 7000) used as elastomeric binders for solid propellants [1]. A solution (approx 2%) of the CTPB sample in CCl$_4$ is prepared, and the magnitude of the 1708-cm^{-1} carboxyl-carbonyl band is measured in a 1-mm cell. Quantitation is achieved either by the calibration-curve method, or by the internal-standard method using the 1435-cm^{-1} methylene band or the 1638-cm^{-1} vinyl band as the standard. The results for the total carboxyl content by this method agree with those of standard

TABLE 8.3. *Examples of Infrared Analysis of Vapors*

Compound	Minimum Detectable Concentration, ppm[a]	Wavelength, μm	Allowable Exposure[b]
Carbon disulfide	0.5	4.54	20
Chloroprene	4	11.4	25
Ethylenediamine	0.4	13.0	10
Hydrogen cyanide	0.4	3.04	10
Nitrobenzene	0.2	11.8	1
Pyridine	0.2	14.2	5
Vinyl chloride	0.3	10.9	1

a. For a 20.25-m cell.
b. Exposure limits for an 8-hr weighted average.

chemical titration procedures to within an error of about ±0.03% carboxyl content, for a carboxyl-content range of about 0.5 to 2.5% in commercial CTPB samples.

High-molecular-weight aliphatic amines are used extensively in many industries. The total primary- and secondary-amine content of aliphatic amines can be determined easily and rapidly by functional-group analysis in the near-infrared [2] using chloroform solvent and 5-cm fused-silica cells. Primary amines have characteristic absorption maxima at 2.02 and 1.55 μm, whereas secondary amines absorb only at 1.55 μm. Quantitation is achieved by the calibration-curve method using a series of standard solutions of primary and secondary amines. Most other methods for the determination of total primary, secondary, or tertiary amine in a mixture are lengthy or inaccurate, or are unsuitable for small samples.

Raman Spectroscopy

Until recently, Raman spectroscopy was used only infrequently as a quantitative method. As was mentioned earlier, low sensitivity of normal Raman scattering is one of the main disadvantages. Obtaining precise values for absolute scattering efficiency for a given situation is extremely difficult since the signal depends on sample alignment, laser power, and collection efficiency. Therefore, it is usually necessary to include a fixed amount of internal standard in both samples and standards. Calibration curves are then constructed based on the ratio of peak intensity of the sample to that of the standard in a manner similar to the one used for the internal-standard method as applied to polarography (Chap. 3) and flame-emission spectroscopy (Chap. 10). For aqueous solutions, it is often convenient to use one of the vibrational bands of the solvent, H_2O, as an internal standard (e.g., the 1640-cm^{-1} bending mode).

Despite the weakness of normal Raman scattering, some compounds are sufficiently strong scatterers to yield relatively sensitive analyses. In one of the earliest applications, Bradley and Frenzel [3] were able to demonstrate a detection limit of 50 ppm of benzene in distilled water using only 5 mW of power from a He-Ne laser at 632.8 nm. More recently, detection limits of 4 to 40 ppm have been obtained for a series of molecular and ionic species in water [4].

To lower the detection limits for the study of pollutants in water, it is necessary to exploit the resonance Raman effect. In one of the earliest applications, Morris [5] demonstrated the potential of resonance Raman spectroscopy for quantitative multicomponent analysis in a study of catecholamines that had been oxidized to the colored aminochromes by ferricyanide. High-quality working curves were generated for concentrations of 10 to 50 μM of adrenaline, noradrenaline, and dopamine using 0.5 M nitrate ion as the internal standard.

Clearly, the continued development of intense laser sources, and further progress on methods for fluorescence rejection, will lead to increased application of resonance Raman spectroscopy for trace-level determination of multicomponent samples. In another application, industrial fabric dyes were studied at levels of 30 to 50 ppb [6]. In addition, it was possible to obtain spectra suitable for definitive qualitative identification at 100 to 200 ppb. It is important to emphasize that quantitative studies using this technique must employ an internal standard. Furthermore, in the case of resonance Raman spectroscopy, it must be remembered that the scattered light as well as the incident beam is in close proximity to an electronic absorption band of the sample. Although the amount of light reaching the sample and reference molecules (H_2O) is the same, the scattered light from each will be affected differently by absorption because the extent of absorption may differ at the two frequencies. Thus, deviations from linearity can occur. Construction of calibration curves can still provide for quantitative measurements, but care must be taken in sample positioning to hold absorption effects constant from sample to sample.

SELECTED BIBLIOGRAPHY

BRAME, E. G., and GRASSELLI, J. G., eds. *Infrared and Raman Spectroscopy*. New York: Marcel Dekker, 1977.

CAREY, P. R. *Biochemical Applications of Raman and Resonance Raman Spectroscopies*. New York: Academic Press, 1982.

COLEMAN, W. F. "Lasers—An Introduction." *J. Chem. Ed.*, 59, 441 (1982).

COLTHRUP, N. B., DALY, L. H., and WIBERLEY, S. E. *Introduction to Infrared and Raman Spectroscopy*, 2nd ed. New York: Academic Press, 1975.

DOLLISH, F. R., FATELEY, W. G., and BENTLEY, F. F. *Characteristic Raman Frequencies of Organic Compounds*. New York: Wiley-Interscience, 1971.

GRIFFITHS, P. R. *Chemical Infrared Fourier Transform Spectroscopy*. New York: John Wiley, 1975.

NAKAMOTO, K. *Infrared and Raman Spectra of Inorganic and Coordination Compounds*, 3rd ed. New York: John Wiley, 1978.

MORRIS, M. D., and WALLAN, D. J. "Resonance Raman Spectroscopy." *Anal. Chem.*, 51, 182A (1979).

OMENETTO, N. *Analytical Laser Spectroscopy*. New York: John Wiley, 1979.

SILVERSTEIN, R. M., BASSLER, G. C., and MORRILL, T. C. *Spectrometric Identification of Organic Compounds*, 3rd ed. New York: John Wiley, 1974.

STROMMEN, D., and NAKAMOTO, K. "Resonance Raman Spectroscopy." *J. Chem. Ed.*, 54, 474 (1977).

REFERENCES

1. A. S. TOMPA, *Anal. Chem.*, 44, 628 (1972).
2. R. B. STAGE, J. B. STANLEY, and P. B. MOSELEY, *J. Amer. Oil Chem. Soc.*, 49, 87 (1972).
3. E. B. BRADLEY and C. A. FRENZEL, *Water Res.*, 4, 125 (1970).
4. K. M. CUNNINGHAM, M. C. GOLDBERG, and E. R. WEINER, *Anal. Chem.*, 49, 70 (1977).
5. M. D. MORRIS, *Anal. Lett.*, 9, 469 (1976).
6. L. VAN HAVERBEKE, P. F. LYNCH, and C. W. BROWN, *Anal. Chem.*, 50, 315 (1978).

PROBLEMS

1. List the upper and lower limits of the near-, mid-, and far-infrared spectral regions in (a) μm; (b) wavenumbers (cm^{-1}).
2. Convert 7000 Å to (a) wavenumbers (cm^{-1}); (b) μm; (c) eV.
3. How many fundamental modes of vibration are predicted for (a) methane, (b) benzene, and (c) acetylene?
4. Single, double, and triple bonds have force constants that are approximately 5, 10, and 15×10^5 dynes/cm. At what frequencies would you expect to find C–C, C=C, and C≡C stretches?
5. The C–H stretch of an alkane occurs at approximately 2900 cm^{-1}. What would be the frequency of the corresponding deuterated analog?

6. The fundamental vibrational frequency of a particular ketone occurs at 1730 cm^{-1}. Determine the absolute wavelengths at which this band will be observed in the Raman spectrum for each of the following excitation lines of an argon-ion laser: 514.5, 488.0, 496.5, 457.9 nm.

7. Given two energy levels that differ in energy by E, the population of the upper and lower states is given by the Boltzmann distribution

$$\frac{\text{Upper-State Population}}{\text{Lower-State Population}} = \frac{N_u}{N_l} = e^{-E/kT}$$

where k = the Boltzmann constant
          T = temperature in kelvins

Considering all other factors to be equal, calculate the ratio of the anti-Stokes to Stokes intensities for each vibration for a molecule that has vibrational transitions at 200, 600, and 1000 cm^{-1}, assuming a temperature of 300 K.

8. Sketch a calibration curve one might expect to obtain for a resonance Raman experiment using a series of solutions containing increasing concentrations of Fe(phenanthroline)$_3^{2+}$ and equal concentrations of internal standard (SO$_4$)$^{2-}$, assuming that the ratio of I complex/I (SO$_4$)$^{2-}$ is plotted versus concentration of the complex and that the 514.5-nm line is used for excitation. The Raman band for the complex occurs at approximately 1460 cm^{-1}, and that for sulfate at about 980 cm^{-1}.

9. Chloroform exhibits Raman bands at 258, 357, 660, and 760 cm^{-1}. Polarized spectra were taken and peak heights measured. For the four peaks, $I_\perp$ was determined to be 30.8, 5.8, 1.3, and 4.7 units; $I_\parallel$ was 40.9, 79.2, 83.2, and 6.0 units. Calculate the depolarization ratios for these bands and indicate whether each is "polarized" or "depolarized."

10. A certain pulsed laser can be operated at either 20 or 50 pulses/sec. Assuming that the average power is 10 W and the pulse width is 10 nsec for each case, calculate the peak power associated with each of the operating frequencies given.

11. An investigator wishes to obtain the infrared spectrum of an aqueous solution of a compound in the region 4000 to 500 cm^{-1}. What sampling arrangement should be used? What change might be made in order to observe those regions obscured by absorption from H$_2$O bands?

CHAPTER 9

Molecular Fluorescence and Phosphorescence

Isiah M. Warner

During the process of absorbing ultraviolet or visible electromagnetic radiation, molecules are elevated to an excited electronic state (Chap. 7). Most molecules will dissipate this excess energy as heat by collision with other molecules. Some molecules will emit some of this excess energy as light of a wavelength different from that of the absorbed radiation. It is this latter process, called *photoluminescence*, that is described in this chapter.

Basically, we can consider *photo*luminescence to be a deexcitation process that occurs after excitation by photons. Other types of luminescence are also characterized by the source of energy used to excite the luminescing molecule. For example, *radio*luminescent molecules derive their excitation energy from high-energy particles, and *chemi*luminescent molecules derive their excitation energy from a chemical reaction.

The discovery of luminescence dates back to as early as the sixteenth century, when the Spanish scientist Nicholas Monardes observed a mysterious blue tinge in water stored in cups made from the wood "lignum nephriticum" [1]. In the seventeenth century, other scientists, including Boyle, gave accounts of similar phenomena [2]. The analytical utility of luminescence was first documented in an 1864 lecture by the English physicist George Stokes. His fundamental paper in 1852 [3] on the subject of luminescence, and subsequent studies, provided the foundation for many of today's concepts. A more detailed historical perspective on luminescence has been provided in an excellent article by T. C. O'Haver [4]. This comprehensive discussion establishes luminescence spectrometry as one of the oldest analytical techniques in use today.

Photoluminescence spectrometry has become one of the most widely used techniques for chemical analysis because of the many associated parameters that can be exploited to achieve additional sensitivity and specificity. An increasing number of publications and numerous monographs [5–10] on the general topic are available.

This chapter will briefly explain the fundamental concepts of photoluminescence and required instrumentation. Then, some applications to specific problems will be described. The reader may wish to consult the referenced monographs for more detailed discussion of the ideas presented.

9.1 PRINCIPLES OF PHOTOLUMINESCENCE

The presentation in this section will focus on some of the basic theory underlying molecular fluorescence and phosphorescence processes, and will include derivations of a few of the most applicable equations. Basically, photoluminescence is an excitation–deexcitation process involving photons. Because there are a number of other competing processes, the rates of all these processes will be of importance.

The Excitation Process

A molecule that has absorbed electromagnetic radiation is in an excited state. It must use some mechanism to eliminate this excess energy. The present discussion covers the competitive processes that lead to elimination of this energy or, more properly, deexcitation.

The molecular *multiplicity*, M, is defined as

$$M = 2S + 1 \tag{9.1}$$

where S = the spin quantum number of the molecule, and is the sum of the net spin of the electrons in the molecule

The applications discussed in this chapter will involve primarily organic molecules. For most organic molecules, $S = 0$ because the molecules have an even number of electrons and, thus, the lowest energy state (ground state) must be one in which all electrons are spin paired [11]. The multiplicity is then computed to be

$$M = 2 \cdot 0 + 1 = 1 \tag{9.2}$$

and is referred to as a *singlet state*. The ground singlet state is designated as S_0, and the first and second excited singlet states are denoted S_1 and S_2, respectively. While a molecule is in an excited state, it is possible for one electron to reverse its spin. Under these conditions, $M = 3$ since $S = +1/2 + 1/2 = 1$. This is termed a *triplet state*, and the lowest-lying triplet is given the symbol T_1. It should be evident from this discussion that a molecule with an even number of electrons cannot have a ground triplet state because the electrons in the lowest energy state have all spins paired. Molecules with one unpaired electron (odd number of electrons) are in a doublet state, as is the case for organic free radicals. Our discussion here will be devoted to excitation and emission processes involving singlet and triplet states of organic molecules.

Qualitatively, the most effective approach to describing the absorption and emission processes is to use a Jablonski energy-level diagram, such as that shown in Figure 9.1. In this type of diagram, we can depict the molecule as having three major contributions to its overall energy: electronic, vibrational, and rotational. The

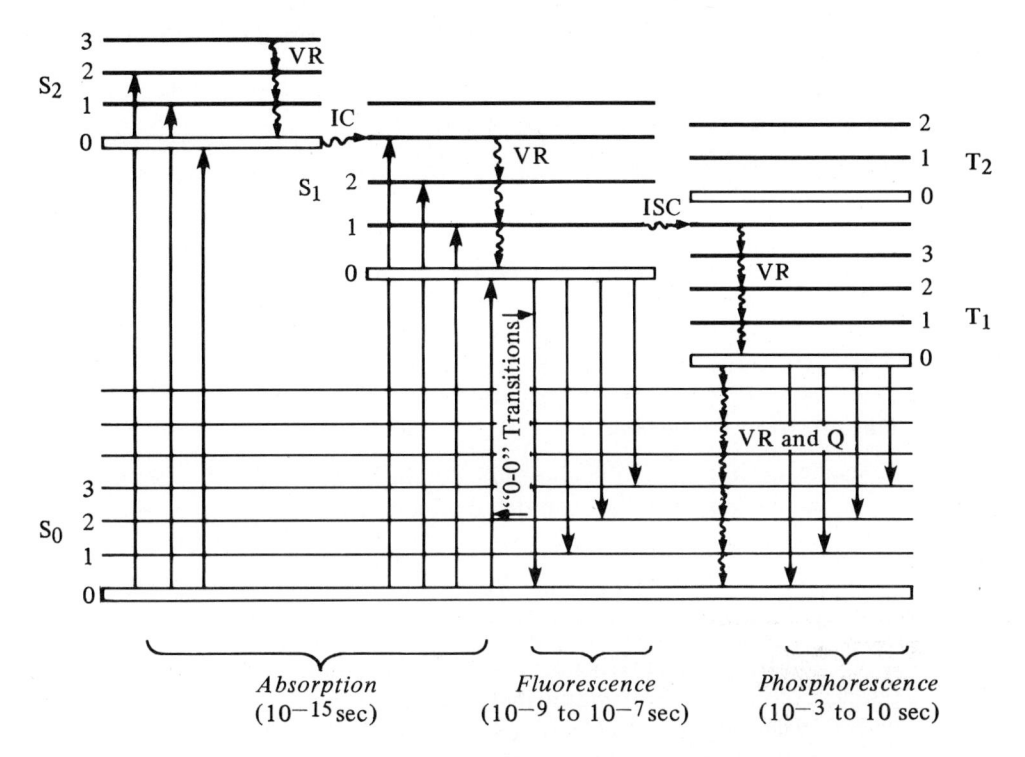

FIGURE 9.1. *Jablonski energy-level diagram depicting absorption and emission processes.*

rotational levels have not been included in the figure because these are not resolved on a conventional spectrometer. Table 9.1 provides an estimation of these contributions for a typical organic molecule, along with the estimated wavelength of electromagnetic radiation necessary for excitation. Although these are approximations which may deviate for some compounds, it is clear from this table that energy requirements for electronic, vibrational, and rotational excitation are very different.

It is generally accepted that electronic absorption of electromagnetic radiation must occur within one wave period of the exciting radiation. If we use the wavelength of light calculated in Table 9.1 for electronic excitation (300 nm), and the speed of light, the time of absorption must be approximately 10^{-15} sec. During this period, if the organic molecule absorbs a photon, it is excited to a higher singlet state (S_1 or

TABLE 9.1. *Approximate Energies and Wavelengths for Various Transitions*

Transition	Approximate Energy, kJ/mole	Approximate Wavelength, nm
Electronic (E_e)	400	3×10^2
Vibrational (E_v)	20	6×10^3
Rotational (E_R)	0.4	3×10^8

SEC. 9.1 Principles of Photoluminescence

S_2 as depicted in Figure 9.1). The absorption process described here is given by the usual Beer-Lambert equation, that is,

$$\log\left(\frac{I_0}{I_T}\right) = \epsilon bc \qquad (9.3)$$

where I_0 = incident intensity of exciting radiation
I_T = transmitted intensity of exciting radiation

(Alternatively, this can be expressed as radiant *power*, P_0 and P_T.) The parameters ϵ, b, and c are respectively the molar absorptivity, cell path length, and molar concentration of the chromophore (see Chap. 7).

Deexcitation Processes

Consider excitation of the molecule to the excited state S_2. In condensed-phase systems, the molecule can rapidly dissipate excess vibrational energy as heat by collision with solvent molecules through a process termed *vibrational relaxation* (VR). In addition, during this deexcitation process the molecule can pass from a low vibrational level of S_2 to an equally energetic vibrational level of the first excited singlet, S_1. This process is called *internal conversion* (IC). The energy-degradation processes of VR and IC occur rapidly (approx. 10^{-12} sec) until the molecule reaches the vibrational levels of S_1. Boltzmann distribution is then rapidly established among the vibrational levels of S_1 and, consequently, the lowest vibrational level ($V = 0$) is most likely to be occupied. Because of this rapid energy loss, emission from excited states higher than the first is rare. Only a few molecules such as azulene and some of its derivatives are found to violate this general rule. For many molecules, once the molecule reaches the first excited singlet, internal conversion to the ground state is a relatively slow process. Thus, emission of a photon from these molecules becomes more competitive with other decay processes. This process of emitting a photon for deexcitation of S_1 to S_0 is termed *fluorescence*. Generally, fluorescence emission occurs very rapidly after excitation (in approx. 10^{-9} to 10^{-7} sec). Consequently, it is not possible for the eye to perceive fluorescence emission after removal of the excitation source.

As can be seen from the energy-level diagram in Figure 9.1, fluorescence and absorption should have at least one electronic transition of the same energy. Because these transitions occur between the zero vibrational levels of S_1 and S_0, they are frequently called the "*0-0*" *transitions*. In practice, there is a slight shift between the "0-0" bands due to the differences in solvent effects on the ground and excited states [8].

As stated earlier, while a molecule is in the excited state, one electron may reverse its spin. This corresponds to a multiplicity of 3, and thus the molecule transfers to a triplet state. This process of nonradiative transfer from the singlet to the triplet state is termed *intersystem crossing* (ISC). The triplet state is lower in energy than the corresponding singlet state [11]. Note, again from the diagram in Figure 9.1, that there is no ground triplet state because we are describing organic molecules with an even number of electrons. Consequently, the lowest energy state corresponds to electrons occupying orbitals with other electrons, and these electrons must be spin paired according to Hund's rule [11]. After transferral to the triplet state, the molecule can rapidly attain the lowest vibrational level of the first excited triplet (T_1) through the VR process. From T_1, the molecule can return to S_0 by emission of a photon. This emission is called *phosphorescence*. Because transitions between states of different multiplicities are "forbidden" [11], phosphorescence is much longer lived than fluorescence (approx. 10^{-3} to 10 sec). Consequently, one can quite often observe "afterglow" when the exciting light is removed from a phosphorescing sample. Because of this relatively long lifetime, radiationless processes can compete more effectively with phosphorescence than with fluorescence. For this reason, phosphorescence must usually be carried out under cryogenic conditions to reduce the competitive nonradiative processes. However, recent studies have shown that it is possible to observe room-temperature phosphorescence if the sample is incorporated into a solid matrix [12–14], treated with micelles [15–17] or mixed with a sensitizer [18, 19].

A deactivation process that will compete with fluorescence is the process of *quenching*. Quenching can occur by a variety of mechanisms [8]. However, the net result in all cases is deactivation of the S_1 or T_1 state through a radiationless process involving interaction with some sort of quencher molecule.

Table 9.2 provides a summary of the excitation and major deexcitation processes discussed above, along with symbols for the rate constants. All of the deexcitation processes are first order or can be described as pseudo first order. Consequently, fluorescence and phosphorescence would be expected to decay exponentially after removal of the excitation source.

Kinetics of Fluorescence and Phosphorescence

The intensity of absorbed light, ΔI, is given by

$$\Delta I = I_0 - I_T \tag{9.4}$$

For convenience, we will refer to ΔI as the rate of absorption. If we excite a sample for a period of time that is long compared to the radiative and radiationless deactivation processes, the number of molecules in the S_1 state will reach a steady state. That is, the rate of producing the S_1 state will exactly equal the rate of deactivation of this

TABLE 9.2. *Summary of Excitation and Deexcitation Processes*

Process	Equation	First-Order Rate Constant
Absorption	$S_0 + h\nu_a \to S_2$ (or S_1)	k_a
Internal conversion/ vibrational relaxation	$S_2 \rightsquigarrow S_1$	k_{IC}
Fluorescence	$S_1 \to S_0 + h\nu_f$	k_f
Intersystem crossing	$S_1 \rightsquigarrow T_1$	k_{ISC}
Phosphorescence	$T_1 \to S_0 + h\nu_p$	k_p
Collisional quenching of S_1	$S_1 + Q \to S_0 + Q +$ Heat	$k_Q[Q]$ (pseudo first order)

state. In mathematical form, we can write

$$\Delta I = (k_{IC} + k_{ISC} + k_f + k_Q[Q])[S_1] \tag{9.5}$$

Here, k_{IC}, k_{ISC}, and k_f are the first-order rate constants for the previously described deactivation processes. The parameter k_Q is the second-order quenching rate constant, which can be combined with the relatively large quencher concentration, [Q], necessary for effective quenching to give a pseudo first-order rate constant, $k_Q[Q]$. The bracketed term $[S_1]$ is the steady-state concentration of S_1 molecules. For convenience, vibrational relaxation has been included in k_{IC}, and it is assumed that vibrational relaxation from S_1 to S_0 is negligible compared with fluorescence.

A certain fraction of the total number of photons absorbed will result in fluorescence emission. We shall call that fraction ϕ_f and will define it as

$$\phi_f = \frac{\text{Total Number of Photons Emitted}}{\text{Total Number of Photons Absorbed}} \tag{9.6}$$

Consequently, the rate of fluorescence can be directly related to the rate of absorption; that is,

$$I_f = \phi_f \Delta I = k_f[S_1] = \phi_f(k_{IC} + k_{ISC} + k_f + k_Q[Q])[S_1] \tag{9.7}$$

Rearranging, we obtain

$$\phi_f = \frac{k_f}{k_{IC} + k_{ISC} + k_f + k_Q[Q]} \tag{9.8}$$

Then, the lifetime of the S_1 state is given by

$$\tau = \frac{1}{k_{IC} + k_{ISC} + k_f + k_Q[Q]} \tag{9.9}$$

If all the processes that compete with fluorescence were absent, we could define the radiative lifetime, τ_r, as

$$\tau_r = \frac{1}{k_f} \tag{9.10}$$

Consequently, Equations 9.8 through 9.10 can be combined to give

$$\phi_f = \frac{\tau}{\tau_r} \tag{9.11}$$

Similar expressions can be derived for phosphorescence, that is,

$$\tau_p = \frac{1}{k_p + k'_{VR} + k_{Qp}[Q_p]} \tag{9.12}$$

and

$$\frac{\phi_p}{\phi_t} = \frac{\tau_p}{\tau_{pR}} \tag{9.13}$$

where k_p = first-order decay constant of T_1 to S_0
k'_{VR} = constant for vibrational relaxation of the T_1 state
$k_{Qp}[Q_p]$ = pseudo first-order rate constant for quenching of the triplet state by an impurity quencher, Q_p

τ_p and τ_{pR} = lifetimes in, respectively, the presence and absence of the competitive radiationless processes

ϕ_t = efficiency of formation of the triplet state

In our derivation, we have neglected the transferral process from the triplet to the singlet state that leads to delayed fluorescence.

Polarization

All of the preceding discussions have related to excitation of the fluorophore with *nonpolarized light*. Nonpolarized light consists of electric vectors vibrating in all planes, and each electric vector has an associated magnetic vector that is perpendicular to the electric vector. For absorption of radiation, we are concerned with only the electric vector. Let us assume that our excitation source is a beam of plane-polarized light, that is, light with an electric vector vibrating in only one plane. Consequently, the greatest amount of absorption occurs when the plane of the electric vector of this polarized light corresponds to the direction of the transition moment of absorption. The emitted fluorescence radiation will be polarized to a degree dependent on the angle between the absorption and emission oscillators. At this point, it is useful to define the "degree of polarization," p, as

$$p = \frac{I_{f\|} - I_{f\perp}}{I_{f\|} + I_{f\perp}} \qquad (9.14)$$

where $I_{f\|}$ and $I_{f\perp}$ = intensities of the components of the fluorescence parallel and perpendicular, respectively, to the direction of the electric vector of the plane-polarized exciting light

For a solution of randomly oriented molecules, p is found to vary from $+1/2$ to $-1/3$ [8]. Weber and Bablouian [20] have indicated that maximum polarization is observed with a dilute solution of the fluorophore in a highly viscous solvent.

Linearity of Fluorescence

Recall from the previous discussion that the intensity of fluorescence is directly related to the amount of absorbed radiation:

$$I_f = \phi_f \Delta I = \phi_f (I_0 - I_T) \qquad (9.15)$$

Using Beer-Lambert's law, we can show that

$$I_T = I_0 \times 10^{-\epsilon bc} \qquad (9.16)$$

Combining these two Equations 19.15 and 19.16, we obtain

$$I_f = \phi_f I_0 (1 - 10^{-\epsilon bc}) \qquad (9.17)$$

For dilute solutions ($\epsilon bc \leqslant 0.01$), Equation 9.17 can be reduced to

$$I_f = 2.303 \phi_f I_0 \epsilon bc \qquad (9.18)$$

This equation implies that, if we assume ϕ_f is wavelength independent, then observation of I_f as a function of exciting wavelengths will give a profile identical to that of the absorption spectrum (correcting for instrumental factors such as variations of I_0 with exciting wavelengths). For this reason, absorption and excitation spectra are often used interchangeably in fluorescence spectroscopy. However, it should be realized that the methods of acquiring normal absorption and excitation spectra are fundamentally different.

For concentration ranges that satisfy Equation 9.18, it is also apparent that I_f is proportional to concentration c. It is also useful to note that the fluorescence yield can be increased by increasing the intensity of the exciting radiation. Of course, there is a limit to how far we can increase I_0 because such an increase will result in increased sample photodegradation. Thus, Equation 9.18 is obviously an important relationship in the use of molecular fluorescence in analytical chemistry.

From the above discussion, we can deduce that all fluorescing molecules can be characterized by two types of spectra: the *excitation spectrum* (fluorescence intensity observed as a function of exciting wavelength at some fixed emission wavelength) and the *emission spectrum* (emission intensity observed as a function of emitted wavelength at a fixed exciting wavelength). These types of spectra will be discussed later in more detail.

Because some energy loss occurs before fluorescence, the fluorescence spectrum will occur at longer wavelengths than does the absorption (excitation) spectrum

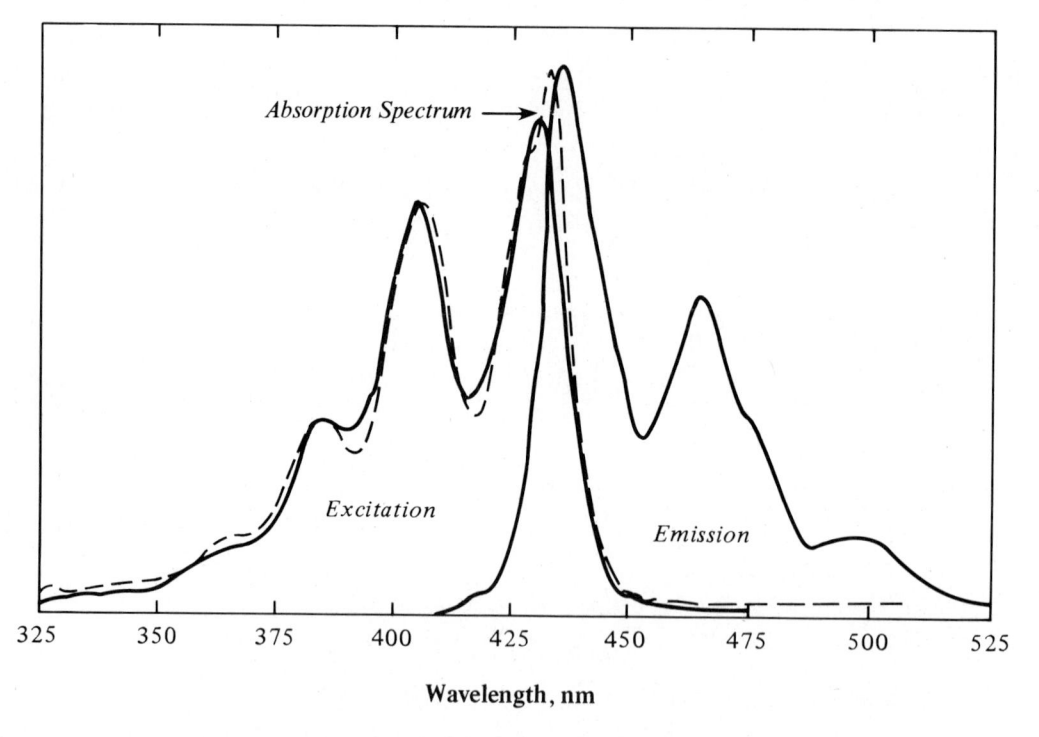

FIGURE 9.2. *Example of mirror-image relationship of the fluorescence and absorption spectra using the polynuclear aromatic compound perylene. (Note: Strict adherence to depiction of the mirror-image relationship would require plots of intensity versus energy, i.e., $1/\lambda$).*

(Stokes' law). Only the longer wavelength band of absorption and the shorter wavelength band of fluorescence (commonly called the "0-0" transition of absorption and emission) will generally overlap. Moreover, since the vibrational spacing in the ground state (S_0) and the first excited singlet (S_1) will often be similar for large molecules, the fluorescence spectrum is often the mirror image of the absorption spectrum (Fig. 9.2). Because phosphorescence emission occurs from a triplet state, there is no mirror relationship with the absorption band of the lowest excited singlet.

Two important properties of emission should now be evident. First, given that emission (phosphorescence and fluorescence) almost always occurs from the first excited state, the emission spectrum is independent of the wavelength of excitation. Second, since the quantum yield of emission is generally independent of wavelength of excitation, it follows that the excitation spectrum is independent of the emission wavelength monitored. These two properties can be schematically represented in a three-dimensional diagram of fluorescence intensity as a function of emission and excitation wavelength. Such a diagram is shown in Figure 9.3 for the fluorescence of the compound perylene. Here, the emission and excitation spectral information is incorporated into a single figure. The multicomponent utility of fluorescence is aptly depicted in such a diagram. One can selectively excite or selectively monitor the emission of certain components. Thus, fluorescence spectroscopy has a much greater potential for multicomponent analysis than does absorption spectroscopy.

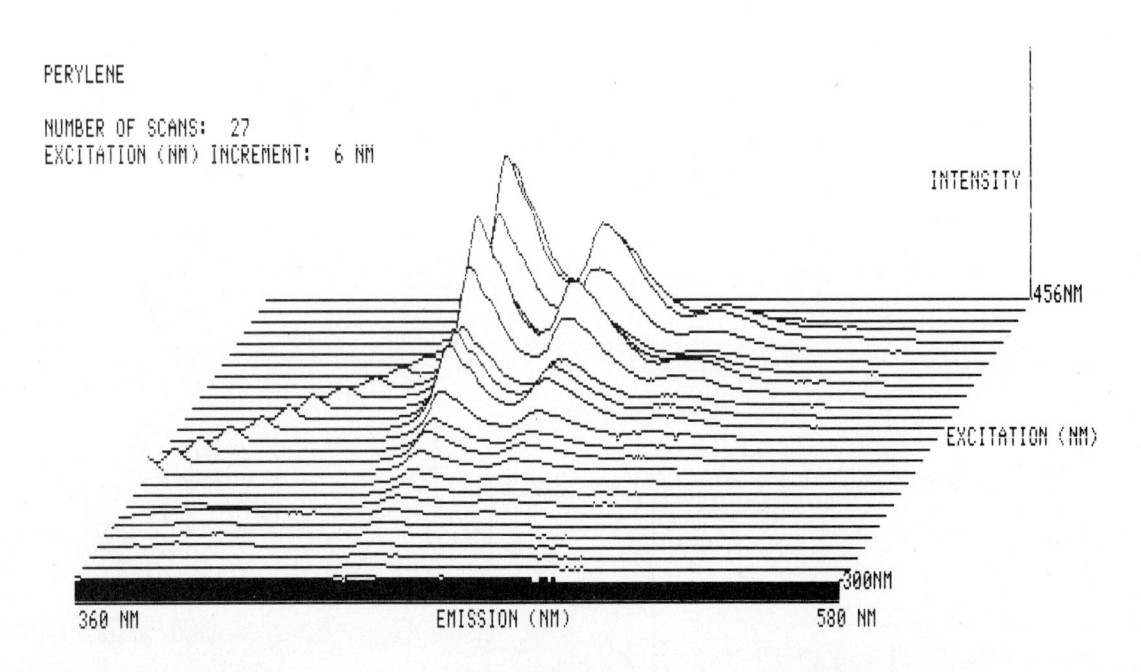

FIGURE 9.3. *Three-dimensional projection (emission-excitation matrix) of perylene. Total of 27 scans, with the excitation wavelength incremented 6 nm each time.*

9.2 FLUORESCENCE AND PHOSPHORESCENCE INSTRUMENTATION

As with most spectral methods, the major components of fluorescence and phosphorescence instrumentation include a source, a monochromator or wavelength-selector system, a sample cell, and a detector. One important difference with these two methods is that they generally require two separate wavelength-selector systems—one for excitation, and a second for emission.

Fluorimeters

Two basic types of fluorimeters are commercially available. One is the filter fluorimeter, which usually operates at fixed excitation and emission wavelengths and is relatively inexpensive. The second is the more expensive grating instrument, which is capable of wavelength scanning and produces better resolution than does the filter instrument. This section will provide a brief description of the design features incorporated into both types of fluorimeters. The abbreviated presentation here is designed to be only somewhat descriptive. More detailed information can be found in other sources [8, 21, 22].

The fluorimeter typically consists of six major components (Fig. 9.4). The first component is the *radiation source*. Because of its relatively broad continuum, extending into the ultraviolet region, the xenon-arc lamp is usually the preferred

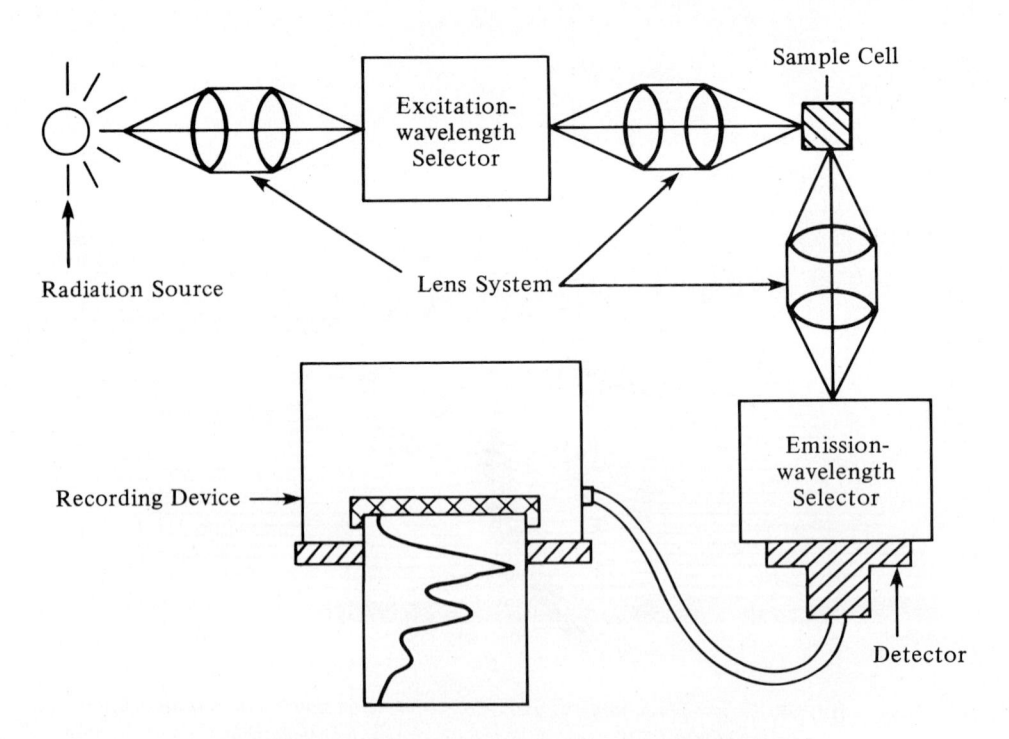

FIGURE 9.4. *Six major components of a basic fluorimeter.*

radiation source for a grating fluorimeter. However, for greater excitation energy at selected wavelengths, the mercury-arc lamp is often used. Because a mercury arc consists of sharp mercury lines superimposed on a continuum, excitation spectra are usually severely distorted with this lamp. Figure 9.5 gives a comparison of the outputs for the xenon- and mercury-arc lamps as provided by the manufacturers. Some fluorimeters use a hybrid xenon/mercury-arc lamp as the source in order to combine the advantages of both types.

The second component of the fluorimeter is the *lensing system* for efficient transfer of the exciting and emitting radiation. Quartz lenses are used in the ultraviolet-wavelength range (200 to 380 nm), and glass lenses can be used in the visible-wavelength range (380 to 700 nm).

Isolation of selected wavelengths from the radiation source is the next necessary part of a fluorimeter. This isolation is provided by the third component, the *excitation-wavelength selector system*. The function of this component is the selection of monochromatic or narrow-band radiation for sample excitation. The excitation-wavelength selector system can be a filter (or series of filters) or a grating monochromator. The filter provides more through-put, but the grating monochromator has the advantage of greater versatility and is usually the preferred system for luminescence research because broad-band excitation will increase the possibility of interferences.

For excitation at 254 nm, a combination of filters is used that transmits only below 300 nm. For example, a 7-54 filter plus a special plastic filter, as shown in Figure 9.6A, could be used for this purpose. Alternatively, because the plastic filter gradually decomposes and must be replaced, a mercury-line interference filter can be

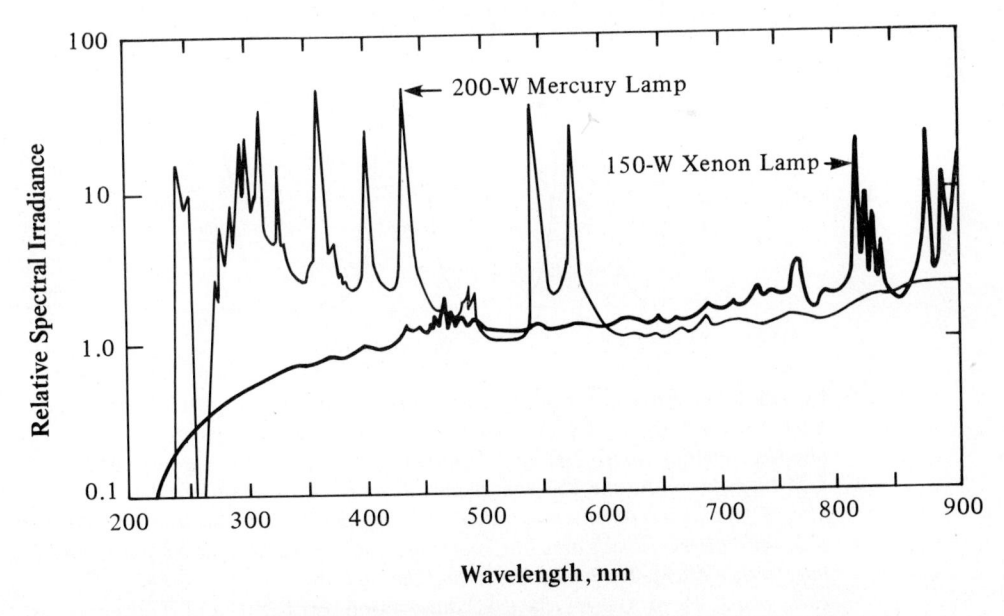

FIGURE 9.5. *Relative spectral output for xenon and mercury lamps. Courtesy of the Oriel Corporation, Stratford, Conn.*

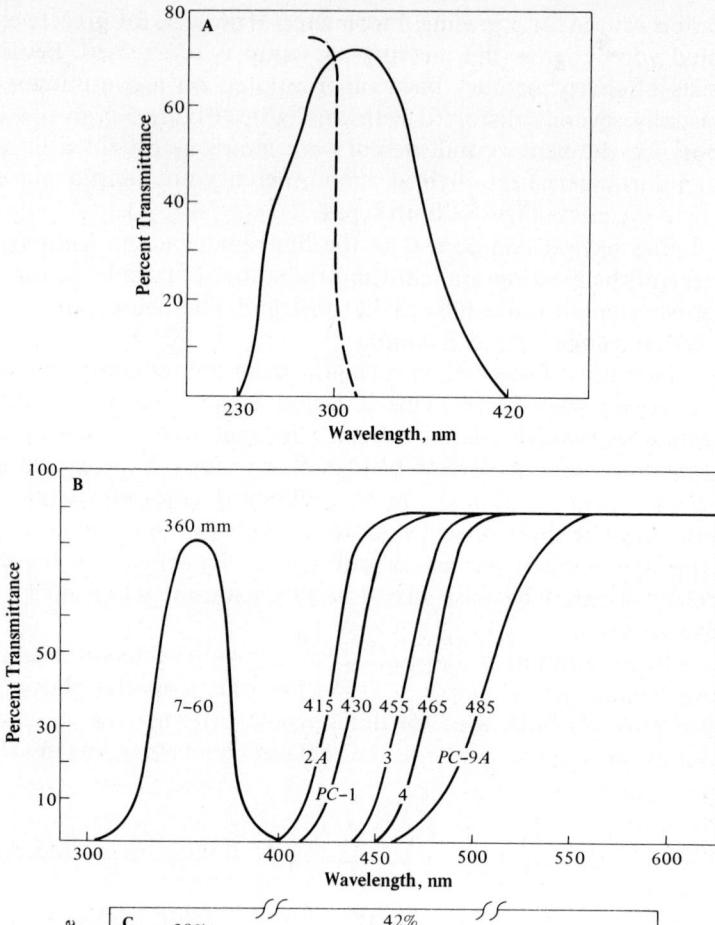

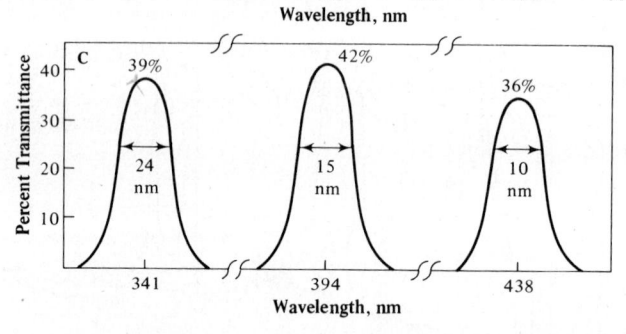

FIGURE 9.6. *Percent transmittance versus wavelength for several types of filters used in fluorescence. A: The* 7-54 *filter (solid line). A plastic filter that absorbs radiation above 300 nm (dotted line) is used in combination with the 7-54 filter to isolate the 254-nm mercury line. B: The 7-60 narrow-bandpass filter (left) and several sharp-cut filters. The nominal cutoff wavelength for the sharp-cut filters is indicated on each curve. C: Several Bausch-and-Lomb interference filters. The bandwidth at half the peak transmittance is given under each peak. From G. H. Schenk,* Absorption of Light and Ultraviolet Radiation, *Boston: Allyn and Bacon, 1973 p. 265, by permission of the publisher.*

used. This is a single, permanent filter for isolating the 254-nm line without transmitting the 313-nm and other lines emitted by the low-pressure mercury arc.

A filter that can be used for excitation at longer wavelengths is the 7-60 filter, whose transmittance is shown in Figure 9.6B. This filter transmits almost all of the intense band of radiation emitted by low-pressure, phosphor-coated mercury sources which are usually used in a filter instrument.

A *sample cell* (cuvet) is the fourth component, and provides sample containment and a uniform illumination surface. These cuvets are typically constructed of quartz to allow passage of ultraviolet radiation. These cells are similar to 1-cm ultraviolet absorption cells, except that all sides are polished because fluorimeters usually use the 90° geometry depicted in Figure 9.4 rather than the 180° geometry common to absorption spectrophotometers.

The fifth component, the *emission-wavelength selector system*, is similar in description to the excitation-wavelength selector system. However, this system, which monitors emitted radiation, is generally placed at an angle of 90° with respect to the excitation axis to minimize interferences from transmitted and scattered exciting light. With regard to filter instruments, the bandwidths of the primary and secondary filters should not overlap. For example, if a 7-60 primary filter is used, then the secondary filter chosen should not transmit below about 400 nm. A good choice would be the 2A secondary filter depicted in Figure 9.6B.

Three types of secondary filters are available: the narrow-bandpass filter, the sharp-cut filter, and the interference filter. Typical examples of all three are shown in Figure 9.6. The most common type used in filter fluorimeters is the sharp-cut filter. Sharp-cut filters are usually made of glass, as are the narrow-pass filters, but they contain chemicals that absorb all ultraviolet and visible radiation up to the wavelength specified. They have a higher transmittance (up to 85%) than do the narrow-pass filters, and have the advantage that they transmit nearly all radiation at wavelengths longer than the specified wavelength. This makes them useful for trace analysis because a much more intense beam of fluorescence emission is allowed to reach the phototube than with narrow-pass or interference filters. For an extensive survey of the transmittance properties of color filters, consult Reference [23].

Interference filters consist of (a) two outer layers of glass on whose inner surfaces a thin semitransparent metallic film has been deposited, and (b) an inner layer of some transparent material, such as quartz, calcium fluoride, or magnesium fluoride. Most radiation striking this filter will exhibit destructive interference, except for the narrow band of radiation that the filter is manufactured to transmit. The bandwidth of the interference filter decreases as the wavelength it is made to transmit increases. Thus, the 341-nm filter (Fig. 9.6C) has a bandwidth of 24 nm, but the 438-nm filter has a bandwidth of only 10 nm.

The advantages of the fluorescence grating over the secondary filter are the same as those of the excitation grating over the primary filter. However, the secondary filter generally transmits a greater fraction of the emitted radiation than does the fluorescence grating (a possible exception would be some newer grating spectrofluorimeters that have wide emission bandwidths). If the secondary filter is a sharp-cut type, then the energy that reaches the detector will be greater than if a narrow-bandpass or interference filter were used. In general, the use of a secondary filter and filter fluorimeters is recommended for trace analysis.

The final component of the fluorimeter system is the *detector*, which is placed at the exit slit of the emission-wavelength selection system. The photomultiplier tube is usually used in most fluorimeters because of its high gain and relatively broad spectral sensitivity to low radiation levels. Figure 9.7 illustrates the spectral response of some common photomultiplier tubes.

Phosphorimeters

A phosphorimeter for phosphorescence measurement is similar in construction to a fluorimeter, with two major differences. First, the sample is usually analyzed at liquid-nitrogen temperatures (77 K) to minimize quenching effects that are usually competitive enough to prevent observation of phosphorescence at room temperature. Second,

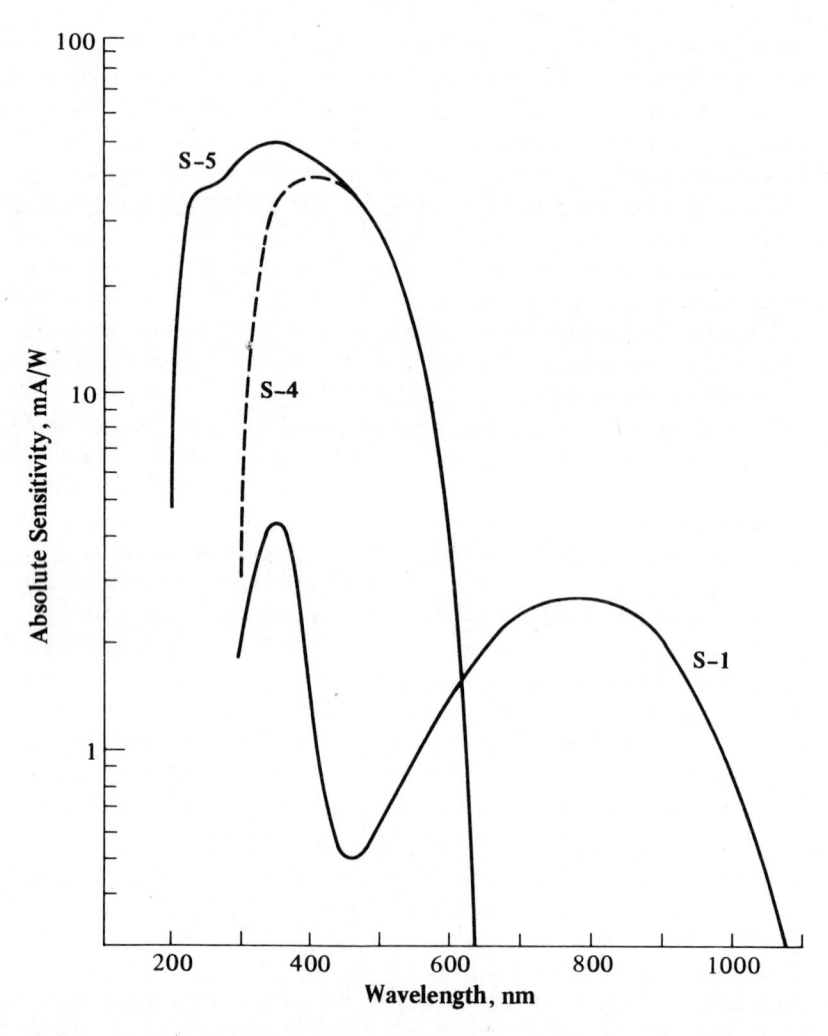

FIGURE 9.7. *Some spectral responses of photomultipliers. S-5 = RCA 1P28; S-4 = RCA 1P21; S-1 = RCA 7102.*

CHAP. 9 Molecular Fluorescence and Phosphorescence

excitation must be "gated" in time to observe phosphorescence emission in the absence of fluorescence emission. In recent years, two approaches to this gating have been emphasized. One approach uses a *rotating can* and a Dewar flask [24] filled with liquid nitrogen to freeze the solution into a glass. Optimum glass formation at liquid-nitrogen temperatures has been achieved using the solvent EPA (ether, isopentane, and ethyl alcohol in a 5:5:2 ratio). The lower part of the Dewar is smaller than the upper part and is made of quartz to permit transmission of the exciting radiation and the phosphorescence emission. The Dewar is placed inside the rotating can, which has two apertures (slits). As a slit moves into line with the excitation beam, the sample is excited. The speed of rotation is such that the short-lived fluorescence ceases before the slit moves into line with the emission detector, so that only phosphorescence is observed.

The second approach to gating the emitted light involves the use of a pulsed excitation source. An example of a commercially available instrument that incorporates this approach is the Perkin-Elmer LS-5 fluorimeter, schematically shown in Figure 9.8. The diagram has been simplified to show only the major components of this system. A phosphorescence sample compartment can be readily interchanged with the standard fluorescence compartment. Samples are maintained under cryogenic conditions using liquid nitrogen and special sample cells. This instrument uses a 9.9-W xenon discharge lamp that is pulsed at line frequency, and which has a width at half peak of less than 10 μsec. The signals for the photomultipliers are gated for the duration of the flash, measured by an A/D converter with an 18-bit dynamic range, and stored by the microprocessor. Shortly before the next flash, the photomultipliers are gated again for the same length of time and the signal is digitized and stored.

By subtracting the values obtained between flashes from those obtained during the flash from both the sample and the reference channels, numbers are obtained that theoretically represent the signal without any contribution from dark current. This system is applicable to any phosphorescent species with a lifetime greater than 20 msec. Other operations such as scale expansion, Savitzky/Golay filtering, integration, and offsetting are performed digitally using the system microprocessor.

Spectral Correction

As has been indicated, the spectral profile of the true excitation spectrum of a molecule should be identical to its absorption profile. In practice, however, the excitation profile is often distorted by instrumental parameters such as source output, monochromator through-put, and detector efficiency. Thus, excitation spectra obtained in the usual nonratio mode, with constant voltage on the photomultiplier, will have lower peaks in the ultraviolet region due to the lower energy from the xenon source. One method for compensating for this apparent loss in signal is to use a second "reference" detector which measures a small fraction of the light emerging from the excitation monochromator [25]. The output of this detector is ratioed with that of the sample detector, giving an output that is almost independent of changes of source intensity with wavelength or time. Excitation spectra obtained in this manner more closely resemble ultraviolet absorption spectra.

However, because the characteristics of the reference detector have not been taken into account, this approach would not provide fully corrected spectra. To

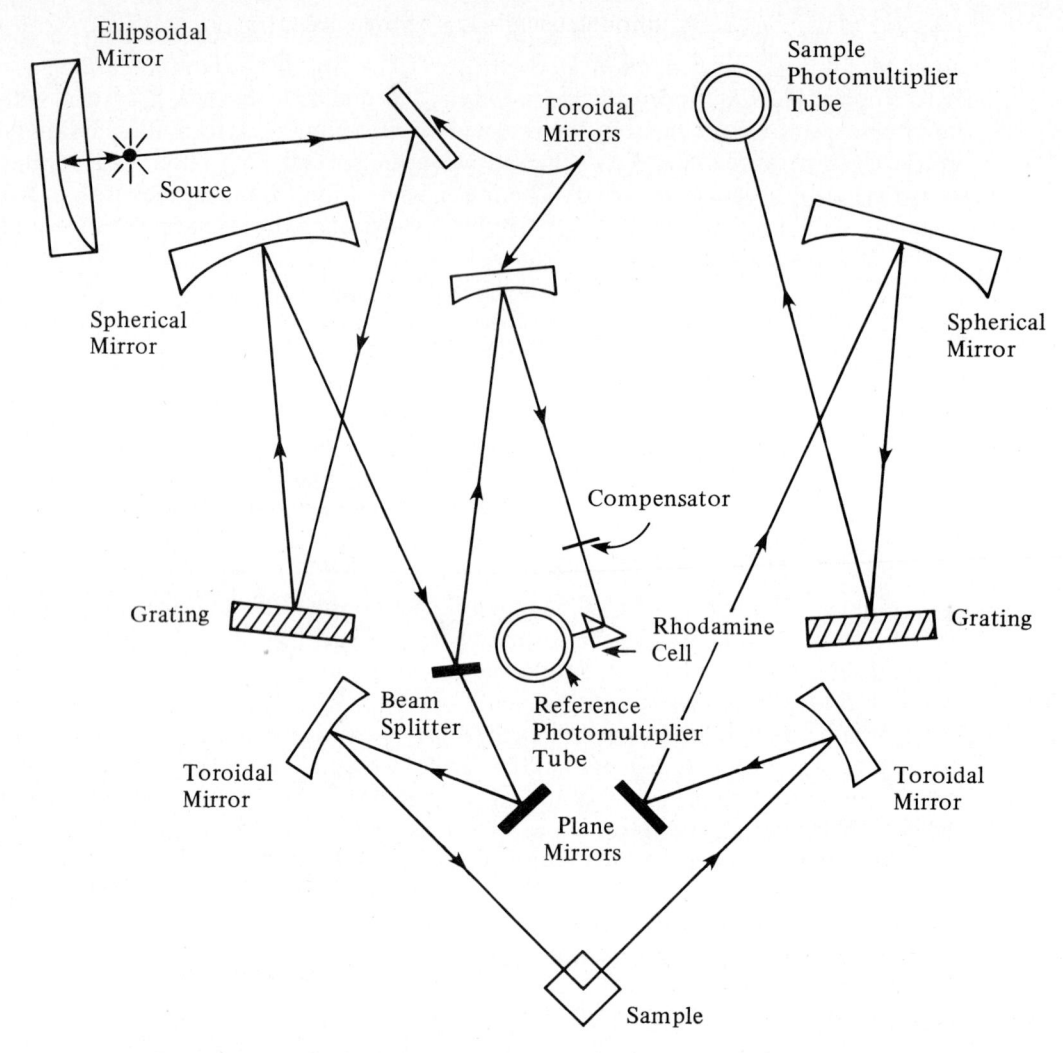

FIGURE 9.8. *Simplified schematic diagram of the Perkin-Elmer LS-5 fluorescence spectrophotometer. Reprinted with permission of the Perkin-Elmer Corporation.*

obtain a corrected excitation spectrum, the reference detector system must have a linear response with the quantum intensity of the incident beam exciting the sample. This is achieved by using a quantum counter (a compound whose fluorescence efficiency is independent of the excitation wavelength) in front of the reference photomultiplier. Such a compensating system is provided in the instrument shown in the optical diagram Figure 9.8. An excitation spectrum acquired on this instrument was previously shown in Figure 9.2. As a comparison, the absorption spectrum (dotted curve) acquired on a conventional absorption spectrophotometer is also displayed. Most of the differences observed in this figure can be attributed to differences in the bandwidths of the two instruments. It should be noted that emission spectra are also instrumentally distorted, although usually less so than excitation spectra. Corrections will usually involve the generation of responsivity curves for the spectral detection system. These and other aspects of spectral correction have recently been discussed in detail [21, 25].

Detection Limit of Fluorescence and Absorption Spectroscopy Compared

One can typically achieve lower detection limits with fluorescence than with absorption spectroscopy. Typically, a good absorption spectrometer can measure the absorbance of a compound in the range of 1.0×10^{-3} absorbance units. If we assume a very efficient absorber with a value of $\epsilon \approx 1 \times 10^5$, we can compute a lower detection limit, C_{DL}, of

$$C_{DL} \cong \frac{1.0 \times 10^{-3}}{1 \times 10^5} = 1 \times 10^{-8}\ M \tag{9.19}$$

for a 1-cm cuvet. On the other hand, detection limits in fluorescence spectroscopy can easily approach concentrations in the range of $1 \times 10^{-12}\ M$ using photon-counting techniques. Consequently, detection limits in fluorescence spectroscopy are generally more than three orders of magnitude better than in absorption spectroscopy.

9.3 APPLICATIONS OF FLUORESCENCE AND PHOSPHORESCENCE

Up to now, we have focused our attention primarily on molecular fluorescence spectroscopy as applied to analysis of organic compounds. However, the sensitivity and selectivity of fluorescence has provided a number of applications of the technique in inorganic analysis. Consequently, the applications discussion here will include both organic and inorganic fluorescence methods of analysis. Many of the compounds cited are also amenable to phosphorescence analysis. However, the routine use of phosphorescence methods of analysis has been circumvented because of the usual requirement of operating under cryogenic conditions.

Organic Compounds

To fluoresce, an organic compound will usually have to be highly conjugated. For example, the structure of vitamin A is as follows:

This compound is known to fluoresce and has a blue green fluorescence with an emission maximum at approximately 500 nm in ethanol. Specifically, there exists a large class of organic compounds that are highly conjugated. These are the polynuclear aromatics. In fact, most of these compounds are good fluorescers. Table 9.3 provides a list of the various classes of organic compounds that are known fluorescers. In general, most of these compounds meet the requirement of extended conjugation.

Fluorescence of Selected Drugs. A number of examples of the use of fluorescence analysis in the determination of organic compounds can be found in the area of drug

TABLE 9.3. *Classes of Organic Compounds Exhibiting Usable Fluorescence*

Class	Best Examples (ϕ_f)	Weak or No Fluorescence	Reference
Hydrocarbons			
1. Aryl-substituted olefins	*trans*-Stilbene	*cis*-Stilbene	[60]
2. Unsubstituted aromatic hydrocarbons	Anthracene (0.2), pyrene (0.3)	Benzene (0.04), biphenyl (P)	[35, 60]
3. Alkyl-substituted hydrocarbons	Toluene (0.1), mesitylene (0.2), 9-methylanthracene (0.3)		[8]
Nitrogen Compounds			
4. Aromatic amines	Aniline (0.1), 2-naphthylamine (0.5)	Nitroanilines (P)	[8, 60]
5. Amino acids	Tyrosine (0.2), tryptophan (0.2)	Phenylalanine (0.04)	[35]
6. "Phenylethylamines"	Amphetamine (0.02)		[61]
7. Heterocyclics	Quinine (0.55)	Pyridine	[35]
Halogen Compounds			
8. Cl-substituted aromatic hydrocarbons	1-Chloronaphthalene (0.06), *p*-chlorotoluene (0.02)	Chlorobenzene (P)	[35]
9. F-substituted aromatic hydrocarbons	Fluorobenzene (0.1), 1-fluoronaphthalene (0.06)		[60]
Oxygen Compounds			
10. Phenols	Phenol (0.2), 2-naphthol (0.3)	Nitrophenols (P)	[35]
11. Phenyl ethers	Anisole (0.3)		
12. Barbiturates	Phenobarbital (0.001)	5,5'-Dialkyl barbiturates	[61, 62]
13. Aromatic acids	Acetylsalicylic acid (0.02)	Benzoic acid (P)	[26]

Note: The quantum efficiency for fluorescence ϕ_f is indicated in parentheses. The designation (P) indicates useful phosphorescence properties.

analysis. Methods have been developed for "phenylethylamines," barbiturates, and aspirin. The phenylethylamines are substituted 1-amino-2-phenylethanes. Amphetamine (2-amino-1-phenylpropane) is the best known of the phenylethylamines. Many of these compounds are excited at 260 to 270 nm, emit at 282 to 300 nm, and can be determined fluorometrically at concentrations as low as 2 mg/L (2 ppm) [26]. Phenylephrine and epinephrine have about the same molar absorptivities as amphetamines, and have quantum efficiencies (ϕ_f) of about 0.08, so they can be determined at concentrations as low as 0.01 ppm.

Several approaches to the fluorometric determination of barbiturates have been published [27, 28]. A general structure for the "enol" tautomer of most common barbiturates is

Most barbiturates are fluorescent in 0.1 M base, but not in acid, because the base extracts protons from the 4-hydroxyl group and the 1-nitrogen to form a fluorescent dianion [28]. Substitution of a methyl group on the 1-nitrogen will result in deactivation by processes other than fluorescence. Most barbiturates can be excited at 255 nm and emit at 405 to 420 nm [27].

For many years, acetylsalicylic acid (ASA) was thought not to fluoresce and was commonly determined by hydrolysis to salicylic acid, followed by fluorometric determination of the salicylic acid. A tedious separation of the salicylic acid from the acetyl derivative was therefore necessary for the determination of both in aspirin tablets. Recently, it was found that ASA does indeed fluoresce in a solvent of 1% acetic acid in chloroform [29]. It is excited at 280 nm and emits at 335 nm, whereas salicylic acid is excited mainly at 308 nm and emits mainly at 450 nm. Therefore, it is possible to determine each in the presence of the other. It is also worth noting that ASA can be determined in the presence of salicylic acid by measuring its phosphorescence [30].

A number of other examples can be cited in the area of organic analysis using fluorescence spectroscopy. However, these will not be discussed here. Instead, the reader is referred to the recent reviews in *Analytical Chemistry* on fluorescence analysis [31–33] for additional information.

Inorganic Compounds

A number of applications of fluorescence spectroscopy to inorganic analysis can be cited. Most of these involve chelates of metal ions. Only a few examples will be considered here. The reader is again referred to supplemental material [34, 35] for additional information.

Simple (Unchelated) Luminescent Ions. An example of a simple ion that luminesces in solution is the uranyl ion, UO_2^{2+}. Another example is the aquated Ce(III) ion. The electron configuration of the latter is $[Xe]4f^1 5d^0$. The observed luminescence involves excitation of a $4f$ electron to a $5d$ orbital, after which luminescence occurs during a $5d \rightarrow 4f$ return transition. Cerium(III) is known to be excited at 254 nm [36]. Five absorption bands in aqueous solution have been assigned to various $4f \rightarrow 5d$ transitions (those at 200, 211, 221.5, 239.5, and 252.5 nm). Cerium(III) has been determined fluorometrically in 0.4 N H_2SO_4 [36] using excitation at 254 nm and measuring the emission at 350 nm. Because Ce(IV) does not fluoresce, Ce(III) can be determined in the presence of Ce(IV).

Of the other lanthanides, Eu(III) chloride, an f^6 ion, and Tb(III) chloride, an f^8 ion, have been reported to fluoresce weakly in dimethylformamide solution [34]. The chloride and sulfate salts of Sm(III), an f^5 ion; of Gd(III), an f^7 ion; and of Dy(III), an f^9 ion, are also reported to luminesce weakly in solution [34]. All five of these lanthanides give rise to weak lines which have been assigned to $f \rightarrow f$ transitions.

The other important ion that is luminescent in solution is the Tl(I) ion. The aquated Tl$^+$ ion can be excited at 215 nm and emits weakly at 370 nm. The $TlCl_4^{3-}$ ion is excited at 240 to 250 nm and emits strongly at 450 nm. A qualitative test for Tl(I) is based on its violet luminescence following addition of 1 M KCl [37]. The fluorometric determination of 10^{-7} M Tl(I) in 3.3 M HCl plus 0.8 M KCl is based on excitation at 250 nm and emission at 430 nm [38].

Chelates of Non-Transition-Metal Ions. This is the largest class of luminescent inorganic systems, and a full discussion is beyond the scope of this chapter. The reader is referred to the monograph of White and Argauer [39] for the many analytical applications of these systems. In general, the diamagnetic ions of the metals in groups IA, IIA, IIB, IIIA, and IIIB, as well as Zr^{4+}, can be determined by measuring the fluorescence of their chelates with aromatic organic ligands. The primary method of exciting these chelates is by promoting the π-π* transitions of the chelated ligand [34]. The ions most frequently measured by fluorescence are those in Group IIIA—Al(III), Ga(III), In(III), and Tl(III)—which form metal chelates with a large number of organic ligands, including 8-hydroxyquinoline, 2,2'-bipyridine, and salicylaldehyde derivatives. These ligands are usually weakly fluorescent when uncomplexed, but intensely fluorescent when complexed by these ions.

The metal ion for which the most methods are available is Al(III). As an example of the kinds of organic ligands that have been employed in fluorescence analysis, some selected organic reagents used for the determination of Al(III) are shown in Table 9.4.

Chelates of Transition Metals. Although many transition-metal ions form stable chelates and complexes with aromatic ligands, relatively few such systems are fluorescent. This is the case for chelated paramagnetic metal ions because the rate of intersystem crossing from the S_1 state to the T_1 state of the aromatic ligand is greatly increased by the unpaired electrons of metal ions. In solution, most T_1 states lose

TABLE 9.4. *Selected Fluorometric Reagents for the Determination of Aluminum*

Organic Chelate	Excitation Wavelength, nm	Emission Wavelength, nm	Concentration Range (Final Solution)
Morin (2',3,4',5,7-Pentahydroxyflavone)	270 (440)	500	4×10^{-6} to 5×10^{-5} M^a; 4×10^{-8} to 2×10^{-7} M^b
Pontachrome Blue Black R (PBBR)	330	635	7×10^{-8} to 6×10^{-7} M^c
8-Hydroxyquinoline	405 (366)	520	Detection limit of 10^{-6} M^d
Acid-Alizarin Garnet R (AAGR or 2,4,2'-Trihydroxyazobenzene-5'-sodium sulfonate)	470	575	$\sim 4 \times 10^{-8}$ to 9×10^{-6} M^e
N-Salicylidene-2-amino-3-hydroxyfluorene (NSAHF)	445	530	$< 3 \times 10^{-8}$ to 3×10^{-7} M^f

a. C. E. White and C. S. Lowe, *Ind. Eng. Chem., Anal. Ed.*, *12*, 229 (1940).
b. F. Will III, *Anal. Chem.*, *33*, 1360 (1961).
c. A. Weissler and C. E. White, *Anal. Chem.*, *18*, 530 (1946).
d. W. T. Rees, *Analyst*, *87*, 202 (1962).
e. C. E. White and R. J. Argauer, *Fluorescence Analysis*, New York: Marcel Dekker, 1970, pp. 55–57.
f. C. E. White, H. C. E. McFarlane, J. Fogt, and B. Fuchs, *Anal. Chem.*, *39*, 367 (1967).

all of their electronic energy by collisional deactivation or by rapid conversion to their S_0 states without emitting a photon. Thus, paramagnetic metal ions such as Fe^{3+}, Co^{2+}, Ni^{2+}, and Cu^{2+} are said to quench the fluorescence of their chelates.

Another phenomenon that operates in these chelates to prevent emission is the heavy-atom effect. Heavy diamagnetic atoms such as Hg^{2+}, Au^+, and Tl^{3+} increase spin-orbital coupling, which in turn increases the rate of intersystem crossing [40]. This effect appears to be most effective with Hg^{2+}, for which no well-documented metal-chelate luminescence has been reported. Certain group VIII d^6 transition-metal ions have been reported to luminesce [35]. When complexed by such strong-field ligands as 1,10-phenanthroline, 2,2-bipyridine, and 2,2,2-terpyridine, Ir(III), Ru(II), Os(II), and Rh(III) form diamagnetic metal chelates. Such chelates exhibit low-energy charge-transfer absorption (d → π^*) and emission (π^* → d) bands [40, 41]. Iridium(III) has actually been determined [41] by measuring luminescence (which appears to be phosphorescence) at room temperature in ethanol-water solution. It is interesting that no Fe(II) chelates have been observed to luminesce, although such chelates exhibit d → π^* absorption bands. Fink and Ohnesorge [42] have postulated that this is the result of a crossover to a $t_{2g}^3 e_g^2 \pi^*$ spin state during the lifetime of the initial excited state of the Fe(II) chelates. Such a spin state is paramagnetic and undergoes rapid intersystem crossing and rapid internal conversion to the ground state before emission can occur.

Multicomponent Analysis

The previous discussion alluded to the possible multicomponent capabilities of fluorescence spectroscopy using the diagram of Figure 9.3. This point can be illustrated further using a similar three-dimensional diagram for a mixture of perylene and anthracene, as shown in Figure 9.9. The initial wavelength of excitation is fixed and the fluorescence spectrum is scanned and stored. The excitation monochromator is then moved to another wavelength of excitation and the fluorescence spectrum is again scanned. This process is repeated until the desired array of fluorescence intensity as a function of various excitation and emission wavelengths is obtained. The data are then processed and plotted. Such a representation of fluorescence intensity as a function of multiple excitation and multiple emission wavelength has been referred to as a "total luminescence spectrum" [43, 44].

It should be noted that, although there are regions of overlap in absorption and fluorescence for both perylene and anthracene, other regions exist where each compound is a sole absorber or a sole emitter. Therein lies the multicomponent capability of fluorescence analysis. For example, to obtain selective information about anthracene, we can (a) obtain the fluorescence spectrum while exciting the mixture at a wavelength where anthracene only absorbs, or (b) obtain the excitation spectrum by using an emission wavelength where anthracene only emits. Figure 9.10 shows the selective excitation and fluorescence spectrum of anthracene obtained from the mixture of perylene and anthracene. A similar manipulation could be used to obtain pure spectra of perylene from the mixture. Sawicki et al. [45] have shown that it is possible to analyze benzo(a)pyrene in the presence of 50 other polynuclear aromatic compounds using this approach of selective excitation and selective fluorescence monitoring.

Fluorescence and Phosphorescence in Chromatography

Fluorescence and phosphorescence measurements have been extensively used to follow the separation of certain luminescent organic compounds in such techniques as liquid chromatography, gas-liquid chromatography, paper chromatography, and thin-layer chromatography (TLC). Such measurements have been found to be very useful in many cases. Recently, fluorescent derivatives that extend this approach to nonfluorescent compounds have been reported [46].

One excellent example of the use of fluorescence in liquid chromatography is the separation and characterization of polynuclear aromatic hydrocarbons in polluted air, reported by Sawicki and coworkers [47]. Column chromatography employing alumina was used to separate the hydrocarbons. Fluorescence, ultraviolet spectrophotometry, and colorimetric tests were used to identify aromatic hydrocarbons in the various fractions.

Fluorescence and phosphorescence have been used by Drushel and Sommers [48] to identify nitrogen compounds in petroleum fractions following gas-chromatographic separations. The nitrogen-rich petroleum fractions investigated were so complex and so small in size that the greater sensitivity of luminescence techniques over other spectroscopic techniques was essential for good analysis.

In paper chromatography, spots of fluorescent organic molecules have been

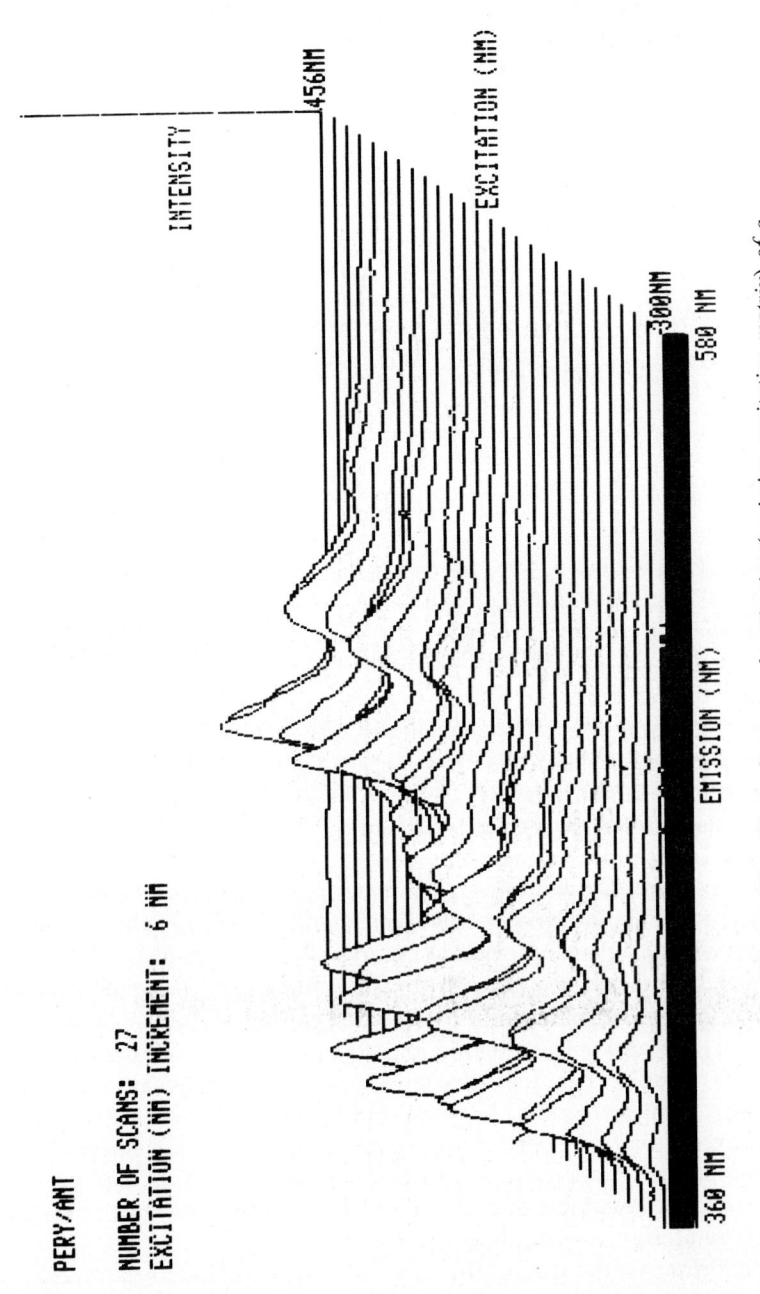

FIGURE 9.9. *Three-dimensional projection (emission-excitation matrix) of a mixture of perylene and anthracene. Total of 27 scans, with the excitation wavelength incremented 6 nm each time.*

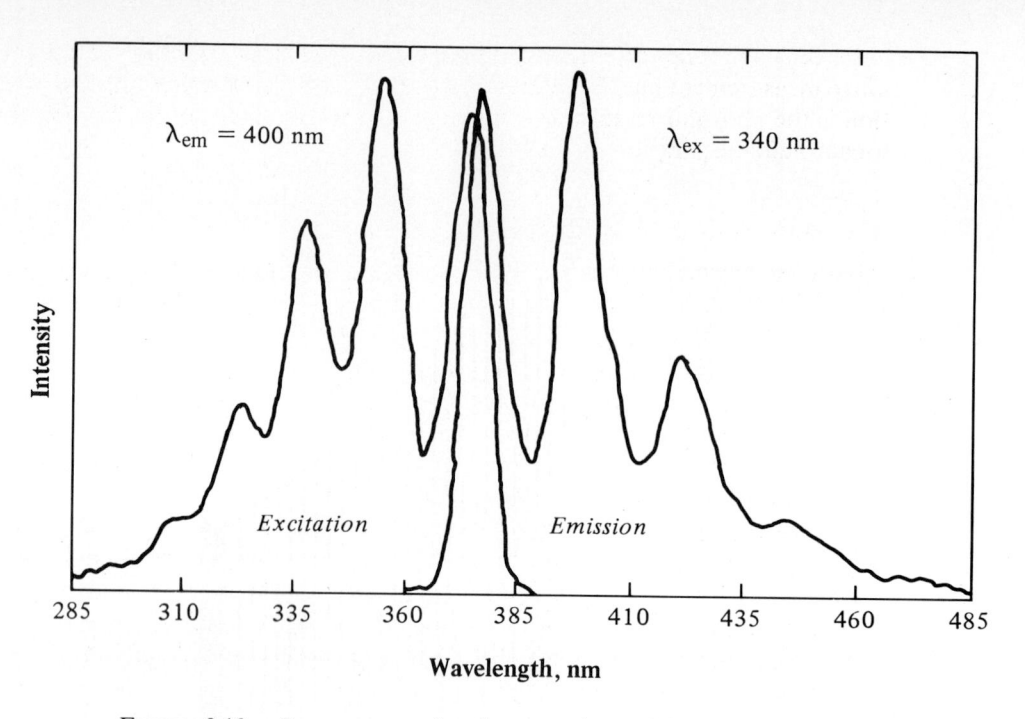

FIGURE 9.10. *Pure spectra of anthracene obtained by selective excitation (λ_{ex} = 340 nm) and selective fluorescence monitoring (λ_{em} = 400 nm) of the mixture displayed in Figure 9.9.*

identified on paper chromatograms by inspecting the paper under either short-wavelength (254-nm) or long-wavelength (300–400-nm) ultraviolet radiation. In most cases, the spots appear blue or violet. Sawicki and Pfaff [49] have shown that phosphorescent organic molecules can also be identified on paper chromatograms.

Luminescence measurements also find general use in thin-layer chromatography (TLC). In the more popular TLC separations, on the other hand, adsorbents or precoated TLC sheets containing certain inorganic phosphors are available that emit intense visible radiation over the entire length of the TLC chromatogram except where spots of the separated organic compounds are present. Most organic compounds quench the luminescence of the phosphor, and the spots appear as dark shadows against the brilliant luminescent background of unquenched phosphor.

Green (522 nm) is the most common luminescent color used on TLC sheets. Typical green phosphors are pure zinc silicate, and calcium silicate with a manganese-lead activator. Most such green phosphors emit only under short-wavelength (254-nm) ultraviolet excitation, thus permitting examination of the TLC sheet for fluorescent organic compounds under long-wavelength excitation. A phosphor of zinc and cadmium sulfides is available that yields an off-white luminescence under both short-wavelength and long-wavelength ultraviolet excitation.

In addition to the qualitative work described above, quantitative measurements on TLC plates are also possible. For example, Lefar and Lewis [50] have measured organic compounds both by the amount of emission on an adsorbent without a phosphor, and by the amount of quenching of a phosphor mixed with an adsorbent. Janchen and Pataki [51] have discussed many examples of direct quantitative measurement on TLC plates of spot luminescence. Another interesting application is the phosphorimetric measurement of nicotine, nornicotine, and anabasine in tobacco after separation by TLC [52].

Applications in Studies of Pollution

Fluorescence and phosphorescence have been used to investigate the pollution of both water and air. Fluorescent compounds, especially, have been used in the study of water flow and water pollution [53]. "Fluorescent tracers" are superior to radioactive tracers because they can be used at concentrations so low (0.001 ppb or even less) that they constitute neither a real contamination nor a health hazard. A common fluorescent dye, the first such dye to be used, is rhodamine B. The fluorescence emission of this compound is independent of pH from pH 5 to 10 and it can be determined at concentrations above 0.01 ppb.

The use of tracers enables industries and cities to control or reduce pollution before it occurs. Measuring the "time of travel" (mean velocity) of rivers and streams and the mixing of those waters into lakes and oceans helps to indicate where to discharge waste, when to discharge it, and at what rate. Tracers have been used, for instance, in San Francisco Bay and in Chesapeake Bay to facilitate proper waste disposal in those waters.

Fluorescence and phosphorescence both find use in the analysis of polluted air for specific chemical pollutants. Fluorescence is especially useful because many aromatic hydrocarbons are intensely fluorescent.

Several polynuclear aromatic hydrocarbons have been labeled as "priority pollutants" by the Environmental Protection Agency, and fluorometry has been used in the determination of these compounds [54]. Fluorometry is more useful than ultraviolet spectrophotometry in pollution analysis, not only because it is more versatile, but also because it can measure the very low concentrations of hydrocarbons found in samples collected from the air. Such analyses are all the more important because some aromatic hydrocarbons are carcinogenic. One of the best-known carcinogenic aromatic hydrocarbons is benzo(a)pyrene, or 3,4-benzopyrene. A very specific method for determining benzo(a)pyrene in the aromatic hydrocarbon fraction of polluted air samples has been developed, using sulfuric acid as a solvent, in which the compound forms a cation with a strong absorption band at 520 nm [54]. A few other aromatic hydrocarbons have weak absorption bands at 520 nm, but none of these emit fluorescent light at 545 nm as does benzo(a)pyrene. To determine benzo(a)pyrene, the sample is excited at 520 nm and the fluorescence emission at 545 nm is measured with the fluorimeter. It has been shown that benzo(a)pyrene can be estimated in artificial mixtures of over 40 similar compounds without separation. This analysis is also unusual in that visible light, not ultraviolet light, is used to excite a molecule.

Medicine

One of the most useful applications of fluorescence is in the routine determination of certain important molecules in body fluids for diagnostic purposes. Some such molecules are naturally fluorescent, but others must be chemically treated to form fluorescent products. For example, the amino acids tyrosine, tryptophan, and phenylalanine are all measured fluorometrically. Both tyrosine and tryptophan possess aromatic rings that absorb intensely and therefore have an intense natural fluorescence. Tyrosine is excited at both 225 and 280 nm, and emits at 303 nm; tryptophan is excited at 220 and 280 nm, and emits at 438 nm [55].

In contrast, phenylalanine possesses a weakly absorbing benzene ring and does not emit fluorescence intensely enough for measurement of trace quantities. The usual analytical methods involve treating it with ninhydrin, Cu(II) ion, and L-leucyl-L-alanine [56] to give a highly fluorescent product. Fluorometric measurement of phenylalanine is useful in testing for phenylketonuria, a hereditary metabolic disorder that causes mental retardation. In one series of tests, adult control samples ran 1.5 mg phenylalanine per 100 mL of blood serum. In contrast, parents of phenylketonuriac children ran 1.9 mg/100 mL, and the phenylketonuriacs themselves ran 30 mg/100 mL. Because phenylketonuriacs cannot convert phenylalanine efficiently to tyrosine, their levels of tyrosine were about half of those in the blood of control samples.

Limitations of Luminescence Measurements

Thus far, we have considered the aspects of luminescence analysis that make it a very useful analytical tool. As with any other analytical tool, fluorescence spectroscopy does have certain limitations.

As was indicated previously, the fluorescence signal is obtained by the measurement of the emitted intensity above background. Because fluorescence measurements are usually not blank compensated as in absorption measurements, the blank can be a serious limitation in achieving optimal detection limits. The major factors that contribute to the luminescence blank are (a) Raman scattering from the solvent, (b) luminescence of the sample holder (cuvet), (c) scattered light from Tyndall and Rayleigh effects, (d) adventitious fluorescence contributed by impurities in the solvent, (e) photodegradation of the sample, and (f) contributions from other fluorescent species in the sample. To some extent, many of these blank problems can be minimized through judicious experimental practice. For example, much of the luminescence from the cuvet can be eliminated by using sample holders made of synthetic silica rather than fused quartz. In addition, sample degradation can be minimized through use of the longer-wavelength absorption band of the analyte, if one is available. These and other aspects of minimizing the luminescence contributed by the sample blank have been discussed in detail elsewhere [8, 21].

Another limitation of luminescence analysis is contributed by synergistic effects, which will usually result in a diminution of the fluorescence intensity of the analyte. Figure 9.11 depicts three processes that contribute to these effects: the "inner-filter" effect, static quenching, and dynamic quenching. The *"inner-filter" effect* (I_a) is caused by the attenuation of the exciting light by excess absorption from the analyte or other molecule, thus decreasing the number of photons available for excitation. In addition, the absorption of the emitted fluorescence light by the analyte or other molecule (thereby preventing detection by the photomultiplier tube) can also be considered an inner-filter effect (I_b). *Static quenching* (II) involves an interaction

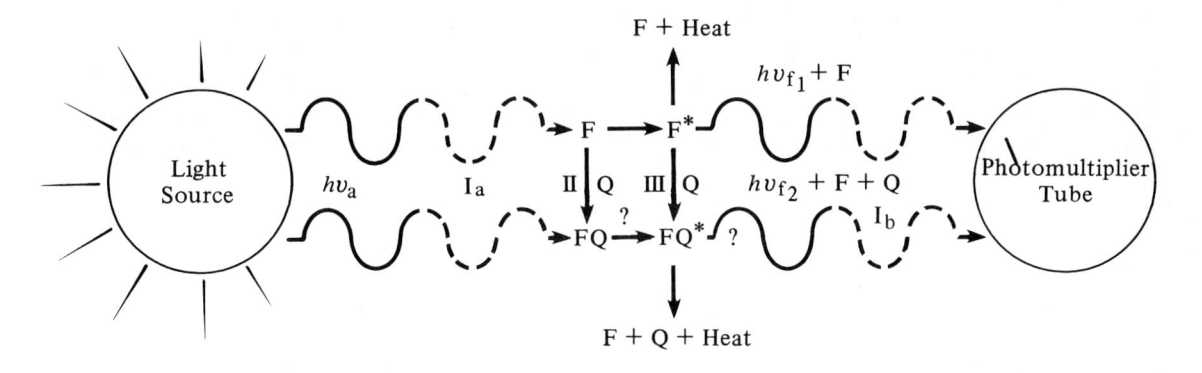

FIGURE 9.11. *Depiction of inner-filter and quenching processes in fluorescence. The fluorescent compound is denoted by F, the quencher by Q. I_a: absorption of exciting light; I_b: absorption of emitted light; II: static quenching; III: dynamic quenching. The question mark (?) indicates processes that do not necessarily occur.*

with the fluorescent substance in the ground state. This type of quenching produces a change in both the excitation and emission spectra. *Dynamic quenching* (III) is a diffusion-controlled process that involves an interaction with the excited state of the fluorescing molecule. This type of quenching produces a change only in the emission spectrum of the fluorescing species.

It is important to distinguish the inner-filter effects from true quenching, both static and dynamic: Quenching effects reduce the true quantum efficiency of the fluorescence process, whereas inner-filter effects will only diminish the fluorescence intensity by lowering the number of excited molecules or by lessening the emitted fluorescence signal. Thus, erroneous interpretation of experimental data can occur if a distinction is not made between these two processes.

Fluorescent compounds can be specifically quenched by one kind of molecule but unaffected by another. For example, bromine will selectively quench the fluorescence of benzopyrene in a binary mixture of benzopyrene and benzofluoranthene. The selective interaction of bromine with benzopyrene has allowed the quantitative determination of each substance in a mixture of the two [57]. Thus, specificity of quenching can sometimes be used for qualitative and quantitative purposes. An example of the latter, the Stern-Volmer relationship, written as

$$\frac{I_f^0}{I_f} = k_{SV}[Q] + 1 \qquad (9.20)$$

has been used extensively in fluorescence spectroscopy. The parameters I_f^0 and I_f are the fluorescence intensities in the absence and presence of the quencher molecule, Q, respectively. The parameter k_{SV} is a proportionality constant often called the Stern-Volmer constant. Thus, a plot of the ratio I_f^0/I_f versus a series of different quencher concentrations will give a straight line of slope k_{SV} and an intercept of unity. Fogarty and Warner [58, 59] have shown that Stern-Volmer quenching may be useful for the deconvolution of mixtures of fluorophores.

SEC. 9.3 Applications of Fluorescence and Phosphorescence

Finally, as indicated earlier, the limitations of fluorescence measurements can often be circumvented though careful experimental practice and the use of the many experimental parameters of fluorescence spectroscopy to achieve greater selectivity. Consequently, the sensitivity and selectivity advantages of fluorescence have led to its wide use in many areas of analytical chemistry.

SELECTED BIBLIOGRAPHY

BIRKS, J. B. *Photophysics of Aromatic Molecules.* New York: Wiley-Interscience, 1970.

BOWEN, E. J., ed. *Luminescence in Chemistry.* London: Van Nostrand, 1968.

GUILBAULT, G. G. *Practical Fluorescence: Theory, Methods, and Practice.* New York: Marcel Dekker, 1973.

HERCULES, D. N., ed. *Fluorescence and Phosphorescence Analysis.* New York: Wiley-Interscience, 1966.

MIELENZ, K. D., ed. *Measurement of Photoluminescence.* Volume 3 of *Optical Radiation Measurement*, edited by F. Grum and C. J. Bartleson. New York: Academic Press, 1982.

PARKER, C. A. *Photoluminescence of Solutions.* New York: Elsevier, 1968.

SCHENK, G. H. *Absorption of Light and Ultraviolet Radiation: Fluorescence and Phosphorescence Emission.* Boston: Allyn and Bacon, 1973.

UDENFRIEND, S. *Fluorescence Assay in Biology and Medicine.* New York: Academic Press, 1962.

REFERENCES

1. N. MONARDES, in Vol. I of *Joyfull Newes of the New Founde Worlde* (translated by J. Frampton), London, 1577. Reprinted by A. A. Knopf, York, 1925.
2. R. BOYLE, *Experiments and Considerations Touching Colors*, London, 1680.
3. G. G. STOKES, *Phil. Trans.*, *142*, 463 (1852).
4. T. C. O'HAVER, *J. Chem. Ed.*, *55*, 423 (1978).
5. S. G. SCHULMAN, *Fluorescence and Phosphorescence Spectroscopy: Physicochemical Principles and Practice*, New York: Pergamon Press, 1977.
6. E. L. WEHRY, *Modern Fluorescence Spectroscopy*, New York: Plenum Press, 1976.
7. J. B. BIRKS, *Photophysics of Aromatic Molecules*, New York: Wiley-Interscience, 1970.
8. C. A. PARKER, *Photoluminescence of Solutions with Applications to Photochemistry and Analytical Chemistry*, New York: American Elsevier, 1968.
9. R. S. BECKER, *Theory and Interpretation of Fluorescence and Phosphorescence*, New York: Wiley-Interscience, 1969.
10. G. G. GUILBAULT, *Practical Fluorescence: Theory, Methods, and Practice.* New York: Marcel Dekker, 1973.
11. B. P. STRAUGHAN and S. WALKER, eds., *Spectroscopy*, Vols. 2 and 3, London: Chapman and Hall, 1976.
12. J. N. MILLER, *Trends Anal. Chem.*, *1*, 31 (1981).
13. E. M. SCHULMAN and C. WALLING, *Science*, *178*, 53 (1972).
14. T. VO DINH, G. L. WALDEN, and J. D. WINEFORDNER, *Anal. Chem.*, *49*, 754 (1980).
15. L. J. CLINE-LOVE, M. SKRILEC, and J. B. HABANTA, *Anal. Chem.*, *52*, 754 (1980).
16. D. W. ARMSTRONG, W. L. HINZE, K. H. BUI, and H. W. SINGH, *Anal. Lett.*, *14*, 1659 (1981).
17. L. J. CLINE-LOVE and M. SKRILEC, *Amer. Lab.*, *13*(3), 103 (1981).
18. J. J. DONKERBROCK, N. J. R. VAN EIKEMA HOMMES, C. GOOIJER, N. H. VELTHORST, and R. W. FREI, *Chromatographia*, *15*, 218 (1982).
19. N. OHTA and H. BABA, *J. Chem. Phys.*, *76*, 1654 (1982).
20. G. WEBER and J. BABLOUIAN, *J. Biol. Chem.*, *241*, 2558 (1966).

21. K. D. Mielenz, ed., *Measurement of Photoluminescence*, Vol. 3 of *Optical Radiation Measurement*, edited by F. Grum and C. J. Bartleson, New York: Academic Press, 1982.
22. A. A. Lamola, ed., *Creation and Detection of the Excited State*, Vol. 1, Part A, New York: Marcel Dekker, 1971.
23. C. W. Sill, *Anal. Chem.*, 33, 1584 (1961).
24. T. C. O'Haver and J. D. Winefordner, *Anal. Chem.*, 38, 682 (1966).
25. J. F. Holland, R. E. Teets, and A. Timnick, *Anal. Chem.*, 45, 145 (1973).
26. C. I. Miles and G. H. Schenk, *Anal. Chem.*, 45, 130 (1973).
27. C. I. Miles and G. H. Schenk, *Anal. Lett.*, 4, 71 (1971); *Anal. Chem.*, 45, 130 (1973).
28. L. A. Gifford, W. P. Hayes, L. A. King, J. N. Miller, D. T. Burns, and J. W. Brides, *Anal. Chim. Acta*, 62, 214 (1972); *Anal. Chem.*, 46, 94 (1974).
29. C. I. Miles and G. H. Schenk, *Anal. Chem.*, 42, 656 (1970).
30. J. D. Winefordner and H. W. Latz, *Anal. Chem.*, 35, 1517 (1963).
31. C. M. O'Donnell and T. N. Solie, *Anal. Chem.*, 50, 189R (1978).
32. E. L. Wehry, *Anal. Chem.*, 52, 75R (1980).
33. E. L. Wehry, *Anal. Chem.*, 54, 131R (1982).
34. F. E. Lytle, *Appl. Spectrosc.*, 24, 319 (1970).
35. G. H. Schenk, *Absorption of Light and Ultraviolet Radiation: Fluorescence and Phosphorescence Emission*, Boston: Allyn and Bacon, 1973, Chap. 4.
36. W. A. Armstrong, D. W. Grant, and W. G. Humphreys, *Anal. Chem.*, 35, 1300 (1963).
37. C. W. Sill and H. E. Peterson, *Anal. Chem.*, 21, 1266 (1949).
38. G. F. Kirkbright, T. S. West, and C. Woodward, *Talanta*, 12, 517 (1965).
39. C. E. White and R. J. Argauer, *Fluorescence Analysis*, New York: Marcel Dekker, 1970.
40. W. E. Ohnesorge, in *Fluorescence and Phosphorescence Analysis*, D. M. Hercules, ed. New York: Wiley-Interscience, 1966, Chap. 4.
41. D. W. Fink and W. E. Ohnesorge, *Anal. Chem.*, 41, 39 (1969).
42. D. W. Fink and W. E. Ohnesorge, *J. Amer. Chem. Soc.*, 91, 4995 (1969).
43. J. T. Brownrigg, A. W. Hornig and H. J. Coleman, Paper 396 presented at the 28th Pittsburgh Conference on Analytical Chemistry and Applied Spectroscopy, March 1977.
44. L. P. Giering, *Ind. Res.*, 20, 134 (1978).
45. E. Sawicki, T. R. Houser, T. W. Stanley, and R. A. Taft, *Int. J. Air Poll.*, 2, 253 (1960).
46. N. Seiler and L. Demisch, in *Handbook of Derivatives for Chromatography*, K. Blan and G. King, eds., London: Heyden, 1978, Chap. 9.
47. E. Sawicki, W. Elbert, T. W. Stanley, T. R. Hauser, and F. T. Fox, *Anal. Chem.*, 32, 810 (1960).
48. H. V. Drushel and A. L. Sommers, *Anal. Chem.*, 38, 10 & 19 (1966).
49. E. Sawicki and J. D. Pfaff, *Anal. Chim. Acta*, 32, 521 (1965).
50. M. S. Lefar and A. D. Lewis, *Anal. Chem.*, 42(3), 79A (1970).
51. D. Janchen and G. Pataki, *J. Chromatogr.*, 33, 391 (1968).
52. J. D. Winefordner and H. A. Moye, *Anal. Chim. Acta*, 32, 278 (1965).
53. G. K. Turner, *Fluorometry Reviews Bulletin on Fluorescent Tracers*, February 1968, Acc. No. 9941, G. K. Turner Associates, Palo Alto, Calif.
54. E. Sawicki, W. Elbert, T. W. Stanley, T. R. Hauser, and F. T. Fox, *Int. J. Air Poll.*, 2, 273 (1960).
55. F. W. J. Teal and G. Weber, *Biochem. J.*, 65, 476 (1957).
56. P. K. Wong, *Clin. Chem.*, 10, 1098 (1964).
57. M. Heros and L. Avry, *Compt. Rend. Acad. Sci. Paris*, 255, 695 (1962).
58. M. P. Fogarty and I. M. Warner, *Anal. Chem.*, 53, 259 (1981).
59. M. P. Fogarty and I. M. Warner, *Appl. Spectrosc.*, 36, 460 (1982).
60. D. N. Hercules, ed., *Fluorescence and Phosphorescence Analysis*, New York: Wiley-Interscience, 1966.
61. G. K. Turner, *Science*, 146, 183 (1964).
62. C. M. Himel and R. T. Mayer, *Anal. Chem.*, 42, 130 (1970).

PROBLEMS

1. Show by calculation that the time of the excitation process must be approximately 10^{-15} sec, using 300-nm light for excitation.
2. Use Equation 9.1 to show that an organic free radical is in a doublet state.
3. Derive Equation 9.13, which relates the efficiency of phosphorescence and the efficiency of formation of the triplet state.
4. Explain qualitatively how it may be possible to obtain information about the Brownian motion and thus the size of a molecule using fluorescence polarization. Would the viscosity of the solvent affect the results? Why or why not?
5. Derive Equation 9.18 starting with Equation 9.15.
6. Derive Equation 9.20 for dynamic quenching using Equation 9.8. What other constants are contained in the Stern-Volmer constant, k_{SV}? Does a similar relationship exist for static quenching? If so, what is the constant in this case?

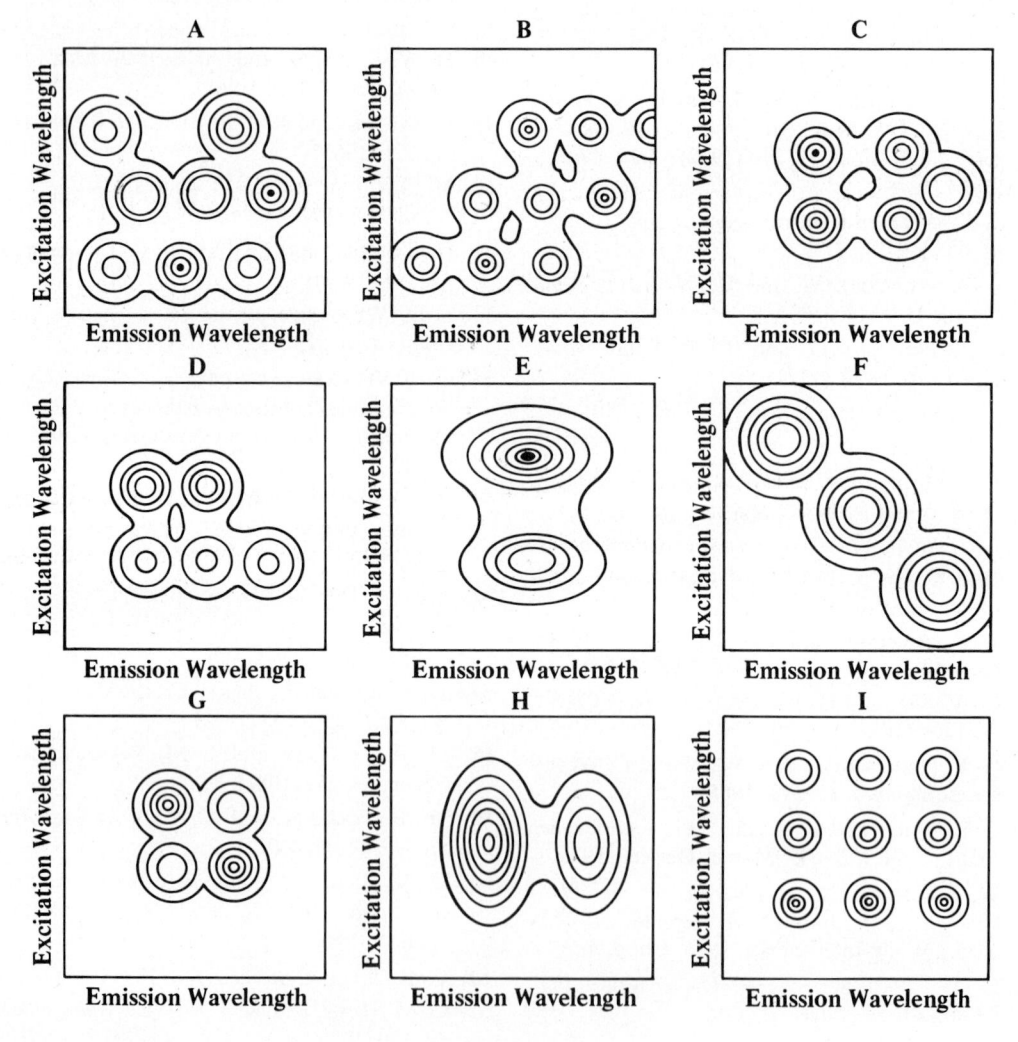

Problem 10

276   CHAP. 9   Molecular Fluorescence and Phosphorescence

7. Would you expect temperature to have an effect on delayed fluorescence? Why or why not?

8. Assuming a diffusion-controlled quenching process and a good absorber and fluorescer, calculate the approximate quencher concentration necessary for 50% dynamic quenching of a fluorophore.

9. Would you expect the quantum efficiency of fluorescence, ϕ_f, to be altered by (a) lowering the temperature? (b) raising the temperature? (c) changing the concentration of the fluorophore? (d) adding a static quencher? (e) adding a dynamic quencher? (f) the solvent viscosity? Explain the reason for your answer.

10. The data displayed as isometric projections in Figures 9.3 and 9.9 can also be represented as contour maps where each contour is an iso-intensity level of fluorescence. Consider the contour maps of fluorescence intensity as a function of multiple excitation and multiple emission wavelengths shown below. Indicate the minimum number of components contributing to each contour map. Assume that the excitation spectrum is independent of the monitored emission wavelength and that the emission spectrum is independent of the excitation wavelength. Why is it not possible to definitively give the exact number of components in each map?

11. The text discussion indicated that fluorescence spectroscopy is typically more than three orders of magnitude more sensitive than absorption spectroscopy. Give an explanation of this greater sensitivity citing the difference in the measurement process of each technique.

12. Compare the slope of a calibration curve for a molecule with a molar absorptivity of 10^5 and a quantum efficiency of 0.01 to that for a molecule with a molar absorptivity of 10^3 and a quantum efficiency of 0.10.

13. Propose two different filter-fluorometric schemes of analysis for the determination of perylene in the presence of anthracene, and another for the determination of anthracene in the presence of perylene. You are not limited to the filters shown in Figure 9.6.

14. A bottle of tonic water is to be analyzed for its quinine content by fluorescence spectrometry, with excitation at 350 nm and emission intensity measured at 450 nm. One milliliter of tonic water is diluted to 100 mL with 0.05 M H_2SO_4; its emission intensity is 8.44 (arbitrary units). A series of quinine standards, in 0.05 M H_2SO_4, is prepared, and the emission intensities are measured (in parentheses): 100 ppm (293), 10.0 ppm (52.3), 1.00 ppm (12.0), 0.100 ppm (1.26), 10 ppb (0.158), 1.0 ppb (0.015). The emission intensity of 0.05 M H_2SO_4 is negligible. Plot the calibration curve for quinine fluorescence, and determine the quinine content of the original tonic-water sample.

15. An organic compound is to be determined by fluorescence spectrometry, with a choice of excitation at 250 nm (with emission at 350 nm) or at 500 nm (with emission at 600 nm). A xenon-arc lamp is to be used as the excitation source of a spectrofluorimeter (see Fig. 9.5) with an S-5 photomultiplier tube as the detector (see Fig. 9.7). The compound has a molar absorptivity of 15,000 at 250 nm and of 4000 at 500 nm. Assume that the quantum efficiencies for fluorescence are the same at the two wavelengths. (a) From this information and that in Figures 9.5 and 9.7, estimate the ratio of the signal obtained with excitation at 250 nm to that with excitation at 500 nm. (b) What other assumptions have to be made?

16. The process of triplet-singlet energy transfer is often referred to as "sensitized phosphorescence." The process involves a donor molecule (D) in the triplet state transferring energy to a singlet acceptor molecule (A). In equation form, we can represent the process as

$$^3D + {}^1A \rightarrow {}^1D + {}^3A$$

This is a spin-allowed process since there is no net change of state. Predict the expected energy relationship between "donor" and "acceptor" triplet energy levels for favored observation of sensitized phosphorescence.

CHAPTER **10**

Flame Emission, Atomic Absorption, and Atomic Fluorescence Spectrometry

Gary Horlick

Among the most common techniques for elemental analysis are flame emission, atomic absorption, and atomic fluorescence spectrometry. All these techniques are based on the radiant emission, absorption, and fluoresence of atomic vapor. The key component of any atomic spectrometric method is the system for generating the atomic vapor (gaseous free atoms or ions) from a sample, that is, the source. Numerous sources have been used to generate atomic vapor, among them, flames, direct-current arcs, alternating-current sparks, electrothermal atomizers, microwave plasmas, radio-frequency plasmas, and lasers. The most widely used sources are the flame and electrothermal atomizers, and it is these sources that will be discussed in this chapter. Spectrometric methods based on arcs, sparks, and plasmas will be covered in Chapter 11.

It is useful to state in broad and somewhat idealized terms the goals for a source.

1. To convert any sample type into gas-phase atomic vapor with little or no sample pretreatment.
2. To do so for all elements at all concentrations.
3. To have identical operating conditions for all elements and samples.
4. To have the analytical signal be a simple function of the concentration of the individual elements; that is, to have no interferences and no matrix effects.
5. To be characterized and understood at a fundamental level.
6. To provide accurate and precise analyses.
7. To have low initial cost and low maintenance and running costs.
8. To be simple and reliable to operate.

Taken as whole, these are rather formidable goals, and no present source combines all these features and characteristics. However, they should be kept in mind as the flame and electrothermal sources are discussed and assessed.

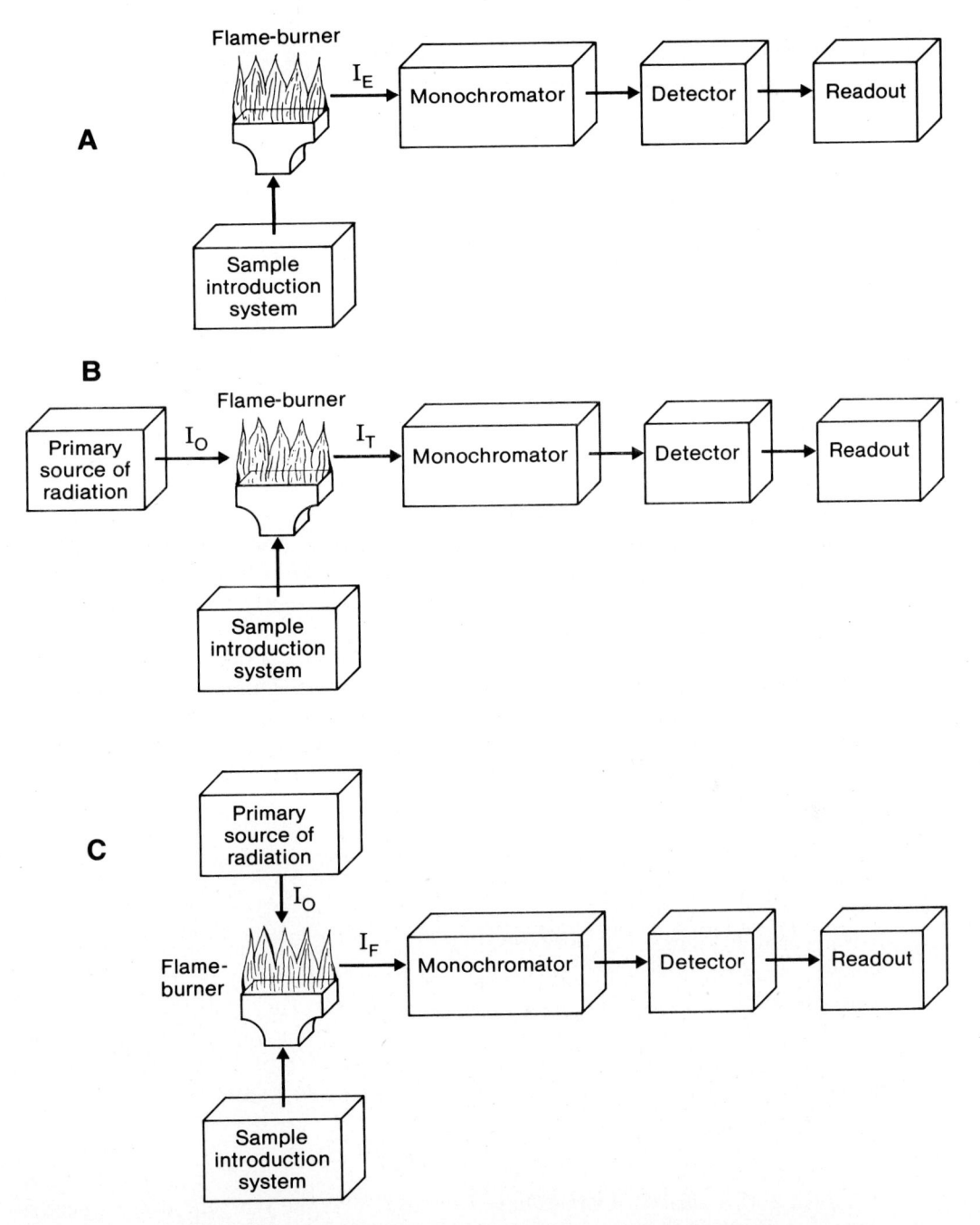

FIGURE 10.1. *Block diagrams of spectrometers. A: Flame emission spectrometer. B: Atomic absorption spectrometer. C: Atomic fluorescence spectrometer.*

With a source such as the flame, three basic measurements can be carried out on the atomic vapor: atomic emission, atomic absorption, or atomic fluorescence. The simplest and historically oldest approach to using the flame for analytical measurements is *atomic emission*, that is, observing the characteristic radiation emitted by the atomic species in the flame. A block diagram of a flame emission spectrometer is shown in Figure 10.1A. Although the flame is the key component, several additional subsystems are required: (a) a sample-introduction system, (b) a source of atomic vapor (the flame), (c) a wavelength-isolation system, (d) a radiation detector, and (e) a readout system.

Since its development as an analytical technique in 1955 by Walsh [1, 2], *atomic absorption spectrometry* has become the most widely used measurement for elemental analysis. A block diagram of an atomic absorption spectrometer is shown in Figure 10.1B. Compared to the flame emission system, one additional key subsystem is required: a primary source of radiation. Light from this primary source is passed through the atomic vapor, and the amount of radiation absorbed by an atomic species is measured.

Even though most of the discussion in this chapter focuses on atomic absorption and flame emission methods, a third technique, *atomic fluorescence spectrometry*, is beginning to find some application to elemental analysis. A block diagram of an atomic fluorescence spectrometer is shown in Figure 10.1C. As with atomic absorption spectrometry, atomic fluorescence spectrometry requires a primary source of radiation. The purpose of the primary source, often in a 90° configuration, is to generate the atomic fluorescence signal from the atomic vapor. This technique will be discussed briefly at the end of the chapter.

For all these techniques, the most common source is the flame and the most common method of sample introduction is solution nebulization. In the next section, the basic atom-formation steps and processes involved in converting an elemental analyte in a solution into analyte atoms in a flame will be outlined. Except for the signal-generation step, all the processes involved are identical for the three methods.

10.1 THE FLAME AS A SOURCE OF ATOMIC VAPOR

As indicated in Figure 10.1, the central two blocks of all the methods represent the sample-introduction system and the source. For the majority of instruments, the source is a flame and samples must be in solution form. The samples are introduced into the flame as an aerosol. The aerosol is normally generated using a pneumatic nebulizer connected to the flame burner by a spray chamber. This very important nebulizer–spray chamber–burner subsystem is shown in Figure 10.2.

Flames

Although a wide variety of flames have been used over the years for atomic spectrometry (see Table 10.1), essentially only two are currently in widespread use for analytical measurements: air-acetylene and nitrous oxide–acetylene. With these two flames, appropriate analytical conditions for most analytes can be achieved for any

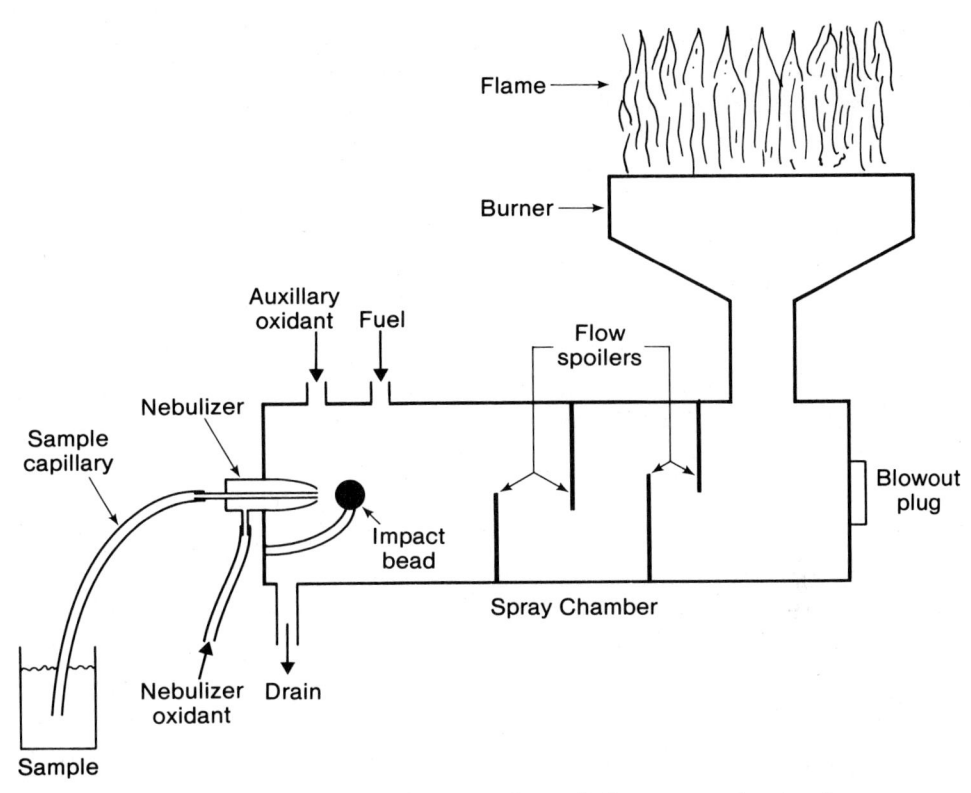

FIGURE 10.2. *Schematic diagram of a nebulizer–spray chamber–burner system.*

TABLE 10.1. *Flames Used in Atomic Emission, Absorption, and Fluorescence Spectrometry*

Oxidant	Fuel	Maximum Temp, °C
Air	Acetylene	2250
Nitrous oxide	Acetylene	2955
Air	Coal gas	1825
Air	Propane	1725
Air	Hydrogen	2045
Entrained air/argon	Hydrogen	1577
Oxygen	Natural gas	2740
Oxygen	Hydrogen	2677
Oxygen	Acetylene	3060
Oxygen/helium	Acetylene	2812
Oxygen	Cyanogen	4500

of the three techniques—emission, absorption, or fluorescence. The choice of flame and operating conditions is dependent on several factors, among the most important, the analyte to be determined, the nature of the sample matrix, and the technique being implemented. A typical breakdown is shown in Table 10.2 for atomic absorption spectrometry. For most elements, the air-acetylene flame is the first choice for atomic absorption determinations. Its lower temperature, when compared to the nitrous oxide–acetylene flame, favors the formation of neutral atoms; and with a fuel-rich flame, oxide formation can be minimized for many elements. However, as an element tends to become more refractory, that is, forms refractory oxides that are difficult to decompose, the higher-temperature nitrous oxide–acetylene flame is recommended. This flame has a zone containing high concentrations of CN and NH species which results in a strongly reducing region inhibiting the formation of refractory oxides. The high-temperature nitrous oxide–acetylene flame may also be useful in eliminating certain vaporization interferences that tend to occur in the lower-temperature air-acetylene flame. On the other hand, a disadvantage of the higher-temperature flame

TABLE 10.2. *Flame Choice and Detection Limits in Atomic Absorption Spectrometry*

	Air-acetylene			Nitrous oxide–Acetylene	
Element	Wavelength, nm	Detection Limit, µg/mL	Element	Wavelength, nm	Detection Limit, µg/mL
Ag	328.1	0.0009	Al	309.3	0.03
As[a]	193.7	0.14	B	249.7	0.7
Au	242.8	0.006	Mo	313.3	0.03
Bi[a]	222.8	0.02	Si	251.6	0.06
Ca	422.7	0.001	Sn	286.3	0.11
Cd[a]	228.8	0.0005	Ti	364.3	0.05
Co	240.7	0.006	V	318.4	0.04
Cr	357.9	0.002	W	255.1	1.2
Cu	324.7	0.001			
Fe	248.3	0.003			
Hg	253.6	0.17			
Li	670.8	0.0005			
Mg	285.2	0.0001			
Mn	279.5	0.001			
Ni	232.0	0.004			
Pb	283.3	0.01			
Pt	265.9	0.04			
Sb[a]	217.6	0.03			
Se[a]	196.0	0.07			
Te[a]	214.3	0.019			
Tl	276.8	0.009			
Zn	213.9	0.0008			

a. Electrodeless discharge lamp (EDL) used as a source; all others use a conventional hollow-cathode lamp.

for atomic absorption measurements is that, for some analytes, there is a marked increase in ionization which reduces the ground-state neutral-atom population on which the atomic absorption signal is dependent. For flame emission measurements, however, the higher-temperature flame is the first choice for almost all elements.

There are some important operational differences between these two flames. For the air-acetylene flame, the common slot burner as illustrated in Figure 10.2 is 10 cm in length. To minimize the danger of flashback, the slot is only 5 cm for the nitrous oxide–acetylene flame. The latter flame has a higher burning velocity than does the air-acetylene flame. This increases the chance of flashback, which occurs when the flame propagates back into the burner and spray chamber. It is unsafe, for example, to directly ignite a nitrous oxide–acetylene flame. Such a flame is established by first igniting an air-acetylene flame, adjusting it to a very fuel-rich condition, and then crossing the air supply over to nitrous oxide. The reverse sequence is necessary to extinguish the flame. In some instruments, completely automatic systems handle these sequences. With properly operated modern instrumentation, the danger of flashback is remote.

The spectral features of the air-acetylene and nitrous oxide–acetylene flames are shown in Figure 10.3. Water is being nebulized in both cases. Both flames have extensive emission from species such as C_2, CH, and OH. The nitrous oxide–acetylene flame has, in addition, very strong emission from CN. The emission intensities of these features are highly dependent on the fuel-to-oxidant ratio, and fuel-rich conditions tend to increase them. For most atomic absorption measurements, the emission or absorption by these species in the flame is not very important; but for atomic emission measurements, great care must be taken to avoid spectral overlap of these complex emissions with analyte emission lines.

While the air-acetylene and nitrous oxide–acetylene flames are the dominant flame systems, some other flames are used in special situations. The low-temperature air-propane and air–natural gas flames are used for the flame emission determination of Na and K. The air-hydrogen flame and the argon-hydrogen-entrained air flame have found some application for atomic absorption determinations of tin, selenium, and arsenic, particularly in combination with hydride-generation techniques. This flame is also frequently recommended for atomic fluorescence measurements. Recently, a helium-oxygen-acetylene flame has been suggested for analytical flame spectroscopy [3], with particular reference to atomic fluorescence spectrometry. Thus, although air-acetylene and nitrous oxide–acetylene flames dominate current applications, many special flames can be usefully tailored for specific features and characteristics for particular measurements. It is interesting to note that even the development of the nitrous oxide–acetylene flame as an analytically useful flame is a relatively recent occurrence [4, 5].

Atom Formation: Steps and Processes

Let us now look at the complete history of analyte atoms from solution to atomic species in the flame. The steps involved are common to all the flame methods. It will become clear that a whole host of chemical, physical, and instrumental parameters affect the relationship between concentration of analyte atoms in solution and the actual readout. The key steps involved are (a) solution transport; (b) nebulization;

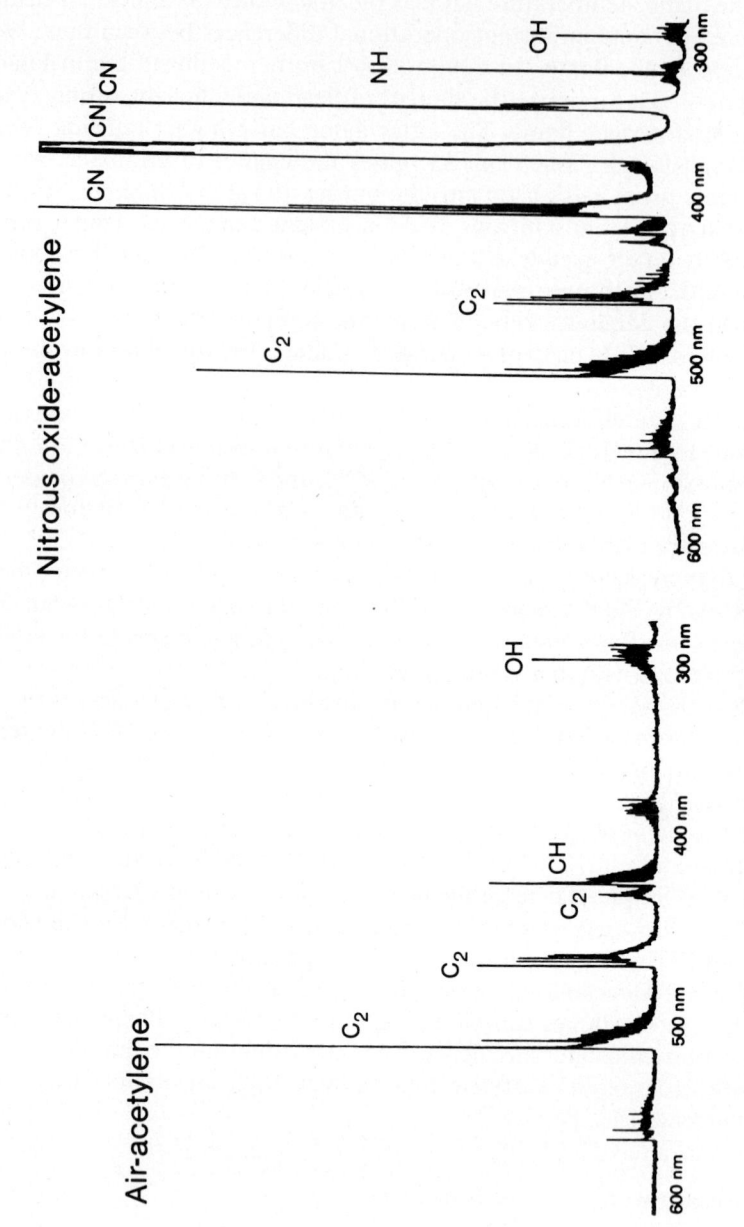

FIGURE 10.3. *Emission spectra of air-acetylene and nitrous oxide–acetylene flames. Courtesy of Varian Instrument Group.*

(c) aerosol transport; (d) desolvation; (e) vaporization; (f) equilibration of vaporized species; and (g) atomic emission, absorption, or fluorescence measurement. These steps are outlined schematically in Figure 10.4, and the first six will be discussed in this section.

Solution Transport. The transport step involves the movement of solution to the nebulizer through a small-bore plastic tube. Even at this apparently simple step, several variables must be considered. An identical and reproducible uptake rate is required for samples and standards. In most cases, the uptake rate is free running in that it depends on the oxidant flow rate to the nebulizer. Thus, this flow rate should be steady and controlled. Solution viscosity and surface-tension changes between samples and standards must be avoided, and a common solvent must be used for all standards and samples. Dusty and turbid solutions will be difficult to transport reproducibly to the nebulizer. Finally, motor-driven peristaltic pumps or syringe drives can be used to ensure accurate and precise delivery of the sample solution to the nebulizer.

Nebulization. The nebulization step is of major importance and a bit of an Achilles' heel of flame spectrometric methods [6]. It involves conversion of the solution into a fine spray or aerosol. Typically, this is accomplished with a concentric pneumatic nebulizer of the general design depicted in Figure 10.5. The ideal goal is to convert the sample solution into usable aerosol with an efficiency of 100%. The problem is that most nebulizers are only about 3 to 15% efficient in this respect. Aerosol, to be truly usable, that is, to have an essentially 100% probability of resulting in free atoms once in the flame, must be fine. Typically, the droplet diameters should be less than 10 μm and preferably approaching 1 to 2 μm. The net result is that about 90% of the sample goes down the drain at this step. This remains an unsolved problem for flame spectrometry, and was one of the reasons for the development of the electrothermal methods discussed later in the chapter.

Aerosol Transport. The third step is aerosol transport to the flame through the spray chamber. The spray chamber and subcomponents within the spray chamber process the aerosol [7, 8]. These subcomponents—impact beads, baffles, impactors, and the geometry of the spray chamber itself—ensure that only the finest aerosol actually reaches the flame, and that all large droplets are condensed and drained off.

Desolvation. The first step once the aerosol reaches the flame is desolvation, that is, conversion of the aerosol into salt particles. The rate of desolvation depends on such factors as droplet size, solvent, time in the flame, velocity of the droplet, flame composition, and flame temperature [9]. As long as only fine aerosol is allowed to reach the flame, incomplete desolvation is seldom a problem.

Vaporization. Vaporization is a critical step in flame spectrometric methods. This step is defined as conversion of a salt particle into vapor in the flame. It is here that many serious interferences occur, which are of significance to all three flame methods. Among the important variables are particle size, particle composition, flame temperature, and time (height) in the flame.

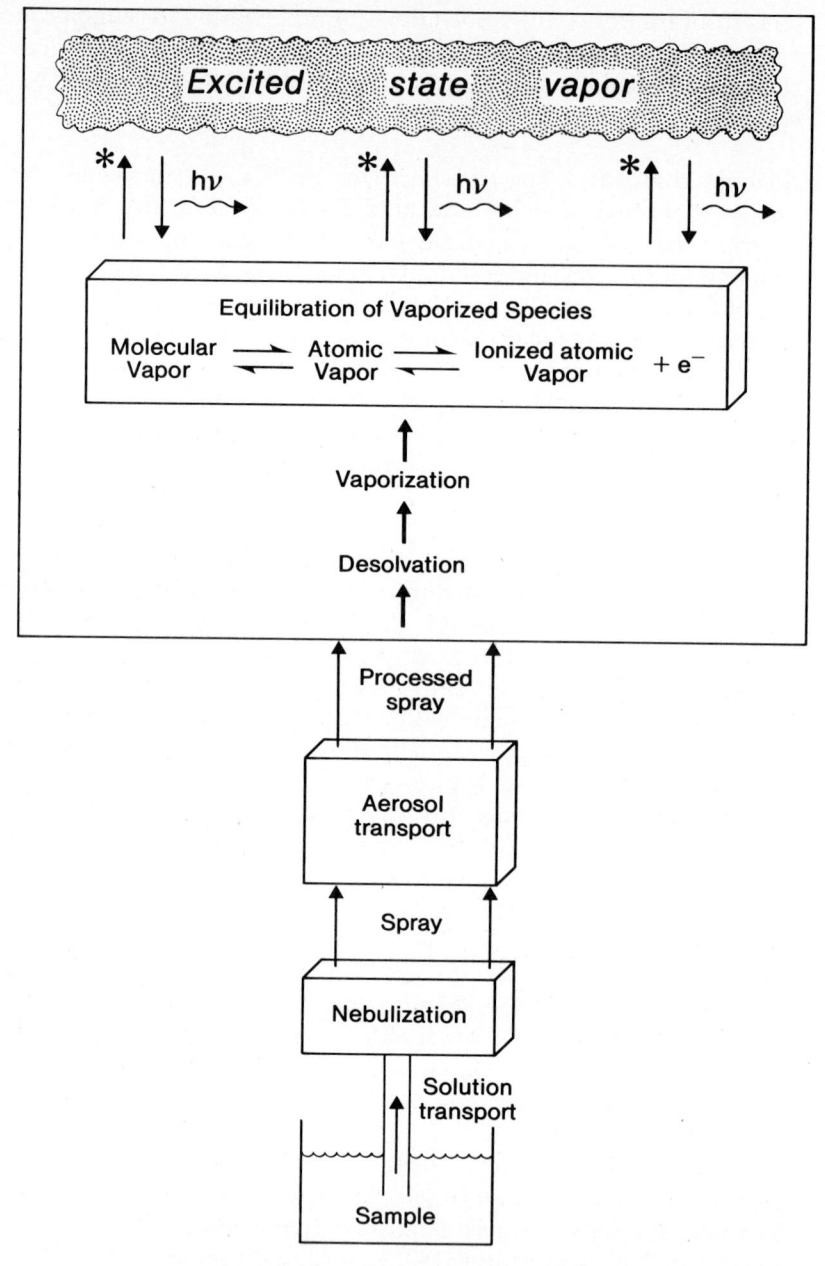

FIGURE 10.4. *Schematic diagram of atom-formation steps and processes.*

The classic, but still important, example of a vaporization interference is that of phosphate on calcium emission [10]. Data illustrating this interference are shown in Figure 10.6A. The leveling off of the suppression of the Ca signal at a specific P/Ca ratio is indicative of compound formation. Also, as more time is allowed for vaporization (higher height in the flame), the interference is less serious. Another classic

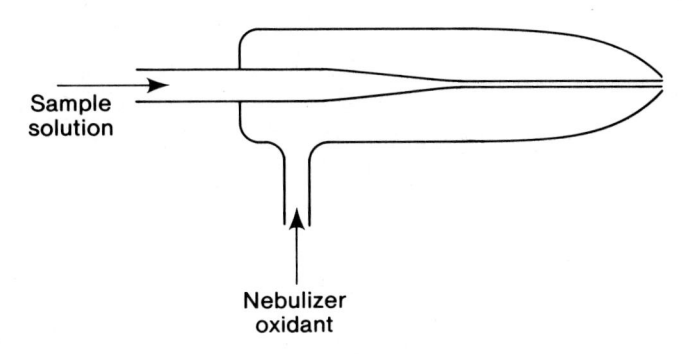

FIGURE 10.5. *Concentric pneumatic nebulizer.*

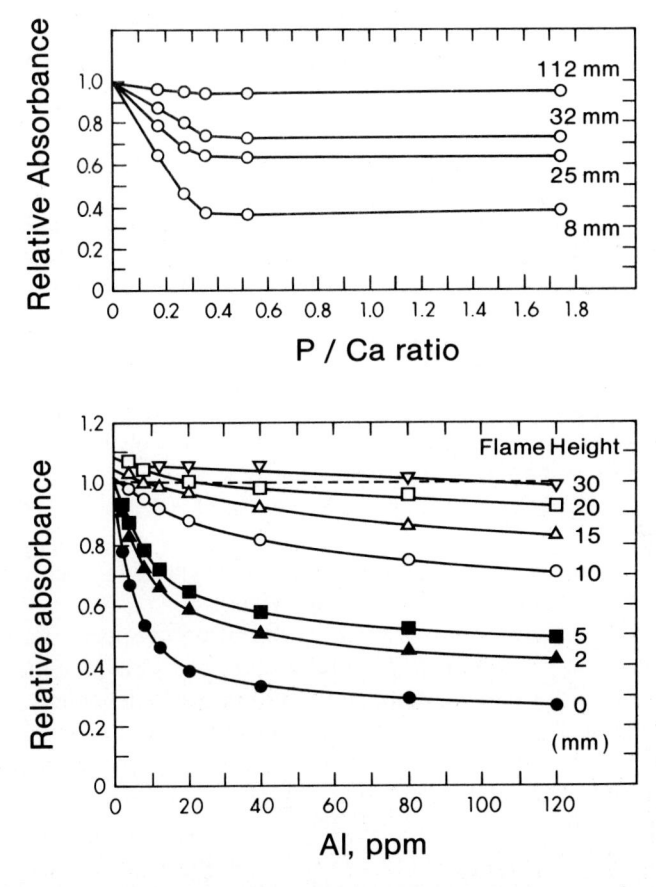

FIGURE 10.6. *Top: Vaporization interferences of phosphate on the atomic absorption signal of calcium. From C. T. J. Alkemade and M. H. Voorhius, Zeit. Anal. Chem., 163, 91 (1958). Used with permission of Springer-Verlag. Bottom: Vaporization interferences of aluminum on the atomic absorption signal of magnesium. Reprinted with permission from W. W. Harrison and W. H. Wadlin, Anal. Chem., 41, 374 (1969). Copyright © 1969 American Chemical Society.*

interference is that of aluminum on magnesium atomic absorption as illustrated in Figure 10.6B. This suppression does not show an obvious "break" and is generally regarded to involve occlusion of Mg in refractory alumina particles. Again, allowing more time for vaporization decreases the interference. In addition, both of these classic interferences can be eliminated by switching to the nitrous oxide–acetylene flame from the air-acetylene flame, indicating that the higher flame temperature is effective in vaporizing the refractory salt particles.

Manipulation of the analyte solution can also alleviate these interferences [11, 12]. The addition of a high concentration of a La or Sr salt as a competing cation can effectively tie up the interferent and displace trace amounts of the analyte element. Protective chelation with EDTA is another way of removing an interferent anion from the analyte ion and effectively eliminating several vaporization interferences. These methods of controlling vaporization interferences remain popular; for example, addition of several thousand ppm of a Sr or La salt to plasma or serum samples is a standard approach in clinical labs for the determination of Mg.

Equilibration of Vaporized Species. The last step before the measurement step is equilibration of the vaporized species. As outlined in Figure 10.4, this equilibrium is between molecular species (often mono-oxide or mono-hydroxide species), neutral atoms, and ions (normally singly ionized). In general, in flame methods free neutral atoms are desired. For absorption and fluorescence methods, these are the primary desired species; whereas for emission, excited neutral atoms are obviously required. In order to minimize the formation of molecular species, reducing flame conditions are desirable as well as a high-temperature flame. But a trade-off is required in the utilization of a high-temperature flame in that ionization will be promoted, diminishing the population of neutral atoms.

An example of this distribution of species is provided by the spectrum shown in Figure 10.7. The spectrum given here is a scan of the emission of "calcium" in a nitrous oxide–acetylene flame. It is clear that calcium exists in several forms in this flame, with CaOH, Ca^0, and Ca^+ emission all being observed. The relative distribution of these species is dependent on flame type, flame–gas ratios, and observation height in the flame.

It is at this step involving equilibration of species that another classic interference effect is operative. This interference is termed *ionization interference*. The relative distribution of atomic and ionic species is dependent on the electron density in the flame. The presence of a high concentration of an easily ionizable element can alter the atom/ion line distribution of another element in the flame. The nature of the ionization interference is shown in Figure 10.8A. The presence of excess Na enhances the neutral-atom emission of K. This occurs because the electrons generated by the ionization of Na in the flame are common to the equilibration of K neutral atoms and ions. The addition of electrons to the flame forces this equilibrium back toward the neutral atom, thereby enhancing neutral-atom emission.

This effect is of equal concern in atomic absorption measurements, as shown in Figure 10.8B for strontium. In the air-acetylene flame, the degree of ionization of Sr is low and excess K has little effect. However, in the higher-temperature nitrous oxide–acetylene flame, excess K significantly enhances the Sr absorption signal. The standard method to control this interference is to swamp the flame with electrons by

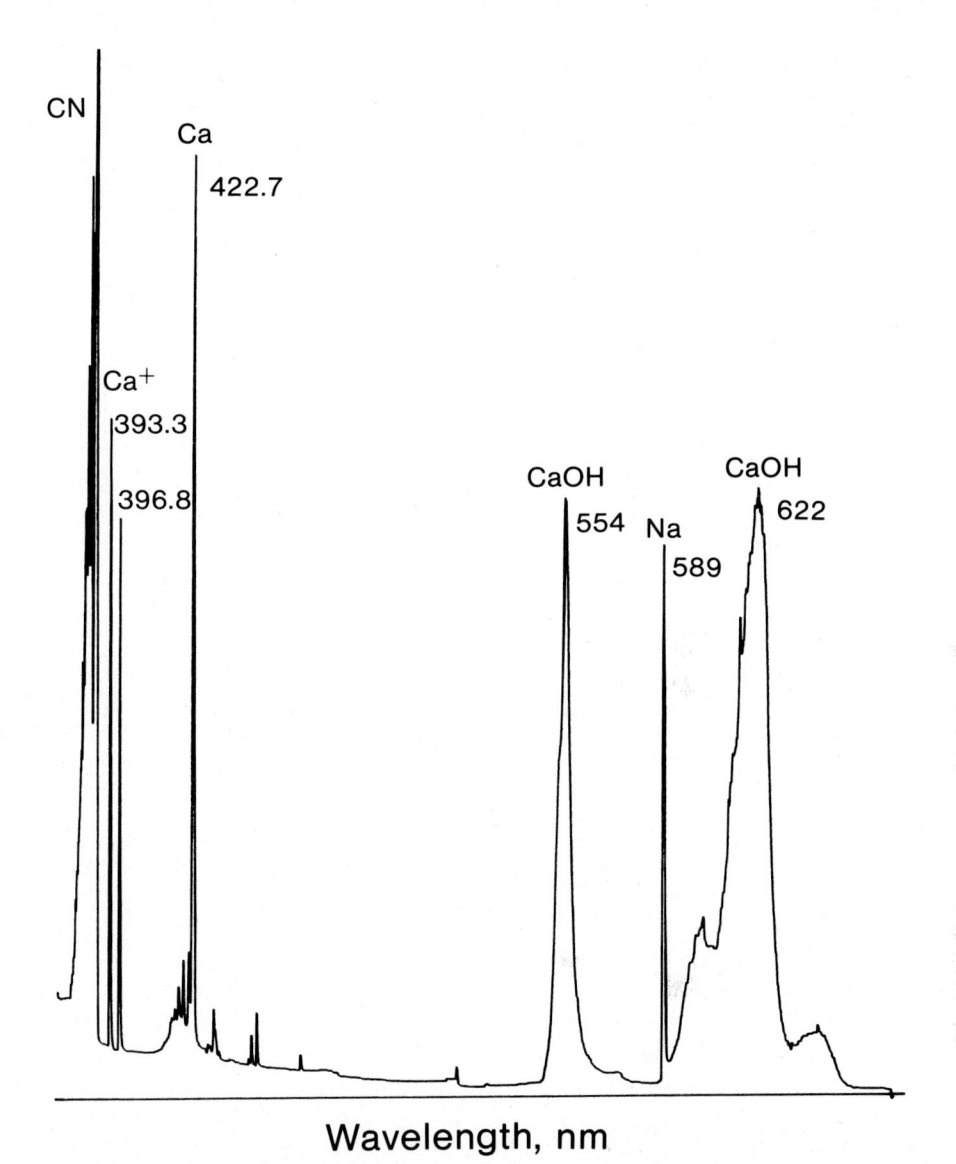

FIGURE 10.7. *Emission spectrum of calcium in a nitrous oxide–acetylene flame.*

the addition of a large excess (1000 ppm) of an easily ionizable element such as cesium. Such a species is known as an *ionization buffer* or *suppressant*.

It is clear that the flame is a complex chemical and physical system. Despite this complexity, it is also, in many ways, one of the most simple and convenient sources for atomic spectrometry.

Having reached the point where we have ground-state atoms, we will now consider in some detail each of the three basic methods—emission, absorption, and fluorescence.

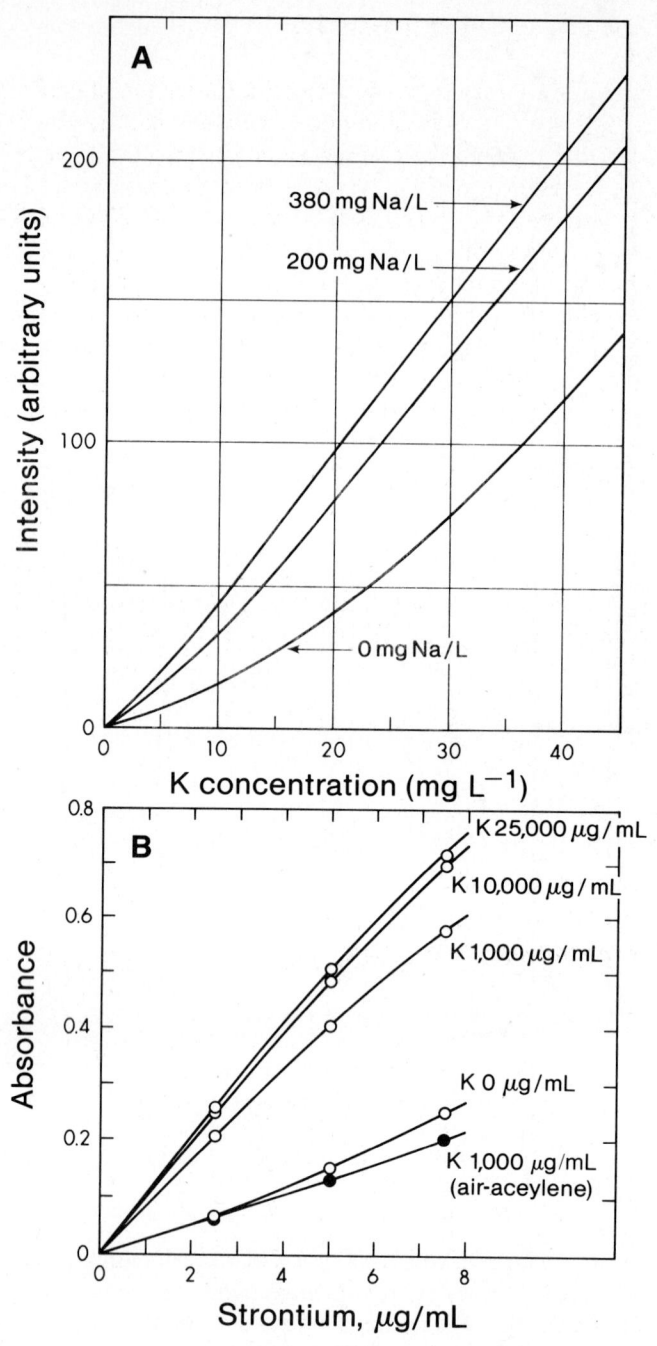

FIGURE 10.8. *Ionization interference effects. A: Sodium on the emission of potassium: From J. Smit, C. T. J. Alkemade, and J. C. M. Verschure,* Biochim. Biophys. Acta, *6, 508 (1951). Used with permission of Elsevier Publishers. B: Potassium on the absorption of strontium. Note that the effect of potassium on the strontium signal is observed only in the case of the nitrous oxide–acetylene flame (open circles) Reprinted with permission from J. A. Bowman and J. B. Willis,* Anal. Chem., *39, 1210 (1967). Copyright © 1967 American Chemical Society.*

10.2 FLAME EMISSION SPECTROMETRY

For emission measurements, the step after formation of neutral, ground-state atoms is the excitation of the atoms and measurement of the characteristic emitted radiation. In flame emission spectrometry, spectra are relatively simple because the flame is not a highly energetic source. The spectral lines observed result from energy transitions involving the outermost electrons. An energy-level diagram (Grotrian diagram) for the sodium atom is shown in Figure 10.9. Sodium has one electron in the $3s$ shell, and this level is designated the ground state. The familiar **D** lines of sodium, the doublet in the yellow region, arise when excited electrons return to this level from the $3p$ level. The doublet results from the two values of the spin quantum number. Transitions from the lowest allowed excited level to the ground state are called *resonance transitions* and the resulting spectral lines are called *resonance lines*. Although several

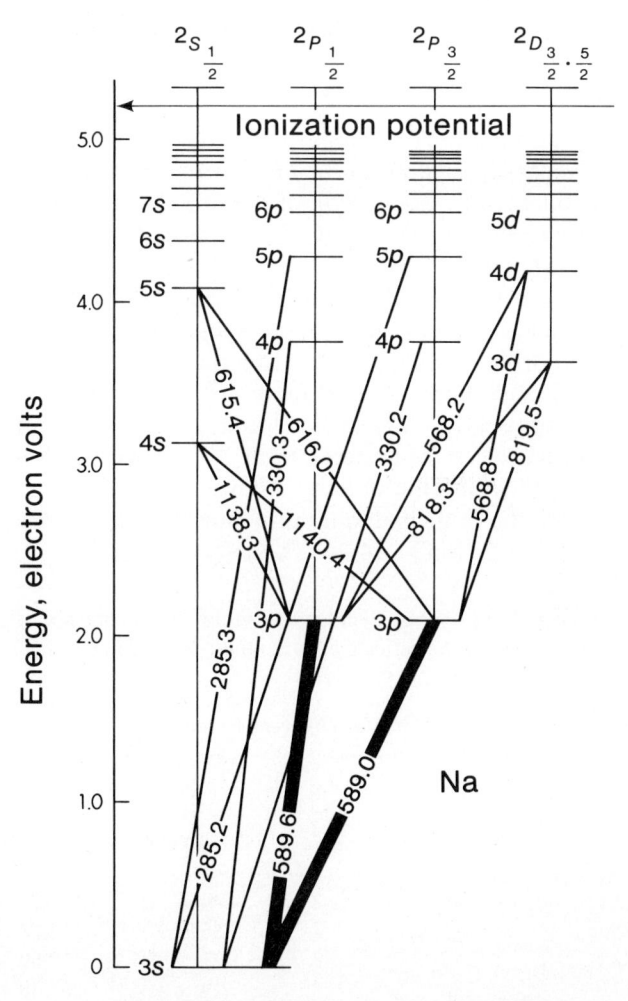

FIGURE 10.9. *Grotrian diagram for sodium.*

other transitions are shown on the energy-level diagram for Na, most are quite weak in conventional flame sources, and Na has one of the simplest flame emission spectra of all the elements. For an element such as Fe, the Grotrian diagram is very complex with literally hundreds of allowed transitions.

It is of interest to know the number of thermally excited atoms relative to the number of ground-state atoms at a given flame temperature. In a quantity of atoms, under the same conditions, the electrons are not all in the same energy level but are statistically distributed among the levels. At a flame temperature T (in K), the ratio of the number of atoms N_u in an excited (upper) state "u" to the number of atoms N_0 in the ground state is given by the Maxwell-Boltzmann expression:

$$\frac{N_u}{N_0} = \frac{g_u}{g_0} e^{-(E_u - E_0)/kT} \tag{10.1}$$

where g_u = statistical weight of the excited state
g_0 = statistical weight of the ground state
E_u = energy of the excited state
E_0 = energy of the ground state
k = the Boltzmann constant

The statistical weights can be regarded as the probability that an electron will reside in a given energy level, and can be obtained from quantum-mechanical calculations. Equation 10.1 permits calculation of the ratio (N_u/N_0) at a given flame temperature when the frequency or wavelength for the transition is known. Some typical values are listed in Table 10.3.

The statistical weights, g, can be calculated from the equation $g = 2J + 1$, where J is the Russell-Saunders coupling and is equal to $(L + S)$ or $(L - S)$. Quantity L is the total orbital angular momentum quantum number, represented by the sharp (S), principal (P), diffuse (D), and fundamental (F) series ($L = 0, 1, 2,$ and 3, respectively); S is spin, $\pm 1/2$. The information is typically supplied in the form of term symbols, which have the general form $N^M L_J$, where N is the principal quantum number and M is the multiplicity. Hence, the transition for the cesium 852.1-nm line, omitting the principal quantum number, N, is $^2S_{1/2} - {}^2P_{3/2}$, and $g_u/g_0 = [2(3/2) + 1]/[2(1/2) + 1] = 4/2 = 2$.

Example 10.1. The 228.8-nm cadmium line corresponds to a $^1S_0 - {}^1S_1$ transition. Calculate the ratio N_u/N_0 in an air-acetylene flame.

TABLE 10.3. *Values of N_u/N_0 for Different Resonance Lines*

Resonance Lines, nm	Excitation Energy, eV	g_u/g_0	N_u/N_0 2000 K	3000 K	4000 K
Cs 852.1	1.46	2	4.44×10^{-4}	7.24×10^{-3}	2.98×10^{-2}
Na 589.0	2.11	2	9.86×10^{-6}	5.88×10^{-4}	4.44×10^{-3}
Ca 422.7	2.93	3	1.21×10^{-7}	3.69×10^{-5}	6.04×10^{-4}
Zn 213.8	5.80	3	7.29×10^{-15}	5.38×10^{-10}	1.48×10^{-6}

Solution: The temperature (Table 10.1) is 2250°C, or 2523 K.

$$\frac{g_u}{g_o} = \frac{2(1)+1}{2(0)+1} = \frac{3}{1}$$

$$v = \frac{c}{\lambda} = \frac{2.998 \times 10^{10} \text{ cm/sec}}{2.288 \times 10^{-5} \text{ cm}} = 1.310 \times 10^{15} \text{ sec}^{-1}$$

$$E_u - E_0 = hv = (6.626 \times 10^{-27} \text{ erg-sec})(1.310 \times 10^{15} \text{ sec}^{-1})$$

$$= 8.682 \times 10^{-12} \text{ erg}$$

$$\frac{N_u}{N_o} = \frac{g_u}{g_o} e^{-(E_u-E_0)/kT}$$

$$= \frac{3}{1} \exp\left[-\frac{8.682 \times 10^{-12} \text{ erg}}{(1.3805 \times 10^{-16} \text{ erg K}^{-1})(2523 \text{ K})}\right]$$

$$= 3e^{-24.93} = 4.5 \times 10^{-11}$$

It is apparent from the data presented in Table 10.3 that for most elements in flames, even the hot nitrous oxide–acetylene flame, the excited-state population is very small compared to the ground-state population. However, if one looks at a table of flame emission detection limits (Table 10.4), they are quite low. It must be kept in mind that even though the ratio of the excited state to the ground state may be small, a very large number of atoms are being introduced into the flame even with ppm levels of analytes. Thus, even if only 0.0001 of the atoms are excited, a large number of photons are emitted. As can be seen from Table 10.4, only a few elements have acceptable detection limits in the air-acetylene flame, and it was the development of the nitrous oxide–acetylene flame that was responsible for bringing some new life to flame emission spectrometry [13].

However, in most measurement situations, flame emission remains a somewhat poor distant cousin to atomic absorption. Perhaps if one does not have a particular hollow-cathode lamp on hand, then one might resort to a flame emission measurement. For elements that exhibit good detection limits, it is a good first choice. Although few spectrometers today are specifically designed for flame emission spectrometry, most high-quality atomic absorption spectrometers can be run in an emission mode. Flame emission can be particularly useful because qualitative scans are possible in order to identify the elemental composition of a sample (see Fig. 10.10). Such information may be useful in clarifying a problem with an atomic absorption determination.

The primary limitation in the use of atomic absorption instrumentation for emission measurement is the resolution capability of the monochromator. Successful emission measurements generally require a smaller bandpass than atomic absorption measurements, in the range 0.02 to 0.04 nm. Atomic absorption measurements can often be made with a bandpass of 5 to 10 times these values. Thus, provided that proper care is taken in the setting of the spectrometer spectral bandpass, that the spectrometer is carefully set on the correct emission line, that one checks for spectral interferences, and that (for most elements) the nitrous oxide–acetylene flame is selected, flame emission can provide a complement to the capability of atomic absorption measurements.

TABLE 10.4. *Flame Choice and Detection Limits in Flame Emission Spectrometry*

	Air-Acetylene			Nitrous Oxide–Acetylene	
Element	Wavelength, nm	Detection Limit, µg/mL	Element	Wavelength, nm	Detection Limit, µg/mL
Ca	422.7	0.005	Ag	328.1	0.02
K	766.4	0.0005	Al	396.1	0.005
Na	589.0	0.0005	Au	267.6	0.5
Rb	780.0	0.001	Ba	553.5	0.001
			Cd	326.1	2
			Co	345.3	0.05
			Cr	425.4	0.005
			Cu	327.4	0.01
			Fe	371.9	0.05
			In	451.1	0.002
			Li	670.7	0.00003
			Mg	285.1	0.005
			Mn	403.1	0.005
			Mo	390.3	0.1
			Ni	341.4	0.03
			Pb	405.7	0.2
			Sn	284.0	0.5
			Sr	460.7	0.0001
			Ti	399.8	0.2
			V	437.9	0.01
			W	400.8	0.5
			Zn	360.1	3

Source: Adapted with permission from E. E. Pickett and S. R. Koirtyohann, *Anal. Chem.*, *41* (14), 28A. Copyright 1969 American Chemical Society.

One of the most common applications of flame emission spectrometry is the determination of the alkali metals, particularly in the clinical laboratory. Instruments designed specifically for this determination may contain simply a single interference filter as the monochromator and a vacuum phototube as the detector. These instruments employ low-temperature flames in which only the most prominent lines of the elements appear. Frequently, a two-filter, two-detector arrangement is used for making automatic internal-standard measurements. Lithium, for example, can be used as an internal standard for sodium measurements. A constant amount of lithium is added to all samples and standards, and the ratio of the intensities of the sodium line and the lithium line is recorded. Such a measurement minimizes the effects of fluctuations in the aspiration rate, flame temperature, and so forth, because the test element and the internal-standard element (if similar enough chemically) should be influenced in the same way, causing the ratio of their spectral intensities to be constant at given concentrations [14].

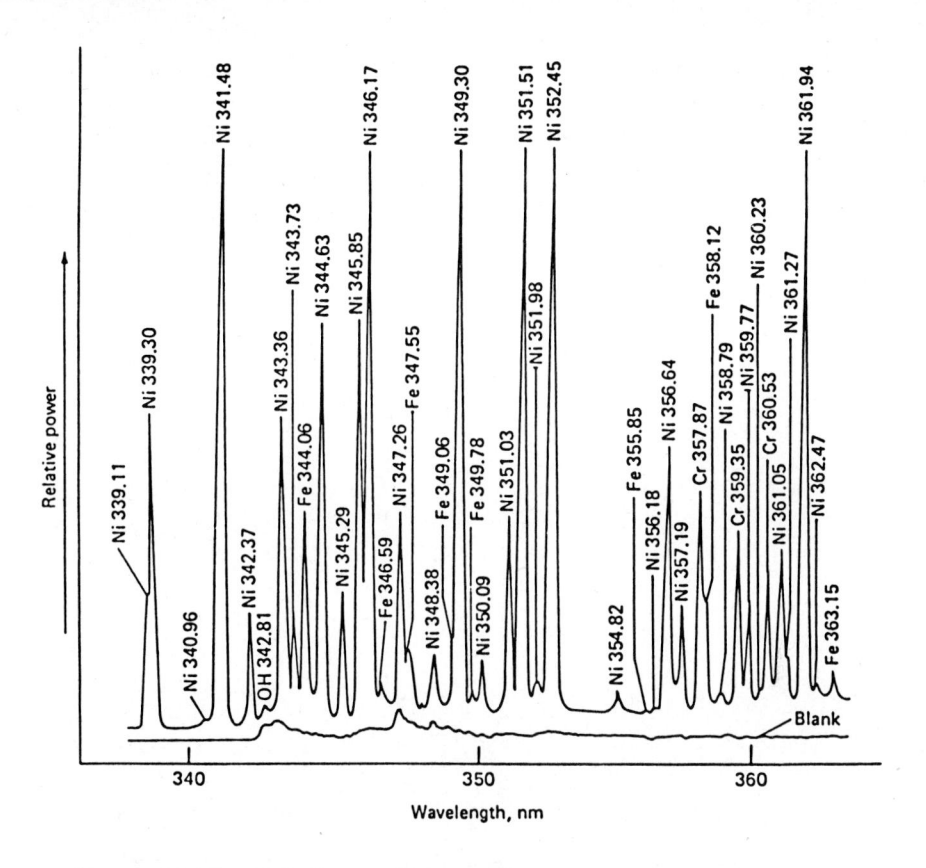

FIGURE 10.10. *Multielement flame emission spectrum. From R. Herrmann and C. T. J. Alkemade,* Chemical Analysis by Flame Photometry, *2nd ed., New York: Wiley-Interscience, 1963, p. 527. Used with permission.*

10.3 ATOMIC ABSORPTION SPECTROMETRY

Atomic absorption spectrometry (AAS) is one of the most important techniques for the analysis and characterization of the elemental composition of materials and samples. It was first proposed as a general-purpose analytical method in 1955 by Walsh [1] in a landmark paper entitled "The Application of Atomic Absorption Spectra to Chemical Analysis." In a more recent paper [2], Walsh indicated that the concept of atomic absorption was slowed over the decades before 1955 by such factors as a lack of photoelectric detection, a misinterpretation of Kirchoff's law, thinking only of continuum sources, and a failure to "avoid being stupid." The key contribution of Walsh was the use of the hollow-cathode lamp as the source. This put the required spectral resolution in the source, and allowed very simple measurement systems to be used. A block diagram of an atomic absorption spectrometer is shown in Figure 10.11.

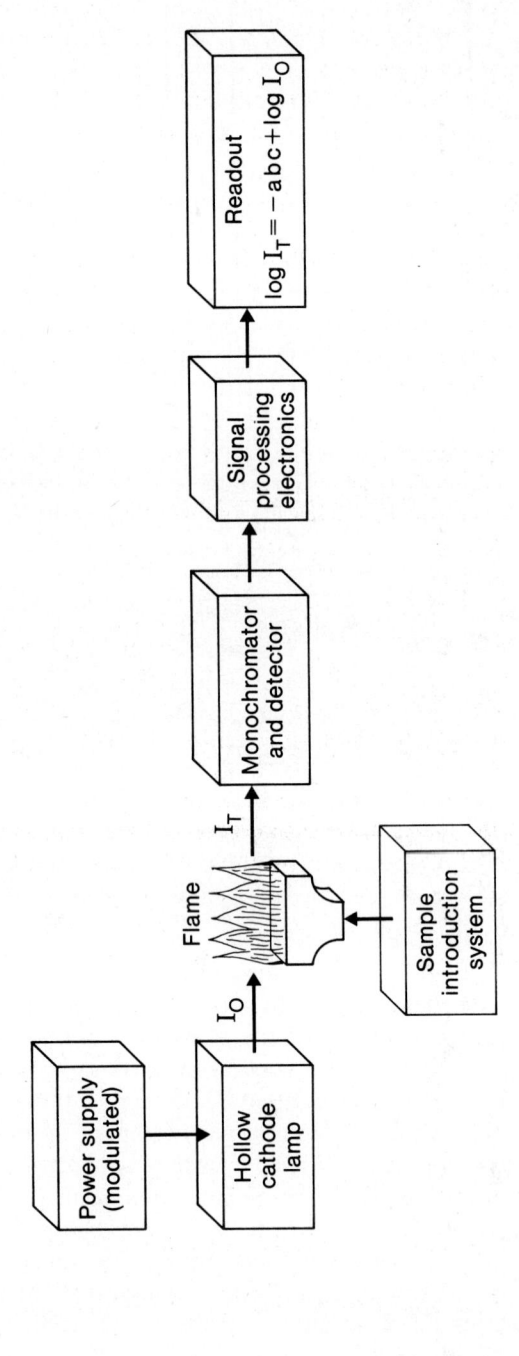

FIGURE 10.11. *Block diagram of an atomic absorption spectrometer.*

Hollow-Cathode Lamps

The choice and development of an appropriate source for an atomic absorption measurement was the key step in the emergence of atomic absorption spectrometry as an analytical method. The requirement perhaps seems simple; as with any absorption measurement, a source of radiation that can be absorbed by the sample is required. For the vast majority of molecular spectrometry absorption measurements, this is readily achieved using the combination of a broad-band source (continuum source) and a monochromator. However, the problem with the atomic absorption measurement is the very narrow (in the range 0.001 to 0.005 nm) absorption profile of atomic species in flames. Thus, starting with a continuum source, a very high resolution and costly monochromator is required to achieve such a bandpass, and then the resulting through-put is likely to be so low that little light is left with which to make the absorption measurement. Walsh's suggestion in 1955 that an element-specific line source be used for the measurement gave birth to atomic absorption spectrometry as a general-purpose analytical technique. The primary source that he suggested be used was the *hollow-cathode lamp*.

A schematic diagram of a hollow-cathode lamp is shown in Figure 10.12. The cathode of the lamp is normally constructed from a single element (or an alloy of the element of interest), and the spectrum emitted by the lamp is the spectrum of the element and the filler gas, most often neon. See the paper by Pillow [15] for a description of the hollow-cathode discharge. If an important line of the element is overlapped by a Ne filler-gas line, argon can be used as an alternative. The lamps are typically run at a current of a few milliamperes, and this requires an applied voltage of 100 to 300 V in series with a 5–10-kΩ resistor. When the lamp is run at relatively low currents—approximately 3 mA—the line width of the atomic emission is about the same as (or narrower than) the atomic absorption profile of the free atoms in the flame. Thus, the hollow-cathode lamp provides an almost ideal source for the atomic absorption measurement, its wavelength exactly matched to that of the analyte and the bandwidth essentially ideal for the atoms in the flame.

The importance of this bandwidth match for the quality of the atomic absorption measurement is illustrated in Figure 10.13. For the purpose of the present discussion, assume that the monochromator bandpass is 0.02 nm. This would require a reasonably high-quality monochromator. As shown in Figure 10.13A for a hollow-cathode lamp source, one can think of the monochromator as being required only to

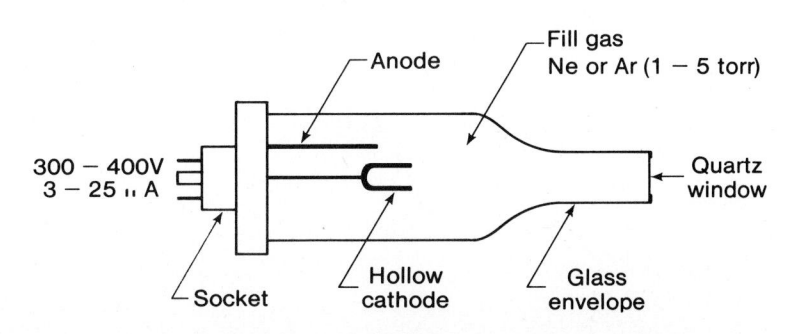

FIGURE 10.12. *Schematic diagram of a hollow-cathode lamp.*

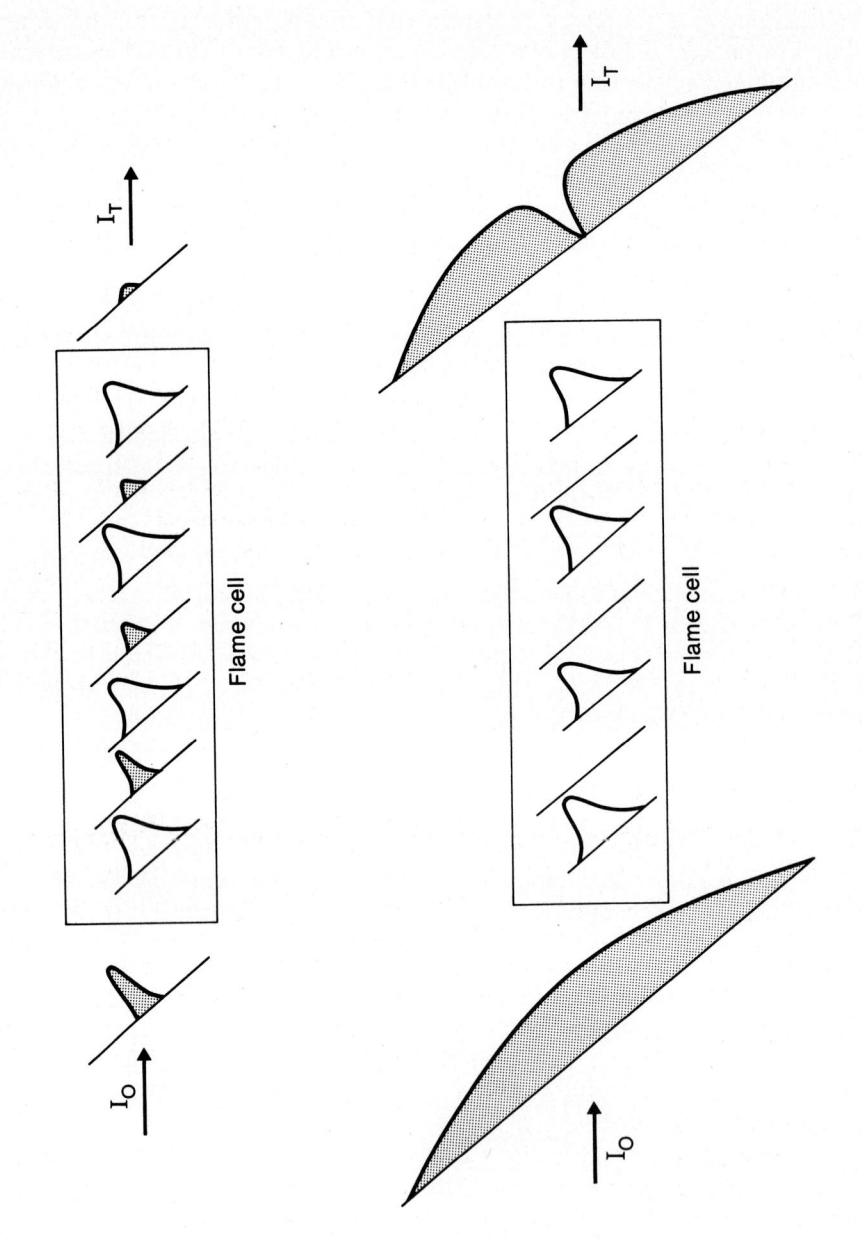

FIGURE 10.13. Schematic diagram of measurement-bandwidth aspects in atomic absorption spectrometry.

"isolate" the line from the lamp. Once this isolation takes place, the required spectral narrowness is built into the line itself, and a sensitive analytical measurement can be carried out. On the other hand, if one started with a continuum source and tried to make the measurement with the same monochromator, the result would be as illustrated in Figure 10.13B. Here, the bandwidth of the source would be the bandpass of the monochromator and even a very strongly absorbing atom concentration in the flame can absorb only a small fraction of the light within the bandpass of the monochromator. The net result is an apparently very low absorbance value, despite the "strong absorbance," due to a classic stray-light effect. This leads to a very insensitive absorption measurement.

It should be mentioned, however, that with a high-resolution echelle spectrometer and an intense continuum source, excellent atomic absorption determinations can be carried out [16].

In addition to their primary lines, most elements have several less sensitive secondary lines that can be used. Such secondary lines can be useful for analyses of solutions for which the concentrations are too large for measurements at the most sensitive line, and dilution of the sample to the operating range of the most sensitive line either cannot be done or is undesirable because of the extra analytical time involved. For example, the normally used (and most sensitive) line for Fe is at 248.3 nm, which provides an optimal working range from about 2 to 10 μg/mL. Three other usable lines for Fe are at 372.0, 386.0, and 392.0 nm; these provide optimal working ranges of about 10, 20, and 390 times higher concentration, respectively.

A number of multielement hollow-cathode lamps are also available. In general, single-element lamps are recommended because they provide the most intense emission of the desired line and the least chance of spectral overlap. However, multielement lamps can be used to back up some elements and to provide a source for elements that are seldom determined.

Electrodeless Discharge Lamps

In addition to hollow-cathode lamps, electrodeless discharge lamps have found some utility as line sources in atomic absorption spectrometry [17]. A small amount of the element or salt of the element or a combination of these is sealed inside a quartz bulb with a small amount of inert gas. This bulb is placed inside a ceramic cylinder on which a helical resonator is coiled. When a radio-frequency field is applied, the energy ionizes the inert gas and excites the metal atoms inside the bulb into emitting their characteristic spectrum. Electrodeless discharge lamps sometimes provide superior performance, and their useful lifetime is longer than that of their respective hollow-cathode lamp. (Note in Table 10.2 that electrodeless discharge lamps were used for several elements.)

To conclude this section, it will prove useful to summarize the consequences to atomic absorption spectrometry as a result of using line sources such as hollow-cathode lamps and electrodeless discharge lamps. The method is very spectrally selective, since the hollow-cathode lamp functions as a "monochromator" set exactly on the analyte line. The actual monochromator serves only to isolate the desired line from nearby lines and does not set the bandpass of the atomic absorption measurement. Because high-resolution monochromators are costly instruments, the net result is that atomic absorption instrumentation can be relatively low in cost and thus has become widely applied. However, use of an element-specific line source does impose some major limitations. Only one element can be determined at a time, making multielement analysis very tedious and time consuming. In other words, atomic absorption is essentially only a quantitative technique and cannot be effectively used for qualitative analysis—that is, identification of the elemental composition of a sample. In addition, measurement of background absorption is considerably complicated. This point will be discussed in more detail in the next section.

10.4 ATOMIC ABSORPTION MEASUREMENTS

Ideally, the transmitted intensity I_t of the hollow-cathode lamp through the flame is governed by the Beer-Lambert law:

$$I_t = I_0(10^{-abc}) \tag{10.2}$$

in which I_0 is the intensity of the source radiation that can be absorbed by the sample atoms, and the other symbols have their usual meanings. In reality, this ideal is not normally achieved because of several factors.

1. The source may not be spectrally pure, leading to a component of the source radiation that cannot be absorbed by the analyte atoms, I_{au}.
2. The analyte atoms in the flame can emit radiation at exactly the same wavelength where the absorption measurement is being made, I_e.
3. The sample matrix may induce scattering losses, I_s.
4. The sample matrix may induce a background absorption loss, I_{ba}.

Therefore, as far as the detection system is concerned, the transmitted intensity is

$$I_t = I_0(10^{-abc}) + I_{au} + I_e - I_{ba} - I_s \tag{10.3}$$

and an effective detection and measurement system must be designed to eliminate or control the effects of the last four terms on the validity of the absorbance values.

The first of these terms, unabsorbed source radiation (I_{au}) is, in fact, handled by using the hollow-cathode lamp as the source. As long as the monochromator bandpass can isolate the analysis line cleanly, this term can be eliminated. In some cases, care must be exercised in setting the monochromator slit width in order to avoid a problem. If any problem should exist with this term, it will be manifested as a stray-light effect and sensitivity is reduced.

The analyte flame emission term is eliminated by *modulation* of the source before it passes through the flame. This is achieved either by mechanical means such as a rotating chopper to interrupt the beam, or by electronic modulation of the hollow-cathode lamp power supply. The measurement system is designed to selectively amplify and synchronously demodulate only this alternating or AC signal from the lamp. The analyte flame emission signal is a DC signal and is not measured. It should be noted that although the emission signal is "not detected," the presence of a very large emission signal can degrade the signal-to-noise ratio of the atomic absorption signal.

Background Correction

To handle the last two terms—scattering losses and background absorption losses—the general area of background correction in atomic absorption analysis must be discussed. The general problem is an absorption caused by the flame or induced by the sample matrix that is independent of the analyte. Typically, such absorption is broad band in nature. If the problem of absorbance originates in the flame, it is not too serious, because the magnitude of the correction can be determined easily by nebulizing a blank solution, and "eliminated" by adjustment of the zero-absorbance setting (100% T). The more serious situation comes from sample matrix–induced problems, such as scattering and molecular absorption. For example, high-salt matrices (NaCl or KI) cause background absorption problems as a result of the presence of molecular species in the flame. In electrothermal atomization, scattering by particulates (smoke) may be serious. The net result, as far as the measurement system is concerned, is that spuriously high absorbance values result.

This problem is not unique to absorption measurements. For emission measurements, a similar problem would be a sample-induced change in background emission. The solution of such a problem for an emission measurement is simple. One can set the monochromator to a wavelength just off the line and measure the background emission while the sample is being nebulized. Should the background change be more complex, a short scan on both sides of the line may be required. However, in atomic absorption, such a straightforward solution is not possible because the source does not emit just off the line, and thus no corresponding absorbance signal can be measured by shifting to an adjacent wavelength. Considerable ingenuity has been required by scientists and manufacturers in order to solve this measurement problem.

Currently, the most common approach is to use a secondary continuum source in the measurement system [18], as depicted in Figure 10.14. A deuterium lamp is frequently used for the continuum source, and the two light beams are combined with a beam splitter. It is important that the two beams, one from the hollow-cathode lamp and one from the deuterium lamp, follow exactly coincident paths through the atom cell; otherwise, proper correction of the background absorbance/scattering will not occur [19]. Each lamp is modulated electronically, but 180° out of phase with the other, and the detection system measures the difference between the two absorbance signals. The hollow-cathode lamp signal is indicative of both the analyte (narrow-line) absorption and the background (broad-band) absorption/scattering. However, the absorbance signal measured with the continuum source is indicative *only* of the broad-band absorption—we have already seen that such a source provides an inadequately sensitive measurement of the atomic absorption signal (see Fig. 10.13). The difference, then, is the net atomic absorption.

An alternative approach to background correction that has met with some success is to measure absorbance near the analytical line using a filler-gas line of the lamp, a nonresonance line of the analyte which is unlikely to be absorbed by the ground-state analyte atoms in the flame, or even a line of another element using a multielement lamp or a second hollow-cathode lamp. In all these cases, it is assumed that the absorbance at the secondary line is due only to the background absorbance/scattering, and the value measured is subtracted from that measured at the analytical line. It is a somewhat more cumbersome approach, and is subject to the problem that the background absorbance may not be exactly the same at the alternate wavelength.

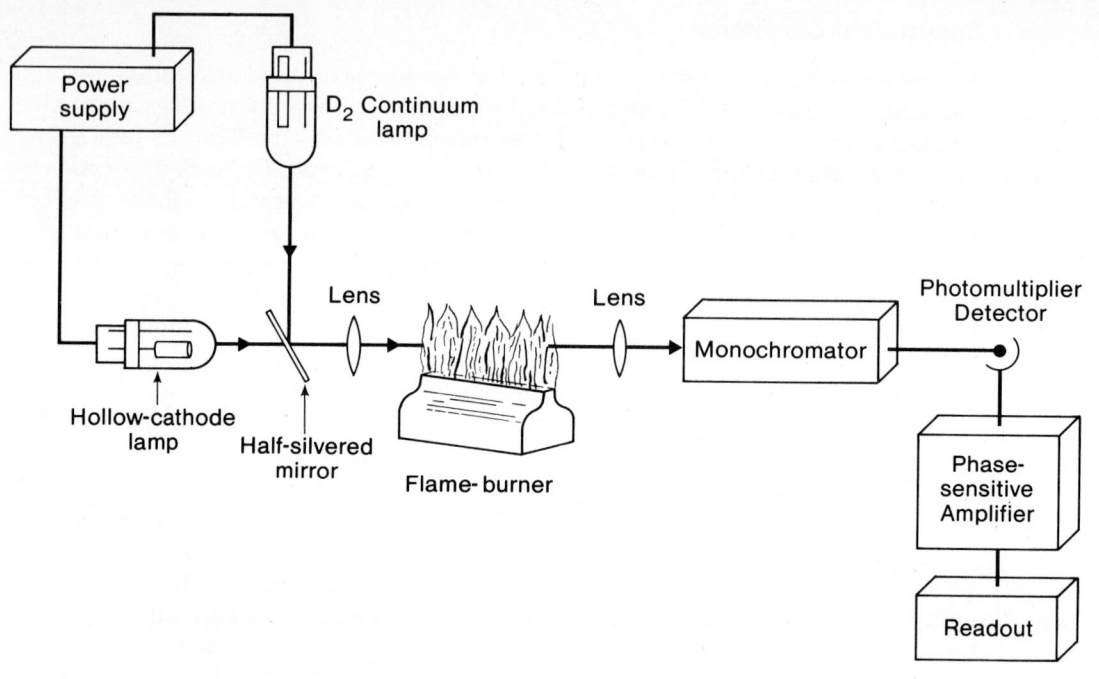

FIGURE 10.14. *Deuterium-lamp background-correction system.*

A rather clever new approach to background correction is to measure "absorbance" while briefly pulsing the hollow-cathode lamp to a high current [20]. This broadens the lamp emission lines so that any absorbance signal measured under these conditions is primarily due to broad-band absorption. Substracting this value from that measured at normal current levels gives a background corrected result. This is an elegant approach in that it does not require any major instrument modifications except to the measurement electronics, and it requires only a single lamp.

Finally, Zeeman techniques can be used to develop elaborate background correction systems. These have been used mostly in conjunction with electrothermal atomization systems and will be discussed in that section dealing with these.

The Analytical Curve

With all the above precautions and subsystems, a reliable measurement of absorption can be obtained and related to concentration by the Beer-Lambert law:

$$\log\left(\frac{I_0}{I_t}\right) = A = abc \tag{10.4}$$

In order to have a readout signal that is linearly related to concentration, the logarithm of the measured signal must be evaluated by the measurement system:

$$\log I_t = -abc + \log I_0 \tag{10.5}$$

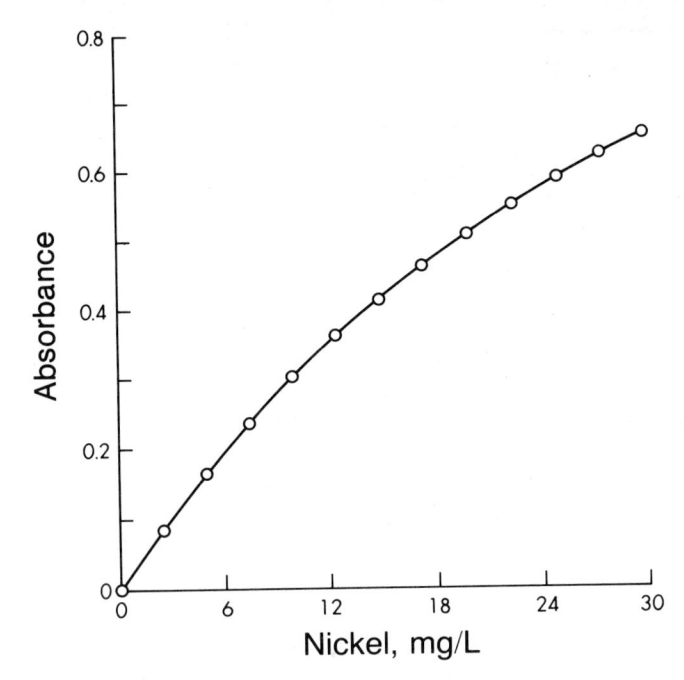

FIGURE 10.15. *Typical atomic absorption analytical curve. Courtesy of the Perkin-Elmer Corporation.*

The log I_0 term is measured while nebulizing the blank, and the readout display is also normally zeroed at this time. The evaluated log I_t signal is then directly related to concentration.

A typical analytical curve is shown in Figure 10.15. Clearly, it is not exactly linear as predicted by Equation 10.5. There are several causes of nonlinear analytical curves in atomic absorption spectrometry, including stray light, line broadening, nonhomogeneities of temperature and spatial structure in the flame, and nonabsorbing lines within the bandpass of the monochromator.

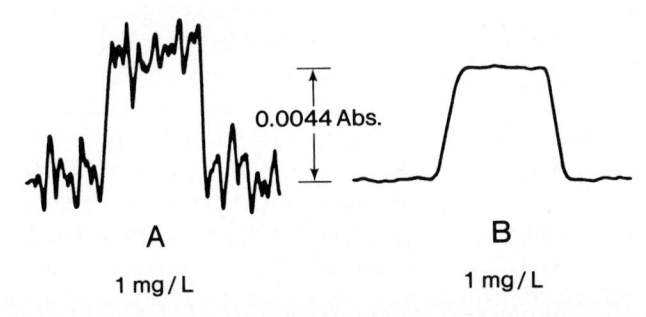

FIGURE 10.16. *Distinction between detection limit and sensitivity. Both signals exhibit the same sensitivity, but the detection limit is superior for case B. Courtesy of the Perkin-Elmer Corporation.*

10.4 Atomic Absorption Measurements

Two figures of merit—sensitivity and detection limit—are often used in conjunction with atomic absorption measurements. *Sensitivity* has been defined as the concentration that gives an absorbance of 0.0044 (1% absorption), but is more clearly stated as simply the slope of the analytical curve in its linear region. Thus, sensitivity is a measure of the size of the absorption signal. *Detection limit*, on the other hand, is defined as the concentration of the element that will produce a signal two or three times the root mean square (rms) value of the baseline noise. In a sense, this is the lowest concentration that can be statistically differentiated from zero (the blank). The distinction between detection limit and sensitivity is illustrated in Figure 10.16. Although sensitivity may have some utility in monitoring instrument performance, detection limit is the better figure of merit.

Thus, we see that there are many aspects to the measurement step, and the simple atomic absorption spectrometer depicted in Figures 10.1 and 10.11 becomes a somewhat complex measurement system if reliable values are to be obtained. Also, the instrumentation discussed here is single beam. A number of commercial instruments contain a double-beam optical system. In such a system, a small part of the hollow-cathode lamp signal is picked off with a beam splitter and shunted around the flame. The readout is the logarithm of the ratio of the signal level in the beam passing through the flame and that of the reference beam. This approach is not fundamentally required for a correct absorbance measurement but it does counteract fluctuations in the lamp intensity and hence provides a more stable baseline.

10.5 ELECTROTHERMAL ATOMIZATION

A key disadvantage of flame systems as sources of atomic vapor is the inefficient use of sample by the nebulizer–spray chamber–burner system. Over the last 10 to 15 years, considerable effort has been directed toward the development of "nonflame" atomization systems for atomic absorption spectrometry. Although not the first and only worker in the field, the early impetus for the development of such systems came from the work of L'vov [21], particularly in his 1969 paper [22]. His systems were furnace-type designs. One of the key goals of such a system is conversion of all the analyte atoms in a particular sample into atomic vapor. This basic approach has now developed to the extent that the atomic absorption literature is replete with both fundamental and applied studies of electrothermal atomizers and related devices.

Over the years, it seems that an almost limitless number of system configurations have been utilized and proposed for electrothermal or flameless atomizers. The basic feature of an electrothermal atomizer (a more accurate designation than "flameless") involves resistive heating of a sample container by passage of an electrical current. A host of subsystems have been used to hold the sample. They include carbon furnaces, carbon rods, carbon cups, carbon braid, nickel cups, and a number of structures such as cups, tubes, wires, and ribbons made out of materials such as tantalum, molybdenum, and tungsten. Two of the more common geometries—a carbon furnace and a carbon tube—are shown in Figure 10.17. See Tables 10.5 and 10.6 for a comparison of detection limits in flames and electrothermal atomizers and typical sensitivities of commercial electrothermal atomizers, respectively.

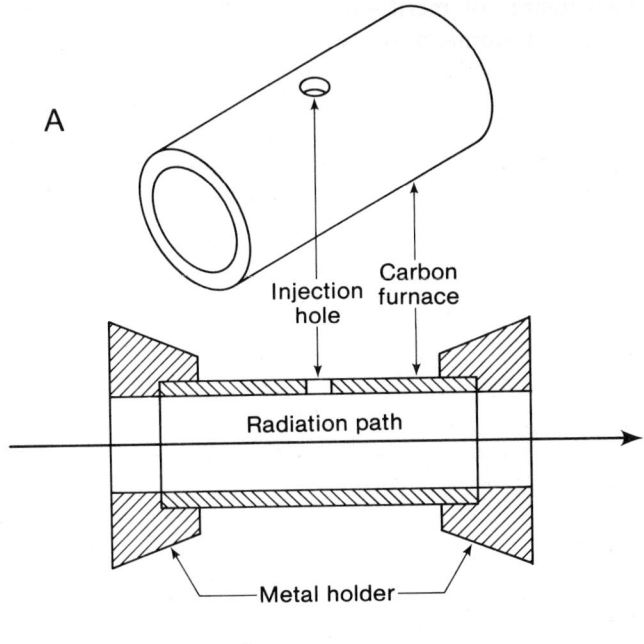

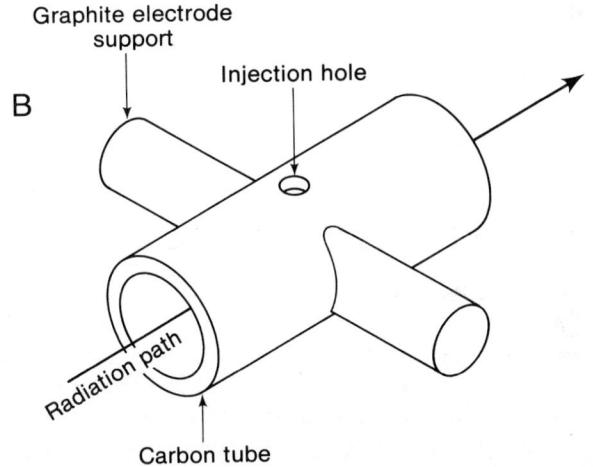

FIGURE 10.17. *Schematic diagrams of furnace (A) and tube-type (B) electrothermal atomizers.*

Basic System Features

Although a large number of systems have been used, the basic features of most of the systems are similar. Most samples are introduced into the sample container as solutions. Sample volumes are small, in the microliter range, typically anywhere from 5 to 100 μL. The sample is normally heated by passage of an electrical current, and often a three-stage heating program is used. The first stage is a *drying cycle*, during which the solvent is vaporized. This is carried out at low temperatures ($< 110°C$).

TABLE 10.5. *Comparison of Detection Limits for Flame and Electrothermal Atomization*[a]

Element	Detection Limit, $\mu g/L$ Electrothermal[b]	Flame	Ratio of Detection Limits Flame : Electrothermal
Ag	0.005	0.9	180
Al	0.01	30	3000
Au	0.1	6	60
Ba	0.04	8	200
Bi	0.1	20	200
Cd	0.003	0.5	170
Co	0.02	6	300
Cr	0.01	2	200
Cu	0.02	1	50
Fe	0.02	3	150
Mn	0.01	1	100
Mo	0.02	30	1500
Ni	0.1	4	40
Pb	0.05	10	200
Pt	0.2	40	200
Si	0.1	60	600
Tl	0.1	9	90
V	0.2	40	200
Zn	0.001	0.8	800

a. Taken from *The Guide to Techniques and Applications of Atomic Spectroscopy*, Perkin-Elmer Corporation, Norwalk, Conn., December 1983. Courtesy of the Perkin-Elmer Corporation.

b. 100-μL sample volume.

The second stage is the *ash cycle*, during which the objective is to char and "burn off" any organic component in the sample. The temperature of the container during this stage may vary widely but is usually in the range 200 to 500°C. Both these stages, drying and ashing, may take from 30 to 90 sec and, clearly, the ash cycle is required only if the sample has an organic matrix. The third stage is the *atomization cycle*, at which time the applied current level is increased significantly in order to reach temperatures in the range 2000 to 3000°C. This is a short cycle, lasting 3 to 5 sec. The analyte is atomized and a transient plume of atomic vapor is generated. Signal measurement will be discussed later in this section.

The above description was purposely semiquantitative in order to emphasize the very large number of empirical variables that must be considered. In fact, a major overall problem in this area is the empirical nature of the method and the lack of fundamental guidelines to aid the choice of analytical conditions. This situation is slowly improving as more basic research is carried out on these methods [23].

Although solution-sample introduction dominates, numerous other methods have been used [24]. These include placement or adsorption of samples on wires that

TABLE 10.6. *Absolute Sensitivities Obtained with Some Commercial Electrothermal Atomizers*

Element	Perkin-Elmer HGA 400[b]	Varian GTA-95[c]	IL 655 CTF[d]
Ag	3	1	0.5
Al	18	7	4
Au	11	4	5
Cd	2	0.3	0.2
Co	17	5	8
Cr	4	2.5	4
Cu	6	2.5	4
Fe	6	2	3
Mg	0.3	0.2	0.07
Mn	3	0.7	1
Mo	9	8	12
Ni	10	10	18
Pb	16	3	4
Pt	90	90	80
Sn	19	22	7
V	30	28	40
Zn	0.7	0.25	0.3

Absolute Sensitivity, pg[a]

a. Absolute sensitivity is defined as the quantity of an element required to produce a 1% absorbance ($A = 0.0044$). Absolute *detection limits* (S/N = 2) average about a factor of 5 (range 2 to 10) smaller.
b. "Analytical Methods for Furnace Atomic Absorption Spectrometry" for the HGA 400 Graphite Furnace, Perkin-Elmer, 1980, 1984. Courtesy of the Perkin-Elmer Corporation.
c. Varian GTA-95 Graphite Tube Atomizer. Courtesy of Varian Associates.
d. Instrumentation Laboratories IL655 CTF (Controlled-Temperature Furnace). Courtesy of Instrumentation Laboratory, Inc.

are then placed or suspended in the furnace, electrochemical preconcentration on wires or on the furnace material itself, and nebulization of a sample onto a heated (low-temperature) atomizer. There has also been increasing use of electrothermal systems for the direct analysis of solid samples. The goal is to eliminate the costly, time-consuming, and contaminating steps necessary to put many matrices into solution. The main problems with the direct introduction of solids are standardization, accuracy, and precision.

As was indicated earlier, a wide range of atomizer geometries and compositions have been utilized and proposed. The most common material is carbon (graphite); and the two most common geometries are closed (the furnace type) and open (the rod or tube type), with the closed types predominating. Considerable effort has been devoted to modifications of the graphite surface for improved performance. Pretreatments include pyrolytic coating and metal (carbidizing) treatment [24] with salt solutions of metals such as tantalum, tungsten, zirconium, molybdenum, and lanthanum. These pretreatments generally make the surface less porous or prevent the sample itself from forming carbides with the atomizer, both of which make the sample more difficult to atomize efficiently and reproducibly.

In the last few years, a technique known as the L'vov platform technique has found wide application [25]. The L'vov platform is a thin pyrolytic graphite plate placed inside the graphite furnace. Volatilization takes place from this surface into a gas that is hotter than the surface. In general, this system has better analytical characteristics than volatilization of the sample from the furnace wall. A schematic representation of this system is given in Figure 10.18.

Performance Characteristics

In the development of electrothermal atomization techniques to date, empiricism has dominated, and only in recent years has progress been made in understanding the complicated processes that occur during electrothermal atomization. In particular, complex time behavior exists during the vaporization of an analyte species; and both the time behavior and the nature of the vaporized species may be highly dependent on the sample matrix. Consider lead, for example. Lead is the element most frequently determined by electrothermal atomization, and chloride is perhaps the main interferent. The problem here is that the lead appears to be lost as the chloride during the atomization step, and thus low results are obtained for lead analyses in this matrix. Using a combination of methodologies such as a short atomization ramp, peak-area measurement, Mo-coated pyroelectric tubes, and the addition of ammonium nitrate as a "matrix modifier," excellent results can be obtained for lead analyses in a high-chloride matrix [26].

This last methodology, the use of a *matrix modifier*, is becoming more common.

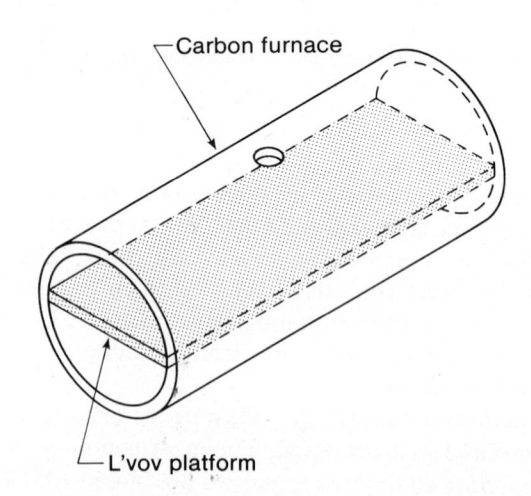

FIGURE 10.18. *Basic geometry of a L'vov platform system.*

In the case of lead, the ammonium nitrate is thought to promote the vaporization of the chloride as ammonium chloride, thus inhibiting the formation of lead chloride. Oxygen gas has also been suggested as a matrix modifier, and numerous other *in situ* chemical pretreatments are being developed and proposed. In particular, one direction involves the use of reagents to stabilize volatilization and to prevent premature vaporization of an analyte during the ashing stage. Most of this work centers on the direct electrothermal atomization of the so-called hydride-forming elements, Sb, As, Se, and Te. Nickel compounds are the most frequently used vaporization-stabilization reagents for these elements [24].

Signal-Measurement Considerations

The crucial difference between the analytical signal in flame-based atomic absorption methods and the electrothermal-based methods is, of course, the transient nature of the atomic vapor in the latter, typically lasting only a few seconds. As obvious as this seems, it took some time before effective and appropriate measurement systems were routinely utilized. The problem was that these atomizers were simply added on to conventional flame-based spectrometers without sufficient care being taken to modify the measurement electronics and readout approaches to correctly respond to, acquire, and measure the transient absorption signal [27]. The desired measurement is the area of the peak-shaped signal, and most modern spectrometers now provide very effective microprocessor-based measurement systems that implement this measurement.

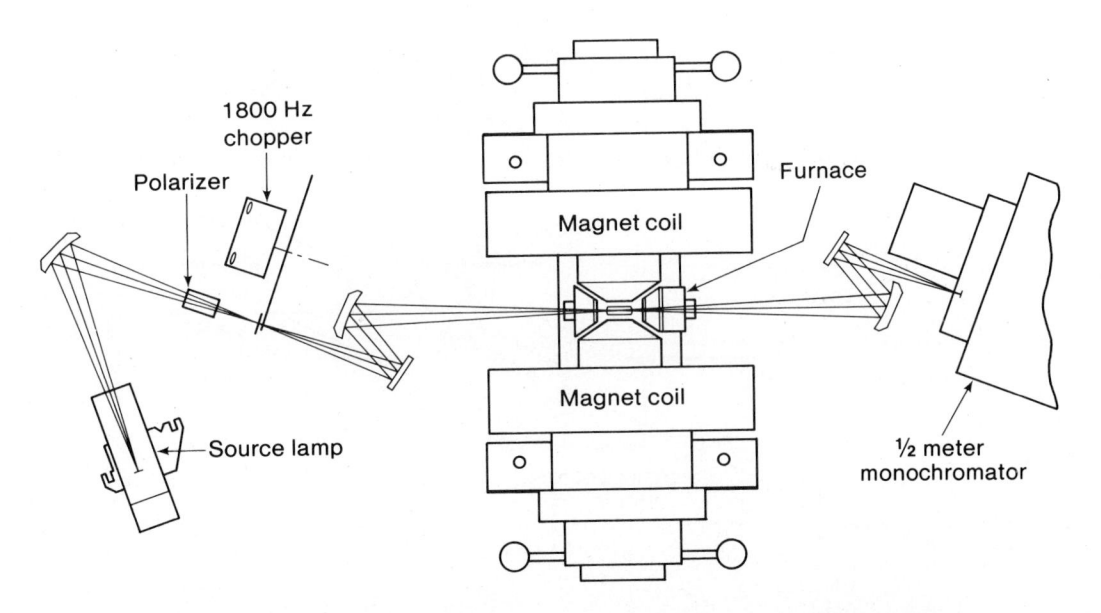

FIGURE 10.19. *Schematic diagram of an analyte-shifted Zeeman-effect background correction system. From F. J. Fernandez, S. A. Myers, and W. Slavin,* Anal. Chem., *52, 741 (1980). Copyright © 1980 by the American Chemical Society. Reproduced with permission.*

The second major problem is that background correction is, in general, considerably more necessary and critical for electrothermal techniques. The deuterium-based method discussed previously can be used. However, there has also been extensive development of Zeeman-based background correction techniques with primary emphasis on their application to electrothermal atomizers.

Zeeman-Effect Background Correction

Systems based on the use of the Zeeman effect are now widely employed for background correction, particularly in conjunction with electrothermal atomizers [28]. Two main approaches have been developed: *source-shifted* and *analyte-shifted* methods. There are some difficulties in running conventional atomic absorption sources in magnetic fields, and analyte-shifted systems are more common. An optical schematic of an analyte-shifted system is shown in Figure 10.19.

In the presence of a strong magnetic field across the atom-vapor cell, the resonance wavelength at which an atom normally absorbs is split into several components.

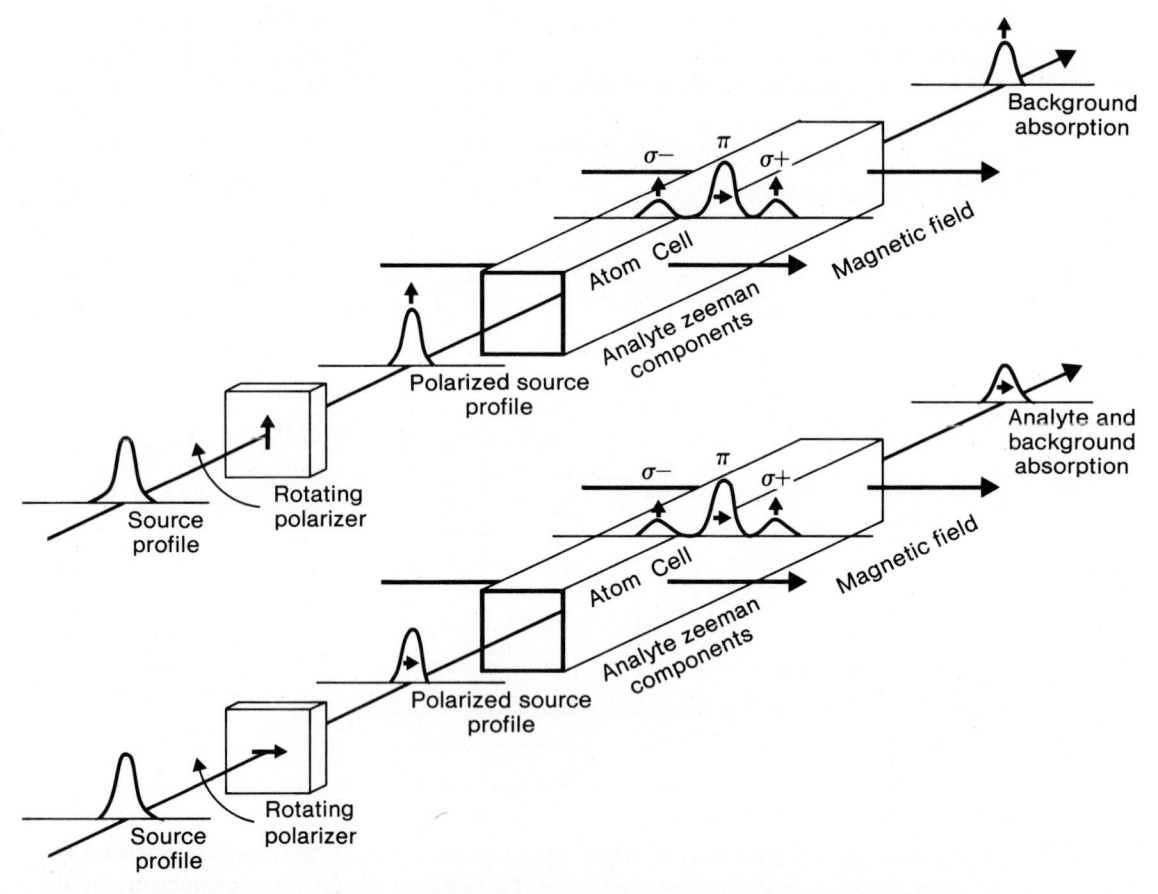

FIGURE 10.20. *Schematic diagram of an analyte-shifted Zeeman-effect background correction process.*

In the simplest case (see Fig. 10.20), the so-called normal Zeeman effect, three components result. These are the *pi* (π) component, corresponding to a change in magnetic quantum number of zero, and two *sigma* (σ) components, corresponding to a ± 1 change in the magnetic quantum number. The sigma components are shifted a few thousandths of a nanometer to either side of the pi component at a field strength of about 10 kilogauss.

The critical difference between these two components is that the pi component can absorb only light that is polarized parallel to the magnetic field, whereas the sigma components can absorb only light that is polarized perpendicular to the magnetic field. The molecular species responsible for background absorption exhibit a negligible Zeeman effect and will absorb both parallel and perpendicularly polarized light equally well. Thus, if the light from the hollow-cathode lamp is alternately polarized parallel and perpendicular to the magnetic field across the analyte vapor, then the difference in absorption between these two polarizations is independent of the background and proportional only to the analyte concentration. Such a system also corrects for scattering losses, because both polarizations of the source will also be equally scattered.

Although somewhat more complicated than the more conventional approaches to background correction, the Zeeman method is more effective in correcting for large background absorptions. Correction occurs at the same wavelength as the analyte absorption and along the exact optical path. It is, however, more complex and costly to implement; and, although used with flames, it is more difficult to engineer such a system around a flame than around an electrothermal atomizer.

10.6 APPLICATIONS

The key strength of any analytical technique is its ability to help solve chemical problems of interest to science, industry, government, and society. Atomic absorption spectrometry is almost unparalleled in the diversity of its application. An excellent and comprehensive source of the latest applications and developments in atomic absorption, atomic fluorescence, and flame spectrometry is The Royal Society of Chemistry series *Annual Reports on Analytical Atomic Spectroscopy*. The latest book is volume 11, covering 1981 [29]. In addition, excellent semiannual bibliographies have been published for a number of years in the *Atomic Absorption Newsletter*, now known as *Atomic Spectroscopy* [30]. Developments in the field are also reviewed every other year in the "Fundamental Review" issue of *Analytical Chemistry* [24, 31]. Literally thousands of applications and developments in technique are reported annually in these publications.

General Considerations

The analytical application of atomic absorption or atomic emission spectroscopy generally involves obtaining the sample in an appropriate solution for measurement and calibrating the instrument properly. Commonly used methods for different materials are described below. Frequently, a releasing agent will have to be added, or a solvent extraction will be required to concentrate the element and increase the sensitivity. Standards should be treated in a similar manner.

Instruments can be calibrated by preparing standard solutions over the concentration range of interest and measuring the absorption or emission of these under the same conditions as sample measurement. At least one standard should be run with each set of samples to determine any correction that should be applied to the calibration curve, because the variables of flame stoichiometry, aspiration rate, and positioning of the burner are difficult to reproduce precisely.

Sometimes, the method of standard additions is used to compensate for chemical and other matrix interference. The principles of this method of calibration, in which the standard is added to an aliquot of the sample, are described in Chapter 2 and elsewhere.

Atomic absorption and atomic emission are used for the analysis of a large variety of materials, containing from trace elements (ppm concentrations) to major ($>1\%$) inorganic constituents. Included are agricultural and biological samples, geological samples, petroleum products, glass and its raw materials, cement, ferrous metals and alloys, water, and air. Because of the general ease—or even lack—of sample preparation and because deaeration of the solution is not required, flame methods have largely replaced polarographic methods for many inorganic analyses. Anodic stripping methods and the more sensitive pulse polarographic techniques are becoming strong competitors, but furnace atomic absorption methods rival the detection limits of these. This last technique can often be used for the direct analysis of small solid samples that can be decomposed or "ashed" at temperatures insufficient to vaporize the test elements present, followed by a higher-temperature atomization step for the analysis itself. Examples are tissue homogenates and leaves.

The primary instrumental disadvantage of atomic absorption techniques is that generally only a single element can be measured at a time. A different light source is required for each element. Atomic emission suffers from a similar disadvantage, although it is fairly simple to change to a different wavelength for the measurement of another element. Simultaneous multielement determinations have been made in recent years by atomic emission spectroscopy using diode-array (Vidicon camera) detectors (described in Chap. 7). The array of diodes effectively serves as multiple detectors (as many as several hundred) which can be arranged to detect several different wavelengths over a range.

Organic or biological samples will usually require destruction by dry ashing or wet digestion with oxidizing acids before flame analysis. Biological fluids (blood or urine) frequently can be aspirated after a simple dilution. For blood analysis, serum or plasma is generally preferred because this fraction of the blood typically contains the clinically significant concentrations of metals. (An exception is blood-lead poisoning, because lead will concentrate in the red cells.) In other cases, the metal may be more concentrated in the red cells, but concentration changes in the serum or plasma are more clinically indicative; examples are potassium, zinc, iron, and magnesium. In these cases, it is critical that blood samples not hemolyze (red cells burst) before separation of the serum or plasma is completed.

Serum is the supernatant obtained from clotted blood. Plasma, the liquid portion of circulating blood, is chemically similar to serum except that it contains in addition fibrinogen, the clotting agent in blood. Plasma is obtained by treating the blood with an anticoagulant such as heparin or oxalate, after which the red cells are separated from the plasma by centrifugation. Oxalate should generally be avoided for metal analysis because many metal oxalates are insoluble; an efficient digestion mixture for biological samples comprises HNO_3, H_2SO_4, and $HClO_4$ in the ratio 3:1:1 by volume, using about 1 mL per gram of wet material.

Metals and alloys can usually be dissolved in acids, whereas materials such as glass will require alkaline or acid fusion. An important consideration in any analysis is matching the matrix of the standards to that of the sample, or else diluting the sample enough to render the physical effects of the matrix harmless. (Chemical effects may still exist.)

One of the most common applications of atomic emission spectroscopy is the determination of the alkali metals, particularly in the clinical laboratory. Blood serum samples need only be diluted with water (or an internal-standard solution) and aspirated.

Use of Organic Solvents

The overall atomization efficiency in a flame is increased by the use of organic solvents. Such an increase may have a variety of causes, including increased rate of aspiration, finer droplets, more efficient evaporation or combustion of the solvent, and so forth. Thus, increased sensitivity would be obtained by adding a miscible organic solvent such as acetone to the solution. (A threefold increase is typical.) The problem is that adding the miscible solvent dilutes the sample solution, which more or less defeats the purpose. Therefore, the technique of solvent extraction (Chap. 20) is usually used to obtain increased sensitivity. The dissolved metal is extracted from the aqueous solution into an immiscible organic solvent in which it is more soluble. The organic phase containing the metal is then aspirated into the flame.

There are a number of advantages to this approach. (a) The test element is separated from the bulk matrix of the sample, frequently eliminating possible interferences. (b) It is obtained in a pure organic solvent, which results in maximum atomization efficiency; a 10-fold signal enhancement can be attained for a given concentration. (c) The test element can be extracted into a smaller volume of organic solvent, with (in many cases) a 10- to 100-fold gain in concentration.

Methyl isobutyl ketone (MIBK) is one of the best materials for solvent extraction and aspiration into a flame. Ammonium 1-pyrrolidinecarbodithioate (APCD) is a commonly used extracting agent because it reacts with a large number of elements in acidic solution (in the literature, this reagent is often referred to as ammonium pyrrolidine dithiocarbamate [APDC]).

When using organic solvents, the initial flame adjustment before aspirating the solvent should generally be very lean because the solvent must be burned as well as the fuel. If the flame is too rich in fuel, the solvent will not be completely burned and the flame will be smoky. The proper flame condition can be adjusted with the solvent aspirating. Solvent should be aspirated between samples because the hot lean flame will tend to heat up the burner.

Lead in Blood

Lead in unclotted blood can be determined by atomic absorption spectroscopy. Five milliliters of heparinized blood are treated with trichloroacetic acid (TCA) to precipitate proteins, which are centrifuged. The pH of the filtrate is adjusted to 3, 1 mL of aqueous APCD is added, and the lead is extracted into 5 mL of MIBK as the Pb(APCD)$_2$ chelate. The organic phase is separated and aspirated into an air-acetylene flame for atomic absorption measurement. Standards are treated in the same manner, and water-saturated MIBK is used to zero the instrument. The detection limit for this procedure is about 0.1 ppm lead in the blood. The upper level of "normal" blood is 0.6 ppm, with most values being in the range 0.3 to 0.4 ppm. Instead of precipitating proteins before solvent extraction, the blood can be hemolyzed with 1 mL of 5% Triton X-100 solution to release the lead from the red cells.

Lead in blood can be determined in microsamples using the Delves microcup sampling procedure. This is a technique similar in operation to furnace atomic absorption. Here, 10 mL of blood is placed in a small nickel crucible, dried on a hotplate at 140°C, and then partially oxidized at 140°C with 20 mL of 30% H$_2$O$_2$ until a dry yellow residue is obtained. The crucible is then mounted on a holder and thrust into a flame where the lead is vaporized into a horizontal nickel tube above the flame (through a hole in the bottom of the tube). The light path is through the tube. A transient signal results (as in furnace atomic absorption), and the peak height of the recorded signal is related to the lead concentration. This technique is generally more precise for blood-lead microanalysis than most other atomic absorption methods (relative standard deviation = $\pm 8\%$ at the 0.4-ppm level). A method of standard additions is used for calibration.

Zinc in Plants

Zinc is an essential trace element in plants. One gram of dried and ground plant material is dry ashed in a silica crucible overnight at 500°C. The ash is treated with 5 mL of 6 M HCl and slowly dried on a steam bath. This operation is repeated with another 5 mL of acid to hydrolyze pyrophosphate and to dehydrate any silica from the sample or the crucible. The residue is taken up in 20 mL of 0.1 M HCl and filtered. This solution can be aspirated directly into an air-acetylene flame for atomic absorption measurement. Standards are prepared in 0.1 M HCl. As always, the blank is prepared in the same way as the sample.

Copper in Seawater

Major constituents of seawater, such as sodium and magnesium, can be determined by appropriate (at least 10-fold) dilution with deionized water and direct aspiration. However, many elements in seawater exist at ppb or lower levels and must be concentrated prior to analysis. Copper (1 to 25 ppb) is one of these. Common procedures include ion-exchange chromatography, solvent extraction, or a combination of these. Copper can be determined by adjusting the pH of seawater (1 L) to 3, adding 5 mL of 2% APCD solution, and extracting the copper with 25 mL of methyl-n-amyl ketone, which has a low solubility in water. The separated ketone layer is aspirated into an air-acetylene flame for atomic absorption measurement.

Beryllium in Airborne Particulate Matter

Particulate matter in a measured volume of air can be collected on a cellulose acetate membrane filter (e.g., Millipore®). The filter is dry ashed in a low-temperature asher. (This device uses oxygen radicals in a radio-frequency plasma for ashing at below 100°C, minimizing losses due to volatility of the test element and retention on crucible walls.) The ash is taken up in dilute HCl and aspirated directly, or the filter can be digested with a mixture of nitric and perchloric acids. For beryllium determination, a nitrous oxide–acetylene flame is used. Results are reported as micrograms per cubic meter of air.

Sodium, Potassium, Magnesium, Manganese, and Calcium in Cement

A 0.5-g sample of cement is decomposed in 4 M HCl and evaporated to dryness, after which the residue is taken up in 4 M HCl. After filtering the solution and diluting it to 100 mL, aliquots are taken to determine each element by atomic absorption spectroscopy using an air-acetylene flame. Standards for sodium and potassium must contain about the same concentration of calcium as the sample solutions. The presence of the high concentrations of calcium in the sample suppresses interference by aluminum or silicon on magnesium absorption. Phosphate, aluminum, and silicon interference on calcium absorption is eliminated by adding 50 mL of 25,000-ppm strontium solution to 5 mL of the stock sample solution and diluting this to 250 mL with water. All standards contain the same concentration of acid or other reagents as the samples.

10.7 ATOMIC FLUORESCENCE SPECTROMETRY

Although atomic fluorescence is not used as frequently as atomic absorption and emission techniques, research in this area is proceeding at a rapid pace. In addition to developments of practical analytical systems and applications, a significant body of research is appearing in which atomic fluorescence, particularly in combination with laser sources, is providing unique fundamental information about atomic vapor systems. Some of the lowest detection limits ever to be reported for atomic spectrometry are now being given in the atomic fluorescence literature, with the measure being number of atoms rather than such a "large" unit as parts per billion or picograms. For example, a detection limit for lead was recently reported as 0.05 pg/mL, equivalent to 250 atoms/cm^3 [32]. Finally, a commercial atomic fluorescence spectrometer (Baird Corporation) is now available, although it is based on use of an inductively coupled plasma as the atom source rather than a flame as has been the case in most research studies.

The application of atomic fluorescence to analytical measurements was primarily developed by Winefordner [33, 34] in the United States and West [35] in England, and several reviews of the method have been published [36–38].

As mentioned before, atomic fluorescence involves excitation of atomic vapor by a radiation source, followed by deactivation by the emission of radiation; the emitted radiation is then measured. This process is not unlike molecular fluorescence spectroscopy, described in Chapter 9. A fluorescent line can be at a wavelength identical to the exciting wavelength, or it can be longer or (very rarely) shorter. There are two main types of fluorescence: resonance and nonresonance. A third type is sensitized fluorescence.

Resonance Fluorescence

Resonance fluorescence occurs when atoms absorb and reemit radiation at the same wavelength. The most common examples correspond to transitions originating in the ground state (resonance transitions). For example, resonance fluorescence is observed for zinc at 213.86 nm, for nickel at 232.00 nm, and for lead at 283.31 nm. Some atoms may have an appreciable population in a low-lying metastable energy level and exhibit resonance fluorescence originating from these levels. The intensity of emitted radiation in this case is generally less than for the more abundant ground-state atoms. Resonance fluorescence originating from ground-state atoms is often accompanied by nonresonance fluorescence having the same upper excitation level.

Nonresonance Fluorescence

Nonresonance fluorescence occurs when the exciting wavelength and the wavelength of the emitted fluorescence line are different. There are two basic types: direct-line fluorescence and stepwise-line fluorescence.

In *direct-line fluorescence*, an atom is excited (usually from the ground state) by a radiation source, and then undergoes a direct radiational transition to a metastable level above the ground state. An example is absorption at the 283.31-nm line by ground-state lead atoms, with subsequent emission at 405.78 nm. As with resonance fluorescence, direct-line fluorescence can be excited by absorption of a nonresonance line, for example tin fluorescence at 333.06 nm.

In *stepwise-line fluorescence*, the upper levels of the exciting and the emitted lines are different. In the normal case, the excited atoms lose part of their energy by collisional deactivation (by collision with flame molecules) and then return to the original (usually ground) state by radiational deactivation. Sodium, for example, is excited at the 330.3-nm line and undergoes stepwise fluorescence to emit a line at 589.0 nm. In a second type of stepwise-line fluorescence, the radiationally excited atom is further excited (thermally) to a higher electronic state and then undergoes radiational deactivation to a metastable state (i.e., the emitted radiation is still longer in wavelength than the exciting radiation).

A less common type of nonresonance fluorescence is *anti-Stokes fluorescence*, or thermally assisted fluorescence, in which the emitted wavelength is shorter than the absorbed wavelength. This occurs when atoms populating an energy level lying near, but above, the ground state are excited to a higher energy level and then undergo a radiational transition to the ground state. Alternatively, this can also occur when a ground-state atom is excited to a certain electronic state by absorbing a photon, subsequently raised to a slightly higher excited state by absorbing thermal energy from the flame, and finally radiationally deactivated to the ground state. The former is a special case of direct-line fluorescence, whereas the latter is a special case of stepwise-line fluorescence. Anti-Stokes fluorescence is always accompanied by resonance fluorescence.

All types of nonresonance fluorescence, particularly direct-line fluorescence, can be analytically useful. Sometimes nonresonance fluorescence is more intense than resonance fluorescence, and it offers the advantage that scattering of the exciting radiation can be eliminated from the fluorescence spectrum by use of a filter or a monochromator. Self-absorption problems (absorption of the emitted radiation by the sample atoms) can also be avoided by measuring fluorescence at a nonresonance line that is not also absorbed.

Sensitized Fluorescence

In sensitized fluorescence, an atom emits radiation after collisional activation by a foreign atom that was excited previously by absorbing resonance radiation, but which has not yet been deactivated again. An example is the sensitized fluorescence of thallium atoms in a gas mixture containing a high pressure of mercury vapor and low pressure of thallium vapor. When irradiated at the 253.65-nm mercury line, the thallium atoms emit at 377.57 and 535.05 nm. This type of fluorescence requires a higher concentration of foreign atoms than can be obtained in flame cells, but presumably it could be observed in nonflame cells.

Atomic Fluorescence Instrumentation

The basic instrumental setup for atomic fluorescence is shown in Figure 10.1C. The source is normally placed at right angles to the atomic vapor source/monochromator axis so that source radiation (except for scattered radiation) does not enter the monochromator. The primary-source radiation is chopped or modulated in order to discriminate against emitted radiation from the atomic vapor source.

A wide range of primary sources have been used in atomic fluorescence measurements. These include hollow-cathode lamps, electrodeless discharge lamps, xenon-arc continuum sources, tunable dye lasers, and inductively coupled plasmas. Pulsed hollow-cathode lamps are used in the commercially available atomic fluorescence spectrometer mentioned above. An intense source is most desirable and, to this end, the high-power tunable dye laser is the most ideal source [39], although dye lasers with the necessary output in the far end of the ultraviolet (eximer pumped dyes) are just becoming available. Thanks to rhodamine 6G, a dye with excellent lasing characteristics in the yellow, Na remains the most frequently used element in fundamental studies.

An important aspect of high-power sources is the potential to saturate the atomic fluorescence signal [40]. The intensity of atomic fluorescence is normally thought of as being proportional to the intensity of the primary source. This is true. However, for very strong sources, a YAG-pumped dye laser for example, the fluorescence signal can be saturated in the sense that the output is no longer proportional to the primary-source intensity. This condition is desirable because the sensitivity is at a maximum, and the atomic fluorescence signal is insensitive to fluctuations in the primary-source intensity. Such dye lasers are very expensive, and thus laser-saturated atomic fluorescence remains a research technique, but a technique that can offer some of the best detection limits in atomic spectrometry.

Most of the work in analytical atomic fluorescence has been carried out using flames as the atomic vapor source. Initially, low-temperature flames were used, but such flames are very prone to vaporization interferences and regular air-acetylene flames provide significantly better overall analytical results. Recent work has shown that the inductively coupled plasma is also a very useful atom cell for atomic fluorescence measurements.

The monochromator/detector end of the experiment may be essentially identical to atomic absorption spectrometry. Of course, the logarithmic transformation is not required, since the fluorescence signal is directly proportional to analyte concentration. At high concentrations, the analytical curve levels off at a maximum fluorescence intensity set by the primary-source intensity. The curve may even bend downward if conditions at high concentration levels result in a decrease in fluorescence efficiency. Detection limits are listed in Table 10.7.

Finally, although atomic fluorescence spectrometry with complex dye laser sources is a sophisticated technique, very simple nondispersive measurement systems can be utilized. If line-specific sources such as hollow-cathode lamps are used, the measurement system can be as simple as an interference filter and a photomultiplier tube [41].

TABLE 10.7 *Comparison of Detection Limits Obtained with Dye Lasers and Conventional Sources in Atomic Fluorescence Flame Spectrometry*

Limits of Detection, µg/mL

	Pulsed Sources			Continuous-Wave Sources	
Element	Laser	Line	Continuum	Line	Continuum Eimac (150 W)
Ag	—	0.004	0.02	0.0001	0.006
Al	0.005	0.07	—	0.1	0.2
Ca	0.005	0.0003	—	0.2	—
Cd	—	0.004	0.03	0.000001	0.01
Co	0.2	0.007	0.1	0.005	0.02
Cr	0.02	0.004	—	0.05	0.01
Fe	0.3	0.008	—	0.008	—
Mg	0.0003	0.001	0.004	0.001	0.0003
Mn	0.01	0.002	0.03	0.006	0.004
Mo	0.3	0.06	—	0.5	0.2
Ni	0.05	0.02	—	0.003	—
Pb	0.03	0.07	—	0.01	—
Sr	0.01	0.01	—	0.03	—
Ti	0.1	—	—	4	0.9
Tl	0.02	—	—	0.008	—
V	0.5	0.2	—	0.07	0.1
Zn	—	0.003	0.1	0.00001	0.006

Source: Adapted with permission from N. Omenetto, *Anal. Chem.*, **48**, 75A. Copyright 1976 American Chemical Society. Values are for analyte in aqueous solutions.

SELECTED BIBLIOGRAPHY

ALKEMADE, C. TH. J., and HERMANN, R. *Fundamentals of Analytical Flame Spectroscopy.* Bristol, England: Adam Hilger, 1979.

CRESSER, M. S. *Solvent Extraction in Flame Spectroscopic Analysis.* London: Butterworths, 1978.

FULLER, C. W. *Electrothermal Atomization for Atomic Absorption Spectrometry.* London: The Chemical Society, 1980.

GAYDON, A. G., and WULFHAND, H. G. *Flames, Their Structure, Radiation, and Temperature,* 4th ed. London: Chapman and Hall, 1978.

KIRKBRIGHT, G. F., and SARGENT, M. *Atomic Absorption and Fluorescence Spectroscopy.* London: Academic Press, 1974.

PRICE, W. J. *Spectrochemical Analysis by Atomic Absorption.* London: Heyden, 1979.

SALVIN, M. *Atomic Absorption Spectroscopy,* 2nd ed. New York: John Wiley, 1979.

VAN LOON, J. C. *Analytical Atomic Absorption Spectroscopy: Selected Methods.* New York: Academic Press, 1980.

WINEFORDNER, J. D., ed. *Spectrochemical Methods of Analysis.* New York: Wiley-Interscience, 1971.

REFERENCES

1. A. WALSH, *Spectrochim. Acta,* 7, 108 (1955).
2. A. WALSH, *Spectrochim. Acta,* 35B, 643 (1980).
3. K. A. SATURDAY and G. M. HIEFTJE, *Anal. Chem.,* 52, 786 (1980).
4. M. D. AMOS and J. B. WILLIS, *Spectrochim. Acta,* 22, 1325 (1966).
5. E. E. PICKETT and S. R. KOIRTYOHANN, *Spectrochim. Acta,* 23B, 235 (1968).
6. R. K. SKOGERBOE and K. W. OLSEN, *Appl. Spectrosc.,* 32, 181 (1978).
7. J. W. NOVAK and R. F. BROWNER, *Anal. Chem.,* 52, 287 (1980).
8. J. W. NOVAK and R. F. BROWNER, *Anal. Chem.,* 52, 792 (1980).
9. G. M. HIEFTJE and H. V. MALMSTADT, *Anal. Chem.,* 40, 1860 (1968).
10. B. SMETS, *Analyst* (London), 105, 482 (1980).
11. J. I. DINNIN, *Anal. Chem.,* 32, 1475 (1960).
12. A. C. WEST and W. D. COOK, *Anal. Chem.,* 32, 1471 (1960).
13. E. E. PICKETT and S. R. KOIRTYOHANN, *Anal. Chem.,* 41(14), 28A (1969).
14. F. J. FELDMAN, *Anal. Chem.,* 42, 719 (1970).
15. M. E. PILLOW, *Spectrochim. Acta,* 36B, 821 (1981).
16. J. M. HARNLY, T. C. O'HAVER, B. BORDEN, and W. R. WOLF, *Anal. Chem.,* 51, 2007 (1979).
17. J. W. NOVAK and R. F. BROWNER, *Anal. Chem.,* 50, 1453 (1978).
18. S. R. KOIRTYOHANN and E. E. PICKETT, *Anal. Chem.,* 37, 601 (1965).
19. D. D. SIEMER, *Appl. Spectrosc.,* 32, 245 (1978).
20. J. J. SOTERA and H. L. KAHN, *Amer. Lab.,* 14(11), 100 (1982).
21. B. V. L'VOV, *Spectrochim. Acta,* 17, 761 (1961).
22. B. V. L'VOV, *Spectrochim. Acta,* 24B, 53 (1969).
23. W. FRECH, J. A. PERSSON, and A. CEDERGREN, *Prog. Anal. At. Spectrosc.,* 3, 279 (1980).
24. G. HORLICK, *Anal. Chem.,* 54, 276R (1982).
25. W. SLAVIN and D. C. MANNING, *Spectrochim. Acta,* 35B, 701 (1980).
26. D. C. MANNING and W. SLAVIN, *Anal. Chem.,* 50, 1234 (1978).
27. D. D. SIEMER and J. M. BALDWIN, *Anal. Chem.,* 52, 295 (1980).
28. S. D. BROWN, *Anal. Chem.,* 49, 1269A (1977).
29. M. S. CRESSER and B. L. SHARP, *Annual Reports on Analytical Spectroscopy,* Vol. 11, London: The Royal Society of Chemistry, 1982.
30. D. M. LAWRENCE, *At. Spectrosc.,* 4, 10 (1983).
31. G. HORLICK, *Anal. Chem.,* 52, 290R (1980).
32. M. A. BOLSHOV, A. V. ZYBIN, V. G. KOLOSHNIKER, and M. V. VASNETSOR, *Spectrochim. Acta.,* 36B, 345 (1981).

33. J. D. Winefordner and T. J. Vickers, *Anal. Chem.*, **36**, 161 (1964).
34. J. D. Winefordner and R. A. Staab, *Anal. Chem.*, **36**, 165 (1964).
35. T. S. West and K. Williams, *Anal. Chem.*, **40**, 335 (1968).
36. J. C. Van Loon, *Anal. Chem.*, **53**, 332A (1981).
37. J. D. Winefordner, *J. Chem. Ed.*, **55**, 72 (1978).
38. T. S. West, *Pure Appl. Chem.*, **50**, 837 (1978).
39. M. S. Epstein, S. Bayer, J. Bradshaw, E. Voigtman, and J. D. Winefordner, *Spectrochim. Acta*, **35B**, 233 (1980).
40. D. R. de Olivares and G. M. Hieftje, *Spectrochim. Acta*, **36B**, 1059 (1981).
41. P. L. Larkins, *Spectrochim. Acta*, **26B**, 477 (1971).

PROBLEMS

1. Why will a nonlinear calibration curve and a loss in sensitivity generally occur in atomic absorption spectroscopy when using a continuum light source rather than a sharp-line source?

2. For the elements listed in Table 10.3, calculate the ratios N_u/N_0 given for the different temperatures.

3. Compare the resolution requirements of a monochromator for atomic absorption measurement and atomic emission measurement.

4. Why is a high-temperature flame, for example the nitrous oxide–acetylene flame, sometimes required in atomic absorption spectroscopy?

5. Why is a high concentration of potassium salt sometimes added to samples and standards in flame-spectroscopic measurements?

6. Why is the light source in atomic absorption instruments usually modulated?

7. Identify and describe the major types of interferences encountered in flame atomic emission and atomic absorption measurements. Discuss how each can be minimized.

8. Why are organic solvents sometimes used in flame-spectroscopic measurements?

9. A 12-ppm solution of lead gives an atomic absorption signal of 30% absorption. What is the atomic absorption sensitivity?

10. A serum sample is analyzed for lithium by atomic emission spectroscopy using the method of standard additions. Three 0.500-mL aliquots of sample are added to 5.00-mL portions of water. To these are added 0, 10.0, and 20.0 μL of standard 0.0500 M LiCl solution. The emission signals (in arbitrary units) are 23.0, 45.3, and 68.0 for the three solutions, respectively. What is the concentration of lithium in the serum sample in ppm (weight/volume)?

11. A sample of an unusual amino-acid analog—L-canavaninosuccinic acid, $C_9H_{16}N_4O_7$—was synthesized and converted to its barium salt to prevent a slow internal cyclization reaction during storage. A barium assay was performed by atomic absorption spectroscopy to obtain an indication of the purity of the compound. To this end, 21.0 mg of the compound was dissolved in dilute HNO_3 and diluted to 100.0 mL; 20.0 mL of the resultant solution was further diluted to 100.0 mL. Five replicate atomic absorption readings were taken on a blank, six standards, and the test sample. The data are as follows:

Sample	Average Meter Reading ± Standard Deviation, arbitrary absorbance units
Blank	0 ± 7
1 ppm Ba	44 ± 3
4 ppm	178 ± 4
10 ppm	483 ± 12
14 ppm	684 ± 21
20 ppm	993 ± 36
30 ppm	1512 ± 88
Test sample	762 ± 18

(a) Plot the calibration curve and determine the percentage of barium by weight in the original compound. (b) Compare this result with the expected Ba content for the "pure" compound, and comment on the purity of the synthesized sample. (c) Calculate the approximate detection limit (S/N = 2) for barium.

12. A sodium solution is analyzed by flame emission using the 589-nm sodium doublet line. In

developing a procedure for the analysis, the analyst notes that a 1-ppm solution of sodium gives an emission reading of 112, whereas the same solution containing 10 ppm potassium gives a reading of 123. In view of the fact that a 10-ppm solution of potassium gives no appreciable reading at 589 nm, give a probable explanation for the enhancing effect of the potassium.

13. List seven goals of an ideal source for an atomic spectrometric method of elemental analysis.

14. In the discussion of flame spectrometry, six steps were considered in the generation of gaseous neutral atoms from analyte dissolved in solution. (a) List the six steps and give a one-sentence definition of each. (b) Briefly describe two interferences that may occur during any one of these steps.

15. List and briefly explain three "consequences" of using hollow-cathode lamps as the primary source in atomic absorption spectrometry.

16. In the discussion of flame spectrometry, six steps were considered in the generation of gaseous neutral atoms from analyte dissolved in solution. Discuss the importance and nature of these steps with respect to the carbon-rod atomizer.

17. Explain why continuum sources have been successfully used for atomic fluorescence measurements but not so successfully for atomic absorbance measurements.

18. Will atomic absorption and atomic fluorescence methods make atomic emission methods obsolete? Discuss and justify your answer.

19. The calcium line at 422.673 nm corresponds to a $^1S_0 \leftarrow {}^1P_1$ spectral transition. The sodium **D** lines at 588.9963 and 589.5930 nm correspond, respectively, to the $^2S_{1/2} \leftarrow {}^2P_{3/2}$ and $^2S_{1/2} \leftarrow {}^2P_{1/2}$ transitions. The potassium doublet at 764.494 and 769.901 corresponds to the same transitions. (a) Calculate the ratio of the statistical weights g_u/g_0 for each of these five transitions. (b) Calculate the fraction of atoms in the upper electronic energy state to that in the ground state in each case for an entrained air/argon-hydrogen flame ($T = 1577°C$), an air-acetylene flame ($T = 2250°C$), and a nitrous oxide–acetylene flame ($T = 2955°C$). (c) Assuming the emission intensities of the lines in the Na and K doublets are directly proportional to the populations in the upper states, calculate the intensity ratios for the two doublets in the three flames.

20. Refer to Figure 10.16, which illustrates the distinction between atomic absorption sensitivity and detection limit. Both signals have the same absorbance, $A = 0.0044$. The rms noise in such signals is often conveniently estimated to be one-fifth of the peak-to-peak noise. (a) For each signal, calculate the ratio of signal to rms noise (S/N_{rms}). (b) Using these values, estimate the detection limits ($S/N = 2$ criterion) for the two cases.

CHAPTER 11

Emission Spectroscopy

Ramon M. Barnes

This chapter describes methods of observing the emission spectra of atoms, ions, and molecules using excitation sources powered by electrical energy. The common names *emission spectroscopy* or *optical emission spectroscopy* are applied to these methods.

In emission spectroscopy, the excitation source transforms the sample from its initial state as solid, liquid, or gas into a plasma of atoms, ions, and molecular radicals that can be electronically excited. The radiative deactivation of these excited states produces light quanta which are sorted by wavelength in a spectrometer or spectrograph; the resulting *emission spectrum* is detected by either photographic or photoelectric means. Many aspects of these measurements resemble those described in Chapter 10 for flame-emission spectroscopy; but electrical arcs or sparks, glow or plasma discharges, or lasers replace the flame as the means of atomization and excitation of the sample. Most of the electrical-discharge excitation sources provide greater energies than do flame sources, and they produce more complex spectra that require spectrometers and spectrographs with better resolution than those used in flame spectroscopy.

Because every element possesses characteristic spectra, emission spectroscopy is applicable in both theory and practice to the entire periodic table. However, the emission spectra for some elements, notably halogens and the noble gases, require more energy to produce than do those for a metallic element, and special excitation conditions must be applied. Normally, the emission spectra of all metals and metalloids in a sample occur simultaneously when the sample is electrically excited.

Emission spectroscopy forms the basis for numerous practical qualitative and quantitative analyses in industrial quality-control and research applications. Quantitative analyses are very rapid once the procedure is defined and the instrument standardized, and analyses are generally simultaneous, multielement ones; the determination of 25 to 35 metals and metalloids in steel or aluminum in a fraction of a minute is common. Furthermore, emission spectroscopy is customarily used to obtain a rapid qualitative or semiquantitative survey of elements contained in unknown materials, because of the ability of electrical excitation sources to atomize and excite samples submitted in many forms—powders, solids, liquids, or gases. A spectrographic qualitative analysis may take no more than 20 min.

This chapter will describe the instrumentation required, the types of samples generally studied, and typical examples of techniques and applications of emission spectroscopy.

11.1 PRINCIPLES AND THEORY

The principles and theory described in Chapter 10 for atomic emission apply also to emission spectra produced by electrical excitation sources. In those sources for which the Maxwell-Boltzmann expression (Eqn. 10.1) describes the distribution of energy levels for an atom or ion, the absolute temperature T represents the equilibrium temperature of the discharge. Since the temperatures of electrical sources are generally higher than those of flames, sufficient energy is available to produce ions. The concentrations of electrons and ions are uniquely related to the temperature and to the composition of the gas (all charged particles originate from thermal ionization of the gaseous form of the elements). The basic concept of thermal ionization introduced by Saha is the application of the law of mass action to ionization. The equilibrium constant for the ionization of atom A to ion A^+, $A \rightleftarrows A^+ + e^-$, can be expressed as a function of the absolute temperature in an expression known as the *Saha relationship* [1]:

$$K_A = \frac{n_e n_{A^+}}{n_A} = \frac{2(2\pi m k T)^{3/2}}{h^3} \frac{Z_{A^+}}{Z_A} e^{-E_A/kT} \tag{11.1}$$

where
- K_A = Saha equilibrium constant
- n_e = number density of electrons (cm^{-3})
- n_{A^+} = number density of ions (cm^{-3})
- n_A = number density of atoms (cm^{-3})
- Z_{A^+} = partition function of the ion
- Z_A = partition function of the atom
- E_A = ionization energy of the atom
- k = Boltzmann constant
- h = Planck's constant
- m = mass of the electron

The first term on the right of Equation 11.1 is the total partition function for the electrons, and the factor of 2 incorporated in it is the state sum for the two possible spin states.

The practical version of the Saha formula (for a total pressure of 1 atmosphere) is given as follows:

$$\frac{n_e n_{A^+}}{n_A} = (4.83 \times 10^{15}) T^{3/2} \frac{Z_{A^+}}{Z_A} 10^{(-5040 V_A/T)} \tag{11.2}$$

When V_A, the ionization potential of the atom, is expressed in electron-volts, kT is given as $0.861 \times 10^{-4}T$ (eV/K), and

$$e^{(-0.4343V_A/0.861 \times 10^{-4}T)} = 10^{(-5040V_A/T)}.$$

Example 11.1. What percentage of Ca ($^1S_0$) is ionized to Ca$^+$ ($^2S_{1/2}$) in a high-current arc with a temperature of 6000 K and an electron density of 1.0×10^{17} cm^{-3}?

Solution: The ionization potential for Ca is 6.09 eV. Therefore,

$$10^{(-5040V_A/T)} = 10^{(-5040 \text{eV}^{-1}\text{K})(6.09\text{eV})/(6000\text{K})} = 10^{-5.12}$$

Only the first term of each of the partition functions is needed:

$$Z(T) = \sum_j g_j e^{-(E_j/kT)} \simeq g_0$$

where g_j = statistical weight of the "j"th level having energy E_j above the ground state

(See Section 10.2 for calculation of statistical weights.)

$$\frac{Z_{A^+}}{Z_A} = \frac{g_{A_0^+}}{g_{A_0}} = \frac{(2J+1)_{A^+}}{(2J+1)_A} = \frac{[2(0+1/2)+1]}{[2(0+0)+1]} = 2$$

$$\frac{n_e n_{A^+}}{n_A} = (4.83 \times 10^{15})(6000 \text{ K})^{3/2}(2)(10^{-5.12}) = 3.4 \times 10^{16}$$

$$\frac{n_{A^+}}{n_A} = (3.4 \times 10^{16})(1.0 \times 10^{-17}) = 0.34 = \textit{degree of ionization} = \alpha$$

The percent ionization is

$$\% = \frac{n_{A^+}(100)}{n_{A^+} + n_A} = \frac{0.34 n_A(100)}{0.34 n_A + n_A} = 25\%$$

In an arc discharge (T = 4000 to 7000 K), at least 0.01 to 0.1% of all particles exist as ions. Electrical discharges require this minimum level of ionization to be conductive and remain self-supporting. The discharges, however, are essentially neutral because the negative charge carried by the free electrons is balanced by the total positive charge of the ions.

Spectral lines from neutral atoms are designated with a Roman numeral I and those from singly charged ions with a Roman numeral II. For example, Ca^0 atom emission is indicated as Ca I and Ca$^+$ ion emission is Ca II.

The absolute emission intensity of a radiating transition can be calculated from the density of the upper-state population. The general form of the equation for emission intensity is

$$I = N \frac{hc}{\lambda} gA \frac{1}{Z_M} e^{(-E_{exc}/kT)} \tag{11.3}$$

where I = emission intensity (erg-sec/cm^3-steradian)
N = number density of the species
h = Planck's constant (erg-sec)
c = velocity of light (cm/sec)
λ = wavelength of the transition (cm)

g = degeneracy of the upper state
A = transition probability (sec^{-1})
Z_M = partition function
E_{exc} = excitation potential (eV when k is in eV/K)

The emission intensity for an atom (I^0) and an ion (I^+) can be obtained by substituting the appropriate values into Equation 11.3. Quantity N then becomes n_{A^+} for the ion or n_A for the atom. Transition probabilities, A, can be obtained from the literature [2], and the partition functions can be calculated with the appropriate number of terms as illustrated in Example 11.1. In practice, the necessity of knowing the densities of upper-state population, N, limits the calculation of absolute emission intensity in emission spectroscopic sources. The number density of species depends on the efficiency of transforming the real sample into an atomic vapor, which changes with operating conditions and is dependent upon the excitation source and method of sample introduction. For research applications, the populations can be estimated or measured, but in spectrochemical analysis the relationship between emission intensity and number density of species is assumed to be proportional to the concentration of analyte in the sample. Therefore, a series of standards with known concentrations is introduced into the excitation source, and the intensities are an empirical function of the concentrations.

For electrical sources in thermal equilibrium, the processes of dissociation, excitation, and ionization can be treated as if the gas mixture or plasma were contained in a furnace at the same temperature. The populations of excited and ionic states are described by the Maxwell-Boltzmann distribution law (Eqn. 10.1), and by the Saha relationship (Eqn. 11.1); velocities of electrons are given by the Maxwell distribution, and dissociation by the general relationships for chemical equilibrium. This condition is known as the *local thermodynamic equilibrium*, LTE.

Some electrical discharges, such as spark discharges, are transient in nature; others, such as glow or microwave discharges, operate at subatmospheric pressures or very high frequency. In these, the equilibria that characterize thermal atmospheric discharges are absent, and so the Boltzmann distribution and Saha relationship are no longer applicable. In practice, this means that one excitation source will produce spectra with absolute intensities different from those produced by another. For instance, the emission spectrum of a spark discharge differs significantly with time from that of an arc; also, the excitation of halogens in a hollow-cathode discharge is more efficient than in an arc.

11.2 INSTRUMENTATION

The measurement of emission spectra requires a number of instrumental components, some common in principle to other spectrometric methods (Chap. 6). An excitation source powered by a suitable generator converts the sample into an emitting discharge plasma. The sample is held or contained in an excitation chamber or stand, and the radiation emitted from the electrical discharge is transferred by suitable optics to a spectrometer or spectrograph, which sorts the radiation according to wavelength.

The readout translates the photoelectric (spectrometric) or photographic (spectrographic) record of the spectrum into an analog or digital display of the wavelengths and intensities. The wavelength region generally encompasses 180 to 900 nm; vacuum instrumentation permits investigation of shorter wavelengths, and many commercial systems do not operate at wavelengths longer than 500 to 600 nm.

Spectrometers

The components of a spectrometer include an entrance slit and at least one exit slit, a dispersive element such as a prism or grating, and optical components such as mirrors or lenses to collimate the entering light and to focus the spectrally resolved wavelengths on the exit slits. The term spectro*meter* implies that the detectors are photoelectric *measuring* devices such as photomultipliers, whereas the name spectro*graph* implies that the detector is an emulsion supported on either a glass or film backing that makes a direct photographic *representation* of a complete spectrum.

A Fastie-Ebert plane grating arrangement which can be used as either a spectrograph or a spectrometer by simply deflecting the spectrum with a large 45° plane mirror is illustrated in Figure 11.1. One large concave mirror is used both to collimate the entering light illuminating the grating and to focus the dispersed light onto the exit focal plane. Since its rediscovery in 1952, the Fastie-Ebert configuration and the closely related Czerny-Turner arrangement with its two small mirrors replacing the single Ebert mirror have become the most popular monochromator-spectrometer-spectrograph instruments in atomic spectroscopy, due to the ease with which their wavelength is scanned and to their excellent optical quality. An optical arrangement of a commercial Czerny-Turner spectrometer with three detectors is represented in Figure 11.2. The spectrum reaching the detector is changed easily by rotating the grating about its vertical axis. When a single photomultiplier is used in a Czerny-Turner *monochromator*, the larger focusing mirror is replaced by a smaller spherical or parabolic one. Scanning monochromators are now available with computer-controlled wavelength selection and data acquisition–management systems to provide rapid sequential multielement analyses.

Another plane grating configuration, using an *echelle* grating rather than a conventional plane (echellette) grating, yields a very high resolution spectrum in a small area, but does require a predispersing grating or prism (Fig. 11.3). The echelle grating produces the primary spectrum in one plane, and the cross-disperser separates the overlapping orders (Eqn. 6.8) in a plane at right angles to the primary spectrum. An echelle spectrometer with a 0.75-m focal length can produce the same resolution at 250 nm as the 3.4-m Fastie-Ebert spectrograph illustrated in Figure 11.1. Although commercial echelle spectrometers use a modified Czerny-Turner arrangement, their convenience for wavelength scanning is sacrificed because both the grating and its predispersing element must be moved simultaneously.

Although plane grating spectrometers (Figs. 11.2 and 11.3) are available, the majority of commercial multichannel spectrometers used for emission spectroscopy employ a concave spherical grating. The imaging characteristics the latter grating exhibits are along the grating foci known as the *Rowland circle* (Fig. 11.4). The diffraction properties of both plane and concave gratings described in Equation 6.8 are identical, although the optical imaging characteristics of the two grating types

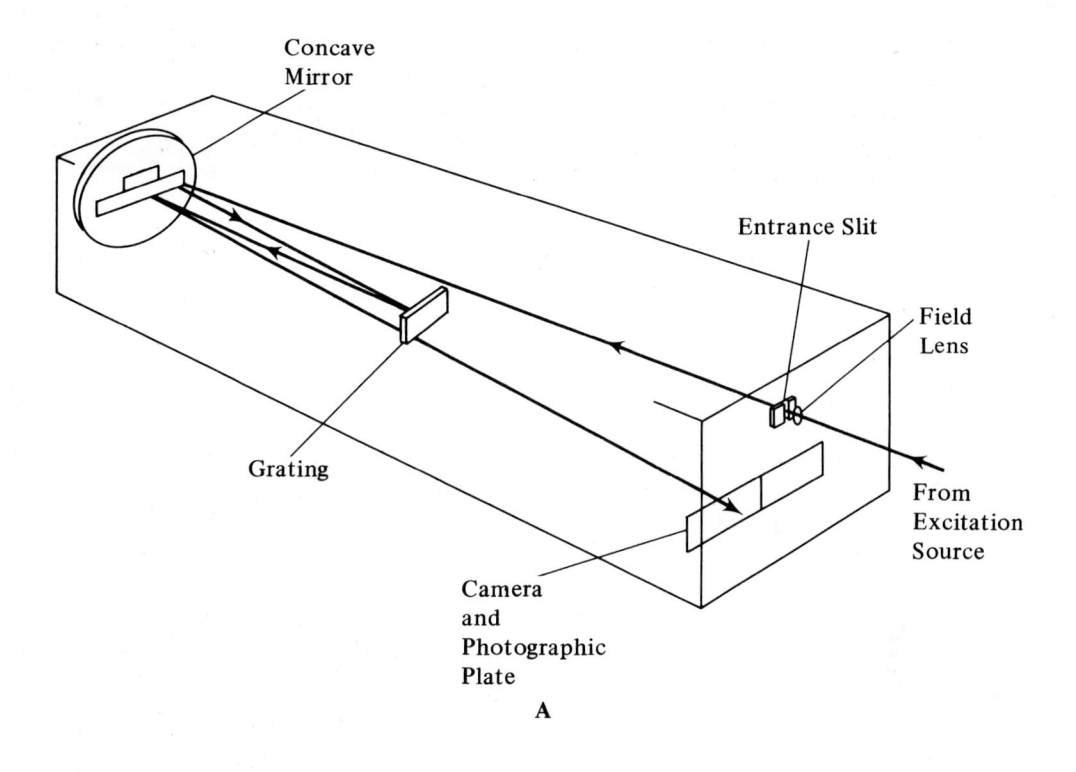

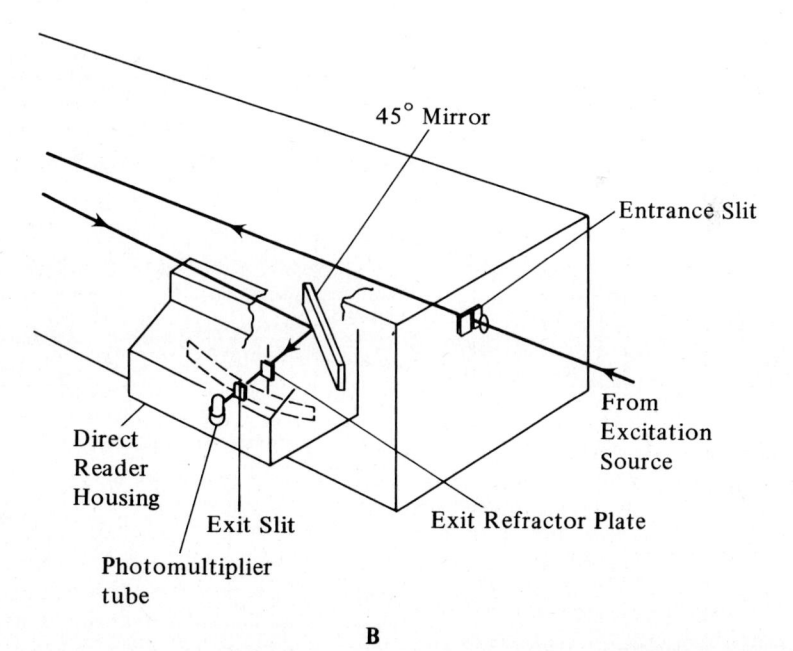

FIGURE 11.1. *Fastie-Ebert plane-grating convertible spectrograph (A) and spectrometer (B). Courtesy of Jarrell Ash Division, Fisher Scientific Co.*

SEC. 11.2 Instrumentation

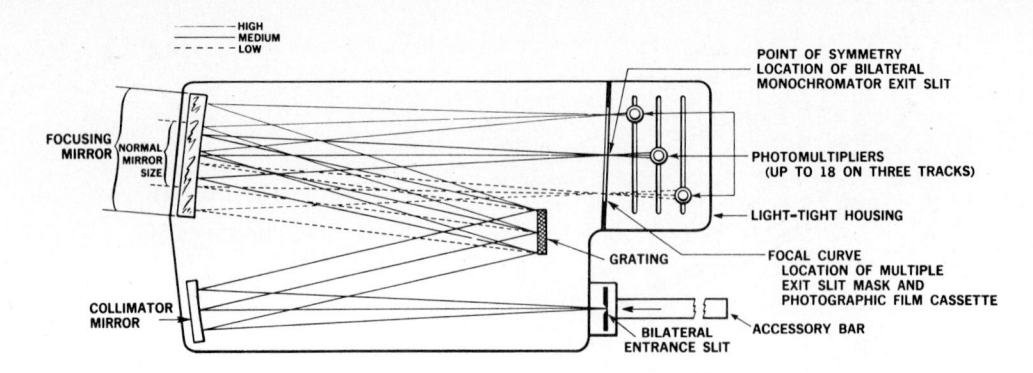

FIGURE 11.2. *Czerny-Turner optical configuration for monochromator, spectrometer, and spectrograph. In the* monochromator *mode, a single exit slit and photomultiplier are located at the point of symmetry on the focal curve, and a normal-size focusing mirror is used. The* spectrometer *arrangement requires a large focusing mirror for the imaging of the spectrum along the focal curve, where a number of exit slits and photomultipliers can be positioned.* Spectrographic *operation is accomplished by inserting a photographic film cassette in the focal curve. The wavelength is changed by rotating the plane grating. Adapted from E. L. Grove, ed.*, Analytical Emission Spectroscopy, *Vol. 1, Part I, New York: Marcel Dekker, 1971. Reprinted by courtesy of Marcel Dekker, Inc.*

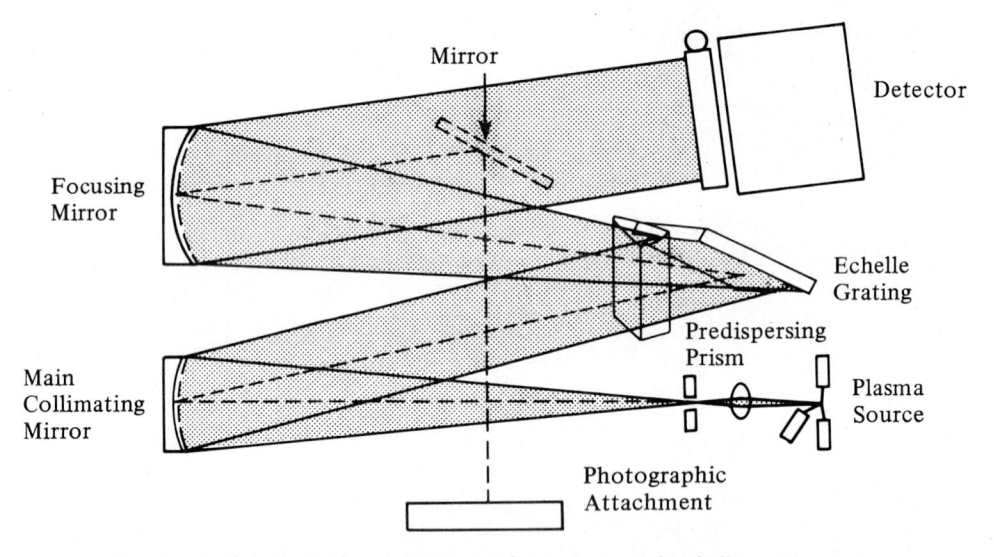

FIGURE 11.3. *Optical arrangement of a commercial echelle spectrometer. An echelle grating is configured as a normal Czerny-Turner spectrometer, and a quartz prism predisperser is positioned so as to refract each order of the grating vertically. Thus, a two-dimensional spectral pattern is imaged at the focal plane. Single or multiple exit slits and photomultiplier detectors are positioned at the focal plane. For photographic recording, the focal plane is diverted by a folding mirror onto a 4-in. by 5-in. camera. Courtesy of Beckman Instruments, Inc.*

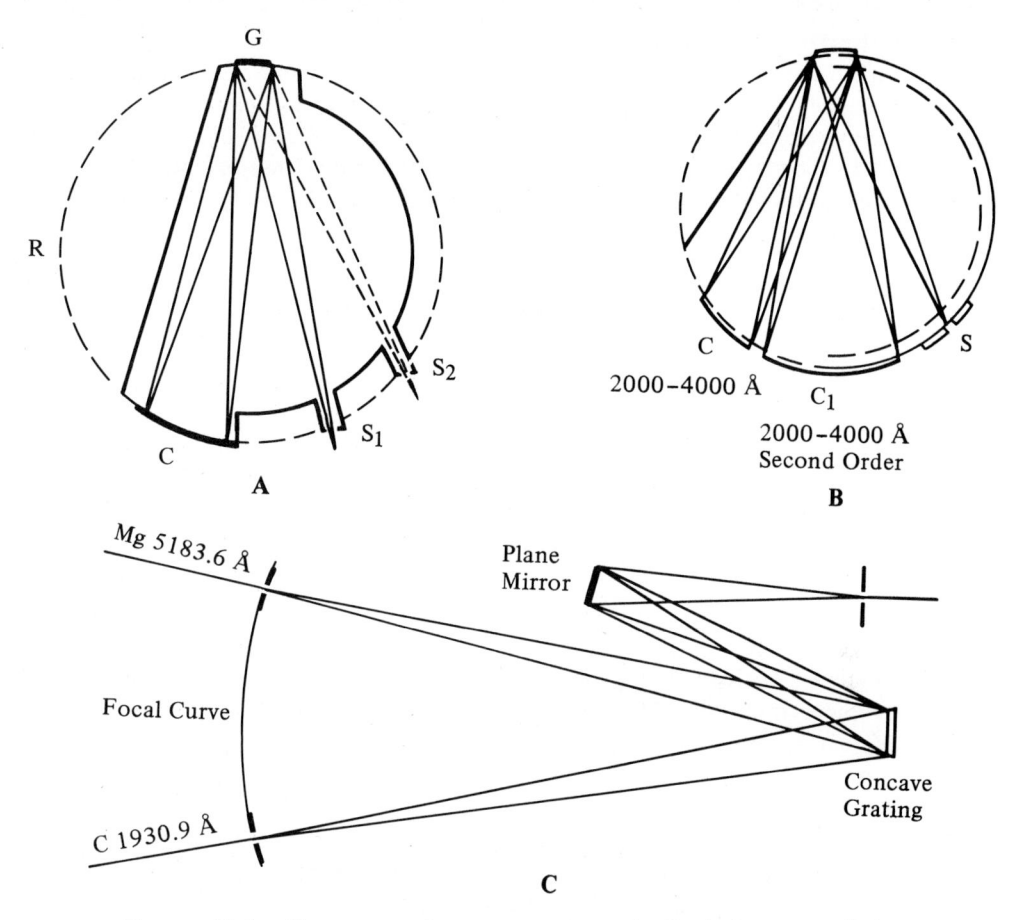

FIGURE 11.4. *Concave grating arrangements. A: Optical system of the modified Abney mounting. Here, R represents the Rowland circle, G is the concave grating, C is the photographic camera, and S_1 and S_2 are fixed entrance-slit and alternative entrance-slit locations. With S_1, the camera covers the wavelength range of 215 to 437 nm when the focal length is 1.5 m. Location S_2 provides longer-wavelength recording or second-order spectra. B: Commercial Paschen-Runge design in which the entrance slit and grating are fixed on the Rowland circle, and the camera, C, is movable on the Rowland circle. C: Modified Paschen-Runge mounting used in a commercial multichannel spectrometer in which the exit slits are located along the Rowland circle, and the entrance-slit position is folded by means of a plane mirror. Adapted from E. L. Grove, ed.*, Analytical Emission Spectroscopy, *Vol. 1, Part I, New York: Marcel Dekker, 1971. Reprinted by courtesy of Marcel Dekker, Inc.*

differ significantly. In the plane grating arrangements like the Fastie-Ebert or Czerny-Turner, a high-quality focused image is obtained because the collimating and focusing mirrors are arranged symmetrically to reduce optical aberrations at the exit slit. For the concave grating configuration, an image from the entrance slit positioned on the Rowland circle is dispersed by the grating, and the vertical focus of each of the wavelengths produced also falls along the Rowland circle where the exit slits (or camera) are located. Thus, for concave grating spectrometers, only one optical element is needed. The major optical disadvantage is that the length of dispersed image is extended, because the horizontal focus falls in a plane removed from the Rowland circle. Concave grating spectrometers are typically constructed so that the grating is fixed and only preselected wavelength ranges are accessible.

The use of photomultipliers in spectrometers requires the accurate positioning of exit slits along the spectrometer's focal curve, allowing the selection of individual spectral lines and groups of lines to detect many elements simultaneously. Emission spectrometers often have room for as many as 90 different exit slits, although only 20 to 35 separate detectors and readouts, called channels, might be used for a particular analysis. For some types of analysis, more than one spectral line from an element can be monitored to provide appropriate concentration ranges for the elements in various samples. Other channels may be used for background detectors or internal reference lines. These instruments are sometimes called *direct-reading spectrometers* or *direct readers.*

Associated with each phototube is an electronic system in which the phototube current is converted to a voltage that is generally collected or integrated on a storage capacitor. The capacitor voltage is read during or at the end of the exposure, and analog devices or digital processors record and compute the concentration of the element detected. Modern spectrometer readout systems operate with digital mini- or microcomputers. The computers control sample handling, excitation conditions, exposure times, and other spectrometer parameters, as well as data acquisition, computation, and presentation. Very rapid, precise readings can be obtained in this manner.

Alternative measurement approaches, including direct photocurrent conversion to a proportional frequency measured by a computer-controlled counting circuit, are practical with low-cost microprocessor and semiconductor products [3]. Although some research efforts have been made to apply semiconductor photodetectors, such as silicon photodiode arrays and vidicon tubes, or a Michelson interferometer for multielement emission spectroscopy [4], the instrumentation has only recently been incorporated into a commercial emission analysis system.

Spectrographs

Spectrographs employ a photographic emulsion to record the entire emission spectrum at one time, so exit slits or multiple detectors are not required. The record is permanent and can be inspected later for more detailed information, which cannot be done with readings obtained from spectrometers. Scanning the entire spectrum with a single-channel spectrometer (monochromator) requires approximately the same time as exposing and processing a spectrogram.

After the photographic emulsion is exposed in the spectrograph, the latent image (the spectrogram) is developed. A dark image appears for each emission line detected (Fig. 11.5). These dark images are generally in the shape of the entrance slit of the spectrograph, and the amount of darkening or blackening is directly related

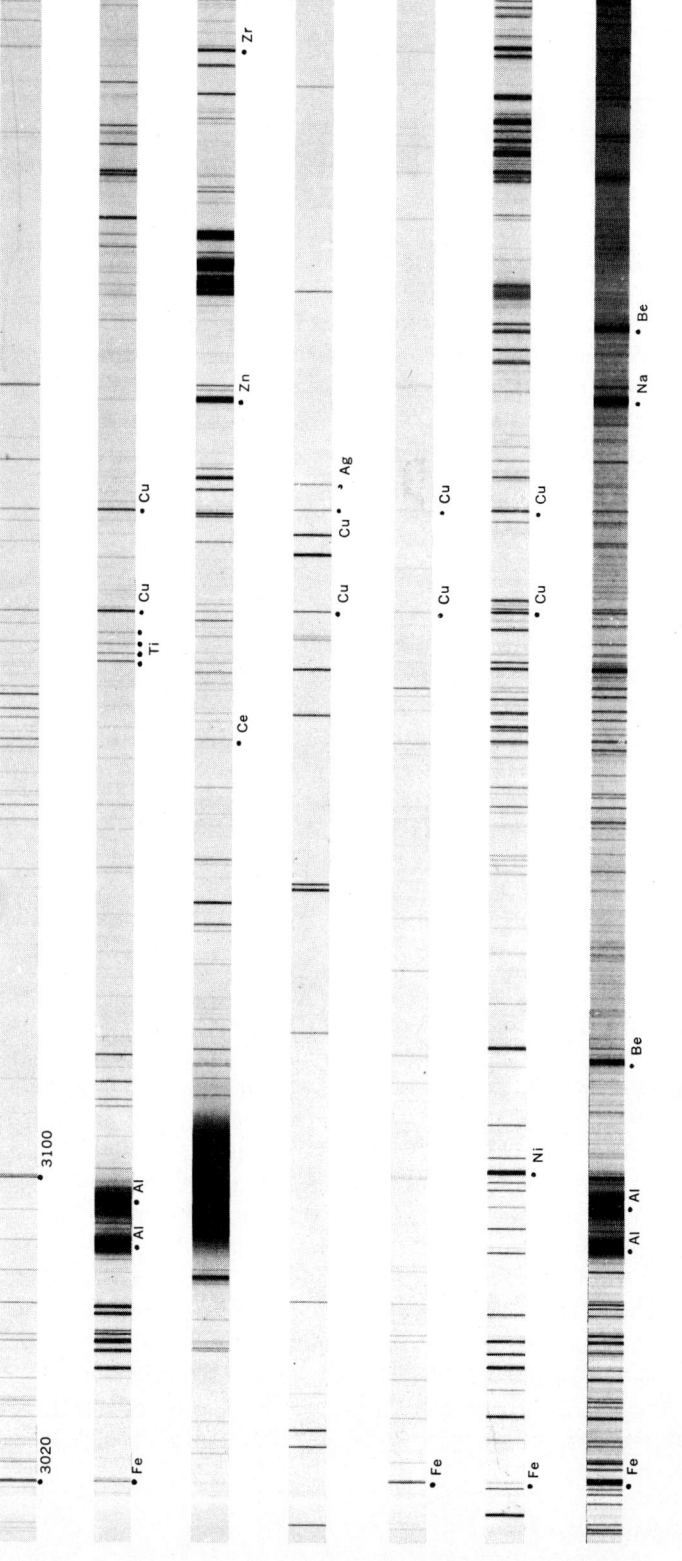

FIGURE 11.5. Portions of typical spectra taken on a 3.4-m Fastie-Ebert spectrograph using a 15,000-line/in. grating giving a dispersion of 0.5 nm/mm at the focal plane. From top down, the materials are iron, aluminum, magnesium, lead, steel, nickel, and beryllium ore. Courtesy of Jarrell Ash Division, Fisher Scientific Co.

to the intensity of the emission signal for each emitting species in the discharge source—that is, to the number of photons striking the emulsion in unit time at each frequency.

To determine the positions and wavelengths of these darkened images or spectral lines, the plate or film is positioned in a comparator or microphotometer which projects a portion of the spectrum onto a viewing screen for direct comparison with a known or standard spectrum. Typically, an iron spectrum will also be recorded on the sample film; this permits alignment of the sample spectrum with the standard spectrum and identification of the wavelengths of the individual spectral lines. To determine the darkening of the spectral lines, the transmittance of the line is measured photometrically with the microphotometer [5]; the radiation from the light source is focused on a line, and the light passing through the line is detected by the photomultiplier tube. A galvanometer or digital readout is obtained. Since the blackening of the photoemulsion is not a linear function of the exposure, the emulsion has to be calibrated; this can be done easily using a laboratory computer [6].

Excitation Sources

A number of electrical excitation sources are available for emission spectroscopy. In most commercial spectrochemical instruments, more than one excitation source is contained in a single power-supply cabinet; a typical combination may include a spark, a direct-current arc, and an alternating-current arc. A list of the various electrical excitation sources, some of their characteristics, their approximate cost, and the types of samples generally required is given in Table 11.1. Because of the actual or potential widespread use in emission spectroscopy, only the arc, spark, and inductively coupled plasma discharges will be described here in detail.

DC Arc. The DC arc is the least complex of the electrical excitation sources discussed here; it consists of a low-voltage (10–50 V), high-current (1–35 A) discharge between a sample electrode and a counter electrode. The DC power supply may consist of no more than a full-wave rectifier and a filter.

The sample, most often prepared as a finely ground powder, is placed in a graphite-cup electrode (see Fig. 11.6); the counter electrode is also fabricated from a graphite rod. The sample electrode is usually (but not always) made the anode, and the counter electrode the cathode. The arc is initiated either by momentarily touching the two electrodes together or, more commonly, by applying an initial high-voltage spark. Once the discharge is formed, a bright, self-sustaining, electrically conducting path is established between the two electrodes. The electrodes heat rapidly and the sample is vaporized into the arc discharge. Graphite electrodes are generally used for routine samples, because graphite does not melt under these conditions and is a good electrical conductor. The anode reaches a temperature of about 4200 K, whereas the cathode is at $\leqslant 3500$ K; the conducting channel between the two electrodes ranges at equilibrium between 4000 and 7000 K, depending on the atmosphere and the composition of the channel plasma. This conducting channel or arc is surrounded by a region of heated gas known as the mantle or envelope. The mantle reaches a temperature of approximately 3000 to 4000 K, but does not serve as a conducting pathway.

TABLE 11.1. *Some Emission Spectroscopy Sources*

Type	Characteristics	Samples Used	Price
DC arc	Continuous, self-maintaining, direct-current discharge with low voltage and high current between sample and counter electrodes	Powders, solids, residues	$6000–$20,000
AC arc	Series of separate discharges individually initiated once during each half-period of supply voltage and extinguished when voltage across the gap falls and becomes too low to sustain the discharge. Current continuous during conduction periods, similar to DC arc	Powders, solids, residues	Not generally manufactured
Spark	Transient discharge reaching high instantaneous current, produced by the discharge of a capacitor between the sample and counter electrodes with a duration of a few hundred microseconds and reoccurring 1–5000 times each second	Solid flats or rods, pressed pellets, liquids, residues	$7000–$20,000
DC-arc plasma jet; gas-stabilized arc	DC-arc discharge formed between nonsample electrodes in flowing-gas streams of argon or helium. Sample introduced separately	Liquids, powders, gases, GLC, HPLC hydride	$7000
Inductively coupled plasma discharge	Electrodeless radio-frequency discharge produced by magnetically induced eddy currents in flowing argon confined in a quartz tube at atmospheric pressure.	Liquids, powders, gases, arc, GLC, HPLC, hydride, spark, electrothermal or laser vaporization	$10,000–$25,000
Microwave plasma discharge	Discharge produced with or without electrodes by microwave fields in stationary or flowing-gas streams in a microwave cavity	Liquids, gases, GLC, hydride, electrothermal vaporization	$3500–$6000
Glow discharge lamp	Low-pressure discharge in an inert gas, characterized by abnormal glow and sputtering of solid-sample cathodes	Solids, residues	$35,000

TABLE 11.1. (*continued*)

Type	Characteristics	Samples Used	Price
Hollow-cathode lamp	Low-pressure discharge in inert gas characterized by normal glow and sputtering of solid cathode	Solids, residues, gases	$3500–$40,000
Laser microprobe	Plasma produced by absorption of laser radiation, forming an emitting vapor cloud. Sometimes supplemented by auxiliary excitation by spark	Conducting and nonconducting solids	$30,000–$40,000

    The sample does not simply "boil out of" the hot electrode, introducing the sample vapor into the arc region. Elements do not all vaporize at the same rate unless some additional steps are taken; the vaporization rate depends on the chemical composition of the sample and on the chemical and physical reactions that occur in the hot electrode. From a very simple point of view, the graphite electrode acts like a high-temperature oven in which each element is vaporized according to its thermal properties. However, factors such as the reduction of oxides, the formation of volatile compounds such as chlorides, alloying, and carbide formation alter these vaporization processes and, hence, the sequential appearance of emissions from the various elements. Refractory materials such as silicates may remain behind in the electrode cup long after other, more volatile, elements have escaped. In some applications, this separation of volatile from nonvolatile elements is deliberately enhanced by the addition of other compounds (the *carrier distillation* method). The determination of trace elements in uranium by addition of a carrier such as AgCl is an example: The trace elements rapidly volatilize along with the AgCl during the first 10 to 20 sec of arcing, but the uranium matrix remains mostly nonvolatilized and, thus, its complex spectrum is greatly diminished in intensity. In other situations, additional compounds are mixed with the sample to promote uniform and simultaneous vaporization. This admixture might include as diluents powdered graphite or salts with an easily ionized cation, such as Li_2CO_3. The presence of lithium in the mixture will also provide a uniform arc temperature during vaporization of the sample. If lithium were not present, the ionization current flow and hence the arc temperature would change as each element or group of elements evaporated from the sample electrode. The arc temperature remains relatively constant in the presence of a sufficient quantity of lithium because the lithium is readily ionized and provides the arc with a constant supply of ions and electrons. In this application, Li_2CO_3 is called a *spectrochemical buffer*.

    The high temperatures of electrodes and arcs result in consumption of the graphite electrodes. When the carbon from the electrode reacts with nitrogen in the ambient air, cyanogen (CN) forms, which emits an intense and complex molecular

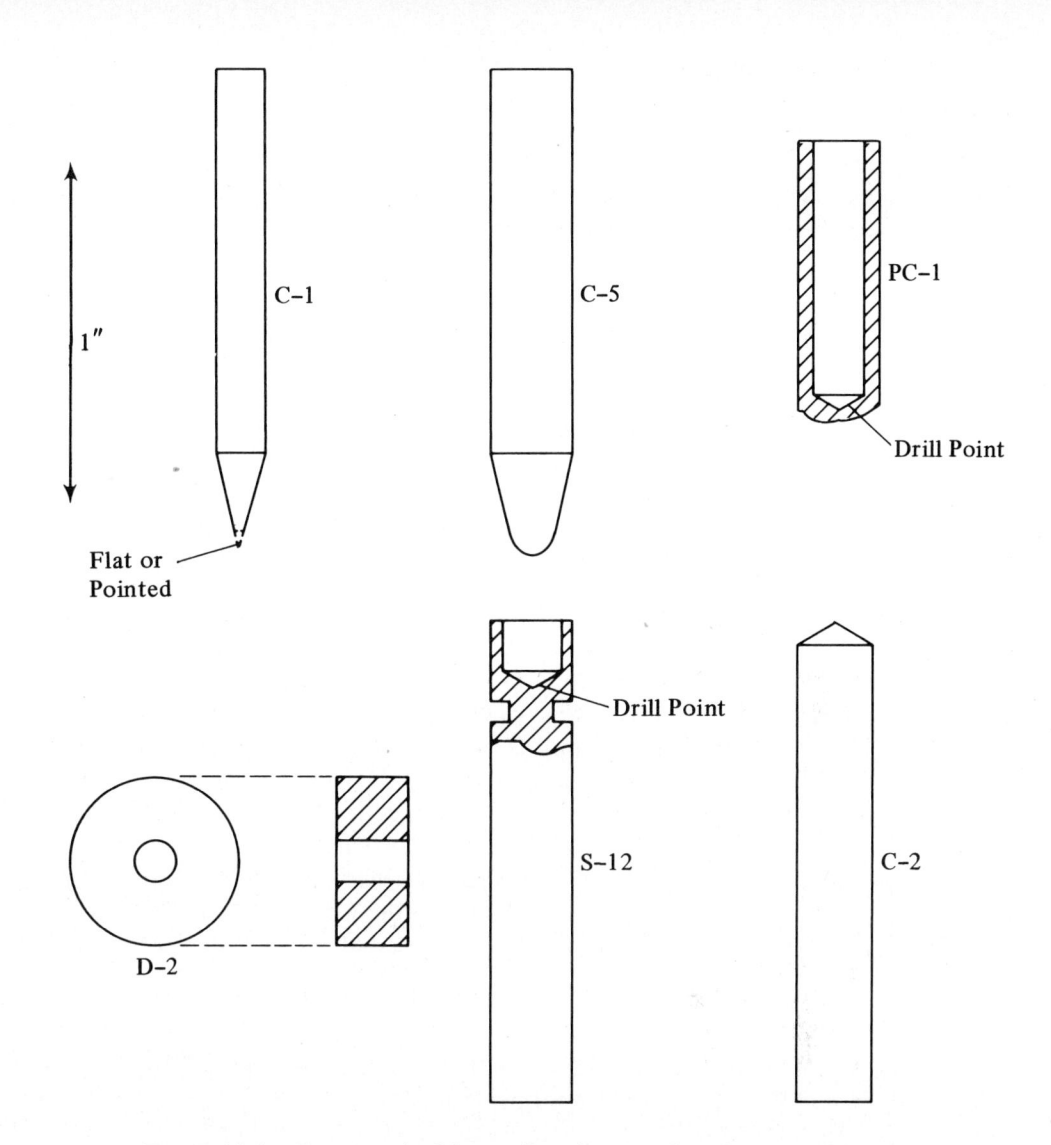

FIGURE 11.6. *Some standard forms of graphite sample and counter electrodes. Counter electrodes represented by C, sample electrodes by S, rotating disk by D, and porous cup by PC. Adapted from* Methods for Emission Spectrochemical Analysis, 6*th ed., Philadelphia: American Society for Testing and Materials, 1971, pp. 105–10, by permission of the publisher. Copyright © 1971 by the American Society for Testing and Materials.*

spectrum that obscures some useful atomic emission spectra. One way to avoid cyanogen bands, and to stabilize the arc discharge at the same time, is to exclude air from the arc by operating it in a closed chamber filled with argon or in a flowing stream (3 to 4 L/min) of mixed argon (70 to 80%) and oxygen (20 to 30%). The latter arrangement is called a *Stallwood jet*.

The evaporation of sample from the electrode and the consumption of the electrode during arcing also cause the arc column to shift its position on the electrode surface. This movement, or wander, of the arc relative to the optical axis of the spectrograph drastically reduces the precision of arc analysis. The Stallwood jet also aids in reducing arc wander.

The spectra of DC-arc discharges are considerably different from those observed for flames and sparks. Because of the high temperature of the arc column, sufficient energy is available to populate higher atomic states. A complex spectrum, generally composed of neutral-atom and ion emission, results.

The high efficiency of atomization and excitation shown by the DC arc, together with the line-rich spectra it produces, make the arc technique valuable for qualitative analysis and multielement quantitative analysis. Large emission signals for relatively small amounts of sample characterize the sensitivity of the DC-arc source.

The DC arc may be an effective emission source for powder samples, but its application for liquids is restricted. A direct-current *plasma jet* (DCP) arrangement is commercially available which extends the DC arc to solution analysis. A DCP is a flowing-gas-stabilized electrical discharge maintained by a DC arc. In the commercial version, two anode jets and one cathode jet form a unique configuration (Fig. 11.7) in which separate arcs are struck between the common cathode and each anode. Tangentially flowing argon introduced into an anode chamber causes a vortex flow around the anode, and a thermal pinch results in an arc column of high current density and higher temperatures than observed in DC-arc discharges. The solution sample is introduced as an aerosol from a nebulizer and spray chamber upward into the junction of the two arc columns, where the analyte emissions are measured. Unlike the DC arc, the DCP does not incorporate the sample in the arc column, and this results in a low background signal. Due to the localized region of analyte emission, the commercial DCP is combined with an echelle spectrometer to provide high resolution (Fig. 11.3).

Spark. The chief characteristic of the spark discharge is its dependence on short times and short-time processes. In routine use, the spark signal is time integrated by a photographic emulsion or by the electronics associated with photoelectric detectors. The spark is typically produced by the discharge of a capacitor between the sample and counter electrodes; Figure 11.8 is a simplified circuit diagram of a typical spark source. A capacitor of 5 to 100 nF is charged to 1 to 30 kV and is then discharged (by the trigger switch) to form a conducting discharge channel or *spark* between the electrodes. There are typically 2 to 30 sparks per cycle of the line signal, or 120 to 1800 sparks per second. The amplitude of the oscillating discharge current flowing through the spark gap decays exponentially with time. The amplitude, frequency, and duration of the oscillation depend on the parameters of the discharge circuit—resistance (R), capacitance (C), and inductance (L)—and on the voltage across the capacitor. For a particular analysis, these various parameters are specified after considerable trial-and-error experimentation to achieve the best quantitative results. Modern design has markedly improved the operation of spark sources [7], and research into the mechanisms of the spark discharge [8, 9] has led to ways of electron-

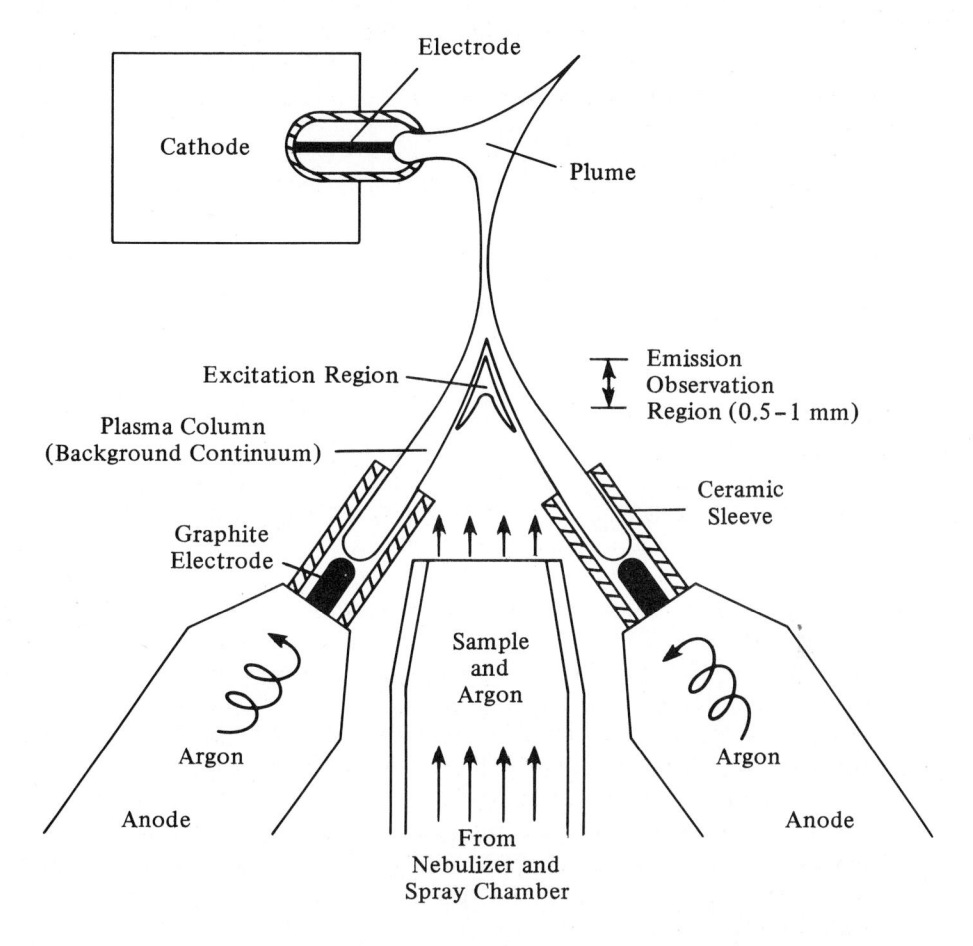

FIGURE 11.7. *Representation of a commercial direct-current plasma jet (DCP) configured with two anodes and a common cathode. The analyte emission is observed in a small region (0.5 to 1 mm) below the junction of the two arc columns. The analyte aerosol is introduced from a nebulizer and spray chamber through a tube located between the two anodes.*

ically shaping the waveform of the spark discharge under computer control [10] for improved analytical results.

Although numerous arrangements of spark source have been developed, two types are most common: the high-voltage source (with or without a diode rectifier) shown in Figure 11.8 and a medium-voltage (1000-V) source.

Events occur very rapidly in the spark discharge, and what appears is substantially different from what happens in an arc. For example, during the first oscillation of the current, which may last only about 10 μsec, the spark-discharge channel is formed and the sample material is vaporized, atomized, ionized, excited, and propelled into the spark gap at velocities of up to 10 km/sec. As the excited atoms and ions travel away from the sample electrode, they emit their characteristic spectra. As the spark current falls, the spark-discharge channel contracts, ions recombine into

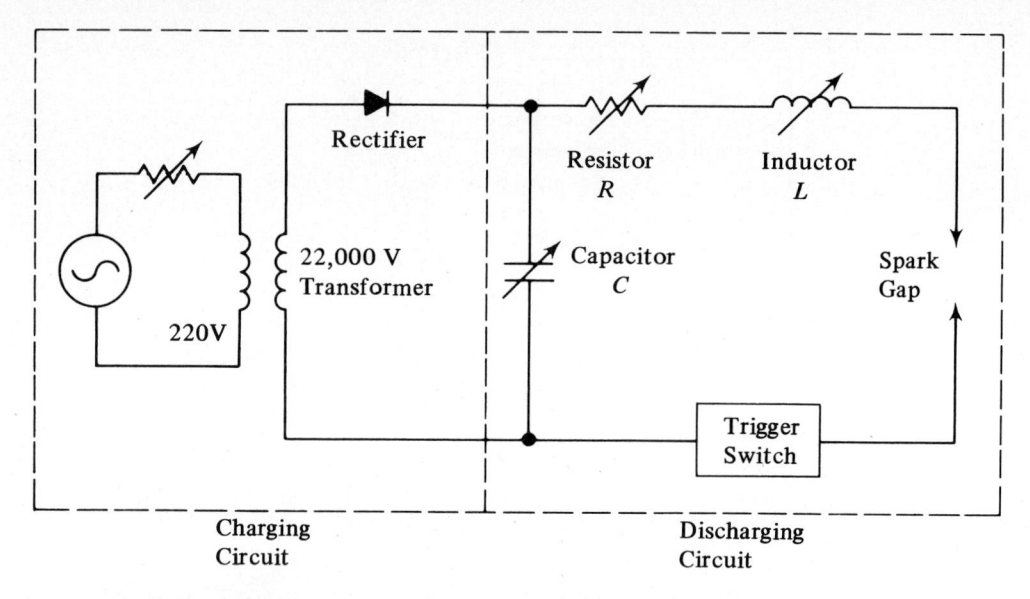

FIGURE 11.8. *Schematic diagram of a spark source for an emission spectrograph.*

excited atoms, and the sample is vapor-deposited on the counter electrode. The process is repeated in subsequent half-cycles until the capacitor voltage is insufficient to maintain the spark discharge. Clearly, the emission spectra from a spark discharge depend not only on the part of the spark gap viewed by the spectrograph (near the anode, near the cathode, or in the middle) but also on the selected time during the spark discharge. Without special equipment, spark-emission spectra are integrated over the total time of all these events.

Although the spark discharge appears complex, the quantitative application of the spark provides a very efficient means for transforming solid samples into emitting atomic vapors. The spark discharge permits fast, precise analysis; in quality-control steel analysis, for example, the time is less than 10 sec. Although the sensitivity is commonly not as high as that obtained with a DC arc, high precision can be obtained without special techniques.

The spark discharge is applied typically to solid samples in the form of rods approximately $\frac{1}{4}$ in. in diameter or cast disks or flats $\frac{1}{2}$ in. thick by 1 to 3 in. in diameter. Powders are sometimes pressed into pellets and treated as solid samples for spark analysis. Liquid samples are analyzed by spark discharges, although plasma sources are replacing the spark. Several different ways of analyzing liquid samples have been developed: the porous-cup electrode (Fig. 11.6); the rotating-disk electrode; and residues dried onto the top of graphite or metal rods, or on graphite disks. The *copper-spark* method uses a dried residue on a quarter-inch copper rod, and the *rotating-platform* method employs a half-inch graphite disk onto which the residue is dried and which is rotated horizontally during sparking. Both methods give good precision, and the copper-spark method is characterized by excellent sensitivity (cf. Table 11.2).

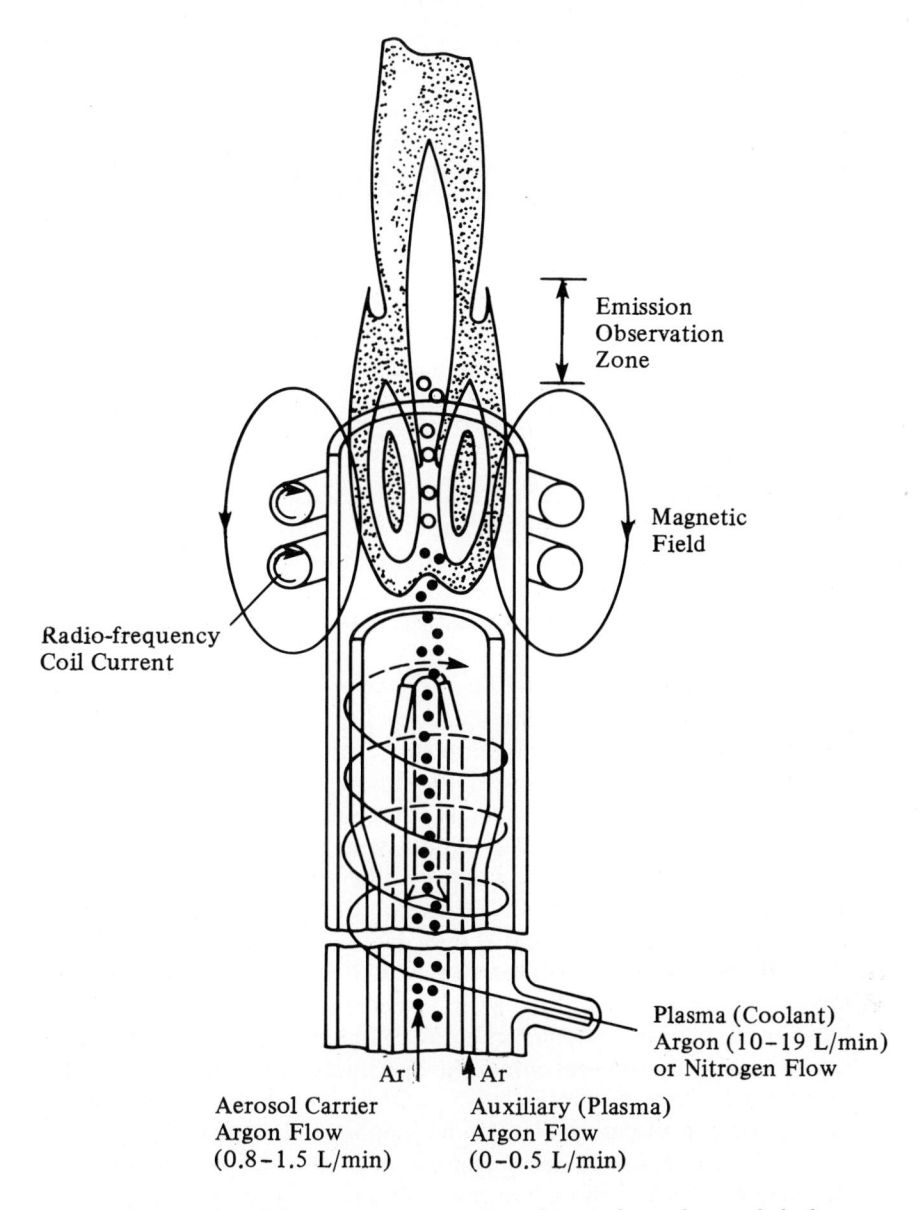

FIGURE 11.9. *Schematic representation of an inductively coupled plasma discharge. From R. M. Barnes*, Wiss. Z. Karl-Marx-Univ. Leipzig, Math.-Naturwiss., *28*(4), 383 (1979), *by permission of the publisher*.

Inductively Coupled Plasma (ICP) Discharge. The arc and spark sources date to the early development of emission spectroscopy in the mid-1800s; the inductively coupled plasma (ICP) discharge is a relatively recent development, and is among the most effective emission spectroscopic sources used today. Commercial ICP systems became available only in 1974, but research on this source has been going on since the early 1960s [11, 12]. More than 15 companies manufacture ICP systems, and more than 5000 different instruments are in operation.

The ICP discharge is caused by the effect of a radio-frequency field on a flowing gas. In Figure 11.9 the discharge is induced without electrode contact in argon flowing upward through a quartz tube inside a copper coil or solenoid. The coil is energized by a radio-frequency generator operating between about 5 and 75 MHz; typical frequencies are 27 and 41 MHz. The radio-frequency signal creates a changing magnetic field H inside the coil in the flowing argon gas.

A changing magnetic field induces a circulating (eddy) current in a conductor, which in turn heats the conductor. At room temperature, argon is not a conductor, but it can be made electrically conductive if heated. To start the ICP discharge, a pilot spark, arc, or Tesla discharge is applied to the argon. This pilot discharge absorbs energy from the changing magnetic field and turns rapidly into a stable discharge plasma that is thermally very hot and spectrally very intense. The equilibrium temperature in the annulus of an ICP discharge operating at 1 to 2 kW input power is about 9000 to 10,000 K.

More than one stream of argon is often used for spectrochemical analysis with the ICP discharge. One argon stream is confined to a volume near the tube walls to protect the quartz from the high-temperature discharge. A second argon stream carries the sample into the center of the discharge to produce an effective pathway through the discharge. If this pathway were not formed, the sample might flow around the hot discharge and be heated less effectively.

Although samples may be injected as powders, gases, or liquids, an arrangement similar to the spray chamber–nebulizer assembly used in flame spectroscopy (Chap. 10) is presently used. A complete nebulizer, spray chamber, and ICP discharge assembly is illustrated in Figure 11.10. Although right-angle and concentric pneumatic nebulizers are widely applied for ICP emission spectroscopy, special-purpose nebulizers (e.g., the clog-free Babington type) and ultrasonic nebulizers have been developed to enhance convenience or efficiency. The aerosol from the pneumatic nebulizers is transported by the central argon flow into the discharge directly, where the solvent is evaporated and the analyte atomized. With ultrasonic nebulization, about 10 times more aerosol reaches the ICP, which could exceed the tolerance of the low-power discharge and cause it to become unstable. Thus, an oven is inserted between the ultrasonic nebulizer spray chamber and the ICP discharge to remove the solvent before the aerosol enters the discharge. This arrangement increases the ICP sensitivity by 3 to 10 times.

The ICP discharge also can be supplied to the analyte as a gas or vapor from a gas chromatograph, hydride generator, or electrothermal vaporizer similar to those designed for atomic absorption spectrophotometry; as a metal aerosol from a spark, arc, or laser; or as an eluant from a liquid chromatograph. These alternative sample-introduction techniques can enhance the analyte sensitivity, reduce the quantity of sample required, or eliminate the need to dissolve solid samples. For example, hydride generation combined with ICP spectroscopy enhances the sensitivity for Se, Te, P, As, Sb, Bi, Ge, Sn, and Pb compared to nebulization of the solution containing these elements [13].

Because of the high temperatures available and the inert atmosphere of the ICP discharge, some of the difficulties found in flame, arc, and spark techniques are not present in the ICP discharge. Chemical interferences caused by the formation of stable compounds in flames (see Chap. 10) are negligible with the ICP discharge, and

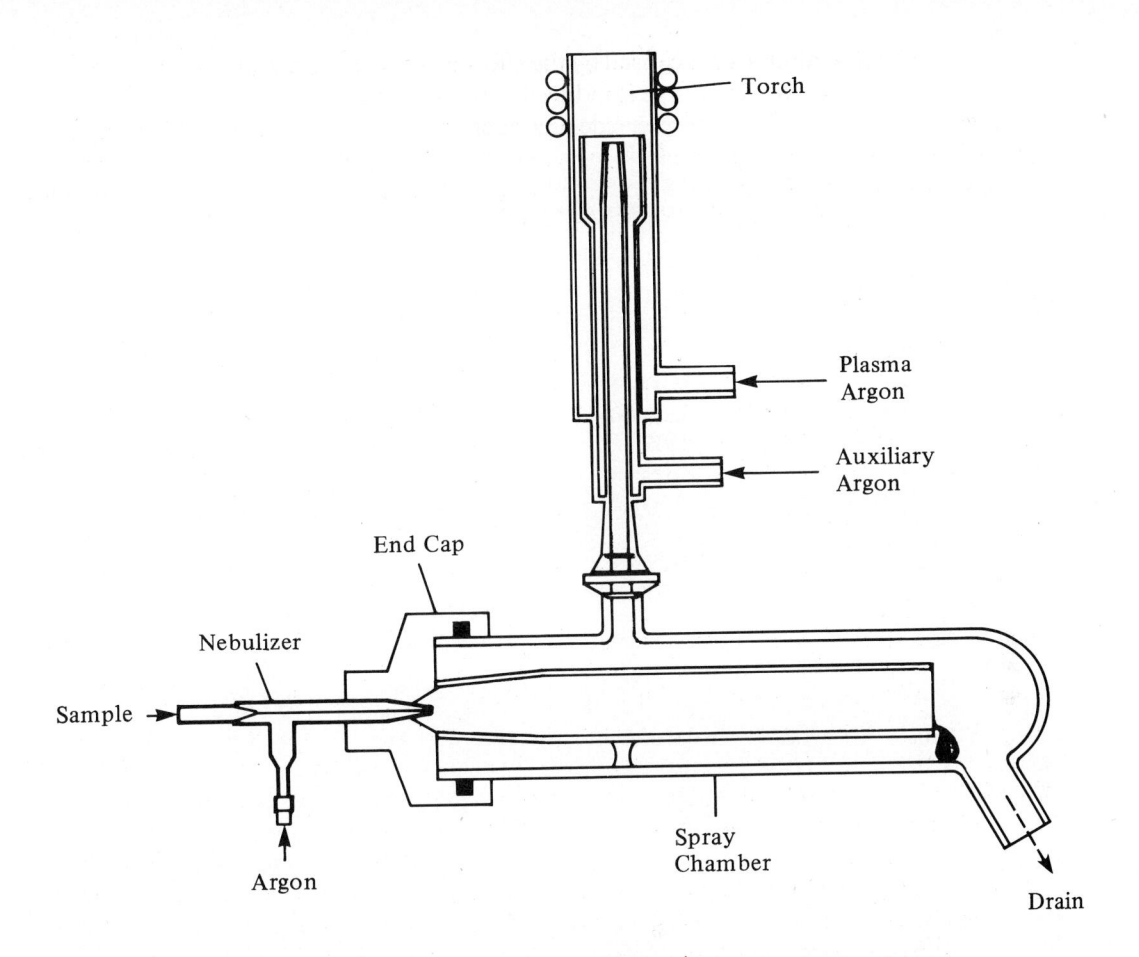

FIGURE 11.10. *Concentric annular nebulizer, spray chamber, and torch apparatus for ICP emission spectroscopy. The sample is drawn into the nebulizer capillary by a controlled flow of argon through the outer jacket. Flow rates are typically 0.8 L/min argon and 1 mL/min analyte solution. The aerosol formed by the nebulizer is passed into a double spray chamber, where the larger droplets collect in the drain. About 2% of the original sample passes into the center tube of the ICP torch and into the ICP discharge. The same spray chamber can be adapted to a variety of nebulizers by replacing the end cap.*

thus releasing agents or special conditions are not needed. All compounds are likely to be atomized completely during their passage through the hot pathway in the center of the discharge. Ionization interferences that occur in excitation sources with high temperatures such as the DC arc are minimal in the ICP plasma. For simultaneous multielement analysis, such interferences can generally be kept to less than 10% under compromise analysis conditions.

Resolving spectral interferences of the matrix or of the emission of concomitant elements from the emission of the analyte is one of the important steps in selecting

TABLE 11.2. *Comparison of Some Experimentally Determined Emission-Spectroscopic Detection Limits ($\mu g/mL$)*

Element	DC Arc[a]	Spark[b]	ICP[c,d]	Element	DC Arc[a]	Spark[b]	ICP[c,d]
Ag	0.0006	0.2	0.004	Nb	5	0.10	0.0002
Al	0.05	0.05	0.00008	Ni	0.02	0.05	0.0001
As	0.1	5	0.002	P	0.15	4	0.015
Au	0.05	0.1	0.04	Pb	0.005	0.1	0.001
B	0.07	0.5	0.0001	Pd	0.02	0.02	0.0008
Ba	0.005	0.02	0.00001	Pt	0.04	0.4	0.08
Be	0.0006	0.0002	0.000003	Rh	0.02	0.05	0.003
Bi	0.03	0.1	0.05	Sb	0.07	2	0.2
Ca	0.01	0.05	0.0000001	Sc	0.2	0.01	0.003
Cd	0.02	1	0.0002	Se	—	—	0.03
Ce	0.02	0.3	0.0004	Si	0.1	0.20	0.01
Co	0.01	0.05	0.003	Sn	0.05	0.30	0.003
Cr	0.01	0.05	0.0008	Sr	0.00003	0.002	0.00003
Cu	0.0003	—	0.0006	Ta	30	0.3	0.03
Fe	0.01	0.5	0.00009	Te	60	4	0.08
Ga	0.02	0.02	0.0002	Th	0.02	0.5	0.003
Ge	0.02	—	0.0005	Ti	0.0001	0.01	0.00003
Hf	1	0.25	0.01	Tl	0.07	0.8	0.2
Hg	0.07	1	0.01	U	—	2	0.03
In	0.03	0.3	0.03	V	0.02	0.02	0.00006
La	0.03	0.02	0.0001	W	0.3	0.4	0.0007
Mg	0.007	0.05	0.000003	Yb	0.0009	0.005	0.00002
Mn	0.003	0.01	0.00002	Zn	0.01	0.5	0.00001
Mo	0.006	0.03	0.0001	Zr	0.004	0.01	0.00006
Na	0.005	0.1	0.00002				

a. V. Svoboda and I. Kleinmann, *Anal. Chem.*, **40**, 1534 (1968).
b. J. P. Faris, *Proc. 6th Conf. Anal. Chem. Nucl. Reactor Tech.*, TID-76655, Gatlingburg, Tenn., 1962.
c. V. A. Fassel and R. N. Kniseley, *Anal. Chem.*, **46**, 1110A, 1155A (1974).
d. P. W. J. M. Boumans and F. J. de Boer, *Proc. Anal. Div. Chem. Soc.*, **12**, 140 (1975).

an appropriate emission wavelength for the analyte element and in applying background correction when developing any new ICP method. However, the continuum background from the hot discharge does not extend appreciably beyond the end of the induction torch, and very high signal-to-background ratios are obtained just a short distance (1 to 3 cm) above the induction coil (Fig. 11.9). Background from the argon continuum and interference from Ar I emission is minimal in this normal analysis region.

The ICP discharge provides a rich spectrum for qualitative, and for sequential or simultaneous multielement quantitative, analysis. The spectrum of an element in the ICP discharge is unlike those obtained in the DC arc, a spark, or a flame; and new wavelength tables giving the spectral-line intensities and spectral interferences have been published [14–16]. Even molecular spectra such as that due to CN are minimized in the ICP discharge or are located in a separate region of the discharge. In addition, the stabilities of the signal intensities observed in the ICP discharge are comparable to those of the flame rather than the arc or spark discharges.

The temperature distribution of the ICP discharge differs from other electronic excitation sources because of the induction-coupling effect. Instead of finding the highest temperature along the axis of the discharge, as in the DC arc, the highest temperatures of the ICP discharge are found off-axis in the induction-coil region.

All of these properties of the ICP discharge provide excellent capabilities for quantitative analysis. Three operating parameters are crucial: input power, plasma gas flow rate, and observation height above the induction coil. These operating conditions can be readily selected so that nearly optimum signal intensities for most elements can be obtained in a single spectroscopic viewing region above the hot discharge. This allows the simultaneous determination of 35 elements, for example, in a single sample without modifying the conditions for each element.

Detection limits for emission spectroscopy with several excitation sources are presented in Table 11.2. (These may be compared with those given in Table 10.4 for flame methods.) Generally, less than 5 mL of sample solution is required for both emission analysis and the flame methods. The ICP discharge has been found to have both the precision of flame methods and the sensitivity of arc methods [17]. (For precise determinations, concentrations should generally be 100 times the detection limits listed.)

Although quantitative analysis with the ICP discharge may be performed with a spectrograph, the ICP discharge is more efficiently used with a spectrometer. One of the major reasons is that the high signal-to-background ratio and high stability provide linear analytical curves (readout signal as a function of concentration) over ranges of 10^4 to 10^5. This linearity exceeds by orders of magnitude that obtained in routine spark and arc analyses, as well as in flame methods. The linearity of the photographic emulsion is insufficient to cover this range, and only photomultipliers have the capability needed.

11.3 QUALITATIVE AND QUANTITATIVE ANALYSES

Emission spectroscopy is widely used for both qualitative and quantitative analysis. The high sensitivity and the possible simultaneous excitation of as many as 72 elements, notably metals and metalloids, make emission spectroscopy especially suited for rapid survey analysis of the elemental content in small samples at the level of 10 μg/g or less. With control over excitation conditions to maintain constant and reliable atomization and excitation, the spectral-line intensities can be used for determining concentrations quantitatively. An analytical curve must be constructed with known standards, and often the ratio of analyte intensity to the intensity of a second element contained in, or added to, the sample (the internal-standard or *internal-reference* method) is used to improve the precision of quantitative analyses. Preparation of standards for arc and spark techniques requires considerable care to match chemical and physical forms to the sample; this is not commonly required for ICP discharge.

Qualitative Analysis

Emission spectroscopy is especially well suited to the identification of elements contained in a sample, because meaningful results are obtained in less than an hour in a single exposure requiring only a few milligrams of sample in almost any form. Conventionally, DC-arc excitation is used for qualitative analysis because of its high sensitivity for metals and metalloids. To perform a DC-arc qualitative analysis, the sample (as a powder, small chunks, chips, filings, residue, or other form) is placed in a graphite-cup electrode (Fig. 11.6), and the electrode arced until the entire sample is vaporized. The spectrum is integrated photographically, providing a permanent record over a comprehensive wavelength range. Generally, several spectra are recorded on one photographic film or plate by moving (racking) this film in the camera between each run (see, e.g., Fig. 11.5). One of the spectra is usually that of iron to allow alignment with master plates. Processing the photoplate takes about 10 min, and the spectrum is compared on a comparator-densitometer with either a master plate (standard spectrum) or a series of spectra of known elements. The master plate, available commercially, contains a standard iron-arc spectrum, a wavelength scale, and wavelength markers for the persistent or most sensitive characteristic lines for each element. A portion of a master plate is shown in Figure 11.11. After aligning the unknown spectrum with the master plate, major line coincidences are identified for characteristic lines on the master. Three lines are generally required for positive identification of an element. An additional aid is that certain elements have characteristic patterns or groupings of spectral lines, which, with a little practice, are rapidly found and used for positive identification. Reference to standard wavelength tables provides additional lines that may be found in the unknown spectrum but not on the master plate.

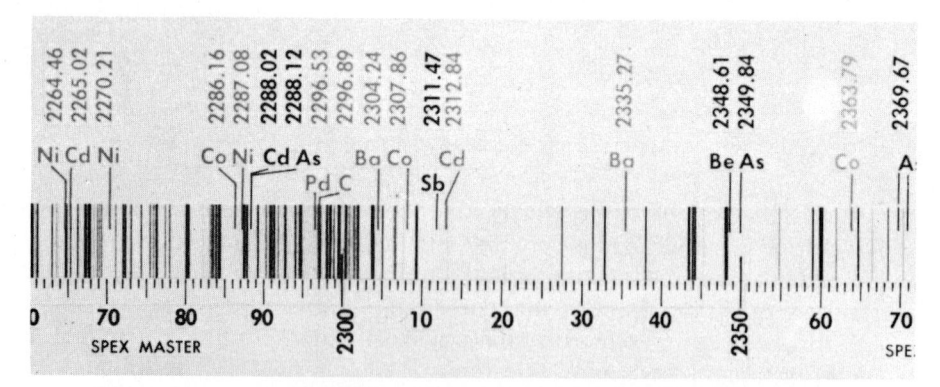

FIGURE 11.11. *Segment of Spex Master plate. In the complete plate, persistent lines of about 70 elements are given with color-coded labeling for ease of identification. Lines are superimposed on an iron spectrum. Units are angstroms. Courtesy of Spex Industries, Metuchen, N.J.*

Qualitative analysis by laser microprobe is helpful for identifying small inclusions or areas in conducting and nonconducting samples. The laser can be focused to sample areas of 10 to 50 μm in diameter.

Often, qualitative analyses are performed with slightly more control over the various experimental conditions to obtain a rough estimate of the concentration range of the elements identified as major, minor, or trace. For better concentration estimates, semiquantitative or quantitative techniques, demanding greater control over parameters, are applied.

Quantitative Analysis

Emission spectroscopy is an important quantitative technique widely applied in many industrial and research laboratories. To achieve a relative concentration error of less than $\pm 10\%$, sample preparation and handling, experimental variables, and operating parameters must be strictly controlled. With conventional arc and spark procedures, relative errors of $\pm 1-5\%$ can be achieved. The development of a routine spectrometric analysis may take months, but once the method is optimized, high-quality quantitative results are obtained rapidly and routinely for large numbers of similar samples.

Fluctuations in electronic excitation sources (described in Sec. 11.2), together with sample irregularities, constitute the major sources of error in emission spectroscopy. Modern spectrometers provide excellent stability and precision, and new sources like the ICP discharge, the glow discharge [18, 19], and the controlled-waveform spark discharge have reduced many of the previous limitations. Photoelectric detection generally provides precision superior to that of photographic methods.

Some other critical considerations in quantitative emission spectroscopy include obtaining a representative sample, treating the sample to provide a suitable form without contamination, and matching standards with samples.

Electrodes. For arc and spark analyses, graphite electrodes are commonly used as sample and counter electrodes; some standardized electrode shapes are presented in Figure 11.6 [20]. The purity of these electrodes must be high, and most suppliers provide a quantitative DC-arc analysis for at least 15 elements with each box of electrodes. Electrodes are generally guaranteed to have a total ash content of less than 1 μg/g, a maximum allowable impurity per element of 2 μg/g, and total maximum impurities of 6 μg/g.

High-purity graphite powder for DC-arc mixtures, or for pressed pellets for spark analysis, is also analyzed and guaranteed by manufacturers. In spark methods, the sample (if conductive) is often one of the two electrodes, and only a counter electrode is needed. Graphite electrodes are common, but metal counter electrodes, especially silver (to permit determination of carbon), are routinely employed in vacuum spark analysis.

Samples. Careful control of sampling and sample preparation is essential in quantitative emission spectroscopy. Even in the routine spark analysis of steel or aluminum, which requires only the grinding or machining of the surface of cast samples, the detailed characteristics of the sample-casting procedures had to be studied extensively during the method development phase.

In arc analyses, the sample may require treatment before analysis; this can contribute contaminants or cause loss of some elements. For example, samples with high carbon content, such as coal, require the removal of the organic portion by ashing in a muffle furnace at elevated temperatures or in a low-temperature oxygen plasma. Coal is ashed in platinum or silica crucibles at 500°C to eliminate the organic portion, but these high temperatures may cause volatilization and loss of some trace elements. Other inorganic materials such as rock, cement, slag, or chemicals need only be dried, ground, and sieved. However, each step can also contaminate the sample. Typically, samples are ground so that known sources of contamination are eliminated. For example, in the analysis of beryllia (BeO) for other elements, samples are ground with a high-purity BeO mortar and pestle. (The use of a mortar and pestle made of tungsten carbide or of alumina will contaminate the specimen with iron and cobalt traces from the tungsten carbide and with aluminum from the alumina.) Sieving with metal screens can also contaminate the sample with traces of the screen material.

Sample contamination must also be taken into account when adding internal-reference and spectrochemical-buffer compounds. High-purity materials used for these special purposes are commercially available, since most laboratory chemicals are not pure enough.

For trace and ultra-trace analysis of many elements, control of contamination or stabilization of the sample, or both, may be the limiting factor in the analytical accuracy [21,22]. At low concentration levels, special care must be taken by the analyst with sample collection, handling and storage, and chemical treatment before the emission analysis. The sample obtained must be representative of the material being examined, and then the collected sample must be maintained or preserved until it is ready for analysis. Contamination during sampling may be a major problem in trace and ultra-trace metal analysis, and a sampling device or apparatus with low potential for metal contamination and a minimum contact time between the sample and the sampler must be employed. Contact between the sample and the storage container also must be planned to minimize contamination and losses. The laboratory environment and apparatus must be maintained contamination free by use of suitable air-purification systems, metal-free furniture, and washing and cleaning procedures. Plastic, especially Teflon (poly-tetrafluoroethylene), or quartz ware is preferred. All emission analyses should be performed with reagent and with analysis blanks (i.e., the technique and analysis procedure concentration determined without the analyte), so as to establish the level of contamination as well as the variability of the blank.

Standards. The preparation of solid and powder standards for quantitative emission analysis procedure concentration determined without the analyte), so as to establish methods. Sometimes the lack of suitable standards hinders the analysis or limits the accuracy obtainable. Standard reference materials are being continually tested and authorized by the National Bureau of Standards [23], and a number of major steel and aluminum companies have developed standard disk samples for emission and x-ray spectrometry. Standards for trace metals in oil are also produced commercially.

In arc- and spark-emission spectroscopy, one of the critical aspects of quantitative analysis is the need to match the standard as closely as possible to the sample.

Dilution of sample and standards by a common matrix in DC-arc methods somewhat reduces the dependence on exact matches.

Internal-Reference Method. In emission spectroscopy, some of the variables in the excitation and processing of spectra can be minimized or eliminated by adopting the internal-reference technique. The technique is based on measuring the ratio of the analyte signal intensity and the reference-line signal intensity. The internal reference is added to both the sample and the standards at the same concentration. The method assumes that as the excitation-source and spectrometer-readout conditions vary, the signal from the internal reference will change in the same way as those for the analyte elements. Thus, a ratio of the line intensities should minimize variations. Barnett et al. [24] have detailed the factors considered in selecting an internal reference. The technique is applied in both photographic and photoelectric detection systems; for photographic emulsions, the internal reference tends to correct for differences in processing and in emulsion properties from one plate to the next. For ICP, the internal reference compensates for changes in nebulization.

In routine industrial analysis, multichannel spectrometers under computer control generally acquire, store, and update analytical curves for numerous elements simultaneously as part of a periodic check on standards. Excellent precision and accuracy can be obtained with these procedures.

In developing an emission-spectroscopic method, especially arc and spark techniques, considerable effort must be devoted to the selection of internal-reference materials and spectral lines. Some arc and all ICP methods exhibit good precision without use of an internal reference.

Analytical Curves. Emission spectroscopy is not an absolute technique, and the intensity response for each analyte element must be calibrated for various known amounts of the element introduced into the excitation source. The analytical curve from the microphotometer represents this calibration. The transmittance of the analyte line is measured along with a nearby background, and these values are transformed into relative intensities using the emulsion calibrations. The selection of photographic emulsion depends on the wavelength range to be covered and on the strengths of the spectral signals. Operating parameters, including exposure conditions and film-developing time and temperature, are selected in preliminary experiments to provide calibration linearity and good signal-to-noise ratio. Once determined, these conditions must then be maintained rigorously throughout the calibration and analysis stages.

Concentration is the independent variable, and the relative intensity or intensity ratio is the dependent variable. Some analytical curves are plotted on logarithmic coordinates, but computer curve-fitting procedures readily allow wide-range rectilinear calibrations, as well as calculation of the concentration values for unknown samples. For very accurate work, emulsion calibrations are repeated for each spectrum by adopting the sample spectrum as the source of the emulsion calibration.

With photoelectric detection, the photocurrent (which is proportional to the analyte emission intensity) is measured directly, is integrated by a capacitor and the capacitor voltage measured, or is converted to a frequency which is counted [3]. The magnitude of the analyte emission is proportional to the analyte concentration and a calibration function can be obtained directly for standard materials. For multielement analysis, calibration is accomplished for each channel with multielement reference standards. Procedures are incorporated into the operation of multichannel spectrometers to test for and correct spectral interferences and background variations with sample matrix, as well as calibration sensitivity and intercept drift with time and environmental conditions. In routine emission analysis, quality-control procedures such as the repeated analysis of control standards (e.g., every 11 samples) ensure continuous monitoring of the calibration function. Modern data-acquisition systems also perform statistical evaluation of the emission data, prepare data reports, and archive recorded information.

Accuracy and Precision. The accuracy and precision required of an atomic-emission spectroscopic method affect the approach used in the analysis as well as the time involved. A qualitative analysis requires a minimum of effort, but as better accuracy and precision are demanded, increasing care is needed. Even if a representative sample has been obtained, errors inherent in the method, human errors, and random errors contribute to inaccuracies. Spectrochemical equipment is largely responsible for the random errors that influence precision, and both method and individual laboratory errors influence the accuracy. In addition, relative precision and accuracy depend on concentration levels. The standard deviation increases with increasing concentration, but the relative standard deviation decreases; the latter may vary from a few percent to less than one percent using photographic detection, depending on the element and the concentration.

Precision is usually improved with a photoelectric detector. Electronic stabilities determined with a stable excitation source have a relative standard deviation ranging over ± 0.03–0.2% (with modern instruments, using 10-sec integrations and 10 runs). Precision may range over ± 0.3–3.0% for homogeneous samples with concentration levels above about 0.5% in spark analysis, or for solution samples with concentration levels about 100 times greater than the limit of detection for the ICP discharge. The precision of DC-arc techniques for determining trace elements in

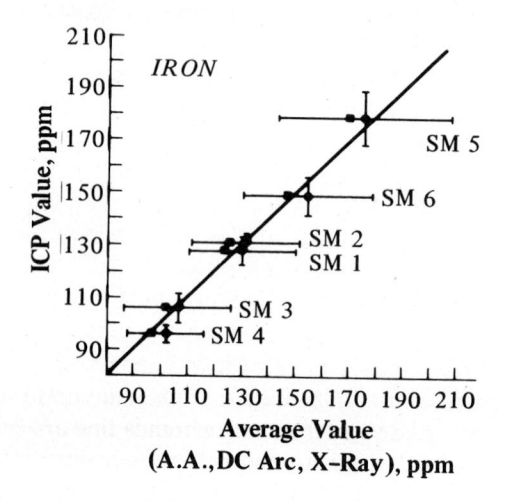

FIGURE 11.12. *Comparison of results for iron in six samples of orchard leaves. The "average value" for each sample is the average of the values obtained by a number of independent laboratories using one or more of the techniques given in parentheses. The standard deviation of these values is given for each sample by the error bars, as well as the standard deviation of the results obtained by ICP emission spectroscopy for 7 dissolutions. A recent atomic absorption value is also given, indicated by* ■. *Adapted from R. H. Scott and A. Strashiem, Anal. Chim. Acta, 76, 71 (1975), by permission of the authors and publisher.*

powdered samples at concentrations greater than 20 µg/g is typically in the range ±2–12% relative standard deviation.

Evaluation of accuracy requires comparing results against standard materials or the results obtained using other independent techniques. The correlation for iron in orchard leaves determined by atomic absorption, DC-arc, and x-ray fluorescence spectroscopy, and ICP emission spectrometry is given in Figure 11.12. The standard deviations are also indicated.

Typically, the accuracy and precision change with the composition of the sample, since the different matrices introduce errors; however, the ICP discharge is particularly free of errors caused by the sample type. For example, determination of some 16 elements in samples as varied as distilled water, steel, blood serum, whole blood, food, and soil showed the detection limits for an element to be within a factor of 2 to 3; the detection limit is proportional to the slope of the analytical curve and the precision (or noise).

11.4 APPLICATIONS

The applications of emission spectroscopy with electrical excitation sources are diverse and extensive. A few examples are presented in this section to illustrate typical analyses. A number of annual and biennial reviews collect and describe new applications as they are published. For example, the *Annual Reports on Analytical Atomic Spectroscopy* contains critical reviews and comprehensive lists of flame and electrical excitation source applications for air, body fluids and tissues, chemicals, foods, metals, minerals, plants, soils, and water [25]. Reviews and extensive surveys of techniques and applications of emission spectroscopy are included in *Progress in Analytical Atomic Spectroscopy* [26], and current literature in emission spectroscopy is examined in the "Fundamental Reviews" issue of *Analytical Chemistry* [27].

Analysis of Metals by Spark Discharge

The determination of 23 elements in aluminum and its alloys by the *point-to-plane* spark technique with an emission spectrometer [28] represents an example of the type of routine quality-control analysis performed on metals and alloys in mills and foundries. Preheated sample molds designed to produce homogeneous castings free of voids or porosity in the regions to be sparked are filled from a sampling ladle containing molten metal taken from the aluminum furnace. After cooling, the disks are transported to the spectroscopy laboratory where an operator machines a smooth surface on the sample. The sample is placed on a Petrey stand that aligns the sample with the entrance slit of the spectrometer. Only predetermined locations on the disk are sparked. A freshly cut graphite counter electrode (C-5 in Fig. 11.6) is positioned 3 mm from the machined surface.

A spark discharge is produced between the flat surface of a chill-cast aluminum sample and the tip of a pointed graphite counter electrode. The emission intensities for 31 different spectral lines and an aluminum internal-reference line are measured simultaneously by 32 photomultiplier tubes positioned behind exit slits. At the end of the 10–15-sec exposure period, the accumulated capacitor potentials for each analytical line relative to the potential for the aluminum internal-reference line are automatically measured and recorded. The unknown values are calculated automatically in terms of percent concentration.

Secondary standards and blank standards of similar metallurgical composition as the unknown samples are used for the principal analytical curve. The averages of 20 results on standards and 20 readings from the blank standards establish the analytical curve. The 20 readings are produced by five separate spark spectra obtained on each of four different occasions.

This overall approach remains basically the same for analysis of steel, brass, zinc, or other metals, although there are specific differences in spectrometers, analyte spectral lines, sample preparation techniques, and excitation conditions.

Metals in Lubricating Oils

The determination of wear metals in the lubricating oils used in aircraft, truck, locomotive, and other engines can provide an excellent indication of the mechanical condition of the engine. In fact, as the presence of certain metals is noticed or their concentrations begin to increase, the parts or components of the engine that are wearing out can be identified and replaced or repaired. This routine program of wear-metal analysis saves tens of millions of dollars annually, and the analysis is one of the largest analytical operations in the world. Tens of thousands of samples are run monthly. The most important wear metals are iron, aluminum, magnesium, copper, and silver. Iron appears as an indicator of more than 80% of all failures detected by wear-metal analysis. Aluminum usually relates to wear of oil pumps, cases, housings, pistons, and cylinder heads, and copper to wear of bronze parts such as bushings and retainers. Silicon is useful as an indicator of lubricant contamination from dust and dirt.

The spark analysis is performed with a rotating graphite-disk electrode (D-2 in Fig. 11.6). The spectra of 10 or more elements in the range 0.1 to 500 μg/mL are determined with a spectrometer during a 45-sec exposure after a 30-sec prespark [29]. For calibration, eight analyses of each of five standards containing the wear metals are used to establish the analytical curves. The samples are agitated in the original container until all sediment is dispersed homogeneously in the oil. The graphite electrode disk is mounted as the cathode on a graphite spindle and positioned in the spark stand. A graphite counter electrode spaced 3 mm above the top of the rotating disk is centered on the optical axis. An aluminum or porcelain boat holding the oil sample is positioned on the spark stand and raised until the disk dips into the oil. The spark is started after the turning disk is evenly coated with the oil sample. Duplicate determinations are made with new electrodes on each sample.

Trace Elements in Airborne Particulate Matter

Emission spectrography has been used extensively for the determination of trace elements in atmospheric particulates, especially in large-scale survey studies in which simultaneous multielement analysis is important [30–32]. Airborne particulate matter is routinely collected by drawing a measured volume of air through filter materials such as fiberglass, asbestos, cellulosic paper, porous plastic, or graphite in the form of disks or electrodes. However, for the determination of trace elements, the chemical composition of the filter is important. For example, glass filters show high concentrations of Ba, Sr, Rb, Zn, Ni, Fe, Ca, As, and other elements. The composition of the filter materials is particularly significant in sampling relatively clean atmospheres because of the low particulate levels collected in reasonable sampling times.

A membrane filter that can be dissolved in acetone, or a spectrochemically pure graphite filter that can be examined directly with a powder DC-arc technique, can provide passable results. After air has been drawn through a previously weighed filter, the membrane filter is dissolved in acetone, then centrifuged. The particulates are collected, dried, and weighed; then a spectroscopic buffer is added composed of one part NaF and one part graphite powder, with 100 μg/g indium oxide and 20,000 μg/g tantalum oxide as internal reference elements. About 35 mg of the final mixture is placed in a graphite electrode and arced at 15 A for 60 sec in a controlled (90% Ar, 10% O_2) atmosphere.

Alternatively, the graphite filter can be a standard porous-cup spectrographic electrode (Fig. 11.6) through which air is drawn. An indium internal-reference solution is dried in the electrode before sampling. When excited in a 28-A DC arc for 20 sec in an argon atmosphere saturated with HCl, this technique gives absolute detection limits between 0.1 and 5 ng for 14 elements.

ICP emission spectroscopy has also been applied for the simultaneous determination of metals in airborne particulates [32]. Due to the high sensitivity of the ICP measurement, large-area sampling filters (i.e., 8 in. by 10 in.) were found to be unnecessary, and a filter area of 3.5 cm^2 was sufficient to provide good results for most elements of interest. Ten elements were determined by ICP emission spectrometry after airborne particulates were collected on glass and quartz fiber filters. Analysis precision, expressed as relative standard deviation, was <1% for Fe, Mn, and Zn; <5% for Cu; and <10% for Cr, Pb, and V. With a filter area of ~20 cm^2, 10 metals including Be, Cd, and Ni could be determined simultaneously with precisions of <10%.

Trace Elements in Plant Material, Soil, and Blood

The routine determination of trace elements in agricultural, geological, and biological samples is of considerable interest [33–37]. Ideally, many trace elements in each sample should be determined simultaneously using a single group of standards. Atomic emission spectroscopy using the ICP source routinely provides that capability.

Inorganic analysis of plant materials for certain trace elements is frequently used in agricultural studies [33]. For example, orchard leaves can be examined for their Fe, Mn, Cu, Al, B, and Zn content [38]. The leaves are first dried and finely ground, then ashed in silica crucibles in a muffle furnace at 500°C. The ash is dissolved in HNO_3 and diluted to a known volume. Once the linearity of response for each element is established by use of standards, calibration is carried out with only one composite standard solution containing all the elements. The ICP discharge results compare well with those obtained by other methods. An example for iron is given in Figure 11.12. Agreement among alternative methods provides a good test for the accuracy of determination. The standard deviations are indicated by the horizontal and vertical lines in the figure. Better precision was obtained for the ICP discharge method.

The sampling and analysis of soils is used extensively in exploring for minerals. The application of the ICP discharge technique for Cu, Zn, Ni, Co, and Pb is rapid and free from certain interference effects common to atomic absorption analysis [39]. After drying and screening, weighed samples are dissolved in a 9:1 mixture of concentrated perchloric and nitric acids.

In health-care programs, knowledge of the concentrations of biologically essential or toxic elements in body fluids is important [34, 37]. Moreover, one must be able to measure accurately small changes in concentration that can be significant with respect to diseases. The ability to determine rapidly and simply several trace elements, some at the ng/mL level, in body fluids such as blood and urine is achieved by the ICP discharge. This analytical system is capable of determining many elements simultaneously, which conserves both sample and time. For biological fluids, only very small volumes (less than 1 mL) are usually available, and sample volumes of 10 to 25 μL can be used with the ICP discharge when employing a single-drop injection technique or electrothermal vaporization [40–44].

A substantial reduction in the absolute detection limit and the volume of sample required for ICP emission spectroscopy can be achieved by combining discrete sample introduction utilizing electrothermal vaporization with the ICP source. For example, Mn and Ni can be determined in 10-μL aliquots of whole-blood or animal-tissue sample solutions [41]; Ni can be determined in freeze-dried human milk [42]; and As, Cr, Cu, Se, and Ni can be determined in a 5-μL aliquot of urine [44].

Element-Specific Detection in Chromatography

Chromatographic methods provide only separation capabilities, and the characterization and identification of resolved components requires further analysis (Chaps. 21 and 22). A significant simplification of the chromatography can occur when a desired compound or compounds contain a particular element not present in the other components of the sample, and a detector is used that responds solely to that element. This results because compounds containing that element need be resolved only from one another and not necessarily from other components of the sample. Element-specific detection for gas and liquid chromatography can be accomplished simply by interfacing an emission spectroscopy source to the output of the chromatograph [45]. These sources include the DCP, ICP, and microwave-induced plasmas (MIPs).

The microwave-excited atmospheric-pressure helium-discharge detector extends the capability of gas chromatography significantly. The limits of detection, linear dynamic range, sensitivity, and selectivity have been measured for more than 20 metals, nonmetals, and halogens. By monitoring the carbon atomic emission wavelength, the MIP can serve as a nonselective universal detector for organic compounds.

With this system, trimethyl-lead chloride and triethyl-lead chloride, for example, are measured in spiked tap water by monitoring the Pb I 405.8-nm line [46]. Because these trialkyl-lead chlorides are very reactive and thermally unstable, a deactivated fused-silica column and a quartz interface are needed. As an alternative to this direct determination of extracted trialkyl-lead chlorides, chemically inert *n*-butylated trialkyl-lead compounds can be generated in the sample by a Grignard reaction, and the compounds collected in a trap prior to chromatographic separation and determination. Detection limits for this approach are 35 ng/mL for the trimethyl-lead chloride and 6 ng/mL for the triethyl-lead compound using the Pb I 283.3-nm line. About 19 ng/mL of the latter have been detected in an industrial plant effluent containing approximately 600 ng/mL of total lead [46]. The analysis requires an element-specific detector, because the background carbon response with a carbon-sensitive detector is high.

SELECTED BIBLIOGRAPHY

AHRENS, L. H., and TAYLOR, S. R. *Spectrochemical Analysis*, 2nd ed. Reading, Mass.: Addison-Wesley, 1961.

BARNES, R. M. *Emission Spectroscopy*. Stroudsburg, Pa.: Dowden, Hutchinson, & Ross, 1975.

BARNES, R. M., ed. *Applications of Inductively Coupled Plasmas to Emission Spectroscopy*. Philadelphia: Franklin Institute Press, 1978.

BARNES, R. M., ed. *Applications of Plasma Emission Spectrochemistry*. Philadelphia: Heyden, 1979.

BARNES, R. M., ed. *Developments in Atomic Plasma Spectrochemical Analysis*. Philadelphia: Heyden-Wiley, 1981.

GROVE, E. L., ed. *Analytical Emission Spectroscopy*, Vol. 1, Parts I and II. New York: Marcel Dekker, 1971 and 1972.

GROVE, E. L., ed. *Applied Atomic Spectroscopy*, Vols. 1 and 2. New York: Plenum Press, 1978.

HARRISON, G. R. *M.I.T. Wavelength Tables*, 2nd ed. Cambridge, Mass.: MIT Press, 1969.

IUPAC. "Nomenclature, Symbols, Units, and Their Usage in Spectrochemical Analysis. I. General Emission Spectroscopy," *Pure Appl. Chem.*, 30, 653–679 (1972); *Spectrochim. Acta*, 33B, 219–39 (1978). "II. Data Interpretation," *Pure Appl. Chem.*, 45, 99–103 (1976); *Spectrochim. Acta*, 33B, 241–45 (1978). "V. Radiation Sources," *Pure Appl. Chem.*, 53, 1913–52 (1981); *Spectrochim. Acta*, 37B, 219–58 (1982).

MIKA, J., and TOROK, T. *Analytical Emission Spectroscopy*. New York: Crane, Russak, 1974.

ROBINSON, J. W., ed. *Handbook of Spectroscopy*, Vol. 1. Cleveland, Ohio: CRC Press, 1974.

SACKS, R. D. "Emission Spectroscopy" in *Treatise on Analytical Chemistry*, 2nd ed., Part I, Vol. 7. New York: John Wiley, 1981, Chap. 6.

SLAVIN, M. *Emission Spectrochemical Analysis*. New York: John Wiley, 1971.

THOMPSON, M., and WALSH, N. *A Handbook of Inductively Coupled Plasma Spectrometry*. New York: Chapman & Hall, 1983.

ZEIDEL, A. N.; PROKOFEV, V. K.; RAISKII, S. M.; SLAVNYI, V. A.; and SCHREIDER, E. Y. *Table of Spectral Lines*. New York: IFI/Plenum Press, 1970.

ZIL'BERSHTEIN, KH. I. *Spectrochemical Analysis of Pure Substances*. New York: Crane, Russak, 1977.

REFERENCES

1. P. W. J. M. BOUMANS, *Theory of Spectrochemical Excitation*, New York: Plenum Press, 1966.
2. C. H. CORLISS and W. R. BOZMAN, *Experimental Transition Probabilities for Spectral Lines of Seventy Elements*, NBS Monograph 53, Washington, D.C.: U.S. Government Printing Office, 1962.
3. G. HORLICK, *Phil. Trans. Roy. Soc.* (London), 305A, 681 (1982).
4. G. HORLICK, R. H. HALL, and W. K. YUEN, in *Fourier Transform Infrared Spectroscopy*, Vol. 3, New York: Academic Press, 1982, pp. 37–81.
5. "Description and Performance of the Microphotometer," in *Methods for Emission Spectro-*

chemical Analysis, 7th ed., Philadelphia: American Society for Testing and Materials, 1982, pp. 278–83, ASTM E 409-81.
6. "Photographic Photometry in Spectrochemical Analysis," in *Methods for Emission Spectrochemical Analysis*, 7th ed., Philadelphia: American Society for Testing and Materials, 1982, pp. 35–65, ASTM E 116-81.
7. J. P. WALTERS, *Appl. Spectrosc.*, 26, 323 (1972).
8. R. J. KLUEPPEL and J. P. WALTERS, *Spectrochim. Acta*, 35B, 431 (1980).
9. P. B. FARNSWORTH and J. P. WALTERS, *Spectrochim. Acta*, 36B, 315 (1982).
10. S. E. MATHEWS and J. P. WALTERS, *Appl. Spectrosc.*, 36, 617 (1982).
11. R. M. BARNES, *CRC Crit. Rev. Anal. Chem.*, 7, 203 (1978).
12. R. M. BARNES, *Phil. Trans. Roy. Soc.* (London), 305A, 499 (1982).
13. M. THOMPSON, B. PAHLAVANPOUR, S. J. WALTON, and G. F. KIRKBRIGHT, *Analyst*, 103, 568, 705 (1978).
14. P. W. J. M. BOUMANS, *Line Coincidence Tables for Inductively Coupled Plasma Atomic Emission Spectrometry*, Vols. 1 and 2, 2nd ed. Oxford: Pergamon Press, 1984.
15. M. L. PARSONS, A. FORESTER, and D. ANDERSON, *An Atlas of Spectral Interferences in ICP Spectroscopy*, New York: Plenum Press, 1980.
16. R. K. WINGE, V. A. FASSEL, V. J. PETERSON, and M. A. FLOYD, *Appl. Spectrosc.*, 36, 210 (1982) and *Inductively Coupled Plasma-Atomic Emission Spectroscopy. An Atlas of Spectral Information*, Amsterdam: Elsevier, 1985.
17. P. W. J. M. BOUMANS and F. J. DE BOER, *Spectrochim. Acta*, 27B, 391 (1972).
18. P. E. WALTERS and H. G. C. HUMAN, *Spectrochim. Acta*, 36B, 585 (1981).
19. J. DURR and B. VANDORPE, *Spectrochim. Acta*, 36B, 139 (1981).
20. "Designation of Shapes and Sizes of Graphite Electrodes," in *Methods for Emission Spectrochemical Analysis*, 7th ed., Philadelphia: American Society for Testing and Materials, 1982, pp. 75–84, ASTM E 130-81.
21. J. R. MOODY, *Phil. Trans. Roy. Soc.* (London), 305A, 669 (1982).
22. M. ZIEF and J. W. MITCHELL, *Contamination Control in Trace Element Analysis*, New York: John Wiley, 1976.
23. R. ALVAREZ, S. D. RASBERRY, and G. A. URIANO, *Anal. Chem.*, 54, 1226A (1982).
24. W. B. BARNETT, V. A. FASSEL, and R. N. KNISELEY, *Spectrochim. Acta*, 23B, 643 (1968).
25. K. W. JACKSON and L. EBDON, eds., *Annual Reports on Analytical Atomic Spectroscopy*, Vol. 13, London: The Royal Society of Chemistry, 1984.
26. C. L. CHAKRABARTI, ed., *Progress in Analytical Atomic Spectroscopy*, Vols. 1 through 8, Oxford: Pergamon Press, 1977–1985.
27. P. N. KELIHER, W. J. BOYKO, J. M. PATTERSON III, and J. W. HERSHEY, *Anal. Chem.*, 56, 133R (1984).
28. "Spectrochemical Analysis of Aluminum and Its Alloys by the Point-to-Plane Technique Using an Optical Emission Spectrometer," in *Methods for Emission Spectrochemical Analysis*, 7th ed., Philadelphia: American Society for Testing and Materials, 1982, pp. 181–91, ASTM E 227-67.
29. "Proposed Spectrochemical Method of Test for Wear Metals in Used Diesel Lubricating Oils by a Rotating-Disk Electrode Technique Using a Direct-Reading Spectrometer," in *Methods for Emission Spectrochemical Analysis*, 6th ed., Philadelphia: American Society for Testing and Materials, 1971, pp. 375–82, D-2-1968.
30. A. SUGIMAE, *Anal. Chem.*, 46, 1123 (1974).
31. J. L. SEELEY and R. K. SKOGERBOE, *Anal. Chem.*, 46, 415 (1974).
32. A. SUGIMAE, *ICP Inf. Newsl.* 6, 619 (1981); 8, 160 (1982).
33. D. J. DAVID, *Prog. Anal. At. Spectrosc.*, 1, 225 (1978).
34. H. T. DELVES, *Prog. Anal. At. Spectrosc.*, 4, 1 (1981).
35. M. A. FLOYD, V. A. FASSEL, and A. P. D'SILVA, *Anal. Chem.*, 52, 2168 (1980).
36. J. N. WALSH and R. A. HOWIE, *Min. Mag.*, 43, 967 (1980).
37. J. M. MERMET and J. HUBERT, *Prog. Anal. At. Spectrosc.*, 5, 1 (1982).

38. R. H. Scott and A. Strasheim, *Anal. Chim. Acta.*, **76**, 71 (1975).
39. R. H. Scott and M. L. Kokot, *Anal. Chim. Acta.*, **75**, 257 (1975).
40. A. Aziz, J. A. C. Broekaert, and F. Leis, *Spectrochim. Acta*, **36B**, 251 (1981).
41. C. Camara Rica, G. F. Kirkbright, and R. D. Snook, *At. Spectrosc.*, **2**, 172 (1981).
42. C. Camara Rica and G. F. Kirkbright, *Sci. Total Environ.*, **22**, 193 (1982).
43. A. Aziz, J. A. C. Broekaert, and F. Leis, *Spectrochim. Acta*, **37B**, 369 (1982).
44. P. Fodor and R. M. Barnes, *Spectrochim. Acta*, **38B**, 229 (1983).
45. P. C. Uden, in *Developments in Atomic Plasma Spectrochemical Analysis*, R. M. Barnes, ed., Philadelphia: Heyden, 1981, pp. 302–320.
46. S. A. Estes, P. C. Uden, and R. M. Barnes, *Anal. Chem.*, **53**, 1336–40 (1981); **54**, 2402–5 (1982).

PROBLEMS

1. Explain why the simultaneous determination of several elements would be more difficult by means of atomic absorption spectrophotometry than by atomic emission spectrometry using either flame, arc, or ICP excitation sources.

2. Alkali and alkaline-earth metals in solution are in the ionic form and appear colorless, but in an excitation source, emission from neutral atoms and ions is observed, perceived as bright colors in the visible region. (a) To go from ions to atoms, these species must acquire one or more electrons. Where do these electrons come from? (b) How do you account for the colorful emission from these metals? (c) What color would you predict each of these metals to show in an arc or ICP discharge?

3. The halogens and gases are conspicuously absent from the lists of elements in Table 11.2. What reasons can you give for the difficulty in determining these elements by arc or spark sources in air? Suggest one or two methods that you would try if you were required to determine halogens, permanent gases, or rare gases by emission spectrometry.

4. Would a DC arc make a very good source for production of neutral atoms for atomic absorption? Explain your answer.

5. If you were given a brass block for analysis and told that the tin and zinc distributions in the block were very heterogeneous, would you or would you not choose a spark point-to-plane technique for analysis? On what grounds do you make your selection? If you chose not to use a spark point-to-plane technique, what alternative spark techniques might you use for the analysis of zinc and tin?

6. A sample of an unknown light-metal alloy was analyzed using the point-to-plane spark technique. By means of a projection comparator, the following wavelengths were identified. What elements are present? (Hint: The CRC *Handbook of Chemistry and Physics* contains lists of wavelengths.) What is the alloy matrix?

236.706 nm	283.307	327.926
251.612	288.158	328.233
252.852	288.958	330.259
255.796	294.920	330.294
256.799	296.116	330.628
259.373	307.399 ←(Internal	332.513
261.020	Reference)	334.502
266.039	317.933	334.557
270.170	318.020	343.823
270.574	322.129	396.153
277.983	324.754	403.076
278.142	327.396	481.053

7. The following figure illustrates the emission signals from Mn in 25-μL aliquots of human whole blood, and of blood to which spikes of a Mn standard solution were added. The whole blood had first been diluted 10-fold with 0.1 M HCl, and an ICP source was used. What is the concentration of Mn in the blood sample?

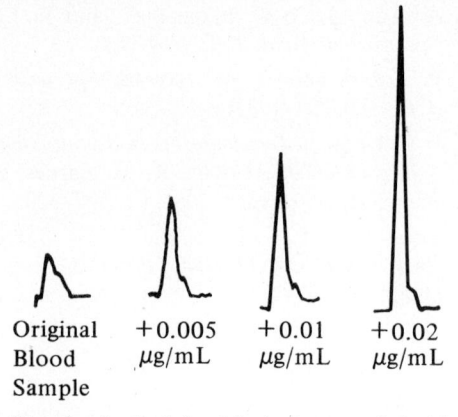

Signals obtained for Mn in human whole blood (10-fold dilution) and from addition standards. Adapted from R. N. Kniseley, V. A. Fassel, and C. C. Butler, Clin. Chem., 19, 807 (1973), by permission of the publisher.

8. The reproducibility of the signals for the Mn 403.0-nm line in Problem 7 is indicated by the recordings in the following figure. What is the precision of these signals expressed as standard deviation and relative standard deviation?

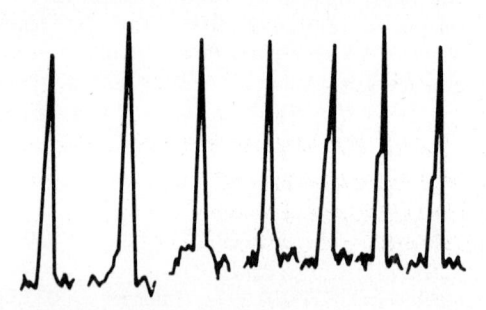

Reproducibility of signals for Mn from 25-μL samples of whole blood (undiluted). Adapted from R. N. Kniseley, V. A. Fassel, and C. C. Butler, Clin. Chem., 19, 807 (1973), by permission of the publisher.

9. Describe the considerations involved in selecting: (a) an internal-reference element and spectral line; (b) a spectrochemical buffer; (c) a matrix diluent.

10. Predict the change in atomic absorption sensitivity when a flame of temperature $T = 2500$ K is replaced by an arc of temperature $T = 5000$ K for the resonance transition of calcium at a wavelength of 422.673 nm. (The transition is $3p^6 4s^2\ {}^1S_0 \leftarrow 3p^6 4s^1\ {}^1P_1^0$, corresponding to $E = 2.93$ eV.) Will ionization make an appreciable contribution?

11. A sample of dolomite was analyzed semiquantitatively for its Si and Na content using a DC arc with a matrix dilution technique. Standards are available (Spex Industries, Inc.) that contain about 50 elements, each at a specified concentration level, mixed with a high-purity spectroscopic graphite powder; for example, one standard, 0.01% of each in a second standard, and so forth. The following data were obtained from a densitometer measurement of seven arcs on an exposed film of the silicon 288.16-nm line and the sodium 330.23-nm line:

Sample	Si Line (% T)	Na Line (% T)
0.0001% standard	>99	>99
0.001% standard	96	92
0.01% standard	66	71
0.1% standard	<1	23
Pure dolomite	<1	<1
1 part dolomite + 9 parts graphite	58	16
1 part dolomite + 99 parts graphite	95	65

Plot a calibration curve (log absorbance versus log concentration) and determine the concentrations of Si and Na in the dolomite sample.

CHAPTER **12**

Nuclear Magnetic Resonance Spectroscopy

S. Sternhell

Like other forms of spectroscopy (e.g., infrared and ultraviolet), *nuclear magnetic resonance spectroscopy* (NMR) deals with the measurement of energy gaps between states of different energy. However, unlike most other forms of spectroscopy, the phenomenon requires the presence of an external magnetic field and concerns nuclei rather than electrons. This is the origin of the terms *nuclear* and *magnetic* in *nuclear magnetic resonance spectroscopy*.

NMR is a powerful technique for structural analysis available to the organic chemist, because it utilizes commonly found elements (in particular, hydrogen and carbon) as "chromophores." With the aid of NMR, it is possible to define the environment of practically all commonly occurring functional groups, as well as of fragments (e.g., hydrogen atoms attached to carbon) that are not otherwise accessible to spectroscopic or analytical techniques. NMR can also be used for quantitative determination of compounds in mixtures and hence for following the progress of chemical reactions. More sophisticated applications often yield kinetic and thermodynamic parameters for certain types of chemical processes; and others, in particular spin-spin coupling, often give accurate information about the relative positions of groups of magnetic nuclei within molecules. The principal limitations of the method are its inherently low sensitivity and its difficult applicability to samples in the solid state.

The phenomenon of nuclear resonance was first observed in 1946 by two teams of physicists—Purcell, Torrey, and Pound at Harvard and Bloch, Hansen, and Packard at Stanford—who shared a Nobel Prize for this discovery. The first observation of the *chemical shift*, the phenomenon on which all chemical applications are based, was made by Knight in 1949, and the first systematic applications to organic chemistry were reported in 1953 by Meyer, Saika, and Gutowsky. The first commercial instruments appeared in about 1956 and, in spite of their high cost, several thousand are presently in use. Relatively inexpensive models are now available that, although not as sophisticated as the more expensive models, are easier to operate and are capable of handling many routine measurements.

12.1 THEORY AND INSTRUMENTATION

Two elementary principles of classical electromagnetism, summarized in Figures 12.1 and 12.2, should be recalled. Atomic nuclei have charge (they contain protons), and some also behave as if they spin. A spinning charge is equivalent to a current in a conductor loop; therefore, nuclei with nonzero spin will generate a magnetic field, that is, will have a *magnetic moment* or a magnetic dipole.

Depending on the shape of the nuclear charge and the number and type of nucleons, the *spin quantum number*, I, can have values 0, 1/2, 1, 3/2, and so on. There

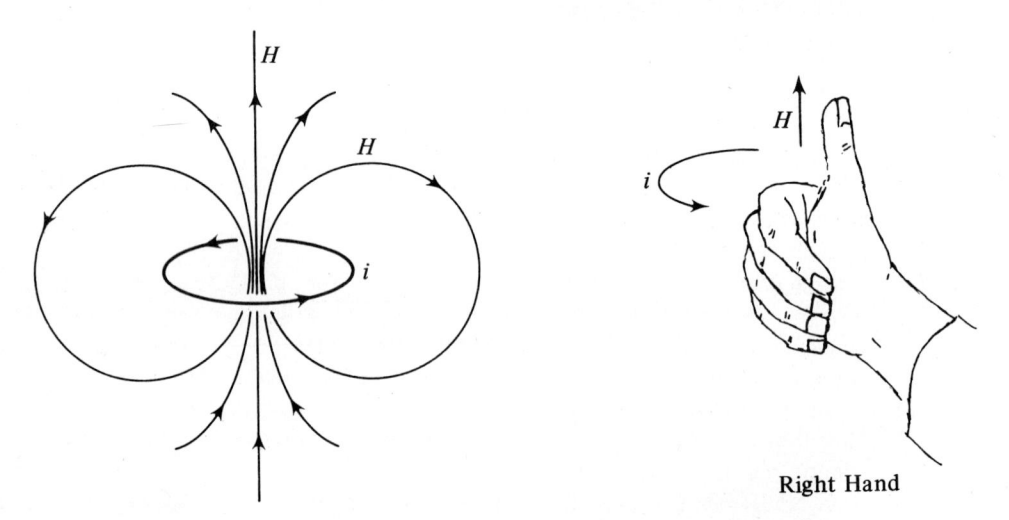

FIGURE 12.1. Right-hand rule. A current i flowing in a conductor loop generates a magnetic field H in the direction shown.

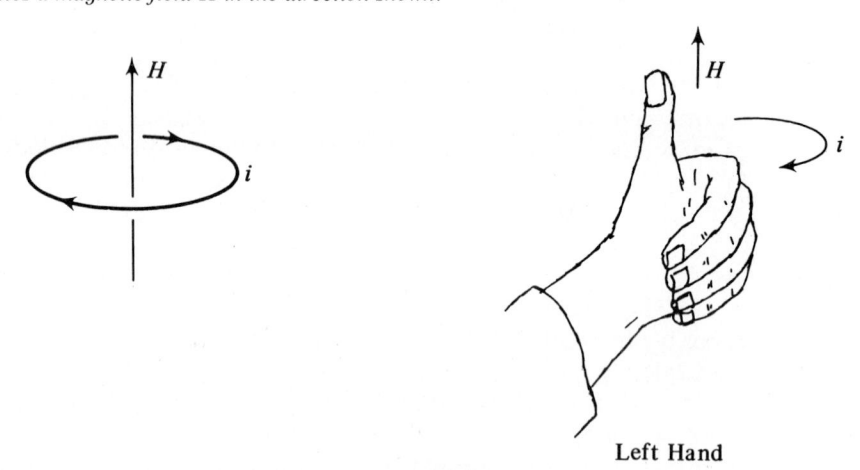

FIGURE 12.2. Left-hand rule. A magnetic field H causes current i to flow in the conductor loop in the direction shown. As given, this rule is incomplete, since relative motion of the field and conductor is necessary in the macroscopic case, but the rule gives the correct direction of the effect in the nuclear case.

are three principal groups of nuclei:

1. $I = 0$ (nonspinning nuclei). These have no magnetic moment and are composed of even numbers of protons and neutrons, for example: $^{12}_{6}C$, $^{16}_{8}O$.
2. $I = 1/2$ (spherical spinning charges). These nuclei have a magnetic moment but no *electric quadrupole*. This group is by far the most important from the chemical point of view. Chemically useful nuclei in this group, in decreasing order of importance, are $^{1}_{1}H$, $^{13}_{6}C$, $^{19}_{9}F$, $^{31}_{15}P$, $^{15}_{7}N$. Of these, the proton (^{1}H) and carbon (^{13}C) account for well over 95% of all NMR observations made.
3. $I > 1/2$ (nonspherical spinning charges). These nuclei have both magnetic dipoles and electric quadrupoles; examples are: $I = 1$: $^{2}_{1}H$, $^{14}_{7}N$; $I = 3/2$: $^{11}_{5}B$, $^{35}_{17}Cl$, $^{37}_{17}Cl$, $^{79}_{35}Br$, $^{81}_{35}Br$, $^{7}_{3}Li$; $I = 2$: $^{36}_{17}Cl$, $^{58}_{27}Co$; $I = 5/2$: $^{25}_{12}Mg$, $^{27}_{13}Al$, $^{17}_{8}O$.

A fundamental quantum law is that: *In a uniform magnetic field, a nucleus of spin I may assume $2I + 1$ orientations.* Thus, for a nucleus of $I = 1/2$ (e.g., the proton), there are $2(1/2) + 1 = 2$ permissible orientations. This makes a nucleus of $I = 1/2$ analogous to a bar magnet in a magnetic field (Fig. 12.3). Since we shall deal exclusively with nuclei of spin $I = 1/2$ in this chapter, and almost exclusively with protons, the bar-magnet analogy will be useful.

As with the bar magnet, the two orientations of the nuclear magnet in the magnetic field (of strength* H_0) have different energies, and it is possible to induce a nuclear transition, analogous to the flipping of the bar magnet, by applying electromagnetic radiation of an appropriate frequency v given by

$$v = \frac{\gamma H_0}{2\pi} \qquad (12.1)$$

where γ = a fundamental constant known as the *gyromagnetic ratio* or *magnetogyric ratio*, and is characteristic of the particular nucleus

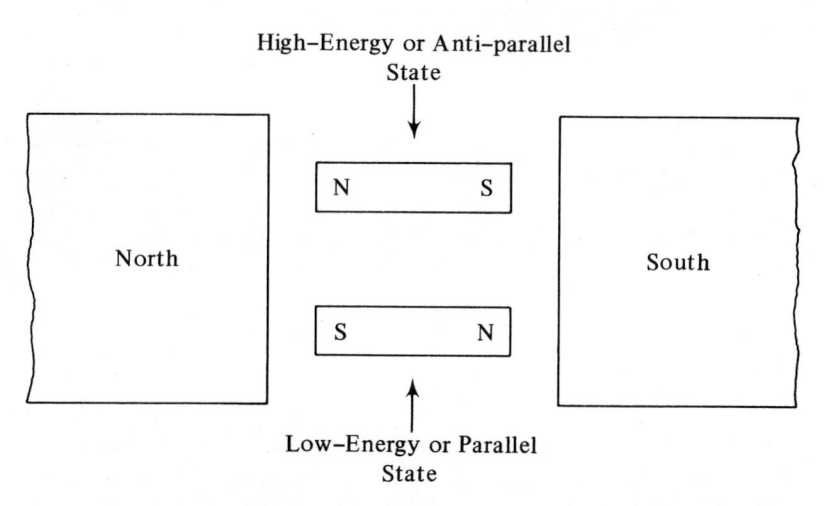

FIGURE 12.3. *Bar-magnet analogy for nuclei with $I = 1/2$.*

---

*The units are *gauss*: 1 gauss is defined as the strength of a magnetic field that induces a voltage of 1 V in a conductor 1 cm long moving at 1 cm/sec.

Note that Equation 12.1 can be reduced to

$$v = \text{Constant} \times H_0 \tag{12.2}$$

Equation 12.1 is known as the *Larmor equation*; it shows that one could observe a nuclear transition (*spin flip*) by keeping the magnetic field constant and varying the applied frequency (or vice versa) until the combination of field strength and irradiating frequency characteristic of the nucleus concerned is reached. This condition is often described as *resonance* and is, of course, the origin of the term "resonance" in *nuclear magnetic resonance*. The term *resonance frequency* is also sometimes used, but it must be remembered that the term would be meaningless without specifying the field strength H_0. Thus, the resonance frequency of the proton is 60 MHz *at 14,092 gauss*. In practice, NMR spectrometers may be capable of varying (or "sweeping") either the frequency or the magnetic field, and one often uses the terms *frequency-sweep spectrometer*, *frequency-sweep spectrum*, *field-sweep spectrometer*, and *field-sweep spectrum*.

The magnitudes of the various constants involved are such that the energy gap corresponding to the spin flip, given by $\Delta E = h\nu$, is very small; 60 MHz corresponds to only 6×10^{-6} kcal/mole. Thus, all NMR frequencies at usable field strengths fall in the radio-frequency (RF) region of the electromagnetic spectrum; the source of the radiation is an RF transmitter, generally a crystal oscillator.

Now consider the details of the energy transfer to a nuclear magnet placed in an external field H_0. The magnet, in either the parallel or the antiparallel orientation (Fig. 12.3), will not remain stationary but will precess (Fig. 12.4) in a magnetic field H_0 with an angular velocity ω_0 given by

$$\omega_0 = \gamma H_0 \tag{12.3}$$

Combining Equations 12.1 and 12.3, we get

$$\omega_0 = 2\pi\nu \tag{12.4}$$

Thus, if one can get ω_0, one can also determine ν, the resonance frequency. This is done as follows. A second magnetic field (H_1) is generated at right angles to H_0 by passing a very high frequency alternating current supplied by an RF oscillator through a coil (the transmitter coil). When the angular component of H_1 matches ω_0, the frequency of this alternating current is equal to ν, and a transition, or spin flip, can occur. The geometry of the arrangement in Figure 12.4 follows from the simple rules of electromagnetism stated at the beginning of the chapter, and the NMR experiment can be seen to amount to "nuclear induction."

The NMR Spectrometer

We can now construct an *NMR spectrometer*. A typical arrangement is shown in Figure 12.5. This diagram represents a *field-sweep, crossed-coil spectrometer*, but the essential features are the same for other types of spectrometers. The crucial parts of any high-resolution (the significance of this term will become apparent later) NMR spectrometer are as follows:

1. The *magnet*, which may be either a permanent magnet, an electromagnet, or a superconducting solenoid ("supercon"), but which must be capable of generating a

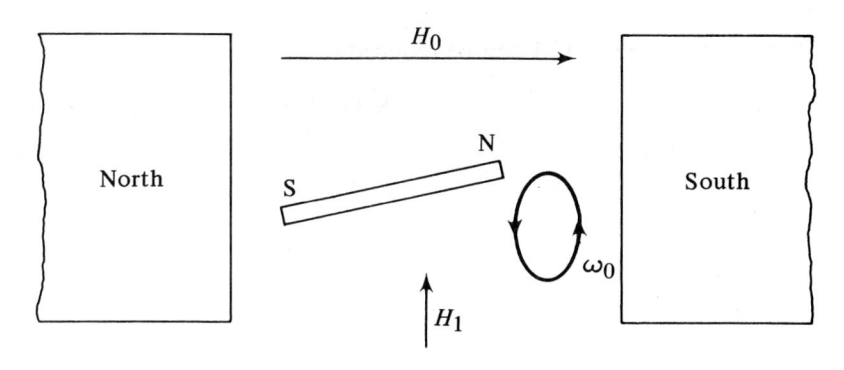

FIGURE 12.4. *Relation between precession and the exciting field H_1.*

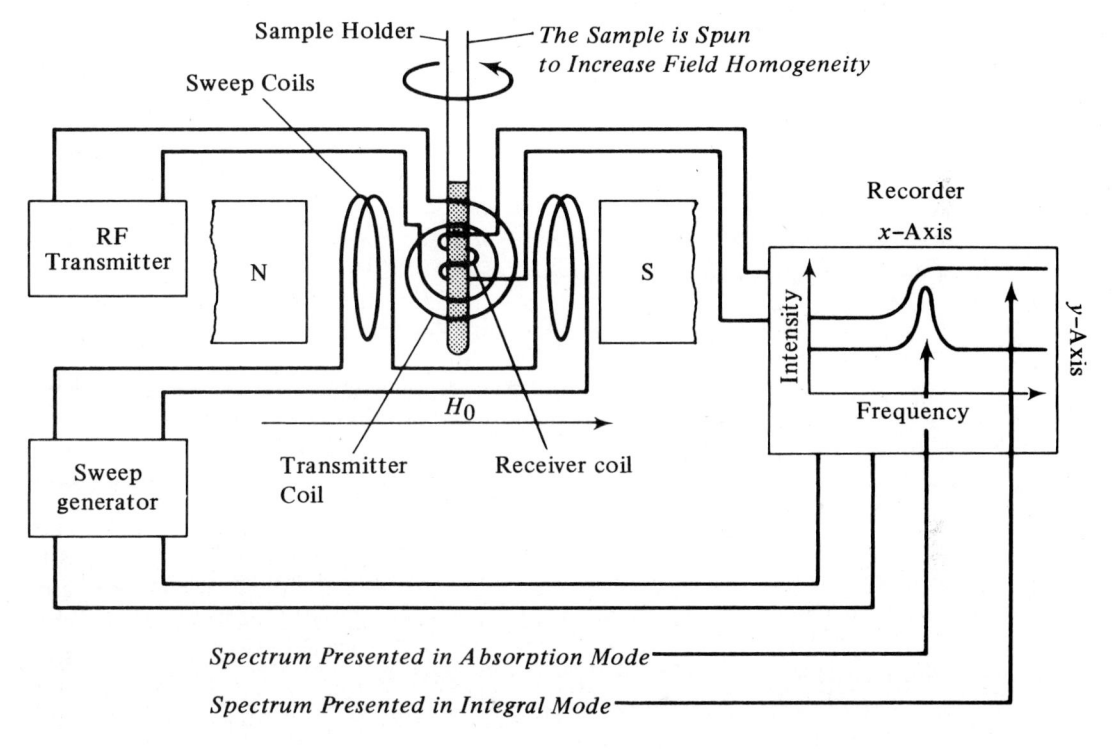

FIGURE 12.5. *Schematic diagram of an NMR spectrometer.*

very strong, very stable, and very homogeneous magnetic field. (These magnetic-field requirements are the principal reasons for the cost and complexity of NMR spectrometers.) To average out small magnetic-field inhomogeneities throughout the sample, the sample tube is roated at several hundred rpm. To obtain high stability, the field is "locked" to the frequency using electronic feedback devices.

2. The *sweep generator*, which is used to vary the magnetic field over a small range by passing a variable direct current through coils that are coaxial with the direction of the main magnetic field H_0.

SEC. 12.1 Theory and Instrumentation

3. The *transmitter coil*, which is placed at right angles to the sweep coils and is used to generate the exciting field H_1.
4. The *receiver coil*, which is placed around the sample holder in the remaining orthogonal plane. A small current is generated in it when the resonance condition is achieved (see nuclear induction, above).

The signal from the receiver coil is suitably amplified and is made to deflect the recorder pen along the *y*-axis while the *x*-axis is synchronized with the sweep generator. Thus, one plots the signal from the receiver coil as a function of the field strength; the intensity of the signal is proportional to the number of nuclei undergoing the transition. This is the *field-sweep experiment*. The *frequency-sweep experiment* is analogous, except that the field is kept constant while the frequency is swept and synchronized with the *x*-axis of the recorder. In either case, one can express the *x*-axis scale in terms of "signal frequency" because, by Equation 12.1, the field strength and frequency for the resonance condition are always directly connected.

For protons, the intensity of the signal is proportional to the area under the absorption curve (Fig. 12.5) and is usually obtained by electronic integration, yielding a step function whose height is a direct measure of the relative intensity of the absorption signal. However, for other nuclei, in particular for ^{13}C, this relationship no longer holds.

The Fourier Transform NMR Spectrometer

There exists another, fundamentally different, method of obtaining an NMR spectrum. Instead of varying the frequency (or the field) until the condition of resonance is reached, a strong pulse of energy is applied over the whole range of frequencies while the field is kept constant. As a result, nuclei are flipped to their upper state from which, over time, they will return (decay) to the lower state. By collecting the thus-induced current as a function of time (this requires a minicomputer), one obtains a time-domain spectrum, which is a generally complex pattern called the *free-induction decay* (FID). Inspection of an FID yields no interpretable information, but a Fourier transformation of an FID, performed on the same dedicated minicomputer, yields a spectrum virtually identical to the regular absorption spectrum. This type of spectroscopy is called *Fourier transform* (FT) *spectroscopy* and has important advantages over the ordinary (*continuous-wave*, or CW) spectroscopy.

The most important advantage of FT spectroscopy concerns an increase in sensitivity as a function of time, because the time needed for collecting a free-induction decay is of the order of seconds, whereas the time needed for a CW scan of the same assemblage of nuclei is of the order of minutes. Thus, during the time taken for a CW scan, the minicomputer of an FT spectrometer can accumulate about 100 scans and add them up in memory. Because coherent signals will add on accumulation, while random noise is as likely to cancel as to add, accumulation will result in a net increase in signal and in sensitivity. The signal-to-noise ratio increases as the square root of the number of scans or pulses. Overall, FT spectroscopy represents about an order of magnitude increase in sensitivity over CW spectroscopy with the same measurement time. It is only the advent of FT spectroscopy early in the 1970s that has enabled ^{13}C NMR to become a routine analytical tool.

There are further advantages to FT spectroscopy. While the signal is stored in the computer as an FID, it may be manipulated before being transformed into an absorption spectrum. The most obvious manipulations result in trade-offs yielding either increased sensitivity, which is critical in ^{13}C NMR, or increased resolution, often very important in ^{1}H NMR. In a more fundamental manner, it is possible to alter the sequence (timing and power) of the energy pulses, as well as to add or subtract FIDs obtained with varying pulse sequences. Such experiments yield data of enormous value, but the principles involved are beyond the scope of this discussion.

The Boltzmann Distribution and Nuclear Relaxation Processes

There are certain consequences of the small size of the energy gap involved in a nuclear-spin transition. The Boltzmann relation gives the populations of nuclear spins in the upper energy state (N_2) and in the lower energy state (N_1) in terms of the energy gap ΔE between them:

$$\frac{N_1}{N_2} = e^{\Delta E/RT} \tag{12.5}$$

When the energy gap is very small, as is the case here, the right side approaches e^0 (unity), and the excess population in the lower energy state, given by $N_1 - N_2$ and called the *Boltzmann excess*, becomes very small. Since an absorption signal can originate only from the Boltzmann excess (typically only 1 nucleus in 100,000 in an NMR experiment), it follows that the method is inherently not very sensitive and that sophisticated signal amplification must be used in an NMR spectrometer. One of the reasons for the use of very high magnetic fields becomes apparent: The Larmor equation shows that high magnetic fields require high resonance frequencies. This in turn widens the energy gap between the spin states ($\Delta E = h\nu$), thereby increasing the Boltzmann excess and the sensitivity of the experiment.

A consequence of the small Boltzmann excess in NMR experiments arises from the general spectroscopic principle that absorption cannot occur unless some mechanism exists for a radiationless transition that can restore the excess population in the lower energy state. This is related to the fact that upward and downward transitions are equally probable on interaction with an appropriate energy quantum. Therefore, excess absorption, that is, observable absorption signals, can originate only from unequal populations, as stated above.

The mechanisms of radiationless transitions from the upper to the lower energy state are particularly critical in NMR because of the small Boltzmann excess. These mechanisms are termed *relaxation* and are characterized by their *relaxation times T*, which are equal to half the time necessary to restore equilibrium by the mechanism considered. Clearly, large values of T indicate inefficient relaxation. Two relaxation mechanisms are important:

1. *Spin-spin* or *transverse relaxation* (characterized by T_2) occurs when the energy is lost by spin exchange, that is, by transmission to neighboring spins. This mechanism operates extremely efficiently in solids, where magnetic nuclei are close together. However, the positive effects of this efficiency are offset by another general

spectroscopic principle, the uncertainty principle, which states that the width of a spectral line is inversely proportional to the time spent in the upper energy state:

$$\text{Line Width} = \frac{\text{Constant}}{\text{Time in Upper State}} \qquad (12.6)$$

Since the spin-spin relaxation mechanism in solids is so efficient, very small values of T_2 result, producing very broad lines. Furthermore, such spin exchanges between identical nuclei average the resonance frequencies of nuclei whose environments are not quite identical, and broaden the spectral lines further (dipolar broadening). For these reasons, solids give spectra with lines about 1000 times too broad to provide information of much chemical interest. Spectra of solids, or *wide-line spectra*, will therefore not be discussed further here, although they are of interest in solid-state physics.

2. *Spin-lattice* or *longitudinal relaxation* (characterized by T_1) occurs when the energy is lost to the "lattice," that is, to any component of the sample, inter- or intramolecular. The lattice contains magnetic nuclei in rapid thermal motion in a magnetic field, generating a variety of electric currents and magnetic dipoles; energy may be lost to them by the nuclear magnets observed, restoring the equilibrium. This mechanism operates with gases, liquids, and solutions and is of just the right efficiency to produce narrow lines, or so called *high-resolution spectra*.

Spin-lattice relaxation times of individual magnetic nuclei (particularly ^{13}C nuclei) can be determined for most molecules and can be correlated with intramolecular or intermolecular mobility.

Interactions between nuclear magnetic dipoles and nuclear electric quadrupoles in nuclei where $I > 1/2$ offer another relaxation mechanism that prevents the observation of the NMR signals from some elements.

Interactions of the nuclear magnet with unpaired electrons (e.g., in free radicals and in atoms of the transition metals) can also result in efficient relaxation. Since an unpaired electron has about 1000 times the strength of a nuclear magnet, line broadening often occurs in solutions containing even small amounts of paramagnetic impurities, which must therefore be rigorously excluded from NMR samples; even dissolved oxygen causes some broadening.

If for some reason the spin-lattice relaxation mechanism is not operating efficiently, as when high viscosity interferes with the thermal movement of the lattice, the signal strength will diminish with time even during the relatively short interval needed to scan the signal, causing the phenomenon of *saturation*. The same phenomenon will occur if the current in the transmitter coil, and therefore the strength of the RF field, is increased to too high a value, flipping the nuclei into their upper states faster than the relaxation processes can restore the equilibrium. The onset of saturation is also accompanied by some line broadening, because it is the nuclei exactly at resonance, and hence in the middle of the signal line, that are saturated first.

With most spectrometers operating under routine conditions, the line widths are controlled by the inhomogeneity of the magnetic field (line widths are generally measured as the width at half-height of a single line and denoted by $W_{1/2}$). In a slightly inhomogeneous field, different parts of the sample will experience slightly different magnetic fields and hence resonate over a range of frequencies, broadening the spectral lines. Line widths of as little as 0.1 Hz are sometimes desirable; line widths in excess of about 1 Hz result in the loss of considerable information. Clearly, this imposes very stringent demands on the magnet as regards homogeneity. A line of 1 Hz width obtained with a spectrometer operating at the equivalent of 100,000,000 Hz requires a homogeneity of better than 1 in 10^8; however, this is routinely available with modern spectrometers.

To summarize: Some nuclei, notably protons, have magnetic moments. "Spin-flip" nuclear magnetic transitions of these nuclei can be observed at frequencies predicted by the Larmor equation, using complicated and expensive apparatus. For ^{1}H, the strength of the signal is directly proportional to the number of nuclei involved.

If this were all NMR had to offer, it would not be considered particularly useful in chemical investigations, since all one achieves is a costly and inconvenient estimate of the total hydrogen, fluorine, etc., content in a sample. In practice, all applications of NMR to chemistry are from three secondary phenomena: the *chemical shift*, the *time dependence* of NMR phenomena, and *spin-spin coupling*. These effects will be considered in the next several sections.

From now on, unless otherwise indicated, the discussion will involve protons and deal with PMR (*proton magnetic resonance*) rather than with NMR, with occasional reference to ^{13}C NMR (CMR). However, the principles are strictly analogous for all magnetic nuclei with $I = 1/2$.

12.2 THE CHEMICAL SHIFT

The statement "protons resonate at 60 MHz at 14,092 gauss" is only an approximation. Actually, protons in organic molecules are found to resonate, at 14,092 gauss, over a frequency range of about 1000 Hz at approximately 60 MHz. The exact frequency at which a proton resonates within this range is related to its chemical environment (hence the term *chemical shift*). The resonance of ^{19}F at 56.54 MHz in the same magnetic field is the closest resonance to that of ^{1}H; this is some 3,500,000 Hz away. It is apparent that the proton chemical-shift range of about 1000 Hz is actually the fine structure of a single line.* To put it pictorially, at a chart width where the chemical-shift range of protons corresponds to about 1 m, the fluorine resonances will turn up 3.5 km away; the ^{13}C range will be found 45 km away.

Since the chemical shift reflects molecular structure, it can be used to determine the structures of unknown compounds; and since carbon and hydrogen are almost universal constituents of organic compounds, the method is very widely applicable. Furthermore, as mentioned before (Fig. 12.5), the intensity of the signal caused by any group of protons (the area under the curve, generally determined by electronic integration) is directly proportional to the number of protons in it. We can therefore determine the environments of hydrogen atoms in an organic molecule and obtain the relative distribution of hydrogens between the various environments. Hydrogen and carbon thus become "chromophores."

---

*The ability to resolve this line defines *high-resolution NMR*.

Measurement of the Chemical Shift

Modern NMR spectrometers can determine resonance frequencies of sharp lines to a precision of better than 0.05 Hz. It would be almost impossible to measure a frequency of, say, 60,000,000 Hz to an absolute accuracy of 0.05 Hz, since this implies an absolute accuracy of 1 part in 10^{10}. Instead, all chemical shifts are measured relative to some standard substance which is added to the sample being investigated; one can then express the chemical shift in terms of displacement, in hertz, from the signal caused by the standard. Because the range of proton chemical shifts at 14,092 gauss is approximately 1000 Hz, measurement to a precision of 0.05 Hz implies an accuracy of 1 part in 10^4—which is realistic, but still requires high stability of the magnetic field over the time necessary to scan the spectrum, and hence an advanced magnet technology.

The standard substance almost universally used is tetramethylsilane (Me_4Si), commonly abbreviated as TMS. This standard was chosen because it gives rise to a single sharp line as a result of the identical environment of all the protons in the symmetrical molecule and because the chemical environment of protons in TMS is such that they resonate at a higher field than practically any other proton. Furthermore, TMS is an inert, low-boiling liquid and can be easily removed from the sample after the spectrum has been run. Therefore, in practice, the procedure is nondestructive. The sample size required for examination by NMR, however, is relatively large, generally at least 10 mg, because of the inherently low sensitivity of the method. For exactly the above reasons, TMS is also used as the internal standard in ^{13}C NMR.

The chemical shift of any proton can be expressed in terms of "hertz from TMS." By convention, the absence of a sign implies "hertz to lower field, or downfield, from TMS," remembering at all times that *field* and *frequency* can be used interchangeably. The chemical shift thus expressed depends on the operating field of the spectrometer (Eqn. 12.1) so that one would have to state: "Proton X resonates at Y hertz from TMS at Z megahertz spectrometer frequency." However, if one divides this value by the spectrometer frequency, one obtains the chemical shift in terms of dimensionless units. In practice, a factor of 10^6 is also introduced to avoid handling very small numbers; so the dimensionless unit turns out to be ppm. Chemical shifts expressed in ppm versus TMS are usually designated as δ. Thus, we write

$$\text{Chemical Shift in ppm } (\delta) = \frac{\text{Chemical Shift in hertz from TMS}}{\text{Spectrometer Frequency in hertz}} \times 10^6 \quad (12.7)$$

The δ scale ranges from 0 to about 12 ppm. Another system sets TMS arbitrarily at 10 and expresses the chemical shifts in terms of τ values, so that

$$\tau = 10 - \delta \quad (12.8)$$

The unit is still ppm, only the scale is different. In other words, with the numbers reading in ppm, we have

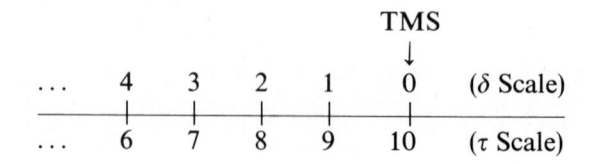

					TMS ↓	
...	4	3	2	1	0	(δ Scale)
...	6	7	8	9	10	(τ Scale)

Chart paper for NMR spectrometers is marked in either, or both, scales and also (nearly always) with a grid scaled in hertz, ensuring maximum confusion for the beginner. (The τ system is no longer recommended usage.)

Physical Causes of the Chemical Shift

The chemical shift occurs because the resonance frequency depends not on the gross field (H_0) between the poles of the magnet of an NMR spectrometer, but on the actual field at the resonating nucleus. Only for the hypothetical case of an isolated proton will the field at the nucleus be equal to the gross field. For all other cases,

$$H_{\text{nucl}} = H_0(1 - \sigma) \tag{12.9}$$

where σ = *shielding constant* for the particular situation

The shielding constant cannot in general be predicted, but the factors governing it, and hence determining the chemical shift, are qualitatively understood.

Consider an isolated hydrogen atom—a proton with its electron. Under the influence of H_0, the 1s electron will circulate in the direction given by the left-hand rule, thus becoming equivalent to a current in a circular loop. This current will generate (by the right-hand rule) a small magnetic field H_e which, in the region of the nucleus, will be in such a direction as to oppose H_0 (Fig. 12.6). The electron is then said to *shield* the proton in a hydrogen atom. Therefore, for a hydrogen atom the gross field H_0 required for resonance at a fixed frequency will be slightly larger than that required for an isolated (unshielded) proton.

Now consider a hydrogen atom bonded to a carbon atom, that is, one existing as a part of a molecule. From simple bonding theory, we know that the electron density about the hydrogen will be reduced because the carbon atom is more electronegative; hence, the shielding effect of the circulating 1s electron (now part of a σ bond) is smaller than that in an isolated hydrogen atom. In other words, a hydrogen atom bonded to carbon is *deshielded*, compared to an isolated hydrogen atom. Clearly, the exact amount of deshielding is related to the electron distribution in the bond joining the hydrogen atom to the rest of the molecule. Thus, through the operation of the inductive mechanism in chemical bonding, one would expect the

FIGURE 12.6. *Shielding of the proton by the electron in an isolated hydrogen atom.*

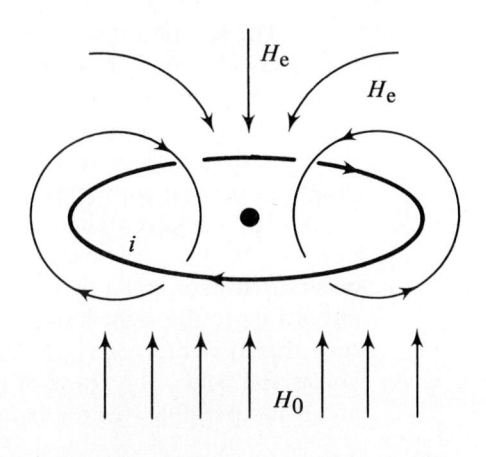

protons of methane to be more shielded than those of methyl chloride, and therefore to resonate at a higher field (closer to TMS). This is, in fact, borne out by experiment:

Compound	$\delta\ ^1H$ (ppm from TMS)
CH_4	0.23
CH_3Cl	3.05 (2.82 ppm downfield of CH_4)
CH_2Cl_2	5.33 (2.28 ppm downfield of CH_3Cl)
$CHCl_3$	7.24 (1.91 ppm downfield of CH_2Cl_2)

This series also shows that the effect of increasing electron withdrawal on the chemical shift of the remaining proton(s) is cumulative, but not strictly additive.

As is typical of all inductive effects, this type of deshielding decreases rapidly with increasing distance from the electronegative atom. Thus, the methyl group of ethyl chloride resonates at $\delta = 1.33$ ppm. In general, factors influencing electron density in the proximity of the proton are reflected in the chemical shift. Electron deficiency is associated with deshielding and therefore results in downfield shifts from TMS.

Because the electron density at a carbon atom is much more directly influenced by the presence of substituents than is the electron density at a hydrogen atom, one would expect more dramatic effects from ^{13}C NMR spectra:

Compound	$\delta\ ^{13}C$ (ppm from TMS)
CH_4	−2.1
CH_3Cl	23.8 (25.9 ppm downfield from CH_4)
CH_2Cl_2	52.9 (29.1 ppm downfield from CH_3Cl)
$CHCl_3$	77.3 (24.4 ppm downfield from CH_2Cl_2)

This is indeed evident from the above series. In addition, the electron distribution about a carbon atom can be profoundly influenced by changes in hybridization, whereas the hydrogen atom is always surrounded by an electron in the same type of orbital. This is shown dramatically by comparing the chemical shifts of protons and carbons in ethane and ethylene:

Compound	$\delta\ ^1H$ (ppm from TMS)	$\delta\ ^{13}C$ (ppm from TMS)
CH_3CH_3	0.86	7.26
$CH_2=CH_2$	5.25	122.1

The second major effect, in addition to deshielding by bonding, that governs the chemical shifts of protons is the influence of *magnetically anisotropic* neighboring groups. A group (which can be a bond or the environment of an atom, but which is here considered to be a collection of electrons) is said to be magnetically anisotropic if the circulation of electrons within it under the influence of a magnetic field depends on its orientation with respect to this field.

For instance, when a molecule of benzene is oriented with respect to the magnetic field H_0 as shown in Figure 12.7, a movement of the delocalized π electrons occurs (left-hand rule), which is known as the *ring current*. This current generates (according to the right-hand rule) a subsidiary magnetic field H_e whose direction is such that it reinforces H_0 at the periphery of the benzene ring while opposing H_0 above and below the plane of the benzene ring. Thus, the aromatic protons, which are at the periphery of the benzene ring, are deshielded, and are found to resonate

FIGURE 12.7. *Induced circulation of π electrons in the benzene ring.*

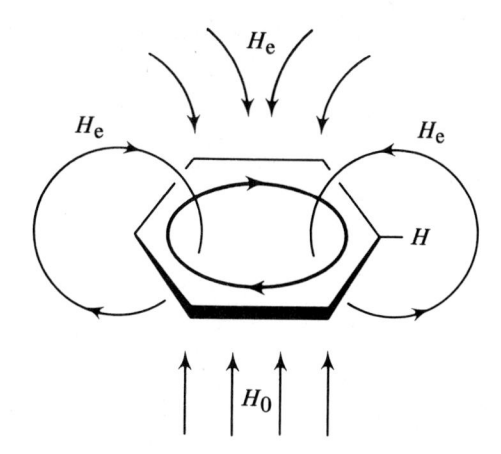

at a field considerably lower than that expected solely on the basis of the electron-density distribution.

When the benzene ring assumes, with respect to the field H_0, an orientation orthogonal to that shown in Figure 12.7, little circulation of electrons takes place, and so the net effect results only from the phenomenon illustrated in the figure. It can be demonstrated, by an extension of such arguments, that magnetically *isotropic* groups will not exert any net shielding effects on neighboring magnetic nuclei, because thermal motion will average all shielding influences.

Magnetically anisotropic groups can be considered to be surrounded by volumes of space in which protons will be shielded (+) or deshielded (−), that is, moved upfield or downfield, respectively. The best-established effects are associated with the groups shown in Figure 12.8.

It can be seen that *both* of the effects discussed above depend on the circulation of electrons in the magnetic field of the spectrometer. Their magnitude will therefore be directly proportional to the spectrometer field, H_0, or the "spectrometer frequency." This is the physical rationalization of the statement made earlier, that the chemical shift expressed in hertz depends on the spectrometer frequency.

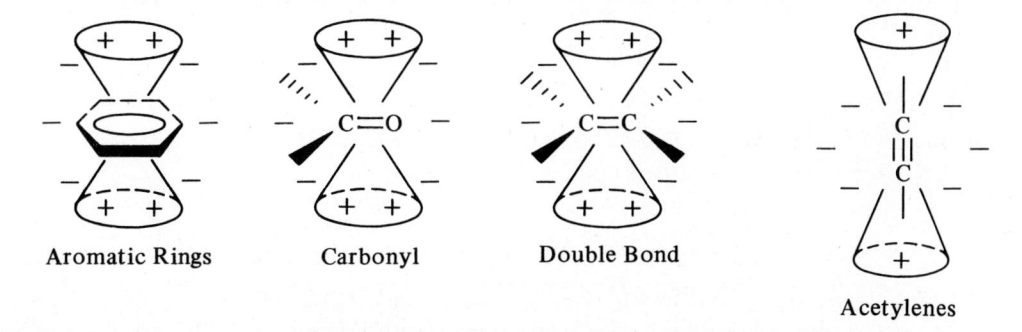

Aromatic Rings Carbonyl Double Bond Acetylenes

FIGURE 12.8. *Shielding zones associated with some magnetically anisotropic groups.*

SEC. 12.2 The Chemical Shift

The chemical shift of any given nucleus depends on the combination of effects, which are (roughly) additive; the effects reinforce or cancel one another. Thus, acetylenic protons are deshielded by the inductive effect (acetylene is acidic) but shielded by the anisotropy of the triple bond; a value of $\delta = 1.80$ ppm results.

Because it is almost impossible to disentangle the various contributory effects, the theory of chemical shifts can be used only as a general guide. For the solution of problems, empirical correlations are almost invariably used. Some of the more fundamental of these are given in Tables 12.1 and 12.2.

The choice of solvent is important. Since the standard (TMS) and the sample are in the same environment, one would expect negligible solvent effects. However, different solvents, which may have different degrees of magnetic anisotropy, will generally interact with various molecules in different ways, and the molecules will on the average be oriented in some preferred manner. Therefore, solutions used for accurate measurements of chemical shifts should be as dilute as possible (preferably less than 10%) to avoid solute-solute interactions, and the solvent should not interact strongly with the sample (as do, e.g., hydrogen-bonding solvents). Carbon tetrachloride is a preferred solvent because it is magnetically isotropic and has no sites for strong interactions. Furthermore, CCl_4 has no protons; so there will be no blanked-out areas in the spectrum. The most commonly used solvent is deuterochloroform ($CDCl_3$), whose dissolving power for most compounds is greater than that of CCl_4 and which is also proton free. Chemical shifts in CCl_4 and $CDCl_3$ are generally very similar. A further advantage of $CDCl_3$ arises from the fact that the signal from the deuterium atom can be used as the "lock" signal. This is particularly useful for ^{13}C spectroscopy, and the signal from the carbon atom in $CDCl_3$ (a triplet, centered at 77 ppm downfield from TMS) provides a secondary internal standard.

TABLE 12.1 *Chemical-Shift Data for Protons (δ Scale)*

Aliphatic protons (cyclic or acyclic, excluding cyclopropane derivatives):
  Methyl (with only H or alkyl substituents on both α and β carbon): 0.9
  Methylene (with only H or alkyl substituents on both α and β carbon): 1.25
  Methine (with only H or alkyl substituents on both α and β carbon): approx. 1.6

Presence of electron-withdrawing substituents on the α carbon (e.g., halogens, —OH, —OR, —O—CO—R, —NH$_2$, —NO$_2$) shifts the proton by 2–4 ppm downfield. Carbonyl groups, C=C, and aromatic rings have a similar but less pronounced effect, the downfield shift being generally about 0.5–1.5 ppm.

Benzylic protons: 2–3. Toluene methyl: 2.34

Acetylenic protons: 2–3

Olefinic protons: 5–7, varying regularly with substitution. Ethylene: 5.30

Aromatic and heterocyclic protons: 6–9. Benzene: 7.27

Aldehydic protons: 9–10

Hydroxylic and amino protons: Anywhere between 1 and 16 ppm, depending on the state of hydrogen bonding (strong hydrogen bonding is deshielding). Signals due to such protons may be easily recognized by shifts with temperature, which alters the degree of hydrogen bonding, and by the facile exchange of the protons with D$_2$O. The latter procedure can be carried out in an NMR sample tube, and the signals due to —OH, —NH$_2$, etc., simply vanish.

TABLE 12.2 *Chemical-Shift Data for* ^{13}C *(δ Scale)*

Aliphatic carbons (cyclic or acyclic, excluding cyclopropane derivatives):
  Methyl (with only H or alkyl substituents on both α and β carbons): 5–30
  Methylene (with only H or alkyl substituents on both α and β carbons): 20–45
  Methine (with only H or alkyl substituents on both α and β carbons): 30–60
  Quaternary carbons (with only H or alkyl substituents on both α and β carbons): 30 50
  CH_3-O- : 50–60

  $CH_3-N\langle$: 15–45

Acetylenic carbons, C≡C: 75–95

Ethylenic carbons, C=C: 105–145. Ethylene: 122

Aromatic carbons: 110–150. Benzene: 128

Heteroaromatic carbons: 105–165

Carbonyl carbons, C=O
  Ketones and aldehydes: 185–225
  Acids, esters, and amides: 160–185

In summary, the phenomenon of chemical shift enables the chemist to obtain some fundamental information about electronegativities, bond anisotropies, and so on. Above all, the ability to observe the chemical shift causes hydrogen and carbon atoms (and to some extent other nuclei) to become *functional groups* that can be qualitatively and quantitatively estimated.

12.3 TIME DEPENDENCE OF NMR PHENOMENA

The time scale of the NMR phenomenon is best understood when it is recalled that NMR transitions occur at the low-frequency end of the electromagnetic spectrum.

Consider two protons situated in different environments. They will give rise to two separate resonances in the NMR spectrum, say Δv hertz apart (Fig. 12.9C). However, if, by one of the mechanisms discussed below, the two protons exchange their environments at a rate *faster* than Δv times per second, only one signal is obtained, at an intermediate frequency (Fig. 12.9A); the two nuclei are *equivalent* on the NMR time scale.

The definition of equivalence is important in NMR. A group of nuclei is defined as *chemically equivalent* if they possess the same chemical shift. Thus, by symmetry, the six protons of the benzene molecule are inherently chemically equivalent. However, the three protons of a methyl group are chemically equivalent only by virtue of the normally fast rotation about the bond joining the methyl group to the rest of the molecule. A group of nuclei is *magnetically equivalent* when they have not only the same chemical shift but also the same spin-spin coupling (see below) to all nuclei outside the group.

NMR spectra are characteristic of exchange rates. Thus, at slow (on the NMR time scale) exchange rates, one can simply observe separate signals for each

FIGURE 12.9. *Effect of exchange rates (k) on the appearance of NMR spectra.*

Fast Exchange
($k \gg \Delta\nu$)

A

Intermediate
Exchange
($k \approx \Delta\nu$)

B

Slow Exchange
($k \ll \Delta\nu$)

C

|←$\Delta\nu$→|

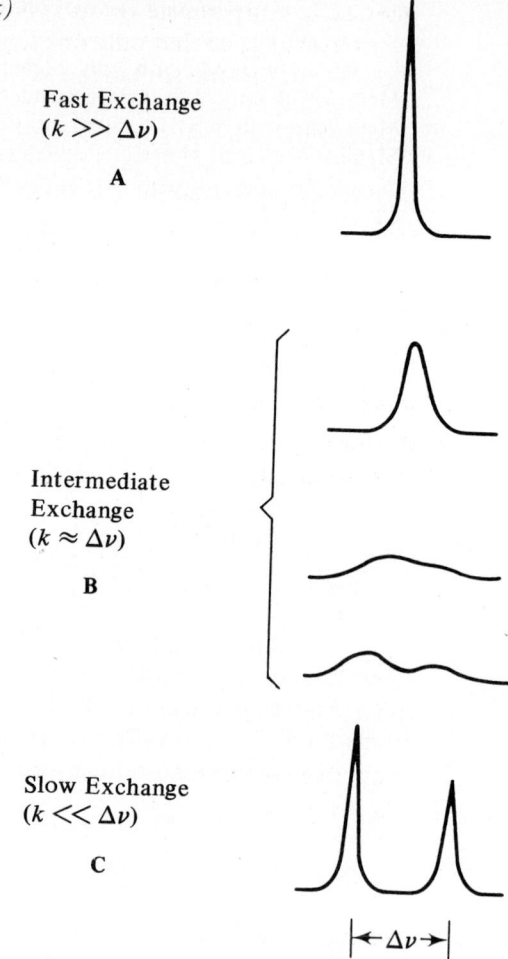

of the environments and estimate the relative populations at each site. At intermediate exchange rates, characteristically broadened spectra are observed (Fig. 12.9B) from which information about the rate of the process taking place can be extracted. At high exchange rates, the single averaged signal occurs at a frequency determined by the relative populations at each site. Given the characteristic frequency of the individual resonances from the slow-exchange case (typically from low-temperature spectra), it is possible to determine the relative populations at two sites from the averaged (typically high-temperature) spectra.

The most common mechanisms for averaging the environments of protons, or of groups of equivalent protons, that can be observed on the NMR time scale are proton exchange, conformational changes, and rotation about partial double bonds. An example of each follows.

1. *Proton exchange.* In dilute solutions in aprotic solvents, the hydroxylic protons of mixtures of ethanol and water give rise to separate signals. However, an

increase in temperature or concentration, or a change in pH, speeds up the prototropic exchange so that only one signal for the –OH protons is observed.

The chemical shift of the –OH protons of just ethanol in an aprotic solvent will also vary with concentration and temperature, because of different degrees of hydrogen bonding. Furthermore, at low exchange rates the –OH signal of ethanol shows splitting due to spin-spin coupling with the methylene protons (see below); but at high exchange rates it gives rise to a singlet because the methylene protons "see" only the average spin state of the –OH protons.

2. *Conformational changes.* At room temperature, the NMR spectrum of cyclohexane consists of a single sharp line, because the rate of conformational inversion between the two equivalent chair forms, which is associated with the interchange between axial and equatorial positions, is fast compared to the difference (in Hz) between the chemical shifts of axial and equatorial protons. At about $-160°C$, this inversion slows down enough to make separate signals for the axial and the equatorial protons observable.

3. *Rotation about partial double bonds.* At low temperatures, the signals caused by the *N*-methyl groups of *N,N*-dimethylformamide

appear as two bands of equal intensity. At higher temperatures, they coalesce to a single band midway between the original signals.

12.4 SPIN-SPIN COUPLING

Many signals in PMR spectra exhibit fine structure because of the splitting of spin-state energy levels of the protons by other magnetic nuclei in the neighborhood.

As an analogy, consider pairs of small bar magnets constrained into a specific spatial relation to each other (as nuclei are in real molecules) and that can assume just two north-south directions (as nuclei of spin $I = 1/2$ do). It is immediately obvious that, for the relative arrangements depicted in Figure 12.10, the antiparallel arrangement **A** has a lower energy than the parallel arrangement **B**.

In the case of the nuclear magnets, the interaction does not take place through space, but through the agency of the binding electrons. The strength of this interaction is expressed by the parameter J, the *coupling constant*, which is related to the degree of splitting of resonances and which is expressed in hertz. The value of J depends only on the electronic and steric relationship between the interacting protons, and hence does not depend on the spectrometer frequency. Thus, obtaining the NMR spectra of the same compound at two frequencies allows one to distinguish between multiple lines caused by protons of different chemical shift and those caused by the splitting of energy levels as a result of spin-spin coupling.

FIGURE 12.10. *Bar-magnet analogy for spin-spin coupling.*

A detectable interaction between protons takes place only across a limited number of bonds. In general, no significant spin-spin coupling is observed between protons separated by more than four σ bonds (or by more than four σ bonds and one π bond).

Two separate problems are involved in obtaining chemically useful information from spin-spin coupling; first, the multiplets in question must be analyzed so as to yield the values of J and the chemical shifts of the interacting protons; and, second, these values must be interpreted in terms of molecular structure.

Analysis of Spin-Spin Multiplets

In principle, any set of multiplets due to a number of interacting protons (or groups of equivalent protons) can be analyzed by computerized quantum-mechanical calculations. Such calculations are always tedious and often difficult; fortunately, the multiplets can be analyzed by direct measurement in a large number of cases. Such spectra are known as *first-order spectra*; we shall deal with the analysis of first-order spectra and with some general features of the more complex *second-order spectra*.

The parameter that determines whether a group of protons (a *spin system*) will give rise to a first- or second-order spectrum is the ratio $\Delta v/J$, of the chemical-shift difference between the relevant protons and the coupling constant J between them, both expressed in hertz. Large $\Delta v/J$ ratios, indicating *weakly coupled* systems, are associated with simple first-order spectra. For a true first-order spectrum, *all* $\Delta v/J$ ratios within the spin system must be large. Quite often a spin system contains sets of both strongly and weakly coupled nuclei, and straightforward application of first-order rules to such cases will lead to errors in analysis. For most purposes, $\Delta v/J \geq 3$ can be considered "large."

Because chemical shifts, and therefore Δv, increase with increasing strength of the magnetic field whereas coupling constants do not, it follows that spectra taken at higher frequencies are easier to interpret. This, together with increased dispersion, is the principal reason that, despite the considerably greater expense, spectrometers operating at ever-higher frequencies are being built. The limits of field strengths obtainable with reasonably sized permanent magnets or iron-core electromagnets have apparently been reached. The latest high-field NMR instruments use superconducting (liquid helium–cooled) solenoids and reach fields corresponding to an operating frequency of up to 600 MHz for protons.

There are certain conventions used in naming spin systems. The letters $A, B, C, D, \ldots$ are used to describe groups of protons whose chemical-shift differences are small compared to the values of their coupling constants, that is, strongly coupled sets. Subscripts are used to give the number of protons in an equivalent group. A break in the alphabetical sequence shows which groups are weakly coupled. For example, writing A$_2$BMXY describes a six-spin system. The two A nuclei and the B nucleus are strongly coupled to each other but only weakly coupled to the nuclei M, X, and Y. The nucleus M is weakly coupled to all other nuclei. The nucleus X is strongly coupled to the nucleus Y but weakly coupled to the other nuclei.

Primes are used to denote protons that are chemically equivalent but are not coupled identically to other protons and therefore are not magnetically equivalent.

The A$_n$X$_m$ system will give rise to a first-order spectrum (A yielding $m + 1$ lines and X yielding $n + 1$ lines, with all spacings equal to J_{AX}), and the A$_n$B$_m$ system will give a complex spectrum. Fortunately, the number of spins in a spin system, and hence its complexity, is limited by the rapid attenuation of J with the number of bonds separating the coupled nuclei.

The rules for interpreting first-order spectra are as follows.

1. When a proton (or a group of magnetically equivalent protons) is spin-spin coupled to n equivalent protons with a coupling constant of J hertz, its NMR signal is split into $n + 1$ lines* separated by J hertz. The relative intensities of the lines are in the ratio of the binomial coefficients of $(x + y)^n$. The true chemical shift of the protons concerned lies at the center of the multiplet.

Splitting by one proton therefore results in a doublet of equal intensity; splitting by two protons results in a triplet of relative component intensities of 1:2:1; splitting by three protons results in a quartet of relative component intensities of 1:3:3:1; splitting by four protons results in a quintet of relative intensities of 1:4:6:4:1; and splitting by six protons results in a septet of relative component intensities of 1:6:15:20:15:6:1. (Compare Table 13.2.)

2. If there are more than two interacting groups of protons (A$_n$M$_m$X$_p$...), the multiplicity of the signal due to the A protons is given by $(m + 1)(p + 1)...$; that is, the part of the spectrum due to nuclei A takes the form of a multiplet of submultiplets. Note that the number n does not enter into the expression. Clearly, the appropriate J-values control splittings.

3. In first-order spectra, equivalent protons appear not to split one another; in other words, the transitions corresponding to such interactions are forbidden, or of zero probability. However, interactions between equivalent protons do take place and the corresponding coupling constants can be obtained from some second-order spectra.

The physical basis of the first-order rules is quite clear. Consider a system of two protons, H$_A$ and H$_X$, and let the two allowed spin states be α (high energy) and β (low energy). Then, for upward transitions of the nucleus H$_A$, we can have

H$_A\beta$ to H$_A\alpha$ with H$_X$ in state α and   H$_A\beta$ to H$_A\alpha$ with H$_X$ in state β

Since the populations of H$_X$ in the α and β states are almost completely equal (recall

---

*More generally, splitting by a nucleus of spin $= I$ gives $2nI + 1$ lines.

the vanishingly small Boltzmann excess discussed previously), the two transitions are of equal probability, and hence H_A will give rise to a symmetrical doublet.

Similarly, for a system of three spins, one H_A and two H_X, we can have the following upward transitions for H_A:

$H_A\beta$ to $H_A\alpha$ with the first H_X in state α and the second in state β
$H_A\beta$ to $H_A\alpha$ with the first H_X in state β and the second in state α
$H_A\beta$ to $H_A\alpha$ with both H_X nuclei in state α
$H_A\beta$ to $H_A\alpha$ with both H_X nuclei in state β

The first two transitions are equivalent (*degenerate*), and hence H_A will give rise to a triplet with the intensity ratio 1:2:1.

However, this sort of reasoning is not the full theoretical treatment for the system; the full treatment merely reduces to this description for cases where $\Delta v/J$ assumes large values, in other words, for first-order spectra.

We shall now deal with the spectral characteristics of some commonly encountered spin systems.

Two-Spin Systems. A two-spin system can, by definition, be either an AX spectrum (i.e., a doublet of equal-intensity lines for H_A centered on the chemical shift δ_A of H_A with a separation of J_{AX}, and an identical doublet centered on δ_X) or a second-order AB spectrum.

The AB spectrum also consists of two doublets whose separations are exactly equal to J_{AB}, but the "inner" lines are more intense than the "outer" lines. The chemical shifts of H_A and H_B are given by the following expression:

$$v_A - v_B = \sqrt{(1-4)(2-3)} \qquad (12.10)$$

where $v_A - v_B$ is the separation of the chemical shifts of H_A and H_B, and the numbers refer to the frequencies of the lines as marked in Figure 12.11. Once $v_A - v_B$ (in Hz)

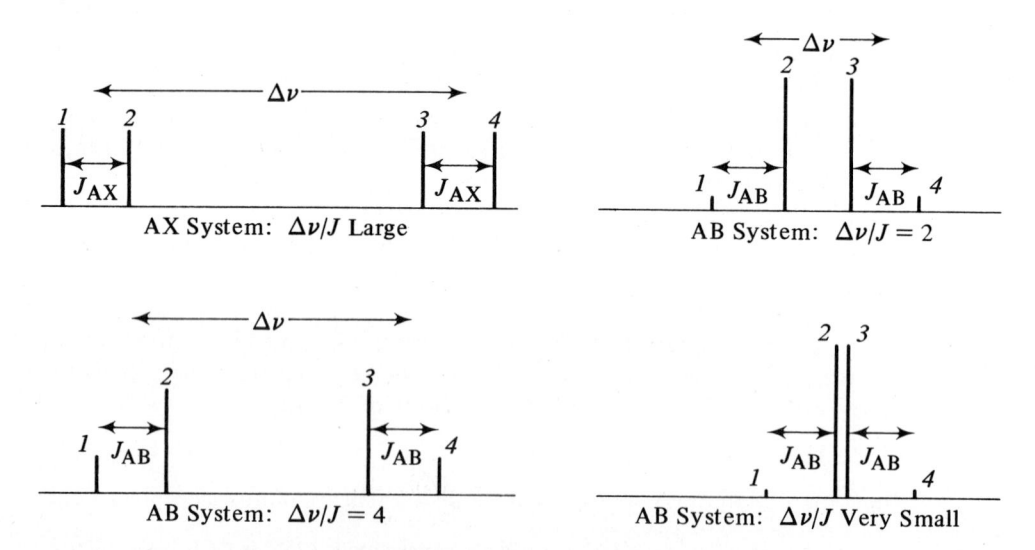

FIGURE 12.11. *Calaulated spectra of two-spin systems.*

has been determined, δ_A and δ_B can be located by measuring from the center of the always perfectly symmetrical AB systems (often referred to as an *AB quartet*).

As δ_A and δ_B become more nearly identical, the intensities of the inner lines increase at the expense of the outer lines until, at the limit of $v_A - v_B = 0$, the transitions corresponding to the outer lines become forbidden and the system reduces to a singlet of two-proton intensity, that is, the trivial A_2 case.

The characteristic of a part of the spectrum "sloping" away from the position of the other part is common to all spectra that are not strictly first order; practically speaking, this means nearly all observable spectra.

Three-Spin Systems. A system of three protons can always be described by no more than six parameters. Thus, the first-order AMX case is described by δ_A, δ_M, δ_X, J_{AM}, J_{AX}, and J_{MX} (obviously, J_{AX} and J_{XA} are the same). In this first-order case, one observes 4 lines for each proton (a doublet-of-doublets) with separations corresponding to the coupling constants with the other nuclei, for a total of 12 lines. Many experimental spectra, such as the one shown in Figure 12.12, approach the ideal AMX case; that is, the directly measured line spacings are very close to the true

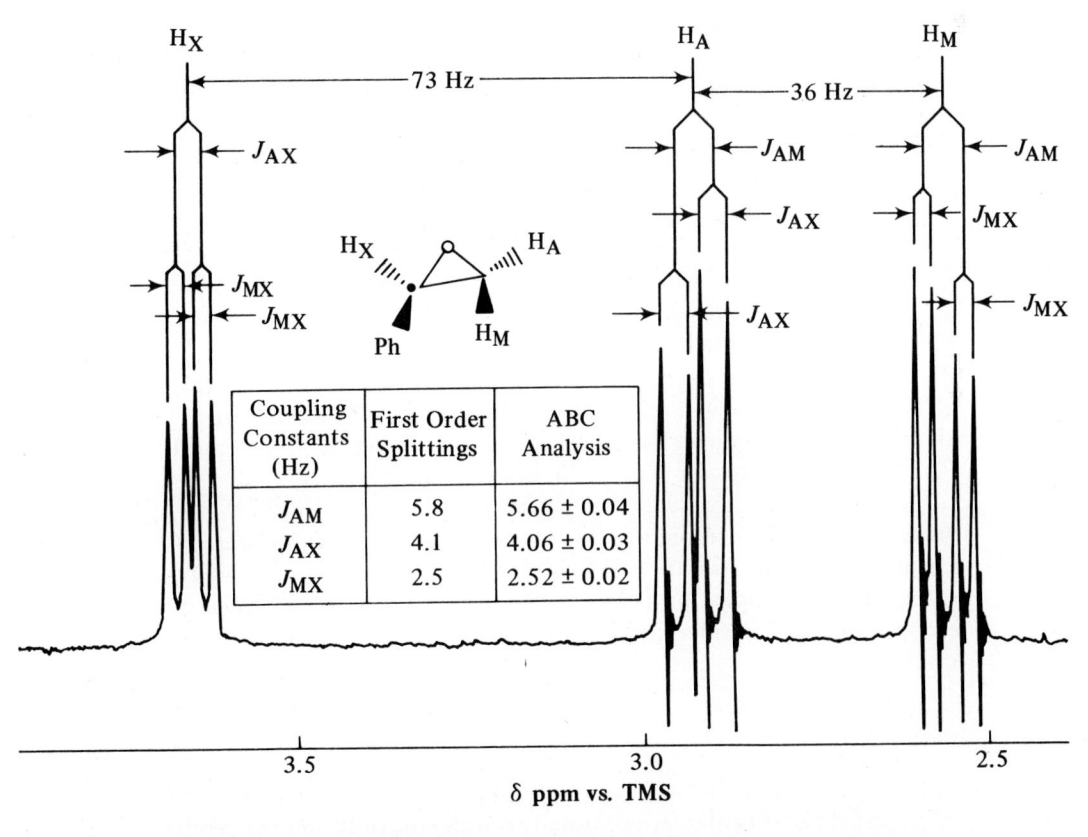

Coupling Constants (Hz)	First Order Splittings	ABC Analysis
J_{AM}	5.8	5.66 ± 0.04
J_{AX}	4.1	4.06 ± 0.03
J_{MX}	2.5	2.52 ± 0.02

FIGURE 12.12. *100-MHz spectrum of styrene oxide (25% in CCl$_4$). The part of the spectrum due to aromatic protons is not shown.*

coupling constants, as computed by the appropriate quantum mechanics ("ABC analysis"). Note that even here the intensities depart from the first-order ideal, where all lines within each doublet-of-doublets should be equal. Thus, the lines in the H_A and H_M multiplets slope toward each other (cf. the AB case above). The lines of the H_X multiplet appear broader because of small additional unresolved coupling to the protons on the phenyl ring.

If all the protons have similar chemical shifts and are coupled—that is, in the ABC case—a spectrum of up to 15 lines results, which is so distorted that it is often not possible to recognize it as such. A very common system is the partially strongly coupled ABX case, which must not be analyzed as an AMX case. Where two of the protons are equivalent, one can get an A_2X or an A_2B system.

By first-order rules, the A_2X system gives rise to a doublet of two-proton intensity and a triplet of one-proton intensity. The A_2B spectrum may have up to nine lines and can be highly asymmetrical.

Common Errors in Analysis of NMR Spectra

The obvious effects associated with second-order spectra (e.g., extra lines, distorted intensity patterns, and unequal spacings) generally preclude any injudicious attempts to analyze such systems by first-order rules. However, in some cases second-order spectra have features that are qualitatively indistinguishable from some features of first-order spectra, and so are often misinterpreted. It must be understood that the three cases discussed below are not physical phenomena; they are simply the result of certain combinations of the chemical-shift and spin-coupling parameters.

Partially Strongly Coupled Spectra. The X portion of an ABX spectrum gives rise to four lines, which are often regularly spaced and appear identical to the X portion of an AMX spectrum. However, the spacings *cannot* be used to obtain the values of J_{AX} and J_{BX}, although the distance between the outer lines does correspond to $J_{AX} + J_{BX}$. When the AB portion of the spectrum can be resolved clearly, no misinterpretation should result, because it is more complex than the AM portion of an AMX spectrum. However, when only the X portion is visible (e.g., when the remainder is hidden by overlapping resonances), the problem is not simple and the possibility of a partially strongly coupled system must be considered.

Virtual Coupling. A resonance due to a proton may be complicated (split) because of a proton which is *not* directly coupled to it, but which is strongly coupled to a proton which *is* coupled to it. This "phenomenon" is really a special case of the trap described above. Consider, for example, a linear system of three protons:

$$-\overset{|}{\underset{H_A}{C}}-\overset{|}{\underset{H_B}{C}}-\overset{|}{\underset{H_C}{C}}-$$

Although protons H_A and H_C are usually not significantly coupled (they are separated by four σ bonds), the resonance due to H_A may not be a simple doublet with spacing J_{AB} if H_B and H_C are strongly coupled, that is, if at the spectrometer frequency used,

$(v_B - v_C)/J_{BC}$ is a small number. Clearly, mistakes are most likely to occur if the B and C portions of the spectrum cannot be discerned, as in the case discussed above.

Deceptive Simplicity. Sometimes the combination of parameters is such that a deceptively simple spectrum results. Consider the spectrum of furan:

$$J_{AB} \neq J_{AB'}$$

The 60-MHz spectrum of this compound gives rise to two triplets, suggesting an A_2X_2 case, whereas symmetry considerations show that it should give rise to an AA'XX' or AA'BB' spectrum. Deceptive spectra should always be suspected when a first-order analysis appears to yield a number of apparently equal coupling constants while structural considerations suggest coupling constants of widely different magnitudes. Thus, the incorrect analysis of the spectrum of furan leads to a postulation of equal *ortho-* and *meta-* coupling constants. More sophisticated analysis shows that this is not the case, as should be expected on structural grounds.

Signs of Coupling Constants

Coupling constants have sign (+ or −) as well as magnitude. By convention, the sign of the coupling constant between two nuclei is taken to be positive if the state with the two spins in an antiparallel orientation is of lower energy. The relative signs of coupling constants cannot be obtained from first-order spectra but may be determined from more strongly coupled spectra and from some multiple-resonance experiments (see below, Special Topics). The absolute signs of coupling constants cannot be obtained from spectral analysis because the reversal of all signs leaves the spectrum unchanged. Many absolute signs have been determined from the NMR spectra of compounds in a nematic (partially oriented) phase.

Interpretation of Spin-Spin Coupling in Terms of Structure

The magnitudes of coupling constants (J) are very characteristic of molecular environment and are especially sensitive to stereochemistry. Furthermore, multiplicities of resonances can give information about the number of neighboring protons. The theory of spin-spin coupling is far too complex to be routinely used and empirical correlation tables are invariably resorted to. The common correlations given below are well established. All coupling constants are quoted as their absolute magnitudes only.

Typical range: 12–18 Hz

The full range is 0 to 22.4 Hz. Double bonds adjacent to the central carbon give larger values (e.g., an aromatic ring or a carbonyl group for R_1). Smaller values are observed when R_1 is a heteroatom.

Vicinal Coupling across Three Single Bonds.

$$H_A-C-C-H_B$$

The magnitude of J_{AB} is dominated by the size of the dihedral angle (ϕ) and is given by the Karplus equation:

$$J_{AB} = J^0(\cos^2\phi) - 0.3 \quad \text{for angles } 0\text{–}90°$$
$$J_{AB} = J^{180}(\cos^2\phi) - 0.3 \quad \text{for angles } 90\text{–}180°$$

(12.11)

The values of the constants J^0 and J^{180} are substituent dependent, with the ranges $J^0 = 9\text{–}12$ and $J^{180} = 14\text{–}16$ covering most situations. Within the variations caused by substituents, those due to the cases

$$H-\overset{|}{\underset{|}{C}}-\overset{|}{\underset{|}{C}}-H, \quad H-\overset{\|}{C}-\overset{|}{\underset{|}{C}}-H, \quad \text{and} \quad H-\overset{\|}{C}-\overset{\|}{C}-H$$

can usually be ignored. Typical values for freely rotating methyl and methylene groups are 6 to 8 Hz. The Karplus relation has obvious importance in determining the stereochemistry of organic compounds and is summarized graphically in Figure 12.13.

Olefinic Systems. Typical values are as follows:

$J_{cis}\ (J_{AB}) = 6\text{–}14$ Hz   cyclohexene 10 Hz

$\underset{H_CR}{\overset{H_BH_A}{C=C}}$ $J_{trans}\ (J_{AC}) = 11\text{–}18$ Hz   cyclopentene 6 Hz

$J_{gem}\ (J_{BC}) = 0\text{–}3$ Hz   cyclobutene 4 Hz

Electronegative substituents (i.e., when R is a heteroatom) lead to smaller values of olefinic coupling constants.

Long-Range Coupling. This is defined as coupling across more than three bonds. Long-range coupling constants are rarely larger than 3 Hz but may be highly charac-

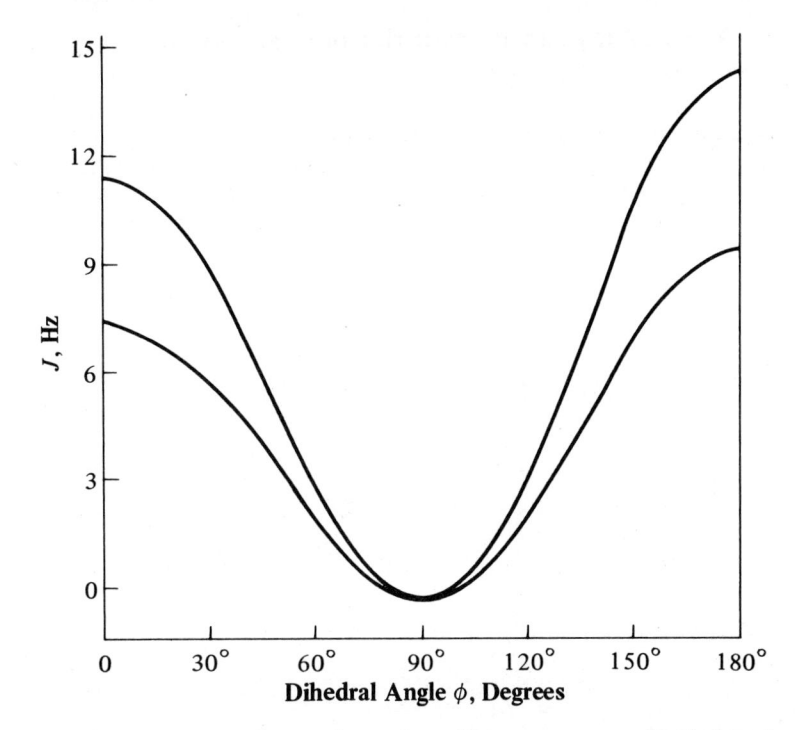

FIGURE 12.13. *Dependence of vicinal coupling constants on dihedral angles.*

teristic of structure. The most common type of long-range interactions is *allylic coupling*, which is due to the protons in H—C—C=C—H, and which is highly dependent on stereochemistry.

Homoallylic coupling, that is, the coupling across five bonds in the fragment H—C—C=C—C—H, takes up a slightly larger range of values and also has a characteristic stereochemical dependence. In general, coupling between two protons separated by four single bonds becomes significant ($J = 1$ to 3 Hz) only if the five atoms of the system H—C—C—C—H take on a planar "W" (or "M") arrangement.

Aromatic Systems. Typical values in benzenoid compounds are:

J_{ortho} = 7–10 Hz
J_{meta} = 1–3 Hz
J_{para} = 0–1 Hz

Similar ranges apply for heterocyclic systems, except that the J_{ortho} involving protons on carbons α to a heteroatom takes on lower values (cf. olefinic systems) and that a ring-size dependence of J_{ortho} (analogous to that in cycloölefins) is also observed. Such influences are cumulative, and so $J_{\alpha,\beta}$ in furans is only 1 to 3 Hz.

SEC. 12.4 Spin-Spin Coupling

12.5 NMR SPECTROSCOPY OF NUCLEI OTHER THAN PROTONS

By accident, almost all the nuclei of interest to the vast majority of chemists have spins $I = 1/2$ and hence do not differ at all from protons in their basic theoretical aspects. However, their usefulness does not warrant the effort of learning any empirical parameters relating to them, so the remarks below do not reflect the amount of data available.

It must be realized that the effect of spin-spin coupling of protons to other magnetic nuclei may be observed in the *proton* spectra, and hence some idea of the magnitudes of coupling between protons and some commonly occurring magnetic nuclei may be useful in interpreting proton spectra.

^{19}F Spectra

Fluorine resonates over a range of some 300 ppm; that is, its chemical shift is more sensitive than that of ^{1}H to the changes of environment. In saturated systems, J_{H-F} for H—C—F (geminal) ranges from 40 to 80 Hz and for H—C—C—F (vicinal) from 0 to 30 Hz. The latter has a Karplus-like dependence on stereochemistry (Eqn. 12.11).

^{31}P Spectra

Phosphorus resonates over about 400 ppm. Parameter J_{P-H} (direct), as in phosphine derivatives, is in the range 200 to 700 Hz, that is, of an entirely different order of magnitude from interproton coupling constants. The J_{H-P} coupling constants for H—C—P, H—C—C—P, and H—C—O—P (as in phosphate esters) vary between 0 and 30 Hz. In phosphate esters, it is generally in the range of 5 to 20 Hz and shows a Karplus-like stereochemical dependence.

^{13}C Spectra

The natural abundance of ^{13}C is only 1%, and thus ^{13}C spectra are difficult to observe in unenriched samples. A further disadvantage is that ^{13}C is a "less good magnet" than a proton. The overall loss of sensitivity compared to ^{1}H is approximately 6000-fold. However, with the advent of FT spectroscopy, it is now possible to obtain ^{13}C spectra routinely.

The basic ^{13}C NMR spectrum is the *fully decoupled* spectrum, that is, one where all spin-spin interactions with protons have been removed by multiple irradiation (see p. 383). Such a spectrum consists simply of a number of singlets corresponding to carbons in different chemical environments. Because the chemical shifts of ^{13}C cover over 200 ppm (Table 12.2), and fall into very characteristic ranges, the structural implications are considerable. Furthermore, by using the "off-resonance decoupling technique," the direct coupling between carbon and protons may be diminished in magnitude, but not completely eliminated, permitting one to distinguish methyl carbons (quartet), methylene carbons (triplet), methine carbons (doublet), and tertiary carbons (singlet), by inspection.

Unfortunately, quantitative applications of CMR spectroscopy are less straightforward than those of PMR because, under the usual experimental conditions, the relative intensities of ^{13}C signals do not accurately reflect their relative abundances. This is due principally to relaxation phenomena and makes proton rather than carbon NMR the method of choice for the analysis of mixtures.

12.6 SPECIAL TOPICS

We will now consider some ways to aid or simplify the NMR experiment and some commercially available instrumentation.

Deuterium Substitution

Substitution of deuterium (D, or $^{2}_{1}H$) for protium ($^{1}_{1}H$) tends to simplify NMR spectra in two ways. First, it removes the part of the spectrum that is due to the replaced proton(s); second, it simplifies the remainder because, although deuterium is magnetic and will split the resonances of the remaining protons, the coupling constants between ^{1}H and ^{2}H are only about one-seventh of the corresponding coupling constants between ^{1}H and ^{1}H. For example, whereas the methyl resonance of ethanol is a triplet (X part of A_2X_3), the methyl resonance of CH_3CD_2OH – is a slightly broadened singlet.

Spin Decoupling

It is possible to introduce one (or more) irradiating radio-frequencies into the transmitter coil of an NMR spectrometer, thus generating one or more perturbing magnetic fields in addition to H_1. Such experiments are known as *double-* (or *multiple-*) *irradiation* experiments and give rise to *double-* (or *multiple-*) *resonance spectra*. *Spin decoupling*, in which the second (and further) fields are relatively strong and are directed at the resonances of nuclei coupled to the nuclei being observed, is a special case and is the most common experiment of this type.

If the resonance from proton A (which is coupled to proton B) is observed while simultaneously proton B is strongly irradiated at its resonant frequency, the normal doublet expected of H_A (half of an AB quartet) will collapse to a singlet. This results from the fact that the second irradiating field causes rapid transitions of H_B between its two spin states, so that H_A experiences only the averaged spin state of H_B and hence no splitting in its energy levels results. Clearly, this phenomenon is related to the time dependence of NMR.

The capacity of spin decoupling in simplifying spectra is obvious. However, strongly coupled protons cannot be decoupled because the introduction of the second RF field perturbs the region near the field and hence makes the resonances impossible to observe.

Shift Reagents

In 1970, Williams, elaborating on the preliminary work of Hinckley, discovered that adding *tris-β-diketonate* lanthanide complexes

$$\text{Lanthanide} \left(\begin{array}{c} O \\ \diagdown \\ O \end{array} \begin{array}{c} R_1 \\ R_2 \\ R_3 \end{array} \right)_3$$

to solutions of substances with lone pairs of electrons available for coordination (e.g., oxygen- and nitrogen-contaning organic compounds) resulted in vastly more dispersed NMR spectra. This phenomenon, called *lanthanide-induced shift* (LIS), arises from the unusual combination of paramagnetic properties in most lanthanides, whereby large local changes in the magnetic fields are produced in the immediate vicinity of the lanthanide. This type of induced shift is known as *pseudo contact shift* and is propagated through space in a normal manner, in contrast to the *contact shifts* found with other paramagnetic species where the effect is transmitted through bonds.

Because lanthanide-induced shifts often have values of up to 20 ppm, the increased dispersion, and hence interpretability, of NMR spectra afforded by this method is enormous. Moreover, pseudo contact shifts diminish regularly with the distance from the paramagnetic center—r in Figure 12.14—and since this can be

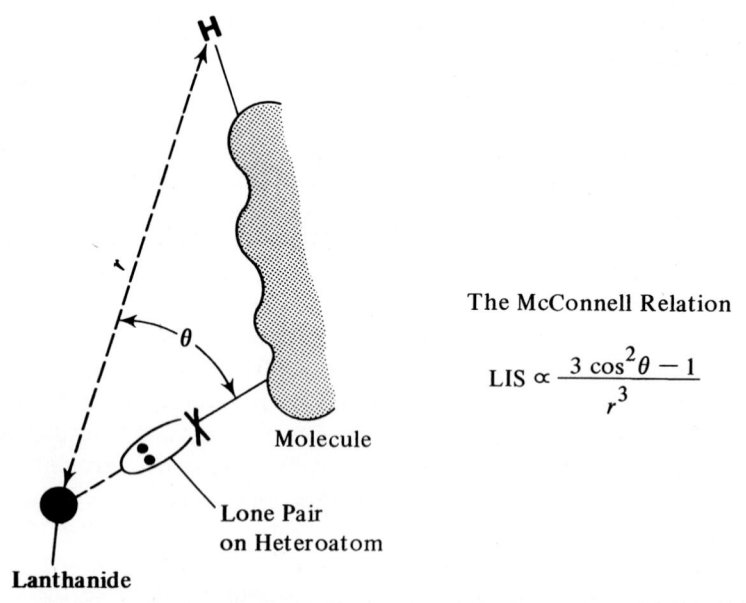

The McConnell Relation

$$\text{LIS} \propto \frac{3\cos^2\theta - 1}{r^3}$$

FIGURE 12.14. *Magnitude of the lanthanide-induced shift (LIS) as a function of molecular geometry. Here, H is a proton on the molecule whose resonance is being shifted by the presence of the lanthanide.*

CHAP. 12 Nuclear Magnetic Resonance Spectroscopy

approximately located (the lanthanide complex is attached to the lone pair), their magnitudes give valuable structural information.

The most commonly used shift reagents are based on europium (downfield shifts) and praseodymium (upfield shifts) and contain either dipivaloyl methane ($R_1 = R_2 = t$-butyl, $R_3 = H$) or fluoroalkyl derivatives of β-diketones as ligands.

Spectra of Compounds Dissolved in Nematic Phases

As already stated, rapid reorientation of molecules is a necessary prerequisite for the observation of high-resolution NMR spectra; solids give rise to "wide-line" spectra of little chemical interest. However, at intermediate rates of molecular reorientation, it is possible to obtain high-resolution NMR spectra without averaging out through-space interactions between magnetic nuclei.

These rates are attainable for molecules dissolved in the nematic phases of liquid crystals; such spectra give information about molecular geometry, in particular about interproton distances, that is not easily available from other physical measurements.

Nematic-phase spectra are considerably more complicated than spectra obtained in liquids or gases. Thus, the spectrum of benzene in nematic phase consists of at least 50 observable lines, whereas benzene in the liquid or gas phase gives rise to only a single sharp line in its high-resolution NMR spectrum.

The Nuclear Overhauser Effect

One of the important ways in which magnetic nuclei may relax from their upper to their lower energy levels is through a dipolar spin-spin interaction with another magnetic nucleus. This process takes place through space, and its efficiency is inversely proportional to the sixth power of the internuclear distance involved. It follows that the most efficient relaxation by that route can take place when two magnetic nuclei are in close spatial proximity in the same molecule.

This phenomenon must not be confused with spin-spin coupling, which is transmitted through bonds and which leads to the splitting of energy levels. Only relaxation times are involved.

Consider now two magnetic nuclei, A and B, which are in sufficiently close spatial proximity to influence each other's relaxation times. If the nucleus A is observed while the nucleus B is simultaneously irradiated, the relaxation process in nucleus A becomes more efficient because nucleus B, which is undergoing rapid up-and-down transitions, becomes effectively a rotating magnetic field. This results in a perturbation of the usual Boltzmann distribution of nuclei A toward the lower state and increases up to 50% the intensity of the signal due to the nucleus A. This enhancement of intensity is known as the *nuclear Overhauser effect* (NOE) and is diagnostic for the presence of magnetic nuclei in close spatial proximity. Structural information can thus be obtained, for example, in connection with *cis-trans* isomerism.

Commercially Available Instruments

High-resolution NMR spectrometers are complex and expensive instruments produced by a very limited number of manufacturers; Varian (United States), Perkin-

Elmer (United Kingdom), Bruker (Germany), and Jeol (Japan) account for virtually all instruments. They range from proton-only routine instruments (e.g., Varian EM360) operating at 60 MHz* and costing approximately $20,000, to flexible research instruments based on superconducting solenoids (e.g., Varian XL-200 and Bruker WM 500) operating at up to 600 MHz and costing up to $500,000. It is widely predicted that extensive use of the current mainstay of NMR instrumentation—spectrometers based on electromagnets and operating at 90 or 100 MHz—will be unlikely to continue, and that NMR instruments will polarize between relatively low-cost units based on permanent magnets and research instruments based on superconducting solenoids.

12.7 ANALYTICAL APPLICATIONS

The vast majority of analytical applications of NMR spectroscopy can be classified under two headings: determination of structures of pure compounds and quantitative determination of mixtures. The monitoring of the progress of reactions is, of course, only a subcategory of the latter class.

In practice, structural determination is not carried out solely by means of NMR spectroscopy, although proton NMR is probably the most important single method available in this area. For this reason, a detailed discussion of structure determination by NMR alone is generally not included even in extensive general texts on NMR spectroscopy and is best considered in conjunction with other major techniques.

Quantitative applications of NMR are generally limited to ^{1}H NMR, and have certain inherent strengths and limitations, which are summarized as follows.

1. NMR spectroscopy is nondestructive; a sample may be recovered completely unchanged after being subjected to an NMR experiment because the energy changes involved are negligible compared to the strengths of chemical bonds. The solvents used are usually easily evaporated, and the cell (a glass test tube) is easily washed out.

2. It is often possible to identify the components of a mixture and to carry out a quantitative analysis in one step; that is, it is not always necessary to carry out precalibration procedures.

3. The results of quantitative analysis by NMR, although not inherently highly accurate, tend to be reasonably good. Thus, even a small number of resonances can lead to a positive identification because chemical shifts can be determined very accurately, and the substances identified can then be quantitatively estimated in a mixture. In addition, the integration of several resonance signals often leads to internal verification.

4. When applicable, quantitative analysis by NMR can be very fast (a typical spectrum takes less than 5 min to run) and convenient (simple sample preparation).

---

*NMR spectrometers are usually described in terms of their operating frequency for protons even if they are not used to obtain proton NMR spectra, rather than in terms of the more logical parameter, the strength of the magnetic field.

5. The principal limitation of NMR spectroscopy as an analytical tool is its inherently poor sensitivity. This is particularly important in examining mixtures, because with pure compounds one may assume that protons resonating at different frequencies are present in ratios of whole numbers whereas the corresponding ratios in mixtures can be determined only to within the accuracy of integration. It is difficult to determine a small amount of impurity, except when it gives rise to well-separated signals; therefore, NMR is very rarely used for this purpose. This insensitivity is not wholly a disadvantage because, by converse reasoning, samples used for structural determination need not be highly purified. As a rule of thumb, a purity of 90% is adequate, and even larger quantities of impurity can be tolerated provided they can be identified.

6. The integrated intensities of resonances in an NMR spectrum give only *relative* abundances of magnetic nuclei in the various environments. This limitation is not serious, since precalibrating the integrator with samples of known composition or introducing internal standards can be used to convert the relative values into absolute numbers. Using an internal standard avoids errors arising from changes of spectrometer response with time, so it is the preferred procedure. Substances suitable for this use should give rise to easily observed resonances (preferably singlets) not overlapping with those being determined, should be chemically inert toward the other components, and should be easy to weigh accurately. The most useful internal standards for nonaqueous systems are 1,3,5-trinitrobenzene and methylene bromide, and for aqueous systems the salts of terephthalic acid.

7. It is obvious that with the aid of an internal standard, it is possible to determine the total weight of hydrogen in a known weight of a pure compound and thus obtain a rapid and nondestructive analysis for this element. Furthermore, by making certain assumptions (e.g., that the resonance of lowest intensity corresponds to a single proton in a molecule or that a sharp singlet near $\delta = 4$ ppm is due to a methoxy group), it is possible to obtain the molecular weight of an unknown substance. If the assumption was wrong, it would typically result in the apparent molecular weight becoming equal to the "equivalent" weight with respect to the fragment whose resonance is considered. This must be less than the true molecular weight and therefore represents a very useful check on molecular weights determined by mass spectrometry (Chap. 16): When the molecular weight obtained by NMR is higher than the m/e ratio of the peak of highest mass in the mass spectrum, the molecular ion is not detectable in the latter. This condition is not uncommon with many compounds (e.g., with many iodo derivatives where the molecular ion cannot be observed even at the lowest practicable electron energy).

8. It cannot be overemphasized that quantitative applications of ^{13}C NMR are much more difficult than those of proton NMR.

Experimental Considerations

The operation of NMR instruments is far too complex to be discussed within the present framework, but the user should know some of the experimental variables, even if he or she does not normally operate the instrument.

Sample Preparation. As mentioned above, very high purity is not normally essential, but certain types of impurities such as paramagnetic substances must be excluded. The presence of solid impurities of any sort in the solution of a substance will degrade the homogeneity of the magnetic field and hence cause line broadening. For this reason, solutions should be filtered before being placed in the NMR sample tube.

Solvent. For all practical purposes, NMR spectra are recorded in solution, although pure ("neat") liquids and even gases can, in principle, also be examined. The solvents must meet certain requirements (Sec. 12.2) and a compromise must often be made between using concentrated solutions (for high sensitivity) and dilute solutions (for measuring chemical shifts uninfluenced by solute-solute interactions). Besides the commonly used carbon tetrachloride, deuterochloroform, and D_2O, a range of deuterated solvents (dimethyl sulfoxide, benzene, pyridine, acetone, dioxane) are commercially available. It must be emphasized that direct comparison of chemical shifts obtained in different solvents is invalid, since solvent-induced changes of up to 0.5 ppm are by no means uncommon.

Computer-Controlled NMR Spectrometers. An NMR spectrometer operating in the FT mode is controlled by a dedicated minicomputer, which is used both for storing the free-induction decay and for performing the Fourier transformation. The experimental variables, set digitally at the start of the experiment, include (a) the size of the data table, N, which corresponds to the number of addresses to be filled and which sets limits to the eventual resolution attainable; (b) acquisition time, AT, generally of the order of 0.5 to 20 sec, which is used to fill these addresses; and (c) the eventual spectral width, SW, of the transformed spectrum in hertz. These three variables are related by the formula

$$N = 2AT \times SW \quad\quad (12.12)$$

Other important variables are the *pulse width*, which is the time (of the order of microseconds) during which the RF power is applied and which thus controls the RF power level; and the delay time, which is the time between pulses (of the order of seconds) and which may be used to allow nuclei to reach magnetic equilibrium or "relax."

Instrumental Variables. Most of these variables serve self-evident purposes (e.g., amplification, noise filtering, phasing of signals, width of sweep, adjustments of field homogeneity), but two are of particular importance in analytical applications. As mentioned before, the strength of the irradiating field H_1 (the amplitude of the RF radiation) may cause saturation when set at too high a value, but high RF field also causes increased sensitivity. For this reason, the RF field is often set at a value that saturates *some* of the resonances over the finite range of relaxation times in a real sample. Deviations from the ideal behavior (the exact correspondence between the number of protons and the height of the integral step) caused by such settings will not lead to error when examining pure substances where the ratios between the various steps can be legitimately rounded off to whole numbers, but may become a source of serious error in quantitative work on mixtures. For this reason, such measurements should be repeated with at least two settings of the RF field strength. Also, since saturation is a function of the duration of exposure of magnetic nuclei to the RF field H_1, as well as the field strength, the parameters governing the scanning velocity (*sweep time*) and RF field strength cannot be considered independently. Furthermore, a pure absorption signal corresponds to only an infinitely slow sweep time, whereas very rapid sweep times will be associated with various distortions.

SELECTED BIBLIOGRAPHY

General Texts

BECKER, E. D. *High Resolution NMR.* New York: Academic Press, 1969.

BOVEY, F. A. *NMR Spectroscopy.* New York: Academic Press, 1969.

EMSLEY, J. W., FEENEY, J., and SUTCLIFFE, L. H. *High Resolution NMR Spectroscopy,* Vols. 1 and 2. Oxford: Pergamon Press, 1965.

GÜNTHER, H. *NMR Spectroscopy.* New York: John Wiley, 1980.

JACKMAN, L. M., and STERNHELL, S. *Applications of NMR Spectroscopy in Organic Chemistry,* 2nd ed. Oxford: Pergamon Press, 1969.

MATHIESON, D. W., ed. *Nuclear Magnetic Resonance for Organic Chemists.* London: Academic Press, 1967.

Analysis of NMR Spectra

ABRAHAM, R. J. *Analysis of High Resolution NMR Spectra.* Amsterdam: Elsevier, 1971.

DETAR, D. F., ed. *Computer Programs for Chemistry,* Vol. 1. New York: W. A. Benjamin, 1968.

ROBERTS, J. D. *An Introduction to Spin-Spin Splitting in High Resolution NMR Spectra.* New York: W. A. Benjamin, 1962.

Carbon-13 NMR Spectroscopy

BREITMAIER, E., and VOELTER, W. ^{13}C *NMR Spectroscopy.* Weinheim: Verlag Chemie, 1974.

CLERC, J. T., PRETSCH, E., and STERNHELL, S. ^{13}C *Kernresonanzspectroscopie.* Frankfurt: Akademische Verlags-Gesellschaft, 1973.

LEVY, G. C., LICHTER, R. L., and NELSON, G. L. *Carbon-13 NMR for Organic Chemists.* New York: Wiley-Interscience, 1980.

STOTHERS, J. B. *Carbon-13 NMR Spectroscopy.* New York: Academic Press, 1972.

Collections of NMR Data

BHACCA, N. S. HOLLIS, D. P., JOHANN, L. F., PIER, E. A., and SHOOLERY, J. N. *High Resolution NMR Spectra Catalog,* Vols. 1 and 2. Palo Alto, Calif.: Varian Associates, 1963.

BREITMAIER, E., HAAS, G., and VOELTER, W. *Atlas of Carbon-13 NMR Data.* London: Heyden, 1979.

BRÜGEL, W. *Handbook of NMR Spectral Parameters.* London: Heyden, 1979.

HOWELL, M. G., KENDE, A.S., and WEBB, J. S., eds. *Formula Index to NMR Literature Data,* Vols. 1 and 2. New York: Plenum Press, 1966.

JOHNSON, L. F., and JANKOWSKI, W. C. *Carbon-13 NMR Spectra.* New York: Wiley-Interscience, 1972.

POUCHERT, C. J., and CAMPBELL, J. R. *The Aldrich Library of NMR Spectra.* Milwaukee: Aldrich Chemical Company, 1974.

Special Topics

BOVEY, F. A. *High Resolution NMR of Macromolecules.* New York: Academic Press, 1972.

CASEY, A. F. *PMR Spectroscopy in Medicinal and Biological Chemistry.* London: Academic Press, 1971.

DWEK, R. A. *NMR in Biochemistry.* Oxford: Clarendon Press, 1973.

FARRAR, T. C., and BECKER, E. D. *Pulse and Fourier Transform NMR.* New York: Academic Press, 1971.

KASLER, F. *Quantitative Analysis by NMR Spectroscopy.* London: Academic Press, 1973.

LEYDEN, D. E., and COX, R. H. *Analytical Applications of NMR.* New York: John Wiley, 1977.

MARTIN, M. L., DELPEUCH, J. J., and MARTIN, G. J. *Practical NMR Spectroscopy.* London: Heyden, 1980.

NOGGLE, J. H., and SCHIRMER, R. E. *The Nuclear Overhauser Effect*. New York: Academic Press, 1971.

Determination of Molecular Structure by Combined Spectroscopic Methods (including NMR).

SILVERSTEIN, R. M., BASSLER, C. G., and MORRILL, T. C. *Spectrometric Identification of Organic Compounds*, 4th ed. New York: Wiley-Interscience, 1981.

SIMON, W., and CLERC, T. *Structural Analysis of Organic Compounds by Spectroscopic Methods*. London: Macdonald, 1971.

WILLIAMS, D. H., and FLEMING, I. *Spectroscopic Methods in Organic Chemistry*, 2nd ed. London: McGraw-Hill, 1973.

PROBLEMS

1. The methyl protons of *n*-propyl alcohol show an absorption peak 528.4 Hz upfield from a benzene external-reference peak, using an RF field of 90 MHz. (a) If the benzene peak occurs at $\delta = 6.73$ ppm (downfield) from the TMS peak, what is the chemical shift of the sample peak relative to TMS? (b) If the applied frequency had been 200 MHz, at what equivalent frequency from the benzene peak would the absorption peak have occurred?

2. Match the following NMR spectra (p. 391) with the following compounds: (a) ethyl bromide; (b) 1,1-dibromoethane; (c) 1,2-dibromo-2-methylpropane; (d) 1,1,2-tribromoethane; (e) ethyl alcohol; (f) *p*-(*t*-butyl)-toluene. The numbers in circles near sets of peaks refer to the relative areas for those peaks.

3. Predict the relative shape of the NMR spectrum for methyl ethyl ketone (2-butanone). Compare it with that for acetone. Include the number of peaks and their relative areas.

4. Predict the relative shapes of the NMR spectra for propane and 1-nitropropane.

5. Using the Larmor equation and the fact that hydrogen resonates at 90 MHz in a field of 21,138 gauss, (a) calculate the gyromagnetic ratio for hydrogen; (b) calculate the resonance frequency for hydrogen in a spectrometer with a magnetic field strength of 23,487 gauss. (c) If the spectrometer described in part (b) is used to obtain ^{13}C spectra, resonance occurs at 25.1 MHz. What is the resonance frequency for ^{13}C in a spectrometer that obtains ^1H signals at 80 MHz?

6. Define, illustrate, or explain each of the following terms or phrases: (a) frequency-sweep spectrometer; (b) spin-lattice relaxation; (c) chemical shift; (d) TMS; (e) ring current; (f) LIS; (g) pulse FT NMR; (h) NOE; (i) off-resonance decoupled ^{13}C spectrum.

7. (a) What is the chemical shift of a proton whose NMR signal is observed at 320 Hz downfield from TMS in a spectrometer whose basic resonance frequency for hydrogen is 90 MHz? (b) What is the chemical-shift difference between two different hydrogens whose NMR signals are observed at 180 and 400 Hz from TMS in a spectrometer operating at 60 MHz? (c) An NMR signal is observed at 7.3 ppm downfield from TMS in a spectrometer operating at 200 MHz. Calculate the position in hertz of that same signal in a spectrometer operating at 400 MHz.

8. Explain why the ^1H NMR spectrum of *N*-methylacetamide shows signals for two different *N*-methyl groups.

9. Sketch the first-order splitting patterns you would expect to observe for the following spin systems (your sketches should be similar to those in Fig. 12.11): (a) AX; (b) A_3X_2; (c) A_3X; (d) AMX with $J_{AM} > J_{MX} > J_{AX}$.

10. Sketch the ^1H NMR spectrum you would expect to observe for each of the following compounds: (a) ethyl chloride; (b) *t*-butyl amine; (c) 1,3,5-trimethylbenzene; (d) methyl methacrylate; (e) 1,1,1-trifluoroethane.

11. The ^1H NMR spectrum of a mixture of toluene and benzene showed two signals: one at 7.3 ppm (integral = 85) and one at 2.2 ppm (integral = 15). From the relative intensities of these signals, calculate the ratio of benzene to toluene in the mixture.

12. Propose structures for the unknown compounds whose ^1H NMR spectra (60 MHz) and molecular formulas are given on p. 392. In each instance, explain your analysis of the spectrum

CH₃−CH₂−OH

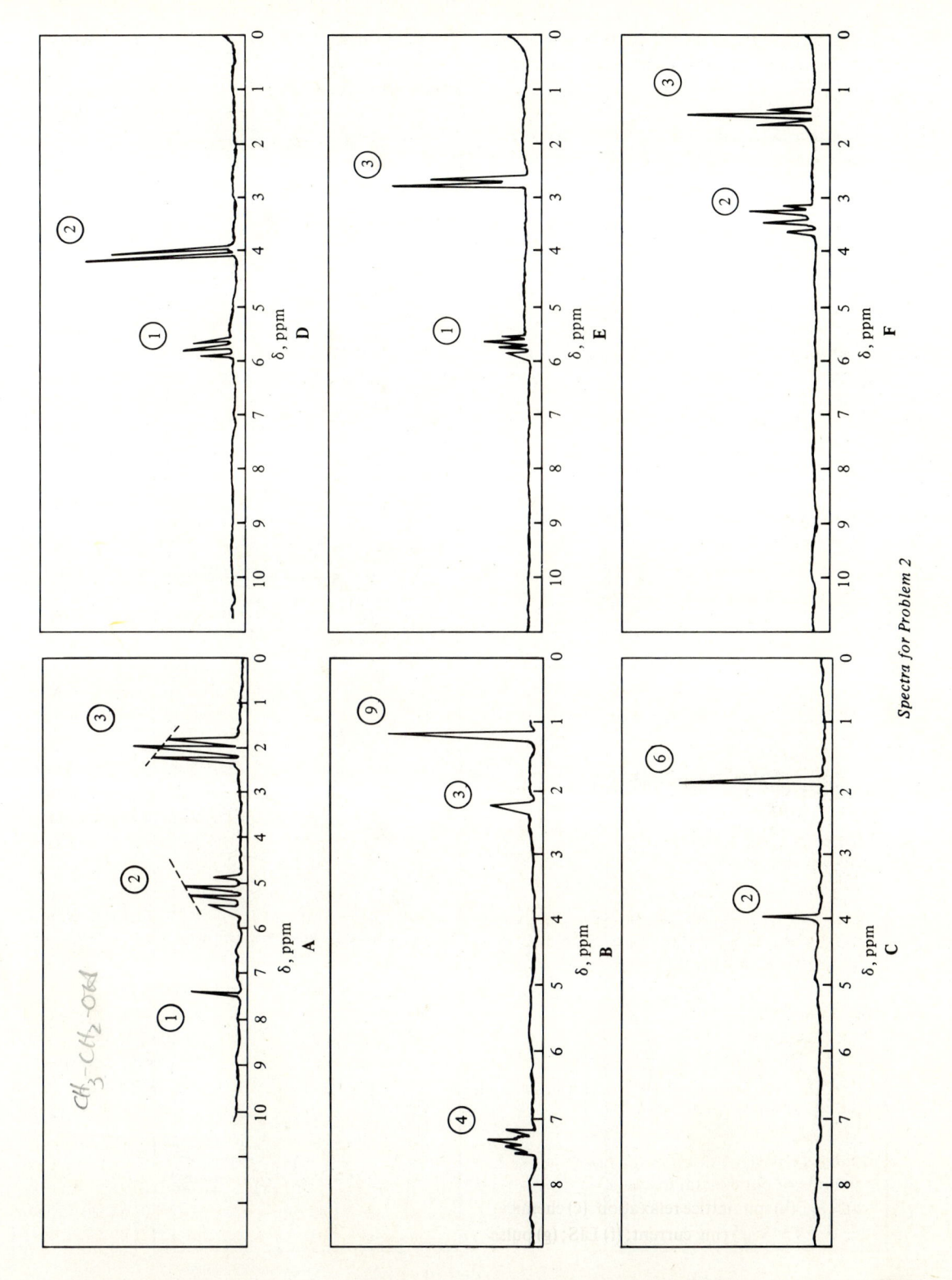

Spectra for Problem 2

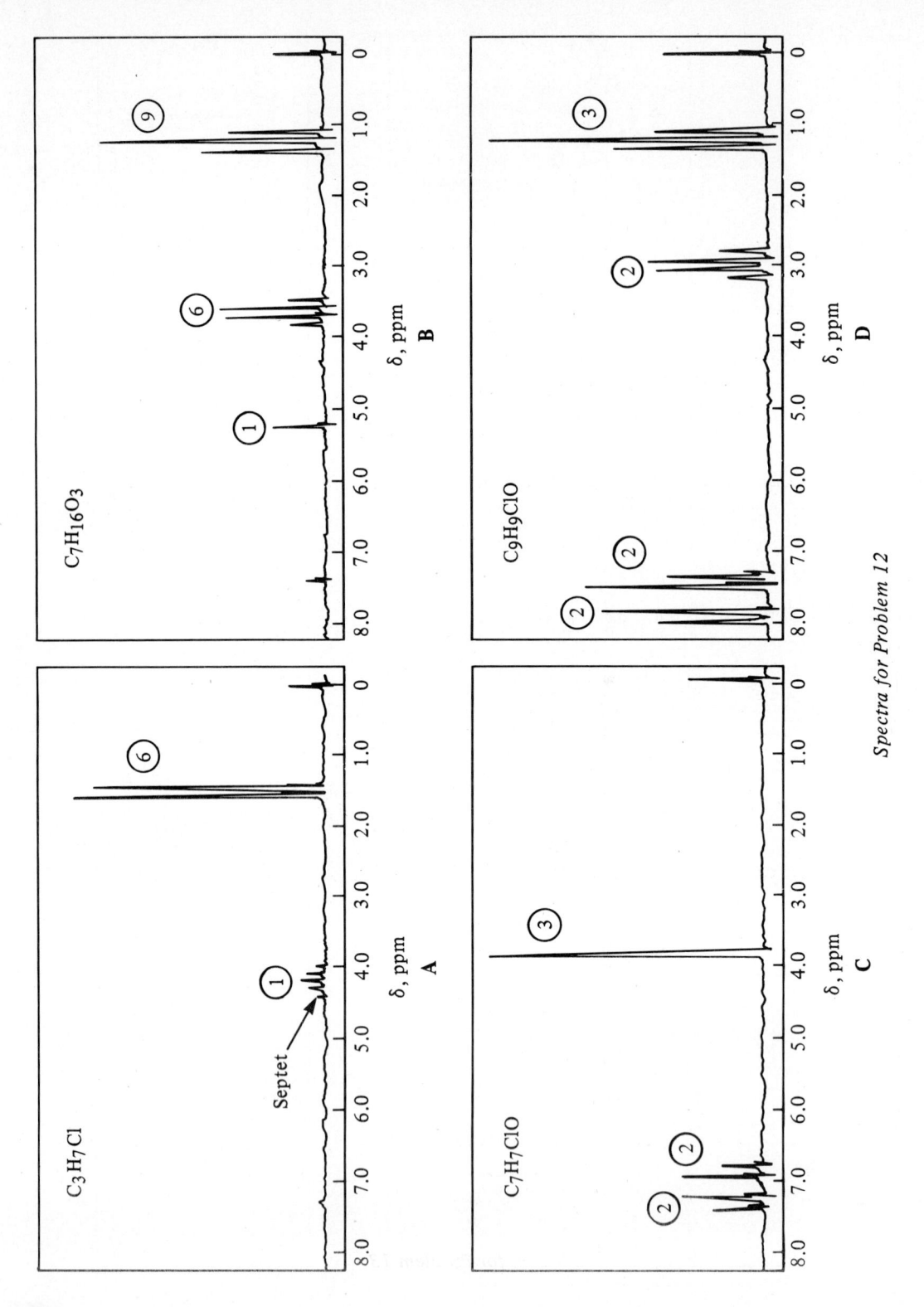

Spectra for Problem 12

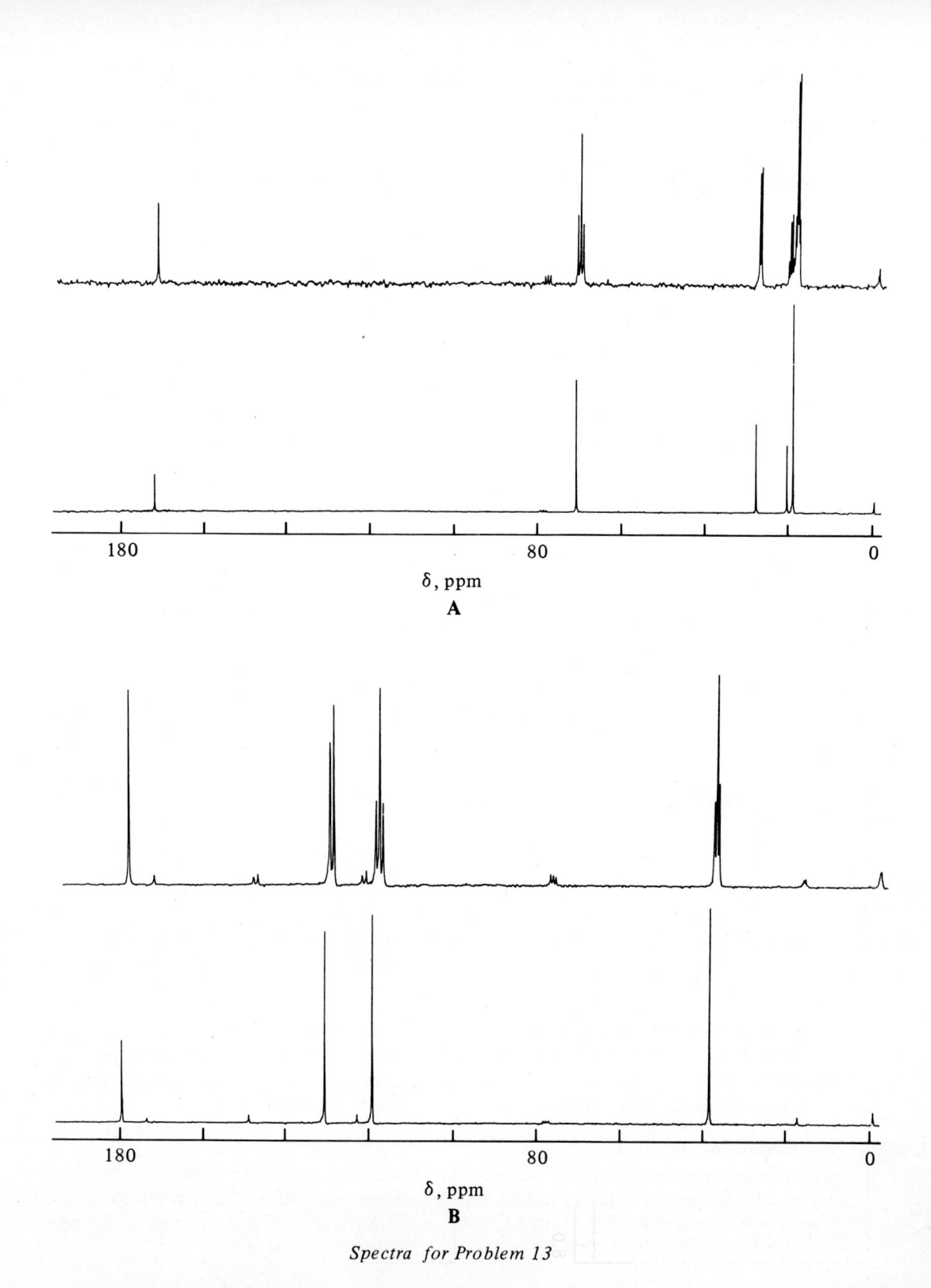

Spectra for Problem 13

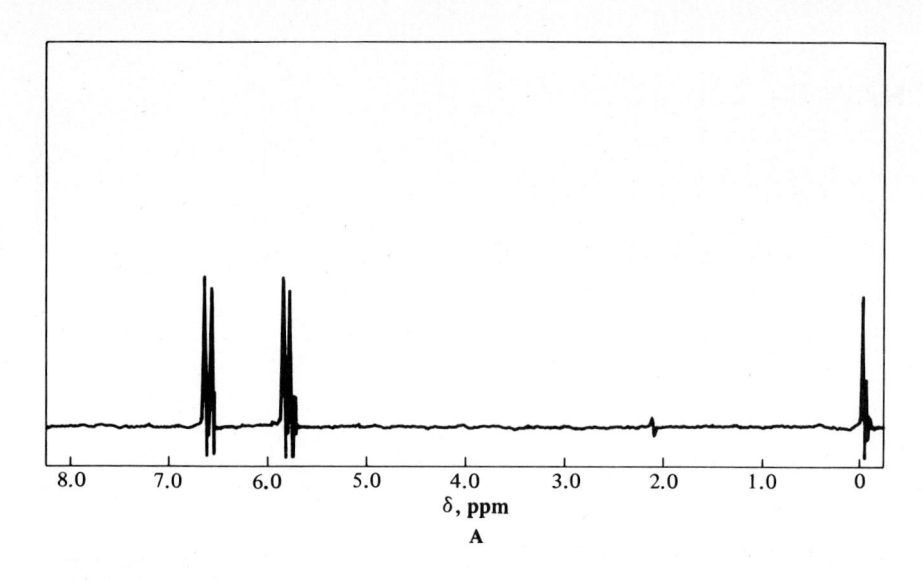

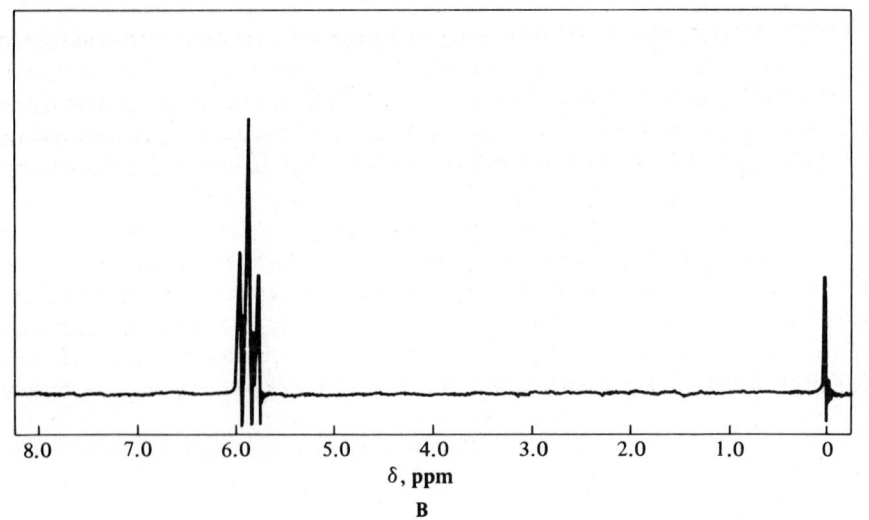

Spectra for Problem 14

and how it leads to the structure you propose. The numbers in circles by sets of peaks refer to the relative areas under the peaks.

13. Propose structures for the compounds whose 50-MHz ^{13}C spectra appear on p. 393. In each case, the lower trace is a fully decoupled spectrum, and the upper is an off-resonance decoupled one. All spectra were run in CDCl$_3$ (77 ppm) and referenced to TMS (0 ppm). One of the spectra shows peaks from a noticeable amount of an isomeric impurity. Can you determine the structure of the impurity?

14. Gas-chromatographic separation of a mixture of halocarbons gave two isomeric compounds whose molecular formula was found to be C$_2$HCl$_3$F$_2$. What are the structures of the two isomers whose 60-MHz NMR spectra are given above? (Hint: Remember that ^{19}F has a spin of 1/2 and splits hydrogen signals as if it were another hydrogen.)

CHAPTER 13

Electron Spin Resonance Spectroscopy

John R. Wasson

Since its discovery by Zavoisky in 1944, *electron spin resonance spectroscopy* (ESR) (also called *electron paramagnetic resonance spectroscopy* [EPR]) has become an essential tool for the study of the structure and dynamics of molecular systems containing one or more unpaired electrons. Such paramagnetic systems can frequently be examined using magnetic susceptibility techniques as well, but these do not provide the detailed information that ESR spectroscopy does. ESR spectroscopy and magnetic susceptibility methods each have their strengths and limitations and often provide complementary information.

The theory of ESR spectroscopy shares much in common with that of nuclear magnetic resonance spectroscopy; however, the magnetic moment of the electron is about 1000 times as large as the nuclear moment and the constants used in NMR theory frequently are different in magnitude and sign. Here, the concern is only with the fundamentals and applications of ESR spectroscopy to chemistry. The texts and specialized monographs cited at the end of the chapter should be consulted for more detailed treatments of the technique.

Types of Materials Studied by ESR

ESR spectroscopy is used to study a wide variety of materials, of which the following is a sample:

1. Inorganic and organic free radicals that possess an odd number of electrons, such as Fremy's radical, $ON(SO_3)_2^{2-}$, and diphenylpicrylhydrazyl (DPPH):

These radicals can be generated by a variety of methods, including pyrolysis and the irradiation of a sample with γ-rays. Free radicals are frequently encountered as intermediates in such chemical reactions as enzyme-substrate reactions. Most free radicals, being unstable, cannot be readily purchased and stored. However, Fremy's radical (derived from Fremy's salt, $K_4[ON(SO_3)_2]_2$), DPPH, and various nitroxide radicals can be obtained from commercial sources.

2. Odd-electron molecules such as NO, NO_2, and ClO_2. Many molecules of this type have been examined by gas-phase ESR techniques.

3. Triplet-state molecules, such as O_2 and S_2. These systems have two unpaired electrons. Optical irradiation of solids and solutions can often permit investigation of photoexcited triplet states, which are important in photochemistry.

4. Transition-metal complexes, organometallic compounds, and catalysts containing metal ions with incomplete *3d*, *4d*, or *5d* electron subshells. The detection of V(IV) (which has the $1s^2 2s^2 2p^6 3s^2 3p^6 3d^1$ configuration) in crude petroleum is one notable application of ESR spectroscopy.

5. Rare-earth and actinide compounds containing incomplete *4f*, *6d*, or *5f* subshells.

6. Impurities in solids, such as semiconductor materials. Odd electrons gained by an acceptor or lost by a donor impurity may be associated with energy bands in crystals.

7. Metals. The electrons in conduction bands of metals can be examined by ESR spectroscopy.

Although ESR spectroscopy can be used to probe the structure of numerous materials, it has certain limitations. Many materials, particularly those containing more than one unpaired electron (e.g., Ni(II) compounds) do not exhibit room-temperature ESR spectra because of large zero-field splitting (discussed later) or unusually large line broadening. These materials are best examined using conventional magnetic susceptibility methods, although on occasion NMR studies are possible. It is also important to be aware that ESR spectroscopy is concerned with a particular electronic state, the ground state (which may be a photoexcited state), and that other electronic states of a system are important only insofar as they become "mixed in" the electronic state being studied via perturbations or structural dynamics.

13.1 THE RESONANCE CONDITION

The electron is a charged particle with angular momentum (orbital and spin) and, as such, it possesses a magnetic moment, μ_e, given by

$$\mu_e = -g\beta J \tag{13.1}$$

Here J (in units of $h/2\pi$, where h = Planck's constant) is the *total angular momentum vector*; g is a dimensionless constant (the *g-value*, g *factor*, or *spectroscopic splitting factor*); and β is a constant, the *Bohr magneton*, which has a value of 9.274×10^{-21}

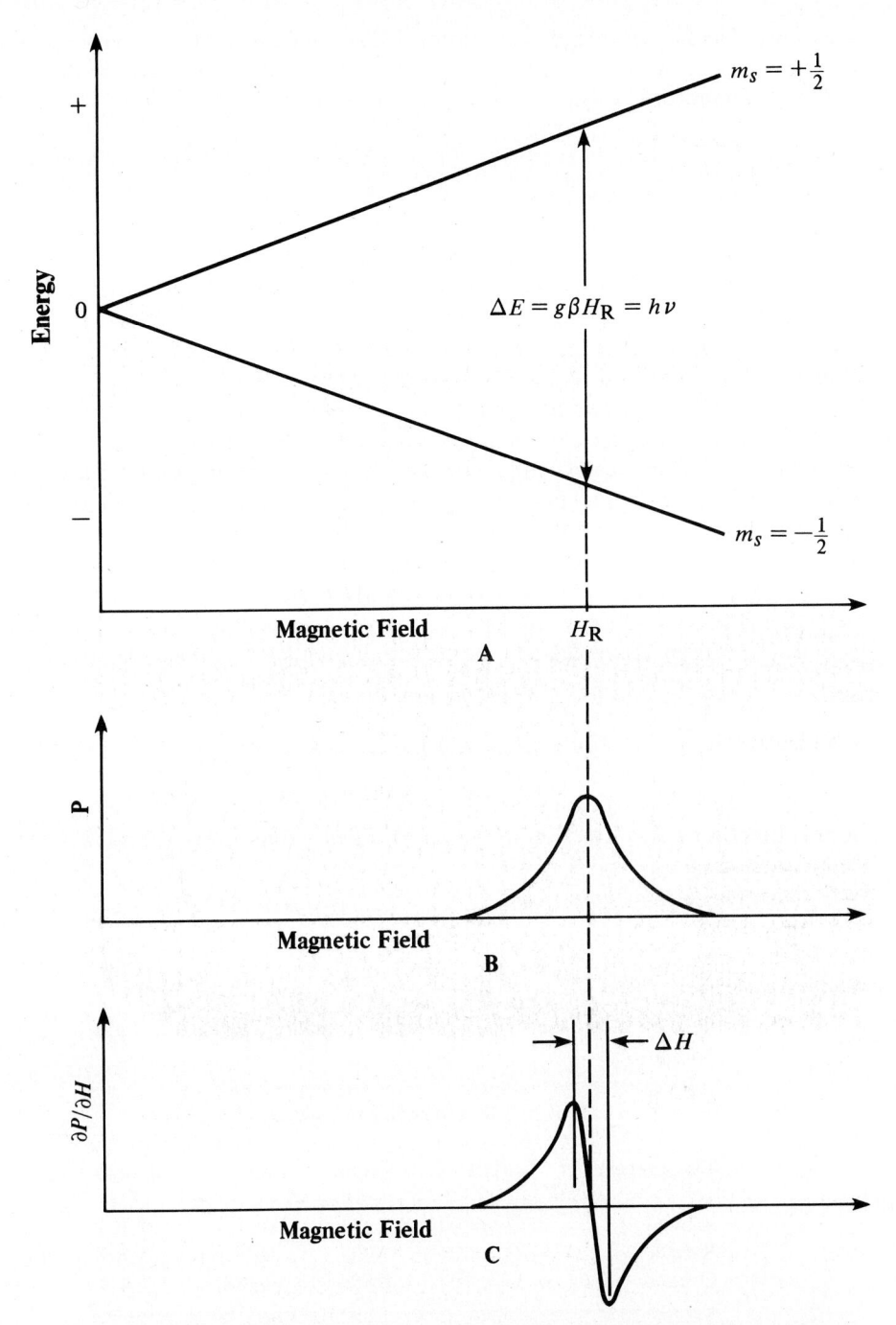

FIGURE 13.1. *Energy levels and spectra in ESR spectroscopy. A: Energy levels for an unpaired electron in a magnetic field. B: ESR absorption peak: RF power (P) absorbed versus magnetic field. C: ESR first-derivative presentation—change of power absorbed per unit change in magnetic field versus magnetic field. ΔH is the peak-to-peak line width. The first derivative spectrum is the usual form obtained using ESR spectrometers, since phase-sensitive crystal detection of the microwave power absorbed by the sample is usually employed.*

erg/gauss. The negative sign in Equation 13.1 is a consequence of negative electronic charge. Neglecting orbital angular momentum and considering only the total spin angular momentum, S, Equation 13.1 can be written as

$$\mu_e = -g\beta S \tag{13.2}$$

The g-value for the free electron, g_e, is 2.0023; for an unpaired electron in an organic molecule or ion, g is generally within a few percent of this value. The approximation made in Equation 13.2 is valid for most discussions of the ESR spectra of the organic free radicals and transition-metal complexes whose orbital angular momentum can be considered to be "quenched." Treating the g-value as an experimental quantity does not harm the present discussion, since deviations of g-values from g_e can be accounted for by introduction of spin-orbital coupling.

Magnetic moments can be detected by their interactions with magnetic fields. In zero field, the magnetic moments of unpaired electrons in a sample are randomly oriented. In the presence of a magnetic field H, electron moments assume orientations with respect to the applied field, giving rise to $2S + 1$ energy states (*Zeeman splitting*). The measurable components of μ_e are $g\beta m_s$, where m_s is the magnetic-spin quantum number, which can take the values $+S, +(S-1), \ldots, -(S-1), -S$. The application of a magnetic field to an $S = 1/2$ (or larger) system is said to remove the *spin degeneracy* (i.e., the equal-energy values of m_s in the absence of an applied magnetic field).

The energy of an electron moment in a magnetic field is given by

$$E = -\mu_e \cdot H \tag{13.3}$$

Upon combining Equations 13.2 and 13.3, the expression

$$E = g\beta H m_s \tag{13.4}$$

results (assuming the direction of the applied field defines the z-axis). When $S = 1/2$, there are two energy levels,

$$E_{m_s = +1/2} = +\tfrac{1}{2}g\beta H \tag{13.5}$$

TABLE 13.1. *Spectrometer Frequencies and g_e, Resonance Field Strength*

Designation	ν (Hz)	λ (cm)	$\bar{\nu}$ (cm^{-1})	g_e (oe)[a]
X-band	~9.5 × 10^9	3.156	0.317	3,390
K-band	~23 × 10^9	1.303	0.767	8,207
Q-band	~35 × 10^9	0.856	1.168	12,489

a. For the purposes of magnetic resonance spectroscopy, oersteds (oe) and gauss (G) are effectively the same and are used interchangeably. The SI system has the tesla as the fundamental unit of magnetic field strength rather than the gauss. Since gauss = 10^{-4} tesla, conversion of units is easy.

and

$$E_{m_s=-1/2} = -\tfrac{1}{2}g\beta H \tag{13.6}$$

whose energy is linearly dependent on H. The separation between these energy levels (Fig. 13.1) at a particular value of the magnetic field, H_R, is

$$\Delta E = +\tfrac{1}{2}g\beta H_R - (-\tfrac{1}{2}g\beta H_R) = g\beta H_R \tag{13.7}$$

In an ESR experiment, an oscillating magnetic field perpendicular to H_R induces transitions between the $m_S = -1/2$ and $m_S = +1/2$ levels, provided the frequency, v, is such that the resonance condition

$$\Delta E = hv = g\beta H_R \tag{13.8}$$

is satisfied. The frequency is held constant and the magnetic field is varied. At a particular value of the magnetic field, H_R, resonance absorption of energy occurs, resulting in a peak in the spectrum (Fig. 13.1B). The frequencies commonly employed in ESR experiments are in the microwave region; these frequencies and magnetic-field strengths for g_e resonance absorption signals are given in Table 13.1.

13.2 ESR INSTRUMENTATION

As in most other types of spectroscopy, the instrumentation used in ESR spectroscopy consists of a source of electromagnetic radiation, a sample holder, and appropriate detection equipment for monitoring the amount of radiation absorbed by the sample. In ESR spectroscopy, a magnetic field provided by an electromagnet is also required. Monochromatic radiation of the various frequencies used in ESR work (Table 13.1) is obtained from *klystrons*, which are electronic oscillators producing microwave energy. Spectrometers operating at X-band (3-cm wavelength) are the ones most commonly employed. The microwave radiation is transmitted along hollow rectangular metal pipes called *waveguides*.

Figure 13.2 gives a block diagram of a simple ESR spectrometer. The sample is placed at the center of the sample cavity, where the magnetic vector is at a maximum. Quartz tubes (approx. 3 mm o.d.) are generally used to contain solid and solution samples. Unlike the NMR technique, the sample tubes are not rotated. The magnetic field is slowly and linearly increased until the resonance condition (Eqn. 13.8) is satisfied, at which point power is absorbed by the sample and a change in current in the detector crystal is monitored. A pair of Helmholtz coils is mounted around the cavity to increase sensitivity. Feeding the coils from an oscillator superimposes a variable amplitude sinusoidal modulation on the slowly varying magnetic field. The signal detected by the phase-sensitive detection system is proportional to the slope of the ESR absorption as the magnetic field passes through resonance. The recorder then presents the first-derivative spectrum. Many spectrometers are also equipped to present second-derivative spectra.

The sample tube must be chosen with careful attention to the physical, chemical, and magnetic properties of each sample. For instance, when the sample is dissolved in a polar solvent with an appreciable dielectric constant, quartz sample

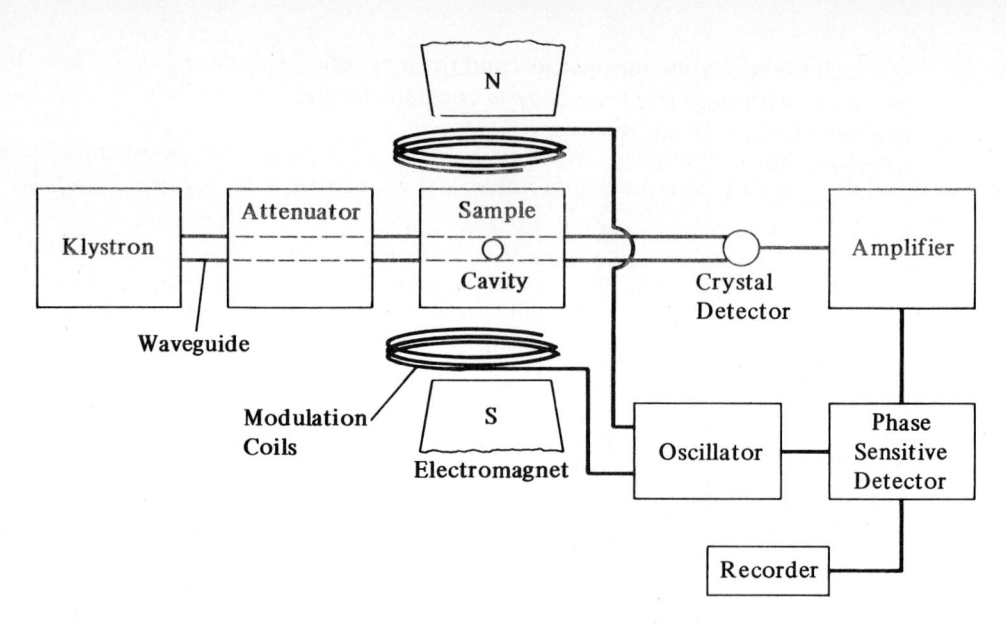

FIGURE 13.2. *Block diagram of a simple ESR spectrometer.*

tubes are usually not suitable and capillaries or thin rectangular cells of glass are used. Again, when studying free radicals with $g \approx g_e$, Pyrex tubing can be used, whereas only quartz can be used for triplet ($S = 1$) compounds because of paramagnetic impurities such as Fe^{3+} in most laboratory-grade glassware. Finger-sized Dewar flasks (of the appropriate materials) and other cryogenic equipment permit ESR spectra to be obtained as a function of temperature.

The spectral sensitivity of ESR depends on a variety of factors, but with a response time of 1 sec, as few as 10^{11} spins (approx. 10^{-12} moles) can be detected with currently available spectrometers. This sort of sensitivity suggests that ESR spectroscopy would be useful for trace analysis. A minimum detectable concentration is perhaps $10^{-9} M$ in samples with very small dielectric loss. For qualitative measurement in aqueous solutions, $10^{-7} M$ is more reasonable, whereas for quantitative measurements the sample concentration should be greater than about $10^{-6} M$. Concentrations greater than about $10^{-4} M$ should not be used in order to avoid spin-spin (exchange) broadening. For high-resolution studies, dissolved oxygen should be removed because it is paramagnetic and can contribute to line broadening.

Unfortunately, ESR spectra are rather more applicable to qualitative and semiquantitative than to quantitative analysis—to answering "What?" rather than "How much?"—because of the variety of instrumental variables in the method and the absence of suitable standards. For example, in the more common type of ESR instrument with a single-sample cavity, a spectrum of the sample is taken, and the standard is then inserted in the same cavity and its spectrum run. Due to instrument drift and changes in other experimental parameters, quantitative measurements have precisions and accuracies of the order of $\pm 25\%$ relative. With instruments with dual resonant cavities, this is improved to about $\pm 10\%$ relative.

In any ESR experiment, it is important to monitor the microwave frequency at which the spectrometer operates and the magnetic-field range swept during the experiment. Although the frequency is constant during an experiment, the frequency available from a given klystron will vary a little with tuning of the instrument. The frequency can be determined using a built-in frequency meter or appropriate transfer oscillators and frequency counters. The magnetic field can be monitored using an NMR gaussmeter or by using samples of known g-value, for instance, the DPPH free radical, for which $g = 2.0036$. The magnetic-field sweep can also be checked using Fremy's radical or oxobis(2,4-pentanedionato)vanadium(IV).

13.3 THERMAL EQUILIBRIUM AND SPIN RELAXATION

For a sample in thermodynamic equilibrium containing N spin systems ($S = 1/2$) in a magnetic field, there is a population difference between the two m_S levels arising from each $S = 1/2$ system; the difference follows the Boltzmann distribution:

$$\frac{N_{m_S = +1/2}}{N_{m_S = -1/2}} = e^{-h\nu/kT} \approx 1 - \frac{h\nu}{kT} \quad \text{where } kT \gg g\beta H \tag{13.9}$$

In this expression, $N_{m_S = \pm 1/2}$ is the number of spins having $m_S = +1/2$ and $m_S = -1/2$. Under normal conditions, there is a slight excess population in the lower ($m_S = -1/2$) level. Absorption of microwave energy by the sample induces transitions from the $m_S = -1/2$ to the $m_S = +1/2$ level. To maintain steady-state conditions, electrons promoted to the excited state must lose energy and return to the lower level; otherwise, saturation would occur and no resonance absorption would be observed. This is similar to the situation in NMR spectroscopy.

Saturation is normally avoided by working at low RF power levels. The electrons lose energy and return to the ground state by two relaxation mechanisms, spin-lattice and spin-spin relaxation, similar to mechanisms encountered in NMR spectroscopy. In spin-lattice relaxation, nonradiative transitions from the $m_S = +1/2$ to the $m_S = -1/2$ state occur because of interactions between the electrons and their surroundings that cause the spin orientation to change. Strong spin-lattice coupling of this type enables the spin system to lose energy to the lattice (surroundings) as rapidly as the oscillating field can supply it; thermal equilibrium values of the spin-state populations are maintained and energy is continuously absorbed as long as the resonance condition is satisfied. Weak spin-lattice coupling leads to saturation at comparatively low microwave power levels, whereas strong coupling can be overcome only by increasing the microwave power. Strong interactions are characterized by short relaxation times giving rise to wide lines (recall the Heisenberg uncertainty principle, which can also be written $\Delta E \Delta t \geq h/2\pi$). If spin-lattice relaxation leads to an m_S-level lifetime of about the period of the microwave radiation, or less, it is impossible to observe ESR spectra, since the microwave-induced transitions are lost among transitions due to relaxation.

The ESR spectra of samples with short relaxation times can be sharpened just by reducing the temperature, since this stabilizes the excited state, thereby lengthening the relaxation time. Spin-lattice relaxation is enhanced (relaxation times shortened) by the presence of energy levels separated from the ground state by the order of kT. This situation is often encountered in paramagnetic ions and radicals where spin-orbital coupling (interaction of electron spin and orbital moments) is relatively large.

Unpaired electrons can interact with other magnetic dipoles in the system. Such interactions do not dissipate energy and hence do not contribute directly to returning the spin systems to equilibrium. However, the spin-lattice transition may be enhanced if the interaction with the magnetic dipoles brings the excess energy to a position for transfer to the lattice. A variety of dipoles are frequently part of an unpaired electron's environment, for example other unpaired electrons, magnetic nuclei of the lattice, and various electronic and impurity dipoles. Since dipolar interactions decrease with the cube of the separation, that is, $E_{di} = \mu_1 \cdot \mu_2 / r_{12}^3$, many of the spin-spin (spin dipolar) interactions can be eliminated by diluting the paramagnetic material of interest into a diamagnetic ($S = 0$, i.e., no unpaired electrons) and (hopefully) isomorphous lattice. The reduction of spin-spin relaxation results in the sharpening of ESR spectra and in improved resolution of g-values, hyperfine structure, and so forth. This will be referred to again later in this chapter.

13.4 ESR SPECTRA OF FREE RADICALS IN SOLUTION

The simplest free radical is the hydrogen atom. In introductory chemistry, the electronic configuration of the ground state is given as $1s^1$ and, at that time, the student is more or less lucidly informed that the spin quantum number, m_S, can take the values $+1/2$ or $-1/2$. From the preceding discussion (Eqns. 13.1 through 13.8), ESR spectroscopy requires these values and configurations. The energy levels and observed ESR spectrum of the hydrogen atom are sketched in Figure 13.3. If only electron spin is considered, the situation depicted in Figure 13.1 would result; however, the nuclear spin of hydrogen ($I = 1/2$) interacts with the electron spin and with the external magnetic field as well. The result of these two interactions (analogous to the interactions producing spin-spin splitting in an NMR spectrum) is that the ESR spectrum consists of two peaks separated by 506.8 gauss with the resonant field H_R centered between them at a strength such that $g_0 = 2.00232$. If A_0, a measure of the *hyperfine coupling* energy, were equal to zero, only a single derivative peak, centered at H_R, would be observed, corresponding to a transition between the dashed levels in Figure 13.3. The components of m_I are shown in Figure 13.3, where it is seen that each spin level is split into two levels by the hyperfine interaction between electron and nuclear spins.

It should be noted that A_0, the isotropic electron spin–nuclear spin hyperfine coupling constant, can be qualitatively related to the amount of time the unpaired electron spends in an s orbital on the nucleus in question; the larger A_0, the greater the probability of finding the electron at the nucleus.

For an ion with a nonzero nuclear moment I, $2I + 1$ lines can be expected centered around g_0 with a hyperfine spacing A_0 between them. For example, for ^{63}Cu(II) (a $3d^9$ ion), $I = 3/2$; four lines are expected and generally observed in the first-derivative spectrum (see Fig. 13.4). The A_0 value is taken as the separation between the two central ESR lines, and the g_0 value is calculated from the magnetic field halfway between those lines. For V(IV) compounds (^{51}V, $I = 7/2$), eight lines

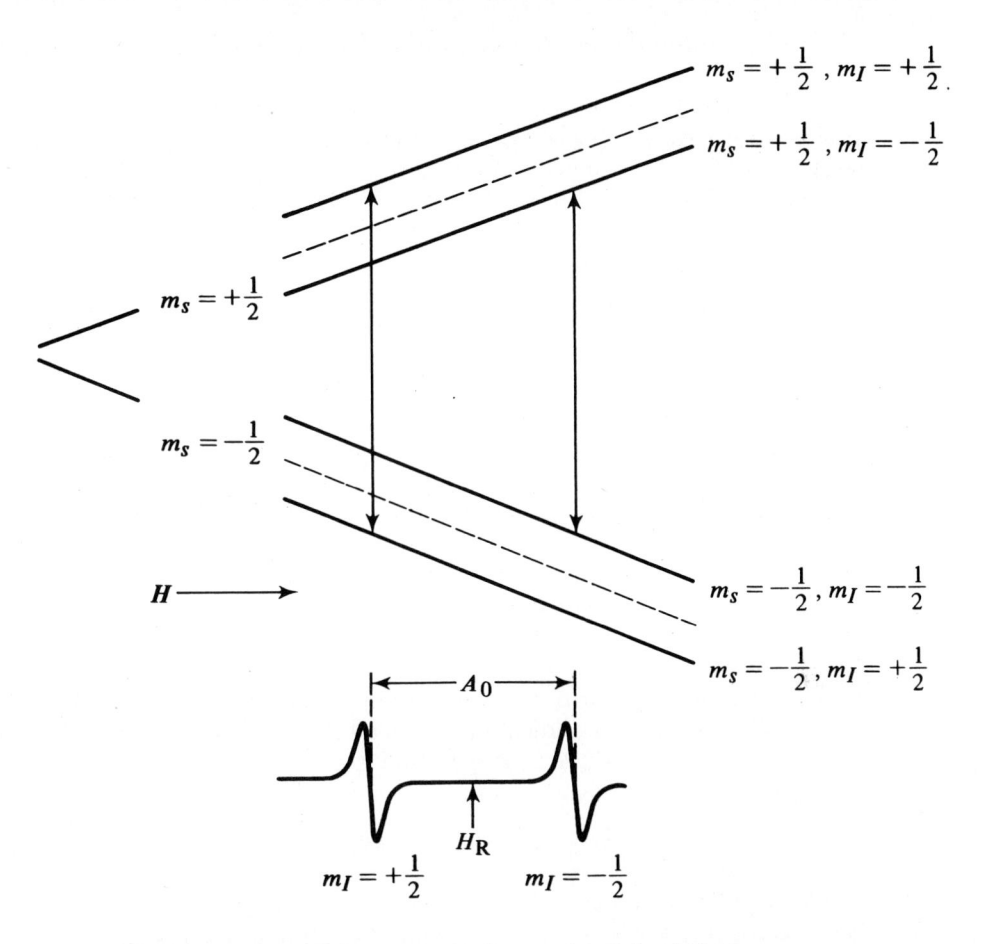

FIGURE 13.3. *ESR spectrum and energy levels for the hydrogen atom.*

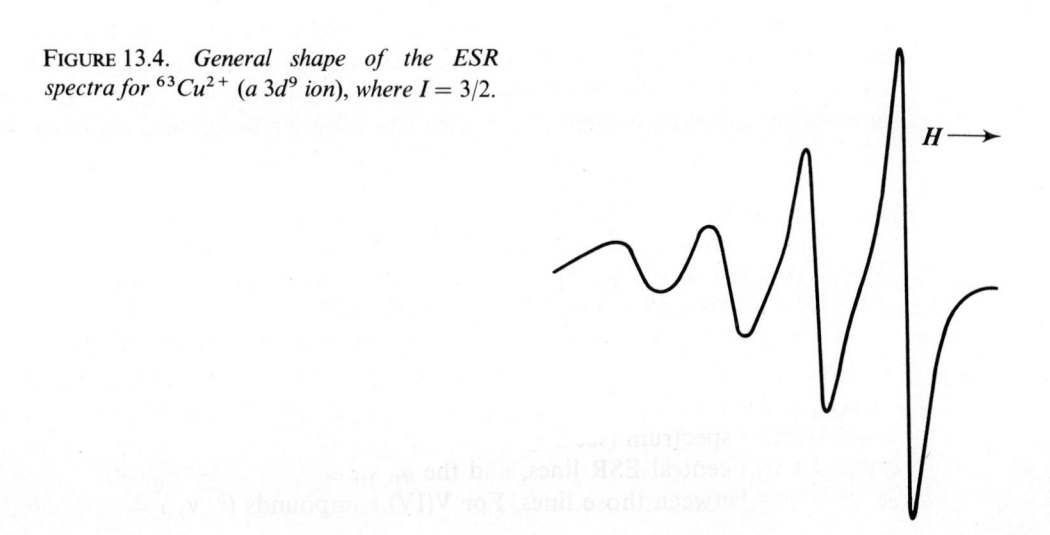

FIGURE 13.4. *General shape of the ESR spectra for $^{63}Cu^{2+}$ (a $3d^9$ ion), where $I = 3/2$.*

are expected for the solution ESR spectra. The selection rules for ESR spectroscopy, as implied in Figure 13.3, are $\Delta m_S = \pm 1$ and $\Delta m_I = 0$.

Organic free radicals in solution generally exhibit hyperfine coupling with several nuclei. When the nuclei are chemically equivalent, the number of lines is given by

$$2n_i I + 1 \tag{13.10}$$

where n_i = number of equivalent nuclei having nuclear spin I

Thus, the methyl radical, H₃C·, exhibits a four-line spectrum from splitting by three equivalent hydrogens. The *p*-benzosemiquinone radical anion (*A*)

(A)   (B)

exhibits a five-line spectrum and the benzene radical anion (*B*) gives a seven-line spectrum. Note that in hydrocarbon radicals the abundance of ^{13}C (1.108%, $I = 1/2$) is so low that satellite peaks due to interaction of the unpaired electron with the ^{13}C atoms are not readily resolved. For hydrocarbon radicals in which the unpaired electron interacts only with hydrogen atoms, Equation 13.10 reduces to $n + 1$.

When an organic radical possesses two or more sets of equivalent atoms, the total number of lines in the ESR spectrum is given by

$$\prod_{i=1}^{j} (2n_i I_i + 1) = (2n_1 I_1 + 1)(2n_2 I_2 + 1) \cdots (2n_j I_j + 1) \tag{13.11}$$

For example, for the hypothetical (H₂N)Ḃ(PF₂) radical where $I = 3$ for ^{10}B, $I = 1$ for ^{14}N, $I = 1/2$ for ^{31}P, and $I = 1/2$ for ^{19}F, a total of 378 lines could be expected if the unpaired electron interacted appreciably with each of the nuclei.

The number of lines in the ESR spectrum of a radical is a clue to the radical's identity. Of further assistance is the intensity pattern in the spectrum. For *n* equivalent nuclei with $I = 1/2$, the $n + 1$ lines have intensities proportional to the binomial expansion of order *n*. Table 13.2 lists the relative intensities up to $n = 5$. The intensity patterns for sets of equivalent nuclei with $I > 1/2$ are handled in a similar, but more

TABLE 13.2. *Relative Intensities of ESR Lines*

Number of Equivalent Atoms with $I = 1/2$ (n)	Relative Intensities of ESR Lines	Number of Lines (n + 1)
1	1 : 1	2
2	1 : 2 : 1	3
3	1 : 3 : 3 : 1	4
4	1 : 4 : 6 : 4 : 1	5
5	1 : 5 : 10 : 10 : 5 : 1	6

complicated, fashion. Figure 13.5 shows the ESR spectrum of the diphenylpicrylhydrazyl (DPPH) free radical in benzene. Under low-resolution conditions, the proton hyperfine coupling is not observed, and the five-line spectrum with the intensity distribution 1:2:3:2:1 results from the interaction of the unpaired electron with two (effectively equivalent) ^{14}N ($I = 1$) nuclei; for nuclei with $I > 1/2$, the intensity distribution is more complex than a simple binomial expansion and is beyond the scope of this text.

Magnetically nonequivalent protons will normally have different splitting constants, and the observed ESR spectra can be analyzed by reconstructing the spectrum from the splitting patterns and intensity distributions expected for each equivalent set of protons. Some examples should serve to illustrate this procedure. The photolysis of hydrogen peroxide in methanol produces the ·CH$_2$OH free radical. The protons bound to carbon and the proton bound to the oxygen atom compose two nonequivalent sets. For ·CH$_2$OH, either a doublet-of-triplets (Fig. 13.6A) or a

FIGURE 13.5. *Low-resolution ESR spectrum of the diphenylpicrylhydrazyl (DPPH) radical in benzene.*

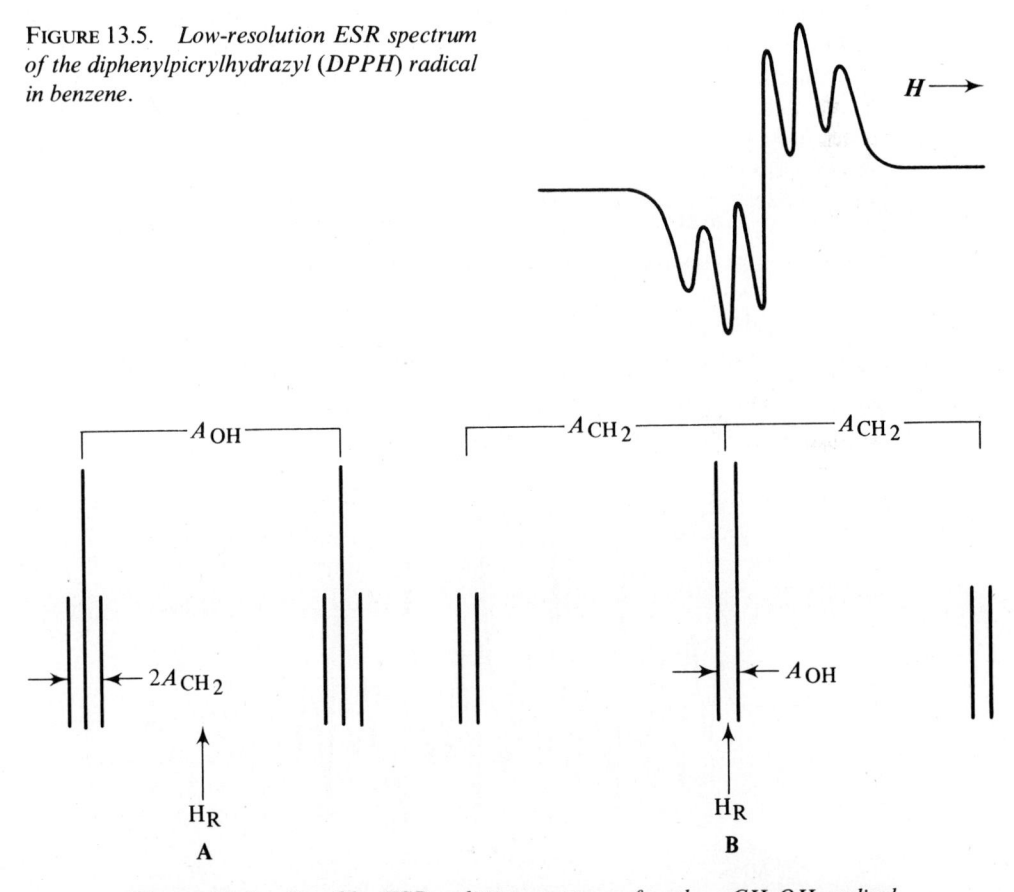

FIGURE 13.6. *Possible ESR splitting patterns for the ·CH$_2$OH radical. A: The hyperfine splitting constant for the OH proton (A_{OH}) is larger than that for the CH$_2$ protons (A_{CH_2}). B: The hyperfine splitting constant for the OH proton is smaller.*

triplet-of-doublets (Fig. 13.6B) could be expected, depending on whether the *A*-value for the OH proton is larger or smaller than the *A*-value for the CH$_2$ protons. Experiments show that a triplet-of-doublets appears, with a 1:2:1 intensity distribution, A_{CH_2} = 17.4 gauss, and A_{OH} = 1.15 gauss in accordance with the stick diagram given in Figure 13.6B [1]. More complicated spectra are analyzed similarly but, as can be imagined, the difficulty of interpretation increases with the increasing number of sets of equivalent protons.

The origin of the proton hyperfine splitting in the ESR spectra of hydrocarbon π radicals can be explained by quantum-mechanical calculations. These results give rise to the well-known McConnell relation [2]:

$$A = \rho Q \tag{13.12}$$

where A = hyperfine coupling constant for a proton in a C–H fragment having an unpaired π-electron density of ρ
         Q = constant having a value of 22.4 gauss

This relation permits unpaired electron distributions in π-electron radicals, such as the benzene radical anion and the *p*-benzosemiquinone radical anion, to be mapped experimentally; while at the same time, the ESR spectra of π-electron radicals can be calculated using simple Hückel molecular orbital theory of the type taught in sophomore-level organic chemistry courses.

Transition-Metal Complexes

Generally, the spectra of transition-metal complexes are associated with the central metal ion. Frequently, *superhyperfine* splitting is also encountered which demonstrates that the unpaired electron of the ligand is delocalized over the metal complex. Figure 13.7 shows the ESR spectrum of VO[S$_2$PC$_2$H$_5$(OCH$_3$)]$_2$ in chloroform solu-

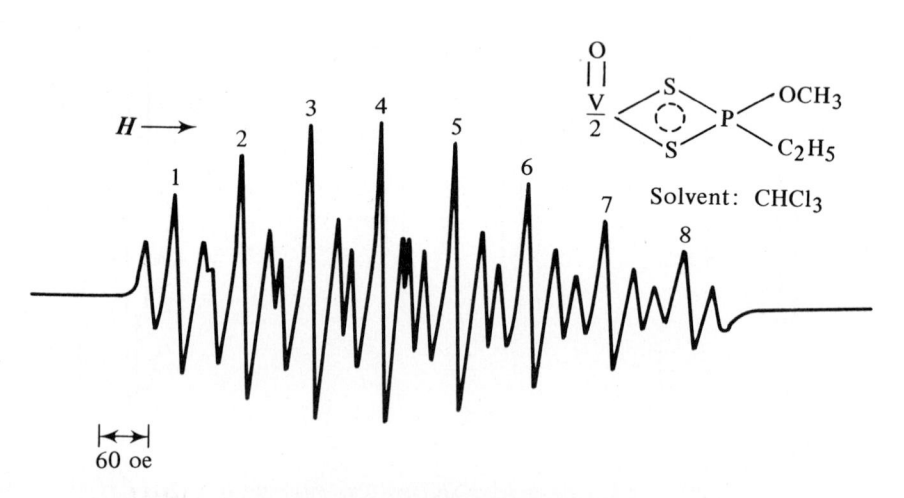

FIGURE 13.7. *ESR spectrum of VO[S$_2$PC$_2$H$_5$(OCH$_3$)]$_2$ in chloroform. The eight vanadium hyperfine lines are numbered; each of these is split into a triplet from interaction with the two equivalent phosphorus nuclei.*

FIGURE 13.8. *Pictorial representation of the orbital interaction in certain oxovanadium(IV) complexes.*

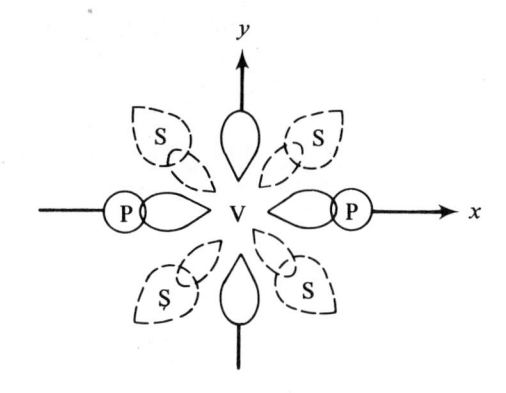

tion. Each of the eight vanadium hyperfine lines is split into three by interaction of the unpaired electron with two equivalent phosphorus nuclei ($I = 1/2$), each line having roughly the 1:2:1 intensity distribution expected. Line overlapping contributes to deviation from the anticipated intensity distribution.

In oxovanadium(IV) complexes ($3d^1$ systems) of this type, it can be shown [3] that the ^{31}P superhyperfine splitting arises from interaction of the ground-state V(IV) $3d_{x^2-y^2}$ orbital with the phosphorus 3s orbitals. This *trans*-annular interaction can be pictorially represented as in Figure 13.8. The $d_{x^2-y^2}$ orbital is not σ bonding with respect to the sulfur atoms, but does possess the correct symmetry to interact directly with phosphorus 3s and 3p orbitals. The essentially isotropic nature of the phosphorus superhyperfine splitting constants indicates that phosphorus 3p orbitals are not appreciably involved in the *trans*-annular interaction. Ligand superhyperfine splitting in the ESR spectra of metal complexes provides detailed information regarding the covalency of metal-ligand bonding. The above example is only one of the many that can be cited.

13.5 ANALYTICAL APPLICATIONS

The chief application of ESR spectroscopy to chemical problems is the identification of the presence and nature of species containing unpaired electrons. Determination of concentrations of electron spins in a given sample is not very precise; in general, only order-of-magnitude estimates of spin concentrations can be obtained. Accurate quantitative analyses can seldom be performed using ESR spectroscopy.

ESR lines usually exhibit shapes very close to those of gaussian or Lorentzian functions. The intensities of ESR lines may be obtained by integration of the full absorption curve, by two consecutive integrations of the first-derivative curve, or by the approximation

$$\text{Intensity} = \text{Derivative Height} \times (\Delta H)^2 \qquad (13.13)$$

where ΔH is defined as in Figure 13.1, which yields the relative intensities of lines in a spectrum.

Applying first principles to measurements of the ESR signal and to all pertinent instrumental parameters, absolute numbers of spins can be determined. However, this is rarely done since the number of variables to be controlled is considerable and the labor involved is extensive. Relative concentrations of species with the same spectral shape (e.g., all Lorentzian-shaped species) and line widths can be determined simply by comparing peak heights of the normal first-derivative curve under identical conditions, that is,

$$\frac{N_1}{N_2} = \frac{h_1}{h_2} \tag{13.14}$$

If the line widths differ, Equation 13.14 can be modified to yield (cf. Eqn. 13.13)

$$\frac{N_1}{N_2} = \frac{h_1}{h_2} \frac{(\Delta H_1)^2}{(\Delta H_2)^2} \tag{13.15}$$

Again, using a standard sample as a reference, intensities of ESR lines obtained by one or more integrations can be compared to yield spin concentrations.

Quantitative Analysis of Metal Ions

In analytical work, the biggest problems are associated with maintaining identical instrumental conditions for both reference compounds and samples, and finding a suitable reference material. The reference sample should be a stable, easily handled material with line-shape, line-width, and power-saturation properties similar to those of the test sample. Reasonably good semiquantitative information can be obtained by preparing calibration curves using standard samples, and comparing them with data for test samples. A good example of this technique is the detection of Cu(II) in seawater at the ppb level [4]. In this example, copper is extracted from seawater with 8-hydroxyquinoline in ethyl acetate. The ESR line intensities of the extracted solutions are then compared to those obtained for standard samples of known copper concentration similarly prepared. The curve of standard line intensity versus copper concentration can lead to a deviation of ± 0.2 ppb in the copper analyses; this means that the relative deviation is about $\pm 8\%$. Whereas a deviation of this magnitude is unacceptable for macroscopic methods, such as gravimetric analysis, it is more palatable with trace analyses. Since this approach to monitoring copper concentration is rather rapid and fairly reliable, its semiquantitative nature can often be tolerated.

Another representative type of analysis by ESR is that of Fe(III) [5]. Iron(III) can be extracted from 1.8 M hydrochloric acid with tributyl phosphate. The ESR intensity of the extracted iron, presumably $FeCl_4^-$, can then be determined relative to the signal from the DPPH radical used as an external standard. The relative intensity is proportional to concentration in the range 10 to 200 ppm of Fe(III) in the extract.

Other metal ions that can be determined quantitatively by ESR include V(IV), Cr(III), Mn(II), and Ti(III).

Spin Labeling of Biological Systems

Spin labels are stable, paramagnetic molecules that, by their structure, easily attach themselves to various biological macromolecular systems such as proteins or cell membranes. Examples of spin labels that can be covalently bonded to specific sites of biological systems include nitroxide derivatives of *N*-ethylmaleimide, which bind specifically to –SH groups, and nitroxide derivatives of iodoacetamide, which bind specifically to methionine, lysine, and arginine residues of amino acids. Noncovalently bonded spin labels that can be incorporated into biological systems include nitroxide derivatives of stearic acid, of phospholipids, and of cholesterol.

Spin labels provide information about the static and dynamic nature of the system, including structure, relative polarity, fluidity, viscosity, conformational changes, phase transitions, and chemical reactions.

In recent years, the systems most often studied by the spin-label method have been biological membranes and various models thereof. For example, the organization of the phospholipid phase of biological membranes has been studied using nitroxide derivatives of stearic acid [6]. One interesting practical application of this type of spin labeling is in the study of disease-state membranes. Intact erythrocyte membranes were shown to be in a more fluid state near the membrane surface in patients suffering from myotonic muscular dystrophy as compared to those from normal controls. For an interesting discussion of the use of spin labels in research on diseases, see the review by Butterfield [7]. A number of the biochemical and other applications of spin labels are discussed in the book by Berliner listed in the Selected Bibliography.

Determination of Surface Area

Because of its specificity for molecules with unpaired electrons, ESR has been used to determine the active surface area of catalysts. The total area of the ESR signal is proportional to the number of unpaired electrons in the sample. A comparison is made of the catalyst, which has had a paramagnetic molecule adsorbed on its surface, with a standard containing a known number of unpaired electrons, usually DPPH, which has 1.53×10^{21} unpaired electrons per gram.

In a study of the surface area of an MnO_2 catalyst [8], ESR measurement of the surface area based on adsorption of DPPH indicated a surface area of 46 m^2/g, whereas ESR measurement of active adsorbed oxygen indicated a surface area of 42 m^2/g. These two values compared favorably with the results from the usual BET method, which gave a surface area of 61 m^2/g. The BET method consists of volumetric or gravimetric measurements of the quantity of gas that will completely cover the surface of the solid with an adsorbed layer.

SELECTED BIBLIOGRAPHY

ABRAGAM, A., and BLEANEY, B. *Electron Paramagnetic Resonance of Transition Ions.* Oxford: Clarendon Press, 1970.

ALGER, R. S. *Electron Paramagnetic Resonance: Techniques and Applications.* New York: Interscience, 1968.

ATHERTON, N. M. *Electron Spin Resonance.* New York: John Wiley, 1973.

BERLINER, L. J., ed. *Spin Labeling II: Theory and Applications.* New York: Academic Press, 1979.

GERSON, F. *High Resolution E.S.R. Spectroscopy.* New York: John Wiley, 1970.

SWARTZ, H. M., BOLTON, J. R., and BORG, D. C., eds. *Biological Applications of Electron Spin Resonance.* New York: Wiley-Interscience, 1972.

WERTZ, J. E., and BOLTON, J.R. *Electron Spin Resonance: Elementary Theory and Practical Applications.* New York: McGraw-Hill, 1972. Probably the best basic text on ESR now available.

REFERENCES

1. R. LIVINGSTON and H. ZELDES, *J. Chem. Phys.*, **44**, 1245 (1966).
2. H. M. MCCONNELL, *J. Chem. Phys.*, **24**, 632, 764 (1956).
3. D. R. LORENZ, D. K. JOHNSON, H. J. STOKLOSA, and J. R. WASSON, *J. Inorg. Nucl. Chem.*, **36**, 1184 (1974).
4. Y. P. VIRMANI and E. J. ZELLER, *Anal. Chem.*, **46**, 324 (1974).
5. T. TAKEUCHI and N. YOSHIKUNI, *Bunseki Kagaku*, **22**, 679 (1973).
6. H. M. MCCONNELL and B. G. MCFARLAND, *Quart. Rev. Biophys.*, **3**, 91 (1970).
7. D. A. BUTTERFIELD, *Biol. Mag. Res.*, **4**, 1 (1982).
8. A. T. T. OEI and J. L. GARNETT, *J. Catal.*, **19**, 176 (1970).

PROBLEMS

1. Calculate the resonant field strengths for substances that have g-values of 2.800 and 1.964 when $v = 9.40, 23.00,$ and 35.00 GHz (1 GHz = 10^9 Hz = 10^9 sec^{-1}), respectively. (Note: Planck's constant = $h = 6.6262 \times 10^{-27}$ erg-sec; Bohr magneton = $\beta = 0.9274 \times 10^{-20}$ erg/gauss.)

2. What are the separations between resonant field strengths for substances having g_1 and g_2 values of 2.045 and 2.160, and of 1.960 and 2.110, when $v = 9.40, 23.00,$ and 35.00 GHz, respectively?

3. What are the values of hv (cm^{-1}/molecule) when $v = 9.40, 23.00,$ and 35.00 GHz? (Note: Planck's constant = $h = 33.3586 \times 10^{-12}$ cm^{-1}-sec.)

4. Evaluate the ratios of populations in $m_S = +1/2$ and $m_S = -1/2$ spin states when $v = 9.40, 23.00,$ and 35.00 GHz and $T = 298$ K (room temperature), 77 K (boiling nitrogen), 20 K (boiling hydrogen), and 4 K (boiling helium). (Note: Boltzmann's constant = $k = 1.3806 \times 10^{-16}$ erg/K.)

5. How many ESR lines can be expected for the 1,2-, 1,3-, and 1,4-difluorobutadiene radical anions? $I = 1/2$ for ^{19}F and ^{1}H.

6. How many ESR lines could be expected for the ^{33}S^{19}F$_6$ radical anion? Radical cation? $I = 3/2$ for ^{33}S, and $I = 1/2$ for ^{19}F.

7. How many ESR lines can be expected for Cu(S$_2$PF$_2$)$_2$ and VO(S$_2$PF$_2$)$_2$ ($I = 3/2$ for Cu; $I = 7/2$ for V; $I = 1/2$ for F) if the metal, phosphorus, and fluorine hyperfine splittings were observable in the solution spectra?

8. How many ESR lines could be expected for the tetrahedral P$_4^-$ radical? $I = 1/2$ for ^{31}P. What would be the theoretically expected relative intensities of the lines?

9. Assuming there is no overlap of the ESR lines, what would be the theoretically expected intensity ratios for the *trans-trans*-1,4-bifluorobutadiene radical anion (see Problem 5)?

10. Calculate the magnitude of the magnetic moment of a free electron with spin angular momentum $S = 1/2$ (neglecting any orbital angular momentum), and the energy of the

electron in magnetic fields of 3.4, 8.2, and 12.5 kilogauss. What would these energies correspond to in terms of calories per mole of electrons?

11. A solution of a stable organic free-radical compound was prepared at $2.5 \times 10^{-5} M$ concentration. This standard solution produced an ESR spectrum with a first-derivative peak height of 60 divisions on a strip-chart recorder, and a line width of 2.5 gauss. A different compound of unknown concentration gave a spectrum of peak height of 75 divisions and a line width of 3.1 gauss under identical experimental conditions. (a) If the standard compound has one unpaired electron per molecule, calculate the number of "spins" per cubic centimeter of the standard solution. (b) If both compounds had Lorentzian-shaped first-derivative spectra, calculate the concentration of the unknown compound assuming it has one unpaired electron per molecule.

12. (a) Given a minimum detectable concentration in ESR of about $10^{-9} M$, calculate the minimum detectable number of spins this represents if one used 3-mm o.d. sample tubes (assume a wall thickness of about 0.3 mm) and filled them about 4 cm high. (b) If there is one unpaired electron per molecule with a molecular weight of 200, what is the absolute detection limit for the compound in grams? (c) For this same compound, calculate the concentration detection limit in nanograms per milliliter (ppb).

CHAPTER 14

X-Ray Spectrometry

Peter A. Pella

Since the discovery of x-rays by Roentgen in 1896, the electromagnetic spectrum between about 0.1 and 200 Å (angstrom units) has been a source of significant contributions to our fundamental knowledge of atomic structure. By 1927, six Nobel Prizes in physics had been awarded for studies of the properties of x-rays and their interaction with matter.

The use of x-rays to probe the elemental composition of matter in the so-called secondary x-ray emission mode has prompted the development of a technique called *x-ray fluorescence spectrometry*. This x-ray technique is easily amenable to almost complete automation, as shown by the growth of commercially available computer-controlled instrumentation, especially within the last decade. These advances permit multielement analysis to be performed, nondestructively, on the order of minutes per specimen. Today, this technique is used extensively in process-control applications in the cement, steel, and mineral-processing industries for major and minor elemental analysis. Ultra-trace analysis in the ppb range is also possible with appropriate sample preconcentration methods.

14.1 INTRODUCTION

X-rays are generated by bombarding matter with high-energy particles such as electrons or protons or with x-ray photons. When an atom is bombarded in this manner, an electron is ejected from one of the inner shells of the atom. This vacancy is immediately filled by an electron from a higher-energy shell, creating a vacancy in that shell which is, in turn, filled by an electron from a yet higher shell. Thus, by a series of transitions, L → K, M → L, N → M, each new vacancy is filled until the excited atom returns to its ground state.

Each electronic transition (apart from radiationless transitions) results in the emission of a characteristic x-ray spectral line whose energy, $h\nu$, is equal to the difference between the binding energies of the two electrons involved in the transition (see Fig. 14.1). Only certain electronic transitions are permitted by quantum-mechanical selection rules, which are described in various textbooks on atomic physics. The energies or wavelengths of the x-ray spectral lines are the basis for qualitative analysis. The lines are designated by symbols such as Ni Kα_1, Fe Kβ_2, Sn Lα_2, and U Mα_1. The symbol of an x-ray line represents the chemical element (here, Ni, Fe, Sn, and U); the notations K, L, and M indicate that the lines originate by the initial removal of an electron from the K, L, and M shells, respectively; a particular line in the series is designated by a Greek letter α, β, etc. (representing the subshell of the

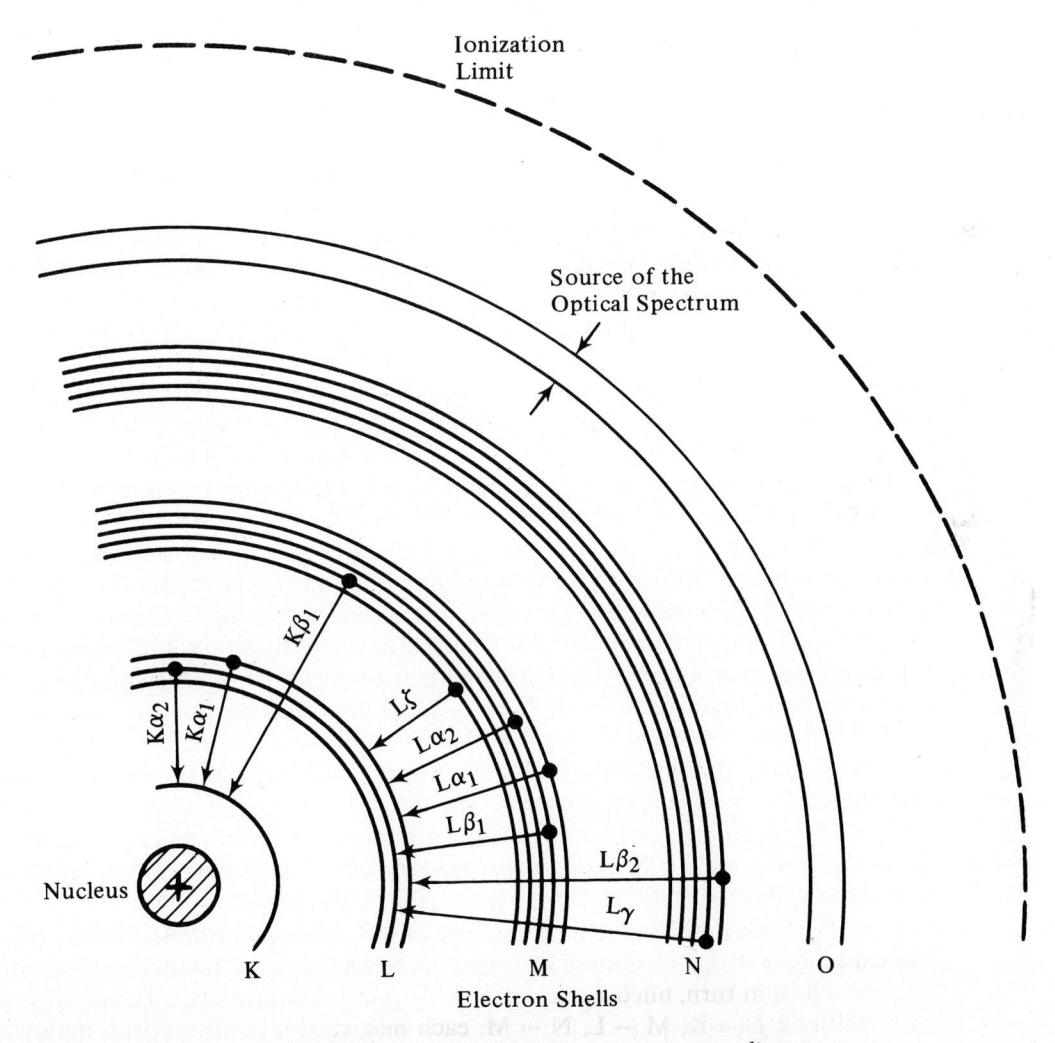

FIGURE 14.1. *Origin of characteristic x-ray lines.*

outer electron involved in the transition), plus a numerical subscript. This numerical subscript indicates the relative strength of each line in a particular series—for example, $K\alpha_1$ is more intense than $K\alpha_2$. Because there are a limited number of possible inner-shell transitions, the x-ray spectrum is much simpler than the complex optical spectrum that results from the removal or transition of valence electrons; in addition, both the intensity and wavelength of x-rays are essentially independent of the chemical and physical state of the excited element.

There are a number of approaches to x-ray analysis. In *x-ray absorption*, the absorption of energetic x-rays that occurs when an electron is removed from its orbital is related to the concentration of the absorbing species. This absorption phenomenon follows the well-known Beer's law. In *x-ray fluorescence*, or *secondary x-ray emission*, the sample is bombarded with an x-ray beam and the reemitted x-radiation, called the characteristic radiation from the analyte elements, is measured.

Crystalline materials, in which the atomic spacing is about the same magnitude as x-ray wavelengths, are capable of diffracting x-rays. This serves as the basis of *x-ray diffraction* analysis; qualitative identification of crystalline materials is readily made from a measurement of the angles of diffraction. X-ray diffraction also serves as a means for isolating x-rays of a particular wavelength in an x-ray spectrometer. This is similar to what is done with a prism or grating in conventional visible or ultraviolet spectrophotometers.

X-ray fluorescence spectrometry is the most widely used x-ray technique for quantitative analysis; this chapter is concerned primarily with this method of analysis and also contains a brief discussion of x-ray absorption.

14.2 INSTRUMENTATION

Until the late 1960s, most x-ray spectrometers were of the wavelength-dispersive type. In wavelength-dispersive spectrometers, wavelengths are separated by Bragg diffraction from a single crystal as shown in Fig. 14.2A. The x-ray tube is usually of high intensity (approx. 3 kW) with a stabilized high-voltage power supply. This is necessary because large losses of characteristic radiation occur due to the relatively low reflectivity of the dispersive crystals. The detector is mounted on a goniometer, which allows the detector to accept one wavelength at a time at the 2θ diffraction angle, and covers a broad range from a few degrees to 150°. Either a proportional or scintillation-counter detector is used, or both can be incorporated in a tandem arrangement. The associated electronics include a DC power supply, scaler-ratemeter, linear amplifier, single-channel analyzer, and recorder. The cost of research-type, single-channel, or sequential-wavelength dispersion x-ray spectrometers with microprocessor or computer-based control of hardware parameters such as choice of crystals, collimator, x-ray tube voltage/current, and so forth, is about $200,000.

In the early 1970s, the development of the solid-state lithium-drifted silicon detector, Si(Li), with a resolution of 165 eV or better, gave rise to another generation of commercially available instrumentation for x-ray fluorescence analysis, at lower cost due to simpler mechanical requirements. Because this semiconductor detector is responsive to the *energies* of the characteristic x-ray radiation, instruments utilizing

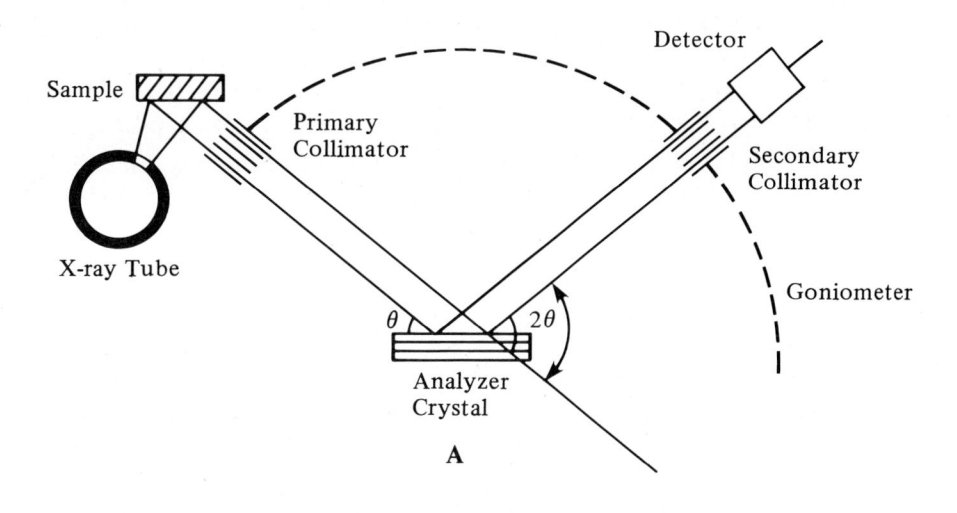

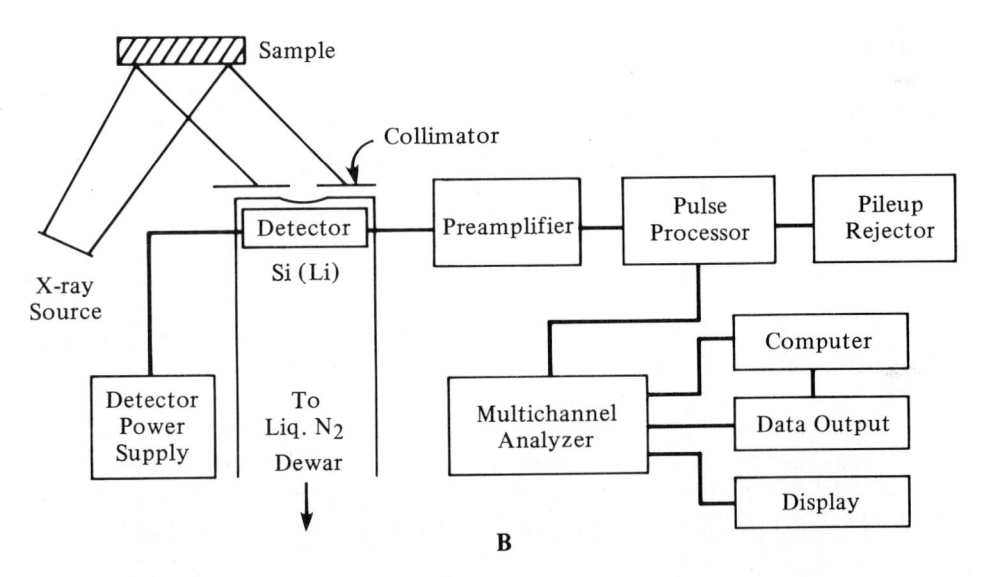

FIGURE 14.2. *X-ray fluorescence spectrometers. A: Wavelength-dispersive spectrometer. B: Energy-dispersive spectrometer.*

these detectors are called energy-dispersive x-ray spectrometers. A schematic diagram of this type of spectrometer is shown in Fig. 14.2B. The energy of a characteristic x-ray produces an electronic pulse-amplitude distribution in the detector. These pulses are sorted and stored according to their respective pulse heights by a multichannel analyzer. Hence, an entire x-ray spectrum from several elements in a specimen can be stored at one time. The detector must, however, operate at liquid-nitrogen temperatures, and a reservoir (Dewar flask) attached to the detector must be filled periodically with liquid nitrogen. The preamplifier, pulse processor, and pileup-rejection circuitry constitute a sophisticated electronic package designed to maintain the quantitative properties of the signals produced in the detector. These are necessary to ensure high performance with respect to energy resolution, data-acquisition times, and signal-to-noise ratio (i.e., detection limit). The x-ray tubes in these systems are usually of much lower power (approx. 500 W) to limit the counting rate in order to avoid x-ray peak shifts and line broadening. However, the recent development of pulsed x-ray tubes has allowed very rapid data-acquisition times without serious loss of resolution. The cost of energy-dispersive instrumentation ranges from $50,000 to $100,000.

Each of the above types of commercial spectrometers has its advantages and disadvantages, especially in particular applications. For example, the resolution of the crystal spectrometer is many times greater than the energy-dispersive system for K and L lines of elements of low to medium atomic number. The converse is true for elements of high atomic number. The choice of system still very much depends on the intended application and the financial resources available.

Generation of X-Rays

For most analytical applications, primary x-rays are produced by bombarding a suitable target with 10–100-keV electrons, as for example in the Coolidge-type x-ray tube illustrated in Figure 14.3. The spectrum resulting from electron excitation consists of a broad band of energies (the *continuum* or *bremsstrahlung*) plus photons of discrete energies that are characteristic of the target element (see Fig. 14.4). The frequency of the characteristic photons is described by Moseley's law:

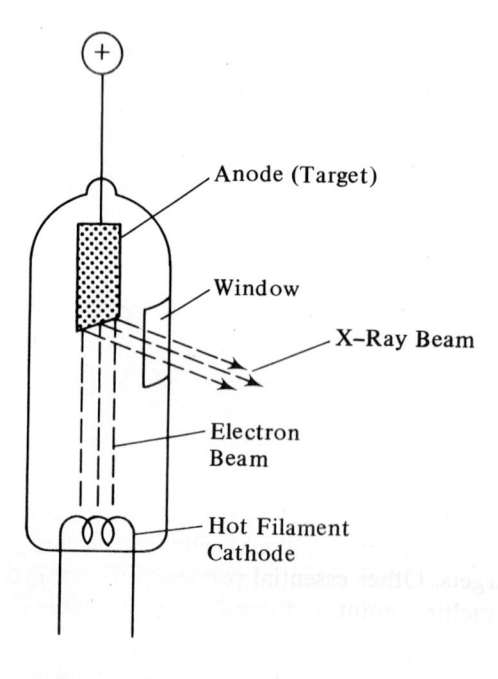

$$\nu = k(Z - 1)^2 \tag{14.1}$$

FIGURE 14.3. *Schematic diagram of a Coolidge high-vacuum x-ray tube.*

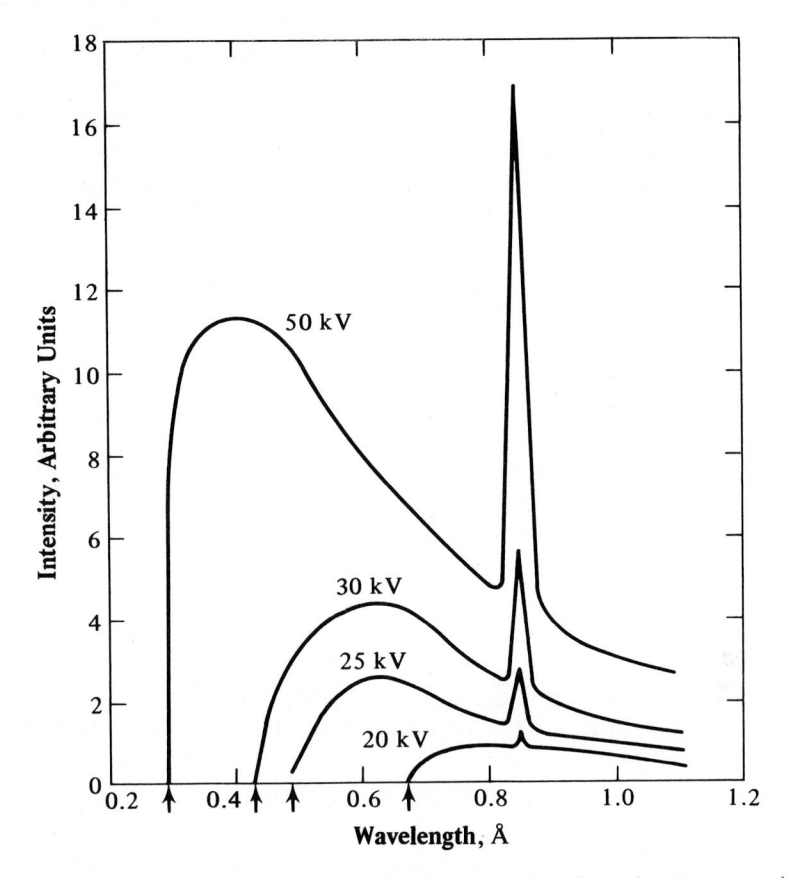

FIGURE 14.4. *Variation in x-ray distribution with voltage (yttrium target). The characteristic Y Kα line is at 0.83 A. The short-wavelength cutoff is indicated by an arrow for each voltage.* From F. J. Welcher, ed., **Standard Methods of Chemical Analysis**, *6th ed., Vol. IIIA, p. 169; Van Nostrand Reinhold Company, Copyright 1966, Litton Educational Publishing, Inc. Used with permission of Wadsworth, Inc.*

where Z = atomic number of the target element
        k = a constant

The integrated intensity (I) of the continuum is related to the current, voltage, and atomic number of the target by

$$I = kiZV^2 \qquad (14.2)$$

where i = x-ray tube current in milliamperes
        V = voltage in kilovolts

Inspection of Equation 14.2 shows that the integrated intensity of the continuum is proportional to the atomic number; therefore, higher-Z elements such as tungsten or platinum are often used as targets. Other essential properties of a target besides high atomic number are a high melting point and good thermal conductivity. The efficiency of primary x-ray production is less than 1%; the other 99+% of the electron energy is transferred to the target as heat.

As indicated in Equation 14.2, the continuum intensity is proportional to the square of the applied voltage. The distribution of the continuum from an yttrium target as a function of applied voltage is shown in Figure 14.4. Note that, as the voltage is increased, the peak of the continuum moves to a higher energy (shorter wavelength). As a first approximation, the wavelength of maximum intensity in the continuum is $3\lambda_0/2$, where λ_0 is the short-wavelength limit (in m) set by the relationship

$$\frac{hc}{\lambda_0} = Ve \tag{14.3}$$

where V = potential in volts
 h = Planck's constant
 c = speed of light in vacuum
 e = charge on the electron

Substitution for the constants in Equation 14.3 results in

$$V = \frac{12.4}{\lambda_0} \tag{14.4}$$

for V in kilovolts and λ_0 in angstroms. At the short-wavelength limit, all the energy of the electron hitting the target is converted to a single photon, with none left over for heat loss or multiple emission.

The applied voltage required to produce the characteristic lines of each spectral series also increases with increasing atomic number. Minimum energies (in keV) for the K, L, and M series x-ray lines are in the following ranges:

K series: 1.1 for ^{11}Na to 115 for ^{92}U
L series: 1.2 for ^{30}Zn to 21.7 for ^{92}U
M series: 0.41 for ^{40}Zr to 5.5 for ^{92}U

Each of these minimum (critical) energies, corresponding to the minimum photon or electron energy that can expel an electron from a given level in the atom, is known as the *absorption edge* of that level for the particular element. Each element has as many absorption edges as it has excitation potentials—one K, three L, five M. For example, the K edge for calcium is 4.038 keV; the three L edges for lead (L_I, L_{II}, and L_{III} corresponding to the different L electrons—see Fig. 14.1) are 15.870, 15.207, and 13.044 keV, respectively.

Example 14.1. Determine the lowest x-ray tube applied voltage that can be used to excite the following characteristic lines: W Lα, Cu Kα, and Pb Kα.

Solution:
$$V \text{ (kV)} = \frac{12.4}{\lambda_{\text{abs edge}} \text{ (Å)}}$$

The $\lambda_{\text{abs edge}}$ values are as follows: W Lα = 1.025 Å, Cu Kα = 1.380 Å, Pb Kα = 0.141 Å. And V (in kV) = 12.1 for W Lα, 8.99 for Cu Kα, and 87.9 for Pb Kα. Note that the K spectrum of lead cannot be excited by a 60-kV high-voltage power supply.

The intensity of a characteristic line excited by electrons is related to the applied voltage by the expression

$$I = ki(V - V_0)^n \quad (14.5)$$

The value of n (which ranges over the interval between 1 and 2) depends on the ratio between the applied voltage V and the critical energy V_0; for voltages less than $4V_0$, the value of n is close to 2.

In x-ray fluorescence spectrometry, the characteristic x-rays of the sample are generated (excited) by both the characteristic and continuum x-rays of the source. An important consideration is the relative excitation of secondary x-rays by the continuum as compared to those generated by primary characteristic x-rays. Although the total intensity of the characteristic primary peak or peaks is low compared to the continuum, the efficiency of excitation by the characteristic lines is significantly greater in many instances because of the higher absorption of these x-rays close to the surface. This higher excitation efficiency is a result of the limiting depth beyond which all secondary (fluorescence) x-rays are absorbed within the sample (the depth being related to the linear absorption coefficient of the sample for the spectral or "analytical" line). The depth of sample analyzed will range from less than 1 mm to 1 cm, depending on the energy of the spectral line and the composition of the sample. Because of this internal absorption, only those incident photons absorbed close to the surface are efficient producers of fluorescence x-rays.

Figure 14.5 shows the relative intensity of the secondary Fe Kα line emitted from an iron sample, using as primary targets in the x-ray tube some of the elements whose atomic numbers are in the range 26 to 47. With an iron target ($Z = 26$), the only source of excitation is the continuum and, therefore, that intensity is considered as unity. With targets of nickel and copper ($Z = 28$ and 29, respectively), the characteristic primary x-rays just exceed the critical energy of the Fe K absorption edge, and so excitation of Fe K radiation is a maximum. When using a high-Z target such as silver ($Z = 47$), the characteristic Ag K radiation is not absorbed near the surface of the iron sample, so the resultant Fe Kα radiation, generated relatively deep in the sample, is absorbed before escaping to the surface.

Methods of Sample Excitation in Energy-Dispersive Instruments

Some commercial energy-dispersive x-ray spectrometers take advantage of the higher excitation efficiency of certain characteristic x-ray lines by incorporating external secondary-target emitters. These provide selectable characteristic x-ray lines for excitation with enhanced sensitivity, as is shown in Figure 14.6A. In this form of excitation, the output spectrum of the x-ray tube impinges on a pure metal foil (the secondary target), generating characteristic x-rays; these in turn are used to excite the analyte x-rays in the specimen. For this so-called monochromatic form of excitation to be efficient, the secondary targets selected must have x-ray energies just above those of the analyte lines of interest. Another advantage of this form of excitation is that the background on the lower-energy side of the scattered radiation from the secondary target is relatively flat, and allows low limits of detection to be attained, at least over a selected energy range. Figure 14.7 is an x-ray spectrum of urban

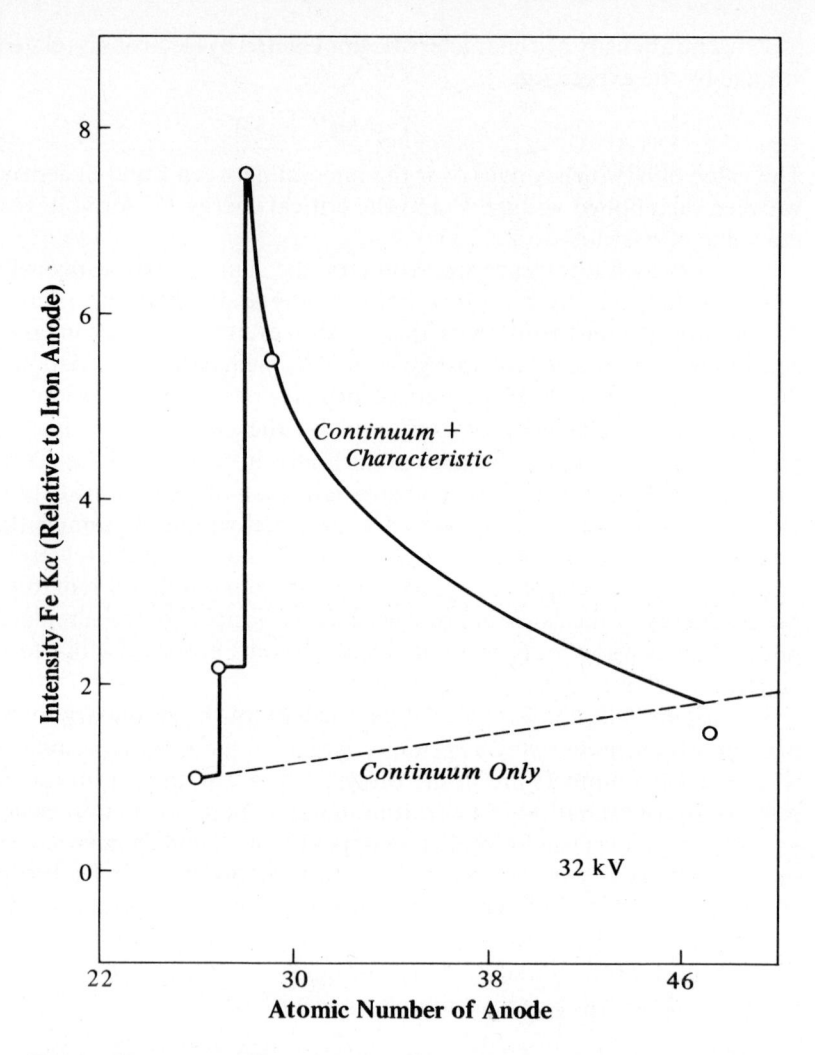

FIGURE 14.5. *Secondary excitation of iron using various x-ray tube targets.*

particulates collected on a membrane filter. A blank filter showing the flat background is shown for comparison. It is important to recognize that an energy x-ray spectrum from a multielement sample can be accompanied by spectral artifacts such as escape and sum peaks. Therefore, caution should be exercised in spectral identification because escape peaks, for example, can fall near or on analyte peaks and can lead to misidentification. Because both escape- and sum-peak energies are predictable in any given spectrum, it is routine practice to identify these peaks during the process of spectral identification.

Figure 14.6B through D illustrates other configurations that are used for excitation in energy-dispersive x-ray spectrometry. Configurations B and C are similar in that the continuum from the x-ray tube provides excitation. The major difference is that the use of a transmission filter, as shown in Figure 14.6B, can make

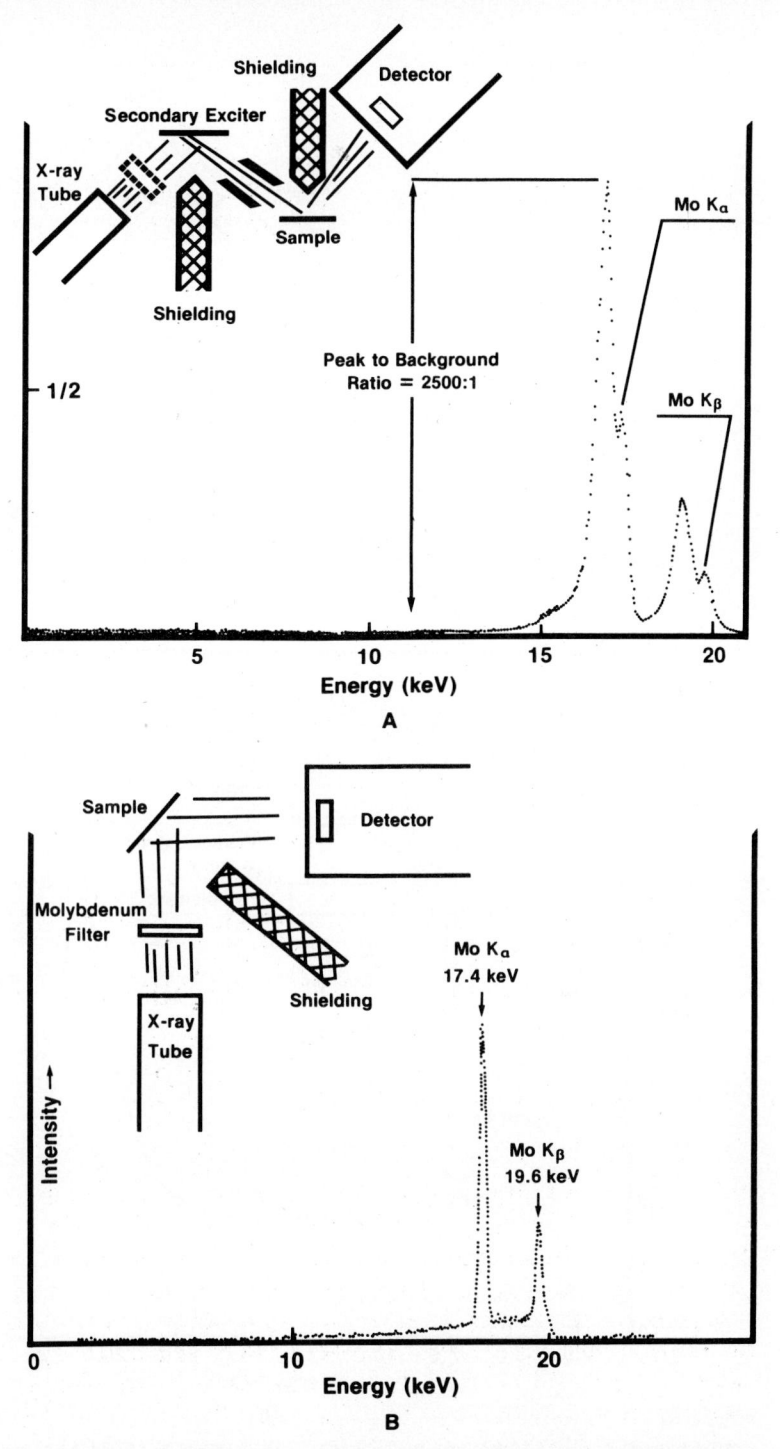

FIGURE 14.6. *Methods of excitation in energy-dispersive x-ray spectrometry. In each figure, the spectrum is that of the x-rays scattered off the sample. A: X-ray tube secondary-target excitation. B: Transmission x-ray tube excitation.*

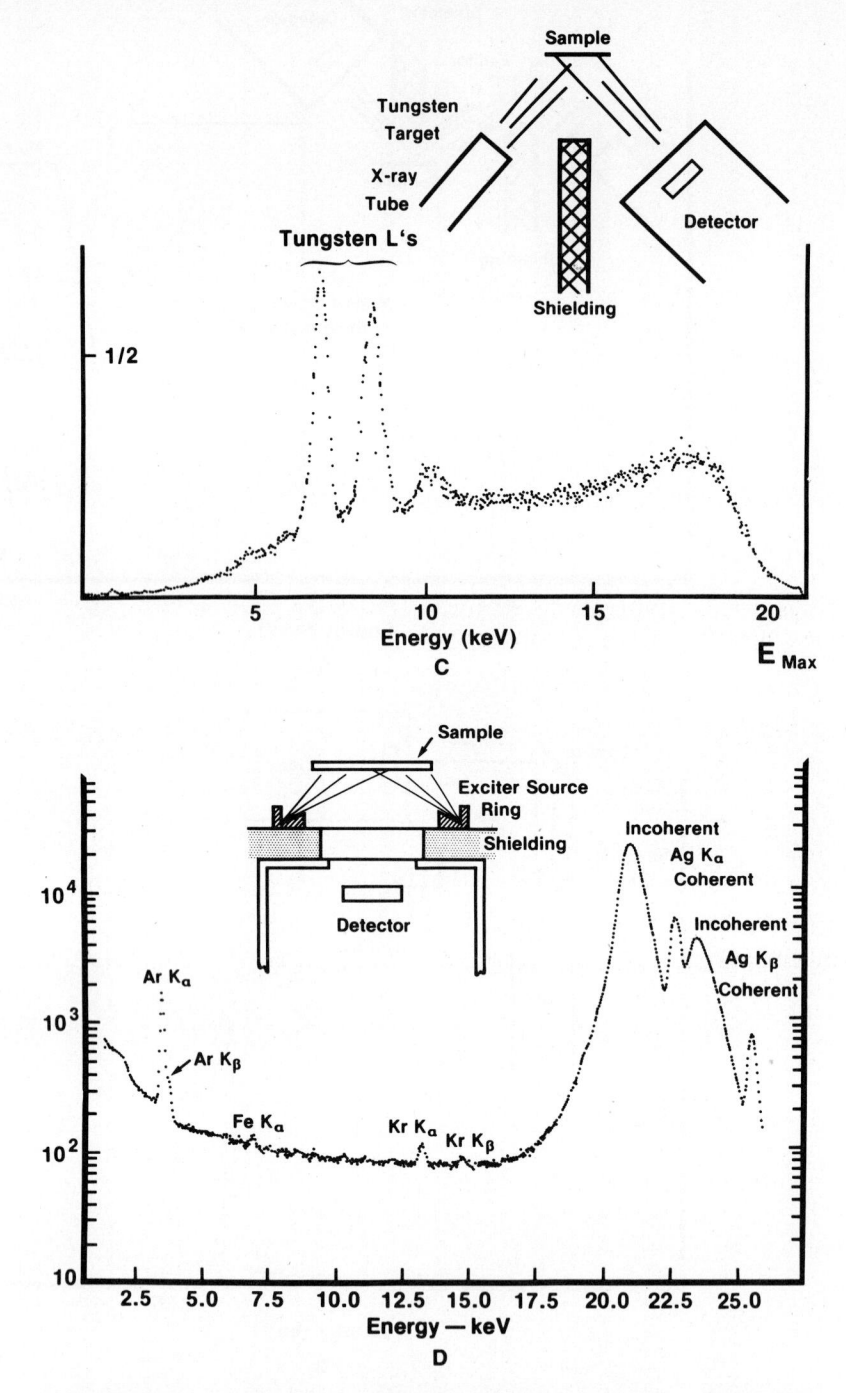

FIGURE 14.6. (continued). C: Bremsstrahlung excitation. D: Radioactive source excitation. The source is ^{109}Cd which decays ($t_{1/2}$ = 450 days) to ^{109}Ag, which produces the Ag x-rays. Ar and Kr peaks originate from the air around the source. Figures courtesy of the Kevex Corporation.

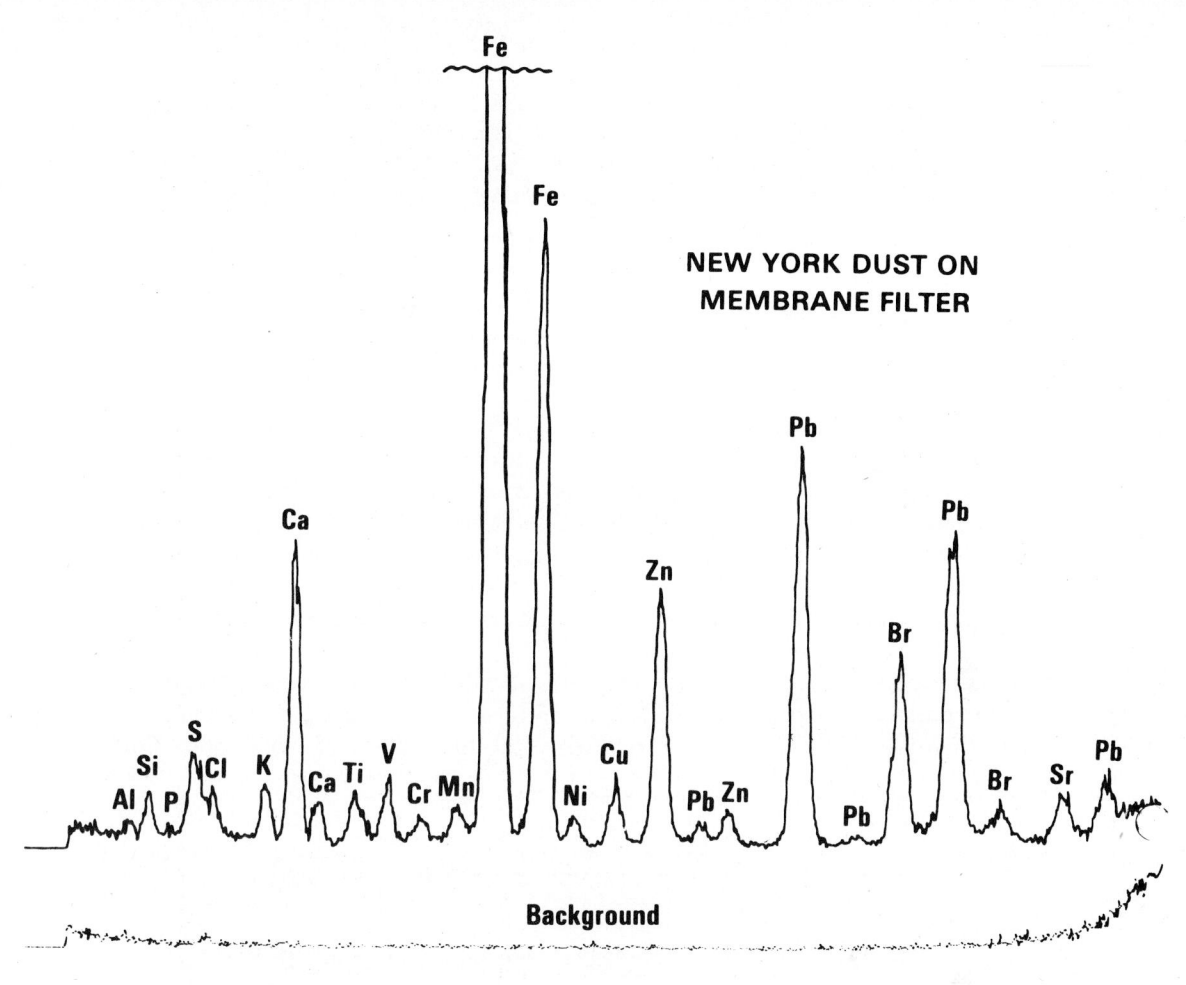

FIGURE 14.7. *Energy-dispersive spectrum of an urban particulate collected on a membrane filter using a Mo secondary-target exciter. (For simplicity, the overlaps of x-ray lines have not been designated.)*

the excitation more selective or can remove x-ray tube target lines that interfere with analyte x-ray peaks. A typical geometry for radioisotopic sources is the annular-ring arrangement shown in Figure 14.6D. A disadvantage of radioisotopic sources is that they have a flux six to eight orders of magnitude lower than high-powered sealed x-ray tubes. Therefore, close coupling of the source, sample, and detector is mandatory.

Example 14.2. For thin specimens, the limit of detection, C_D can be defined as the smallest mass loading (in $\mu g/cm^2$) that can be detected reliably with 95% confidence. Assuming the measurement errors are random only, C_D can be expressed as

$$C_D = 3.29(N_b)^{-1/2}(\text{BEC}) \tag{14.6}$$

where N_b = total background count in t sec
BEC = background equivalent concentration

If 10 measurements of the blank filter for the Pb Lα line gave an average count rate of 3.2 counts per second (cps), calculate the limit of detection for Pb if 100-sec and 1000-sec counting times are used. (The calibration function is 9.2 cps per $\mu g/cm^2$, as measured from a thin-film standard containing Pb. Thus, BEC = 3.2/9.2 = 0.35 $\mu g/cm^2$.)

Solution: For 100 sec, $C_D = 3.29(320)^{-1/2}(0.35) = 0.064$ $\mu g/cm^2$.
For 1000 sec, $C_D = 3.29(3200)^{-1/2}(0.35) = 0.020$ $\mu g/cm^2$.

Notice how the limit of detection can be improved (decreased) by increasing the counting time. Why? (*Answer:* The experimental estimate of the precision of the blank is increased. The detection limit is proportional to $t^{-1/2}$.)

Radioactive sources are used especially in portable units in applications encompassing a limited range of elements. The detector in such instruments is usually a scintillation counter. Portable analyzers employ the technique of balanced filters to isolate the analyte line of interest. These filters consist of two thin metallic foils with K absorption edges on the low- and high-energy sides of the x-ray line of interest. For example, as shown in Figure 14.8, the Sn Kα line just exceeds the K edge of the palladium filter, and therefore is absorbed, whereas the Sn Kα x-rays are readily transmitted by the silver foil. The thicknesses of the two filters are carefully controlled so that their absorption characteristics are virtually the same for all radiation, except in the narrow pass-band. The difference in measured intensity, using first the silver and then the palladium filter, is related to the tin content of the sample. Portable

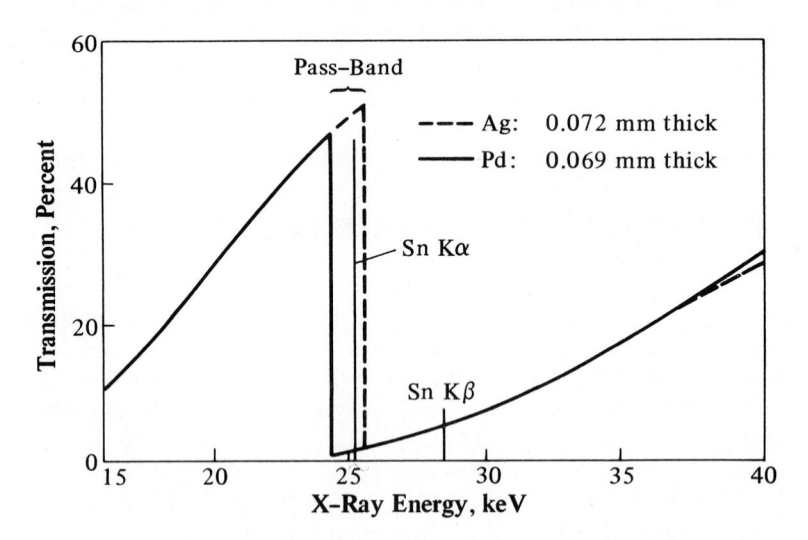

FIGURE 14.8. *Balanced filters for isolating Sn Kα radiation. The transmission and absorption filters are made of silver and palladium, respectively. From S. H. V. Bowie, A. G. Darnby, and J. R. Rhodes,* Trans. Inst. Min. Met. 74, 36 (1964–65), *by kind permission of J. R. Rhodes, Columbia Scientific Industries, P. O. Box 9908, Austin, Texas.*

analyzers complement the conventional x-ray spectrometer for process control and field applications, but are not competitive in the laboratory.

Radioactive Sources

Radioisotopes commonly used for energy-dispersive analysis are listed in Table 14.1. Sealed sources of all types are now commercially available. The price of a source depends on the cost of the radioisotope and the complexity of the instrument design.

Alpha emitters such as polonium-210 and curium-242 are used to excite emission of low-energy x-rays. They offer the advantage of large signal-to-background ratios, but the thickness of sample analyzed is extremely small. The alpha emitters are health hazards and generally have very limited application in energy-dispersive analysis.

Beta emitters are used to generate continuum plus characteristic x-rays that collectively excite the sample. These beta sources may be a thin layer of the isotope on a suitable target, or a mechanical or chemical mixture of isotope and target.

Isotopes that decay by K electron capture emit essentially monoenergetic x-radiation; for example, ^{109}Cd and ^{55}Fe emit Ag K and Mn K x-rays, respectively. Large signal-to-background ratios can be achieved with these "monoenergetic" sources. Many analysts recommend the use of a K-capture isotope whose characteristic radiation just exceeds the absorption edge of the element being determined. This criterion is applicable to wavelength dispersion. However, for energy dispersion one must allow for the poorer resolution of the analytical system, and so the criterion must be modified to resolve the characteristic x-ray emitted by the sample from the incoherent Compton radiation (see Chap. 19) and the coherent scattered incident radiation. To achieve this resolution, the primary x-rays must be either significantly higher in energy than the K or L absorption edge of the element being determined, or so close to the edge that the incoherent scatter falls below the energy of the analytical line.

TABLE 14.1. *Radioisotopes Commonly Used in Energy-Dispersive X-Ray Analysis*

Radioisotope	Half-Life, yr	Principal Mode of Decay	Principal Radiation
^{55}Fe	2.7	Electron capture	Mn K x-rays
^{109}Cd	1.3	Electron capture	Ag K x-rays, 88 keV gamma
^{125}I	0.16	Electron capture	Te K x-rays, 35 keV gamma
^{3}H/Zr	12.3	β emission	Bremsstrahlung 3–12 keV
^{147}Pm/Al	2.6	β emission	Bremsstrahlung 10–50 keV
^{241}Am	458.0	α emission	Np L x-rays, 26 and 60 keV gamma
^{153}Gd	0.65	Electron capture	Eu K x-rays, 97 and 103 keV gamma
^{57}Co	0.74	Electron capture	Fe K x-rays, 14, 122, and 136 keV gamma
^{242}Cm	17.6	Electron capture	Pu L x-rays
^{236}Pu	86.4	Electron capture	U L x-rays

Dispersion of X-Rays

As indicated earlier, there are two general approaches to sorting x-rays of different wavelengths (different energies) in a spectrometer: by *wavelength dispersion*, in which the x-rays are identified after the original beam has been spread out by an analyzing crystal; and by *energy dispersion*, in which the detected x-rays are sorted into 100 to 1000 energy groups by means of a high-resolution semiconductor detector coupled to a multichannel analyzer. Here, we will consider the former.

Chapter 6 described the use of diffraction gratings to disperse electromagnetic radiation in the ultraviolet to infrared regions. The radiation is diffracted if the spacings between lines on the grating are of the same magnitude as the wavelength of the radiation. This condition would be very difficult to meet for x-rays, because their wavelengths are only a few angstroms at most. The spacings of atoms in crystals, however, are small enough to diffract x-rays; this has led to the widespread use of x-rays for determining crystal structures and, conversely, to the use of crystals to disperse x-rays in spectrometers.

The condition for diffraction of x-rays can be described by considering the diffraction of a monochromatic beam impinging on a row (plane) of atoms (or ions) as shown in Figure 14.9. Lines AD and CD are perpendicular to the incident and diffracted beams, respectively. The beam diffracted from plane P_2 must travel a distance $\overline{ABC}$ further than that diffracted from plane P_1. Since angles ADB and BDC are equal to the angle of incidence and diffraction θ,

$$\overline{AB} = d \sin \theta \qquad (14.7)$$

or

$$\overline{ABC} = 2d \sin \theta \qquad (14.8)$$

where d = distance between planes in the crystal

Those waves out of phase after diffraction will interfere destructively and not be observed. Only those in phase will reinforce and be observed. This occurs when $\overline{ABC}$ is an exact multiple of the wavelength of the incident beam λ. Hence,

$$n\lambda = 2d \sin \theta \qquad (14.9)$$

where n = *order* of the diffraction pattern

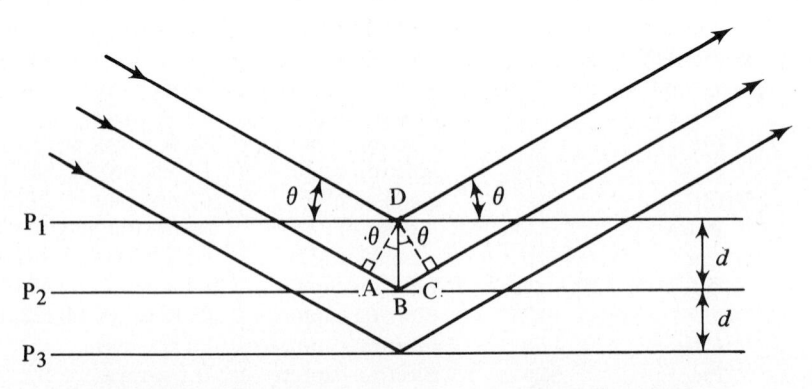

FIGURE 14.9. *Diffraction of x-rays from the planes of a crystal.*

When n is unity, the diffracted radiation is called first order. Higher orders of diffracted radiation fall off in intensity. Equation 14.9 is known as *Bragg's law*, and may be compared to Equation 6.8 for the diffraction of radiation from a grating. For x-rays of a given wavelength, then, diffraction will be observed at only certain values of θ (determined by d). These θ-values can be determined by rotating the crystal and measuring the angle of diffracted x-rays with respect to the angle of the incident beam. The d spacings of a given plane can therefore be calculated. Or, if the d spacing of a crystal is known, the angle at which a given x-ray will be diffracted (in a spectrometer, for instance) can be calculated.

Example 14.3. The first-order diffraction of the Mo Kα 0.712 Å line from a plane of calcium fluoride is observed as strong radiation at a 2θ angle of 12.96°. What is the distance between planes?

Solution: $n\lambda = 2d \sin \theta$
$1(0.712) = 2d \sin(6.48)$
$d = 3.15$ Å

Bragg originally used this relationship to determine the wavelengths of x-rays. He diffracted x-rays from a sodium chloride crystal and used as the d-value that calculated from the density of the crystal and Avogadro's number. Once λ was determined, then x-ray diffraction could be used to determine d spacings in other crystals, and x-ray crystallography (the study of crystals) was born.

There are two types of x-ray optical systems used in wavelength-dispersive spectrometers. They are the *nonfocusing* (flat-crystal) and *focusing* (curved-crystal) systems. In flat-crystal optics, the x-ray beam is collimated by a closely spaced series of parallel metal foils (see Fig. 14.2A), where the beam divergence is limited by the spacing between the foils and the lengths of the foils. In curved-crystal optics, either a small region of the sample is excited using a pinhole collimator to limit the size of the primary x-ray beam, or a large area is irradiated and a divergence slit serves as the "optical source" of secondary x-rays. With focusing optics, a cylindrical concave crystal focuses the diffracted x-rays onto a circle having a radius equal to one-half the radius of curvature of the crystal. The crystal has been mechanically bent to a radius $2r$, then ground to the radius r of the focusing circle. The x-ray source, crystal surface, and detector slit all lie on the focusing circle. Focusing and nonfocusing optics give essentially the same intensity and line-to-background ratio for large samples, whereas for small samples, focusing optics are at least an order of magnitude more efficient.

Analyzer crystals must satisfy the Bragg relationship for the analytical line without exceeding the maximum 2θ angle available with commercial instrumentation, (approx. 150°). In general, the analyzing crystal should be composed of elements of low atomic number to avoid high background from fluorescence x-rays generated in the analyzing crystal itself. The dispersion of x-rays is inversely related to the d spacing and to the Bragg angle; so

$$\frac{d\theta}{d\lambda} = \frac{n}{2d \cos \theta} \qquad (14.10)$$

Therefore, crystals of low d spacing (high dispersion) are used to disperse partially overlapping x-ray spectral lines. The characteristics of some typical crystals are given in Table 14.2.

> *Example 14.4.* The wavelengths of two analyte lines are 1.240 and 1.261 Å. Compare their angular dispersions using LiF$_{200}$ ($2d = 4.028$ Å) and LiF$_{220}$ ($2d = 2.848$ Å), assuming first-order reflection. (Note: LiF$_{200}$ and LiF$_{220}$ crystals are cut in such a way that the reflecting planes [hkl] are 200 and 220, respectively.)
>
> *Solution:* First, solve for the diffraction angles for each crystal from Bragg's law (see Example 14.3):
>
> For LiF$_{200}$: 1.240 Å diffracts at 17.93°, 1.261 Å at 18.24°.
> For LiF$_{220}$: 1.240 Å diffracts at 25.81°, 1.261 Å at 26.28°.

Then,

$$\text{LiF}_{200}: \frac{d\theta}{d\lambda} = \frac{1}{(4.028)(0.951)} = 0.261 \text{ at } \theta = 17.93°$$

$$\frac{d\theta}{d\lambda} = \frac{1}{(4.028)(0.949)} = 0.261 \text{ at } \theta = 18.24°$$

$$\text{LiF}_{220}: \frac{d\theta}{d\lambda} = \frac{1}{(2.848)(0.900)} = 0.390 \text{ at } \theta = 25.81°$$

$$\frac{d\theta}{d\lambda} = \frac{1}{(2.848)(0.897)} = 0.392 \text{ at } \theta = 26.28°$$

Detectors

Three types of detectors are used in wavelength-dispersive x-ray analysis: Geiger, proportional, and scintillation detectors. (They detect ionizing radiation, of which x-rays are an example.) Geiger tubes are simple to operate and do not require highly stabilized electronic circuitry. However, they are of limited use because of two principal disadvantages: First, the counting loss (nonlinear response) is significant,

TABLE 14.2. *Properties of Analyzing Crystals*

Crystal	Reflecting Plane (hkl)	2d, Å	Comments
Lithium fluoride	200	4.03	Optimum crystal for all wavelengths less than 3 Å
Silicon	111	6.27	Suppresses even-ordered reflections, $n = 2, 4, 6, \ldots$
Pentaerythritol	002	8.74	Optimum for atomic numbers 13–17
Mica	002	19.93	Used primarily in curved-crystal optics for long-wavelength x-rays
Potassium hydrogen phthalate	002	26.63	Used for atomic numbers 6–12
Barium stearate	—	100	Used with curved- or flat-crystal optics for wavelengths greater than 20 Å

even at moderate x-ray intensities, because of the relatively long dead-time between counts; second, discrimination of x-ray energies is not possible.

The spectral sensitivity of proportional detectors is similar to that of Geiger counters, but proportional detectors can be used at relatively high counting rates (high x-ray intensities), since the dead-time is small. Also, the output voltage of this type of detector is proportional to the energy of the incident x-ray photon, so that direct energy discrimination is possible without the need for a dispersing medium. For x-rays longer than 2 Å, the windows of these detectors are very thin organic films of mylar, formvar, or nitrocellulose, all of which have a high transmittance for long-wavelength x-rays. Since these windows are porous, the counting gas is passed in a continuous stream through the detector (*flow-proportional detectors*).

The most generally useful detector in conventional x-ray spectrometry is the scintillation counter, which incorporates a very low dead-time and excellent sensitivity for x-rays of wavelength less than 2 Å. Since the output voltage is proportional to the energy of the incident x-ray photon, electronic pulse-amplitude discrimination can be used to reject x-rays whose energies are sufficiently different from those of the spectral lines being measured. The resolving power of the scintillation counter is approximately one-half to one-third that of the proportional detector. However, a high degree of energy resolution is not generally required in wavelength-dispersion applications, since the discrimination is against higher-order x-rays whose energies are multiple integer values of the desired radiation.

Dead-time refers to the time interval during which the detector is insensitive and gives no response to a successive incoming x-ray photon. It is usually used to characterize the performance of the overall detection system and can be determined experimentally. For example, the measured count rate will be lower than the "true" count rate, especially above 50,000 cps for most systems. From experimental data, the dead-time of a system can be calculated from the equation

$$I_\tau = \frac{I_m}{1 - I_m \cdot t_d} \tag{14.11}$$

where I_τ = "true" count rate
I_m = measured count rate
t_d = dead-time in seconds

Example 14.5. From experimental data, the dead-time of an x-ray system is 3.0 μsec. What relative error does this represent at a measured count rate of 20,000 cps?

Solution:
$$I_\tau = \frac{I_m}{1 - I_m \cdot t_d}$$
$$= \frac{20{,}000}{1 - (20{,}000)(3.0 \times 10^{-6})} = \frac{20{,}000}{1 - 0.060} = 21{,}280 \text{ cps}$$

$$\text{Relative Error} = \frac{I_\tau - I_m}{I_\tau} \times 100\% = 6.0\%$$

Solid-state semiconductor detectors are excellent for energy resolution. These (and other) detectors are described in detail in Chapter 19.

In Figure 14.10, the energy resolution of proportional, scintillation, and semiconductor detectors is compared to wavelength-diffraction resolution. The energy resolution of the three types of detectors has approximately the same slope; that is, it is proportional to the square root of the photon energy. The resolving power of the semiconductor Si(Li) detector is approximately a factor of 10 better than that of the scintillation counter and a factor of 3 better than the gas-proportional detector.

The Si(Li) detector resolution is conventionally specified at the full width at half maximum (FWHM) of the Mn Kα line of 5.9 keV. State-of-the-art detectors may have resolutions as low as 145 eV, but typically closer to 160 eV in the 3–20-keV range. For lower energies, the resolution provided by a LiF-crystal (diffraction) spectrometer is a factor of 5 to 10 superior to the best semiconductor detector. In the spectrum in Figure 14.7, taken with a Si(Li) detector, several spectral lines are overlapping other lines not labeled in the figure. The characteristic lines that overlap are as follows: S Kα and Pb M; K Kβ and Ca Kα; Ti Kβ and V Kα; V Kβ and Ca Kα; Cr Kβ and Mn Kα; Mn Kβ and Fe Kα. These lines can be resolved with the LiF-crystal spectrometer.

However, a powerful technique called *spectrum stripping* is now part of the software available with energy-dispersive spectrometers. One procedure allows

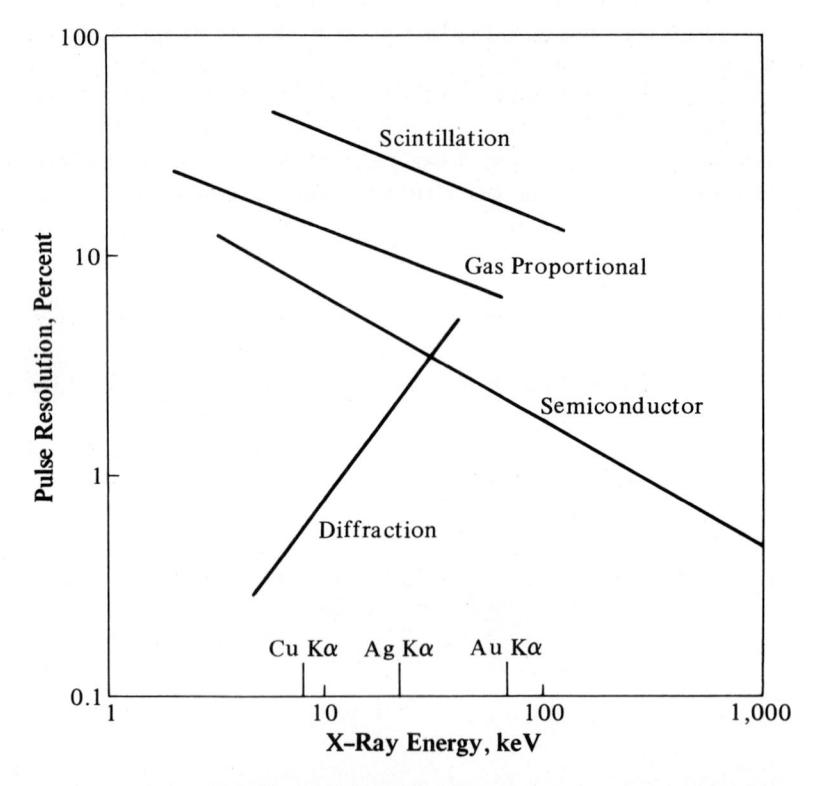

FIGURE 14.10. *Resolution of energy-proportional detectors and of crystal diffraction as a function of photon energy.*

reference spectra from noninterfering analytes to be subtracted from the spectrum of interest to remove the interfering peak. With increasing photon energy, the resolution of a semiconductor detector becomes more favorable and, at high energy, provides superior resolution, because high-reflectivity crystals of very low d spacing are not available.

14.3 PREPARATION OF SAMPLES FOR X-RAY FLUORESCENCE ANALYSIS

X-ray fluorescence spectrometry is applied to virtually every type of elemental determination encountered by control and research laboratories. The types of samples analyzed include ashes, ores, minerals, ceramics, metals and alloys, and films and coatings. The samples may be in the form of powders, solutions, rods, sheets, films, or particulates. The method is, in general, nondestructive; the sample can be retained for other analyses.

Solid samples used in x-ray fluorescence analysis include metallurgical specimens, briquetted powders, and borax disks. For quantitative analysis, both standards and unknowns must be in the same matrix and subjected to the same preparation for reliable results. Because of the limited escape depth of secondary x-rays, particularly in the long-wavelength region, the surface layers must be representative of the entire sample.

Surface preparation is an essential step in achieving good quantitative results. For metallurgical samples, the surface is generally prepared by grinding, followed by polishing. (Etching techniques are not used, because some elements may be preferentially removed from the surface layer.) Particle size and particle-size distribution are very important variables in the analysis of powdered samples. Powdered samples are processed either by adding a small amount of binder and forming into a briquet using high pressure, or by fusing with a suitable flux such as sodium tetraborate and casting the molten sample into a glass disk. Fusion methods are generally preferable to briquetting, since the fused sample is homogeneous on a scale of micrometers, eliminating particle-size effects. Also, the fluxing reagent serves as a diluent for the sample, thereby reducing interelement effects.

Solution samples are held in plastic or metallic containers with thin mylar windows transparent to x-rays. The most convenient procedure is to use cuplike containers with 0.006- or 0.02-mm mylar windows.

Samples in the gaseous or vapor state require cells with x-ray–transparent windows that can withstand high pressure differentials. Some type of pressure regulation must be used to maintain the number of atoms in the x-ray beam at a specified level.

Preparation of Samples for Trace Analysis

Trace analyses are conveniently classified into two types: minor or trace constituents in large samples (e.g., gold in low-grade ores), and major constituents of very small samples (e.g., titanium in a flake of paint). Samples of the first type can be converted to the second by physical or chemical concentration of the desired elements. Limits of detectability for the first type range from 0.1 to 100 ppm, depending on the elements being determined, the overall sample composition, and the complexity of the x-ray spectra. This is a concentration detection limit. Limits of detectability with the second type of sample are expressed in micrograms (generally about 0.01 to 1 μg) and may represent ppm, ppb, or lower concentrations in the original sample, depending on the sensitivity for the element and the size of the starting sample. The minimum amount detected is an absolute detection limit. Good examples of these sample types that are of environmental concern are aqueous discharges from industrial plants (type 1) and particulates collected on filter disks from air samples (type 2).

For trace analysis, it is desirable to prepare the sample such that matrix effects are negligible, and the lowest possible detection limits can be obtained. Analytes that can be deposited in a thin layer on suitable substrates can meet these requirements. Preconcentration has been accomplished through the use of ion-exchange-resin-loaded filter papers, or by coprecipitating the analytes with chelating agents and then collecting on filters. One drawback of the ion-exchange filters is their limited loading capacity. This does not present a problem when other sample constituents do not contribute to the ionic loading, for example, when determining trace metals in fresh water. However, in samples such as seawater, the alkali and alkaline-earth metals exceed the trace-element concentrations by 10^9. Therefore, prior separation of the trace metals from the matrix is necessary. Chelating resins such as Chelex-100 have been used for this purpose. These techniques have been successfully applied to a wide variety of environmental and biological samples. Using 1-L samples of seawater, Ni, Mn, Zn, Cu, and Pb have been determined in the 2–5-ppb range. Another advantage of such techniques is that standards prepared with resin-loaded papers have a very long shelf life and are easily stored.

When the separation-collection procedure is not quantitative, a radiotracer of the element being determined can be added to the initial sample to serve as a collection monitor. For example, in the determination of gold in low-grade ores, radioactive ^{195}Au is added to the ore prior to dissolution and subsequent collection on chelate-resin–loaded paper. The intensity of the Pt K x-rays from the K-capture decay of the ^{195}Au is a linear function of the fraction of gold collected on the resin-loaded paper. Radiotracers of ^{203}Hg^{2+} and CH$_3$^{203}Hg$^+$ are used to determine the collection efficiency for inorganic and organic mercury in aqueous effluents. The actual chemical form in which the mercury is present, as well as the total mercury content, is an important parameter in environmental studies.

In the analysis of particulates in air, the samples are collected by passing measured volumes of air through filter paper (or some other suitable filter medium). One method of standardization is to pipet small volumes of solution containing known amounts of the elements to be determined onto the same filter medium. Depending on the energy of the x-ray line and the size of the particulates, there may be systematic errors in the analysis using standards prepared from solutions. Another approach to standardization is to disperse known quantities of elements in the form of finely divided powders onto paper or glass-fiber filters. Problems with this approach include variation in particle size between standards and unknowns, and quantitative collection of the standard powders by the filter.

14.4 QUANTITATIVE X-RAY FLUORESCENCE ANALYSIS

Quantitative analysis is achieved by comparing x-ray intensities from unknowns to those from primary or secondary standards. Depending on the degree of similarity between unknown and standard, small or large correction factors may be necessary.

The measured intensity depends both on the concentration of the element being determined and on the overall composition of the sample. For example, the intensity of the Ag Kα line from various silver ores depends on both the total silver content and the matrix composition (Fig. 14.11). These samples represent matrices containing various amounts of lead, barium, and transition elements, plus minerals such as quartz and feldspar. Therefore, the analyst must have adequate knowledge regarding the composition of the unknowns, or use some technique that will minimize the compositional dependence. This dependence, commonly called the *matrix effect*, is the principal source of systematic error in quantitative x-ray analyses.

Seven general procedures are available to reduce or correct for the matrix dependence: comparison standards, thin films, dilution, internal standards, standard addition, scatter correction, and mathematical correction procedures. Each of these is discussed below. One essential requirement for both samples and standards in all

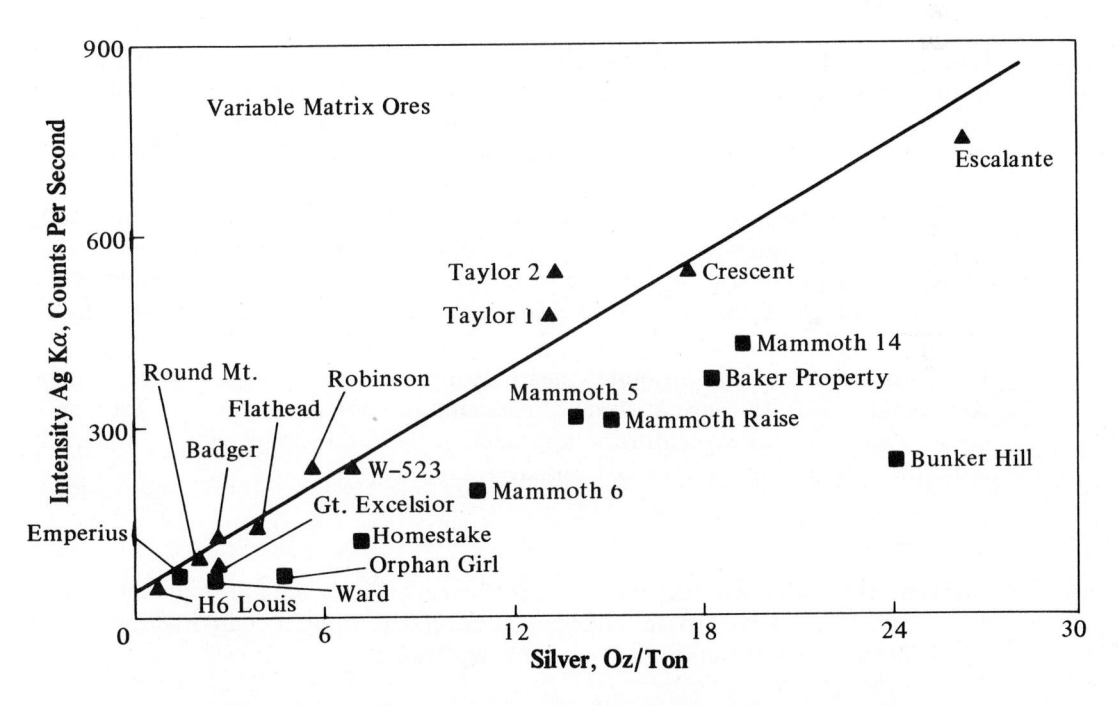

FIGURE 14.11. *Variation in Ag Kα intensity with mineralogical composition. ▲ = low-Z matrix; ■ = high-Z matrix. The line is a calibration curve derived from silver standards prepared in a silica matrix. From P. G. Burkhalter, Anal. Chem., 43, 10 (1971), by permission of the publisher. Copyright © 1971 by the American Chemical Society.*

methods (other than those using thin films) is homogeneity on a micrometer scale because of the low penetration of incident x-ray photons. Other practical factors to be considered for quantitative results include surface preparation and particle-size distribution.

Comparison Standards

The most widely used method of calibration is the comparison of intensities from unknowns with those from standards of similar composition, so-called in-type analysis. Obviously, this direct comparison requires knowledge regarding the probable composition of the sample (in industrial-control analyses, for example, the approximate composition of the sample is frequently known). When unknowns and standards are very close in composition, a simple ratio of intensities and composition is adequate, for example

$$\frac{I_u}{I_s} = \frac{c_u}{c_s} \tag{14.12}$$

where the subscripts "u" and "s" refer to unknowns and standards, respectively.

Thin Films

Interelement effects are small or negligible in thin-film samples, because neither the primary nor the secondary x-rays are strongly absorbed. Thus, the intensities of the secondary x-rays are directly proportional to the amount of the element present; matrix effects are minimized by this technique. Because standards and unknowns are prepared in a similar manner, linear comparison of intensities is valid.

To prepare thin-film samples, the desired elements must usually be chemically or physically separated from the host compound. The elements being determined are collected in a physical form suitable for x-ray analysis using such methods as ion exchange, solvent extraction, or precipitation. Metallic ions, for example, may be collected on resin-loaded paper, which also serves as the mechanical support in the x-ray spectrometer. The absolute sensitivity for elements isolated from the host compound is 0.01 to 1 μg; analysis of elements present in trace concentrations is possible with this preconcentration approach, if the total sample size is sufficiently large. (See the section on trace analysis, above.)

Dilution

One general method of reducing the matrix effect is to dilute standards and unknowns in a common "solvent." With sufficient dilution, the small weight fraction of the solute makes its contribution to the matrix negligible:

$$\lim_{W_A \to 0} \mu_{\text{sample}} = \mu_{\text{solvent}} \tag{14.13}$$

where W_A = weight fraction of the sample
μ = averaged linear absorption coefficient for the sample or solvent at a given wavelength

As a result of dilution, the element or elements being determined are now present as minor constituents; so the intensity-to-concentration relationship is linear and thus minimizes the matrix effects.

Dilution may be achieved by dissolving the sample in an inorganic or organic solvent (e.g., water or chloroform), or by fusing it with a flux (e.g., borax, carbonate, or pyrosulfate). Another approach is to add a strong absorber of x-rays, such as lanthanum oxide, to the flux to minimize the degree of dilution required to reduce matrix effects, that is, to swamp out the effects of a variable sample matrix. A typical application of the La_2O_3-flux approach is determining copper in a series of ore samples having a variable iron concentration. The dilution plus the strong absorber minimizes the absorption of Cu Kα by the iron.

Internal Standards

In the internal-standardization method, a known concentration of a reference element is added to the sample being analyzed. For x-ray spectrometry, the reference element should have a characteristic radiation that will be excited and absorbed to a similar extent as the characteristic radiation of the element being determined. Therefore, the internal standard is generally an element one atomic number higher or lower than the element being determined. In some instances, it may be necessary to use an element of much higher atomic number and to make use of its L or M radiation, for example Br Kα (11.9 keV) and Au Lβ_1 (11.4 keV).

Regardless of the reference line used, the matrix may affect the relative intensities of the reference and analytical lines in one of the following ways:

1. The matrix will have a slightly higher absorption coefficient for the longer-wavelength line.
2. Another element in the matrix will have an absorption edge between the reference line and the analytical line of interest.
3. An emission line from the matrix will preferentially excite the element of lower atomic number.

Classes of samples representing each of these situations are shown graphically in Figure 14.12. Matrix M is considered to be the third element, and the symbols X and R represent the element being determined (the unknown) and the reference element, respectively. If the matrix element emits line L_1 or L_3 or has absorption edge E_1 or E_3, there is no significant preferential absorption or enhancement of either the unknown or reference emission line. In contrast, if the matrix element has an absorption edge at E_2, then X Kα is preferentially absorbed by the matrix element and the internal standard is not applicable. Another source of systematic error occurs when the matrix element emits a strong line of wavelength L_2, whereby the reference element, R, is preferentially excited (by absorption at RE). The original sample, of course, must not contain appreciable amounts of the reference element.

Within these limitations, the internal-standard method has enjoyed wide application in mineral and ore processing. It is essential that the reference element be intimately mixed with the sample on a micrometer scale. Various procedures have been used to achieve this blending, such as grinding the sample and reference with an abrasive such as silicon carbide, or by fusion using borax, carbonate, or pyrosulfate as a flux.

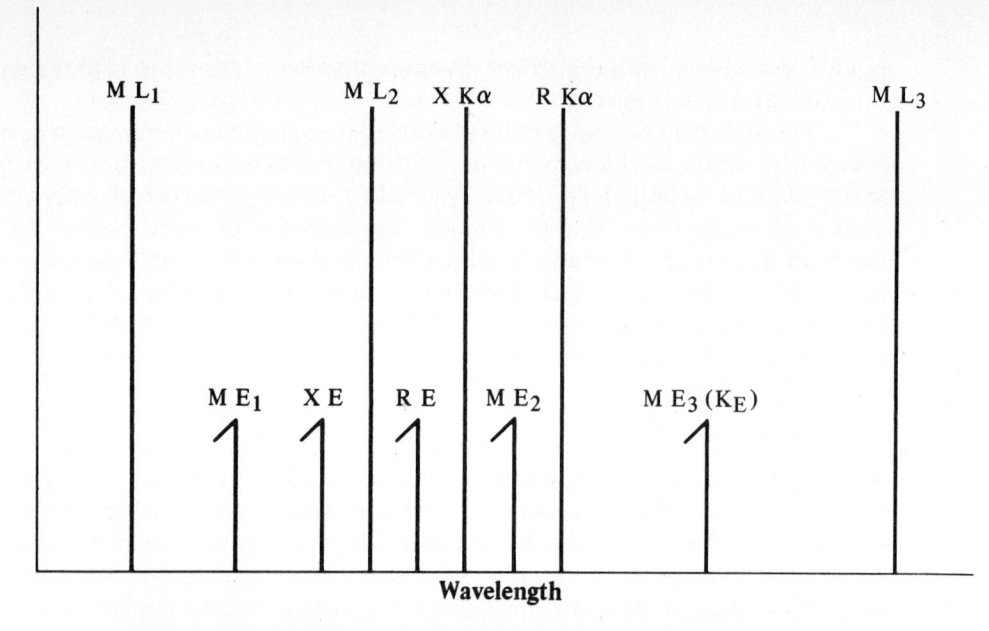

FIGURE 14.12. *Possible interfering lines and absorption edges when using internal standards. M is the matrix element, X is the element being determined, R is the reference element, E_1 represents an absorption edge, K_E is the K absorption edge, and vertical lines represent emission wavelengths.*

Standard Addition

The standard addition method is similar to the internal-standard method, except that the element being determined serves as the reference standard itself. Analysis is accomplished by measuring the intensity of the characteristic spectral line before and after the addition of a known amount of the element being determined. It is assumed that a linear relationship exists between line intensity and concentration (over a limited concentration range), and that the matrix is not significantly altered by the addition. In general, the addition method is limited to the determination of trace and minor elements. All the comments regarding sample preparation for internal standards apply directly to the standard addition approach.

Scatter Correction

Scattered intensities from the sample, such as a coherently (elastic) or an incoherently (inelastic or Compton) scattered target line, can be used to correct for matrix effects because the analyte line and scattered line are both affected by the total mass absorption coefficient of the specimen. These methods are particularly advantageous for specimens having a low effective atomic number, such as mineral samples where the ratio of analyte to scattered intensity is relatively constant.

Mathematical Correction Methods

In the last 10 years, considerable progress has been made in the development of mathematical procedures to correct for matrix effects in thick specimens. Of particular interest has been the *fundamental-parameter method* of Criss and coworkers [1], who developed a comprehensive user-oriented computer program (NRLXRF) at the Naval Research Laboratory. Fundamental-parameter formulas are based on physical-mathematical models of the actual analysis experiment. The equations are derived from first principles and contain physical constants and parameters including absorption coefficients, fluorescence yields, form of x-ray-tube excitation spectrum, and geometric factors. These can be used to predict the x-ray intensity emitted by an element in any prescribed specimen composition. For this reason, it is most convenient to use these formulas to simulate "theoretical intensities" for arbitrary compositions. Standards containing the analyte element are then used to rescale the theoretical intensities so that they are on the same basis as measured intensities. In this manner, it is possible to perform a multielement analysis with a minimum number of standards. In the analysis of alloys, it has been shown that using just pure-element standards results in relative errors of 5% or less for concentrations ranging from a few percent to 95%. With just one multielement standard, the relative error can be decreased to about 1 to 2%.

The intensities of characteristic x-rays produced within a specimen depend on the concentrations or weight fractions of the elements, and on the mass absorption coefficients of these elements for the primary radiation. Also, absorption of the characteristic x-rays within the specimen depends on the mass absorption coefficients of the matrix elements for the fluorescence radiation. These effects can be summarized by reference to Figure 14.13. For convenience, the analyte intensity from the specimen is expressed relative to the 100% analyte (i.e., the relative intensity), and is plotted versus its weight fraction in the specimen.

A linear calibration curve, such as curve **A** in Figure 14.13, is obtained over a wide range of compositions only when the absorption coefficients of the elements in the specimen for the primary x-rays from the excitation source and for the characteristic x-rays are nearly constant over the same range of compositions. The shape of curve **B** can arise when the absorption by the matrix elements in the specimen of either the primary or the analyte fluorescence radiation (*or both*) is greater than the absorption by the analyte alone. If the matrix absorbs the primary or the fluorescence radiation emitted by the analyte (*or both*) to a lesser extent than does the pure analyte, then curve **C** is obtained. This is sometimes referred to as a negative absorption effect. Note that this would lead to a *positive* interference if uncorrected: The quantity of analyte element determined to be present would be higher than that actually present. A good example is the effect of silicon on the element iron as the analyte.

Both curves **B** and **C** in Figure 14.13 can be described by an equation of a hyperbola for a two-component system as follows:

$$R = \frac{c}{1 + \alpha(1 - c)} \tag{14.14}$$

where R = a relative intensity
         c = concentration (mass fraction) of the analyte element
         α = a constant determined from fundamental-parameter calculations or experimentally with reference materials

Alpha, α, is also called the *influence coefficient* or interelement-effect coefficient. The value of α is a real number whose magnitude is related to the displacement of the

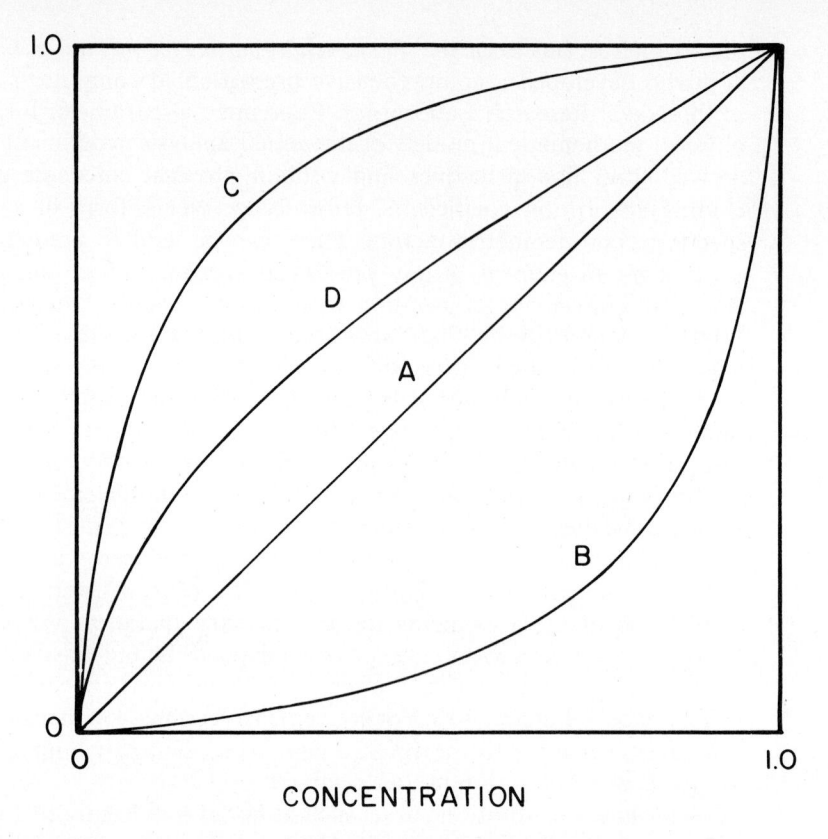

FIGURE 14.13. *Relative x-ray fluorescence intensity (i.e., x-ray intensity from the analyte in the specimen relative to that from the pure analyte) versus concentrations of the analyte in the specimen. A: Linear calibration. B: Preferential absorption by the matrix. C: Preferential absorption by the analyte. D: Secondary fluorescence effect.*

hyperbola from the linear calibration curve. Referring to Figure 14.13, $\alpha = 0$ for curve A, α is positive for curve B, and α is negative for curve C.

Equation 14.14 can be rewritten for a two-component system A and B in a form used by LaChance-Traill [2]:

$$R_A = \frac{c_A}{1 + \alpha_{AB} c_B} \quad (14.15)$$

where R_A = relative intensity of analyte A
c_A and c_B = concentrations of A and B, respectively, in the specimen
α_{AB} = influence coefficient, with the subscript denoting the influence of matrix element B on analyte A.

To determine α_{AB} experimentally, it is necessary to have reference materials in which A and B are present at different concentration ratios, preferably over a wide range.

Equation 14.15 can be rewritten to solve for α_{AB}:

$$\alpha_{AB} = \frac{(c_A/R_A) - 1}{c_B} \quad (14.16)$$

From reference materials, R_A is measured; c_A and c_B are known; and α_{AB} is computed. In practice, the "best value" of α_{AB} can be obtained by plotting $[(c_A/R_A) - 1]$ versus c_B. The slope of this line is α_{AB}.

Another interelement effect can exist (see curve **D** in Fig. 14.13) when the characteristic fluorescence radiation emitted by a matrix element in the specimen enhances the analyte x-ray-emitted intensity. Curve **D**, however, does not rigorously follow the hyperbolic model, because the analyte line intensity increases with increasing concentration of both the analyte (excited element) and excitant element. The phenomenon is also referred to as a *secondary fluorescence effect*.

The above hyperbolic model can also be extended to multicomponent samples as follows:

$$R_i = \frac{c_i}{1 + \sum \alpha_{ij} c_j} \quad (14.17)$$

where "i" and "j" = analyte and matrix-element indices, respectively

For example, in a ternary system consisting of elements *A*, *B*, and *C*, three equations can be written,

$$R_A = \frac{c_A}{1 + \alpha_{AB} c_B + \alpha_{AC} c_C} \quad (14.18)$$

$$R_B = \frac{c_B}{1 + \alpha_{BA} c_A + \alpha_{BC} c_C} \quad (14.19)$$

$$R_C = \frac{c_C}{1 + \alpha_{CA} c_A + \alpha_{CB} c_B} \quad (14.20)$$

for each of the analyte elements taken in turn. Thus, it can be seen that six interelement coefficients need to be determined for a ternary system. For experimental determination of reliable coefficients using regression methods, it is desirable to use as many reference materials as are available that encompass a fairly wide composition range for each of the analyte elements. Since in practice it is seldom possible to have an adequate number of standards, the current trend is to calculate coefficients using fundamental-parameter models. After the coefficients have been determined, Equations 14.18 through 14.20 are solved for the unknown concentrations in the specimen, usually by means of an iterative process with a computer.

Other workers, including Rasberry and Heinrich [3], Claisse and Quintin [4], and Tertian and Vié Le Sage [5], have proposed other models to account for both absorption and enhancement effects in thick samples. These models are not within the scope of this chapter; but, basically, other terms and coefficients are introduced in order to overcome the limitations of the purely hyperbolic model, especially for enhancement effects.

Accuracy and Precision

The accuracy and precision of the x-ray fluorescence method are related to the care taken in sample preparation, to the stability of the instrument, and ultimately to counting statistics.* With modern instrumentation, minor or major constituents may give counting rates of 10,000 cps or higher above background, so that a relative precision of 0.1% can be achieved in 100-sec counting times. The accuracy of the analyses is related to the similarity of composition between the unknowns and the standards or to the analyst's ability to apply appropriate correction factors. In routine industrial-control applications, a relative analytical accuracy of 1% is not unusual, whereas for many unknown samples the relative accuracy may be 5 to 10% or poorer. X-ray fluorescence methods can be applied for concentrations ranging from several ppm in favorable cases to essentially 100% by weight.

14.5 SPECIAL TOPICS AND OTHER X-RAY METHODS

Coating and Film-Thickness Determinations

An important application of x-ray spectrometry in the metallurgical industry is measuring the thickness of coatings or films. The thickness can be calculated by determining either the intensity I_s of the characteristic radiation emitted from the substrate (after being attenuated by the coating) or by measuring the intensity I_f of a characteristic line emitted from the coating (see Fig. 14.14). These measurements are correlated to data obtained with coatings of known thickness. Virtually all coating-thickness monitors used in the metallurgical industries are based on such x-ray-intensity measurements. The methods are rapid, nondestructive, and can be applied to a variety of coatings and thicknesses; measurements are not affected by physical properties such as hardness or magnetism, and only to a minor or negligible extent by oils used to protect metal surfaces. Compared to wavelength dispersion, the energy-dispersive analyzers using radioisotopic sources of x-rays or gamma rays have the advantages of compactness, reduced maintenance, stability, and lower price, and can be used with much lower radiation fluxes.

A typical "thickness gauge" of this type will measure an area of 40 cm² with a repeatability of 1%. The measurement ranges for tin and zinc gauges using different sources are given in Table 14.3. Errors due to changes in hardness, chemical composition, and thickness are less than 0.3%.

One of the first important applications of x-rays to the measurement of the thickness of coatings was in the control of tin plate on steel, using attenuation of the Fe Kα line. The incident radiation may be either polychromatic or monochromatic (the latter makes isolation of the signal from background easier). However, it is essential that the primary radiation be of sufficient energy to excite the substrate. In the example of tin on steel, x-rays or gamma rays in the range 10 to 25 keV will

---

*The counting error is inversely proportional to the square root of the number of photons counted for each analytical line being measured.

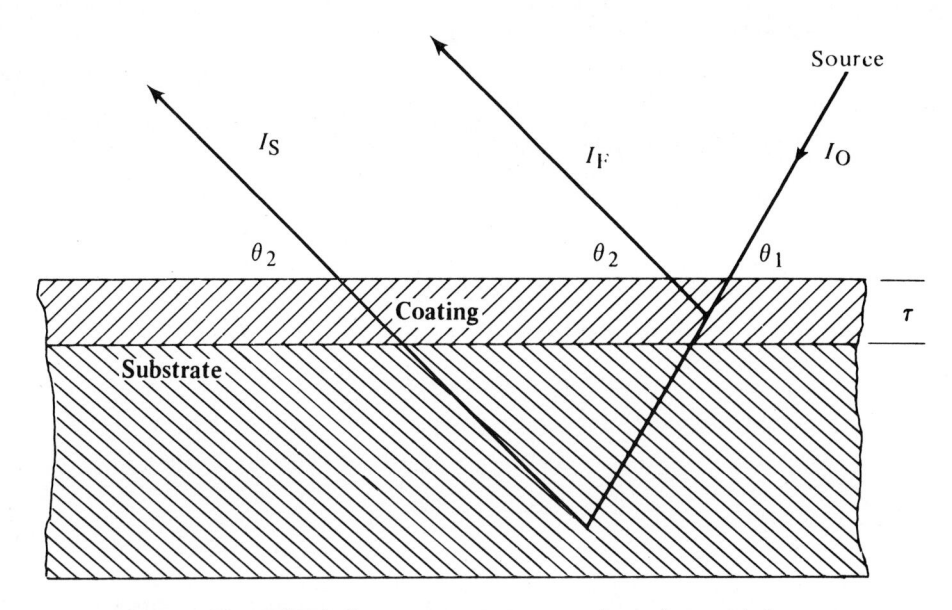

FIGURE 14.14. *Film-thickness measurements using characteristic x-rays emitted by the coating or the substrate.*

TABLE 14.3. *Typical Instrument Specifications for Determining Tin and Zinc Plating Thicknesses*

	Electrolytic Tin	Hot-Dipped Zinc	Electrolytic Zinc
Measurement range (one side of sample)	1–15 g/m²	60–450 g/m²	5–30 g/m²
Approximate plating thickness	0.1–2 μm	8–60 μm	0.7–4 μm
Preferred source and energy	²⁴¹Am 17 keV Secondary γ-rays	²⁴¹Am 60 keV γ-rays	²⁴¹Am 17 keV Secondary γ-rays

Source: Adapted from J. F. Cameron and C. G. Clayton, *Radioisotope Instruments*, New York: Pergamon Press, 1971, by permission of the publisher.

excite Fe K lines without exciting the characteristic Sn K lines. Referring to Figure 14.14, the coating thickness τ is related to the intensity I_s from the substrate by

$$\frac{I_s}{I_{s(\tau=0)}} = e^{-(\mu_1 \csc\theta_1 + \mu_2 \csc\theta_2)\tau} \tag{14.21}$$

where $I_{s(\tau=0)}$ = intensity when the coating thickness equals zero
μ_1 = linear absorption coefficient (in cm⁻¹) of the coating for the primary radiation
μ_2 = linear absorption coefficient of the coating for the secondary radiation
τ = coating thickness in centimeters

Example 14.6. Calculate the thickness (τ) in micrometers of a tin coating ($\rho = 7.3$ g/cm^3) deposited uniformly on an iron substrate, using Equation 14.21. Use the following parameters: Cu Kα (1.54 Å) as the monochromatic source; Fe Kα (1.93 Å) as the secondary radiation; $\theta_1 = \theta_2 = 45°$; and the measured value $I_s/I_{s(\tau=0)} = 0.50$.

Solution: The magnitude of τ will depend on the values of the *mass absorption coefficient*, μ_m, which equals μ/ρ, where ρ is the density of the material. The following values were taken from the CRC *Handbook of Chemistry and Physics*: (1) μ_m for tin = 247 at 1.54 Å; (2) μ_m for tin = 470 at 1.93 Å.

$$\mu_1 = \mu_{(m, 1.54\text{Å})}\rho = 247(7.3) = 1800 \text{ cm}^{-1}$$

$$\mu_2 = \mu_{(m, 1.93\text{Å})}\rho = 470(7.3) = 3430 \text{ cm}^{-1}$$

Substituting these values, and noting that csc 45° = $2^{1/2}$ = 1.41 :

$$0.50 = e^{-[1800(1.41) + 3400(1.41)]\tau} = e^{-7400\tau}$$

$$-0.69 = -7400\tau$$

$$\tau = 9.36 \times 10^{-5} \text{ cm} = 0.94 \text{ } \mu\text{m}$$

An important point from the problem is that input parameters will vary 2 to 10% or more; therefore, the reliability of the calculated value is related proportionally.

The range of thicknesses over which this method can be used is determined primarily by the average value of μ_1 and μ_2. The higher the value of μ, the smaller the range of thicknesses that can be measured. In the case of electroplated tin on steel, the usual range is 0.1 to 2.5 μm.

Most determinations of coating thickness are based on measurements of characteristic radiation from the coating. The thickness can be considered to fall into one of three regions: linear, exponential, and infinite thickness. In the linear region, absorption of the incident and fluorescent radiation by the very thin coating is negligible; therefore, the intensity is linearly related to the coating thickness. For films of intermediate thickness, the intensity is an exponential function of thickness. As the thickness approaches infinity, the intensity becomes constant.

A typical example of the use of characteristic radiation is measuring the thickness of nickel coatings on steel. The function of the nickel is to provide a thin, but strong, bond between an enamel or porcelain coating and the steel substrate. Current practice in the enamel-coating industry is to cut samples from the sheet and submit them to the x-ray laboratory. With a portable analyzer, however, semiskilled personnel can obtain satisfactory results directly on the sheets.

Energy-dispersive x-ray spectrometry has received considerable attention for the determination of lead on interior walls of houses in the inner cities of the United States (ingestion of lead-bearing paints is a significant health hazard to young children). The highly penetrating Pb K radiation is used for these measurements, since the lower-energy Pb L lines may be absorbed in overlayers of nonlead paints, giving a false negative reading.

Positive-Ion Excitation

The use of highly energetic positive ions for the generation of low- and medium-energy characteristic x-rays from samples is receiving considerable attention for applications in surface-structure analysis and in the environmental sciences. The important features of positive-ion excitation are the high sensitivity for low-Z elements, the small depth of sample analyzed, and the relatively low level of continuous radiation scattered from the sample. Generally, protons in the range 100 keV to 5 MeV are used; however, alpha particles and higher-Z positive ions have some unique advantages.

In surface studies, particularly in the fields of corrosion and thin-film technology, low-energy x-ray spectra generated by 100–300-keV protons are used in conjunction with Auger and photoelectron spectroscopy (Chap. 15) to analyze changes in the surface caused by implantation, oxidation, diffusion, and so on. Fractions of a monolayer of various elements on the surface can be determined by this analytical technique. The sampling depth is of the order of 1 to 5 monolayers using protons for excitation of low-energy x-rays. Depth profiles can be obtained by sputtering successive monolayers and analyzing the new surface. The beam diameter can be reduced to approximately 1 mm if spatial resolution is required.

For trace applications, microgram amounts of sample are supported on a very thin substrate. The detection limits for most elements using a 1.5-MeV proton beam at 5 μA current is of the order of 10^{-9} to 10^{-12} g, the detection limit being a function of the thickness of the support material. Trace applications include examination of water residues, biological specimens, tissue sections, and air particulates.

Electron-Probe Microanalysis

In the early 1940s, Hillier of the Radio Corporation of America conceived the idea of using a focused electron beam for localized x-ray spectroscopic analysis. Several years later, the first practical electron-probe x-ray spectrometer was designed by Castaing at the University of Paris. This microanalyzer proved to be a major breakthrough, since the instrument made possible the nondestructive analysis of micrometer-sized volumes.

In *electron-probe microanalysis* (EPMA), an electron beam of moderate energy, 10 to 50 keV, is focused on the sample at the place where elemental composition is to be determined. The atoms in a minute volume, one-half to several micrometers in diameter and one or more micrometers in depth, are excited by the incident electrons and, upon returning to the ground state, emit x-rays characteristic of the excited elements. The actual volume of sample analyzed depends on such variables as the diameter and energy of the electron beam, the diffusion of electrons in the sample, and the path length of the scattered primary and secondary x-rays.

Until the late 1960s, commercial EPMA instruments consisted of four major components: an electron optical system for producing a stable electron beam of 1 μm (or smaller) diameter, light optics for viewing the microscopic area under investigation, a precision stage for accurately locating and translating the sample under the electron beam, and one or more focusing x-ray spectrometers for measuring the characteristic x-rays (see Fig. 14.15). More recently, EPMA has been augmented by *scanning electron microscopy* (SEM). In the scanning electron microscope, the sample image is obtained by either backscattered or transmitted electrons, and the elemental composition is determined by energy-dispersive x-ray methods.

Quantitative analysis is accomplished by the use of standards similar in composition to the sample being analyzed, or by using pure elements as standards and

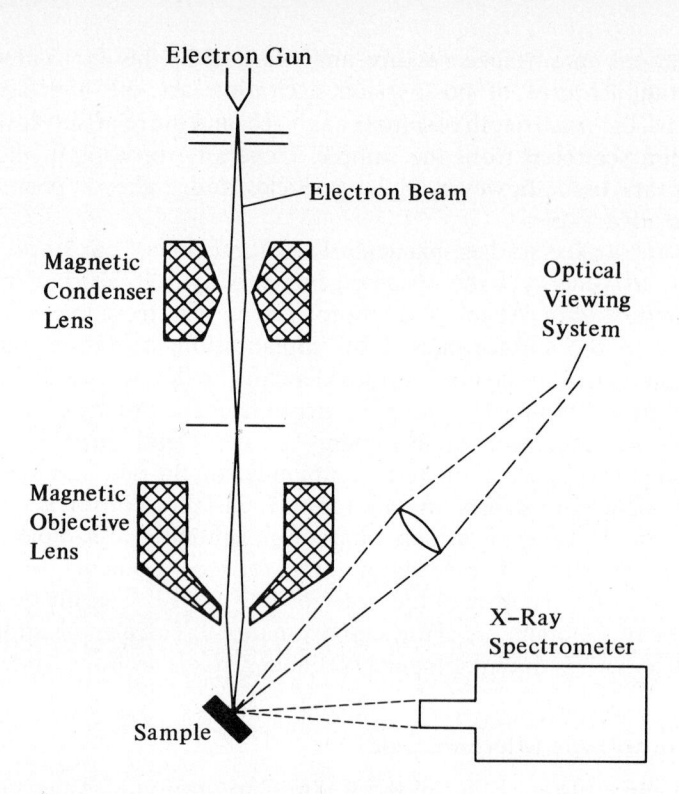

FIGURE 14.15. *Electron optics for the excitation of x-rays in a micrometer-sized volume.*

applying theoretical and semiempirical corrections for differences in atomic number (Z), absorption (A), and secondary fluorescence (F) (i.e., the so-called ZAF method). The mathematical methods using pure-element standards have received the greatest attention because of the difficulty in obtaining multielement standards that are homogeneous on a submicrometer scale. Computer programs are available for performing the complex corrections. Since the development of SEM, it has been realized that qualitative or semiquantitative analysis is adequate for many practical applications. Therefore, pictorial displays of concentration across the sample surface, obtained by sweeping the electron beam across the sample, are generally adequate, as compared to the older technique of point-by-point analysis.

For most elements, the sensitivity of the EPMA method is approximately 0.01 to 0.1 weight-percent with a relative accuracy of 1 to 5%, depending on the correction procedure, the availability of suitable standards, and the reliability of the parameters used in the calculations. Boron ($Z = 5$) and all higher-Z elements can be determined. Although the EPMA method is not applicable to trace elements in a homogeneous sample, its absolute detection limit—10^{-12} to 10^{-16} g—is very impressive.

The EPMA or modified SEM methods have been applied extensively to virtually all fields of science. Applications include the qualitative identification of the composition of small inclusions, the determination of diffusion rates of elements in solids, phase-equilibria studies, the determination of the thickness and uniformity of films, and the qualitative and quantitative analysis of minute samples such as airborne particulates. Specific applications include the study of the metallic phases in meteorites, the depletion of uranium in the grain boundaries of iron alloys, the diffusion of chromium during oxidation of steels, and inorganic inclusions in oil shale.

Essentially all types of samples can be analyzed by this technique. Ideally, samples should be good electrical conductors to avoid charge buildup; however, a very thin conducting film, for example gold or carbon, can be deposited on nonconductors such as ceramics, glasses, and biological samples.

X-Ray Absorption Spectrometry

The intensity of an x-ray beam is attenuated during passage through matter by a dual process: *photoelectric absorption* and *x-ray scattering*. The amount of attenuation can serve as the basis of quantitative analysis in specific cases. In photoelectric absorption, virtually all the energy of the incident x-ray quantum is converted into the kinetic energy of the photoelectron ejected; this is the process that results in the emission of characteristic (fluorescence) x-rays. The extent of attenuation is governed by the linear absorption coefficient μ, which is similar to *absorptivity* in molecular absorption (Beer's law). For a given element, this is equal to the sum of the photoelectric absorption coefficient τ and the scattering coefficient σ. Photoelectric absorption predominates, except for low values of Z and high values of λ. By application of Beer's law, x-ray transmittance (P/P_0) can be used to determine either composition or thickness:

$$\frac{P}{P_0} = e^{-\mu_m \rho b} \tag{14.22}$$

where μ_m = mass absorption coefficient, in square centimeters per gram
ρ = density of the sample in grams per cubic centimeter
b = thickness in centimeters

Values for the *mass absorption coefficient*, which is equal to μ/ρ, are listed in standard reference tables. As a first approximation, μ_m is proportional to $Z^4 \lambda^3$ up to the K or L absorption edge; thus, the absorption coefficient is dependent on both the element composition and the wavelength of the x-ray (see Fig. 14.16). As mentioned before, there can be one K absorption edge, three L edges, and five M edges.

X-ray absorption measurements have been used for two types of analytical applications. The principal application has been for process monitoring in which the sample composition is essentially constant except for a variable amount of the element of interest. Examples are tetraethyl lead in gasoline or sulfur in fuel oil, where intense continuous radiation provides adequate signal for instantaneous process control. This approach, using a polychromatic source, has essentially no specificity, and thus the sample parameters are very limited. Unknown samples can be analyzed by measurement of two monochromatic x-ray line intensities, one on each side of a characteristic absorption edge of the element being determined. This absorption-edge technique is then specific for the element being determined.

Quantitative calculations when there is more than one absorber are based on the additive absorption of different elements. The mass absorption coefficient of a

FIGURE 14.16. *Relationship of the mass absorption coefficient and wavelength.*

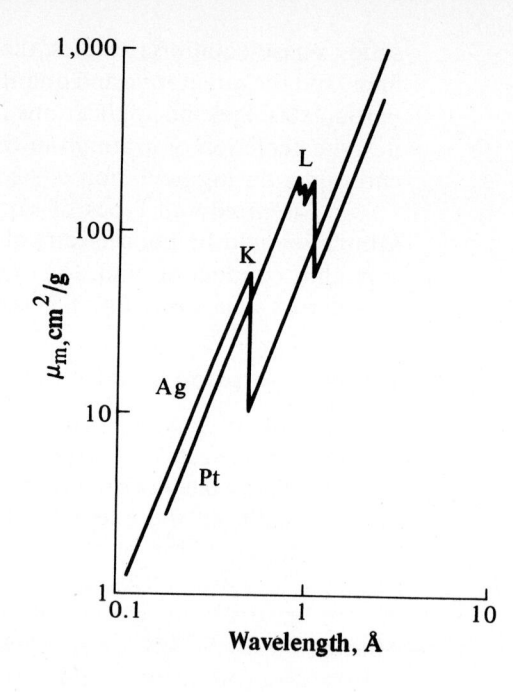

sample containing different elements is given by

$$\mu_{m(s)} = \sum_{i=1}^{j} \mu_{m(i)} W_i \tag{14.23}$$

where $\mu_{m(i)}$ = mass absorption coefficients (in cm^2/g) of the individual elements $m(i)$ at the given wavelength
W_i = weight fractions of the elements

Strictly speaking, this equation holds only for monochromatic radiation. The sensitivity for a given element depends greatly on the sample composition. Very small concentrations of elements with large mass-absorption coefficients can be determined in samples where the bulk matrix has a low mass-absorption coefficient. Thus, for instance, 10^{-10} g of phosphorus can be determined in biological tissues. Classical examples of quantitative x-ray absorption analysis are the determination of lead tetraethyl in gasoline and sulfur in petroleum.

Example 14.7. A 3.00-mL gasoline sample is placed in a cell of 0.250 cm thickness. The sample, which contains a small amount of lead tetraethyl in *n*-octane and has a density of 0.720 g/cm^3, absorbs 75.0% of the Cu Kα line from an x-ray source. If the mass absorption coefficients ($\mu_m = \mu/\rho$) for the Cu Kα line for lead, carbon, and hydrogen are 230, 4.52, and 0.48 cm^2/g, respectively, what is the percentage of lead tetraethyl in the sample?

Solution: We must first calculate the mass absorption coefficients for the two compounds present and then for the sample. Then, we can calculate the weight fraction of lead tetraethyl. The formula weight of lead tetraethyl, $Pb(CH_2CH_3)_4$, is 323.4, so

$$W_{Pb} = \frac{207.2}{323.4} = 0.6407$$

$$W_C = \frac{8(12.011)}{323.4} = 0.2971$$

$$W_H = \frac{20(1.0080)}{323.4} = 0.06234$$

Therefore, for lead tetraethyl,

$$\mu_{m(LTE)} = 230(0.641) + 4.52(0.297) + 0.48(0.062)$$
$$= 149 \text{ cm}^2/\text{g}$$

The formula weight of n-octane, $CH_3(CH_2)_6CH_3$, is 114.23, and thus

$$W_C = \frac{8(12.011)}{114.23} = 0.8412$$

$$W_H = \frac{18(1.0080)}{114.23} = 0.1588$$

Therefore, for octane,

$$\mu_{m(O)} = 4.52(0.841) + 0.48(0.159) = 3.88 \text{ cm}^2/\text{g}$$

We can calculate the mass absorption coefficient for the sample, $\mu_{m(S)}$, from Equation 14.22:

$$\log \frac{100}{25.0} = \frac{\mu_{m(S)}(0.720)(0.250)}{2.303}$$

$$\mu_{m(S)} = 7.70 \text{ cm}^2/\text{g}$$

Let W be the weight fraction of lead tetraethyl in the sample. Therefore,

$$\mu_{m(S)} = \mu_{m(LTE)}W + \mu_{m(O)}(1 - W)$$
$$7.70 = 149W + 3.88(1 - W)$$
$$W = 0.0264$$

Thus, the sample contains 2.64% lead tetraethyl.

Mass absorption coefficients at different wavelengths for the various elements are available in numerous handbooks such as those in the Selected Bibliography at the end of this chapter. As in conventional spectrophotometry (Chap. 7), the optimum absorbance range for x-ray absorption analysis is about 0.1 to 1.

Although specific examples like those mentioned above are well suited for x-ray absorption analysis, with good selectivity using the absorption-edge technique, in general the method does not offer significant advantages over x-ray fluorescence analysis. Since the latter technique is easier to apply, it is more widely used.

SELECTED BIBLIOGRAPHY

Books

Advances in X-Ray Analysis. Proceedings of the Annual Conference on Applications of X-Ray Analysis, University of Denver, Vols. 1–26. New York: Plenum Press, 1958–1982.

BERTIN, E. P. *Principles and Practices of X-Ray Spectrometric Analysis*, 2nd ed. New York: Plenum Press, 1978.

BIRKS, L. S. *Electron Probe Microanalysis*. New York: Interscience, 1963.

GOLDSTEIN, J. I., NEWBURY, D. E., ECHLIN, P., JOY, D. C., FIORI, C., and LIFSHIN, E. *Scanning Electron Microscopy and X-Ray Microanalysis*. New York: Plenum Press, 1981.

HEINRICH, K. F. J. *Electron Beam X-Ray Microanalysis*. New York: Van Nostrand Reinhold, 1981.

JENKINS, R. *An Introduction to X-Ray Spectrometry*. New York: Heyden, 1974.

JENKINS, R., and DEVRIES, B. *Worked Examples in X-Ray Spectrometry*. New York: Springer-Verlag, 1970. *An excellent set of problems for students.*

JENKINS, R., and DEVRIES, J. L. *Practical X-Ray Spectrometry*. Phillips Technical Library. New York: Springer-Verlag, 1969.

JENKINS, R., GOULD, R. W., and GEDCKE, D. *Quantitative X-Ray Spectrometry*. New York: Marcel Dekker, 1981.

LIEBHAFSKY, H. A., PFEIFFER, H. G., WINSLOW, E. H., and ZEMANY, P. D. *X-Rays, Electrons, and Analytical Chemistry*. New York: Wiley-Interscience, 1972.

WOLDSETH, R. *X-Ray Energy Spectrometry*. Burlingame, Calif.: Kevex Corporation, 1973.

Tables of Wavelengths, 2θ Angles, Photon Energies, and Mass Absorption Coefficients.

BEARDEN, J. A. *X-Ray Wavelengths*. U.S. Atomic Energy Commission Report, N.Y.O. 10586, 1964.

FINE, S., and HENDEE, C. F. "A Table of X-Ray K and L Emission and Critical-Absorption Energies for All the Elements." *Nucleonics*, *13*(3), 36 (1955).

STAINER, H. M. *X-Ray Absorption Coefficients: A Literature Survey*. Bureau of Mines Information Circular 8166, 1963.

WHITE, E. W., GIBBS, G. V., JOHNSON, G. G., JR., and ZECHMAN, G. A., JR. *X-Ray Emission Line Wavelength and Two-Theta Tables*. ASTM Series 37. Philadelphia: American Society for Testing and Materials, 1965.

SWITZER, G., AXELROD, J. M., LINDBERG, M. L., and LARSEN, E. S. *Tables of Spacing for Angle 2θ, Cu Kα, Cu Kα_1, Cu Kα_2, Fe Kα, Fe Kα_1, Fe Kα_2*. U.S. Geological Survey Circular 29. Washington, D.C.: U.S. Department of the Interior, 1948.

Sources of Standards

Catalog of Standard Reference Materials. Special Publication 260. Washington, D.C.: National Bureau of Standards, 1984–85.

Standard References Materials and Meaningful Measurements. Sixth Materials Research Symposium, National Bureau of Standards, Washington, D.C.

Report on Available Standard Samples and Related Materials for Spectrochemical Analysis. Special Technical Publication 58-D. Philadelphia: American Society for Testing and Materials, 1960.

REFERENCES

1. J. W. CRISS, L. S. BIRKS, and J. V. GILFRICH, *Anal. Chem.*, **50**, 33 (1978).
2. G. R. LA CHANCE and R. TRAILL, *J. Can. Spect.*, **11**(1), 43 (1966).
3. S. D. RASBERRY and K. F. J. HEINRICH, *Anal. Chem.*, **46**, 81 (1974).
4. F. CLAISSE and M. QUINTIN, *J. Can. Spect.*, **12**, 129 (1967).
5. R. TERTIAN and R. VIÉ LE SAGE, *X-Ray Spectrom.*, **6**, 123 (1977).

PROBLEMS

1. You have been asked to determine low concentrations of tantalum and niobium by wavelength dispersive XRF in an ore concentrate in which the tantalum-to-niobium ratio is 1:50. Discuss methods to achieve maximum fluorescence signal from tantalum with minimum interference from niobium.

2. Using wavelength dispersion, which detector would you use for the x-ray fluorescence measure of Si Kα, U Mα, Cu Kα, and I Kα?

3. For the determination of 1 to 10 weight-percent lead in an ore concentrate, what elements and characteristic lines can be used as internal standards for x-ray fluorescence?

4. An analytical laboratory was requested to determine potassium (present as KCl) in a silica matrix. Suggest some possible approaches to the problem.

5. Discuss methods for determining the thickness of a zinc coating on an iron substrate; include discussion on the preparation of standards and on calibration.

6. Outline a method for calculating the energies of the Kα and Lα lines of plutonium ($Z = 94$).

7. Discuss methods for extending the limits of detectability of x-ray fluorescence spectroscopy down to the ppm-ppb range. Also include comments regarding possible sources of systematic errors.

8. Describe the operation of a Coolidge tube.

9. What is an absorption edge?

10. Describe the principles of x-ray emission, x-ray absorption, and x-ray fluorescence analysis. Distinguish between each technique with respect to instrumentation requirements.

11. An analyst, using an x-ray fluorescence spectrometer with a molybdenum target source and a sodium chloride analyzer crystal, wishes to determine nickel in a meteorite sample. At what 2θ value would the analyst look for the nickel peak?

12. Using a tungsten x-ray tube and a LiF analyzing crystal ($d = 2.01$ Å), a very strong x-ray fluorescence peak for a pure but unknown metal was observed at $2\theta = 69.36°$. Calculate the wavelength of the fluorescence radiation and identify the metal.

13. An x-ray tube with a copper target is operated at 50 kV. What will be the cutoff wavelength for the continuous radiation?

14. Calculate the transmittance of a monochromatic x-ray beam from a copper target (Kα line) passing through a 1-cm³ sample of carbon tetrachloride in benzene (1% by weight) contained in a cell with a cross-sectional area of 9.80 cm². The density of the solution is 0.880 g/cm³.

15. A wavelength-dispersive spectrum of a quartz sample using a Cr-target x-ray tube exhibited two peaks, at 109.21 and 101.16° (2θ). The peak at 109.21° was about three times more intense than the peak at 101.16°. Assuming the peaks are the principal K lines of silicon, what crystal was used in the spectrometer?

16. One liter of seawater is processed for the removal of the alkali and alkaline earths. The separated metals nickel, manganese, and zinc are then quantitatively collected on SA-2 paper (effective area = 14.5 cm²). X-ray intensities have been measured and corrected for background and are: $I_{Ni} = 5.0$ cps; $I_{Mn} = 4.0$ cps; $I_{Zn} = 6.0$ cps. SA-2 standards used for spectrometer calibration gave the following data after correction for background: Ni = 59 cps μg^{-1} cm²; Mn = 27 cps μg^{-1} cm²; Zn = 17.4 cps μg^{-1} cm². Compute the concentrations of these metals in seawater in ppb.

17. An x-ray spectrum of an austenitic steel containing primarily Fe, Cr, and Ni taken with a Mo secondary-target energy-dispersive x-ray spectrometer shows x-ray peaks at the following energies (in keV): 1 = 3.68; 2 = 4.66; 3 = 5.42; 4 = 5.94; 5 = 6.40; 6 = 7.05; 8 = 7.47; 9 = 8.26. Identify these peaks in the spectrum.

18. Calculate the weight fractions (W) of Fe and Ni in a binary alloy having relative x-ray intensities of $R_{Fe} = 0.3172$ and $R_{Ni} = 0.5483$. Use the following equations and the data below from a set of Fe-Ni standard binary alloys.

For *absorption*:
$$W_{Ni} = R_{Ni}[1 + \alpha_{NiFe} W_{Fe}]$$
For *enhancement*:
$$W_{Fe} = R_{Fe}\left[1 + \beta_{FeNi}\left(\frac{W_{Ni}}{1 + W_{Fe}}\right)\right]$$

where α_{NiFe} and β_{FeNi} are the influence coefficients.

Standard Number	W_{Fe}	W_{Ni}	R_{Fe}	R_{Ni}
1	0.0462	0.9516	0.0789	0.8782
2	0.0659	0.9322	0.1104	0.8321
3	0.1018	0.8964	0.1621	0.7595
4	0.3067	0.6911	0.4007	0.4515
5	0.3431	0.6552	0.4373	0.4073
6	0.5100	0.4820	0.5907	0.2553
7	0.6315	0.3599	0.6958	0.1720
8	0.9549	0.0329	0.9659	0.0125

CHAPTER **15**

Electron Spectroscopy

Lo I Yin
Isidore Adler

Recent years have seen a spectacular development in instrumental techniques, and one of the fastest-growing areas is electron spectroscopy. Electron spectroscopy is a technique for studying the energy distribution of electrons ejected from a material that has been irradiated with a source of ionizing radiation such as x-rays, ultraviolet light, or electrons.

It is convenient to distinguish among the various electron spectroscopies on the basis of the excitation sources used. When x-ray radiation is employed, the technique is commonly called ESCA [1] (for *electron spectroscopy* for *chemical analysis*); it is also sometimes called *x-ray photoelectron spectroscopy* (XPS). When ultraviolet excitation is used, the method is generally called *photoelectron spectroscopy* (PES) or *ultraviolet photoelectron spectroscopy* (UPS). A third variation, in which electrons are generally used as the ionizing radiation, is commonly referred to as *Auger spectroscopy*.

Of the three types of electron spectroscopy, ESCA has perhaps been the most widely used for chemical studies. For this reason, this chapter will concentrate most heavily on ESCA. Although simple quantitative chemical analysis can be performed by ESCA, this probably represents the least effective use of this powerful tool, which provides quantitative information about such basic parameters as binding energies, charges, valence states, and so on, which involve the atom as a function of its chemical environment.

15.1 PRINCIPLES OF ELECTRON SPECTROSCOPY

A unique quality of ESCA is that it permits direct probing of the valence and core electrons, where much chemical information is contained. This is achieved mainly through the photoelectric process: When any material is bombarded by photons with

Photoelectron Production *Auger Electron Production*

M_5 —— $3d_{5/2}$

M_4 —— $3d_{3/2}$

M_3 —— $3p_{3/2}$

M_2 —— $3p_{1/2}$

M_1 —— $3s_{1/2}$

KL_2L_3 Auger Electron

L_3 —— $2p_{3/2}$

L_2 —— $2p_{1/2}$

L_1 —— $2s_{1/2}$

$h\nu$ — K or 1s Photoelectron

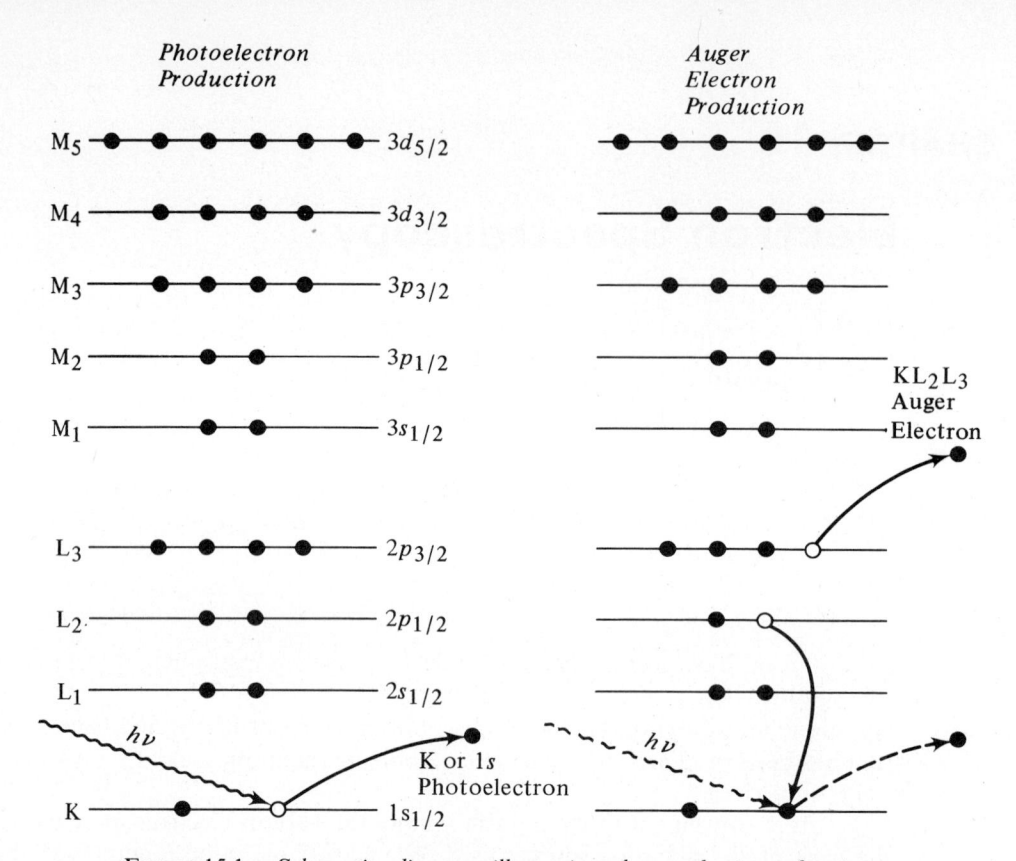

K —— $1s_{1/2}$

FIGURE 15.1. *Schematic diagram illustrating the production of primary photoelectrons and Auger (secondary) electrons in an atom. In primary-photoelectron production, an atom absorbs an x-ray photon which causes the ejection of a core electron (in this case, a K-shell electron). An Auger electron is produced when, after the primary photoelectron has been ejected from the inner shell (indicated by the dashed line), an electron from a higher shell (in this case, from the L shell) fills the orbital vacancy. The excess energy, instead of being emitted as a secondary x-ray, is simultaneously transferred to another electron (the Auger electron), which is ejected from the atom; in this instance, the Auger electron is also from the L shell.*

energy greater than the binding energy of an electron in a given atomic shell or subshell, there is a finite probability that the incident photon will be absorbed by the atom and an atomic electron either promoted to an unoccupied level or ejected as a photoelectron. Figure 15.1 illustrates schematically the process of photoelectron production by absorption of a photon. The probability of photoelectric absorption depends on the energy of the incident photon and the atomic number of the element being irradiated.

The kinetic energy of the photoelectron is, to a first approximation (ignoring solid-state and relaxation effects), given by

$$E_p = h\nu - E_b \quad (15.1)$$

where E_p = kinetic energy of the photoelectron
hv = energy of the incident photon
E_b = binding energy of the electron in its particular shell

It is clear from Equation 15.1 that, if the incident photons are "monoenergetic," the photoelectrons ejected from a given shell will also be monoenergetic. Thus, at a given energy of the incident photons, the photoelectron spectrum of a material reflects the various occupied electronic levels and bands in the material.

Because the energies of the various electronic levels are usually different for different materials, photoelectron spectra are characteristic of the material. It is important to emphasize that the photoelectrons possess characteristic energies as they leave the atom, but that only a relatively small fraction of the electrons emerge from the target material with their energies undisturbed, since energy is lost by a variety of mechanisms.

Figure 15.2 illustrates the excitation of a solid sample by x-ray photons. Although the x-rays may penetrate deeply into the sample to produce photoelectrons, most of these electrons lose energy in numerous inelastic collisions; only the atoms residing in the top few monolayers give rise to undistorted photoelectron spectra. Thus, a typical spectrum resulting from a group of initially monoenergetic photoelectrons will consist of a single peak due to the "undisturbed" electrons (i.e., those that are directly ejected), plus a large continuum on the low-kinetic-energy side of the peak. Since the typical escape depth in ESCA and Auger spectroscopy is only about 3 to 50 Å, these are truly techniques for surface analysis.

The electron continuum, however, does not begin at the photoelectron peak but rather at a discrete distance from the peak. The reason for this is that an electron emerging from a solid loses its kinetic energy in quantized amounts by exciting plasma oscillations (plasmons). For example, Figure 15.3A shows a wide-range scan (170 to 1480 eV) of a germanium sample. The x-axis is the kinetic energy of the electrons (E_p). For the purpose of display, the electron counts (intensity) are compressed into a logarithmic scale on the y-axis. We see the photoelectrons from the various L and M shells as well as the LMM Auger-electron lines.* We also note a small, broad shoulder at the same distance (approx. 17 eV) from the low-energy side of each peak. This is the plasma-loss peak, which is common to all electrons (whether photoelectrons or Auger electrons) emerging from the solid sample. In Figure 15.3B, two of the more prominent electron lines are shown in detail, with electron intensity displayed linearly on the y-axis. In this case, the plasma-loss peaks are situated far enough from the main peak to allow the precise positions of the various electron lines to be identified unambiguously. However, plasma-loss peaks can occasionally cause confusion in the identification of fine structure associated with some electron lines.

Because the positions of photoelectron lines can be precisely determined from these spectra, and the photoelectron energies are characteristic of the atomic levels of a given element, these energies can be used as the basis for elemental identification.

---

*An LMM Auger electron is emitted by the following mechanism: A vacancy, initially created in the *L shell* by the photoejection of an electron, is filled by the fall of an *M-shell* electron, accompanied by the ejection of another *M-shell* electron from the atom. The second M-shell electron is the LMM Auger electron.

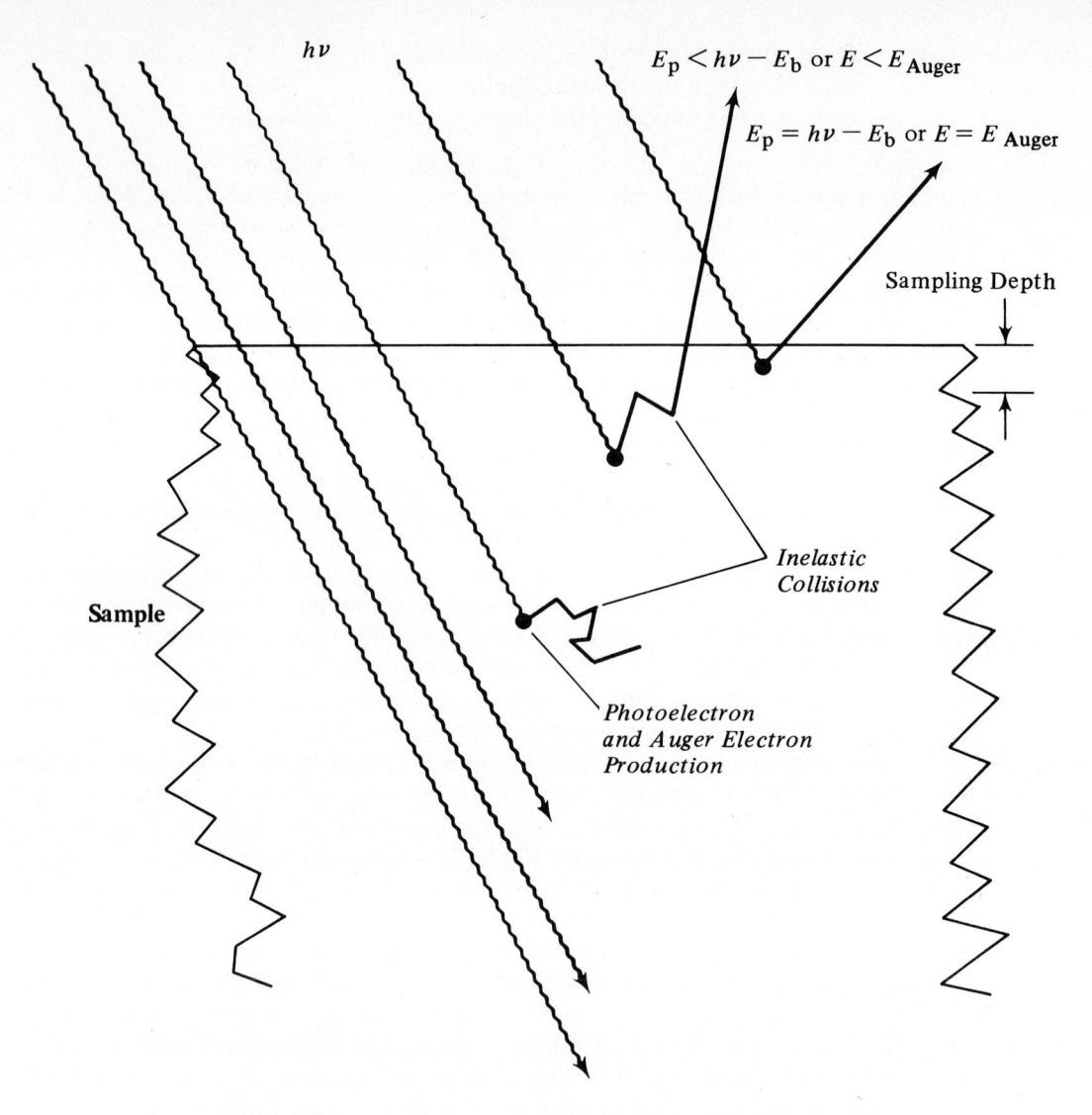

FIGURE 15.2. *Production and escape characteristics of photoelectrons and Auger electrons in a solid sample. Because most electrons lose energy by inelastic collisions, the effective sampling depth is about 3 to 50 Å. E_p is the kinetic energy of the ejected photoelectron and E_b is its binding energy; hν is the (monoenergetic) energy of the incident x-ray photon. In the case of Auger electrons, E_{Auger} is its original kinetic energy and E is that seen by the spectrometer.*

On closer examination, one finds that the kinetic energy of the "undisturbed" emerging photoelectrons is not entirely constant for a given shell in a given element, even for a monoenergetic photon source: There are variations as the chemical environment of the atom changes. As the outer electrons participate in forming chemical bonds, the net charge on the atom changes, which in turn affects the binding energy

454   CHAP. 15   Electron Spectroscopy

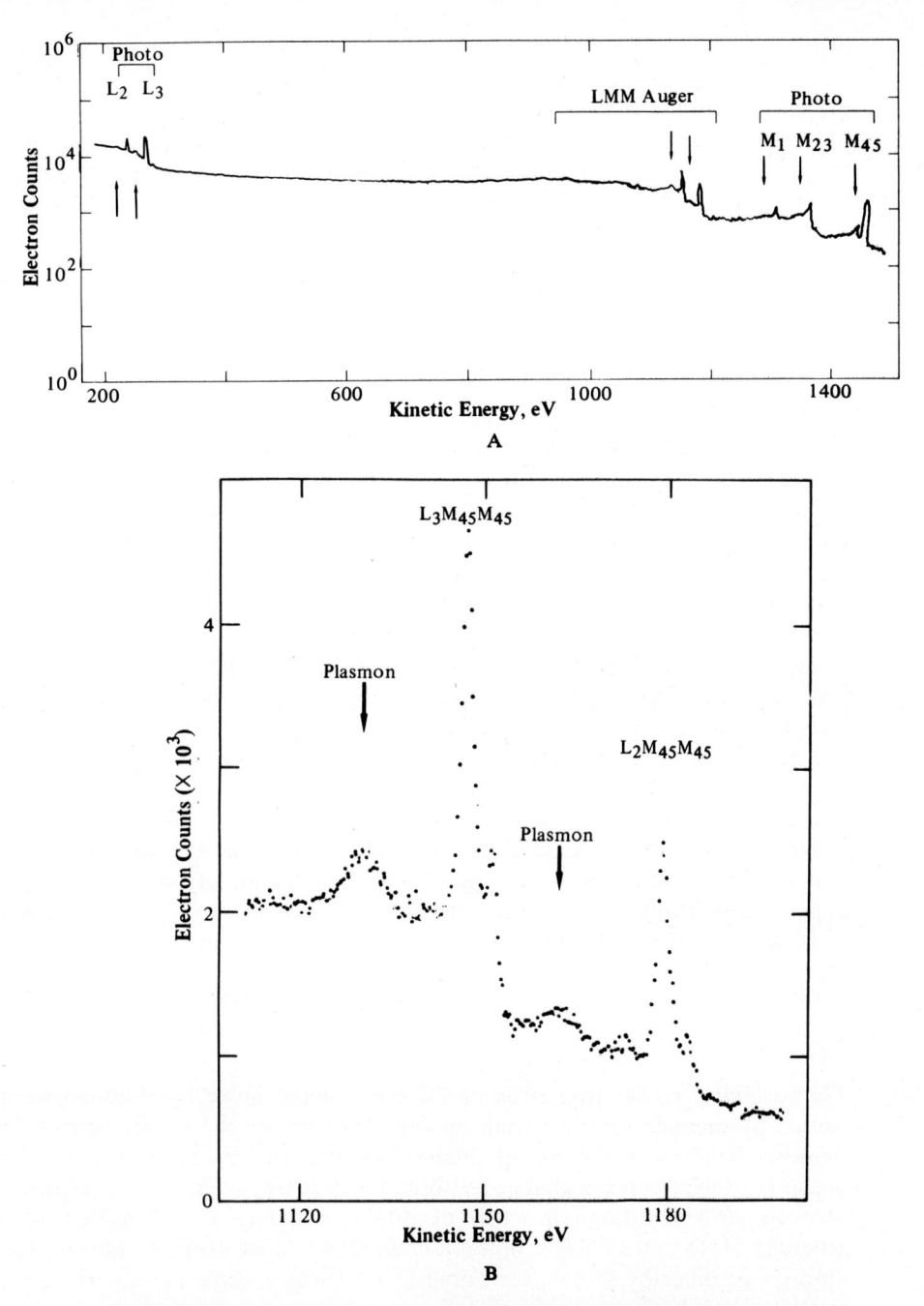

FIGURE 15.3. *A: Al-K$\alpha_{1,2}$ x-ray excited electron spectrum of germanium. Photoelectron peaks from various atomic shells are labeled "photo." Note the presence of a plasma-loss shoulder, indicated by an arrow, about 17 eV on the low-energy side of each prominent electron line. B: Expanded display of the Auger portion of the germanium spectrum showing plasma-loss ("plasmon") peaks adjacent to the main peaks.*

of the core electrons. For example, in electropositive elements (e.g., metals), the outer-shell electrons move away from the nucleus as bonds are formed, after which the core electrons become more strongly bound because the atom now has a net positive charge. As a consequence, the kinetic energies of the photoelectrons shift toward lower values relative to the uncombined element (see Eqn. 15.1). On the other hand, the electronegative elements show a net increase in negative charge as they form chemical bonds, so the ejected photoelectrons emerge with higher kinetic energies, reflecting the decreased binding energies the core electrons have to overcome to escape.

This is a highly simplified picture of the effects of chemical bonding on photoelectron spectra. There are a variety of other factors that affect the binding energy of the electron and the kinetic energy of the ejected photoelectron: for example, relaxation effects, stereochemistry, crystal structure, and lattice energy, to name a few. The electron senses the total of all these effects; their proportionate contributions are very difficult to assign.

An excellent discussion of the various factors involved in binding energies is given in a paper [2] dealing with a series of metal-dithiene compounds of the type $[M(S_2C_2R_2)_n]^z$, where M is one of a large variety of transition metals; n is generally 2 or 3; R is one of a variety of substituents such as CN, C_6H_5, or CH_3; and z is 0, -1, or -2. The problem presented by these compounds is the possible oxidation states of the metal necessary to satisfy stoichiometrically the possible charges of the complex. For instance, the nickel species in both $[(C_2H_5)_4N]_2\{Ni[S_2C_2(CN)_2]_2\}$ and $[(C_2H_5)_4N]\{Ni[S_2C_2(CN)_2]_2\}$ have identical ligands, have square-planar geometry, and are diamagnetic. Tests using ESCA showed that the binding energies of the Ni electrons were identical in both. By contrast, the binding energies of the sulfur ($2p$) electrons decreased as the negative charge on the complex increased. This evidence shows clearly that the "oxidation state" of the Ni remains relatively constant and that the increased negative charge of the complex mainly resides on the anionic ligands, specifically on the sulfur atoms. Some investigators use such observations to deduce the chemical environment around an atom, whereas others use theoretical calculations to predict the binding energies.

Resolution

The resolving power needed for ESCA is defined more by binding-energy shifts caused by chemical changes than by overlaps between the binding-energy values of elements. For a normal range of chemical phenomena, the binding energy for a particular photoelectron can change by about 0.1 to 10 eV, whereas the energy separation between corresponding electrons of different elements is considerably greater, generally 50 to 100 eV. (The binding energies of $1s$ electrons in boron and carbon atoms, e.g., differ by 96 eV. See Table 15.1.) There is some energy overlap between electrons in *different* orbitals of different elements—the scandium $2p$ electrons at 407 and 402 eV are very close in energy to the nitrogen $1s$ electron at 399 eV—but these overlaps are not usually significant. The chemical changes, being relatively small (<10 eV) compared to the usual kinetic energies of photoelectrons (500 to 1500 eV), require resolutions of up to about $\pm 0.01\%$ for adequate measurements. These requirements will be discussed in more detail in the section on instrumentation.

Table 15.1. *Selected ESCA Binding Energies*

Atomic Number	Element	Binding Energy (E_b), eV	Type of Electron
3	Li	55	$1s$ ↓
4	Be	111	
5	B	188	
6	C	284	
7	N	399	
8	O	532	
9	F	686	
10	Ne	867	$1s$
11	Na	1072; 63	$1s$; $2s$
12	Mg	89	$2s$
13	Al	74; 73	$2p_{1/2}$; $2p_{3/2}$
14	Si	100; 99	
15	P	136; 135	
16	S	165; 164	
17	Cl	202; 200	
19	K	297; 294	
20	Ca	350; 347	
21	Sc	407; 402	
22	Ti	461; 455	
23	V	520; 513	
24	Cr	584; 575	
25	Mn	652; 641	
26	Fe	723; 710	
27	Co	794; 779	
28	Ni	872; 855	
29	Cu	951; 931	
30	Zn	1044; 1021	$2p_{1/2}$; $2p_{3/2}$
32	Ge	129; 122	$3p_{1/2}$; $3p_{3/2}$
47	Ag	373; 367	$3d_{3/2}$; $3d_{5/2}$
78	Pt	74; 70	$4f_{5/2}$; $4f_{7/2}$
79	Au	87; 83	$4f_{5/2}$; $4f_{7/2}$

Source: These values are taken from Appendix 1 of K. Siegbahn, C. Nordling, A. Fahlman, R. Nordberg, K. Hamrin, J. Hedman, G. Johansson, T. Bergmark, S. Karlsson, I. Lindgren, and B. Lindberg, *ESCA: Atomic, Molecular, and Solid-State Structure Studied by Means of Electron Spectroscopy*, Uppsala, Sweden: Almquist and Wiksells, 1967, by permission of the senior authors.

Auger Spectra

The nature of atomic deexcitation processes is such that photoelectrons are accompanied by x-ray emission or by Auger-electron emission. For the lighter elements and for the outer atomic shells of the medium and heavy elements, Auger-electron emission is the predominant mode by which an atom deexcites. Moreover, when high-energy electrons rather than x-rays are used to bombard a sample, Auger-electron emission predominates.

When a vacancy is produced in an inner shell by photoelectron ejection, the filling of this vacancy by an electron from a higher-energy shell is followed either by the emission of a fluorescence x-ray photon or by the simultaneous ejection of another outer-shell electron—the Auger or secondary electron (Fig. 15.1). Like the fluorescence x-rays, these electrons have characteristic energies for each atomic shell and each element. However, the Auger spectrum is generally more complex than the x-ray spectrum. For instance, two prominent x-ray peaks, $L\alpha$ and $L\beta$, are observed with initial vacancies created in the L_3 ($2p_{3/2}$) and L_2 ($2p_{1/2}$) shells of copper (Ll and $L\eta$ are extremely weak). In contrast, Figure 15.4 shows the corresponding Auger spectrum of Cu with initial vacancies in the L_2 and L_3 shells. There are many more groups of prominent peaks, as well as fine structures in the peaks themselves. Most of these Auger spectral lines have been cataloged for various elements and can be identified by reference to these tables. Like photoelectrons, the Auger electrons are sensitive to the chemical environment, which may produce some ambiguities in identification; so the tables must be used with care.

Some characteristics of Auger lines can be used to distinguish them from photoelectron lines. This is important since photoelectron lines are frequently accompanied by Auger lines. One useful fact is that the position (energy) of the Auger lines is always independent of the energy of the exciting photon because the Auger process occurs after the atom is ionized. For example, the kinetic energy of a KL_2L_3 Auger electron is given by

$$E_{\text{Auger}} = E_K - E_{L_2} - E'_{L_3} \qquad (15.2)$$

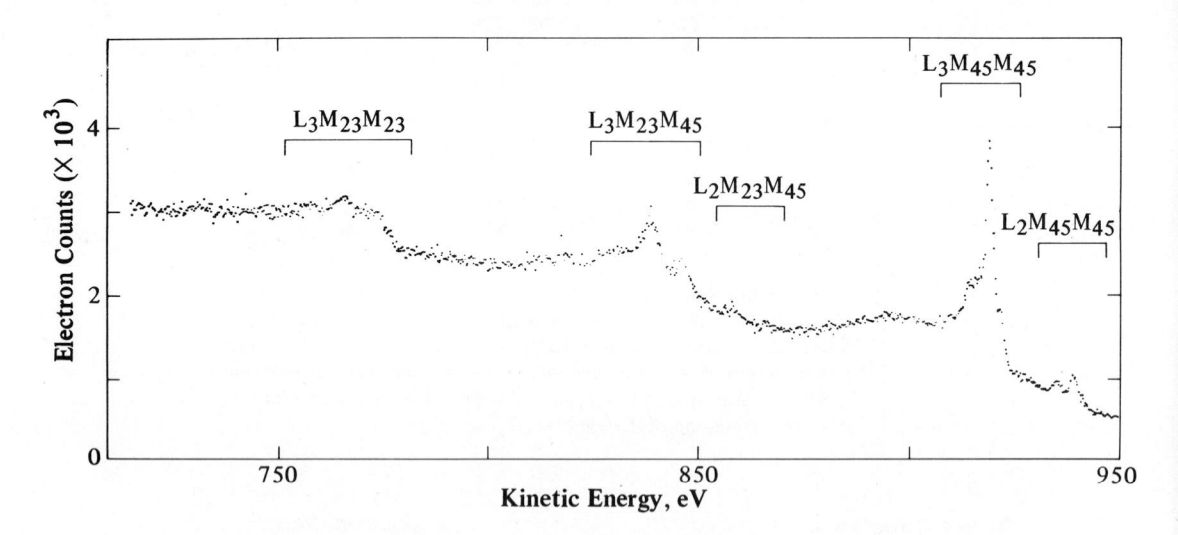

FIGURE 15.4. *Detailed $L_{2,3}MM$ Auger spectrum of copper. These lines are due only to vacancies in the L_2 and L_3 shells of copper. Note the abundant fine structure in the spectrum. The separation between a particular $L_3M_iM_j$ group and the corresponding $L_2M_iM_j$ group is equal to the binding-energy difference between L_3 and L_2 shells.*

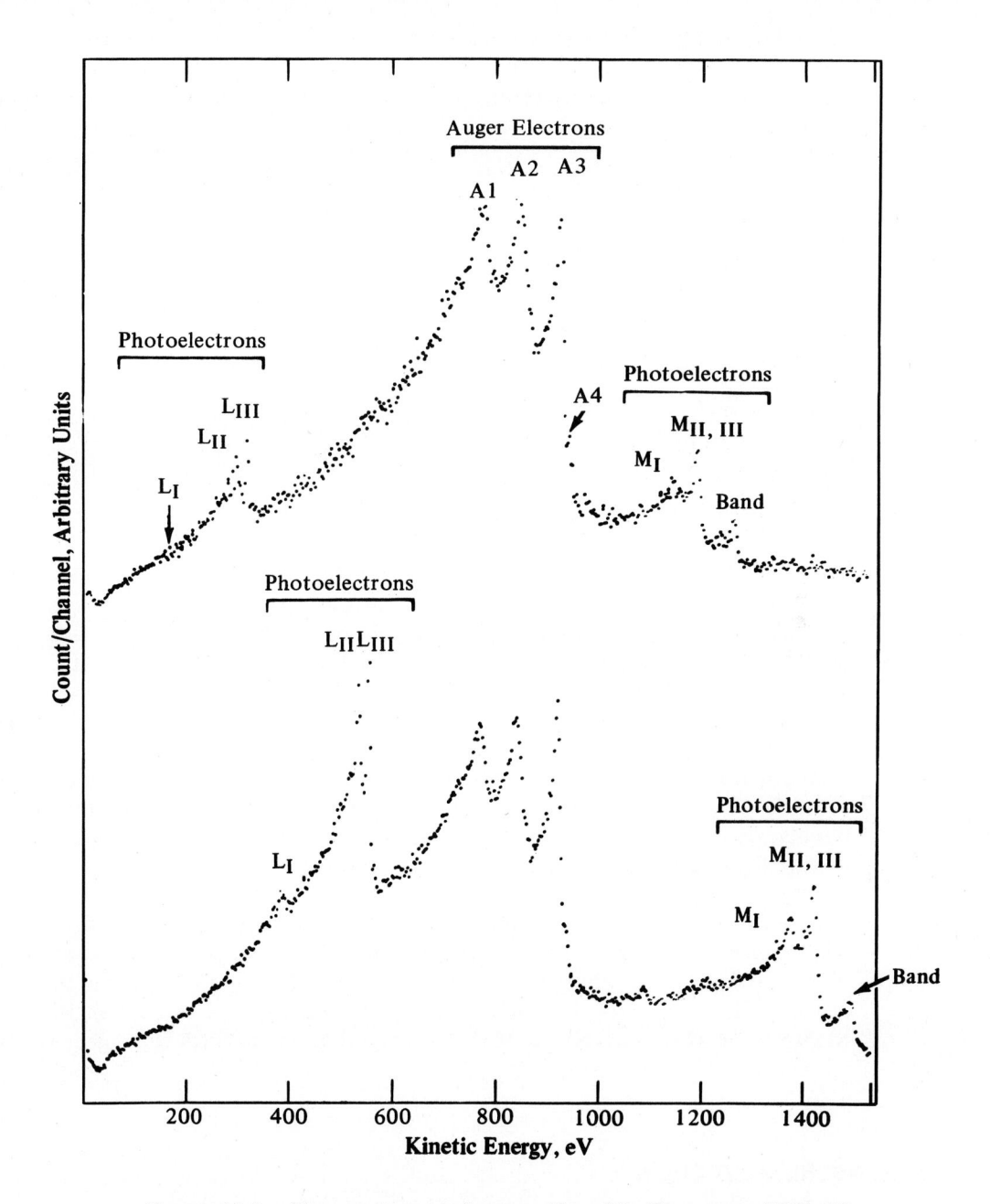

FIGURE 15.5. *Electron spectra of copper. Top: Mg Kα$_{1,2}$ x-ray (1254 eV) excitation. Bottom: Al Kα$_{1,2}$ x-ray (1487 eV) excitation. Note that, whereas the kinetic energies of the photoelectrons are proportional to the energies of the incident x-rays, those of the Auger electrons are independent. "Band" refers to the conduction band. The details of the Auger portion of the spectra are shown in Figure* 15.4. *From L. Yin, E. Yellin, and I. Adler,* J. Appl. Phys., *43*, 3595 (1971); *by permission of the authors and the American Institute of Physics.*

where E_K and E_{L_2} = binding energies of the K and L_2 shells, respectively, of a *neutral atom*

E'_{L_3} = binding energy of an electron in the L_3 shell of an *ion* having a single vacancy in the L shell

A wide-range scan of the electron spectrum of copper is shown in Figure 15.5. The top spectrum was obtained by using Mg K$\alpha_{1,2}$ x-rays (1254 eV) as an excitation source, the bottom by using Al K$\alpha_{1,2}$ x-rays (1487 eV). Whereas the kinetic energies of the groups of peaks labeled "photoelectrons" increase with the incident x-ray energies (Eqn. 15.1), the energies of the group labeled "Auger electrons" remain unchanged. This method for distinguishing between Auger and photoelectron lines can now be used with relative ease. The new x-ray sources being marketed by several companies are dual-target sources, aluminum and magnesium; so the target can be changed without breaking the vacuum or taking the x-ray tube apart.

In some instances, there are other methods of distinguishing Auger electrons from photoelectrons. In the case of KLL Auger electrons, there is a single group of lines which one identifies from appropriate tables, keeping in mind a possible discrepancy of a few electron-volts caused by chemical shifts. For outer-shell Auger electrons, there are groups of lines corresponding to vacancies produced in subshells. Although these groups may contain complex features, their component lines are separated by energies corresponding to the difference in binding energies. For example, the L_3MM group of Auger lines is separated from a comparable group of L_2MM Auger lines by the difference in binding energy between the L_3 and L_2 shells (Fig. 15.4). Furthermore, this difference is usually independent of the chemical environment.

Another practical generalization is that, for any given group of Auger lines from a major shell, the most intense group results from vacancies in the outermost subshell. This follows quite naturally, because in a given major shell the electron population is greatest in the outermost subshells. Therefore, roughly speaking, the probability of producing a vacancy by x-ray absorption is also highest for the outermost subshell. Furthermore, after photoionization, physical processes occur that reorganize the vacancy distribution among the subshells. Such reshuffling always results in increased vacancies in the outermost subshell. Thus, for example, one would predict (and actually find) that for the L-shell Auger groups, the intensities are as follows: L_3MM > L_2MM > L_1MM; similarly, for the M-shell groups, M_5NN > M_4NN > M_3NN > M_2NN > M_1NN; and so forth for higher shells.

15.2 INSTRUMENTATION

As in many fields, ESCA instrumentation was initially designed and built by pioneering investigators and thus was found in very few laboratories. Today, ESCA instrumentation may be purchased commercially. There are excellent instruments available that offer relative ease of operation, high sensitivity, and good resolution.

In this section, we shall review the principles underlying the instrumentation, and briefly describe in general terms some of the state-of-the-art devices that are commercially available. Although the emphasis in this chapter is on ESCA, mention should be made of other modes of operation now available in commercial equipment.

Figure 15.6 is a schematic diagram of a basic ESCA system. The sample is bombarded by low-energy monoenergetic x-rays which produce characteristic photoelectrons or Auger electrons. The electron energies are then sorted by means of an analyzer so that only electrons of a selected energy reach the detector. The resulting signals are then amplified and sent to a readout system.

Excitation Sources

There are alternative methods of excitation used in ESCA, and modern commercial devices provide them as a routine matter. For the practice of XPS, x-radiation is needed, whereas if Auger spectroscopy is the method of choice, an electron gun is used. It should be added that with x-ray excitation both photoelectron and Auger spectra appear, whereas with electron excitation only the Auger spectrum is available.

As implied in Equation 15.1, the exciting source must be monoenergetic. The important question in ESCA applications is, How monoenergetic? This is a relative

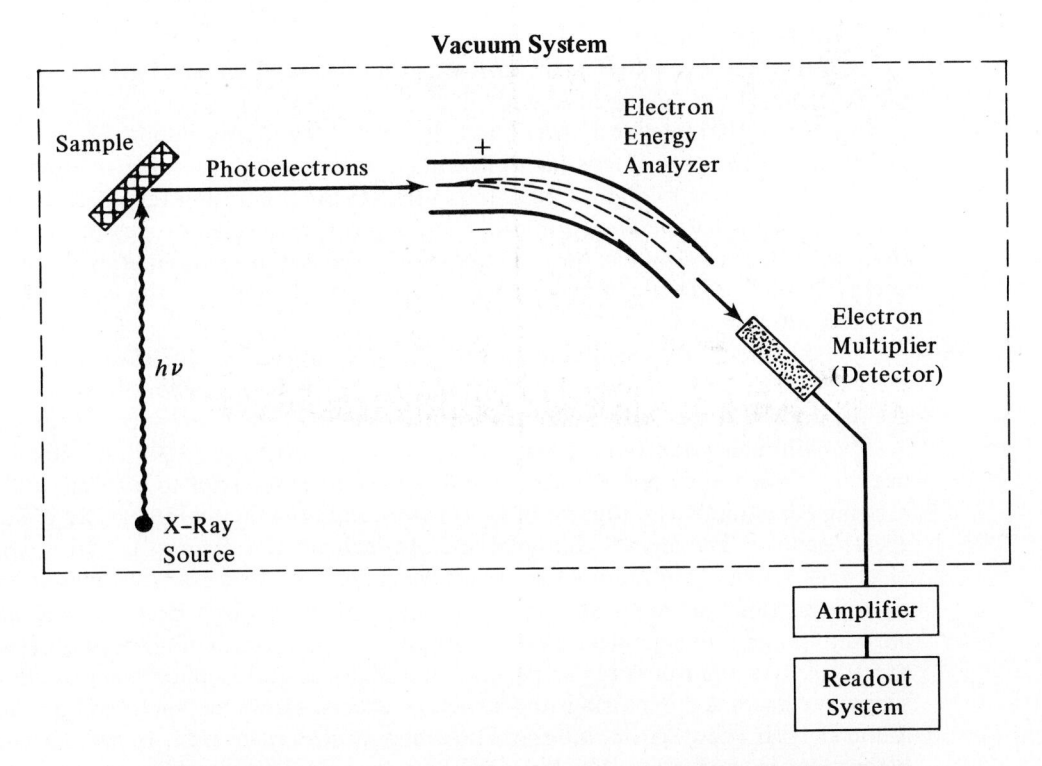

FIGURE 15.6. *Schematic diagram of an electron spectrometer. The electrostatic analyzer "sorts" or spreads out the photoelectrons, Auger electrons, and other secondary electrons of various energies so that only monoenergetic electrons reach the detector. Here, the energy analyzer serves much the same function as does a monochromator in optical spectroscopy.*

term. Because the demands of ESCA require measurement of line shifts of the order of 0.1 to 1 eV, the energy spread of the exciting radiation should be of comparable or lesser magnitude. Furthermore, the exciting source should be of high intensity to deal with the inefficiencies of the production of x-ray photoelectrons of undegraded energy.

The best way of obtaining such a source is to make use of the characteristic x-ray lines from an x-ray anode. These lines have a finite energy spread whose width decreases with decreasing atomic number. As a general rule, K x-ray lines are preferable as excitation lines because the L spectra are inherently more complex. Using these criteria (as well as the practical considerations below), the characteristic Al or Mg K lines (1487 eV or 1254 eV, respectively) were chosen for use in ESCA. The elements heavier than Al or Mg produce x-rays with higher efficiency, but the lines are inherently too broad and the $K\alpha_1$ and $K\alpha_2$ components begin to become distinct; so the radiation is no longer monoenergetic. The elements lighter than Mg or Al are either gases, or (if solid) the width of the K line begins to increase because the x-rays are now coming from transitions of electrons in the valence bands. In the case of Mg or Al, the $K\alpha_1$ and $K\alpha_2$ lines are so close in energy that for practical purposes they can be considered as a single line; the total energy spread is about 0.7 and 0.9 eV for Mg and Al, respectively. Power dissipation of the order of 1 kW or more is required to produce adequate electron intensities.

A recent development for producing monoenergetic x-rays involves a crystal monochromator in conjunction with an Al x-ray source to further reduce the energy spread of the characteristic x-rays and the background continuum. Because of the greatly reduced x-ray intensities following the use of the monochromator, it has been necessary to develop an ingenious method for processing the electron output from the spectrometer. The solution is analogous to a parallel-processing procedure that permits a band of electron energies to be measured simultaneously. The total energy spread of the exciting x-rays and the spectrometer is of the order of 0.5 eV for such instruments.

Another readily available monochromatic source uses the resonance lines of He I at 21.2 eV or He II at 40.8 eV, which have line widths of about 0.005 eV. These are the major sources used in vacuum ultraviolet photoelectron spectroscopy (UPS).

Although good photon sources with energies between 1000 and 40 eV have been developed, they are still not generally available. We refer to a recent and very exciting development in the use of *synchrotron radiation* in a wide variety of surface spectroscopies. The special characteristic of synchrotron radiation is that it consists of a very intense continuum that is highly polarized. It is possible to select very narrow portions of the x-radiation continuum by means of a Bragg crystal monochromator and thus obtain a very narrow band of x-rays for exciting the photoelectron spectrum. It is then possible to perform the high-resolution photoelectron spectroscopy necessary for exploring fine-structure effects. However, such capabilities are found only in a few special laboratories where accelerators exist. Experiments must be brought to the facility, which is used on a time-sharing basis.

Electron Spectrometer

Because of the commercial availability of very adequate electron spectrometers (the heart of the ESCA instrument), it will not be necessary to furnish here the detailed theory of operation. The discussion will concern itself only with the principles of operation.

Three types of electron spectrometers are presently in use: magnetic, electrostatic, and retarding grid types. Of these, the first two are focusing instruments whereas the last is not. Historically, the magnetic spectrometer was the first type, used notably in the work of Siegbahn and coworkers [1]. Today, the most common type of spectrometer in photoelectron spectroscopy, whether x-ray or ultraviolet, is the electrostatic type. The third variety (retarding grid) is used mainly with electron excitation, usually in conjunction with the LEED (*low-energy electron diffraction*) apparatus for studying Auger spectra.

The magnetic spectrometer is a momentum analyzer—the momentum of the electron is proportional to the applied magnetic field. In such instruments, the percent momentum resolution ($\Delta mv/mv$) is an instrumental constant. The momentum resolution achieved by Siegbahn et al. is about 0.01%. Instruments of this type are usually very large and very sensitive to stray magnetic fields. It is usually necessary to cancel any extraneous magnetic fields by the use of large Helmholtz coils and to place the spectrometer in iron-free rooms. These are essentially custom-built research instruments with low efficiency and are found in very few laboratories.

Electrostatic instruments come in a variety of forms: complete concentric hemispheres, hemispheric and spherical sectors, coaxial cylinders, and so forth. In such analyzers, the kinetic energy of the electron is proportional to the applied potential between the two conducting surfaces. Owing to the spherical or axial symmetry of such spectrometers, most have double-focusing properties. In general, the percent energy resolution ($\Delta E/E$) is a constant of the instrument; resolutions of the order of 0.05% have been achieved. Because the percent resolution is a constant, the absolute resolution ΔE is directly proportional to the energy E of the electron. In other words, at low electron energies the absolute resolution is better (ΔE is smaller). Consequently, there are two approaches to achieving good resolution in electrostatic analyzers: (a) building an instrument with a high enough resolution that the absolute resolution ΔE is small enough even at high kinetic energies to contribute little to the measured line width; and (b) reducing the energy of the electron entering the spectrometer to a constant value so that ΔE becomes small and constant. This is done by the simple expedient of applying a retarding potential, or by using a retarding lens which reduces the initial kinetic energy of the electron to the desired value prior to the electron entering the spectrometer. At present, most commercial ESCA instruments employ the latter technique and offer resolutions of the order of 0.1 eV. Like the magnetic analyzer, the electrostatic analyzer is sensitive to stray magnetic fields, and some form of magnetic shielding is usually required.

The third type of spectrometer, which uses retarding grids and electronic differentiation for energy analysis, is rather uncommon in conventional photoelectron spectroscopy. The reader is referred for further details to texts on LEED and Auger spectroscopy.

Detection Systems

As a general rule, the electron energies and intensities measured in ESCA are both generally low because of the various factors discussed above. The low electron energies dictate the use of windowless detectors, and the low intensities dictate the use of pulse-counting techniques; most of the available ESCA instruments employ both methods. The low counting rates also make automated data acquisition and analysis attractive; thus, many commercial instruments offer on-line computers as part of the entire ESCA system. It is worthy of note that detectors are undergoing remarkable improvements, particularly the continuous-surface multipliers and microchannel plate detectors. One commercial supplier is now offering position-sensitive electron detectors that will improve sensitivity and resolution in focal-plane spectrometers by factors of 20 to 100.

Sample Systems

Because ESCA can be applied to a great variety of problems, sample-handling capabilities should have enough flexibility that analysis can be performed on solids, liquids, or gases. Modern commercial instruments include a variety of sample-handling methods for dealing with bulk materials, films or foils, powders, wire and fibers, and so forth. In addition, there are capabilities for either heating or cooling the samples.

Vacuum System. It is essential to have strict control over the sample surface because the response in ESCA is entirely determined by the surface. A minimum requirement is a "very clean vacuum," that is, a vacuum relatively free of vapors that can be adsorbed on the surface of the sample and thus distort and contaminate the observed spectra. As a rule of thumb, a vacuum of 10^{-8} torr or better is necessary for general purposes. Because surface contamination is very rapid and the surface in poorer vacuums may be contaminated to an unknown extent, a vacuum in the range of 10^{-10} torr is often required. In fact, since most surfaces are already contaminated by the time the sample is introduced into the sample chamber, it is also very desirable to have auxiliary surface-cleaning capabilities in the sample chamber itself, such as sample-heating and ion-sputtering devices (e.g., an argon-ion gun), to clean the samples in place so they can be quickly studied before the surface becomes contaminated again.

Surface Charging. The ESCA investigator working on nonconducting samples must be constantly aware of the problem of surface charging. Surface-charge buildup will affect the values of the observed kinetic energies of photoelectrons from which chemical information is subsequently obtained. Surface charge occurs as a consequence of the ejection of electrons from the sample by the incident x-rays. The surface of a nonconductor becomes electron deficient (positively charged) because there are not enough charge carriers to neutralize the deficiency, as would happen in a conductor. The electrons then must leave a surface that attracts them and thus experience a retardation and consequent loss of kinetic energy. Sample charging is not a simple effect but is sensitive to the geometry and environment of the sample and its container. The amount of charging will thus vary from instrument to instrument, even for the same sample.

There are various techniques for dealing with sample charging—some instrumental, some involving special sample preparation, and some involving both. Instrumental techniques include using an electron gun to flood the sample with low-energy electrons and designing special geometries to surround the sample with a cage that discharges it. Among sample-preparation methods are such techniques as depositing on the nonconducting sample a very thin conducting film of a noble metal (e.g., gold or platinum) that has strong photoelectron lines with well-established energies. The deposited film must be thin enough not to obscure the sample of interest underneath, should not have electron lines with energies similar to those of the sample, and must not react with the sample. Other methods to minimize sample charging include depositing a thin film of nonconducting sample onto a conducting substrate and incorporating the nonconducting sample into a conductive wire mesh.

The phenomenon of charging is important only when absolute-energy values are required; provided the degree of charging is constant or quickly reaches an equilibrium value, the relative energies are unaffected.

Another parameter that affects the measured electron energies and which must be taken into account in determining the electron-binding energies is the *work function*. The kinetic energy the spectrometer sees is not necessarily the energy the electron has as it leaves the atom; every electron that escapes the surface of the sample must overcome a surface potential known as the work function, usually of the order of a few electron-volts. Although work functions differ from sample to sample, it has been shown that only the work function of the entrance material to the spectrometer needs to be accounted for in the determination of absolute kinetic energies. Because the work function of a particular spectrometer is an instrumental constant, it can readily be determined either by direct measurement or by reference to some standard electron energy. For example, the well-known 83.8-eV binding energy of the $4f_{7/2}$ level of gold is often used to calibrate the spectrometer work function. The work function can then be subtracted from the right side of Equation 15.1 to give

$$E_p = h\nu - E_b - E_{wf} \qquad (15.3)$$

15.3 APPLICATIONS

In current applications, the somewhat limited usefulness of ESCA in the area of simple quantitative elemental analysis does not detract from its potential in a myriad other chemical studies. To demonstrate the principles of the technique, we will show how to go about interpreting ESCA data in a few examples. Some of these examples are from our own work and were selected merely for convenience. Although typical, they are by no means exhaustive; they serve only to highlight some of the current applications of ESCA.

Elemental Analysis

In general, the application of ESCA for simple quantitative elemental analysis is somewhat limited. There have been such studies, such as that illustrated in Figure 15.7. Shown is a quantitative determination of layer thickness of an amide lubricant on polyethylene. Both nitrogen and oxygen were used as flags, or indicators. However, examination of the literature discloses very few examples of classical quantitative elemental analysis. ESCA is an extraordinarily sensitive surface technique involving

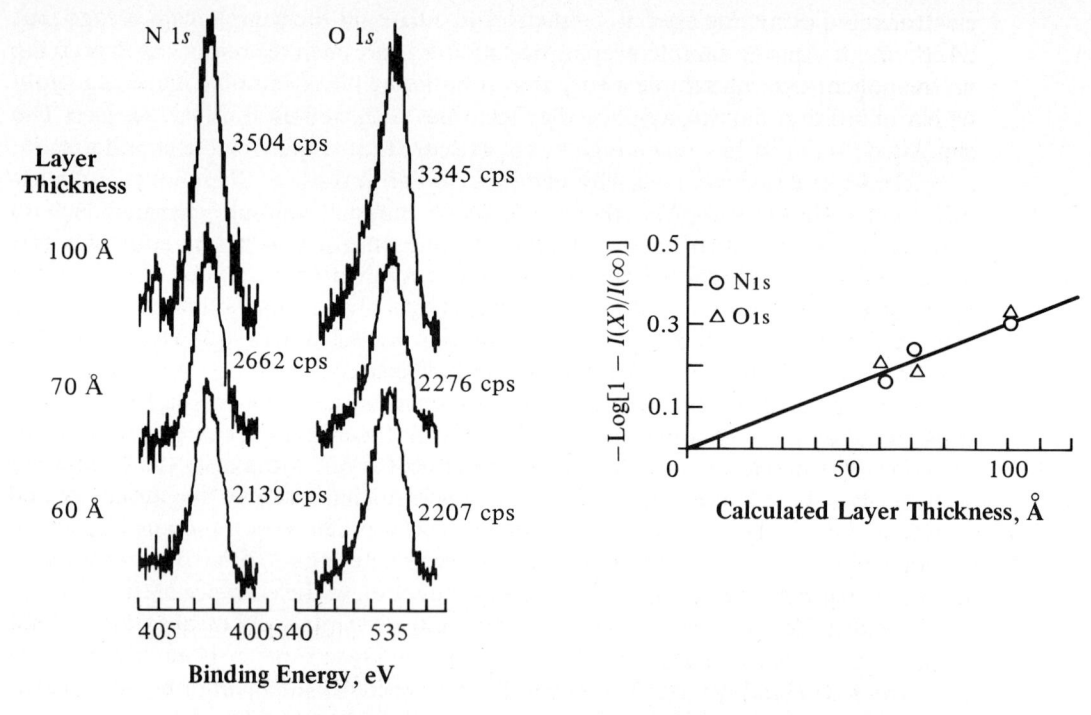

FIGURE 15.7. *Quantitative measurement of the film thickness of an amide lubricant on polyethylene. A: ESCA spectra of the N 1s and O 1s photoelectron peaks for several film thicknesses. B: Plot for determining the layer thickness. Circles are for N 1s at 402 eV; triangles are for O 1s at 535 eV; I(X) is the intensity (cps) of a peak for a layer thickness X, and I(∞) is the limiting intensity for a very ("infinitely") thick film. Reprinted with permission from W. M. Riggs and R. G. Brimer,* Chemtech, 5, 652(1975). *Copyright* © 1975 *American Chemical Society.*

the top 20 or so angstroms; in this sense, almost vanishingly small amounts of an element, about 0.001 monolayer, can be detected. To attempt an elemental analysis of a sample, however, immediately presents the analyst with the question of how representative the surface is of the rest of the sample, particularly in view of the possibility of surface contamination. Sample preparation is critical and must contend with a wide variety of surface phenomena such as adsorption and chemisorption, oxidation, and mechanical contamination, as well as more subtle phenomena that will be brought out in greater detail below. One important point is that both ESCA and Auger spectroscopy are essentially nondestructive techniques.

In summary, the use of ESCA for elemental analysis is primarily limited to the qualitative identification of surface elements, and to monitoring the presence or absence of these elements. Nevertheless, progress is being made in the application of both ESCA and Auger spectroscopy to quantitative analysis.

Valence States

Studies of chemical bonding, charge distribution, and valence state are perhaps the best-established applications of ESCA at present, and account for the bulk of the published papers in this area. In contrast to the heretofore more classical techniques, which are essentially inferential in character, ESCA is able to directly probe both the valence electrons, which actually participate in bonding, and the core electrons, which are directly influenced by the behavior of the valence electrons. It is this capability of ESCA that has led to its rapid growth; it is perhaps the most powerful and direct tool for these types of studies.

The transition metals are excellent examples of elements capable of various valence states, some of which are stable. In a systematic study of Fe_2O_3 and FeF_2 under argon-ion bombardment [3], the Fe $2p_{3/2}$ photoelectron peak was examined after various periods of ion bombardment (Fig. 15.8). With extended periods of bombardment, the peak shifts toward higher kinetic (lower binding) energy until its position coincides with that for metallic iron. In other words, Fe^{2+} and Fe^{3+} are being reduced to their metallic state under argon-ion bombardment. This is a clear example of how one can study surface reactions in a dynamic or time-dependent way. The spectra also show what is assumed to be FeO as an intermediate product in the case of Fe_2O_3 samples. It is well known that FeO is unstable in air and would normally be prepared only with extreme difficulty. There is no way of knowing whether the observed FeO is stoichiometric. This kind of chemical reduction under ion bombardment is fairly common among the $3d$ transition-metal compounds.

The effect of surface contamination is shown by the fact that initially the position of the photoelectron line from the "metallic iron" actually is indistinguishable from that for Fe_2O_3. This is caused by a surface layer of oxidized iron that is subsequently sputtered away by the ion bombardment.

The examples cited above demonstrate a need for caution in the use of ion-sputtering as a cleaning technique in the preparation of sample surfaces for ESCA analysis; the ion-sputtering process itself can produce chemical changes on the sample surface.

Stereochemistry

Because ESCA can directly probe the electronic structures of substances ranging from free atoms to solids, it is useful in a host of related fields such as stereochemistry, geochemistry, crystallography, and atomic and solid-state physics. Among the phenomena lending themselves to study are stereostructures, band structures, paramagnetism, atomic lifetimes, and Auger transitions.

An example of the use of ESCA to study the effect of steric arrangements concerns the binding energies of the core electrons of Ni in nickel compounds [4]. Some 70 compounds containing Ni in all of its known oxidation states were examined. Among other things, the results indicate that when Ni is bonded to the same ligand under different geometries, the binding energy of the Ni-ion electrons increases in the order: planar < tetrahedral < octahedral. This is not surprising because, for a given type of ligand, the nickel-to-donor distances increase in the same order; thus, since the valence electrons in the octahedral case are farthest removed from the Ni core, the binding energy of the remaining electrons is greatest. This relationship will not necessarily hold if the ligand is varied.

There is also a direct correlation between Ni $2p$ binding energy and the estimated charge on Ni for some simple Ni(II) compounds.

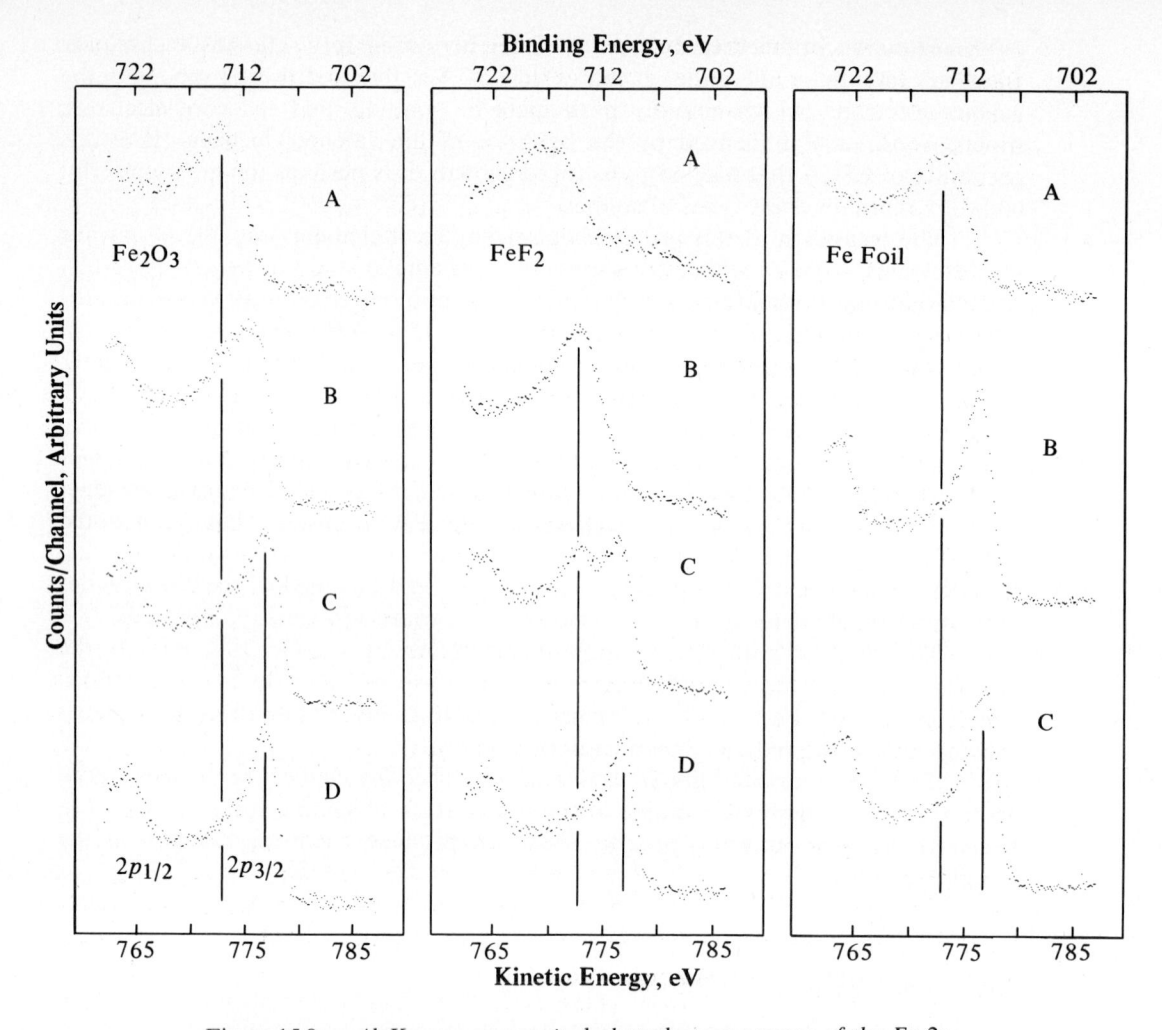

Figure 15.8. Al Kα$_{1,2}$ x-ray excited photoelectron spectra of the Fe 2p$_{3/2}$ (L$_3$) level in Fe$_2$O$_3$, FeF$_2$, and Fe foil, showing chemical reduction induced with argon-ion sputtering. Sputtering was performed at a pressure of 25 microns, a voltage of 1.5 kV, and a current density of 0.2 mA/cm^2. Fe$_2$O$_3$ sequence: A, prior to ion sputtering; B, after 12 min; C, after 72 min; D, after 120 min of sputtering. FeF$_2$ sequence: A, prior to sputtering; B, after 15 min; C, after 120 min; D, after 210 min of sputtering. Fe-foil sequence: A, prior to sputtering; B, after 60 min; C, after 160 min of sputtering. From L. Yin, S. Ghose, and I. Adler, Appl. Spectrosc., 26, 355 (1972); by permission of the publisher.

Surface Studies

One of the more important practical applications of ESCA is in the study of surface phenomena. Such areas as the direct study of surface reactions, diffusion processes under preselected conditions, the study of dopants in solids, surface-catalysis phenomena, sputtering processes, and gas-surface interactions are open to investigation.

Some contributions in this area have been made by the LEED method, with the instruments modified to measure the energies of the emerging Auger electrons. Because LEED is based on excitation of the sample by electrons (typically, samples are bombarded with a focused beam of electrons in the energy range up to 5 keV), these studies are limited to the use of Auger spectra. Furthermore, electron excitation requires electron differentiation of the energy spectra to minimize the high background of scattered electrons. Therefore, Auger spectra have the shapes shown in Figure 15.9 rather than the hump-shaped peaks of ESCA spectra. The degraded resolution, the complexity of Auger spectra, and the uncertainty of Auger chemical shifts limit such surface studies to the identification of elements and the monitoring of the presence or absence of surface constituents, rather than their chemical states or activities. Nevertheless, a great deal of surface information can be extracted even from the simple elemental identification of the presence, absence, or change of surface constituents.

An excellent example of surface studies is given by Weber [5] using ion sputtering to determine the in-depth composition of thin films on selected substrates. Figure 15.9 shows the effect of sputtering and surface thickness on sequentially observed Auger spectra. Note that the unit on the y-axis is $dN(E)/dE$, the first derivative of the number of electron counts, rather than $N(E)$. The top spectrum is that of a 150-Å nichrome film deposited on a silicon substrate. The Ni and Cr from the nichrome as well as the surface oxygen are visible, but not the substrate. In the middle spectrum, after 100 Å has been sputtered away, an Auger line from the silicon substrate appears while the oxygen line intensity drops greatly. In the bottom spectrum, after 200 Å of material has been removed, the Si from the substrate shows very strongly, whereas the Ni and Cr peaks have almost disappeared. Auger peaks for argon from the sputtering source have also begun to appear. The amplitudes of the various Auger peaks are shown as a function of material sputtered in Figure 15.10. By contrast, these same amplitude profiles are shown for a heat-treated film in Figure 15.11. It is clear that in the latter case, the first 75 Å of the surface consists of Cr and O, whereas the Ni has now diffused to the interface between the film and the silicon substrate.

Other examples of surface phenomena that have been studied involve grain boundaries, surface diffusion, corrosion, and so on.

For the most part, the behavior of surface catalysts has not been well understood; as a result, their development is still something of an art. ESCA offers perhaps the best potential for studying such phenomena because, since the suspected mechanisms probably involve some form of charge transfer, there should be an associated change in binding energy.

One such example is the study of the binding energy of platinum in a series of complexes [6]. The ESCA binding-energy data show that the coordination of "neutral ligands" can lead to a considerable transfer of charge from the metal to the ligands, confirming that the catalytic behavior of these platinum complexes is indeed related to charge transfer. Other typical examples have been given by Kelly and Tyler [7] and Larsson et al. [8].

The Scanning Auger Microprobe (SAM)

One significant advantage of Auger spectroscopy with electron bombardment is that the electron beam can be focused on a very small area of the sample surface.

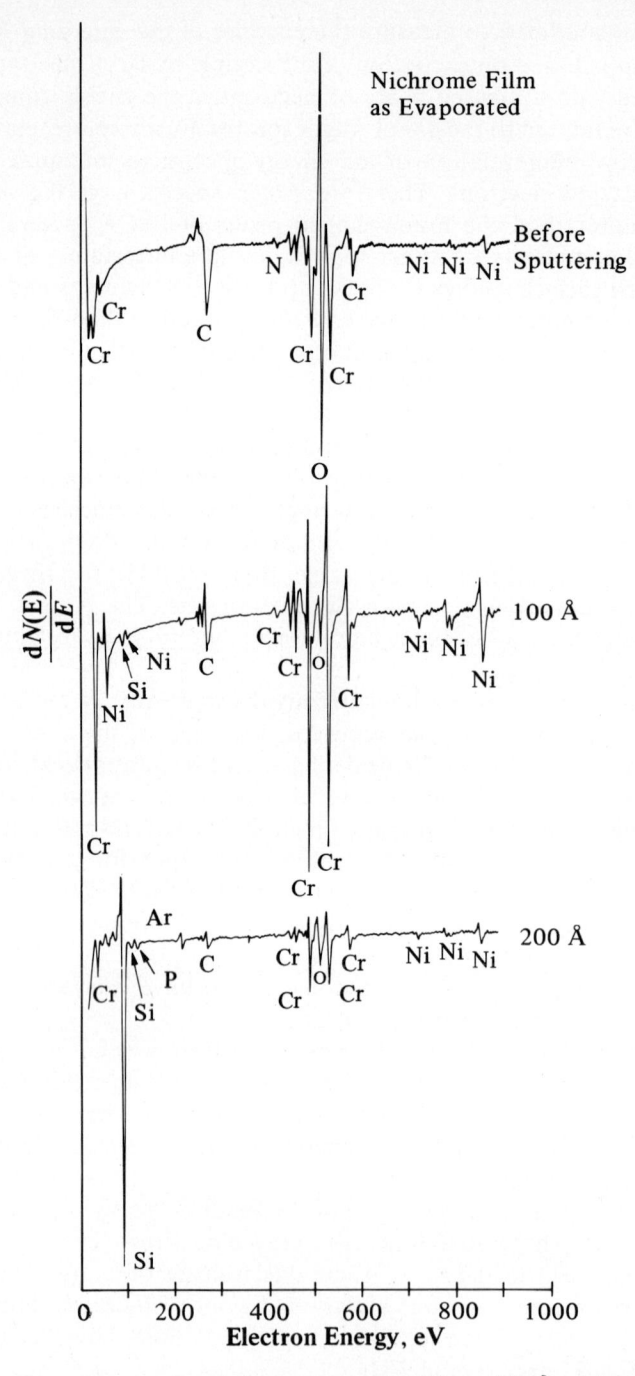

Figure 15.9. *Electron-excited Auger spectra from a 150-Å nichrome film deposited on a silicon substrate. The spectra represent a profile of the film from the surface down to a depth of 200 Å as various amounts of the material have been sputtered off. Note that the unit on the ordinate is the derivative of the electron intensity, dN(E)/dE, rather than the intensity, N(E). From R. E. Weber*, J. Crystal Growth, *17*, 342 (1972), *by permission of the author and the North-Holland Publishing Company.*

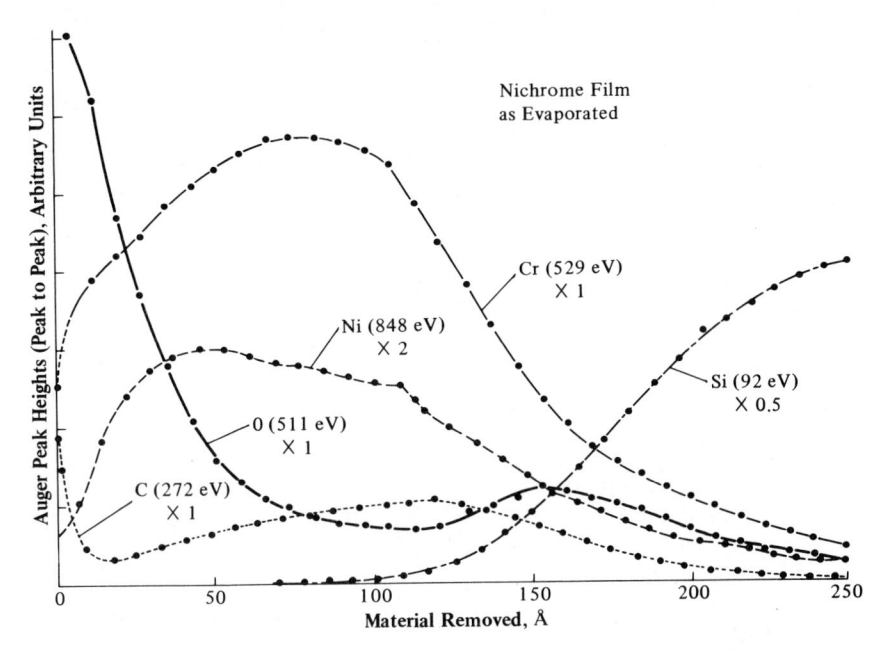

FIGURE 15.10. *Amplitudes of the various Auger peaks from a nichrome film on a silicon substrate as a function of the amount of material sputter-etched from the film. From R. E. Weber*, J. Crystal Growth, *17*, 342 (1972), *by permission of the author and the North-Holland Publishing Company.*

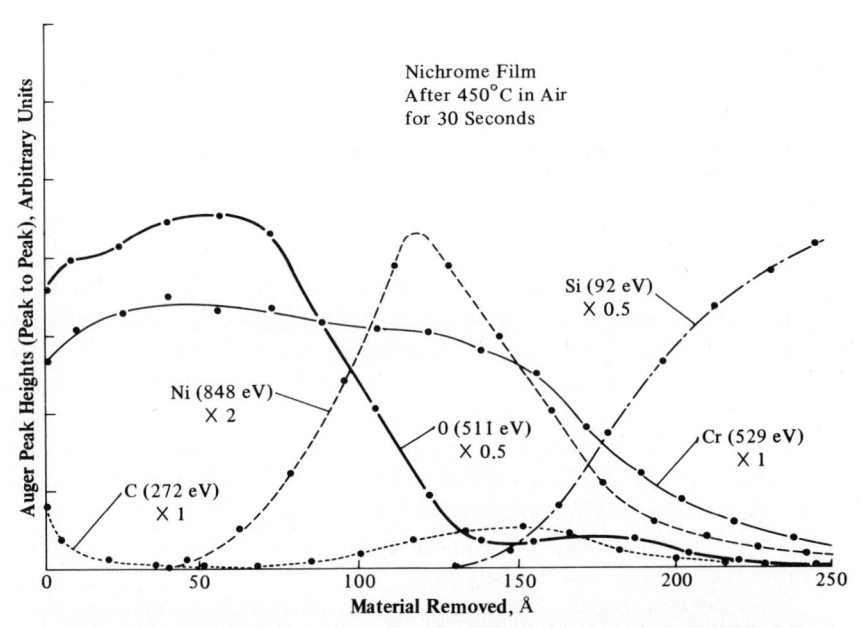

FIGURE 15.11. *Amplitudes of the Auger peaks from a heat-treated nichrome film as a function of the amount of material sputter-etched from the film. From R. E. Weber*, J. Crystal Growth, *17*, 342 (1972). *Reproduced with the permission of the author and the North-Holland Publishing Company.*

Typical beam diameters have been as little as 25 μm, although the use of Auger spectroscopy in conjunction with the scanning electron microscope allows the use of beam diameters below 5 μm. At present, ESCA photon-beam diameters are limited to about a millimeter or so. With scanning capability and small beam diameters, one can now perform lateral or even two-dimensional characterization of a surface, with a spatial resolution down to about 0.2 μm. For a two-dimensional semiquantitative analysis of a surface, the spectrometer is locked onto the energy of an Auger peak characteristic of a particular element. The electron beam is scanned systematically across the surface of the sample, and the position of the probe beam is correlated with the x-y position of the light beam on an oscilloscope screen. Finally, the light intensity of the oscilloscope signal is made proportional to the intensity of the Auger signal. The net result is an oscilloscope picture of the sample surface magnified perhaps 400×, with the bright areas on the screen corresponding to the "image" of the element analyzed on the surface. One can then "lock" on to the Auger peak of a second element, get its "picture," and so on. By this sequential multielement capability, the presence of several elements at specific sites can be ascertained.

Examples of this type of application abound in the literature. Segregation of impurities in metals and alloys often occurs at grain boundaries. One study of the embrittlement and stress failure of a tungsten sample revealed a nearly uniform distribution of phosphorus across the sample, except for certain grains that appeared to be completely free of it; this complete absence of phosphorus could be interpreted as related to cleavage failure at these points.

Molecular and Valence-Band Energy Levels

Studies of molecular structure and molecular energy levels by ultraviolet photoelectron spectroscopy are being conducted by a large number of investigators. Such studies are fundamental in nature and to a large extent theoretical. To provide meaningful data free of solid-state effects, the samples are usually gaseous. Furthermore, in order to see the closely spaced molecular orbitals and final-state vibrational structures, very high instrumental resolution (a few meV) and (usually) ultraviolet excitation are required. This subject is thoroughly covered in review articles and texts [9, 10].

Photoelectron spectroscopy provides a useful (because direct) tool for studying the valence-band structure of solids. It is unlike soft x-ray emission spectroscopy, where one must contend with transitions (to inner shells) that are constrained by selection rules and where one must take into account the character of the shell to which the transitions occur. In photoelectron spectroscopy, any of the occupied states in the band can be examined by ejecting the band photoelectrons. Thus, the photoelectron spectral shape essentially reflects the structure of the occupied band itself. Consequently, ESCA has been widely used to study the band structures of metals, alloys, and compounds. These data in turn are compared with density-of-state or molecular orbital calculations.

Because band structures are of interest to investigators in a wide variety of disciplines besides chemistry (e.g., solid-state physics, metallurgy, materials science, geochemistry, and crystallography), this area of application of ESCA is growing rapidly. This is especially so with the improved vacuum and resolution of second-generation instruments.

In principle, ultraviolet-excited photoelectron spectroscopy would be ideally suited to studies of valence-band structure because of its extremely high resolution. However, ultraviolet photoelectron spectra may not truly represent the band structure being probed; because the energy of the ultraviolet source is so low, it tends to induce valence-electron transitions and thereby distort the intensity distribution of the resulting photoelectron spectrum. X-ray excitation, on the other hand, although providing poorer resolution, is essentially free of such distortions. Thus, the two types of photoelectron spectra complement each other and together provide a more complete picture of the valence-band structure.

SELECTED BIBLIOGRAPHY

CARLSON, T. A. *Photoelectron and Auger Spectroscopy.* New York: Plenum Press, 1975. *A text on ESCA that covers theory and practice.*

HERCULES, D. M. *ESCA and Auger Spectroscopy.* Washington, D.C.: American Chemical Society, Audio Course Division. *An excellent review of ESCA and Auger spectroscopy consisting of six cassettes and 152 pages of written text.*

HERCULES, D. M., and HERCULES, S. H. *J. Chem. Educ.,* **61**, 402, 483, *and* 592 (1984). *Three articles on the analytical chemistry of surfaces, including ESCA and Auger spectroscopies.*

SIEGBAHN, K., NORDLING, C., FAHLMAN, A., NORDBERG, R., HAMRIN, K., HEDMAN, J., JOHANSSON, G., BERGMARK, T., KARLSSON, S., LINDGREN, I., and LINDBERG, B. *ESCA: Atomic, Molecular, and Solid State Structure Studied by Means of Electron Spectroscopy.* Uppsala, Sweden: Almquist and Wiksells, 1967. *This book, one of the first on electron spectroscopy, is now something of a classic and is still worth reading.*

WAGNER, C. D., RIGGS, W. M., DAVIS, L. E., MOULDER, J. F., and MUILENBERG, G. E., eds. *Handbook of X-Ray Photoelectron Spectroscopy.* Perkin-Elmer, Physical Electronics Division, 1979. *A very useful handbook that is both a guide and a reference book for the practice of ESCA.*

REFERENCES

1. K. SIEGBAHN, C. NORDLING, A. FAHLMAN, R. NORDBERG, K. HAMRIN, J. HEDMAN, G. JOHANSSON, T. BERGMARK, S. KARLSSON, I. LINDGREN, and B. LINDBERG, *ESCA: Atomic Molecular, and Solid-State Structure Studied by Means of Electron Spectroscopy,* Uppsala: Almquist and Wiksells, 1967.
2. S. O. GRIM, L. J. MATIENZO, and W. E. SWARTZ, JR., *Inorg. Chem.,* **13**, 447 (1974).
3. L. YIN, S. GHOSE, and I. ADLER, *Appl. Spectrosc.,* **26**, 355 (1972).
4. L. J. MATIENZO, L. YIN, S. O. GRIM, and W. E. SWARTZ, JR., *Inorg. Chem.,* **12**, 2762 (1973).
5. R. E. WEBER, *J. Crystal Growth,* **17**, 342 (1972).
6. C.D. COOK, K. Y. WAN, U. GELIUS, K. HAMRIN, G. JOHANSSON, E. OLSON, H. SIEGBAHN, C. NORDLING, and K. SIEGBAHN, *J. Amer. Chem. Soc.,* **93**, 1904 (1971).
7. M. A. KELLY and C. E. TYLER, *Hewlett-Packard Journal,* July 1973, pp. 1–14.
8. R. LARSSON, B. FOLKESSON, and G. SCHÖN, *Chemica Scripta,* **3**, 88 (1973).
9. K. SIEGBAHN, C. NORDLING, G. JOHANSSON, J. HEDMAN, P. F. HEDEN, K. HAMRIN, U. GELIUS, T. BERGMARK, L. O. WERME, R. MANNE, and Y. BAER, *ESCA Applied to Free Molecules,* Amsterdam: North-Holland, 1969.

10. D. W. Turner, C. Baker, A. D. Baker, and C. R. Brundle, *Molecular Photoelectron Spectroscopy*, London: Wiley-Interscience, 1970.

PROBLEMS

1. Identify the elements in the sample that gives the ESCA spectrum pictured below.

2. Identify the elements in an organic compound that has the ESCA spectrum shown at the foot of this page.

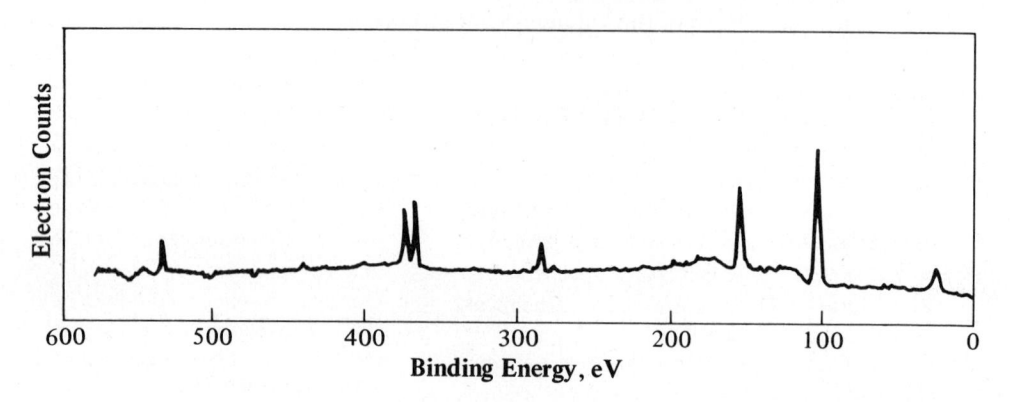

Problem 1

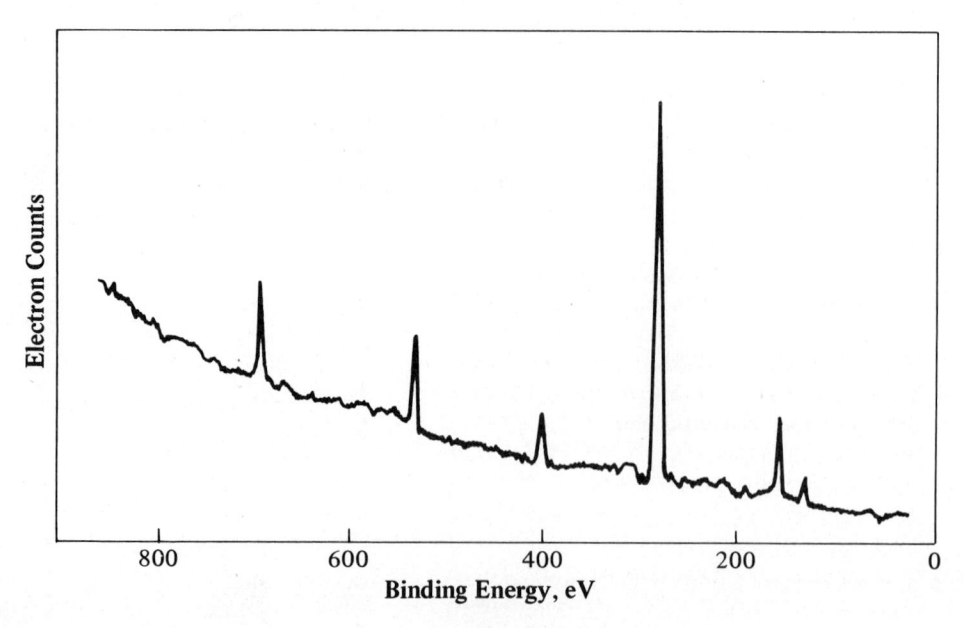

Problem 2

3. Determine the binding energies ($2p_{3/2}$ electrons) for iron in the compounds whose spectra are shown in Figure 15.8. Compare the shifts in binding energy caused by the differing chemical environments with the binding-energy differences between elements near iron in the periodic table.

4. What are the advantages and disadvantages of ESCA as a method for analysis of surfaces? Compare ESCA with Auger spectroscopy.

5. In an attempt to solve an analytical problem in the design and production of a semiconductor device, you are trying to determine the approximate detection limit for silver vacuum-deposited on a silicon substrate. Given that the beam-probe diameter in ESCA is about 1 mm, that the analysis depth is about 20 Å, and that ESCA can detect about 0.001 of a monolayer of an element, answer the following questions. (a) What is the absolute "detection limit" for Ag in number of atoms? (b) How many grams does this correspond to? (c) If the Ag were homogeneously distributed in the Si substrate, what would be the concentration detection limit in grams per cubic centimeter and in ppm? The densities of elemental Ag and Si are 10.5 and 2.33 g/cm^3; the atomic radii are 1.44 and 1.32 Å; and the atomic weights are 107.87 and 28.086 g/mole, respectively.

6. The ESCA spectrum of an inorganic compound is taken using Al Kα radiation. The photoelectron energy (E_p) of the $4f_{7/2}$ level for gold, a thin layer of which has been deposited on the sample, is measured at 1353 eV. (a) At what photoelectron energy should you look for the carbon 1s photoelectron peak to determine whether the surface of the sample has been contaminated with oil from the vacuum pump? (b) With the same spectrometer, at what photoelectron energy should the carbon 1s peak be if Mg Kα radiation were used?

7. An ESCA spectrum of a gaseous mixture of CO, CO_2, and CH_4 is taken. Prominent peaks are noted at binding energies of 290.1, 295.8, 297.9, 540.1, and 541.3 eV. Assign the observed peaks to the element and compound responsible.

8. The surface of an aluminum sample was thoroughly cleaned by abrasion and immediately put into the sample chamber of an ESCA spectrometer. Two prominent peaks in the spectrum occurred at binding energies of 72.3 and 75.0 eV, whose relative intensities were 15.2 and 5.1 (arbitrary) units. After a week's exposure to laboratory air, the same sample was rerun under the same conditions; the two peaks were again observed (at 72.2 and 74.5 eV), although the intensities were now 6.2 and 12.3 units. Explain.

9. ESCA can be used for the quantitative analysis of MoO_3/MoO_2 mixtures, compounds for which an instrumental method was not previously available [W. E. Swartz, Jr., and D. M. Hercules, *Anal. Chem.*, **43**, 1774 (1971)]. This is done by analyzing the Mo $3d_{3/2}$-$3d_{5/2}$ region of the electron spectrum. Molybdenum(VI) in MoO_3 has a peak at 235.6 eV for $3d_{3/2}$ and at 232.5 eV for $3d_{5/2}$; Mo(IV) in MoO_2 has corresponding peaks at 233.9 and 230.9 eV. Thus, the electron counts at 235.6 eV are due primarily to MoO_3, and those at 230.9 to MoO_2. The following data were obtained on standard MoO_3/MoO_2 samples:

% MoO_2	$N_{235.6} : N_{230.9}$
100	1.30 ± 0.10
95	1.41 ± 0.20
90	1.70 ± 0.10
80	2.10 ± 0.20
75	2.30 ± 0.20
60	2.95 ± 0.05
55	2.99 ± 0.20
50	3.31 ± 0.10
45	3.46 ± 0.06
40	3.60 ± 0.17
25	4.17 ± 0.05
20	4.53 ± 0.20
10	5.00 ± 0.30
5	5.05 ± 0.20
0	5.07 ± 0.05

Three unknowns gave the following count ratios: 1.94 ± 0.15, 2.55 ± 0.15, and 4.08 ± 0.10. Determine the percentage of MoO_2 in the three unknowns.

10. Calculate the binding energies of the L and M photoelectrons of Mg from the spectra shown in Figure 15.5. Comment on the differences in the corresponding binding energies between the two spectra. Which of these peaks corresponds to the $2p_{1/2}$ and $2p_{3/2}$ Cu electrons listed in Table 15.1?

CHAPTER 16

Mass Spectrometry

| Michael L. Gross

The practice of mass spectrometry is carried out with rather sophisticated instruments (mass spectrometers) that produce, separate, and detect both positive and negative gas-phase ions. Because samples are typically neutral in charge, they first must be ionized in the spectrometer. Ionization of molecular substances is often followed by a series of spontaneous competitive decomposition or fragmentation reactions that produce additional ions. The ion masses (more correctly, their mass-to-charge ratios) and their relative abundances are displayed in a *mass spectrum*. Most compounds produce unique or distinctive patterns; so most substances can be identified by their mass spectra.

Mass spectrometry is noteworthy among modern structural tools because the information gathered is of a chemical nature. The signals produced by a spectrometer are the direct result of chemical reactions (ionization and fragmentation) rather than energy-state changes which typify most other spectroscopic tools. To apply mass spectrometry intelligently, it is important to understand how this chemical information is produced. This chapter is intended to introduce the reader to the instrumentation necessary to carry out and identify these chemical reactions and to the procedure used in interpreting the information obtained.

Mass spectrometry is one of the oldest instrumental methods used in chemical analysis. The most important contribution of early mass spectrometry was the discovery of nonradioactive isotopes. The first study was reported by J. J. Thomson in 1913 [1] on the gaseous element neon. He showed that neon consists of two isotopes, ^{20}Ne and ^{22}Ne.

Major instrumental improvements followed closely just after World War I. F. W. Aston [2] in England constructed a more elaborate mass spectrograph which was used to verify the neon work and to identify other isotopes. The first mass spectrometer that used electrical rather than photographic detection was built in the United States by A. J. Dempster [3]. Whereas the mass spectrograph was better suited to measuring the exact masses of the isotopes, the spectrometer excelled in determinations of isotopic abundances. By the mid-1930s, most of the stable isotopes had been identified, and exact mass measurements established the idea that atomic masses are not whole numbers, a fact crucial to the understanding of nuclear chemistry. It was not until 1942, however, that the first commercial mass spectrometer was put into use, for petroleum analysis at the Atlantic Refining Corporation.

More general application of mass spectrometry as an analytical tool in the organic and biochemical areas was not developed until the 1960s. Within a few years, a number of chemists in the United States (McLafferty, Bieman, and Djerassi) and in England (Beynon) demonstrated that mass spectrometry could be used to elucidate molecular structure for a wide variety of substances. Since then, the technique has become a standard addition to most research and analytical laboratories.

16.1 INSTRUMENTATION IN MASS SPECTROMETRY

To obtain a mass spectrum, the sample must be vaporized, ionized, and then (provided the substance is molecular) allowed to fragment or decompose. The various ions must then be separated according to their mass-to-charge ratios (m/z-values) and finally detected. The instrumentation necessary to accomplish these requirements has four major components: (a) inlet systems for vaporization, (b) a source that serves to ionize and then detain the ions for a short period of time (usually about 1 μsec) so that fragmentation may occur, (c) a method of mass analysis, and (d) a detection scheme.

Inlet Systems

A generally useful inlet system must be able to vaporize molecules of quite low vapor pressure such as high-molecular-weight organic and organometallic compounds. Because many substances of interest to chemists do not have large equilibrium vapor pressures at room temperature, the inlet must operate at low pressure (10^{-4} to 10^{-7} torr) and high temperature (up to 300°C). Actually, the full development of mass spectrometry in the organic and inorganic areas had to await the development of these inlet systems. The problems involved with leak-tight high-temperature vacuum systems with many remotely operated valves are not trivial, and these problems have been overcome only in the last 20 to 25 years.

Most analytical mass spectrometers have two inlet systems: a batch inlet for gases and liquids and for solids of moderately high vapor pressure, and a direct inlet for high-molecular-weight nonvolatile solids and for thermally unstable compounds. A typical design for a batch inlet is shown in Figure 16.1.

A small quantity of solid or liquid sample (approx. 10 to 100 μg) is introduced via the detachable sample tube into the reservoir. The sample is maintained in the gaseous state by the low background pressure of the inlet (10^{-5} to 10^{-6} torr) and the high temperature of the surrounding oven. It should be obvious that a sample whose vapor pressure at the oven temperature is less than the background pressure of the inlet cannot be admitted to the spectrometer using this inlet. In fact, it is

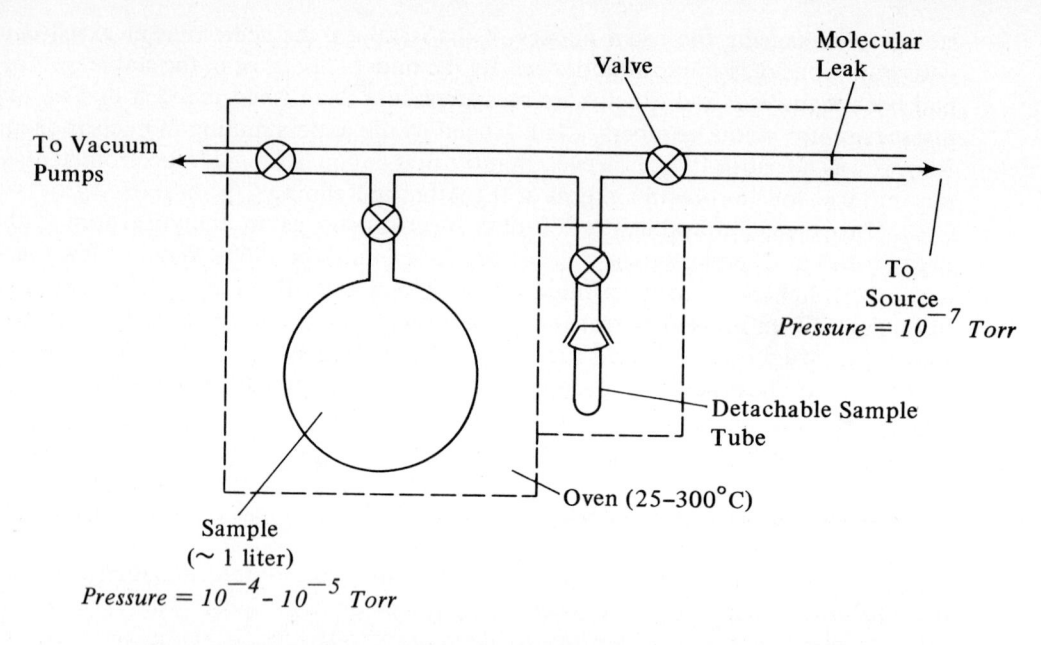

```
                    Valve         Molecular
                                  Leak
To Vacuum
Pumps
                                         To
                                         Source
                                         Pressure = $10^{-7}$ Torr

                                  Detachable Sample
                                  Tube
                          Oven (25–300°C)
Sample
(~ 1 liter)
Pressure = $10^{-4}$ – $10^{-5}$ Torr
```

FIGURE 16.1. *Schematic diagram of a typical batch-inlet system. This inlet is used for gases, volatile liquids, and volatile solids.*

preferable that the sample have a vapor pressure of 10^{-2} to 10^{-3} torr at the operating temperature so that a steady stream of the vapor can be admitted to the source through the leak—a glass or metal diaphragm containing a pinhole—for ionization. The flow is often molecular, which means the rate of effusion is inversely proportional to the square root of the molecular weight. This is the case when the opening is very small (10 times smaller than the mean free path of the gas particles). Since the lower-weight molecules pass more quickly through the leak, the inlet reservoir slowly becomes enriched in the higher-weight molecules. (This is not a serious problem unless very precise data are needed [e.g., in isotope-ratio work], because leak rates are usually quite slow.)

Solid samples that do not have a vapor pressure high enough to evaporate under conditions of the batch inlet, or that are thermally sensitive, are admitted directly to the source. Usually, the sample is placed in a small cup and introduced into the source through a vacuum lock. The sample cup can be cooled (e.g., with liquid nitrogen) or heated by infrared radiation or by thermal contact with a hot metal block surrounding the container. Using this technique, it is no longer necessary to fill the sample reservoir with vapor, and thus smaller sample sizes (as low as 1 ng) and substances with lower vapor pressures can be readily admitted. The direct inlet provides a dramatic increase in the versatility of mass spectrometry. No other analytical tool can produce as much information on such small quantities of complex organic or organometallic compounds. Mass spectra can be obtained for such diverse and nonvolatile samples as steroids, carbohydrates, dinucleotides, and low-molecular-weight polymers.

In many cases, nonvolatile substances are converted into more volatile derivatives prior to mass spectral analysis. Examples include trimethylsilyl derivatives of alcohols or molecules containing sugar groups, ester derivatives of acids, and volatile chelates of trace-metal ions. Usually, a suitable volatile derivative can be synthesized by well-established and relatively simple procedures.

Another method for sample introduction is the gas chromatograph (GC), discussed in Chapter 22. Components separated by a GC using packed columns can be admitted to the source of a mass spectrometer after enriching the eluent vapor (i.e., separating the helium carrier gas from the sample vapor with a molecular separator). The most common design (Fig. 16.2) is a jet separator made entirely of glass to prevent adsorption or decomposition of the sample molecules (a problem with hot metal). As the helium and sample pass through the constriction (jet), located below the vacuum-pump outlet, they expand into the low-pressure, glass envelope. Because helium diffuses much more rapidly than larger organic molecules, it is preferentially pumped away. The trajectory of the sample molecules is less affected, and most continue directly into the tube leading to the mass spectrometer source. The entire jet assembly is contained in an oven, and the temperature is controlled to be slightly greater than that of the GC column.

Capillary columns involve much slower flow rates than packed columns (1 to 2 mL/min compared to 30 to 50 mL/min). As a result, an interface is often not required. In fact, the new, fused-silica capillary columns can be threaded directly from the GC oven into the source ionizing chamber, permitting 100% efficiency.

Because the width of capillary GC peaks is of the order of a few seconds, a fast-scanning mass analyzer is necessary. The combination of gas chromatography and mass spectrometry (usually given the acronym GCMS) is perhaps the most versatile and sensitive tool in mixture analysis and is often used in petroleum, environmental, and biochemical research.

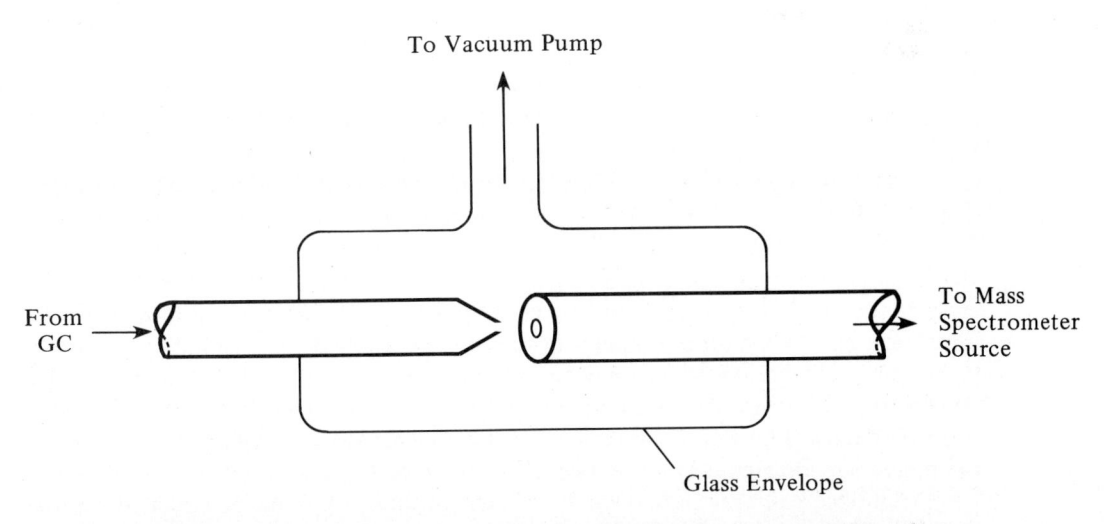

FIGURE 16.2. *Jet-separator interface connecting a gas chromatograph and a mass spectrometer.*

Electron-Impact Source

The ion source—the region where the sample ionizes and fragments—is the heart of the mass spectrometer. One may consider the source as a rather sophisticated chemical reactor initiating a series of characteristic degradation reactions (fragmentation) of an ionized sample. The decompositions take place in a very short time (usually 1 µsec), so a mass spectrum can be obtained very rapidly; as will be seen, some instruments can produce up to 1000 spectra per second.

The most common method of ionization is by electron impact; that is, a high-energy electron beam dislodges an electron from a sample molecule to produce a positive ion:

$$M + e^- \rightarrow M^+ + 2e^- \tag{16.1}$$

where
M = molecule under study
M^+ = *molecular* or *parent ion*

The beam produces M^+ in a variety of energy states. Some molecular ions are produced with rather large amounts of internal energy (rotational, vibrational, and electronic) that is dissipated by fragmentation reactions; for instance,

$$M^+ \diagup^{M_1^+ \rightarrow M_3^+}_{M_2^+ \rightarrow M_4^+, \text{ etc.}} \tag{16.2}$$

where $M_1^+, M_2^+, \ldots$ are lower-mass ions. Other molecular ions resist decomposition because they are formed with insufficient energy for fragmentation. It should be noted that most fragmentation processes are endothermic, and thus low-energy molecular ions will not fragment in the source and will be detected at the molecular mass.

A schematic of a typical electron-impact source is given in Figure 16.3. The device consists of an *electron gun* that accelerates and focuses electrons emitted by a thin, red-hot filament usually made of rhenium or tungsten. The electrons are accelerated by placing a negative bias of 70 V on the filament, producing a beam with a gaussian distribution of kinetic energies around a maximum at 70 eV (1 eV = 23.06 kcal/mole). Because most covalent molecules have ionization potentials of around 10 eV (see Table 16.1), 70-eV electrons are sufficient to dislodge an electron from one of the higher-energy molecular orbitals and produce a molecular ion with a distribution of internal energies. Molecular ions in higher-energy states can then decompose, producing various fragment ions. Since fragmentation patterns do not change significantly above 25 to 30 eV of ionizing energy, 70 eV is simply an arbitrary choice, and 50 eV or 80 eV would be equally acceptable. It is possible to vary this ionization energy by simply changing the voltage applied to the filament. A useful technique is to obtain a spectrum at a low voltage, such as 15 eV; this spectrum will be considerably simplified because of decreased fragmentation. In addition, ionization potentials of complex molecules can be measured by studying the decrease in the molecular-ion signal as a function of ionizing voltage.

The electron beam is collected on a target usually operated at a positive voltage (100 V) with respect to the filament. The target and filament are incorporated into the proper electronic circuitry to ensure a constant current or flow of electrons through

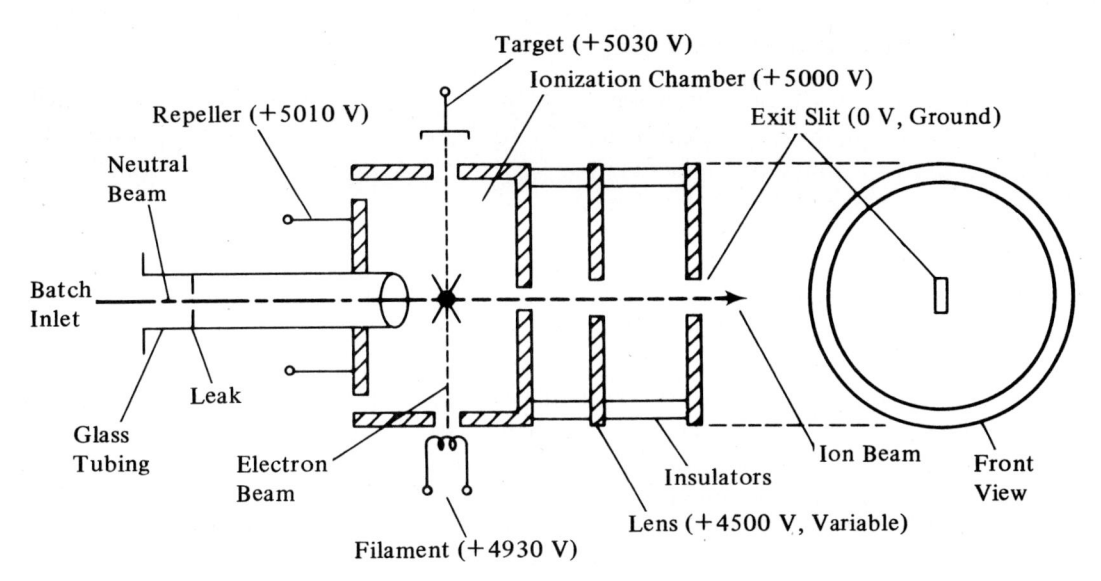

FIGURE 16.3. *Typical electron-impact source. The source is mounted on a frame and inserted into the flight tube of the mass spectrometer. The voltages in parentheses are typical values for the component parts of a spectrometer operating at 5000 V accelerating potential. Note that the target is at $+100$ V with respect to the filament, and that the filament is at -70 V with respect to the ionization chamber.*

TABLE 16.1. *Ionization Potentials of Some Common Organic Molecules*

Compound	Ionization Potential, eV
Methane	12.98
n-Hexane	10.17
Benzene	9.25
Naphthalene	8.12
Ethanol	10.48
Ethylamine	8.86
Acetone	9.69
Acetic acid	10.35

Source: J. L. Franklin, J. G. Dillard, H. M. Rosenstock, J. T. Herron, K. Draxl, and F. H. Field, *Nat. Stand. Ref. Data Ser. Nat. Bur. Stand.*, **26** (1969).

the ionization region. The regularity of the electron beam is important to achieve a constant number of ionizing events per unit time. (Under normal conditions, only about 1 neutral molecule out of every 1000 introduced into the source is ionized; the remainder are pumped away by the large-capacity diffusion pumps located outside the source.)

The second important feature of the source is the *ion gun*, which accelerates all the molecular and fragment ions out of the source into a mass-analysis sector (the region in which sorting of ions occurs according to m/z value). Accelerating voltages are unique for each spectrometer and are in the range 1000 to 10,000 V. If a spectrometer operates with 5000 V accelerating potential, the voltages applied to the various components are as shown in Figure 16.3.

The repellers are charged to a slightly higher voltage than is the chamber in order to push the positive ions into the ion gun. The lens plates are two semicircular disks to which is applied a variable voltage to focus the ion beam. Thus, the ions are formed in a positive field whose strength decreases in the direction of the exit slit and therefore accelerates positive sample ions in that direction. As was mentioned previously, the residence time of the ions in the ionization chamber is about 1 μsec. Since the time is short and the pressure low (10^{-6} to 10^{-7} torr), each ion acts as an independent entity and will remain as a molecular ion or as a fragment, depending on the amount of internal energy imparted to it by the electron beam.

To review: A steady stream of neutral molecules is drawn into the source from an inlet system and ionized where the stream of molecules intercepts the electron beam. The positive ions created are constantly drawn out with the ion gun, and the remaining neutral molecules are steadily pumped away. Thus, the source operates on a steady-state principle: constant input of neutral molecules and output of ions (and leftover neutral molecules).

The electron-impact source is the workhorse of analytical mass spectrometry. It is efficient, durable, and capable of producing a steady, intense beam of positive ions. Like all instruments, it must be periodically disassembled and cleaned, in addition to being equipped with a new filament. However, a source that is well cared for may operate for 6 months or more.

Spark Source

Of course, the electron-impact source cannot be used if nonvolatile inorganic samples such as metal alloys or ionic residues are to be analyzed. These substances can be investigated using a different kind of ionization chamber called a spark source, similar to the excitation sources used in emission spectroscopy (Chap. 11). The other parts of the spectrometer can be the same as in a general-purpose instrument; however, a Mattauch-Herzog double-focusing instrument is preferred (Fig. 16.7, below), because the spark source produces ions with a wide spread of kinetic energies. The entire device is known as a *spark-source mass spectrometer* (SSMS).

Ions are produced by applying a pulsed radio-frequency voltage of approximately 30 kV to a pair of electrodes mounted directly behind the ion gun, that is, about where the electron beam is located in an electron-impact source (see Fig. 16.3). The electrodes may be made of the sample itself if it is an electrical conductor or, for a nonconducting specimen, from a mixture of graphite and sample mixed and pressed into an electrode. The high-voltage spark causes localized heating of the electrode with simple vaporization as atoms or simple ions. As in electron impact, the ions are accelerated through the ion gun and then mass analyzed.

There are a number of advantages of SSMS that recommend it as a general-purpose tool for trace-element analysis in a variety of samples. First, the method has uniformly high sensitivity for almost all the elements; as little as 1 ppb can be detected. Second, extremely complex samples can be submitted to elemental analysis; as many as 60 different elements have been determined simultaneously in a given sample. Third, the information is relatively simple—only the mass-to-charge ratios of the elements are observed. Complications can arise if the element has a large number of isotopes or if the probability of forming multicharged ions is large; but nonetheless, the mass spectrum is much simpler than the spectrum obtained in emission spectroscopy. Fourth, the response of the instrument is linear over a wide range of concentrations of a given element in the sample; therefore, it is not necessary to use a wide range of standards to calibrate the measurements.

The detection systems used in SSMS are either photographic plates or electron multipliers (discussed below). The former has the advantage of being an integrating detector and is used for rapid monitoring of the elemental composition of a complex sample. If accurate data are required, electrical recording is preferred.

Thus, SSMS is an extremely powerful technique for routine elemental analysis of complex nonvolatile samples. The chief disadvantages are the high cost of the equipment and the fact that a skilled technician is needed to operate the instrument. Some specific applications will be discussed in Section 16.4.

Mass Analysis by Magnetic Sectors

A number of methods of mass analysis can be used in mass spectrometry. The most common type involves a magnetic sector. Once outside the source, the ion beam moves down a straight, evacuated tube toward a curved region placed between the poles of a magnet (see Fig. 16.4). This region is called the magnetic sector and its purpose is to disperse the ions in curved trajectories that depend on the m/z of the ion. Low-mass ions (beam 1 in the figure) are deflected most, and the heavier-mass ions (beam 3) the least. The symbol z refers to the number of unit charges on the ion; q is the actual charge.

The kinetic energy of the ions leaving the source is given by the product of the ion charge q and the accelerating voltage V

$$\frac{mv^2}{2} = Vq \qquad (16.3)$$

where m = mass of the ion
v = its velocity

In the magnetic field, the ions experience a centripetal force of Bqv, where B is the field strength that causes deflection. This force must be balanced by the centrifugal force of the ions, mv^2/r, where r is the radius of curvature. Therefore,

$$\frac{mv^2}{r} = Bqv \qquad (16.4)$$

or

$$v = \frac{Bqr}{m} \qquad (16.5)$$

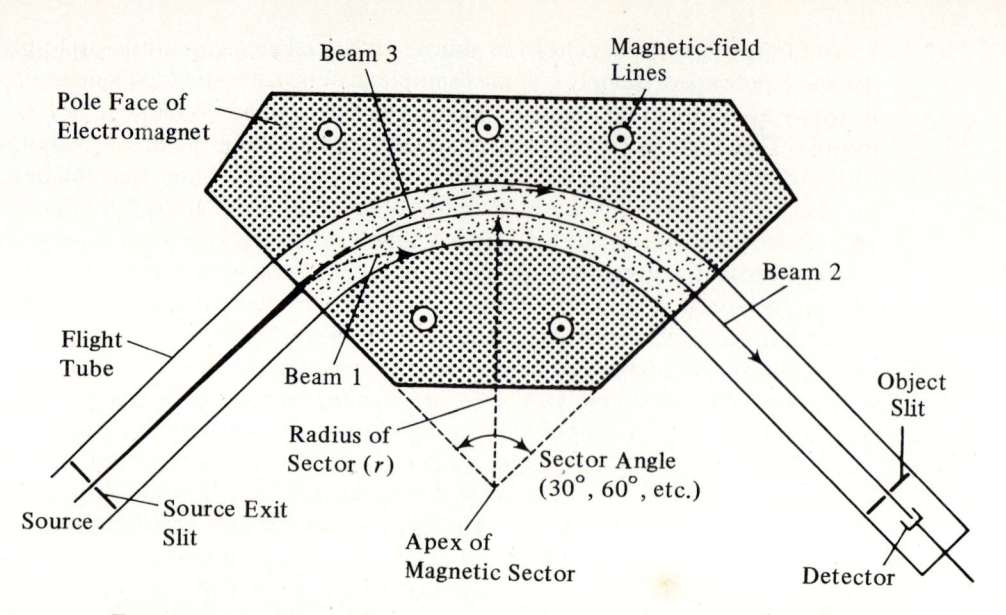

FIGURE 16.4. Mass analysis in a sector magnetic field. Poles of the electromagnet are located above and below the plane of the page. The magnetic field B points perpendicularly out of the plane of the page. Note that the two slits and the apex of the sector are colinear.

If the ion velocity is substituted into Equation 16.3, one obtains

$$\frac{m}{2}\left(\frac{Bqr}{m}\right)^2 = Vq \tag{16.6}$$

or

$$m/z = \frac{B^2 r^2}{2V} \tag{16.7}$$

This equation may be rewritten in a form where m is in atomic mass units, z is the number of charges ($+1$, $+2$, etc.), B is in gauss, r is in centimeters, and V is in volts:

$$m/z = \frac{B^2 r^2}{20,740 V} \tag{16.8}$$

Example 16.1. To illustrate the use of the above equations in the design of a mass spectrometer, we will determine the radius of trajectory of a monopositive ion of mass 100 in a magnetic field of 12 kilogauss if the accelerating potential is 6000 V. Using MKS units (common in electricity and magnetism): 12 kilogauss = 1.2 tesla; $m = 0.100$ kg/6.02×10^{23} molecules = 1.66×10^{-25} kg/molecule; $q = 1.6 \times 10^{-19}$ coulomb (the charge of an electron or unit positive charge).

Solution: Rearranging Equation 16.7 for r, we obtain

$$r = \left(\frac{2Vm}{qB^2}\right)^{1/2} = \left(\frac{2 \times 6000 \times 1.66 \times 10^{-25}}{1.6 \times 10^{-19} \times (1.2)^2}\right)^{1/2} = 0.093 \text{ m}$$

The radius of deflection necessary for an ion to impinge on the detector is determined by the curvature built into the flight tube, and is therefore a constant. A scan of mass spectrum is accomplished either by keeping B constant and decreasing the accelerating voltage so that ever-increasing ion masses are brought to focus (i.e., are given a deflection equal to the radius of curvature of the flight tube), or by increasing B at constant V to accomplish the same result. Older instruments that used permanent magnets were scanned by the first method; however, most modern instruments are equipped with electromagnets and are scanned by increasing the electric current in the magnet coils.

Mass spectrometers with only a sector magnetic field for mass analysis are known as single-focusing instruments. A well-designed single-focusing spectrometer may have resolution as high as 5000. In mass spectrometry, resolution R is defined as

$$R = \frac{m}{\Delta m} \tag{16.9}$$

where Δm = mass difference between two resolved or separated peaks
m = nominal mass at which the peaks occur

A resolution of 5000 would indicate that $m/z = 5000$ would be resolved from $m/z = 5001$ (or $m/z = 50.00$ from $m/z = 50.01$). A resolution of 500 is sufficient for many applications in organic chemistry, and low-cost instruments offering such resolution may be adequate to solve numerous problems.

Double-Focusing Mass Spectrometers

If resolution greater than about 5000 is required, a double-focusing mass spectrometer is necessary. Two factors that limit resolution in the single-focusing instruments are the angular divergence and spread in kinetic energy of the ion beam as it leaves the ion gun (see Fig. 16.5). The various ions in the beam always have a small spread of kinetic energies (of the order of a few eV) because they are formed in different regions of the ionization chamber and, therefore, experience different total accelerations. In addition, the neutral molecules enter the source with a Boltzmann distribution of thermal energies, which must be added to the ionization and acceleration energies to obtain the total kinetic energy.

To correct for these aberrations, an *electrostatic analyzer* or sector is introduced, usually before the magnetic sector. This device consists of two cylindrical

FIGURE 16.5. *Angular divergent beam at the exit of an electron-impact source.*

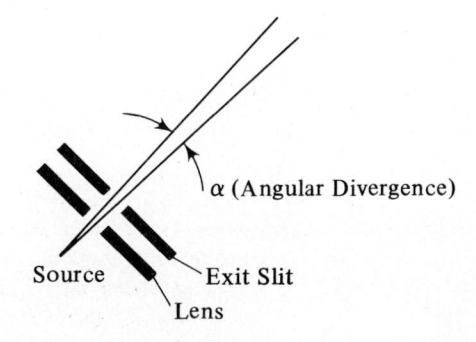

electrodes; a positive voltage is applied to the outer one and a negative voltage of equal magnitude to the inner one (see Fig. 16.6). The radius of curvature of an ion beam through this sector is determined by the kinetic energy of the beam for a constant voltage; the higher the kinetic energy, the greater the radius. Ultimately, high-energy ions will be deflected so little that they will impinge on the positive electrode. Ions of low kinetic energy are discharged on the negative electrode. Thus, the electrostatic analyzer serves as a kinetic-energy analyzer.

More specifically, the electrostatic analyzer serves to sort out ions of equal kinetic energy and bring them to a common focus. Thus, a beam emanating from a single point (the source) is brought to focus at many points, each representing a common kinetic energy (only two are shown in Fig. 16.6). In turn, the magnetic-field shape can be designed, by proper machining of the pole faces, to refocus the separate beams at one point for each mass-to-charge ratio.

Resolution as high as 150,000 with mass-measuring accuracy of 0.3 ppm can be achieved with one commercial double-focusing spectrometer, and a resolution of 20,000 to 50,000 is not uncommon. Thus, the exact weight of a compound of nominal molecular weight 600 could be measured to approximately ± 0.0002 mass unit using the 150,000 resolution instrument. This accuracy allows unambiguous assignment of the elemental composition (chemical formula) of the sample ion and consequently of the neutral sample.

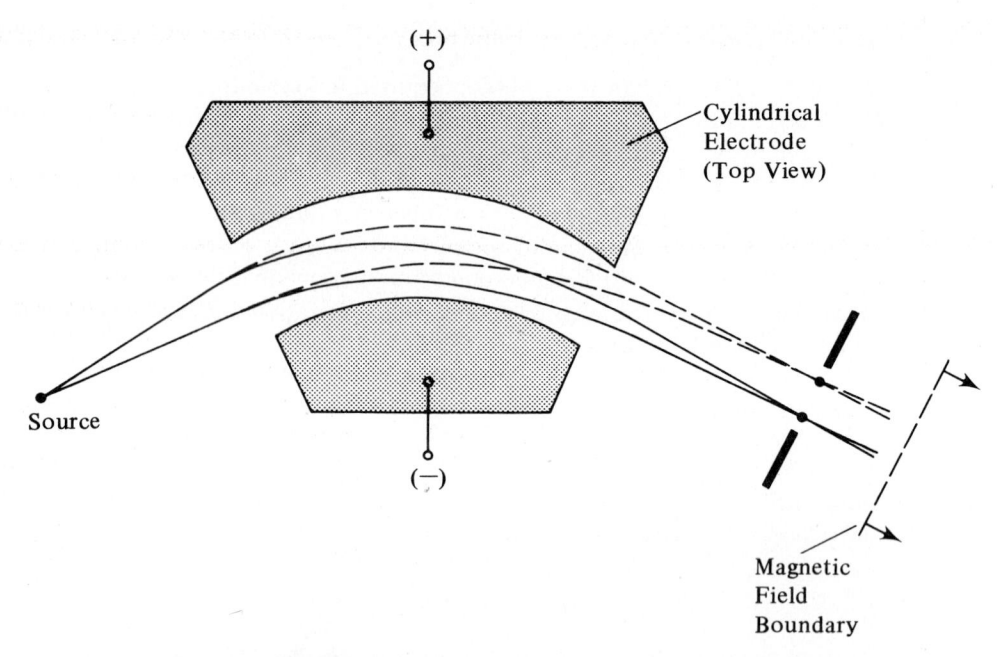

FIGURE 16.6. *Focusing of a divergent beam of two slightly different kinetic energies (solid and dashed lines) by an electrostatic analyzer. The electrostatic analyzer is followed by a sector magnet; see Figures 16.4 and 16.7. So-called reverse-geometry instruments are also double focusing. In these, the magnet sector precedes the electrostatic analyzer.*

Two designs are prevalent for double-focusing or high-resolution mass spectrometers: (a) Nier-Johnson and (b) Mattauch-Herzog (both shown in Fig. 16.7). The Nier-Johnson design operates only with an electrical detector. The Mattauch-Herzog design uses either a photographic-film detector at the focal plane or an electrical detector placed at one point of the plane. The advantage of a photoplate is that it is an integrating detector and can give a reliable spectrum even when evaporation of the sample is discontinuous or sporadic. It is also useful for very small samples because the operator does not have to wait for the magnetic field to be scanned; instead, the entire spectrum is exposed at once.

The disadvantage of film detection is that the plate must be developed and the lines identified to obtain the mass spectrum. Also, intensity data suffer in accuracy (at best, ion abundances can be measured to 10% relative error). Identification of line position and intensity is done with a microdensitometer (as in x-ray crystallography or emission spectroscopy); high-resolution measurements can be made in this manner. The densitometer is usually interfaced with a computer.

Using a Nier-Johnson design, the exact (high-resolution) mass measurement is obtained by peak matching. The exact mass of an unknown peak is determined by a high-precision measurement of the changes in accelerating voltage and ESA voltage that are necessary to superimpose the unknown peak on a peak of known mass produced by a mass standard introduced with the unknown sample. Typical mass standards are perfluorinated hydrocarbons or amines. Since $m/q = B^2 r^2 / 2V$,

$$\frac{m_s}{m_{std}} = \frac{V_{std}}{V_s} \qquad (16.10)$$

where V_{std} = accelerating voltage needed to focus the standard ions at constant B and r
          V_s = voltage needed to focus the sample ions at constant B and r

These measurements are very time consuming and are generally used to obtain exact mass measurements for only a few important ions in an unknown spectrum.

A more convenient and expeditious means of mass measurement with either design is to interface an electronic detector with an on-line computer that acquires and stores all the data, both m/z-values and intensity data, while the spectrum is being scanned. After identifying the m/z ratios of the mass standard, the computer calculates the exact masses of all the unknown ions from the scanning time between standard and unknown and, within a few minutes, prints on a teletype the exact masses and intensities of all the peaks in the mass spectrum. This is possibly the most elegant technique in mass spectrometry, because it provides the analyst with exact masses, which can be used to determine the elemental compositions of all peaks in a mass spectrum.

For example, a molecular mass of 150.0681 ± 0.0003 (2-ppm accuracy) is unique for the composition $C_9H_{10}O_2$ and rules out other samples of nominal mass 150 such as $C_5H_{10}O_5$ (m/z = 150.0528), $C_7H_6N_2O_2$ (m/z = 150.0429), or $C_9H_{14}N_2$ (m/z = 150.1157). Similar arguments can be made for fragment ions. Certainly, sample identification using this technique is greatly facilitated compared to that using a low-resolution spectrum, which yields nominal masses only. As might be expected,

Mattauch–Herzog

SOURCE

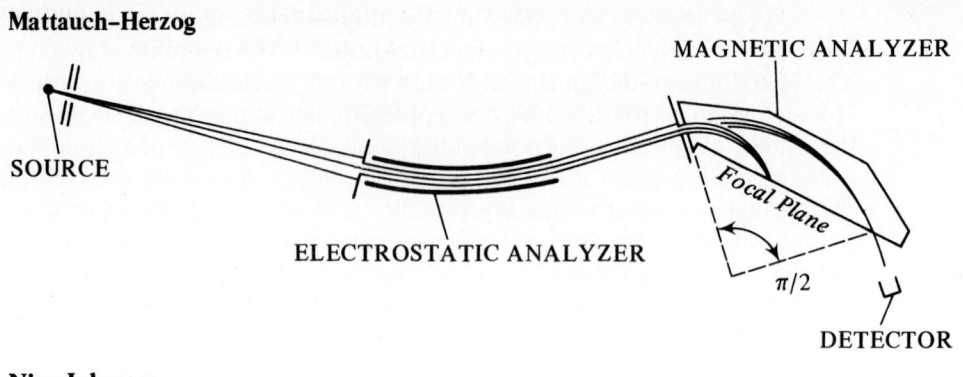

ELECTROSTATIC ANALYZER

MAGNETIC ANALYZER

Focal Plane

$\pi/2$

DETECTOR

Nier–Johnson

Alpha Slit

x-y Lens

ELECTROSTATIC ANALYZER

Beta Slit (Variable)

Monitor Collector

38 cm Radius

Hexapoles

Y and Z Deflectors

Defining Slit (Variable)

Focus

Ion Chamber

SOURCE

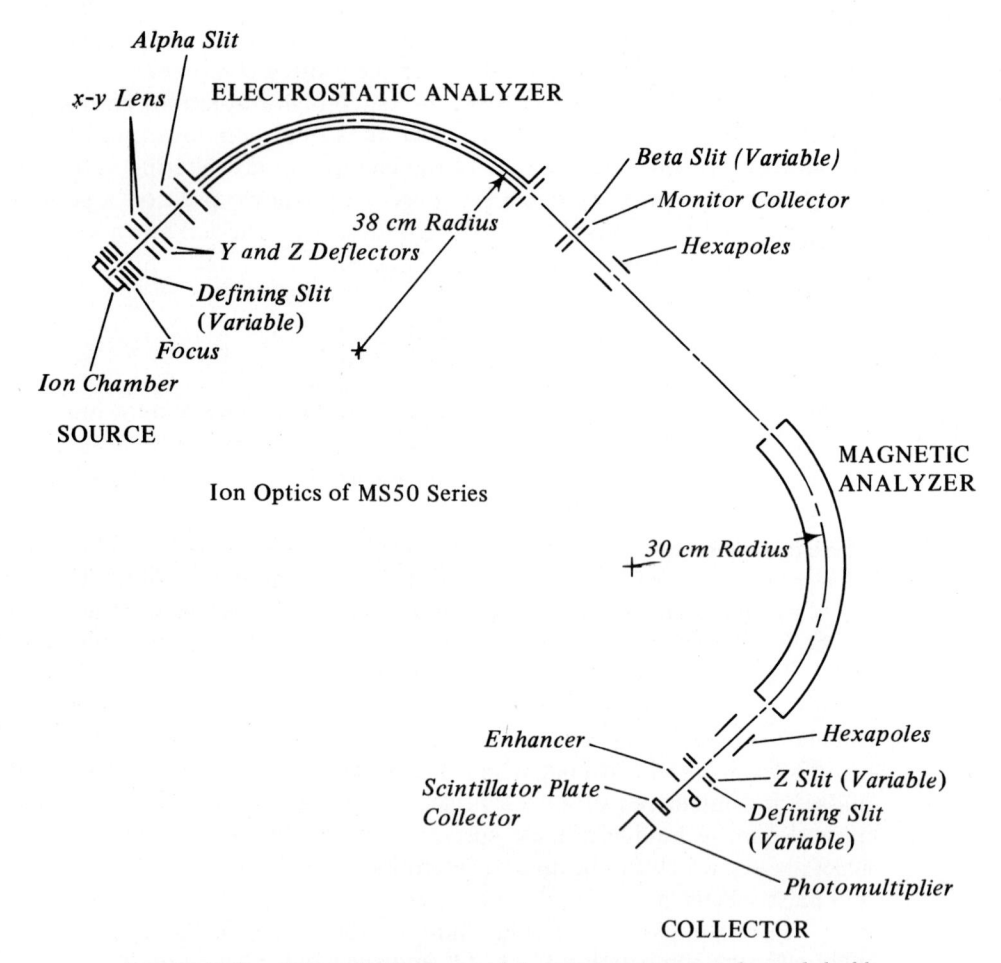

Ion Optics of MS50 Series

MAGNETIC ANALYZER

30 cm Radius

Enhancer

Hexapoles

Scintillator Plate Collector

Z Slit (Variable)

Defining Slit (Variable)

Photomultiplier

COLLECTOR

FIGURE 16.7. *Schematic diagrams of the two most commonly used double-focusing mass spectrometers. The Nier-Johnson is really a modified design capable of resolution of 150,000 because of the added hexapole lenses. From S. Evans and R. Graham, Advan. Mass Spectrom., 6, 429 (1974), by permission of Applied Science Publishers, Ripple Road, Barking, Essex, England.*

the high-resolution mass spectrometer equipped with computer is quite expensive ($250,000 to $500,000), and complicated to operate and maintain.

Note that the exact mass is that of the precise isotopic species involved—e.g., $^{12}C_9{}^1H_{10}{}^{16}O_2$—not the normal molecular weight. For $C_9H_{10}O_2$, the latter would be 150.77.

Time-of-Flight Mass Analysis

Time-of-flight (TOF) mass spectrometers are equipped with a modified electron-impact source and a long, straight flight tube. Different masses are distinguished by their different arrival times at the detector located at the end of the tube. Since the kinetic energy of the ions after acceleration is given by Equation 16.3, we can write

$$v = \left(\frac{2Vq}{m}\right)^{1/2} \tag{16.11}$$

Every ion has its own unique velocity, inversely proportional to the square root of the mass. If the ions are accelerated into a long flight tube of length L, the time necessary for an ion to reach the end of the tube is

$$t = \frac{L}{v} \tag{16.12}$$

The difference in time, Δt, that separates one ion (ion 1) from another ion (ion 2), is

$$\Delta t = L\left(\frac{1}{v_1} - \frac{1}{v_2}\right) \tag{16.13}$$

$$\Delta t = L\frac{\sqrt{m_1} - \sqrt{m_2}}{\sqrt{2Vq}} \tag{16.14}$$

and depends on the difference in the square roots of the masses.

The instrument is operated in a pulsed mode because continuous ionization and acceleration would lead to a continuous output at the detector with intractable overlapping of various masses. A typical sequence of events for pulsed operation is as follows. (a) The electron gun is turned on for about 10^{-9} sec to form a packet of ions. (b) The accelerating voltage is turned on for about 10^{-4} sec to draw the ions out into the flight tube. (c) All power shuts off for the rest of the millisecond pulse interval, allowing the ion packet to "coast" unhindered down the flight tube. (d) The electron gun turns on again, forming a fresh packet of ions. The spectrum is recorded by bringing the amplified signals from the detector to the vertical deflection plates of a storage oscilloscope. The horizontal axis of the scope is a time base and starts when the accelerating voltage is activated (see Fig. 16.8). With this instrument, as many as 1000 spectra per second can be obtained.

The advantage of a TOF mass spectrometer is its rapidity in scanning a spectrum. The device is extremely useful in monitoring fast gas-phase kinetics, flash photolysis, and shock-tube experiments, and can be used in GCMS applications. It can be used in routine analytical applications, and the best designs have a resolution of around 500. Recently TOFs have become popular for high-mass ion analysis because of their unlimited upper mass limit.

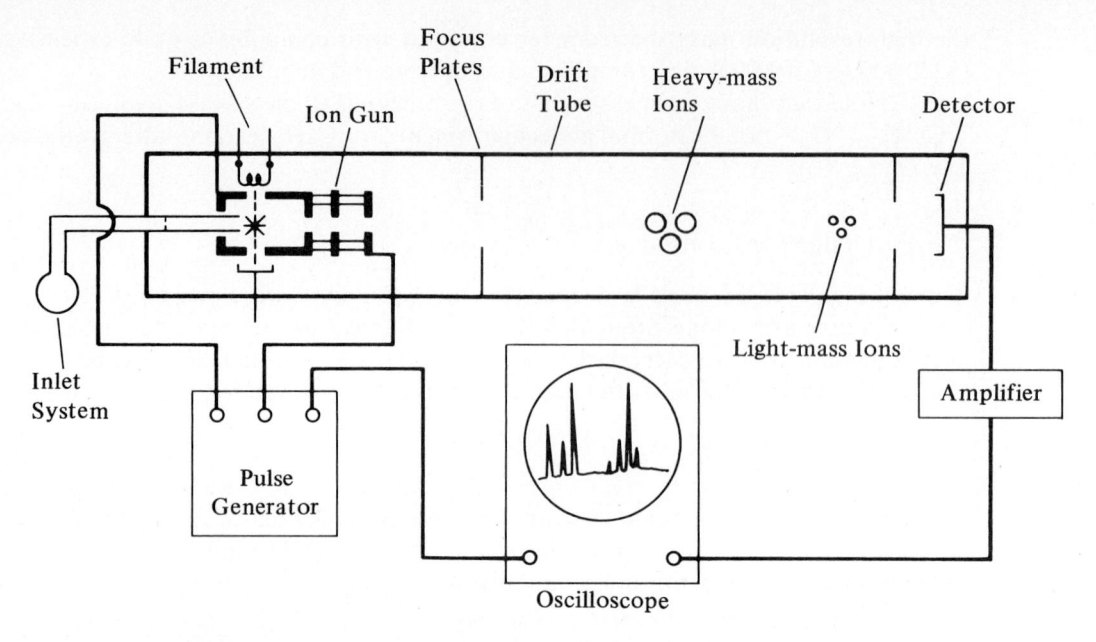

FIGURE 16.8. *Schematic diagram of a time-of-flight mass spectrometer.*

Quadrupole Mass Analyzers

Another means of accomplishing mass analysis with the use of magnetic fields is a *path-stability mass spectrometer* (often called a *mass filter*). In these devices, an ion beam from a conventional source is injected into a dynamic arrangement of electromagnetic fields. Certain ions will take a "stable" path through the analyzer and be collected; others will describe "unstable" paths and be filtered out. The quadrupole is one example of this type of mass spectrometer and has become quite popular in recent years, especially in the area of GCMS.

The quadrupole mass analyzer consists of four poles arranged as shown in Figure 16.9. Ions are injected along the z-axis into a radio-frequency field formed by application of a DC voltage U and an RF voltage $V \cos \omega t$ to the four electrodes. The voltage of the positive electrode is $+(U + V \cos \omega t)$, and that of the negative electrode is $-(U + V \cos \omega t)$. Because V is larger than U, the opposite poles change polarity at twice the RF frequency. The polarity changes are 180° out of phase (i.e., when the vertical poles are positive, the horizontal poles are negative). Solution of the differential equations of motion for the ions is rather complicated; it is sufficient here to point out that two types of solution are obtained, representing either a stable path or an unstable path. Both paths involve oscillation about the z-axis because of the alternating field, but as can be seen from Figure 16.9, the ions (ideally of one mass only) in a stable path will pass through whereas all others will take the unstable course and be discharged on collision with the poles (they are filtered out). As U and V are varied while the U/V ratio is kept constant, ions of one mass after another will take the stable path and be collected on a detector located at the end of the mass analyzer.

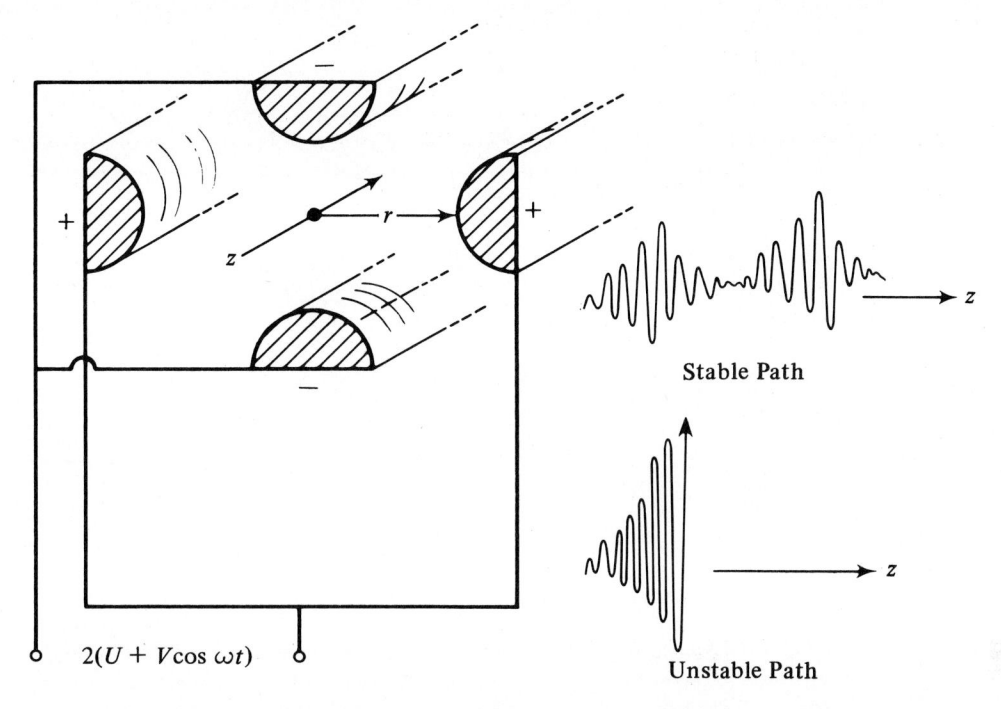

FIGURE 16.9. *Quandrupole mass analyzer. Ions are injected from the electron-impact source along the z-axis. Ideally, the poles are hyperbolic in cross section; in practice, circular poles are used. Pole length is approximately 20 cm, and r about 2 cm.*

Quadrupoles have a number of distinctive advantages. First, the path does not depend on the kinetic energy or the angular divergence of the incoming ions; therefore, these instruments have high transmission. Second, they are relatively inexpensive and compact. Third, a complete scan can be achieved very rapidly since only a change in voltage is required. As a result of these advantages, such instruments are often used in the GCMS combination (where rapid scanning is a requirement) and in space or satellite work. They perform fairly well as routine analytical instruments; for some designs, an upper mass limit of approximately 2000 atomic mass units with resolution of 1400 to 1600 can be achieved.

Ion Cyclotron Resonance Spectrometry and Fourier Transform Mass Spectrometry

A relatively new and important branch of mass spectrometry is ion cyclotron resonance (ICR) spectrometry and Fourier transform mass spectrometry (FTMS) [4–6]. Both are based on the same principles. Ions are formed by a short (10-msec) pulse of electrons from a heated filament, and trapped in a cubic cell like that illustrated in Figure 16.10. Diffusion of the ions from the cell is prevented by applying appropriate voltages to the six metal plates as shown. Because the ions are not accelerated (they

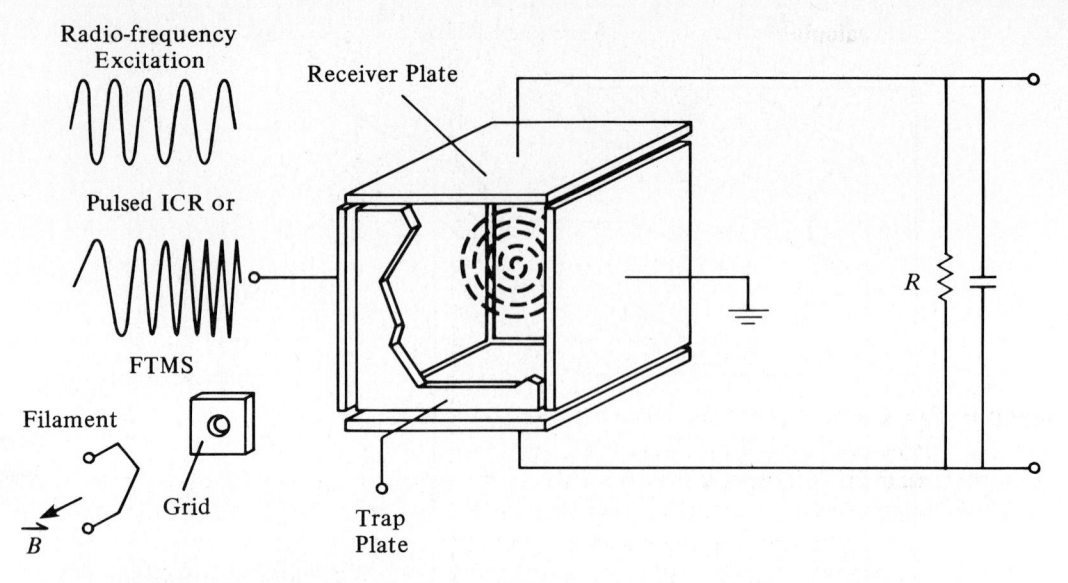

FIGURE 16.10. *Diagram of the cell used in pulsed ICR and in Fourier transform mass spectrometry. Pulsed ICR uses a single-frequency RF excitation, whereas a scanned frequency is used in FTMS. A voltage of about* 1 *to* 5 *V is applied to trap positive ions. The grid is used for pulsing the electron beam. Note the direction of the applied magnetic field B, perpendicular to the front and back plates of the cell.*

have thermal energies, approximately), they travel in small circular orbits (about 0.1 mm in diameter) due to the magnetic-field force. They experience a centripetal force of Bqm, which is balanced by mv^2/r (as noted earlier). Solving for v/r, which is frequency ω expressed as radians/sec, we obtain

$$\omega = \frac{qB}{m} \quad (16.15)$$

If an alternating electric field of frequency ω is applied to the side plate of the cell, (see Fig. 16.10), ions of mass m will absorb energy and begin to spiral (a resonance condition) as denoted by the dashed-line trajectory. Typical frequencies are in the RF range (hundreds of kHz). If the electric field is turned off (actually, a pulse of RF energy is used), the ions remain in a "parking orbit," as shown by the solid-line trajectory, until the motion is disrupted by collisions with background gas. The motion induces "image currents" of the same frequency as the resonant ion in the circuit composed of the top and bottom plates. To understand this, consider the conduction electrons in the metal. As the positive ions, which move together in a packet (coherent motion), approach the top plate, the electrons in the circuit are attracted to the ion packet. The electrons will then move in the opposite direction as the ions approach the opposite or bottom plate. This motion constitutes an alternating current of frequency ω, which produces an alternating voltage of the same frequency across the shunt resistor R (Ohm's law). Measurement of the frequency, which can be done with high precision and accuracy, coupled with knowledge of the magnetic field B, allows us to calculate the m/z of the resonant ions. A mass spectrum can be obtained by scanning either the excitation frequency or the magnetic field B. Signals will be observed when the cell contains ions that have a natural cyclotron frequency equal to the excitation frequency, that is, when Equation 16.15 is satisfied.

FTMS is conceptually a relatively simple adaptation of the pulsed ICR experiment described above. Instead of exciting with a single-frequency pulse, a rapidly scanned pulse (1 to 2 msec) of frequencies is admitted to the cell, exciting all ions having cyclotron frequencies in the scanned frequency band. The signal produced is a composite, consisting of all frequencies of the resonant ions summed together. The constituent frequencies are sorted out by a Fourier transformation of the composite signal (see discussions of other Fourier spectrometers in this text).

Both pulsed ICR and FTMS are invaluable for studies of the rates and equilibria of reactions of gas-phase ions and neutral molecules. Rates are measured by inserting a variable delay between the ionizing pulse and the excitation pulse. Knowing the pressure (density) of the neutral molecules and measuring the time dependence of the ion abundances allow us to calculate a rate constant. Investigations of these reactions increase our understanding of the fundamentals of chemical kinetics and the structures and properties of ions in the absence of solvent. Inherent acidities, basicities, stabilities, and so on, of ions have been measured using this technique. Conventional mass spectrometers can also be used for these measurements, but higher pressures (0.01 to 1 Torr compared to 10^{-7} to 10^{-6} torr) are required because of the short residence times.

The chief advantage of FTMS is that all ions are detected (a full mass spectrum is obtained) simultaneously for a single ionizing pulse (Fellgett advantage). To obtain a full spectrum by pulsed ICR, one ion is detected after another by making a slow scan. The fast scanning of FTMS has been shown to be useful in GCMS applications.

Compared to conventional mass spectrometry, FTMS can produce a mass spectrum at very high mass resolution if the transient decay of the signal is observed for a long period of time (e.g., 1 sec). In fact, resolving powers of greater than 1,000,000 at $m/z = 100$ have been obtained using superconducting magnets, which is approximately a factor of 10 greater than those obtainable with a double-focusing mass spectrometer. The high resolution should prove to be useful in assigning exact masses and examining high-molecular-weight materials (mol. wt. > 1000).

Methods of Ion Detection

The most useful and sensitive method of detection is to focus the mass-analyzed beam of ions on an *electron multiplier*. (The design of electron multipliers is much the same as that of a photomultiplier tube in ultraviolet-visible spectroscopy [see Chap. 7].) The tube current is amplified and sent to a strip-chart recorder, which often contains, instead of a pen, a mirrored galvanometer that reflects high-intensity light onto photographic paper. The mirror method allows faster scanning than a pen trace; it also allows simultaneous recording of the spectrum at a number of different sensitivities using several galvanometers.

Another, less common, method of detection involves a photographic film placed along the focal plane of a Mattauch-Herzog instrument.

16.2 INTERPRETATION OF A MASS SPECTRUM

Certainly the most important applications of mass spectrometry are the identification of complex molecules and the elucidation of their structures. It might be expected that a specific molecule would give a unique fragmentation pattern that would distinguish it from all other substances. This expectation is often realized, but not always. As in other forms of spectroscopy, the analyst must be able to interpret the pattern observed; this requires considerable skill and experience. The following pages will present a discussion of the types of ions found in a mass spectrum in order to illustrate some rather basic procedures in interpretation. More thorough approaches can be found in specialized monographs. See, for example, those by Hill, McLafferty, and Reed at the end of this chapter.

Mass spectrometry is an extremely information-rich technique, producing many signals or peaks for a single substance. Its chief advantage over other information-rich tools (e.g., NMR, infrared, and x-ray spectroscopy) is its sensitivity: Useful spectra can be obtained for samples as small as 1 ng. However, a complete understanding of all the fragmentation mechanisms has not yet been achieved.

Assignment of the Molecular Ion

As previously discussed, the molecular ion has a mass that corresponds to the molecular mass of the neutral sample. Because one electron has been removed, it is a radical cation, symbolized by $M^{\ddot{+}}$ or often just M^+. Most substances produce a recognizable molecular ion, although there are important exceptions. High-molecular-weight hydrocarbons, aliphatic alcohols, ethers, and amines produce only a small number of molecular ions because of extensive fragmentation. Polyfunctional compounds such as carbohydrates and polyamines often do not yield a molecular ion upon electron impact. On the other hand, molecules possessing an aromatic ring often give abundant molecular ions, presumably because of their ability to delocalize positive charge.

The first problem in dealing with an unknown spectrum is the identification of the molecular ion. Some simple rules are helpful. First, M^+ should have the highest mass, ignoring isotopic contributions. Second, the molecular mass will be an even mass number if it contains an even number (0, 2, 4, ...) of nitrogen atoms, and will be an odd mass number otherwise; this is known as the "nitrogen rule." Some examples are: benzene, C_6H_6, $M^+ = 78$; ethanol, C_2H_5OH, $M^+ = 46$; cholesterol, $C_{27}H_{46}O$, $M^+ = 386$; dimethyl hydrazine, $CH_3NHNHCH_3$, $M^+ = 60$; methylamine, CH_3NH_2, $M^+ = 31$; and pyridine, C_5H_5N, $M^+ = 79$. A final test of a correct M^+ assignment is that no illogical losses should be found. Seldom do organic molecules lose more than four hydrogen atoms, to give $(M - 4)$ fragments. The next reasonable fragmentations of molecular ions are losses of a methyl group $(M - 15)$, NH_2 or O $(M - 16)$, OH or NH_3 $(M - 17)$, H_2O $(M - 18)$, F $(M - 19)$, HF $(M - 20)$, and C_2H_2 $(M - 26)$. Thus, if a tentative molecular ion has lost 4 to 14 or 21 to 25 mass units, either the assignment of M^+ is incorrect or the spectrum is of a mixture.

Elemental Composition of the Molecular Ion

In the spectrum of methane (CH_4), a small peak located at $m/z = 17$ has an intensity 1.1% that of the M^+ peak at $m/z = 16$. The signal at $m/z = 17$ arises because carbon consists of two naturally occurring stable isotopes: ^{12}C and ^{13}C. Assigning the value 100% to the quantity of ^{12}C (an incorrect, but useful, procedure), we find that ^{13}C is 1.1%. Thus, $m/z = 17$ in methane is $^{13}CH_4{}^+$. A molecule that contains six carbon atoms, such as benzene (C_6H_6), will have M^+ at $m/z = 78$ and $^{13}CC_5H_6$ at $m/z = 79$, but then the intensity at 79 is 6.6% (1.1% × 6), because the probability of finding one ^{13}C is six times greater.

The analyst can make use of the natural abundance of ^{13}C to assign the number of carbon atoms in M^+. For example, if M^+ is 100% and $(M + 1)^+$ is 7.7%, M^+ contains 7 carbons. Often, M^+ is not the largest peak in a mass spectrum and therefore is not assigned an intensity of 100% (the largest peak in a spectrum is usually arbitrarily assigned an intensity of 100% and all other peaks are measured relative to this). In this case, a useful formula is

$$\text{Number of Carbon Atoms} = \left(\frac{M+1}{M}\right) \bigg/ 0.011 \quad (16.16)$$

where $M + 1$ and M are the intensities of the respective peaks. This procedure works fairly well if M^+ contains 10 or fewer carbon atoms. A relative error of 10% in the measurement of $M + 1$ or impurities in the spectrum make the number of carbon atoms at best a maximum rather than exact number. If $M^+ = 100\%$ and $M + 1 = 17.8\%$, the maximum number of carbon atoms, is 16 (1.1% × 16 = 17.6%); although the molecule may contain 15 carbon atoms, it cannot contain 17.

Other elements have isotopic contributions helpful in determining how many atoms of that element are contained in M^+. Table 16.2 gives some examples. The halogens are noteworthy: An M^+ that contains one Cl must have an $M + 2$ peak with at least 1/3 the abundance of M^+, and a bromine-containing molecule will have nearly a 1:1 ratio of $M:(M + 2)$.

Using the data in Table 16.2, the chemist can predict the pattern at M, $M + 1$, and $M + 2$ for a suspected compound and compare it with experiment. For example,

TABLE 16.2. *Relative Isotopic Abundances of Some Common Elements*

Element	M Mass	M Percentage	M+1 Mass	M+1 Percentage	M+2 Mass	M+2 Percentage
H	1	100	2	0.015	—	—
C	12	100	13	1.08	—	—
N	14	100	15	0.36	—	—
O	16	100	17	0.04	18	0.20
S	32	100	33	0.80	34	4.4
Cl	35	100	—	—	37	32.5
Br	79	100	—	—	81	98.0

with 4-chloropyridine, one would expect the following:

Ion	Composition	m/z	Abundance (Calculated)
M^+ =	$C_5H_4N^{35}Cl$	113	100%
$M + 1$ =	$^{13}CC_4H_4N^{35}Cl$ $C_5H_4^{15}N^{35}Cl$	114	$5(1.1) + 1(0.4) = 5.9\%$
$M + 2$ =	$C_5H_4N^{37}Cl$	115	$1(32.5) = 32.5\%$
$M + 3$ =	$^{13}CC_4H_4N^{37}Cl$ $C_5H_4^{15}N^{37}Cl$	116	$0.325[5(1.1) + 1(0.4)] = 1.9\%$

If the observed pattern at $m/z = 113–116$ agrees with the calculation, the evidence is strong that M^+ is C_5H_4NCl.

Most other elements have distinctive isotopic compositions which can be obtained by referring to a handbook. These patterns are useful for confirming the presence of these elements in M^+.

The procedures outlined above for estimating the elemental composition of molecular ions can be applied to fragment ions as well. For example, a common ion in hydrocarbons or molecules with large alkyl substituents is $C_3H_7^+$ ($m/z = 43$). Molecules containing acetyl groups will give a CH_3CO^+ ion, also at $m/z = 43$. The $m/z = 44$ in the former case will be 3.3% of the intensity at 43, whereas in the latter $m/z = 44$ will be 2.2%. A cautionary note must be interjected: To apply the rules, one must be certain that the F + 1 and F + 2 masses (where F is a fragment mass) consist only of isotopic contribution to F and not of other fragment ions.

Fragment Ions from Simple Cleavage Reactions

After ionization, most molecules fragment by the simple loss of a portion of the molecule in the form of a free radical. For example, isobutane can readily lose a methyl radical to form the propyl ion ($m/z = 43$):

$$\begin{array}{c} \overline{CH_3} \\ | \\ CH_3CHCH_3 \end{array}\Bigg]^{\ddot{+}} \rightarrow CH_3\overset{+}{C}HCH_3 + \dot{C}H_3 \qquad (16.17)$$

The fragment ions formed in these reactions are often gas-phase carbonium ions, the same ions observed as intermediates in certain solution reactions of organic compounds. Many mass spectral fragmentations produce the thermodynamically most stable carbonium ions. Notice that the propyl ion has an *odd* mass number ($m/z = 43$), which is typical for all simple cleavage ions possessing an even number of nitrogen atoms. (Molecular ions with an even number of nitrogen atoms have *even* masses, by the nitrogen rule described above.) Although a complete treatment of simple cleavage reactions is not possible here, a brief introduction in terms of structural types will be presented [7–9].

Aliphatic Hydrocarbons. The mass spectra of two isomeric hydrocarbons are given in Figure 16.11. Straight-chain hydrocarbons and molecules containing large *n*-alkyl groups typically give a pattern similar to that shown in Figure 16.11A. The data are reported in bar-graph form with the largest peak (called the *base peak*) assigned an abundance of 100%. The right-hand axis is expressed as a percentage

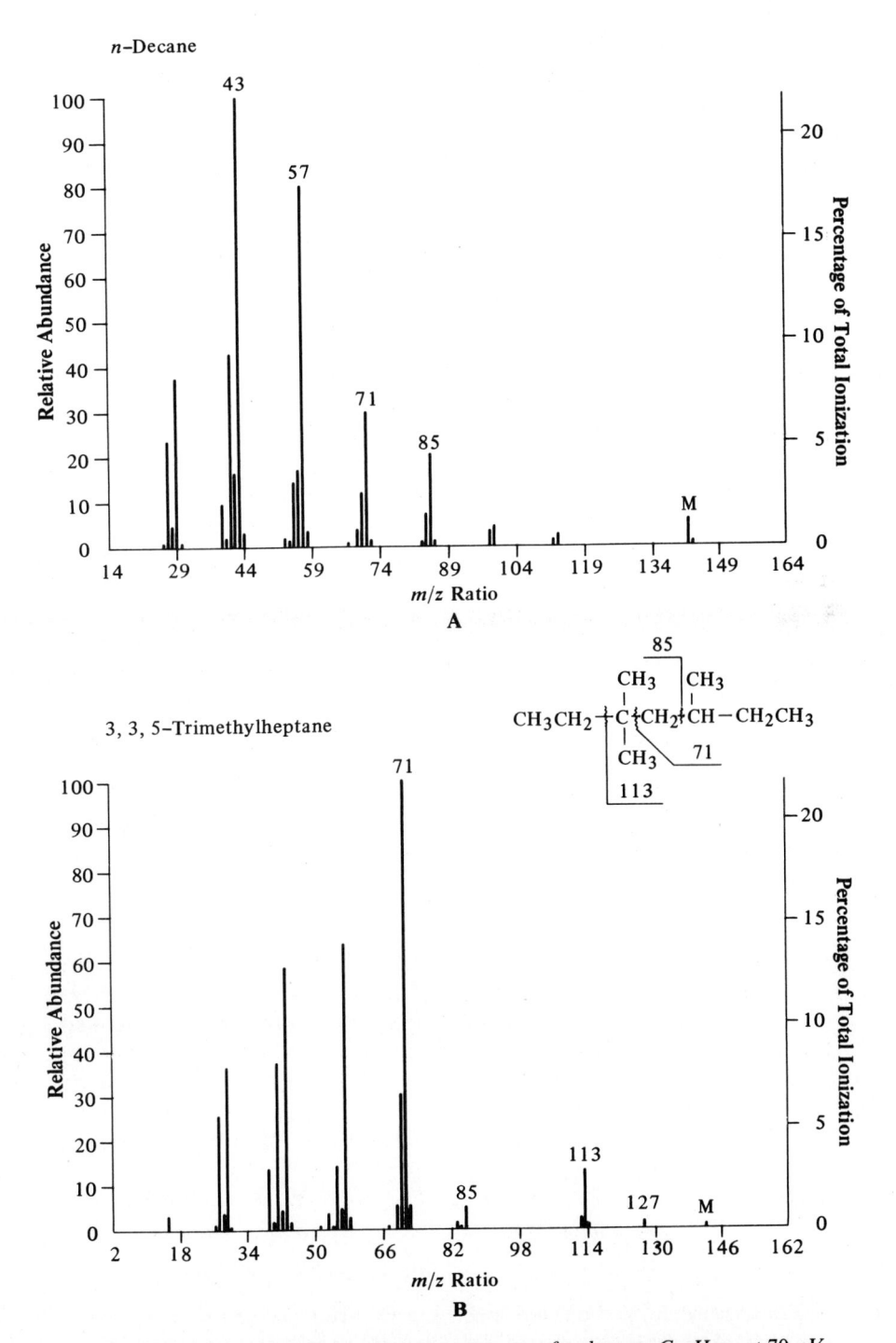

FIGURE 16.11. *A: Electron-impact spectrum of n-decane, $C_{10}H_{22}$, at 70 eV ionizing energy. B: Electron-impact spectrum of 3,3,5-trimethylheptane, an isomer of n-decane. Notice the change from the "normal" alkyl pattern.*

of all ion intensities summed together. Another common method of presenting mass spectral data is in tabular form—a list of all m/z values and their relative abundances (again normalized to the most abundant ion).

Various alkyl ions such as $C_3H_5^+$, $C_3H_7^+$, $C_4H_7^+$, and $C_4H_9^+$ ($m/z = 41, 43, 55,$ and 57, respectively) dominate hydrocarbon spectra, but these do not come from initial simple cleavages of M^+, but instead from subsequent decomposition of the initially formed ions. Branching in an alkyl chain can be detected by slight perturbations of the straight-chain spectrum caused by preferential cleavages at the branch points (see Fig. 16.11B). Thus, the most significant fragments are not the high-abundance ions at low mass, but rather the low-abundance high-mass ions formed by simple cleavage at branch points.

Saturated Aliphatic Compounds Containing Heteroatoms. A great variety of organic matter falls in this classification—for example, alcohols, ethers, mercaptans, amines, and halides. Two types of simple cleavage reactions may occur that are initiated or directed by the presence of the heteroatom (O, S, N, X, etc.), as exemplified by the following reactions of ethyl ether:

$$CH_3CH_2OCH_2CH_3^{+\cdot} \rightarrow \cdot CH_3 + [CH_3CH_2\overset{+}{O}CH_2 \leftrightarrow CH_3CH_2\overset{+}{O}=CH_2] \quad (16.18)$$
$$m/z = 59 \text{ (51\% Rel. Abund.)}$$

$$CH_3CH_2OCH_2CH_3^{+\cdot} \rightarrow CH_3CH_2O\cdot + \overset{+}{C}H_2CH_3 \quad (16.19)$$
$$m/z = 29 \text{ (40\% Rel. Abund.)}$$

Heteroatoms that can stabilize the positive charge by resonance prefer Path 16.18. Other, more electronegative, heteroatoms prefer Path 16.19. Aliphatic amines fragment almost exclusively by Path 16.18; alcohols, ethers, and thio compounds fragment by both pathways; and halogen compounds preferentially lose the $X\cdot$ (Path 16.19). Thus, 2-aminopropane undergoes loss of both hydrogen and a methyl group to give $m/z = 58$ and 44, respectively; whereas 2-bromopropane loses bromine almost exclusively:

$$\begin{array}{c}\overline{CH_3}\\|\\CH_3-CH-NH_2\end{array}\Bigg]^{+\cdot} \longrightarrow \begin{array}{c} CH_3-CH=\overset{+}{N}H_2 + \cdot CH_3 \\ m/z = 44 \text{ (100\% Rel. Abund.)} \\ CH_3 \\ | \\ CH_3-C=\overset{+}{N}H_2 + \cdot H \\ m/z = 58 \text{ (10\% Rel. Abund.)} \end{array} \quad (16.20)$$

$$\begin{array}{c}\overline{CH_3}\\|\\CH_3-CH-Br\end{array}\Bigg]^{+\cdot} \longrightarrow \begin{array}{c} CH_3 \\ |_+ \\ CH_3CH \end{array} + Br\cdot \quad (16.21)$$
$$m/z = 43 \text{ (100\% Rel. Abund.)}$$

Try to rationalize the three spectra in Figure 16.12 using the rules just discussed.

Notice that when there is a choice between the loss of various radicals (hydrogen, methyl, ethyl, etc.), the larger group is preferred, as in Reaction 16.20. This is generally true of all simple cleavage reactions at 70 eV of ionizing energy.

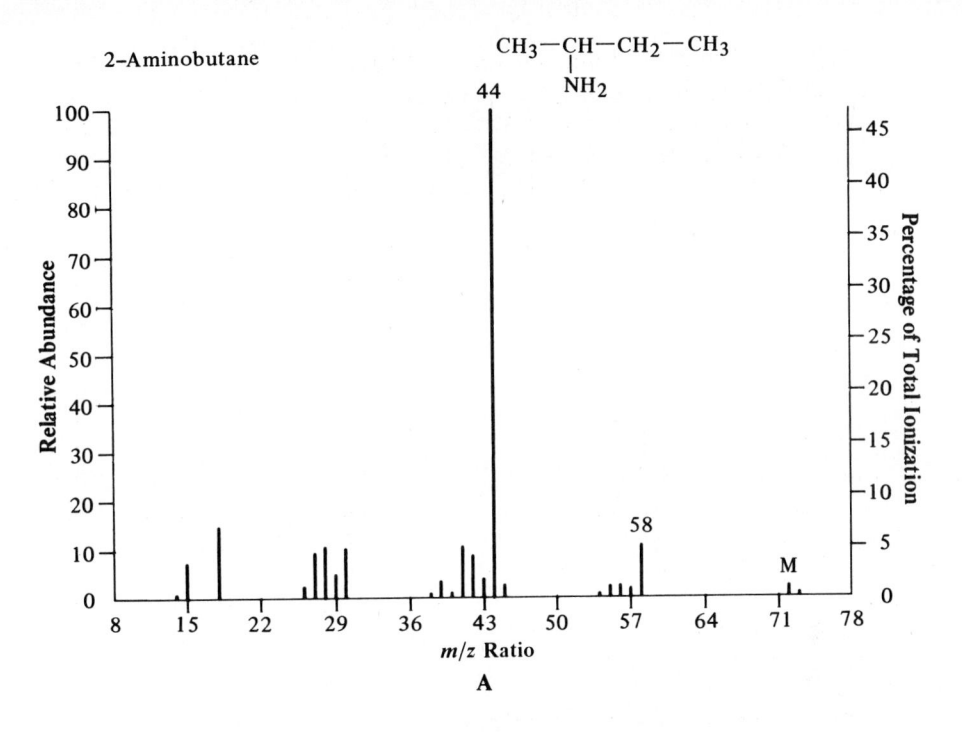

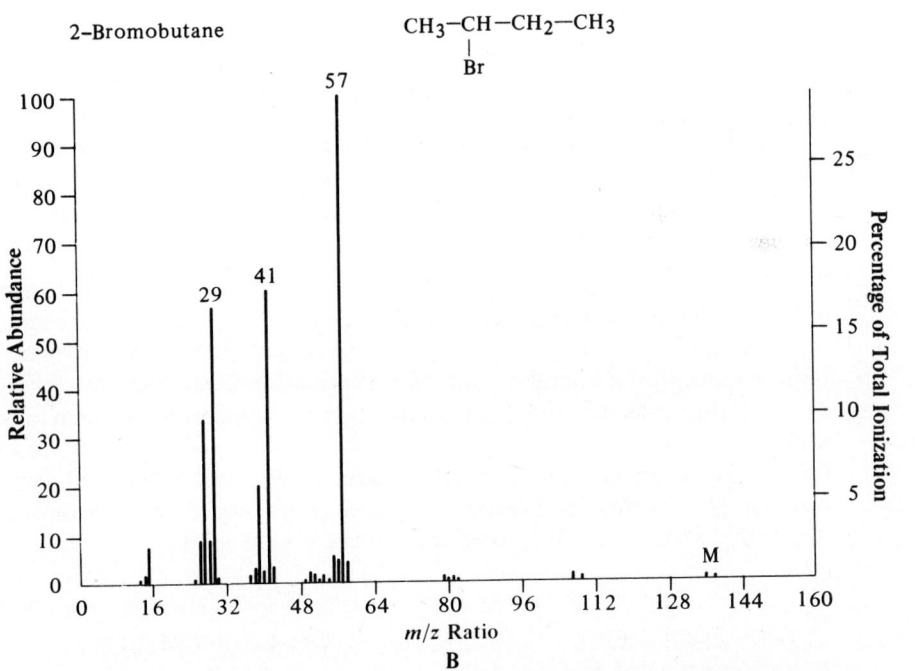

FIGURE 16.12. *Mass spectra of a series of sec-butyl compounds. Major features of the spectra can be interpreted in terms of simple initial cleavage reactions. All are at 70 eV ionizing energy.*

499

2-Butanethiol

$$CH_3-CH-CH_2-CH_3$$
$$|$$
$$SH$$

C

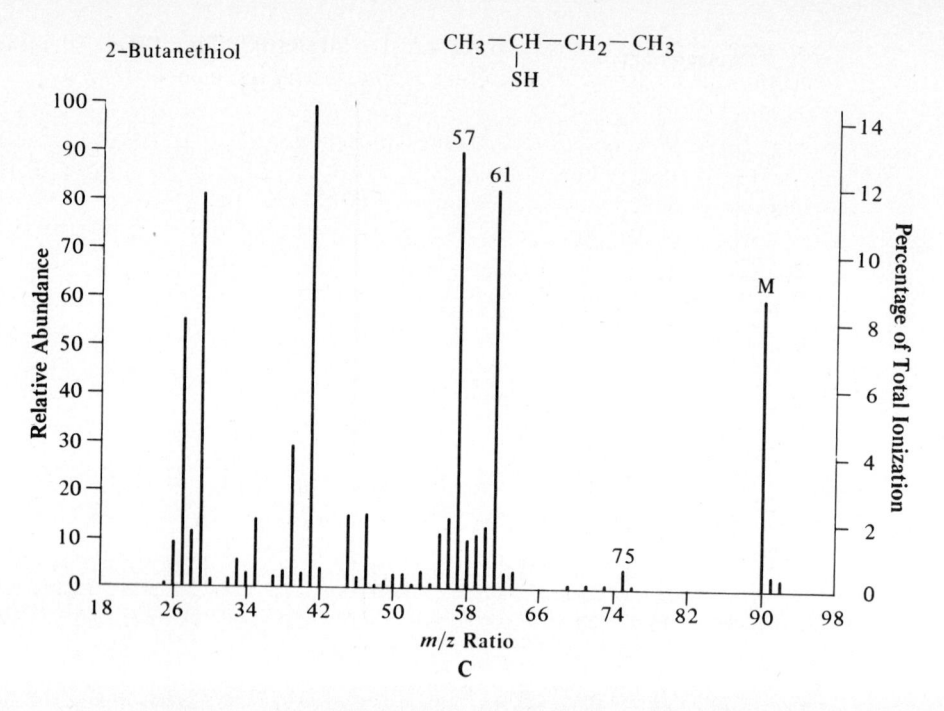

FIGURE 16.12. *(continued)*

Alkenes and Doubly Bonded Heteroatoms. One might expect that a double bond in a long hydrocarbon chain, such as a fatty acid, could be located by mass spectrometry because formation of allyl ions would be preferred:

$$R-CH_2-CH=\overline{CH_2}]^{+\cdot} \xrightarrow{-H} R-\overset{+}{C}H-CH=CH_2 \longrightarrow \text{etc.}$$
$$\phantom{R-CH_2-CH=\overline{CH_2}]^{+\cdot}}\xrightarrow{-R} \overset{+}{C}H_2-CH=CH_2 + \text{Other Fragments} \tag{16.22}$$

This possibility is not realized because the double bond rearranges or migrates after ionization but prior to fragmentation. As a result, isomeric olefins tend to give nearly identical spectra. Examples of this kind constitute a serious drawback of electron-impact ionization if complete sample identification is required.

If the molecule contains a double bond to a heteroatom, such as in a carbonyl group, the simple cleavage reaction occurs adjacent to this group and locating the functional group is often straightforward. Here, double-bond migration is not a problem. An example is 2-butanone, which gives a base peak corresponding to loss of an ethyl group, but also experiences some loss of methyl:

$$CH_3\overset{\overset{\displaystyle O}{\|}}{C}CH_2CH_3^{+\cdot} \longrightarrow \begin{cases} CH_3\overset{\overset{\displaystyle O}{\|}}{C}^+ + CH_3\dot{C}H_2 \\ m/z = 43\ (100\%\ \text{Rel. Abund.}) \\ CH_3CH_2\overset{\overset{\displaystyle O}{\|}}{C}^+ + \cdot CH_3 \\ m/z = 57\ (7\%\ \text{Rel. Abund.}) \end{cases} \tag{16.23}$$

Note again that loss of the larger alkyl is preferred. This type of fragmentation is found in most carbonyl compounds (acids, aldehydes, esters, etc.).

Aromatic Compounds. The spectrum of benzene (Fig. 16.13) is archetypal of unsubstituted aromatic compounds. Usually, one of the most intense peaks is M^+, and the fragmentation pattern is quite simple (compare with Fig. 16.11). The reason is that aromatic compounds are readily able to stabilize a positive charge by delocalization. The only possible single cleavage in benzene is loss of H to give $C_6H_5^+$ ($m/z = 77$). Ring opening and cleavage give rise to $C_4H_4^+$ ($m/z = 52$) and $C_3H_3^+$ ($m/z = 39$).

Substituted aromatics such as shown in the following reactions fragment preferentially by loss of R to give the stable benzyl ion, which is known to rearrange to the symmetrical tropylium ion (A):

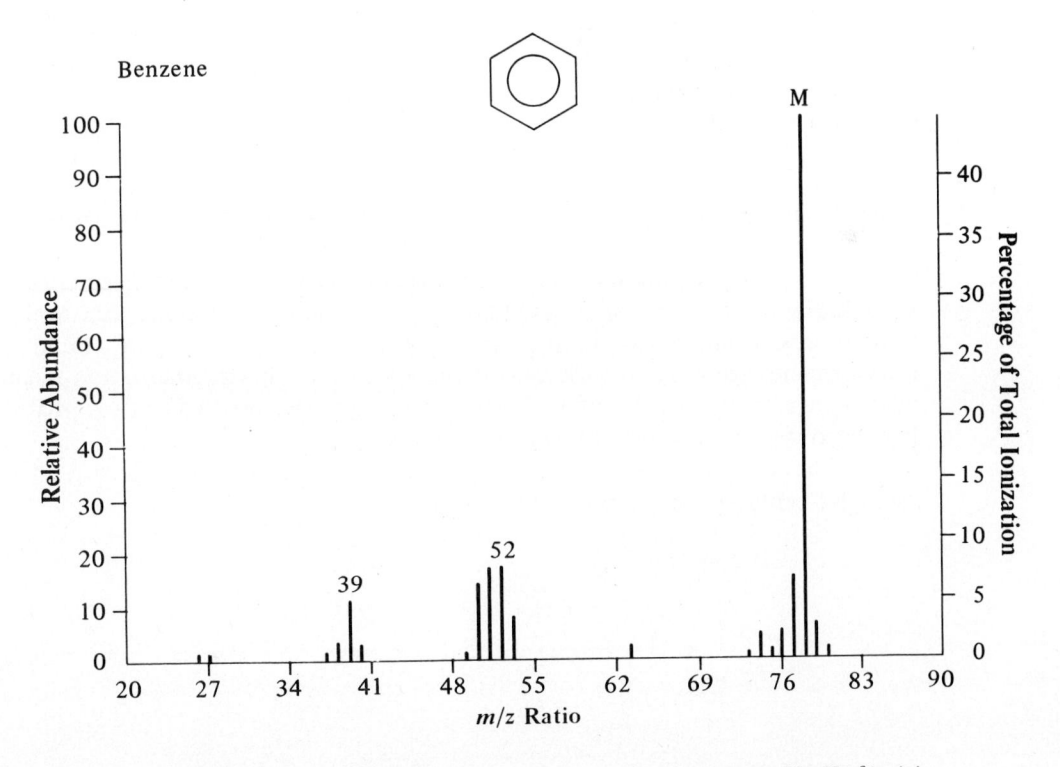

$$\underset{X}{\underset{|}{C_6H_4}}-CH_2R \quad \overset{+\cdot}{\longrightarrow} \quad \underset{-R\cdot}{\longrightarrow} \quad \underset{X}{\underset{|}{C_6H_4}}-\overset{+}{C}H_2 \quad \longrightarrow \quad \underset{(A)}{\text{tropylium}}-X \longrightarrow \text{Fragments} \tag{16.24}$$

FIGURE 16.13. *Electron-impact mass spectrum of benzene at 70 eV of ionizing energy.*

The spectra of *ortho*, *meta*-, and *para*-substituted compounds are often identical because the substituent location is lost in the 7-membered ring ion. Except for specific *ortho*-disubstituted compounds, the chemist cannot use mass spectrometry for assigning ring-substituted isomers in various aromatics.

Fragment Ions from Rearrangements of M⁺

Since rearrangements may alter the original skeleton of a molecule, one might think that they are troublesome in mass spectral interpretation. Actually, a number of fragmentations of M⁺ involving rearrangement are analytically very useful. Most of these processes occur by loss of a neutral molecule, rather than a radical, and are found at even masses in a mass spectrum (if the number of nitrogen atoms is even). Thus, a rather abundant even-mass ion at the higher-mass end of the spectrum should be singled out for special attention. Only a few examples will be considered here to illustrate the interpretive procedure.

The McLafferty Rearrangement. In many compounds containing a doubly bonded heteroatom (C=X), a hydrogen will transfer to X from the third carbon down the chain from C=X, with the loss of an olefin. The process is illustrated for 2-hexanone:

$$\begin{array}{c}\text{[structure of 2-hexanone with H transfer]} \longrightarrow \text{[enol ion]} + C_3H_6 \\ m/z = 58 \end{array} \qquad (16.25)$$

The transfer is highly specific, involving only the hydrogens on the carbon shown. This fragmentation is known as the *McLafferty rearrangement*. Notice that the fragment ion has an even mass number ($m/z = 58$), which is typical for rearrangements of this kind. Because most initial fragmentations are cleavage reactions that yield odd mass numbers (unless the number of nitrogen atoms is odd), highly abundant rearrangements are easy to pick out of the spectrum. The fragmentation occurs in many other carbonyl compounds (acids, esters, amides, aldehydes, etc.), provided a hydrogen is situated three atoms from the carbonyl.

The analytical utility of the McLafferty rearrangement is illustrated for 3-methyl-2-pentanone, an isomer of 2-hexanone:

$$\begin{array}{c}\text{[structure of 3-methyl-2-pentanone with H transfer]} \longrightarrow \text{[enol ion]} + C_2H_4 \\ m/z = 72 \end{array} \qquad (16.26)$$

The molecular ion is again of the correct structure to give a McLafferty rearrangement, but this time to yield $m/z = 72$, indicating that the methyl substituent is in position 3. Rather subtle differences in molecular structure, such as the position of the branch point, can often be uncovered using information from this rearrangement.

Rearrangements in Aromatic Compounds. An important fragmentation in various substituted aromatic compounds involves the transfer of a side-chain hydrogen atom to X (where X = CH_2, O, S) or to the aromatic ring:

$$\text{Ph-X(CH}_2)_n\text{CH}_3]^{+\cdot} \longrightarrow \text{Ph-XH}]^{+\cdot} \quad \text{or} \quad \text{Ph(X)(H)(H)}]^{+\cdot} \quad (16.27)$$

$[n = 1, 2, \ldots]$

Thus, the base peak in the spectrum of phenyl ethyl ether is $m/z = 94$ (C_6H_5OH); and an important peak (55% relative abundance) in the spectrum of butyl benzene is $m/z = 92$ (C_7H_8). Isotopic labeling studies with deuterium show that hydrogen transfer occurs from any one of the four carbon atoms in phenyl butyl ether, suggesting a nonspecific rearrangement. Nevertheless, the fragmentation is analytically useful— only phenyl ethyl ether, of the other $C_8H_{10}O$ isomers shown, gives loss of C_2H_4.

Only two types of rearrangement involving M^+ processes have been discussed here. Many others are highly specific and quite useful in elucidating molecular structure; others, less specific, can still yield information (e.g., the side-chain rearrangement above). Rearrangements invariably involve the loss of small neutral molecules such as olefins, H_2O (in alcohols), small alcohols (in some esters), acids, carbon monoxide, formaldehyde, and so forth. They are readily identified in the high-mass region of the spectrum as even-mass-number ions (for an even number of nitrogen atoms) and are usually useful in mass spectral interpretation.

Further Fragmentation Reactions

Enough energy is often deposited in the molecular ion by the ionization process to again decompose the initially formed fragments, producing secondary, tertiary, etc., ions at lower masses. The mass spectra for complicated molecules often contain very abundant fragment ions of low mass, which are products of these consecutive decompositions. In the hydrocarbon spectra of Figure 16.11, notice the abundant peaks around $m/z = 29, 43$, and 57. These are not initially formed fragments, but rather arise by successive rearrangement reactions. For example, $m/z = 43$ ($C_3H_7^+$) probably originates by the loss of C_2H_4 from $C_5H_{11}^+$, of C_3H_6 from $C_6H_{13}^+$, or of other neutral olefins from higher-mass primary ions. These rearrangements are less useful for interpretation than the initial fragmentations and may even be misleading to the inexperienced analyst. There is a strong tendency, for instance, to incorrectly interpret abundant $C_3H_7^+$ as indicating a branched propyl group, although in some instances this may be the case. (An abundant $m/z = 43$ [$C_3H_7^+$] would be significant if the abundances at $m/z = 57$ [$C_4H_9^+$] and $m/z = 29$ [$C_2H_5^+$] were very small.)

The extensive fragmentation that often occurs in complex molecules can be attenuated so as to emphasize the initial cleavages. This is done by lowering the ionizing energy to 15 to 20 eV or by using other methods of ionization (e.g., chemical or field), which are discussed later.

Multiply Charged Ions

Besides the singly charged ions that dominate a mass spectrum, some doubly charged fragments can be found. For example, a weak peak at $m/z = 38.5$ (77/2) in the spectrum of benzene is $C_6H_5^{2+}$. Gas-phase metal ions from organometallic compounds or from volatile metals such as mercury are often found in $+2$ or even $+3$ states. Usually, multiply charged ions are of low abundance and not very useful in interpretation.

Metastable Ions

In all mass spectrometers, fragmentation continues to occur outside the source (i.e., after full acceleration). If, in a sector instrument, an ion M_1^+ decomposes to M_2^+ prior to entering the magnetic field [$M_1^+ \rightarrow M_2^+ + (M_1 - M_2)$], the M_2^+ will no longer have the same kinetic energy as the "normal" ions and, therefore, will not be mass analyzed at M_2. Instead, its kinetic energy is $M_2(M_2/M_1)$ because of energy conservation, and the ion will be analyzed at a lower apparent mass $m^* = M_2^2/M_1$.

Metastables are usually identified by broad low-abundance peaks occurring at fractional masses. For example, in acetophenone, one might postulate that a methyl group is lost to form the $m/z = 105$ ion, which then loses CO to form $m/z = 77$. An alternative route is the direct, one-step loss of CH_3CO:

$$\text{PhCOCH}_3^{+\cdot} \xrightarrow{-CH_3^{\cdot}} \text{PhCO}^+ \xrightarrow{-CO} C_6H_5^+ \quad (16.28)$$
$$m/z = 120 \qquad m/z = 105 \qquad m/z = 77$$
$$\xrightarrow{-CH_3CO^{\cdot}}$$

Both processes are verified by the observation of two metastable peaks (designated by * in Reaction 16.28) at $(77)^2/105 = m/z = 56.5$ and at $(77)^2/120 = m/z = 49.4$. Metastable peaks, then, are invaluable aids in determining which decompositions take place to give the observed mass spectral patterns.

Negative Ions

Negative ions can be investigated by reversing the polarity of the accelerating voltage and the magnetic field. Usually they are formed by one of two processes: (a) capture by the neutral molecule of an electron from the ionizing beam to form M^-; or (b) ion-pair production ($AB + e^- \rightarrow A^+ + B^- + e^-$), which yields fragmentary negative ions. The abundances of negative ions are 10 to 1000 times less than those of positive ions in an electron-impact source. However, high-abundance negative ions can be produced by chemical ionization and by fast atom bombardment (see Sec. 16.4).

16.3 ANALYTICAL APPLICATIONS OF ELECTRON-IMPACT MASS SPECTROMETRY

Identification and Structural Elucidation of Compounds

It should be clear from the last section that a mass spectrum yields a wealth of information from very small samples (10^{-6} to 10^{-9} g). If the mass spectrum has been previously reported, identification of the unknown is accomplished by checking for a match. However, caution must be used because spectra obtained with different instruments using different temperatures and source conditions will not match exactly; small quantitative differences in relative abundances arise because of different residence times in the source and because certain instrument designs may discriminate against low- or high-mass ions. Checking for a match can be done rapidly using a computer with a data file of previously determined spectra. Compilations are available on magnetic tape or in book form (see the Selected Bibliography at the end of the chapter). Ideally, the file spectra should be determined using the same instrument under the same conditions; proof of identity is then a peak-to-peak match of the unknown spectrum with a reference.

The use of mass spectra to identify dangerous drugs, both for diagnosis in a hospital setting and in forensics, is an important application. Applications in other fields include identifying pollutants (environmental work), natural products (biochemistry), flavor components (the food industry), or hydrocarbons (the petroleum industry). Often, the analyst begins with a complex mixture which can be separated by a gas chromatograph on-line with a mass spectrometer.

Proof of structure for new compounds is more difficult since the mechanisms of mass spectral fragmentations are not well enough understood to be used to predict the entire spectrum of a postulated structure from basic principles. Using mass-spectral information together with infrared, ultraviolet-visible, and nuclear magnetic resonance data is a powerful approach. High-resolution mass spectrometry expedites structural studies by providing the formulas (elemental compositions) of M^+ and the fragment ions.

Analysis of Mixtures

Mass spectrometry made important contributions in the analysis of petroleum mixtures during World War II. However, gas chromatography is now preferred because of the convenience and simplicity inherent in a chromatographic procedure. Mass spectra of different hydrocarbons contain many identical peaks, and it is difficult to find one peak characteristic of each component; the reverse is often the case in GC. Nevertheless, the percent composition of a mixture can be obtained by quantitative mass spectrometry using a series of simultaneous equations, much as is done in analyzing mixtures by ultraviolet or visible spectrophotometry (see Chap. 7).

Mass spectral analysis of simple mixtures may be used in one-time experiments for which the setup and calibration of a gas or liquid chromatograph are too time consuming, even though the mass spectrometer must also be calibrated. The convenience of mass spectral methods for gaseous mixtures recommends this approach, especially if the appropriate gas-handling apparatus is not readily available for GC. However, GC is often preferred for routine work.

With complex mixtures, it may be necessary to know only the types of compounds present. For example, it is important in the petroleum area to determine the approximate concentrations of saturated hydrocarbons, alkenes and cycloalkanes, and aromatics or substituted aromatics in some mixture. The lower-mass series of ions are useful in this pursuit. Alkanes yield abundant fragments at $C_2H_5^+$, $C_3H_7^+$, $C_4H_9^+$, ... ; alkenes and saturated cyclic alkanes at $C_2H_3^+$, $C_3H_5^+$, $C_4H_7^+$, ... ; and aromatics at $C_6H_5^+$, $C_7H_7^+$, $C_8H_9^+$. The sum of the intensities of the fragment peaks for any of these series is proportional to the concentration of each type of hydrocarbon.

For complex mixtures, which often occur in biochemical and environmental problems, the mass spectrometer can serve as a highly specific detector for a gas chromatograph. For example, the organic extract of polluted waters may contain hundreds of organic compounds that cannot be perfectly separated by one pass through a GC, even using a capillary column. Thus, what appears to be a single peak by GC may actually be a mixture of a number of components. Each component may have unique M^+ or fragment peaks. To make use of this fact, mass spectra are rapidly obtained at many points across the GC peak and the data stored in an on-line computer. The abundance of certain ions can then be plotted versus time; each ion is specific for one compound in the unresolved GC peak (see Fig. 16.14).

The procedure can be made approximately 100 to 1000 times more sensitive by rapid switching between selected peaks characteristic of the substance to be quantitated. This procedure, called *multiple ion monitoring* (MID) can be accomplished by rapid switching of the accelerating voltage of a magnet sector at constant B (with a double-focusing instrument, both the accelerating and ESA voltages must be switched in tandem) or U and V of a quadrupole. Since certain masses are preselected, the analyst must know what masses are of interest. GCMS in the MID mode can be used to monitor chemical substances at the part-per-trillion level in 1-g samples of soil, water, and biological fluids and tissues [10].

In inorganic chemistry, mixtures of metal ions in solution can be analyzed by electron-impact mass spectrometry. First, the metal ions are complexed with an organic ligand (usually various substituted acetylacetonates) to form volatile metal

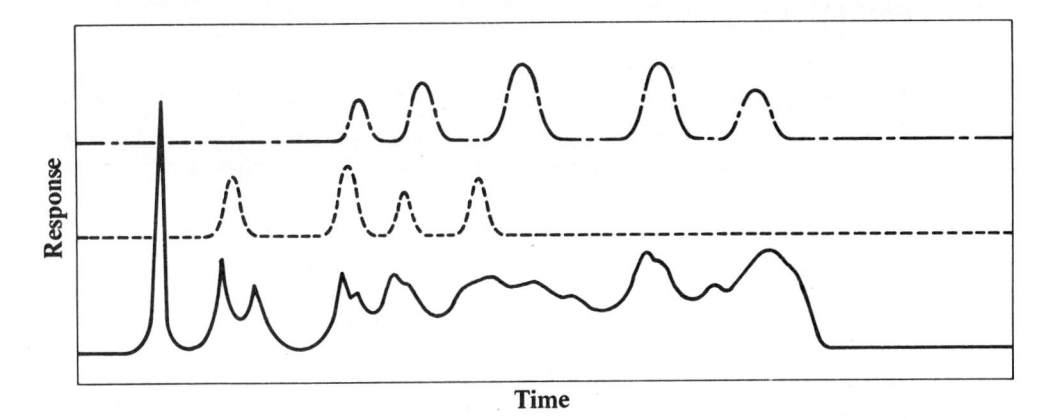

FIGURE 16.14. *Comparison of a total gas-liquid chromatogram obtained from the output of the GC detector or the total ion-current monitor of the mass spectrometer, and two hypothetical chromatograms obtained by monitoring specific peaks in the mass spectra. Solid line = total chromatogram; dashed line = chromatogram made by monitoring one specific ion; broken solid line = chromatogram made by monitoring another specific ion.*

chelates. If many metal ions are anticipated, the mixture is separated by GC and the separated fractions are identified by mass spectrometry. Simple mixtures can be analyzed directly using the mass spectrometer. Because of the high sensitivity of mass spectrometry, trace analysis is possible.

Biochemical Applications

The need for sophisticated analytical tools in the biochemical and health-science areas continues to grow each year. Mass spectrometry is ideally suited for many problems because of its high sensitivity. For instance, the action of drugs in living systems can be better understood if the drug metabolites are isolated and identified. A urine or blood specimen will contain metabolites in trace quantities, and a mass spectrum can be obtained after extraction, concentration, and separation.

Many compounds of biochemical interest have been thoroughly investigated by mass spectrometry. The mechanisms of fragmentation of alkaloids, steroids, and terpenes are fairly well understood [11]. Mass spectral studies have been reported for amino acids, carbohydrates, and various lipids [12]. However, some molecules of biochemical importance (e.g., proteins and nucleic acids) are highly polar, thermally sensitive, and of high molecular weight. Thus, it is impossible to vaporize these substances in the source of a mass spectrometer. Prior chemical conversion to various volatile derivatives is helpful, provided the molecular weight is not too high. In this way, small polypeptides (containing 10 to 12 amino acids) have been analyzed, and the amino-acid sequence determined by mass spectrometry. Spectra can also be obtained of mono- and dinucleotides after converting the OH-groups to $OSi(CH_3)_3$ to break the hydrogen bonding.

Isotope-Abundance Studies

Mass spectrometry was originally developed to quantitatively identify and analyze the natural abundances of stable isotopes. The determination of isotopic abundances is still important, but for different reasons. Isotopic labeling of molecules is quite important in studies of chemical mechanisms and kinetics, both in the organic and biochemical areas. Prior to studies of this kind, the extent of labeling must be determined, and mass spectrometry is usually the method of choice. For example, the amounts of benzene-d_5, -d_4, etc., in benzene-d_6 can be determined by measuring the abundances of $C_6D_6^+$ ($m/z = 84$), $C_6D_5H^+$ ($m/z = 83$), $C_6D_4H_2^+$ ($m/z = 82$), etc. The best procedure is to lower the ionizing energy so that no peaks corresponding to loss of H or D interfere [13].

Sometimes the position of an isotopic label (^{2}H, ^{13}C, ^{15}N, etc.) can also be determined. The analyst must know whether the compounds undergo prior scrambling reactions before attempting studies of this nature. It would be folly to search for the position of a deuterium in an olefin by mass spectrometry because of the tendency for double-bond migrations to occur in the ionized molecule. However, for compounds giving no scrambling, or only well-understood rearrangements, mass spectrometry can be used to locate the label.

Precise isotope ratios are also necessary in studies of kinetic isotope effects, isotope-dilution studies, and dating work.

The isotope-dilution technique involves "spiking" the sample with the element of interest; the added element, however, has an appreciable difference in isotopic abundance ratios from the natural abundance. As an example, consider a trace analysis of organic bromide, say bromobenzene. If the sample is spiked with 1 μg of C_6H_5Br that contains only bromine-81, and the observed ratio of ^{79}Br:^{81}Br changes from 1:1 (natural abundance) to 1:2, then the original sample must contain 2 μg of bromobenzene. The method is sensitive to trace amounts of particular elements and has been used for determination of carbon, nitrogen, oxygen, and sulfur in organic and biochemical samples, and for analysis of metals in geological specimens.

As one example of dating work, the age of rocks can be determined by measuring the ratio of argon-40 to argon-36 using the mass spectrometer. Argon-40 is a product of potassium-40 decay with a half-life of 1.3×10^9 years. The exact amount of argon-40 is obtained by measuring the ratio of argon-36 to argon-40 with argon-38 added as a tracer. In this way, ages of meteorite and geological samples have been determined in the one-million- to one-billion-year range.

If high precision (1 part in 10,000) is required, an isotope-ratio mass spectrometer is used. These instruments are normal magnetic-sector instruments with dual inlets and dual collectors. The ion containing one isotope is focused on one collector (^{40}Ar), and an adjoining detector collects the other peak (^{36}Ar). The signals are accurately compared using precision resistors and null detection.

Another timely example is tracing the origin of nitrate in ground and surface waters. At high concentration, this substance is toxic, and can be dealt with only if its source can be identified. Different sources (animal waste, fertilizers, natural rocks) have different ^{15}N/^{14}N ratios; however, the differences are so small that precise ratios requiring an isotope-ratio instrument are needed.

Thermodynamic Studies

An ionization-efficiency curve is a plot of the decrease in intensity of a certain ion (molecular ion or fragment) as the ionizing energy is lowered. There is usually a particular "threshold" energy below which a negligible number of ions appear; this is the ionization potential.

In one method, the energy of the electron beam is varied by changes in the voltage applied to the filament; a more elegant way is to use an intense light source and a monochromator (photoionization). Because most molecules have ionization potentials around 10 eV (124 nm), a rather costly and complicated vacuum monochromator is required to select the ionizing wavelength. A mass spectrometer is still used to monitor ion intensities.

From ion-efficiency curves, the ionization potential of M and the "splitting" potentials of the fragments can be obtained and, in turn, bond-dissociation energies and heats of formation of gas-phase ions can be obtained from the ionization potentials [14].

16.4 OTHER METHODS OF VAPORIZATION AND IONIZATION

Chemical- and Field-Ionization Mass Spectrometry

Certainly, the molecular weight of the unknown is one of the most important pieces of information to gain from a mass spectrum, but the M^+ ion for certain types of compounds is often absent or of low intensity. Another related drawback to mass spectrometry is consecutive ion fragmentation in complex molecules, which results in low-mass ions carrying a large share of the total intensity (see Fig. 16.11). Thus, the more analytically important molecular ion and primary fragments (those formed in the initial fragmentation of M^+) are of low abundance or even missing in the spectrum. A major advance is the development of *chemical-* (CIMS) [15, 16] and *field-ionization* (FIMS) [17, 18] techniques, which are gentler ionization procedures. The result is enhancement of the abundance of ions containing information on molecular weight and initial fragmentation.

Instead of ionizing with an energetic electron beam, *chemical ionization* occurs via ion-molecule reactions. The ion, often referred to as a reagent ion, reacts with a sample molecule by transferring a proton or by abstracting a H^- or an electron, which imparts a $+1$ charge to the sample molecule.

Typically, the source of a conventional mass spectrometer is redesigned to operate at a higher pressure (1 to 10 torr). In FTMS, the pressure is kept at 10^{-6} torr, and a long delay (200 msec) is inserted between the ionization and excitation (start of mass analysis) pulse. Methane is admitted and ionized to produce CH_4^+ and CH_3^+. These react to form CH_5^+ and $C_2H_5^+$ as follows:

$$CH_4^+ + CH_4 \rightarrow CH_5^+ + CH_3 \qquad (16.29)$$

$$CH_3^+ + CH_4 \rightarrow C_2H_5^+ + H_2 \qquad (16.30)$$

The CH_5^+ and $C_2H_5^+$ do not react further with the neutral methane, but once a small amount of sample (XH) is admitted to the source (1 part in 1000 parts methane), the sample molecules are ionized by proton and hydride-ion transfers:

$$CH_5^+ + XH \rightarrow XH_2^+ + CH_4 \qquad (16.31)$$

$$C_2H_5^+ + XH \rightarrow X^+ + C_2H_6 \qquad (16.32)$$

The XH_2^+ and X^+ may then fragment, giving a mass spectrum. No molecular ion per se is observed, but the molecular weight is readily obtained from the M + H or M − H ions.

Other reagent ions, which are weaker gas-phase acids than CH_5^+, can be used to further simplify the spectrum. Examples include $C_4H_9^+$ (from isobutane), NH_4^+ (ammonia), or H_3O^+ (water). These acids also ionize by proton transfer, but the energies are somewhat less and the fragmentation of XH_2^+ is minimized. Figure 16.15 shows the electron-impact and chemical-ionization spectra of ephedrine, a biologically important amine. Notice the striking absence of M^+ by electron impact and the abundant M + 1 in the chemical-ionization spectrum. The fragmentation is also simplified by chemical ionization.

Field ionization employs a sector mass analyzer with a modified source. A small wire or sharp-edged anode is mounted at the input end of the ion gun as shown in Figure 16.16, and a very large electric field (10^5 V/cm) is applied between the anode and cathode. (The anode is first activated so that its surface contains many sharp points [carbon whiskers] by filling the chamber with benzonitrile or a similar compound with a low hydrogen-to-carbon ratio at 0.04 to 0.1 torr and applying 10,000 V for several hours.) The electric field is then sufficient to remove an electron from gaseous molecules in the field. It is thought that the high field so distorts the potential-energy surfaces of the sample molecule that an electron is quantum mechanically tunneled through the energy barrier to the anode. The practical consequence of this is that the molecular ion formed in this manner is not excited, and very little fragmentation results. Almost every compound gives an abundant M^+ or $(M + 1)^+$ by field ionization, and considerably simplified fragmentation patterns. The electron-impact and field-ionization spectra of ribose are compared in Figure 16.17.

Chemical and field ionization are important complements to electron-impact mass spectrometry because they emphasize the molecular-ion region producing small numbers of primary fragmentations and almost no secondary and further fragmentations. By contrast, electron impact gives a wealth of fragmentations at the expense of the molecular ion and of primary fragments. In most cases, the analyst can combine this information to identify the sample. As a result, chemical and field ionization are especially useful for analysis of complex molecules, and in direct analysis of mixtures because of the simple spectra that are produced. Chemical ionization is potentially more sensitive than electron impact, for two reasons. First, the ionization of the sample is concentrated in a few ions rather than dispersed over hundreds of ions as in the electron-impact ionization of complex molecules. Second, the efficiency of ionization by electron impact is approximately 0.1%; that is, only one out of every thousand sample molecules is ionized. The un-ionized sample molecules are lost to the vacuum pumps. Because chemical ionization takes place by ion-molecule reactions, ionization efficiency could be raised significantly by increasing the reagent gas

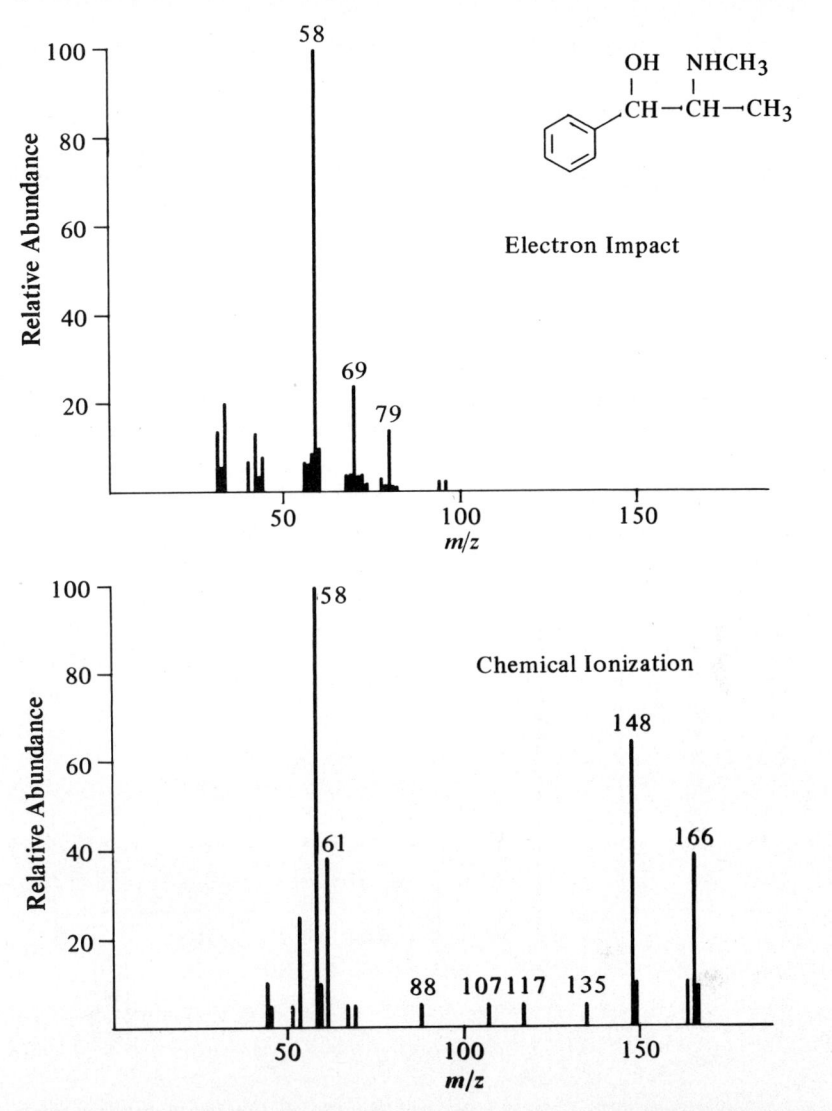

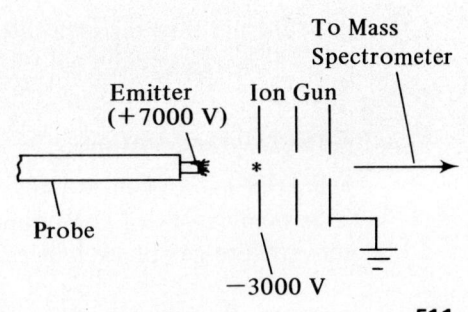

FIGURE 16.15. *Comparison of the electron-impact and chemical-ionization mass spectra of ephedrine.* From H. M. Fales, H. A. Lloyd, and G. A. W. Milne, J. Amer. Chem. Soc., **92**, 1590 (1970), *by permission of the publisher. Copyright © 1970 by the American Chemical Society.*

FIGURE 16.16. *Schematic diagram of a typical field-ionization source. This source is simply substituted for the conventional electron-impact source. The remainder of the mass spectrometer is the same. Combined electron-impact/field-ionization sources have been used.*

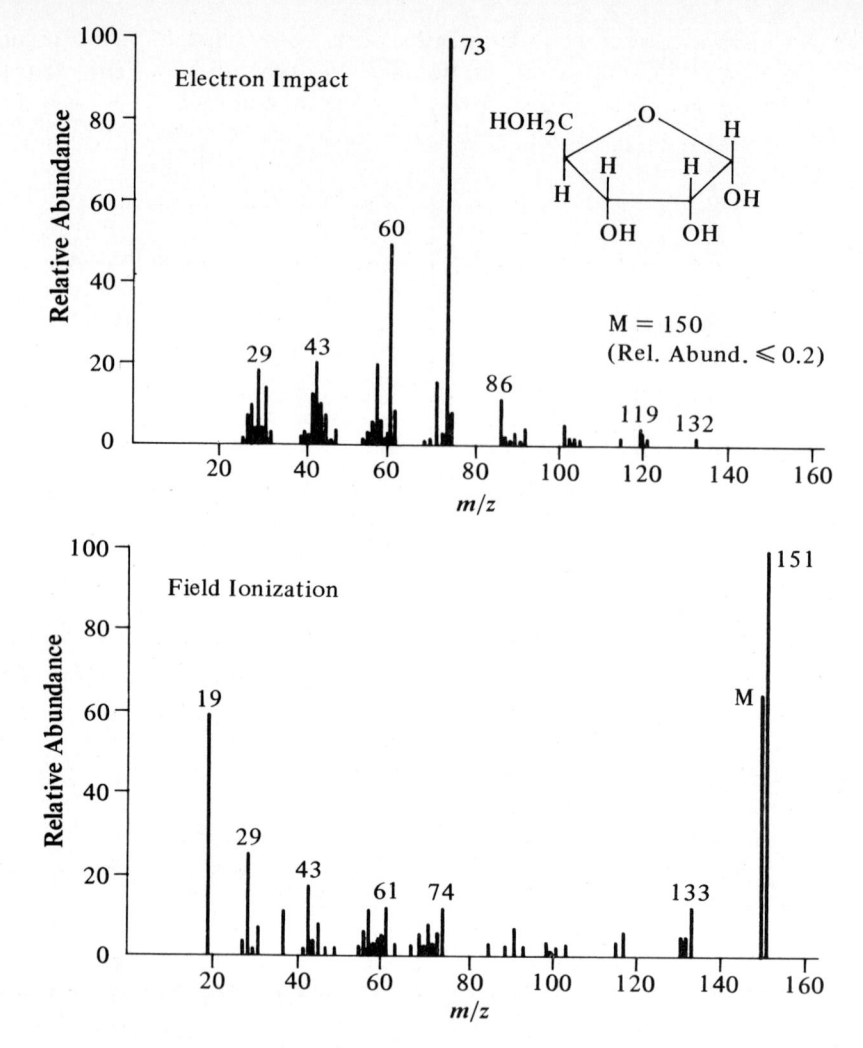

FIGURE 16.17. Comparison of the electron-impact and field-ionization mass spectra of d-ribose. The molecular ion is found at m/z = 150. The abundant peak at m/z = 151 is formed by a protonation ion-molecule reaction occurring near the emitter. From H. D. Beckey, Field Ionization Mass Spectrometry, Braunschweig-Vieweg, 1971; New York: Pergamon Press, 1971, p. 284, by permission of the publisher.

pressure and thus the concentration of reagent ions, and by extending ion residence times as is a possibility with FTMS.

Desorption Ionization

A goal of mass spectroscopists has always been to make the technique applicable to as many types of compounds as possible. The requirement that the sample be converted to gas-phase ions can be the chief drawback for thermally sensitive solids (e.g., carbohydrates, peptides, nucleic acids, organometallics) and organic salts (e.g., quaternary ammonium salts, sulfonates, phosphates). New methods of ionization in which the sample ions desorb directly from the solid state to the gaseous state without decomposition have been developed recently, and they have revolutionized the field of mass spectrometry [19].

One method is *thermal desorption mass spectrometry*. The sample is placed on a thin wire and heated very rapidly—hundreds of degrees per second. At high temperature, desorption or evaporation is kinetically more favorable than decomposition. This method can be used to convert organic salts to the gas phase; for example, the sodium salt of a sulfonic acid will desorb as RSO_3^-. Instead of using a heated wire, the sample can be desorbed by rapid heating using a pulse of infrared radiation from a CO_2 laser ($\lambda = 10$ μm) or a neodymium YAG laser (1 μm). Time-of-flight and Fourier transform mass analyzers are ideally suited to these experiments because an entire spectrum can be obtained with a single shot of the laser.

High electric fields can also be used to facilitate desorption. *Field desorption mass spectrometry* involves coating the field-ionization emitter (discussed above) with the nonvolatile sample [18]. Heating of the wire, coupled with application of the high field, causes solid-state ions to desorb directly into the gas phase.

The most important desorption methods involve particle bombardment of the sample either in the solid state or in solution using a nonvolatile solvent such as glycerol. Particle-bombardment methods owe their existence to the development of californium-252 *plasma desorption mass spectrometry* [20]. Californium-252 is radioactive and decomposes into two particles, each of approximately half the nuclear mass, which travel in opposite directions at very high kinetic energies (millions of electron-volts). One fission fragment passes through a metal foil coated with a thin film of sample, causing the sample to desorb directly as ions. The other fragment is used to trigger a TOF mass analyzer. Nucleic-acid oligomers and complicated organic toxins having molecular masses as high as 23,000 have been analyzed using this technique. Its chief drawbacks are the low ion yield and long ion counting (many hours) required to obtain a spectrum, and the low mass resolution of the TOF analyzer. Increasing the flux of fission fragments to speed up the experiment causes sample damage.

Lower-energy and higher-flux particle beams (1000 to 10,000 eV) have recently been used with conventional double-focusing mass spectrometers to obtain better mass resolution (up to 10,000) and to shorten the time of the experiment. Sample damage is avoided by the simple expedient of dissolving the sample in glycerol (or similar liquid). This important discovery [21] has permitted particle-bombardment sources to be installed on conventional double-focusing and quadrupole instruments. The technique is now readily available to chemists who have access to a mass spectrometry facility.

This lower-energy particle bombardment has taken two forms. The first, *fast-atom bombardment* (FAB), uses a discharged beam of argon or xenon atoms traveling at 6000 eV to bombard the sample, which is dissolved in glycerol and placed on the metal tip of a probe admitted to the back of the ion source (Fig. 16.18). The second form of the technique is called *organic* or *molecular secondary-ion mass spectrometry* (SIMS; see Sec. 16.5). Here, fast ions (Ar^+, Xe^+, Cs^+, etc.) are used to bombard the sample, which is either coated on a silver grid or dissolved in glycerol as for FAB.

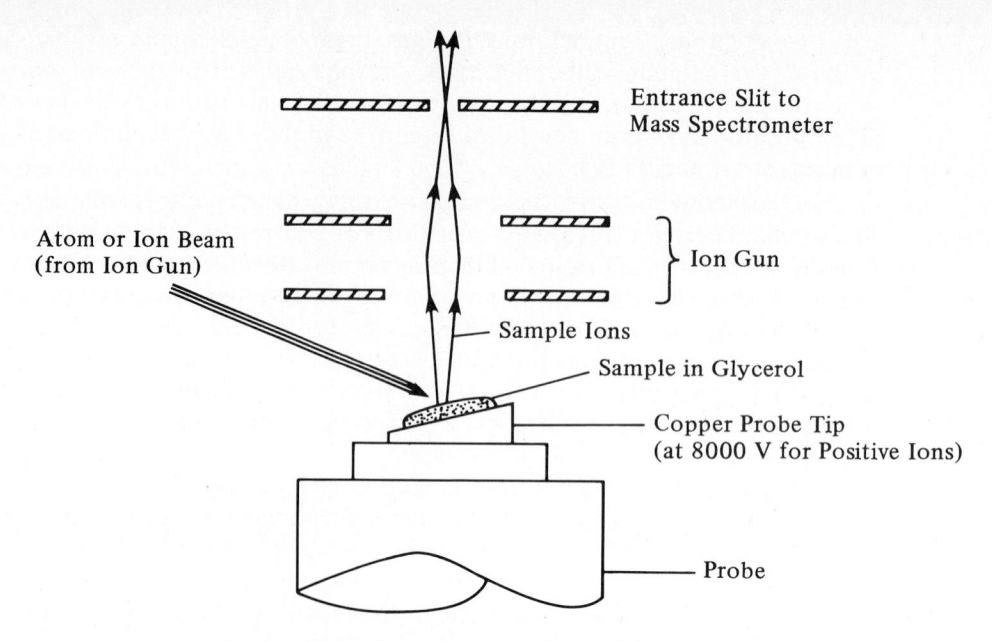

FIGURE 16.18. *Particle-bombardment desorption source.*

Although the californium-252 technique still holds the record for high mass determination (>12,000), materials such as insulin (mass approx. 5000) have been analyzed using FAB mass spectrometry. It is expected that important advances in biochemical and medical research will now be possible because of these new particle-bombardment desorption methods.

Spark-Source Mass Spectrometry

As mentioned previously, mass spectrometry can be applied to nonvolatile inorganic substances by using a spark source. One important analytical application is the analysis of fossil fuels, fly ash, and coal dust for trace metals. Because many metals such as mercury, cadmium, arsenic, beryllium, and lead are environmental hazards, it is important to determine their concentration in coal for evaluation of suitable fuels. Another useful application is the analysis of semiconductors for trace elements that affect the electrical properties of the material.

In the biological area, *spark-source mass spectrometry* (SSMS) is an ideal tool for trace-elemental analysis. First, the sample is ashed by strong heating or by a microwave discharge in oxygen to remove the organic material. The residue is then mixed with pure graphite and sparked in the spectrometer. The presence of elements at ppb levels has been verified in samples such as plant leaves, human and animal tissue, bones, and plasma. A major advantage is that multielement analysis is possible, although the precision may be less than in other trace-element techniques such as anodic stripping voltammetry or atomic absorption spectroscopy.

16.5 OTHER INSTRUMENTATION DEVELOPMENTS

Surface Analysis

Analysis of solid surfaces in terms of elemental composition is an important problem in modern research. Mass spectrometry, as *secondary-ion spectrometry* (SIMS) or *ion-probe analysis* [22], can play a useful role. The procedure involves bombarding the sample surface with a fast-moving (5–20-keV) ion beam. This beam is produced in a highly efficient discharge source called a duoplasmatron (see Fig. 16.19). The primary ions are focused by a series of lenses to obtain a narrow, well-defined beam. The impacting ions dislodge ("sputter") atoms and ions from the sample as gas-phase species. The secondary ions are drawn into the mass spectrometer for mass analysis. Figure 16.19 shows the use of a double-focusing spectrometer to perform an energy

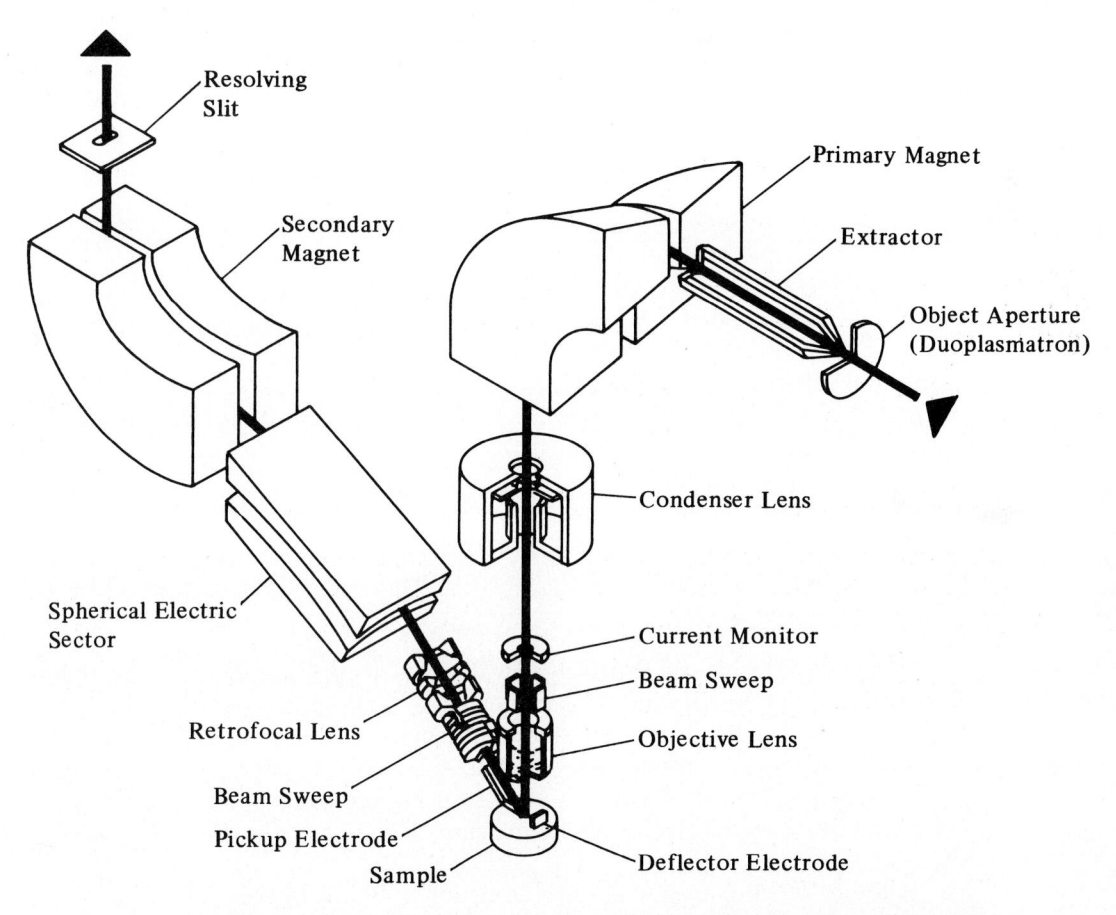

FIGURE 16.19. *Schematic diagram of a secondary-ion mass spectrometer. This particular design is called the ion microprobe because very small areas can be investigated with it.* From Chemical and Engineering News, *August* 19, 1968, *p. 30, by permission of the American Chemical Society, copyright owner.*

analysis on the secondary beam before a mass analysis. Suitable primary ions can be either inert (Ar^+, N_2^+) or reactive (O_2^+, O^+).

When the primary-ion beam impacts the surface, energy transfer occurs to break lattice bonds and vaporize ions or atoms. Some of the atoms may possess sufficient internal excitation to eject an electron, forming additional gas-phase ions. The majority of the secondary ions are singly charged and monoatomic. One mechanism for ion loss is neutralization at the surface before the gas phase is entered. This can be minimized if the surface is a nonconducting matrix, such as a metal oxide, or if the primary-ion beam is O_2^+, which reacts at the surface to form a less strongly conducting oxide region. Thus, an oxide-coated metal surface bombarded with Ar^+ will release more ions than a pure metal surface.

This phenomenon brings up a disadvantage of the technique. The ion yield or ionization efficiency of a surface depends on the nature of the element and on the chemical composition at the bombardment location. The yield of an ion from one element relative to another will change as the composition and conditions vary. The results may be complicated because a change in ion output as the surface is sputtered does not necessarily mean a change in the relative elemental composition. Another problem occurs because the sputtering may be nonuniform, that is, certain elements are removed more efficiently than others. This effect can be minimized by scanning the ion beam across the surface.

Nevertheless, there are many important advantages of SIMS:

1. The technique is highly sensitive, permitting detection of ppb concentrations.
2. Surface areas between 1 mm² and 1 μm² can be investigated, depending on the width of the primary-ion beam. The narrower limits are approached with instrumentation called an *ion microprobe*.
3. Like SSMS, the technique is applicable to all elements and is especially useful for light elements (H through Na), which are not amenable to techniques such as the electron microprobe (Chap. 14).
4. Changes in isotopic composition can be examined because the procedure is mass spectral in nature. Again, this is not true for electron microprobes.
5. Three-dimensional analysis is possible with depth resolution of approximately 50 to 100 Å.

Although SIMS is still relatively new in terms of applications, it is clear that it is extremely useful for studies of metallurgical, geological, and semiconductor surfaces, and for samples from corrosion and metal-catalyst studies. There may be biochemical and organic applications as well. A number of units are commercially available, including instruments with double-focusing and quadrupole mass analyzers for the secondary beam. However, the expense is high, ranging from $250,000 to $500,000.

Another technique that permits in-depth profiling of solid surfaces is *ion-scattering spectrometry* (ISS). In this method, a sample is mounted in a vacuum chamber and bombarded with noble-gas ions at 0.5 to 3 keV. The primary ions scattered at 90° pass into an electrostatic analyzer identical in principle to those used in double-focusing mass spectrometers. By scanning the voltage to the electrostatic analyzer, velocity analysis is accomplished.

It can be shown that the energy loss of the primary beam at the first monolayer of surface is determined by the mass of the surface atom responsible for the scattering. The relevant equation is

$$\frac{E_1}{E_0} = \frac{m_s - m_0}{m_s + m_0} \qquad (16.33)$$

where m_0 = mass of the primary beam
E_0 = kinetic energy of the primary beam
E_1 = energy of the ions scattered at 90°
m_s = mass of the surface atom

It is clear that if E_1 is measured and m_0 and E_0 are known, the mass of the surface atom, m_s, can be determined and the element identified. The method involves an indirect mass measurement. The number of ions scattered at a certain intensity is proportional to the amount of the element present.

This technique possesses many of the advantages of SIMS, including studies of isotopic abundances. The method is not as extensively evaluated, however; nevertheless, it appears to be capable of major-component analysis down to the 0.1 part-per-thousand range.

Mass spectrometry principles and techniques have been used in other kinds of surface studies in which sample atoms are sputtered by interaction with a laser beam or by radio-frequency glow discharges. These approaches are more highly specialized, but it should be clear that mass spectrometry is an important tool in surface chemistry. The reader should compare SIMS and ISS with other surface-analytical techniques such as ESCA, Auger spectroscopy, electron microprobe, and low-energy electron diffraction (see Chaps. 14 and 15).

Tandem Mass Spectrometry

An alternative to linking a gas or liquid chromatograph to a mass spectrometer for analysis of mixtures is to couple two mass spectrometers in series (MS/MS) [23]. This concept is shown schematically in Figure 16.20. Any of the ion sources can be used. The first mass spectrometer (MS-I) can be a quadrupole [24], magnet-sector, or double-focusing instrument [25]. It is used to separate the ion beam of interest from other ions (originating from other components in a mixture, for example). To obtain a "mass spectrum" of the separated ions, some method of activation is required. This is usually accomplished by inserting a collision cell after MS-I. The ions are activated by collisions with inert gas molecules at a pressure of 10^{-3} to 10^{-4} torr and decompose to give spectra very similar to mass spectra produced by direct electron ionization. MS-II is then used to scan the spectrum.

Three important applications have been developed. First, MS/MS can be used for rapid mixture analysis of targeted compounds. MS-I serves the purpose of the gas chromatograph in GCMS, and MS-II acts as the mass spectrometer. The advantage of MS/MS is that the analysis can be conducted very rapidly because the entire mixture is admitted to the ion source at once. The analyst is not required to wait 10 to 60 min for the targeted component to emerge from the GC column. The second application is its use as a mass analyzer for ions produced by chemical ionization or

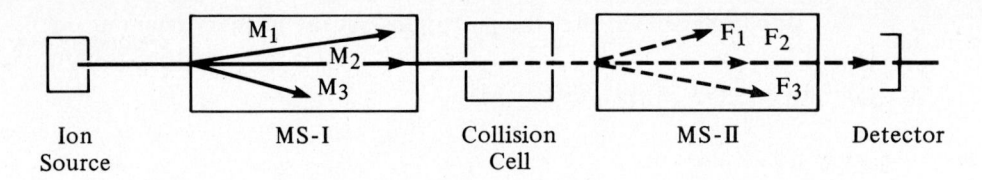

FIGURE 16.20. *Schematic diagram of a mass spectrometer/mass spectrometer (MS/MS). M_1, M_2, M_3 are various molecular ions; F_1, F_2, F_3 are fragments of M_2.*

desorption ionization. Because these methods tend to be "soft," often little fragmentation is observed. With MS/MS, the molecular mass (and formula if MS-I is double focusing) is obtained with MS-I; and structural information, produced by the collisional activation, is gathered with MS-II. Finally, the technique is very useful in fundamental studies of ion chemistry. The ion of interest, which can be a fragment or a product of an ion-neutral reaction, is "isolated" with MS-I and "analyzed" with MS-II.

SELECTED BIBLIOGRAPHY

General

BIEMANN, K. *Mass Spectrometry: Organic Chemical Applications.* New York: McGraw-Hill, 1962.

GROSS, M. L., ed. *High Performance Mass Spectrometry: Chemical Applications.* American Chemical Society Symposium Series 70. Washington, D.C.: American Chemical Society, 1978.

LEVSEN, K. *Fundamental Aspects of Organic Mass Spectrometry.* New York: Verlag Chemie, 1978.

MIDDLEDITCH, B. S. *Practical Mass Spectrometry: Contemporary Introduction.* New York: Plenum Press, 1979.

SCHLUNEGGER, U. P. *Advanced Mass Spectrometry: Applications in Organic and Analytical Chemistry.* Oxford: Pergamon Press, 1980.

WATSON, J. T. *Introduction to Mass Spectrometry: Biomedical, Environmental, and Forensic Applications.* New York: Raven Press, 1976.

Instrumentation in Mass Spectrometry

BEYNON, J. H. *Mass Spectrometry and Its Applications to Organic Chemistry.* Amsterdam: Elsevier, 1960.

DAWSON, P. H., ed. *Quadrupole Mass Spectrometry and Its Applications.* Amsterdam: Elsevier, 1976.

KISER, R. W. *Introduction to Mass Spectrometry and Its Applications.* Englewood Cliffs, N.J.: Prentice-Hall, 1965.

ROBOZ, J. *Introduction to Mass Spectrometry: Instrumentation and Techniques.* New York: Interscience, 1968.

Compilations of Standard Spectra

Catalog of Mass Spectral Data. API Research Project 44. Pittsburgh, Pa.: Carnegie Institute of Technology.

CORNU, A., and MASSOT, R. *Compilation of Mass Spectral Data.* London: Heyden, 1966.

HELLER, S. R., and MILNE, G. W. A. *EPA/NIH Mass Spectral Data Base*, Vols. 1 through 4. Nat. Stand. Ref. Data Ser., Nat. Bur. Stand. (U.S.), 63.

Index of Mass Spectra Data. Philadelphia,: American Society for Testing and Materials, 1969.

STENHAGEN, E., ABRAHAMSSON, S., and MCLAFFERTY, F. W., eds. *Atlas of Mass Spectral Data.* New York: John Wiley, 1969.

STENHAGEN, E., ABRAHAMSSON, S., and MCLAFFERTY, F. W., eds. *Registry of Mass Spectral Data.* New York: John Wiley, 1974.

Interpretation of Mass Spectra

HILL, H. D. *Introduction to Mass Spectrometry.* London: Heyden, 1966.

MCLAFFERTY, F. W. *Interpretation of Mass Spectra*, 3rd ed. Mill Valley, Calif.: University Science Books, 1980.

REED, R. I. *Application of Mass Spectrometry to Organic Chemistry.* London: Academic Press, 1966.

REFERENCES

1. J. J. THOMSON, *Rays of Positive Electricity and Their Application to Chemical Analyses*, London: Longmans, Green, and Co., 1913.
2. F. W. ASTON, *Phil. Mag.*, *38*, 707, 709 (1919).
3. A. J. DEMPSTER, *Phys. Rev.*, *11*, 316 (1918).
4. J. D. BALDESCHWIELER and S. S. WOODGATE, *Acc. Chem. Res.*, *4*, 114 (1971).
5. C. L. WILKINS and M. L. GROSS, *Anal. Chem.*, *53*, 1661A (1981).
6. J. L. BEAUCHAMP, *Ann. Rev. Phys. Chem.*, *22*, 527 (1971).
7. H. BUDZIKIEWICZ, C. DJERASSI, and D. H. WILLIAMS, *Mass Spectrometry of Organic Compounds*, San Francisco: Holden-Day, 1967.
8. J. H. BEYNON, R. A. SAUNDERS, and A. E. WILLIAMS, *The Mass Spectra of Organic Molecules*, Amsterdam: Elsevier, 1968.
9. Q. N. PORTER and J. BALDAS, *Mass Spectrometry of Heterocyclic Compounds*, New York: Wiley-Interscience, 1971.
10. M. L. GROSS, *J. Chem. Ed.*, *59*, 921 (1982).
11. H. BUDZIKIEWICZ, C. DJERASSI, and D. H. WILLIAMS, *Structure Elucidation of Natural Products by Mass Spectrometry*, Vol. 1, *Alkaloids*; Vol. 2, *Steroids, Terpenoids, Sugars, and Miscellaneous Natural Products*, San Francisco: Holden-Day, 1964.
12. G. R. Waller, ed., *Biochemical Applications of Mass Spectrometry*, New York: Wiley-Interscience, 1972.
13. K. BIEMAN, *Mass Spectrometry: Organic Chemical Applications*, New York: McGraw-Hill, 1962, pp. 204–50.
14. R. W. KISER, *Introduction to Mass Spectrometry and Its Applications*, Englewood Cliffs, N.J.: Prentice-Hall, 1965, pp. 162–206.
15. F. H. FIELD, *Acc. Chem. Res.*, *1*, 42 (1968).
16. B. MUNSON, *Anal. Chem.*, *43*(13), 28A (1971).
17. H. D. BECKEY, *Field Ionization Mass Spectrometry*, Oxford: Pergamon Press, 1971.
18. W. D. Reynolds, *Anal. Chem.*, *51*, 283A (1979).
19. K. L. BUSCH and R. G. COOKS, *Science*, *218*, 247 (1982).
20. R. D. MACFARLANE and D. F. TORGERSON, *Science*, *191*, 920 (1976).
21. M. BARBER, R. S. BORDOLI, R. D. SEDGWICK, and A. N. TYLER, *Nature*, *293*, 270 (1981); K. L. RINEHART, JR., *Science*, *218*, 254 (1982).
22. C. A. EVANS, JR., *Anal. Chem.*, *44*(13), 67A (1972).
23. R. G. COOKS and G. L. GLISH, *Chem. Eng. News*, *59* (November 30, 1981), pp. 40–52.
24. R. A. YOST and C. G. ENKE, *Anal. Chem.*, *51*, 1251A (1979).
25. M. L. GROSS, E. K. CHESS, P. A. LYON, F. W. CROW, S. EVANS, and H. TUDGE, *Intl. J. Mass Spectrom. Ion Phys.*, *42*, 243 (1982); F. W. MCLAFFERTY, P. J. TODD, D. C. MCGILVERY, and M. A. BALDWIN, *J. Amer. Chem. Soc.*, *102*, 3360 (1980).

PROBLEMS

1. A sector instrument is designed to operate with a radius of 30.00 cm and an accelerating voltage of 3000 V. Calculate the magnetic field (in gauss) necessary to focus the M^+ of methane. What would be the radius of trajectory for CH_3^+ under these conditions?

2. Repeat the calculation of Problem 1 for the necessary field to focus the M^+ of naphthalene $(C_{10}H_8^+)$. What would be the radius of curvature for the $M - 1$ ion in naphthalene? Comment on the resolution necessary to separate M and $M - 1$ in methane and naphthalene.

3. A sector mass spectrometer is built to scan to $m/z = 500$ with the maximum field of the electromagnet at 6000 gauss and an accelerating voltage of 3000 V. To examine the molecular ion of a compound of molecular mass 850, what modification in the accelerating-voltage power supply is required?

4. Draw a schematic diagram of an electron-impact source showing what voltages should be applied to the various components using an accelerating voltage of 1765 V, an ionizing energy of 50 eV, a repeller voltage of 10 V, and a target voltage of 80 V.

5. An unknown compound gives a nominal molecular mass of 220. A mixture of the unknown and perfluorotributylamine $[(C_4F_9)_3N]$ is admitted to a double-focusing mass spectrometer with an accelerating voltage of 5000 V. With $C_4F_9^+$ in focus ($m/z = 218.9856$), the accelerating and electrostatic-analyzer voltages are reduced to exactly 99.463% of the originals to bring the unknown peak to an identical focus. What is the exact mass of $m/z = 220$? From the following data, what is the elemental composition of $m/z = 220$?

Compound	m/z
$C_{17}H_{16}$	220.1251
$C_{10}H_{20}O_5$	220.1311
$C_{13}H_{20}N_2O$	220.1576
$C_{11}H_{24}O_4$	220.1674
$C_{15}H_{24}O$	220.1827

6. The exact mass of CO is 27.9949, and that of C_2H_4 is 28.0313. What resolution is necessary to just separate CO^+ and $C_2H_4^+$ found in a mixture of carbon monoxide and ethylene? Compare this requirement with that necessary to separate $C_{20}H_{40}^+$ and $C_{19}H_{36}O^+$, both nominally at $m/z = 280$.

7. The mass spectrum of benzene is obtained on a time-of-flight mass spectrometer of length 100 cm with an accelerating voltage of 3000 V. Calculate the time required for $C_2H_2^+$ (one of the low-mass ions), $C_6H_5^+$, and $C_6H_6^+$ to reach the detector.

8. An electric field of what frequency (in kHz) is necessary to observe the $m/z = 78$ ion from benzene in an ICR cell in a magnetic field of 7800 gauss (0.78 tesla)?

9. Calculate the $M + 1$, $M + 2$, etc., abundances for the molecular ions listed below. To approximate the probability of finding two ^{13}C atoms in a molecule, the following relation can be used:

$$\frac{M + 2}{M} = \frac{(1.1N)^2}{200}$$

where $(M + 2)/M$ is expressed in percent and N is the number of carbon atoms. Consider the abundance of M^+ to be 100%. (a) C_6H_6; (b) $C_2H_4O_2$; (c) $C_2H_8N_2$; (d) C_3H_7Cl; (e) C_4H_4S; (f) $C_{16}H_{34}$.

10. Mercury as an environmental pollutant is sometimes found as dimethyl mercury. Look up the natural abundance of the mercury isotopes and calculate the pattern to be expected in the molecular-ion region.

11. The methyl group in acetophenone can be deuterated by refluxing a mixture of acetophenone and D_2O with a little base catalyst:

$$PhC(O)CH_3 + \tfrac{3}{2}D_2O \underset{}{\overset{OD^-}{\rightleftharpoons}} PhC(O)CD_3 + \tfrac{3}{2}H_2O$$

At 15 eV of ionizing energy, no $M - 1$ is detected for unlabeled acetophenone, and $M + 1$ is 8.8%. Calculate the percentage of $-d_3$, $-d_2$, and $-d_1$ after one exchange, if the following abundances are found in the mass spectrum. Remember to correct for ^{13}C.

m/z	Rel. Abund.
124	8.75
123	100.00
122	7.01
121	4.42
120	—

12. Postulate a structure for the compound with the following mass spectrum. After you have settled on a structure, account for the ions found in the spectrum.

m/z	Rel. Abund.	m/z	Rel. Abund.
25	0.10	61	0.74
26	0.36	62	0.64
27	0.77	63	0.51
28	0.14	64	0.07
35	0.12	72	0.35
36	0.30	73	2.1
37	2.0	74	4.3
37.5	0.84	75	4.6
38	5.71	76	3.4
38.5	0.33	77	45.2
39	1.10	78	3.0
40	0.05	79	0.06
47	0.18	84	0.70
48	0.19	85	0.89
49	1.5	86	0.89
50	9.6	87	0.32
51	12.0	88	0.22
52	1.0	97	0.26
53	0.03	99	0.07
54	0.05	108	0.11
54.5	0.08	109	0.04
55	0.34	110	0.10
55.5	0.07	111	0.78
56	3.91	112	100.00
56.5	0.28	113	6.9
57	1.25	114	32.9
57.5	0.09	115	2.1
60	0.47	116	0.06

13. The following mass spectrum was obtained for an amine comtaining four carbon atoms. What is the compound?

m/z	Rel. Abund.	m/z	Rel. Abund.
15	1.9	43	3.1
27	0.75	44	100.0
28	4.2	45	2.8
29	9.1	56	2.3
30	2.9	57	1.6
31	4.1	58	10.0
32	0.39	59	0.41
33	1.1	71	0.39
39	2.0	72	2.3
40	0.75	73	1.2
41	9.4	74	0.07
42	6.0		

14. The first evidence to prove the existence of a certain inorganic compound was the mass spectrum taken on a time-of-flight mass spectrometer. The following data were measured from a photograph of the oscilloscope display, and as such the M + 1 peaks were too weak to measure. Identify the compound.

m/z	Rel. Abund.	m/z	Rel. Abund.
111	100	128	6
112	10	129	15
113	100	130	7
114	12	144	33
127	15	146	33

15. The following compounds are isomeric C-6 ketones. Complete identification should be possible by considering carbonyl-directed cleavages and the McLafferty rearrangement (or lack of it). Identify the compound using these processes.

	Ketone 1		Ketone 2	
m/z	Rel. Abund.	m/z	Rel. Abund.
27	14.8	27	49.4
28	4.4	28	10.9
29	33.7	29	68.3
30	0.8	30	1.7
39	7.7	39	12.8
40	1.0	40	1.6
41	26.2	41	23.0
42	3.8	42	8.0
43	100	43	100
44	2.2	44	3.3
55	2.7	55	3.1
56	9.0	56	1.9
57	27.4	57	76.1
58	1.3	58	2.5
71	0.9	71	45.6
72	17.1	72	2.5
73	0.8	73	—
85	2.4	85	2.0
100	3.5	100	19.9
101	0.2	101	1.3

16. The following spectrum is of a compound containing C, H, and O. Also present in the molecule is an aromatic ring. Identify the material.

m/z	Rel. Intensity	m/z	Rel. Intensity
26	0.10	67	0.35
27	4.0	74	0.40

m/z	Rel. Intensity	m/z	Rel. Intensity
28	0.2	75	0.35
39	6.0	76	0.39
40	1.4	77	7.2
41	4.8	78	0.80
42	0.3	79	0.81
43	4.1	93	0.70
44	0.10	94	100.00
49	0.05	95	6.6
50	1.3	96	0.42
51	4.6	107	2.0
52	0.43	108	0.40
55	1.1	121	0.20
56	0.06	122	0.03
62	0.34	135	0.20
63	1.2	136	25.0
64	0.50	137	2.5
65	4.3	138	0.18
66	5.9		

17. What resolution is necessary to distinguish between molecular oxygen and sulfur in a mass spectrometer?

18. The mass spectrum of methyl alcohol has peaks at $m/z = 15, 28, 29, 30, 31$, and 32. A broad low-intensity, metastable peak was found at $m/z = 27.13$. Determine the mother and daughter ions.

19. The decomposition of ions with the elemental composition $C_2H_5O^+$ has been studied by Shannon and McLafferty [*J. Amer. Chem. Soc.*, **88**, 5021 (1966)]. (a) Metastable peaks caused by the following decompositions were observed:

    (1) $C_2H_5O^+ \rightarrow H_3O^+ + C_2H_2$
    (2) $C_2H_5O^+ \rightarrow CHO^+ + CH_4$.

    At what m/z values would the metastable peaks caused by these decompositions be found? (b) Of the structural formulas $HOCH_2CH_2Y$, $CH_3CH(OH)Y$, CH_3OCH_2Y, and CH_3CH_2OY, one was found to decompose by route (1) one hundred times less often than any of the other three structures did. Predict which structure it was.

20. The mass spectra of two different trimethylpentanes showed the following relative abundances [H. W. Washburn et al., *Ind. Eng. Chem., Anal. Ed.*, **17**, 75 (1945)]:

m/z	Rel. Abund. (1)	Rel. Abund. (2)
43	20	50
57	80	9
71	1	40
99	5	0.1
114	0.02	0.3

Of the two isomers, 2,2,4- and 2,3,4-trimethylpentane, which is more likely to produce the abundances given in column (1)? Write the structures of the ions most likely responsible for the m/z values found.

21. From the following table of mass spectral data, deduce the probable structure of the unknown $C_xH_yN_z$:

m/z	Rel. Abund.	m/z	Rel. Abund.
15	3.7	43	2.7
27	3.3	44	0.29
28	4.1	55	2.0
29	3.6	56	2.7
30	6.2	57	5.6
31	0.10	58	100
39	4.8	59	3.6
40	1.4	60	0.05
41	18	73	0.41
42	11	74	0.02

22. The mass spectrum of an unknown compound had the following relative intensities for the M ($m/z = 86$), M + 1, and M + 2 peaks, respectively: 18.5, 1.15, and 0.074 (percentage of base peak). From the following partial list of isotopic abundance ratios, determine the molecular formula of the unknown:

Formula	Isotope-Abundance Ratios (M = 100%) M + 1	M + 2
$C_4H_6O_2$	4.50	0.48
C_4H_8NO	4.87	0.30
$C_4H_{10}N_2$	5.25	0.11
$C_5H_{10}O$	5.60	0.33
$C_5H_{12}N$	5.98	0.15
C_6H_{14}	6.71	0.19

23. Determine the m/z order in which the following gases will appear in the mass spectrum of a mixture containing them: C_2H_4, CO, N_2

24. Compare and contrast ESCA, Auger spectroscopy, ISS, and SIMS as methods for surface analysis.

CHAPTER 17

Thermal and Calorimetric Methods of Analysis

Neil Jespersen

In *thermal* methods of analysis, the temperature is manipulated to produce the measured parameter. Thermogravimetry (TG), differential thermal analysis (DTA), and differential scanning calorimetry (DSC) are the three major methods that use temperature change as the independent variable. In *calorimetric* methods, by contrast, temperature is the dependent variable. Examples of calorimetric methods are thermometric titration (TT) and direct-injection enthalpimetry (DIE). These five methods will be discussed primarily from an analytical point of view. Each technique has its unique characteristics and capabilities; for this reason, the major aspects of each method are considered individually.

17.1 GENERAL CHARACTERISTICS OF THERMAL METHODS

Thermogravimetry involves measuring the mass of a sample as its temperature is increased. A plot of mass versus temperature permits evaluation of thermal stabilities, rates of reaction, reaction processes, and sample composition.

Differential thermal analysis is the monitoring of the difference in temperature between a sample and a reference compound as a function of temperature. These data can be used to study heats of reaction, kinetics, heat capacities, phase transitions, thermal stabilities, sample composition and purity, critical points, and phase diagrams.

Measurement of the differential power (heat input) necessary to keep a sample and a reference substance isothermal as temperature is changed (scanned) linearly is the basis of *differential scanning calorimetry*. DSC is used to study the same effects as DTA. Initially, DSC provided a great deal of improvement over DTA; but, owing to continued improvements in instrumentation, DTA and DSC can currently provide data of nearly equal quality.

Thermometric titrations involve monitoring of temperature change as a function of the volume of titrant added. From this, concentrations may be evaluated as in normal titrimetry; ΔH^0, ΔG^0, and ΔS^0 for the reaction may also be calculated under appropriate conditions.

Direct-injection enthalpimetry data are similar to data from TT. However, titration is replaced by a virtually instantaneous injection of reagent, and temperature is monitored as a function of time. As a result, more rapid analysis is possible. Heats of reaction can be readily deduced, and kinetics may be studied in favorable situations.

General Thermodynamic Relationships

Since thermal analyses are usually run under conditions of constant pressure, the underlying thermodynamic equation is the Gibbs-Helmholtz expression:

$$\Delta G^0 = \Delta H^0 - T \Delta S^0 \qquad (17.1)$$

where
G = free energy of the system
H = enthalpy of the system
S = entropy of the system
T = temperature in kelvins

The general chemical reaction

$$aA + bB \rightarrow cC + dD \qquad (17.2)$$

is spontaneous as written if ΔG is negative, is at equilibrium if $\Delta G = 0$, and does not proceed if ΔG is positive.* Thermal analysis involves the monitoring of spontaneous reactions.

Methods involving temperature change as the independent variable (TG, DTA, and DSC) take advantage of the $T \Delta S$ term in Equation 17.1. Differentiating the Gibbs-Helmholtz equation with respect to temperature, one obtains†

$$\frac{d(\Delta G)}{dT} = -\Delta S \qquad (17.3)$$

This shows how to move from a stable situation (ΔG positive) to one where reaction will occur. If ΔS is positive, an increase in temperature will eventually cause ΔG to become negative, whereas if ΔS is negative, decreasing the temperature will achieve the desired spontaneous reaction.‡ Once the reaction is made to occur, each of the three methods may be used to detect the process, often yielding different and complementary information.

---

*It is important to note that in some cases the reaction may not proceed under certain conditions even though ΔG is negative. A mixture of oxygen and hydrogen is a case in point. Such metastable states, although of considerable interest, are outside the scope of this discussion.

†Equation 17.3 is a gross oversimplification: It assumes that neither ΔS nor ΔH varies with temperature. This is not so. However, the possible variations of ΔH and ΔS will usually affect only the temperature at which ΔG becomes negative, not the fact that it will eventually do so.

‡Once the spontaneous reaction starts, it proceeds at a rate dependent upon the kinetic and microscopic characteristics of the sample.

The second group of methods (TT and DIE) involves the creation of a spontaneously reacting mixture by combining two or more chemical species. Then,

$$\Delta T = \frac{-\Delta H n_p}{C'_p} \tag{17.4}$$

where n_p = number of moles of product formed
C'_p = heat capacity (cal/deg) for the entire system

Since the amount of product (n_p) may be equilibrium controlled (as in TT) or kinetically controlled (as in DIE), Equation 17.4 may be expanded in various ways to calculate the parameters listed previously.

These relationships have been presented in their simplest forms; later discussions will illustrate the complexities involved. For instance, the thermodynamic terms calculated are not usually standard-state terms (e.g., ΔH calculated is not ΔH^0). The additional effort necessary to extrapolate to standard states (e.g., infinite dilution) is usually excessive for analytical situations, and in some cases the quality of the data does not warrant such treatment. References will be given to treatments of data or theory beyond the scope of this text that may interest those who wish to become more deeply involved with methods of thermal analysis.

17.2 THERMOGRAVIMETRY

Thermogravimetry (TG) involves continuously measuring the mass of a sample as a function of its temperature. Plots of mass versus temperature are called *thermogravimetric curves* or *TG curves*. The use of other names is discouraged.

General Considerations

Suitable samples for TG are solids that undergo one of the two general types of reaction

$$\text{Reactant(s)} \rightarrow \text{Product(s)} + \text{Gas}$$

$$\text{Gas} + \text{Reactant(s)} \rightarrow \text{Product(s)}$$

The first process involves a mass loss, whereas the second involves a mass gain. Processes occurring without change in mass (e.g., the melting of a sample) obviously cannot be studied by TG.

A simple thermogram, for the dehydration of copper sulfate pentahydrate, is shown in Figure 17.1. There are two points of major interest to the analytical chemist. First is the general shape of the thermogram and the particular temperatures at which changes in mass occur. From this information, individual compounds may be identified under given conditions. Unfortunately, the reproducibility of the temperatures at which mass changes occur is severely affected by many experimental conditions, as will be discussed later. For this reason, the obvious qualitative analytical capabilities of TG have yet to be fully realized.

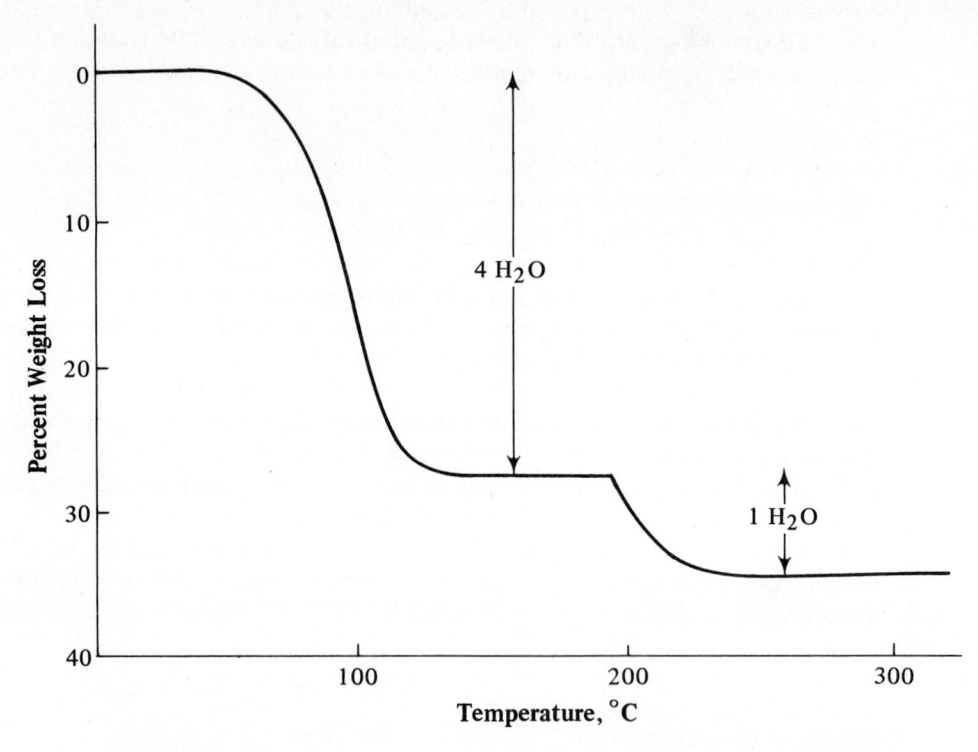

FIGURE 17.1. *Thermogravimetric curve for the dehydration of copper sulfate hydrate.*

The second major feature of the curve (the magnitudes of the mass changes observed) has found much more use, because it is independent of the many factors that affect the shape of the thermogram. Mass changes are directly related to the specific stoichiometries of the reactions occurring, independent of the temperature. As a consequence, precise quantitative analysis of samples whose qualitative composition is known can be made, or the composition of novel compounds can be deduced.

Instrumentation

Thermogravimetric instrumentation should include several basic components to provide the flexibility necessary for the production of useful analytical data. These components are: (a) a balance, (b) a heating device, (c) a unit for temperature measurement and control, (d) a means for automatically recording the mass and temperature changes, and (e) a system to control the atmosphere around the sample.

Balances. Balances must remain precise and accurate continuously under extreme temperature and atmospheric conditions, and should deliver a signal suitable for continuous recording. These requirements may be met in many ways; the two books by Duval and Wendlandt (see the Selected Bibliography) consider at least 10 different commercial thermobalances that perform satisfactorily. The basic characteristics of such balances will be illustrated by describing one, the Cahn Electrobalance (Ventron Instruments Corp., Cahn Div., Paramount, Calif; see Fig. 17.2). It operates as a null-type device by providing an electrical force to restore the beam to a predetermined position as the mass of the sample changes. As the beam is moved off-balance, a shutter affixed to the beam changes the amount of light reaching a phototube, which causes a restoring force to be generated by passing a current through an electromagnet that serves as the pivot for the balance beam. A permanent magnet above and below the pivot point supplies magnetic attraction to the electromagnet. The force necessary to restore the beam is proportional to the current passed through the electromagnet, which is recorded. Additional features of the balance are two loops for hanging sample pans (with loop A affording 2.5 times the sensitivity of loop B) and controls that adjust the range of the recorder and set the initial balance of the instrument (mass dials).

Null and other types of balances are available that have the capability of measuring the mass of a sample between 100 and 0.02 g to a precision of 0.01 to 1%. As will be shown later, the smaller the sample, the better the results often will be.

Heating Devices. The sample can be heated by resistance heaters, by infrared or microwave radiation, or by heat transfer from hot liquids or gases. Resistance heaters are the most common.

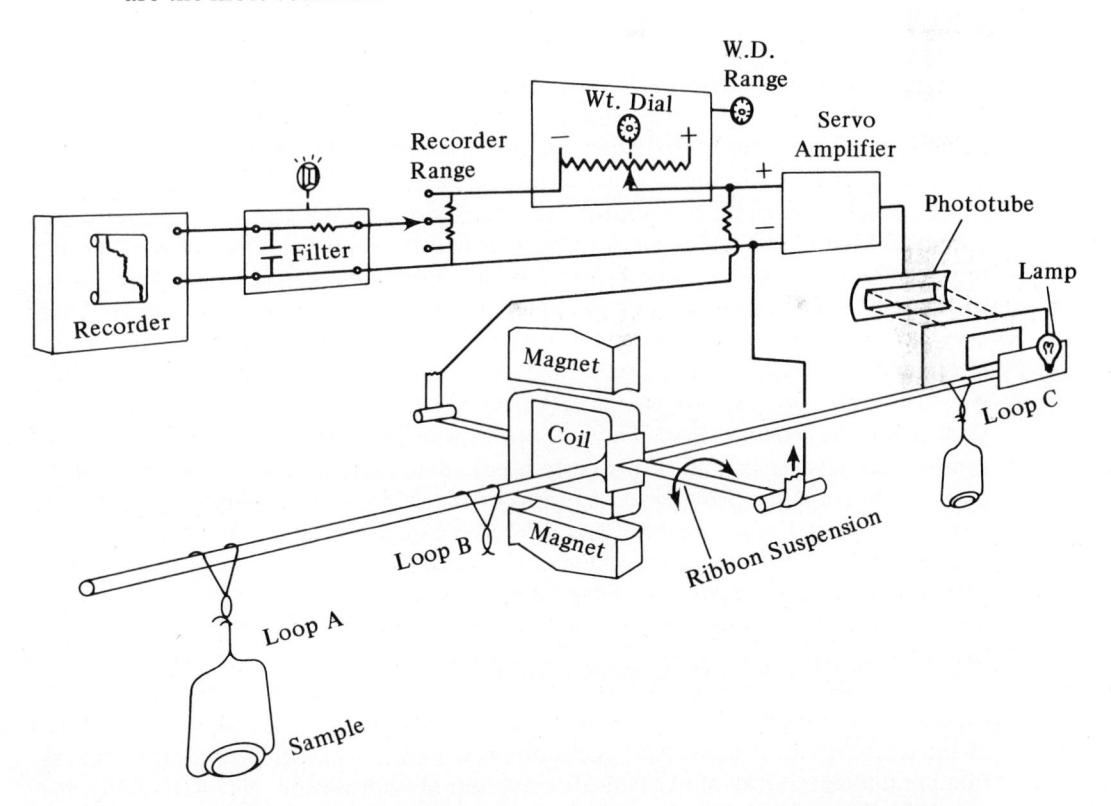

Figure 17.2. Schematic diagram of the Cahn Electrobalance. Courtesy of Cahn Instruments Div., Ventron Instruments Corp.

Furnaces should be designed so that the sample is heated uniformly and symmetrically. The furnace must also be designed so that the heat generated is localized on the sample. Heat flow into the environment is an inconvenience to the operator, and heat flow to the balance results in serious errors. These problems are minimized by cooling the exterior of the oven, and by locating the oven as far as possible from the balance itself. Convection currents can be minimized by designing the sample holder to provide the least possible resistance to air flow, and by installing appropriate baffles. Apparent mass changes are always observed at the start of an experiment because of convection currents. Best results are obtained when these currents are held constant after start-up.

Temperature Measurement and Control. Temperature-sensing devices are usually thermocouples placed as close to the sample as possible. Thermocouples are inexpensive, rugged, and fairly linear in their response to temperature changes. Platinum resistance thermometers are also used in this application. The electromagnetic force generated by the thermocouple may be used to drive one axis of an *x-y* recorder, or a feedback circuit to the heater may be used to obtain a programmed linear heating rate. In the latter case, the time axis of a strip-chart recorder is proportional to temperature. Instruments that depend on a linear increase in power to the heater often have severely nonlinear temperature increases because of heat losses to the environment.

Recording the Signal. Electrical signals from the balance (see Fig. 17.2) and from the measuring thermocouple (after amplification) are fed into a recording potentiometer. If a strip-chart recorder is used, then the time-base axis is also the sample-temperature axis; an event marker may be used to indicate the temperature in increments of 10, 25, 50, or 100°C in order to monitor the linearity of heating. The *x-y* recorder is also used, but this method of recording suffers from the disadvantage that any nonlinearity in the heating rate will not be observed. Newer instruments may incorporate microprocessors or microcomputers to control the experiment and acquire data. These may be programmed for data analysis, reducing the effort and subjectivity of data treatment.

Controlling the Atmosphere. The composition of the atmosphere surrounding the sample can have large and (if properly used) advantageous effects. For this reason, most TG instruments provide some means of altering this atmosphere; in most cases, a static or flowing atmosphere of any desired composition can be provided. In addition, TG determinations can be done in a vacuum (many systems can achieve pressures of 10^{-3} torr or less), or at elevated pressures.

Theory and Experimental Considerations

Figure 17.1 is a typical TG curve. In virtually all TG analysis, the mass is monitored as the temperature is increased; accurate measurements under decreasing temperature are difficult and tend to yield little additional information. As discussed earlier, a reaction occurs when ΔG for the process becomes zero or negative, and its start is indicated when the mass deviates from the initial plateau. When the reaction stops, a new plateau is reached. The temperatures at which the reaction appears to start (T_i) and end (T_f), as well as the shape of the curve, depend on many factors. Some of these are heating rate, heat of reaction, furnace atmosphere, amount of sample, nature of sample container, particle size, and packing of sample.

The major use of TG is in the precise determination of mass changes for several sequential reactions. Necessarily, each reaction involving a change in mass must begin with and be followed by a plateau in order to distinguish sequential events. Two processes may occur more or less simultaneously, resulting in shoulders or complete merging of the decomposition reactions. This situation can be avoided by proper use of the variables listed above. A rather simplified discussion of these variables is presented to illustrate the possibilities.

Heating Rate. In any heating process, there is always a difference in temperature between the sample and the oven. The magnitude of this *thermal lag* is roughly proportional to the rate at which the sample is heated. As a result, if a change in mass of a sample always occurs at a temperature T_i^0, then the observed T_i (measured outside the sample) will always be greater than T_i^0. Because of the thermal lag, the difference between T_i and T_i^0 will increase as the heating rate is increased. Temperature T_f is the temperature at which the end of mass change occurs. This value will depend on the heating rate in a similar fashion. In addition, the temperature range of mass loss ($T_f - T_i$) tends to increase with the heating rate. Figure 17.3 shows the effect of a much slower heating rate for the same sample as in Figure 17.1.

Heat of Reaction. Since this is an intrinsic property of the material studied, it cannot be altered; however, its effects can be modified. An endothermic process will show a larger thermal lag than an exothermic one; the latter will sometimes cause the sample to be hotter than the observed temperature. Obviously, T_f is greatly affected whereas T_i is not. If T_f increases greatly, other processes may be obscured. This increase occurs only in endothermic reactions and may be minimized by heating the sample at a low rate so that the heat absorbed by the reaction can be replaced by heat flow from the oven. Decreasing the temperature range of mass loss ($T_f - T_i$) increases the probability of obtaining usable plateaus.

Furnace Atmosphere. This is perhaps the most useful variable in altering TG curves. The basic feature is that, by providing an atmosphere rich in the reaction product, decomposition is delayed to higher temperatures. Conversely, in an inert atmosphere or vacuum, the reaction will proceed at lower temperature. Simultaneous reactions can be separated by the choice of atmosphere, provided different gases are liberated.

In addition to moving the temperatures of the decompositions, it is possible to alter the reaction that occurs. A notable example of this is found in the heating of organic samples: *Oxidation* will occur in the presence of oxygen, whereas *pyrolysis* will occur if oxygen is excluded.

Nature of Sample Container. The container is important in that (a) the material used to hold the sample may catalyze an entirely unexpected reaction and (b) the container may trap some of the gases generated. The former may be evaluated by analyzing the reaction products. In the latter case, irreproducible mass changes can occur

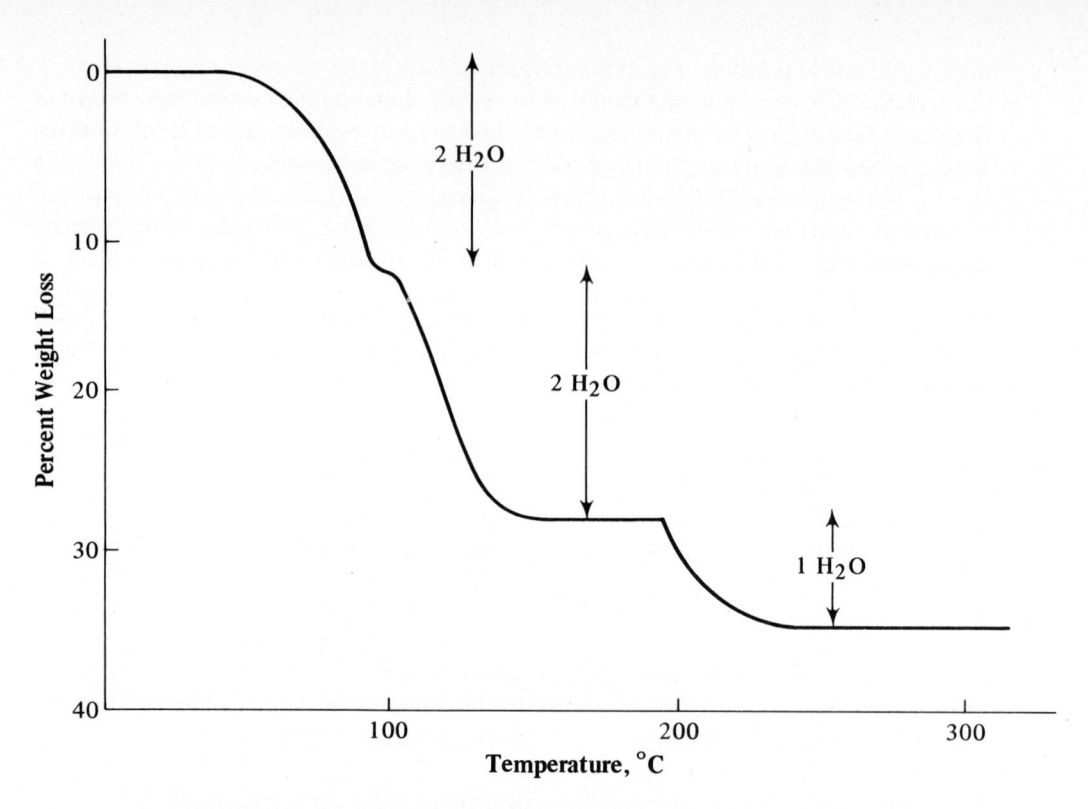

FIGURE 17.3. *Increased resolution of thermogravimetric curves by lower heating rate (compare to Fig. 17.1).*

because of adsorption of the gas, or the curve can be displaced by a self-generated atmosphere around the sample. Metallic sample holders reduce the chances of adsorption, and self-generated atmospheres can be removed by flowing an inert gas past the sample during analysis.

Physical Characteristics of Sample. The amount, particle size, and packing of the sample generally affect its thermal homogeneity. Large systems have large temperature gradients whereby the outer portions may be reacting while the inner portions remain cool and stable. The result is a greater thermal lag for larger samples, with the effects noted above. Smaller particle size tends to decrease thermal lag, but tighter packing increases it. Nonuniformity of particle size and packing leads to irreproducible curves.

Depending on the processes studied, it may be necessary to have very good control over some or all of these variables. Generally, to distinguish between reactions that occur within 50 to 100°C of one another, very stringent control is needed. If long plateaus separate the decompositions, control is less critical.

In addition to gravimetric analysis, TG has also been used to elucidate the kinetics of decomposition reactions. This involves analyzing the shape of the TG curve. In general, the rate of reaction at any measured temperature is proportional to the slope of the curve, but a number of uncertainties sometimes make these analyses of questionable value. Freeman and Carroll [*J. Phys. Chem.*, 62, 389 (1958)] describe the most popular of the kinetics analysis methods, and Clarke et al. [*Chem. Comm.*, 266 (1969)] present the major objections to kinetics analysis by TG.

Analytical Calculations

Under controlled and reproducible conditions, quantitative data can be extracted from the relevant TG curves. Most commonly, the mass change observed is related to sample purity or composition.

Example 17.1. A mixture of CaO and $CaCO_3$ is analyzed. The thermogram shows one reaction between 500 and 900°C, where the mass of the sample decreases from 125.3 to 95.4 mg. What is the percentage of $CaCO_3$ in the sample?

Solution: $CaCO_3 \rightarrow CaO + CO_2$

$CaO \rightarrow$ No Reaction

$$\text{mmoles } CO_2 = \frac{\text{mg Lost}}{\text{Mol. Wt. } CO_2} = \frac{29.9}{44.0} = 0.680$$

mmoles CO_2 = mmoles $CaCO_3$

mg $CaCO_3$ = (mmoles $CaCO_3$)(Mol. Wt. $CaCO_3$) = (0.680)(100.1)
= 68.0 mg

$$\% \; CaCO_3 = \left(\frac{68.0}{125.3}\right)(100\%) = 54.3$$

Example 17.2. A pure compound may be either MgO, $MgCO_3$, or MgC_2O_4. A thermogram of the substance shows a loss of 91.0 mg from a total of 175.0 mg used for analysis. What is the formula of the compound?

The relevant possible reactions are

$$MgO \rightarrow \text{No Reaction}$$
$$MgCO_3 \rightarrow MgO + CO_2$$
$$MgC_2O_4 \rightarrow MgO + CO_2 + CO$$

Solution: $\% \text{ Mass Loss Sample} = \left(\frac{91.0}{175.0}\right)(100\%) = 52.0$

$$\% \text{ Mass Loss if } MgCO_3 = \left(\frac{44.01}{84.31}\right)(100\%) = 52.2$$

$$\% \text{ Mass Loss if } MgC_2O_4 = \left(\frac{44.01 + 28.00}{112.3}\right)(100\%) = 64.1$$

If the preparation was pure, the compound present is $MgCO_3$.

Applications

Books by Duval, Vallet, and Wendlandt (see the Selected Bibliography) contain a good summary of TG work. A few interesting analytical applications are given below.

Mixtures of divalent-cation oxalates can be analyzed successfully with a high degree of precision. A mixture of calcium, strontium, and barium oxalate monohydrates will lose all its water of hydration between 100 and 250°C; the three anhydrous oxalates will decompose simultaneously to the carbonates between 360 and 500°C; and the carbonates will in turn decompose to the oxides in the following order: calcium (620 to 860°C), strontium (860 to 1100°C), and barium (1100°C and up). In addition to the rather common oxalates, the precipitates formed with other organic precipitating agents have been studied, including those of the very similar lanthanide metals. Examples of precipitating agents are cupferron and neocupferron, and significant differences in the decomposition curves of their chelates may be needed for analysis of mixtures.

Direct analysis of solid materials eliminates the precipitation steps referred to above. Clays and soils can be evaluated by TG to determine water content, carbonate content, and organic-matter content.

Thermograms can be used to compare the stabilities of similar compounds (e.g., of metal carbonates by studying the thermal decomposition to their respective oxides). Qualitatively, the higher the decomposition temperature, the more positive is the ΔG value at room temperature and the greater the stability.* The example given earlier of the dehydration of copper sulfate pentahydrate is interesting in that one of the five water molecules is not equivalent to the other four. This is confirmed by x-ray crystallography, which shows that four water molecules surround the Cu(II), whereas the more tightly bound water molecule is hydrogen-bonded to two neighboring sulfate ions.

17.3 DIFFERENTIAL THERMAL ANALYSIS

In *thermal analysis* (TA), the temperature of a sample is monitored while heat is supplied at a uniform rate. In the more sophisticated and sensitive method of *differential thermal analysis* (DTA), the difference in temperature between a sample and a reference is measured as the sample and reference are heated.

General Considerations

Differences in temperature between the sample and an inert reference substance will be observed when changes that involve a finite heat of reaction, such as chemical reactions, phase changes, or structural changes, occur in the sample. If ΔH is positive (endothermic reaction), the temperature of the sample will lag behind that of the reference. If ΔH is negative (exothermic reaction), the temperature of the sample will exceed that of the reference. Figure 17.4 shows TA and DTA curves for these two cases. DTA is more widely applicable than TG because it is not limited to reactions involving a change in mass. On the other hand, a reaction with a ΔH of zero will not be observed. However, if a measurable change in heat capacity accompanies the process,

---

*This assumes a constant ΔS for the reactions compared, and is most valid when a homologous series of compounds is studied.

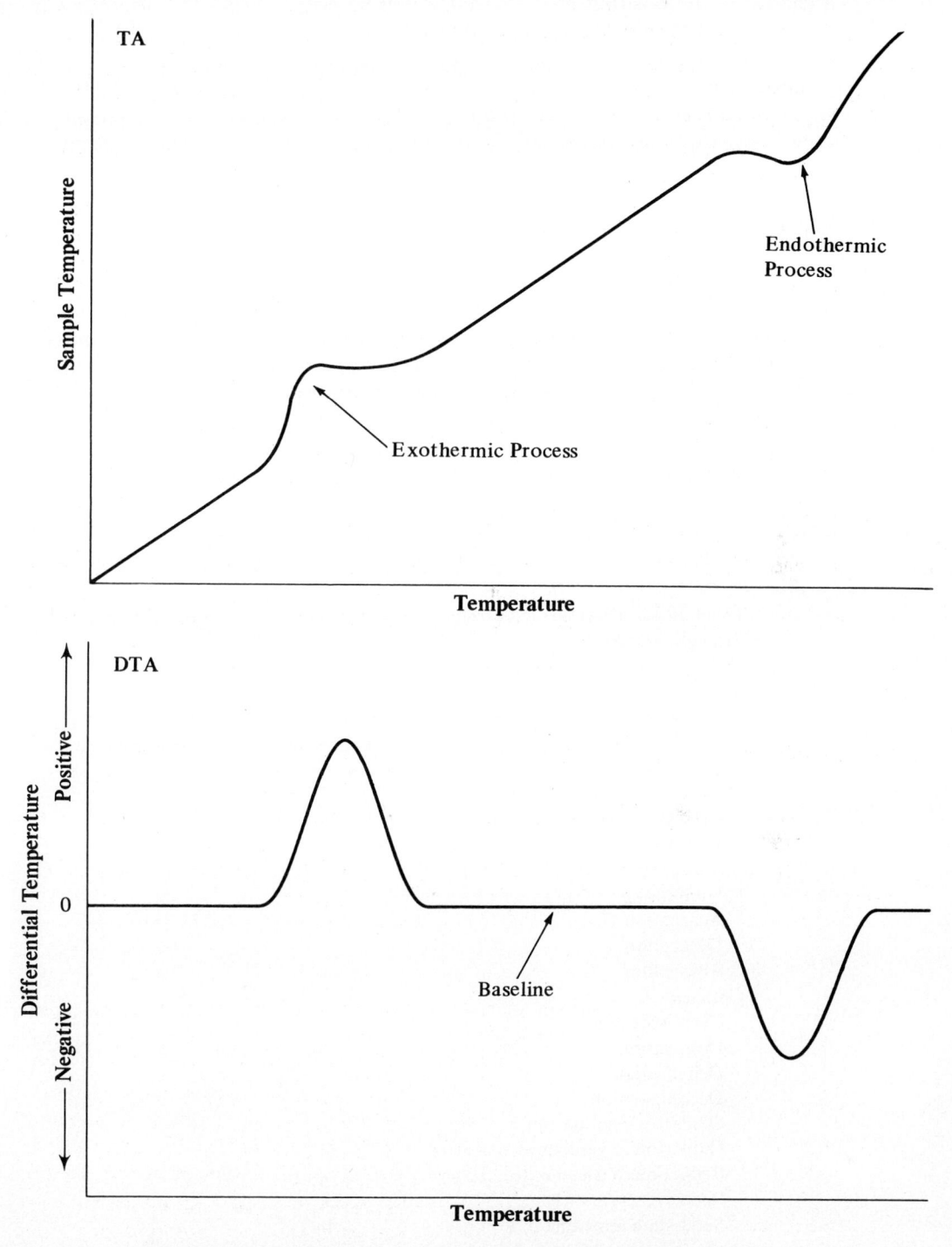

FIGURE 17.4. *Comparison of thermal analysis (TA) and differential thermal analysis (DTA) curves, illustrating exothermic and endothermic peaks.*

a change in the position of the baseline will be noted. Table 17.1 lists the various reaction types observable and the expected nature of ΔH.

DTA heating curves are useful both qualitatively and quantitatively. The positions and shapes of the peaks can be used to determine the composition of the sample. (Sadtler publishes an index of DTA curves similar to the tabulations of optical spectra.) The area under the peak is proportional to the heat of reaction and the amount of material present, and thus permits quantitative analysis. In addition, the shape of the heating curve can be used in evaluating the kinetics of the reaction under carefully controlled conditions.

Instrumentation

Implementing DTA requires the following components: (a) a circuit for measuring differences in temperature, (b) a heating device and temperature-control unit, (c) an amplifying and recording apparatus, and (d) an atmospheric-control device.

Temperature Measurement. Thermocouples are by far the most reliable devices for monitoring temperature in DTA. A typical arrangement is shown in Figure 17.5.

One of the major considerations in DTA is obtaining valid readings of the actual temperature of the sample and reference materials conveniently and reproducibly. As in TG, thermal equilibrium is of utmost importance. There is always a

TABLE 17.1. *Processes Observable Using DTA, and the Heats of Reaction Typically Observed*

Phenomenon	Heat of Reaction Exothermic	Heat of Reaction Endothermic
Physical		
Crystalline transition	×	×
Fusion	—	×
Vaporization	—	×
Sublimation	—	×
Adsorption	×	—
Desorption	—	×
Absorption	—	×
Chemical		
Chemisorption	×	—
Desolvation	—	×
Dehydration	—	×
Decomposition	×	×
Oxidative degradation	×	—
Oxidation in gaseous atmosphere	×	—
Reduction in gaseous atmosphere	—	×
Redox reactions	×	×
Solid-state reactions	×	×

Source: Adapted from S. J. Gordon, *J. Chem. Ed.*, 40, A87 (1963), by permission of the publisher.

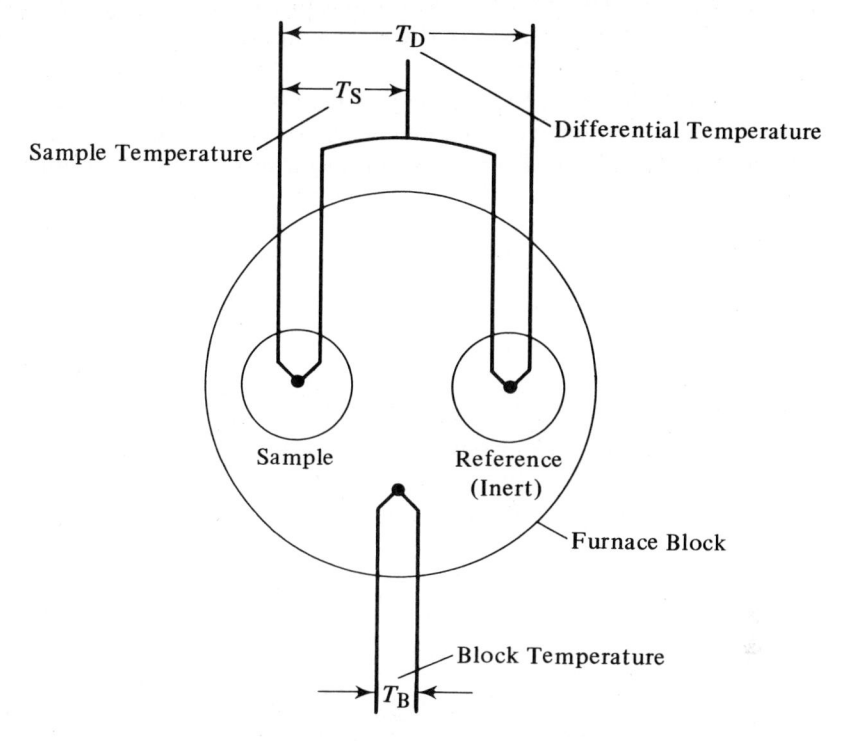

FIGURE 17.5. *Schematic of the thermocouple arrangement in the DTA cell. Adapted from S. J. Gordon, J. Chem. Ed., 40, A87 (1967), by permission of the journal editor.*

definite temperature difference between the outer and inner portions of the sample; indeed, reactions often occur at the surface of the sample while the interior remains unreacted. This effect is minimized by using as small a sample as possible with uniform particle size and packing. Depending on the instrument used, the thermocouple may be embedded in the sample or, at the other extreme, may simply be in direct contact with the sample holder. In any case, the thermocouple must be precisely positioned for every experiment. To obtain the best results, the reference and sample thermocouples should be matched in temperature response and the geometric arrangement of the sample and reference thermocouple should be perfectly symmetrical within the oven.

Heating and Temperature Control. The heating and temperature-control units are very similar to those used in TG. Ovens should be constructed to avoid electrical interference with the thermocouples. To reduce this possibility even further, most instruments have an inner metallic chamber for the sample and reference, to act as an electrical shield and to minimize thermal fluctuations. Needless to say, convection currents are no longer important, and atmospheric control is much more easily achieved—particularly a flowing atmosphere.

Recording the Signal. Amplifying the signal and recording the results are done in much the same way as in TG, but replacing the signal for mass changes with the potential difference between the thermocouples. Figure 17.6 shows a schematic of a typical DTA apparatus.

The similarities of TG and DTA are obviously great, at least instrumentally. As a consequence, many commercial instruments are designed to perform both types of analysis; the heating device, temperature-control unit, atmospheric control, and recording device are essentially used in common and are contained in a single control unit, only the thermobalance and DTA sample compartments being separate. As with TG, most modern instruments are equipped with digital data-acquisition and data-treatment systems. In addition, many commercial instruments are set up to run simultaneous TG and DTA experiments.

Theory and Experimental Considerations

The equation for heat flow from the environment to the sample or vice versa is given by Newton's law of cooling:

$$\frac{dQ}{dt} = k(T_s - T_e) \tag{17.5}$$

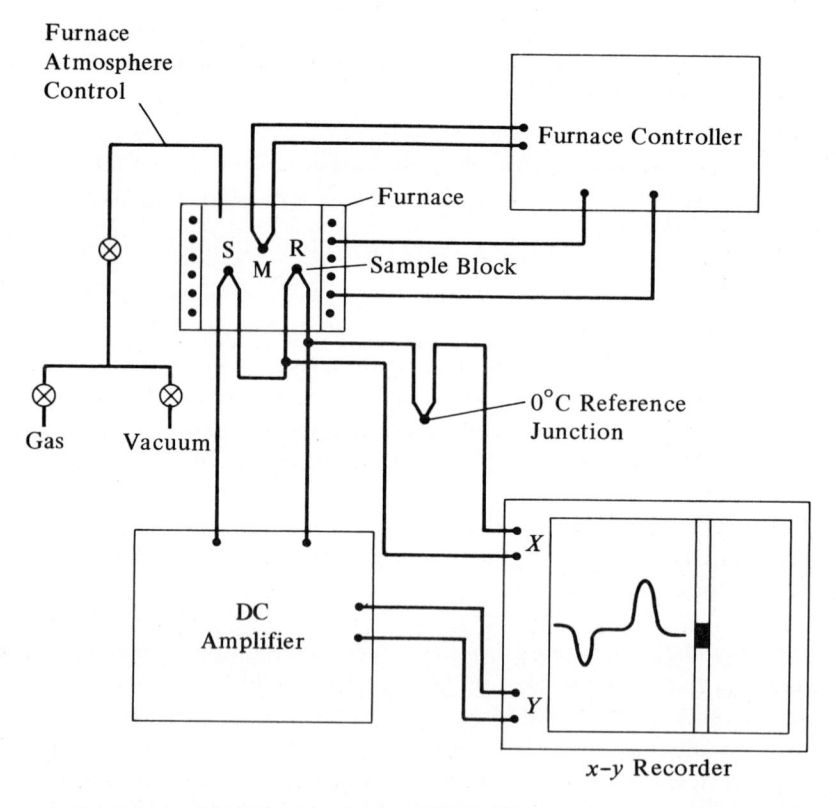

Figure 17.6. *Schematic of the entire DTA setup. S, R, and M represent the sample, reference, and furnace-monitoring thermocouples, respectively. Adapted from W. W. Wendlandt,* Thermal Methods of Analysis, *New York: John Wiley, 1964, by permission of the publisher.*

where k = a constant related to the thermal conductivity of the system
T_s = temperature of the sample
T_e = temperature of the environment

The flow of heat (Q) is from the environment to the sample if dQ/dt is negative; this is always the case in a heating curve, in the baseline regions. The equation for the rate of heat production for chemical processes is

$$\frac{dQ}{dt} = (-\Delta H)\frac{dn_p}{dt} \tag{17.6}$$

where n_p = number of moles of product formed

The net heat change in the sample is given by the sum of Equations 17.5 and 17.6. For an exothermic process, the start of the reaction causes an increase in the rate of gain of heat (Eqn. 17.6) but the rate of transfer of heat from the environment is decreased [in Eqn. 17.5, $(T_s - T_e)$ tends to decrease]. When Equation 17.5 is equal to Equation 17.6, the maximum (or minimum) of the DTA peak is observed. If the constant k in Equation 17.5 is known, the exact point at which the reaction ends can be calculated. Usually, however, the magnitude of k is not known and all that can be said is that the reaction ends at some point after the maximum. This behavior may be compared to that in TG, where the reaction unambiguously ceases when a new plateau is reached.

DTA peak areas depend mainly on the amount of material, the heat of reaction, and the thermal flow to or from the sample. These are related by the equation

$$A = \frac{-m\Delta H}{gk} \tag{17.7}$$

where g = a constant related to the geometry of the sample
k = a constant related to the thermal conductivity
m = amount of reactive component in the sample in moles

Although the constants g and k can be evaulated experimentally, they are usually combined into a simple empirical conversion factor, k', to give

$$A = k'm(-\Delta H) \tag{17.8}$$

The parameters discussed below also have some effect upon the observed areas, but the proportionality given by Equation 17.8 holds well under controlled experimental conditions.

As in TG, the maintenance of thermal equilibrium throughout the DTA system becomes the overriding consideration in obtaining reproducible results. As noted previously, the placement of the thermocouple becomes critical; slight displacements in the position of the thermocouple can contribute to irreproducible results. Placing the thermocouple outside the sample cell in good thermal contact with it ensures consistent positioning, while simplifying the experimental procedure.

The magnitude of the difference in temperature between the exterior and the interior of the sample depends on two factors: the rate of heating (as in TG), and the thermal conductivity of the sample and sample holder. Thus, a metal sample (which has a high thermal conductivity) is close to isothermal, even at high rates of heating.

An apparent solution to the problem of thermal equilibrium is to increase the thermal conductivity of the sample appreciably (e.g., by mixing it with a diluent of high conductivity). This approach has limitations, however: Equation 17.5 shows that the heat produced or absorbed by the reaction will then be partially or completely compensated for by heat flow to or from the environment.

The best solution is to use a cell whose thermal conductivity is low relative to that of the sample. This minimizes the effect described by Equation 17.5 so that a large proportion of the effects described by Equation 17.6 can be measured. Therefore, ceramic sample holders are used in reactions where ΔH is relatively small, and more convenient metallic sample holders are used in reactions involving great amounts of heat. Best results are obtained with well-powdered samples, and uniformity of results is enhanced by consistently using samples of the same particle size and density of packing.

Table 17.2 summarizes the major factors that affect the shape and size of the DTA heating curve. The effects of the atmosphere around the sample are precisely the same as those in TG and may be either a serious problem or a tool to use to analytical advantage.

Reference Materials. The subject of reference materials is important and often neglected. The major requirements are that the reference material be inert over the temperature range of the analysis, that it not react with the sample holder or thermocouples, and that its thermal conductivity match that of the sample. The last item is important since changes in thermal conductivity or heat capacity with temperature result in sloping baselines. Table 17.3 lists some of the more common reference materials.

TABLE 17.2. *Some Common Factors That Influence DTA Heating Curves*

Factor	Effect	Correction or Control
Heating rate	Changes in peak size and position	Use low heating rate
Sample size	Changes in peak size and position	Decrease size or lower heating rate
Thermocouple placement	Irreproducible curves	Use same location for each run
Sample particle size	Irreproducible and erratic curves	Use small, uniform particle size
Thermal conductivity of sample	Changes in peak position	Mix with thermally conductive diluent or lower heating rate
Thermal conductivity of cell	Changes in peak area	Decrease thermal conductivity to increase peak area
Reaction with atmosphere	Changes in peak size and position	Control carefully (can be used advantageously)
Sample packing	Irreproducible curves	Control carefully (affects thermal conductivity)
Diluent	Changes in heat capacity and thermal conductivity	Choose carefully (can be used advantageously)

TABLE 17.3. *Common Diluents and Reference Materials for DTA*

Compound	Approximate Temperature Limit, °C	Reactivity
Silicon carbide	2000	May be a catalyst
Glass beads	1500	Inert
Alumina	2000	Reacts with halogenated compounds
Iron	1500	Crystal change at ~700°C
Iron(III) oxide	1000	Crystal change at 680°C
Silicone oils	1000	Inert
Graphite	3500	Inert (in O_2-free atmosphere)

Diluents. The materials listed in Table 17.3 can also be used as diluents. Naturally, the diluent must also be inert in the presence of the sample.

One purpose of using a diluent has already been discussed: It permits the thermal conductivities of sample and reference to be matched. In addition, it may be used to maintain a constant sample size while the amount of the reacting component is varied; this will decrease the influence of many of the factors listed in Table 17.2. A diluent can also be used where the sample is so small that weighing it directly is inconvenient.

Analytical Calculations

Equation 17.8 predicts a direct proportionality between peak area and mass. Hence, for quantitative analysis the peak area (A) of a sample of known mass (m_k) is compared to that for an unknown sample run under identical conditions:

$$m_{unk} = m_k \left(\frac{A_{unk}}{A_k} \right) \tag{17.9}$$

Similarly, the heat of reaction may be determined by comparison with a sample of known ΔH. Particular caution has to be exercised in determining heats of reaction and sample masses. The constant in Equation 17.8 that relates peak area and ΔH varies with temperature and can cause significant errors. As a result, the known sample should react at the same temperature as the unknown, and the peak areas of the known and unknown should be roughly equal.

Example 17.3. Compound A has a molecular weight of 98.4 and a heat of fusion of 1.63 kcal/mole. Compound B has a molecular weight of 64.3 and melts at approximately the same temperature as compound A. Samples of 500 mg of each yield DTA peak areas of 60.0 and 45.0 cm² for A and B, respectively. What is the heat of fusion of compound B?

Solution: From Equation 17.8,

$$\Delta H_B = \Delta H_A \left(\frac{A_B}{A_A} \right) \left(\frac{m_A}{m_B} \right)$$

Quantities m_A and m_B must be expressed in molar units to compare different compounds:

$$\Delta H_B = 1.63 \text{ kcal/mole} \left(\frac{45.0 \text{ cm}^2}{60.0 \text{ cm}^2}\right)\left(\frac{500/98.4}{500/64.3}\right) = 0.799 \text{ kcal/mole}$$

Calculation of the area under the peak of the heating curve (see Fig. 17.7) can be subject to ambiguity, since, more often than not, the initial and final baselines do not coincide—the thermal conductivity or heat capacity has changed as a result of the reaction. A method for rapidly estimating the area of interest is illustrated in Figure 17.7. Both baselines are extended to a perpendicular line drawn from the maximum of the curve, and the areas under the two halves of the curve are determined and summed to give the total area.

The rates and activation energies of the reactions observed in DTA can be calculated from observed changes in DTA curves as the heating rate is changed; the essential data are the rate of heating b and the temperature at the curve maximum T_{max} (in K). The applicable equation is

$$\frac{d[\ln(b/T_{max}^2)]}{d(1/T_{max})} = -\frac{E^*}{R} \quad (17.10)$$

where E^* = activation energy of the reaction
R = the gas constant

A plot of $\ln(b/T_{max}^2)$ versus $1/T_{max}$ yields a line of slope $-E^*/R$ [1].

The order of the reaction process (first order, second order, etc.) can be determined from the asymmetry of the DTA curve [1] (see Fig. 17.8). The asymmetry of the peak is simply x/y, and the reaction order is estimated from

$$\text{Reaction Order} = 1.26(x/y)^{1/2} \quad (17.11)$$

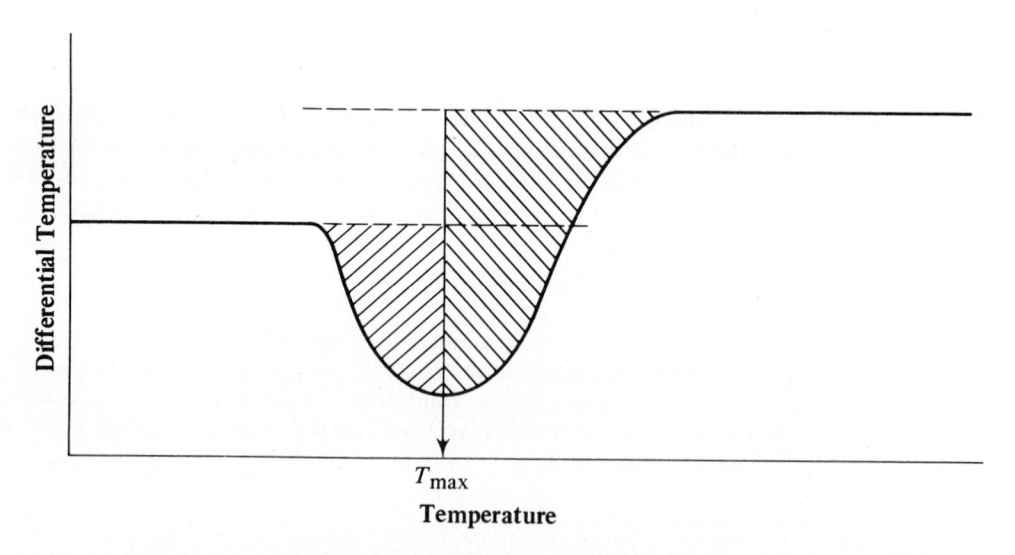

FIGURE 17.7. *Illustration depicting the calculation of DTA peak areas. The displacement of the baseline indicates a change in heat capacity.*

FIGURE 17.8. *Parameters used to calculate the DTA peak asymmetry. Adapted from H. E. Kissinger,* Anal. Chem., *29*, 1702 (1957), *by permission of the publisher. Copyright* © *1975 by the American Chemical Society.*

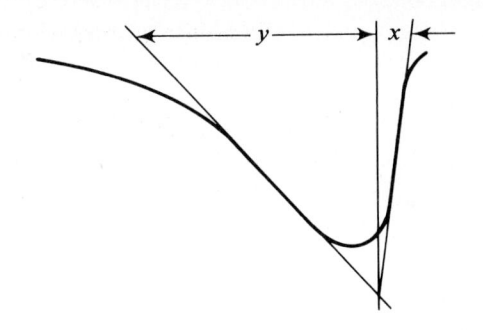

There are many more complete treatments, and interested readers are referred to the Selected Bibliography at the end of this chapter for more detail. The problems associated with kinetic analysis by TG are also characteristic of DTA. In particular, the microscopic state of the solid sample often has a great effect on the apparent rate of reaction.

Heat-Capacity Measurements

Heat-capacity estimates can be made using DTA. In the ideal system, with identical reference and sample materials, the "true" baseline of the instrument is obtained. When a reference and sample of different heat capacities are used, the baseline will not be the same; and an estimate of the heat capacity of an unknown can be obtained by comparing the baseline shift with that for a sample of known heat capacity. A shift in the baseline is almost always observed after a DTA peak because of change in heat capacity of the sample. In addition, some reactions such as the glass transition of polymers yield virtually no DTA peak, but there is a rather sharp shift in the baseline at the transition temperature.

Applications

The general reference books of MacKenzie and Wendlandt summarize a large number of DTA studies and analytical applications. A few of these are outlined below.

Polymer analysis is perhaps the most common application of DTA. Under carefully controlled conditions, the shape of the heating curve indicates both the type of polymer and the method used to prepare it; consequently, not only can the polymer be identified, but often (when production processes differ) the particular manufacturer as well. The "crystallinity" of a polymer determines its physical properties to a great extent. In DTA, there are commonly two peaks, one for the reaction of the crystalline part of the sample and the other for that of the noncrystalline part (these two peaks often overlap). The magnitudes of these peaks can be used to evaluate the percent crystallinity. A significant advantage of DTA in this application is that the untreated polymer can be studied, thus avoiding possible changes caused by pretreatment (e.g., dissolution or grinding) of the sample.

Fuels (e.g., coal) can be evaluated rapidly to determine the source and BTU rating. As with TG, clays and soils can be analyzed using DTA.

Some of the most interesting analyses are of biological materials. Heating curves of such materials (e.g., plant leaves, cell cultures) give characteristic plots; indeed, cell cultures of the same strain of bacteria yield different heating curves, depending on the growth medium. In addition, the calorific value of organic material and foods can be evaluated using DTA.

17.4 DIFFERENTIAL SCANNING CALORIMETRY

General Considerations

In DTA, reactions are observed by measuring the deviation of the sample temperature from the temperature of the reference material. This deviation causes thermal fluxes (Eqn. 17.5) which complicate the theoretical description of the curves and decrease the sensitivity. It would be advantageous to keep the sample and reference at the same temperature and to measure the rate of heat flow into each that is necessary to maintain the constant temperature. This is achieved by placing separate heating elements in the sample and reference chambers; the rate of heating by these elements can be controlled and measured as desired. This is the basis of *differential scanning calorimetry* (DSC).

DSC plots are graphs of the differential rate of heating (in cal/sec) versus temperature (see Fig. 17.9). The area under the peak is directly proportional to the heat evolved or absorbed by the reaction, and the height of the curve is directly proportional to the rate of reaction. Although a proportionality constant similar to k' in Equation 17.8 exists, it is an electrical conversion factor rather than one based on

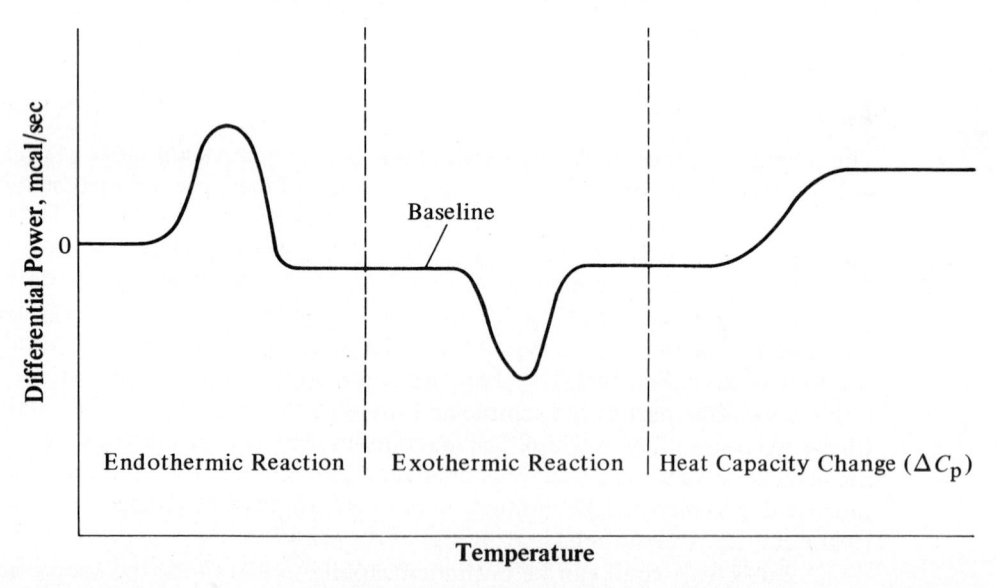

FIGURE 17.9. *Ideal representation of the three processes observable via DSC.*

sample characteristics. That k' is largely independent of temperature is a major advantage of DSC over DTA.

Instrumentation

Figure 17.10 illustrates the circuitry of a differential scanning calorimeter. There are two separate heating circuits, the average-heating controller and the differential-heating circuit. In the average-temperature controller, the temperatures of the sample and reference are measured and averaged and the heat output of the average heater is automatically adjusted so that the average temperature of the sample and reference increases at a linear rate. The differential-temperature controller monitors the difference in temperature between the sample and reference and automatically

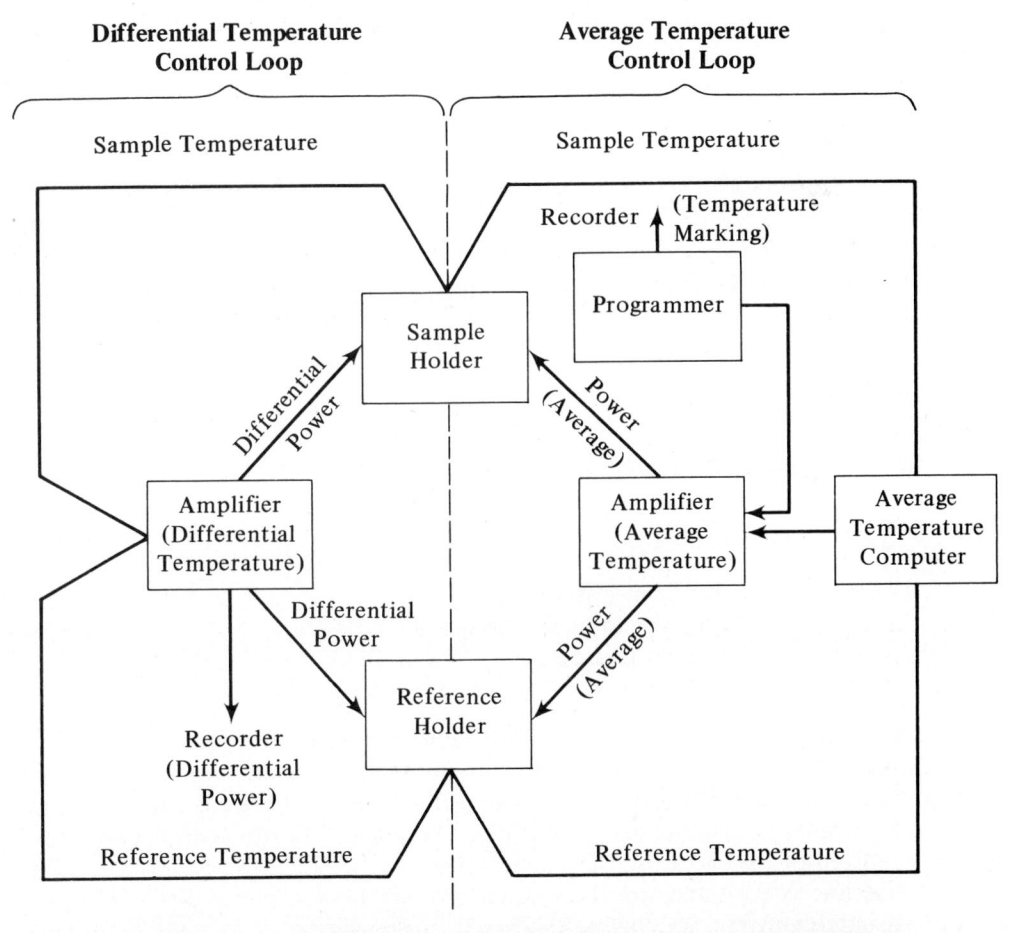

FIGURE 17.10. *Schematic diagram of the DSC apparatus. Adapted from E. S. Watson, M. J. O'Neill, J. Justin, and N. Brenner*, Anal. Chem., *36*, 1233 (1964), *by permission of the publisher. Copyright* © *1964 by the American Chemical Society.*

SEC. 17.4 Differential Scanning Calorimetry 543

adjusts the power to either the reference or sample chambers to keep the temperatures equal. The temperature of the sample is put on the x-axis (time) of a strip-chart recorder, and the difference in power supplied to the two differential heaters is displayed on the y-axis. The power difference is calibrated in terms of calories per unit time.

A simple differential scanning calorimeter can be constructed to monitor endothermic reactions, since heat can be added to the sample compartment without affecting the heating rate. If the process is exothermic, however, the mere addition of heat to the reference will, while keeping the sample and reference isothermal, make the rate of heating nonlinear. The circuit in Figure 17.10 avoids this problem: When an exothermic process occurs, the average-temperature circuit decreases the rate of heating of both the reference and sample equally. In the sample compartment, this decrease in the rate of heating is compensated for by the heat of the reaction, while the differential heater in the reference compartment compensates for the decreased heating by the average heater.

Samples for analysis range in size from 1 to 100 mg and are sealed in a foil or metallic container for direct contact with the heaters and temperature sensors. The sample and reference compartments are well isolated to avoid flow of heat from one to the other, and heat flow to the environment is equalized by careful choice of material and geometry in the compartments. A wide range of heating rates (0.5 to 80°C/min) can be used, and instruments are generally sensitive enough to detect heat evolution or absorption at a rate of less than 1 mcal/sec. Electrical signals are amplified and recorded as in TG and DTA, with microprocessors becoming more frequently used for data acquisition and treatment. The use of sealed sample containers eliminates atmospheric considerations in most cases. Temperatures are monitored using platinum resistance devices.

Experimental Considerations

The amount of heat generated by the differential heaters per unit time is written as

$$P = \frac{dQ}{dt} = i^2 R \tag{17.12}$$

where P = power in watts
Q = quantity of heat in joules
i = current in amperes
R = resistance in ohms

Chemical reactions liberate or absorb heat according to Equation 17.6. Thus, when ΔH is positive (endothermic reaction), the sample heater is energized and a positive signal is obtained; when ΔH is negative, the reference heater is energized and a negative signal is obtained (see Fig. 17.9). The integral of the peak is equal to the heat evolved or absorbed by the reacting sample. In DSC, the reaction ends when the baseline is reestablished. This situation may be contrasted with DTA, where the reaction may end well before the baseline is reached.

Not only is DSC sensitive to processes where there is a finite ΔH, but it is also very sensitive to differences in the heat capacities of the sample and reference. If the sample has a greater heat capacity than the reference, the sample differential heater will be operating even in the baseline region, giving a positive signal; similarly, a higher heat capacity for the reference will yield a negative baseline. A change in the heat capacity of either the sample or reference will be seen as a displacement of the baseline. The difference between the actual baseline and the zero of the instrument (in cal/sec), divided by the heating rate (in °C/sec), is equal to the difference in heat capacities (in cal/°C) between the sample and reference systems.

If the heat capacity of the reference is known, then the heat capacity of the sample can be determined over a wide range of temperatures. There is a great interest in this type of application; for example, changes in structure of many large polymers have a very small ΔH (virtually undetectable by DTA), but a ΔC_p quantitatively measurable by DSC.

The factors in Table 17.2, which have detrimental effects on DTA curves, have minimal effects on DSC curves. In particular, measurements obtained from the total area under the curve (calculation of ΔH and sample mass) are not affected. However, these factors can still have an effect on the rate of reaction and any values calculated from these rates, particularly if large thermal gradients are allowed to develop in the sample or reference.

The rate at which heat is evolved in exothermic reactions must be taken into account in DSC; rapid exothermic reactions may cause the rate of temperature increase of the sample to be greater than the programmed rate of heating, even when both the average and differential heaters are off. A similar problem sometimes occurs with endothermic processes, where a rapid endothermic reaction may cool the sample so severely that the combined maximum heating of the two heaters cannot maintain a linear heating rate and isothermal conditions. Both these situations can be easily rectified by adjusting the heating rate or the sample size.

Analytical Calculations

Virtually every chemical process involves a change in the heat capacity of the sample. When measured by DSC such changes produce a curve similar to Figure 17.7 (except with a y-axis in cal/sec). The area under the DSC curve is determined in the same manner as in DTA. This area is proportional to the amount of heat evolved or absorbed by the reaction, and the heat of reaction is obtained by dividing this by the moles of sample used. If the heat of reaction is known, the moles of sample present can be calculated from essentially the same equation (i.e., the integral of Eqn. 17.6). All determinations should be preceded by an analysis of a standard sample of known mass and ΔH in order to calibrate the particular instrument used.

Processes where ΔH is zero yield no area for the curve (see the heat-capacity change in Fig. 17.9). In this case, the change in the *specific heat* is determined from

$$\Delta C_p \text{ (cal/°C-g)} = \frac{\Delta \text{Baseline}}{mb} \qquad (17.13)$$

where m = mass of the sample
b = heating rate

Applications

Because of the great similarity between DSC and DTA, the analyses previously described and referred to for DTA are amenable to DSC studies.

The unique feature of DSC is the determination of heat capacities (specific heats). As noted, the differential power (in cal/sec) divided by the heating rate (in °C/sec) yields the difference in heat capacities between the sample and the reference (in the baseline regions only). A change in heat capacity is seen by a shift in the baseline. A sharp increase in the baseline of the plot is typical of glass transitions in polymers. By comparing the heat capacity of the sample with the known heat capacity of the standard, the absolute heat capacity of the sample can be calculated. The specific heat is then calculated by dividing the absolute heat capacity of the sample by its mass.

17.5 THERMOMETRIC TITRATION AND DIRECT-INJECTION ENTHALPIMETRY

General Considerations

Early thermometric measurements were laborious and time consuming and not very sensitive, although many excellent thermometric studies were reported up to the early 1950s, when thermometric titration became suitably automated for routine analytical use. Around that time, the thermistor temperature sensor, the constant-delivery pump, and sophisticated thermostatic control were introduced in rapid succession.

Thermometric titration (TT) and *direct-injection enthalpimetry* (DIE) are both calorimetric techniques; the heat evolved or absorbed serves as an indicator of the progress of the reaction. TT and DIE are used for routine analysis and in fundamental research involving the chemical equilibrium, reaction kinetics, and thermochemistry of processes not readily studied by other methods.

TT plots are characteristically graphs of temperature change versus titrant added. DIE yields plots of temperature versus the time following injection of a titrant. Both are illustrated in Figure 17.11. The methods are discussed together in this section because of the many similarities between them.

In TT, temperature changes occur only when titration is in progress and when there is sample reactant present. As a consequence, the start and endpoint of a titration are readily observed, and the number of moles titrated is calculated as in regular titrimetry. By determining the heat capacity of the system under study, heats of reaction can be readily determined. In addition, equilibrium constants can be evaluated under the appropriate conditions.

In DIE, the titrant at the same temperature as the sample is injected rapidly into the sample and the data are obtained as a temperature-versus-time curve. No endpoint is determined, but the magnitude of the temperature change is proportional to the concentration. Also, one can generate kinetic curves to evaluate slow reactions. The speed of analysis is enhanced, and processes with equilibria unfavorable for titration are readily studied by using a large excess of one reactant.

Both TT and DIE are subject to the same restrictions as classical calorimetry. If TT is used simply as an endpoint-detection method, the restrictions also apply, but to a much lesser degree.

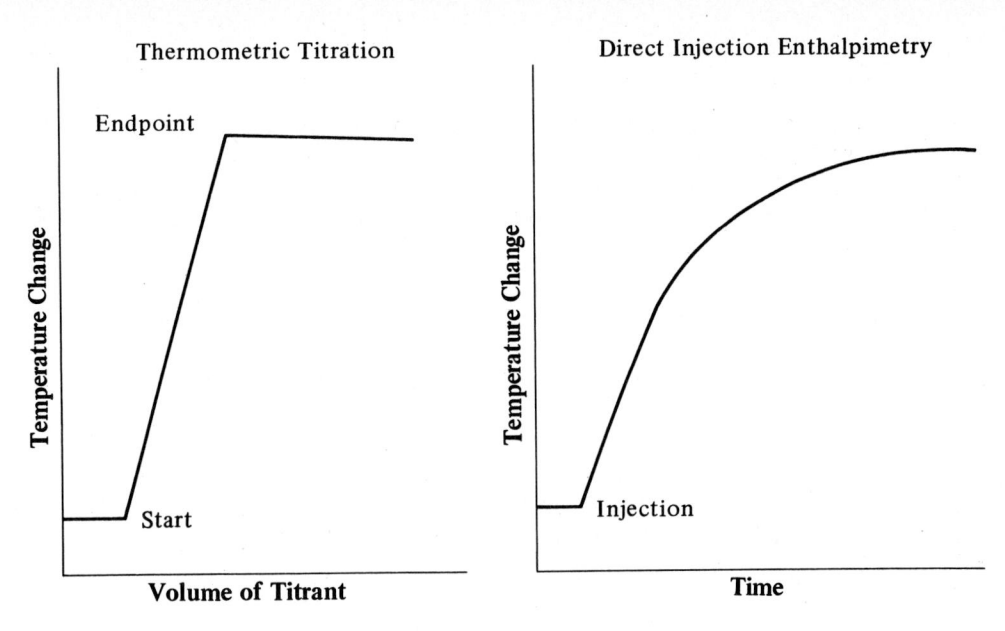

FIGURE 17.11. *Characteristic curves obtained from TT and DIE.*

Instrumentation

Thermometric titrations can be implemented with a buret, a Dewar flask, and a Beckman thermometer, as in early studies. In modern instrumentation, however, automated electronic methods are used to obtain and often to evaluate the data. A typical TT setup consists of (a) a constant-delivery pump, (b) a temperature-control system, (c) an adiabatic cell, (d) a calibration unit, (e) an electronic temperature-sensing system, and (f) an amplifying and data-processing system.

Delivery Pump. A constant-delivery pump permits the time axis of a strip-chart recorder to be used as the volume-of-titrant axis (with a simple conversion factor). Typically, a syringe driven by a synchronous motor (that drives a carriage or screw) is used, and solutions can be delivered at constant rates ranging down to a few microliters per minute. Because of their variable flow rates, the more common peristaltic pumps are not often used for thermometric titrations.

In DIE, the syringe is rapidly emptied at the start of the experiment to deliver the titrant virtually instantaneously into the sample cell.

Temperature-Control Requirements. The temperature control needed in the TT apparatus depends on the results desired. It is often possible to obtain useful endpoints in titrations simply by bringing both the sample and titrant to room temperature. However, for precise calorimetric results, the titrant and sample must be as close to the same temperature as possible. This is the main purpose of the thermostat.

Currently, using modern temperature controllers, it is possible to maintain the temperature of the system at a wide range of set temperatures to a precision of $\pm 0.001\,°C$ or less. As a rule of thumb, the temperature change caused by the reaction observed must be at least as great as the temperature difference between the titrant and sample (i.e., the precision of the thermostat). Consequently, letting the random heat flows of the environment control the temperature of the apparatus is sufficient only for highly exothermic or endothermic processes.

Adiabatic Cell. The "adiabatic" cells used for TT and DIE have widely varying designs. They range from an insulated beaker to a Dewar flask to the highly elegant and efficient Dewar-type cell of Christensen et al. [2] (Fig. 17.12). All are designed to minimize the heat transfer from the cell to the environment, thus maximizing the temperature change observed. When only titration endpoints are of interest, the simplest cell suffices; but if quantities such as heats of reaction, equilibrium constants, or kinetic parameters are sought, better cells are necessary.

These cells may be evaluated in terms of their *heat-leak modulus*, which is defined by Newton's law of cooling:

$$\frac{dT}{dt} = -C(T_c - T_e) \tag{17.14}$$

where C = heat-leak modulus
 T_c = temperature of the cell
 T_e = temperature of the environment

The Christensen cell has the very low heat-leak modulus of 1.1×10^{-3} min^{-1}. Another factor is the mass of the cell and its contribution to the overall heat capacity and response; the better cells have thin walls to minimize the heat capacity and maximize the speed of response to temperature changes.

Figure 17.12. *Efficient adiabatic cell used for TT and DIE. Adapted from J. J. Christensen, R. M. Izatt, and L. D. Hansen, Rev. Sci. Instr., 36, 779 (1965), by permission of the senior author.*

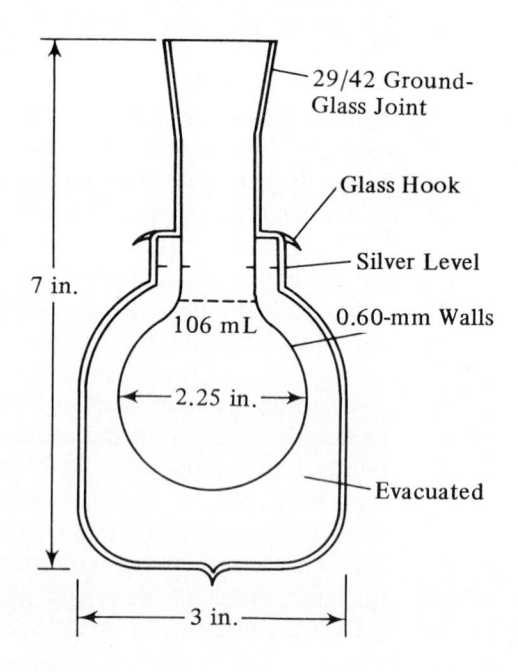

Whereas it is relatively easy to maintain the titrant at any particular temperature by having it in good thermal contact with the thermostat bath, the cell temperature is not controlled easily in this manner, and an external means is usually employed to bring the cell quickly to thermal equilibrium. This is done using the calibration heater.

Calibration Unit. The calibration circuitry has two purposes: to determine the heat capacity of the system, and to control the temperature in the cell itself.

The heat evolved or absorbed is calculated from the temperature change using the relation

$$\Delta Q = \Delta T C'_p \tag{17.15}$$

Here, C'_p is the heat capacity of the system, readily measured as the amount of heat necessary to raise the cell temperature a known amount by electrical-resistance heating:

$$C'_p(\text{J}/°\text{C}) = \frac{i^2 R t}{\Delta T} (\text{W-sec}/°\text{C}) \tag{17.16}$$

where t = time of heating
i = current (measured)
R = resistance (known)

Division by 4.184 J/cal permits conversion to calories per °C. The voltages V_s and V_h across a standard resistor (R_s) and the heater (R_h) connected in series are the measurements of interest, since $i^2 R_h = V_s V_h / R_s$. The heater can be used to advantage in hastening the cell to thermal equilibrium with the thermostat bath and titrant by heating the cell and its contents to a temperature very close to that at which the thermostat is set.

Temperature-Sensing System. Temperature sensing is the heart of the TT technique. The principal temperature sensors used are thermistors. A thermistor is a temperature-sensitive semiconductor whose resistance obeys the equation

$$R_T = A e^{B/T} \tag{17.17}$$

where A and B are constants whose values depend on the nature of the thermistor. In general, thermistors decrease in resistance as the temperature increases, by 3 to 6% per °C. If the thermistor is incorporated into a Wheatstone bridge, the off-balance potential caused by the change in R_T can be recorded on a strip-chart recorder; when the input potential of the Wheatstone bridge is about 1 V, the output voltage changes by approximately 10 mV/°C (thermocouples produce a voltage of about 10 μV for a similar temperature change).

Neither the thermistor response nor the Wheatstone bridge readout is linear with temperature. However, it has been found that the nonlinearity is unimportant over a small enough temperature range (less than 0.1°C). Other factors that make thermistors ideal for TT and DIE are their small size, fast response to temperature change, and (when encapsulated with glass) inertness to most chemicals.

Amplification and Recording. Although the thermistor is already as sensitive as a thousand-junction thermocouple would be, it is often advantageous to amplify the signals obtained. Using a DC amplifier, it is possible to obtain good signals for temperature changes of the order of 10^{-4}°C or less. An AC Wheatstone bridge with a lock-in amplifier can detect temperature changes of the order of 10^{-6}°C [3].

The two most popular data-acquisition systems are the strip-chart recorder and the digital data-storage system. The latter is used when a great deal of data processing is anticipated.

Experimental Considerations

Figure 17.13 represents an idealized TT curve. Region 1 is the baseline. Ideally horizontal, in practice it has a finite slope as a result of frictional heat added by stirring, resistive heat added by the thermistor (Eqn. 17.12), and the transfer of heat from the cell to the thermostat (Eqn. 17.14). If frictional and resistance heating is constant and equal to W, then the slope in region 1 is

$$\frac{dT}{dt} = -C(T_c - T_e) + W \qquad (17.18)$$

The slope in region 2 is due to the same effects, plus the following: the temperature change generated by the reaction, the heat of dilution of the reactants (ΔH_D), and

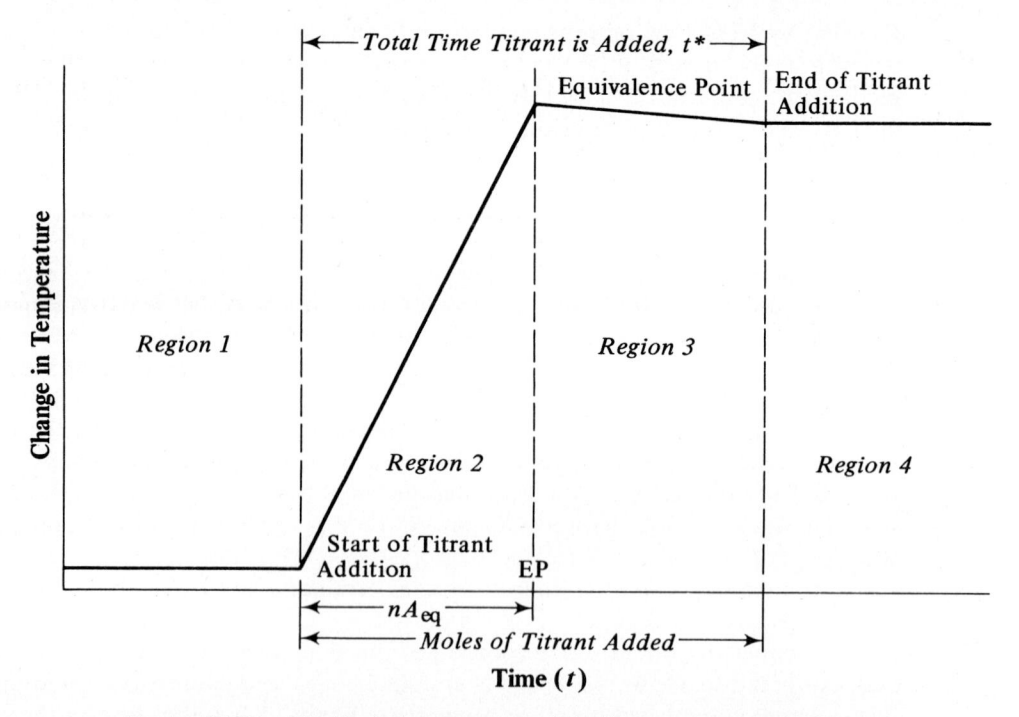

FIGURE 17.13. *Idealized representation of the four major regions of the TT curve. With a constant-delivery pump, the x-axis can be in units of time or moles of titrant.*

the difference in temperature between titrant and sample after the start of titration (ΔT_R). This may be expressed as

$$\frac{dT}{dt} = -C(T_c - T_e) + W + \left(-\frac{\Delta H}{C'_p}\right)\left(\frac{dn_p}{dt}\right) + \frac{\Delta H_D}{C'_p} + \Delta T_R k \qquad (17.19)$$

where $k = $ a constant

In region 3, where the equivalence point has been passed, the slope of the curve is described by

$$\frac{dT}{dt} = -C(T_c - T_e) + W + \frac{\Delta H_D}{C'_p} + \Delta T_R \qquad (17.20)$$

When no further titrant is added (region 4), the slope again obeys Equation 17.18. (There is usually some rounding at the equivalence point in real titrations.)

These expressions can be combined to obtain the heat of reaction ΔH and, if the production of product (n_p) is equilibrium controlled, the equilibrium constant of the reaction can also be calculated [4].

In DIE, temperature-versus-time curves indicate the progress of the reaction. Kinetic processes can be evaluated (usually for reactions having half-reaction times greater than 5 sec), as can heats of reaction from the total temperature change. The general equation for a DIE curve is

$$\frac{dT}{dt} = \frac{-\Delta H}{C'_p}\left(\frac{dn_p}{dt}\right) + W \qquad (17.21)$$

where dn_p/dt is governed by the appropriate rate expression. An obvious advantage of this type of kinetic analysis is that the system is not disturbed (and is continuously monitored) in obtaining the data.

Equations 17.18 through 17.21 are straightforward, but often it is not necessary to solve them rigorously to obtain useful and reliable results (see the following section). In applying the equations, however, there are a few points to consider. First, the heat capacity of the system in TT is continuously changing in regions 2 and 3; thus, C'_p in Equation 17.19 is not truly a constant. As the volume of the system increases during the titration, the change in temperature per unit time decreases, even though the amount of heat released remains constant; this can cause a large error if not corrected [5]. The correction is made by replotting the entire curve in terms of Q (cal of heat evolved)—instead of T—versus volume of titrant added. To obtain Q, the change in temperature at each point on the curve is multiplied by the heat capacity (C_p) at that point. The heat capacity at every point in regions 2 and 3 can be determined by measuring C_p in regions 1 and 4 of the curve, obtaining C_{p1} and C_{p4}. At any point t in regions 2 and 3, the heat capacity is given by

$$C_{pt} = C_{p1} + \left(\frac{t}{t^*}\right)(C_{p4} - C_{p1}) \qquad (17.22)$$

where t^* is the total time interval over which titrant is added (this assumes a linear increase in heat capacity during the course of titration). The curve of heat evolved (Q) versus volume of titrant is called an *enthalpogram*. This procedure is tedious, and can be avoided by using a titrant that is 100 or more times as concentrated as the sample [6]; the term $(C_{p4} - C_{p1})$ in Equation 17.22 then approaches zero.

DIE curves are equivalent to enthalpograms, since the heat capacity does not change significantly once the reactants are mixed.

Analytical Calculations

When the concentration of either the titrant or the sample is known, the volume added to reach the endpoint yields the concentration of the unknown. Heats of reaction are obtained from the rigorous solution of Equations 17.18 through 17.20 [4].

Example 17.4. A thermometric titration of acid A with base B was performed, and a curve similar to that in Figure 17.13 was obtained. The slopes of the four regions of the curve were 1.0×10^{-5}, 8.0×10^{-4}, -1.0×10^{-5}, and -0.5×10^{-5} °C/sec, respectively. The overall temperature change was 0.100°C. Prior to the experiment, the heat capacity of the cell was determined to be 1.000 cal/°C. The titration rate was 6.0×10^{-8} moles B per second. In addition, it was found that under identical experimental conditions, the titration of B into pure water gave a slope of 2.0×10^{-5} °C/sec. Use these data to calculate the heat of reaction.

Solution: Using Equation 17.18, the slope in region 1 corresponds to the value W, since $T_c = T_e$. The same equation applies to region 4, but in this case the constant C can be evaluated since W and $(T_c - T_e)$ are known. The value of C is 1.5×10^{-4} sec^{-1}. The heat-of-dilution factor in Equation 17.20 is the difference between the slope of the titration of B into pure water and the slope in region 1:

$$\frac{\Delta H_D}{C'_p} = (2.0 \times 10^{-5}) - (1.0 \times 10^{-5}) = 1.0 \times 10^{-5} \text{ °C/sec}$$

Now it is possible to solve Equation 17.20 for T_R, giving a value of -1.5×10^{-4} sec^{-1}. Finally, all of the above information is used in Equation 17.19, to obtain

$$8.1 \times 10^{-4} = \frac{-\Delta H}{C'_p}\left(\frac{dn_p}{dt}\right)$$

The term dn_p/dt is equal to the titration rate, and C'_p was given as 1.000 cal/°C. The heat of reaction is then: $\Delta H = (-8.1 \times 10^{-4})/(6.0 \times 10^{-8}) = -13.5 \times 10^3$ cal/mole $= -13.5$ kcal/mole.

In the above data, the terms in units of °C can be replaced by any other value proportional to the temperature (e.g., recorder deflection); similarly, the terms in units of seconds can be replaced by any measure that is directly proportional to time. Equation 17.19 can also be integrated and then evaluated. This requires a knowledge of the variation of T_c and C'_p with time in region 2.

Instead of using the equations, heats of reactions can be estimated graphically to a few percent if the change in temperature, the heat capacity at the midpoint of the titration curve, and the moles of product formed (not the molarity) are known. For example, consider the titration curve shown in Figure 17.13. The temperature change at the midpoint of the reaction is obtained by extrapolating regions 1 and 3 into region 2. The change in temperature at the midpoint of titrant addition is measured from these extrapolations. At this point, the heat capacity is the average of the heat capacities in regions 1 and 4. The moles of product can be calculated from the titration rate, the equivalence-point time, and the molarity of the titrant.

Determination of Equilibrium Constants

Estimates of equilibrium constants can be made from TT curves when there is distinct curvature near the equivalence point. Taking the general reaction

$$A + B \rightleftarrows AB$$

at the equivalence point, the analytical concentrations (A) and (B) can be determined. The equilibrium concentrations [A] and [B] are then calculated as

$$[AB] = \frac{h}{h_t}(A) \tag{17.23}$$

$$[A] = (A) - [AB] \tag{17.24}$$

$$[B] = (B) - [AB] \tag{17.25}$$

These concentrations can then be combined to obtain the equilibrium constant. Figure 17.14 illustrates the meaning of h and h_t. As the equilibrium constant increases, the degree of curvature decreases and h approaches h_t.

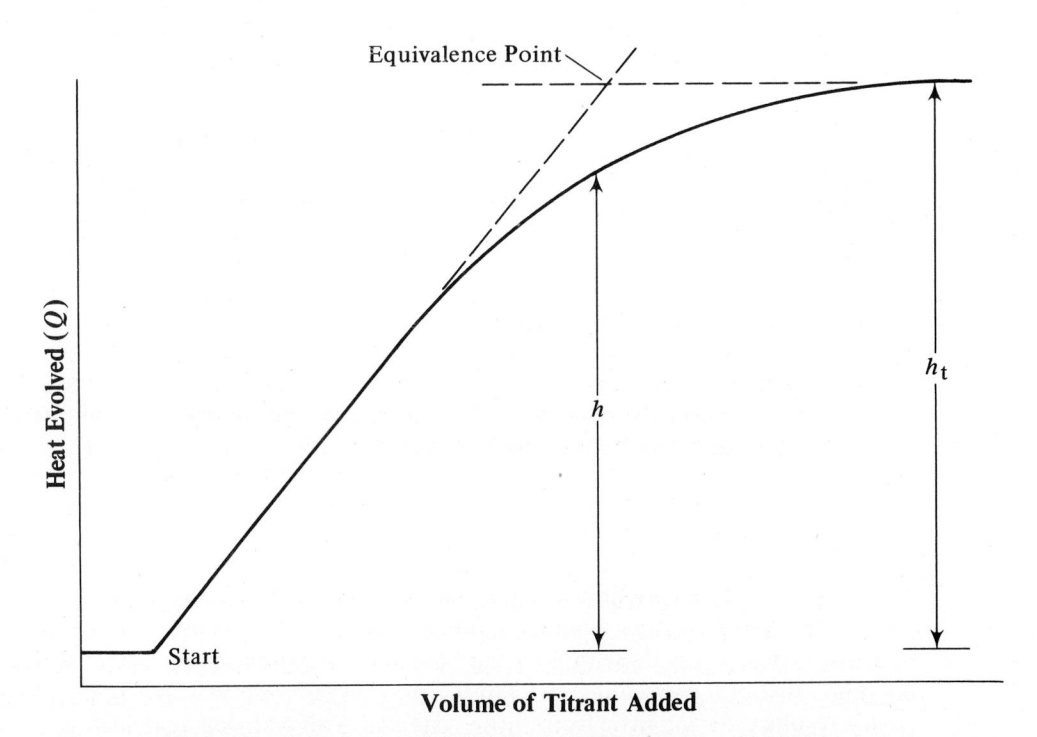

FIGURE 17.14. *Equilibrium curvature of enthalpograms, illustrating parameters h and h_t used to estimate equilibrium constants.*

Example 17.5. Figure 17.14 is a titration curve where chemical equilibrium causes curvature near the endpoint. An estimate of the equilibrium constant is made by measuring h and h_t as diagrammed. The concentration of the sample must be known. Assume that the reaction of a metal (M) with a ligand (L) takes place to form the complex ML. The sample concentration is given to be 0.0100 M initially. Calculate the stability constant of the complex.

Solution: At the endpoint, the following relationships hold:

$$C_L = 0.0100 = [ML] + [L]$$
$$C_M = 0.0100 = [ML] + [M]$$

Then,

$$[ML] = C_M\left(\frac{h}{h_t}\right) = C_L\left(\frac{h}{h_t}\right)$$

$$[L] = [M] = C_M - [ML] = C_L - [ML]$$

The stability constant is

$$K_s = \frac{[ML]}{[M][L]} = \frac{C_M(h/h_t)}{(C_M - [ML])^2}$$

If h and h_t are measured to be 77.8 and 87.3, respectively, then

$$K_s = \frac{(0.0100)(77.8/87.3)}{[0.0100 - 0.0100(77.8/87.3)]^2} = 7.53 \times 10^3$$

The units of h and h_t cancel in the above expressions, and any appropriate and convenient measure of their relative magnitudes may be used (e.g., °C, cm, mV). This example has demonstrated the calculation at the equivalence point. Similar calculations can be done wherever the actual curve departs significantly from the extrapolated curve. An estimate of the precision of the results can be obtained by performing these calculations at several points.

Once the heat of reaction and equilibrium constant have been determined, the entropy change, ΔS, of the reaction can be calculated from

$$-RT \ln K_{eq} = \Delta H - T\Delta S \tag{17.26}$$

Often, titrations can be done with samples of 10^{-2} M or less. When dilute solutions are used, it may be safely assumed that the measured thermodynamic parameters are essentially the same as in the standard state of infinite dilution (i.e., $\Delta H_{mean} \approx \Delta H^0$, etc.).

Applications

Some of the analyses possible using TT or DIE are listed in Table 17.4.

An almost classical example of the advantages of TT is in the titration of boric acid ($K_a = 6.4 \times 10^{-10}$) with a strong base. This titration is impossible by classical methods without pretreating the sample. However, as shown in Figure 17.15, TT yields results differing little from those obtained with a strong acid such as HCl, the reason being that the values of ΔH for the two processes are essentially the same, whereas those for ΔG and ΔS are significantly different.

TABLE 17.4. *Reaction Types Amenable to Thermometric Enthalpy Titration*

Reaction Type	ΔH_r, kcal/mole[a]	Precision
Neutralization		
Strong acid + strong base	−13.5	±0.1%
Weak acid + strong base	−13.5 to −4	±0.1–5%
Weak base + strong acid	−13.5 to −4	±0.1–5%
Polyprotic or basic systems	−14 to −4	±0.1–10%
Oxidation-Reduction		
Inorganic[b]	−40 to −10	±0.1%
Organic[b]	−40 to 0	±0.1–10%
Complexation		
EDTA[b]	−15 to +10	±0.1–5%
Other[b]	−20 to +20	±0.1–10%
Precipitation		
Inorganic[b]	−20 to −5	±0.1–10%
Organic[b]	−20 to −5	±0.1–10%
Heats of Mixing		
Dilution of inorganic ions	−15 to +10	—
Organic solvents	−15 to +15	—

a. ΔH_r is the approximate overall heat of the process; 1 cal = 4.184 J.
b. These processes may have slow kinetics which severely affect the results.

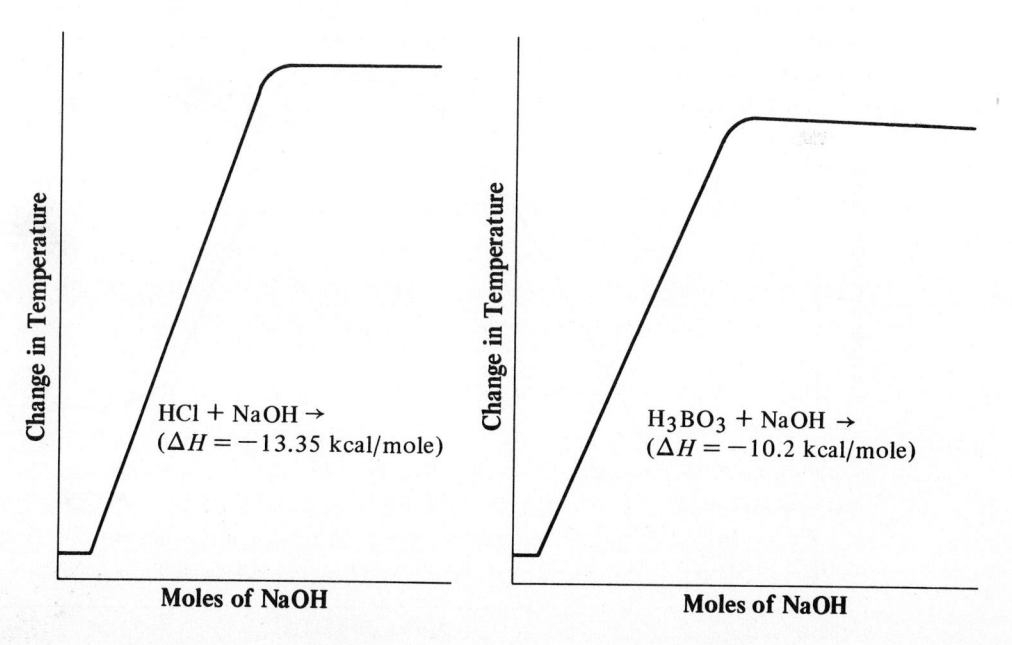

FIGURE 17.15. *Comparison of the titration of H_3BO_3 and HCl with NaOH, observed thermometrically.*

Analysis of mixtures is possible when the two species have different equilibrium constants and heats of reaction with the titrant. This occurs, for example, in the titration of a mixture of calcium and magnesium with EDTA. Calcium ($K_f = 10^{11}$) reacts first and exothermically ($\Delta H = -5.7$ kcal/mole); magnesium ($K_f = 10^{9.1}$) reacts second and endothermically ($\Delta H = +5.5$ kcal/mole). This titration is illustrated in Figure 17.16.

It is also possible to titrate biochemical species. Antibodies have been titrated with antigen, and enzyme-substrate mixtures have been titrated with appropriate coenzymes. Proteins are readily titrated with acid or base, or precipitated with phosphotungstic acid, yielding very informative TT curves [7, 8].

DIE permits rapid analysis. For instance, SO_2 and CO_2 in air [8] can be determined by injecting air samples into concentrated KOH. Sharp temperature changes or pulses indicate the presence of reactants, and the magnitude of the temperature change gives the concentration. Analysis time is very short (approx. 3 min) for this type of determination, and the precision and accuracy, although not as good as with TT, are acceptable.

Kinetic studies of the hydrolysis of organic nitrates and of esters have been made by DIE. Enzyme reactions have also been studied. In all cases, a continuous recording of the process is obtained, and the system is not disturbed by sampling. This can be important in sensitive biochemical reactions.

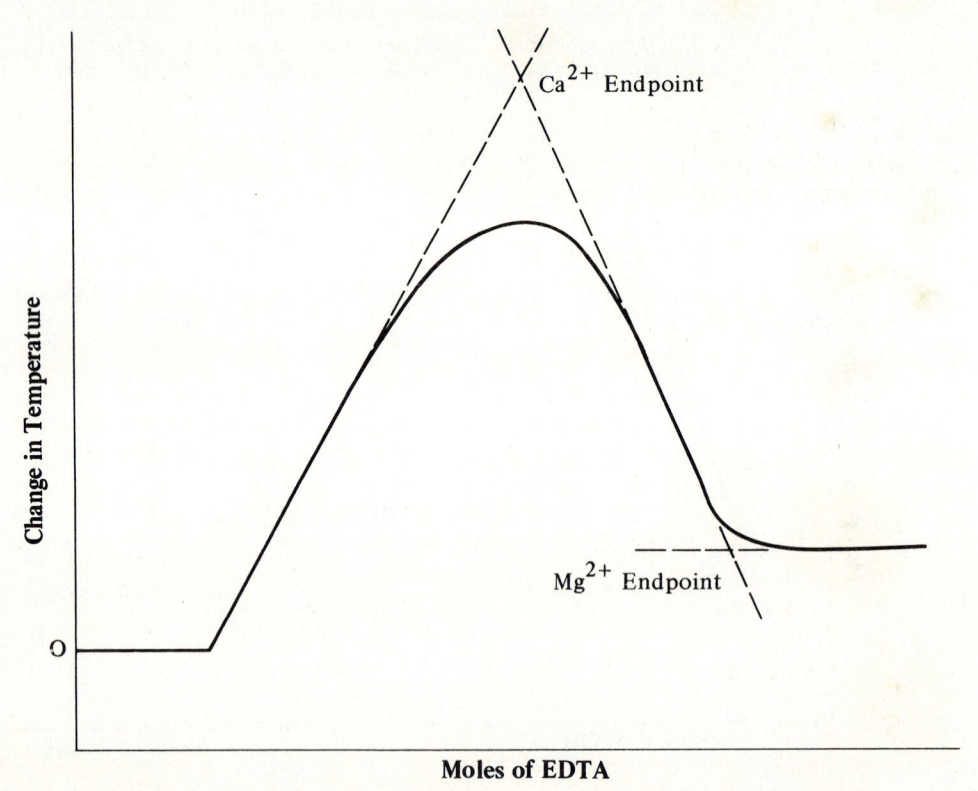

Figure 17.16. *Thermometric titration of a mixture of Ca^{2+} and Mg^{2+} with EDTA.*

SELECTED BIBLIOGRAPHY

BARTHEL, J. *Thermometric Titrations*. New York: John Wiley, 1975.

DUVAL, C. *Inorganic Thermogravimetric Analysis*, 2nd ed. Amsterdam: Elsevier, 1963.

GRIME, J. K. "Enthalpimetry—A Change of Emphasis," *Trends Anal. Chem.*, *1*, 22 (1981).

JESPERSEN, N. D., ed. *Biochemical and Clinical Applications of Thermometric and Thermal Analysis*. Amsterdam: Elsevier, 1982.

JORDAN, J., GRIME, J. K., WAUGH, D. H., MILLER, C. D., CULLIS, M. M., and LOHR, D. "Enthalpimetric Analysis," *Anal. Chem.*, *48*, 427A (1976).

MACKENZIE, R. C. *Differential Thermal Analysis*, Vol. 1. New York: Academic Press, 1970.

VALLET, P. *Thermogravimetrie*. Paris: Gauthier-Villars, 1972.

VAUGHAN, G. A. *Thermometric and Enthalpimetric Titrimetry*. London: Van Nostrand Reinhold, 1973.

WENDLANDT, W. W. *Thermal Methods of Analysis*, 2nd ed. New York: Interscience, 1974.

REFERENCES

1. H. E. KISSINGER, *Anal. Chem.*, *29*, 1702 (1957).
2. J. J. CHRISTENSEN, R. M. IZATT, and L. D. HANSEN, *Rev. Sci. Instr.*, *36*, 779 (1965).
3. E. B. SMITH, C. S. BARNES, and P. W. CARR, *Anal. Chem.*, *44*, 1663 (1972).
4. J. J. CHRISTENSEN, J. RUCKMAN, D. J. EATOUGH, and R. M. IZATT, *Thermochim. Acta*, *3*, 203, 219, 233 (1972).
5. N. D. JESPERSEN and J. JORDAN, *Anal. Lett.*, *3*, 323 (1970).
6. P. W. CARR, *Thermochim. Acta*, *3*, 427 (1972).
7. E. B. SMITH and P. W. CARR, *Anal. Chem.*, *45*, 169 (1973).
8. P. G. ZAMBONIN and J. JORDAN, *Anal. Chem.*, *41*, 437 (1969).

PROBLEMS

1. Define the following terms: (a) diluent; (b) thermal conductivity; (c) heat capacity; (d) pyrolysis; (e) thermogram; (f) dynamic atmosphere.

2. Differentiate between temperature and heat.

3. Melting of any compound is always endothermic. Why?

4. (a) Show mathematically the direction in which the reaction temperatures will move (i) when the atmosphere contains the gaseous product, and (ii) when the atmosphere contains the gaseous reactant. (b) Show how the stoichiometric coefficient of the gaseous reactant can be deduced by varying the atmosphere.

5. If the heat capacity of a sample changes with temperature, what does this mean in terms of ΔG^0, ΔH^0, and ΔS^0?

6. List some of the factors that influence (a) TG curves, (b) DTA curves, and (c) DSC curves, indicating which are most important for the various techniques.

7. What is a self-generated atmosphere? What effect may it have on DTA and TG?

8. A TG curve shows that a compound will not start to decompose until the temperature reaches 150°C. However, upon storage overnight at 140°C, this same compound decomposes totally. Suggest why.

9. Comment on the advisability of using DTA for routine determinations of melting and boiling points.

10. The DSC curve for an inorganic complex exhibits an endotherm at 375°C. A TG trace shows no weight loss at this temperature. What transition may be occurring?

11. Why is DTA more sensitive than TA?

12. On heating, a sample of bismuth gives an endothermic DTA peak at 270°C, Whereas on cool-

ing, it gives an exothermic peak at 257°C. What is occurring?

13. Describe what is meant by the crystalline and glass character of a polymer. Why are thermal methods well suited to the analysis of polymers?

14. Suggest three analyses that can be done by DTA or DSC that would be more difficult or impossible by other methods.

15. Describe some problems involved in the determination of the BTU content of coal by DSC or DTA.

16. Computers are becoming integral parts of all instrumentation. Suggest the three best ways in which a computer could aid in thermal analysis and justify your choices.

17. Why is the TT curve for boric acid so sharp (Fig. 17.15), although the ionization constant indicates that it should be otherwise? (Hint: There is another chemical process taking place.)

18. Ignoring the heat capacity of the adiabatic cell, show that ΔT is independent of the sample volume for a given thermometric titration. What effect does the cell have?

19. If a reaction is slow, what sort of error (positive or negative) can be expected in calculating unknown concentrations in a thermometric titration?

20. Suggest three analyses that can be done by TT or DIE that would be more difficult or impossible by other methods.

21. Sketch a family of TG curves illustrating the effect of heating rate.

22. Figure 17.1 is a typical TG curve. What would an ideal curve look like? (*Hint*: See footnotes.)

23. Using data in handbooks, calculate the TG curve for the thermal decomposition of 100 mg of urea (see Merck Index). Assume all processes are quantitative.

24. From the information in the text, draw a fully labeled diagram of the TG curve obtained by heating to 1200°C a mixture of 50 mg of $CaC_2O_4 \cdot H_2O$ and 50 mg of $BaC_2O_4 \cdot H_2O$. Calculate the magnitude of all weight losses.

25. Sketch a DTA curve for the process in Problem 24. Assume no additional reactions.

26. Phase diagrams can be deduced from DTA curves. A simple phase diagram is given below. Sketch the heating curves obtained when the mole percent of solid A is 0, 25, 50, 75, and 100%. Assume the heat of fusion of solid A is twice that of solid B.

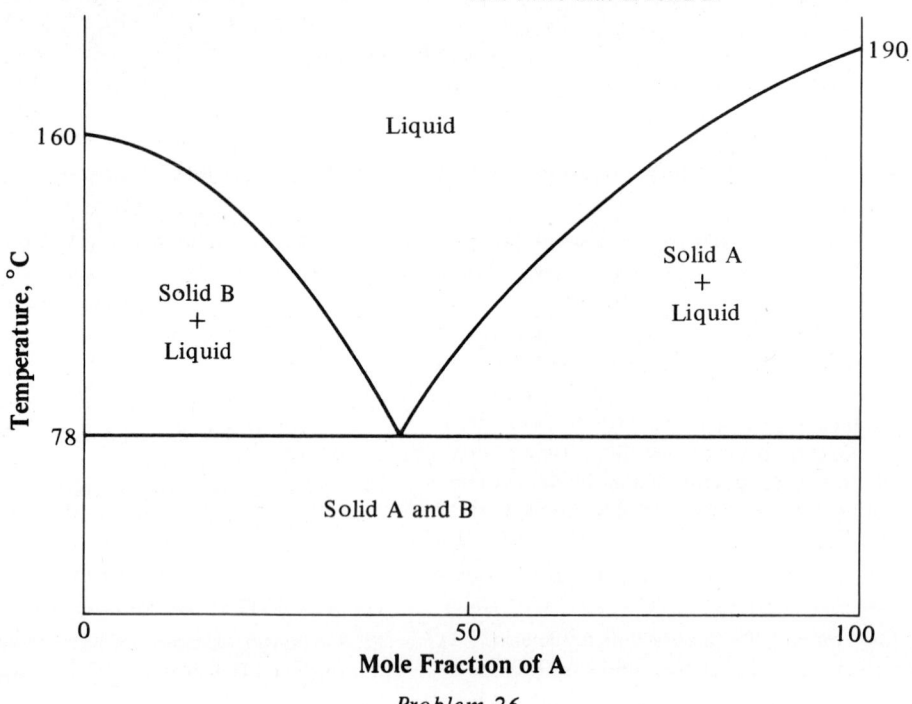

Problem 26

27. The heat of ionization of tris-(hydroxymethyl)-aminomethane (TRIS) is +11.45 kcal/mole. (a) Sketch the TT curve of 50.0 mL of 0.01 M TRIS with 1 M HCl. (b) Sketch the titration curve of the resulting cation with 1 M NaOH. (c) Estimate the temperature change for each titration.

28. A hydrate of Na_2HPO_4 weighing 45.0 mg decreases to a weight of 35.7 mg after heating to 150°C. Calculate the number of waters of hydration.

29. A compound composed of copper(II), ammonia, and chloride is subjected to TG analysis. A 25-mg sample of this compound decreases in weight to 14.2 mg. What formula may be assumed if the loss is all ammonia?

30. Determine the area of the peak in Figure 17.7 by (a) planimetry, (b) counting squares on transparent graph paper, (c) the cut-and-weigh method (copy the curve first), (d) triangulation, and (e) any other method.

31. Figure 17.16 looks like a DSC or DTA curve. Assuming that it is, determine its area by the methods suggested in Problem 30.

32. A 15.4-mg sample of a polymer undergoes a transition wherein the DSC baseline shifts from 4.22 to 8.80 mcal/sec at a heating rate of 10.0°C/min. What is the change in heat capacity of the sample? If its original heat capacity was 2.73 cal/°C-g, what is the new heat capacity?

33. The heat of fusion of naphthalene is 4.63 kcal/mole at 80°C, and the DTA peak observed using 100 mg of sample is 36.3 cm². Water has a heat of fusion of 1.43 kcal/mole at 0°C. What should be the peak area for 100 mg of ice under the same conditions? From experimental considerations, would it be expected to be slightly larger or smaller? Why?

34. Silver nitrate is thermally stable up to 473°C, at which point NO_2 and O_2 are gradually lost, leaving a residue of metallic silver at about 608°C. On the other hand, $Cu(NO_3)_2$ decomposes below 470°C in two steps to CuO, which is stable up to at least 950°C. Suggest a method for determining the percent composition of the alloy formed by heating an unknown mixture of silver and copper nitrates above 950°C.

35. Using the data in Figure 17.15, determine the heat of ionization of boric acid.

36. (a) Calculate the heat capacity of a TT system (cal/°C system) given the following data: $\Delta T = 0.0235°C$; $V_s = 1.234$; $V_h = 1.876$; $t = 64.3$ sec; $R_s = 10.003$. (b) What is the resistance of the heater R_h? (c) Is the answer too high or too low if the heater leads have an appreciable resistance? Why?

37. Calculate K, ΔG, ΔH, and ΔS at 25°C for the reaction $M + L \rightleftarrows ML$, given the following data and the tabular material below: sample concentration (L) = 1.00×10^{-2} M; sample volume = 100.0 mL; titrant (M) = 1.00 M; titration rate = 0.0400 mL/sec.

Time, sec	Heat Evolved, cal
5.0	1.95
10.0	3.87
15.0	5.73
20.0	7.42
25.0	8.68
30.0	9.30
35.0	9.56
40.0	9.69
50.0	9.89
60.0	9.97
70.0	10.0
80.0	10.0

Assume all appropriate corrections have been made. Sketch the titration curve.

38. Using the data in Problem 37, calculate the equilibrium constant, assuming the product is ML_2. Assume the sample concentration (L) = 0.0200 M.

39. For the data in Problem 37, calculate K at 15, 20, 25, 30, and 35 sec assuming the stoichiometry first in Problem 37 and then in Problem 38. Based on your calculations, which stoichiometry is correct?

CHAPTER **18**

Kinetic Methods

Horacio A. Mottola
Harry B. Mark, Jr.

Chemical analysis, the branch of analytical chemistry based on chemical reactions, can be divided into two approaches to measurement: *thermodynamic* and *kinetic*. The rate profile shown in Figure 18.1 is exhibited by all types of chemical reactions regardless of their complexity. The *kinetic* and *equilibrium* regions illustrated reflect the two complementary approaches with which modern analytical chemists are confronted: (a) signal measurements made in systems at equilibrium (thermodynamic approach), and (b) signal measurements made under dynamic conditions in systems approaching equilibrium (kinetic approach). The thermodynamic approach involves changing the equilibrium conditions of the system to render thermodynamically unfavorable all reactions except the one of analytical interest. The kinetic approach, on the other hand, involves adjusting, or simply taking advantage of, the difference in reaction rates of the components of the mixture in order to measure the reactions of the desired species.

The concept of using reaction-rate parameters to determine the initial analytical concentration of reactants dates back over a half of a century to the early literature in biochemistry, radiochemistry, and gas-phase diffusion; furthermore, among all the analyses performed in all the laboratories around the world, the number carried out by kinetic-based methods probably exceeds that carried out by thermodynamic methods and direct instrumental measurement combined. This comes as a surprise at first, until one considers the large numbers of enzymatic and other determinations done on multichannel autoanalyzers used in clinical laboratories. Most of these rapid automated instruments use kinetic methods.

The expanded use of automated continuous-flow sample processing in clinical and industrial laboratories is also responsible for the increasing role that kinetics plays in contemporary analytical chemistry. In these continuous-flow procedures, detection occurs in the unsegmented continuous-flow stream while the system is attaining equilibrium by a physical or chemical process or both; thus, they are kinetic based. Recent developments in instrumentation (particularly microprocessor-based instruments) have contributed to a decrease in the dominance of equilibrium-based methodology. Consequently, a chapter on kinetic methods is appropriate in this text.

Figure 18.1. *Reaction-rate versus time profile.* R_f and R_r *represent the forward and reverse reaction rates, respectively. Region A is the* kinetic *region; region B is the* equilibrium *region, in which* $R_f = R_r$.

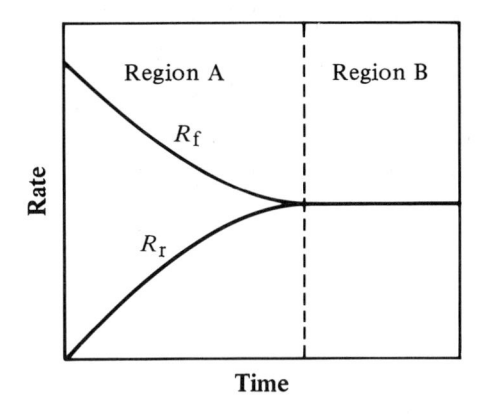

18.1 TYPES OF KINETIC METHODS

Table 18.1 presents three classification schemes that characterize the field of kinetics and its approaches. These classifications reflect the large variety of chemical situations, sample processing, and measurement and data treatment that make kinetic-based methodology a rich analytical approach with a unique nomenclature. Some classifications involve particular areas of application or author preference. Pardue [1], for instance, has classified the kinetic methods used in clinical chemistry into two categories: *fixed-sensor signal* and *variable-sensor signal*. Each of these two groups is subdivided into *one-point*, *two-point*, and *multipoint* methods, which are subdivided further according to what types of blanks are used, what variables are measured, and how the collected data are used in computations.

Kinetic-based methods involving catalyzed reactions are by far the most commonly used. This is in large part due to the extensive application of enzyme-catalyzed reactions in clinical (analytical) chemistry. Moreover, nonenzymatic catalytic methods have been extensively recorded in the literature on kinetic methods of determination; redox systems and catalysts of transition-metal ions dominate this area. The relevance of enzymatic methods to clinical chemistry becomes obvious when one realizes enzyme (activity) determinations constitute about 20 to 25% of the workload in clinical laboratories, and that relevant biological species such as glucose, urea, creatinine, cholesterol, uric acid, and bilirubin are routinely determined in biological fluids by enzyme-catalyzed kinetic approaches. For substrate determinations, enzymes are remarkable analytical reagents because of their high selectivity (occasional specificity) and their capacity for self-regeneration via the catalytic cycle. The popularity of nonenzymatic catalytic methods results from the low limits of detection possible, the extensive use of well-accepted monitoring approaches (e.g., photometric), and the almost endless availability of *indicator reactions* for catalyst determinations. In catalytic methods, the main reaction in which the monitored species participates is called the indicator reaction.

Everyday application of uncatalyzed reaction-rate methods (for either single species or multispecies determination) is less common, and is limited to rather specific, specialized cases. The same is true for heterogeneous systems involving elec-

TABLE 18.1. *Classifications of Kinetic-Based Methods*

A. Classification Based on Chemistry of Reactions Employed

   Homogeneous Systems
   A1. Catalytic methods
      • Enzymatic methods employing soluble enzyme preparations
      • Nonenzymatic methods (mainly catalysis of redox reactions by transition-metal ions)
   A2. Uncatalyzed reaction-rate methods
      • Single-component determinations
      • Multicomponent determinations (differential reaction-rate methods)
   A3. Chemiluminescence-based methods

   Heterogeneous Systems
   A4. Kinetic methods based on electrode reactions
   A5. Enzymatic methods employing immobilized enzyme preparations

B. Classification Based on Methods of Mixing Reactant Solutions
   B1. Batch methods (discrete sampling)
   B2. Stopped-flow methods
   B3. Continuous-flow mixing
   B4. Catalytic titrations and catalytic endpoint detection
   B5. Centrifugal mixing

C. Classification Based on Measurement Approach
   C1. Initial rate (fixed and variable time)
   C2. Integral methods (fixed and variable time)
   C3. Derivative methods
   C4. Methods based on kinetic plots (method of the tangents)
   C5. Multipoint methods (delta and regression methods)
   C6. Measurement of length of induction periods

trode reactions. The use of immobilized enzyme systems is on the increase, and chemiluminescence is also gaining in popularity. Measurement approaches used in chemiluminescent methods are kinetic in nature; they are sometimes classified as catalytic methods although the so-called catalyst is irreversibly converted into inactive products of higher oxidation states, and a catalytic cycle is difficult to envision.

Continuous-flow sample processing continues to gain acceptance and is slowly replacing the time-honored discrete sampling approach. [Different aspects of this type of sample processing and of centrifugal analyzers are discussed in the chapter on automation (Chap. 25)]. Stopped-flow mixing has been used mainly for the kinetic and mechanistic study of fast reactions, but its rather limited application as a direct determination approach has been extended as a result of its implementation in conjunction with unsegmented-flow sample processing.

The increasing role of digital computers in performing computational tasks makes multipoint methods more attractive than techniques based on a small number of data points. From an analytical viewpoint, such approaches offer two main advantages: (a) less experimental work is required to seek discriminating variables, and (b) the use of a large number of data points greatly reduces errors.

18.2 MEASUREMENT OF REACTION RATES

Chemical reaction rates cover a very wide range of velocities. Some reactions, such as the neutralization of a strong acid with a strong base, are so rapid that they appear to reach equilibrium instantaneously; whereas others, such as the (noncatalyzed) reaction between oxygen and hydrogen at room temperature, are so slow that no reaction can be detected.

To determine the initial concentration of a desired species by kinetic-based methods, the rate of the chemical reaction must be measured by monitoring the concentration of at least one of the reactants or products as a function of time. Chemical methods (titration) or physical methods (spectrophotometry or conductivity) can be used. If chemical methods are used, the reaction must be quite slow, or else quenching methods must be utilized (when applicable). Continuous measurement of the reaction rate is possible by physical, but not by chemical, methods; the reaction rates observable are limited only by the response times of the instruments.

Reactions with half-times, or half-lives, larger than about 10 sec are considered to be *slow*, whereas those with half-times smaller than 10 sec are considered to be *fast*. The methods and experimental limitations for measuring these types are briefly discussed below.

Slow Reactions

The rates of slow chemical reactions in solution can generally be studied by quite simple and conventional methods. The reactants are mixed in some vessel, and the progress of the reaction is followed by titrating aliquots of the mixture or by measuring, at known times, a physical property of the solution such as optical absorption or voltammetric diffusion current.

The speed of initial mixing of the components in the vessel places a limit on the minimum half-time that can be measured in this way. If the mixing is accomplished by simple stirring devices such as magnetic stirring bars, the mixing time is a few seconds, and reactions with half-times smaller than 10 sec are difficult to measure with acceptable accuracy. On the other hand, the kinetics of reactions with long half-times can be determined, but such determinations take a long time and are, therefore, undesirable for analytical purposes. Two hours is arbitrarily considered the longest acceptable time for routine analysis.

If the reaction is over in more than 2 hr or less than 10 sec, several simple techniques can be used to adjust the rate so that the half-time will lie within the desired range. These are: (a) changing the temperature of the reaction system, (b) changing the concentration of the reactants, and (c) changing the solvent medium or ionic strength of the solution. For reactions that are over in 10 sec or less, stopped-flow mixing offers an attractive alternative, although this means of mixing is not compatible with simple, unsophisticated instrumentation.

Change in Temperature. The relationship between temperature and the rate constant k of a chemical reaction is given by the Arrhenius equation

$$\frac{d(\ln k)}{dT} = \frac{E^*}{RT^2} \quad (18.1)$$

or (in its integrated form)

$$k = Ae^{-E^*/RT} \tag{18.2}$$

where E^* = activation energy for the reaction
R = universal gas constant
A = frequency factor

For a large number of homogeneous reactions, the rate constant increases two or three times for each 10°C rise in temperature.

Reactions that are fast at room temperature can be slowed by cooling to allow measurement by "convenient" or "conventional" physical or chemical methods. For example, the conductometric analysis of a mixture of certain aliphatic aldehydes by differential reaction rates can be carried out at 0 to 5°C, whereas at room temperature the reactions are too fast. On the other hand, slow reactions can be made to proceed considerably faster by raising the temperature. For instance, fructose reacts with anthrone at room temperature in the presence of glucose and within 10 min develops a highly colored product. Glucose does not react appreciably even over a long period of time, but by elevating the temperature of the reaction mixture to 100°C after the fructose has reacted, the glucose can be made to react to completion in a few minutes and can then be determined.

Change in Concentration. Reactions with very large rate constants can be measured simply by using low concentrations of reactants, provided methods sensitive enough to measure the small changes in concentration are available. For example, the bromination of N,N-diethyl-m-toluidine is extremely fast in aqueous solution, but since 10^{-8} M bromine can be estimated from the redox potential of a platinum electrode, the concentration of the free amine can be reduced to 10^{-8} M and the reaction can then be readily followed. Spectrophotometric methods can be used to measure extremely small concentrations of highly colored compounds. A reaction involving color change and a large rate constant occurs between ferrous ion and cobaltioxalate ion in aqueous solution:

$$Fe^{2+} + [Co(C_2O_4)_3]^{3-} = Fe^{3+} + [Co(C_2O_4)_3]^{4-} \tag{18.3}$$

High concentrations of the reacting species can be employed to speed up reactions with small rate constants, although changes in activity coefficients may hamper the calculation of initial concentration when high concentrations of reactants are used.

Change in Solvent or Ionic Strength. The rate constant of a chemical reaction can be altered considerably by changing the solvent or by adding a salt to the reaction medium, partly as a consequence of changes in the dielectric constant of the solvent or the ionic strength of the solution. More specifically, as the dielectric constant increases, (a) the rate of a reaction between two ions of the same sign increases, (b) the rate of a reaction between two ions of the opposite sign decreases, (c) the rate of a reaction between two neutral species that form a polar product increases, and (d) the rate of reaction between an ion and a neutral molecule is not significantly changed. As the ionic strength of the medium increases, (a) the rate of reaction between two ions of the same sign increases, (b) the rate of reaction between two ions of opposite sign decreases, (c) the rate of reaction between two neutral species that form a polar product changes only slightly, and (d) the rate of reaction between an ion and a neutral molecule changes only slightly.

Fast Reactions

Although the kinetic methods discussed in this chapter have been applied mostly to slow reactions, many are also applicable to fast reactions, provided sufficient accuracy in the measurement of the reaction rate can be achieved. Special techniques to measure fast reactions have become more accurate in recent years, and practical analytical applications of them are now being devised.

For the most part, experimental methods for studying fast reactions can be classified into four groups: *mixing*, *relaxation*, *periodic*, and *continuous*. The approximate upper limit of reaction rates that can be measured by each of these techniques depends on the mixing time or, in the case of relaxation and periodic methods, on the displacement time, which is the time required to bring the system to a suitable nonequilibrium condition.

Mixing methods, the experimental methods most commonly used in kinetic studies and analytical applications of fast reactions, involve the rapid mixing of reacting species that were initially separated. They are of special interest because they are the only methods that do not rely on displacing an established equilibrium. Hence, reactions that are virtually irreversible under conditions of interest can be studied; it is for this reason that mixing methods are also the most applicable to pseudo-first-order reactions.

In the *continuous-flow* method, the reactants flow in separate continuous streams that meet in a mixing chamber and then pass along an observation tube or chamber with detection devices placed at appropriate points along its length. The detection devices, which measure the composition of the flowing sample, may be optical, thermal, chemical, or electrical, or may apply any other method appropriate to a rapidly moving sample. Reactions with half-times of the order of 10^{-3} sec can be observed by this method.

The *stopped-flow* method employs a pair of driven syringes to force the reactants into a mixing chamber and then into the observation cell. As soon as the mixed solution reaches the observation cell, the flow is stopped so that changes in the measured parameter can be observed without interference from artifacts arising from flow and turbulence.

Figure 18.2 compares, in terms of half-time of reaction and rate constants, the mixing capabilities of conventional (simple means of mixing) and stopped-flow mixing. The mixing speed of stopped-flow mixing is comparable to that of continuous flow mixing, but stopped-flow methods are more routinely used. The continuous-flow mixing referred to here operates under conditions of *turbulent* flow, and should not be confused with unsegmented continuous-flow sample processing (as in flow-injection techniques), in which *laminar* flow predominates.

A typical stopped-flow setup is illustrated in Figure 18.3. The stopped-flow technique requires only about 100 to 500 μL of solution for a complete run, has dead-times as low as 0.5 msec, permits extending observations to minutes, but requires fast

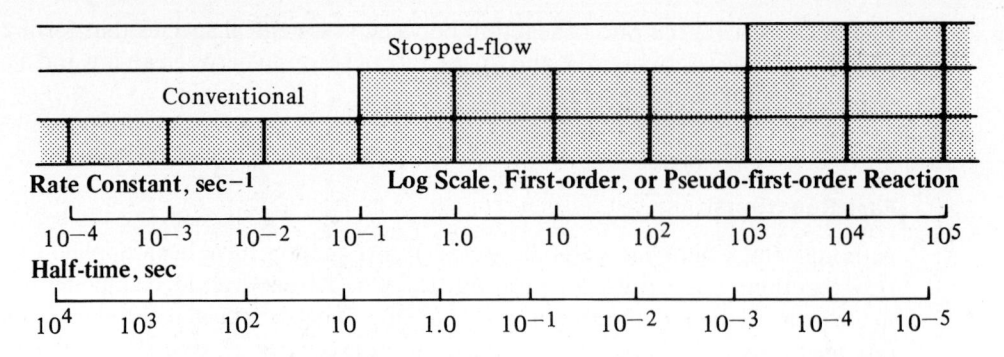

FIGURE 18.2. *Comparison of the mixing capabilities of conventional mixing methods and stopped-flow mixing.*

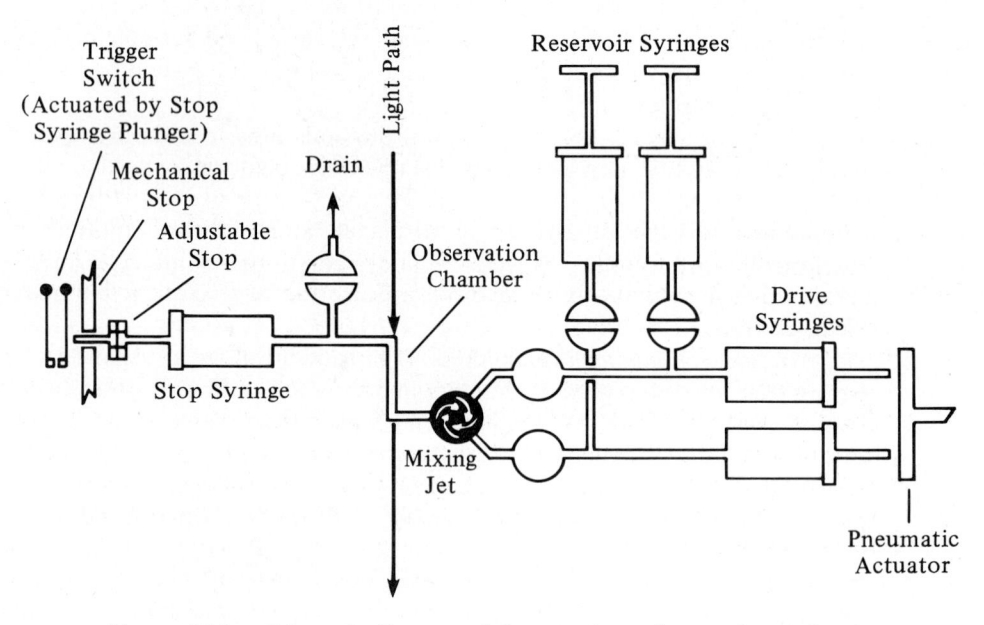

FIGURE 18.3. *Schematic diagram of the experimental setup for performing stopped-flow kinetics with optical detection. Reprinted with permission of Dionex Corporation.*

detection and rapid readouts. Continuous-flow methods allow the use of slow-responding detectors because, at a particular point in the flow stream, the "age" of the solution is constant. However, continuous-flow operation requires large reaction volumes; in addition, the whole length of the flow tube must be scanned by the detection system. Continuous-flow methods are slightly faster, with dead-times as low as 100 to 200 μsec because of absence of the mechanical problems associated with abruptly stopping the flow.

Stopped-flow methods require a very rapid and efficient mixing of the separate reagent streams that will form the final solution. A diagram of a typical *mixing jet*,

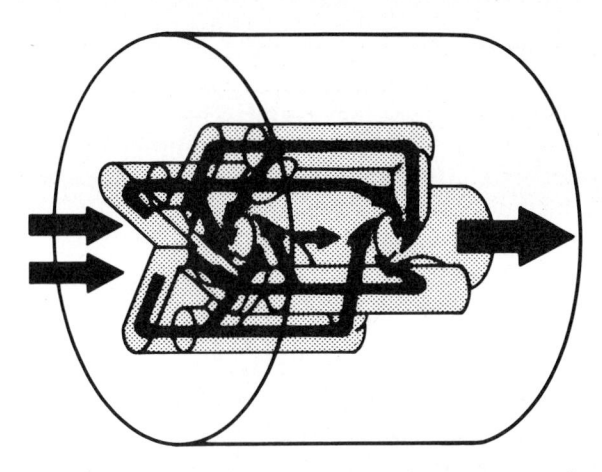

FIGURE 18.4. *Flow diagram of the tangential mixing jet used in the experimental setup for stopped-flow kinetics shown in Figure 18.3. The paths for solution flow are indicated by heavy lines. Separate reagent streams enter from the left, and the (mixed) solution exits to the right to the observation chamber. Reprinted with permission of Dionex Corporation.*

such as that shown in the middle of Figure 18.3, is illustrated in Figure 18.4. Tangential ports in the mixing jet create turbulent flow that intimately intermixes the two reactant solutions, which are forced into the mixing jet by pneumatically actuated drive syringes. The freshly mixed solutions enter the observation chamber, forcing the spent fluid from the previous experiment out and into a stop syringe. When this syringe plunger reaches the mechanical stop, all flow is instantaneously stopped and a trigger switch initiates data collection by an oscilloscope, strip-chart recorder, or computer. At the end of the measurement, the stop syringe is manually purged, making the instrument ready for the next experiment. The reactants are stored in the reservoir syringes and transferred, as needed, to the drive syringes with the help of four rotary valves, which eliminate the possibility of cross-contamination during filling.

18.3 MATHEMATICAL BASIS OF KINETIC METHODS

The past few years have seen the development of many methods for calculating the initial concentration of the species of interest from reaction-rate data. These methods involve, in general, manipulating and rearranging the differential or integral forms of the classical reaction-rate equations to put them in a form convenient for calculating the initial concentrations of the unknown reactants. Such methods can be classified into two main categories: methods for a single species and methods for the simultaneous (*in situ*) analysis of mixtures. The methods within each of these two categories can be subdivided according to the kinetic order of the reactions employed: pseudo-zero-order or initial-rate methods, first-order and pseudo-first-order methods, and second-order methods.

Understanding the operational characteristics of analytical determinations in the kinetic approach requires a background in the mathematical formulation and manipulation of chemical rate expressions. This background can be kept within rather simple terms and should not lead to the assumption that kinetic methods are necessarily "too mathematically involved." The chemical reactions employed in virtually all of these methods (except those involving catalytic reactions or radiochemical decay) are bimolecular ones of the type

$$A + R \underset{k_b}{\overset{k_f}{\rightleftharpoons}} P \tag{18.4}$$

where A = species of analytical interest
R = added reagent
P = product (or products)
k_f = forward rate constant
k_b = backward rate constant

The general differential rate expressions then have the form

$$-\frac{d[A]_t}{dt} = -\frac{d[R]_t}{dt} = \frac{d[P]_t}{dt} = k_f[A]_t[R]_t - k_b[P]_t \tag{18.5}$$

(Quantities $[A]_t$, $[R]_t$, and $[P]_t$ represent the concentrations of species A, R, and P, respectively, at any time t.) Thus, the nomenclature "zero," "first," and "second order" refers to the experimental conditions under which the rate measurements are made or to the relative concentrations of the reactants A and R.

18.4 RATE CONSIDERATIONS FOR THE DETERMINATION OF SINGLE SPECIES

Pseudo-Zero-Order Conditions (Initial-Rate Methods)

If the rate data are taken only during the initial 1 to 2% completion of the total reaction, then the concentrations of A and R remain virtually unchanged and equal to the initial concentrations ($[A]_0$ and $[R]_0$, respectively); and the reverse reaction can be ignored, since only a negligible amount of product is formed. Thus, Equation 18.5 simplifies to a pseudo-zero-order form:

$$\frac{d[P]_t}{dt} \approx k_f[A]_0[R]_0 \approx \text{Constant} \tag{18.6}$$

Expression 18.6 provides the basis for initial-rate measurements since the constant value of the rate (*initial rate*) depends on $[A]_0$ or $[R]_0$. Hence, if for instance $[R]_0$ = constant, the initial rate plotted against $[A]_0$ should yield a straight line useful as a calibration plot for the determination of $[A]_0$.

Initial-rate measurements are frequently recommended because the reaction has proceeded only a small fraction to completion and the back reaction does not contribute appreciably to the overall reaction rate. Complications resulting from side reactions are also less during the initial time periods of any reaction.

The estimation of chemical concentration is accomplished instrumentally by measurement of some physical parameter functionally related to such concentrations. Commonly, the monitoring of these physical parameters results in an electrical signal detected as a current or a voltage level. Three different approaches are used to extract initial-rate information: (a) the *derivative* or *slope* approach, the most straightforward, in which one obtains the derivative of the electrical signal by electronically differentiating the signal from the detector; (b) the *fixed-time* or *constant-time* approach, and (c) the *variable-time* approach. The fixed-time and variable-time approaches result from the use of the integral forms of the rate equations, but under conditions such that Equation 18.6 is approximately valid. The fixed-time method involves the measurement of a small signal change, $\Delta[\text{Signal}]$ or $\Delta \mathscr{S}$, at a finite but short Δt close to $t = 0$. Application of the variable-time approach involves measuring the Δt necessary for a finite but small $\Delta[R]$ or $\Delta[A]$ close to $[R]_0$ or $[A]_0$. In both cases, the ratio $\Delta[\text{signal}]/\Delta t$ gives an approximate estimate of the initial rate.

First-Order and Pseudo-First-Order Conditions

A first-order irreversible reaction of a reactant A to form the product P can be written as

$$A \xrightarrow{k_A} xP \tag{18.7}$$

where k_A = rate constant
x = number describing the stoichiometry of the reaction

The rate of disappearance of A as a function of time is

$$-\frac{d[A]_t}{dt} = k_A[A]_t \tag{18.8}$$

where $[A]_t$ = concentration of A at any time t

Integrating Equation 18.8 yields a relationship between $[A]_t$ and the initial concentration, $[A]_0$, which is the quantity to be measured in a kinetic-based determination:

$$[A]_t = [A]_0 e^{-k_A t} \tag{18.9}$$

Substituting Equation 18.9 into Equation 18.8 defines the rate of the reaction in terms of $[A]_0$:

$$-\frac{d[A]_t}{dt} = k_A[A]_0 e^{-k_A t} \tag{18.10}$$

Equation 18.10 is the basis for the *derivative* approach to rate-based analysis, which involves directly measuring the reaction rate at a specific time or times and relating this to $[A]_0$. Equation 18.9 is the basis for the two different *integral* approaches to kinetic determinations: the *fixed-time* and the *variable-time* procedures.

It should be noted that, if Reaction 18.4 is run under such conditions that the initial concentration of one of the reactants (either A or R) is very large compared to that of the other, then the concentration of that reactant will remain virtually unchanged as the reaction proceeds to equilibrium and can be considered equal to the initial concentration. Also, the reverse reaction can usually be neglected since the large excess of one of the reactants drives the reaction to virtual completion. Under these conditions, the reaction is pseudo first order and the rate expression takes the form (for R in excess)

$$-\frac{d[A]_t}{dt} \approx k_f[R]_0[A]_t = k_f'[A]_t \qquad (18.11)$$

A completely analogous expression can be written for $-d[R]_t/dt$ for the case where A is in large excess; this similarity permits mathematically treating pseudo-first-order reactions in the same manner as true first-order reactions.

It is obvious from the discussion above that any kinetic-based analytical procedure must take into account the degree of approximation made in the various rate equations with respect to the period of measurement, the relative initial concentrations of reactants, and, in some cases, the reversibility of the reactions. Care must be taken, for example, in using a pseudo-first-order method when the initial concentration of the unknown varies over several orders of magnitude; the error introduced in assuming the validity of the pseudo-first-order approximation of Equation 18.11 is a function of $[A]_0$. Although the reaction mechanisms and rate equations for enzymatic and other catalyzed reactions in general are somewhat more complex, similar assumptions and simplifications (and, therefore, restrictions in validity) apply to the rate-measurement techniques employed in the analytical use of these systems. In this respect, it should be noted that a very common procedure in analytical applications of kinetics is to select experimental conditions to achieve pseudo-first-order rates.

Catalytic Rate Considerations: Nonenzymatic Catalytic Methods

The rate of any catalyzed reaction can be shown to be directly or nearly proportional to the initial concentration of catalyst as a result of occurrence of catalytic cycles. This relationship is more likely to apply if certain conditions are met, such as keeping constant all variables affecting the rate (temperature, ionic strength, solvent used). The concentrations of reactants, other than those of the catalyst and the species whose change in concentration is monitored, must be such that they affect the rate pseudo zero order. The monitored reaction species is adjusted to first-order dependence. When all these requirements are met, and for the generalized case,

$$R + X \xrightarrow{C} P \qquad (18.12)$$

where R = species monitored
X = other reactant(s) species
C = catalyst

Because all catalyzed reactions can proceed in the absence of catalyst (the catalyzed reaction proceeds simultaneously with the uncatalyzed one), the expression for the general case of Reaction 18.12 can be written

$$-\frac{d[R]}{dt} = [R](k_u + k_c[C]_0) \qquad (18.13)$$

where k_u = rate constant for the uncatalyzed reaction (plus some concentration terms)
k_c = rate constant for the catalyzed reaction (plus some concentration terms)

Successful catalytic methods require that $k_u[R] \ll k_c[R][C]_0$ because $k_u[R]$ characterizes the rate of the uncatalyzed reaction and determines the limits of detection for given procedures. Equation 18.13 is valid for rates estimated near initial reaction time, or under conditions where side reactions or the back reaction (for the indicator reaction) do not affect the rate of the catalyzed reaction. At any given time t, then,

$$-\frac{\Delta[R]}{\Delta t} \propto [C]_0 \qquad (18.14)$$

or, if Δt is constant,

$$-\Delta[R] \propto [C]_0 \qquad (18.15)$$

Expressions 18.14 and 18.15 reflect the proportionality between the concentration of catalyst and the rate in direct catalytic determinations. It is worth noting that the general mechanism of catalyzed reactions (excluding chain reactions) can be described by

$$R + C \underset{k_{-1}}{\overset{k_1}{\rightleftharpoons}} [CR] \qquad (18.16)$$

$$[CR] + X \xrightarrow{k_2} P + C \qquad (18.17)$$

and mathematical treatment of this mechanism as a preequilibrium case (Reaction 18.17 being rate determining) or a steady-state case (Reaction 18.16 being rate determining) reduces to relationships of the basic forms characterized by Expressions 18.14 and 18.15.

Expression 18.14 indicates directly that initial-rate measurements are possible for catalyst determination. It has been shown [2] that the variable-time approach is preferred in the determination of catalysts (including enzymes). The variable-time procedure requires the rearrangement and integration of Equation 18.13 between times t_1 and t_2 to give

$$-\ln\frac{[R]_2}{[R]_1} = (t_2 - t_1)(k_u + k_c[C]_0) \qquad (18.18)$$

If $[R]_1$ and $[R]_2$ are kept constant, as reference-signal values between runs, then

$$-\ln\frac{[R]_2}{[R]_1} = \kappa = \text{Constant} \qquad (18.19)$$

and

$$[C]_0 = \frac{\kappa}{\Delta t k_c} - \frac{k_u}{k_c} \qquad (18.20)$$

where $\Delta t = (t_2 - t_1)$. The analytical value of Equation 18.20 resides in the fact that a plot of $1/\Delta t$ versus $[C]_0$ is a straight line defining a calibration curve for determination of the catalyst by the variable-time procedure, in which the time Δt required for a fixed change in the species monitored is measured.

When the concentration is measured instrumentally, it is related to the magnitude of some electrical signal $\mathscr{S}$ produced in the detector or sensor portion of the instrument. When $\mathscr{S}$ is linearly related to the concentration of the product—for example, in conductance or amperometric measurements—$\Delta\mathscr{S} = v\Delta[R]$ and $d\mathscr{S} = vd[R]$, where v is the porportionality constant or *transfer function* in electrical units per concentration unit.

Often, an instrumental method is used in which $\mathscr{S}$ is not a linear function of [R]. For example, in optical absorption or potentiometric measurements, the output of a photomultiplier or an electrode is a logarithmic function of concentration. In such cases, the direct instrumental response can be written in a general form as

$$\mathscr{S} = f([R]) \tag{18.21}$$

where $f[R]$ = an arbitrary function

The mathematics becomes more complicated and, in general, nonlinear calibration curves result.

Catalytic Rate Considerations: Enzymatic Methods

Enzyme-catalyzed reactions are used analytically to determine both enzyme activities [E] and substrate concentrations [S], and are very important in clinical diagnoses. The usual Michaelis-Menten mechanism for enzymatic reactions is

$$E + S \underset{k_{-1}}{\overset{k_1}{\rightleftharpoons}} E \cdot S \overset{k_2}{\longrightarrow} P + E \tag{18.22}$$

In this equation, E·S is the intermediate enzyme-substrate complex. A steady-state treatment of this reaction mechanism gives the rate law

$$\frac{-d[S]_t}{dt} = \frac{d[P]_t}{dt} = \frac{k_2[E]_0[S]_t}{K_M + [S]_t} \tag{18.23}$$

The enzyme concentration appears only as the initial concentration in Equation 18.23, since the enzyme is cyclically regenerated during the reaction. Quantity K_M, the so-called Michaelis constant, is equal to $(k_{-1} + k_2)/k_1$. Equation 18.23 is the basis for the derivative techniques for determining both $[E]_0$ and $[S]_0$. If the concentration of product is monitored by a linear-response sensor, the resulting electrical signal at a given time is

$$\frac{d\mathscr{S}_t}{dt} = \frac{vk_2[E]_0[S]_t}{K_M + [S]_t} \tag{18.24}$$

If the substrate concentration is large compared to K_M, then the reaction is pseudo zero order and the rate of change of the signal with time is directly proportional to $[E]_0$. Under conditions where $[S]_t \ll K_M$, the rate of change of the signal is also directly proportional to $[S]_t$ and, when initial reaction rates are measured, $[S]_t \approx [S]_0$.

Integrating Equation 18.23 between two substrate concentrations, $[S]_1$ and $[S]_2$ (the concentrations at times t_1 and t_2, respectively), yields

$$-K_M \ln\left(\frac{[S]_2}{[S]_1}\right) - \Delta[S] = k_2[E]_0(t_2 - t_1) \tag{18.25}$$

which is the basis of the integral methods. The fixed-time procedure is, at least from a theoretical viewpoint, better for pseudo-first-order reactions and for the determination of substrates in enzymatic reactions. As stated earlier, however, for rate determination of enzyme activity (or the concentration of other catalysts), the variable-time approach is preferred. In general, if the monitored signal is nonlinear with concentration, the variable-time approach is superior to the fixed-time approach. A compromise is necessary, however, because the error introduced by signal nonlinearity is decreased while the uncertainty is increased by the use of the variable-time procedure with a pseudo-first-order reaction. It is clear from the discussion above that the two approaches for implementation of the integral methods are complementary, and that the type of reaction and the characteristics of the signal-concentration response of the detection system play important roles in dictating what approach to take [2].

The major advantages of using initial-rate data were mentioned earlier in the chapter. Application of the theory of propagation of errors has suggested, however, that instantaneous rate measurement (derivative approach) at a time $t = 1/k$ (k is the first-order or pseudo-first-order rate constant) yields reaction rates that are essentially independent of changes in parameters affecting the value of the rate constant; and the precision of the determination should be expected to improve [3]. The effect of temperature on the H_2O_2 oxidation of iodide ion supports this idea that fixed-time measurements are best performed at $t = 1/k$. However, the mechanism of many enzyme-catalyzed reactions becomes complex as the reaction proceeds, and initial-rate measurements may be required for practical reasons. This explains the preponderance of discussions of enzymatic determinations based on initial reaction rates.

Figure 18.5 shows the effect of substrate concentration and enzyme concentration (activity) on the initial rate (IR) of an enzyme-catalyzed reaction. The portions

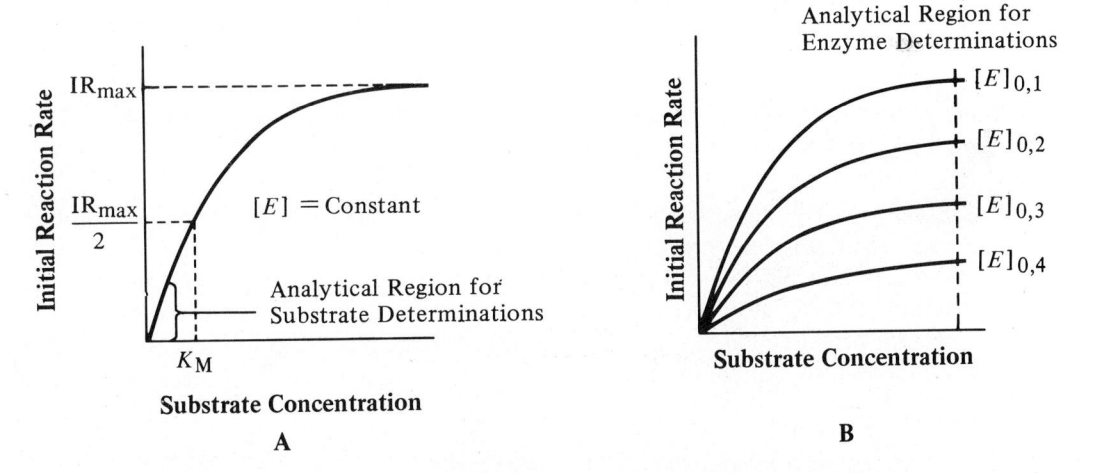

FIGURE 18.5. *Initial rate (IR) of an enzyme-catalyzed reaction as a function of the initial substrate concentration. [E] is the enzyme concentration and $[E]_{0,1} > [E]_{0,2} > [E]_{0,3} > [E]_{0,4}$; K_M is the Michaelis constant for the reaction.*

designated "analytical regions" represent situations in which the initial rate is directly proportional to the substrate concentration (Fig. 18.5A) or to the enzyme concentration (Fig. 18.5B). Referring to Equation 18.23, if $K_M \gg [S]$, we have (because [E] is constant)

$$IR = (k')[S]_0 \tag{18.26}$$

and, if $[S] \gg K_M$,

$$IR = (k'')[E]_0 \tag{18.27}$$

where k' and k'' are constants. These two equations provide direct relationships for the construction of calibration curves for the determination of substrate concentration (Eqn. 18.26) or of enzyme activity (Eqn. 18.27).

Enzyme activity is a rate concept in itself since it is defined as *the amount of enzyme that will catalyze the transformation of 1 micromole of substrate per minute or, where more than one bond of each substrate molecule is attacked, 1 microequivalent of the group concerned per minute, under defined conditions.* This amount is called a unit (U) or an international unit (IU). The temperature should be stated, and it is generally suggested that, where practical, it be 30°C. Other conditions (pH, substrate concentration, etc.) should be optimal. The concentration of an enzyme in solution should be expressed as units per milliliter. The so-called specific activity is expressed as units of enzyme per milligram of protein. For the convenience of expressing results in whole numbers, the concentration may be multiplied by 1000 to yield international units per liter. The specific activity can differ from one preparation of an enzyme to another; that is, the same molar concentration of an enzyme can yield different activities.

A common practice in the enzymatic determination of substrates is the coupling of enzyme-catalyzed reactions. A typical example is the determination of glucose by use of glucose oxidase. The H_2O_2 produced in the main reaction is coupled with a peroxidase-catalyzed oxidation in a second reaction:

$$\beta\text{-D-Glucose} + O_2 \xrightarrow{\text{glucose oxidase}} \text{Gluconic Acid} + H_2O_2 \tag{18.28}$$

$$H_2O_2 + \text{Dye (Reduced Form)} \xrightarrow{\text{peroxidase}} H_2O + \text{Dye (Oxidized Form)} \tag{18.29}$$

A variety of dyes have been used in the coupled reaction, among them *o*-tolidine, *o*-dianisidine, and leuco bases.

The oxygen reduced in Reaction 18.28, or the H_2O_2 produced in the same reaction, can be monitored electrochemically (amperometric detection), and this detection principle has been used in several enzyme-electrode configurations and in continuous-flow sample processing.

18.5 RATE CONSIDERATIONS FOR DIFFERENTIAL REACTION-RATE METHODS

The term *differential* has no mathematical connotation in its use here; it only implies the possibility of differentiating chemical species (via rate measurements) without prior separation. Often, the reaction rates of closely related components of a mixture with a common reagent are similar, and the rates cannot be sufficiently separated by either a thermodynamic or a kinetic masking technique to permit the faster- or slower-reacting component to be neglected. When this specific situation occurs, *differential reaction-rate methods* can be used for analyzing the mixtures without resorting to separation techniques. In contrast to catalytically based methods, differential rate techniques are not aimed primarily at determining low concentrations of materials in solution; most of the reactions used are uncatalyzed.

Consider the irreversible bimolecular reactions of a binary mixture of A and B with a common reagent R:

$$A + R \xrightarrow{k_A} P \qquad (18.30)$$

$$B + R \xrightarrow{k_B} P' \qquad (18.31)$$

The range of concentrations of reactant and reagent for which general differential reaction-rate methods have been developed is illustrated in Figure 18.6. When the concentration of common reagent is very large with respect to the total concentration of A and B, the reaction proceeds by pseudo-first-order kinetics (region I), and several general methods are available. The rates of change of the concentrations of either the product or the total reactants (total A and B) are monitored as a function of time.

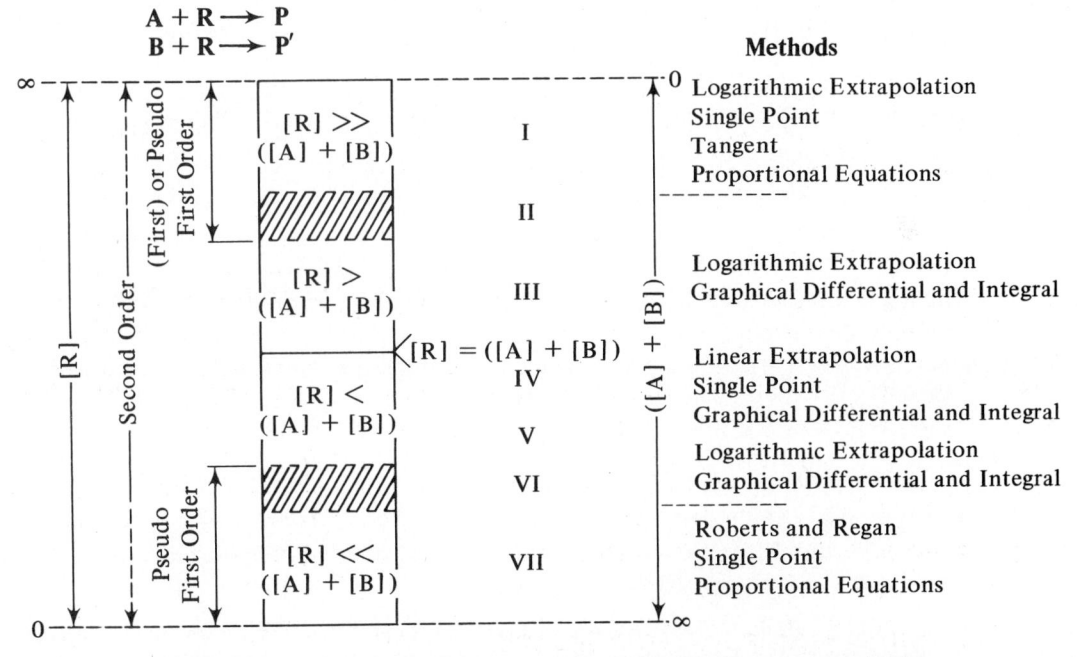

FIGURE 18.6. *General analytical techniques applicable to second-order reactions. From H. B. Mark, Jr., G. A. Rechnitz, and R. A. Grienke,* Kinetics in Analytical Chemistry, *New York: Wiley-Interscience, 1968, by permission of John Wiley and Sons. Copyright* © *1968 by John Wiley and Sons.*

As [R] becomes less than 50 times the total concentration of reactants ([A] + [B]), pseudo-first-order kinetics are no longer valid (region II); however, as [R] approaches ([A] + [B]) in magnitude, simple second-order kinetic treatments of the rates of reaction can be used. Regions III, IV (special cases where [R] = ([A] + [B])), and V of Figure 18.6 represent the concentration ranges for which the second-order treatment is employed. General methods based on second-order kinetics have been developed in which either [R], ([A] + [B]), or [P] can be followed.

As [R] decreases further, the kinetics again approach pseudo-first-order rates (region VI), but this time with respect to R. Because [R] ≪ ([A] + [B]) (region VII), a pseudo-first-order rate again applies, and general differential reaction-rate methods have been developed for this situation. There are also differential methods based on measurements of *initial* reaction rates, where the kinetics become pseudo zero order.

Theoretical treatments for analysis based on second-order kinetics are considerably more involved than those for first-order or pseudo-first-order processes. Therefore, whenever possible, the conditions of a bimolecular reaction are adjusted so that the reaction follows pseudo-first-order kinetics; a 50-fold or greater excess of reagent (region I) or reactants (region VII) is necessary. However, there are systems for which pseudo-first-order conditions cannot be employed, for example, with a large excess of either reagents or reactants, the reactions of interest may be too fast for practical measurements.

In a more general view, differential rate methods fall into two major categories: those based on *graphical* computation and those based on *mathematical* computation. There are features common to each approach, however, and in some cases the classification must be based on which one predominates. Discussed in detail below are the two most commonly used first-order differential kinetic methods. These are the *logarithmic extrapolation method* (basically a graphical approach) and the *method of proportional equations* (a mathematical computational approach). Table 18.2 lists salient advantages and disadvantages of both approaches.

Logarithmic Extrapolation Method

The logarithmic extrapolation method is suitable for reactions that are first order or pseudo first order with respect to the reactants, that is, in region I of Figure 18.6, where $[R]_0 \gg ([A]_0 + [B]_0)$. Consider two competing irreversible reactions of the type

$$A \xrightarrow{k_A} P \qquad (18.32)$$

$$B \xrightarrow{k_B} P \qquad (18.33)$$

in which A and B react to form a common product P, whose concentration at any time t is given by the expression

$$[P]_\infty - [P]_t = [A]_t + [B]_t = [A]_0 e^{-k_A t} + [B]_0 e^{-k_B t} \qquad (18.34)$$

After A has reacted essentially to completion ($[A]_t \approx 0$) in the case where $k_A \gg k_B$, one can take the logarithm of both sides of Equation 18.34 and obtain

$$\ln([P]_\infty - [P]_t) = \ln([A]_t + [B]_t) = -k_B t + \ln[B]_0 \qquad (18.35)$$

TABLE 18.2. *Advantages and Disadvantages of the Graphical Logarithmic Extrapolation Methods and the Method of Proportional Equations*

A. *Graphical Extrapolation Methods.* These were the first differential rate methods to be widely used and are the most frequently mentioned in the literature.

Advantages

1. Since these methods depend on plotting the logarithm of the total reactant concentration against time, there is no need to determine rate constants.
2. Temperature is not a critical variable.
3. The plotting procedure usually minimizes small errors since the "best" straight line is drawn through several points.
4. Since the methods are not restricted to constant processes following first-order kinetics, they can be applied in some cases where synergism is observed.
5. The first-order method can be used for the determination of three components in a mixture.

Disadvantages

1. The faster-reacting component must be about 99% consumed before useful data can be obtained.
2. The total initial concentration of reactants must be known. Its determination often requires following the reaction to completion.
3. If continuous monitoring of the reaction is not possible, a rather large number of samples have to be withdrawn from the mixture to follow the progress of the reaction.

B. *Method of Proportional Equations.* This is the most flexible approach to reaction-rate determination of closely related components.

Advantages

1. Requires generally shorter times than the other approaches.
2. Even if the mixture reacts by complex kinetics, can be applied if the proportionality constants can be determined.
3. A priori knowledge of the total initial concentration of reactants is not required.
4. Values of ratios of rate constants $\cong 4$ suffice.
5. Easily adaptable to automation, being ideal for fast reactions and for initial-rate methods.
6. Readily adaptable to the determination of more than two components in a mixture. Preferably used in combination with a minicomputer since the formulation and solution of the necessary simultaneous equations is then a simple task.

Disadvantages

1. The monitored property must be additive; hence, is not applicable in case of synergism.
2. Rate constants must be measured carefully.

Source: Adapted from H. A. Mottola and H. L. Pardue, *Crit. Rev. Anal. Chem.*, 4, 229 (1975). Copyright CRC Press, Inc., Boca Raton, Fla.

Thus, a plot of $\ln([A]_t + [B]_t)$ or $\ln([P]_\infty - [P]_t)$ versus time yields a straight line with a slope of $-k_B$ and an intercept (at $t = 0$) of $\ln[B]_0$. The value of $[A]_0$ may then be obtained by subtracting $[B]_0$ from the total initial concentration of the mixture; the latter can be determined by independent methods or can be calculated from $[P]_\infty$, provided the reaction mechanism does not change during the final stages. A typical reaction-rate curve of this type is illustrated in Figure 18.7. Because of its simplicity, this method is one of the most widely used differential kinetic techniques; it gives somewhat greater accuracy than do the other methods for mixtures in which the ratio of rate constants is relatively large.

Method of Proportional Equations

The method of proportional equations is based on the principle of constant fractional life (usually called *half-life*), which applies to a species undergoing reaction in such a way that, after any time interval, a constant fraction of the amount left unreacted at the end of the previous interval has reacted (or a constant fraction remains unreacted), irrespective of the initial concentration. This property is associated with first-order or pseudo-first-order reactions—such as radioactive decay—for which half-lives are often quoted as a measure of reaction rate. This property of constant fractional life also applies to more complex reactions—such as successive and parallel reaction sequences—involving first-order reactions. The initial concentration of a species reacting with constant fractional life is directly proportional to the amount of product formed at any given time.

For Reaction 18.7,

$$[P]_t = x([A]_0 - [A]_t) = x[A]_0[1 - e^{-k_A t}] = G_A[A]_0 \qquad (18.36)$$

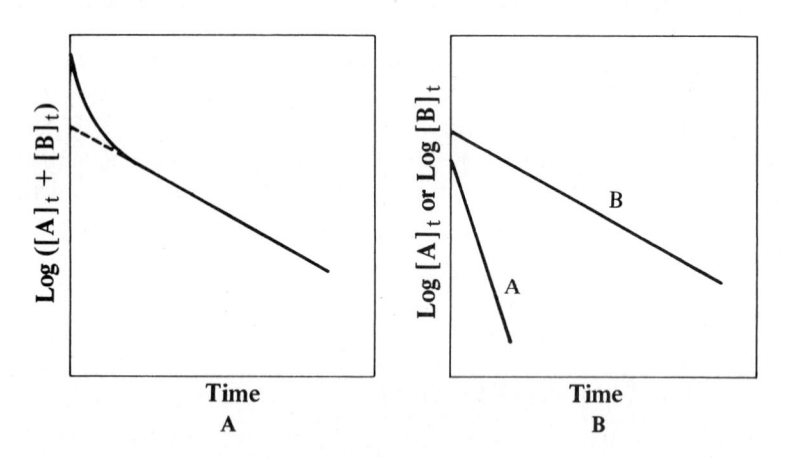

FIGURE 18.7. *Logarithmic extrapolation method for a mixture of species A and B reacting by first-order kinetics. A: Rate data obtained for the mixture. B: Rate data obtained for each component separately.* From H. B. Mark, Jr., G. A. Rechnitz, and R. A. Grienke, Kinetics in Analytical Chemistry, New York: Wiley-Interscience, 1968, by permission of John Wiley and Sons. Copyright © 1968 by John Wiley and Sons.

Therefore, the concentration of P at any given time is directly proportional to the initial concentration of A. The proportionality constant is a function only of the stoichiometry, the reaction time, and the rate constant.

Instead of the actual concentration, any parameter directly proportional to $[P]_t$, such as absorbance of light, electrical conductivity of the solution, voltammetric diffusion current, or volume of reagent required for a titration, can be measured. Then, since $\mathscr{S} = v[P]$,

$$\mathscr{S}_t = K_A[A]_0 \tag{18.37}$$

where $\quad K_A = vG_A$
$\quad\quad\quad v =$ proportionality constant

Consider the analysis of a mixture of two similar species, A and B. If B also reacts by first-order kinetics to produce P (not necessarily with the same stoichiometry as A), the same treatment given above for A can be applied. For the reaction of B alone,

$$[P]_t = G_B[B]_0 \tag{18.38}$$

where $\quad G_B = x_B[1 - e^{-k_B t}]$ and is a constant at any specified time t

If the reactions of A and B are independent of each other, then

$$[P]_{t_1} = G_{A_1}[A]_0 + G_{B_1}[B]_0 \tag{18.39}$$

$$[P]_{t_2} = G_{A_2}[A]_0 + G_{B_2}[B]_0 \tag{18.40}$$

The numerical values of constants G_A and G_B at times t_1 and t_2 are determined by measuring the amount of P produced by known amounts of pure A and pure B after times t_1 and t_2. Alternatively, the constants can be calculated by substituting known reaction-rate constants (k_A and k_B), stoichiometries, and times into the equations for G_A and G_B. Usually, the first procedure is preferred because it minimizes the influence of the numerous experimental variables.

Thus, the analysis of a two-component mixture is accomplished by measuring the concentration of P at times t_1 and t_2. These data are then substituted into Equations 18.39 and 18.40, which are solved simultaneously to yield the concentrations $[A]_0$ and $[B]_0$. This method can also be applied to situations in which two species react to form different products, that is,

$$A \rightarrow nC \tag{18.41}$$

$$B \rightarrow mD \tag{18.42}$$

provided [C] and [D] are both directly proportional to the same parameter, the instrumental signal $\mathscr{S}$. This may be the case, for instance, in analyzing a mixture of two organic compounds containing the same functional group. If the two reactions proceed independently, one can write

$$\mathscr{S}_t = K_A[A]_0 + K_B[B]_0 \tag{18.43}$$

The initial concentrations of the two species can be found by determining $\mathscr{S}_t$ at two reaction times and solving the resulting simultaneous equations.

This method can also be used for mixtures containing more than two reacting species. For a series of compounds (A, B, ..., N) that react with constant fractional lives to yield products directly proportional to $\mathscr{S}$, a series of n equations analogous to Equation 18.43 can be written for n different reaction times and, in theory, can be solved for the initial concentration of each species.

18.6 INSTRUMENTATION

The accuracy and precision of the experimental measurement is, of course, important to both kinetic-based and equilibrium-based analytical methods. However, a few special factors—instrumental and experimental—are of critical importance in kinetic-based techniques; these are discussed below with regard to spectrophotometric monitoring. The block diagram of a typical rate-measuring system is shown in Figure 18.8.

Instrument Stability

High-frequency noise in an instrument used to measure reaction rates can generally be eliminated or minimized by simple electronic filtering, since the frequency of the noise is usually very high compared to the rate of change of the signal. However, low-frequency noise or drift, which comes mainly from the reaction monitor and which varies at about the same rate as the reaction rate itself, presents a much more difficult problem. In this case, it is necessary to design the instrument in such a way as to eliminate these sources of drift, because there is no way of electronically separating the reaction signal from the random drift once the data reach the rate-computation system.

Light-Source Stabilization. The effect of light-source variations can be minimized by several methods. In most conventional spectrophotometers, a double-beam

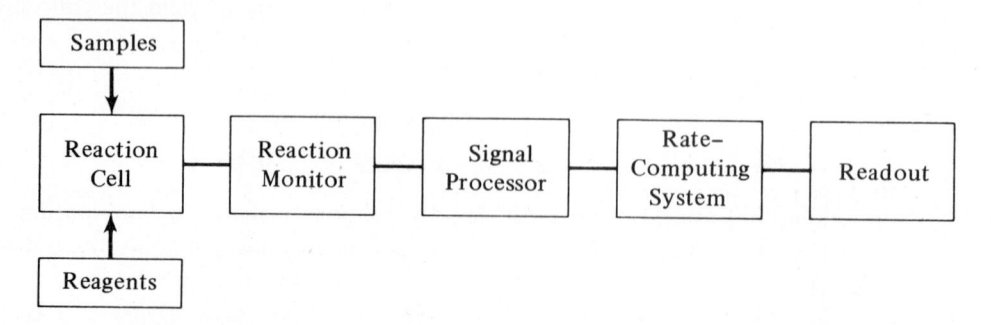

FIGURE 18.8. *Schematic diagram of a typical rate-measuring instrument. Adapted from J. S. Mattson, H. B. Mark, Jr., and H. C. MacDonald, Jr., eds.,* Spectroscopy and Kinetics, *Vol. 3, New York: Marcel Dekker, 1973, by permission of the publisher.*

systems can partially cancel fluctuations by comparing reference and sample beam intensities. However, since instruments used for kinetic methods are not often called upon for spectral scanning, most designers have discarded the double-beam approach and have developed less expensive electronic stabilization techniques for light-source regulation.

The simplest technique for stabilizing the light source against short-term fluctuations is to regulate the AC line voltage, since line-voltage fluctuations can cause changes in light intensity comparable to those caused by the measured reaction rate. Circuits that regulate the lamp power-supply voltage or current have also been used. However, this regulation is not sufficient; a more elaborate electronic regulator has been designed that compensates for changes in light-bulb resistance, and is thus superior to circuits that regulate only the applied voltage or the applied current.

Variations in light intensity can result from heat flow around the light bulb itself; the use of baffling to eliminate convection currents across the optical path has been shown to give satisfactory results. Such a stabilized light-source spectrophotometer can provide drift stability of better than 0.003 absorbance unit (AU) per hour, a noise level that produced a rate error of less than 0.001 AU per minute, and a photometric accuracy of 0.01 AU at 1.0 AU and 0.001 AU near zero absorbance.

One disadvantage of such control systems is that they control the electrical input to the lamp rather than the actual light intensity. Several systems have been described using optical feedback techniques to directly control the light intensity. Usually, in these techniques the light intensity is monitored with a second photodetector, and a signal from this detector is fed back to the lamp power supply.

Detector Noise and Drift. Transducers for modern spectrophotometric systems are usually vacuum phototubes or photomultiplier tubes. Noise and drift from the transducer can be quite troublesome in kinetic methods, and special care is normally taken to ensure low-noise operation.

There is some controversy about which detector provides the highest signal-to-noise ratio, the phototube or the photomultiplier. If comparisons are made at the same *light level*, the photomultiplier is capable of a larger signal-to-noise ratio, because its internal amplification is so high (often about 10^6) that the Johnson noise of the load resistor is insignificant.* On the other hand, when a phototube is used at low light levels, Johnson noise becomes the limiting factor and external amplification does not help. If comparisons are made at the same *anode current*, the phototube has the larger signal-to-noise ratio because the light level has to be much higher to obtain an equivalent output current. Thus, in systems where the operator can control the light level, such as with absorption spectrophotometers, it is recommended that intense light sources be used with a phototube detector. In this respect, the use of "fast optics" (optical systems characterized by low aperture ratios) offers the possibility of high light throughput. Grating systems rated at aperture ratios of about f/3.5 are particularly attractive for such purposes.

As is true in any measurement, if the input signal can be made high enough that little or no amplification is needed, higher signal-to-noise ratios can be obtained than with amplification. For systems where the light level cannot be controlled, such as in

---

*Johnson or thermal noise is produced by random thermal motion in resistive circuit elements.

emission spectroscopy, or where the light level is very low, the photomultiplier will give superior results.

Another important consideration in detectors is power-supply stability. For phototubes, regulation is not critical because the current-versus-voltage characteristic is essentially flat in the usual operating region. For photomultipliers, however, the gain is highly dependent on the power-supply voltage. As a rule of thumb, the power-supply stability should be at least an order of magnitude greater than the desired stability in gain.

Silicon photovoltaic detector/amplifier combinations are generating more interest because they offer flat noise spectrum to DC without hysteresis or memory effects as a characteristic of spectral stability.

Temperature Control

The rate of chemical reaction is considerably more sensitive to temperature variation than is the position of equilibrium (provided the formation constant is very large and the reaction can be considered "quantitative"). Thus, temperature control is critical in reaction-rate methods.

Two factors in temperature control must be considered. The accuracy of temperature control in the jacket of the reaction cell is the initial consideration, of course. However, since chemical reactions are either exothermic or endothermic, it has been shown that rapid temperature exchange and equilibration of the reaction solution with the cell jacket is also extremely important in obtaining good data.

Rise-Time of the Instrument

As mentioned above, high-frequency noise can often be eliminated by simple electronic filtering. However, caution is necessary and the investigator should be very familiar with the actual effective rise-time of the measuring instrument over all operating ranges. Obviously, a fixed time constant is not applicable over a large range of initial concentrations, since the reaction rate at any time is a function of initial concentration. Thus, it is necessary to quantitatively evaluate the rise-time of the instrument under all conditions of damping employed and to compare those results with the maximum reaction rates measured under each setting of the filter time-constant.

Linearity of Transducer Response

Most transducers converting chemical concentration into an electrical signal have a nonlinear response; for example, electrode potential and optical transmission are not directly proportional to concentration. In general, this nonlinearity is easily and simply corrected in equilibrium analytical measurements. However, it is considerably more difficult to instrumentally correct the response-versus-concentration function in reaction-rate methods, and often the correction itself can introduce significant errors in the analytical results. For example, the simple nonlinear feedback elements employed in log-response operational-amplifier circuits are not sufficiently accurate in transforming transmittance into absorbance to be used for many analytical purposes.

As mentioned earlier, the variable-time approach can be used advantageously in the case of nonlinear response, since the measured reaction rate in this procedure is linearly proportional to the initial concentration of the species of interest even though the actual transducer response is not proportional to concentration. The time required to reach a fixed concentration level is the parameter measured and, thus, linearity of the overall response-versus-concentration curve is not necessary. The point here is that in some cases the instrument and not the chemical reaction can dictate the method used.

Data-Acquisition Considerations

Most reaction-rate methods involve only a small number of actual data points (from one to about four) in the calculation of the initial concentrations. Clearly, this approach throws away a considerable amount of data that could be used advantageously. Studies examining parameters affecting the accuracy of mixture analysis by pseudo-first-order reaction methods, have shown that continuous data utilization is superior in several cases. In the early development of reaction-rate methods, the procedures involved chemical reactions that were not suitable for the continuous automatic measurement of reaction-rate curves. Also, calculations at that time attempted to limit the number of data points taken and to predetermine the optimum times, concentrations, and so on, for taking this minimal amount of data. However, recent advances in electronic circuitry and computer technology have had a tremendous influence on the design of kinetic-analysis instrumentation. These advances have also strongly influenced differential rate methods in both principles and approach. Several groups have designed instruments with built-in computation systems allowing continuous analysis of the reaction-rate curve over the entire reaction; in these instruments, data are processed using both ensemble-averaging and smoothing routines in real time. Experimental results and detailed error analysis have shown conclusively that this approach to data acquisition, reduction, and display leads to much greater accuracy and precision in the analytical results. In fact, good results can be obtained from fast differential rate determinations in which the usual finite or minimal data-point methods fail completely.

Automation of Operations

It is obvious that automated control of solution mixing, measurement sequences, and so on, will minimize the time-measurement errors arising from manual solution handling and measurement control, and will hence increase the accuracy of a given reaction-rate procedure.

As the cost of high-speed digital computers decreases, the applications of these small computers as built-in (on-line) units in chemical instrumentation for data reduction, system control, data acquisition, and experimental optimization in real time have increased sharply. Systems have been described in which the computer not only handles the data acquisition and sample manipulation, but actively takes part in all stages of the experiment by examining the data in real time and making decisions to optimize the experimental parameters and variables while the experiment is running. This is probably the most important single improvement in kinetic-based analysis; it will greatly expand both the routine and the specialized applications of the technique.

The block diagram for a completely computer-automated rate-measurement system is presented in Figure 18.9. By preliminary treatments (e.g., dissolution, dilution, filtration, ion exchange), the sample and reagent solutions are prepared as

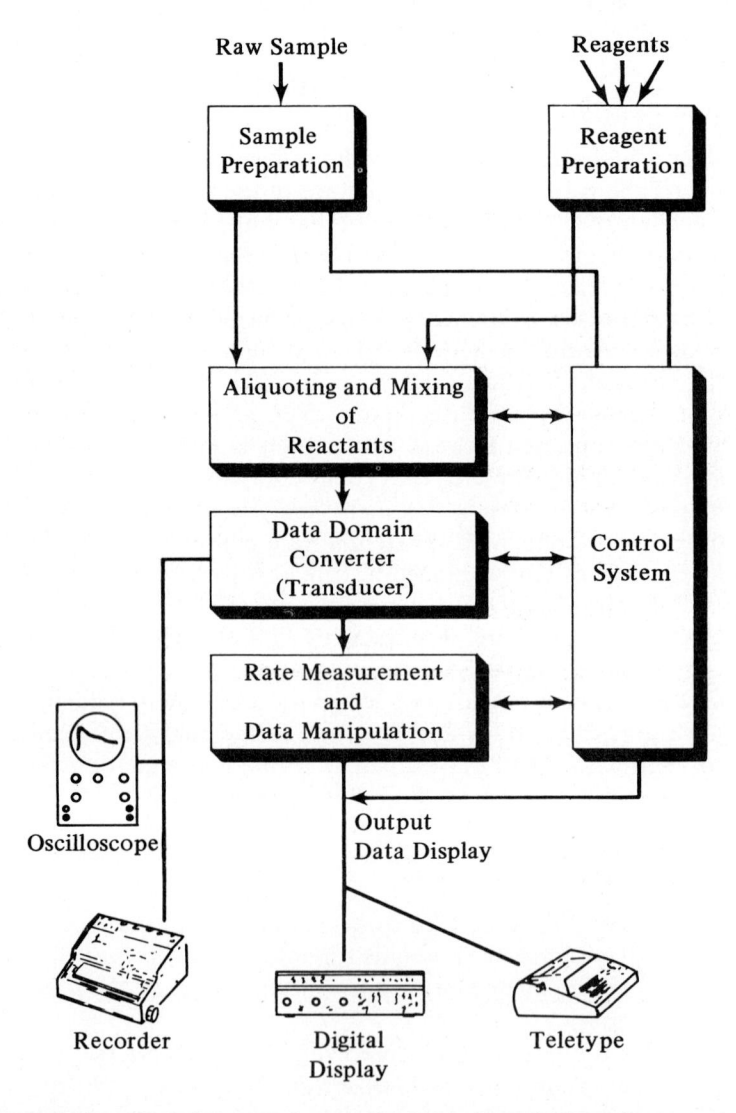

FIGURE 18.9. *Block diagram of a completely automated system for reaction-rate methods. Reprinted with permission from H. V. Malmstadt, E. A. Cordos, and C. J. Delaney*, Anal. Chem., *44*(12), 26A (1972). *Copyright 1972 by the American Chemical Society.*

required for the specific procedure being used. Predetermined volumes of sample and reagents are then introduced, mixed, and transported to the reaction cell. The control and rate-measurement systems can be hardwired for specific applications, or they can be incorporated in a minicomputer-interfaced system that can provide, through software, much versatility in control of the measurement sequence and the processing and readout of data. Readout is visually displayed with digital lights or printed out on a serial teletype or a high-speed parallel printer. When desired, a servo recorder or storage oscilloscope can display the parameter-versus-time and rate curves. Totally automated units can analyze as many as 1000 samples per hour.

18.7 ANALYTICAL APPLICATIONS OF KINETIC-BASED METHODS

Catalytic Determinations

Catalytic methods using enzyme-catalyzed reactions are applied mainly for the determination of substrate species in biological fluids and for determination of enzyme activities. Industrial applications are less frequent, although a wealth of analytical procedures could be implemented using enzyme-catalyzed reactions for analyses of food products, agricultural materials, industrial waters, and some pharmaceuticals. Nonenzymatic catalytic methods have been successfully applied to the determination of trace substances in high-purity materials, biological samples, environmental samples, geological samples, and industrial and natural waters.

Table 18.3 gives a summary of substrates and enzymes generally determined in clinical laboratories by kinetic methods based on enzyme-catalyzed reactions. Table 18.4 lists some nonenzymatic catalytic determinations which have been selected from the literature of the past 10 years or so. These are typical of the variety of samples for which chemical species have been determined by catalytic methods; and the chemistry and analytical approaches are representative of those encountered in these types of applications.

TABLE 18.3. *Substrates and Enzymes Commonly Determined in Clinical Chemistry by Kinetic Methods Based on Enzyme-Catalyzed Reactions*

Substrates (Enzyme used in the determination)
    Ethanol (alcohol dehydrogenase), determination of blood alcohol
    Glucose (glucose oxidase)
    Galactose (galactose oxidase)
    Urea (urease)
    Uric acid (uricase)
Enzymes (Substrate + other reactant(s) used in the determination)
    Acetylcholinesterase (acetylcholine + H_2O)
    Acid phosphatase (sodium thymolphthalein monophosphate, pH 10.1)
    Creatinine phosphokinase (creatinine phosphate + ADP)
    Glutamic-oxaloacetic transaminase (L-aspartate + 2-oxoglutarate)
    Glutamic-pyruvic transaminase (L-alanine + 2-oxoglutarate)
    Lactate dehydrogenase (pyruvate + NADH)

TABLE 18.4. *Selected Nonenzymatic Catalytic Determinations*

Biological and Pharmaceutical Samples

Copper in human blood serum [4] by a flow-injection procedure (closed-loop configuration). The indicator reaction is that of Fe(III) with thiosulfate. The copper catalyst is removed from the recirculating reagent by on-line controlled-potential electrolysis after detection. Electrochemical removal of the catalyst is paralleled by reoxidation of Fe(II) to Fe(III), permitting maintenance of a constant level of the monitored species (the red $Fe(H_2O)_5SCN^{2+}$ complex). The limit of detection is 0.25 µg/mL, with a determination frequency of 325 injections per hour.

Iron in human skin, guinea pig kidney, lung and spleen [5] by use of the indicator reaction involving the H_2O_2 oxidation of *p*-phenetidine in the presence of 1,10-phenanthroline as activator. Because of the low limit of detection and good sensitivity, this catalytic method permits working with samples 10 to 20 times smaller than those required in atomic absorption procedures.

Molybdenum in blood and urine [6] by means of its catalytic effect on the reduction of Se(IV) to Se(0) by Sn(II) in hydrochloric acid medium. The method requires dry ashing of the sample, dissolution in 6 *M* HCl, extraction with pentyl acetate, back extraction of the molybdenum into an aqueous phase, addition of acid, and catalytic determination by the fixed-time procedure (10-min reaction time). The red colloidal selenium, stabilized with gum arabic, is measured at 390 nm.

Enantiomeric iodinated thyronines in blood serum [7] by a postcolumn reaction system in a high-performance liquid chromatograph (HPLC). Direct ultraviolet detection after HPLC is not sufficiently sensitive for plasma levels of thyroid hormones, but the use of the catalytic effect of iodide on the reaction between chloramine-T and *N,N'*-tetramethyldiaminodiphenylmethane permits detecting as little as 3 nmol/L of isomeric tetraiodothyronines in serum. The color produced in the reaction is monitored photometrically at 600 nm.

Mercury in pharmaceutical ointments [8] by a titrimetric procedure based on a precipitation reaction involving iodide (the titrant). The endpoint of the titration is catalytically determined by the effect of excess iodide on the Ce(IV)-As(III) or the Ce(IV)-Sb(III) reaction. Amperometry and constant-current potentiometry are used for monitoring the reaction. Sample preparation prior to the titration is required.

Agricultural and Food-Product Samples

Calcium and magnesium in sterilized milk [9] by an EDTA back-titration using Mn(II) as a catalytic titrant. When in excess, Mn(II) triggers the auto-oxidation of 1,4-dihydroxyphthalimide dithiosemicarbazone to detectable levels.

Zinc in fruit juices [10] by a method based on restoring the catalytic activity of apocarbonic anhydrase.

Molybdenum in plant materials [11] by its catalytic effect on the I^- reaction with H_2O_2. Continuous-flow determination is employed after sample preparation.

Water Samples

Copper in river waters [12] by its catalytic effect on the oxidation of *p*-hydrazinobenzenesulfonic acid by H_2O_2. Fixed-time determination (20-min reaction) is used with monitoring at 454 nm of the oxidation product coupled with *m*-phenylenediamine (formation of a yellow azo dye).

Iron in drinking water [13] by a catalytic automated procedure (continuous-flow sample processing) based on the iron catalysis of the H_2O_2 oxidation of *p*-phenetidine in the presence of 1,10-phenanthroline as activator.

TABLE 18.4. *(continued)*

Petroleum Samples

Acids (including weak acids such as phenol) in straight-run and air-blown petroleum bitumens [14] by nonaqueous titrimetry and catalytic-thermometric endpoint detection.

Vanadium in crude oils [15] after sample preparation using the enhancement of the V(V)-catalyzed aerial oxidation of sodium 4,8-diamino-1,5-dihydroxyanthraquinone-2,6-disulfonate by Fe(III). The fluorescent oxidation product is monitored after 60 sec irradiation for initial-rate measurements.

The determinations of protein-bound iodine (PBI) and blood urea nitrogen (BUN) are representative of applications of catalytic and enzymatic-catalytic methods, respectively. The determination of PBI requires protein precipitation with trichloroacetic acid, ashing or acid digestion of the precipitated protein material, and final measurement of the catalytic effect of I^- on the classical Ce(IV)-As(III) reaction. A fixed-time procedure (absorbance measurement at 420 nm after 20 min reaction) is generally used. The enzymatic determination of BUN requires the preparation of a protein-free filtrate by addition of tungstic acid to precipitate the protein. The filtrate is incubated with the enzyme urease at pH 6.8, and the NH_3 produced by the enzymatic reaction is measured colorimetrically by means of Nessler's reagent or with an ion-selective electrode.

Besides direct determination of catalysts and substrates, *inhibitors* or *activators* of a catalytic reaction are sometimes also determined. The determination of inhibitors to catalysis by transition-metal ions has permitted the indirect determination of such species as aminopolycarboxylic acids and proteins. These materials complex the metal-ion catalyst, lowering its effective concentration and slowing the rate of the catalytic reaction. Also, such determinations form the basis of a unique method of endpoint detection involving: (a) a titration reaction in which a catalytic titrant is added to the solution containing the sample to react rapidly and stoichiometrically with the species of interest; and (b) an indicator reaction, which involves the monitored species and occurs at a noticeable rate only if an excess of titrant (catalyst) is present in the system. These have been referred to in the literature as *catalymetric* or *catalytic titrations*, but a more appropriate term would be *catalytic endpoint* detection.

The inhibition or slowing down of enzyme-catalyzed reactions has also been used, primarily for the determination of metal ions.

Differential Reaction-Rate Determinations

The method of proportional equations (MPE), when applicable, is the method of choice; this is reflected in the increased use of this procedure in recent years. Almost any experimental variable or situation can be utilized to formulate the simultaneous equations to be used for resolution of the components simultaneously determined. Two distinctive areas of application can be singled out from the contributions in the past 20 years or so. The first one is the use of metal(ligand)-exchange reactions for the simultaneous determination of metal-ion species which normally are not determined by catalytic procedures, and the second one is in the resolution of chemical species of pharmaceutical interest.

The rapid nature of metal-ion exchange reactions requires the application of stopped-flow mixing. The use of digital computers for the processing of data allows the simultaneous determination of n species by use of a number of measurements, m, such that $m \gg n$. In resolving binary and ternary mixtures of alkaline earth metals, for instance, a linear least-squares treatment of 200 data points, taken at regularly spaced time intervals, yielded precision and accuracy 5 to 10 times greater than those obtained when only 30 points were used in a similar manner [16].

A typical system of ligand(metal)-exchange reactions is illustrated by

$$M(CDTA) + H^+ \rightarrow (HCDTA)^{3-} + M \tag{18.44}$$

$$(HCDTA)^{3-} + M' \text{ (or } xH^+\text{)} \rightarrow M'(CDTA)^{2-} + H^+ \text{ (or } H_xCDTA^{4-x}\text{)} \tag{18.45}$$

Reaction 18.44 is the rate-determining step with a characteristic rate for each metal, and reaction 18.45 is rapid. The M' is added as a scavenger for $(HCDTA)^{3-}$ to force the dissociation step in Reaction 18.45, as well as to permit spectrophotometric monitoring of the extent of reaction. Copper(II) and lead(II) have been used as M'; excess acid also may be used in Reaction 18.45. Here, CDTA stands for the aminopolycarboxylic acid anion *trans*-1,2-diaminocyclohexane-$N,N,N'N'$-tetraacetate.

Figure 18.10 depicts values for the rate constant for Reaction 18.44, k_H^{MCDTA}, and suggests the use of an overall scheme for the determination of 28 metal ions, with half-lives adjustable to less than 10 min in 0.1 M acid. In a buffer solution and in an excess of scavenger, each M(CDTA) complex reacts with H^+ independently of all others, which makes the reaction particularly useful for differential rate determinations of binary and ternary mixtures of metal ions.

The simultaneous determination of ampicillin and its pro-drugs pivampicillin and bacampicillin can be cited as an example of differential determinations with relevance to the pharmaceutical and medical fields [17]. Ampicillin pro-drugs have been used recently because they improve the oral absorption of the parent ampicillin. The structure of ampicillin (D-α-aminobenzylpenicillin) is shown here, along with the substituents used in the ester preparations.

R = H (Ampicillin)
R = $CH_2-OOC-C(CH_3)_3$ (Pivampicillin)
R = $CH-O-C-O-C_2H_5$ (Bacampicillin)
     | ||
     CH_3 O

R = CH   C = O (Talampicillin)

The imidazole-catalyzed rearrangement of penicillins into the corresponding acids, which yield the stable Hg(II) mercaptides in presence of $HgCl_2$, has been used for the spectrophotometric determination of penicillins. The ampicillin esters exhibit faster reaction rates than the parent compound in the imidazole method, providing the

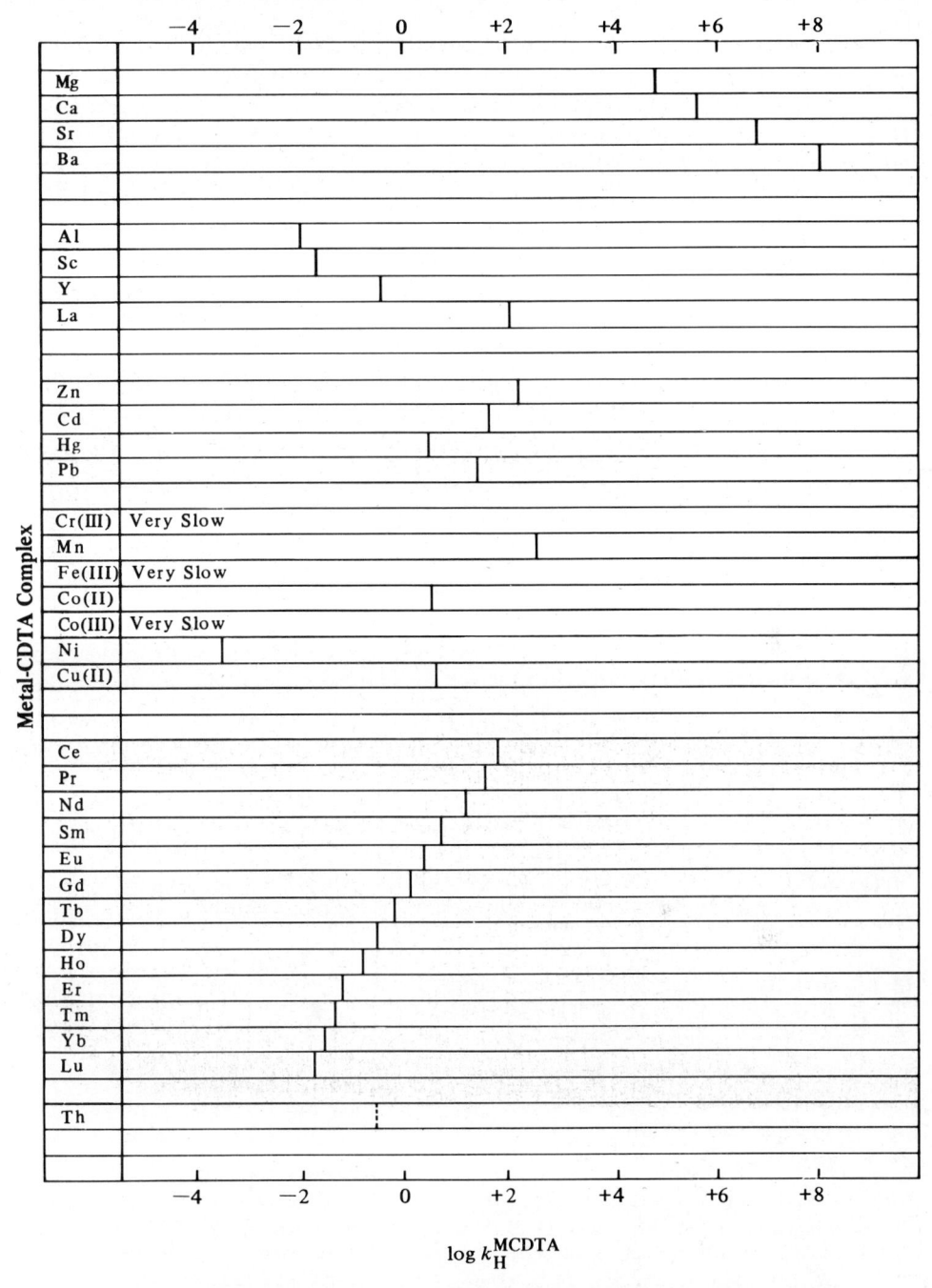

FIGURE 18.10. Experimental rate constants for the reaction of H^+ with a number of metal-CDTA complexes at 25°C. The half-life of each reaction can be calculated by using the relation $\log t_{1/2}$ (in sec) $= pH + pK_H^{MCDTA} - 0.16$. Taken from D. W. Margerum, J. B. Pausch, G. A. Nyssen, and G. F. Smith, Anal. Chem., 41, 233 (1969). Copyright © 1969 by the American Chemical Society. Reproduced by permission.

basis for a differential rate determination, even though the method under equilibrium conditions cannot distinguish between the esters and the free-acid forms. The method of proportional equations (measurement of two different reaction times) makes possible determinations of binary mixtures of ampicillin and pivampicillin or ampicillin and bacampicillin with errors of about 2 to 3%. This approach has been used for analysis of pharmaceutical tablets, and in studies to assess the stability of ampicillin esters. The major advantage of the differential reaction-rate method is its ability to distinguish between the pro-drug and the parent compound in a simple manner. A similar application is the determination of carbenicillin mixed with its pro-drugs carindacillin and carfecillin, whose structures are as follows [18]:

The method used in this case is based on the different rates of degradation of carbenicillin and its α-carboxy ester derivatives in acidic aqueous solutions and on spectrophotometric determination by the imidazole method. Logarithmic extrapolation allows stability studies and the determination of binary mixtures.

SELECTED BIBLIOGRAPHY

Monographs

GUILBAULT, G. G. *Enzymatic Methods of Analysis.* Oxford: Pergamon Press, 1970.

MARK, H. B., JR., RECHNITZ, G. A., and GREINKE, R. A. *Kinetics in Analytical Chemistry.* New York: Wiley-Interscience, 1968.

YATSIMIRSKII, K. B. *Kinetic Methods of Analysis.* Oxford: Pergamon Press, 1966.

Book Chapters

CROUCH, S. R. In *Computers in Chemistry and Instrumentation,* J. S. MATTSON, H. B. MARK, JR., and H. C. MACDONALD, JR., eds., Vol. 3. New York: Marcel Dekker, 1972, Chap. 3. *Excellent source for mathematical, instrumental, and computerization considerations in rate measurement.*

MALMSTADT, H. V., KROTTINGER, D., and MCCRACKEN, M. S. In *Topics in Automatic Chemical Analysis,* J. K. FOREMAN and P. B. STOCKWELL, eds., Vol. 1. Great Britain: Ellis Horwood, 1979, Chapter 4. *Review on automated reaction-rate methods with details on preparation and handling of reactants, and on measurement of chemical reactions by electronic devices and some commercially available automated systems.*

RIDDER, G. M., and MARGERUM, D. W. In *Essays on Analytical Chemistry (in Memory of Prof. Anders Ringbom),* E. WANNINEN, ed. Oxford: Pergamon Press, 1977, pp. 515–28. *Improved techniques to acquire, to calculate, and to test data sets for differential reaction-rate methods with emphasis on the use of the entire response curve and minicomputer manipulations.*

Review Articles

MOTTOLA, H. A. "Catalytic and Differential Reaction Rate Methods," *CRC Crit. Rev. Anal. Chem.,* **4,** 229 (1975).

MULLER, H. "Catalytic Methods," *CRC Crit. Rev. Anal. Chem.*, **13**, 313 (1982).

MOTTOLA, H. A., and MARK, H. B., JR. *Anal. Chem.*, **54**, 62R (1982). *Part of the biannual reviews on kinetic determinations and some other kinetic aspects of analytical chemistry.*

MARK, H. B., JR. *Talanta*, **20**, 257 (1973). *Development and publication of new kinetic-based methods.*

REFERENCES

1. H. PARDUE, *Clin. Chem.*, **23**, 2189 (1977).
2. J. D. INGLE, JR., and S. R. CROUCH, *Anal. Chem.*, **43**, 697 (1971).
3. F. J. HOLLER, R. K. CALHOUN, and S. F. MCCLANAHAN, *Anal. Chem.*, **54**, 755 (1982).
4. S. M. RAMASAMY and H. A. MOTTOLA, *Anal. Chim. Acta*, **127**, 39 (1981).
5. A. A. ALEXIEV, V. RACHINA, and P. R. BONTCHEV, *Anal. Biochem.*, **99**, 28 (1979).
6. G. D. CHRISTIAN and G. J. PATRIARCHE, *Anal. Lett.*, **12**(B1), 11 (1979).
7. E. P. LANKMAYR, B. MAICHIN, and G. KNAPP, *Fresenius' Z. Anal. Chem.*, **301**, 187 (1980); *J. Chromatogr.*, **224**, 239 (1981).
8. F. F. GAAL and B. F. ABRAMOVIC, *Talanta*, **27**, 733 (1980).
9. M. TERNERO, D. PEREZ-BENDITO, and M. VALCARCEL, *Microchem. J.*, **26**, 61 (1981).
10. K. KOBAYASHI, K. FUJIWARA, H. HARAGUSHI, and K. FUWA, *Bull. Chem. Soc. Jap.*, **52**, 1932 (1979).
11. B. F. QUIN and P. H. WOODS, *Analyst* (London), **104**, 552 (1979).
12. S. NAKANO, H. ENOKI, and T. KAWASHIMA, *Chem. Lett.*, **1980**, 1173.
13. H. SCHURIG and H. MULLER, *Acta Hydrochim. Hydrobiol.*, **7**, 281 (1979).
14. E. J. GREENHOW and A. NADJAFI, *Anal. Chim. Acta*, **109**, 129 (1979).
15. A. NAVAS, M. SANTIAGO, F. GRASES, J. J. LASERNA, and F. GARCIA SANCHEZ, *Talanta*, **29**, 615 (1982).
16. B. G. WILLIS, W. H. WOODRUFF, J. M. FRYSINGER, D. W. MARGERUM, and H. L. PARDUE, *Anal. Chem.*, **42**, 1350 (1970).
17. H. BUNDGAARD, *Arch. Pharm. Chemi, Sci. Ed.*, **7**, 81 (1979).
18. H. BUNDGAARD, *Arch. Pharm. Chemi, Sci. Ed.*, **7**, 95 (1979).

PROBLEMS

1. The kinetics of the hypothetical reaction

$$2A + 2B \rightleftarrows 2C + D$$

have been studied by initial-rate measurements at 25°C for mixtures of various compositions. None of the solutions contained C or D initially. The rate law can be expressed in the form:

$$-\frac{d[A]}{dt} = k[A]^a[B]^b[C]^c[D]^d$$

Consider the following data:

	Run 1	Run 2	Run 3
A, $\times 10^3$ M	1.0	2.0	2.0
B, $\times 10^3$ M	1.0	1.0	2.0
Initial rate $\times 10^3$ (M/h)	1.0	2.0	4.0

With this information and the knowledge that a trace amount of a transition metal ion, X, catalyzes the reaction (order with respect to the metal ion is 1), calculate the minimum value for the rate constant of the catalyzed and uncatalyzed reactions if 10^{-8} M X must be determined in a 1-min monitoring of the appearance of C at 420 nm (all other species do not absorb light at this wavelength). The molar absorptivity for C at 420 nm is 1017 cm²/mmole, and a relative concentration error of 10% is tolerable in the spectrophotometric measurement of C. Assume equal order dependence for reactants and products in the catalyzed and uncatalyzed paths and a 1.0-cm cell. (For expressions to account for the measurement of C with a relative error of 10%, see E. B. Sandell, *Colorimetric Determination of Traces of Metal*, 3rd ed.,

New York: Interscience, 1959, p. 98.) What will be the value of [C] measured after 1 min reaction?

2. The reaction between peroxydisulfate and iodide ions

$$S_2O_8^{2-} + 3I^- \rightleftarrows 2SO_4^{2-} + I_3^-$$

can be kinetically described by

$$\frac{d[S_2O_8^{2-}]}{dt} = -k[S_2O_8^{2-}][I^-]$$

with $k = 2.2 \times 10^{-3}$ M^{-1} sec^{-1}. What length of time will be needed to react 10% of the peroxydisulfate in a solution that is 0.0100 M in iodide and 0.000500 M in peroxydisulfate ion?

3. Iridium(III) is a catalyst for the Hg_2^{2+} oxidation by cerium(IV). The mechanism

$$Ce(IV) + Ir(III) \underset{k_{-1}}{\overset{k_1}{\rightleftarrows}} [Ce(IV)Ir(III)]$$

$$[Ce(IV)Ir(III)] \underset{k_{-2}}{\overset{k_2}{\rightleftarrows}} Ce(III) + Ir(IV)$$

$$Ir(IV) + [Hg(I)] \xrightarrow{k_3} Ir(III) + [Hg(II)]$$

is assumed to operate as a result of the following rate law:

$$\text{Rate} = \frac{k_2 k_3 [Hg(I)][Ir]_0}{k_3 [Hg(I)] + k_{-2}[Ce(III)]}$$

If $k_3 > k_2$ and $k_2 = k_{-2}$, which of the following approaches would you use to determine iridium, and why: (a) initial-rate measurement; (b) fixed-time (measurement of Ce(IV) absorbance after 50% reaction); (c) method of tangents (determination of the slope of first-order kinetic plots following the reaction for two half-lives)?

4. When a new enzyme is isolated and purified, an attempt is made to determine its specificity (selectivity). This is done by characterizing k_2 and K_M by measuring initial rates. A simple way to estimate k_2 and K_M consists in plotting $1/IR$ (IR = initial rate) versus $1/[S]_0$. This plot gives an intercept on the ordinate equal to $1/(k_2[E]_0)$ and a slope of $K_M/(k_2[E]_0)$. Derive the equation that justifies the plot.

5. Briefly discuss the analytical relevance of (a) k_2 and (b) K_M in enzymatic determinations.

6. Strictly speaking, all enzyme-catalyzed reactions are reversible. The simplest representation of the overall reaction takes into account the formation of a single central complex:

$$E + S \underset{k_{-1}}{\overset{k_1}{\rightleftarrows}} [ES] \underset{k_{-2}}{\overset{k_2}{\rightleftarrows}} E + P$$

Referring to this simplified model, derive the expression for the net velocity (forward), taking into account the effect of product formation on the rate.

7. Consider the method of proportional equations for a binary mixture. Does the relative concentration error for either one of the components to be determined depend on the values of the physical parameter being measured? Justify your answer mathematically.

8. With reference to H. Steinhart, *Anal. Chem.*, *51*, 1012 (1979) on the determination of tryptophan in foods and feedstuffs, indicate to what type of kinetic procedure the method discussed belongs.

9. Y. Onoue et al. [*Anal. Chim. Acta*, 106, 67 (1979)] suggested the simultaneous determination of 8-quinolinol and 5,7-dihalo-8-quinolinols by exploiting the difference in decay of their fluorescence. The plot illustrated here was used for the simultaneous determination of 8-quinolinol and 5,7-dibromo-8-quinolinol. Given the following: curve 1: mixture; curve 2: 1.0×10^{-5} M pure 5,7-dibromo-8-quinolinol; curve 3: 4.0×10^{-7} M 8-quinolinol; curve 4:

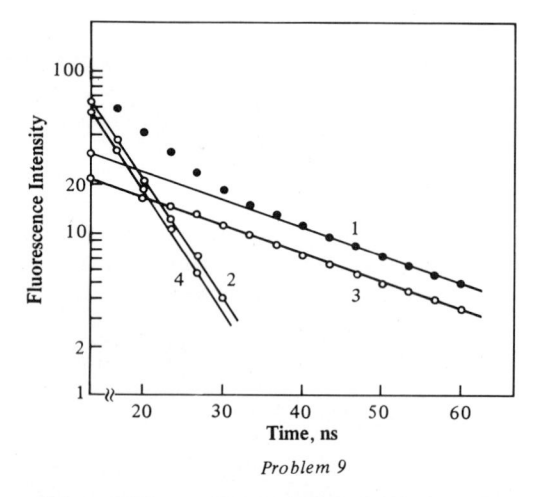

Problem 9

From: Y. Onoue, K. Morishige, K. Hiraki, and Y. Niskikawa, *Anal. Chim. Acta*, *106*, 67 (1979). With permission.

obtained by subtracting the contribution of 8-quinolinol from curve 1. Provide the concentration of each component in the mixture.

10. The ligand exchange reaction

(1) $\dfrac{\text{NiL}^{2-}}{\text{NiHL}^-} + \text{Zn}^{2+} \xrightleftharpoons{k_{\text{Zn}}^{\text{NiL}}} \dfrac{\text{ZnL}^{2-}}{\text{ZnHL}^-} + \text{Ni}^{2+}$

(with L^{4-} = ethylenediamine-N,N,N',N'-tetraacetate ion) is catalyzed by traces of copper since

(2) $\text{NiL}^{2-} + \text{Cu}^{2+} \xrightarrow{k_{\text{Cu}}^{\text{NiL}}} \text{CuL}^{2-} + \text{Ni}^{2+}$

(3) $\text{CuL}^{2-} + \text{Zn}^{2+} \rightleftharpoons \text{ZnL}^{2-} + \text{Cu}^{2+}$

Equation (2) represents a reaction 6500 times as fast as that represented by Equation (1). The attack of Zn^{2+} is also faster than Reaction (1). Knowing that ionic strength = 1.25 M, temperature = 25.0°C, $[\text{NiL}]_{\text{total}} = 0.0155$ M, pH = 5.00, and using the following data:

$[\text{Cu}^{2+}]_0$, M	K^*, min^{-1}
1.34×10^{-5}	7.5×10^{-5}
5.36×10^{-5}	11.0×10^{-5}
1.34×10^{-4}	18.5×10^{-5}

where K^* is the observed rate coefficient dependent on catalyst concentration, calculate the values of $k_{\text{Cu}}^{\text{NiL}}$ and $k_{\text{Zn}}^{\text{NiL}}$. [Reference: D. W, Margerum and T. J. Bydalek, *Inorg. Chem.*, 1, 852 (1962)].

11. Discuss the special instrumentation requirements imposed on spectrophotometric measurements as applied to kinetic-based methods.

12. In what applications are kinetic-based methods commonly used?

13. K. A. Connors [*Anal. Chem.*, 48, 87 (1976)] introduced a graphical treatment of data for the kinetic resolution of binary and ternary mixtures of closely related species. By application of this approach, Connors determined simultaneously each component in a mixture of cyanobenzoate, chlorobenzoate, and benzoate esters. Summarize the treatment used by Connors to resolve this mixture.

CHAPTER **19**

Radiochemical Methods of Analysis

William D. Ehmann
Morteza Janghorbani

Radiochemical methods of analysis employ radioactivity, with or without chemical manipulations, to obtain qualitative or quantitative information about the composition of materials. This information may concern the nature and quantity of elements or the specific chemical form of the component of interest. For example, qualitative and quantitative determinations of elements present in river waters can be readily accomplished; on the other hand, radiochemical methods can be used to determine the quantity of vitamin B_{12} (which contains an atom of cobalt) in a mixture of similar organic compounds. The fundamental difference between this method of analysis and all others is that, in this method, one either induces radioactivity in the sample or adds a radioactive substance to the sample.

Radiochemical methods of analysis are used in a wide range of analytical applications. Not only can these methods be used to obtain information regarding the nature and quantities of substances present in materials of interest, but radioactive elements can also be used as tracers to study various physicochemical processes. Radioactive substances can be used to follow the movement of elements or of specific compounds in soils and plants, the absorption of elements in the body, and the self-diffusion of lead atoms in metallic lead, among other applications. Although these tracer applications are of great practical value, the present chapter will be concerned only with applying radioactivity to determining the presence and quantity of elements and compounds in various materials—that is, the use of radioactivity in chemical analysis.

19.1 FUNDAMENTALS OF RADIOACTIVITY

In this section, we will discuss only those fundamental aspects of radioactivity directly relevant to radiochemical analysis. For a more detailed treatment, the Selected Bibliography at the end of the chapter should be consulted.

All atomic nuclei are made up of protons and neutrons (known collectively as *nucleons*); the only exception is the lightest hydrogen nucleus, which consists of a single proton. The *atomic number* (Z) of an atom is the number of protons present in its nucleus (also the number of electrons in the neutral atom). The sum of protons (Z) and neutrons (N) in a nucleus is termed the *mass number* (A). The mass number should not be confused with the atomic or nuclidic mass, which is the mass of the atom relative to that of a ^{12}C atom (which is, by definition, exactly 12.000... *atomic mass units*, amu).

All atoms whose nuclei contain the same number of protons (and are thus atoms of the same element) have virtually the same chemical properties, because these properties are determined by the structure of the orbital-electron cloud. But atoms of the same element may have a different number of neutrons and, therefore, a different mass number. Atoms having the same Z and a different A are called *isotopes*. On the earth, with few exceptions, the abundance of each isotope is in a fixed ratio to that of the other isotopes of the same element; the relative abundance of any isotope (usually expressed in units of atom percent) is called its *isotopic abundance*. Each element has two types of isotopes: (a) *stable isotopes*, and (b) *radioactive isotopes*. Stable isotopes are isotopes whose nuclei have not been observed to undergo spontaneous radioactive disintegration. The nuclei of radioactive isotopes, on the other hand, undergo spontaneous disintegration and eventually become stable isotopes of some element. Radioactive isotopes disintegrate by emitting *electromagnetic radiation* (x-rays or gamma rays),* by emitting *elementary particles* (α, β, n, p, or e^-), or by undergoing *fission* (breaking up into smaller nuclei). These isotopes are either *artificial* (manmade; e.g., ^{60}Co) or *natural* (found in nature; e.g., ^{40}K).

Whether or not a given isotope is stable depends on the particular number of protons and neutrons present in its nucleus and on the state of excitation of the nucleus. If one plots the atomic numbers of all stable isotopes as a function of the number of neutrons present in their nuclei, one obtains the graph shown in Figure 19.1. To form stable light elements, approximately equal numbers of neutrons and protons are required; however, for the heavier elements, considerably more neutrons are needed. This deviation from $Z/N = 1$ occurs because protons are charged particles and repel each other according to Coulomb's law. Since the nucleus is held together by strong, short-range binding forces between nucleons, additional forces provided by added neutrons are needed to dilute the increased coulombic repulsion forces.

Nuclei that do not fall on this curve are unstable and disintegrate, emitting the appropriate particles or radiation until the final product nuclei are on the curve. If the unstable *nuclide*† is on the proton-rich side of the curve, it disintegrates by emitting a *positron*,‡ or by a related process known as *electron capture*. The *daughter* nuclide will then have one fewer proton and one more neutron than the *parent* nuclide. If, on the other hand, the radionuclide is on the neutron-rich side of the curve, it

---

*X-rays are electromagnetic radiations resulting from an orbital-electron rearrangement in the atom, whereas gamma rays are emitted directly by the nucleus. Only the point of origin distinguishes one from the other.

†*Nuclide* is a general term referring to an atomic species containing Z protons and N neutrons; if the nuclide is radioactive, it is called a radionuclide.

‡A positron is a particle having the mass of an electron but a unit positive charge. Its symbol is β^+.

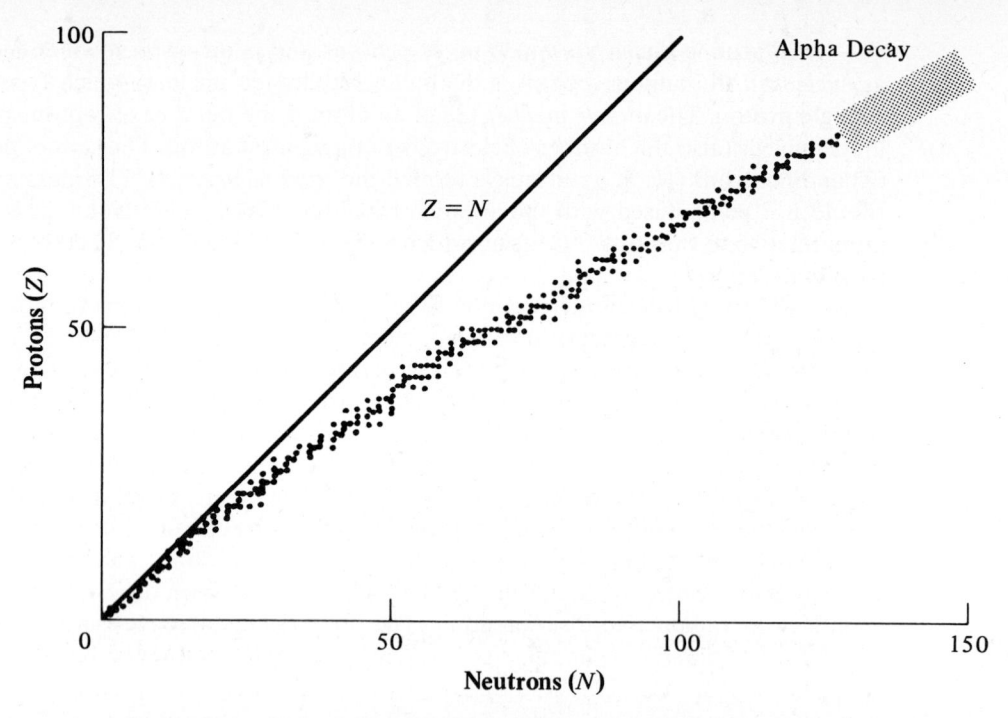

FIGURE 19.1. *Line of beta stability. Each point corresponds to a stable isotope.*

emits a *negatron** and the daughter nuclide will contain one fewer neutron and one more proton than the parent nuclide. For heavy nuclides, emission of an *alpha particle* (the nucleus of an ordinary helium atom) is also very common. The daughter product of an alpha decay will contain two fewer protons and two fewer neutrons than does the parent nuclide.

Rates of Disintegration

Not all radionuclides disintegrate at the same rate. The disintegration of any radionuclide is a first-order process and follows

$$-\frac{dN}{dt} = \lambda N \qquad (19.1)$$

where N = number of radioactive atoms of a specific radionuclide present at time t
λ = *nuclear-decay constant* (in sec^{-1})

The left side of Equation 19.1, dN/dt, is the number of disintegrations taking place per unit time. Each radionuclide has its own characteristic nuclear-decay constant. The minus sign indicates that the decay results in a decrease in N. Equation 19.1 indicates that the rate of decay of any radionuclide is directly proportional to the

---

*A negatron is an ordinary negative electron emitted from the nucleus. Its symbol is β^-. It is also often called a *beta ray* or *beta particle*.

number of those atoms present at that time. One can use this equation to calculate N, the number of radioactive atoms present at any time t, by separating variables and integrating to give

$$N = N_0 e^{-\lambda(t - t_0)} \tag{19.2}$$

Here, N_0 is the number of radioactive atoms present at time t_0, which is the reference point in time. For convenience, t_0 is generally taken to be zero time, in which case Equation 19.2 simplifies to

$$N = N_0 e^{-\lambda t} \tag{19.3}$$

Equation 19.3 states that if, at time $t = 0$, there are N_0 radioactive atoms having a nuclear-decay constant λ, then at any later time t there will be N radioactive atoms remaining, and $(N_0 - N)$ radioactive atoms will have undergone radioactive decay during the time t.

A convenient measure of how fast a radionuclide disintegrates is the *half-life* $(t_{1/2})$ of the radionuclide. This parameter is defined as the length of time required for one-half of a statistically large number of radioactive atoms to undergo radioactive decay. Thus, if there are N_0 radioactive atoms at time $t = 0$, one half-life later there will be $N_0/2$ radioactive atoms of the original radionuclide remaining. From Equation 19.3, it can be shown that the relationship between the half-life and the decay constant of a radionuclide is

$$t_{1/2} = \frac{0.693}{\lambda} \tag{19.4}$$

Since each radionuclide has its own characteristic decay constant, its half-life has a definite value. Half-lives of radionuclides vary over an extremely large range. For instance, one radioactive isotope of boron has a half-life of about 3×10^{-19} sec, whereas the half-life of naturally occurring bismuth is greater than 2×10^{18} yr.

From a knowledge of the half-life of any radionuclide, one can accurately predict the relative amount of that nuclide at any time later than or prior to some reference time. This fact is extensively used in determining the ages of materials by methods of radioactive dating. More important, as far as this chapter is concerned, Equation 19.1 can be used to determine the amount of radioactive material present in a sample at any given time. If the sample contained a single radionuclide, its amount could be measured by simply measuring the absolute activity of the sample with an appropriate detection system (see Sec. 19.3). Combining Equations 19.1 and 19.3,

$$A = -\frac{dN}{dt} = \lambda N_0 e^{-\lambda t} \tag{19.5}$$

One can measure A, the *absolute activity*,* experimentally and then calculate N_0, given knowledge of the decay constant λ and the decay time t.

---

*Activity is the measure of the number of specific particles or radiations emitted per unit time (commonly, per second). In contrast, the absolute activity (A) of a given radionuclide in units of disintegrations per second (dps) or disintegrations per minute (dpm) is the total number of disintegrations taking place per unit time without regard to the distribution of emitted particles or radiations. For example, one could measure either the positron or the negatron activity from a ^{64}Cu source (a radionuclide that decays by both modes) or the absolute activity of the source, which is independent of the mode of decay. The fraction of decays occurring by either path is described by the *branching ratio*.

SEC. 19.1 Fundamentals of Radioactivity

If the sample contains more than one radionuclide, the method of activity measurement can still be used with the following modifications. The total (gross) activity of a sample containing several independently decaying radionuclides is

$$A_{\text{total}} = \lambda_1 N_1 + \lambda_2 N_2 + \cdots$$

or

$$A_{\text{total}} = \lambda_1 (N_0)_1 e^{-\lambda_1 t} + \lambda_2 (N_0)_2 e^{-\lambda_2 t} + \cdots \tag{19.6}$$

Inspection of Equation 19.6 shows that, if one knows the nature of the radionuclides present in the sample, one should be able to calculate their abundances $[(N_0)_1, (N_0)_2,$ etc.$]$, by making the appropriate number of activity measurements as a function of time.

For example, if the sample contains only two independent radionuclides, two independent measurements will permit application of the two following simultaneous equations with two unknowns, $(N_0)_1$ and $(N_0)_2$:

$$(A_{\text{total}})_{t_1} = \lambda_1 (N_0)_1 e^{-\lambda_1 t_1} + \lambda_2 (N_0)_2 e^{-\lambda_2 t_1} \tag{19.7}$$

$$(A_{\text{total}})_{t_2} = \lambda_1 (N_0)_1 e^{-\lambda_1 t_2} + \lambda_2 (N_0)_2 e^{-\lambda_2 t_2} \tag{19.8}$$

In general, it should be possible to calculate the individual abundances of all radioisotopes present in the sample by making the appropriate number of activity measurements, provided that two conditions are met: (a) the half-lives of the sample constituents are sufficiently different, and (b) the half-lives are such that at least one of the components has significantly changed its activity during the time interval between consecutive measurements.

A general method for distinguishing among the radioactive constituents in a sample is to plot the logarithm of total observed activity as a function of time. First, consider a single-component system. Taking the logarithm of Equation 19.5 yields

$$\log\left(-\frac{dN}{dt}\right) = \log(\lambda N_0) - 0.43 \lambda t \tag{19.9}$$

Therefore, if $\log(-dN/dt)$ is plotted as a function of time, one obtains a straight line whose slope, -0.43λ, is a measure of the half-life of the radionuclide and whose intercept, $\log(\lambda N_0)$, at $t = 0$ is a measure of N_0. In practice, one plots the observed activity on the logarithmic ordinate of semilog graph paper as a function of time, which is plotted on the linear abscissa. If there were more than one radioactive component present in the sample, the observed activity at any time would be the sum of the activities of all components. A plot of total activity on semilog graph paper as a function of time would then appear as a curve. However, if the half-lives of the components are sufficiently different and if the time of the experiment is long compared to at least one of the half-lives involved, one can graphically resolve this composite curve into its component straight lines. By extrapolating these straight lines to $t = 0$, one can then calculate the abundance of each of the radionuclides. Figure 19.2 shows such a plot for a two-component system.

Normally, the experiment is conducted for a sufficient length of time that all but the longest-lived component have decayed essentially completely, and the longest-lived component, therefore, produces a straight-line plot. One then fits the best

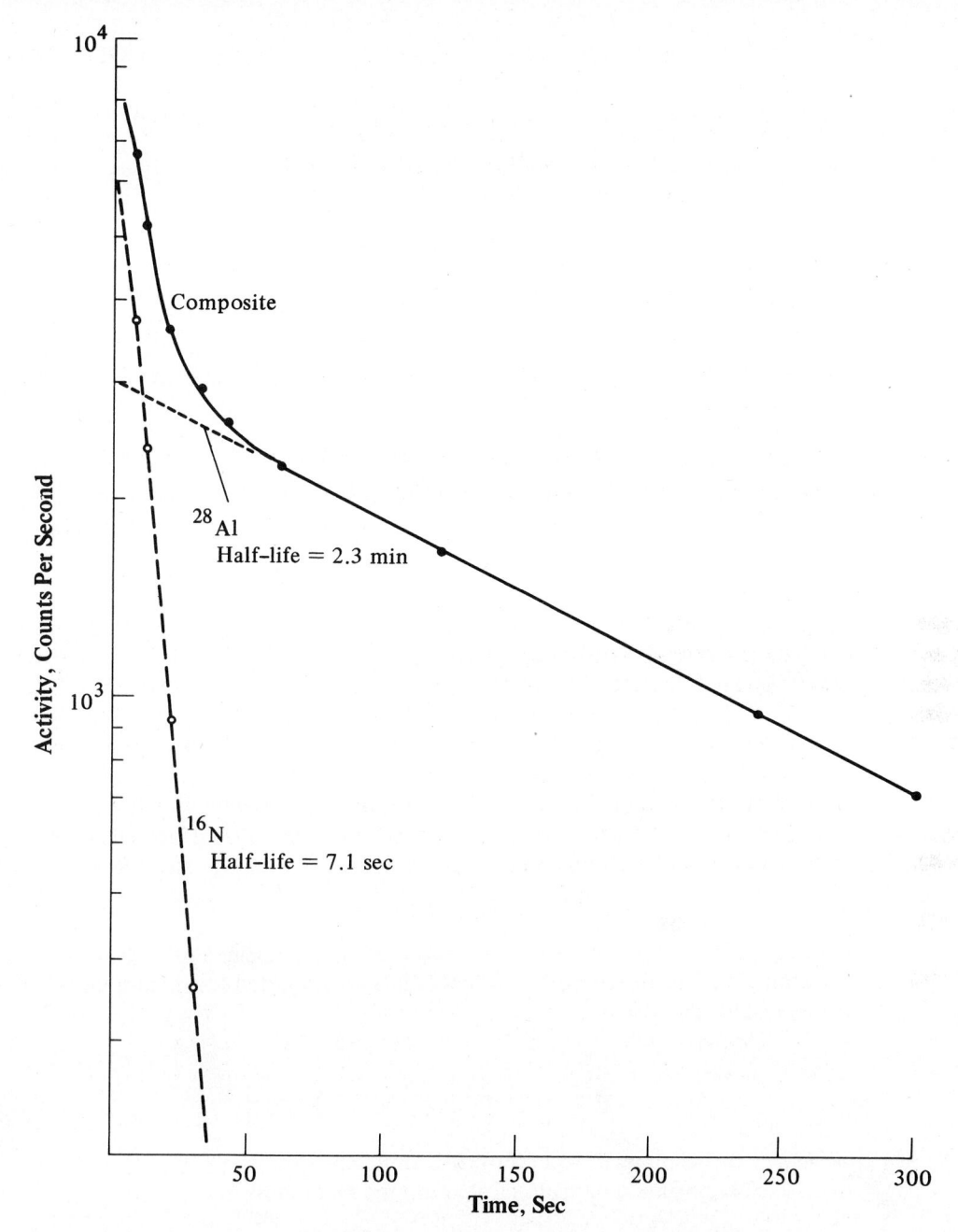

FIGURE 19.2. *Two-component decay curve for* ^{28}Al *and* ^{16}N.

straight line to this linear portion (computer line-fitting routines are commonly used) and extrapolates it to the origin. Point-by-point subtraction of this fitted straight line from this composite curve leaves another composite curve with one fewer member than the original one and whose end portion is also a straight line. Repetition of this *stripping* technique will give straight lines for all components, from which one can calculate the quantities $(N_0)_1$, $(N_0)_2$, etc. Ordinarily, the stripping technique is applied to mixtures of three or fewer components.

In general, the great advantage of the decay curve and stripping method over the method of direct calculation using simultaneous equations is that these plots often reveal the presence of unsuspected contaminant radionuclides in the sample. In addition, a large number of points are used to characterize each straight line in the decay curve and stripping method. Therefore, the effect of random errors associated with single determinations is lessened. The advantage of the method of simultaneous equations lies in the simplicity of the experiment and subsequent mathematical manipulations, but the method assumes a priori knowledge of all sample components.

For the resolution of a decay curve of a two-component system whose half-lives are not greatly different, another simple method is often used. Writing Equation 19.6 for a two-component system and multiplying both sides by $e^{\lambda_1 t}$ yields

$$A_{\text{total}}(e^{\lambda_1 t}) = \lambda_1 (N_0)_1 + \lambda_2 (N_0)_2 e^{(\lambda_1 - \lambda_2)t} \tag{19.10}$$

If one plots $A_{\text{total}} \exp(\lambda_1 t)$ as a function of $\exp(\lambda_1 - \lambda_2)t$, a straight line is obtained whose intercept and slope may be used to directly calculate $(N_0)_1$ and $(N_0)_2$.

19.2 NUCLEAR REACTIONS AND TYPES OF RADIOACTIVE DECAY

By convention, a nuclide is specified as $^A_Z X$, where X is the chemical symbol for the element, the superscript A denotes the mass number, and the subscript Z the atomic number of the nuclide. When particles or radiations interact with a nucleus, nuclear reactions occur. For example, a neutron may interact with the nucleus of the nuclide $^{27}_{13}\text{Al}$ according to the following equation written in the style of a chemical reaction:

$$^{27}_{13}\text{Al} + ^1_0 n \rightarrow ^{28}_{13}\text{Al} + Q \tag{19.11}$$

This equation states that a neutron is added to the nucleus of $^{27}_{13}\text{Al}$ to produce the nuclide $^{28}_{13}\text{Al}$ and the *energy Q*, which is largely imparted to an emitted gamma ray in this particular reaction.

Quantity Q (MeV) may be calculated as follows:

$$Q = 931 \left[\sum_{i=1}^{n} m_{r_i} - \sum_{i=1}^{k} m_{p_i} \right] \tag{19.12}$$

where m_{r_i} = mass of reactant r_i in atomic mass units
m_{p_i} = mass of product p_i in atomic mass units

The constant 931 is the proportionality constant between mass in atomic mass units and energy in million electron-volts (MeV). If the value of Q is positive, energy is produced and the reaction will proceed spontaneously; such a reaction is said to be *exoergic*. If Q is negative, energy is required for the reaction to proceed, and the reaction is said to be *endoergic*. The energy required to initiate endoergic reactions is often supplied in the form of kinetic energy of the incident particle. The energy released in exoergic reactions may be in the form of electromagnetic radiation (called gamma radiation), or in the form of kinetic energy given to emitted particles and the product nucleus.

An abbreviated form is usually used for writing nuclear reactions. The reaction given in Expression 19.11 can be written as

$$^{27}_{13}\text{Al}\,(n, \gamma)\,^{28}_{13}\text{Al} \qquad (19.13)$$

where the actual emitted particle, in this case the gamma ray, is formally denoted. This reaction of $^{27}_{13}\text{Al}$ is called an (n, γ) reaction. This is understood to mean that a neutron is absorbed by a nucleus of $^{27}_{13}\text{Al}$ and gamma radiation is emitted, resulting in the formation of a product nucleus $^{28}_{13}\text{Al}$. The product nucleus of a nuclear reaction can be either stable or radioactive. If the product nuclide is radioactive, it will eventually decay to a different nuclide. The most common modes of decay are emission of alpha particles, beta particles, and gamma rays; other particles or radiations can also be emitted in radioactive decay, but they are of little analytical utility and will not be discussed here. Radioactive decay may involve a single-step transformation or may proceed through a series of steps. An example of the former is

$$^{28}_{13}\text{Al} \rightarrow {}^{28}_{14}\text{Si} + \beta^- \qquad (19.14)$$

The nuclide $^{28}_{14}\text{Si}$ is stable with respect to further nuclear decay. The latter type of decay scheme is exemplified by

$$^{47}_{20}\text{Ca} \rightarrow {}^{47}_{21}\text{Sc} + \beta^- \\ \downarrow \\ {}^{47}_{22}\text{Ti} + \beta^- \qquad (19.15)$$

The final product nuclide, $^{47}_{22}\text{Ti}$, is stable.

Beta Decay

There are three forms of beta decay. One is called *negatron* (β^-) *emission*. Negatrons are ordinary electrons that are emitted from nuclei as the result of a nuclear transformation. Negatron decay is illustrated by the symbolic expression

$$^A_Z\text{X} \rightarrow {}^A_{Z+1}\text{Y} + \beta^- + \bar{\nu} \qquad (19.16)$$

where $\bar{\nu}$ is an *antineutrino*, an elusive particle of zero or near-zero mass, that is of no practical analytical interest. The daughter nuclide Y formed by this transformation may initially exist in an excited state or in its ground state. If negatron emission results in an excited daughter state, deexcitation usually follows promptly, usually with the emission of one or several gamma rays. The ejection of high-energy orbital electrons is an alternative to gamma-ray emission; this is known as *internal conversion*. The radionuclides ^3H (tritium) and ^{14}C are examples of pure negatron emitters that decay directly to the ground state of their daughters without the emission of gamma radiation. Many other negatron emitters do have accompanying gamma radiations which may be detected in preference to measuring the short-ranged negatrons, which may be absorbed within the sample.

The second beta-decay process is *positron* (β^+) *emission*. Positron decay is illustrated by the symbolic expression

$$^A_Z\text{X} \rightarrow {}^A_{Z-1}\text{Y} + \beta^+ + \nu \qquad (19.17)$$

where v is a *neutrino*—again a particle of no analytical utility with essentially no mass. The radionuclides $^{22}_{11}$Na and $^{65}_{30}$Zn are examples of positron emitters commonly used in radioanalytical work. Again, the daughter nuclide may be formed in an excited state or in its ground state. Hence, gamma radiation may also accompany this mode of decay.

The third beta-decay process is *electron capture*. In this process, the nucleus captures an orbital electron (usually a K electron), which creates an orbital vacancy. This vacancy is then filled by electrons from higher energy levels. The energy difference between the electronic-energy levels involved is released in the form of x-rays, or by the emission of low-energy *Auger* or *secondary electrons* (see Chap. 15) from orbitals of higher electronic energy. The symbolic expression for this process is similar to that for positron decay, but no positrons are emitted from the nucleus. Gamma radiation may also accompany this type of beta decay. The x-rays emitted in electron-capture decay are sometimes useful for analytical determinations. An example of a radionuclide decaying by electron capture is

$$^{7}_{4}\text{Be} + e^- \rightarrow {^{7}_{3}}\text{Li} + \text{x-Rays} + v \qquad (19.18)$$

The energy distribution of beta particles emitted in negatron or positron decay is continuous (Fig. 19.3). The maximum energy associated with the distribution is called E_{max} and is characteristic of the particular nuclear transformation. At energies less than this, part of the energy resides in the neutrino or antineutrino emitted with

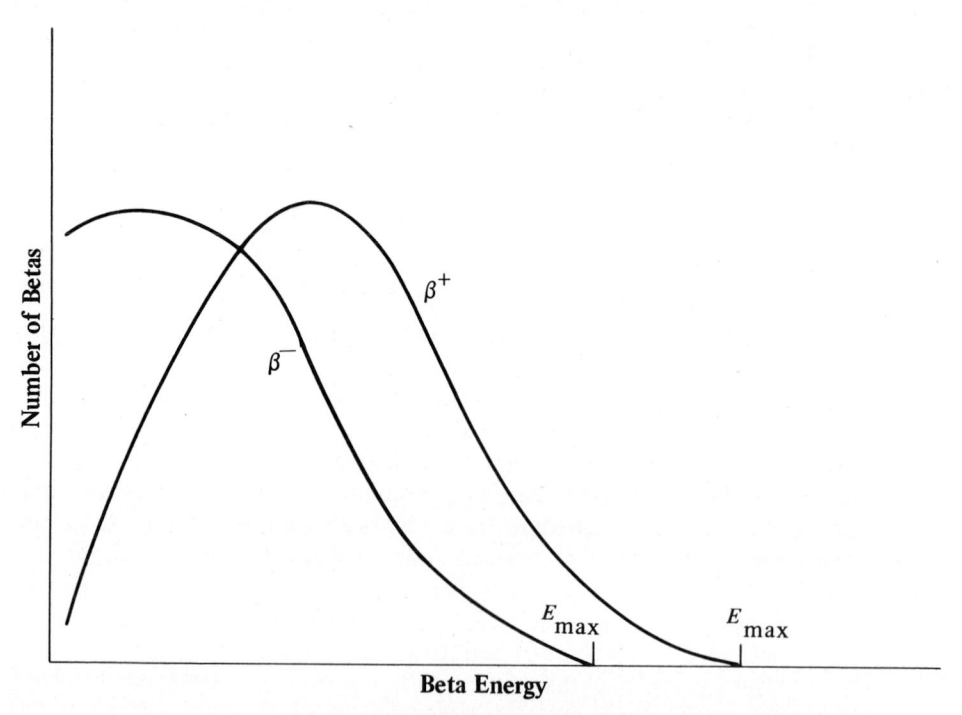

FIGURE 19.3. *Spectra of negatrons and positrons emitted during beta decay. E_{max} corresponds to the total energy of the transition.*

the beta particle; the sum of the two energies is equal to the characteristic maximum energy.

Beta particles generally have a short range in matter. For instance, beta particles from a transformation having $E_{max} = 1.0$ MeV are completely stopped by a 1.5-mm thickness of aluminum foil. From an analytical point of view, this is a very important observation. When beta rays are being counted, sample thickness becomes an extremely important parameter and must be controlled accurately (i.e., all samples must have equal thicknesses), or appropriate corrections must be applied.

Negatrons commonly interact with matter in three ways: excitation, ionization, and bremsstrahlung production. As negatrons pass through matter, they may interact with a bound electron and either eject the electron, forming an *ion pair* of the ejected electron and the ionized atom, or simply raise the electron to one of its excited atomic- or molecular-energy levels. The incident negatron may lose part or all of its energy in each collision. On the other hand, high-energy beta particles may strike the target nucleus itself; the sudden deceleration in the strong electric field surrounding the nucleus generates an "electromagnetic shock wave," called *bremsstrahlung radiation*, carrying away part of the kinetic energy of the electron.

Interactions of positrons with matter are similar to those encountered with negatrons. However, after dissipating most of its energy, the positron undergoes a process known as *annihilation*. The positron interacts with an electron in the vicinity of an atom, and the masses of both particles are converted to energy according to Einstein's equation $E = mc^2$. Two gamma-ray photons, each having an energy equal to 0.511 MeV, are produced simultaneously. The detection of 0.511-MeV gamma rays usually indicates that a sample contains a positron-emitting radionuclide.

Alpha Decay

Alpha decay is common among nuclei with high atomic number and among several rare-earth elements. This type of decay is illustrated by

$$^{A}_{Z}X \rightarrow {^{A-4}_{Z-2}}Y + {^{4}_{2}}He \qquad (19.19)$$

An example of a radionuclide decaying by alpha emission is $^{210}_{84}Po$.

The emitted alpha particles have discrete energies. In passing through matter, they interact chiefly with electrons, dissociating molecules and exciting or ionizing molecules and atoms. The range of alpha particles in matter is much shorter than that of beta particles of similar energies. For example, a beam of 1.0-MeV alpha particles is stopped completely by a 3–4-μm thickness of aluminum foil (compared to about 1.5 mm for beta particles). This difference in penetrability is due to the lower velocity and greater charge of the alpha particle as compared to the beta particle. As in the case of beta decay, gamma radiation may accompany alpha decay, if the immediate product nuclide is formed in an excited state.

Electromagnetic Radiation

Electromagnetic radiation (photons) emitted from the nucleus is called gamma radiation. This radiation has neither charge nor mass, although its energy can be converted to an equivalent mass by using the energy-mass equation ($E = mc^2$). Since gamma rays carry only energy, the emitting nuclide does not change in mass number or atomic number, thus preserving its chemical identity. Gamma rays emitted from any radionuclide have discrete energies characteristic of the different nuclear-energy states of that nuclide. Therefore, they can be used to "fingerprint" the materials from which they are emitted. The technique used to measure the numbers and energies of gamma rays emitted by radionuclides is called *gamma-ray spectrometry.*

The interaction of gamma radiation with matter is much more complex than the interaction of charged particles. There are three modes of interaction; the extent of each depends on the nature of the material and the energy of the radiation. In the first mode, called the *photoelectric effect*, the incoming gamma ray interacts with one of the orbital electrons in an absorber atom, ejecting that electron. The electron carries away kinetic energy equal to the energy of the gamma ray less the binding energy of the electron to the absorber atom. The ejected electron may then interact with other electrons, causing secondary ionization. The important characteristic of this process is the fact that essentially all of the energy of the gamma ray is given up in a single primary interaction. This process is most important for gamma rays of low energy (up to 1 MeV) and target materials with high atomic number.

The second process of gamma-ray interaction is called *Compton scattering.* In this process, the incoming gamma ray interacts with either a bound or free electron, losing only part of its energy in the encounter. If an electron is ejected from a bound state, its kinetic energy will be equal to the energy given up by the gamma ray less the binding energy of the electron. In this process, any one gamma ray may interact with many electrons. Of course, a gamma ray that has lost only part of its energy by this process may then undergo a photoelectric interaction, or simply escape from the absorber or detector.

The third process of gamma-ray interaction is *pair production.* This process occurs only when gamma rays have an energy equal to or greater than 1.02 MeV. In this process, the gamma-ray photon interacts with the absorber to produce an electron-positron pair. This is the reverse of the positron annihilation process discussed earlier. The reason for the minimum energy requirement is that 1.02 MeV is the energy equivalent of the two electron masses that must be created. The excess energy of the gamma ray $(E_\gamma - 1.02 \text{ MeV})$ appears largely as kinetic energy given to the positron and the electron. The two particles may then interact further with other atomic electrons, causing secondary ionization.

Electromagnetic radiation penetrates much deeper into matter than do charged particles. Attenuation of gamma rays in matter follows an exponential law, similar to the Beer-Lambert law for absorption of visible light. To decrease the intensity of a 1.0-MeV parallel beam of gamma rays to one-half its original value requires a 4-cm thickness of aluminum (compared to 3–4-μm for alpha and 1.5-mm for beta radiation). In general, light materials are very ineffective in stopping gamma rays. Common shielding materials used are lead and high-density concrete.

19.3 DETECTION OF NUCLEAR RADIATION

To detect nuclear radiation, one ordinarily uses a transducer capable of converting the energy of the radiation into an electrical signal, usually a voltage pulse. Depending on the nature of the application, a satisfactory transducer must meet one or both of the following criteria: (a) There must be strict proportionality (preferably one-to-one) between the number of photons or particles interacting with the detector and the number of voltage pulses generated, and (b) there should be a strictly known relationship (preferably linear) between the energy that the radiation dissipates in the detector and the amplitude of the voltage pulse (called *pulse height*, PH) generated by the transducer.

Three types of radiation detectors are in common use: the *gas-ionization detector*, the *scintillation detector*, and the *solid-state* (or *semiconductor*) *detector*. Generally, the type used depends on the specific application. Gas-ionization detectors are commonly used for inexpensive detection of charged particles, scintillation detectors for beta- and gamma-ray detection, and solid-state detectors for x-ray and gamma-ray detection. The operation and properties of these detectors will be briefly described.

Gas-Ionization Detectors

A *gas-ionization detector* consists of two electrodes at different potentials and a (nonconducting) gas between them. The radiation produces ion pairs in the gas; the ions are then collected by the electrodes, yielding a voltage pulse, which is measured. Figure 19.4 shows a typical cylindrical detector together with its associated measurement system. The detector proper is made of a cylindrical conducting material with an electrically isolated wire located on the central axis of the cylinder. A very thin window made of Mylar, aluminum, or beryllium separates the radioactive source from the main volume of the detector (or the source can be placed directly inside the chamber). Either the chamber can be permanently filled with some appropriate gas

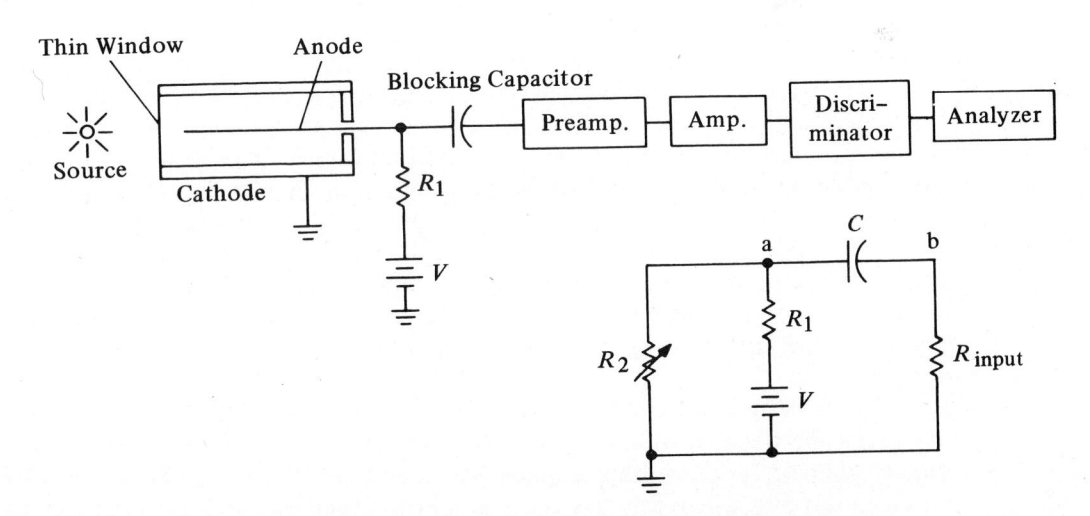

FIGURE 19.4. *Gas-ionization detector system. Top: Block diagram illustrating system components. Bottom: Simplified electrical analog.*

SEC. 19.3 Detection of Nuclear Radiation

(e.g., 90% Ar + 10% CH$_4$), or a continuous flow of the gas can be sent through the chamber.

The detector operates as follows: As long as there is no ionizing radiation present, the filter gas acts as a very large resistor (R_2) and allows virtually no current to pass through the tube. The voltage appearing at point "a" is then equal to the bias voltage V of the battery or regulated power supply. Capacitor C acts to block this voltage from reaching point "b." Therefore, the voltage seen by the input of the amplifier (R_{input}) is zero. Now, assume that a single ionizing particle enters the detector volume and produces a number of ion pairs. Since there is an electrostatic potential between the two electrodes, the positive ions are attracted toward the cathode and the electrons toward the anode. The electrons that reach the anode pass through the external circuit (made of R_1 and V), while an equal number of electrons flow from the negative terminal of the battery to the cathode to neutralize the positive ions that reach the cathode. This is analogous to a sudden drop in resistance of the filler gas (R_2 of Fig. 19.4). Thus, the voltage at point "a" will suddenly drop from V to v:

$$v = V \frac{R_2}{R_1 + R_2} \qquad (19.20)$$

Here, R_2, the effective resistance of the filler gas, depends on the number of ion pairs produced. The overall effect is a sudden change in voltage at point "a" in the form of a sharp spike. This spike appears at the input of the preamplifier as a negative voltage pulse. The amplitude of this spike is the PH.

Three types of ionization detectors, differing by the type of filler gas and the magnitude of the bias voltage V, are generally recognized: *ion chambers*, *proportional counters*, and *Geiger-Müller counters*. Each of these is best suited for a specific application. A plot of PH as a function of bias voltage has the general features shown in Figure 19.5. (It is important to note at this stage that PH is proportional to the number of ion pairs actually reaching the electrodes and not necessarily to how many were produced inside the detector, since some ion pairs may combine on their way and not reach the electrode—a phenomenon called ion recombination.)

Depending on their mode of production, two types of ion pairs are recognized: *primary* ion pairs and *secondary* ion pairs. The former are ion pairs produced from the direct interaction of radiation with the filler gas, whereas the latter are pairs produced by the energetic electrons of the primary ion pairs. The number of primary ion pairs produced per incoming particle depends on the amount of energy dissipated inside the detector volume. If all of the energy of the incoming particle is dissipated inside the detector, the number of primary ion pairs will then be directly proportional to the energy of the incoming radiation. The number of secondary ion pairs produced depends on the kinetic energy of each primary electron and on the voltage applied to the detector.

To understand the curve in Figure 19.5, assume that a single particle of radiation enters the detector volume, dissipating all of its energy and producing 1000 primary ion pairs. If the voltage applied is smaller than V_1, only a fraction of these primary ion pairs will reach the electrodes. The remainder will undergo ion recombination. The higher the applied voltage (for $V < V_1$), the larger the number of primary ion pairs reaching the electrodes and the larger the PH. This region of the

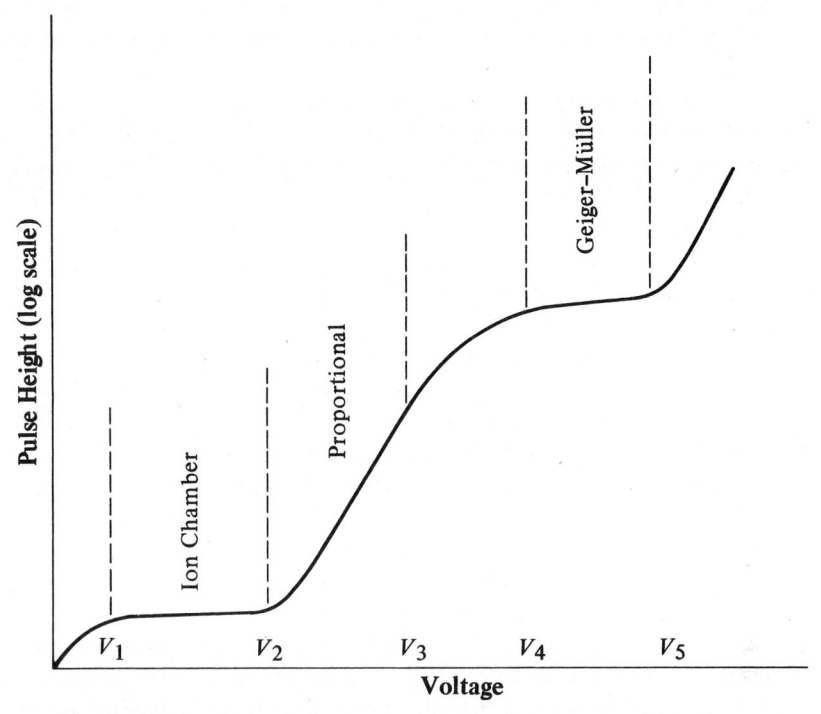

FIGURE 19.5. *Pulse height (PH) as a function of applied bias voltage for a gas-ionization detector.*

curve is of limited practical value. For voltages between V_1 and V_2, the kinetic energy of the primary ion pairs is sufficient for almost all of them to reach the electrodes, but is not enough to produce secondary ion pairs. In this region, PH is independent of applied voltage but is directly proportional to the energy dissipated inside the detector. This is the region used for ion-chamber detectors. Many types of filler gases (even air) may be used. Since there are no secondary ion pairs produced, PH is extremely small (the voltage pulse would be approximately 0.001 μV).

If the applied voltage is increased beyond V_2, the primary electrons will be subject to a relatively large electrostatic field and will acquire enough kinetic energy to, in turn, ionize the filler gas (typically mixtures of argon and methane), producing secondary ion pairs. In the region between V_2 and V_3, the total number of ion pairs reaching the electrodes, and thus PH, is proportional to both the applied voltage and the energy dissipated in the detector. This region is called the *proportional region* and is extensively used for measuring alpha and beta radiation. One important feature of this region is that PH is generally in the range of 1 to 100 mV and, therefore, is relatively easy to measure. Also, since PH is proportional to the energy dissipated in the detector, not only can the number of incoming particles per unit time be measured, but also their energies. Another very important feature of proportional counters is that of *detector dead-time*—only a short time (a few microseconds) passes between the entry of radiation and the time when all ion pairs are collected at the electrodes and current stops flowing. Therefore, radioactive sources with fairly high disintegration rates can be measured without correcting for detector dead-time losses.

SEC. 19.3 **Detection of Nuclear Radiation**

When the applied voltage exceeds V_3, the electrostatic field becomes so large that a chain reaction results from the interaction of any ionizing particles, resulting in loss of proportionality between energy dissipated and PH. This proportionality is lost first gradually (region between V_3 and V_4) and then completely ($V_4 < V < V_5$)—the detector discharges throughout its entire volume. For the region between V_4 and V_5, PH becomes independent of the energy dissipated in the detector by the incoming particle, but may increase slightly with increasing applied voltage. This is called the *Geiger-Müller* region, and detectors operating in this region are referred to as *GM counters*. Filler gases are typically mixtures of helium with small amounts (2 to 3%) of a volatile hydrocarbon or elemental halogen gases added. These counters produce large voltage pulses (a fraction of a volt to a few volts) and require little external amplification. Therefore, GM counting systems are often inexpensive. The basic limitations of GM counters are their large dead-time (a few hundred microseconds per pulse) and their lack of energy discrimination. These counters are widely used for portable radiation monitors and for counting gross beta activity in tracer experiments.

At voltages greater than V_5, the filler gas itself begins to ionize because of the large voltages involved, and produces a continuous discharge whether or not there is any ionizing radiation present. When using either proportional or GM counters, it is important to realize that the magnitude of the bias voltage may affect the observed count rate. Therefore, before using the counter, one must establish the proper bias voltage by counting an essentially constant activity source with different bias voltages (Fig. 19.6). The proper operating voltage is that corresponding to the midpoint of the plateau. With a proportional counter, different plateaus are obtained for alpha and beta particles and, hence, discrimination is possible.

Scintillation Detectors

It was pointed out earlier that interaction of radiation with matter can result in the production of ion pairs. The electron from the ion pair can in turn produce secondary

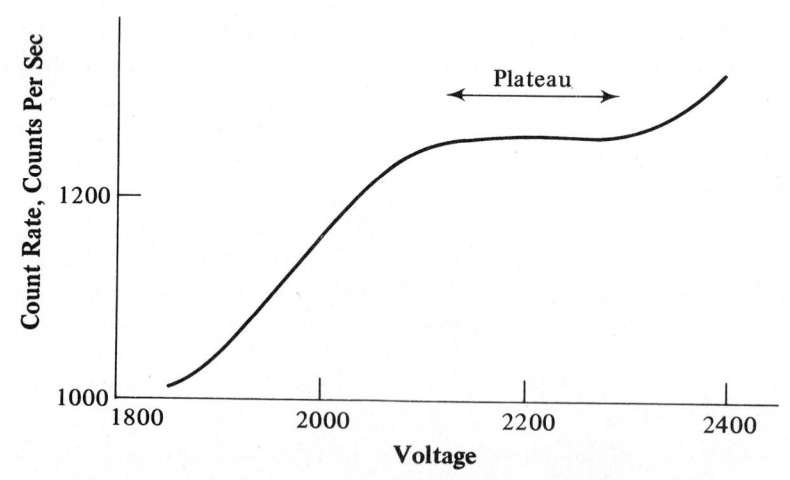

FIGURE 19.6. *Voltage plateau for a neutron proportional counter. The proper operating voltage is at the midpoint of the plateau.*

ionization, which is important in gas-ionization detectors. These secondary electrons can also electronically excite the atoms in the detection medium, which in turn can emit light quanta. *Fluors* or *phosphors* are materials in which such a sequence of events occurs. The number of light quanta emitted is proportional to the energy of the radiation absorbed; so these materials can be used as radiation detectors permitting energy discrimination. Detectors of this type are called *scintillation detectors*. Typical materials used as scintillators are thallium-doped NaI crystals, anthracene crystals, and certain organic compounds such as *p*-terphenyl dissolved in organic solvents. NaI(Tl) crystals are particularly efficient for gamma-ray detection because they contain a high-Z material (iodine) and have a relatively high density.

Liquid scintillators (organic scintillators dissolved in an appropriate solvent) are generally used for detecting beta particles. They are particularly useful for low-energy beta emitters such as ^{3}H or ^{14}C; these are used widely as tracers in biochemistry and organic chemistry. The sample is commonly dissolved in the solvent along with the scintillator. Special counters are required because the pulses generated by tritium, ^{3}H, are of nearly the same amplitude as the background pulses from thermionic emission in the photomultiplier detector at room temperature; therefore, the photomultiplier and amplifier electronics are cooled to $-10°$C. Some newer liquid scintillation counters employ two photomultiplier detectors mounted at 180° with the sample cavity between them. Only pulses detected simultaneously by both detectors are recorded (*coincidence detection*). Hence, background pulses generated by thermionic emission in the individual detectors are discriminated against without the need for cooling.

A scintillation detector generally consists of a fluor placed in close contact with a photomultiplier tube. The flashes of light emitted from the fluor enter the photomultiplier, generating a large current pulse from each primary scintillation event. The current pulse is then converted to a voltage pulse, which is amplified and analyzed. The amplitude of this pulse is the PH, and is proportional to the energy originally deposited in the fluor by the radiation.

As discussed previously, gamma rays have discrete, well-defined energies. Each incoming gamma ray interacts with the detector material by one or more of the three processes discussed earlier, leaving part or all of its energy inside the detector. Since each gamma ray (of the same energy) may deposit a different fraction of its initial energy, the output voltage pulses will not have identical PHs, although the PH distribution will be characteristic.

If the material of interest emits gamma rays having more than one characteristic energy, the PH distribution (gamma-ray spectrum) will be quite complex. To analyze such a spectrum, a multichannel analyzer (MCA) is used with the scintillation detector. The MCA receives the voltage pulses, classifies each pulse according to its PH, and stores all the pulses of equal PH in the same memory location (channel). For example, a 400-channel analyzer can be calibrated to store a 10-V pulse in channel 400 and a 5-V pulse in channel 200. Any gamma ray giving rise to a PH of 7.5 V will be stored in channel 300, since the relationship between PH (gamma-ray energy) and channel number for a MCA is (generally) linear.

If one places a monoenergetic source of gamma radiation in front of a NaI(Tl) detector and uses a MCA to analyze the PH distribution of the detector, a spectrum similar to that in Figure 19.7 is obtained. Note that the total number of counts

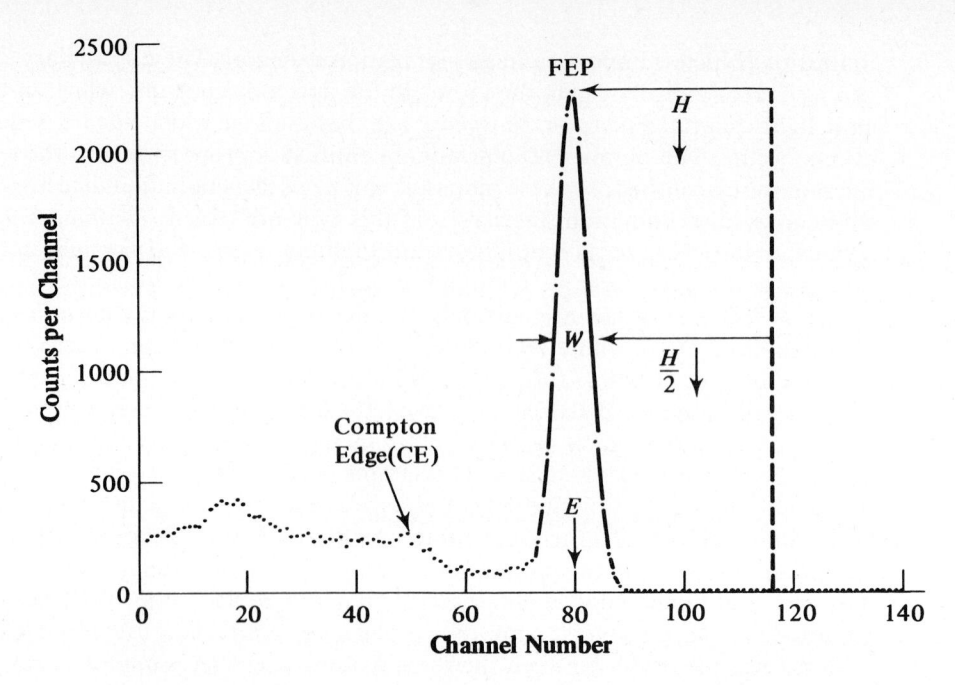

FIGURE 19.7. *Gamma-ray spectrum of a* ^{137}Cs *source, using a* 4" × 4" *NaI(Tl) scintillation detector. The energy resolution of the detector is measured by the width* (W) *of the full-energy peak* (FEP) *at one-half the maximum height* (H/2) *of the FEP.*

registered in each channel during the entire counting time is plotted as a function of channel number (which is proportional to energy). The actual gamma-ray spectrum is a continuum (the *Compton continuum*) with a peak (the *full-energy peak*), even if the source emits only gamma rays with a single energy. The Compton continuum, starting from low energies and ending at the point CE, is caused by incomplete deposition of the gamma-ray energy following a Compton interaction with the detector crystal. The edge of the Compton continuum is called the *Compton edge* (CE) and corresponds to the maximum energy that a gamma ray can transfer to an electron following a Compton interaction. The degraded gamma-ray produced in the Compton interaction may either escape the crystal or undergo further interaction. If all the energy of the primary gamma ray is eventually deposited in the crystal, the event is recorded in the *full-energy peak* (FEP), together with the events from photoelectric-effect interactions. If the degraded gamma ray escapes the crystal, the event is recorded in the Compton continuum.

For analytical purposes, the most important feature of the spectrum is the FEP. This peak includes all primary gamma-ray interactions that occur by the photoelectric effect, as well as those interactions caused by Compton scattering and pair production in which the energies of the secondary radiations or particles are completely dissipated in the detector crystal. Although the gamma-ray energy is well defined, the FEP always has a certain width that is a function of both the type of detector and the energy of the FEP. The energy corresponding to the channel at the center of the FEP is essentially the energy of the primary gamma ray that interacted with the crystal. The area under this peak is related to the *activity* of the source at the beginning of the counting time (t_c) by

$$\text{FEP Area} = \xi \int_0^{t_c} A_0 e^{-\lambda t}\, dt = \frac{\xi A_0}{\lambda}(1 - e^{-\lambda t_c}) \qquad (19.21)$$

where ξ is the overall efficiency of the counting system and includes factors for the counting geometry and FEP detection probability; it is a constant for any fixed experimental setup. Quantity A_0 is the activity of the specific gamma ray emitted by the radioactive source at $t_c = 0$.

The width (W) of the FEP at the point corresponding to one-half its height (Fig. 19.7) is called the *full width at half maximum* (FWHM). The ratio of W to E_γ (both in units of energy) expressed as a percentage is called the *resolution* (R) of the detector:

$$R = \frac{W}{E_\gamma} 100\% \qquad (19.22)$$

For a typical NaI(Tl) detector, R is about 8% for the 1.332-MeV FEP of ^{60}Co; for the newer Ge(Li) semiconductor detectors, R may be as low as 0.2% for ^{60}Co.

If a source emits gamma rays with different energies, a composite spectrum results. Figure 19.8 shows a composite spectrum for $^{54}_{25}$Mn, $^{60}_{27}$Co, and $^{137}_{55}$Cs. In

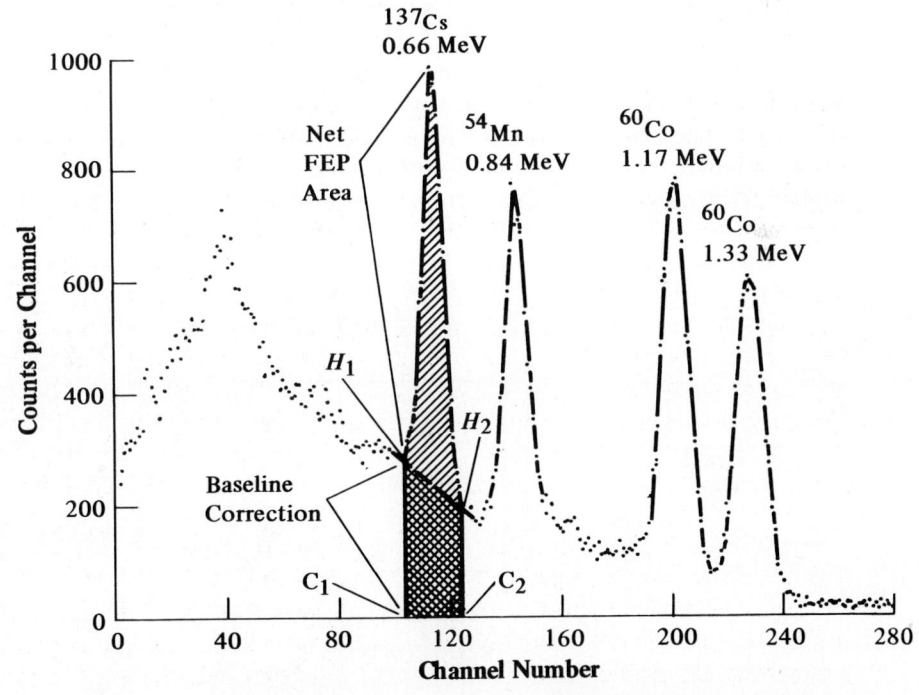

FIGURE 19.8. *Composite gamma-ray spectrum of* ^{137}Cs, ^{54}Mn, *and* ^{60}Co. *The net area of the full-energy peak (FEP), as obtained by baseline subtraction, is proportional to the activity of the radionuclide.*

any quantitative analysis using composite gamma-ray spectra, one needs to measure the area under each FEP. To do this, one must first select the portion of the FEP that lies above the Compton background from higher-energy peaks. This is commonly done by locating the left channel (C_1) just before the FEP appears to rise above the background and the right channel (C_2) at the point where the FEP disappears into the background. (This selection is sometimes difficult because of statistical variations, especially if the FEP of interest is small compared to the underlying Compton background.) One then calculates the net area of the FEP by subtracting from the total area the average background count per channel multiplied by the number of channels:

$$\text{Net Area (counts)} = \left(\frac{\text{Total Area}}{C_1 \rightarrow C_2}\right) - \left(\frac{H_1 + H_2}{2}\right)(C_2 - C_1 + 1) \quad (19.23)$$

It is often necessary to determine the statistical error involved in this calculation, following the procedures given in Section 19.5. If the background-count distributions around C_1 and C_2 appear to be horizontal, it is often advisable to take averages of several channels preceding the FEP and several channels following the FEP in order to obtain a more representative baseline correction. There are several more sophisticated methods of baseline correction, but their discussion is beyond the scope of this chapter.

Solid-State (or Semiconductor) Detectors

A *solid-state* (or *semiconductor*) *detector* (SSD) operates on the same principle as a gas-ionization detector, but using a solid semiconductor instead of a filler gas. A SSD is a block of some semiconductor material (commonly Ge or Si) into which has been incorporated a minute quantity of a group IIIA element such as gallium (Fig. 19.9). The doped block, having a lower density of free electrons than the pure semiconductor, is called a *p*-type semiconductor. (An *n*-type semiconductor has a higher density of free electrons than the pure semiconductor.) A very thin layer (a few micrometers) of Li is then diffused into one surface of the block. If a reverse bias is

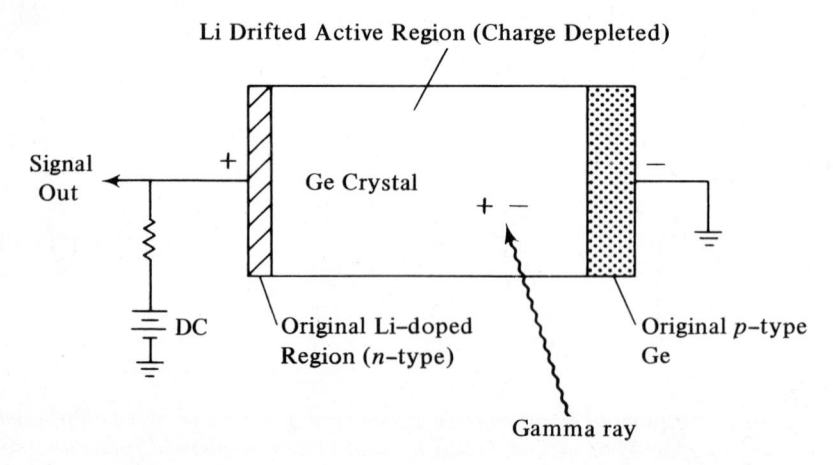

FIGURE 19.9. *Schematic diagram of a Ge(Li) detector.*

applied between the p side ($-$ bias) and the lithium side ($+$ bias) of the block, it will develop a *charge-depleted region*. This region has a very high effective resistance; by cooling the block to the temperature of liquid nitrogen, the current flow is decreased even further. The depletion region constitutes the effective volume of the detector.

If a gamma ray enters the depletion region of the detector, it may interact and form an ion-electron pair, which may in turn cause secondary ionization on their way to the lithium side of the detector (the positive electrode). These electrons charge up the detector (the two electrodes of the detector separated by its dielectric form a capacitor) and produce a voltage pulse across the electrodes. The magnitude of this

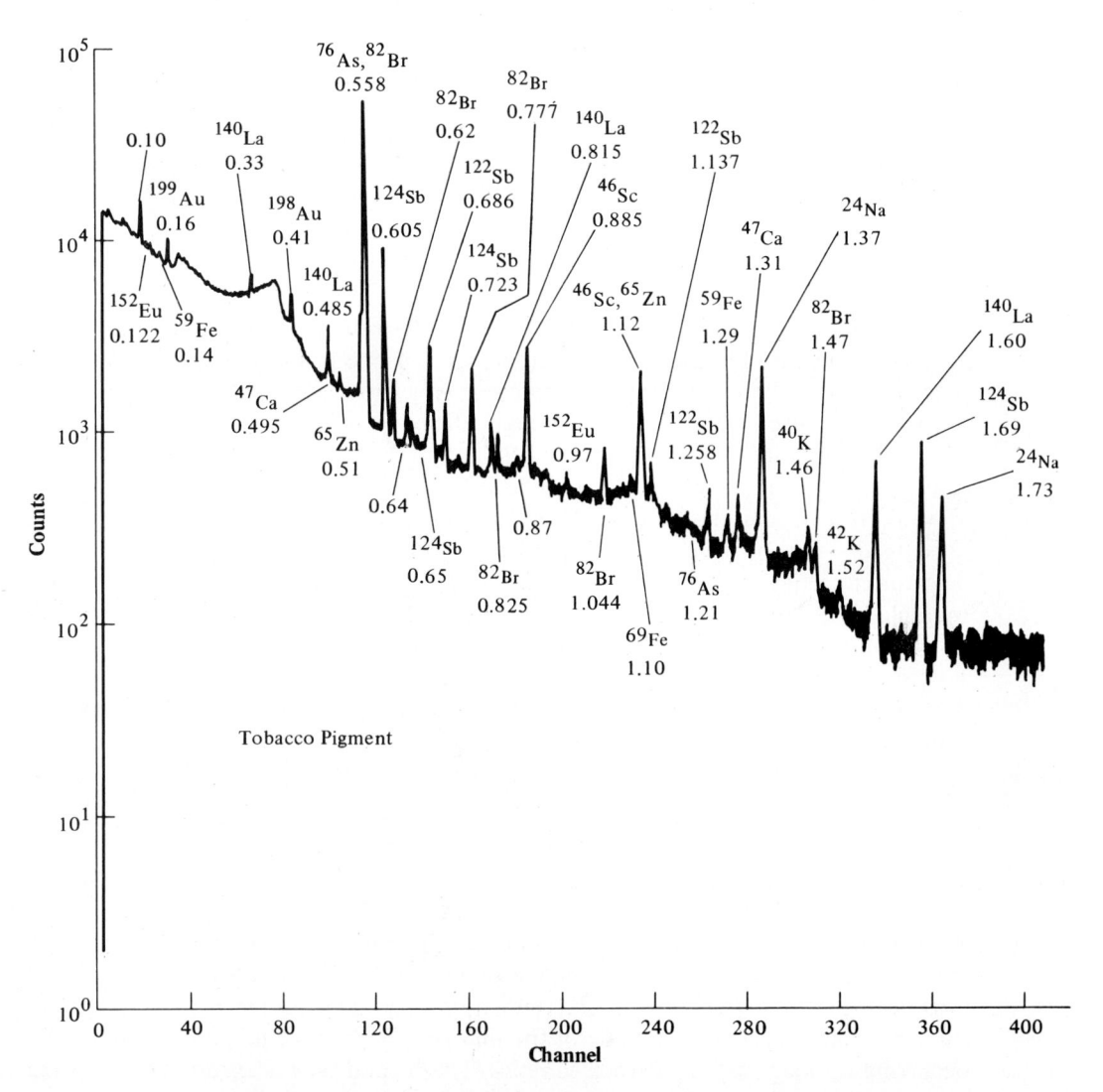

FIGURE 19.10. *High-resolution gamma-ray spectrum obtained with a 35-cm^3 Ge(Li) detector and a reactor-irradiated tobacco extract. Note that the number of counts is on a logarithmic scale.*

pulse is

$$V = \frac{Q}{C} \tag{19.24}$$

where Q = total charge collected on the electrodes
C = detector capacitance

This voltage pulse is very small (of the order of microvolts) and is amplified by a very sensitive preamplifier before it can be processed by conventional amplifiers and MCAs.

    The two important parameters characterizing gamma-ray detectors are *resolution* and *efficiency*. The resolution of a detector measures its capability to separate (resolve) adjacent gamma rays and becomes very important when unwanted gamma rays are present near the FEP of interest. Efficiency, on the other hand, measures only the fraction of the incoming gamma rays of any given energy that contribute to the corresponding FEP. Ideally, the detector has as good a resolution (low R) and as large an efficiency (ξ) as possible. However, with present-day gamma-ray detectors, these two properties are conflicting—detectors with high efficiency have poor resolving power and vice versa. Therefore, the type of detector chosen depends on the requirements of the analysis. For example, if a sample contains many gamma-emitting radionuclides, one must select a detector with good resolution and accept its lower inherent efficiency. On the other hand, if one is analyzing a region of the spectrum where there are no spectral interferences and the sample has a low activity, one would choose a detector having a high efficiency.

    The fundamental advantage of a SSD over a scintillation detector is its superior resolution. The efficiency of a Ge(Li) SSD is often expressed relative to that of a 3″ × 3″ NaI(Tl) scintillation detector for a point source of ^{60}Co 1.332-MeV gamma rays at a distance of 25 cm from the detectors. Measured in this way, a typical Ge(Li) detector efficiency is only 10%. Therefore, although the resolution may be improved by a factor of 40, the efficiency is decreased by a factor of 10. The superior energy resolution of a Ge(Li) detector is well illustrated by the spectrum in Figure 19.10. This spectrum was obtained by counting a reactor-irradiated extract from tobacco with a Ge(Li) detector. Gamma-ray peaks from at least a dozen elements can be distinguished.

19.4 ACTIVATION ANALYSIS

Radiochemical methods of analysis can be grouped according to whether one measures the natural radioactivity present in the sample or uses some means of introducing radioactivity into an otherwise nonradioactive sample in order to analyze for some component. An example of the first type is the determination of naturally occurring radioactive ^{40}K in rock samples. The second type is exemplified by using labeled KI*O$_3$ (I* denoting a radioisotope of iodine) to determine the concentration of SO$_2$ in air by the radiorelease method. This chapter deals almost exclusively with the use of radioactivity to analyze otherwise nonradioactive substances.

Radiochemical methods that induce radioactivity in a nonradioactive sample are further divided into two general categories: those that induce radioactivity in the components of the sample to be analyzed by some sort of bombardment process (Sec. 19.4), and those that make a sample radioactive by adding a radionuclide (Sec. 19.5). In this section, we will consider the first category. In all such methods, the sample is bombarded with nuclear radiations or particles (neutrons, protons, gamma rays, etc.), and the radiations emitted from the sample are measured, either simultaneously or subsequently. This general class of radiochemical methods is referred to as *activation analysis*.

Figure 19.11 presents in a schematic manner several different types of activation analysis used by radioanalytical chemists. If the sample or target $^A_Z E$ is bombarded with neutrons, the method is called *neutron activation analysis* (NAA). If radiochemical separations are required following the irradiation and prior to counting the sample, the method is termed *radiochemical neutron activation analysis* (RNAA). If the technique requires no chemical separations, the term *instrumental neutron activation analysis* (INAA) is used. The neutrons most commonly used in NAA are thermal-energy neutrons (approx. 0.04 eV) produced by nuclear-fission reactors. Nuclear reactors also produce neutrons of somewhat higher energies known as epithermal, resonance, and reactor fast neutrons. These higher-energy neutrons may have different interaction probabilities and may induce different nuclear reactions than reactor thermal neutrons. In *epithermal neutron activation analysis* (ENAA), samples are encased in a material that absorbs thermal neutrons (typically cadmium or a material containing boron) and allows the higher-energy neutrons to pass through and activate the sample. Because reactions and reaction probabilities vary with neutron energy, additional selectivity may be obtained by using ENAA for specific types of samples. Another type of NAA uses fast (14-MeV) neutrons produced by an accelerator (*fast neutron activation analysis*, FNAA).

Other types of activation analysis use charged particles (*charged particle activation analysis*, CPAA) from an ion accelerator such as a cyclotron or gamma rays

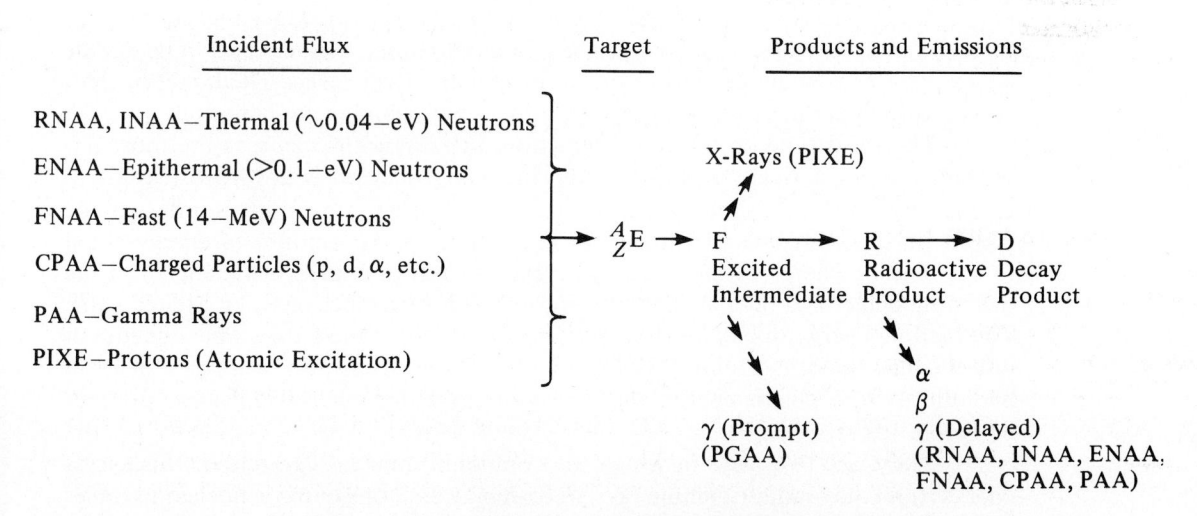

Incident Flux	Target	Products and Emissions
RNAA, INAA—Thermal (~0.04–eV) Neutrons		X-Rays (PIXE)
ENAA—Epithermal (>0.1–eV) Neutrons		
FNAA—Fast (14–MeV) Neutrons	$^A_Z E \rightarrow$ F $\rightarrow$ R $\rightarrow$ D
CPAA—Charged Particles (p, d, α, etc.)		Excited Radioactive Decay
PAA—Gamma Rays		Intermediate Product Product
PIXE—Protons (Atomic Excitation)		
		γ (Prompt) α, β, γ (Delayed)
		(PGAA) (RNAA, INAA, ENAA, FNAA, CPAA, PAA)

FIGURE 19.11. *Methods of activation analysis.*

produced by an electron linear accelerator (*photon activation analysis*, **PAA**) to induce the analytical nuclear reactions. *Proton induced x-ray emission analysis* (**PIXE**) employs 1–5-MeV protons from a Van de Graaff electrostatic generator to excite atomic x-ray emissions, which can be used for determination of elemental abundances in specially prepared matrices. The prompt gamma rays emitted in the course of thermal neutron (n, γ) reactions may also be used for analytical purposes (*prompt gamma activation analysis*, **PGAA**), rather than counting the "delayed" particles or radiations emitted by the radioactive nuclear reaction product nuclides. CPAA, PAA, PIXE, and PGAA are not so widely applied as NAA with either thermal or fast neutrons. Hence, the following discussion will deal primarily with the more widely employed techniques of RNAA and INAA.

Principles of Neutron Activation Analysis

Neutron activation analysis involves bombarding the sample with neutrons and measuring the radioactivity induced in the sample (commonly using gamma-ray spectrometry). In order to understand the principles of this method, some pertinent properties of neutrons and their interactions with matter will first be discussed.

Neutrons are nuclear particles with unit mass number and neutral charge; they are commonly produced as a result of nuclear reactions or nuclear fission, and interact with matter almost exclusively by collisions with nuclei. A neutron interacts with the nucleus of an atom in several ways. It can undergo *elastic scattering*, whereby the neutron collides with the target nucleus and is scattered (similar to a moving billiard ball striking another [stationary] ball). Depending on the size of the target nucleus and the angle of collision, a varying amount of the kinetic energy of the neutron is lost in adding kinetic energy to the target nucleus. If the target nucleus has a low mass (hydrogen, deuterium, carbon, etc.), a considerable fraction of the energy of the incident neutron may be lost in the collision. This is why low-mass materials (H_2O, D_2O, etc.) are used to reduce the kinetic energy of fast neutrons produced by fission in nuclear reactors—a process known as *thermalization*.

A neutron also undergoes *inelastic scattering* with a target nucleus. In this case, the neutron scatters off the nucleus of a target atom, transfers part of its kinetic energy, and excites the nucleus to one of its higher energy levels. The target nucleus can then dissipate this excess energy by emitting electromagnetic radiation.

The third type of neutron interaction, the *capture reaction*, is the most important one for activation analysis. The incoming neutron is absorbed (captured) by the target nucleus, forming a new nuclide with the same atomic number as the parent nuclide, but one unit higher in mass number. An amount of energy equal to the binding energy of the neutron in that nucleus plus the kinetic energy of the incoming neutron is then available to raise the product nucleus to an excited state. The binding energy differs for different nuclides; but, for the most stable nuclides of intermediate mass, it is about 8 MeV/nucleon. Thus, even if the captured neutron had almost zero kinetic energy, the excess energy of the *compound nucleus* is about 8 MeV.

There are two ways in which the compound nucleus can release this excess energy: (a) It may radiate gamma rays, or (b) it may emit one or more nuclear particles (neutrons, protons, or alpha particles). Which of these two processes predominates depends on the total excitation energy of the compound nucleus. If sufficient energy is available, more than one reaction can take place.

In order to determine whether a given nuclear reaction can occur, the energy balance for the complete reaction must be calculated. If the overall reaction produces energy (Q is positive), the reaction proceeds spontaneously. Consider the nuclear reaction

$$^{27}_{13}\text{Al} + ^{1}_{0}\text{n} \rightarrow ^{28}_{13}\text{Al} + \gamma + Q \tag{19.25}$$

Note that this is the same reaction as written in Expression 19.11, except that we have formally written in the emitted gamma ray. Remember that the total (excess) nuclear energy Q that we will now be calculating for this process is largely imparted to the gamma ray. Using Equation 19.12,

$$\sum m_r = 26.981535 + 1.008665 = 27.990200 \text{ amu}$$

$$\sum m_p = 27.981908 = 27.981908 \text{ amu}$$

$$Q = 931(27.990200 - 27.981908) = +7.7 \text{ MeV}$$

The positive value of Q indicates that the reaction will proceed with neutrons having nearly zero kinetic energy (thermal neutrons have an average energy of approximately 0.04 eV).

For the reaction

$$^{27}_{13}\text{Al} + ^{1}_{0}\text{n} \rightarrow ^{26}_{13}\text{Al} + 2^{1}_{0}\text{n} + Q \tag{19.26}$$

$Q = -13.1$ MeV. Since the value of Q is negative, this reaction cannot take place without the input of energy. The needed energy must be supplied by the kinetic energy of the incoming neutron. The minimum amount of kinetic energy that the incoming neutron must provide for the above reaction is somewhat more than the calculated 13.1 MeV, because part of the kinetic energy of the incoming neutron is merely transferred to the target nucleus ($^{27}_{13}\text{Al}$) to produce a moving product nucleus, according to the principle of conservation of momentum. The *laboratory threshold energy*, E_T, required to initiate the reaction may be calculated by means of

$$E_T = Q\left(\frac{m_a + m_n}{m_a}\right) \tag{19.27}$$

where m_a = mass of the target nuclide
m_n = mass of the neutron

Therefore,

$$E_T = -13.1\left(\frac{26.981535 + 1.008665}{26.981535}\right) = -13.5 \text{ MeV}$$

For the above reaction to occur, the incoming neutron must have at least 13.5 MeV kinetic energy.

Now consider the slightly more complicated case

$$^{27}_{13}\text{Al} + ^{1}_{0}\text{n} \rightarrow ^{27}_{12}\text{Mg} + ^{1}_{1}\text{p} + Q \tag{19.28}$$

where $Q = -1.8$ MeV and $E_T = -1.9$ MeV. Quantity E_T is the minimum energy required for the reaction; but, once the proton is created, it has a low probability of

SEC. 19.4 Activation Analysis

leaving the nucleus because it is a charged particle. To increase the probability of leaving the nucleus, it must have enough energy to overcome the *coulombic barrier*. For the case of the emission of a neutron, there is no coulombic barrier and, therefore, once the neutron is created it can leave the nucleus. The minimum kinetic energy that a charged particle must have in order to overcome the coulombic barrier and leave the nucleus is determined by the following equation:

$$E_c = -1.44 \frac{Z_a Z_b}{r_s} \qquad (19.29)$$

where E_c = coulombic-barrier energy (in MeV)
Z_a = atomic number of the product nuclide
Z_b = atomic number of the emitted particle
$r_s = r_a + r_b$
r_a = radius of the product nucleus
r_b = radius of the emitted particle

The various radii are calculated using the empirical expression

$$r \approx 1.5 A^{1/3} \qquad (19.30)$$

where A = mass number
r = radius in Fermis (1 Fermi = 10^{-13} cm)

In this example,

$$E_c = -1.44 \left(\frac{12(1)}{1.5(27^{1/3} + 1^{1/3})} \right) = -2.9 \text{ MeV}$$

Therefore, at least 1.9 MeV is needed to create the proton and an additional 2.9 MeV for it to overcome the coulombic barrier. For the above reaction to take place with high probability, the incident neutron must have a minimum kinetic energy of 1.9 + 2.9 = 4.8 MeV. Endoergic reactions are also called *threshold reactions*.

The probability that a nuclear reaction will occur is measured by a quantity called the *reaction cross section*. The most common unit of cross section is the *barn* (1 barn = 1 × 10^{-24} cm²). If the energetics are favorable for more than one reaction, then each reaction has a specific reaction cross section and proceeds independendently of other reactions.

The rate at which a nuclear reaction proceeds depends on three parameters: the number of target atoms present, the reaction cross section, and the number of neutrons incident per unit area of the target material per unit time. This relationship is expressed by

$$R = N\phi\sigma \qquad (19.31)$$

where R = reaction rate (in sec^{-1})
N = number of target nuclei present
ϕ = neutron flux density (in n/(cm²-sec))
σ = reaction cross section (in cm²)

The magnitude of the reaction cross section depends on the nature of the target nuclide and on the energy of the incident neutrons. With thermal (low-energy) neutrons, (n, γ) reactions generally have large cross sections, although there are some exceptions. Threshold reactions, of course, cannot take place with thermal neutrons to any appreciable extent.

Sources of Neutrons

Three sources of neutrons are commonly used in activation analysis: nuclear reactors, isotopic sources, and accelerators.

Nuclear Reactors. A nuclear reactor generates neutrons by the process of *fission*. Although the actual workings of nuclear reactors are quite complicated, the principles, for the present purpose, can be understood by considering a $^{235}_{92}$U-fueled nuclear reactor. Upon capturing a neutron, a $^{235}_{92}$U nucleus breaks up into several lighter nuclei and produces more neutrons:

$$^{235}_{92}\text{U} + ^{1}_{0}\text{n} \rightarrow ^{A_1}_{Z_1}\text{X} + ^{A_2}_{Z_2}\text{X}' + k^{1}_{0}\text{n} + Q \qquad (19.32)$$

where
$$A_1 + A_2 + k = 236$$
$$Z_1 + Z_2 = 92$$

The *average* value of k is 2.5. The fact that each nucleus of $^{235}_{92}$U produces more neutrons than it requires for fission is responsible for the copious production of neutrons by nuclear reactors.

The fission neutrons produced in nuclear reactors have a continuous kinetic-energy spectrum, mostly in the range of 1 to 10 MeV. Since (n, γ) reactions are of more widespread analytical use, fission neutrons must be slowed to thermal energies by passing them through H_2O, D_2O, or graphite, which act as *moderators*. Depending on the type of nuclear reactor and the irradiation position in the reactor, the neutron spectrum may vary widely. Therefore, both (n, γ) and threshold reactions can occur in samples placed in nuclear reactors. Threshold reactions may produce interferences, of which the experimenter should be aware.

Isotopic Sources of Neutrons. Nuclear reactors are the only sources of copious quantities of neutrons. A typical research reactor might have a useful flux density of 10^{11} to 10^{14} n/(cm^2-sec). However, moderate flux densities of neutrons can be obtained from isotopic sources of neutrons at relatively low cost and with minimal space and maintenace requirements.

Isotopic neutron sources are of two general types. The first is a manmade radionuclide that undergoes spontaneous fission and produces neutrons. $^{252}_{98}$Cf is a radionuclide commonly used for this purpose; a 1-mg $^{252}_{98}$Cf source will produce 2.34×10^9 n/sec. The neutron spectrum of this source is similar to that of reactor neutrons and, therefore, for practical applications the source is placed in a moderator or "thermalizer." The useful thermal-neutron flux density available in a typical facility is about 3×10^7 n/(cm^2-sec).

The second type of isotopic neutron source consists of a radionuclide emitting intense alpha or gamma radiation, mixed with the element beryllium; one of the following reactions takes place:

$$^{9}_{4}\text{Be} + ^{4}_{2}\text{He} \rightarrow ^{12}_{6}\text{C} + ^{1}_{0}\text{n} \qquad (19.33)$$

$$^{9}_{4}\text{Be} + \gamma \rightarrow 2\,^{4}_{2}\text{He} + ^{1}_{0}\text{n} \qquad (19.34)$$

These neutrons also have a "fast," continuous spectral distribution and are usually slowed (moderated) by placing the source in a hydrogen-rich medium, such as water or paraffin.

Accelerators. The accelerator most commonly used for the production of neutrons is the Cockcroft-Walton neutron generator. A schematic diagram of this generator is given in Figure 19.12. Deuterium molecules are ionized in the ion-source bottle, accelerated in an electrostatic field of 100 to 200 kV, and focused on a target containing tritium (^3_1H). The following nuclear reaction takes place:

$$^3_1H + {}^2_1H \rightarrow {}^4_2He + {}^1_0n + Q \tag{19.35}$$

where $Q \approx +14$ MeV. The neutrons produced are, therefore, nearly monoenergetic at 14 MeV.

These neutrons are capable of inducing many threshold reactions. For example, consider the reaction

$$^{16}_8O + {}^1_0n \rightarrow {}^{16}_7N + {}^1_1H + Q \tag{19.36}$$

for which $Q = -9.6$ MeV, $E_T = -10.2$ MeV, and $E_c = -1.9$ MeV. Therefore, the minimum kinetic energy of neutrons required for the above reaction must be 12.1 MeV. This method is widely used for the determination of oxygen, an element that is difficult to determine by other analytical techniques. Benchtop-sized sealed-tube neutron generators are commonly employed.

Theory of Instrumental Neutron Activation Analysis (INAA)

The procedure in INAA is as follows:

1. The sample is exposed to neutrons for a known length of time, t_i.

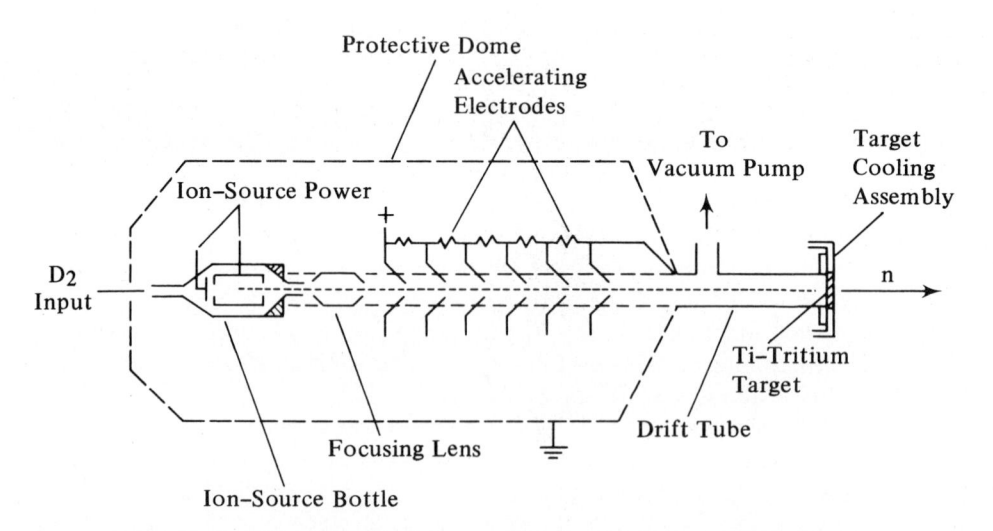

FIGURE 19.12. *Schematic diagram of a Cockcroft-Walton 14-MeV neutron generator.*

2. It is then transported to the counting station and allowed to *cool* or decay for a definite length of time, t_d.
3. The gamma-ray spectrum is acquired for counting time, t_c.
4. The area under the FEP of interest is calculated.

This procedure is repeated for another sample (the standard) containing a known amount of the element of interest. From the weight of the element in the standard, the relative FEP areas of the sample and standard, the relative neutron fluxes used for irradiating the sample and standard, and the times involved, the amount of the element in the sample is calculated.

Assume that the weight of the element present in the unknown sample is W_u grams and that one irradiates the sample for $t_{i(u)}$ sec, allows it to decay for $t_{d(u)}$ sec, and counts the emission for $t_{c(u)}$ sec. The activity of the radionuclide of interest at the end of the irradiation is given by

$$A_u^0 = N\phi_u\sigma(1 - e^{-\lambda t_{i(u)}})$$

$$= 6.02 \times 10^{23} \frac{I}{M} \phi_u\sigma(1 - e^{-\lambda t_{i(u)}})W_u \qquad (19.37)$$

where I = isotopic abundance of the element
M = atomic mass of the element
ϕ_u = neutron flux density (in n/(cm^2-sec))
N = number of atoms of the target present

Most tables list I in units of atom percent, in which case M should be the atomic mass of the element, not the mass of the individual isotope.

When acquisition of the spectrum begins, the activity will be

$$A_u = A_u^0 e^{-\lambda t_{d(u)}} \qquad (19.38)$$

The detector will, of course, detect only a fraction of this activity. Furthermore, only those events that register in the FEP are of interest. These factors are accounted for by the *detector photopeak efficiency* ξ. The *count rate* registered by the detection system in the FEP at the instant that counting starts is therefore given by

$$CR_u = \xi A_u^0 e^{-\lambda t_{d(u)}} \qquad (19.39)$$

The analyzer will integrate the count rate for the period of time t_c. At the end of this time, *total counts* registered in the FEP, excluding any background effects, will be

$$C_u = \xi A_u^0 e^{-\lambda t_{d(u)}} \int_0^{t_{c(u)}} e^{-\lambda t}\, dt$$

$$= \frac{1}{\lambda} \xi A_u^0 e^{-\lambda t_{d(u)}}(1 - e^{-\lambda t_{c(u)}}) \qquad (19.40)$$

Combining Equations 19.37 and 19.40 results in

$$C_u = \frac{\xi}{\lambda} \frac{I}{M} \phi_u \sigma(1 - e^{-\lambda t_{i(u)}})e^{-\lambda t_{d(u)}}(1 - e^{-\lambda t_{c(u)}})W_u(6.02 \times 10^{23}) \qquad (19.41)$$

If the exact values of the parameters in the above equation were known, W_u could be calculated directly. However, because of uncertainties in the numerical values of

ξ, ϕ, and σ, it is more convenient to use a comparative method whereby one also irradiates a sample of known content of the element of interest. Then

$$\frac{C_u}{C_s} = \frac{\xi_u \phi_u (1 - e^{-\lambda t_{i(u)}}) e^{-\lambda t_{d(u)}} (1 - e^{-\lambda t_{c(u)}}) W_u}{\xi_s \phi_s (1 - e^{-\lambda t_{i(s)}}) e^{-\lambda t_{d(s)}} (1 - e^{-\lambda t_{c(s)}}) W_s} \quad (19.42)$$

Several parameters having the same values for the sample and standard (I, M, and σ) have been canceled out. It is quite a simple matter with present-day solid-state electronics to accurately control the various times involved. If each corresponding time is the same for both the sample and standard, and if both are counted with the same detection system, then

$$W_u = \frac{C_u}{C_s} \frac{\phi_s}{\phi_u} W_s \quad (19.43)$$

Equation 19.43 is the working equation commonly used in neutron activation analysis. When employing nuclear reactors or an isotopic source, the values of ϕ_u and ϕ_s may also be the same, and a further simplification results. However, when using accelerator-generated neutrons, this is not easily done, and the values of ϕ_u and ϕ_s (or the ratio ϕ_s/ϕ_u) must be determined experimentally.

Capabilities and Limitations of Neutron Activation Analysis

Neutron activation analysis is a method for determining the elemental contents of substances. Its fundamental limitation is its inability to distinguish among different chemical forms or oxidation states of an element. Like most analytical methods, this technique also suffers from possible interferences and matrix effects. Three types of interferences may occur.

Type I Interferences. These arise from nuclear reactions in the other elements present in the sample that produce the same radionuclide as the one measured. For example, in determining Al in rocks by reactor irradiation employing the reaction $^{27}_{13}$Al (n, γ) $^{28}_{13}$Al, a possible interference is $^{28}_{14}$Si (n, p) $^{28}_{13}$Al.

Type II Interferences. These are caused by the release of secondary nuclear particles from a primary reaction. For instance, when determining nitrogen in protein products with a neutron generator, the reaction employed may be $^{14}_{7}$N (n, 2n) $^{13}_{7}$N. If the sample is packaged in polyethylene containers, the incident neutrons may collide with the hydrogen atoms present in the container material, producing energetic protons which may in turn react with carbon in the sample according to the reaction $^{13}_{6}$C (p, n) $^{13}_{7}$N. This type of interference is generally of limited significance, because the flux density of protons produced is much less than that of the primary neutrons.

Type III Interferences. These are caused by the inability of some detectors to resolve closely similar gamma-ray energies. For example, when determining the Al content of a material by the reaction $^{27}_{13}$Al (n, p) $^{27}_{12}$Mg, one employs the 0.842-MeV gamma ray emitted by $^{27}_{12}$Mg. Iron, if present in the sample, will undergo the reaction $^{56}_{26}$Fe (n, p) $^{56}_{25}$Mn, which emits 0.847-MeV gamma rays. If a NaI(Tl) detector is used to detect the 0.842-MeV gamma rays, the two gamma rays cannot be resolved.

Sometimes type III interferences involve radionuclides with half-lives different from those of the desired elements, and can be resolved by the decay-curve method discussed previously.

One of the most important advantages of INAA over many other methods of analysis is that it is essentially nondestructive. Very often a complete analysis can be performed without appreciably altering the physical or chemical nature of the sample. This is important for several reasons. First, it may be imperative to preserve the sample, such as in forensic analysis where the sample is needed as evidence in a courtroom, or in the analysis of lunar samples or works of art. Second, nondestructive analysis involves minimum sample manipulation and, therefore, a trace sample is not contaminated by reagents and containers as in conventional destructive wet-chemical techniques.

Neutron activation analysis has a high degree of sensitivity for the majority of elements. Trace-level determinations are routinely performed with reactors and can, in certain favorable cases, be performed with the other types of neutron sources. A very important advantage of neutron activation analysis over many other analytical methods is that simultaneous analyses of multicomponent systems are easy to perform; many routine procedures are available to determine more than a dozen elements in a single small sample.

Practical Considerations in Neutron Activation Analysis

Figure 19.13 shows the block diagram of a complete NAA facility. As in any other analytical method, each step may introduce both random and determinate errors, degrading the overall precision and accuracy of the results. We will briefly examine each step and point out a few of the most important factors that should be kept in mind.

The sample may be solid, liquid, or gas, although the first two forms are most commonly used. Problems associated with sampling are the same as in any other method of analysis. Once the sample is secured, it is packaged in an appropriate container. An important point to keep in mind is that the sample and standards should be as similar as possible in matrix composition. For example, when rocks are being analyzed, the sample is generally pulverized to a fine powder, and the standards are also preferably made from finely powdered standard rocks (e.g., those provided by the U.S. Geological Survey). An alternative standard could be prepared by evaporating an aliquot of a standard solution of the element on a matrix of high-purity SiO_2.

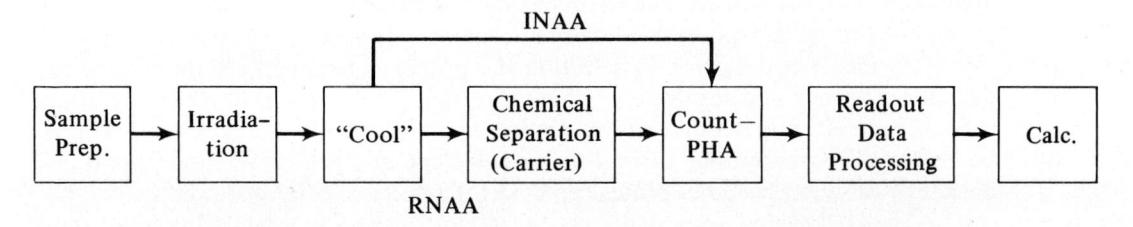

FIGURE 19.13. *Block diagram of a typical neutron-activation-analysis experiment.*

The size of the packaged samples should be as close to that of the standards as possible, so that self-absorption of the neutron flux does not introduce errors. When heterogeneous materials are being analyzed, complete mixing is very important. When organic materials are irradiated, decomposition of the sample may occur; this is especially serious when high fluxes of neutrons are employed, since considerable heat may be generated inside the sample. The same problem occurs in the irradiation of aqueous solutions, where the buildup of pressure inside the container must be allowed for. Heat-sealed quartz vials are often used for reactor irradiations. If pressure buildups are anticipated, the vials may be cooled to liquid-nitrogen temperatures before they are opened.

The irradiation assembly must provide the same neutron flux for both sample and standard, or appropriate correction factors must be determined. In addition to variations in the absolute magnitude of the thermal-neutron flux as a function of the position inside the nuclear reactor, the ratio of thermal-neutron and fast-neutron fluxes changes appreciably with position. This may result in serious errors caused by unwanted threshold reactions. When 14-MeV generators are used, time-dependent variations in the neutron flux may also become significant.

The optimum irradiation time is a very important factor in activation analysis. The decision is based on the specific nature of the sample and the type of information desired. Generally, two factors are considered. First, longer irradiation times increase the activity produced. However, Equation 19.37 shows that the factor $[1 - \exp(-\lambda t_i)]$ approaches unity as t_i becomes large with respect to the half-life of the product radionuclide. Therefore, irradiation times in excess of 3 to 5 half-lives of the product desired result in little additional activity. Second, the longer the irradiation time, the greater will be the induced activities due to long-lived radionuclides that may interfere with the specific determination. (Of course, the higher the overall activity of the sample, the higher the health hazard and the more care must be used in handling it.) In general, irradiation times of approximately 3 half-lives of the product, but rarely more than one week total time, are used for conventional activation analysis.

It is often desirable to allow the irradiated specimen to decay for a period of time (to cool) before counting. A suitable decay period permits short-lived interfering activities to decay and, again, lessens the health hazard.

After the cooling period, either the sample is counted directly (INNA) or some chemical manipulation is performed before counting (RNAA). In RNAA, a stable *carrier* for the element to be determined may be added to the sample after irradiation. The carrier is equilibrated with the element in the sample (often by fusing it with Na_2O_2, or treating it with strong acid). Then the element of interest is separated along with the carrier. The chemical yield of the separation is determined from the amount of carrier recovered, and this correction is applied to the measured activity.

As mentioned earlier, in selecting the proper detector, the criteria used are detector efficiency and resolution. Where sensitivity is the overriding consideration, a NaI(Tl) detector is the detector of choice. If there are interferences, RNAA must be employed to eliminate them. In multielement analyses of complex matrices, detector resolution becomes critical and Ge(Li) detectors should be used.

The electronic components needed for processing the detector signals have evolved into standardized modular units, and are relatively simple to select. One important factor when using MCAs is the analyzer dead-time. Typically, a MCA receives a pulse, digitizes that pulse, and stores it in the proper memory channel. During this time, the analyzer is *dead* to any incoming pulses. If the sample has a high activity level, this could result in an appreciable loss of counts. Most modern analyzers keep track of this *dead-time* and automatically lengthen the counting time to compensate for its effect. This internal correction works best when the counting time is short compared to the half-life of the radionuclide of interest.

Figure 19.14 compares NAA with other commonly used analytical techniques. NAA, with its important advantage of high sensitivity for many elements and its

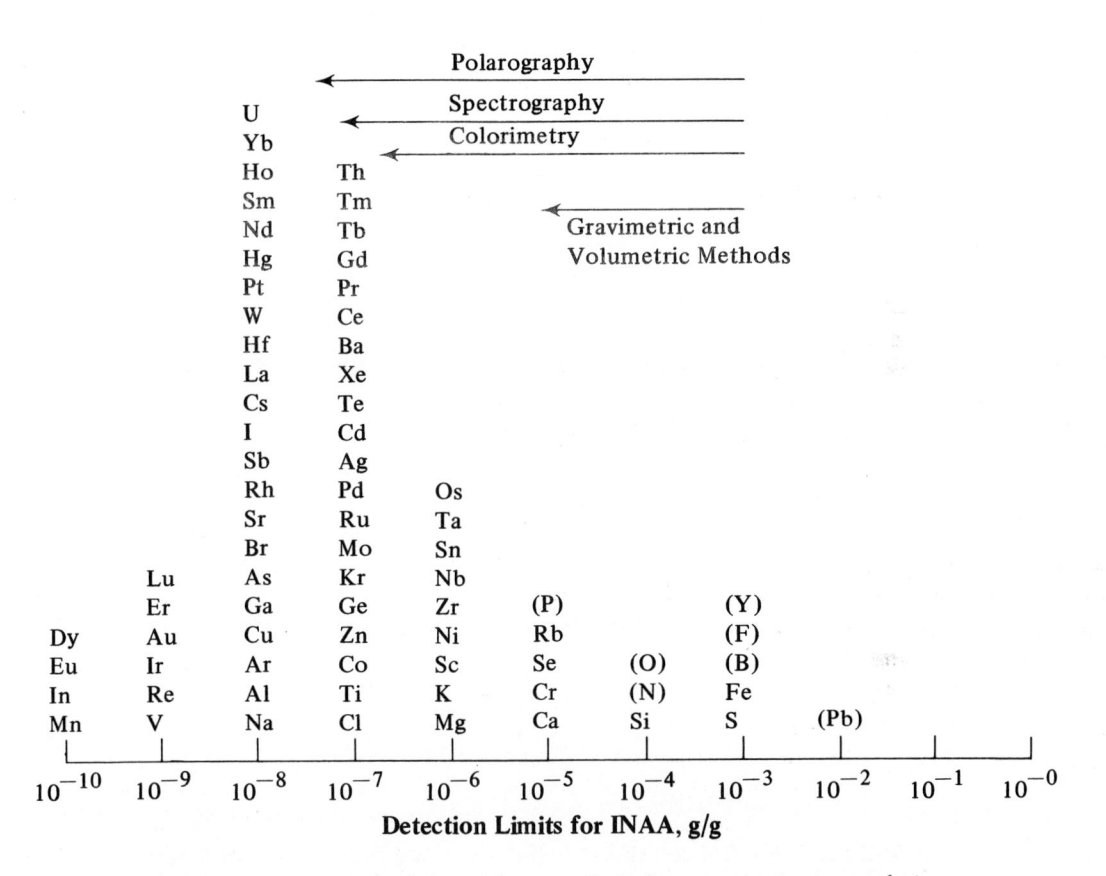

FIGURE 19.14. *Detection limits for instrumental neutron activation analysis (INAA) compared to the usual sensitivity ranges for several commonly used analytical techniques. Elements listed in parentheses are determined by 14-MeV neutron activation with a flux density of 2×10^8 n/(cm^2-sec) for a maximum irradiation time of 5 min, followed by NaI(Tl) gamma-ray spectrometry. Detection limits for all other elements are based on reactor irradiations of 1 hr or less at a flux density of 10^{13} n/(cm^2-sec), followed by gamma-ray spectrometry using a 40-cm^3 Ge(Li) detector. Sensitivities for many elements could be improved by several orders of magnitude by using longer irradiation times, positions of higher flux density, or radiochemical separations. For example, Ir has been determined in rocks at levels well below 10^{-12} g/g (0.001 ppb) by long, high-flux reactor irradiations.*

inherent freedom from problems of reagent and laboratory contamination, is often the benchmark technique against which other trace-element techniques are measured. The increasing availability of inexpensive, easily housed, isotopic neutron sources and sealed-tube neutron generators can now put a "reagent bottle" of neutrons in even the most modest analytical laboratory.

Let us consider an example of an activation-analysis calculation:

Example 19.1. We wish to determine the element vanadium in a 1.00-g sample of petroleum. We anticipate the vanadium content to be about 100 ppm by weight. A ^{252}Cf neutron source with a useful flux density of 2×10^7 n/(cm^2-sec) is available, along with a detector system with a total FEP efficiency of 10.0% for the ^{52}V 1.43-MeV gamma ray. The 1.43-MeV gamma rays are emitted in 100% of the ^{52}V disintegrations. Calculate the number of counts that would be obtained for ^{52}V if the sample is irradiated to saturation and "cooled" for 2.00 min prior to counting for a period of 10.0 min.

Nuclear Data

Atomic mass of V = 50.94 amu
Natural abundance of ^{51}V = 99.76%
Cross section of ^{51}V = 4.9 barns
Half-life of ^{52}V = 3.75 min
The nuclear reaction is ^{51}V (n, γ) ^{52}V

Solution: Combining Equations 19.37 and 19.40 and noting that if the sample is irradiated to saturation (a time long with respect to the half-live of ^{52}V), the term $(1 - e^{-\lambda t_1}) \sim 1$, then

$$\text{Counts} = \frac{\xi}{\lambda} \frac{W}{M} I\phi\sigma e^{-\lambda t_d}(1 - e^{-\lambda t_c})(6.02 \times 10^{23})$$

$$= \left(\frac{0.100}{0.693/[3.75 \text{ min } (60 \text{ sec/min})]}\right)\left(\frac{100 \times 10^{-6} \text{ g}}{50.94 \text{ g/g atom}}\right)$$

$$\times [6.0 \times 10^{23} \text{ atom/(g-atom)}](0.9976)[2 \times 10^7 \text{ n/(cm}^2\text{-sec)}]$$

$$\times (4.9 \times 10^{-24} \text{ cm}^2)(e^{-0.693(2.00 \text{ min})/3.75 \text{ min}})$$

$$\times (1 - e^{-0.693(10.0 \text{ min})/3.75 \text{ min}})$$

$$= 2180$$

Assuming no interferences, this calculation shows that one could determine V in this sample at the 100-ppm level with a relative statistical error due to counting of about 2% (see Sec. 19.6). Indeed, this method has been commonly employed by the petrochemical industry for the determination of vanadium in petroleum and its products.

Well over 15,000 papers dealing with activation analysis have appeared in the literature. Most of these (99%) have been published since 1955. Some of the more interesting applications have been determining potentially toxic trace elements in natural waters and environmental samples; authenticating paintings and other objects of art; and studying impurities in semiconductor materials, trace elements in plant and animal metabolism, and trace-element abundances in terrestrial rocks, meteorites, and lunar samples. In the analyses of lunar samples, more than twice as many trace-element determinations have been reported by activation analysis than by any other technique. In fact, the activation-analysis determinations on these rare samples probably exceed those by all other techniques combined.

Neutron activation analysis employing 14-MeV generators has been widely employed in the direct determination of oxygen in rocks and of nitrogen in food grains and explosives. Charged-particle activation analysis is useful in the analysis of thin films or coatings on metals.

Activation analysis is not without its own unique problems. However, for the determination of elements at the sub-ppm level, it is certainly the technique against which other methods must be compared. Accuracy and precision of the order of a few percent are readily attainable at the nanogram level for many elements. High sensitivity, multielement capability, and freedom from reagent and laboratory contamination problems are the major advantages offered.

19.5 METHODS INVOLVING ADDITION OF RADIONUCLIDE

The second general category of radiochemical analysis involves adding a radioactive substance to the sample, manipulating the sample by chemical or physical means, measuring the radioactivity, and ultimately calculating the amount of the component of interest. This category includes *direct* and *inverse isotope dilution analysis*, *radiorelease methods of analysis*, and *radioimmunoassay*.

Direct Isotope Dilution Analysis

In the method of activation analysis, radioactivity is induced in the sample to be analyzed. In the method of *direct isotope dilution analysis* (DIDA), a radioactive form of the component of interest is added to the sample. The component is then exhaustively purified without regard to quantitative recovery and a fraction of the pure component isolated. The amount and activity of the isolated component are measured and the quantity present in the original sample is calculated using that information.

Theory of DIDA. Consider a complex sample of W grams containing W_1 grams of the component of interest. To this sample is added W_1^* grams of a radioactive form of the component with a total activity A_1. W_2 grams of the pure component is then isolated; it contains both the active and the inactive forms and has an activity of A_2. The *specific activity* SA_1 of the radioactive ("spike") material before it is mixed with the sample is defined as

$$SA_1 = \frac{A_1}{W_1^*} \qquad (19.44)$$

and SA_2, the specific activity of the *recovered* component, as

$$SA_2 = \frac{A_2}{W_2} \qquad (19.45)$$

Quantity SA_2 will remain constant regardless of how much of the pure component was isolated, since it is activity per unit weight of recovered component. One can then write the following balance sheet:

	Weight of Component	Specific Activity
Before mixing	W_1 (inactive form)	0
	W_1^* (active form)	SA_1
After mixing but before purification	$W_1 + W_1^*$ (mixture)	SA_2
After purification	$f(W_1 + W_1^*) = W_2$ (isolated component)	SA_2

Note that f is the fraction of the component isolated and is unknown. Also, note that the specific activity of the sample-spike mixture remains the same before and after chemical isolation of the component. It follows that

$$\left[\frac{A_1}{W_1^* + W_1}\right] = \left[\frac{A_2}{W_2}\right] \tag{19.46}$$

and solving this equation for W_1 gives

$$W_1 = \left(\frac{A_1}{A_2}\right) W_2 - W_1^* \tag{19.47}$$

Since W_1^* (the weight of the component in the spike added to the sample) is known from the listed concentration of the standard spike solution, and A_1, A_2, and W_2 are determined experimentally, the amount of the component in the original sample can be easily calculated without a knowledge of f, the fraction of the compound actually isolated. Equation 19.47 can also be rewritten to include the specific-activity terms as follows:

$$W_1 = W_1^* \left(\frac{SA_1}{SA_2} - 1\right) \tag{19.48}$$

Example 19.2. A synthetic mixture containing vitamin B_{12} and a number of very similar chemical compounds was produced in a pharmaceutical laboratory and is to be analyzed for its B_{12} content by isotope dilution analysis. Vitamin B_{12} contains a cobalt atom, and radiolabeled B_{12} may be prepared using radioactive ^{60}Co. The labeled B_{12} spike solution available was known to contain 1500 cpm activity per milliliter from the labeled B_{12} and had a specific activity of 125 cpm per milligram of B_{12}. Exactly 5.00 mL of the spike solution was added to a solution containing 1.512 g of the synthetic mixture. After mixing, a series of chemical separations were performed to obtain a sample of pure B_{12} from the mixture. The chemical yield of the separation process was unknown, but the pure B_{12} isolated weighed 125 mg and yielded a counting rate of 2750 cpm. Calculate the percentage of B_{12} in the original synthetic mixture.

Solution:

5.00 mL Spike Solution × 1500 cpm/mL = 7500 cpm in Spike Added

$$\frac{7500 \text{ cpm Spike Added}}{125 \text{ cpm/mg } B_{12} \text{ in Spike}} = 60.0 \text{ mg Labeled } B_{12} \text{ in Spike Added}$$

Therefore, using Equation 19.47,

$$W_1 = \left(\frac{7500 \text{ cpm}}{2750 \text{ cpm}} \times 125 \text{ mg}\right) - 60.0 \text{ mg} = 281 \text{ mg B}_{12} \text{ in Mixture}$$

and

$$\% \text{ B}_{12} \text{ in Original Synthetic Mixture} = \frac{281 \text{ mg B}_{12}}{1512 \text{ mg Sample}} \times 100 = 18.6\%$$

Advantages and Limitations of DIDA. In wet-chemical analyses, exhaustive multistep purification procedures are often required to obtain the component in a highly pure form, and a quantitative yield is almost impossible to achieve. The main advantage of DIDA is that no quantitative separation of the component of interest is necessary. The instrumentation required is usually quite simple, since measurements of gross activity with simple counting systems are sufficient. The separated component must be in highly pure form; and once the pure component is obtained, its quantity must be accurately measured, or deduced from stoichiometric considerations. The separated component must also have a high enough level of activity to minimize statistical counting error. (This is usually not a serious limitation, since the activity of the initial labeled compound can often be adjusted to compensate for a low efficiency in the purification step.) The weight W_1^* should not be much larger than W_1, and tracer solutions of high specific activity are ordinarily used.

A very important effect, which could become either an advantage or a disadvantage, is inherent in the fundamental requirement of the method: Both the active and inactive forms behave identically in the subsequent purification steps. This means either that the labeled component must be in the same chemical form as the inactive component, or that the mixture must be treated chemically to convert both forms into the same chemical compound. This situation can, of course, be of great advantage if one is trying to distinguish among different chemical forms of a given element. For instance, a solution containing both Cr^{3+} and $Cr_2O_7^{2-}$ can be analyzed for $Cr_2O_7^{2-}$ by adding $^{51}Cr_2O_7^{2-}$ tracer and excess NaOH, after which Ba^{2+} is added to precipitate $BaCrO_4$.

Inverse Isotope Dilution Analysis

In DIDA, a radioactive form of the component of interest is added to the sample and the quantity of the inactive form initially present is determined. In some instances, one may wish to determine the amount of a radioactive substance in the sample. A method similar in principle to DIDA can then be used wherein a quantity of an inactive form of the component of interest is added to the sample, the sample is purified without regard to quantitative recovery, and the amount of the recovered component and its activity are measured. From this information, the quantity of the radioactive substance initially present in the sample is calculated. This method is referred to as *inverse isotope dilution analysis* (IIDA). Equation 19.47 is used and solved for W_1^*, the unknown amount of the radioactive substance in the sample.

Advantages and Limitations of IIDA. The main advantage of this method is that one can determine the quantity of a specific radioactive component of a sample without comparing it with a known radioactive standard. The method also avoids preparing standards with the same matrix as the sample in order to ensure equivalent counting efficiencies. However, the method cannot be applied if spectral interferences prevent the specific measurement of A_1. Furthermore, the method is applicable only when W_1 does not differ greatly from W_1^*. In the case of trace analysis, the method offers the advantage of not requiring a quantitative separation of the component of interest. The method of IIDA has not been applied as widely as has DIDA.

Radiorelease Methods of Analysis

The *radiorelease* methods are based on the chemical reaction of the constituent of interest with a radiolabeled reagent. The labeled component is then released either as a gas or in some readily extractable form. From a measurement of the amount of radioactivity released and the stoichiometry of the reaction, the quantity of the constituent of interest is determined. Consider the determination of SO_2 in air by this method. If air is passed through a basic solution of KI^*O_3, the following reaction takes place:

$$5SO_2 + 2KI^*O_3 + 4H_2O \rightarrow K_2SO_4 + 4H_2SO_4 + I_2^* \qquad (19.49)$$

The solution is then acidified and the liberated I_2^* is extracted into chloroform. The chloroform phase is separated and counted for its I_2^* content. From the stoichiometry of the reaction and the quantity of liberated I_2^*, the content of SO_2 in air can be determined.

Advantages and Limitations of Radiorelease Methods. The chief advantage of this method of analysis is its sensitivity, since highly active radioreagents are available. For instance, a micromole of I_2^* may easily have 10^7 dpm activity. However, the method is chemical in nature and suffers from all limitations inherent in the particular chemical reaction involved. In the above example, any other substance that can reduce KI^*O_3 to I_2^* will, of course, interfere with the determination (oxides of nitrogen are potential interferences). Furthermore, at trace levels quantitative extraction of the released species becomes critical.

Radioimmunoassay Methods of Analysis

The method of *radioimmunoassay* is used for direct measurement of concentrations of substances that possess antigenic properties. Examples of such substances are insulin, adrenocorticotropic hormone (ACTH), gastrin, glucagon, and others. The method is based on competition between a radiolabeled antigen (Ag*) and the unlabeled antigen (to be assayed) for a specific antibody (Ab). Its major features relate to its capability for high sensitivity and direct concentration measurement (as opposed to the measurement of biological response) of these substances. In order to understand the principles behind this method, it is necessary to define some terms used in connection with this technique.

Antigen. Any substance that is capable of inducing formation of antibodies in living organisms. Antigens could be soluble molecules such as insulin or various toxins, or insoluble bodies such as bacterial cells.

Antibody. A protein (immunoglobulin molecule) with a specific amino-acid sequence so that it reacts with and inactivates a specific antigen. For radioimmunoassay purposes, specific antibodies are produced from the blood of experimental animals such as rabbits by injection of an antibody-producing substance. They are then separated from other blood constituents and preserved for later use.

The procedure is illustrated as follows. A small amount of high-specific-activity antigen (Ag*), labeled usually with radioiodine (^{125}I or ^{131}I), is mixed with the unknown sample containing unlabeled but otherwise identical antigen (Ag) to be assayed. To this sample is then added a suitable antibody (antiserum), which specifically inactivates the antigen of interest. The following reaction takes place:

$$\underset{\text{antigen}}{(Ag^* + Ag)} + \underset{\text{antibody}}{Ab} \rightleftharpoons \underset{\text{antigen-antibody complex}}{(Ag^* + Ag)\cdot Ab} \tag{19.50}$$

$$K = \frac{[(Ag^* + Ag)\cdot Ab]}{[(Ag^* + Ag)][Ab]} \tag{19.51}$$

Under a given set of experimental conditions, the amount of radioactivity incorporated in the antigen-antibody complex is inversely related to the concentration of the unlabeled antigen (unknown) in the test solution, due to direct competition for a limited number of binding sites. Thus, by comparing the level of radioactivity in the antigen-antibody complex with a standard curve prepared by addition of variable amounts of the antigen of interest to the Ag*-Ab system, one can determine the amount of the antibody in the unknown.

For this method to be successful, a number of important factors must be taken into account. Some of these are briefly described here, and the interested reader is referred to the Selected Bibliography for additional reading.

1. *Radiolabeling of the antigen.* The antigen of interest must be labeled with a suitable radioisotope. The most commonly used isotopes are ^{125}I and ^{131}I, which are labeled on the tyrosine residue of the antigen.

2. *Specificity of the antibody.* A suitable antibody must be available to react with the antigen of interest. This is usually prepared by previous injection of an antigen into an animal such as a rabbit. The antibody must be specific to the antigen of interest and not be affected by other constituents of the test sample.

The amount of antibody added to the test system must be such that the total amount of the antigen is always in excess. Otherwise, no competition would occur between the labeled antigen and the antigen in the test sample.

3. *Experimental conditions.* The equilibrium concentrations involved are dependent on solution pH and temperature, and so these conditions must be carefully controlled between the standard and the unknown assays.

Because this technique is based on the measurement of the radioactivity of the antigen-antibody complex, a suitable method must be used to separate the complex from the nonreacted antigen. Otherwise, the labeled noncomplexed antigen would be an interferent. Several methods including immunoprecipitation, electrophoresis, adsorption chromatography, ion-exchange chromatography, or paper chromatography can be used for this purpose.

19.6 STATISTICAL CONSIDERATIONS IN RADIOCHEMICAL ANALYSIS

In reporting the results of any analysis, two important parameters are the accuracy and the precision of the data. Accuracy is a measure of how close the reported data are to the true values. Precision, on the other hand, is only a measure of how closely one can expect to reproduce the reported data, if the experiment is repeated. Good precision does not necessarily imply accurate results. A discussion of these factors can be found in most books on quantitative analysis. However, since radioactive counting follows a different distribution law than do most other analytical manipulations, calculations of the precision of radiochemical methods require a knowledge of this distribution. In this section, the distribution law involved will be presented, and the calculations compared to those of most other types of analysis.

It is generally accepted that random errors arising in various analytical operations, except radioactive counting, follow a gaussian distribution, described by the distribution function (see Fig. 19.15)

$$p_{(x,\mu,\sigma)} = \frac{1}{\sigma\sqrt{2\pi}} e^{-(x-\mu)^2/2\sigma^2} \qquad (19.52)$$

where μ = the true mean or the "central tendency" of the population*
x = an individual measurement
σ = *standard deviation* of the distribution function
p = probability of obtaining the value x in a single trial

Practically speaking, σ is a measure of the broadness of the curve in Figure 19.15; a total of 68.3% of the entire area under the curve falls within $\pm 1\sigma$ of the value of the mean (which, in this case, is the same as the most probable value). Gaussian distributions are often called "normal" distributions because of the frequency with which they occur. There is really nothing "abnormal" about other types of distributions.

In practice, a finite number (n) of measurements are made, all of which would be identical were it not for the random errors involved in each measurement. From this finite (and often small) set of measurements, one estimates the most probable value of the function by averaging the experimental results. The question is: If another single measurement were made, how close would this last measurement be to the mean of the previous measurements? The answer, of course, is that this depends on the broadness of the distribution function for these measurements, as well as on the degree of confidence to be placed on the answer. For this purpose, one calculates the *standard deviation of the individual sample determination*, using

$$s_x = \sqrt{\frac{\Sigma(x_i - \bar{x})^2}{n-1}} \qquad (19.53)$$

where $\bar{x}$ = *average* or *mean* of the n individual measurements, $x_1, x_2, \ldots, x_i$

---

*The true mean of a population is the quantity obtained if a given measurement is repeated an infinite number of times and the results averaged, assuming no determinate errors.

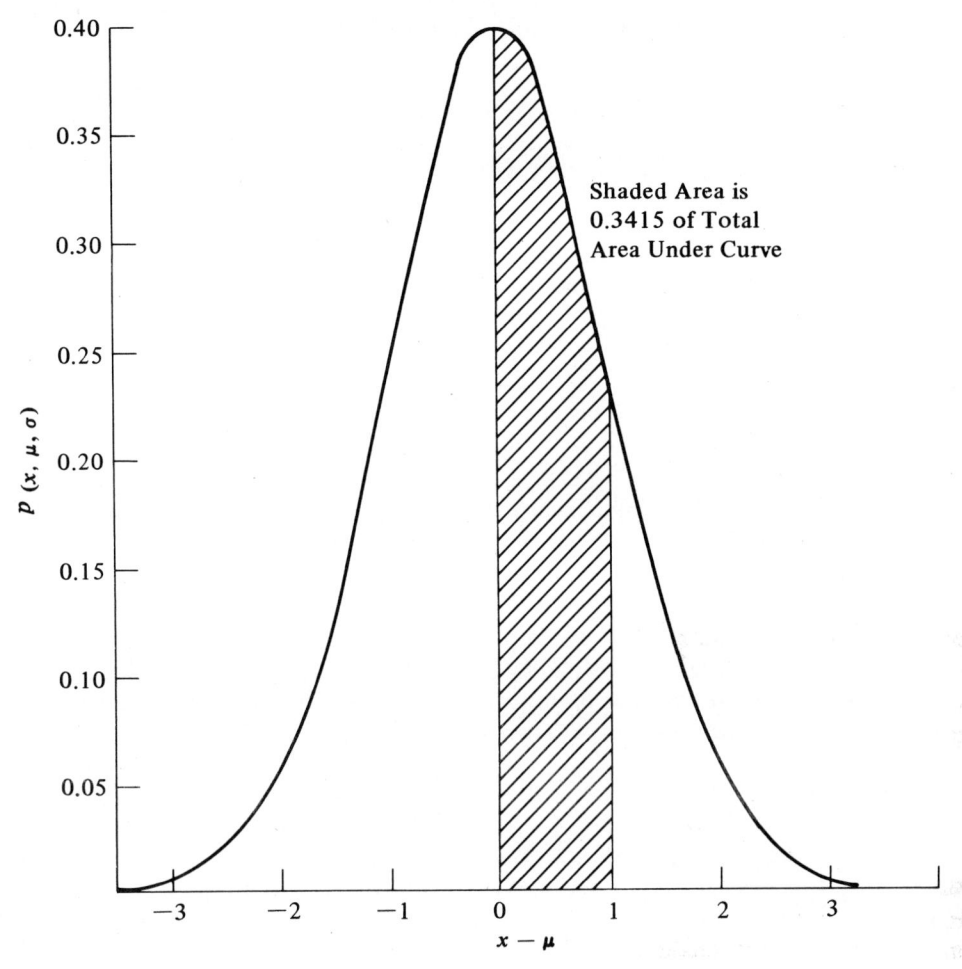

FIGURE 19.15. *Probability curve: the gaussian or "normal" distribution curve for $\sigma = 1.0$.*

One would thus expect any other individual measurement to fall within $\mu \pm 1s_x$ in 68.3% of the measurements and within $\mu \pm 2s_x$ in 95.5% of such measurements. As can be seen, the greater the degree of confidence, the wider the range of possible values.

One can now ask, how close would one come to the first mean if one makes another set of measurements and calculates a new mean? To answer this, calculate the *standard deviation of the mean* using the following formula:

$$s_{\bar{x}} = \frac{s_x}{\sqrt{n}} \tag{19.54}$$

In contrast to most analytical operations, radioactive counting does not, in general, follow the normal distribution law. It follows the *Poisson distribution law*,

SEC. 19.6 **Statistical Considerations in Radiochemical Analysis** 633 an asymmetric distribution function described by

$$p(x, \mu) = \frac{\mu^x}{x!} e^{-\mu} \quad (19.55)$$

Figure 19.16 shows a typical Poisson distribution function. A very important distinction between this function and the normal distribution function is that, in order to characterize the latter, we must know both μ and σ because the broadness of the normal distribution is independent of its mean. In contrast, the Poisson distribution curve is completely characterized by its mean alone. The broadness of the distribution is a function of the mean, and is given by

$$\sigma_x = \sqrt{\mu} \quad (19.56)$$

for an infinite number of measurements. For a finite set of measurements,

$$s_x = \sqrt{\bar{x}} \quad (19.57)$$

Thus, from only a single measurement (an estimate of the mean), the standard deviation can be estimated to be

$$s_x = \sqrt{x} \quad (19.58)$$

The standard deviation of the mean is, as before, given by Equation 19.54. As the magnitude of μ (or its estimator $\bar{x}$) increases, the portion of the Poisson distribution curve close to its mean becomes more symmetrical, and resembles more closely the normal distribution curve. Therefore, only for large values of $\bar{x}$ can one assume that

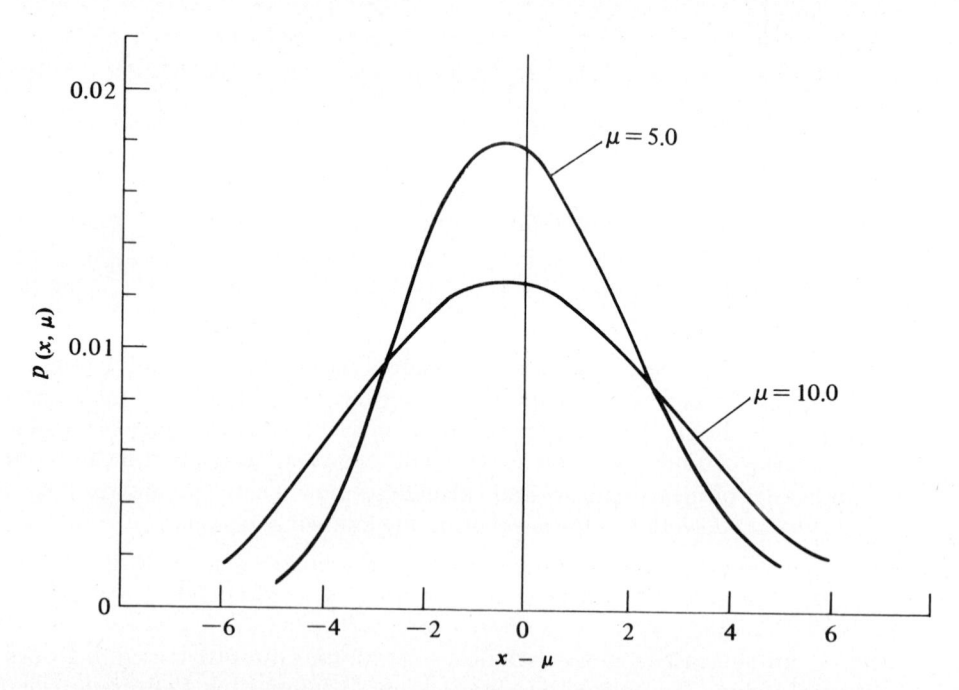

FIGURE 19.16. *Probability curves: Poisson distribution curves for* $\mu = 5.0$ *and* $\mu = 10.0$.

634 CHAP. 19 Radiochemical Methods of Analysis

the statistics of radioactive counting follow the normal distribution law, and then only as an approximation.

Propagation of Errors

Most radiochemical procedures consist of a number of measurement steps. In each step, random errors can occur which will contribute in different degrees to the uncertainty of the final result. One can therefore estimate the uncertainty of the final result, given estimates of the uncertainty of each step. Assume that two measurements or two steps with numerical values $A \pm \sigma_A$ and $B \pm \sigma_B$ yield the result $C \pm \sigma_C$. If $C = A + B$, or $C = A - B$, then

$$\sigma_C = \sqrt{\sigma_A^2 + \sigma_B^2} \qquad (19.59)$$

If $C = AB$ or $C = A/B$, then

$$\sigma_C = C\sqrt{\left[\frac{\sigma_A}{A}\right]^2 + \left[\frac{\sigma_B}{B}\right]^2} \qquad (19.60)$$

In combining the uncertainties, each step is assumed to have the same error distribution function, so that σ_C also has the same function. This is generally true for most analytical operations. However, when combining measurements of radioactivity, one cannot always obtain a meaningful error distribution in the manner indicated above (this is analogous to adding apples to oranges!). However, if the specific Poisson distribution being considered does approximate a normal distribution, it is approximately correct to pool the standard deviations in this way. In addition, if one or more steps contribute predominantly to the uncertainty of the complete analysis, one can safely calculate error based on the uncertainty of only those steps.

SELECTED BIBLIOGRAPHY

BRUNE, D., FORKMAN, B., and PERSSON, B. *Nuclear Analytical Chemistry.* Deerfield Beach, Fla.: Verlag Chemie, 1984. *A good introduction to all aspects of nuclear methodology as applied to analytical chemistry.*

CHAPMAN, D. I. "Radioimmunoassay," *Chemistry in Britain, 15,* 439 (1979). *A good, easy-to-read article on the principles of radioimmunoassay.*

CHOPPIN, G. R., and RYDBERG, J. *Nuclear Chemistry Theory and Applications.* Oxford: Pergamon Press, 1980. *This book presents the basic theories of nuclear chemistry and contains extensive discussions of radioanalytical applications and topics related to nuclear power.*

EHMANN, W. D. "Nondestructive Techniques in Activation Analysis," *Fortsch. Chem. Forsch., 14*(1), 49 (1970). *This is a general article on the technique of nondestructive neutron activation analysis with major emphasis on 14-MeV neutrons. Practical aspects of this area as well as some advanced developments are discussed.*

ERDTMANN, G. *Neutron Activation Tables.* Weinheim: Verlag Chemie, 1976. *A compilation of data useful in thermal, epithermal, and 14-MeV neutron activation analysis.*

FRIEDLANDER, G., KENNEDY, J. W., MACIAS, E. S., and MILLER, J. M. *Nuclear and Radiochemistry,* 3rd ed. New York: John Wiley, 1981. *Standard senior-level textbook with the emphasis on the principles of nuclear chemistry and radio-*

chemistry, and containing a brief treatment of radioanalytical applications.

KRUGER, P. *Principles of Activation Analysis.* New York: Wiley-Interscience, 1971. *A senior-level textbook covering various aspects of neutron activation analysis in fair depth.*

KRUGERS, J., ed. *Instrumentation in Applied Nuclear Chemistry.* New York: Plenum Press, 1973. *A comprehensive, in-depth treatment of all aspects of nuclear instrumentation.*

TÖLGYESSY, J., BRAUN, T., and KYRS, M. *Isotope Dilution Analysis.* Oxford: Pergamon, 1972. *The only comprehensive book on the subject of isotope dilution analysis.*

WALKER, F. W., MILLER, D. G., and FEINER, F. *Chart of the Nuclides*, 13th ed. General Electric Company, 175 Curtner Avenue, San Jose, CA 95125. *A compilation of nuclear data in a paperback foldout chart format. The chart is revised periodically; this edition is dated 1984.*

WANG, C. H., WILLIS, D. L., and LOVELAND, W. D. *Radiotracer Methodology in the Biological, Environmental, and Physical Sciences.* Englewood Cliffs, N.J. Prentice-Hall, 1975. *A senior-level textbook with an emphasis on the use of tracers, radioanalytical instrumentation, and the design of radiotracer experiments.*

PROBLEMS

1. A ^{252}Cf isotopic neutron source is available that provides a flux density of 2.0×10^7 n/(cm^2-sec) at the irradiation positions. The aluminum content of 1.0-g alloy samples is to be determined by the ^{27}Al (n, γ) ^{28}Al reaction; a counting rate of at least 100 cpm at the start of the counting period is required for these determinations. Using the following data, calculate the minimum aluminum content that could be determined in these alloys by this technique: (1) ^{28}Al is a negatron and gamma-ray emitter that emits one 1.78-MeV gamma ray per disintegration. (2) Half-life of ^{28}Al = 2.3 min. (3) Cross section of ^{27}Al for the (n, γ) reaction = 0.24 barn. The natural isotopic abundance of ^{27}Al is 100%, and its atomic mass is 26.98. (4) The overall efficiency of the counting system is 10%, based on the 1.78-MeV gamma-ray FEP. (5) The sample is irradiated to saturation (the time of irradiation is long with respect to the half-life of ^{28}Al) and "cooled" 2 min prior to the start of counting.

2. The radionuclide $^{16}_{7}$N emits a high-energy gamma ray at 6.13 MeV. A gamma-ray spectrum of a $^{16}_{7}$N source exhibits peaks at 5.62, 5.11, and 0.511 MeV in addition to the FEP at 6.13 MeV and the usual Compton distribution. Assuming these features are not due to primary gamma rays emitted in the decay of $^{16}_{7}$N, how would you explain the existence of these three extra peaks?

3. A sample containing both Mn and Fe was irradiated with neutrons in a nuclear reactor for a period of 50 hr. The radionuclides ^{56}Mn and ^{59}Fe were formed by (n, γ) reactions on 100% natural-isotopic-abundance ^{55}Mn and 0.33% natural-isotopic-abundance ^{58}Fe. At the end of the irradiation, the absolute activity ratio, ^{56}Mn activity/^{59}Fe activity, was observed to be 10^5. Using the following data, calculate the weight ratio (grams Mn/grams Fe) of the elements in the sample. (1) Cross section for (n, γ) reactions: ^{55}Mn = 13.3 barns; ^{58}Fe = 1.2 barns. (2) Atomic masses: Mn = 54.94 amu; Fe = 55.85 amu. (3) Half-lives: ^{56}Mn = 2.57 hr; ^{59}Fe = 45 days.

4. A sample of seawater (density approx. 1.00 g/mL) is to be analyzed for its I$^-$ content. A 5-mL aliquot is placed in an electrolytic cell, to which is added 5 mL of a mixture of 0.1 M sodium acetate and 0.1 M acetic acid as supporting electrolyte, plus 1.0 mL of a solution of K^{129}I containing 1.0 μg of iodide with a specific activity of 312,000 cpm per microgram I$^-$. An identical cell containing only the supporting electrolyte plus 1.0 mL of K^{129}I is also prepared. The two cells are placed in series, and a potential of -100 mV with respect to a saturated calomel electrode is imposed on two silver electrodes acting as the anodes for each cell. After 10 min, the activity of the two anodes is measured to be 20,800 and 104,000 cpm.

Assuming that the current efficiencies of the two cells for deposition of I⁻ are the same, calculate the I⁻ concentration in the seawater in ppm.

5. A sample containing N_0 atoms of $^{38}_{17}Cl$ (half-life = 37.3 min) was received at 1200 hours for beta counting. It was placed in a liquid scintillation counter with 100% efficiency at 1230 hours and counted for exactly 60 min. The accumulated counts were 100,000. Calculate N_0.

6. A 50-mL solution of 1.00 mM Ag⁺ is titrated with a 1.0 mM solution of $K_2{}^{51}CrO_4$ (half-life = 27.8 days) with an activity of 1.00×10^7 dps/mole. Plot the anticipated radiometric titration curve.

7. With reference to a *Chart of the Nuclides*, discuss the various neutron-activation-analysis techniques that might be used to determine Ni at the milligram level. What neutron sources could be used? What nuclear reactions would be involved? What are the relative advantages and disadvantages of the various approaches to this determination?

8. Rapid analyses for Co in steel are often done by isotope dilution analysis. Assume that a 1.00-g sample of steel is dissolved in acid and that exactly 2 mL of a spike solution of ⁶⁰Co is added to the solution. The spike solution has a concentration of 3 mg of Co per milliliter and a specific activity of 1.50×10^4 dpm per milligram Co. Two electrodes are immersed in the solution and a small amount of Co_2O_3 plated out on the anode. The weight increase of the anode is determined to be 12.5 mg, and its activity is 2500 dpm. Calculate the percentage of Co in the steel sample.

9. A 2.00-min background count for a given counter yielded 3600 counts. A radioactive sample was counted for 2.00 min with the same counter and a total of 6400 counts were recorded. Calculate the background-corrected counting rate for the sample and its standard deviation, s_x. Express the counting rate in counts per minute.

10. Sample A, sample B, and background alone were each counted for 10 min with a given counter. The observed counting rates were 110, 205, and 44 cpm, respectively. Calculate the ratio of the activity of sample A to that of sample B and determine the standard deviation of this ratio.

11. Carbon-14, a β emitter with a half-life of 5720 yr, is produced in the atmosphere by the reaction $^{14}N (n, p) {}^{14}C$, the neutrons coming from cosmic rays. The dating is based on the assumption that the amount of ^{14}C in the atmosphere (the ratio of $^{14}C/^{12}C$) remains constant over thousands of years. A living species incorporates this same ratio in all its carbon-containing molecules. When it dies, the incorporation of ^{14}C ceases, and the ^{14}C decays. The $^{14}C/^{12}C$ ratio determined from a wood sample taken from a dugout canoe at the bottom of a lake was found to be one-tenth the ratio determined from a wood sample less than 1 yr old. How old is the dugout canoe?

12. One gram of pure radium emits 3.70×10^{10} dps (1 curie). (a) How many atoms of radium are decaying each second? (b) How long will it take before half of the radium atoms have disintegrated?

13. Radon-222, the first decay product of ^{226}Ra, is an alpha-particle emitter with a 3.82-day half-life. A sample of ^{222}Rn gas was found to have an activity of 2.22×10^6 dpm. (a) What is its activity in microcuries? (b) What is the decay constant in sec⁻¹? (c) How many atoms of ^{222}Rn does the sample contain? (d) How many grams? (1 curie = 3.70×10^{10} dps.)

14. Sodium-24 is a beta-particle emitter with a half-life of 14.8 hr. Calculate the activity in curies of a 20-mg sample of NaCl enriched with ^{24}Na so that it contains 1 atomic percent of ^{24}Na. (1 curie = 3.70×10^{10} dps.)

15. The neutron-activation-analysis limit of detection for arsenic is listed as 2×10^{-10} g for 1-h irradiation in a neutron flux of 10^{13} n/(cm²-sec). Using the value of 4.3×10^{-24} cm² as the cross section of the arsenic nucleus for neutron capture, calculate the disintegrations per second expected from ^{76}As after irradiation for this period of time.

16. For any given isotope, what period of neutron irradiation is required to raise the observed activity to one-half the maximum activity?

17. Derive the relationship between the average lifetime of a radioactive atom and its half-life.

18. Human growth hormone (HGH, mol. wt. 21,500) is to be labeled with a radioactive

Problems   **637**

isotope of iodine for use in radioimmunoassays. If one atom of ^{125}I or ^{131}I is incorporated into each molecule of HGH, calculate the resultant specific activity (in microcuries per microgram HGH) of the labeled protein. Which of the two isotopes of iodine would be preferred for this purpose?

19. Derive the alternative form of the isotope dilution equation (Eqn. 19.48) from the form given in Equation 19.47.

CHAPTER 20

Fractionation Processes: Solvent Extraction

Henry Freiser

Great strides have been made in the development of highly selective analytical methods. However, the analytical chemist is called upon nowadays to deal with increasingly complex samples; as a result, separation steps can be necessary even with highly selective instrumental methods such as neutron activation or atomic absorption. Furthermore, separation of a component of interest can also concentrate it, which effectively increases the sensitivity of the analytical technique ultimately used. Although separations procedures are often not, strictly speaking, instrumental techniques, they frequently constitute an integral part of an instrumental procedure and, in fact, may be incorporated into an instrumental design. Because of their importance, solvent extraction and chromatographic techniques are covered in the next three chapters.

Chromatography, a multistage counter-current process involving a bulk fluid mobile phase and a dispersed condensed phase, has grown in importance as analysts have been faced with increasingly complex separations problems. Major advances in natural-products microanalysis, amino-acid and other biochemical analyses, and petroleum chemistry have occurred as successive development of adsorption chromatography, gas chromatography, ion-exchange chromatography, high-performance liquid chromatography, gel-permeation chromatography, and ion chromatography has occurred.

20.1 PHASE PROCESSES

One of the most powerful approaches to separations involves pairs of phases in which the component of interest transfers from one phase to the other more readily than do interferring substances. For all phase-distribution equilibria, the classical phase rule of Gibbs is applicable and useful. The phase rule

$$P + V = C + 2 \qquad (20.1)$$

relates the number of independent variables (degrees of freedom) V needed to describe a system of C components with a number of phases P that can coexist in equilibrium with one another. Note that a *component* is not the same as a *chemical species*; a two-component mixture of NaCl and H$_2$O contains the following species: H$^+$, OH$^-$, Na$^+$, Cl$^-$, H$_2$O, and NaCl.

The degrees of freedom include both temperature and pressure. Hence, in systems containing only condensed phases (liquids or solids) whose properties are only slightly affected by pressure changes, the phase rule reduces to

$$P + V = C + 1 \tag{20.2}$$

It is useful to classify phase-separation processes according to the following criteria:

1. The states of the phases involved (solid, liquid, or gas)
2. Whether the phase is in bulk or spread thin as on a surface
3. The manner in which the two phases are brought into contact (batch, multistage, or counter current)

Bulk and "thin" phases can be distinguished by the fact that, in the latter, the phase involved spreads out over a relatively large area. Thus, both distillation and gas-liquid chromatography (GC) are separations involving a gas and a liquid phase, but in the latter the liquid phase is spread out as a thin layer on a largely inert, solid supporting material, in the form of a column. Similarly, solvent extraction and liquid-partition chromatography (either paper or column) involve two liquid phases, but, in the latter, one of the liquid phases is present as a supported thin layer. In these examples, the mode of contacting the phases can also be different. In a simple distillation process, a batch of the mixture is heated in the boiler, and the distillate consists of the more volatile components. In contrast, in GC the gas mixture moves in a *counter-current* manner to the immobilized liquid layer, ensuring that the increasingly depleted, mobile gas phase encounters a fresh, clean portion of the immobilized liquid phase. In counter-current processes, a component comes—or almost comes—to equilibrium between two phases many times. It is possible to carry out separations using pairs of *bulk* phases which undergo counter-current contact. Thus, fractional distillation, in which a packed distillation column and reflux head are used, involves counter-current contact.

The focus of this chapter will be the chemistry of solvent extraction and its use as a separation process, particularly altering the chemical parameters of an extraction system in order to bring about the desired separation in a single step. The following two chapters will describe in more detail the principles and applications of chromatographic processes (in which a large number of equilibrium steps occur), in contrast to the "batch" solvent-extraction processes.

20.2 GENERAL PRINCIPLES AND TERMINOLOGY OF SOLVENT EXTRACTION

Solvent extraction enjoys a favored position among separation techniques because of its ease, simplicity, speed, and wide scope. Separation by extraction can usually be accomplished in a few minutes using a simple pear-shaped separatory funnel, or an ordinary stoppered centrifuge tube and the solvent layers separated by centrifugation. It is applicable both to trace-level impurities and to major constituents. Furthermore, inorganic constituents are often separated in a form suitable for direct analysis by spectrophotometric, atomic absorption, radiochemical, or other methods.

In solvent extraction, a solute of interest transfers from one solvent into a second solvent that is essentially immiscible with the first. The extent of transfer can be varied from negligible to essentially total extraction through control of the experimental conditions.

All solvent-extraction procedures can be described in terms of three aspects, or steps.

1. *The distribution of the solute*, called the extractable complex or species, *between the two immiscible solvents*. This step can be quantitatively described by Nernst's distribution law, which states that the ratio of the concentrations of a solute distributing between two essentially immiscible solvents at constant temperature is a constant, provided the solute is not involved in chemical interactions in either solvent phase (other than solvation). That is,

$$K_D = \frac{[A]_o}{[A]} \qquad (20.3)$$

where $[A]_o$ = molar concentration of solute A in organic solvent
   $[A]$ = concentration of solute A in aqueous solution
   K_D = distribution constant or distribution coefficient of A

2. *Chemical interactions in the aqueous phase or formation of the extractable complex*. Inasmuch as most of the substances of interest, particularly metal ions, are not usually encountered in a form that can be directly extracted into an organic solvent, chemical transformations to produce an extractable species are of primary importance in solvent-extraction processes. For organic compounds, the pH of the aqueous phase is often the primary controlling factor. At pHs where the organic compound exists in a neutral, noncharged form, the species will usually extract quite efficiently into a nonaqueous solvent, whereas the charged, protonated (or deprotonated) form will stay in the aqueous layer.

3. *Chemical interactions in the organic phase*, such as self-association or mixed-ligand-complex formation. Such chemical interactions do not invalidate the Nernst distribution law (Eqn. 20.3), but the extraction cannot be quantitatively described by that simple equation. It becomes necessary to know how each of the contributing reactions affects the extent of extraction.

The extent of extraction is described in terms of the *distribution ratio, D*, given by

$$D_A = \frac{C_{A(O)}}{C_A} \qquad (20.4)$$

where $C_{A(O)}$ = *total analytical concentration* of component A (in whatever chemical form) in the organic phase
   C_A = total analytical concentration of A in the aqueous phase

If the substance does not undergo chemical reactions in either phase, then D_A reduces to K_D.

Another important way of expressing the extent of extraction is by the *fraction extracted F*, which is

$$F_A = \frac{C_{A(0)}V_o}{C_{A(0)}V_o + C_A V} = \frac{D_A R_V}{D_A R_V + 1} \tag{20.5}$$

where V_o = volume of the organic phase
V = volume of the aqueous phase
R_V = phase-volume ratio, V_o/V

The *percent extraction* is simply $100F$. The fraction remaining in the aqueous phase, G_A, is

$$G_A = \frac{1}{D_A R_V + 1} = 1 - F_A \tag{20.6}$$

Equation 20.5 shows that, for a given value of D, the extent of extraction can be increased by increasing the phase-volume ratio—that is, by increasing the volume of the organic phase. Another way of increasing the fraction extracted is to extract several times using only part of the total volume of the organic solvent at a step. With $D_A R_V = 10$, the fraction extracted in a single batch extraction is about 0.90. Two extractions with $R_V/2$ increases the fraction extracted to about 0.97; three extractions with $R_V/3$ increases it to 0.99—essentially quantitative extraction.

Consider two substances, A and B, present in a solution. Initially, the concentration ratio is C_A/C_B; after extraction, the concentration ratio in the organic phase will be $C_A F_A/C_B F_B$, where F_A and F_B are the corresponding fractions extracted. The ratio F_A/F_B (the factor by which the initial concentration ratio is changed by the separation) is a measure of the *separation* of the two substances. A corollary measure of separation is G_A/G_B, the change in the ratio of concentrations remaining in the aqueous phase.

Two substances whose distribution ratios differ by a constant factor will be most effectively separated if the product $D_A D_B$ is unity. For instance, consider a pair of substances whose distribution ratios D_A and D_B are 10^3 and 10^1, respectively. If these substances were present in equal quantity, then a single extraction would remove 99.9% of the first and 90% of the second. However, if the two distribution ratios were 10^1 and 10^{-1} (again differing by a factor of 100), the respective fractions extracted would be 90% of A and 10% of B, a much more effective separation.

Classification of Extraction Systems: Organic and Inorganic

The following classification refers essentially to inorganic systems, particularly those involving metal ions. Many organic compounds, of course, are extractable without any significant chemical reaction occurring—for instance, alcohols, ethers, carboxyl compounds, and so on. Systematic changes in the extraction of such compounds by various solvents can be related to the degree of hydrogen bonding and to other, less specific, interactions of the organic compounds, as well as to their molecular weights.

Most metal salts are soluble in water but not in organic solvents, particularly hydrocarbons and chlorinated hydrocarbons. This solubility is caused by the high dielectric constant of water and by the ability of water to coordinate with ions, especially metal ions, so that the hydrated salt more nearly resembles the solvent. To form a metal complex that can be extracted by an organic solvent, it is necessary to replace the coordinated water around the metal ion by groups, or ligands, that will form an uncharged species compatible with the low-dielectric-constant organic solvent.

Extractable metal species can be formed in a great variety of ways. A classification system of metal extractions is therefore very useful, particularly as a guide to understanding the hundreds of different extraction systems now in use. Methods for forming an extractable species include the following.

1. *Simple* (monodentate) *coordination* alone, as with $GeCl_4$.

2. *Heteropoly acids*, a class of coordination complexes in which the central ion is complex rather than monatomic, as with phosphomolybdic acid, $H_3PO_3 \cdot 12MoO_3$.

3. *Chelation* (polydentate coordination) alone, as with Al(8-quinolinate)$_3$ or Cu(diethyldithiocarbamate)$_2$.

4. *Ion association* alone, as with $Cs^+,(C_6H_5)_4B^-$ (the comma is used to indicate association between the two ions).

Combinations of the above can be used, such as

5. *Simple coordination and ion association*, as with ("Onium")$^+$,$FeCl_4^-$. "Onium" stands for one of the following cation types: hydrated hydronium ion, $(H_2O)_3H^+$, a rather labile cation requiring stabilization by solvation with an oxygen-containing solvent; a substituted ammonium ion, $R_nNH_{(4-n)}^+$, where R is an alkyl or aralkyl group and n may vary from 1 to 3; a substituted phosphonium ion, R_4P^+; stibonium ion, R_4Sb^+; sulfonium ion; and other ions of this sort, including cationic dyes such as Rhodamine B.

6. *Chelation and ion association* with either positively or negatively charged metal chelates, such as Cu(2,9-dimethyl-1,10-phenanthroline)$_2^+$,ClO_4^- or $3(n\text{-}C_4H_9NH_3^+)$,Co(nitroso-R-salt)$_3^{3-}$.

7. *Simple coordination and chelation*, such as in Zn(8-quinolinol · pyridine). This category is of significance for coordinatively unsaturated metal chelates—those with a monoprotic bidentate reagent in which the coordination number of the metal is greater than twice its valence.

Clearly, a thorough understanding of solvent extraction of metals presupposes a deep knowledge of coordination chemistry.

Methods of Extraction

Generally, one has a choice of three methods of extraction: batch extraction, continuous extraction, and counter-current extraction. The choice is usually determined by the distribution coefficient of the substance extracted and, in the case where separation is desired, by the closeness of the various distribution coefficients involved. Because of the speed and simplicity of batch extraction, this method is preferred when applicable.

Batch Extraction. When experimental conditions can be adjusted so that the fraction extracted is 0.99 or higher ($DR_V \geq 100$), then a single or batch extraction will transfer the bulk of the desired substance to the organic phase. Most analytical extractions fall into this category. The usual apparatus for a batch extraction is a separatory funnel such as the Squibb pear-shaped funnel, although many special types of funnels have been designed [1].

Even with DR_V equal to only 10, two successive batch extractions will transfer 99% of the material to the organic phase. If one chooses as a criterion of separation that substance A be at least 99% extracted and substance B be no more than 1% extracted, then, from Equation 20.5, one must have $D_A R_V > 100$ and $D_B R_V < 0.01$ for a single (batch) extraction.

Continuous Extraction. For relatively small DR_V values, even multiple batch extraction cannot conveniently or economically be used—too much organic solvent is required. Continuous extraction using volatile solvents can be carried out in an apparatus in which the solvent is distilled from an extract-collection flask, condensed, contacted with the aqueous phase, and returned to the extract-collection flask in a continuous fashion.

Counter-Current Distributions. A special multiple-contact extraction is needed to effect the separation of two substances whose D values are very similar. In principle, counter-current distribution (CCD) could be carried out in a series of separatory funnels, each containing an identical lower phase. The mixture is introduced into the upper phase in the first funnel. After equilibration, the upper phase (containing the substance of interest) is transferred to the second funnel, and a new portion of upper phase (devoid of sample) is introduced into the first funnel. After both funnels are equilibrated, the upper phase of each is moved on to the next funnel, and a fresh portion of upper phase is again added to the first funnel. This process is repeated as many times as there are funnels, or more, collecting the upper phases as "elution fractions." With automated CCD equipment, several hundred transfers can be conveniently accomplished, which will permit the separation of two solutes whose D_A/D_B ratio is less than 2.

It can be shown that the distribution ratio D of a solute in a CCD process is related to the concentration in the various separatory funnels or stages by the binomial expansion

$$(F + G)^n = 1 \tag{20.7}$$

where n is the number of stages in the CCD process and F and G are given by Equations 20.5 and 20.6, respectively.

The fraction $T_{n(r)}$ of the solute present in the rth stage for n transfers can be calculated from

$$T_{n(r)} = \frac{n!}{r!(n-r)!} \frac{(DR_V)^r}{(1 + DR_V)^n} = \frac{n!}{r!(n-r)!} F^r G^{n-r} \tag{20.8}$$

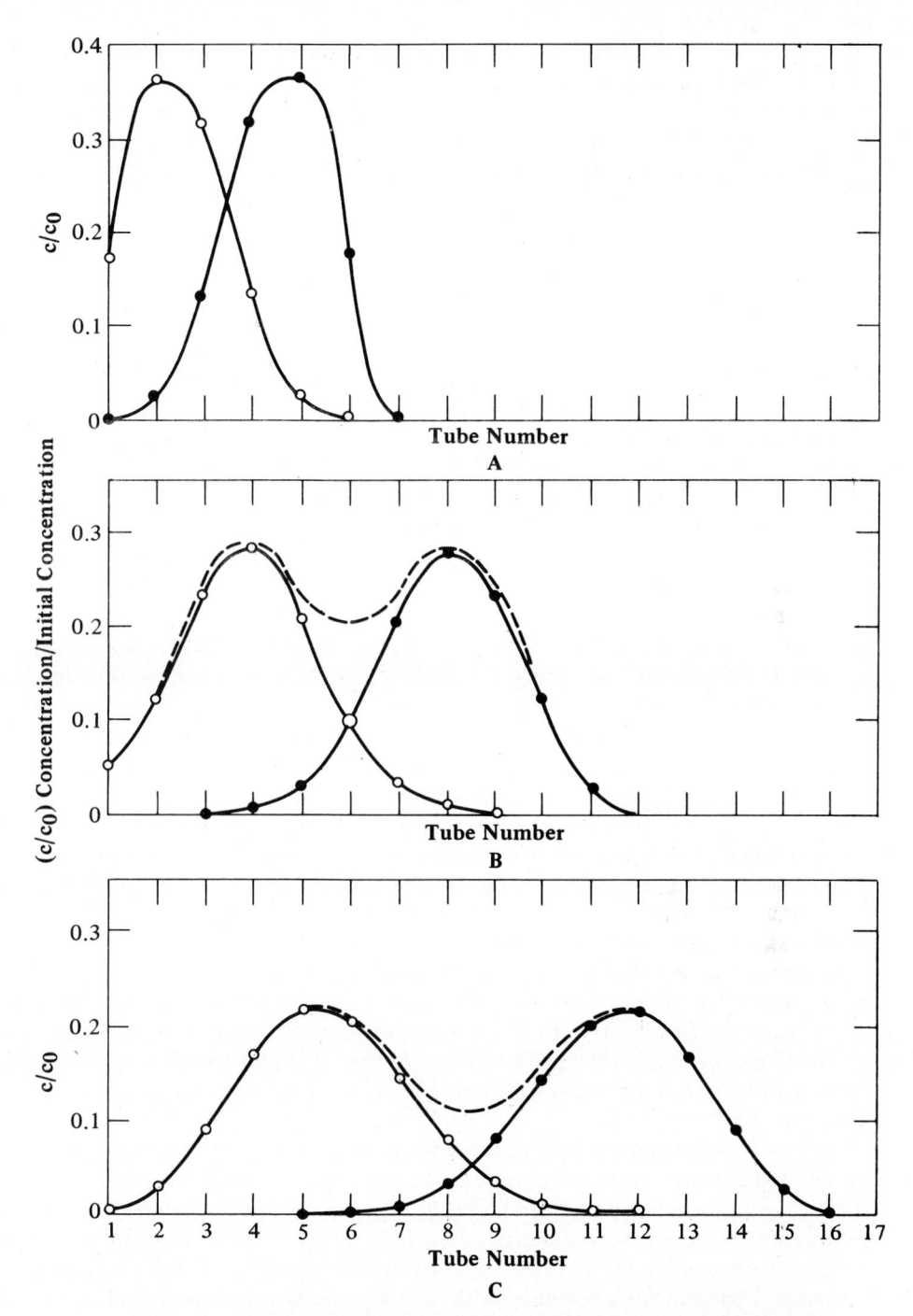

FIGURE 20.1. *Distribution of two solutes in counter-current distribution tubes after different numbers of transfers. The distribution ratio for the left peak is 7/3 and that for the right peak is 3/7. A: After 5 transfers ($n = 5$). B: After 10 transfers ($n = 10$). C: After 15 transfers ($n = 15$). From H. Purnell,* Gas Chromatography, *New York: John Wiley, 1962, p. 95, by permission of the publisher.*

The distribution of two solutes with differing distribution ratios in the tubes after different numbers of transfers is illustrated in Figure 20.1. As the number of transfers is increased, the solute is spread through a larger number of tubes. The separating ability, however, increases with increased number of transfers. After the run is complete, each tube is analyzed for the solute. In a preparative run, the contents of the tubes containing a particular solute are combined.

20.3 EXPERIMENTAL TECHNIQUES

Selection of a particular extraction method from the large number of methods available involves considering the behavior of interfering substances that might be present, as well as that of the substance of interest. Another important factor is how the species in question is to be analytically determined after extraction. Some of the chelate systems (e.g., dithizone chelates) are strongly enough colored to provide the basis for a spectrophotometric determination. If the extract is to be aspirated into the flame of an atomic absorption apparatus, however, a dithizone solution is not as desirable as a nonbenzenoid reagent, because of its behavior in the flame.

The problem is greatly simplified by referring to the literature, from which one can choose a method on the basis of similarity or even exact matching of separation problems. One generally proceeds in a new situation by following published precedents, but a better understanding of the design of an extraction procedure can be obtained from a careful study of the *basis* of previous work.

Choice of Solvent

Solvents differ in polarity, density, and ability to participate in complex formation. Generally, it is more convenient to use a solvent denser than water when the substance of interest is being extracted and a less dense solvent when interferences are extracted away from the substance of interest. In the former case, this is because the (denser) phase containing the extracted substance of interest can be conveniently drained from a separatory funnel, leaving behind the original solution; this can then be easily reextracted if necessary. In the latter case, the (less dense) solvent containing the interferences can be left behind in the funnel. If multiple extractions are necessary to remove the interferences, however, it may be more convenient to use a solvent denser than water.

Ion-association complexes in which one of the ions is strongly solvated, such as the hydrated hydronium ion encountered in extracting chloride complexes from HCl solutions [e.g., $(H_2O)_3H^+, FeCl_4^-)$], can be most effectively extracted with oxygen-containing solvents such as alcohols, esters, ketones, and ethers. Such solvents increase extractability significantly over that obtainable with hydrocarbon (or chlorinated hydrocarbon) solvents with coordinatively unsaturated chelates—that is, those in which the coordination number of the metal ion is greater than twice its oxidation state (e.g., $ZnOx_2$, where Ox refers to the 8-quinolinate [oxine] anion).

On the other hand, ion-association complexes involving quaternary ammonium, phosphonium, or arsonium ions, and coordinatively saturated chelates, can be readily extracted into hydrocarbons as well as into oxygenated solvents. In such cases, the principle of "like dissolves like," as expressed by the Hildebrand "solubility parameter" δ (defined as the heat of vaporization of 1 cm^3 of a liquid), offers a guide to extractability. Using this approach, the following expression can be derived:

$$2.3RT \log K_D = V_s(\delta_o - \delta_w)(\delta_o + \delta_w - 2\delta_s) \quad (20.9)$$

where V_s = molar volume of the solute
δ_o = solubility parameter of the organic solvent
δ_w = solubility parameter of the aqueous phase
δ_s = solubility parameter of the distributing solute

A plot of K_D versus δ_o is parabolic, with a maximum K_D for an organic solvent whose δ_0 matches δ_s. Simply expressed, in the absence of specific chemical interactions, a substance will be most extractable in a solvent whose δ-value most closely matches its own. Thus, 8-quinolinine ($\delta = 10$) is more extractable into benzene ($\delta = 9.2$) than into CCl$_4$ ($\delta = 8.6$), and more into CCl$_4$ than into heptane ($\delta = 7.4$). The application of this principle is limited by the lack of known δ-values for many extractable species.

It must not be assumed that the best solvent to use is always the one that gives the highest extractability, because a poorer solvent is often more selective for separations.

Stripping and Backwashing

Occasionally, it is of advantage to remove (*strip*) the extracted solute from the organic phase into which it has been extracted as part of the analytical procedure. This is done by shaking the organic phase with a fresh portion of aqueous solution containing acids or other reagents that will decompose the extractable complex in the organic phase. The charged metal ion will then be extracted preferentially into the new aqueous solution.

The technique of *backwashing* also involves contacting the organic extract with a fresh aqueous phase. Here, the combined organic phases from multiple extraction of the original aqueous phase, which contain almost all the desired element and some of the impurities, are shaken with small portions of a fresh aqueous phase containing the same reagent concentrations initially present. Under these conditions, most of the desired element remains in the organic phase while the bulk of the impurities are back-extracted (backwashed) into the aqueous phase because of their lower distribution ratios.

Treatment of Emulsions

After two immiscible liquids have been shaken, a sharp phase boundary should rapidly reappear; therefore, emulsions should not be allowed to form. The tendency to form emulsions decreases with increasing interfacial tension. In liquids of relatively high mutual solubility or that contain surfactants, the interfacial tension is low and the tendency to form emulsions is correspondingly high. Low-viscosity solvents and solvents with densities significantly different from that of water are also helpful in avoiding emulsions. With systems that tend to form emulsions, repeated inversion of the two phases rather than vigorous shaking is called for. In an extreme case, using a continuous extractor rather than a separatory funnel is often successful. The tendency to form emulsions can be reduced by adding neutral salts or an anti-emulsion agent.

20.4 IMPORTANT EXPERIMENTAL VARIABLES

In addition to such important factors as the choice of organic solvent, the avoidance of emulsions, and the actual method of extraction—batch, continuous, or countercurrent—there are other important and easily controlled experimental variables. In the extraction of metals, the most important and critical of these are the pH of the aqueous solution and the use of masking agents. These two factors are often primarily responsible for the specificity (degree of separation) of an extraction method.

Chelate Extraction Systems

Many chelating extractants are weak acids and can be represented as HR. For a chelate extraction process,

$$M^{n+} + nHR\,(\text{Org}) \rightleftharpoons MR_n\,(\text{Org}) + nH^+ \qquad (20.10)$$

the distribution ratio, D_M, is given by

$$D_M = \frac{C_{M(0)}}{C_M} = \frac{[MR_n]_0}{[M^{n+}]/\alpha_M} = \frac{[MR_n]_0}{[M^{n+}]}\alpha_M \qquad (20.11)$$

In this equation, $C_{M(0)} = [MR_n]_0$, since there is only one metal-containing species in the organic phase. In the aqueous phase, there may be many metal-containing species in addition to M^{n+}, but these can be accounted for by using α_M, the fraction of the total metal concentration actually present as M^{n+},

$$\alpha_M = \frac{[M^{n+}]}{C_M} \qquad (20.12)$$

The formation of the chelate MR_n is expressed by the formation constant

$$\beta_n = \frac{[MR_n]}{[M^{n+}][R^-]^n} \qquad (20.13)$$

and the formation of the anion R^- from HR is quantitatively given by the acid-dissociation constant

$$K_a = \frac{[H^+][R^-]}{[HR]} \qquad (20.14)$$

Incorporating these expressions as well as those for the distribution of the reagent

HR and chelate MR_n (their distribution coefficients),

$$K_{D_R} = \frac{[HR]_0}{[HR]} \tag{20.15}$$

$$K_{D_C} = \frac{[MR_n]_0}{[MR_n]} \tag{20.16}$$

it can be seen that

$$D_M = \frac{[MR_n]_0}{[M^{n+}]} \alpha_M = \beta_n \frac{[MR_n]_0}{[MR_n]} [R^-]^n \alpha_M = \beta_n \frac{K_{D_C} K_a^n}{K_{D_R}^n} \frac{[HR]_0^n}{[H^+]^n} \alpha_M \tag{20.17}$$

The combination of constants (K_{DC}, K_a, K_{DR}) in Equation 20.17 is called the *overall extraction constant*, K_{ex}. Representative values of K_{ex} are listed in Table 20.1. The $pH_{1/2}$ values are explained below.

Equation 20.17 shows that the value of D_M increases with increasing concentration of the reagent in the organic phase, and decreases with increasing hydrogen-ion concentration in the aqueous phase. Control of pH is therefore important in chelate extractions. Inasmuch as the extractions of different metal ions with a given reagent are characterized by different extraction constants, the extraction curves (percentage extracted vs. pH) will be similar in shape, but displaced in pH. Figure 20.2 shows a typical set of extraction curves for various metal dithizonates. Note that, whereas the curves of all the divalent metal ions are parallel, those for Ag^+ and Tl^+ are less steep, because $n = 1$ in Equation 20.17. From the curve, it can be concluded that at pH = 2, Hg^{2+} is 100% extracted; Ag^+, Cu^{2+}, and Bi^{3+} are fractionally extracted; and the other ions listed are not extracted at all. It would be simple to separate Hg^{2+} from Sn^{2+}, Pb^{2+}, Zn^{2+}, Tl^+, and Cd^{2+} in a mixture by extracting with dithizone at pH = 2, but difficult to separate Hg^{2+} from Ag^+, Cu^{2+}, or Bi^{3+}.

TABLE 20.1. *Values of Extraction Constants, K_{ex} and $pH_{1/2}$, in Selected Metal-Chelate Systems*

	Extractant			
	8-Quinolinol (0.10 M in $CHCl_3$)		Dithizone (10^{-4} M in CCl_4)	
Metal Ion	Log K_{ex}	$pH_{1/2}$	Log K_{ex}	$pH_{1/2}$
Ag^+	—	6.5	7.18	−3.2
Al^{3+}	−5.22	2.87	Not extracted	
Ca^{2+}	−17.9	10.4	Not extracted	
Cd^{2+}	—	4.65	2.14	2.9
Cu^{2+}	1.77	1.51	10.53	−1.3
Fe^{3+}	4.11	1.00	Not extracted	
Pb^{2+}	−8.04	5.04	0.44	3.8
Zn^{2+}	—	3.30	2.3	2.8

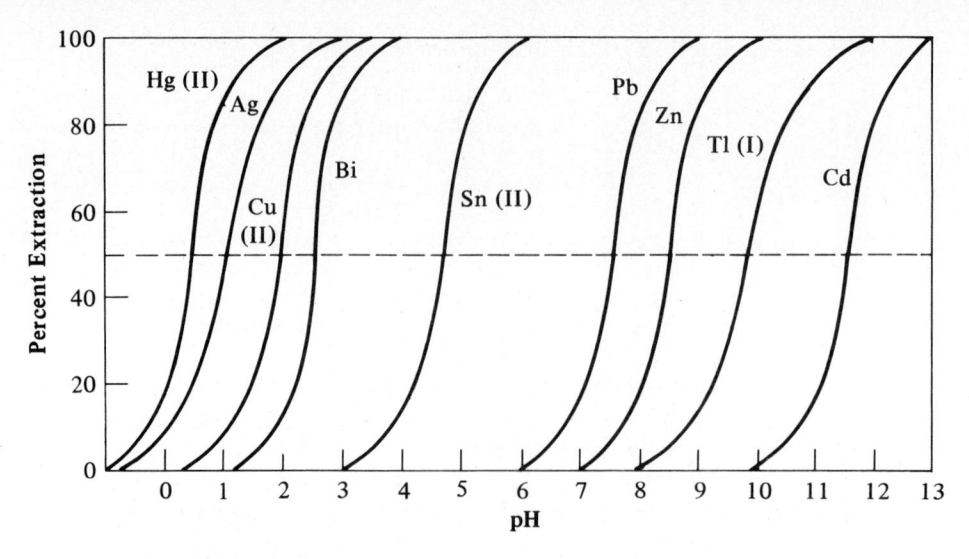

FIGURE 20.2. *Qualitative extraction curves for metal dithizonates. From G. H. Morrison and H. Freiser, in C. L. Wilson and D. Wilson, eds.,* Comprehensive Analytical Chemistry, *Vol. 1A, Amsterdam: Elsevier, 1959, by permission of the publisher.*

Since Bi^{3+} is only 10% extracted at this pH, backwashing several times with fresh aqueous (pH = 2) portions would quantitatively remove Bi^{3+} (90% of the remaining amount each time) from the extract without appreciably affecting the extracted Hg.

One useful way to condense extraction information from curves such as those in Figure 20.2 or from expressions such as Equation 20.17 is to specify the $pH_{1/2}$ value for the metal ion, obtained with a particular concentration of the reagent. The $pH_{1/2}$ is the pH value at which half the metal is extracted into the organic solvent (i.e., when $D = 1$). Thus, from Figure 20.2, the $pH_{1/2}$ values for the dithizonates are 0.3 for Hg, 1.0 for Ag, 1.9 for Cu, 2.5 for Bi, 4.7 for Sn, 7.4 for Pb, 8.5 for Zn, 9.7 for Tl, and 11.6 for Cd. For a single batch extraction, a minimum of 3 units difference in $pH_{1/2}$ is required to permit the quantitative separation of two metal ions; however, as mentioned above, a smaller difference suffices if backwashing is used.

Masking Agents. The factor α_M in Equation 20.17, which represents the fraction of the total metal concentration in the aqueous phase that is in the form of the simple hydrated metal ion, points to the importance of masking agents in improving the selectivity of extraction. Masking agents are competing complexing agents that form charged water-soluble complexes. Their effectiveness in preventing reaction of a metal ion with an extracting agent increases with increasing formation constant of the masking complex, increasing concentration of the masking agent, and, for the many masking agents that are bases, with increasing pH. Some representative values of α_M are listed in Table 20.2 for different masking agents.

As an illustration of masking, consider a mixture of Ag^+ and Cu^{2+} from which Ag^+ is to be selectively extracted. It can be seen from Figure 20.2 that Ag^+

TABLE 20.2. *Values of Masking Factor* ($-Log\,\alpha_M$ *from Eqn. 20.12) for Representative Metal Ions and Masking Agents at Various pH Values*

Metal Ion[a]	Masking Agent	pH 2	pH 5	pH 8	pH 10
Ag^+	EDTA	0	0.5	3.7	5.5
	NH_3	0	0.1	4.6	7.2
	CN^-	4.7	10.7	16.7	19.0
Al^{3+}	EDTA	1.8	3.2	14.5	18.3
	OH^-	0	0.4	9.3	17.3
	F^-	10.0	14.5	14.5	17.3
Ca^{2+}	EDTA	0	3.2	7.1	8.9
	Citrate	0	1.8	2.5	2.5
Cd^{2+}	EDTA	1.8	7.9	12.2	14.0
	NH_3	0	0	2.3	6.7
	CN^-	0	0.7	10.1	14.5
Cu^{2+}	EDTA	4.6	10.7	15.0	16.8
	NH_3	0	0	3.6	8.2
Fe^{3+}	EDTA	10.3	17.2	22.0	26.4
	OH^-	0	3.7	9.7	13.7
	F^-	5.7	8.9	9.8	13.7
Pb^{2+}	EDTA	4.2	10.2	14.4	16.2
	OH^-	0	0	0.5	2.7
	Citrate	1.0	4.2	4.2	5.3
Zn^{2+}	EDTA	2.8	8.8	12.9	14.7
	NH_3	0	0	0.4	0.7
	CN^-	0	0	7.5	2.3

a. Al^{3+}, Ca^{2+}, and Pb^{2+} are not masked by NH_3 or CN^-. Cu^{2+} is very strongly masked by CN^-, >20. Fe^{3+} is very strongly masked by CN^-, but is not masked by NH_3.

can be quantitatively extracted at pH = 3; but Cu^{2+} is also appreciably extracted. In the presence of 0.1 M EDTA, the value of log α_{Cu} = -6.6 (estimated from Table 20.2), which displaces the extraction curve of copper dithizonate to the right, increasing $pH_{1/2}$ by 3.3 units (see Eqn. 20.17). Because the value of log α_{Ag} under these conditions is about -0.2, EDTA has little effect on the extraction curve of silver dithizonate. Hence, in the presence of 0.1 M EDTA at pH = 3, Ag^+ will be selectively extracted from Cu^{2+}.

Similarly, the use of cyanide as a masking agent will permit the selective extraction of Al^{3+} by 8-quinolinol in the presence of such transition-metal ions as Cu^{2+} and Fe^{3+}, as well as Ag^+; Al^{3+} does not form a CN^- complex, whereas the other metals form strong complexes. Other examples of successful masking can be predicted with the help of Table 20.2. Masking such as this is also useful in improving the selectivity of ion-exchange separations.

Kinetics of Extraction. Kinetic factors may be important in all types of extraction, but they are most frequently observed with chelate extraction systems. Extraction equilibrium can usually be achieved in 1 or 2 min of shaking because mass-transfer rates are reasonably rapid; however, the formation of an extractable complex is sometimes slow enough to affect the course of the extraction, particularly with certain metal chelates. For example, most substitution reactions of Cr^{3+} are very slow; thus, although Cr^{3+} forms stable chelates, it is rarely extracted in the usual chelate extraction procedure. Less dramatic, but analytically useful, is the difference in the speed of formation of various metal dithizonates, which makes it possible to separate Hg^{2+} from Cu^{2+}, and Zn^{2+} from Ni^{2+}, by using shaking times of no longer than 1 min.

Ion-Association Extraction Systems

As with chelate systems, ion-association extraction equilibria involve a number of reactions. An example is the extraction of Fe^{3+} from HCl solutions into ether:

$$Fe^{3+} + 4Cl^- \rightleftharpoons FeCl_4^- \tag{20.18}$$

$$H(H_2O)_4^+ + FeCl_4^- \rightleftharpoons H_9O_4^+, FeCl_4^- \tag{20.19}$$

$$H_9O_4^+, FeCl_4^- \rightleftharpoons H_9O_4^+, FeCl_4^- \text{ (Ether)} \tag{20.20}$$

The importance of chloride and of acid in the overall extraction is evident. About 6 M HCl is required for optimum extraction of iron. Ether, an oxygen-containing solvent, is needed to stabilize the $H_9O_4^+$ ion. If a $(C_4H_9)_4N^+$ salt is added, then the iron can be extracted out of a much less acidic solution, provided the chloride concentration is about 6 M; and, more significantly, it would be possible to use benzene, CCl_4, or $CHCl_3$ for the extraction as well as oxygen-containing solvents.

In many ion-association extraction systems, high concentrations of electrolyte are effective in increasing the extent of extraction. The addition of such salts, referred to as *salting-out* agents, serves two purposes. The first, and more obvious, is to aid the direct formation of the complex by the mass-action effect—the formation of a chloro or nitrato complex, for instance, is promoted by increasing the concentration of Cl^- or NO_3^-. Second, as the salt concentration increases, the concentration of "free" (uncomplexed) water decreases because the ions require a certain amount of water for hydration. This decreases the solubility of the complex in the aqueous phase. Because Li^+ is more strongly hydrated than K^+, $LiNO_3$ is a much better salting-out agent than KNO_3 for nitrate extraction systems, even though equimolar solutions supply the same nitrate concentration.

20.5 EXTRACTION SYSTEMS AND EXAMPLES

In this section, we will describe the properties of a few representative organic compounds that are used for the extraction of metal ions, and also present a few examples of simple solvent-extraction procedures that have been used to separate the analyte of interest prior to its determination by some instrumental method.

Metal-Extraction Systems

Sodium Diethyldithiocarbamate. This reagent is a white crystalline compound, readily soluble in water, that is an effective extractant for about two dozen metals (Cu, Fe, Co, Bi, Ni, U, Cr, Te, Se, Ag, Hg, As, Sb, Sn, Pb, Cd, Mo, Mn, V, Zn, In, Ga, Tl, W, Re, Os, Nb), most of which are the "hydrogen sulfide–group" metal ions. The selectivity of the reagent can be improved by the use of masking agents. Thus, with EDTA, Cu may be extracted from a steel sample. With a combination of EDTA and cyanide, bismuth can be selectively removed.

Sodium Diethyldithiocarbamate

$$(C_2H_5)_2N-C\begin{smallmatrix}\nearrow S \\ \searrow S^-, Na^+\end{smallmatrix}$$

Although the reagent itself does not absorb in the visible region, a number of its transition-metal complexes are colored, making spectrophotometric determination possible. Diethyldithiocarbamates extracted into an ester or ketone solvent are widely employed as a means of preconcentration of metal ions in atomic absorption spectrometry.

Diphenylthiocarbazone (Dithizone). Dithizone is a black-violet crystalline compound, essentially insoluble in water and readily soluble in many organic solvents. It forms highly colored chelates with Ag, Au, Bi, Cd, Co, Cu, Fe, Hg, In, Ni, Pb, Pd, Pt, Te, and Zn that provide the basis for very sensitive extraction spectrophotometric determinations.

Dithizone

8-Quinolinol (Oxine). Oxine, a white crystalline compound, insoluble in water but soluble in many organic solvents, forms extractable chelates with almost 60 metal ions. Selectivity is achieved by pH control and masking. Most metal oxinates can be determined spectrophotometrically; those with some nontransition metals such as Al and Zn fluoresce.

8 – Quinolinol

Outline of Illustrative Extraction Procedures

Several specific extraction procedures are outlined below to illustrate the principles discussed in previous sections. For a working method, more detailed procedures can be found in the Selected Bibliography at the end of the chapter. The usual precautions peculiar to trace-element determinations (e.g., impurities and solution stability) must also be carefully observed.

Extracting Cadmium with Dithizone. It is possible to separate Cd^{2+} from Pb^{2+} or Zn^{2+} by using a highly alkaline solution during extraction; and from Ag^+, Hg^{2+}, Ni^{2+}, Co^{2+}, and Cu^{2+} by stripping the Cd^{2+} at a pH of 2 where the other dithizonates are stable.

A solution containing up to 50 μg Cd^{2+} is treated with tartrate (to avoid precipitating the hydroxide), and made basic with an excess of 25% KOH. This is now shaken with successive 5-mL portions of dithizone in $CHCl_3$ until the aqueous layer remains yellowish brown (indicating excess dithizone). The combined chloroform extracts are then shaken for 2 min with an aqueous solution buffered at pH = 2, which will strip the Cd^{2+} quantitatively. To remove small amounts of Cu^{2+} and Hg^{2+} that may accompany the Cd^{2+}, the aqueous solution (pH = 2) is reextracted with a fresh portion of dithizone in $CHCl_3$; the Cd^{2+} will remain in the aqueous solution.

Extracting Methyl Mercury from Seafood. A very interesting example of the use of simple solvent extraction in sample preparation is provided by considering the procedure originally developed by Westöö for the gas-chromatographic determination of methyl mercury in seafood [2]. To a homogenized fish sample is added concentrated HCl and NaCl solutions, and about an equivalent volume of benzene. The strongly acid aqueous condition serves to completely displace the methyl mercuric ion, which is normally very tightly bound to the sulfhydryl groups in proteins, into solution. The high concentration of chloride ion largely converts the methyl mercuric ion to the neutral, molecular, methyl mercuric chloride form, which is efficiently extracted into the nonpolar benzene layer. The methyl mercury is then back-extracted into an aqueous cysteine solution to remove it from nonpolar organic compounds that may also have been extracted into the benzene layer and might later serve as an interference in the determination step (chlorinated pesticides would be particularly troublesome). The cysteine solution is then acidified with HCl and the resultant CH_3HgCl extracted again into benzene. The methyl mercury level is finally determined by gas chromatography with an electron-capture detector, which is a very sensitive method for chlorinated compounds. The methyl mercury found must be corrected for the overall extraction efficiency of the three sequential extractions, which runs quite high—about 80 to 90% overall. This extraction procedure very nicely separates the methyl mercury from the fish matrix and from interfering compounds, and the sensitivity of the electron-capture detector permits determinations as low as about 0.05 μg CH_3Hg^+ per gram of seafood.

Extracting Quinine from Urine. A fairly simple procedure for extracting quinine from urine, followed by determination with a scanning spectrofluorimeter, was reported by Mulé and Hushin [3]. This is of interest because screening for quinine is a fairly common method for effective surveillance of heroin abuse within a narcotic control, treatment, or aftercare program. Quinine is a common diluent of illegal heroin samples, and quinine can be detected in urine for over a week after ingestion of only a few hundred milligrams.

A few milliliters of urine are adjusted to pH 9 or 10 to convert the quinine to its noncharged molecular form, which is extracted by shaking the urine with 3:1 chloroform-isopropanol. The nonaqueous layer is back-extracted with 0.05 M H_2SO_4, and the aqueous extract analyzed directly. This straightforward procedure has a high extraction efficiency, about 95%, and serves to largely separate the quinine from interfering fluorescent compounds in urine. The overall method is capable of reliably detecting quinine down to about 0.10 μg per milliliter of urine.

Liquid Chromatography

As one proceeds from a single-stage batch separation to the multistage countercurrent distribution process, separating two components whose D values are relatively close together becomes easier. Chromatography can be viewed as a logical extension of counter-current distribution (L. C. Craig first developed CCD as a simple form of liquid chromatography) in which discrete stages or tubes are replaced by immobilized solvent supported on granules of carrier either in a two-dimensional (paper or thin-layer chromatography) or three-dimensional (column) bed. After the mixture of solutes is added to the bed, the eluant, a solvent of constant or uniformly varying composition (the latter used in *gradient elution*), is passed over or through the bed. This *elution* may be continued just long enough to separate the components on the bed—"developing the chromatogram." (This is common practice in paper or thin-layer chromatography.) In column chromatography, it is more customary to continue the flow of eluant until all of the components have appeared in the eluate flowing out of the column. The distribution of solutes along the column is similar to the distribution illustrated in Figure 20.1, where tube number represents the distance along the column. Regardless of the particular chromatographic process used (i.e., adsorption, partition, etc.), the separation of components is related to differences in their distribution ratios between the immobile and mobile phases. In ordinary *elution chromatography*, the eluting solvent has a low D-value, and so the components move at large and varying elution volumes. On the other hand, if a component with a very high D-value is part of the eluant, it will displace the solute components. In *displacement chromatography*, therefore, the solute components appear one after the other in much more tightly spaced bands than in elution chromatography. An interesting variation involves using the sample solution itself as a developing solvent. In this case, called *frontal development*, the solutes do separate in the order of their D-values, but appear in the eluate as fronts rather than bands, in which the least retained solute appears first, followed by a mixture of the first solute and the next more strongly retained component and so on, until finally all of the components of the original solution appear together, as in the original solution.

The next two chapters will treat chromatography in more detail.

SELECTED BIBLIOGRAPHY

Analytical Chemistry Fundamental Reviews. The reviews published in April of even-numbered years include comprehensive surveys of newly published extraction procedures, and many references.

GIDDINGS, J. C. "Principles of Chemical Separations." In I. M. Kolthoff, P.J. Elving, and F. H. Stross, eds., *Treatise on Analytical Chemistry*, 2nd ed., Part I, Vol. 5. New York: Wiley-Interscience, 1982, pp. 449–504.

IRVING, H. M. N. H. "Liquid-Liquid Extraction." In I. M. Kolthoff, P. J. Elving, and F. H. Stross, eds., *Treatise on Analytical Chemistry*, 2nd ed., Part I, Vol. 5. New York: Wiley-Interscience, 1982, pp. 505–84.

MORRISON, G. H., and FREISER, H. *Solvent Extraction in Analytical Chemistry.* New York: John Wiley, 1957.

ROTHBART, H. L., and BARFORD, R. A. "Countercurrent Distribution." In I. M. Kolthoff, P. J. Elving, and F. H. Stross, eds., *Treatise on Analytical Chemistry*, 2nd ed., Part I, Vol. 5. New York: Wiley-Interscience, 1982, pp. 585–652.

SEKINE, T., and HASEGAWA, Y. *Solvent Extraction Chemistry: Fundamentals and Applications.* New York: Marcel Dekker, 1977.

STARY, J. *The Solvent Extraction of Metal Chelates.* New York: Macmillan, 1964.

ZOLOTOV, Yu. A. *Extraction of Chelate Compounds.* Ann Arbor, Mich.: Ann Arbor-Humphrey Science Publishers, 1970.

REFERENCES

1. G. H. MORRISON and H. FREISER, *Solvent Extraction in Analytical Chemistry*, New York: John Wiley, 1957.
2. G. WESTÖÖ, *Acta Chem. Scand.*, *20*, 2131 (1966); *21*, 1790 (1967); *22*, 2277 (1968).
3. S. J. MULÉ and P. H. HUSHIN, *Anal. Chem.*, *43*, 708 (1971).

PROBLEMS

1. Name the categories of extractable complexes used in metal-extraction procedures and illustrate each by specific examples.
2. How does Hildebrand's theory of regular solutions apply to the role of the organic solvent in extraction processes?
3. What are masking agents? How can one tell whether a substance that masks effectively in one situation will do so in another?
4. What are ion-association complexes? Under which conditions will they form and what properties must they have to be useful in extraction?
5. What factors affect the value of an oxygen-containing solvent used in extracting ion-association complexes?
6. Relate D, the distribution ratio, to $\%\,E$, percent extraction, as a function of R_V (the phase-volume ratio V_o/V). For what purposes is D a more (or less) appropriate criterion than $\%\,E$?
7. When can the ratio of D-values for two substances A and B, D_A/D_B, serve as a satisfactory means of evaluating the separation of A and B? Why is D_A/D_B not used as a "separation index"?
8. What are batch, continuous, and countercurrent extraction processes and when would each be used?
9. Develop an algebraic expression with which to define the influence of each of the following factors on the extraction of metal ions using chelating extractants: (a) the reagent concentration; (b) the metal-ion concentration; (c) the pH; (d) the presence (and concentration) of masking agents; (e) the nature of the organic solvent; (f) the ionic strength of the aqueous phase.

10. What is meant by K_{ex} (extraction constant) and pH_0 (or $pH_{1/2}$), and how can each be used to calculate the D (or $\% E$) value under various conditions? How different must the K_{ex} or pH_0 values be for a pair of metals to get a separation of 100/1? Does the amount of difference depend on the valence of the metal ions? Are other characteristics relevant as well? If so, which ones? Does the efficiency of separation depend on the value of K_{ex} or on the ratio of the two constants?

11. What are the optimal conditions for using a dithizone solution in $CHCl_3$ to separate Bi(III) from a solution containing a 100-fold excess of zinc?

CHAPTER 21

Solid- and Liquid-Phase Chromatography

| Ronald E. Majors

Chromatography is the general name given to the methods by which two or more compounds in a mixture are physically separated by distributing themselves between two phases: (a) a *stationary phase*, which can be a solid or a liquid supported on a solid; and (b) a *mobile phase*, either a gas or a liquid, which flows continuously around the stationary phase. The separation of individual components results primarily from differences in their affinity for the stationary phase.

In *liquid chromatography* (LC) the flowing or mobile phase is a liquid, whereas in *gas chromatography* (GC) it is a gas. *Gas-solid chromatography* (GSC) is the specific term used when the stationary phase is a solid; in *gas-liquid chromatography* (GLC), the stationary phase is a liquid spread over the surface of a solid support. The present chapter is concerned with LC; the following chapter deals with GC in its various forms.

21.1 INTRODUCTION

Chromatography was discovered and named in 1906 by Michael Tswett, a Russian botanist, when he was attempting to separate colored leaf pigments by passing a solution containing them through a column packed with adsorbent chalk particles. The individual pigments passed down the column at different rates and were separated from each other. The separated pigments were easily distinguished as colored bands—hence, *chroma* ("color") + *graphy* ("writing").

The next major development was that of *liquid-liquid* (*partition*) *chromatography* (LLC) by Martin and Synge in 1941. Instead of only a solid adsorbent, they used a stationary liquid phase spread over the surface of the adsorbent and immiscible with the mobile phase. The sample components partitioned themselves between the two liquid phases according to their solubilities. For this work, Martin and Synge received the Nobel Prize in chemistry in 1952.

In the early days of column chromatography, reliable identification of small quantities of separated substances was difficult; so *paper chromatography* (PC) was developed. In this "planar" technique, separations are achieved on sheets of filter paper, mainly through partition. Appreciation of the full advantages of planar chromatography then led to *thin-layer chromatography* (TLC), in which separations are carried out on thin layers of adsorbent supported on plates of glass or some other rigid material. TLC gained popularity after the classic work by Stahl in 1958 standardizing the techniques and materials used. To aid or enhance the separation of ionic compounds by PC or TLC, an electric field can be applied across the paper or plate. The resulting techniques are referred to as *paper and thin-layer electrophoresis*, respectively.

The most recently developed chromatographic technique—GC—was first described by Martin and James in 1952 and has become the most sophisticated and widely used of all chromatographic methods, particularly for mixtures of gases or for volatile liquids and solids. Separation times of a matter of minutes have become commonplace even for very complex mixtures. The combination of high resolution, speed of analysis, and sensitive detection have made GC a routine technique used in almost every chemical laboratory.

In recent years, interest has renewed in closed-column LC because of new instrumentation, new column packings, and a better understanding of chromatographic theory. *High-performance liquid chromatography* (HPLC) is rapidly becoming as widely used as GC and is the preferred technique for the rapid separation of nonvolatile or thermally unstable samples.

21.2 BASIC PRINCIPLES OF LIQUID CHROMATOGRAPHY

To illustrate the basic principles of liquid chromatography, we shall consider a hypothetical separation of a three-component sample in a closed column. The stationary phase (*packing*) consists of solid porous particles (normally small—less than 150 μm in diameter) contained inside a long narrow tube, the *column*.

Figure 21.1 demonstrates the chromatographic process. A small volume of sample solution is injected at the column inlet (Fig. 21.1A). The mobile solvent phase moves the sample through the column packing (Fig. 21.1B). The individual components undergo sorption and desorption on the packing, thereby slowing their motion in varying degrees depending on their affinity for the packing. Each component X is distributed between the stationary phase (s) and the mobile phase (m) as it passes down the column. According to

$$X_m \leftrightharpoons X_s \qquad (21.1)$$

the corresponding *distribution coefficient* for component X is given by

$$K_X = \frac{[X]_s}{[X]_m} \qquad (21.2)$$

A large value of K_X indicates that the component favors the stationary phase and moves slowly through the column, whereas for small values of K_X the component favors the mobile phase and moves quickly through the column.

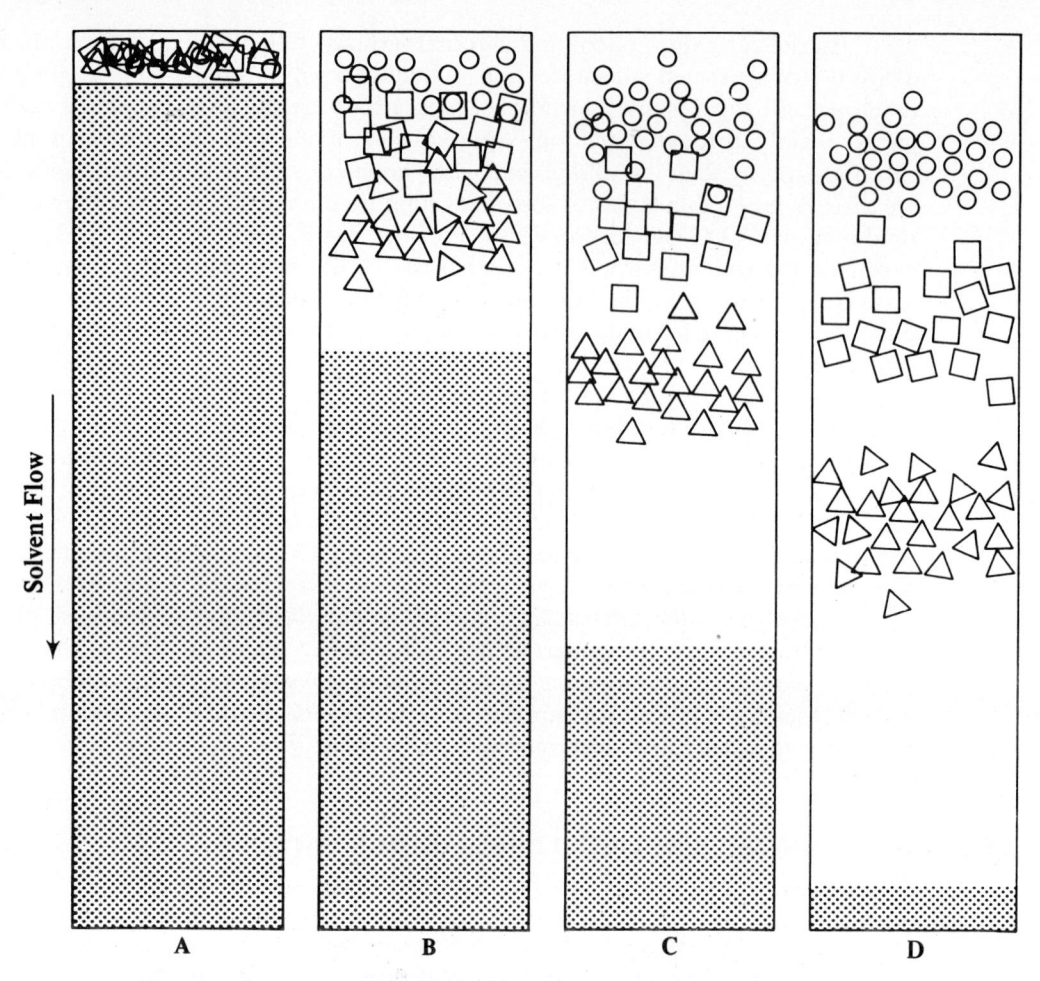

FIGURE 21.1. *Hypothetical separation of a three-component mixture: Component A:* △; *Component B:* □; *Component C:* ○. *The dotted area represents the original solvent in the column, which is being "displaced" during elution.*

The different relative speeds of the components separate them along the column (Fig. 21.1C). In elution chromatography, the separated components are moved down the entire length of the column by the mobile phase (Fig. 21.1D). If one measures the concentration of each component as it exits from the column and plots it as a function of the volume of mobile phase passed through the column, a *chromatogram* results. Individual volumes of mobile phase can be collected and the solute concentration in each measured externally (e.g., spectrophotometrically), but normally the column effluent is monitored continuously by a detector that measures some physical or chemical property of the solute or of the mobile phase. A chromatogram for the hypothetical three-component sample is depicted in Figure 21.2.

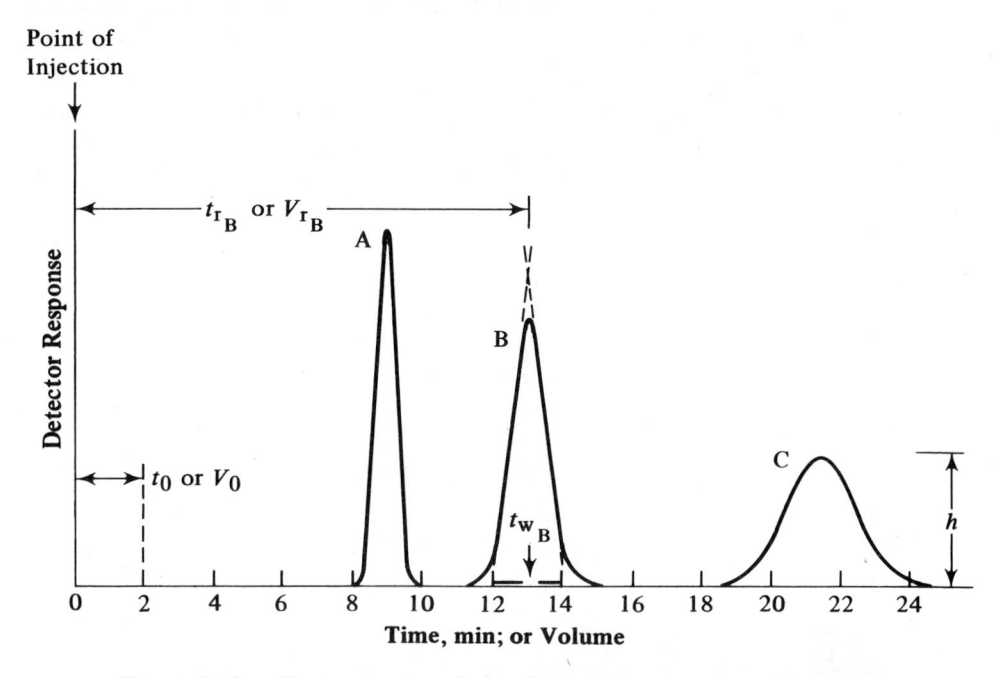

FIGURE 21.2. *Chromatogram of the three-component mixture of Figure 21.1.* t_0 = *time for solvent to traverse the column;* t_{r_B} = *retention time of substance B;* t_{w_B} = *peak basewidth of substance B;* h = *peak height. Units can also be given in terms of volume rather than time:* V_0, V_{r_B}, V_{w_B}, *and so forth.*

Several types of LC are distinguished by their predominant mechanism of separation. The stationary phase governs the separation mode. The various modes will be briefly outlined here; each will be dealt with in somewhat greater detail in Sections 21.6 and 21.7. Since the solute molecule usually has some affinity for the stationary phase, it transfers from the mobile phase to the stationary phase, setting up the equilibrium described by Reaction 21.1. This alternating process of solute mass transfer, depicted in Figure 21.3A, eventually leads to some degree of separation.

Adsorption Chromatography

Adsorption chromatography (Fig. 21.3B), often referred to as *liquid-solid chromatography* (LSC), is based on interactions between the solute and fixed active sites on a solid adsorbent used as the stationary phase. The adsorbent can be packed in a column, spread on a plate, or impregnated into a porous paper. The adsorbent is generally an active, porous solid with a large surface area, such as silica gel, alumina, or charcoal. The active sites, such as the surface silanol groups of silica gel, generally interact with the polar functional groups of the compounds to be separated. The nonpolar (e.g., hydrocarbon) portion of a molecule has only a minor influence on the separation. Thus, LSC is well suited for separating classes of compounds (e.g., separating alcohols from aromatic hydrocarbons).

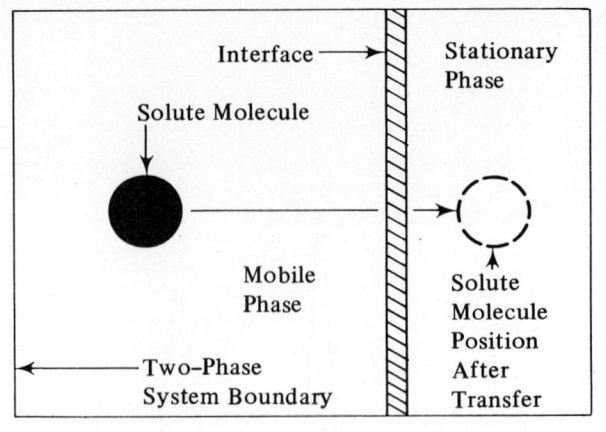

A. Transfer of Solute to a Generalized Stationary Phase

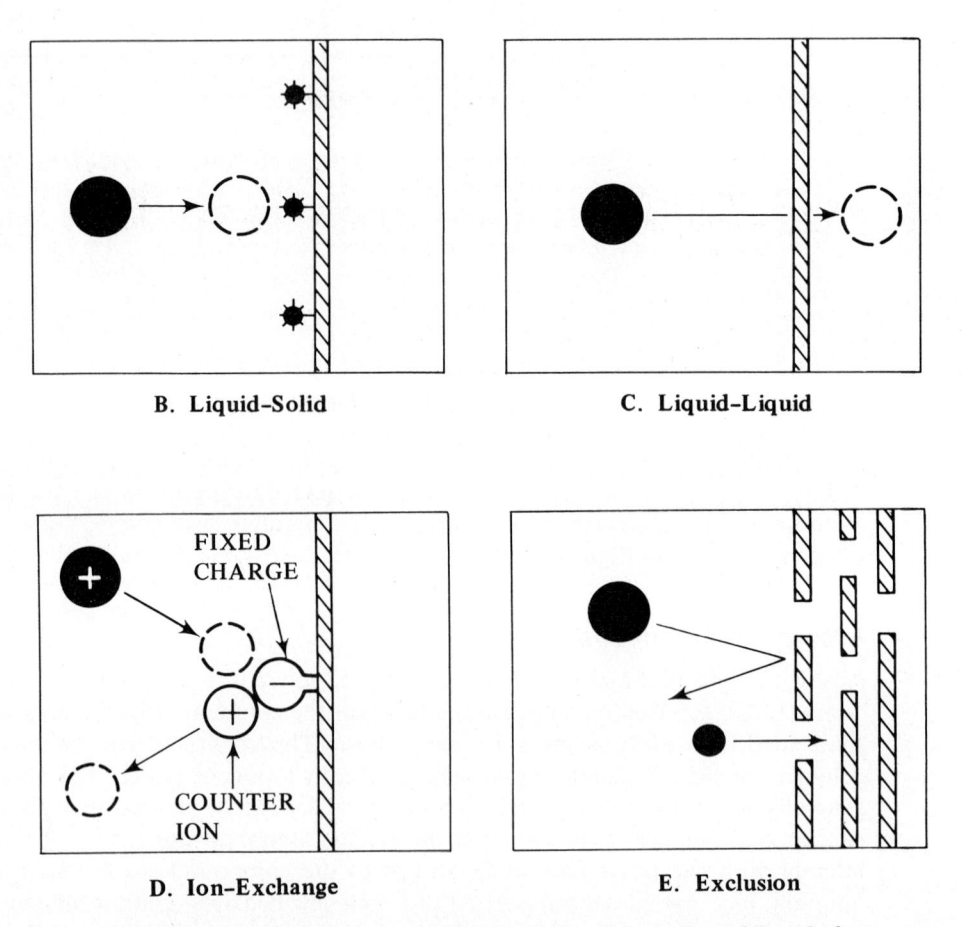

B. Liquid–Solid

C. Liquid–Liquid

D. Ion-Exchange

E. Exclusion

FIGURE 21.3. *Schematic representation of the four modes of liquid chromatography. Courtesy of Varian Associates.*

Partition Chromatography

In partition chromatography (Fig. 21.3C), also referred to as liquid-liquid chromatography, the solute molecules distribute themselves between two immiscible liquid phases, the stationary phase and the mobile phase, according to their relative solubilities. The stationary phase is uniformly spread on an inert support—a porous or nonporous particulate solid or porous paper (paper chromatography). To avoid mixing of the two phases, the two partitioning liquids must differ greatly in polarity. If the stationary liquid is polar (e.g., ethylene glycol) and the mobile phase is nonpolar (e.g., hexane), then polar components are retained more strongly; this is the usual mode of operation. On the other hand, if the stationary liquid is nonpolar (e.g., decane) and the mobile phase polar (e.g., water), polar components favor the mobile phase and elute faster. The latter technique (which has a reversed polarity) is referred to as *reverse-phase* LLC. Because of the subtle effects of solubility differences, LLC is well suited for separating homologs and isomers.

Most often, the stationary phase in LLC is chemically bonded to the support material rather than mechanically applied to it. This is referred to as *bonded-phase chromatography* (BPC). The separation mechanism of this technique is not clear, but both partition and adsorption mechanisms may be involved, depending on experimental conditions. In HPLC, the use of BPC dominates all other modes.

Ion-Exchange Chromatography

Ion-exchange chromatography, depicted in Figure 21.3D, is based on the affinity of ions in solution for oppositely charged ions on the stationary phase. Ion-exchange packings consist of a porous solid phase, usually a resin, onto which ionic groups are chemically bonded. The mobile phase is usually a buffered aqueous solution containing a counter ion whose charge is opposite to that of the surface groups—that is, it has the same charge as the solute—but which is in charge equilibrium with the resin in the form of an ion pair. Competition between the solute and the counter ion for the ionic site governs chromatographic retention. Ion-exchange chromatography has found wide application in inorganic chemistry for separating metallic ions, and in biological systems for separating water-soluble ionic compounds such as proteins, nucleotides, and amino acids.

Size-Exclusion Chromatography

The mechanism of size-exclusion chromatography, also referred to as *gel-permeation* or *gel-filtration chromatography*, is shown in Figure 21.3E. Here, the stationary phase should be chemically inert. Size-exclusion chromatography involves selectively diffusing solute molecules into and out of mobile phase–filled pores in a three-dimensional network, which may be a gel or a porous inorganic solid. The degree of retention depends on the size of the solvated solute molecule relative to the size of the pore. Small molecules will permeate the smaller pores, intermediate-sized molecules will permeate only part of the pores and be excluded from others, and the very large molecules will be completely excluded. The larger molecules will travel faster through the stationary phase and elute from the column first. Thus, size-exclusion chromatography is especially useful in separating high-molecular-weight organic compounds (e.g., polymers) and biopolymers from smaller molecules.

Uses of Liquid Chromatography

Liquid chromatography is most applicable to nonvolatile compounds such as ionic compounds or polymers, thermally unstable compounds such as explosives, and labile compounds such as many biological substances. For volatile compounds, GC is the preferred technique. However, approximately 80% of the known organic compounds are nonvolatile enough to be handled by LC. Liquid chromatography is a relatively "gentle" technique in the sense that many separations can be carried out at ambient temperature, provided the sample can be dissolved in a suitable solvent. It suffices to say that if the sample can be dissolved, it can be analyzed by LC. Sample sizes can range from nanograms to grams; the only limitation on the low end is finding a suitable detector for picogram quantities. As new detectors are developed, detection limits for HPLC will be lowered into the picogram range.

In a given chromatographic system, the volume of mobile phase at which a particular component elutes is usually constant and a characteristic of that component. Thus, the *retention volume* V_r for a chromatographic peak or spot can be used for its *qualitative identification*.

Sometimes one or more substances may elute at the same V_r; in those cases, cross-correlation techniques may allow a more positive identification. In these techniques, a sample is run using two or more different chromatographic systems or conditions; it is unlikely, although still possible, that two substances will give the same elution behavior in more than one system. Also, the use of response *ratios* for two different detection principles (e.g., ultraviolet-visible spectroscopy and fluorescence) or absorbances at two different wavelengths can lead to further confirmation of peak identity. On-line spectroscopic investigations can be carried out. Examples are LC mass spectrometry, LC Fourier transform infrared spectroscopy, and LC ultraviolet-visible absorbance spectroscopy. Positive identification is best accomplished by trapping the peak of interest and subjecting it to off-line mass-spectral, infrared, NMR, or some other appropriate analysis.

Liquid chromatography can also be used for the quantitative analysis of the separated compounds. In column chromatography, the detector response is normally proportional to the mass or concentration of the sample in the effluent. Thus, the area under a chromatographic peak is useful for quantitative analysis; in Figure 21.2, the darkened area under peak C represents the peak area of that component. The peak height (distance h in the figure) can also be used. In modern HPLC, digital integrators or computer-based data systems are used that automatically measure peak areas (or heights) and calculate the concentration (or mass) of individual peaks based on responses for injected standards.

In TLC or PC, the area of the spot is related to the amount of substance. The separated component can also be eluted from the plate or paper and measured externally by another technique (e.g., spectrophotometry).

Since many sample components are completely separated, the eluted fractions from LC can be used for preparing pure materials in milligram to gram quantities for further use in experimentation, or for studying the molecular structure of the compounds. The technique of preparative LC is widely employed; automatic fraction collectors are used to collect the effluent corresponding to the peaks of interest.

21.3 THEORY RELATED TO PRACTICE

Theoretical considerations are a useful guide to the practical design and operation of the chromatographic experiment. The object of chromatography is separation—or rather, separation in a reasonable time.

Retention

To achieve separation, one must first have retention. Earlier, the thermodynamic distribution coefficient K_X was defined as a measure of the degree of retention for compound X. The *capacity factor* k'_X is a more practical quantity that can be determined directly from the chromatogram. It is given by

$$k'_X = \frac{\text{Total Moles of X in Stationary Phase}}{\text{Total Moles of X in Mobile Phase}} = \left(\frac{V_s}{V_m}\right)\frac{[X]_s}{[X]_m} = \left(\frac{V_s}{V_m}\right)K_X \quad (21.3)$$

where V_s = volume of the stationary phase within the column
V_m = volume of the mobile phase within the column

The fundamental equation for any chromatographic process, relating the *retention volume* V_r to other quantities, is

$$V_r = V_m(1 + k'_X) = V_m + V_s K_X \quad (21.4)$$

The value of V_r can be obtained from the chromatogram, since $V_r = Ft_r$, where F is the *flow rate* (in mL/min) and t_r is the peak *retention time* (Fig. 21.2). Similarly, V_m, termed the *void volume* (also *dead volume* or *interstitial volume*) is equal to Ft_0, where t_0 is the time required for an injected solvent molecule or any other nonretained compound to traverse the column. Note that V_m is the total volume of mobile phase in the column at any given time. Substituting for V_m and V_r in Equation 21.4, and rearranging, produces a more useful expression for the capacity factor:

$$k'_X = \frac{t_r - t_0}{t_0} \quad (21.5)$$

Example 21.1. (a) Using the hypothetical chromatogram in Figure 21.2, calculate the capacity factors of peaks B and C. (b) For a flow rate of 2.0 mL/min, calculate the retention volume and k'-value for peak A.

Solution:
(a) By Equation 21.5,

$$k'_B = \frac{t_r - t_0}{t_0} = \frac{13.0 - 2.0}{2.0}$$

$$= 5.5$$

$$k'_C = 9.8$$

(b) $$k'_A = \frac{t_r - t_0}{t_0} = \frac{9.3 - 2.0}{2.0} = 3.6_5$$

$$V_m = F(t_0) = 2.0(2.0) = 4.0 \text{ mL}$$

$$V_r = V_m(1 + k'_A) = 4.0(1 + 3.6_5) = 18.6 \text{ mL} \approx 19 \text{ mL}$$

In TLC or PC, the degree of retention for a compound is its R_f-value, defined as the ratio of the distance the solute has moved to the distance the solvent front has moved. Figure 21.4 shows how one measures the R_f-value; normally, the center of the solute spot is used for calculating the distance a.

Column Efficiency

Column efficiency describes the rate of band broadening as the solute travels through the column or across the plate or paper. As illustrated in Figure 21.1, all molecules do not move at the same speed. Dispersion of molecules generally results in a gaussian profile. The center of the profile or elution band—that is, the k'-value of each component—represents the average rate of travel of a solute molecule. Small deviations from the mean value are brought about by the finite rate of solute mass transfer between the mobile and stationary phases, the different flow paths through the stationary phase caused by irregular packing in the bed, and axial (*longitudinal*) diffusion in the direction of flow.

The quantitative measure of efficiency is the number of *theoretical plates N*

Figure 21.4. *Measuring the R_f value from a paper or thin-layer chromatogram. R_f for component 2 = a/b.*

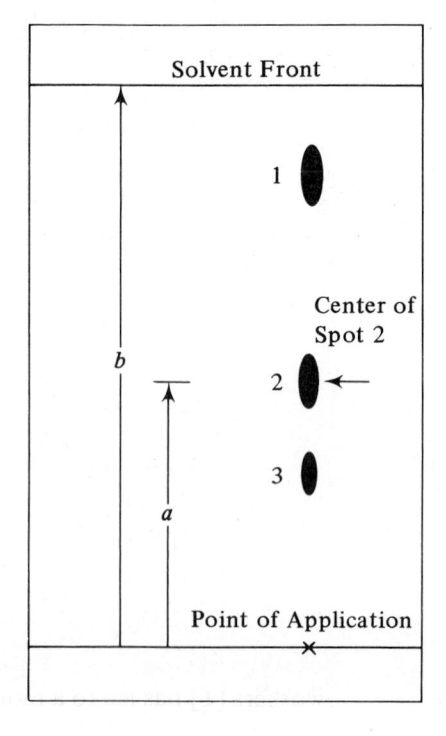

calculated from the chromatogram by using

$$N = 16\left(\frac{t_r}{t_w}\right)^2 \tag{21.6}$$

where t_w = peak width measured in the same units as the retention time

The peak width is obtained from the intersection of the baseline with the tangents drawn through the inflection points on the sides of each peak (see Fig. 21.2). The theoretical-plate model, a carryover from distillation theory, assumes a column to be made up of a series of plates. At each plate, one equilibrium distribution of solute between the mobile phase and stationary phase occurs. Thus, the higher the value of N, the more chance there is for separation to occur (i.e., the better the separating power of the column).

Another useful parameter for column efficiency is the *height equivalent to a theoretical plate*, HETP (the H-value). The following simple relationship shows that H has the units of length:

$$H = \frac{L}{N} \tag{21.7}$$

where L = column length

Thus, a column or plate with a low H-value is better than one with a high value. Values of H less than 1 to 3 mm are commonplace in GC, and in modern HPLC are one to two orders of magnitude lower.

> *Example 21.2.* (a) From the chromatogram of Figure 21.2, determine the number of theoretical plates for peak C. (b) Assuming a 25-cm column, calculate the value of H in millimeters for peak C. (c) How would doubling the column length affect the peak width of peak C, provided other parameters were kept constant?
>
> *Solution:*
> (a) By Equation 21.6,
>
> $$N = 16\left(\frac{t_r}{t_w}\right)^2 = 16\left(\frac{21.5}{4.1}\right)^2 = 440 \text{ plates}$$
>
> (b) By Equation 21.7,
>
> $$H = \frac{L}{N} = \frac{250 \text{ mm}}{440 \text{ plates}} = 0.57 \text{ mm}$$
>
> (c) By Equations 21.6 and 21.7,
>
> $$N = 16\left(\frac{t_r}{t_w}\right)^2 \quad \text{and} \quad \frac{L}{H} = 16\left(\frac{t_r}{t_w}\right)^2$$
>
> Therefore, $t_r/t_w \propto \sqrt{L}$. Since $t_r \propto L$, then $t_w \propto \sqrt{L}$, and so t_w increases by $\sqrt{2} = 1.41$.

Band-Broadening Contributions to H

The development of the generalized nonequilibrium theory by Giddings and co-workers [1] has led to a more thorough understanding of the factors contributing to band spreading in chromatography and hence to the design of better chromatographic columns. In the nonequilibrium theory of chromatography, the movement of the solute through the column is treated as a random walk—that is, the progress of a molecule through the column is a succession of random stops and starts about a mean equilibrium concentration. In this dynamic nonequilibrium, represented in Figure 21.5, mass transfer of the solute into the stationary phase results in a lag behind the equilibrium concentration (band center): When it desorbs and transfers into the mobile phase, the solute moves more rapidly than the band center. Thus, dispersion increases with the number of transfers and decreases as the velocity of the mobile phase decreases (closer approach to equilibrium).

Solute mass transfer in the stationary phase (H_{sp}) or in the "stagnant" mobile phase contained in the pores of the column packing (H_{sm}) are a source of band broadening (Fig. 21.6). The mass-transfer rate can be increased by (a) decreasing the mean diffusion path through which the solute must pass (i.e., decreasing the pore depth or the particle size), (b) increasing the rate of solute diffusion by decreasing the viscosity of the media through which it passes, (c) decreasing the thickness of the stationary phase so that the molecule can diffuse into and out of it very rapidly, or (d) lowering the k'-value of the molecule so that the molecule spends less time in the stationary phase.

Mass transfer of solute in the mobile phase (H_{mp}) also contributes to band spreading. Complex flow patterns arise from the flow of the mobile phase through a packed bed of particles. These patterns are difficult to describe quantitatively, but this form of band dispersion is minimized by homogeneous packing of the bed with uniform particles, by small interparticle channels (less convective mixing between particles), and by the use of mobile phases of low viscosity.

Another source of band dispersion is *eddy diffusion* (H_{ed}). This term describes the irregular flow through the packed particles in a column. The solute proceeds

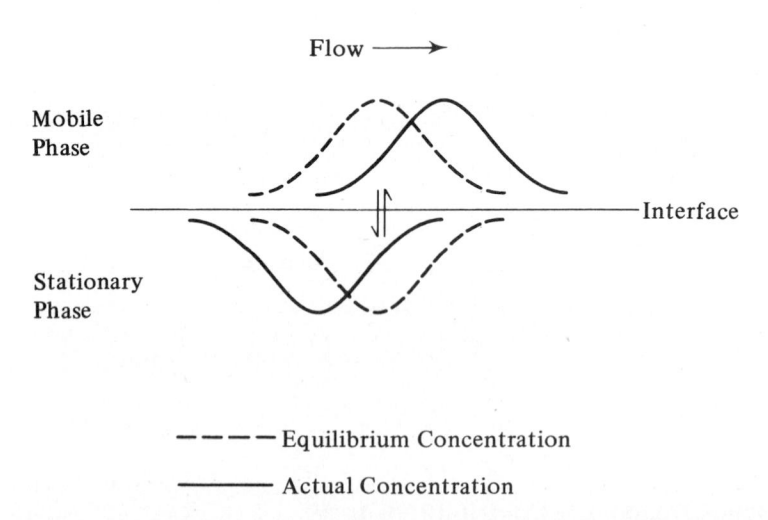

FIGURE 21.5. *Illustration of the influence of local nonequilibrium on band dispersion. Dashed lines: equilibrium concentration profile; solid lines: actual concentration profile.*

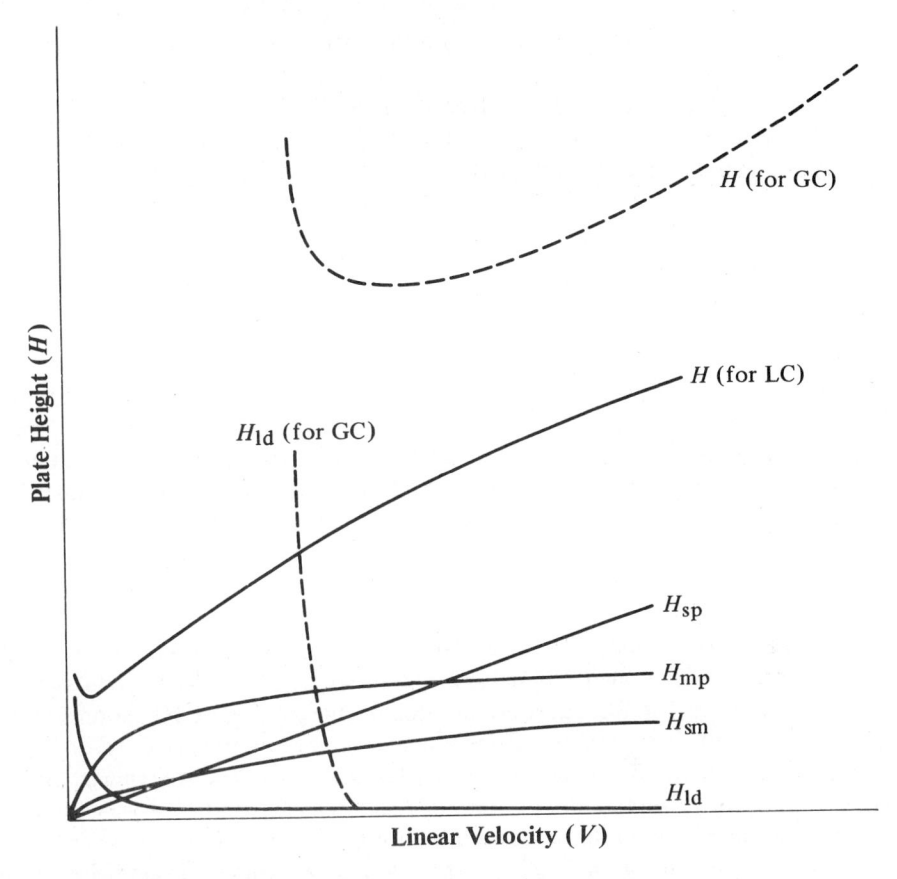

FIGURE 21.6. *Plot illustrating the flow-velocity dependence of overall plate height H and various contributions to H (Eqn. 21.8).*

through the channels between the particles by many interconnected paths that differ in their tortuousness and degree of constriction. Because of the many possible paths, solute molecules arrive at the column exit at different times. Longitudinal molecular diffusion, H_{1d} (i.e., random diffusion in and against the direction of mobile-phase flow) is minimal in LC, although it is an important source of band broadening in gas chromatography. Quantity H_{1d} becomes appreciable only at very low flow rates (Fig. 21.6), because the solute diffusion coefficients in liquids are 10^5 smaller than they are in gases. Thus, these diffusional contributions to H are rarely observed.

The individual band-broadening contributions to H can be described mathematically by

$$H = \frac{1}{(1/H_{ed} + 1/H_{mp})} + H_{1d} + H_{sm} + H_{sp} \quad (21.8)$$

where $H_{ed} = C_e d_p$ (eddy diffusion)
$H_{mp} = C_m d_p^2 v / D_m$ (mobile-phase mass transfer)
$H_{1d} = C_d D_m / v$ (longitudinal diffusion)

$H_{sm} = C_{sm}d_p{}^2v/D_m$ (stagnant-to-mobile-phase mass transfer)
$H_{sp} = C_s d_f{}^2 v/D_s$ (stationary-phase mass transfer)
d_p = particle diameter
D_m = solute diffusion coefficient in the mobile phase
D_s = solute diffusion coefficient in the stationary phase
v = linear velocity of the mobile phase
d_f = thickness of the stationary-phase film (mainly for LLC)

When the stationary phase is the same as the support or particle (as in LSC), d_f is replaced by d_p. Quantities C_e, C_m, C_d, C_{sm}, and C_s are coefficients whose characteristics are given in Reference [1].

Equation 21.8 is a slightly modified version of the Van Deemter equation—Chapter 22, Equation 22.10—used in GC. Because of the so-called coupling between the A term (eddy diffusion) and the C_{liq} term of Equation 22.10, a different relationship between H and v is observed in LC than in GC. The fundamental difference between GC and LC is the great difference in sample diffusion rates in liquids and in gases: D_m is 10^4 to 10^6 times greater in GC. Thus, the H_{1d} contribution in LC is much smaller than in GC, whereas the H_{mp} term is larger.

A pictorial representation of the various contributions to H at different mobile-phase velocities is given in Figure 21.6. For comparison, a similar curve is presented for a typical GC column. In practice, H-versus-v relationships are determined experimentally. These curves are then used for optimizing operating conditions. For best efficiency, one prefers to use a velocity (proportional to flow rate) near the minimum of the H-versus-v plot. Although this is often done in GC, in LC one normally works above the minimum since the latter occurs at very low velocities and separation times would be intolerably long.

From Equation 21.8, one can deduce experimental conditions that minimize the value of H. It can be seen that H decreases with decreasing particle diameters (d_p), linear velocities (v), thickness of films of the stationary phase (d_f), viscosity of the mobile phases and elevated temperatures (to increase D_m and D_s), and uniform packing of the particles (to decrease eddy diffusion).

Care must be taken to keep additional band broadening outside the column to a minimum. Extra-column contributions to band broadening, such as mixing or poor sample-introduction technique, will increase the H-value. Because of the slow diffusion of samples in liquid phases, extra-column volumes are more detrimental to efficiency in LC than in GC; hence, great effort should be exercised to keep those volumes between the point of injection and the top of the column to a minimum. Likewise, the volume between the column exit and the detector, and the volume of the detector itself, should be minimized. In TLC or PC, the applied spot should be kept as small as possible.

Resolution and Its Optimization

The degree of separation is referred to as *resolution*, R. Figure 21.2 and the equation

$$R = 2 \frac{t_{r_B} - t_{r_A}}{t_{w_A} + t_{w_B}} \tag{21.9}$$

are helpful in discussing the characteristics of chromatographic peaks that determine R. If the two components A and B of the chromatogram of Figure 21.2 are examined, resolution is determined by (a) the distance between the peak maxima and (b) the peak or bandwidths. The separation between peaks is related to the *selectivity factor* α, sometimes called relative retention, by

$$\alpha = \frac{t_{r_B} - t_0}{t_{r_A} - t_0} = \frac{k'_B}{k'_A} \tag{21.10}$$

Selectivity refers to the capability of a chromatographic system to distinguish between two components, and is a thermodynamic quantity governed by the relative solute distributions between the mobile phase and the stationary phase. Selectivity in chromatography is very difficult to predict, but one can often use possible molecular interactions between the solute and stationary phase, such as hydrogen-bonding or acid-base relationships, to roughly predict α-values. As α approaches 1, separation becomes exceedingly difficult. To modify selectivity, one must change the stationary phase, the mobile phase, or both.

The bandwidth is related to the efficiency of the chromatographic process, which was discussed in the previous section. Unlike selectivity, efficiency is a kinetic phenomenon and can be increased by better column design as well as by the other factors discussed previously. Figure 21.7 illustrates the influence of selectivity and efficiency on chromatographic resolution. Figure 21.7A depicts a two-component separation displaying poor resolution. By decreasing peak width (i.e., more theoretical plates), resolution can be increased without affecting selectivity (Fig. 21.7B). Note that the two peaks in Figure 21.7B have the same t_r's as those in Figure 21.7A. On the other hand, Figure 21.7C indicates that better resolution results from increasing the distance between peak maxima (i.e., improving selectivity), even without increasing efficiency. Poor resolution caused by a low capacity factor (despite adequate column efficiency and selectivity) is depicted in Figure 21.7D. Better resolution can be obtained by increasing V_s (Eqn. 21.3)—for instance, by using a longer column or an adsorbent with a greater surface area.

In chromatography, a value of $R = 1$ is considered the minimum value for quantitative separation. For a two-component separation, this corresponds to a 2% contamination of each band by the other. When $R = 1.5$, the separation of the two bands is considered complete (cross-contamination less than 1%). Such a *baseline separation* is often required in preparative chromatography when pure substances are desired. As the value of R approaches zero, it becomes more difficult to discern separate peaks.

Example 21.3. (a) For the chromatogram shown in Figure 21.2, determine the resolution between peaks B and C. (b) Determine the value of α for the same peaks. (c) Two peaks with similar retention times generally have similar widths; for retained peaks with a width of 2 mL, what must be the difference in retention volumes to increase the resolution to 2?

Solution:

(a) $$R = \frac{2(t_{r_C} - t_{r_B})}{t_{w_B} + t_{w_C}} = \frac{2(21.5 - 13.0)}{2.1 + 4.1} = 2.7$$

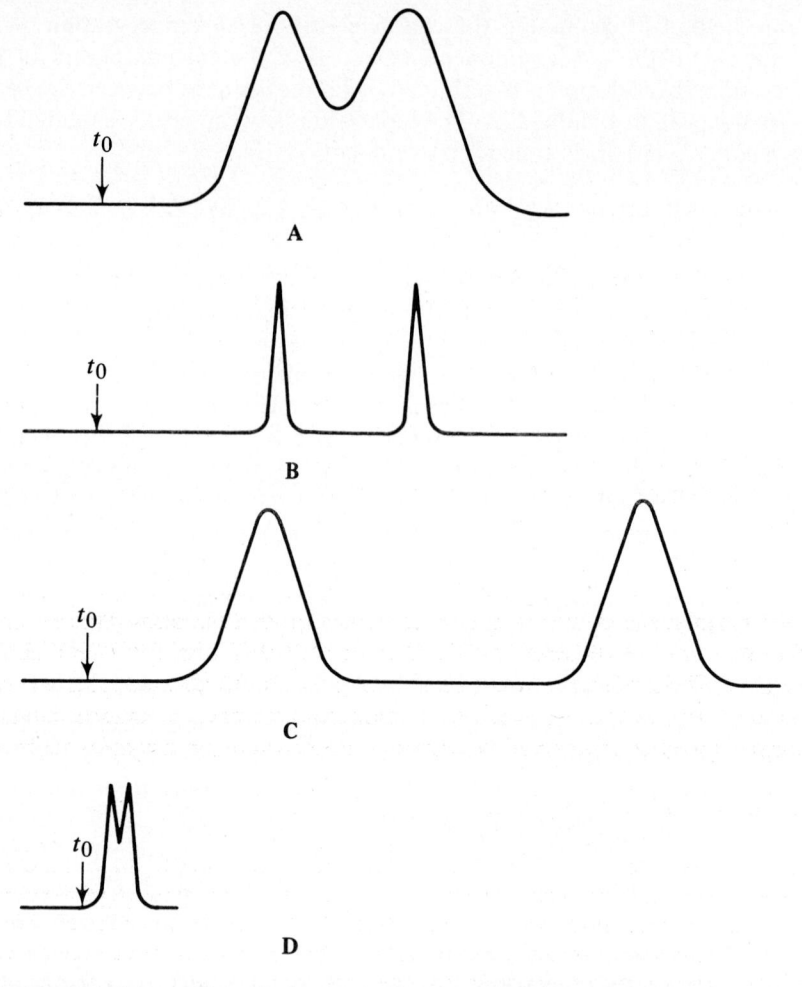

FIGURE 21.7. *Effect of selectivity, efficiency, and capacity factor on resolution. A: Poor resolution. B: Good resolution due to column efficiency. C: Good resolution due to column selectivity. D: Poor resolution due to low capacity factor despite adequate column efficiency and selectivity. Courtesy of Varian Associates.*

(b) $$\alpha = \frac{t_{r_C} - t_0}{t_{r_B} - t_0} = \frac{21.5 - 2.0}{13.0 - 2.0} = 1.8$$

(c) $$R = 2\frac{(t_{r_2} - t_{r_1})}{t_{w_1} + t_{w_2}} = 2\frac{(t_{r_2} - t_{r_1})}{2t_{w_1}} = \frac{t_{r_2} - t_{r_1}}{t_{w_1}}$$

$$2 = \frac{t_{r_2} - t_{r_1}}{t_{w_1}} = \frac{t_{r_2} - t_{r_1}}{2 \text{ mL}} \quad \text{and} \quad t_{r_2} - t_{r_1} = 4 \text{ mL}$$

An alternative equation for the resolution, which can be derived from Equation 21.10, namely

672 CHAP. 21 Solid- and Liquid-Phase Chromatography

$$R = \frac{1}{4}\sqrt{N}\underbrace{\left(\frac{\alpha-1}{\alpha}\right)}_{b}\underbrace{\left(\frac{k'}{1+k'}\right)}_{c} \quad \text{(21.11)}$$

$\underbrace{\phantom{\frac{1}{4}\sqrt{N}}}_{a}$

shows that resolution is a function of three separate factors: the efficiency term a, the selectivity term b, and the capacity-factor term c (see Eqn. 22.12 in the next chapter); these parameters a, b, and c can be adjusted independently.

It is evident from Equation 21.11 that R approaches zero (resolution is lost) as N or k' approaches zero or as α approaches 1. An increase in α, N, or k' favors better resolution, but, according to Equation 21.5, a large value of k' corresponds to a long separation time. In practice, one does not optimize all factors simultaneously, but selects a column with a high plate number, then optimizes the k'-value (usually between 2 and 5) by modifying the mobile-phase composition. If resolution is not adequate, then either N or α can be increased. If α is close to 1, an increase in N may give only a modest increase in R. In that case, it is better to change α by changing the mobile phase or the stationary phase. This often time-consuming trial-and-error process of selecting and optimizing mobile and stationary phases makes the practice of LC something of an art.

In an attempt to ease the burden in the development and optimization of an analytical HPLC method, more sophisticated optimization schemes are being developed. These approaches generally use established methods such as SIMPLEX, multilevel factor analyses, or response-surface methodology. Modern HPLC instruments with microcomputers can be programmed to handle these mathematical approaches, and there is hope for the future development of "smart" chromatographs.

Example 21.4. On silica gel an unretained peak, benzene, gave $V_m = 2.0$ mL, and *o*-diaminobenzene (ODB) and *m*-diaminobenzene (MDB) displayed k'-values of 10 and 12, respectively. For a 100-cm column, the resolution between the isomers was 1.2. (a) Determine the length of column necessary to achieve complete baseline resolution. (b) With the new column from part (a), what will be the void volume for benzene? (c) If the k'-value for MDB is increased to 18 while k' for ODB remains the same, what length of column would be necessary to resolve them to baseline? Note that in Equation 21.11, the k'-value generally refers to the last eluting peak. (d) If the original column efficiency were tripled by using smaller particles, would the two isomers be baseline resolved?

Solution:

(a) By Equations 21.7 and 21.11,

$$\frac{R_2}{R_1} = \frac{\sqrt{N_2}}{\sqrt{N_1}} = \frac{\sqrt{L_2}}{\sqrt{L_1}}$$

$$\sqrt{L_2} = \frac{R_2\sqrt{L_1}}{R_1} = \frac{1.5\sqrt{100}}{1.2}$$

$$L_2 = 160 \text{ cm}$$

(b)
$$\frac{V_{m_2}}{V_{m_1}} = \frac{L_2}{L_1}$$

$$V_{m_2} = \frac{V_{m_1}L_2}{L_1} = \frac{2.0(160)}{100} = 3.2 \text{ mL}$$

(c) By Equations 21.10 and 21.11,

$$\alpha_1 = \frac{k'_{1(\text{MDB})}}{k'_{1(\text{ODB})}} = \frac{12}{10} = 1.2 \quad \text{and} \quad \alpha_2 = \frac{k'_{2(\text{MDB})}}{k'_{2(\text{ODB})}} = \frac{18}{10} = 1.8$$

$$\frac{R_2}{R_1} = \frac{\sqrt{L_2}}{\sqrt{L_1}} \frac{\left(\frac{\alpha_2 - 1}{\alpha_2}\right)\left[\frac{k'_{2(\text{MDB})}}{1 + k'_{2(\text{MDB})}}\right]}{\left(\frac{\alpha_1 - 1}{\alpha_1}\right)\left[\frac{k'_{1(\text{MDB})}}{1 + k'_{1(\text{MDB})}}\right]}$$

$$\frac{1.5}{1.2} = \frac{\sqrt{L_2}}{\sqrt{100}} \frac{\left(\frac{1.8 - 1}{1.8}\right)\left(\frac{18}{1 + 18}\right)}{\left(\frac{1.2 - 1}{1.2}\right)\left(\frac{12}{1 + 12}\right)}$$

$$L_2 = 21 \text{ cm}$$

(d) $$\frac{R_2}{R_1} = \frac{\sqrt{N_2}}{\sqrt{N_1}} = \frac{\sqrt{3}}{\sqrt{1}}$$

$$R_2 = R_1\sqrt{3} = 1.2(1.7) = 2.0$$

Yes, $R_2 > 1.5$.

Note that this example illustrates that improvement in the selectivity factor α is a more powerful way to improve resolution than by merely increasing L.

Sample Capacity

The earlier discussion involving the distribution coefficient was based on the assumption that the coefficient is linear with respect to sample concentration (i.e., a linear sorption isotherm). Nonlinear isotherms, however, are sometimes encountered with the sample sizes employed in practical applications of chromatography, particularly in adsorption systems. The *sample capacity* of the stationary phase is an important consideration in practical applications. The sample capacity corresponds to the amount of sample that can be sorbed onto a particular stationary phase before overloading occurs. Exceeding the sample capacity results in unsymmetrical peak shapes, change in retention times, and loss of resolution. Sample capacity is generally expressed in milligrams of sample per gram of stationary phase. It is proportional to V_s, the volume of available stationary phase (e.g., adsorbent surface area in LSC or bonded-phase loading in BPC). For porous LSC adsorbents, typical sample capacities are in the range 2 to 5 mg/g. Sample capacity should not be confused with the capacity factor k' defined earlier. Sample capacity is most important in preparative LC, because in this technique sample yield is the parameter of interest.

The Chromatographic Compromise

The relationships among sample capacity, speed, and resolution can be represented by the triangular diagram in Figure 21.8. For a particular LC system, any one of these attributes can be improved at the expense of the other two, or any two can be

FIGURE 21.8. *Relationship of resolution, speed, and capacity. Courtesy of Varian Associates.*

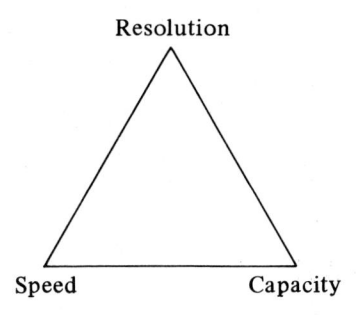

improved at the expense of the other one. The chromatographer must always compromise. In analytical LC, speed and resolution are the desired characteristics; sample capacity is usually unimportant, provided a detectable amount of sample is separated. In preparative LC, capacity is the main objective, provided the resolution is consistent with purity requirements; speed is usually sacrificed.

21.4 PAPER AND THIN-LAYER CHROMATOGRAPHY

Experiments in LC can be performed with equipment ranging from simple laboratory glassware to complex and expensive automated chromatographs. The simplest technique—PC—requires a filter paper, a pipet to apply a sample, a closed jar, and the necessary solvents. On the other hand, HPLC can require high-pressure pumps with electronic programmers, columns with specially prepared microparticles, and highly sensitive, flow-through, microvolume detectors.

Paper Chromatography

The basic technique of paper chromatography (PC) is quite simple. A sheet of cellulose filter paper, such as Whatman No. 1, serves as the separation medium. For one-dimensional PC, the paper is cut into strips about 5 cm wide and 20 cm long; for two dimensional PC (below), a 20-cm by 20-cm sheet is commonly used. The papers come in various porosities (fine, medium, coarse); the porosity determines the rate of movement of the developing solvent. Low-porosity paper gives slow solvent movement but good resolution. Thick papers, which have increased sample capacity, are available for preparative separations.

Preparation. Prior to use, the paper strips are stored under conditions of controlled humidity. Since the predominant mechanism is partition between sorbed water and the mobile phase, the amount of water in the cellulose fibers governs the paper's separating characteristics. The paper may also be impregnated with another stationary phase by dipping and careful drying.

The sample, dissolved in a volatile solvent, is applied to the paper as a drop or "spot" by means of a syringe or micropipet. To minimize band spreading, the spot should be restricted to about 2 mm in diameter. Sample sizes are normally

10 to 50 µg, and the total quantity of the sample should not exceed 500 µg. Larger spots and larger sample sizes lead to poorer separations. Figure 21.9 illustrates the correct manner of sample application. Several samples or standards can be applied as separate spots across the bottom of the paper. The solvent is removed by evaporation, often by means of a hair dryer or heat gun. If the sample is too dilute, several drops may be applied to concentrate it; the solvent should be evaporated between each application. For preparative PC, multiple spots or bands of sample are applied across the bottom of the paper. The paper is now ready to be developed.

Operation. The separation takes place inside a closed container, usually glass, as shown in Figure 21.10. Within the chamber, the paper can be supported so that the solvent flows upward (ascending PC), downward (descending PC), or horizontally. The airtight container ensures that the paper and developing solvent vapors are in equilibrium; for reproducible R_f-values, the paper is usually preequilibrated with solvent vapor for 1 to 3 hr before development begins. For ascending PC, development begins by placing the bottom edge of the suspended paper (but not the spots) into the mobile phase, which ascends through the fibers by capillary action (Fig. 21.10A). In the descending method, the spotted edge of the paper strip is immersed in a trough near the top of the chamber, containing the solvent (Fig. 21.10B). The downward flow of solvent, caused by both capillary action and gravity, moves the solvent farther than in the ascending method. For this reason, the descending PC method is often preferred.

In ascending PC, the paper is supported by means of a clip or hook, or wrapped into a cylinder as depicted in Figure 21.10A. In descending PC, the top edge of

FIGURE 21.9. *Application of sample to paper in PC (or TLC). A: Applying the sample to paper. B: Drying the spots on the paper. From* An Introduction to Chromatography *by D. Abbott and R. S. Andrews. (Boston: Houghton Mifflin Company, 1965.) Reprinted by permission.*

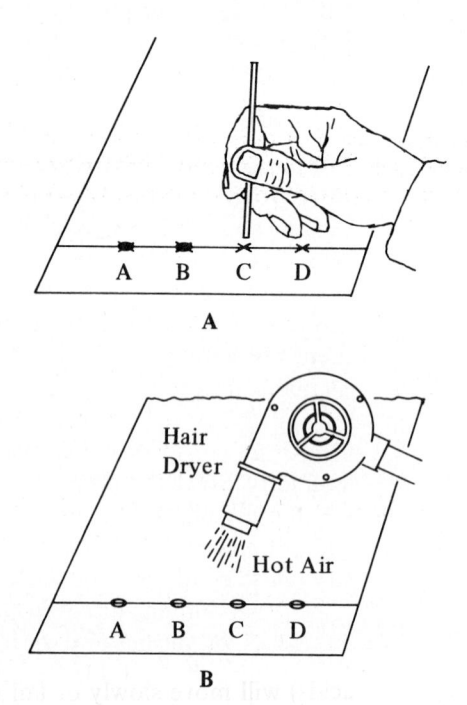

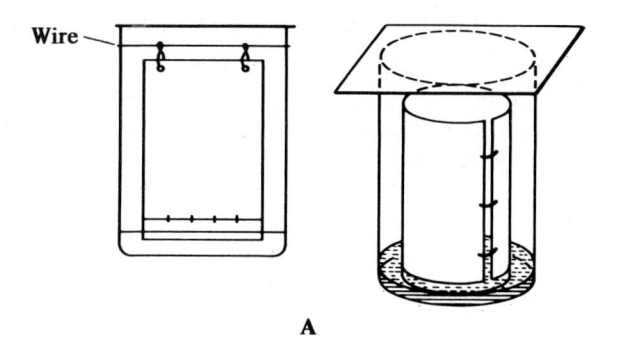

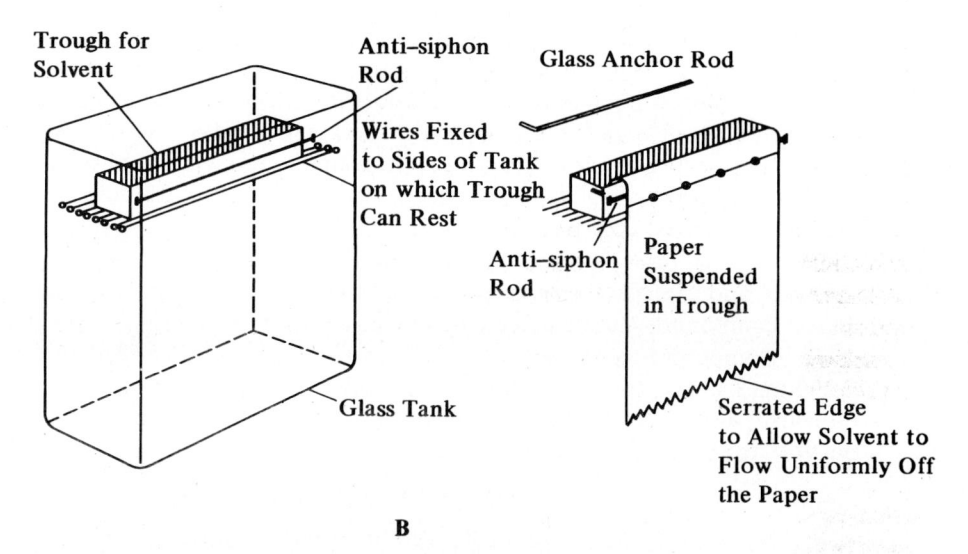

FIGURE 21.10. *Developing chambers for paper chromatography. A: Ascending development. B: Descending development. Note that similar chambers are used in TLC. From* An Introduction to Chromatography *by D. Abbott and R. S. Andrews. (Boston: Houghton Mifflin Company, 1965.) Reprinted by permission.*

the paper is held down by a glass rod or strip. In the horizontal (or radial) method, a circular paper is used and the sample is applied in its center. After the sample dries, solvent is applied at the center and spreads out radially, carrying the sample with it. The main advantage of the radial method lies in its simplicity and its economical use of paper and solvents.

The mobile phase used for development depends on the nature of the substances to be separated. The sample should be only sparingly soluble in the solvent; if it is too soluble, distribution coefficients will strongly favor the mobile phase, components will move with the solvent front, and poor resolution will result. A single organic solvent (e.g., *n*-butanol) saturated with water may serve as a mobile phase. Many popular solvents are of this type, but several useful partition solvents incorporate only small amounts of water. Very polar compounds (phenols, sugars, amino acids) will move slowly or fail to separate in these binary systems. Often, including another component in the mixtures (an acid, a base, or a complexing agent) results in a better separation of polar compounds. Organic solvent mixtures are frequently employed. One disadvantage of solvent mixtures, however, is that multicomponent solvents themselves can partition along the paper, resulting in solvent "bands" which may affect component separation.

The development time should be sufficient to separate the components of interest. The rate of solvent movement depends on factors such as the porosity of the paper; the surface tension, viscosity, and volatility of the solvent; and the ambient temperature. Reasonable R_f-values for good resolution are about 0.4 to 0.8; typical separation times for modern PC papers are in the range 2 to 4 hr.

Detection. After separation, the solvent front is marked and the sheet is dried. The separated compounds are then detected in a variety of ways, chemical or physical. If they are colored, detection presents little or no problem; usually, however, the substances are colorless. In the latter case, the sheet is sprayed with or dipped in a chemical reagent to produce a colored product. There are a large variety of such visualization reagents [2] for various classes of compounds. For example, amino acids (colorless) are easily detected as a pale blue-violet product by treating the paper with a 0.2% solution of colorless ninhydrin. A number of unsaturated organic compounds fluoresce and can readily be detected under an ultraviolet lamp, and labeled (radioactive) compounds can be detected using a radiation counter.

Compounds are identified by their R_f-values, as described in Section 21.3. Sometimes a reference substance, chemically similar to the sample, is run simultaneously alongside the sample and relative migration rates are obtained by comparison. In this case, the R_X-value defined by

$$R_X = \frac{\text{Distance Moved by Substance}}{\text{Distance Moved by Standard Substance X}} \qquad (21.12)$$

is used. This procedure is especially useful in descending PC, in which the solvent is allowed to run off the bottom end of the paper to increase the migration distances.

In two-dimensional PC, a single sample is applied near one corner of the paper and the paper developed. The paper is then removed from the tank, dried, turned 90°, and developed in a second solvent. This procedure effectively increases the distance of migration but, more important, can separate the unresolved components, because the second solvent can have characteristics different from the first.

Thin-Layer Chromatography

Thin-layer chromatography (TLC), like paper chromatography, is performed on an open bed—a glass, aluminum, or plastic plate of dimensions not unlike those mentioned for PC, covered with a porous solid powder comprised of small particles about 5 to 40 μm in diameter. Commonly used phases include silica gel, alumina, cellulose, polyamides, and ion-exchange resins. To promote adhesion and to give better mechanical strength to the TLC plate, a binder such as calcium sulfate (5 to 10% by weight) is mixed with the powder.

Preparation. In preparing the plate, the powder is first usually made into a slurry with water (or other solvent) and then spread on the plate. For plate-to-plate reproducibility, the layer must be very uniform. The slurry can be spread manually using a spatula or another plate, or (more reproducibly and conveniently) with a special apparatus. A "moving spreader" can be used in which the cleaned plates are held in a frame and the applicator containing the slurry is passed over them, depositing the slurry as a thin film. The thickness can be varied by adjusting flanges at the base of the applicator. For analytical work the layer is 0.2 to 0.3 mm thick, whereas for preparative TLC the thickness varies from 2 to 10 mm. After spreading is completed, the plate is dried in air and activated by heating it at 110°C for a short period of time. Ready-made plates are available commercially; they are more expensive than homemade plates, but they are undoubtedly convenient and provide good reproducibility.

Modern TLC plates with reversed bonded-phase coatings have proven to be quite useful. They generally do not require as much preequilibration as the silica-gel plates. Such plates can be used to quickly study a series of mobile phases for possible use in reversed-phase HPLC.

Operation. After activation, the sample application and developing procedures are carried out almost exactly as in PC. Normally, sample sizes range from 10 to 100 μg per spot for analytical TLC, but in preparative TLC when samples are applied as bands, up to 100 mg can be used with a 20-cm by 20-cm plate. Spots should be 2 to 5 mm in diameter. Ascending or descending development and one- or multidimensional techniques can be used. Because of differences in capillary action and possibly in solvent heat of absorption, development times are usually faster in TLC than in PC. Depending on the mobile phase and the particle size of the adsorbent, a typical time is 20 to 30 min for a 10-cm distance, whereas for a high-porosity paper developed under similar conditions, development might take 2 hr. On thin layers, spots often remain compact; on filter paper, they tend to spread somewhat because of the paper's fibrous structure. Thus, resolution is better in TLC and smaller amounts of substances can be separated and identified than in PC.

Detection. In TLC, since the supports, such as silica and alumina, are chemically more inert than paper, more strongly reactive reagents can be used to locate the separated substances. Concentrated sulfuric acid sprayed onto a silica plate makes organic substances visible as charred spots after the plate is heated in an oven. Selective color-forming reagents or iodine vapor is also used. Viewing the plate under ultraviolet radiation can reveal fluorescent substances, or an immobile fluorescent compound can be added initially to the preparation slurry. In the latter case, separated substances show up as dark spots against a fluorescing background when the plate is viewed under ultraviolet, because of their quenching effect.

Values of R_f are more difficult to reproduce in TLC than in PC because there are more experimental variables. R_f-values are influenced by the following factors:

1. The nature of the adsorbent (its chemical nature, particle size, surface area, and binder)
2. The nature of the mobile phase (its purity, precision of mixing, moisture content, and volatility)

3. The activity of the adsorbent and its thickness and uniformity
4. The temperature of the apparatus
5. The amount of sample used
6. The vapor-pressure equilibrium between the plate and the development-chamber atmosphere

A comparison of TLC and PC for the separation of nucleotides is given in Figure 21.11. A one-dimensional development using identical conditions shows the superiority of cellulose-layer TLC over PC for separating various mixtures (a different mixture for each vertical column of spots). Note that, in the same development distance, the isomeric 2'- and 3'-nucleotides (spots 1 to 4) were only partially resolved by PC but were fully resolved by TLC, whereas spots 7 to 9 were completely resolved only by TLC. The reduced degree of spot diffusion in TLC can be readily observed. Standardization of procedure is of considerable importance. For details on TLC standardization, the reader is referred to the book by Stahl [3].

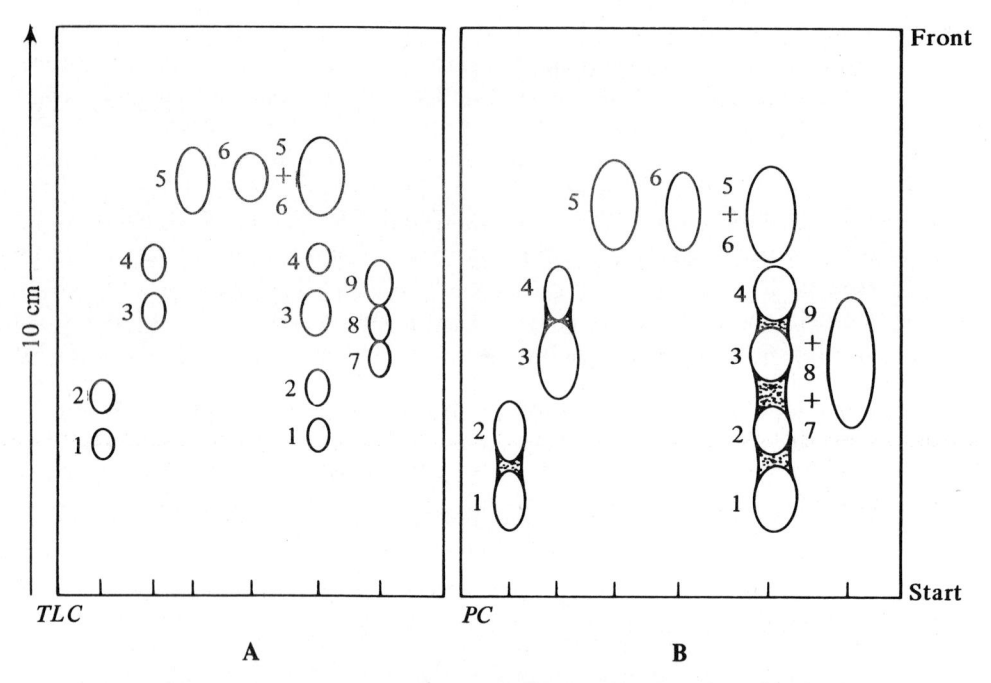

FIGURE 21.11. *Comparative TLC and PC separation of nucleotides. A: Cellulose thin-layer chromatogram—development distance 10 cm in 91 min. B: Paper chromatogram run under identical conditions—development distance 10 cm in 134 min; paper Schleicher and Schull 2043b. The solvent used for both was saturated ammonium sulfate/1 M sodium acetate/isopropanol (80:18:2). Each vertical column of spots corresponds to separate mixtures separated. Samples:* (1) 3'-AMP; (2) 2'-AMP; (3) 3'-GMP; (4) 2'-GMP; (5) 2'- and 3'-CMP; (6) 2'- and 3'-UMP; (7) 5'-AMP; (8) 5'-ADP; (9) 5'-ATP. (A = adenosine, G = guanine, C = cytidine, U = uridine, M = mono-, D = di, T = tri-, P = phosphate.) *From K. Randerath,* Biochem. Biophys. Res. Comm., *6, 452 (1961–62), by permission of Academic Press.*

Quantitative Aspects of TLC and PC

Compared to TLC or PC, quantitation is carried out more easily in column chromatography. A good deal of care is required to obtain reproducible and accurate quantitative results in TLC or PC. It is imperative that chromatographic conditions be well standardized. Standards and samples must be applied to the paper (or plate) in spots of similar size and at similar concentrations; solvents must be prepared, the chamber brought to equilibrium, and so forth, in the same manner. The locating reagent must be applied in a reproducible way. After a developed chromatogram has been obtained, the separated substance can be measured directly on the paper (or plate), or it can be removed from the paper and measured by some other means. The following measurement techniques are used.

1. *Visual comparison of spots.* Samples and reference solutions containing known amounts are run on the same sheet, and the relative areas of the unknown and the standards estimated by eye.

2. *Physical measurements of colored spots.* Transmission reflectance or fluorescence measurements on a strip or plate are made using a spectrophotometer. Scanning photodensitometers are devices that measure spot intensity by reflectance and display the result on a recorder. Visual methods are accurate to 5 to 10%; scanning methods, to 3 to 5%.

3. *Radioactive measurements.* For radioactive substances included in the sample, one may scan the strip with an automatic scanning device.

4. *Spot-area measurement.* The area of a spot is proportional to the amount of substance. The spot area can be determined by using transparent graph paper and counting the squares covered by the spot. Standards are run under the same conditions, and a calibration curve of area versus standard weight applied is obtained.

5. *Spot removal.* The spot may be removed from the paper by cutting or from a plate by scraping off the adsorbent containing the spot. The substance can be eluted or extracted from the strip or plate, and then handled as any other sample solution (e.g., measured by spectrophotometry, polarography, etc.).

21.5 COLUMN LIQUID CHROMATOGRAPHY

Earlier, the basic principles of column chromatography were outlined. Although classical open-column LC is still a widely used technique, modern high-performance liquid chromatography (HPLC)—also called *high-speed liquid chromatography* (HSLC)—is quickly becoming the standard technique for column separation. There is no difference in the basic mechanism involved; only the apparatus used and the practice of the technique are different. Relative to classical LC, the main advantages of HPLC are increased speed, resolution, and sensitivity, and its convenience for quantitative analysis.

When particles are packed into a column, they offer a restriction to solvent flow. The longer the column and the smaller the particles, the greater the restriction.

If flow is forced through the column, it generates a back pressure. The relation of this column back pressure ΔP to the other chromatographic variables is given by

$$\Delta P = \frac{\phi \eta L v}{d_p^2} \tag{21.13}$$

where ϕ = a dimensional structural constant, approximately 600 sec²/cm for packed beds (column resistance factor)
η = viscosity of the mobile phase
L = column length
v = linear velocity of the mobile phase
d_p = average diameter of the particles

If the chromatographic variables are expressed in the cgs-mks system, then ΔP has units of kilograms per square centimeters or of atmospheres. Often, the pressure is expressed in units of pounds per square inch above gravity, or psig.

High-Performance Liquid Chromatography

The difference between classical LC and HPLC can be explored by referring to Figures 21.12 and 21.13. For classical LC, large porous particles with $d_p = 100$ to 250 µm (Fig. 21.12A) are packed into columns with internal diameters of 1 to 5 cm (Fig. 21.13A). Little pressure is required to permit slow solvent flow between these large particles. Normally, a small head of liquid in the column above the surface of the packing or, in some cases, a reservoir container connected to and placed above the column acts as the constant-pressure source. Pressure drops are of the order of 0.1 to 1 atmosphere. Flow rates are very slow (approximately 0.1 mL/min or less), and separation times very long. If attempts are made to speed up the solvent velocity, say by pumping, then, according to Equation 21.8, column efficiency (already low) and resolution will decrease because of mass-transfer limitations in the deep pores (i.e., large values of H_{sm} or H_{sp}), and large interparticle channels (i.e., large values of H_{ed} and H_{mp}). H-versus-v curves for such packings give steep slopes; hence the need for low flow rates. Because of their large surface areas or high ion-exchange capacities,

A. Large Porous   B. Pellicular   C. Micro–

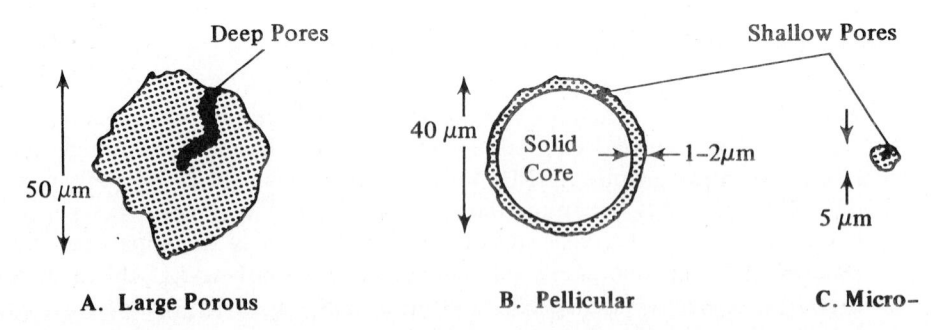

FIGURE 21.12. *Types of particles used in liquid chromatography. A: Large porous particle (d_p = 50 to 250 µm). B: Pellicular particle (d_p = 37 to 50 µm). C: Porous microparticle (d_p = 5 to 10 µm).*

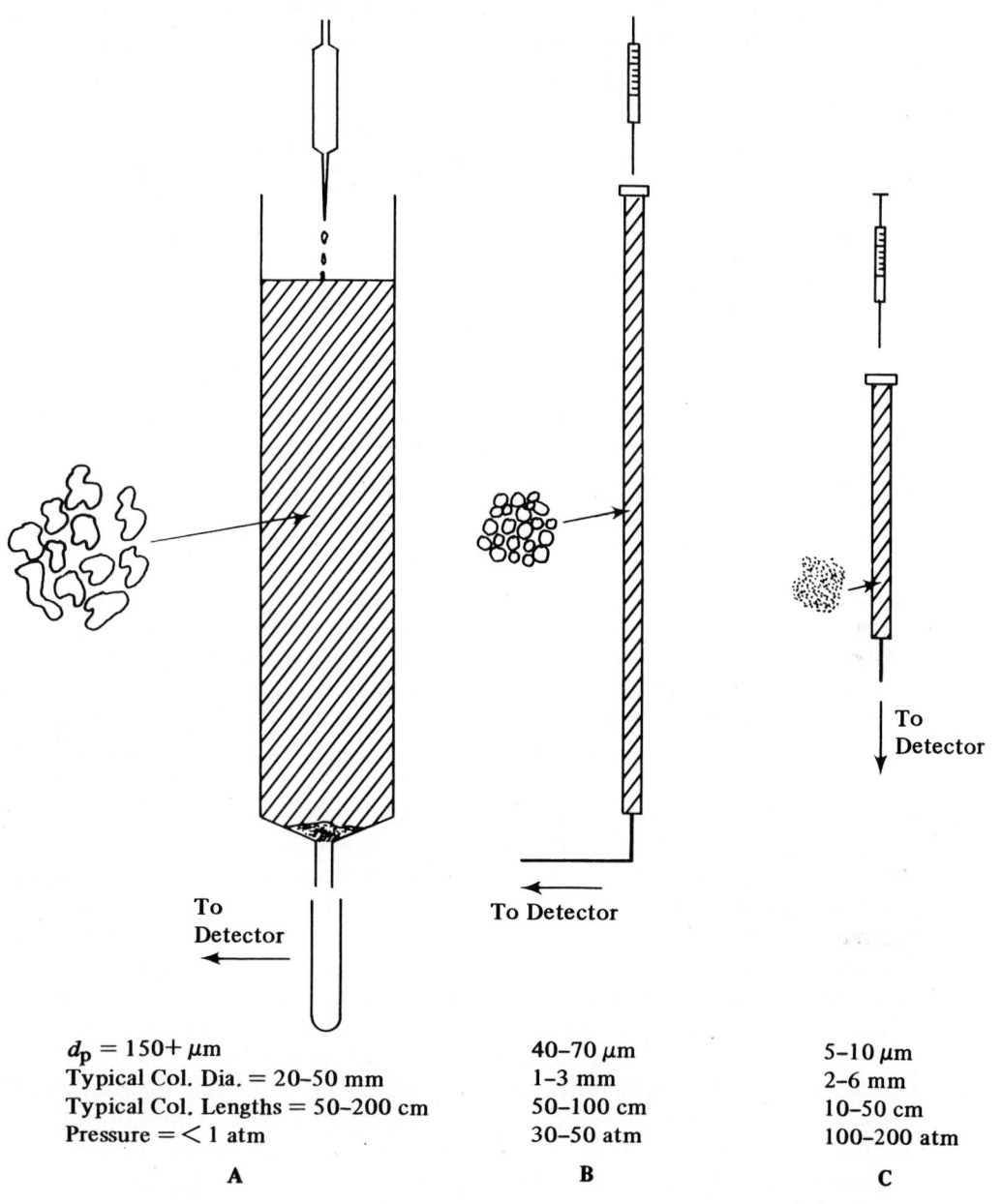

d_p = 150+ μm   40–70 μm   5–10 μm
Typical Col. Dia. = 20–50 mm   1–3 mm   2–6 mm
Typical Col. Lengths = 50–200 cm   50–100 cm   10–50 cm
Pressure = < 1 atm   30–50 atm   100–200 atm

      A   B   C

FIGURE 21.13. *Comparison of columns used in liquid chromatography. A: Classical open-column chromatography with large porous particle packings. B: HPLC with pellicular packings. C: HPLC with microparticulate packings.*

though, the large porous packings exhibit large sample capacities, important in preparative chromatography. Large porous packings are available at a nominal price and can be packed into columns by simple procedures.

    Although an increase in column efficiency with a decrease in d_p was predicted very early in the development of column chromatography, only during the 1960s were column packings available that permitted application of the theory. These packing materials (in the range $30 \leq d_p \leq 75$ μm), when packed into narrow columns, gave rise to larger back pressures than the classical LC columns (Fig. 21.13B). Thus, to assist in mobile-phase flow through the column, high-pressure pumps were required. This was the advent of HPLC. For analytical HPLC, flow rates of 0.5 to 5 mL/min became typical, and pressure drops up to 300 atmospheres were obtained. On the other hand, column efficiency was increased 10- to 100-fold compared to classical LC, and separation times were decreased.

These improvements resulted from the development of pellicular packings in the late 1960s. These spherical packings consist of a solid, nonporous core (usually a glass bead approximately 40 μm in diameter) and a thin, porous outer shell, as depicted in Figure 21.12B. The outer shell, normally 1 to 3 μm thick, may be silica gel, alumina, resin, or polyamide. Because of their dense solid cores, pellicular particles are easily packed into columns. Compared to a porous particle of equivalent diameter, stationary-phase mass transfer in this thin shell (the value of H_{sm} or H_{sp} of Eqn. 21.8) is greatly improved. However, because of the thin shell, V_s is significantly reduced and sample capacity is 0.05 to 0.1 that of the totally porous packings. Therefore, pellicular packings are less useful in preparative LC. In the 1980s, the pellicular packings have given way to the microparticulates, discussed next. Pellicular packings do find use as quickly repacked material for *guard columns*, which are sample-impurity filters placed between the injector and the analytical column.

A decrease in d_p below 30 μm for porous particles leads to further improvements in efficiency. Ion-exchange chromatographers were among the first to recognize the advantages of using very small spherical particles—in the 10-μm range. Ion-exchange resins were synthesized and then separated into narrow size fractions by sedimentation or elutriation procedures. For the other modes of LC, commercial quantities of microparticles in narrow size distributions, and the technology to pack them, were unavailable until the early 1970s. Through the use of air-centrifugal particle classifiers, narrow size distributions of adsorbents are now commercially available. In addition, high-pressure slurry-packing procedures have been developed. Microparticles in the range 5 to 15 μm (Fig. 21.12C) are commonly used for producing highly efficient HPLC columns. The increased efficiency (an extra factor of 10) is caused by improvements in the stationary-phase and mobile-phase mass-transfer terms of Equation 21.8. Thus, separation times and H-values 0.001 to 0.01 those of classical LC are obtained routinely. In addition, unlike the pellicular packings, since their surface areas (or ion-exchange capacities) are the same as those of the large porous packings, the microparticles provide high sample capacity. Unfortunately, these microparticles are more expensive than large porous particles and require rather specialized packing techniques.

Because efficiencies for microparticulate columns are very high (optimum H-values of 0.01 to 0.03 mm), only short columns (15 to 25 cm) are required for analytical HPLC, as can be seen in Figure 21.13C. The use of 5-μm particles should yield increased column back pressures compared to those produced by the larger porous or pellicular particles, as suggested by Equation 21.13. However, these short columns exhibit moderate back pressures (less than 200 atmospheres) when used at flow rates of 1 to 2 mL/min with nonviscous mobile phases. But, for more difficult separations which may require tens of thousands of theoretical plates, long columns (50 to 100 cm) with the smallest available particles and high-pressure solvent feed are sometimes required. Likewise, according to Equation 21.13, high flow rates or viscous mobile phases result in increased column back pressure. In these cases, pressure drops of several hundred atmospheres may be encountered.

The overall influence of a reduction in particle size on the efficiency of a packed column is illustrated in Figure 21.14. Here, curves of H versus v are plotted for 6.1 μm $\leq d_p \leq$ 44.7 μm for six different particle sizes of a porous silica gel. A standard test solute (N,N'-diethyl-p-phenylazoaniline) and the same mobile phase were used for all columns. Columns were packed by a high-pressure slurry technique. Note the significant decrease in H as the particle size is reduced to 6.1 μm. For comparison, an H-versus-v curve for Corasil®, a pellicular silica with an average d_p of 42.5 μm, is included. Porous silicas below 20 μm in d_p would be expected to give greater efficiency than 40-μm pellicular packings. Such H-versus-d_p relationships hold for all LC modes, including TLC.

LC Columns

In open-column chromatography, a glass, metal, or plastic tube with a tapered outlet (Fig. 21.13A) is used as the column. To contain the packing, a wad of glass wool or a

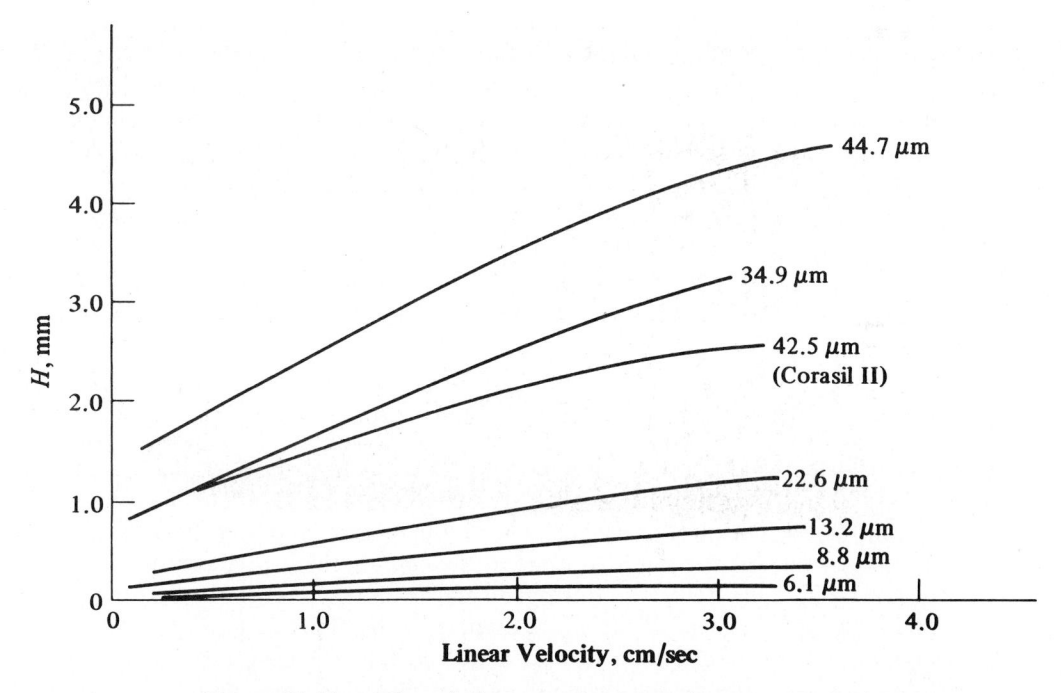

FIGURE 21.14. *Effect of velocity on plate height for porous silica gels of various particle diameters. Test solute: N,N'-diethyl-p-phenyl azoaniline ($k' = 1.2$). Mobile phase: 90/9.9/0.125 parts by volume of hexane/methylene chloride/isopropanol. Included for comparison is the same test solute run on Corasil II, a pellicular silica gel of d_p = 37 to 50 μm. From R. E. Majors*, J. Chromatog. Sci., *11*, 92 (1973), *by permission of the publisher.*

porous metal frit is placed at the bottom of the column, and the solid particles are poured into the top in increments until the column is full. It is important that the column be tightly packed, with no voids. Although gravity can be used to force the liquid through the packed column, a pump is used for smaller-diameter packings—less than 100 μm. Glass columns are useful for low-pressure and classical open-column work, but stainless-steel columns and compression-type fittings are usually required for high pressures—above 70 atmospheres. Glass-lined stainless-steel and fused-silica microcolumns are available for special applications requiring extremely inert surfaces. The column temperature can be controlled by placing the column in an oven or a thermal contact block, or by using a water jacket.

In HPLC, a recent trend has been toward use of columns with smaller internal diameters (i.d.'s). An effect of column diameter is illustrated in Figure 21.15. There are several advantages in the use of smaller-diameter, often termed *microbore*, columns. First, if the same amount of sample is injected onto a smaller-i.d. column, the resulting narrow peak widths give an increased peak concentration—provided, of course, that the column is not overloaded. The increase, in turn, gives greater sensitivity at the detector. This enhancement effect is most important in trace analysis, where the total amount of sample is often limited. For instance, compare the 20-fold increase in peak height for naphthalene for the 1-mm- versus the 4.6-mm-i.d. column of equal length (Fig. 21.15). Second, solvent consumption is reduced. To maintain the same linear velocity v, the flow rate must be reduced in proportion to the relative cross-sectional areas of the columns (see Eqn. 21.13). Third, certain detectors will have enhanced capabilities because of the lower flow rates and smaller sample sizes. Electrochemical detectors, for example, can be built with very small cell volumes, and the electrochemical efficiency can still be maintained. The direct coupling of mass spectrometers to the smaller-i.d. columns requires a flow rate lower than 50 μL/min to avoid overpressuring the ion source. Even flame-based gas-chromatographic detectors have been interfaced to microcolumns.

The general schematic of a high-pressure liquid chromatograph is depicted in Figure 21.16. Each component of the liquid chromatograph has its role in carrying out a successful separation.

Pump and Hydraulics

The most popular type of pump used in HPLC is the reciprocating type, shown in Figure 21.17. Such a pump consists of a piston in a low-volume chamber (typically 70 to 100 μL) that alternately sucks in mobile phase from a reservoir (low-pressure side), then delivers the pump volume to the column (high-pressure side). To prevent backflow into the reservoir during displacement or from the pressurized mobile phase during filling, check valves are used, typically of the ball-and-seat type depicted in Figure 21.17. Because no flow is delivered to the high-pressure side during the filling stroke, multiple out-of-phase pistons are sometimes used to minimize flow disruptions. Likewise, pulse dampers, which are the hydraulic equivalent of electrical capacitors, serve to minimize pump pulsation. Pulse-free flow is important because detectors that are flow sensitive will show noisy baselines if the flow fluctuates or has pulses. Noisy or drifting baselines make it more difficult to accurately measure chromatographic peaks, especially at trace levels.

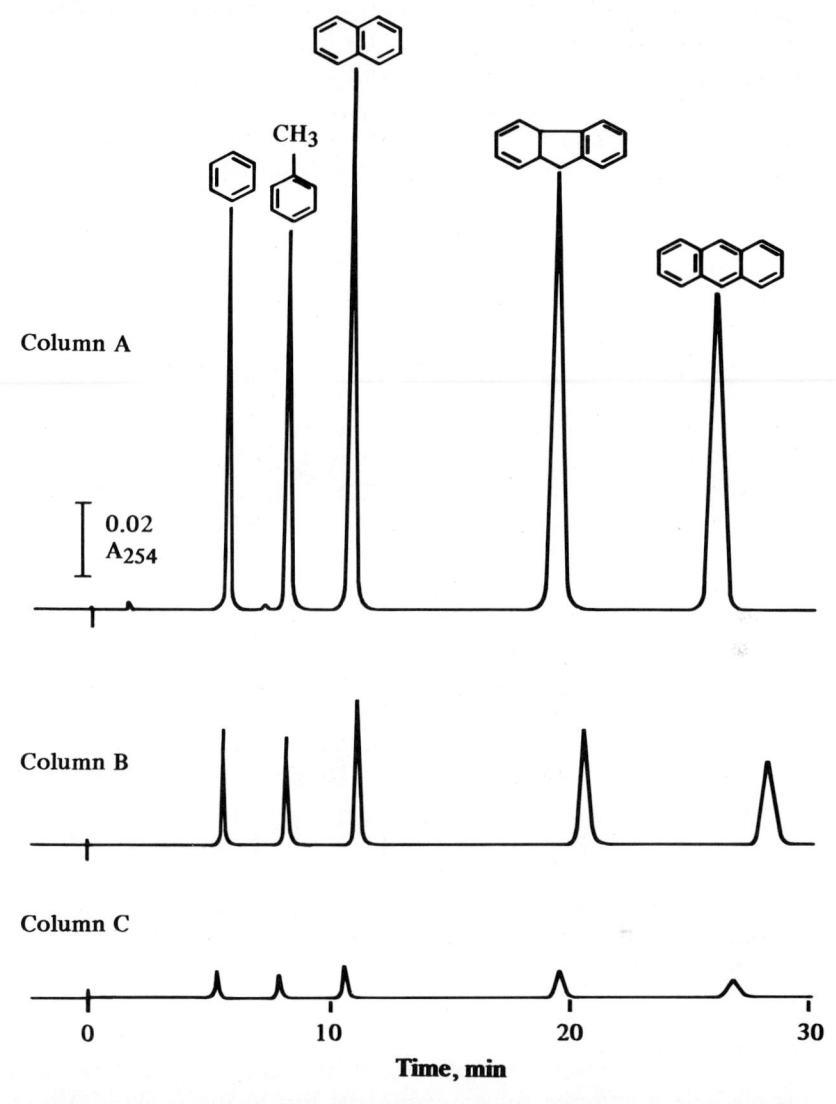

FIGURE 21.15. *Detectability differences for microbore and conventional LC columns. Sample is 4 µg anthracene in 1 µL. Mobile phase: 55% acetonitrile/ water. Ultraviolet detector (254 nm) with submicroliter flow cell. Column packing: SP-C_{18}-5 in 15-cm columns. Column A: 1 mm i.d. and 50 µL/min solvent flow. Column B: 2.1 mm and 200 µL/min. Column C: 4.6 mm and 1000 µL/min. Courtesy of Varian Associates.*

Other common devices in the high-pressure side of the hydraulic system are pressure-readout devices, either gauges or electronic pressure transducers. Mechanical or electrical flow controllers strive to keep the flow rate constant despite downstream column-pressure changes that occur when gradient elution (see Section 21.6) is carried out. These pressure changes are a result of differences in solvent viscosity, as described in Equation 21.13.

SEC. 21.5 Column Liquid Chromatography

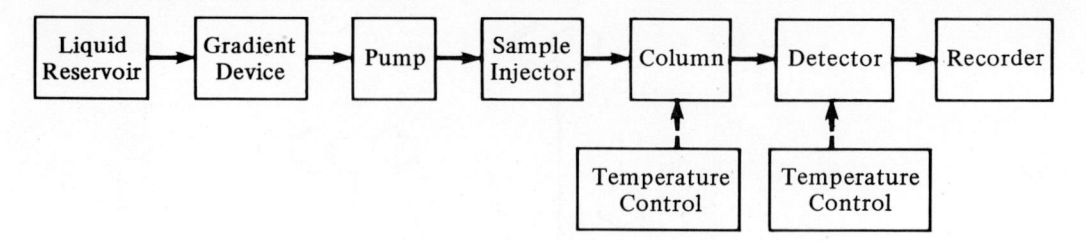

FIGURE 21.16. *Block diagram of a liquid chromatograph. Courtesy of Varian Associates.*

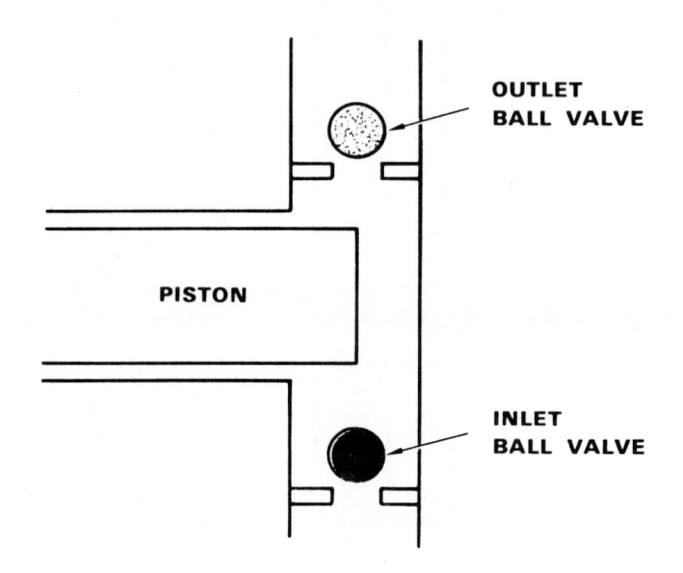

FIGURE 21.17. *Schematic diagram of a reciprocating pump. Solvent flow is in an upward direction. Courtesy of Varian Associates.*

Sample Injector

In open-column chromatography, the sample can be injected by pipeting the sample, dissolved in a suitable solvent, onto the top of the packing. For the forced-flow systems in HPLC, a sample injector is placed between the pump and the column as shown in Figure 21.16. The most commonly used sample injector in HPLC is the *loop injector*. A six-port, high-pressure, external loop injector and its basic operating principle are illustrated in Figure 21.18. In the load position, sample is forced through the loop by a syringe. Once the loop is completely filled, the valve core is rotated either manually or by an automatic air-operated or electrically operated actuator. The chromatographic pump then forces mobile phase through the loop and displaces the load to the column. To refill the loop, the core is rotated in the opposite direction.

For automatic injection of repetitive samples, autosamplers are used. These devices hold a number of samples (up to 100), which are usually placed in small sealed vials. In most cases, an autosampler will permit transfer of samples from the vials to an automatic loop injector, which is operated automatically by an actuator.

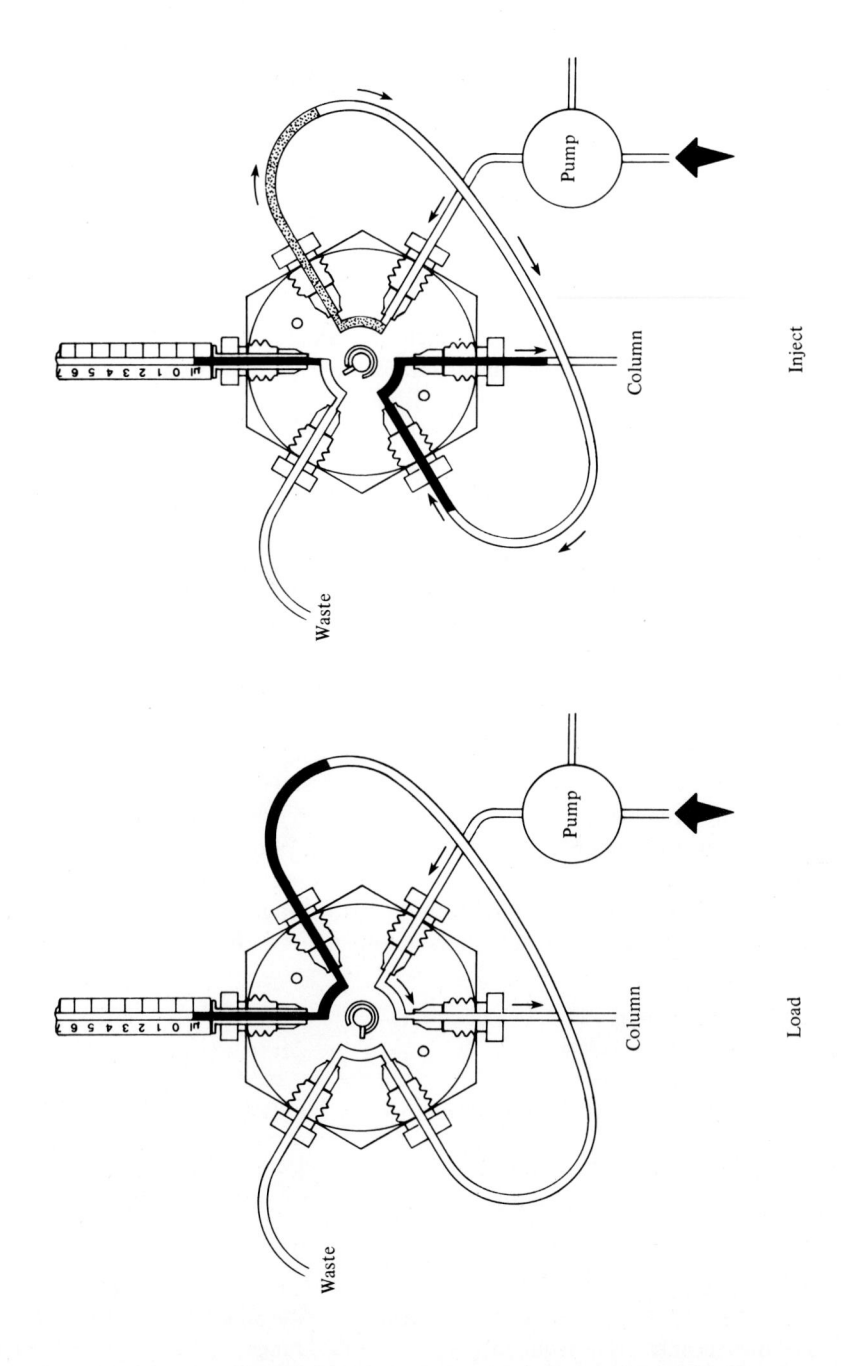

FIGURE 21.18. *Six-port rotary sample-injection valve for HPLC. The arrows indicate the path of solvent flow. Courtesy of Varian Associates.*

689

Detectors

To measure the substances eluting from the column, fractions of mobile phase can be collected and the concentration of the separated components measured externally—for example, with a spectrophotometer or a pH meter. An automatic fraction collector that collects a defined volume of column effluent in test tubes is sometimes employed. However, it is more convenient to use a continuous-detection device at the exit of the column. The detector can be selective (i.e., it detects ultraviolet-absorbing or fluorescent compounds only) or universal (i.e., it detects all components). In some cases, two detectors placed in series are used to gain additional information. The detector output is usually displayed on a strip-chart recorder or some other data-acquisition device. For quantitative analysis, an integrator or a minicomputer is useful to automatically measure peak areas.

To minimize broadening of the narrow peaks often obtained in HPLC, modern detectors measure a small volume, usually 15 μL or less. For microbore columns, detectors volumes may be below a microliter.

The continuous detectors most often used in LC are based on ultraviolet or visible absorption, fluorescence, electrochemistry, and differential refractometry, although a number of others based on other physical principles are commercially available.

Table 21.1 compares the most popular LC detectors according to several performance criteria. At the present time, LC detectors are generally less sensitive than GC detectors, many of which can detect picograms of material under optimum conditions. In LC, there is no equivalent to the flame-ionization detector (see Chap. 22) in GC—that is, a sensitive, universal, gradient-compatible detector. Nevertheless, selective LC detectors such as fluorescence and electrochemical detectors approach the sensitivity of GC detectors.

Most LC detectors provide only limited structural information. However, spectrophotometers with optics optimized for micro flow cells, can be used to obtain a stop-flow ultraviolet-visible absorption spectrum of an LC peak trapped in a flow cell. Photodiode array–based spectrophotometric detectors can give an on-the-fly spectrum. The on-line coupling of liquid chromatographs with mass or Fourier transform infrared spectrometers offers sophisticated (but expensive) detection or

TABLE 21.1. *Comparison of LC Detectors*

Factor	Ultraviolet (UV)	Refractive Index (RI)	Fluorescence	Electrochemical
Specificity	Selective	Universal	Selective	Selective
Detection limit	10^{-10} g/mL	10^{-7} g/mL	10^{-11} g/mL	10^{-11} g/mL
Gradient compatible	Yes	No	Yes	No
Major limitations	Non-ultraviolet active solvents only	Low sensitivity, precise temperature control required	Limited dynamic range	Compound adsorption, no electroactive solvents

identification methods. With these detection schemes, the choices of LC mobile phases are often limited.

Table 21.2 reflects the current usage of LC detectors [4] based on literature publications. The optical detectors predominate, with absorbance detectors being the most widely used. Absorbance detectors are generally the easiest to use, are applicable to a wide variety of compounds containing chromophores, and are compatible with many popular nonabsorbing LC mobile phases.

UV Absorbance Detectors. A schematic of a simple ultraviolet detector with a flow-through cell is presented in Figure 21.19. The light source is a low-pressure mercury lamp with an intense emission line at 254 nm. The light is collimated by a quartz lens, passes through the reference and sample cells, is filtered to remove unwanted radiation, and is sensed by a dual photocell or photodiode. Typically, a flow-through cell of this kind has a 1-mm diameter and 10-mm length with an 8-μL volume.

In addition to Hg, other fixed-wavelength detectors employ interchangeable monochromatic source lamps such as Zn (206 nm) and Cd (214 nm). Deuterium lamps provide a continuum output over the ultraviolet and into the visible. Filter photometers that use interference or bandpass filters to give the desired wavelengths, or spectrophotometers that employ monochromators to give a narrower bandpass of radiation, use deuterium lamps as sources. The most sophisticated and costly ultraviolet-visible detectors are those utilizing photodiode arrays that provide continuous spectra of the LC effluent.

A *photodiode array* (PDA) is a solid-state imaging device containing many individual photodiodes that can simultaneously measure many wavelengths. The optical system of one such PDA LC detector is shown in Figure 21.20. This system uses "reverse optics"; that is, the entire output of the source lamp passes through the flow cell prior to the monochromator, which then disperses the light into its individual wavelengths. Since these rapid-scanning detectors produce many spectra during a normal chromatographic run, computers are required to store and process the voluminous amounts of data produced. The stored spectra can be retrieved and

TABLE 21.2. *HPLC Detector Usage*

Type	Percent Usage
Absorbance	70.7
Ultraviolet, fixed λ	(28.0)
Ultraviolet, filter	(8.7)
Spectrophotometric	(34.0)
Fluorescence	15.0
Refractive index	5.5
Electrochemical	4.4
Other	4.4

Source: R. E. Majors, H. G. Barth, and C. H. Lochmüller, *Anal. Chem.*, *54*, 323R (1982).

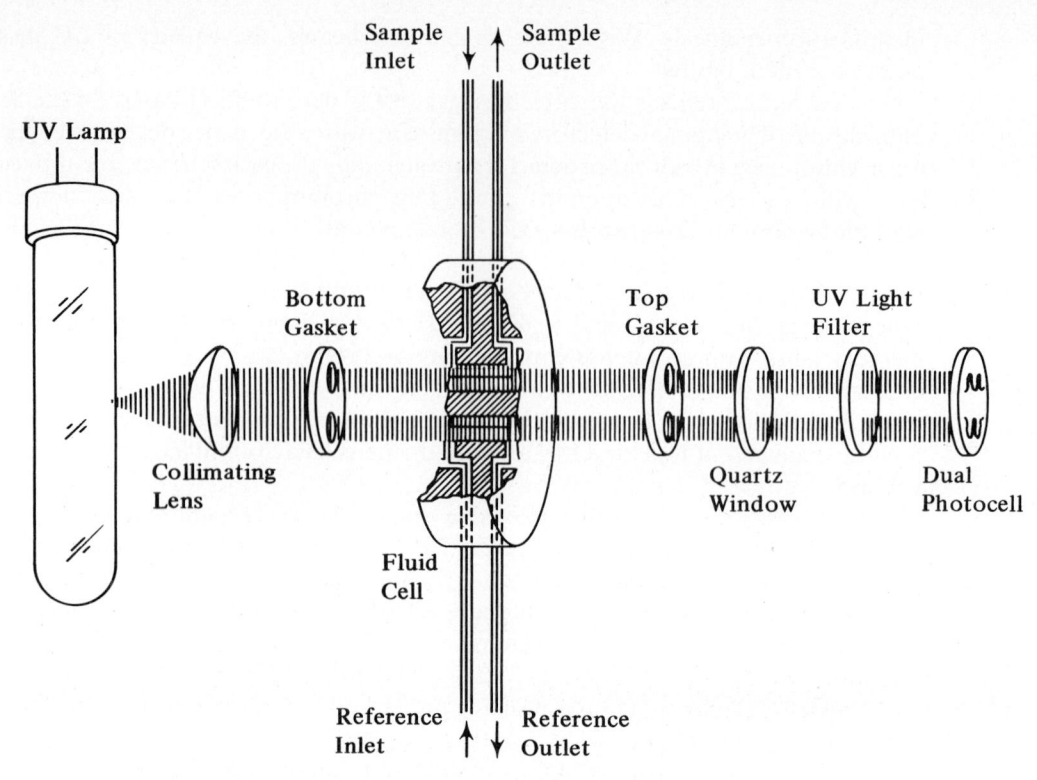

FIGURE 21.19. *Expanded optical schematic of an ultraviolet detector. Courtesy of Varian Associates.*

compared to standard spectra side-by-side or overlaid on a CRT screen for peak confirmation or identity. Three-dimensional plots of absorbance versus wavelength versus time, such as that shown in Figure 21.21, can aid the chromatographer in the detection and possible identification of impurities.

For absorbance detectors, the normal laws of spectrophotometry hold, and so absorbance is linear with concentration. Modern ultraviolet detectors can resolve absorbance differences as small as 0.00001 absorbance unit, equivalent to sub-ppb detection limits in favorable cases.

Fluorescence Detectors. Both monochromator and filter-type fluorescence detectors are useful for naturally fluorescing classes of compounds (see Table 9.3). For example, the specificity and sensitivity of fluorescence detection permit the analysis of illicit samples for LSD without preliminary sample cleanup, with a detection limit of about 9 pg. Fluorescence detection can offer as much as two to three times the sensitivity of absorbance detection.

Such higher sensitivity can be further exploited by taking advantage of selective chemical reactions that can be used to convert nonfluorescent or weakly fluorescent compounds into strongly fluorescent species. The technique of *postcolumn reaction detection* can be carried out on-line using the experimental setup depicted in

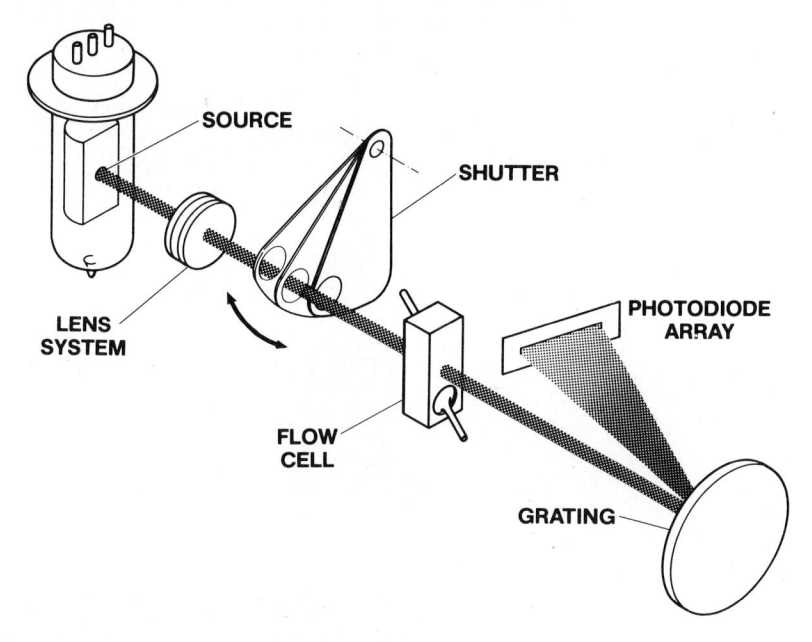

FIGURE 21.20. *Optical system of a photodiode-array LC detector. Courtesy of the Hewlett-Packard Company.*

Figure 21.22, where a derivatizing reagent is continuously added to the column effluent. When a solute that can react with the reagent elutes from the column, the chemical reaction occurs. Usually, the chemical reaction needs a finite time to go to completion. Therefore, a reaction coil or packed-bed reactor is placed between the tee and the detector.

An example of postcolumn reaction detection is the use of orthophthaldehyde as a reactant in the determination of several important biogenic amines: cadaverine (1,5-diaminopentane); putrescine (1,4-diaminobutane); spermidine, the N-(3-aminopropyl) derivative of putrescine; and spermine, the N,N'-bis derivative. These amines are widespread in tissues and appear to occur in association with nucleic acids, and are thus suspected as possible participants in mechanisms regulating protein synthesis or cell division. They are thus studied as biological markers in studies of cancer and organ malfunctions. Saturated organic amines cannot be detected by ultraviolet, fluorescence, or refractive detection at the levels ordinarily found in urine or blood plasma. With postcolumn reaction, however, parts-per-trillion sensitivity can be achieved. Precolumn chemical derivatization can also be used for detection at trace levels. For example, reaction of o-phthaldehyde/2-mercaptoethanol with as many as 25 amino acids followed by LC separation results in detection limits of about 50 femtomoles (10^{-15} mole) of an amino acid. Both compound selectivity and increased sensitivity can be obtained using derivatization.

Electrochemical Detectors. Electrochemical (EC) detection has become popular because it can be applied to trace analysis of several classes of organic compounds in

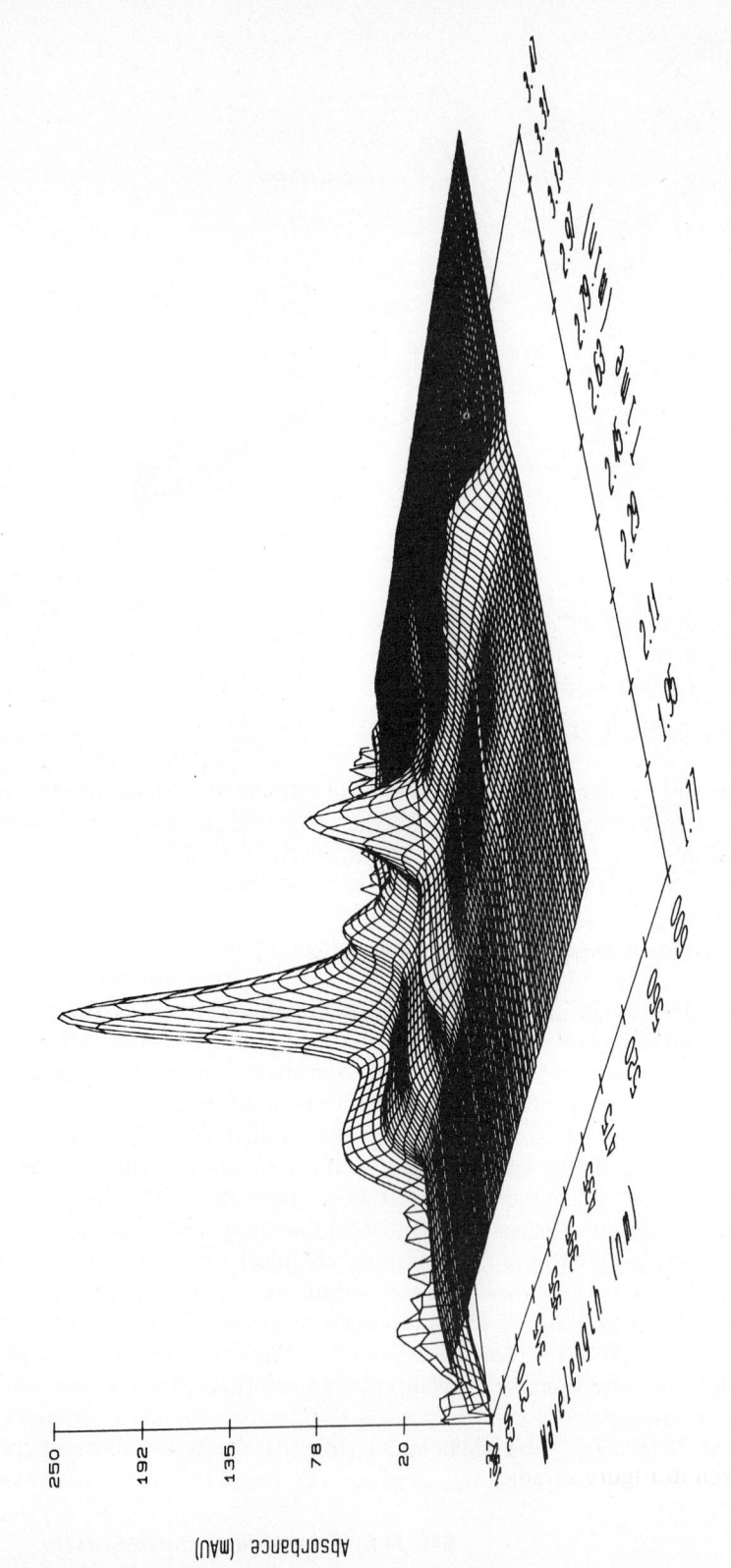

FIGURE 21.21. Typical three-dimensional plot (absorbance versus wavelength verses time) for a liquid chromatograph. Separation of hydroxylcobalamine (peak 1, at about 2.5 min) and cyanocobalamine (peak 2, at about 3.0 min). Courtesy of the Hewlett-Packard Company.

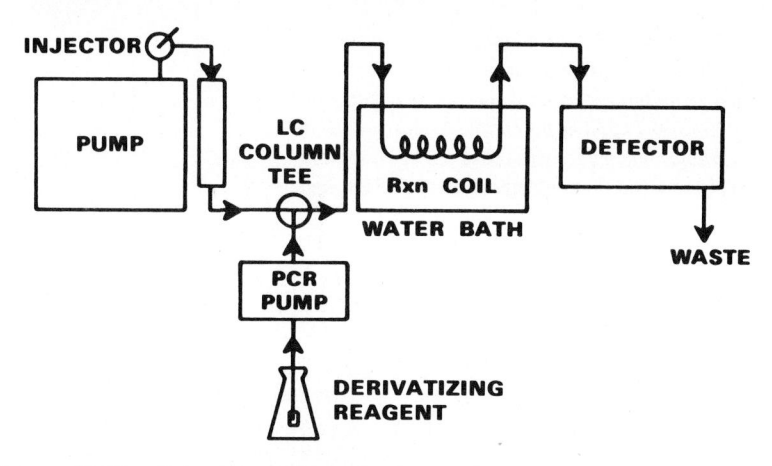

FIGURE 21.22. *Schematic diagram of a postcolumn reaction system for liquid chromatography.*

water or body fluids. Chapters 1 through 5 described the principles and many applications of electrochemistry. Here, it is sufficient to say that all of these techniques can be or have been applied to HPLC detection in combination with microliter-volume flow-through electrochemical cells. By far the most popular use of EC detection has been in the analysis of catecholamines and neural transmitters such as serotonin, norepinephrine, dopamine, and L-dopa in urine and tissue. These compounds are important in the pathways of tyrosine metabolism and as chemical neurotransmitters. Measurement of their levels can provide guidance in the diagnosis of certain metabolic disorders. Figure 21.23 shows a chromatogram of trace amounts of several catecholamines in a urine sample using an amperometric detector with a glassy-carbon working electrode. Sub-picomole detection limits can often be obtained with liquid chromatography-electrochemistry (LC-EC).

Mass Spectrometric Detectors. The use of on-line gas chromatography-mass spectrometry (GC-MS) revolutionized peak identification and confirmation for GC and has opened the way for a similar revolution in analytical detection techniques for LC. At first glance, the techniques of LC and MS would appear to be completely incompatible, as can be seen in Table 21.3 [5]. However, with some compromises by both the liquid chromatographer and the mass spectroscopist, the marriage of these techniques can be successfully accomplished.

The most popular liquid chromatography-mass spectrometry (LC-MS) interface techniques are: (a) direct liquid inlet (DLI), where a small portion of the LC effluent is directed into the ion source; (b) moving-belt method, where LC effluent is deposited onto a moving belt, the solvent is evaporated, and the sample is volatilized from the belt into the ion source; and (c) thermospray ionization, where ions are created when aqueous buffered mobile phase is passed through a heated stainless-steel capillary, creating a supersonic jet of vapor with subsequent evaporation of the mobile phase from charged liquid droplets. A schematic of a thermospray LC-MS interface is given in Figure 21.24.

FIGURE 21.23. *Chromatogram of two catecholamines (norepinephrine and epinephrine) in a urine sample using amperometric detection at constant applied potential. A glassy-carbon working electrode was used. The retention times for the two compounds are 5.28 and 8.38 min, respectively. Courtesy of Varian Associates.*

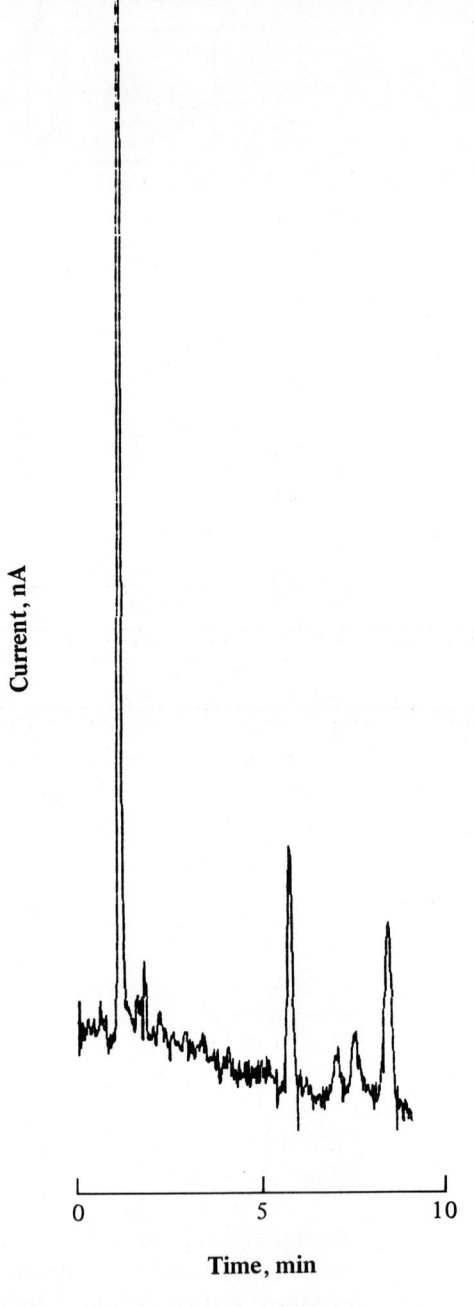

The LC-MS technique can provide structural and molecular-weight information. In the selected ion-monitoring mode, where a single mass-to-charge value is monitored, the technique is very sensitive, approaching sub-nanogram sensitivity. The biggest problem in LC-MS is the LC mobile phase. For example, 1 mL/min of hexane, methanol, or water generates, respectively, 180, 350, and 1250 mL/min of

TABLE 21.3. *Incompatible Factors in LC-MS*

HPLC	MS
Liquid-phase operation	Gas-phase operation
25–50°C operation	100–350°C operation
Almost no sample limitation	Some volatility desired
No MW limitation	Has MW limitations
Relatively inexpensive	Expensive
Uses inorganic buffers	Cannot tolerate inorganic buffers
Conventional flow rates produce 550 mL/min gas at STP	Accepts 10 mL/min at STP

Source: R. E. Majors, *LC Magazine*, *1*, 488 (1983). Reprinted with permission.

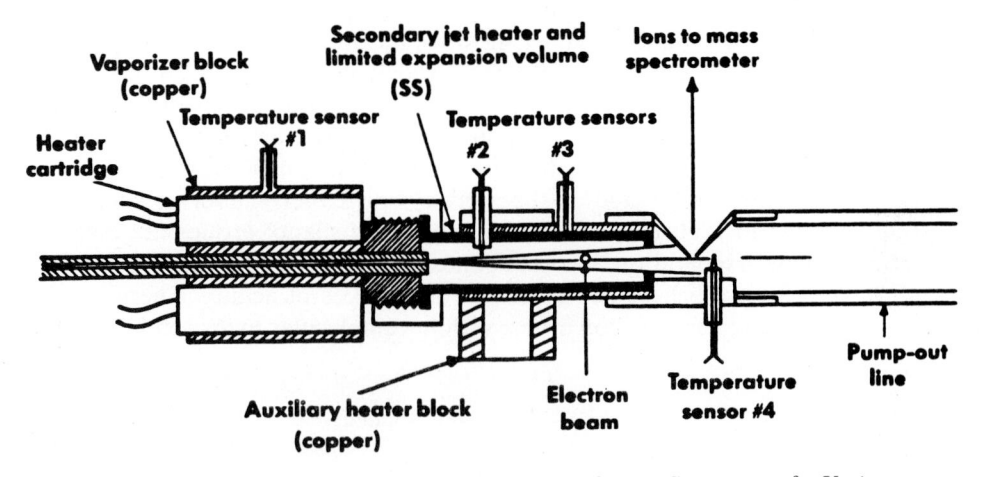

FIGURE 21.24. Thermospray LC-MS interface. Courtesy of Varian Associates.

gas at standard temperature and pressure. Since the MS ionization source must operate under high vacuum, this amount of vapor must be removed without removing a substantial amount of the solute. For DLI interfaces to accomplish vapor removal, large rotary vacuum or cryogenic pumps are needed. Also, the use of microbore LC with optimum flow rates in the tens of microliters per minute range cuts down on the volume of vapor without affecting column efficiency. The use of nonvolatile inorganic salts is generally precluded for all of the interfaces. However, volatile compounds such as trifluoroacetic acid or ammonium acetate can be used in LC mobile phases.

21.6 USES AND APPLICATIONS OF ADSORPTION CHROMATOGRAPHY

Selecting the correct chromatographic technique for a particular mixture is sometimes difficult. We will discuss the properties of the various types of stationary phases that serve to distinguish one mode of LC from another.

Influence of the Stationary Phase in LSC

Adsorption or liquid-solid chromatography (LSC), the oldest chromatographic method, is the most widely used of all modes. Thin-layer or column chromatography is used by most laboratories that use liquid-chromatographic techniques, often as a screening method to select the best experimental conditions for LC. Adsorbents are porous solids with specific surface areas ranging from 50 to 1000 m^2/g.

The most popular adsorbents for TLC are silica, alumina, diatomaceous earth, kieselguhr, and polyamides; silica dominates HPLC, with alumina being used occasionally. Silica gel is also widely used as the base material for chemically bonded phases employed in the other LC modes. Most adsorbents come in particle sizes to suit the needs of the various kinds of chromatography, are specially made, and can be purchased commercially. For TLC, particle sizes of 20 to 40 μm are frequently used, whereas for open columns the particles are larger (100 to 150 μm), and for HPLC they are smaller (down to 3 μm).

To illustrate the manner in which adsorbents separate compounds, consider the surface characteristics of the most widely used adsorbent, silica gel. Chromatographic-grade silica gels are prepared by reacting sodium silicate with a mineral acid, such as hydrochloric acid. Polymerization occurs, and a three-dimensional array of SiO$_4$ tetrahedra results. This polysilicic acid, when dehydrated, forms a stable porous solid, terminated at the surface with either silanol or siloxane bonds, as illustrated in Figure 21.25. The slightly acidic silanol groups are considered to be important in separation; siloxane bonds, to have little or no influence. Silanol groups themselves are believed to have varying degrees of acidity. The most acidic ones, located on adjacent silicon atoms with intramolecular hydrogen bonding, often lead

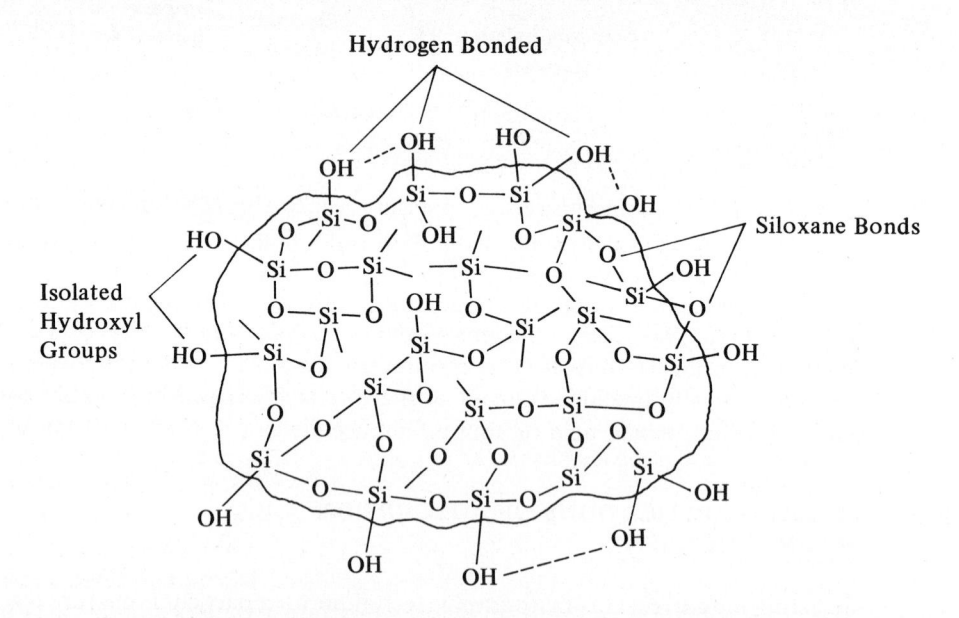

FIGURE 21.25. *Structure of silica gel depicting the various types of bonds and silanol groups present.*

to undesirable chromatographic effects, such as chemisorption and peak tailing. Often a polar modifier, such as water, is added to the adsorbent in order to deactivate the strongest adsorption sites.

Interactions between the adsorbent surface and the solute can vary from nonspecific ones (e.g., dispersion or van der Waals' forces) to specific ones (electrostatic interactions involving permanent dipoles or electron donor–acceptor interactions, such as hydrogen bonding). Retention on silica gel or alumina is governed mainly by interactions with the polar functional groups of the solute. Thus, compounds of different chemical types (e.g., hydrocarbons and alcohols) are easily separated by LSC. Weak dispersive interactions with the hydrocarbon (especially aliphatic) portion of the solute allows little or no differentiation among homologs or other mixtures differing only in the extent of aliphatic substitution. The relative positions of the functional groups in the solute molecule, and the number and spatial arrangement of surface adsorption sites, lead to a geometric specificity that makes LSC unique in its ability to separate polyfunctional compounds, especially positional isomers. For instance, *cis-trans* pairs or substituted aromatic isomers can be separated. Such a separation, by HPLC, of the three isomers of nitroaniline is shown in Figure 21.26. Note that the first to elute is the *ortho-* isomer, which is intramolecularly hydrogen bonded and thus has less (intermolecular) interaction with the surface, whereas the *para-* isomer is more likely to react intermolecularly and is the most strongly retained.

Influence of the Mobile Phase in LSC: Gradient Elution

Actually, the interactions in LSC involve a competition between the solute molecules (X) and the molecules (S) of the mobile phase for the adsorption sites. This equilibrium is illustrated by

$$X_m + nS_{ads} \rightleftharpoons X_{ads} + nS_m \tag{21.14}$$

where
- X_m = solute molecules in the mobile phase
- X_{ads} = solute molecules in the adsorbed state
- S_{ads} = mobile-phase molecules adsorbed on the surface site
- S_m = solvent molecules in the free mobile phase
- n = number of adsorbed solvent molecules displaced by the adsorption of one molecule of X

Thus, stronger adsorption of the mobile phase decreases adsorption of the solute. Solvents can be classed according to their strength of adsorption. Such a quantitative classification is referred to as an *eluotropic series*. Table 21.4 is an abbreviated eluotropic series specifically for alumina as the adsorbent, but qualitatively this series holds for other polar adsorbents as well [6].

An eluotropic series can be used to find an optimum solvent strength for a particular separation. Using a solvent of constant composition is called *isocratic elution*. If an isocratic solvent is too strong (the k-values for the solutes are too small), a weaker solvent is substituted. On the other hand, if the initial solvent is too weak (the k'-values are too large), a stronger solvent is selected. This trial-and-error approach to finding the optimum solvent can sometimes be done more rapidly by TLC than by column chromatography. With the advent of microprocessor- and

FIGURE 21.26. *LSC separation of nitroaniline isomers on 10-μm alumina*: (1) *o-nitroaniline*; (2) *m-nitroaniline*; (3) *p-nitroaniline. Column: Micropak® Al-10. Packing: 10-μm alumina, type T. Dimensions: 15 cm × 2.4 mm. Mobile phase: 40% CH_2Cl_2 in hexane. Flow rate: 100 mL/hr. Sample size: 1 μg of each isomer. Detector: 254-nm ultraviolet absorption. From R. E. Majors, Anal. Chem., 45, 757 (1973), by permission of the publisher. Copyright © 1973 by the American Chemical Society.*

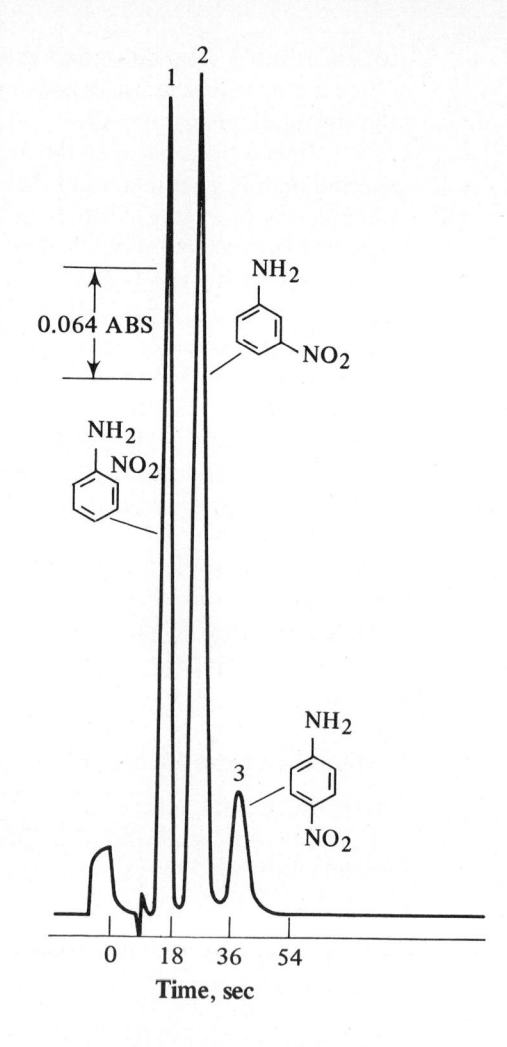

computer-controlled HPLC instruments, solvent-optimization schemes are being perfected to aid in unattended methods development. Decision-making programs can be used to optimize resolution or time by automatically manipulating mobile-phase composition as well as temperature, and other chromatographic parameters.

Binary solvent mixtures may also be used to find an optimum value of the *solvent-strength* parameter ϵ^0. For example, a mixture of isoöctane ($\epsilon^0 = 0.01$) and methylene chloride ($\epsilon^0 = 0.42$) can be found (see Table 21.4) with an isocratic solvent strength similar to that of carbon tetrachloride ($\epsilon^0 = 0.18$). However, the relationship between binary composition and solvent strength is not necessarily linear, owing to the solvent-solvent and preferential solvent-surface interactions. Additional selectivity in LSC can be achieved through secondary solvent effects. Such effects are produced by solvent mixtures that display equivalent values of ϵ^0 but that, because of various solvation interactions such as hydrogen-bonding ability, basicity, and so forth, also give rise to variations in relative retention (i.e., selectivity). Ternary

TABLE 21.4. *Eluotropic Series for Alumina*

Solvent	Solvent-Strength Parameter (ϵ^0)
n-Pentane	0.00
Isoöctane	0.01
Cyclohexane	0.04
Carbon tetrachloride	0.18
Xylene	0.26
Toluene	0.29
Benzene	0.32
Ethyl ether	0.38
Chloroform	0.40
Methylene chloride	0.42
Tetrahydrofuran	0.45
Acetone	0.56
Methyl acetate	0.60
Aniline	0.62
Acetonitrile	0.65
i-Propanol, *n*-propanol	0.82
Ethanol	0.88
Methanol	0.95
Ethylene glycol	1.11
Acetic acid	Large

Source: L. R. Snyder, *J. Chromatogr.*, *16*, 55 (1964), by permission of the author and the North-Holland Publishing Company.

and quaternary solvent mixtures can also be used to generate additional selectivity and retention effects.

In all forms of chromatography, one must be aware of the *general elution problem* (illustrated in Fig. 21.27) when dealing with isocratic solvent systems and multicomponent samples with widely differing k'-values. If a strong isocratic mobile phase is selected that will adequately elute strongly retained compounds, then the weakly retained ones will be eluted too quickly and will be poorly separated (Fig. 21.27A). Conversely, if a weak mobile phase is chosen, so that weakly retained sample components will be retained and separated, then very strongly retained solutes may not be eluted at all—or only very slowly (Fig. 21.27B)—and possibly with the peaks so broadened as to be undetectable. No single isocratic solvent can be found that will be effective for such a mixture of components with widely varying k'-values. To handle this kind of sample, the rates of band migration must be changed during the chromatographic run.

In GC, the general elution problem is solved by temperature programming and to a lesser extent by flow programming (see Chap. 22). In LC, the most common technique is called *solvent programming* or *gradient elution*. Here, elution is begun with a weak solvent and the solvent strength is increased with time. The changes are made either stepwise or continuously. The overall effect is to elute successively the more strongly retained substances and at the same time to reduce peak broadening. The k'-values, and hence the analysis time, can be decreased by as much as 10^6 using solvent programming. Figure 21.27C demonstrates how solvent programming provides a solution to the general elution problem for the compounds shown.

In HPLC, since columns are reusable, the stationary phase must be returned to its initial condition at the conclusion of a solvent program so that, if necessary,

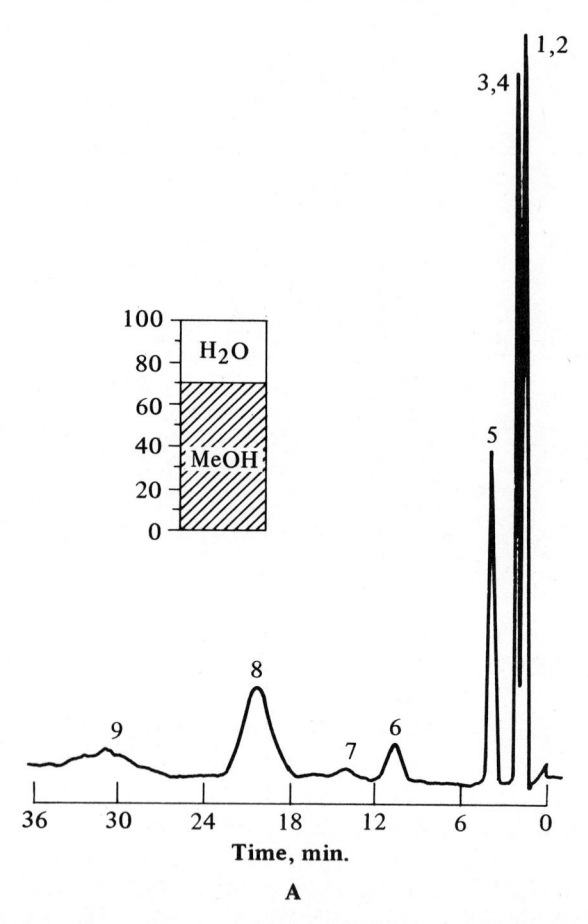

FIGURE 21.27. *Illustration of the general elution problem: separation of commercial flame retardants (brominated aromatics). Column: Permaphase®-ODS (DuPont), 1 m × 2.4 mm. Solvent flow rate: 1 mL/min. A and B are isocratic elutions. C is a solvent-programmed run: 40% methanol in water to 100% methanol at 3%/min. Compounds: (1) Bisphenol A; (2) Firemaster LV-723P; (3) Firemaster BP4A (Tetrabromobisphenol A); (4) p-Dibromobenzene; (5) 4,4'-Dibromobiphenyl; (6) Hexabromobenzene; (7) Hexabromobenzene impurity; (8) Firemaster BP-6 (Hexabromodiphenyl); (9) Firemaster BP-6 impurity (probably isomer). (Firemasters are products of the Michigan Chemical Company.) Chromatograms courtesy of Varian Associates.*

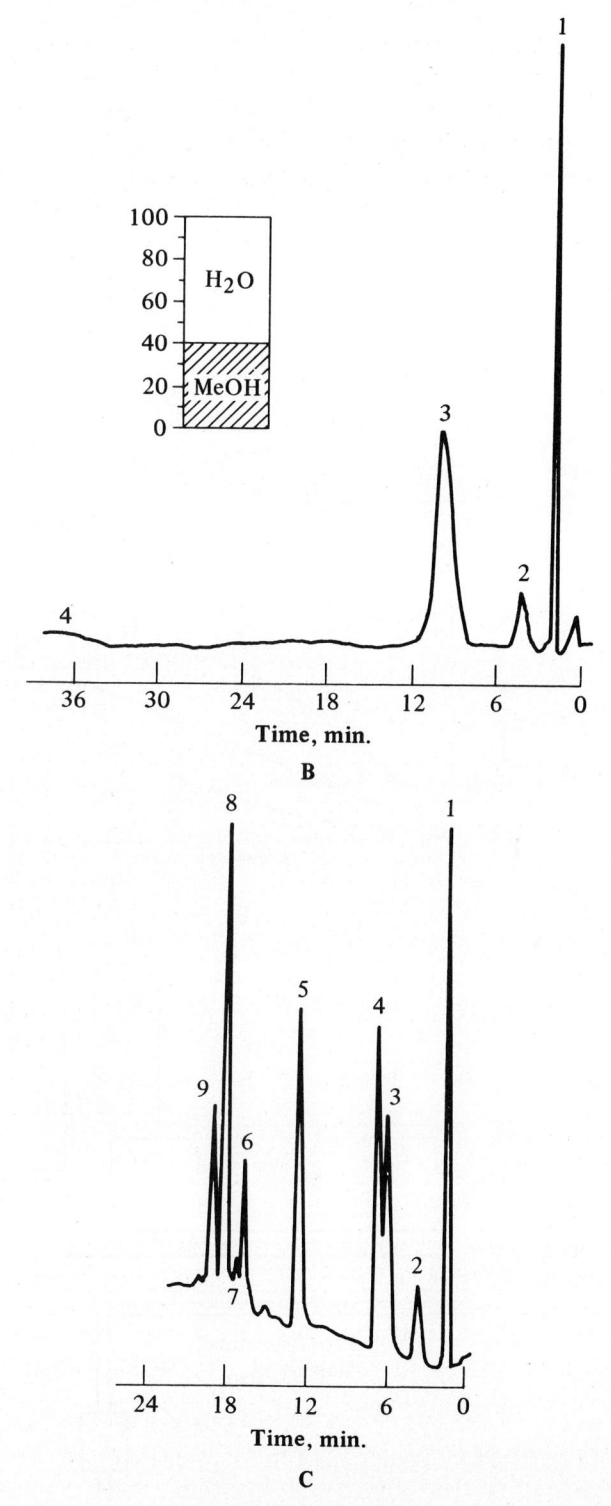

FIGURE 21.27 *(continued)*

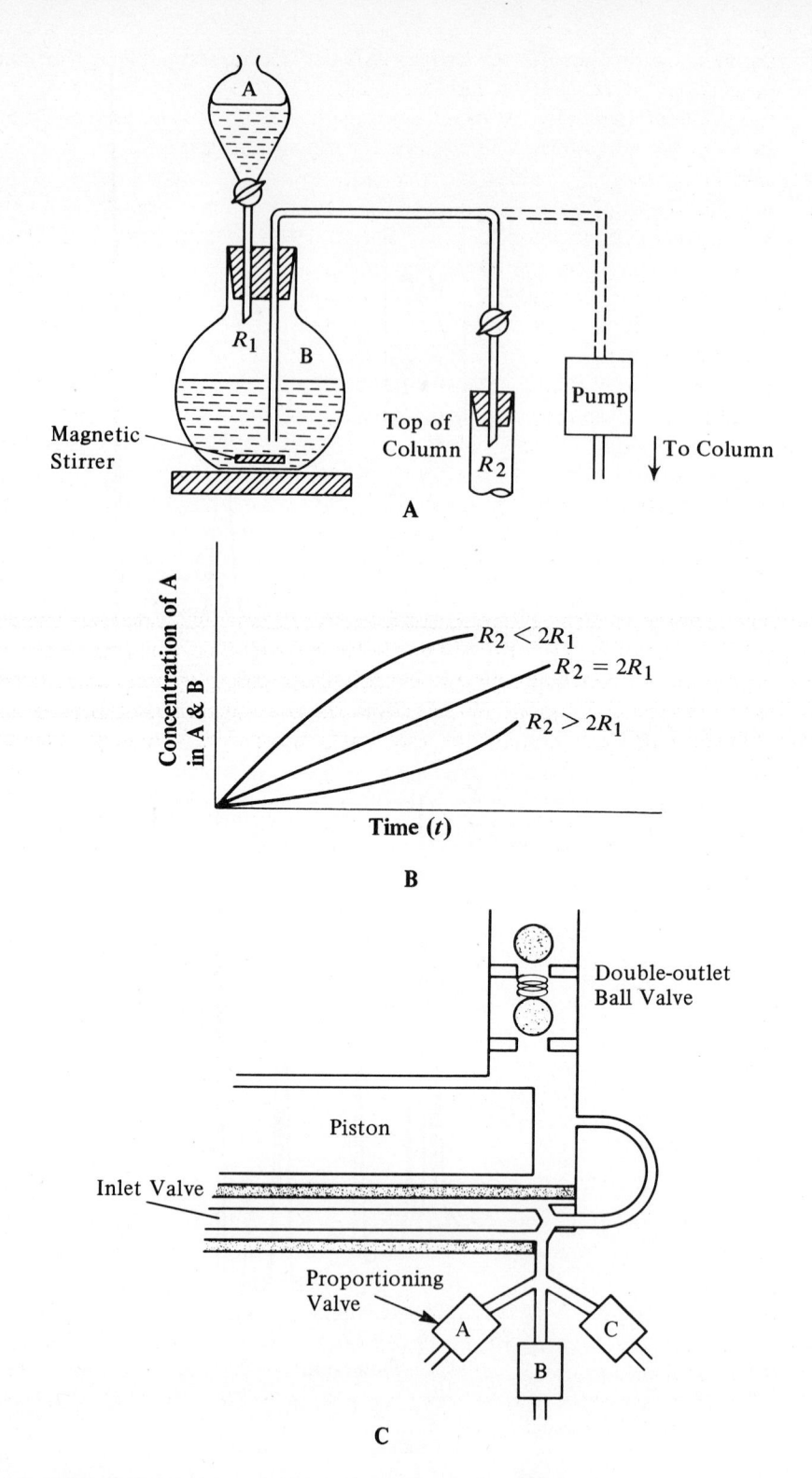

A

Magnetic Stirrer

R_1, B, R_2, Top of Column, Pump, To Column

B

Concentration of A in A & B

$R_2 < 2R_1$
$R_2 = 2R_1$
$R_2 > 2R_1$

Time (t)

C

Double-outlet Ball Valve

Piston

Inlet Valve

Proportioning Valve

A, B, C

704   CHAP. 21   Solid- and Liquid-Phase Chromatography

another sample can be run under equivalent conditions. This process is called *regeneration*. In adsorption chromatography, regeneration is accomplished by an instantaneous return to the initial composition of the solvent, then prolonged washing to remove the stronger solvent. The time required depends on V_s, the amount of stationary phase. An alternative, and in most cases more rapid, regeneration technique is to run a "negative" solvent program to remove the stronger solvent more gradually. At the end of the negative program, the column is usually ready for another injection.

In column chromatography, solvent gradients can be formed stepwise by adding successively stronger solvents one at a time to the top of the column. A commercial apparatus is available to automatically select from 2 to 12 successive solvents, each at predetermined times. Continuous binary gradients can be formed by using a simple apparatus equipped with a magnetic stirrer (see Fig. 21.28A). The pure weaker solvent (B) is placed in the mixing flask and the stronger solvent (A) is placed in the reservoir. Solvent A is permitted to flow into the mixing flask and the mixture of A and B is delivered to the column, often by means of a reciprocating pump, as shown by the dashed lines in Figure 21.28A. Equations are available for calculating the composition of the mixture in the flask at any given time. The shape (profile) of the gradient can be varied, as shown in Figure 21.28B; both linear and nonlinear profiles are useful in LC.

Homogeneous mixing is required to provide smooth and reproducible gradient profiles. In HPLC, gradients can be formed on the low-pressure or on the high-pressure side of the pump. Low-pressure gradient formation is somewhat analogous to the technique illustrated in Figure 21.28A. However, as illustrated in Figure 21.28C, high-speed proportioning valves, controlled by a microprocessor-based

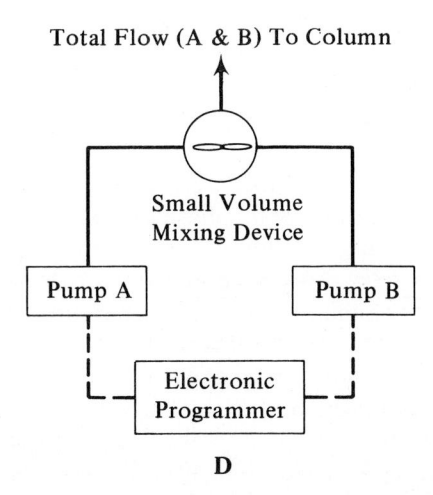

FIGURE 21.28. *Gradient devices used in liquid chromatography. A: Simple mixing device for gradient elution, gradient formed on low-pressure side of pump. R_1 is the flow rate of A into the flask. R_2 is the flow rate of A + B to column. B: Gradient profiles obtained from A. C: Modern single pump with low-pressure, high-speed, three-solvent proportioning valves. D: Schematic for two-pump gradient chromatograph. Gradient formed on high-pressure side of pump.*

programmer, are used to meter solvents in the pump's piston chamber during the fill stroke. Ternary or quaternary solvent mixtures can be prepared in this manner. Only a single pump is required for gradient formation. High-pressure gradient formation, depicted in Figure 21.28D, involves multiple pumps, each of which is independently controlled by an electronic programmer. The ratio of A to B can be varied easily, and gradient profiles changed reproducibly. With both single- and multiple-pump systems, stepwise gradients, negative gradients, and isocratic mobile phases can be generated conveniently.

In TLC, although continuous solvent programming is feasible, it is seldom used because it is experimentally inconvenient. Stepwise development or two-dimensional TLC is used instead. Programmed multiple development with the same solvent system has also been used.

21.7 USES AND APPLICATIONS OF PARTITION AND BONDED-PHASE CHROMATOGRAPHY

Partition or liquid-liquid chromatography (LLC) is similar to solvent extraction. In fact, solvent-extraction data can be used to predict partition coefficients for LLC. The resolving power and speed of LLC are considerably greater than those of solvent extraction, however, since the equivalent of several thousand partitions takes place as the sample components move down a column. Liquid-liquid chromatography is generally better suited to analytically separating complex mixtures, whereas extraction is used more for large-scale preparative separations or for separating relatively simple mixtures.

Selecting Stationary and Mobile Phases in LLC

The stationary and mobile phases are selected so as to have little or no mutual solubility. Therefore, they generally are quite different in their solvent properties. For example, referring to Table 21.4, one might choose water as the stationary phase and pentane as the mobile phase for normal LLC. However, water does have a finite (though very slight) solubility in pentane. If pentane, used as a mobile phase, is allowed to flow over a water-coated support long enough, it will slowly remove the water and change the nature of the separation mechanism. For this reason, the mobile phase must be presaturated with the stationary phase before it enters the column (or plate). Presaturation can be done by stirring the two phases together until equilibration takes place; but, in LC, it is more conveniently done by placing a *precolumn* before the injector and the chromatographic column. The precolumn should contain a high-surface-area packing, such as silica gel, coated with a high percentage (say 30 to 40% by weight) of the stationary phase used in the analytical column. As the solvent passes through the highly dispersed stationary phase, the solvent becomes saturated with it and will not remove it from the analytical column. In column chromatography, LLC has given way to the use of the more stable chemically bonded phases.

Liquid-liquid chromatography is limited to compounds with comparatively low values of K (or k'), because the stationary phase must be a good solvent for the sample but a poor solvent for the mobile phase. In practice, increasing solvent strength in order to elute compounds with high K- (or k'-) values will increase the solubility of the stationary phase and remove the stationary phase from its support. When the solvent strength is high enough to dissolve an appreciable amount of stationary phase, presaturation is made difficult. Needless to say, in conventional LLC solvent programming is ruled out.

Even with its limitations, LLC is a very useful technique because it can resolve minute differences in the solubility of the solute. Many solvent pairs are available, and the choice of the proper ones allows great selectivity to be achieved. Classification of solvents in terms of their ability to undergo different types of intermolecular interactions has greatly improved selection of useful solvent pairs. A scheme based on the Hildebrand solubility parameter (δ) has been described [7, 8]—common chromatographic solvents are classified quantitatively in terms of parameters such as dispersion interactions, dipole interactions, and proton donor-acceptor ability. By matching the properties of the particular solute (e.g., its proton-acceptor ability) with one or more of these individual parameters, one can estimate K-values and vary δ until the desired separation is obtained. The K-value of a solute is related to the ratio of the solute's concentrations (more correctly, its activities) in the two partitioning phases; selecting partitioning solvents with quite different δ-values (polarity) magnifies these differences in solubility. Because the activity coefficients of members of a homologous series vary with the molecular size of the compound, members of a homologous series can be separated by LLC, whereas in LSC there is little discrimination between successive members of a homologous series.

Both PC and TLC have been carried out with coated liquid phases. The paper strip (or thin-layer plate) is impregnated with stationary phase, either neat or dissolved in a volatile solvent. In the latter case, the solvent should be allowed to evaporate slowly to ensure homogeneous distribution of the stationary phase. Normal development is carried out, but some care is required to locate the separated spots since the coated liquid phase may interfere with detection.

Substances only very sparingly soluble in water are not separated by ordinary PC since they move with the solvent front. If the paper is impregnated with silicone oil or paraffin and a highly polar solvent is used as the mobile phase, such samples are more easily separated. This technique is referred to as reversed-phase PC. In the same way, reversed-phase TLC can be carried out.

Bonded-Phase Chromatography

The severe limitations of conventional LLC—that is, finding immiscible solvent pairs, presaturing the mobile phase to avoid removal of coated stationary phase, and the impossibility of using gradient elution to solve the general elution problem—have been overcome by the use of chemically bonded stationary phases. *Bonded-phase chromatography* (BPC) now dominates in use all modes of HPLC.

Microparticulate silica gel is the base material used for the synthesis of almost all chemically bonded phases. Recall from Figure 21.25 that silica gel contains the reactive silanol groups. These groups can be silanized by reaction with organochloro- or organoalkoxysilanes to form the stable siloxane bond. The ≡Si—O—Si—C bond is stable under most conditions used in LC but is attacked by hydrolysis under basic conditions (pH > 7). Siloxane phases have been prepared by reaction under anhydrous conditions to yield monomeric phases or by polymerization under controlled humidity to yield polymeric phases. The monomeric phases generally provide better efficiency owing to more rapid mass transfer of solute, whereas the polymeric phases provide higher sample capacity. Both types of phase can be used with gradient elution, a major advantage of BPC. Figure 21.27C depicts a solvent-programmed separation of several brominated aromatic compounds using a water-methanol gradient and a reversed-phase octadecylsilane (hydrophobic) bonded phase. A conventionally coated LLC phase could not have been used in such a solvent system; the stationary phase would have been dissolved in, and removed by, the mobile phase.

To simplify this discussion of BPC, two main techniques can be classified, based on the relative polarities of the stationary and mobile phases: (a) Normal-phase BPC, and (b) reversed-phase BPC. *Normal-phase BPC* is used when the stationary phase (e.g., aminopropyl) is more polar (as evidenced by the predominant functional group) than the mobile phase (e.g., hexane). In some cases, solute elution is similar to that observed for LSC on silica gel. Thus, nonpolar compounds prefer the mobile phase, exhibit low retention, and elute first. Polar compounds interact with the polar stationary phase, are retained, and elute later.

Reversed-phase BPC is used when the stationary phase is nonpolar (e.g., octadecylsilane) and the mobile phase is polar (e.g., water-methanol). Solute-elution order is often the reverse of that observed with normal-phase BPC. Polar substances prefer the polar mobile phase and elute first, whereas nonpolar compounds are retained more strongly. The technique is ideally suited to substances insoluble or only sparingly soluble in water but soluble in alcohols or other water-miscible organic solvents. Because many organic compounds show this solubility behavior, reversed-phase BPC is the most widely used mode of HPLC, accounting for about 60% of the published applications.

Normal-Phase BPC. By variation of the nature of the polar functional group on the organic side chain, quite different selectivities can be imparted relative to the unbonded silica packing. Table 21.5 lists the more polar commercially available

TABLE 21.5. *Polar Bonded Phase Structures*

Diol	—$(CH_2)_3OCH_2CHCH_2$ with OH, OH
Cyano	—$(CH_2)_3C{\equiv}N$
Amino	—$(CH_2)_nNH_2$ $n = 3$ or 4
Dimethylamino	—$(CH_2)_3N(CH_3)_2$
Diamino	—$(CH_2)_3NH(CH_2)_2NH_2$

siloxane-bonded packings for normal BPC work. Compared to silica gel, the polar bonded phases have the advantage of responding to solvent composition changes more rapidly. They also give less tailing since, during bonding, the most acidic silanol groups are replaced by less polar functionalities such as cyano or amino. The former advantage is most useful in gradient-elution work because column regeneration is frequently a time-consuming step in LSC.

Although normal-phase BPC can replace LSC on silica gel in many applications, several unique separations have been demonstrated that make it the preferred technique. One such example is the class separation of polynuclear aromatic hydrocarbons (PAH) with alkyl side chains. This addresses an important area in the characterization of fuel sources and trace analysis of these potential carcinogenic and mutagenic species in air, waste water, and industrial-process streams. Reversed-phase techniques, although useful for separation of PAHs based on their hydrophobic character, cannot separate these compounds according to the number of condensed rings. Using a diamino column and a heptane mobile phase, however, makes possible the separation by number of condensed rings for aromatics with up to four rings. Other unique separations for which normal-phase BPC is well suited include the class separation of alkanes and lipids, and the separation of saccharides, steroids, and fat-soluble vitamins. Of course, compounds that are unstable in aqueous solution must be separated by normal-phase techniques.

Reversed-Phase (RP) BPC. In its regularly practiced form, RP-BPC utilizes a hydrophobic stationary phase usually possessing an octadecyl or octyl silane functional group, and a polar mobile phase usually water mixed with a miscible organic solvent, such as methanol or acetonitrile. Table 21.6 lists the more popular commercially available siloxane-bonded packings for RP-BPC work.

The popularity of RP-BPC is undoubtedly due to the following:

1. Nonionic, ionic, and ionizable compounds can often be separated, sometimes at the same time, using a single column and mobile phase.
2. Bonded-phase columns are relatively stable provided certain precautions, especially pH control, are taken.
3. The predominant mobile phase, water, is inexpensive and plentiful.
4. The most frequently used organic modifier, methanol, can be obtained at a reasonable price and of sufficient purity in most places in the world.

TABLE 21.6. *Popular Commercially Available Siloxane Bonded-Phase Packings for Reversed-Phase Work*

—n-$C_{18}H_{37}$

—n-C_8H_{17}

—n-C_6H_{13}

—$(CH_2)_3C{\equiv}N$

—phenyl

5. The elution order is often predictable because retention time usually increases as the hydrophobic character of the solute increases.
6. Columns equilibrate rapidly, thereby permitting faster method development and sample turnaround after gradient elution.

Because RP-BPC is in such widespread use, a number of subclassifications have developed based on the predominant mechanism being used to control the separation. Table 21.7 takes a simplistic look at these various forms of RP-BPC. *Regular* RP-BPC refers to use of a hydrophobic packing and water plus a water-miscible organic modifier as the mobile phase (see Fig. 21.27). *Ionization control* is the general technique whereby, for an un-ionized or weakly ionized species, one can change the retention characteristics by the addition of buffer salts, by pH changes, or by other means. *Ion suppression*, a subcategory of ionization control, describes the techniques whereby, for a weakly acidic solute (e.g., acetic acid), a low concentration (1 to 2% vol/vol) of a stronger acid is added to keep the weak acid in its un-ionized form, resulting in better peak shape; the same may be achieved for weak bases by the addition of ammonium carbonate or very dilute ammonia.

Reversed-phase ion-pair chromatography (RP-IPC) is used for separating ionic or ionizable compounds through the formation of neutral ion pairs in solution by using counter ions containing an ionic functional group of an opposite charge. For example, consider the analysis of sulfonic acid dyes, which are completely ionic

TABLE 21.7. *Reversed-Phase BPC Techniques*

Technique	Uses	Typical Mobile Phase
Regular	General	(A) Water plus water-miscible organic solvent
Ionization control	Ionizable compounds	(B) As in (A), plus buffer salt
Ion suppression	Weak acids or bases	(C) As in (A), plus acid or base for weak acids; as in (A), plus acid (e.g., phosphoric, perchloric) for weak bases; as in (A), plus base (e.g., carbonate, dilute ammonia)
Ion pair	Strong or weak acids and bases; cations and anions	(D) For cations, as in (B), plus alkyl sulfonate or sulfate (e.g., sodium heptanesulfonate); for anions, as in (B), plus tetraalkylammonium salt (e.g., tetrabutylammonium chloride)
Complexation	Metals, chelates, stereoisomers	(E) As in (B), plus metal chelates, chiral reagents, silver ion (for olefins), ligands
Nonaqueous reversed phase (NARP)	Very nonpolar compounds (e.g., lipids, triglycerides, long-chain fatty acids)	(F) Acetonitrile or methanol plus tetrahydrofuran or methylene chloride

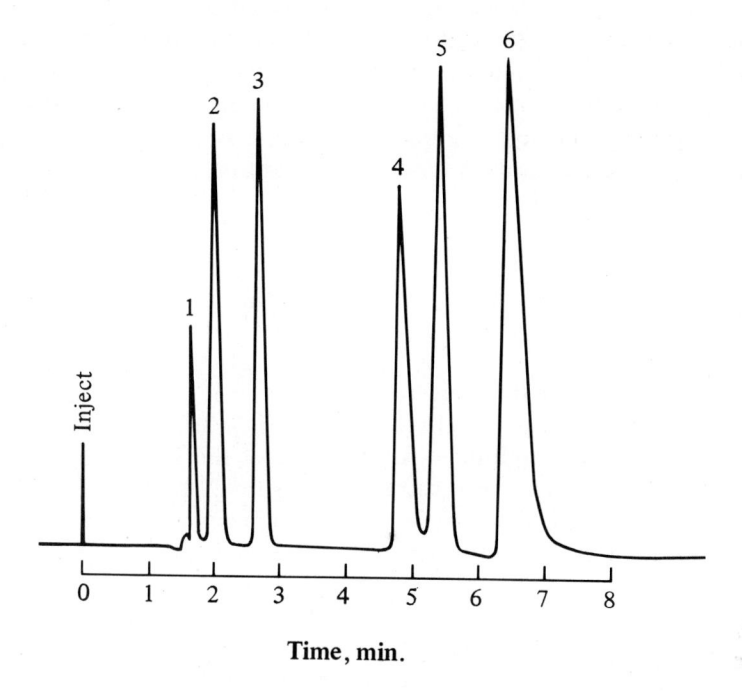

FIGURE 21.29. *Separation of several antihistamines and decongestants by reverse-phase ion-pair chromatography. Courtesy of Varian Associates.*

711

at almost all pH values. If one attempts to inject such compounds onto a silica-based bonded reversed-phase column using a water-methanol mobile phase, very poor peak shape may likely result. This behavior is caused by interaction between the ionic functional group and residual weakly acidic silanol groups on the silica surface, and also by mixed separation modes. The addition of a tetraalkylammonium compound to the mobile phase causes a neutral ion pair to be formed, which provides better peak symmetry.

The actual mechanism of RP-IPC is more complex, involving more than simple electrostatic interactions. Possible ion-pair formation in solution or dynamic formation of an ion exchanger on the surface of the stationary phase have been proposed; but the ion-interaction model that also takes into account adsorptive forces seems more realistic. Nevertheless, the RP-IPC technique has become competitive with ion-exchange chromatography (Sec. 21.8) for the liquid-chromatographic separation of ionic compounds. An example of the separation of several antihistamines and decongestants is illustrated in Figure 21.29. In this separation, pentanesulfonic acid added to the mobile phase interacts with the nitrogenous groups of these drugs, resulting in a separation based on ionic interactions.

The use of *complexation* by the addition of other additives to the mobile phase allows specific chemical equilibria to govern the separation selectivity. For example, olefins can form charge-transfer complexes with silver ions added to the aqueous mobile phase. Relative to compounds not possessing olefinic bonds, the retention times of olefins are altered. This can provide better chromatographic resolution.

21.8 ION-EXCHANGE CHROMATOGRAPHY

Ion-exchange chromatography is generally applicable to ionic compounds, to ionizable compounds such as organic acids or bases, and to compounds that can interact with ionic groups (e.g., chelates or ligands). Ion-exchange chromatography is carried out with stationary phases having charge-bearing functional groups. The mobile phase usually contains a counter ion, opposite in charge to the surface ionic group, in equilibrium with the resin in the form of an ion pair. The presence of a solute ion of the same ionic charge sets up an equilibrium as follows:

$$\text{Cation exchange: } X^+ + R^-Y^+ \rightleftharpoons Y^+ + R^-X^+ \qquad (21.15)$$

$$\text{Anion exchange: } X^- + R^+Y^- \rightleftharpoons Y^- + R^+X^- \qquad (21.16)$$

where   X = sample ion
        Y = mobile-phase ion (counter ion)
        R = ionic site on the exchanger

Competition between the sample ion and the counter ion for the fixed ionic site is very similar to the competition between solute and solvent for adsorption sites in LSC. In fact, sometimes ion exchange is referred to as adsorption chromatography involving electrostatic interactions. However, as the nature of the stationary and mobile phases, as well as the samples handled, are quite unlike those used for LSC, we prefer to classify ion-exchange chromatography separately.

Stationary Phases

The stationary phases used in ion exchange may be naturally occurring inorganic solids such as sodium aluminosilicate and clays such as montmorillonite, or synthetic ones such as zirconium phosphate. More often, though, they are resins prepared by the copolymerization of styrene and divinylbenzene. The amount of divinylbenzene used in the synthesis controls the extent of cross-linking in the resin. High cross-linking decreases the solubility of polystyrene and improves the structural rigidity, required for high-pressure use in HPLC. However, high cross-linking also decreases the porosity required for good mass transfer. Low-cross-linked resins have a tendency to "swell" by absorption of mobile phase. The amount of cross-linking expressed as percent divinylbenzene varies from 2 to 12%, with 8% being an average value.

The resin-based column packings give rise to lower column efficiency than is observed for the silica-based packings used in the other LC modes. For this reason, ionogenic groups have been chemically bonded onto microparticulate silica gel. These silica-based ion exchangers display better solute mass-transfer characteristics and can be used at ambient temperature. Resin-based columns frequently must be used at higher temperatures (e.g., 60 to 80°C) to improve their efficiency. On the other hand, resin columns have a greater pH range (0 to 14 in many cases) than silica (1 to 7.5), show longer lifetimes, and can be reused if they become contaminated.

Usually, the ionic groups are added to the resin, by chemical reaction, following cross linking. Cationic or anionic resins are classified as strong or weak, depending on the acidic or basic strength of the functional group. Strong cation exchangers normally have sulfonic ($-SO_3H$) functionality (incorporated by sulfonation of the resin), whereas weak ones have carboxyl ($-COOH$) functionality. Strong anion exchangers have tetraalkylammonium groups—for instance, $-CH_2-N(CH_3)_3{}^+Cl^-$ (incorporated by chloromethylation followed by treatment with trimethylamine)—whereas weakly basic ones might have $-NH_3{}^+Cl^-$ or $-NHR_2{}^+Cl^-$ functional groups. Resins of intermediate strength are also available.

Another stationary-phase variable that influences ion-exchange behavior is the pore size. For the separation of large molecules such as proteins and oligonucleotides, wide-pore packings are required so that these molecules can diffuse into the pore and interact with the ionogenic groups. Wide-pore silica-based packings with average pore sizes in the range 300 to 500 Å are available and are becoming the preferred types for separation of large biomolecules.

Exchange Capacity

The number and strength of fixed ionic groups on the solid govern the exchange capacity. Since the ion-exchange capacity affects solute retention, exchangers of high capacity are most often used for separating complex mixtures, where increased retention improves resolution. The capacities of weakly acidic and basic resins show a marked dependence on pH; generally, these resins have a small range of maximum capacity, dependent on the pK of the functional group. The strongly acidic and basic resins have a much wider range of maximum capacity and are generally more widely used. Weak ion exchangers are most often used in separating strongly basic, strongly acidic, or multifunctional ionic substances which are often firmly retained on the strong ion exchangers; they have been used for separating such substances as proteins, peptides, and sulfonates. For porous strong resins, capacities are on the order of 3 to 10 meq/g. For microparticulate silica-based packings, the exchange capacities are a factor of 5 to 10 lower. For pellicular strong resins, the capacity is much lower (5 to 50 μeq/g).

Influences on Distribution Coefficients and Selectivity

Ion-exchange chromatography involves more variables than other forms of chromatography. Distribution coefficients and selectivities are functions of pH, solute charge and radius, resin porosity, ionic strength and type of buffer, type of solvent, temperature, and so forth. The number of experimental variables makes ion-exchange chromatography a very versatile technique, since each may be used to effect a better separation, but a difficult one because of the time needed to optimize a separation. When polystyrene-divinylbenzene resins are used, organic ions (especially aromatic ones) are sorbed both by ionic forces with exchange groups and by interactions with the resin matrix itself. For example, because of "solvent" effects of the resin matrix, phenols are more strongly retained in anion exchange than their weak ionization would suggest. Even nonionic compounds can be separated on resins, probably by a partition mechanism. In these cases, the presence of a buffer decreases the solubility of the compound in the mobile phase, thus increasing the affinity of the compound for the resin. This form of "salting-out" chromatography is used to separate alcohols in order of increasing molecular weight.

Electrically neutral species that form a complex with ions can be separated by the exchange process. A well-known example is separating sugars through the adducts formed with the borate buffer used to elute them. Ligands can be separated through their interaction with metallic ions sorbed by the resin.

Cellulose powder, chemically modified to contain ion-exchange groups, is also used in both thin-layer and column chromatography. Because of its lack of rigidity, application is limited to low-pressure columns. Sheets of modified cellulose are available for use as exchangers in PC, and ion-exchange resins and inorganic ion exchangers have been impregnated into cellulose strips. Liquid ion exchangers, such as trioctylamine and bis-(2-ethylhexyl)-phosphoric acid (which are immiscible with aqueous solutions) can be coated onto a support, as in LLC.

Uses of Ion-Exchange Chromatography

Ion-exchange chromatography is used most often in inorganic chemistry and in biochemistry, the latter often to deal with water-soluble polar compounds such as proteins and amino acids. Metallic ions can be separated by cation exchange using the characteristic charge-to-radius ratio of the hydrated ions. Under comparable conditions, tetravalent ions are generally retained more than monovalent ions. Within a particular series of ions carrying the same charge, there is also a range of selectivity. As a rough guide, resin affinity decreases as the radius of the hydrated ion increases.

The development of ion-exchange methods for separating lanthanides and various fission products was instrumental in the development of atomic reactors. Tables of distribution coefficients as a function of pH for almost every cation in the periodic table, and for many synthetic resins, came out of this monumental work. More recently, the extension of ion-exchange methods into the biochemical field has aided in the structural elucidation of proteins and nucleic acids. Figure 21.30 depicts the separation of amino acids, one of the most used areas of ion-exchange chromatography. The amino acids are detected as they are eluted by their reaction with ninhydrin in a postcolumn reactor. The colored product is measured in a flow-through colorimetric detector.

Ion Chromatography

Ion chromatography (IC) is an ion-exchange technique that uses, most popularly, a low-capacity column combined with a conductivity detector to provide for the analysis of inorganic or organic ions, usually in trace amounts. Its most frequent practical application is the determination of trace anions in aqueous solution. The low-capacity column allows the use of a buffer with a low ionic strength. There are two forms of IC practiced today: (a) *suppressed* or *dual-column* IC, and (b) *nonsuppressed* or *single-column* IC.

The schematic of Figure 21.31 represents the normal practice of the suppressed technique. The separation of two simple anions, chloride and bromide, is illustrated. The technique employs basic mobile-phase buffers (e.g., salts of weak carbonate acids) to separate the anions in the separator column. The mobile phase then flows through a high-capacity cation-exchange column called the suppressor column. This column, being in the hydrogen form, converts the mobile phase into a low-conducting weak

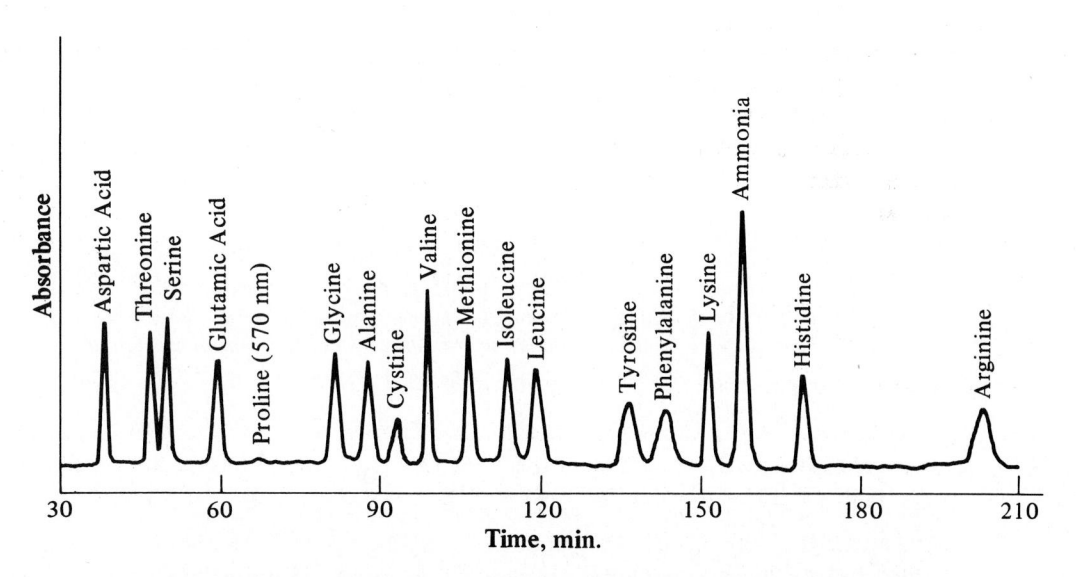

FIGURE 21.30. *Separation of amino acids by ion-exchange chromatography. Resin: Durrum® DC-1A, a sulfonated polystyrene-divinylbenzene cation exchanger of $d_p = 8$ μm. 10-nanomole calibration mixture. Flow rate: 70 mL/hr. From James R. Benson, Durrum Resin Report No. 5, April 1973, Durrum Chemical Corp., by permission of Durrum Chemical Corp.*

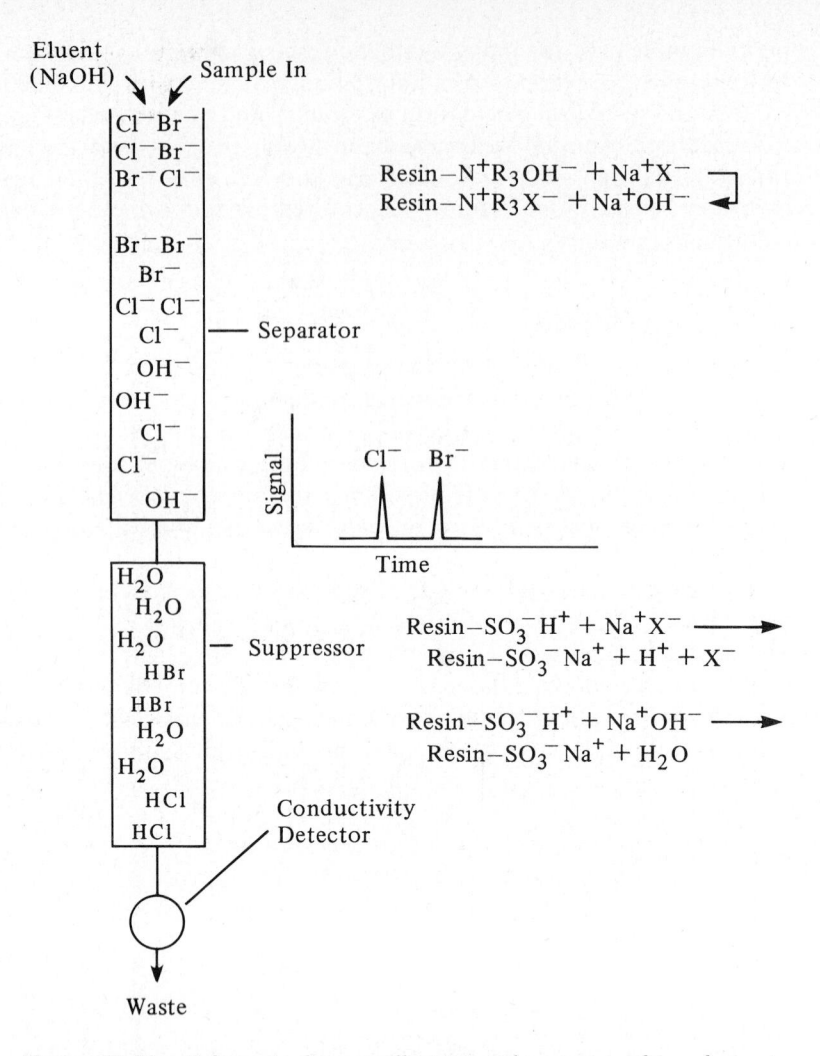

FIGURE 21.31. Schematic diagram illustrating the process of ion chromatography of anions. The upper (separator) column actually separates Cl^- and Br^-. The lower (suppressor) column is a strong-acid resin, and exchanges H^+ for other cations, thereby eliminating the large conductivity from the dilute NaOH eluent. The final conductivity, therefore, is due only to H^+Br^- and H^+Cl^-. Courtesy of Varian Associates.

acid. The sample anions (Cl^- and Br^-) are converted to highly conducting, fully ionized acids by the suppressor column. The advantage of the dual-column technique is that the total ionic content of the mobile phase is reduced, allowing the measurement of sample anions in a weakly conducting buffer. The disadvantages are that the suppressor column must occasionally be regenerated, and some peak broadening (with resulting decrease in resolution) is encountered due to the presence of the suppressor column.

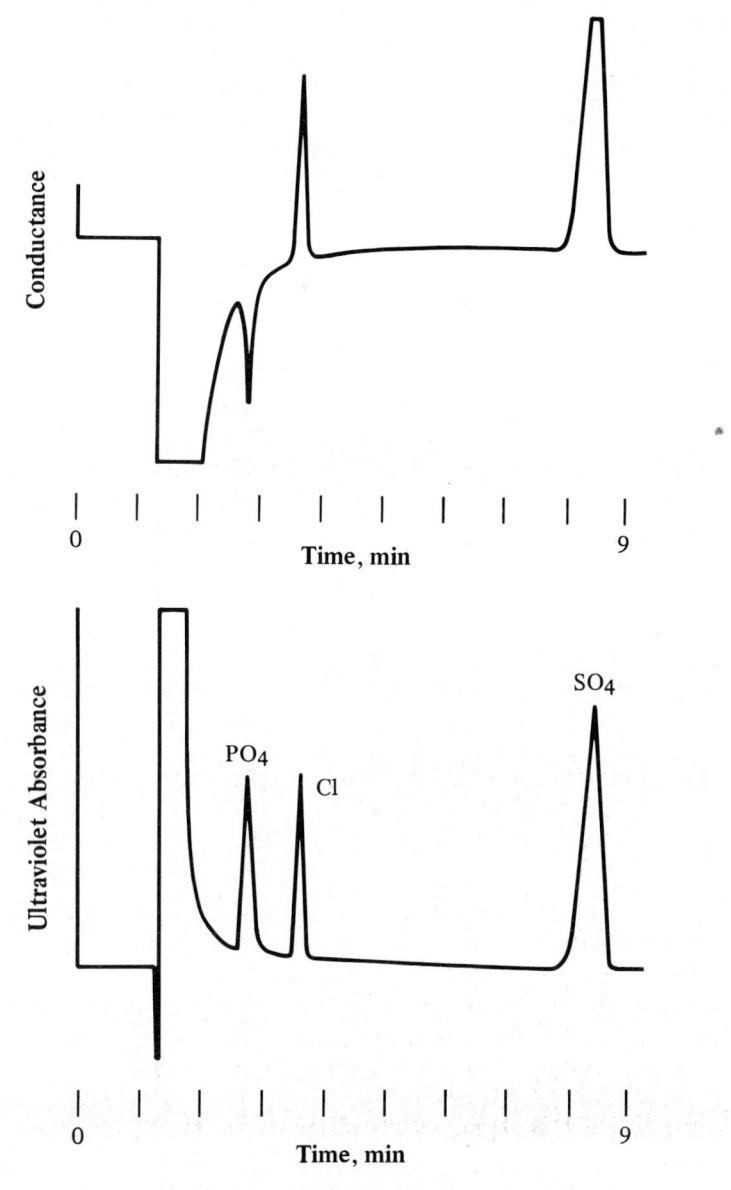

FIGURE 21.32. *Ion-chromatographic separation of several anions in a fertilizer sample, using both conductivity (upper) and ultraviolet (lower) detection. Each hashmark represents 1 min. Courtesy of Varian Associates.*

In the nonsuppressed IC technique, the suppressor column is absent, and the mobile phase passes from the anion column directly into the conductivity detector. The advantages of this technique are its simplicity and the absence of additional band broadening. The main disadvantage is the difficulty associated with measuring small amounts of sample ions in a highly conducting medium.

SEC. 21.8 Ion-Exchange Chromatography

Ion chromatography has found widespread application for the trace analysis of anions and cations in samples ranging from groundwater to body fluids. Spectroscopic techniques such as atomic absorption or flame-emission spectrometry are usually preferred for the trace analysis of cations. However, no analytical technique can rival IC for the convenient trace analysis of anions. Figure 21.32 illustrates the determination of small amounts of anions in a fertilizer sample using the single-column (nonsuppressed) technique. The negative-going peak for phosphate in the conductivity mode is due to the protonation of the phosphate, resulting in lower conductance than the eluting medium itself.

21.9 SIZE-EXCLUSION CHROMATOGRAPHY

Size-exclusion chromatography, also called gel chromatography, is the predominant method used for separating and characterizing substances of high molecular weight. The process is almost always carried out in a column, but it also has been performed on a thin layer. Column packing materials with pores of different (controlled) sizes are generally used. The materials can be soft gels, semirigid gels, or rigid materials. The soft and semirigid gels can change their pore sizes, depending on the solvent used as a mobile phase. The soft gels, of the polydextran or agarose type, can swell to many times their dry volume, whereas the semirigid gels of the polyvinylacetate or polystyrene type swell to 1.1 to 1.8 times their dry volume. Rigid materials, such as porous glass or porous silica beads, have fixed pore sizes and do not swell at all.

General Considerations

To understand how size exclusion differs from the other forms of chromatography, refer to Equation 21.4. In this context, V_m and V_s are referred to as the *void volume* and the *total pore volume*, respectively. The distribution coefficient K_X depends on the molecular weight of the sample and on the pore size of the packing. The equilibrium established in exclusion chromatography is described by Equation 21.1; K_X is defined by Equation 21.2. In a true permeation process, assuming all pores to be accessible to a small solute molecule, $X_s = X_m$ and $K_X = 1$. If none of the pores is available to a large solute molecule (i.e., it is excluded), then $X_s = 0$ and $K_X = 0$. Intermediate-sized molecules have access to various portions of the pore volume; for them, $0 < K_X < 1$. Unlike in other forms of LC, all sample molecules elute between the excluded volume V_m and the total permeation volume V_t. Note that V_r of Equation 21.4 is then equal to V_t. If $K_X > 1$, another mechanism of sorption is present and the process is not strictly exclusion.

Selecting the pore size of the packing depends on the size of the solute molecules to be separated as well as on their overall geometric shape. Often, the samples have a wide variation of solute sizes (i.e., molecular weights), and one pore size is insufficient to separate all molecular species. Some are completely excluded from the pores ($K_X = 0$) and elute as a single peak at V_m, whereas others may permeate all the pores and elute as a single peak at V_t. Others will selectively permeate part of the pores and elute at various values of V_r. A calibration curve (Fig. 21.33) is usually

Calibration Curve for Exclusion Chromatography

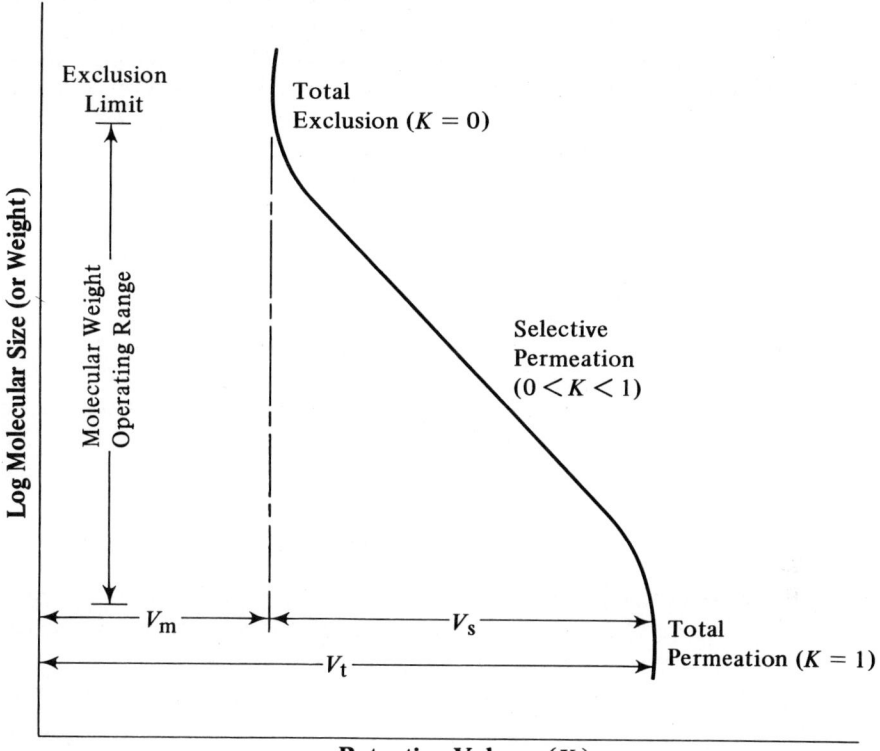

Size-Exclusion Chromatogram

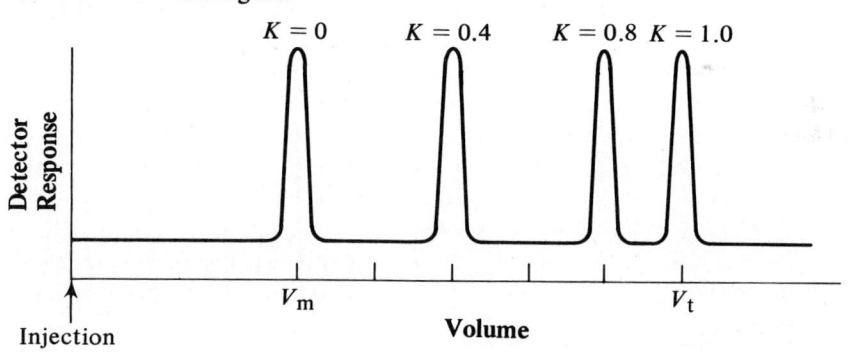

FIGURE 21.33. *Calibration curve and chromatogram for size-exclusion chromatography. Courtesy of Varian Associates.*

plotted as log(mol. wt.) versus V_r. Each exclusion packing of a different average pore size will have its own calibration curve. Note that neither the exclusion limit nor the molecular-weight range is sharply defined. Lack of precise definition occurs because the pores of the packings do not have a narrow distribution, and the distribution of the pores governs the slope of the calibration curve. If the pore distribution is wide, the curve will have a steep slope. Thus, the molecular-weight operating range will be large, but the column will provide less discrimination (resolution) of species of close molecular sizes. If the pore distribution is narrow, the curve will be flatter, the molecular-weight operating range smaller, but the resolution of closely sized molecules will be increased.

Uses of Size-Exclusion Chromatography

In exclusion chromatography, columns of different molecular-weight operating ranges are used to separate components in a sample of wide molecular-weight distribution. The columns are usually defined in terms of their molecular-weight exclusion limits. As many as eight columns, each covering a different molecular-weight range, are connected in series; each set of columns will have its own calibration curve. Calibration curves are determined by injecting standard samples of known molecular weight, and V_r is determined for each. With nonaqueous mobile phases, polystyrenes with

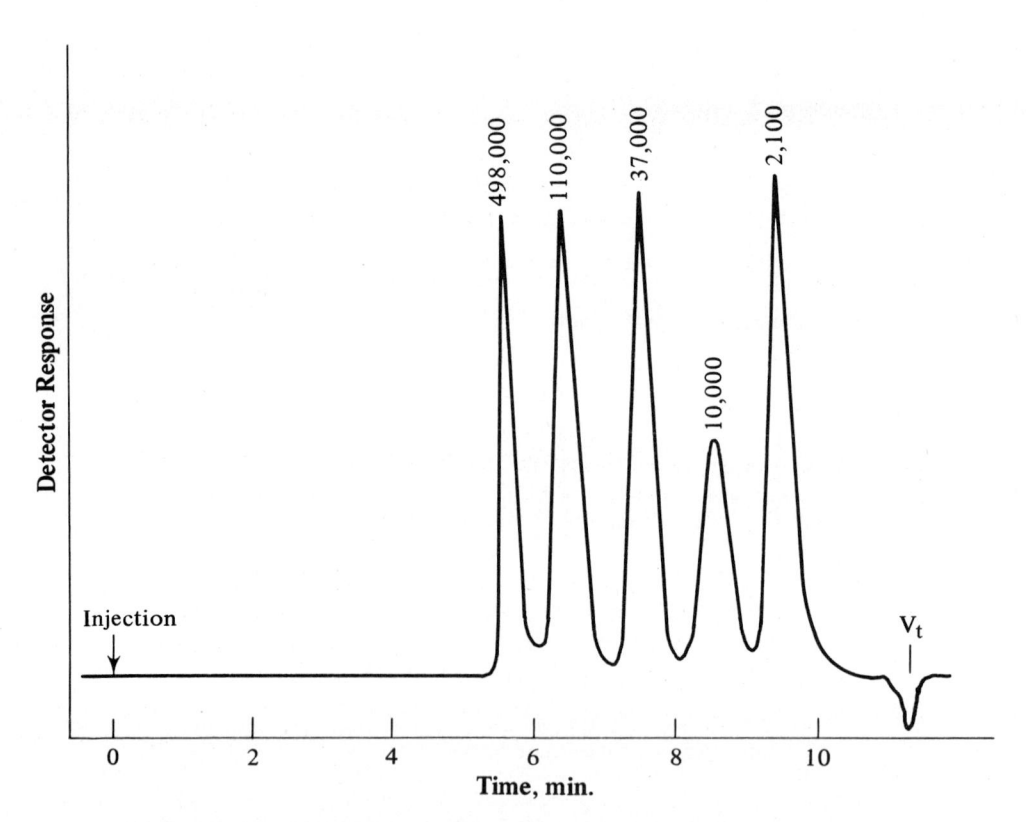

FIGURE 21.34. *High-speed size-exclusion chromatogram of polystyrene samples with various molecular weights. Column: Toyo Soda TSK® G 4000H8 Gel. Mobile phase: Tetrahydrofuran. Flow rate: 1.7 mL/min. Numbers on peaks represent molecular weights of fractions separated. Courtesy of Toyo Soda Manufacturing Ltd.*

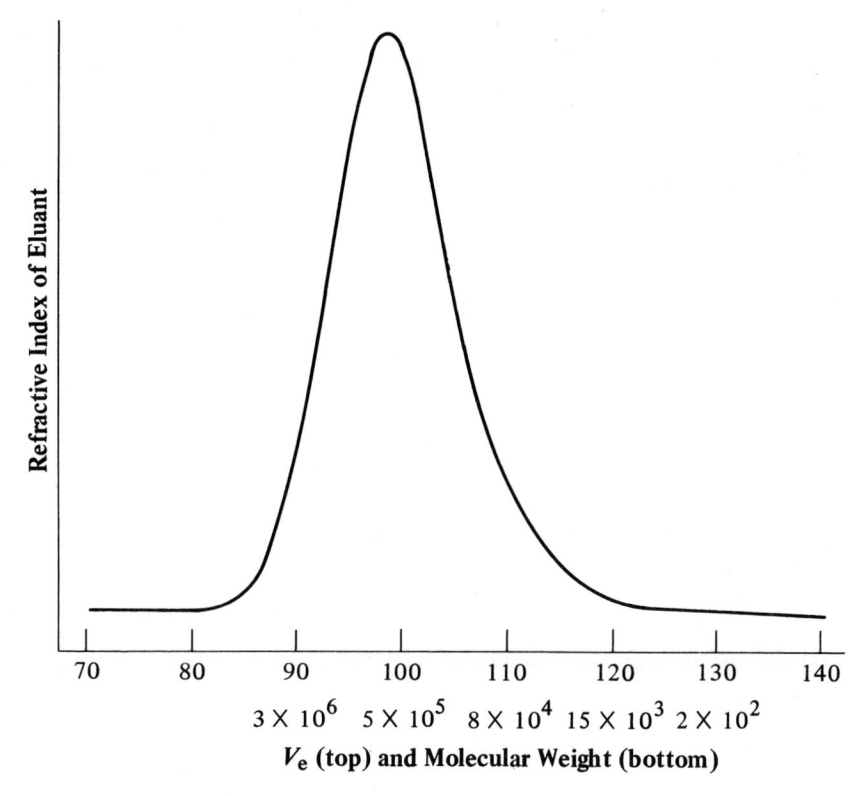

FIGURE 21.35. *Typical chromatogram of a polymer sample (polyvinyl chloride) by exclusion chromatography.* From L. R. Snyder and J. J. Kirkland, Modern Liquid Chromatography Slide Book, *Washington, D.C.: American Chemical Society*, 1973, Fig. 1.7, by permission of the publisher. Copyright © 1973 by the American Chemical Society.

known narrow molecular-weight ranges are used; with aqueous mobile phases, soluble dextrans or polyethyleneoxides are employed.

To illustrate the results obtainable in size-exclusion chromatography, Figure 21.34 depicts an HPLC separation of polystyrene standards in a column packed with cross-linked polystyrene particles (average pore size 260 Å) of 10-μm d_p, using tetrahydrofuran as a mobile phase. Unlike mobile phases in other forms of chromatography, the mobile phase in exclusion chromatography serves only to dissolve the sample and transport it through the column; it does not interact with the column packing. Note that in Figure 21.34, the polymers elute in the order of decreasing molecular weight, which is the reverse of that normally encountered in chromatography.

Size-exclusion chromatography is used not only for separating sample molecules, but also (in organic chemistry) to determine the average molecular weight and the molecular-weight distribution of polymers. Polymers are not separate, unique chemical entities, but are comprised of a continuous distribution of molecular weights. In these cases, the chromatogram shows a single broad peak, as depicted in Figure 21.35. From such a chromatogram, the polymer chemist can obtain molecular-weight and molecular-weight-distribution data and relate these to the physical properties of the polymeric materials (e.g., rigidity, tensile strength, stability). Thus, optimum polymerization conditions can be established.

Size-exclusion chromatography is also widely used to separate biological compounds, which are often water soluble and of high molecular weight. Proteins, nucleic acids, enzymes, and polysaccharides are routinely examined by exclusion chromatography in aqueous solution. Until recently, most applications were performed on soft gels of the crossed-linked polydextran variety, such as Sephadex®, in open-column chromatography. These soft gels are limited by their compressibility to low column-inlet pressures. Rigid hydrophilic gels and controlled-pore-size glasses, which can withstand high pressures, are now in routine use in many laboratories. The most popular are silica-based packings of 5- and 10-μm average particle sizes. Because bare silica gel itself possesses acidic silanol groups which could interact with the ionic or ionizable functional groups of biomolecules, hydrophilic phases are chemically bonded to its surface. Short-chain organic phases with moderately polar functional groups such as diol are generally used because too great an organic character results in undesirable interactions with the hydrophobic (hydrocarbonlike) portions of biomolecules such as proteins or polynucleotides. An illustration of one of these aqueous size-exclusion chromatographies can be seen in Figure 21.35.

21.10 TECHNIQUES RELATED TO LIQUID CHROMATOGRAPHY

A number of peripheral techniques closely allied to the basic chromatographic methods are worthy of mention.

Zone Electrophoresis

In electrophoresis, separation depends on the differences in the electrical properties of the components in a mixture. Although not, in principle, a chromatographic technique, electrophoresis is especially helpful in separating ionic compounds difficult to separate by PC alone. The principles of operation are depicted in Figure 21.36. Each end of a supported filter paper or cellulose-acetate strip is dipped into a vessel containing a buffer solution, which also acts as an electrolyte. The strip is carefully moistened with buffer solution. The sample is placed at some point on the paper strip. Then electrodes are dipped into the two vessels and connected to a high-voltage DC source for a predetermined period of time, permitting a constant current to flow through the electrolyte-moistened paper. Any substance that bears an electrical charge will migrate along the paper, the direction and rate being governed by the sign and magnitude of the charge on the ion and by the mobility of the ion. The current is switched off before the substances reach the two vessels. The paper is removed and dried, and the spots or zones are located in the same way as in paper chromatograms.

The migration rate of each substance depends on several factors, such as the voltage applied, the structure and charge of the ion, and the type and pH of the buffer employed. The migration of each mixture component is independent of that

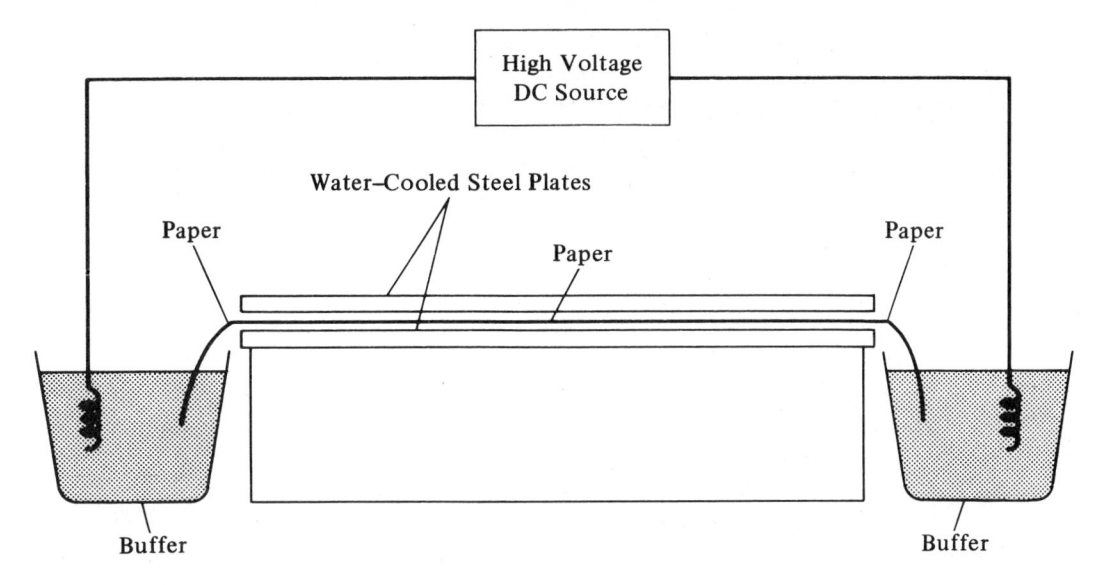

FIGURE 21.36. *Experimental setup for high-voltage electrophoresis.*

of the other substances present. Since there is no solvent front, a known substance is normally used as a standard.

The primary applications of the technique are in the fields of biology and clinical chemistry, for separating amino acids and proteins. Because of the presence of acidic and basic ionic groups in amino acids, proteins, and other biological macromolecules, pH has a profound effect on migration. In acidic solution, amino acids are positively charged, whereas in basic media they are negatively charged. At a certain pH value, the net charge is zero and the amino acid exists as a zwitterion. This point is called its *isoionic* or *isoelectric point*; provided other ionic interactions are absent, there should be zero electrophoretic mobility at this point.

In some cases, compounds that bear no electrical charge themselves can be separated because they can form complexes with ions present in the buffer solution. These separations are probably based on the principle of electroosmotic buffer flow and, in that respect, are similar to separations in ordinary PC. An example is the separation of borate complexes of sugars in a sodium-borate buffer.

Voltages applied are expressed in terms of volts per centimeter. In low-voltage electrophoresis, the applied voltage is up to about 500 V. High-voltage electrophoresis, in which voltages up to several thousand volts are used, is a technique well suited to the high-speed separation of low-molecular-weight substances; it is less suited to separating macromolecules, presumably because of their lower ionic mobilities.

In addition to filter paper and cellulose-acetate membranes, other supports can be used. Gels such as agar, starch, and polyacrylamide are supported in the form of a slab or block on a special rack. The technique is termed *gel electrophoresis*. The gels with varying pore sizes exert a slight molecular-sieve effect. In fact, specially prepared polyacrylamide-gel slabs with a continuous gradient of pore size can be used to improve resolution, since the migration of a particular macromolecule will be retarded when the pore limit is reached. *Thin-layer electrophoresis,* usually used for polar compounds such as phenols or amines, uses a thin-layer silica or alumina plate to take advantage of adsorption effects.

Two-dimensional combined techniques using electrophoresis in one direction and chromatography in the other can sometimes give a better separation than either alone.

Affinity Chromatography

Affinity chromatography is a selective "filtration" technique for macromolecules utilizing highly specific, reversible biochemical reactions. A ligand with a high specificity for the component of interest is attached to a support particle, much like a chemically bonded phase. The ligand is an immobilized enzyme or an antigen, and the macromolecule of interest is an enzyme inhibitor or antibody in a complex biological sample. The immobilized enzyme (or antigen) is placed in a column and the sample introduced in a suitable buffer, the pH and ionic strength of which favor the selective interaction. Elution with the buffer removes the unwanted substances from the column, then the buffer is changed to one that reverses the enzyme-inhibitor interaction and the inhibitor is eluted. A small number of specific systems have been developed, and they have proved to be very successful. Depending on the system of interest, each ligand must be individually chemically bonded. Some intermediate packings onto which a ligand can attach are available, and this technique should continue to become even more useful in the future.

Desalting and Deionization

Often, isolating pure substances from biological systems requires the elimination of samples containing large amounts of inorganic salts. These samples can be desalted by one of several techniques. Exclusion chromatography is quite useful if the component of interest has a higher molecular weight than the salts. Most salts are of low molecular weight and, provided the correct pore-size packing is selected, are retained at the total permeation volume V_t. The higher-molecular-weight sample passes through the column and is collected before the salt elutes. This technique is commonly used for desalting proteins that have been purified by salting out.

A second technique for desalting involves the use of a hydrophobic packing, such as polystyrene-based resins like Amberlite XAD-2®. Provided the component to be collected is organic in character, the sample will pass through the column, selectively concentrating the organic species but allowing the salt to pass through unretained. Afterward, an organic solvent is used to elute the organic compounds. Organic pollutants can be selectively concentrated from large volumes of water using this approach.

Alternatively, an ion-exchange resin can be used to adsorb ionic species, allowing uncharged organic compounds or nonelectrolytes to pass through. For example, if the contaminant is sodium chloride, a cationic resin in the hydrogen form could be used. Passing a salt sample through the column will exchange Na^+ for H^+. The hydrogen chloride can, in turn, be removed from the effluent by evaporation, or the effluent can be passed through an anion-exchange column in the OH^- form, thereby removing the Cl⁻ ions. Of course, this classic technique has been used for many years for deionizing water.

SELECTED BIBLIOGRAPHY

General

HEFTMANN, E. *Chromatography*, 3rd ed. New York: Van Nostrand Reinhold, 1974.

KARGER, B. L., SNYDER, L. R., and HORVATH, C. *An Introduction to Separation Science*. New York: John Wiley, 1973.

MILLER, J. M. *Separation Methods in Chemical Analysis*. New York: John Wiley, 1975.

MORRIS, C. J. O. R., and MORRIS, P. *Separation Methods in Biochemistry*. New York: Wiley-Interscience, 1964.

STOCK, R., and RICE, C. B. F. *Chromatographic Methods*, 3rd ed. London: Chapman and Hall, 1974.

High-Performance Liquid Chromatography

JOHNSON, E. L., and STEVENSON, R. L. *Basic Liquid Chromatography*. Palo Alto, Calif.: Varian Associates, 1978. *A paperback coverage of practical aspects of HPLC.*

SNYDER, L. R., and KIRKLAND, J. J. *Introduction to Modern Liquid Chromatography*, 2nd ed. New York: John Wiley, 1979.

Ion Exchange

BIDLINGMEYER, B. A. *J. Chromatog. Sci.*, *18*, 525 (1980).

FRITZ, J. S., GJERDE, D. T., and POHLANDT, C. *Ion Chromatography*. Heidelberg: Dr. Alfred Hüthig Verlag, 1982.

HELFFERICH, F. *Ion Exchange*. New York: McGraw-Hill, 1962.

RIEMAN, W., and WALTON, H. F. *Ion Exchange in Analytical Chemistry*. New York: Pergamon Press, 1970.

Paper and Thin-Layer Chromatography

BLOCK, R. J., DURRUM, E. L., and ZWEIG, G. *A Manual of Paper Chromatography and Paper Electrophoresis*, 2nd ed. New York: Academic Press, 1958.

KIRCHNER, J. G. *Thin-Layer Chromatography*. New York: Wiley-Interscience, 1976.

YAU, W. W., KIRKLAND, J. J., and BLY, D. D. *Modern Size-Exclusion Liquid Chromatography*. New York: John Wiley, 1979.

Electrophoresis

BIER, M., ed. *Electrophoresis*, Vol. 1 (1959); Vol. 2 (1967). New York: Academic Press.

SHAW, D. J. *Electrophoresis*. New York: Academic Press, 1969.

REFERENCES

1. J. C. GIDDINGS, *Dynamics of Chromatography*, Part 1, *Principles and Theory*, New York: Marcel Dekker, 1965. *A complete theoretical account of chromatographic principles.*
2. I. M. HAIS and K. MACEK, *Paper Chromatography*, New York: Academic Press, 1963.
3. E. STAHL, *Thin-Layer Chromatography: A Laboratory Handbook*, 2nd ed., Heidelberg: Springer-Verlag, 1969.
4. R. E. MAJORS, H. G. BARTH, and C. H. LOCHMÜLLER, *Anal. Chem.*, 54, 323R (1982).
5. R. E. MAJORS, *LC Magazine*, *1*, 488 (1983).
6. L. R. SNYDER, *Principles of Adsorption Chromatography*, New York: Marcel Dekker, 1968.
7. L. R. SNYDER and J. J. KIRKLAND, *Introduction to Modern Liquid Chromatography*, 2nd ed., New York: John Wiley, 1979.
8. R. A. KELLER, B. L. KARGER, and L. R. SNYDER, in R. Stock and S. G. Perry, eds., *Gas Chromatography 1970*, Institute of Petroleum, 1971.

PROBLEMS

1. The R_f-value of a solute can be expressed as the probability of finding it in the mobile phase at any given instant, expressed as the mole fraction of a solute in the mobile phase. (a) Derive a simple equation relating the solute R_f-value to its k'-value. (b) From this equation, determine the k'-value of estradiol, a steroid whose R_f-value on a silica-gel plate was 0.3. (c) Relative to estradiol, estriol showed a selectivity factor $\alpha = 1.2$ on a silica-gel column. Would you expect the two steroids to be separated on a silica-gel thin-layer plate, assuming spots are distinguishable only if their R_f-values are at least 0.02 unit apart?

2. Because of refractive-index effects, an unretained solvent used to dissolve the sample—if different from the chromatographic mobile phase—often deflects the baseline when passing through an ultraviolet detector cell. This indicates the void volume or the void time. Consider the chromatogram in Figure 21.26. (a) Determine the capacity factors for each nitroaniline isomer. (b) Determine the selectivity factor for the m- and p-substituted isomers relative to the o-nitroaniline.

3. (a) From the TLC chromatogram of Figure 21.11, determine the R_f-values for 3'-GMP and 2'-GMP. (b) Using an equation similar to that derived in Problem 1(a), determine the selectivity factor α between the two nucleotides.

4. In Figure 21.14, H was found to be proportional to $d_p^{1.8}$ at constant v. (a) If the average d_p of a silica gel were reduced from 20 μm (where $H = 0.3$ mm) to 2 μm, how many theoretical plates would one obtain from a 25-cm column? (b) Would this column be able to give a baseline separation between geometric isomers whose α-value is 1.05 and k'-value is close to 10?

5. The polystyrene peaks in Figure 21.34 were obtained by using standards with narrow molecular-weight distributions. (a) Assuming that the polystyrene with the highest molecular weight was totally excluded, construct a calibration curve like that depicted in Figure 21.33. (b) An unknown polybutadiene gave a peak maximum $V_r = 17$ mL. Determine its average molecular weight based on polystyrene. (c) Estimate the lower molecular-weight limit of the operating range (i.e., the molecular weight below which no separation will occur).

6. Porous silica has a packing density of 0.55 g/mL, whereas pellicular silica, because of its glass-bead core, has a packing density of 3.0 g/mL. Surface areas measured by nitrogen adsorption are typically 400 m²/g for porous silica and only 10 m²/g for pellicular silica. The sample capacity for a steroid on porous silica was found to be 2 mg/g. (a) For a preparative LC column of dimensions 50 cm by 0.8 mm, determine the maximum amount of sample that can be injected into the column filled with porous silica. (b) Assuming the sample capacity is proportional to the surface area, what sample size can be injected into the same column packed with pellicular silica? (c) How long would the latter column have to be to have the same sample capacity as the porous silica from part (a)? (d) Assume 200 mg of pure steroid was needed for further experimentation. If a "heart-cutting" (i.e., collecting the middle two-thirds on the LC peak) technique was used, how many injections would be required on the porous-silica column?

7. Linear velocity can be determined from a chromatogram by $v = L/t_0$. For a chromatographic separation, t_r for the last eluting peak is considered to be the separation time. (a) Derive an expression relating separation time to v and k'. (b) Using the equation from part (a), calculate the total separation time on a 100-cm column exhibiting an H-value of 0.4 mm for the last eluting peak ($k' = 24$). The flow rate was 2 mL/min, and the column void volume was determined to be 4 mL.

8. Some HPLC systems are limited to an operating pressure of 250 atmospheres. A 100-cm column containing 40-μm particles of ion-exchange resin had a pressure drop of 10 atmospheres at $v = 1.0$ cm/sec. The total time required for the analysis was 30 min. The column produced 300 theoretical plates, but the separation of interest required 8000 plates. The number of plates can be increased by (1) using a longer column, (2) using a lower linear velocity (i.e., flow rate), or (3) using a smaller particle size in the packing. Assuming the relationship

726   CHAP. 21   Solid- and Liquid-Phase Chromatography

$H = A(v^{0.6})d_p^{1.8}$ holds for the above system (A is a constant), which of the three options would be the most advantageous to pursue, and why?

9. In reversed-phase chromatography, there appears to be a logarithmic relationship between the capacity factor and the number of units in a homologous series. (a) Based on the chromatogram in Figure 21.15, column A, estimate the capacity factor for the 4-ring fused aromatic naphthacene. Note that the void time t_0 is about 1.57 min. (b) How could the retention time be decreased for this compound?

10. During gradient elution using binary mixtures of mobile phases such as ethanol and water, the viscosity changes nonlinearly. In fact, at 45% methanol in water, it reaches a maximum of 2.0 centipoises (25°C). If 6000 psig is the maximum operating pressure of a liquid chromatograph, what would be the maximum linear velocity that could be used for a 30-cm-long × 4-mm-i.d. column packed with 10-μm particles?

11. Assuming 60% of the volume of a column 25 cm in length by 0.40 cm in internal diameter is occupied by solid packing particles, calculate the expected retention volume for peak C of Figure 21.2. (Hint: Treat the column as a cylinder.)

12. In Figure 21.11A, assume that the 3'-GMP spot represents 10 μg of nucleotide standard. The TLC separation of the mixture 1–6 can be found on the plate (vertical column 5). (a) What would be the estimated amount of 2'-GMP in this mixture? (Hint: Treat the spot as an ellipse.) (b) Calculate the R_f-value for 5'-ATP.

13. Two different columns were available for a separation of enantiomers. One column displayed 20,000 theoretical plates, a selectivity of 1.25, and a k'-value of 3 for the second of the two enantiomers. The second column gave relatively narrow peaks (baseline t_w = 180 sec), an exceedingly long retention time of 2 hr (t_0 = 3.5 min), but a selectivity value of 1.05. (a) Which column would give the best overall resolution? (b) Keeping other parameters the same, would doubling the number of theoretical plates (N) for column 1 or doubling the selectivity (α) for column 2 change the situation?

14. The peaks for naphazoline and phenacetin in Figure 21.29 just barely overlap. Calculate the resolution between the two compounds.

15. In Figure 21.15, the retention times for anthracene are 25.7, 28.3, and 26.8 min for columns A, B, and C, respectively. The peak widths t_w at baseline are constant at 1.0 min. The void time for column A is 1.6 min. (a) Assuming the extra-column void volume is 20 μL, calculate the void volume of column A. (b) Assuming the same percentage of within-column void, calculate the total void volume for columns B and C. (c) Calculate the k' for anthracene for each column.

16. Estimate the volume of gas that would be produced at standard temperature and pressure from the solvent used for each of the three columns in Figure 21.15 during the total 30-min analysis time. Assume acetonitrile-water is an ideal mixture.

CHAPTER 22

Gas Chromatography

Charles H. Lochmüller

It was noted in the previous chapter that, in 1952, A. J. P. Martin and R. L. M. Synge received the Nobel Prize for the discovery of partition chromatography. In 1941, these authors had also written [1] that,

> The mobile phase need not be a liquid but may be a vapour. We show below that the efficiency of contact between the phases (theoretical plates per unit length of column) is far greater in the chromatogram than in ordinary distillation or extraction columns. Very refined separation of volatile substances should therefore be possible in a column in which a permanent gas is made to flow over a gel impregnated with a nonvolatile solvent in which the substances to be separated approximately obey Raoults' law.

Despite this clear and unequivocal prediction of gas-liquid partition chromatography, no one seized the opportunity until ten years later, when Martin and James [2] demonstrated its great potential. Gas chromatography has grown phenomenally in the last quarter century, and its use is now routine in many aspects of experimental chemistry, including analysis. The aim of this chapter is to give the beginning student a practical understanding and a physical model of gas chromatography, sufficient to carry out elementary experiments and to provide a basis for further reading.

Gas chromatography (GC) is one type of partition chromatography; it is similar in many ways to other techniques of this kind, such as HPLC (high-performance liquid chromatography), paper chromatography, and so on. The distinguishing features are that the mobile phase is a gas and that the motion of the component bands, in the direction of "chromatographic development," involves the forced diffusion of the respective substances in their vapor phases. Many of the differences between, say, HPLC and GC are due to the physical properties of the mobile phase—for instance, its viscosity, acidity, basicity, and compressibility. The basis for differential zone migration remains the same: Two components will migrate at different rates in the same chromatographic system if their distribution constants are different. Here, the main emphasis will be on gas-liquid chromatography, but gas-solid techniques (which have some advantages) will also be mentioned.

22.1 THE THERMODYNAMICS OF GAS CHROMATOGRAPHY

Gas chromatography involves the same two types of phenomena as any chromatographic method: first, static or equilibrium processes that can be described thermodynamically; second, dynamic or flux processes (including mass transport) that must be described kinetically. A rudimentary understanding of both statics and dynamics as they apply to gas chromatography should help the student to understand the potential and limitations of this technique and to improve his or her attack on a problem in analysis. In this section, the static aspects are considered.

Principles

The concept of retention volume in chromatography was discussed in Chapter 21. One can distinguish between the retention volume V_r (or retention time t_r) and the *adjusted retention volume* V'_r (or adjusted retention time t'_r):

$$V'_r = V_r - V_0$$

or
(22.1)

$$t'_r = t_r - t_0$$

where V_0 = elution volume of an unretained species
t_0 = elution time of the species

Quantities V'_r and t'_r are also called the *elution* volume or time. The relationship between t_r and V_r is

$$V_r = t_r F_c \tag{22.2}$$

where F_c = adjusted flow rate

The adjusted flow rate is most commonly calculated by first measuring the volumetric flow rate F_0 at the exit port of the gas chromatograph by use of a simple soap-bubble flowmeter such as that illustrated in Figure 22.1. Quantity F_0 must then be corrected for the vapor pressure of water, and back to the column temperature, T_c.

Consider a chromatographic column. When the mobile phase is a liquid, the carrier velocity (v) is not a strong function of the axial position in the column because, in general, liquids are not very compressible. Gases are, however, quite compressible, and so a correction is applied to the adjusted volume V'_r to obtain the *net volume* V_n. It can be shown that the average carrier velocity in the column $\bar{v}$ is related to the outlet velocity v_0 by a correction factor j, given by

$$j = \frac{3}{2}\left[\frac{(P_i/P_o)^2 - 1}{(P_i/P_o)^3 - 1}\right] \tag{22.3}$$

where P_i = inlet pressure
P_o = outlet pressure

Now

$$V_n = jV'_r \tag{22.4}$$

FIGURE 22.1. *Bubble-type flowmeter for volumetric flow measurements. A soap-bubble film is swept past the indices V_0 and V_f and the transit time is measured. ΔV = the volume of tube between V_0 and V_f; Δt = the transit time; F_0 = the volumetric flow-rate; F_c = the adjusted flow rate; T_c = the column temperature (K); T_A = the ambient temperature (K); P_A = the ambient pressure; P_{H_2O} = the vapor pressure of water at the ambient temperature.*

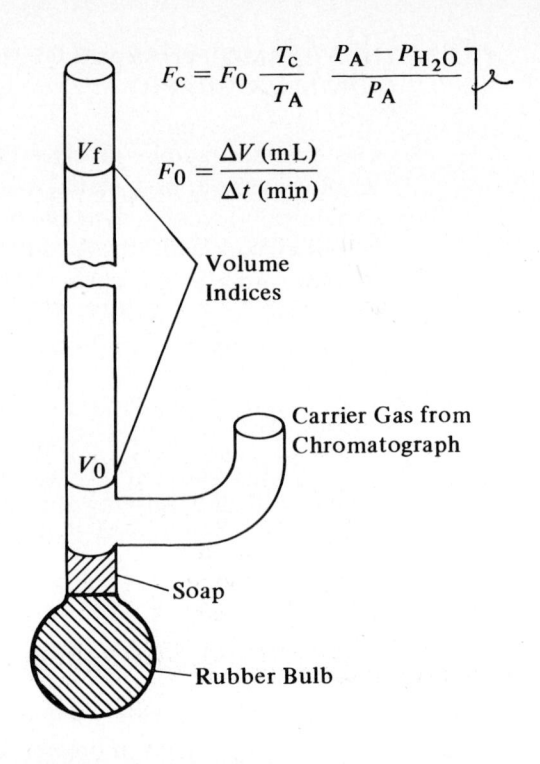

$$F_c = F_0 \frac{T_c}{T_A} \left[\frac{P_A - P_{H_2O}}{P_A} \right]$$

$$F_0 = \frac{\Delta V \text{ (mL)}}{\Delta t \text{ (min)}}$$

Volume Indices

Carrier Gas from Chromatograph

Soap

Rubber Bulb

and $\bar{v} = jv_0$; and since v and F are related (assuming a column of constant cross section), $\bar{F} = jF_0$. When the sample (generally a mixture of solutes) enters the column containing the stationary phase (solvent in GLC), it rapidly distributes itself between the vapor and condensed or solution phases. This is described by a distribution coefficient K, given by

$$K = \frac{c_s}{c_m} = \left(\frac{n_s}{V_s}\right)\left(\frac{V_m}{n_m}\right) \tag{22.5}$$

where c = concentration of the solute in a phase (stationary or *mobile*)
n = number of moles of solute in a phase
V = total volume of a phase in the column

In conventional gas-chromatographic experiments (i.e., at low pressures and with inert carrier gases such as helium), nonideal behavior of the solute in the gas phase contributes to only about 1 to 3% of the observed distribution coefficient.

The net retention volume increases or decreases with increasing or decreasing mass of stationary phase W_s, and the specific retention volume V_g is defined as

$$V_g = \frac{V_n}{W_s} \frac{273}{T_c} \tag{22.6}$$

The factor $273/T_c$ corrects V_g to the reference temperature of 273 K. It can be shown that, for ideal behavior, the specific retention volume is given by

$$V_g = \frac{273R}{\gamma_\infty P^0 M W_s} \tag{22.7}$$

where R = the gas constant
γ_∞ = activity coefficient of the solute at infinite dilution in the stationary phase
P^0 = saturation vapor pressure of pure solute at a given temperature
MW_s = molecular weight of the stationary-phase material

Thus, the specific retention volume depends on only two factors in a given solvent (MW_s is assumed constant): the saturation vapor pressure of the solute P^0 and the activity coefficient of the solute γ_∞.

Influence of Temperature and Volatility on Retention

Consider the practical significance of the above result. First, other things being equal, the greater P^0 is, the smaller V_g will be (Eqn. 22.7), and the shorter the retention time will be (Eqn. 22.2). The value of P^0 increases with increasing temperature; thus, retention time decreases as the temperature is increased. One can also shorten retention times by converting the solutes of interest into more volatile derivatives, thus increasing P^0. For example, amino acids converted to the volatile ester amides have smaller overall retention times than the amino acids themselves:

$$\underset{\displaystyle \overset{R}{\underset{O}{H_2N-C-C-OH}}}{} \longrightarrow \underset{\displaystyle \overset{R}{\underset{O=C}{H-N-C-C-OR'}}}{\underset{CF_3}{}}$$

The CF_3 amide is more volatile than the CH_3 analog owing to the greater electronegativity of fluorine: The tendency for the $>C=O$ group to accept hydrogen bonds is lessened.

The Separation Factor

Separation of several substances occurs when the respective values of either K or V_g are different. A measure of this difference is the *selectivity factor* (or *separation factor*) α (see Eqn. 21.10). For two substances indicated by subscripts 1 and 2, we have

$$\alpha = \frac{V_{g_2}}{V_{g_1}} = \frac{k'_2}{k'_1} = \frac{K_2}{K_1} = \frac{t'_{r_2}}{t'_{r_1}} \tag{22.8}$$

For instance, if solute 1 is a hydrocarbon and solute 2 an alcohol, a polar stationary phase such as the polyethylene glycols or the polar silicones (Table 22.1) will increase the retention of the alcohol because of the strong specific interactions involving the $-OH$ group. This is an example of stationary-phase selectivity.

Extreme selectivity is required for the direct chromatographic resolution of enantiomers. Such molecules differ only in their "handedness"; stationary phases must be designed to take advantage of this single distinguishing property. The enantiomers of amino acids are resolved as their ester-*N*-trifluoro-acetamides by using the

TABLE 22.1. *A Tabulation of Commercial Stationary Phases and Their Skeletal Structures*

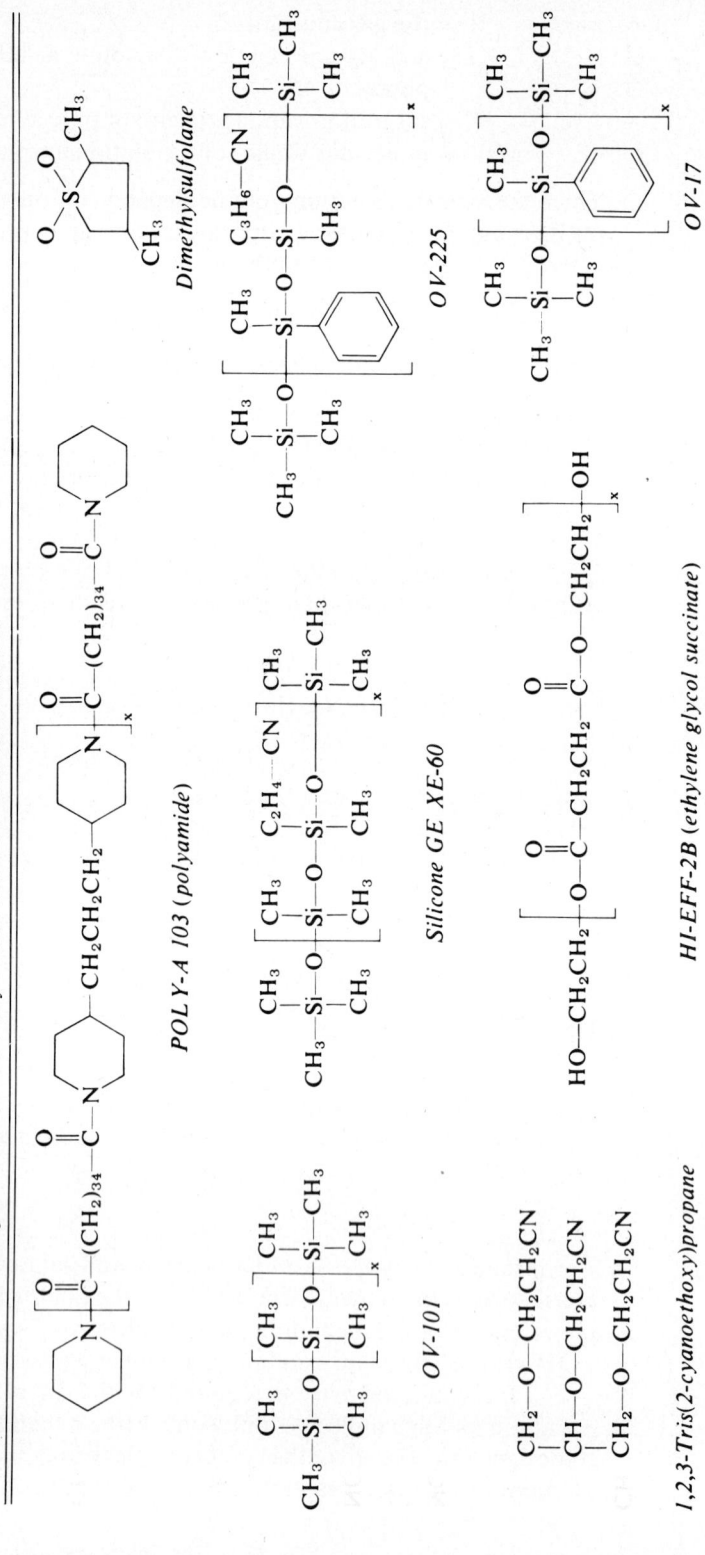

Source: Courtesy of Applied Science Laboratories, State College, Pennsylvania, Catalog 17 (1974), p. 12.

stationary phase *N*-trifluoroacetyl-L,L-valylvaline cyclohexyl ester. There the separation mechanism involves the formation of transient diastereomeric complexes of R-S and S-S configuration.

At an intermediate level of selectivity, adding certain metal ions, especially Ag^+, to a stationary phase significantly increases the retention of alkenes and the α-values for geometrical (*cis-trans*) isomers. The mechanism probably involves π complexes formed between Ag^+ and the unsaturated C=C double bond.

To achieve good separation, α needs to be either large or small in magnitude compared to unity. Thus, one desires a large difference in P^0 (by selective derivatization or varying the temperature) or a large difference in γ_∞ (Eqn. 22.7). The latter can often be achieved by invoking the rule that "like dissolves like"; but, at times finding a suitably selective liquid stationary phase can require a great deal of experience, since there are systems whose interaction mechanisms are not known.

For chemically dissimilar solutes, the enthalpies of vaporization are often quite different, and so α can depend strongly on temperature. For example, at low temperatures, compound 1 will elute before compound 2; therefore, the separation factor as defined in Equation 22.8 is greater than 1.0. But, as the temperature is increased, the difference in retention time (specific retention volumes V_{g_1} and V_{g_2}) becomes zero. At even higher temperature, compound 2 can elute before compound 1, and α becomes less than 1.0 and continues to decrease with increasing temperature. Such a reversal of elution order is not uncommon for dissimilar molecular species, and changing the column temperature is sometimes one way of "moving" an interferent peak away from the analyte peak to permit more accurate analyses.

22.2 THE DYNAMICS OF GAS CHROMATOGRAPHY

The successful separation of two substances depends not only on the separation factor α but also on the quality of the column in terms of performance or efficiency. The latter is described in terms of the *height equivalent to a theoretical plate* (*HETP*) or *plate height* (*H*). The plate height (more precisely, the number of plates) for a given solute on the column is related to the variance σ^2 of the chromatographic zone. This total variance is the sum of many contributing factors, but three general areas of variance production can be identified—the inlet system, the column, and the detector —and

$$\sigma^2_{total} = \sigma^2_{in} + \sigma^2_{col} + \sigma^2_{det} \qquad (22.9)$$

The standard deviation σ of a chromatographic zone with (approximately) a gaussian profile is related to the width of the zone at the inflection points. As in any error process, the square of the standard deviation (the variance) accumulates and hence Equation 22.9 follows. It is useful to keep in mind that in column-chromatographic methods, σ is a volume element and that increasing extra-column "dead volume" (for example) will increase σ^2_{total} and decrease the total number of plates.

Broadening Factors Affecting Column Performance

The analyst should understand the physical basis for zone broadening in gas chromatography in order to properly select the right column system for the job to be done. In the following discussion, the term σ_{col}^2 of Equation 22.9 will be expanded into separate factors, each factor adding to the plate height H. The broadening factors to be considered are:

1. Finite rate of diffusion of the solute vapor in the mobile (gas) phase along the length of the column.
2. Noninstantaneous equilibration of the solute vapor with the stationary solvent phase.
3. Factors that depend on the geometry of the column packing.

This is not a complete list, but it is sufficient for the present purpose.

It is important that the reader have a clear mental picture of an actual chromatographic column before proceeding in this section. Packed columns are similar to those discussed in Chapter 21. Open tubular columns are practically unique to gas chromatography, and the reader may wish to go forward to the section on columns before continuing here. These columns are open tubes coated on the inside with a film of stationary phase, the advantage being that very long columns (large numbers of theoretical plates) can be attained with a small pressure drop.

The overall plate height arising from the above factors is described by the *Van Deemter equation* as a function of carrier-gas velocity v:

$$H = A + \frac{B}{v} + (C_{liq} + C_{gas})v \qquad (22.10)$$

where
$$A = 2\lambda d_p$$
$$B = 2\gamma D_g$$
$$C_{liq} = \frac{8}{\pi^2} \frac{k'}{(1+k')^2} \frac{d_f^2}{D_l}$$
$$C_{gas} = \frac{d_p^2}{D_g}$$

The A term is the eddy diffusion contribution and has the form $2\lambda d_p$, where d_p is the diameter of the particles packed in the column and λ is a geometric factor indicating how uniformly the column is packed. Variable A represents the distance a flowing stream moves before its velocity is seriously changed by the packing; A is independent of the velocity of the gas.

The B term is the longitudinal or molecular-diffusion contribution, which is a function of the diffusion coefficient D_g of the solute vapor in the carrier gas, and a correction factor γ, which accounts for the tortuosity of the gas channels in the column. As a limiting condition, γ approaches 1 for large particles; that is, the flow of the gas through the column approaches a straight-line path. If a sample could be placed on the column as a zone of infinitesimal width at a time $t_0 = 0$, diffusion would cause the zone to become wider and less concentrated as time goes by, even with no flow. Diffusion is much more important in gas chromatography than in liquid chromatography (H_{ld} in Eqn. 21.8) because the diffusion coefficients of solutes are approximately 10^5 greater in gases than in liquids.

The C terms represent the rate of mass transfer—the finite time required to establish equilibrium between the two phases. The C_{liq} term is a function of the capacity (partition) factor k' (Chap. 21), the film thickness d_f, and the interdiffusion constant D_l of the solute in the liquid stationary phase. The C_{gas} term is related to d_p^2/D_g in packed columns or r^2/D_g in open tubular columns. The C_{liq} and C_{gas} terms are often combined into a single C term, and represent the kinetic lag in attaining equilibrium between phases as well as transverse diffusion within the mobile phase itself. The B and C terms depend on the carrier-gas velocity in opposite manners; B decreases with increased velocity whereas C increases.

The resulting equation and its components are shown in Figure 22.2. The sum curve is quasi-hyperbolic, exhibiting a minimum value of H (or $HETP$) at an optimum carrier velocity (v_{opt}); $H_{min} = A + 2\sqrt{BC}$ and $v_{opt} = \sqrt{B/C}$. Equation 22.10 can be compared with Equation 21.8 for liquid chromatography. Also, as in Equation 21.6 used in liquid chromatography, $N = 16(t_r/t_w)^2$ in gas chromatography.

The Van Deemter equation contains important information for the practical analyst planning to use gas chromatography. The A term suggests that the use of smaller-sized supporting particles will result in a smaller H. In practice, particles with a mesh size of greater than 100 to 120 (i.e., smaller particles) are not used, since decreasing d_p increases the pressure drop in the columns.* Narrow mesh ranges produce a more uniform packing geometry, also resulting in a smaller H. Efficiency increases with decreasing column radius, but it is difficult to pack columns less than 3 mm i.d. Diffusion in the gas phase will be reduced by using carrier gases of higher molecular weight or by operating the column at increased pressure. The combined C factor suggests that the plate height depends on the partition ratio, and therefore that the observed plate height is different for different compounds. Because small ratios between the volume of gas and the volume of liquid lead to small values of H, as much stationary phase as possible should be used. When the available surface area of the support has been covered, however, the film thickness d_f increases (and along with it H, since it is proportional to d_f^2), thus degrading column performance—an important consideration in scaling up from lightly loaded analytical columns to heavily loaded "preparative" columns. The stationary phase should not be viscous at the column temperature; high viscosity means small D_l, which increases H. A compromise must be made between capacity and sample size: The amount of stationary phase must be large enough, and the operating temperature low enough, to produce significant retention times, but not so large or low as to broaden zones excessively by producing large mass-transfer effects.

The Resolution Factor

The separation efficiency of a particular pair of components is described by the resolution obtained, as determined by the peak-to-peak separation α and the average

---

*Supports are usually sized by screening through standard ASTM screens. Mesh numbers refer to the number of openings per linear inch. Particles that will pass through 60 mesh, but not through 80 mesh, are referred to as 60/80 mesh.

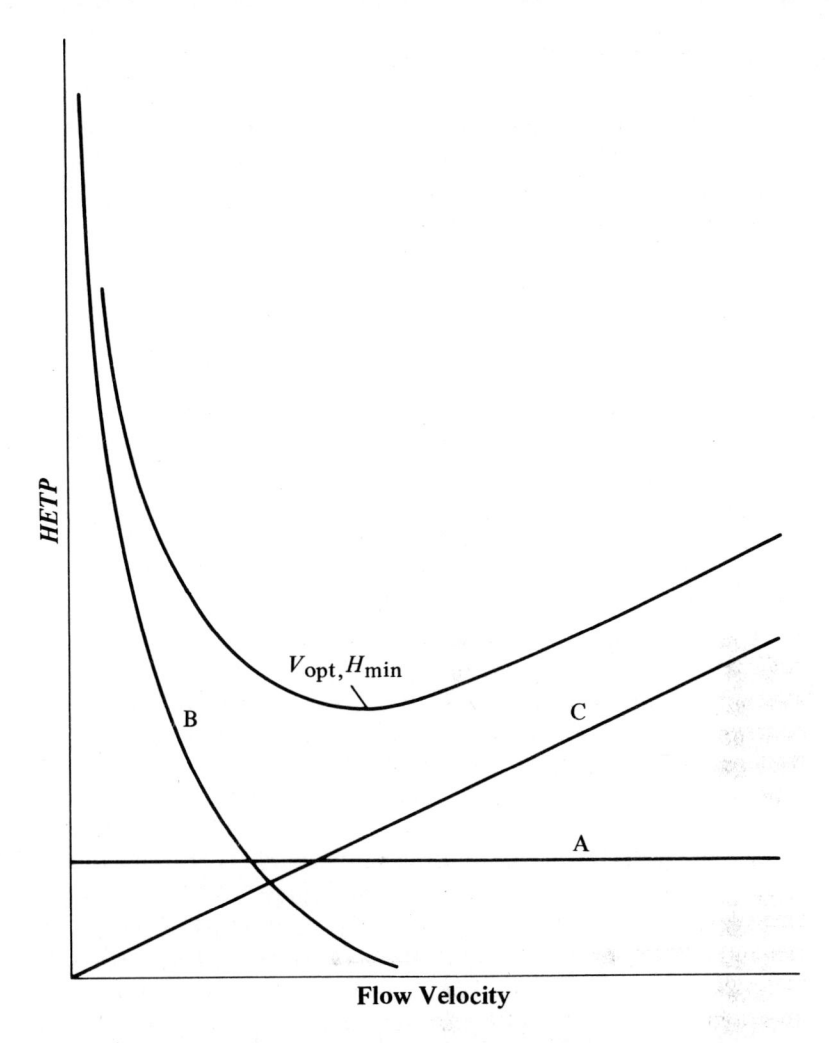

FIGURE 22.2. Plot of HETP (height equivalent to a theoretical plate) against flow velocity, illustrating the contributions of the following factors to plate height and the position of optimum velocity for minimum HETP: (A) eddy diffusion, (B) ordinary diffusion, and (C) combined gas and liquid resistance to mass transfer.

peak width at the baseline V_w, given in the same units as the retention volumn. A resolution factor R for substances 1 and 2 can be defined as follows:

$$R = \frac{V_{r_2} - V_{r_1}}{0.5(V_{w_1} + V_{w_2})} = \frac{V_{r_2} - V_{r_1}}{4\sigma} \tag{22.11}$$

This equation is similar to Equation 21.9 and either equation applies—that is, either volumes or times can be used (see Eqn. 22.1). $R = 1$ corresponds to reasonably good separation, since there are 2σ units between zone centers and thus only 2% of each zone overlaps the other. At $R = 1.5$, there is a *baseline separation* with a zone overlap of less than 1%. The effective resolution also depends on the relative concentrations of the solutes.

Equation 22.11 does not show the relationship between experimental variables and the quality of a given separation. A more fundamental equation was given in the previous chapter (Eq. 21.11):

$$R = \left(\frac{\sqrt{N}}{4}\right)\left(\frac{\alpha - 1}{\alpha}\right)\left(\frac{k'}{1 + k'}\right) \tag{22.12}$$

This shows the relationship between the number of theoretical plates N, the separation factor α, and the capacity factor k' for the second solute in determining R. The analyst is often interested in R but more often in N_{req}, the number of plates required to give a certain R-value. For $R = 1$, Equation 22.12 can be rearranged to yield

$$N_{req} = 16\left(\frac{\alpha}{\alpha - 1}\right)^2\left(\frac{k' + 1}{k'}\right)^2 \tag{22.13}$$

Usually, $k' = 2-3$ is optimum. The effect of k' on resolution is important only for fast-moving peaks with packed columns where k'-values range from 2 to 200.

The whole advantage of open tubular columns is lost if k' has a fractional value (the usual range of k'-values being 0.2 to 20). As an example, consider the case of a vapor with $k' = 0.2$ on an open tubular column and $k' = 2$ on a packed column. The packed column would require 1/16 the number of plates to give the same resolution. Many examples of the performance of open tubular columns show capacity ratios even smaller than 0.2, so that the resolution is poorer than could be obtained on a short packed column of only a few thousand plates. Actually, it is often convenient to speak in terms of *effective* plates, N_{eff}, when comparing columns, where $N_{eff} = N(k'/1 + k')^2$. For example, open tubular columns (because of their greater length) have much larger values of N than do packed columns, but the values of N_{eff} are often comparable. In other words, the plate heights of open tubular columns are longer than those of properly prepared packed columns.

Optimizing Speed in Chromatographic Analysis

An analyst seeks to obtain chromatographic results in the minimum possible time; thus, he or she must compromise between larger k'-values and shorter times of analysis. This is simple if the initial value of R is greater than 1.5, since increasing the carrier velocity will shorten the analysis without degrading the results. Since $v = 1/t_m$ and $H = L/N$ (where L = total length of the column), then $t = [NH(1 + k')]/v$; combining this with the expression for the resolution yields a relationship between analysis time and resolution,

$$t = 16R^2\left(\frac{\alpha}{\alpha - 1}\right)^2\frac{(k' + 1)^3}{(k')^2}\frac{H}{v} \tag{22.14}$$

which indicates that if double the resolution is required (k', α, H, and v being constant), then the analysis time is increased by a factor of 4. If k' is either very small or very large, then t tends toward very large values. The minimum in the expression occurs at $k' = 2$. This corresponds to a retention time for the second peak that is three times that of an unretained species (*air peak* or *methane time*).

The value of R has a practical significance that depends on the relative concentrations of adjacent zones. For example, quantitative analysis is always possible for $R = 1.5$ regardless of relative concentrations; but for $R = 0.6$, it is difficult to detect two peaks at concentration ratios smaller than 1/8. An approach that takes into account the degree of zone overlap is that of Glueckauf [3].

22.3 GAS-CHROMATOGRAPHIC INSTRUMENTATION

Gas-chromatographic instrumentation differs very little from that used for other forms of column chromatography (see Fig. 22.3). A gas chromatograph consists of (a) a source of carrier gas, the flow rate of which can be fixed at a desired magnitude within the range provided; (b) an inlet that can be heated (25 to 500°C); (c) a column in a thermostatted air bath (25 to 400°C); and (d) a detector suitable for vapor-phase samples. The high temperatures are needed to vaporize the solutes of interest and

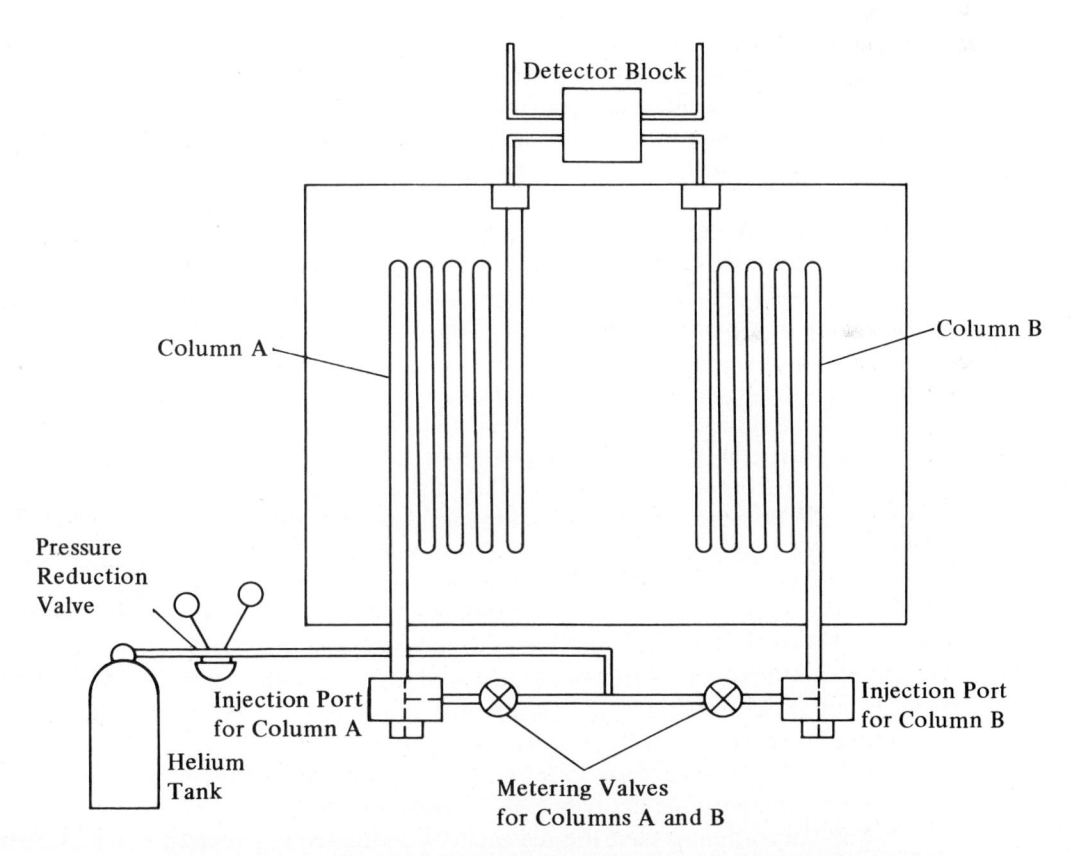

FIGURE 22.3. *Block diagram of a dual-column gas chromatograph showing essential parts. Courtesy of Gow-Mac Instrument Co., Madison, New Jersey.*

maintain them in the gas phase. Because the distribution coefficient depends on the temperature, the latter is controlled to between ± 0.1 and $\pm 0.01\,°C$ (depending on the precision desired in the measured retention times). The inlet and detector are generally maintained at a temperature approximately 10% (in °C) above that of the column (in any case, above 100°C for flame-ionization detectors; see later) to ensure rapid volatilization of the sample and to prevent condensation. The temperature of the column is usually set at least 25°C higher than the boiling point of the solute. (This is not, of course, an absolute requirement, since it is only necessary that a substance have a reasonably high vapor pressure at the operating temperature.)

Columns for Gas Chromatography

The most commonly used gas-chromatographic column consists of a tube filled with solid particles of fairly uniform size; the particles are coated with the liquid stationary phase. Perhaps the most commonly used support is marine diatomite (e.g., Johns-Manville Chromosorb®). The choice of tubing material depends on the experiment. Aluminum and copper are commonly used, but may have chromatographically and catalytically active oxide films that make them undesirable for sensitive compounds (e.g., steroids); in such cases, stainless steel or glass is used (the latter is more inert, but is less conveniently manipulated).

Open tubular columns consist of 25–100-m lengths of 0.3–0.6-mm-i.d. steel, glass, or quartz (fused silica) tubing coated on the inside wall with a film of stationary phase. These are called wall-coated open tubular (WCOT) columns. The fused-silica types have become fairly popular because of the relative inertness of the inner surface, which results in reduced tailing of more polar solutes, and because of the high mechanical flexibility of such columns, although this contributes nothing to the chromatographic characteristics of the column. The flexibility is the result of the polymer coating on the outside of the silica tubing, which excludes moisture and prevents hydration and cracking of the otherwise thin wall. There are problems with coating polar stationary-phase films onto silica columns, but coatings can be prepared by forming a "bonded phase" in place using chemistries analogous to those used for preparation of "bulk" bonded phases for liquid chromatography.

Capillary gas chromatography was proposed and demonstrated in the first decade of gas chromatography. It was then virtually ignored until the 1970s, when a combination of patent maturation and the need to separate more and more complex mixtures (especially for environmental and biomedical studies) spurred commercial exploitation. The improvement in commercial instrumentation followed the research of the earlier workers and little or no fundamental advances were involved. Most of the changes centered on (a) reduction of dead volumes in inlet and detector designs; and (b) better, more uniform, temperature-control ovens and improved linearity in temperature programming. The vast majority of capillary applications are temperature programmed with injection of the sample onto a relatively cold column. Injection onto a cold column tends to focus the injection band at the column head and reduces the apparent inlet contribution to band broadening.

Capillary columns have limited capacity compared to packed columns, that is, less stationary phase and therefore less absolute retention. As a result, many examples of capillary gas chromatography involve the determination of minor-level, rather than trace-level, components in the solutions injected. (Through sample-preparation procedures, trace components in the actual analytical sample can still be determined by appropriate extraction and concentration methods, but the components are in relatively high concentration in the injected solution.) Because of the low capacity of WCOT columns, and for other reasons, surface-coated open tubular (SCOT) columns have been introduced. These columns are internally coated with finely divided metal oxide, graphite, or alumino-silicate before the stationary phase is applied. As a result, they have somewhat increased capacity because of the larger surface area presented.

The major advantage of capillary columns is not in plate height, which is generally larger than with well-packed columns, but in the number of plates achievable with a relatively small pressure drop. For example, if 20,000 theoretical plates is a good upper limit for packed columns, then open tubular columns can have 75,000 to 150,000 plates. Capillary gas chromatography is a technique complementary to the use of packed columns. The latter are to be preferred when available resolution is adequate and the highest quantitative precision is desired.

Chromatographic Support Materials

The function of a chromatographic support is to hold the stationary phase. One useful type of support is provided by the marine diatomites, which are the skeletons of tiny unicellular algae (diatoms) and consist chiefly of amorphous hydrated silica with traces of metal-oxide impurities. This material has the advantages of high porosity and large surface area. Some properties of a variety of diatomite supports are given in Table 22.2. Chromosorb P, for example, is prepared from one particular grade of firebrick and is a pink (hence P), calcined diatomite that is relatively hard

TABLE 22.2. *Properties of Some Diatomite Supports*

Properties	Chromosorb® A	Chromosorb® G	Chromosorb® P	Chromosorb® W
Color	Pink,	Oyster White,	Pink,	White,
Type	Flux-Calcined	Flux-Calcined	Calcined	Flux-Calcined
Density, g/cm^3				
(i) Loose weight	0.40	0.47	0.38	0.18
(ii) Packed	0.48	0.58	0.47	0.24
Surface area, m^2/g	2.7	0.5	4.0	1.0
Surface area, m^2/cm^3	1.3	0.29	1.88	0.29
Maximum liquid-phase loading	25%	5%	30%	15%
pH	7.1	8.5	6.5	8.5
Handling characteristics	Good	Good	Good	Slightly Friable

Source: Courtesy of Johns-Manville Corporation.

and not easily friable. It is used mainly with solutes of low to moderate polarity (e.g., hydrocarbons). It is a relatively good adsorbent, a quality that can be an interference. If there were no liquid phase at all, the support would act as an adsorbent, and gas-solid chromatography could be carried out. The effect of placing a thin film of liquid on an active adsorbent is to moderate the gas-solid activity, but not to eliminate it. It has been shown that even 20% by weight liquid loading does not eliminate this activity. Several techniques are used to reduce the activity—for instance, acid washing, and "silanizing" the active silica sites with dimethyldichlorosilane to displace the hydrogen. The effect of these treatments on chromatograms is shown in Figure 22.4. The choice of a support for a given analysis is as important as the choice

Figure 22.4. *Effect of treatment on support activity (Chromosorb® P). 1. Nonacid-washed; 2. acid washed; 3. acid washed, dimethyldichlorosilanized. Solutes: (A) ethanol, (B) methylethyl ketone, (C) benzene, (D) cyclohexane. Conditions: 6' × 1/4" column, 60/80 mesh support, 75 mL/min He flow, 100°C (no liquid coating). Courtesy of Johns-Manville Corporation.*

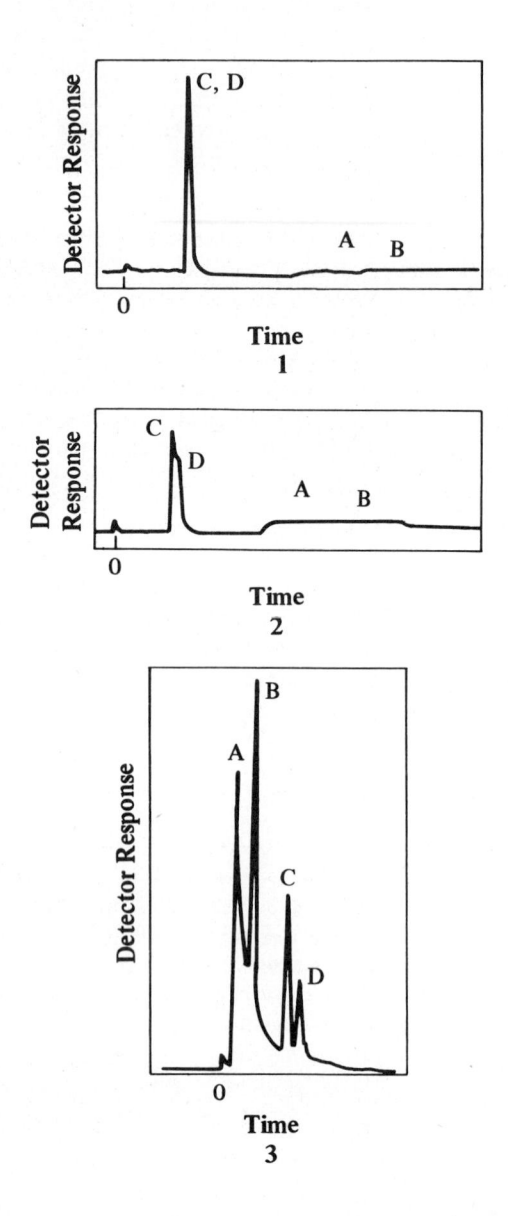

of a stationary phase; for instance, if retention is partly due to solution in the stationary phase and partly to adsorption on the support, then the retention time will vary with the size of the sample. Some compounds (e.g., sterols) may actually decompose on the column if a poor choice of support has been made.

Stationary Phases

The selection of the stationary phase is also important. Several guides are available in which the types of solute and of stationary phase are correlated (see the Selected Bibliography). Although such information is useful, the phases listed are not necessarily the best ones; the lists merely indicate that a given stationary phase has been used with some success for a given class of compounds. The situation is complicated by the fact that about a thousand stationary phases have been reported in the literature. Many attempts have been made to glean a list of standard phases from these reported phase materials, many of which give substantially the same separations and possess unique stability. Many researchers have developed various retention-index methods to indicate the preferred phases for a given separation. Such studies involve investigating specific solute-solvent interactions fundamental to our understanding of stationary-phase selectivity. (Note particularly the recent work of Rohrschneider and of McReynolds [4,5].) Table 22.3 contains a list of 12 stationary phases and the compound classes that can be separated by each. The list was compiled [6] by applying the following criteria: Phases should be (a) well tested, (b) readily available, (c) stable over a wide range of temperatures, and (d) cover a wide polarity range. Of the several hundred phases studied, this list is presented as a preliminary guide, but of course the infinite number of possible combinations of solutes will not always be separated using phases from this list. Squalene is a hydrocarbon; SE-30 is a methylsilicone rubber with hydrocarbonlike properties. These are selective for nonpolar species, but will also separate alcohols with sufficiently different boiling points. Carbowax 20M is moderately polar, and is selective for hydrogen-bonded species such as alcohols. Many synthesis laboratories do 90% of their work with SE-30 and Carbowax 20M. As often as not, a significant difference of vapor pressure exists between starting material and product, and little selectivity is required. The result is that two phases of markedly different polarity may serve quite well.

Preparing the Column

The three most critical steps in the actual preparation of a gas-chromatographic column are: (a) coating the support, (b) packing the column, and (c) curing or condition after packing but before use. Highly efficient columns give lower limits of detection and shorter analysis times; a common goal, though rarely achieved, is a column with more than 3000 plates/m.

There are some guidelines, however, by which columns with more than 2000 plates/m can be prepared consistently. Uniformity of particle size is achieved by using a narrow range of mesh size, by carefully removing any *fines* (particles of very small size), and by not producing more fines by rough handling in the coating and packing processes. For analytical columns of 2 to 3 mm i.d., a mesh size of 100 to 120 is desirable.

TABLE 22.3. *A Proposed List of Standard Stationary Phases*

Preferred Phase	Structure	Temp Limit, °C	Uses
Squalene	2,6,10,15,19,23-Hexamethyltetracosane	150	Hydrocarbons, gases
SE-30	Polydimethyl siloxane	350	Gases, hydrocarbons, aldehydes, ketones (b.p. separ.)
OV-3	Polyphenyl methyl dimethyl siloxane	350	Alcohols, fatty acids, esters, aromatics
OV-7	Polyphenyl methyl dimethyl siloxane	350	Aromatics, heterocyclics
DC-710	Polymethyl phenyl siloxane	300	Aromatics (similar to OV-17)
OV-22	Polyphenyl methyl diphenyl siloxane	350	Alcohols, aromatics
QF-1	Polytrifluoropropyl methyl siloxane	250	Alcohols, amino acids, steroids, nitrogen compounds
XE-30	Polycyanomethyl siloxane	275	Drugs, alkaloids, halogenated cmpds.
Carbowax 20M	Polyethylene glycol	250	Alcohols, esters, pesticides, essential oils
DEG adipate	Diethylene glycol adipate	200	Fatty acids, esters, pesticides
DEG succinate	Diethylene glycol succinate	200	Steroids, amino acids, alcohols
TCEP	*Tris*-cyano ethoxy propane	175	Alcohols, steroids, pesticides

Source: Adapted from J. Leary, J. Justice, S. Tsuge, S. Lowry, and T. L. Isenhour, *J. Chromatog. Sci.*, *11*, 201 (1973), by permission of the senior author and the publisher.

Coating the support can be carried out by many methods; the least desirable is using a rotary evaporator in the drying step, which tends to agitate the packing and produce fines. A reliable method is as follows:

1. Dissolve the liquid phase in a solvent in a flask. (A suitable solvent can be found in any of the suppliers' catalogs. Select one that does not boil under the vacuum to be used.)
2. Add the cooled solution to the support, swirling gently to ensure wetting.
3. Stopper, and apply vacuum to remove air from the pores of the support. When no more air bubbles escape, seal the vacuum and hold for 5 min.
4. Release the vacuum, transfer the support to either a glass funnel with a coarse-porosity frit in the neck or to a fluidized-bed dryer, and immediately suck the solution off.

5. When the solution ceases to drip out, fluidize the bed of coated packing and dry with a gentle flow of hot nitrogen gas. (The packing is dry and ready for filling into the column when no odor of solvent remains.)

The coated packing is transferred, a little at a time, into the column with the aid of vacuum and gentle tapping. Electric vibrators are popular, but their use will degrade column performance by producing fines. The key factors are uniformity of the packing material and gentle tapping. To condition the packing, heat the column slowly to the upper working limit and maintain this temperature, along with a small flow of carrier gas, for several hours. This distributes the liquid evenly over the surface of the support. Many of the modern polysiloxanes do not require extensive conditioning. The most important consideration in obtaining good column performance and long life is to avoid overheating.

Column Inlets

As in liquid chromatography, the gas-chromatography sample is introduced to the column through a specially designed inlet, generally by injecting it in nanoliter amounts through a rubber septum with a microliter syringe. The inlet should be hot enough to flash evaporate the sample, and large enough in volume to allow the sample vapor to expand without blowing back through the septum. Two types of inlet are shown in Figures 22.5 and 22.6, which are illustrative of the component design of modern inlets.

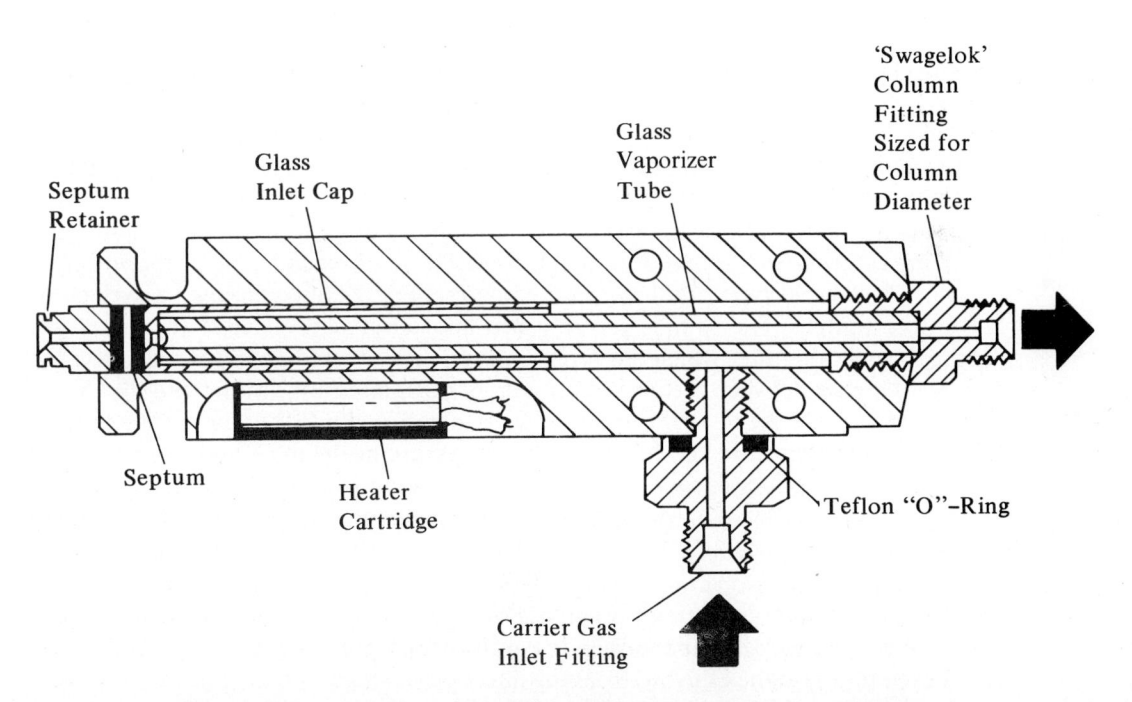

FIGURE 22.5. *Diagram of glass-lined flash-evaporation inlet. Courtesy of Hamilton Co.*

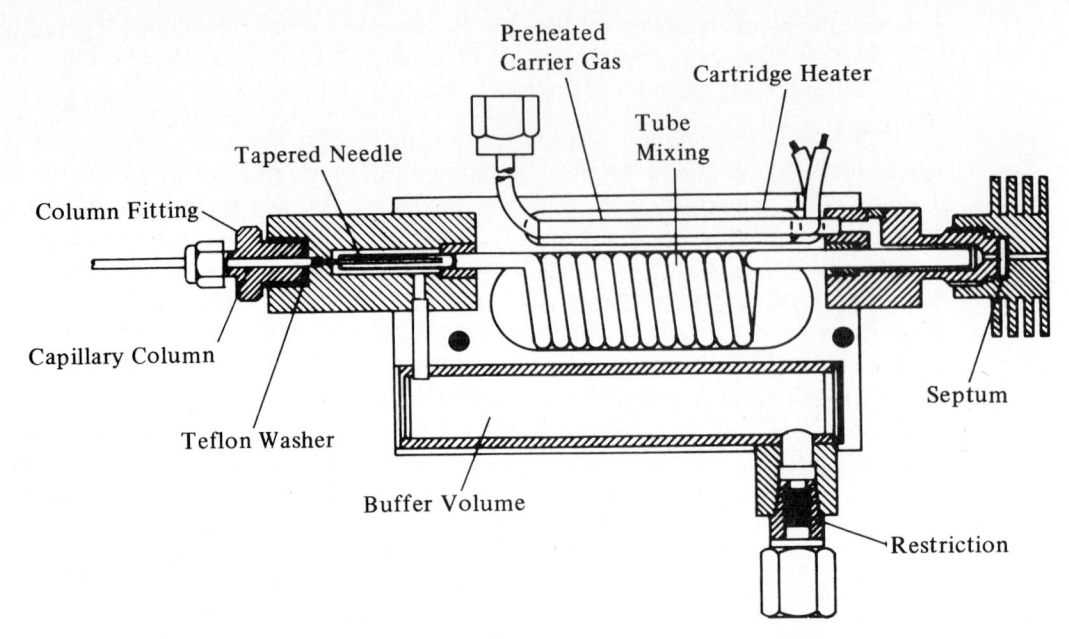

FIGURE 22.6. *Diagram of an inlet splitter for open tubular or small-diameter packed columns. Courtesy of Hamilton Co.*

The first (most common) type is the *flash-vaporization inlet* (Fig. 22.5), an arrangement of concentric glass tubes that washes the septum area with a high-velocity, preheated, carrier-gas stream, thus preventing blowback. The gas passes to the column through a 1- or 2.5-mm glass vaporizer tube heated by a cartridge heater mounted in the body of the inlet. A glass lining is preferred for samples that might decompose in an all-metal inlet. A slight modification in the position of the septum makes it possible to inject samples almost directly onto the column, for studying compounds that are too thermally labile to withstand flash evaporation.

The second type of inlet, called a *splitter* (Fig. 22.6), again uses a syringe to inject the sample. The sample is vaporized and effectively mixed with carrier gas in the mixing tube, after which the vapor passes over a tapered hollow needle. Because of the difference between the inside diameter of the mixing-tube outlet and that of the tapered needle, a fraction of the total sample is introduced into the column as a narrow zone. The splitting ratio is a function of the pressure in the inlet, and can be varied. Such a sample splitter is used with capillary columns and high-resolution columns because of their relatively low capacity. Problems involving fractionation of sample components at the needle split point exist with splitter-type inlets. A large variety of mechanical and fluidic variations have been introduced to improve the quantitative precision of capillary chromatography. Two such approaches are the *splitless* and *direct* injection modes. In the former, the entire evaporated sample is kept in the inlet for a fixed time and then flushed from the inlet by carrier gas. Direct injection involves placing the sample into the capillary column itself. Improvements in mechanical design make this otherwise difficult task somewhat easier than was originally the case. Direct injection, of course, has the advantage of placing all the sample onto the column and, thus, offers the best quantitative precision.

Detectors

The three most common detectors in gas chromatography are those using thermal conductivity, flame ionization, and electron capture. The first is also the oldest; it measures heat conductivity, which is different for different gases. The second and third types respond to changes in electron currents; the electrons are produced in a flame by burning the sample, or by exposing the sample to a radioactive source.

Thermal Conductivity Detector. The thermal conductivity detector (TCD) is a simple universal detector (see Fig. 22.7) that produces a large signal requiring no amplification. The detector cell has either two or four filaments arranged in a Wheatstone bridge circuit (Fig. 22.8). In the four-filament model, two filaments in opposite arms of the bridge are surrounded by carrier gas flowing in a reference stream, the

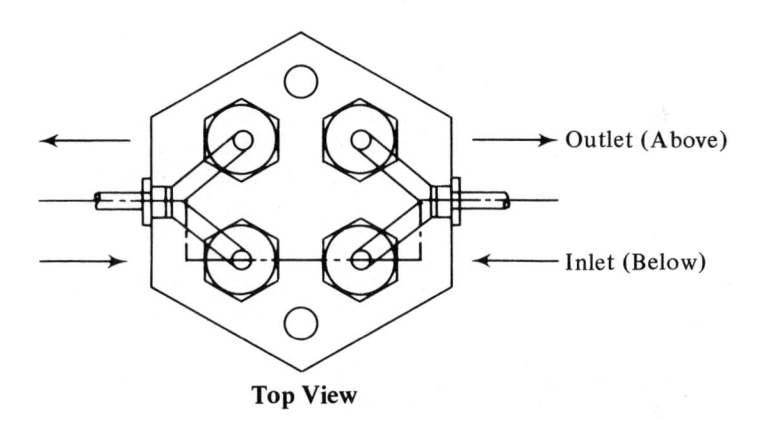

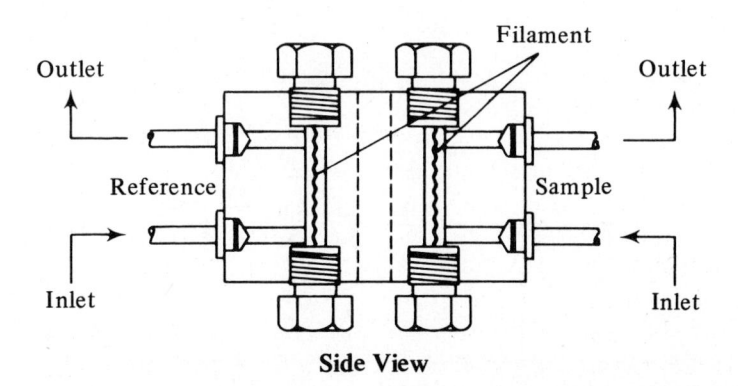

FIGURE 22.7. *Cross section of a typical four-wire conductivity cell. Courtesy of Gow-Mac Instrument Co.*

other pair by carrier gas flowing out of the column. When the bridge is balanced, no signal appears across points 1 and 2 in Figure 22.8. Since the temperature of the filaments is proportional to the rate at which heat is transported to the cell body by the gas, and since resistance is proportional to temperature, a change in the heat conductivity of the gas will produce an output signal at points 1 and 2. Most organic vapors have low thermal conductivities (λ) compared to hydrogen or helium ($\lambda_{acetone} = 2.37 \times 10^{-5}$, $\lambda_{H_2} = 41.6 \times 10^{-5}$, $\lambda_{He} = 34.80 \times 10^{-5}$ at 0°C). For this reason, helium is widely used as a carrier gas; but if one were interested in analyzing the noble gases, nitrogen might be the carrier of choice. The TCD is reliable, simple, nondestructive, and moderately sensitive; it responds to essentially all compounds, and is widely used in preparative work. Since it is nondestructive, the solutes can be collected (e.g., in a dry ice–acetone bath) for further examination by other means, such as infrared spectroscopy. Relative responses vary widely and are frequently nonlinear with concentration, so quantitative analyses require careful calibration. The analyst is rarely justified in merely taking peak ratios as an accurate indication of relative amounts. This detector has a concentration detection limit of about $5-10 \times 10^{-6}$ g/mL of eluant gas, and a dynamic (working) range of about 10^5.

Flame-Ionization Detector. The flame-ionization detector (FID) has a wide linear range and high sensitivity, and is quite reliable. It consists of a hydrogen-air flame polarized in an electrostatic field (Fig. 22.9). The flame ignites and ionizes the combustible sample components as the carrier gas passes into it, after which the ions (primarily carbon compounds) are collected at the electrodes, producing a current. The FID does not respond fully to oxygenated carbons such as carbonyls, carboxylic acids,

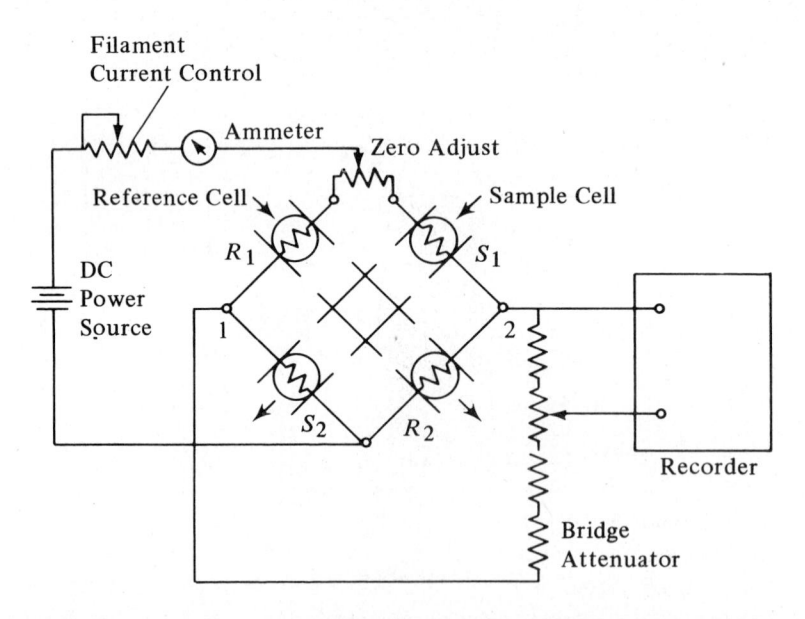

FIGURE 22.8. *Typical bridge configuration for thermal conductivity detection. Courtesy of Gow-Mac Instrument Co.*

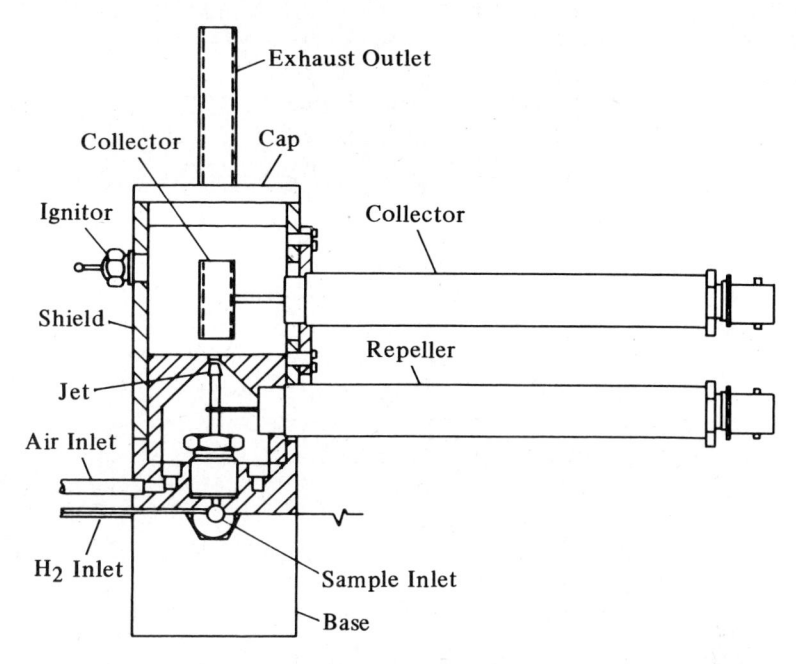

FIGURE 22.9. *Flame-ionization detector. Courtesy of Gow-Mac Instrument Co.*

or their sulfur analogs (e.g., cyclohexane and cyclohexanone have different response factors). However, it does not respond at all to water or to the permanent gases (N_2, O_2, CO_2, etc.), making it ideally suited for trace analysis in aqueous solutions and atmospheric samples. Response is proportional to the number of carbon atoms, but diminishes with increasing substitution by halogens, amines, hydroxyl groups, or any electron-capturing species. The limit of detection is $1-5 \times 10^{-9}$ g/mL of sample gas, with a dynamic range of 10^8. Sample collection is possible if the column effluent is split into two streams.

Electron-Capture Detector. The electron-capture detector (ECD) takes advantage of the affinity of certain functional groups for free electrons (the reason for loss of sensitivity in the FID). The principle is almost identical to flow-through proportional counting of a radioactive source. The carrier gas is passed through a cell containing a beta source (e^- for nuclear decay), which ionizes the carrier gas. The source can be a Pt foil saturated with $^3H_2$, but a ^{63}Ni foil is used more frequently because of its higher temperature stability. Some typical carrier gases are He-CH_4, N_2-CH_4, and Ar-CH_4. The beta particles ionize the carrier molecules and produce electrons, which migrate to the anode (Fig. 22.10) under an applied potential of 1 to 100 V. An electron-capturing species eluting from the column will react with the electrons to form an ion or neutral molecule, which is swept from the cell. The net result is a reduction in the number of electrons found at steady state or a drop in the *standing current*. A "peak" in ECD detection is therefore actually a detector-current "valley," since the maximum current is found in the absence of capturing species. Response is very

FIGURE 22.10. *"Pin-cup"* design for electron-capture detection, in cross section.

nonlinear, but a linear range of $0.5-1 \times 10^3$ can be achieved by pulsing the polarizing voltage. The pulse duration is long enough for electron collection, but not for ion collection. The limit of detection is about 1×10^{-12} g.

The major advantage of the ECD, however, is its selectivity. The ECD is insensitive to amines, alcohols, and hydrocarbons, but very sensitive to halogens, anhydrides, peroxides, ketenes, nitro groups, and so forth, with selectivity ratios of $10^5:1$ being not uncommon in practice. Table 22.4 lists electron-capturing compounds

TABLE 22.4. *Electron Absorption Coefficients of Various Compounds and Classes of Compounds for Thermal Electrons*

Electron Absorption Coefficient[a]	Compounds and Classes	Electrophores
0.01	Aliphatic saturated, ethenoid, ethinoid, and diene hydrocarbons; benzene; cyclopentadiene	None
0.01–0.1	Aliphatic ethers and esters; naphthalene	None
0.1–1.0	Aliphatic alcohols, ketones, aldehydes, amines, nitriles; monofluoro- and chloro- compounds	—OH —NH$_2$ >CO —CN Halogens
1.0–10	Enols; oxalate esters; stilbene; azobenzene; acetophenone; dichloro-, hexafluoro-, and monobromo- compounds	—CH=C—OH —CO—CO— Halogens
10–100	Anthracene; anhydrides; benzaldehyde; trichloro- compounds; acyl chlorides	—CO—O—CO— Phenyl—CO— Halogens
100–1000	Azulene; cyclooctatetrene; cinnamaldehyde; benzophenone; monoiodo-, dibromo, trichloro-, and tetrachloro- compounds; mononitro- compounds	Halogens NO$_2$ Phenyl—CH=CH—CO—
1000–10,000	Quinones; 1,2-diketones; fumarate esters; pyruvate esters; diiodo-, tribromo-, polychloro-, and polyfluoro- compounds; dinitro compounds	—CO—CO— —CO—CH=CH—CO— Quinone structure Halogens NO$_2$

a. Values are relative to the absorption coefficient of chlorobenzene, which is arbitrarily taken to be unity.
Source: From J. E. Lovelock and N. L. Gregory, in N. Brenner, J. E. Callen, and M. D. Weiss, eds., *Gas Chromatography*, New York: Academic Press, 1962, by permission of the publisher.

and relative sensitivities. It is not uncommon for FID and ECD to be combined, displaying the response of both detectors on the same chart using a two-pen recorder. The many applications of ECD include analyzing pesticides (e.g., aldrin, dieldrin, DDT, lindane) and organometallics (e.g., lead alkyls) and tracing SF_6 in flue and stack gases.

Several classes of compounds can be reacted with derivatizing reagents containing Cl or F, not only to make the product more volatile and less likely to exhibit tailing, but also to provide high electron-capture sensitivity. For example, it is fairly common to prepare the methyl esters of carboxylic acids using various reagents such as BF_3/methanol prior to chromatography. Compounds BCl_3/2,2,2-trifluoroethanol or BCl_3/2-chloroethanol can be used to esterify acids to provide good electron-capture sensitivity, Trifluoroacetic, pentafluoroproprionic, and heptafluorobutyric anhydrides can be used to derivatize carbohydrates, amino acids, alcohols, phenols, and amines. Pentafluorobenzyl bromide has been used for phenols; and condensation of a number of amines with pentafluorobenzaldehyde will produce stable imines for subsequent analysis.

Other Specific Detectors. The most elegant of the specific detectors used for gas chromatography is the mass spectrometer. (This application of mass spectrometry was discussed in Chapter 16.) Infrared spectrometry has become a practical detection method, now that rapid-scan infrared systems have made collecting samples unnecessary. In fact, part of the impetus behind developing both these techniques was the difficulty in collecting gas-chromatographic fractions; vapor samples entering cold traps from detectors at elevated temperatures tend to form aerosol fogs that do not condense on the trap walls, but are swept out by the carrier gas instead. However, at high concentrations, enough material can be collected to run remote infrared, mass spectral, or nuclear magnetic resonance spectra. The concentrations here are in the range of 10^{-3} g/sec, as compared to 10^{-12} to 10^{-9} g/sec with the directly coupled methods.

If the detector is of the FID type, flame optical emission or absorption can also be used. Commercial detectors are available that use essentially nondispersive or filter analyzers coupled to a FID. Phosphorus, sulfur, and nitrogen are commonly detected by this method. A hollow-cathode light source makes possible the detection of many organometallic compounds by atomic absorption.

22.4 QUALITATIVE AND QUANTITATIVE ANALYSIS

Qualitative Analysis

Gas chromatographic retention times are most frequently determined from the positions of peak maxima, although the thermodynamically meaningful (and analytically preferred) value is the position of the peak center of gravity. Under carefully controlled conditions, values of t_r are reproducible to better than $\pm 0.1\%$; however, agreement between retention times of a standard and an unknown peak is not conclusive evidence that they arise from the same substance. To produce more conclusive evidence, the retention times can be varied by changing the operating conditions (e.g., column material, flow rate, and temperature); if the two peaks move identically, they probably represent the same material. Most analyses are carried out under isothermal conditions, but it is possible to program the temperature to change at some predictable rate (°C/min). The latter technique can be especially valuable in separating mixtures of substances with widely varying vapor pressures, but determining a suitable programming rate can be a tedious trial-and-error process. An example of temperature programming is given in Figure 22.11. When a lower initial temperature is used, the germane (GeH_4) and arsine (AsH_3) peaks are more fully resolved, and when programming is done to higher temperatures, the stannane (SnH_4) and stibine (SbH_3) are eluted sooner and with less broadening.

It is a common practice for qualitative analysis to be based on measurements of t_r; this is especially true in those laboratories that run standards with each analysis. Nevertheless, t_r is not the ideal parameter for identification purposes because it is a function of temperature, flow rate, and liquid-phase volume. (Indeed, the liquid-phase volume is continuously changing with time because of evaporation; even the chemical composition of the liquid phase can vary under the conditions of the experiment.) What is needed, then, is a parameter that is independent of all these factors. A very successful, but not perfect, solution is the Kováts index system, which relates the

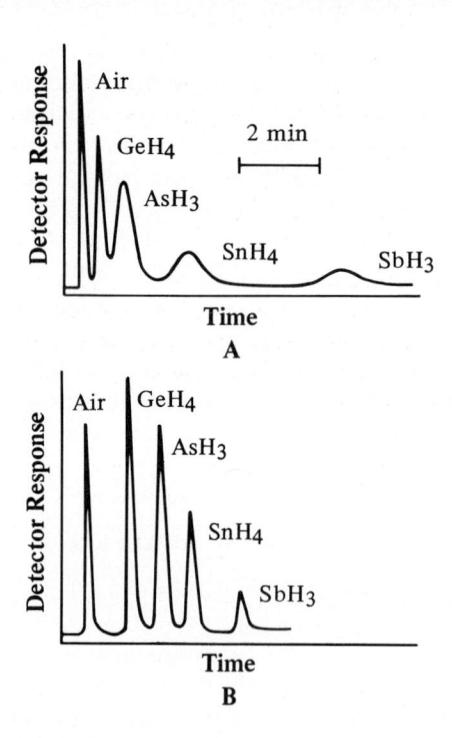

FIGURE 22.11. *Example of temperature programming to improve separations. A: Isothermal temperature, 85°C. B: Temperature programmed 8°C/min, 75° to 120°C. From R. D. Kadeg and G. D. Christian,* Anal. Chim. Acta, *88, 117 (1977), by permission of the publisher.*

retention volume V_r (or the retention time t_r) of the unknown compound to that of n-hydrocarbons eluting before and after it. To each of a series of paraffins is attached an index I, given by

$$I = 100n \qquad (22.15)$$

where n = carbon number of a given paraffin

The retention index of an unknown is calculated from the relationship

$$I = 100\left[\frac{\log V_n^u - \log V_n^x}{\log V_n^{x+1} - \log V_n^x}\right] + 100x \qquad (22.16)$$

where x = carbon number of the compound eluted before the unknown
V_n^u = net retention volume of the unknown
V_n^x = net retention volume of the hydrocarbon eluted before the unknown
V_n^{x+1} = net retention volume of the hydrocarbon eluted after the unknown
$x + 1$ = carbon number of the compound eluted after the unknown

(Here, x and $x + 1$ are the bracketing hydrocarbon carbon numbers.)

This method is based on a linear relationship between log V_r and carbon number in a homologous series and essentially places any species on its appropriate place on the plot. The basic assumption is that, under isothermal conditions, variations in the retention of hydrocarbons will be reflected in the retention of all other species; hence, if the flow rate changes or if the stationary-phase volume is reduced, all observed V_r-values will change, but the value of I for a given species will not. A problem arises only when the chemical composition of the phase changes with time because of polymerization or oxidation. For example, a polyglycol polymer stationary phase might change in polarity after long heating and exposure to an active catalytic support surface, so that changes in its retention of alcohols will not be mirrored by changes in its retention of paraffins. The retention of the hydrocarbons on this phase is mainly a function of vapor pressure, with almost a uniform activity-coefficient contribution and no directed (i.e., hydrogen-bonding) interactions, whereas the retention of alcohols depends not only on vapor-pressure differences but also on directed solute-solvent interactions. A loss of hydrogen-bonding capacity will influence the retention of alcohol markedly, but may not affect that of the paraffins. Some workers use homologous series of analogous compounds—alcohols with alcohols, ketones with ketones, and so on—in an attempt to avoid this problem. The use of computers in gas chromatography is strongly encouraged, since they make the calculation of retention indices a trivial operation. Of course, just as a single t_r measurement (even with standards) does not conclusively identify a given substance, neither does a single index value. Several columns should be used and index values calculated for each as cross-references to known materials.

Quantitative Analysis

If the analyst is willing to assume that peak shape is not a function of solute concentration, he or she can carry out quantitative analyses by establishing standard curves of peak height versus concentration. The quantitative information in a gas chromatogram is found in the peak areas and not in the peak heights, since zone shape is a function of many different variables, notably injection rate. Peak areas are obtained by conventional methods, such as triangulation, mechanical or electronic integration by analog devices on the recorder, and summation in digital recorders (analog-to-digital conversion); the ease and accuracy of measurement increase in the order of the methods mentioned, as does the cost. The relative response factors for the different species are important to determine, so as to normalize the areas measured to a common base for comparison. Quantitative analysis is generally carried out with reference to standard calibration curves; in practice, an error of $\pm 2\%$ is quite reasonable. Quantitative analysis also depends on using a recording method that has the correct frequency response for the signal observed.

22.5 APPLICATIONS OF GAS CHROMATOGRAPHY

Applications of gas chromatography are best illustrated by the actual chromatograms themselves. The examples presented here were chosen to illustrate the utility of the technique in many areas of scientific endeavor, and because they combine many of the ideas presented in this chapter. They do not necessarily illustrate the method of choice for a given analysis.

Industrial Environmental Analysis

A good example of a gas-chromatographic analysis in the area of air quality is in the study of coke-oven emissions. By its very nature, the coke process can be expected to produce polycyclic organic matter; this is of concern because many compounds in this class are carcinogenic. Such compounds as benz[a]anthracene, benzo[a]pyrene, and benz[c]acridine are of particular importance.

The chromatograms shown in Figure 22.12 were obtained using a 10-ft by 0.125-in. (o.d.) column packed with 2% SE-30 on Chromosorb G. Temperature programming was utilized (175 to 275°C at 4°C/min, then held for 15 min). Pertinent fractions were trapped and analyzed remotely by ultraviolet spectrometry. Recoveries (grams found/grams present) were about 86% for benzo[a]pyrene. Actual coke-oven samples were collected by ambient air-filter techniques and extracted with cyclohexane. Detection was by FID in a split-effluent stream.

Forensic Analysis for Drugs of Abuse

One of the major applications of gas chromatography is in finding legal evidence of the presence of illicit material. Two examples are given here. The first is the analysis of "street-quality" heroin; the second is an analysis for amphetamine and related materials in biological fluids.

Analysis for the heroin content of illicit heroin, which is often "cut" with quinine hydrochloride, can be accomplished directly using gas-liquid chromatography. In this example, an internal standard (cholesterol) is added to improve both qualitative and quantitative accuracy. The chromatogram is isothermal (235°C) with a 6-ft by 0.25-in. column packed with 3% OV-1 on 80/100-mesh Chromosorb W. The

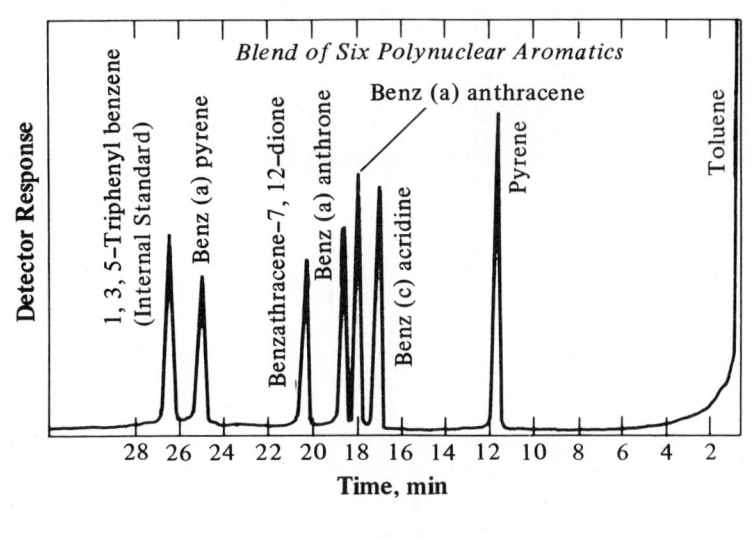

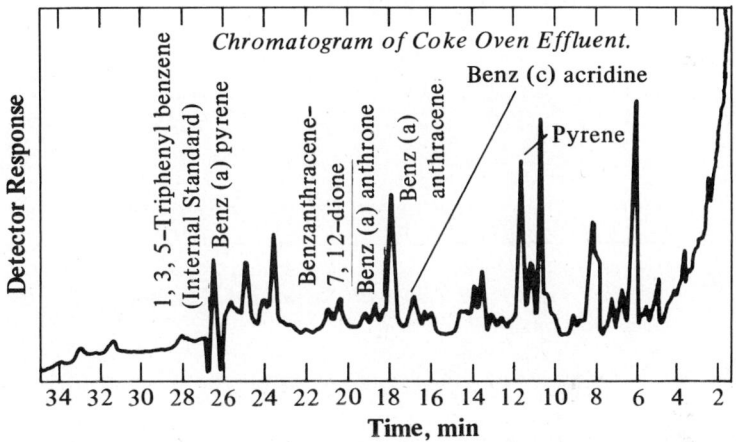

FIGURE 22.12. *Chromatograms of a standard mixture of polycyclic aromatic hydrocarbons and of a coke-oven emission. From T. D. Searl, F. J. Cassidy, W. H. King, and R. A. Brown,* Anal. Chem., *42, 954 (1970), by permission of the senior author and the publisher. Copyright © 1970 by the American Chemical Society.*

peak area is determined using a digital integrator. Figure 22.13 shows a typical analysis. Heroin content is calculated by the following relationships:

$$c_s = \frac{A_s}{A_{std}} \left(\frac{A_{i.s./std}}{A_{i.s./s}} \right) c_{std} \tag{22.17}$$

$$\% \text{ Heroin} = \frac{c_s}{S}(100\%) \tag{22.18}$$

where A_s = area count of sample
A_{std} = area count of standard

SEC. 22.5 Applications of Gas Chromatography 755

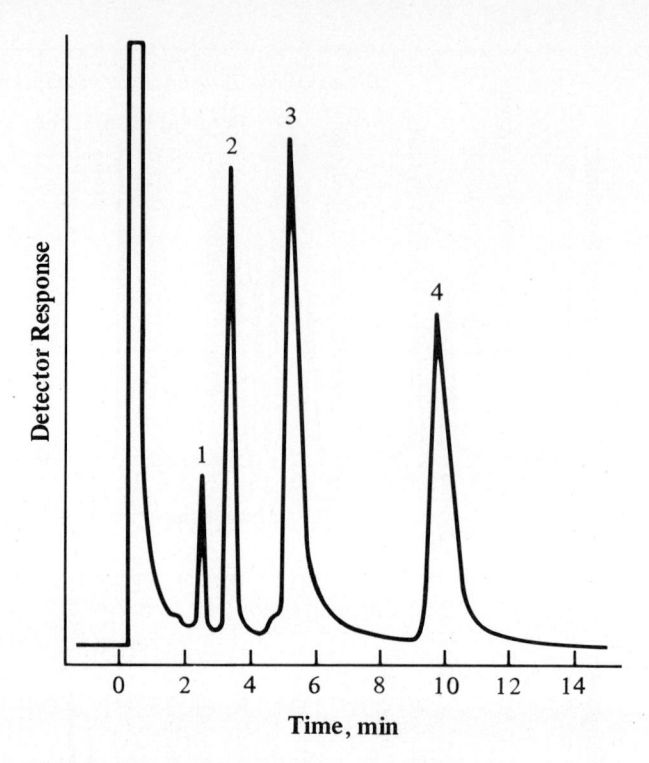

FIGURE 22.13. *Chromatogram of heroin preparation showing internal standard. Compounds:* 1. *Acetylcodeine or O^6-monoacetylmorphine;* 2. *Heroin hydrochloride;* 3. *Quinine hydrochloride;* 4. *Cholesterol internal standard. From P. De Zan and J. Fasenello*, J. Chromatog. Sci., *10*, 333 (1972), *by permission of the publisher.*

$A_{\text{i.s./std}}$ = area count of the internal standard in standard solution
$A_{\text{i.s./s}}$ = area count of the internal standard in sample solution
c_{std} = concentration of the standard
c_s = concentration of the sample (heroin)
S = weight of the sample

An analysis of amphetamine and related compounds is easily achieved in standard solutions by derivatization to the *N*-trifluoroacetamide. The situation is much different in the world of "real" samples. Biological fluids are very complex mixtures of materials. In urine, such materials as amphetamine can be extracted, a derivative made, and the latter analyzed without much problem; at high dilution in blood, the problem becomes more complicated. In this example, the following procedure yielded the highest recoveries (98%) at the 2.5×10^{-8} g/mL of blood level: (a) extracting the substance with benzene in coated glassware, (b) scavenging the amine with a volatile amine (diethylamine) in a HCl salt-formation step, and (c) derivativization to the *N*-trifluoroacetamide. Chromatographic conditions were similar to that of the previous example. A typical chromatogram is given in Figure 22.14.

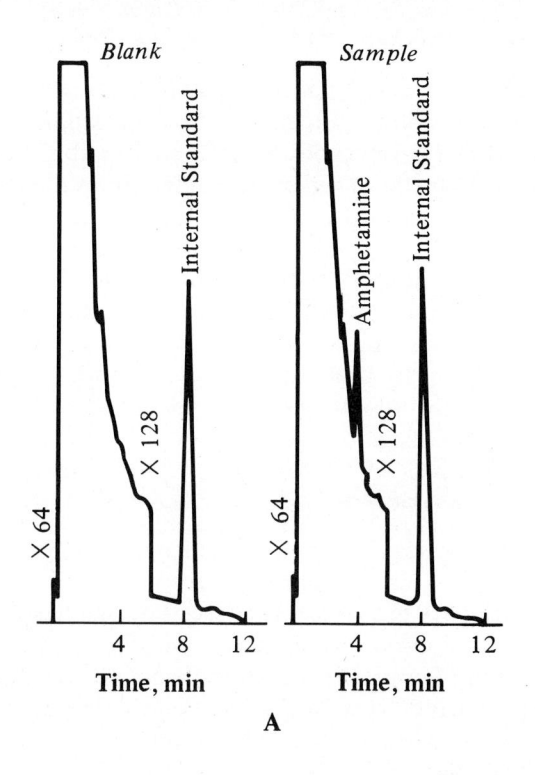

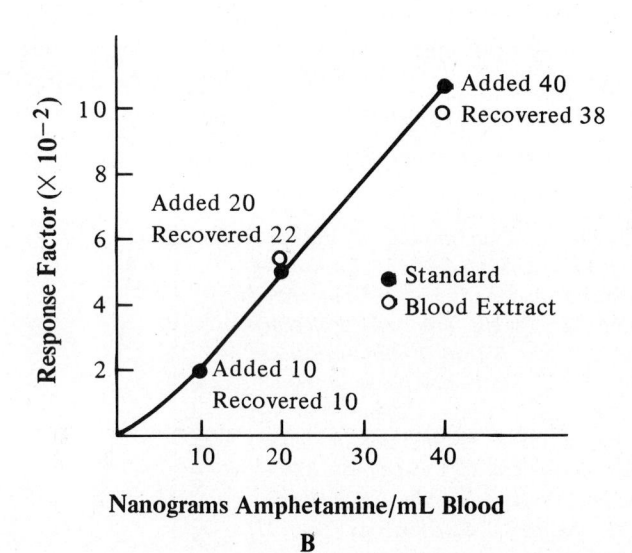

FIGURE 22.14. *Typical chromatogram of amphetamine as determined in an extract of human whole blood. A: Chromatogram. B: Calibration curve illustrating linearity, sensitivity, and recovery from whole blood. From J. E. O'Brien, W. Zazulan, V. Abbey, and O. Hinsvark, J. Chromatog. Sci., 10, 336 (1972), by permission of the publisher.*

Pharmacological Studies

This example of the recovery of a drug and its metabolites is important because it illustrates electron-capture detection. The drug studied is 7-chloro-1,3-dihydro-5-(2'-chlorophenyl)-2H-1,4-benzodiazepin-2-one. In this study, the intent was to recover the intact drug and to recover and identify its metabolites in blood and urine. Figure 22.15 shows a typical chromatogram of a diethyl-ether extract of blood. The detection limit for the drug was 0.002 μg/mL of blood; the very low levels of the drug expected in human-subject studies required gas chromatography with electron-capture detection (^{63}Ni sources) as opposed to more common spectrophotometric procedures. The chromatogram is relatively simple, because of the selectivity of the detection system; the ECD is essentially responding only to the chlorinated compounds. The only metabolite positively identified was Lorazepam, the 3-hydroxy analog of the parent drug.

Resolution of Enantiomers

The separation of optical isomers is a challenge for separation science. There are two common approaches for gas-chromatographic separation of enantiomers: direct, using chiral stationary phases; and indirect, through the prior formation of diastereotopic isomers by reaction of the analyte with an optically resolved reagent. In the example shown in Figure 22.16, an optically active silicone-polymer stationary phase coated on a glass capillary column was used to separate the enantiomers of a series of amides. This is a particularly good example of the effective use of enhanced selectivity from the column type, combined with the high number of theoretical plates possible with capillary columns. In fact, the selectivity illustrated in this example is so high as to make possible separation of these solutes using a packed column. In a "real" sample, however, the capillary column would offer the advantage of better resolution of the peaks of interest from background interferences.

FIGURE 22.15. *Chromatogram of diethyl-ether extract of blood showing a drug and its metabolites. Compounds: 1. Lorazepam (an identified metabolite); 2. Unidentified metabolite; 3. Parent drug; 4. Reference standard. Column: 4 ft by 4 mm borosilicate glass, 3% OV-17 stationary phase on 60/80-mesh diatomite. Argon-methane (90:10) carrier gas. Column: Isothermal at 240°C. Injection port, 280°C; detector, 325°C. Adapted from J. A. F. de Silva, I. Bekersky, and C. V. Puglisi, J. Chromatogr. Sci., 11, 547 (1973), by permission of the publisher.*

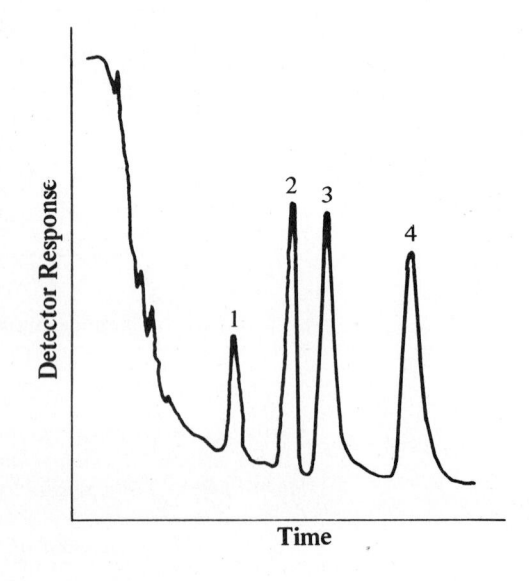

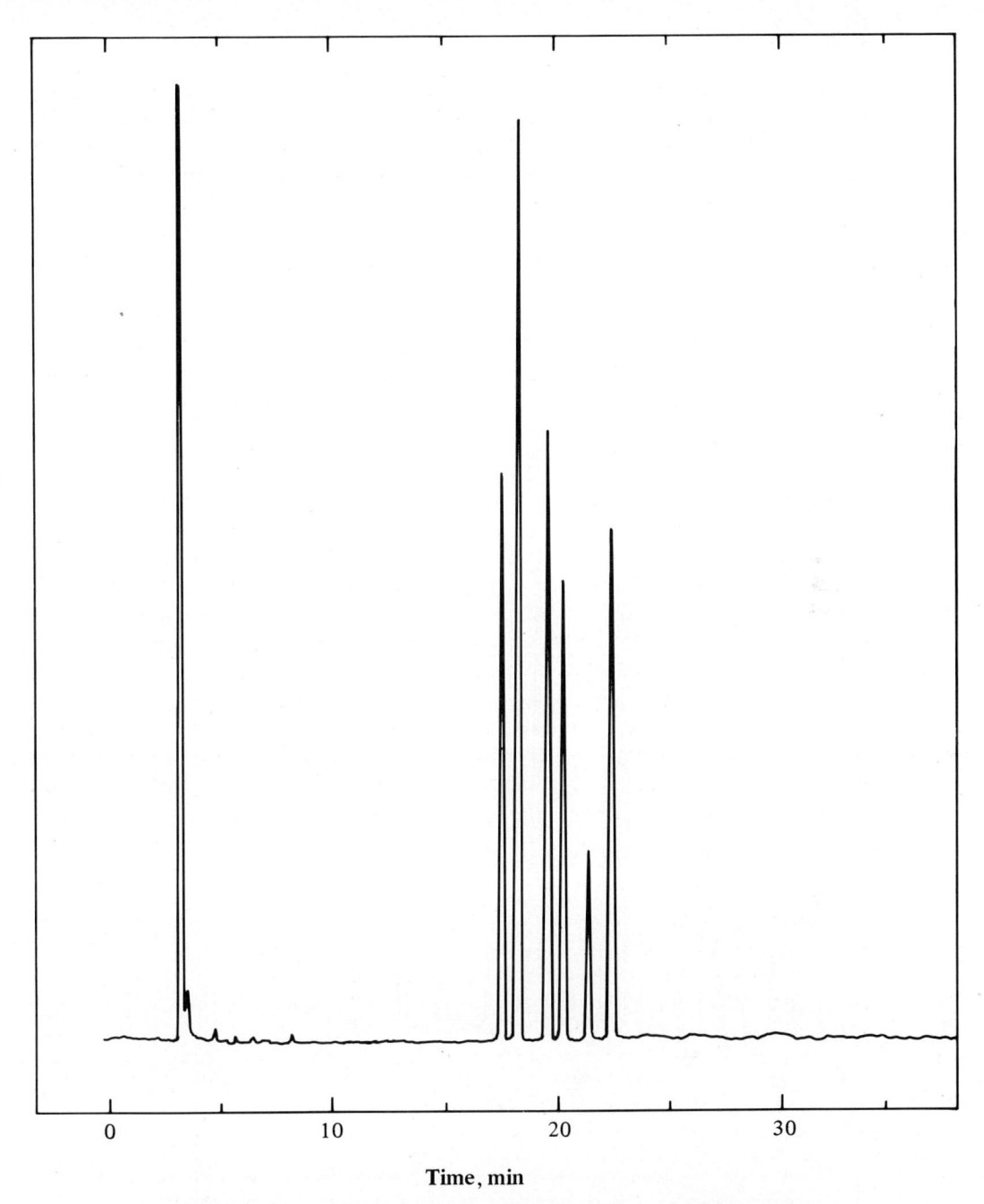

FIGURE 22.16. *Separation of the trifluoroacetyl, pentafluoropropyl, and heptafluorobutyl amides of α-methylbenzylamine on a 40-m wall-coated open tubular (WCOT) column coated with 32% chiral peptide/68% methyl silicone polymer. Column at 125°C; injector at 220°C; detector (FID) at 270°C. Splitter at 40:1 ratio; detector operated at 4×10^{-12} A full scale. Similar to C. H. Lochmüller and J. V. Hinshaw, Jr., J. Chromatogr., 202, 363 (1980). The peak on the left, at about 3.1 min, is that for methane, giving the void time t_0.*

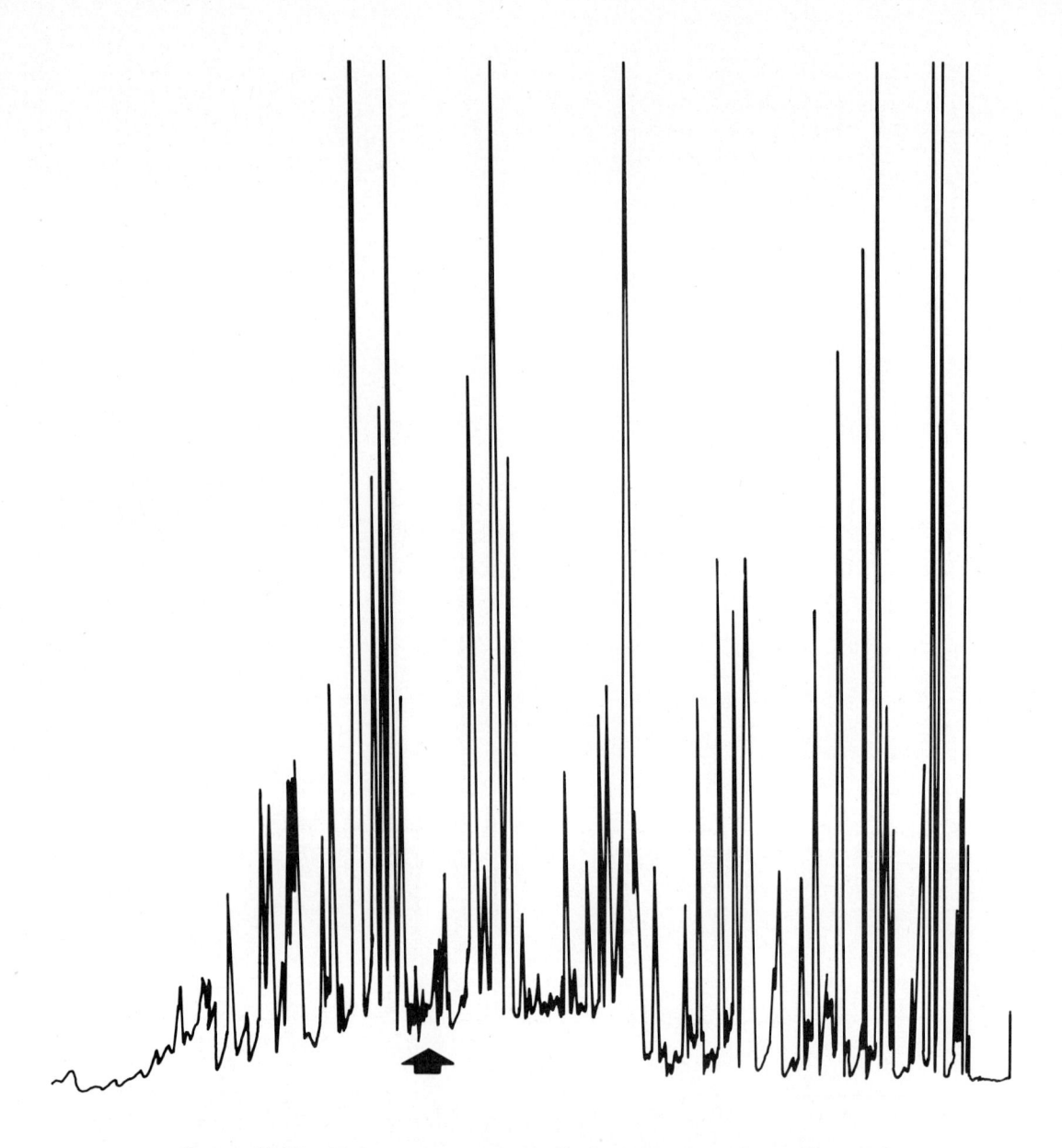

FIGURE 22.17. *Temperature-programmed chromatogram of a commercial premium-grade gasoline, illustrating the separation of major components and the capabilities of a thermal conductivity detector of optimized design. A 42-m by 0.5-mm SCOT column coated with SE-30 was held isothermal at 30°C for 1 min after injection, then ramped at 35°C/min to 190°C. Detector current, 200 mA; attenuation of 256. Reproduced from C. H. Lochmüller, B. M. Gordon, A. E. Lawson, and R. J. Mathieu,* Journal of Chromatographic Science., **16**, 523 (1978), *with permission of Preston Publications, Inc.*

FIGURE 22.18. *Example of a tandem gas-liquid chromatogram. The vertical lines in chromatogram A indicate the "cut" of eluting species which was shunted to a second column to produce chromatogram B. Courtesy of Siemens Corporation.*

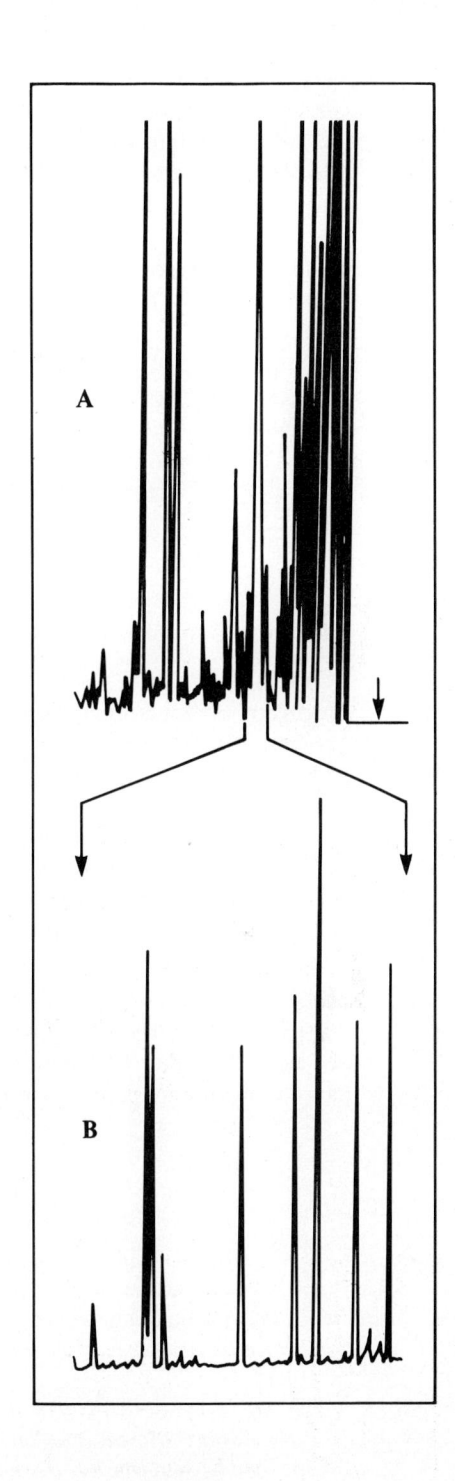

Effect of Detector Design

The most common detector for capillary work is the FID. This is true for several reasons; but most important, it is fairly easy to eliminate the effect of detector contributions to band broadening in the FID through the use of "make-up" gas (additional gas flow near the end of the column in the inlet which sweeps the eluting zone rapidly into the flame), and by running the tip of the capillary column to just below the flame itself. The other advantage of the FID is its response to *mass* directly, rather than to *concentration*. In concentration-response detectors, the use of make-up gas reduces the response because of dilution. To make effective use of concentration detectors (ECD, TCD, FPD, etc.) with capillary columns, it is necessary to reduce the volume of such detectors to reduce band broadening. Careful studies of the effect of detector volume and flow-path design can yield practical detectors that rival the FID in terms of minimal contribution to band broadening. The example shown in Figure 22.17 represents a temperature-programmed separation of the components of gasoline using a TCD of optimized design. Note that the very sharp peaks typical of capillary columns are hardly broadened by use of a TCD.

Tandem Capillary Separations

Capillary columns do not provide the ultimate solution to the "general elution problem": That is, there is no guarantee that all peaks of interest will be resolved in a reasonable time on a single column, even with complex temperature programming. The two chromatograms presented in Figure 22.18 illustrate two points. (a) Through proper engineering, it is possible to fluidically switch parts of a developing chromatogram from one column to a second column in a separate oven, even with capillary columns, without significant degradation of peak shape. The solutes eluting in the one small section of chromatogram A were shunted to a second capillary column in series, and the peak shape is still excellent in chromatogram B. (b) Single peaks, even in capillary chromatography, can be complex mixtures. Note that in chromatogram A, the material rechromatographed appears to consist of one major component (one large peak) with perhaps four minor components (peaks). Chromatogram B shows at least 27 peaks, 9 of them of appreciable size. As is generally the case, multidimensional chromatography provides better resolution.

SELECTED BIBLIOGRAPHY

GIDDINGS, J. C. *Dynamics of Chromatography*. New York: Marcel Dekker, 1965. *This classic text provides insightful discussions of fundamental processes involved in the chromatographic process.*

GROB, R. L., ed. *Modern Practice of Gas Chromatography*. New York: Marcel Dekker, 1977. *A thorough and comprehensive treatment of basic theory and practice.*

HARRIS, W. E., and HABGOOD, H. W. *Programmed Temperature Gas Chromatography*. New York: John Wiley, 1966. *A thorough and clearly written reference for this subject area.*

JENNINGS, W. *Gas Chromatography with Glass Capillary Columns*, 2nd ed. New York: Academic Press, 1980. *A concise treatment of all topics associated with the use of narrow-bore open tubular columns.*

KARGER, B. A., SNYDER, L. R., and HORVATH, C. *An Introduction to Separation Science.* New York: Wiley-Interscience, 1973. *This text organizes the various separation techniques into a unified theme and is strongly recommended to those individuals whose work involves separations.*

LITTLEWOOD, A. B. *Gas Chromatography: Principles, Techniques, and Applications,* 2nd ed. New York: Academic Press, 1970.

MCFADDEN, W. *Techniques of Combined Gas Chromatography/Mass Spectrometry.* New York: Wiley-Interscience, 1973. *An authoritative reference for this important combination, most useful for its excellent treatment of mass spectrometry and interfacing details.*

MCNAIR, H. M., and BONELLI, E. J. *Basic Gas Chromatography.* Varian Aerograph, 2700 Mitchell Drive, Walnut Creek, CA. 94598. *A very good "first" book on practical aspects.*

PERRY, T. A. *Introduction to Analytical Gas Chromatography.* Chromatographic Science Series, Vol. 14. New York: Marcel Dekker, 1981. *This excellent text gives a very readable account of the theory and practice of modern gas chromatography. Highly recommended as an introductory text.*

YANCY, J. A., ed. *Guide to Stationary Phases for Gas Chromatography.* Analabs, Inc., North Haven, Connecticut. *A correlated listing of stationary phases and solute types, updated on a regular basis.*

REFERENCES

1. A. J. P. MARTIN and R. L. M. SYNGE, *Biochem. J.*, **35**, 1358 (1941).
2. A. T. JAMES and A. J. P. MARTIN, *Biochem. J. Proc.*, **48**, vii (1951); *Analyst*, **77**, 915 (1952).
3. E. GLUECKAUF, *Trans. Faraday Soc.*, **51**, 34 (1955).
4. L. ROHRSCHNEIDER, *Z. Anal. Chem.*, **170**, 256 (1959).
5. W. O. MCREYNOLDS, *J. Chromatog. Sci.*, **8**, 685 (1970).
6. J. LEARY, J. JUSTICE, S. TSUGE, S. LOWRY, and T. L. ISENHOUR, *J. Chromatog. Sci.*, **11**, 201 (1973).

PROBLEMS

1. Predict the effect of changing from He to N_2 carrier gas on (a) retention volume, and (b) *HETP*. What might be the effects of using a supercritical vapor such as CO_2?

2. Compare the following two methods of scaling up an analytical separation to preparative levels: (1) keeping column size (volume and length) constant and increasing the percent liquid loading; and (2) increasing column diameter, keeping the length and the percent liquid loading constant.

3. The following data were obtained on a nonpolar column for *n*-butylacetate (retention time in mm of chart paper): *n*-heptane, 174 mm; *n*-octane, 373.4 mm; *n*-butylacetate, 310 mm. (a) Calculate the Kováts index *I* for *n*-butylacetate. (b) Does the value of *I*/100 have any physical significance?

4. A column of support coated with an ester of low volatility is operated at 23 lb/in.^{2} and 40°C. The following results were obtained:

Compound	t_r, min
Methane + air	1.8
Ethane	2.4
Propane	3.6
Propylene	4.3
Isobutane	5.5
Butane	7.5
Isobutylene	8.6
Trans-2-butene	10.6
Cis-2-butene	12.3
Isopentane	13.6
Pentane	18.2

Construct a plot of log V'_r versus carbon number, including all compounds significantly retained. What can be seen in the results?

5. Derive the expressions for optimum velocity and minimum plate height using the simple Van Deemter expression. What are the dominant factors in each?

6. Name the three major nonequilibrium effects that lead to band broadening in continuous-flow systems such as a gas-chromatographic column.

7. The adjusted retention times (t'_r) in minutes for a series of compounds were determined carefully on a nonpolar column: *n*-pentane, 2.8; *n*-hexane, 5.3; *n*-heptane, 13.7; *n*-octane, 29.3; toluene, 16.5; cyclohexane, 12.4. Calculate the Kováts index for toluene and for cyclohexane.

8. Gas-reduction valves used on helium tanks in gas chromatography commonly give the pressure in units of pounds per square inch above atmospheric pressure. (a) Calculate the actual inlet pressure in millimeters of mercury (torr) for 20, 40, 60, and 80 psig if ambient pressure is 740 mm Hg and normal atmospheric pressure (760 mm) is 14.696 psi. (b) Calculate the compressibility factor j for each of these cases.

9. Uncorrected flow rates (F_0) measured on a bubble-type flowmeter must be corrected for temperature and pressure (see Fig. 22.1). Calculate F_c for $F_0 = 23.8$, 40.2, and 51.9 mL/min. The ambient pressure and temperature are 750 mm Hg and 25°C, respectively; the column temperature is 110°C.

10. The following data were obtained on a 1/8-in. by 10-ft column of 15% SE-30 (by weight) on Chromosorb W. Inlet pressure = 60.0 psig; ambient pressure = 740 mm Hg; column temperature = 116°C; ambient temperature = 24°C; uncorrected He flow rate = 26.4 mL/min on a bubble-type flowmeter. A total of 1.09 g of SE-30 was contained in the colum

Compound	t'_r, min	Peak Width (W), min
Ether	1.78	0.31
Hexane	6.78	0.84
Ethylbenzene	18.14	1.64

For each compound, calculate (a) the adjusted retention volume, (b) the net retention volume, and (c) the specific retention volume (at 0°C). Calculate (d) the resolution between ether and hexane, and (e) the resolution between hexane and ethylbenzene.

11. The following chromatographic data were obtained on the column used in Problem 10 with 2-μL injections of heptane:

F_0, mL/min	t_0 ("Air Peak"), min	t_r, min	Peak Height (Chart) Divisions	Peak Width (W), min
121.2	1.38	4.49	50.2	0.30
91.3	1.69	5.37	60.3	0.34
72.8	1.94	6.17	64.9	0.38
63.7	2.09	6.62	67.6	0.42
51.2	2.42	7.62	69.2	0.49
40.9	2.78	8.83	79.4	0.63
32.7	3.30	10.31	78.5	0.76
27.4	3.74	11.69	76.8	0.90
18.1	5.03	15.84	72.3	1.47

Ambient temperature and pressure were 25.0°C and 740 mm Hg, respectively; inlet pressure was 60.5 psig; column temperature was 109°C. For each flow rate, calculate F_c, t'_r, V'_r, V_n, N, H, and the (triangulated) area A of the peak. Plot N and H versus F_c and estimate the optimum flow rate. Plot the peak height and peak area versus F_c and comment on the effect of flow rate on these.

12. What number of plates would be required to effect a separation with less than 1% contamination ($R = 1.5$) for $\alpha = 1.10$ in (a) a packed column with a capacity factor of 50; (b) a packed column with $k' = 5$; (c) an open tubular column $k' = 0.5$. (e) Calculate the effective number of

plates in each case. (f) What column lengths would be necessary for a good packed column of $H = 0.2$ mm and an open tubular column of $H = 10$ mm?

13. The chromatogram shown in Figure 22.16 illustrates the separation of six solutes—three enantiomeric pairs of compounds. Each series of three (the right-and left-handed series) is a homologous series in the number of fluorine-substituted carbons. (a) Identify the sets of D and L isomers by fitting the peaks to plots of $\log t_r'$ versus number of carbons in the homologous series. (b) Calculate the α-values for each pair of enantiomers. (c) Calculate the α-values between members of the homologous series of the same handedness, and compare these to the values for pairs of enantiomers. (d) The width of the peaks at baseline t_w is about 0.30 min for each peak. Calculate the resolution R between the two trifluoroacetyl enantiomers, and between the trifluoroacetyl and pentafluoropropyl derivatives of the same handedness. Use Equation 22.11 or its equivalent, Equation 21.9. (e) Calculate N, H, and k' for the first enantiomeric pair. (f) Calculate the resolution between the first enantiomeric pair using equation 22.12, and compare it to that calculated in part (d).

CHAPTER 23

Introduction to Analog Circuits and Devices

F. James Holler

The rate of growth of electronics, instrumentation, and microcomputer technology is nearly incomprehensible. Only a short time ago, computers were expensive, cumbersome devices rarely seen in chemical laboratories; but in the last decade, the advent of the inexpensive mass-produced microcomputer with its collection of associated peripheral devices has moved computers from the rare to the commonplace. Microcomputers are now found in most laboratory instruments, including even balances and pH meters.

The proliferation of automated, computerized instrumentation dictates that chemists develop an understanding of the advantages and limitations that characterize modern electronic devices. Although it is clearly impossible, and perhaps undesirable, for all chemists to attain knowledge of electronics at the circuit-design level, the development of high-function integrated-circuit modules allows a conceptually simple "top-down" approach to the study of electronics. Using such an approach, it is possible to carry out very sophisticated physicochemical measurements quite simply, by correctly connecting several function modules or integrated circuits. There is little need for detailed knowledge of the internal design of individual modules. In addition to facilitating the learning process, this approach aids in diagnosing malfunctioning instruments and in the intelligent application of instrumental systems to the solution of chemical problems.

The purpose of this chapter and the following one on use of computers in analytical chemistry is *not* to provide the level of discussion and coverage of topics that one would encounter in a full-semester electronics course. Rather, our objective is only to include enough material for the reader to attain a very basic and rudimentary knowledge of electronics in order to understand and appreciate the functioning of the various measurement systems and methods discussed elsewhere in the text. For this reason, the topics discussed will be very selective. Furthermore, in the interests of space and simplicity, discussion will be largely restricted to consideration of DC signals, with only occasional inclusion of AC.

The structure of the next two chapters is as follows. We will begin with a top-down overview of data domains to provide a philosophical framework in which the entire measurement process is approached. We will then do a complete about-face, discuss small, discrete circuit components and very simple circuits, and then continue building up to more and more complicated and sophisticated circuits and systems. Finally, we will arrive back at the grand overview.

23.1 DATA DOMAINS

Electronic measurement is aided by a wide variety of devices that convert information from one form to another. In order to investigate how electronic instruments function, it is important to understand the way in which electrical information is *encoded*, or transformed from one system of information to another, as a characteristic of *electrical signals*—that is, as voltage, current, charge, or variations in these quantities. The various modes of encoding electrical data are called *data domains*. Enke [1] has developed a classification scheme based on this concept that greatly simplifies the analysis of instrumental systems and promotes understanding of the measurement process. As shown in the data-domains map of Figure 23.1, data domains can be broadly classified into nonelectrical domains and electrical domains.

Nonelectrical Domains

The measurement process begins and ends in nonelectrical domains. The physical and chemical characteristics that may be of interest in a particular experiment reside in these data domains. Among these characteristics are length, density, chemical composition, intensity of light, and pressure. Clearly, a measurement can be made entirely in nonelectrical domains. For instance, the determination of the length of an object with a vernier caliper is a comparison of that object with the scale on the caliper. The information representing the length in standard units is encoded by the experimenter directly from the scale. The ultimate objective in this case, as in all measurements, is that the encoded number be in some way proportional to the chemical or physical characteristic of interest.

Electrical Domains

The modes of encoding information as electrical quantities can be subdivided into *analog domains*, *time domains*, and *digital domains*, as is shown in the bottom half of the circular map in Figure 23.1. Note that the digital domain spans both electrical and nonelectrical domains because written numbers or numbers on any type of display convey digital information, and they can also be encoded electrically.

Any measurement process can be represented as a series of *interdomain conversions*. For example, Figure 23.2 illustrates the measurement of the pH of a solution and, in a general way, the data-domain conversions necessary to arrive at a number expressing the pH. The information begins in the chemistry of the solution. It is encoded into an electrical domain by a special type of interdomain converter called an

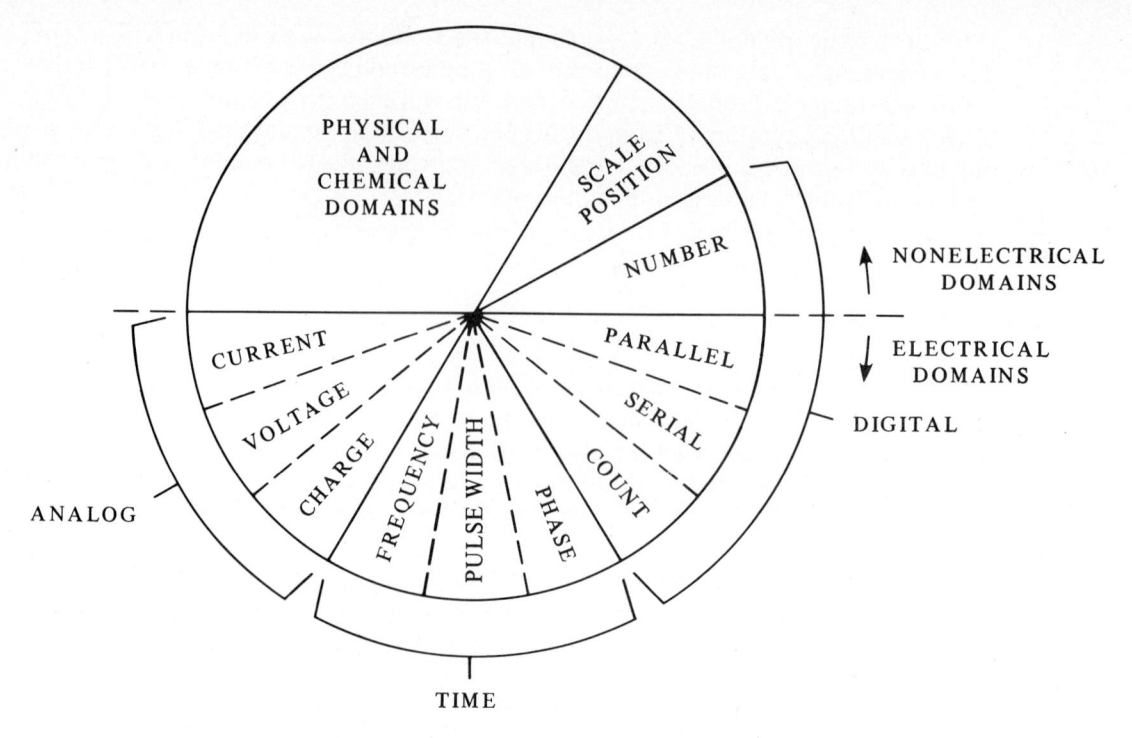

FIGURE 23.1. *Data-domains map. A number of domains in which data can be encoded are displayed in the circular data-domains map. Nonelectrical domains appear in the upper semicircle, and electrical domains compose the lower semicircle.*

input transducer. Voltmeters, alphanumeric displays, electric motors, cathode ray tubes (CRTs), and many other devices that serve to convert data from electrical to nonelectrical domains are called *output transducers.* The voltage output of the glass electrode is converted by the voltmeter to a scale position that can be read directly as a number representing the pH of the solution, provided that the meter face has been properly calibrated in pH units. A modern pH meter can contain, internally, a number of data-domain converters to process the information provided by the glass electrode, and to present it directly in digital form. However, discussion of these conversions must be deferred until we have developed the appropriate circuitry for such systems.

Information in *analog domains* is encoded as the magnitude of one of the electrical quantities voltage, current, charge, or power and is presented to the outside world by an output transducer. These quantities are continuous in both amplitude and time, as shown by the analog outputs of Figure 23.3. Magnitudes of analog quantities can be measured continuously or they can be sampled at specific points in time dictated by the needs of a particular experiment. Although the data of Figure 23.3 are recorded as a function of time, any variable such as wavelength, magnetic-field strength, or temperature can be used as the independent variable, under the appropriate circumstances. The correlation of two analog signals resulting from corresponding measured physical or chemical properties is important in a wide variety of

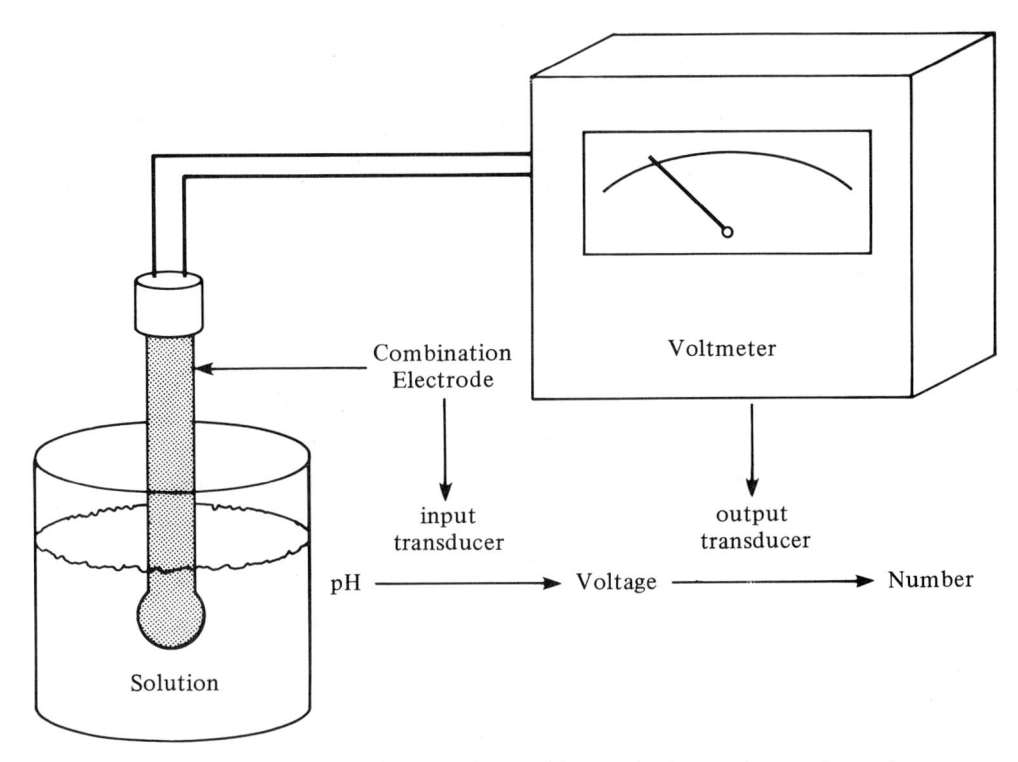

FIGURE 23.2. *Data domains and devices in a pH measurement. A chemical property of the solution is converted to an electrical signal by the glass-electrode input transducer. The output transducer converts this electrical signal into an interpretable number (pH) in the nonelectrical domain.*

instrumental techniques such as nuclear magnetic resonance spectroscopy, infrared spectroscopy, differential thermal analysis, and so on.

Analog signals are particularly susceptible to electrical noise that results from interactions within measurement circuits or from other electrical devices in the immediate vicinity. Such induced signals are obviously undesirable because they bear no relationship to the information of interest.

Information is stored in *time domains* as the time relationship of signal fluctuations, rather than in their amplitudes. Figure 23.4 illustrates three different time-domain signals. The horizontal dashed lines represent an arbitrarily chosen threshold that is used to decide whether a signal is HI (above the threshold) or LO (below the threshold). The time relationships between transitions (HI → LO or LO → HI) contain the information. The number of HI-LO transitions per unit time is called the *frequency*. As a practical example, in Raman spectroscopy the frequency of arrival of photons at the photocathode of a photomultiplier is directly related to the intensity of scattered light. The time between consecutive LO → HI transitions is called the *period*, and the time between a LO → HI and a HI → LO transition is called the *pulse width*. Voltage-to-frequency converters and frequency-to-voltage converters are examples of devices that are used to convert analog-domain signals to the time

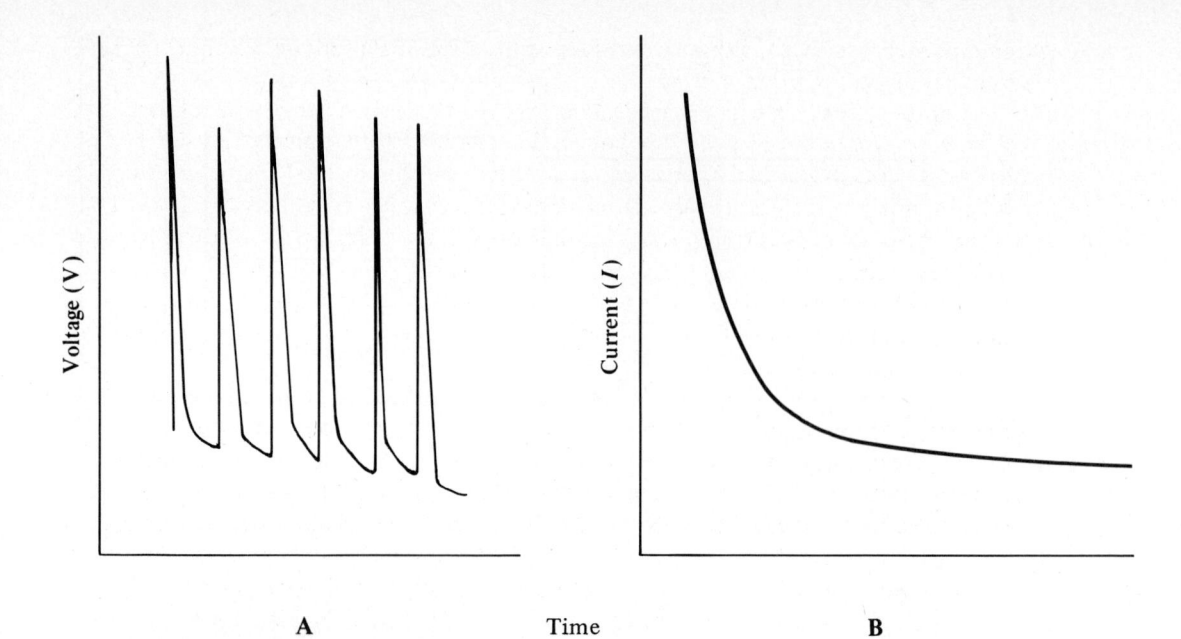

FIGURE 23.3. *Variation of analog quantities with time. A: Recorder trace of the voltage of a glass electrode in a flowing stream as samples of different pH pass the transducer. B: Continuous plot of photocurrent from a photomultiplier tube in the observation cell of a stopped-flow apparatus following rapid mixing of Fe^{3+} and SCN^-.*

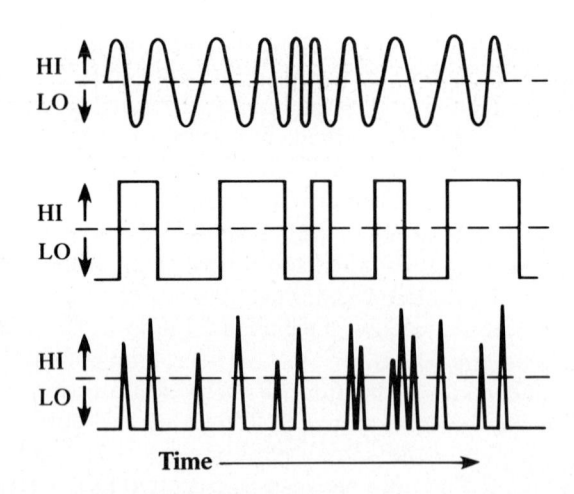

FIGURE 23.4. *Time-domain signals. Information is contained in the time relationship of transitions between LO and HI levels. These relationships can be encoded as frequency, pulse width, or period.*

domain, and vice versa. We will consider such devices in more detail after we have discussed operational amplifiers.

Data are encoded in the *digital domain* in a two-level scheme. The information can be represented by the state of a light bulb or light-emitting diode (ON-OFF), the state of a switch (ON-OFF), the state of a logic-level signal (HI-LO), or the state of any mechanical or electrical device that has two distinct positions or conditions. Several methods of encoding digital signals are shown in Figure 23.5. As with time-domain signals, a practical definition is made to distinguish between HI and LO. In the common transistor-transistor logic (TTL) family of digital circuits, for example, LO is represented by any voltage level between 0.0 V and $+0.4$ V, and HI by voltages between $+2.4$ V and $+4.5$ V.

The fundamental unit of information in the digital domain is the *bit*, which is one piece of HI-LO data. Numbers can be transmitted *serially* as a string of LO → HI → LO transitions on a single line by simply counting the number of cycles of the signal, as shown in Figure 23.5A. A more efficient method of encoding data is to simply slice time into equally spaced intervals and note the state of the signal for each slice. Each consecutive slice is assigned a value corresponding to increasing powers of 2. Slice zero is assigned a value of $2^0 = 1$, slice one is assigned $2^1 = 2$, slice two $2^2 = 4$, and so on for as many bits as are necessary to encode the data at hand. The powers of 2

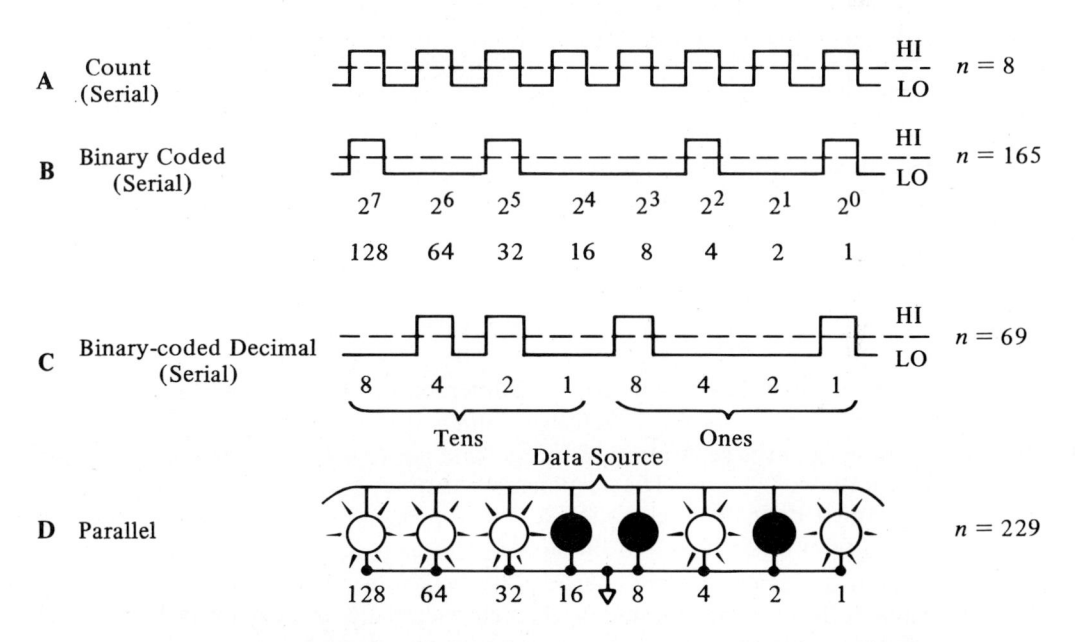

FIGURE 23.5. *Digital-domain signals. A:* The count serial signal *encodes data as the number of HI-LO transitions in the train of pulses. B:* The binary serial signal *is encoded as a binary number with either a 1 or a 0 in each slice of time. C:* The binary-coded-decimal *(BCD)* serial signal *is interpreted by grouping the time slices into decimal digits, each of which contains four slices. D:* Parallel signals *are transmitted simultaneously over several signal sources—eight lights in this case.*

corresponding to each time slice containing a HI are then summed to obtain the final number. In the example of Figure 23.5B, $n = 2^7 + 2^5 + 2^2 + 2^0 = 128 + 32 + 4 + 1 = 165$. With 8 bits or time slices available, as illustrated, the numbers 0 to 255 can be represented; 16 bits can represent numbers from 0 to 65535.

A slightly less efficient but more convenient method of encoding data for those who have not acquired the knack of thinking in binary is the *binary-coded-decimal* (BCD) scheme depicted in Figure 23.5C. Time slices are taken in groups of four, with each group of 4 bits capable of representing 10 states, the integers 0 to 9. Each group of four is assigned a power of 10 so that, for example, 8 bits can represent the integers from 0 to 99, 16 bits the numbers from 0 to 9999. Although the use of BCD-coded data represents a loss of efficiency of a factor of 6.5535 for 16 bits of information, the loss is more than offset by the convenience realized when the output transducer presents data to a human observer.

Serial data transmission is convenient for data sent over relatively long distances, and when the speed of the transmission is not critical. When high speed is important, however, *parallel data transmission* (Fig. 23.5D) allows any number of bits to be conveyed from one point to another in the same time as that required to serially transmit a single bit. Thus, for short distances and high speeds, parallel transmission is preferred. These considerations are especially important in computer applications. One last point: Digital data need no interdomain conversions to be converted to a number. They require only interpretation (decoding) and display, either electronically or manually, as we have illustrated for some of the data of Figure 23.5.

In the sections that follow, we shall examine a few of the basic principles of DC circuits, AC circuits, semiconductors, power-control devices, operational amplifiers, and some data-domain converters. Whenever possible, we will attempt to explain the operation of these devices in terms of the data-domains concept and to locate them in the framework just discussed.

23.2 ELECTRICAL QUANTITIES AND BASIC CIRCUITS

Electrical quantities result from the interaction of charge with its surroundings. *Current* is the rate of flow of charge, *voltage* is the potential energy per unit charge resulting from the separation of charge, and *power* is the rate of work resulting from the motion of charge. The response of various materials to electrical quantities constitutes their characteristic electrical properties. For example, the intrinsic resistance of a conductor to the flow of charge is its *resistivity*, and the ability of a material or device to store charge is its *capacitance*. The idea, then, is to combine available devices and manipulate their characteristics in such a way as to accomplish a given task, that is, to get data into the proper domain.*

---

*In this and the next chapter, we will use the uppercase symbols I, Q, and V to represent DC current, charge, and voltage, respectively; lowercase symbols i, q, and v will represent AC quantities. This is fairly common practice in electronics.

Ohm's Law

We begin our discussion with the simplest imaginable circuit, shown in Figure 23.6, which consists of a battery connected to a light bulb through a switch. The battery of voltage V is a charge-storage device, electrochemical in nature, that is capable of delivering current to a load. In this case, the filament of the light bulb is the load of resistance R. When the switch is closed and the circuit is completed, a current I is generated in the loop, and the bulb lights. The relationship among V, I, and R is given by Ohm's law as

$$V = IR \tag{23.1}$$

As indicated in Chapter 5, Section 5.1, the conductance G is the reciprocal of the resistance; thus,

$$I = \frac{V}{R} = VG \tag{23.2}$$

If the voltage of the battery is 3 volts (V) and the resistance of the bulb is 15 ohms (Ω), then the current (which is the same in all parts of the series circuit) is $I = 3\,\text{V}/15\,\Omega = 0.2$ ampere (A).

The electrochemical energy of the battery is released as heat and light in the bulb, and the power dissipated can be calculated from the known variables. We can calculate the work performed on the light bulb from the amount of charge Q that passes through the potential difference V across the bulb. Thus, since $W = QV$ and $I = dQ/dt$ (or Q/t for a constant current), we have

$$W = QV = ItV \tag{23.3}$$

for work expressed in joules (J). Power is defined as work per unit time, and so the power P dissipated in the bulb is

$$P = \frac{W}{t} = \frac{ItV}{t} = IV \tag{23.4}$$

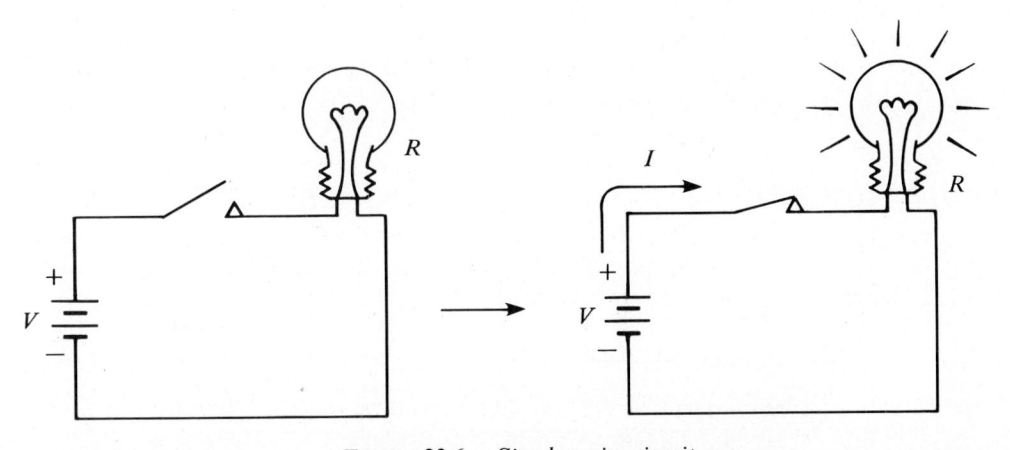

FIGURE 23.6. *Simple series circuit.*

for power expressed in watts (W). Finally, by rearranging Ohm's law and substituting for I or V in Equation 23.4, we obtain

$$P = I^2 R \quad \text{and} \quad P = \frac{V^2}{R} \quad (23.5)$$

The practical result of this last equation is that the power dissipation of a given load can be calculated at any instant in time if either the instantaneous current in the load or the voltage across the load is known. As a practical matter, such information can be used to prevent the thermal destruction of circuit components because manufacturers' literature specifies power ratings for most components. Thus, a simple power calculation can often prevent such mishaps.

Series Resistors

When resistors in a circuit are arranged in series as shown in Figure 23.7A, the resistances are additive, and the total resistance R_S is

$$R_S = \sum R_j = R_1 + R_2 + R_3 \quad (23.6)$$

The circuit shown can be rewritten as the *equivalent circuit* of Figure 23.7B. Of course, this equivalent circuit is a simple series circuit, and the current is $I = V/R_S = 3\,\text{V}/30\,\Omega = 0.1\,\text{A}$. Once the current is known, the voltage difference, or "IR drop," across each of the three resistors can be calculated. Thus, $IR_1 = (0.1\,\text{A})(5\,\Omega) = 0.5\,\text{V}$, $IR_2 = (0.1\,\text{A})(10\,\Omega) = 1.0\,\text{V}$, and $IR_3 = (0.1\,\text{A})(15\,\Omega) = 1.5\,\text{V}$. Note that the sign of the voltage change across each of the resistors is truly an IR drop; that is, the sign is opposed to the applied voltage as indicated by the $+$ and $-$ shown on the resistors. This makes sense because the work done by the battery on each successive resistor around the loop causes a decrease in the potential energy available. The sum of the IR drops around the loop is 3 V and is opposite in sign to the applied voltage. This is confirmation of *Kirchhoff's voltage law*, which states that the sum of the voltages around a closed loop in a circuit is zero. Moving clockwise around the circuit of Figure 23.7A, we have

$$+V - IR_1 - IR_2 - IR_3 = 0 \quad (23.7)$$

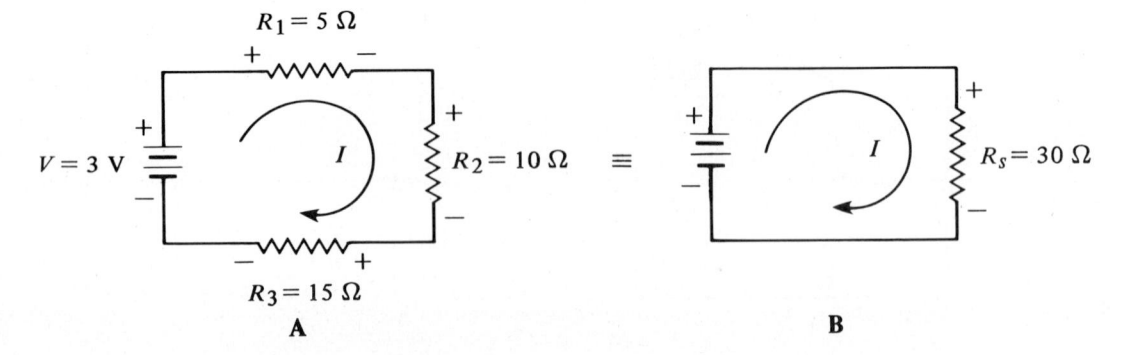

FIGURE 23.7. *A*: Series-resistance circuit. *B*: Its equivalent circuit.

The Voltage Divider

A slightly different arrangement of series resistors is depicted in the circuit of Figure 23.8A. The voltage source V is applied across $R_S = 900\text{ k}\Omega + 90\text{ k}\Omega + 10\text{ k}\Omega = 1\text{ M}\Omega$, and thus the current in the loop is given by $I = V/R_S$. The voltage selected when the switch is at position 2 is of course $V_2 = V - IR_1$, so that if we substitute for I and R_1, we have

$$V_2 = V - \frac{V}{R_S}(R_S - R_2 - R_3) = V\left(1 - \frac{R_S - R_2 - R_3}{R_S}\right) \tag{23.8}$$

and, finally,

$$V_2 = V\left(\frac{R_2 + R_3}{R_S}\right) \tag{23.9}$$

In a similar fashion, we can show that

$$V_3 = V\left(\frac{R_3}{R_S}\right) \tag{23.10}$$

For the values given for the resistors, we calculate $V_2 = V(90\text{ k}\Omega + 10\text{ k}\Omega)/1\text{ M}\Omega = V/10$ and $V_3 = V(10\text{ k}\Omega)/1\text{ M}\Omega = V/100$. The values for R_1, R_2, and R_3 were chosen deliberately to provide a *decade voltage divider*, that is, a divider that provides $(1/10)^n$ fraction of a voltage source, operated simply by rotating a switch to the proper junction between resistors. However, *any* fraction of the voltage can be selected by choosing the proper combination of resistors so that the ratio b of the selected resistance to the total divider resistance is equal to the desired fraction of the voltage.

A useful variation on this idea is the continuously variable voltage divider of Figure 23.8B, in which the resistor chain is replaced by a variable resistor called a *potentiometer*. In its simplest form, the potentiometer can be a length of nickel-chromium alloy wire (nichrome) attached to a meter stick. A sliding contact or wiper

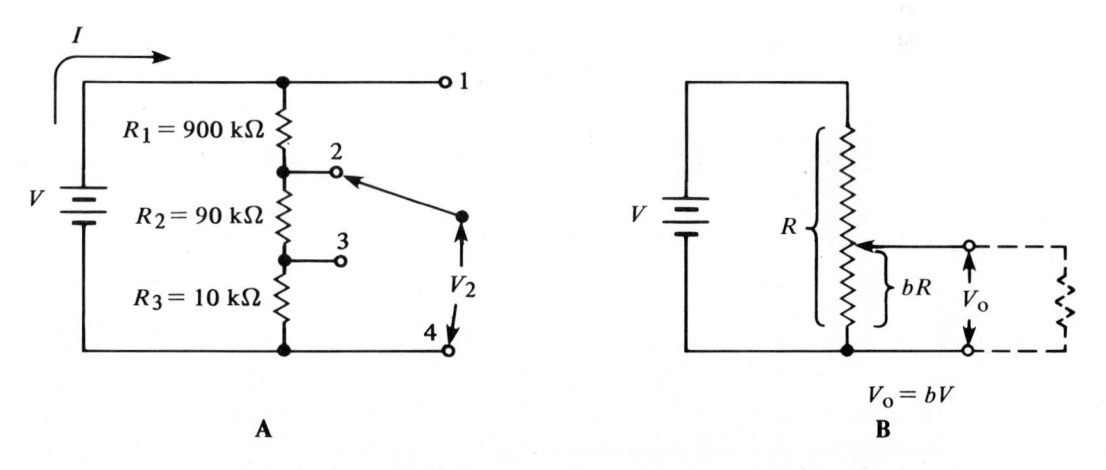

FIGURE 23.8. Voltage dividers. A: Decade voltage divider. B: Continuously variable divider. The dashed portion indicates a load, R_L, attached to the output of the voltage divider.

is moved to a position on the scale corresponding to the desired fraction of the total resistance, which in turn yields the same fraction of the total voltage applied to the ends of the wire. This arrangement, although reasonably accurate and precise, is not convenient owing to its size. Therefore, the resistance wire is normally wound in a helix and enclosed in a small cylindrical package with a shaft that can be rotated to give the desired resistance. Such helical potentiometers, or helipots, have been used for many years whenever high-resolution selectable voltages and resistances are desired. The technology of producing precision resistors and variable potentiometers has progressed dramatically, making available high-precision, high-linearity potentiometers that are no larger than a pencil eraser.

The voltage divider is one of the most used and least appreciated circuits in all of electronics. It has wide applicability if its limitations are recognized and taken into account; in particular, the output of a voltage divider cannot drive a load and maintain the expected divider voltage. If a resistive load, R_L, is attached to the output of the divider shown in Figure 23.8B, a current path is created in parallel with bR, which causes a decrease in the voltage across R_L. Later, when we discuss operational amplifiers, we will consider an elegant way to "unload" a voltage divider.

Parallel Resistors

When two or more resistors are connected in parallel as illustrated in Figure 23.9A, the voltage source must supply current through all of the parallel resistances, and the sum of the individual currents—I_1, I_2, and I_3 in this case—must be equal to the total current I supplied by the battery. This is a statement of *Kirchhoff's current law*, which requires that the algebraic sum of all currents *into* a node or junction point be zero. If we apply the law at point "a," we have $I - I_1 - (I_2 + I_3) = 0$, and if we apply it at point "b," we have $I_1 + (I_2 + I_3) - I = 0$, which of course are identical results. Armed with this knowledge and the fact that the potential differences across all three resistors are identical, we can prove that the parallel equivalent resistance R_P (Fig. 23.9B) is given by

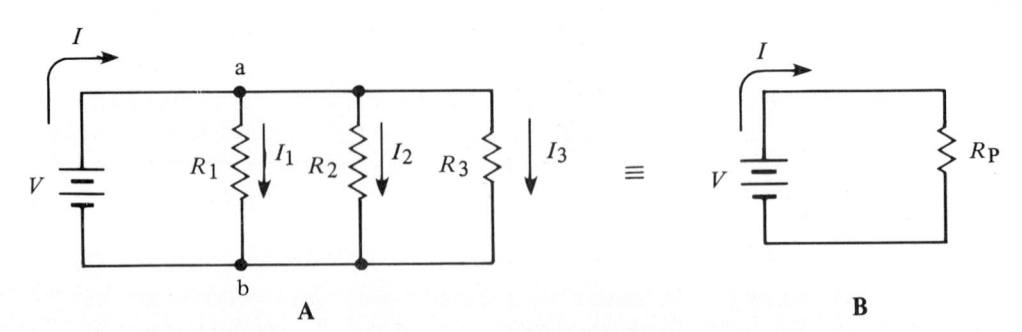

$$\frac{1}{R_P} = \sum \frac{1}{R_j} = \frac{1}{R_1} + \frac{1}{R_2} + \frac{1}{R_3} \tag{23.11}$$

FIGURE 23.9. *A*: Parallel-resistance circuit. *B*: Its equivalent circuit.

or, alternatively, in terms of conductances,

$$G_P = \sum G_j = G_1 + G_2 + G_3 \qquad (23.12)$$

Another useful result is obtained if we consider the special case of a pair of parallel resistors:

$$\frac{I_1}{I} = \frac{R_2}{R_1 + R_2} \quad \text{and} \quad \frac{I_2}{I} = \frac{R_1}{R_1 + R_2} \qquad (23.13)$$

That is, the fraction of the total current in one resistor is given by the ratio of the value of the other resistance to the total resistance. This is the so-called current-splitting expression, which is useful when a certain fraction of the total current must be "shunted" around a conductor. As we will see, such situations often arise in circuit networks designed to allow fixed-range meters to be used for other ranges.

Meters and the Measurement of DC Quantities

Until the last few years, most DC electrical measurements were made with the familiar D'Arsonval moving-coil meter. With the advent of low-cost, high-accuracy, high-precision digital voltmeters (DVMs), the moving-coil meter has become nearly obsolete. With this in mind, we will use the DVM as a state-of-the-art measurement tool and leave the discussion of the moving-coil meter to the books listed in the Selected Bibliography at the end of the chapter.

Modern DVMs are usually constructed from a single circuit, a battery or other power supply, a liquid-crystal display (LCD), and a few external components as shown in Figure 23.10A. They are inexpensive ($20 to $30), small, and can be easily incorporated into virtually any measurement system. The input resistance of these devices is extremely high (10^{11} to 10^{12} Ω), and thus for the purposes of our discussion, the DVM can be considered to be a nearly ideal "black box" voltage-measurement device or output transducer.

DVMs generally have a fixed input-voltage range, typically 0 to 199.9 mV for unipolar models and -199.9 to $+199.9$ mV for bipolar models. This limits their utility for wide-range voltage measurements, and so it is convenient to add a decade voltage divider to the input of the DVM as shown in Figure 23.10B. When the rotary switch is in position "a," the entire input voltage is connected to the DVM, so the range is simply the intrinsic range of the DVM. When the switch is in position "b," one-tenth of the input voltage is applied, the range is 10 times the range of the DVM, or ± 1.999 V, and so on for ranges of ± 19.99 V (position "c"), ± 199.9 V (position "d"), and ± 1999 V (position "e"). The convenience of the decade voltage divider is offset to some extent by the fact that it decreases the input resistance of the voltmeter to 10 MΩ, the total resistance of the divider. This value is still rather large and should not be a problem in most applications, but the potential for loading a high-resistance voltage source must be recognized.

An analogous circuit can be used to adapt the DVM for DC-current measurements as illustrated in Figure 23.10C. The idea here is to connect the current source in series with a small precision resistor, to measure the resulting IR drop across the resistor, and to calculate the current from Ohm's law. For the example shown, the

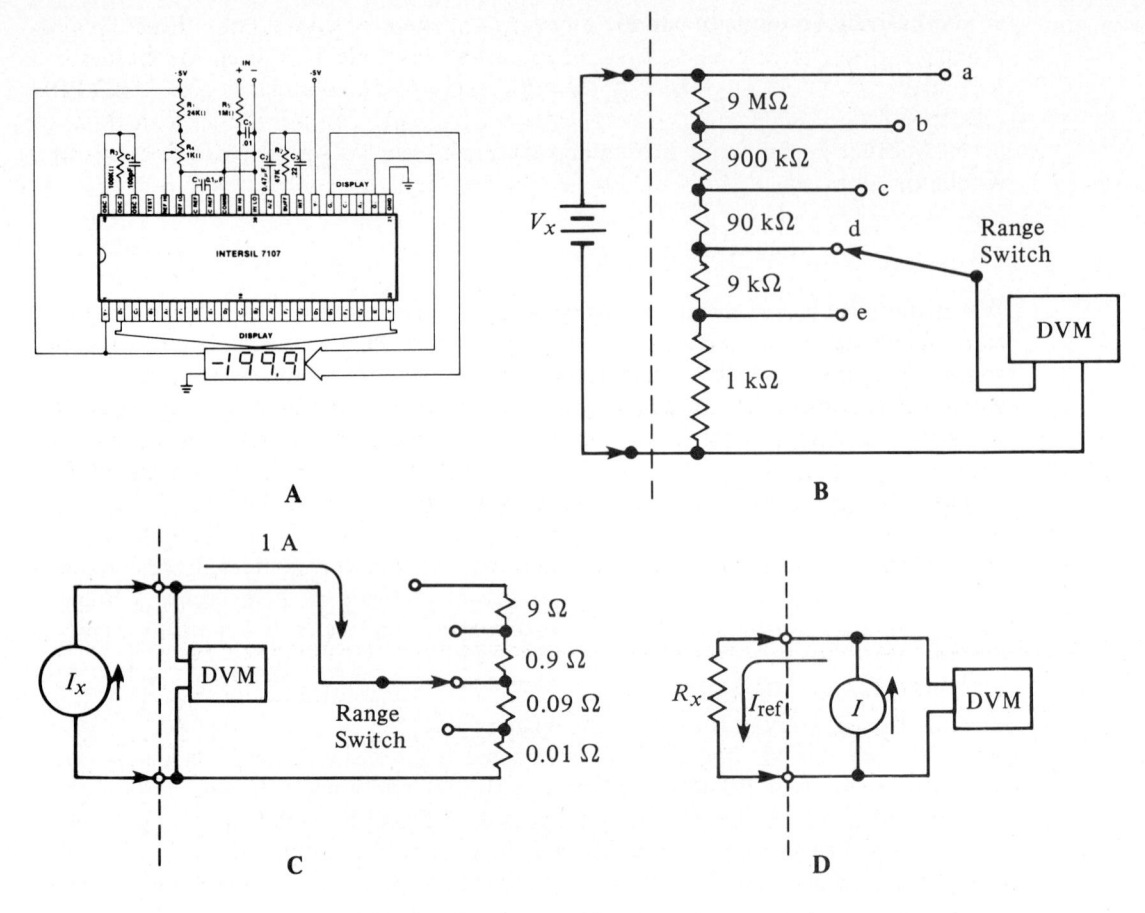

FIGURE 23.10. *Digital voltmeter. A: Commercial integrated-circuit digital voltmeter. From Intersil, Inc., 3 1/2 Digit A to D Converter Evaluation Kits, Cupertino, CA, 1981. Copyright 1981, Intersil, Inc. B: Multirange digital voltmeter; V_x is an unknown voltage to be measured. C: Multirange digital current meter; I_x is an unknown current. D: Resistance measurement with a digital voltmeter; R_x is an unknown resistance.*

1-A current source is passed through the bottom pair of resistors in the chain, whose total resistance is 0.1 Ω, so the voltage drop across the resistors is $V = IR =$ (1 A)(0.1 Ω) = 100 mV. To make the DVM read directly in amperes, we need only move the decimal point on the meter face. This is easily accomplished using a second rotary switch that is mechanically linked to the range switch.

Resistances can be measured using a DVM configured as shown in Figure 23.10D. A constant-current source is connected in series with the unknown resistance R_x, and the measured potential difference gives the resistance directly from $R_x = V/I$. The current source is constructed so as to produce a convenient amount of current such as 1 mA; so that if a 1-kΩ resistor is measured, exactly 1.000 V is read on the meter, a value that can be interpreted as 1.000 kΩ.

Many different manufacturers offer packaged versions of the three circuits discussed above. In addition, these meters usually include provisions for measuring AC current and voltage. Such devices are called *digital multimeters* (DMMs). They sell for as little as $75 for a battery-powered, portable, $3\frac{1}{2}$-digit model with $\pm 0.5\%$ accuracy on the DC volts function. Very high accuracy, high-resolution, microprocessor-controlled models can cost as much as several thousand dollars.

Equivalent Circuits

We have shown how series and parallel combinations of resistors can be drawn as simplified equivalent circuits. In fact, any combination of resistors and voltage sources with a voltage output is equivalent to a battery or a voltage source in series with a single resistor, called the *Thévinin equivalent circuit*. The equivalent resistance, R_{th}, is the resistance of the circuit when all voltage sources are set to zero; and the equivalent voltage, V_{th}, is the "no-load" voltage—the voltage at the output of the circuit as measured by a very high resistance voltmeter.

As an example, consider the simple voltage divider of Figure 23.11A. If we short circuit the voltage source, R_{th} is the parallel combination of R_1 and R_2, or

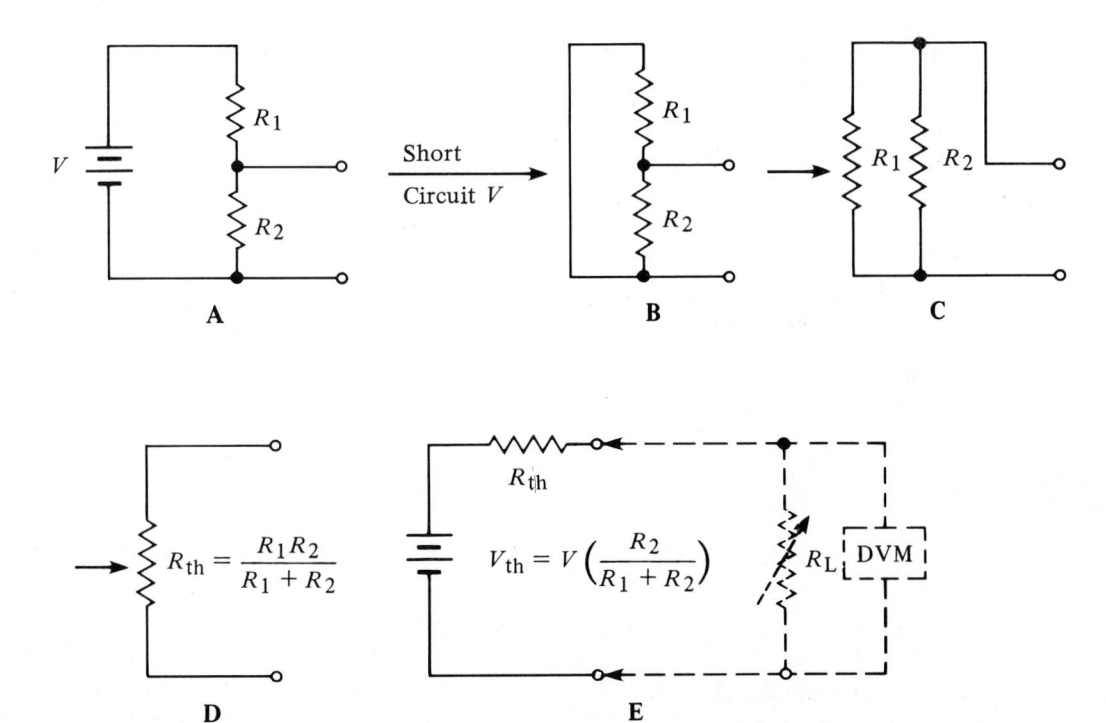

FIGURE 23.11. *Thévinen equivalent circuit. A: A voltage divider attached to a voltage source. B: The voltage source is set to zero. C: The circuit is redrawn to emphasize that it is now two resistors in parallel. D: The circuit equivalent R_{th} is determined. E: Experimental determination of R_{th} by attaching a variable load resistor and a DVM.*

$R_{th} = R_P = R_1R_2/(R_1 + R_2)$. The no-load voltage is calculated from the divider expression (Eqn. 23.13) to be $V_{th} = V[R_2/(R_1 + R_2)]$. The resulting circuit, shown in Figure 23.11E, is called the Thévenin equivalent circuit. The value of R_{th} can be measured experimentally by attaching a variable load resistor such as R_L in Figure 23.11E (dashed lines), and changing the resistance until the measured voltage is one-half the no-load voltage. This is a useful technique for obtaining the characteristic output resistance of any circuit.

23.3 AC QUANTITIES AND MEASUREMENTS

Up to this point in our discussion, we have dealt exclusively with DC quantities. However, *all* signals contain fundamental AC or time-varying components in spite of the best efforts to eliminate them. In fact, as we will see in Chapter 24, many of the most sophisticated measurement techniques rely on the time-domain characteristics of signals for their success, and AC measurements are often used to avoid fundamental noise sources inherent in DC transducers and signals. Some knowledge of AC circuits is therefore indispensable in understanding and applying modern data-acquisition and signal-processing technology.

Consider the periodic signals of Figure 23.12. These signals are characterized by the time duration of consecutive repetitive waveforms t, which is called the *period*. The *frequency* f is the number of cycles per unit time. Because there is one cycle per period, the relationship between f and t is given by

$$f = \frac{1}{t} \tag{23.14}$$

The units of frequency are hertz (Hz) or $\sec^{-1}$.

As illustrated in Figure 23.12, periodic waveforms are also characterized by their amplitudes, which can be in either the current or the voltage domain. There are several ways to represent amplitudes, two of which are shown: the *peak-to-peak voltage*, V_{p-p}, and the *peak voltage*, V_p. The amplitude of a sinusoidal waveform at any time t is given by

$$v = V_p \sin 2\pi f t = V_p \sin \omega t \tag{23.15}$$

or

$$i = I_p \sin 2\pi f t = I_p \sin \omega t \tag{23.16}$$

where ω = *angular frequency* in radians/second ($=2\pi f$).

Note the use of lowercase i and v for time-varying quantities, whereas uppercase variables are used for DC or time-invariant quantities. The sine, square, and triangle waveforms are centered about zero, and thus their negative and positive peak values are equal. For the sawtooth and pulse waveforms presented, $V_{p-p} = V_p$ because the signals are unipolar.

A third useful measure of amplitude is the *root mean square* (rms) voltage or current. This measure is defined such that 1 A of AC current produces the same

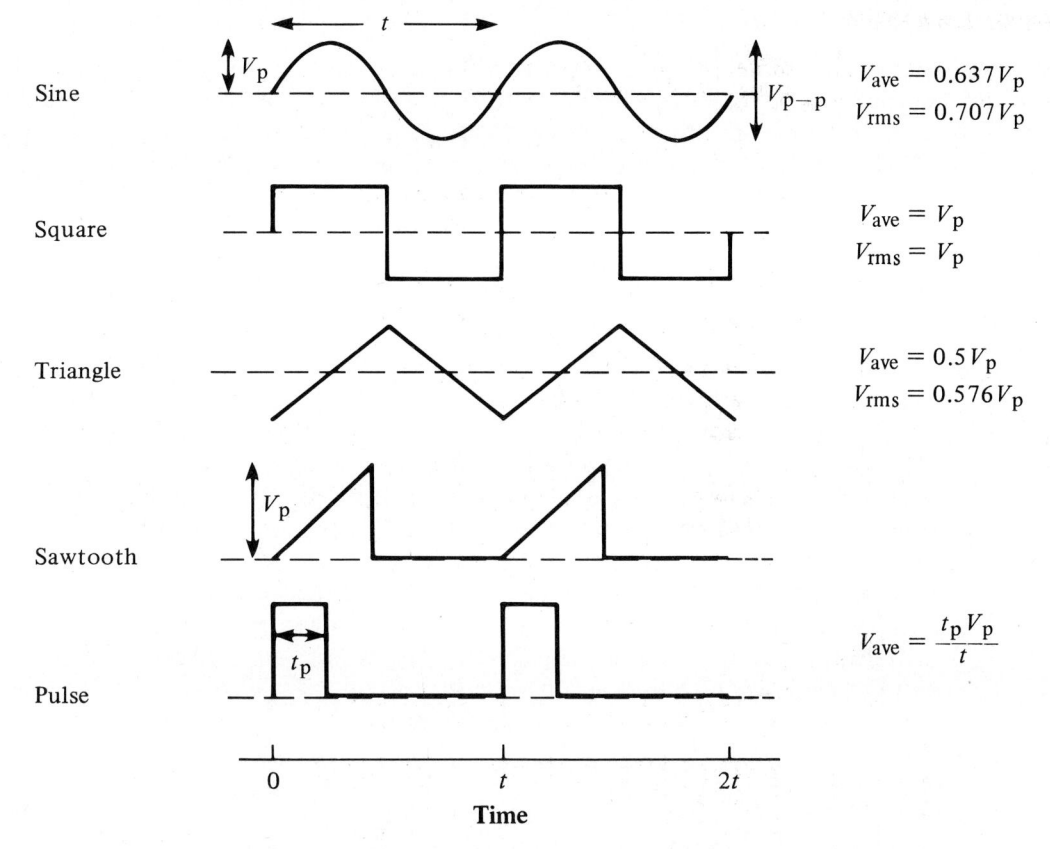

FIGURE 23.12. *Examples of periodic signals. The period of the signal, t, is the same in each of the five signals.*

amount of heat as would 1 A of DC current in the same resistor. The definition can be shown to lead to $V_{rms} = 0.707 V_p$ and $I_{rms} = 0.707 I_p$ for a sinusoidal waveform.

Finally, it is sometimes convenient to express amplitudes as the average absolute value of the magnitude over a complete cycle of the wave, for $V_{ave} = 0.637 V_p$ and $I_{ave} = 0.637 I_p$ for sinusoids. For reasons of economy, most AC meters are average-reading meters, so they give accurate readings only for sinusoidal AC signals. In order for such devices to be used for other waveforms, the appropriate correction factor must be applied.

In the section that follows, a brief introduction to the analysis of AC circuits with sinusoidal signal sources is presented. This analysis might at first seem to be a very specific tool that has little utility for other circuits and waveforms. However, *any* waveform can be synthesized from the sum of sinusoids that have appropriate amplitude, frequency, and phase relationship to one another. It is also true that any waveform can be analyzed and the constituent sinusoids determined by *Fourier analysis*, as will be discussed in Chapter 24.

Reactive Circuits

When a sinusoidal voltage is applied across a pure resistor, the resulting instantaneous current is given by Ohm's law: $i = v/R$ or $I_p = V_p/R$. Other circuit elements such as capacitors and inductors do not respond instantaneously to time-varying signals, however. So for AC circuits, a new term called *impedance*, Z, is defined that takes this into account. Impedance is exactly analogous to resistance in DC circuits and is given by $Z = V_p/I_p$, and hence the units of impedance are ohms. As the frequency of the sinusoid approaches zero—that is, as the signal becomes DC—the impedance becomes simply the resistance. Series and parallel impedances are additive in exactly the same manner as are resistances (see Eqns. 23.6 and 23.11).

Capacitance

The simplest capacitor comprises two parallel conducting plates separated by a dielectric such as air. If the capacitor is connected to a DC voltage source V, work is performed on the capacitor as charge is removed from one plate and placed on the other. A finite period of time is required for the capacitor to charge to voltage V. It is the inability of a capacitor to instantaneously change its voltage that is of interest in AC circuits. The "capacity" of a capacitor to store charge is called the *capacitance*, C. The charge Q (coulombs) on a capacitor C (farads) with an applied voltage V (volts) is given by $Q = CV$ (Eqn. 1.1).

If a sinusoidally varying voltage is applied to a capacitor, the current required to charge the capacitor is given by

$$i = \frac{dq}{dt} = C\frac{dv}{dt} \tag{23.17}$$

By differentiating Equation 23.15, we have

$$\frac{dv}{dt} = \frac{d(V_p \sin \omega t)}{dt} = \omega V_p \cos \omega t \tag{23.18}$$

and so

$$i = \omega C V_p \cos \omega t \tag{23.19}$$

The maximum value of the cosine function occurs at $t = 0$, and thus $I_p = \omega C V_p$. Finally, the impedance of the capacitor, called the *capacitive reactance*, X_C, is given by

$$Z = X_C = \frac{V_p}{I_p} = \frac{1}{\omega C} = \frac{1}{2\pi f C} \tag{23.20}$$

It should be clear from Equation 23.20 that, at low frequency, the quantity $1/\omega C$ is large and capacitors tend to behave as open circuits. Conversely, as the frequency becomes very high, the capacitive reactance becomes small and capacitors act as short circuits. Equation 23.19 also indicates that the current in a capacitor is a cosine waveform whereas the voltage is a sine waveform. This means that there is a *phase difference* between the current and the voltage of $\pi/2$ or $90°$. Hence, the current through a capacitor is said to "lead" the voltage drop across the capacitor by $90°$.

In a similar fashion, it can be shown that the *inductive reactance* of an inductor of inductance L (henries) is given by $X_L = \omega L$. An inductor usually consists of a coil of copper wire, often surrounding a soft iron core. These devices resist changes in current such that there is a 90° phase difference between the current and the voltage of an inductor; but, in this case, the voltage leads the current. The inductive reactance is zero as $\omega \rightarrow 0$ and is directly proportional to the frequency. An inductor can be considered sort of an "inverse capacitor."

Series RC Circuit

A very useful circuit in a wide variety of applications is the series RC circuit shown in Figure 23.13A. An AC voltage source of peak voltage V_i is impressed upon a series combination of a resistor R and a capacitor C. The total impedance of the circuit is given by the vector sum of the resistance and the capacitive reactance as depicted in Figure 23.13B, so that $Z^2 = X_C^2 + R^2$. We can obtain the voltage across the capacitor if we realize that this circuit is an AC voltage divider. Hence, the peak voltage across the capacitor, V_o, is given by

$$V_o = \left(\frac{X_C}{Z}\right) V_i = \left(\frac{X_C}{\sqrt{X_C^2 + R^2}}\right) V_i \qquad (23.21)$$

It is useful to consider how V_o/V_i, the *transfer function*, varies as a function of frequency. At low frequency, $X_C \gg R$, and thus $V_o/V_i \rightarrow 1$; and at high frequency,

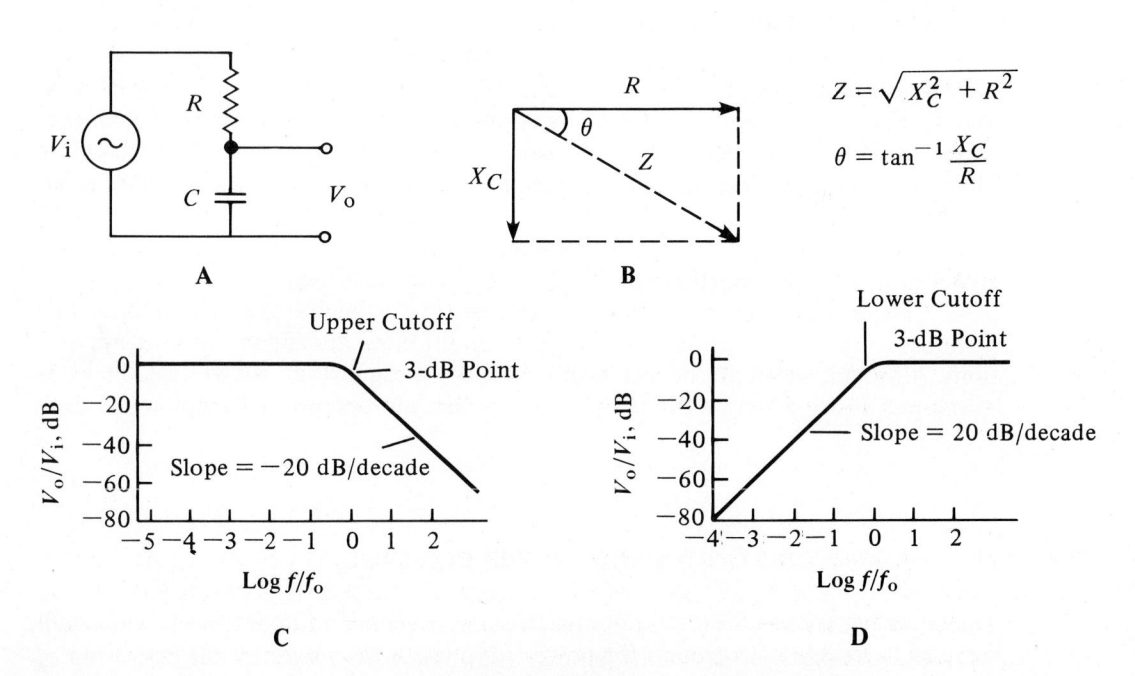

FIGURE 23.13. Series RC circuit. A: Schematic diagram. B: Vector diagram showing the phase and magnitude relationships among R, X_c, and Z. C: Bode diagram for a low-pass filter. D: Bode diagram for a high-pass filter.

$X_C \ll R$, so $V_o/V_i \to 0$. The intermediate frequency, at which $X_C = R$, is defined as the *upper cutoff frequency*, ω_o. Thus, we have

$$\frac{1}{\omega_o C} = R \quad \text{or} \quad \omega_o = \frac{1}{RC} \quad \text{and} \quad f_o = \frac{1}{2\pi RC} \tag{23.22}$$

At this frequency, the transfer function is

$$\frac{V_o}{V_i} = \frac{X_C}{\sqrt{X_C^2 + R^2}} = \frac{R}{\sqrt{2R^2}} = \frac{1}{\sqrt{2}} = 0.707 \tag{23.23}$$

These facts are easily visualized with the aid of the plot of Figure 23.13C, in which the transfer function in decibels (dB) is presented as a function of $\log f/f_o$. This figure is called a *Bode plot*. The decibel results from the definition of *power gain*, which is $10 \log(P_o/P_i)$. From the relationship $P = V^2/R$ (Eqn. 23.5), we see that for voltage ratios such as the transfer function, the voltage gain of the circuit is $20 \log(V_o/V_i)$. The point on the Bode plot corresponding to the upper cutoff frequency is called the *3-dB point*, because at this point $20 \log 0.707 = -3$ dB. At frequencies beyond the 3-dB point, the slope of the plot reaches a limiting value of -20 dB per decade of frequency; thus, the transfer function is said to "roll off" at -20 dB/decade.

The circuit of Figure 23.13A is called a *low-pass filter* because signals of frequencies lower than the cutoff frequency pass through essentially unchanged, and those of frequencies higher than the cutoff frequency are attenuated, or "filtered out". Low-pass filters are often used at the input of DVMs and other DC-measurement devices to reduce undesirable high-frequency noise.

If the roles of the resistor and capacitor of the low-pass filter are reversed by taking V_o across the resistor, we obtain a *high-pass filter*. The transfer function of the high-pass filter is given by $V_o/V_i = R/(R^2 + X_C^2)^{1/2}$. At low frequency, X_C is very large and V_o/V_i is very small. As the frequency increases past the *lower cutoff frequency* (Fig. 23.13D), $f_o = 1/\omega RC$, $V_o/V_i \to 1$, and $20 \log V_o/V_i \to 0$ as with the low-pass filter. Obviously, the high-pass filter is used when DC and low frequencies are undesirable, and high-frequency signals are of interest. This device is useful at the inputs of AC-measurement devices such as an oscilloscope in order to block the DC portion of the signal so that the AC portion can be observed and measured.

There are many configurations of resistors, capacitors, and inductors that have been used in a wide variety of signal-conditioning and signal-filtering applications. Although we shall not specifically consider these further, we will use the ideas developed above when we discuss power supplies and operational-amplifier circuits.

23.4 SEMICONDUCTOR DEVICES AND POWER CONTROL

The following section presents a brief discussion of semiconductor diodes, which will serve as necessary background for power supplies. A discussion of the operation of transistors, simple amplifiers, and difference amplifiers will then follow, preparing us for the development of operational amplifiers in Section 23.5.

Semiconductor Basics

In pure crystalline silicon, each atom is surrounded by four other silicon atoms so that the four bonding electrons on a silicon atom are paired to form four covalent bonds. If a group V element such as arsenic is added homogeneously to pure silicon at a concentration of about 1 ppm, the properties of the material change dramatically. Each arsenic atom is surrounded by four silicon atoms, but because arsenic has one more electron in its outermost energy level, the extra electrons can enhance the electrical conductivity of the material upon thermal excitation. In this example, arsenic is referred to as the *dopant*, and the resulting material is called an *n-type semiconductor* because the majority charge carriers move in the direction of *n*egative charge flow.

In a similar way, it is possible to add a small amount of a group III element, such as gallium, to pure silicon. Because gallium has one fewer bonding electron than silicon, a vacancy, or *hole*, is produced for each dopant atom added. Electrons in the crystal lattice adjacent to a hole can move into the hole, leaving an electron vacancy behind. This results in the motion of holes in the direction opposite to that of electron flow. These materials are called *p-type* because the mobile charge carriers are holes that migrate in the direction of *p*ositive charge flow.

Diodes

It is possible to fabricate metallurgically a device that qualitatively appears as a piece of p-type material joined to a piece of n-type material with wires attached to both sides as shown in Figure 23.14A. The region between the p-type and n-type materials is termed a *pn junction*, and the resulting device is called a *diode*. In the diagram, open circles represent holes and filled circles represent electrons. The symbol for a diode, depicted in Figure 23.14B, is reminiscent of an arrow pointing from the p-type to the n-type material.

A qualitative understanding of the operation of a semiconductor diode can be achieved by considering Figure 23.14C. The upper circuit applies a positive voltage or *bias* to the anode of the diode, and the diode is said to be *forward biased*. The schematic of the diode itself suggests that both the holes in the p material as well as the electrons in the n material are repelled toward the pn junction. But their motion will be in the opposite directions. As electrons and holes cross the junction, they combine with holes and electrons normally present, and their net migration across the boundary results in a large current through the diode.

When the anode of the diode is made negative, the diode is said to be *reverse biased*. Holes in the p material are attracted toward the negative potential and electrons are attracted toward the more positive potential. This results in a region at the pn junction depleted of mobile charge carriers, and thus little current passes through the diode. This behavior results in the ideal characteristic *I-V* curve of Figure 23.14D, in which the diode has essentially zero resistance (slope = ∞) when forward biased, and infinite resistance (slope = 0) when reverse biased. This naive picture of diode operation is exactly analogous to the operation of a liquid flow check valve. In fact, the symbol for a diode is often used in flow-system schematics, and some manufacturers use the term *fluid diode* to describe check valves.

○ = Hole
● = Electron

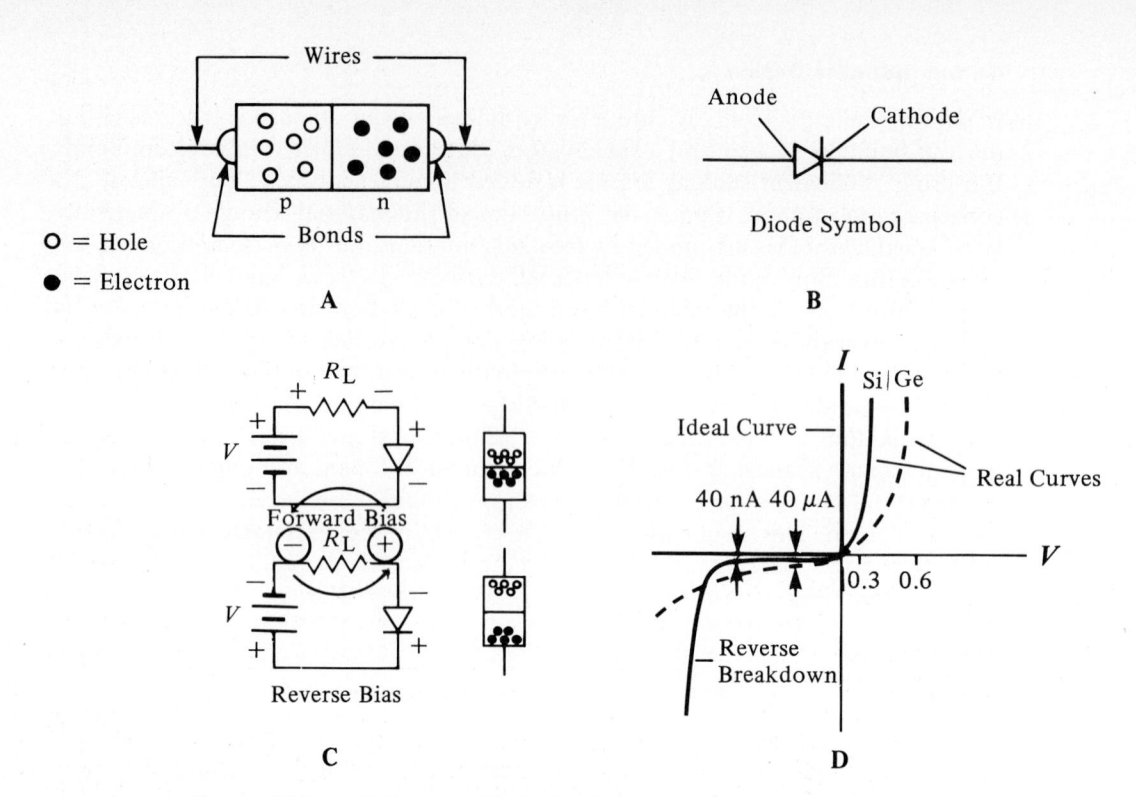

FIGURE 23.14. *Semiconductor diode. A: Schematic of a pn junction. B: Diode symbol. C: Biasing a diode. D: Comparison of the characteristic curve of an ideal diode (heavy solid line) with those of a silicon and a germanium diode.*

For a more complete theoretical picture of semiconductor devices, the reader is referred to the Selected Bibliography. However, it is important for us to consider the practicalities of real diodes, which do not exhibit the ideal characteristic curve of Figure 23.14D. The behavior of a real diode is described by the Shockley equation,

$$I = I_i(e^{Q_e V/kT} - 1) \qquad (23.24)$$

where I_i = temperature-dependent intrinsic current of the pn junction
Q_e = charge on the electron
V = voltage applied across the diode
k = Boltzmann constant in joules per kelvin
T = temperature in kelvins

At constant temperature, a diode is a circuit element whose current increases exponentially with the applied voltage, a fact that will serve us well when we consider exponential and logarithmic amplifiers.

The intrinsic current I_i is different for germanium and silicon diodes, and this results in somewhat different characteristic curves, as the plot of Figure 23.14D shows. A silicon diode carrying substantial current exhibits a voltage drop of approximately

0.3 V, whereas the voltage drop across a germanium diode is about 0.6 V under similar conditions. The small reverse-bias currents are also substantially different.

From Figure 23.14D, it is apparent that the reverse-bias voltage cannot increase without limit. As reverse-bias current begins to pass through a diode, the temperature of the junction increases. Hence, the intrinsic current increases, which results in increased temperature, and so on. This is known as *avalanche breakdown*, and if it proceeds unchecked, the pn junction will be destroyed. Diodes that are specially fabricated to have a sharp breakdown curve and to withstand the resulting temperature increase are called *zener diodes*. The nearly vertical *I-V* curve of reverse-biased zener diodes is used to advantage in voltage reference circuits and in power supplies.

Power Supplies

All of the voltage sources mentioned thus far have been represented as batteries. Although batteries provide relatively stable DC voltages, they are inconvenient because they discharge readily, and their voltages gradually change as they do so. The advent of semiconductor diodes, transistors, and integrated-circuit voltage regulators has made the construction of precision DC power supplies that use ordinary house current almost trivial.

A block diagram of a basic power supply is shown in Figure 23.15. Most modern power supplies consist of a *transformer* to convert 115-V AC house current (Fig. 23.15A) to an AC voltage level consistent with the desired DC output (Fig. 23.15B), a *rectifier* to convert the bipolar AC voltage to unipolar AC (Fig. 23.15C), a

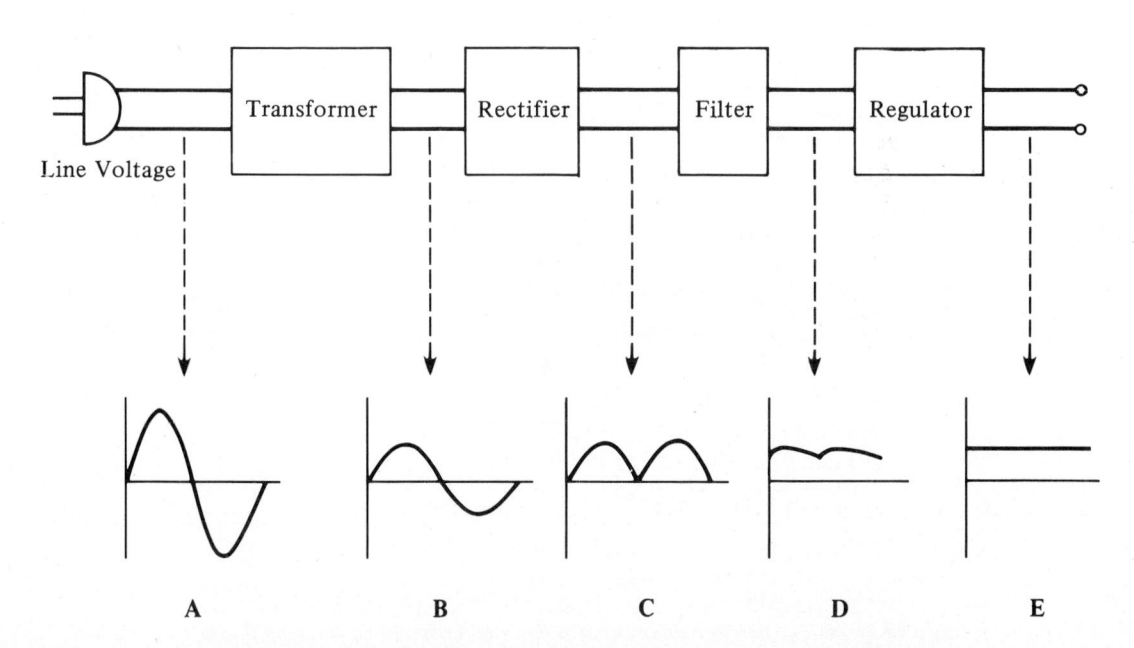

FIGURE 23.15. *Block diagram of a power supply and waveforms at various points in the circuit.*

filter to smooth the AC signal (Fig. 23.15D), and a *regulator* to remove most of the remaining AC signal and maintain a fixed DC voltage (Fig. 23.15E). We shall consider the components of the basic power supply in turn.

Line voltage, which is usually 110 to 120 V AC, is converted to a different AC voltage by a transformer. A transformer consists of a number of coils of wire (the primary) wound on a core of high magnetic permeability, and a second set of coils (the secondary) wound on the same core. An AC current in the primary induces an AC current in the secondary. The ratio of the voltage in the secondary to that in the primary, V_S/V_P, is directly proportional to the ratio of the number of coils in the secondary to that in the primary, N_S/N_P. For example, the transformer in Figure 23.16 has a larger number of windings in the secondary than in the primary, and thus produces a large voltage relative to the input voltage across the primary.

The bridge rectifier depicted in Figure 23.16 is one of the common configurations of germanium rectifier diodes used to convert bipolar signals to unipolar. During the positive half-cycle of a sinusoidal AC waveform, point "a" in the figure becomes positive with respect to point "b." Diode 1 is forward biased and provides a current path, denoted by the dashed line, into the positive node (+), while diode 3 is also forward biased to provide a path out of the negative node (−). Diodes 2 and 4 are reverse biased and act as open circuits to the current. During the negative half-cycle of the waveform, the roles of the diodes are reversed: Point "b" becomes positive relative to point "a," and thus diodes 2 and 4 are forward biased and diodes 1 and 3 are reverse biased. The current path indicated by the dotted lines out of the negative node and into the positive node results. Even though the current in the secondary of the transformer flows in opposite directions during the two halves of the cycle, the bridge rectifier directs the current in the same direction in the load R_L, as shown by *i*.

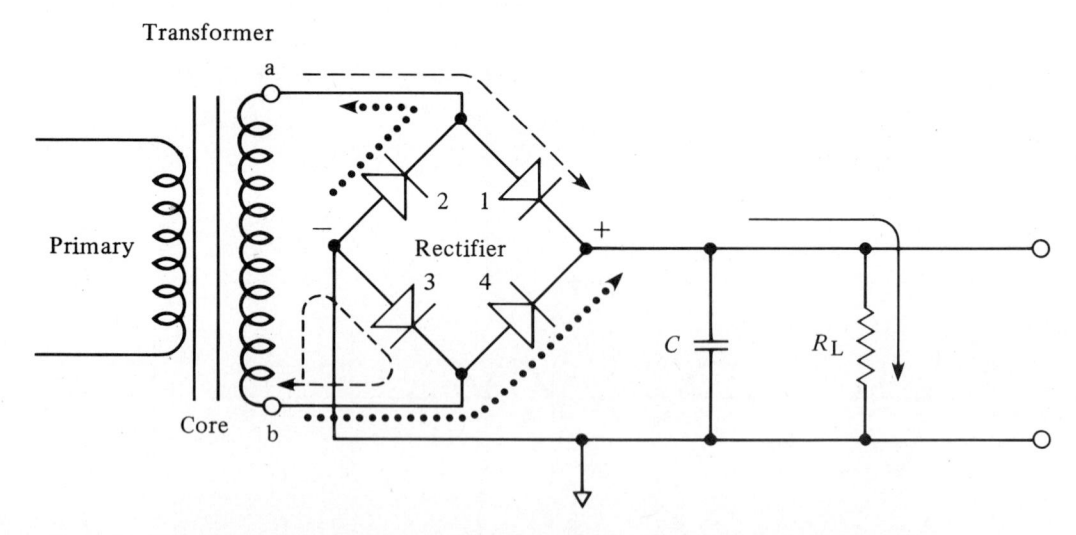

FIGURE 23.16. *Schematic diagram of a basic unregulated power supply. The dashed lines indicate the direction of current flow during the positive half-cycle of the AC voltage input; the dotted lines during the negative half-cycle.*

Without the filter capacitor C in the circuit, the voltage waveform across R_L exhibits the shape indicated in Figure 23.15C. This type of behavior is called *full-wave rectification*.

The filter capacitor is added to store charge during the portion of the waveform when the current through the rectifier is large so that the current in the load can be maintained during the portions of the waveform when the current through the bridge is low. The remaining AC component of the output of the power supply illustrated in Figure 23.15D is called *ripple*. The ripple factor, r, is defined as the ratio of the rms AC component to the average DC voltage:

$$r = \frac{I_{AC}}{I_{DC}} = \frac{V_{AC}}{V_{DC}} = \frac{1}{2\sqrt{3}fCR_L} \tag{23.25}$$

The average DC voltage is given by

$$V_{DC} = 1.4 V_{rms}\left(1 - \frac{1}{2fCR_L}\right) \tag{23.26}$$

where V_{rms} = rms voltage of the transformer secondary
       f = frequency of the full-wave rectified signal

For power-line signals, the frequency is 2×60 Hz = 120 Hz. As an example, consider the circuit of Figure 23.16 if $V_{rms} = 3.7$ V, $C = 470\,\mu$F, and $R = 200\,\Omega$. By applying Equations 23.25 and 23.26, $V_{DC} = 1.4(3.7\,\text{V})\{1 - [1/(2 \times 120\,\text{Hz} \times 4.70 \times 10^{-4}\,\text{F} \times 200\,\Omega]\} = 4.95$ V, and $r = 0.025$ or 2.5%. The ripple factor may seem acceptably low, but in many applications this 120-Hz component superimposed on the power-supply voltage is simply too large. Additionally, if the load varies, both V_{DC} and r change as well. For instance, if $R_L = 100\,\Omega$ in the previous example, $r = 5.1$% and $V_{DC} = 4.72$ V, which are substantially different from the previous values. It is usually necessary to regulate the voltage in the interest of accurate and precise measurements, and because many modern integrated circuits require highly stable power supplies for reliable operation.

The circuit of Figure 23.17A shows how a zener diode can be used to provide a

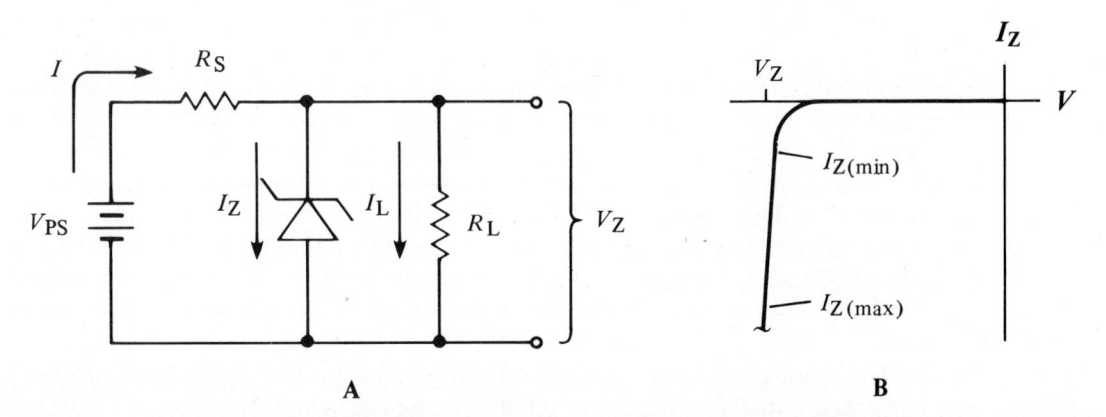

FIGURE 23.17. *A: Zener-regulated power supply. B: Reverse-bias characteristic of a zener diode.*

shunt regulator. As mentioned above, the nearly vertical reverse-bias breakdown curve (Fig. 23.17B) provides a wide range of currents over which the voltage across the diode is essentially constant. In the circuit shown, $I = I_Z + I_L$ is nearly constant, and the output voltage $V_Z = V_{PS} - IR_S$. If R_L decreases, I_L increases and I_Z decreases to maintain constant V_Z. If R_L increases, I_L decreases and I_Z increases. The series resistor R_S is chosen to be consistent with the maximum expected value of I, and to maintain I_Z between permissible limits. The maximum I_Z is determined from the power rating of the zener diode. The zener diode can respond very rapidly to regulate the 120-Hz ripple of a typical power supply. With careful design, zener-regulated voltage sources are stable to $\pm 0.01\%$ of the output voltage.

Transistors

It is impossible in limited space to treat more than a very small fraction of the tremendous number of available transistor devices. Our approach will be to examine two devices, the *npn bipolar-junction transistor* and the *n-channel field-effect transistor*, which will serve as background for the study of a number of more complex devices.

The npn bipolar transistor, depicted in Figure 23.18, is fabricated as a very thin layer of p-type material sandwiched between two layers of n-type material; hence the designation npn. Wires are bonded to the three regions of the device for electrical connection. The back-to-back pn junctions of the transistor suggest the diode model shown in the figure. This model is useful for demonstrating biasing in transistors, but it falls short in helping us to understand how transistors work because (a) in a real transistor, there is no wire separating the two halves of the p region of the transistor; and (b) real transistors are not actually symmetrical as shown. In fact, the center p-type region, called the *base* (B), is lightly doped and extremely thin. The upper n-type region, called the *collector* (C), is also lightly doped, but the lower n-type region, the *emitter* (E), is heavily doped to keep its resistance low. The emitter is designated by the small arrow in the direction of the base-emitter pn junction as shown in the circuit symbol.

The operation of the npn transistor can be understood by considering the biasing circuit of Figure 23.18B. The battery V_E applies a forward bias across the base-emitter junction, but the battery V_C reverse biases the collector-base junction. It is instructive to consider the flow of electrons through the transistor. Electrons move across the forward-biased base-emitter junction (in the direction opposite that of the current) into the base. At this point, electrons can recombine with holes in the p-type material of the base, or they can be swept on to the collector under the influence of the bias V_C. As a result of the thinness of the base region, the majority of electrons pass on to the collector, which results in the current I_C in the external circuit. Only a small fraction of the electrons recombine with holes to produce I_B. Thus, the ratio $I_C/I_E = \alpha$ is usually 0.95 to 0.99 for typical commercial transistors. The *current gain*, β, is defined as the ratio I_C/I_B, which for the typical values of α can be shown to yield values of 19 to 99.

The main point to be gained from this discussion is that bipolar transistors are current amplifiers of gain β. A relatively small expenditure of current into the base of an npn transistor results in a large current in the collector. A myriad useful circuits utilize this property of transistors in a variety of configurations in audio

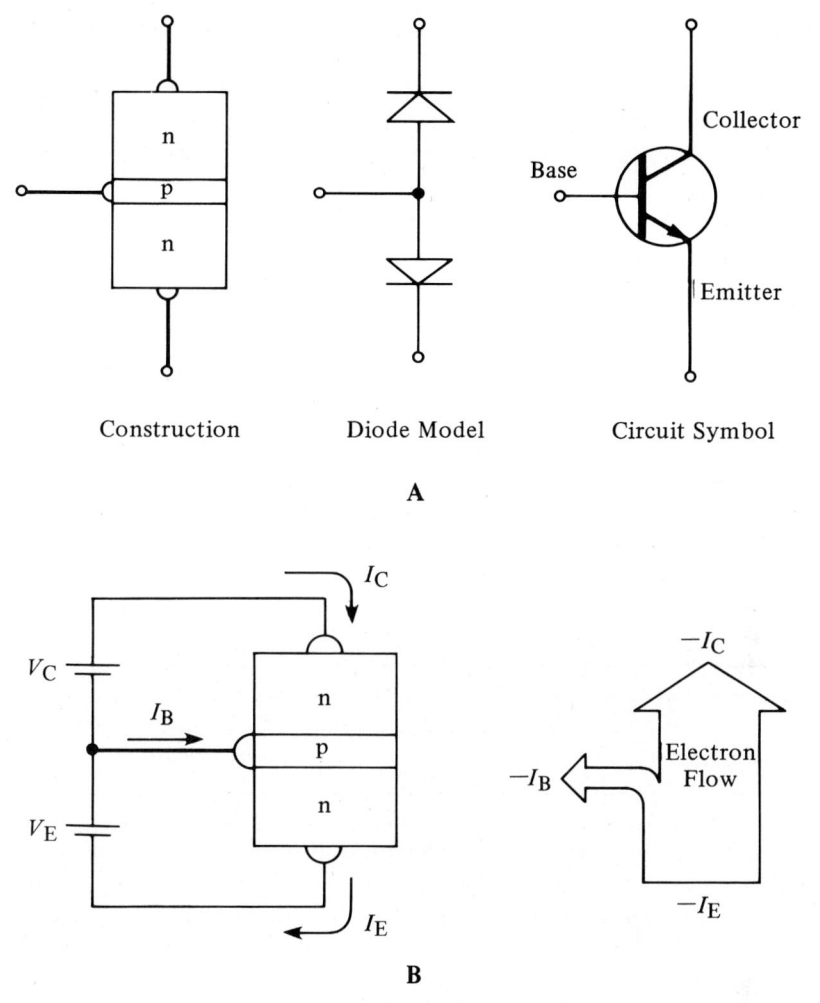

FIGURE 23.18. *A: npn bipolar-junction transistor. B: Biasing circuit for an npn transistor and diagram showing the flow of electrons and of current.*

amplifiers, power amplifiers, power supplies, operational amplifiers, and so forth. One such configuration is the common-emitter (CE) circuit of Figure 23.19A, so called because the emitter is connected directly to circuit common (usually the same as earth ground).

Characteristic curves for the common-emitter configuration are shown in Figure 23.19B. These are often supplied by manufacturers, or they can be generated for a given transistor by varying the collector-to-emitter voltage V_{CE} and measuring I_C with a current meter. When this is repeated for various constant values of $I_B = V_B/R_B$, the family of curves shown in the figure results.

The application of the circuit is best understood by a graphical device known as the *load line*. The load line is determined by applying Kirchhoff's voltage law to the loop containing V_C, R_L, and V_{CE}. Thus, $V_C - I_C R_L - V_{CE} = 0$; or dividing by R_L

SEC. 23.4 Semiconductor Devices and Power Control

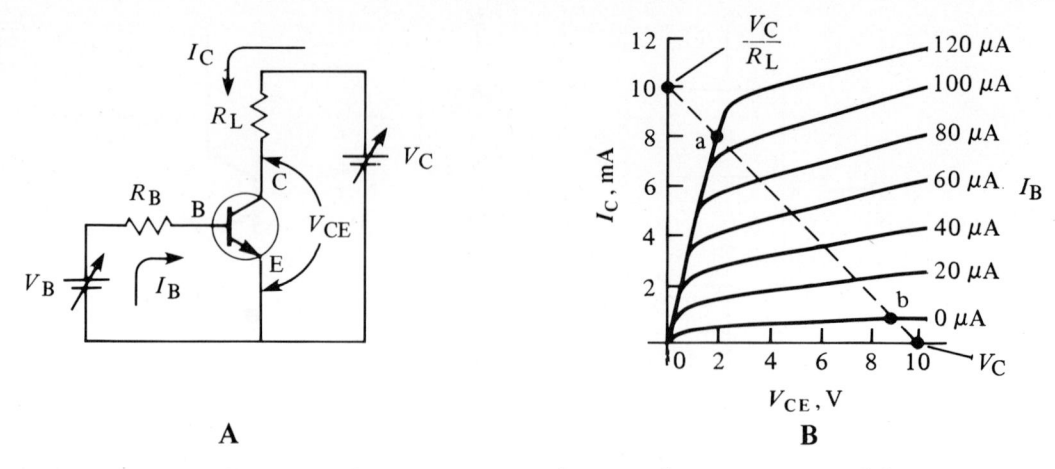

FIGURE 23.19. *A: Common-emitter configuration for a transistor amplifier. B: Characteristic curves and load line for the circuit in A.*

and rearranging, $I_C = -(1/R_L)V_{CE} + V_C/R_L$. The maximum value of I_C that the transistor can pass occurs if $V_{CE} = 0$ and thus $I_C = V_C/R_L$. The minimum value is of course $I_B = 0$, which would occur if the transistor could be shut off completely. In this case, the entire voltage drop V_C would appear across the transistor as V_{CE}. These two extremes determine the ends of the dashed load line shown superimposed on the family of curves. The slope of the line, as suggested by the equation, is $-1/R_L$. In this case, $V_C = 10$ V, $V_C/R_L = 10$ mA, and $R_L = 10$ V/10 mA $= 1$ kΩ.

Using the load line for this circuit, the collector current and V_{CE} can be estimated for any reasonable value of I_B. It should be mentioned that in practice, $I_C = V_C/R_L$ cannot be attained; that is, once I_B exceeds the value indicated at point "a" (approx. 105 μA), the transistor is in *saturation*. This results from the finite resistance of the transistor when the transistor is completely turned on. Likewise, when $I_B = 0$, a small current indicated by point "b" still passes through the transistor. Clearly, in order for a linear relationship to exist between I_B and i_C, the operating region of the transistor must be kept between points "a" and "b."

Digital transistor circuits such as the TTL family are designed to operate at saturation because they need only be either ON or OFF. Thus, the linear operating range is of little importance in these circuits.

Field-Effect Transistors

A second type of transistor that plays a major role in modern circuitry is the *field-effect transistor* (FET). There are several types of FETs, but we will consider only *junction field-effect transistors* (JFETs) here. A JFET is made of a single channel of either n-type or p-type material with contacts at each end, called the *drain* (D) and *source* (S), as shown in Figure 23.20A. The *gate* (G) is constructed by providing pn junctions along the channel and connecting them together as shown. The device illustrated is an n-channel JFET, but of course p-channel JFETs are available as well.

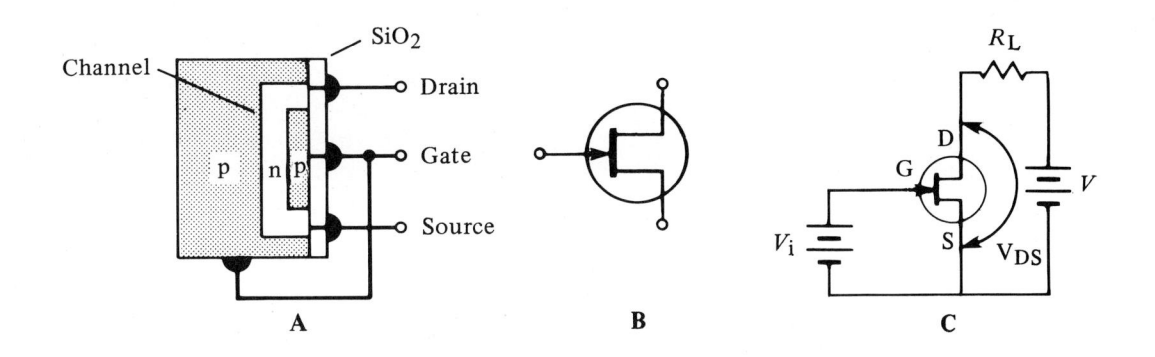

FIGURE 23.20. *Junction field-effect transistor (JFET). A: Construction. B: Circuit symbol. C: Biasing circuit.*

The conductivity of the channel between the drain and source is dependent on the concentration of charge carriers, which can be controlled by changing the thickness of the depletion region at the pn junctions of the device. For example, if a reverse bias is placed across the gate-source junction, the depletion region increases in size, dramatically increasing the effective resistance of the channel. If the bias is made positive, the channel resistance decreases accordingly. This means that if the JFET is biased properly, it can act as a voltage-dependent resistor. In addition, the input resistance of the gate is extremely high (approx. $10^8 \, \Omega$); so it does not load the signal source, and it provides excellent isolation characteristics between the control potential and the output signal.

The simplified biasing circuit of Figure 23.20C shows how the JFET can be used for voltage amplification. Any change in the input voltage V_i is reflected in a corresponding change in V_{DS}. The gain of the circuit is $A = \Delta V_{DS}/\Delta V_i$. Other pertinent circuit parameters can be obtained by a load-line analysis similar to that demonstrated for the bipolar-junction transistor.

Difference Amplifiers

The capability of amplifying small signal differences superimposed on large but constant signals is useful in a variety of applications. The input stage of an operational amplifier illustrates such an application. The circuit of Figure 23.21 is a JFET difference amplifier whose transfer function is given by

$$V_o = V_{o_1} - V_{o_2} = -A_1 \left(\frac{V_1 - V_2}{2} \right) - A_2 \left(\frac{V_1 - V_2}{2} \right) \quad (23.27)$$

where V_1 and V_2 = two input voltages of the differential amplifier
A_1 and A_2 = voltage gains of the JFETs Q_1 and Q_2
V_o = output voltage of the circuit

If the two JFETs are matched, that is, if $A_1 = A_2 = A$, then $V_o = -A(V_1 - V_2)$.

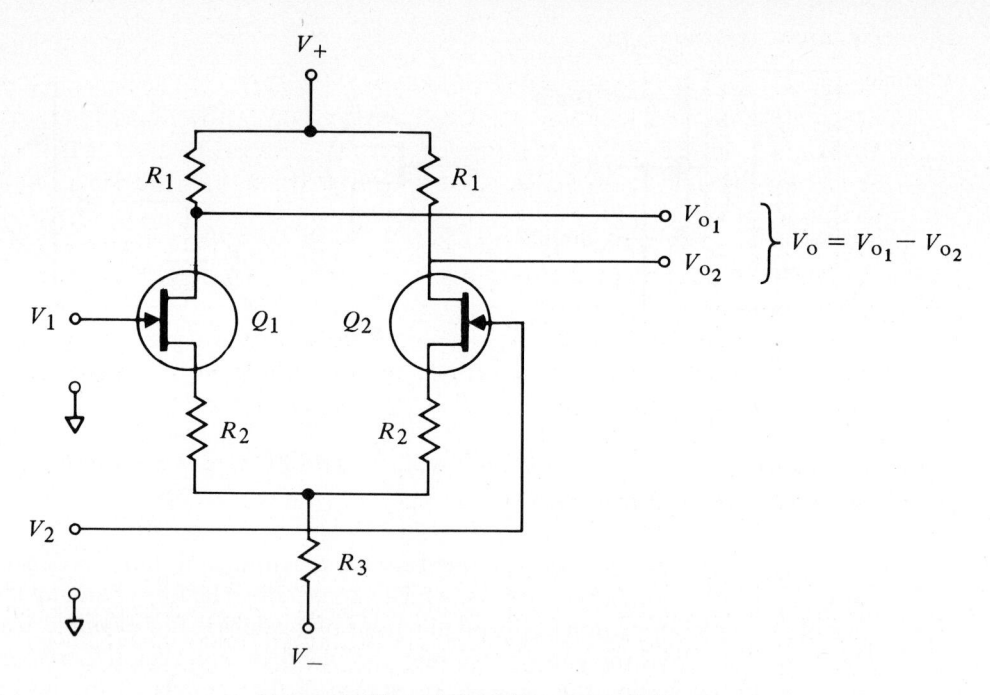

FIGURE 23.21. *JFET difference amplifier.*

Although a detailed analysis of this circuit is beyond the scope of the chapter, the circuit operation can be described qualitatively as follows. Suppose that initially $V_1 = V_2$. If V_1 increases and V_2 remains fixed, Q_1 begins to conduct to a greater extent than does Q_2, and V_{o_1} decreases with respect to V_{o_2}. Alternatively, if V_2 increases with respect to V_1, then Q_2 conducts to a greater extent than does Q_1, and V_{o_1} increases with respect to V_{o_2}. This action results in the inversion indicated by the negative sign in the transfer function. The total current in Q_1 and Q_2 is held constant by making R_3 very large, which results in a linear transfer function. The high input resistance of this circuit and the circuit's ability to subtract out *common-mode signals*, that is, signals present at both inputs, make it extremely useful in many different types of signal-conditioning circuits.

23.5 OPERATIONAL AMPLIFIERS: PRINCIPLES AND APPLICATIONS

Most modern analog signal-conditioning circuits owe their success to the class of integrated circuits known as *operational amplifiers* (op-amps or OAs). Op-amps are ubiquitous. Open any instrument or piece of electronic equipment, or scan any instrument schematic, and you will most likely find one or more op-amps. This fact, coupled with the ease with which relatively complex functions can be carried out with op-amp circuitry, emphasizes the importance of having a basic understanding of their principles of operation.

Op-Amp Basics

The circuit symbol for an op-amp is shown in Figure 23.22A. The op-amp is powered by supply voltages $+V_{ps}$ and $-V_{ps}$, which are typically $+15$ and -15 V, respectively. Op-amps have two inputs, the inverting input V_- and the noninverting input V_+, and a single output, V_o. Note that all three of these voltages are measured with respect to circuit common ($\triangledown$), which is nearly always equal to earth ground potential ($\bot$). From this point on, we will indicate a single circuit common for each of the circuits that are presented and assume that all voltages are measured with respect to the circuit common.

The op-amp is a high-gain amplifier consisting of three stages: an input stage that is very similar to the difference amplifier of Figure 23.21, a gain stage, and an output stage. The *transfer function* of the op-amp is given by

$$V_o = A(V_+ - V_-) = -AV_s \qquad (23.28)$$

where $A =$ open-loop gain of the op-amp
 $V_s = V_- - V_+$

This transfer function is shown graphically in Figure 23.22B. The *open-loop gain A* is the slope of the transfer function in the narrow linear range of the curve. This is the intrinsic gain of the amplifier when there is no circuit element connecting the output of the amplifier back to one of the inputs. For the typical values of V_s and V_o given in the figure, $A \cong 2 \times 10^5$. Note that outside the ± 10-V linear output range of the op-amp, the transistors internal to the device saturate, and the output reaches one of its limits, $+V_L$ or $-V_L$. These voltages are somewhat smaller in magnitude than V_{ps} and represent the practical upper limit of the magnitude of output voltages.

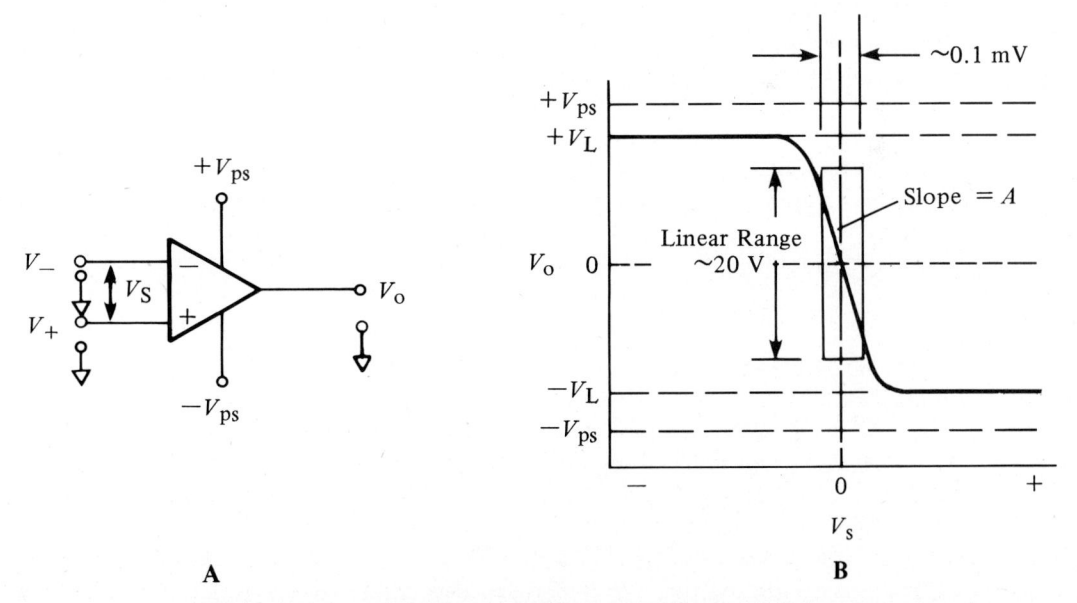

FIGURE 23.22. *Operational amplifier. A: Circuit symbol and associated input and output voltages. B: Transfer function.*

Op-amps are specifically designed to operate in the linear region in conjunction with networks of passive components connected to provide negative *feedback* from the output to the inverting input. As we shall see, the overall circuit gain is independent of the open-loop gain of the op-amp, and the transfer function of the circuit depends on the nature of the feedback network.

For many useful signal-conditioning circuits, we can assume that op-amps possess ideal characteristics:

1. The input resistance is infinite.
2. The output resistance is zero.
3. The open-loop gain A is infinite.
4. The output voltage is zero when the input voltages are zero.
5. The bandwidth is infinite.

Although no op-amp is truly ideal, the characteristics of many integrated-circuit op-amps approach the ideal, and for this reason our analysis will be valid for a majority of applications.

Voltage Follower

Our first use of negative feedback is illustrated by the op-amp voltage follower of Figure 23.23. This circuit is constructed by simply connecting the output to the inverting input so that $V_- = V_o$ and $V_+ = V_i$. If we substitute for V_- and V_+ in Equation 23.28, we have $V_o = A(V_i - V_o)$. Solving this equation for V_o gives

$$V_o = \frac{AV_i}{A+1} \qquad (23.29)$$

For very large A, the transfer function becomes $V_o = V_i$; that is, the output voltage follows the input voltage.

In this application and all others involving negative feedback, the action of the amplifier is to provide an output voltage sufficient to keep $V_s = V_- - V_+$ negligibly small. The actual value of V_s is the difference between V_o and V_i, which is the error in the transfer function. For an open-loop gain of 2×10^5, the error is one part in 2×10^5, a value considerably below the resolution of most DVMs.

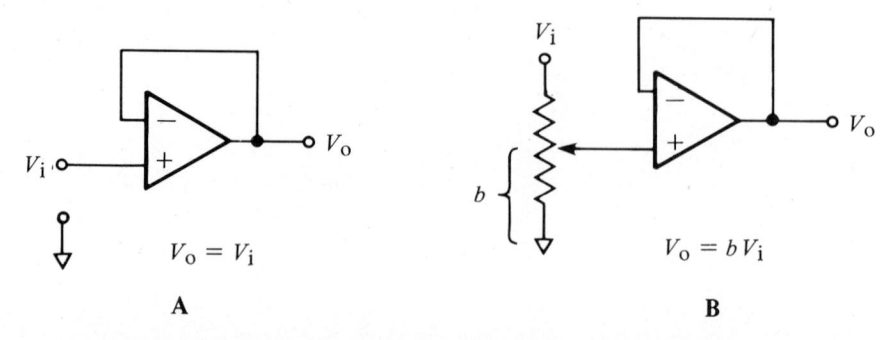

FIGURE 23.23. *A: Voltage follower. B: Unloading a voltage divider with a voltage follower.* b = *divider fraction.*

The voltage follower is extremely useful as a buffer amplifier because the input presents essentially no load to a voltage source ("infinite" input resistance) and the output is capable of driving substantial loads by virtue of the low output resistance of the op-amp (zero in the ideal case). As suggested earlier, a voltage follower can be connected to the output of a voltage divider to unload it. As shown in Figure 23.23B, the input of the op-amp presents essentially no load to the divider, and the op-amp output can drive a substantial load of 10 to 200 mA.

Follower with Gain

For applications in which low-level signals must be amplified, a variation on the voltage follower is desirable. Only a fraction of the output voltage is fed back to the inverting input of an op-amp, as illustrated in Figure 23.24A, so that $V_- = bV_o$. Because $V_i = V_+$, on substitution into Equation 23.28 as before, we have

$$V_o = \frac{AV_i}{1 + bA} \quad \text{or} \quad V_o \cong V_i\left(\frac{1}{b}\right) = V_i\left(\frac{R_1 + R_2}{R_2}\right) \tag{23.30}$$

when A is very large. Thus, the circuit gain, $1/b$ or $(R_1 + R_2)/R_2$, is independent of the open-loop gain. It can be arbitrarily selected (within limits) by choosing appropriate values of R_1 and R_2. In practice, a linear helical potentiometer with a calibrated dial can be used for both R_1 and R_2 with the wiper connected to the inverting input of the op-amp to adjust the gain. The nominal gain of the circuit can be dialed directly on a 1000-division dial as $D = 1000(1/\text{Gain}) = 1000b$.

Current Follower

A very useful op-amp circuit is the current follower, depicted in Figure 23.25A. In this circuit, a resistor R_f is connected from the output to the inverting input at point S to provide negative feedback. The noninverting input is connected directly to ground.

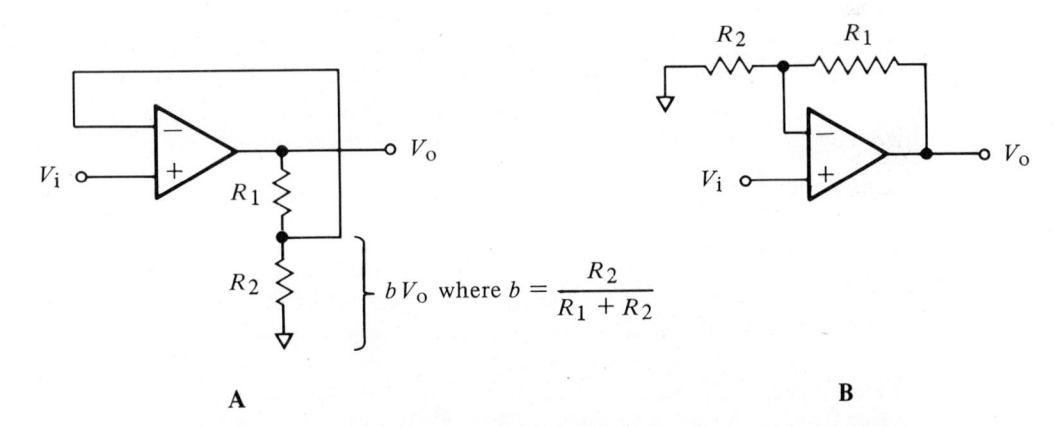

FIGURE 23.24. Follower with gain. A: Drawn to emphasize the voltage divider at the output of the op-amp. The fraction b is fed back to the inverting input. B: Circuit redrawn in the conventional manner.

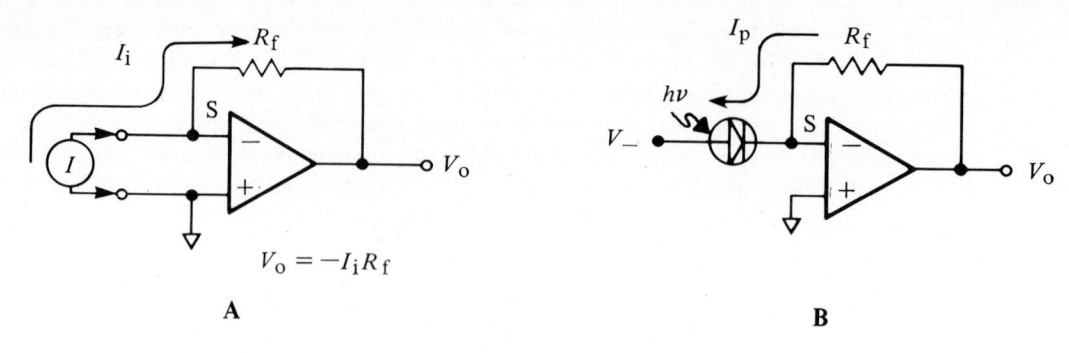

FIGURE 23.25. *Current follower, or current-to-voltage converter. A: Basic circuit. B: Application to the output current of a photodiode.*

Because V_s is very nearly 0 V, point S is kept at ground potential as well, and thus is termed a *virtual ground*. When a current source I_i is connected to point S, essentially all of the current must pass through R_f to the output. As you recall, the inverting input at point S draws no current due to its extremely high resistance. Since one end of R_f is at virtual ground and the other end is attached to the output of the op-amp, which is at potential V_o, then the IR drop across R_f must be

$$V_o = -I_i R_f \tag{23.31}$$

The negative sign indicates that the current passes out of the virtual ground at point S through R_f to the output. Thus, a current *into* point S produces a negative V_o, and a current *out of* point S produces a positive voltage. If the proper value of R_f, is chosen, a voltmeter attached to the output can be made to read current directly. For example, if $R_f = 1$ MΩ, the voltmeter indicates 1 μA per volt.

A typical application of the current follower in making data-domain conversions is illustrated in Figure 23.25B, in which a reverse-biased photodiode is connected to point S. When photons strike the pn junction of the photodiode, electron-hole pairs are created and photocurrent I_p is generated out of point S and through the diode as shown. This results in a positive voltage at the output of the op-amp, which is directly proportional to the light intensity incident on the photodiode. In addition, the current follower is often used in electrochemical instrumentation to monitor currents without placing a load on the electrodes themselves.

Op-Amp Inverting and Summing Amplifiers

When the current source of Figure 23.25A is replaced by a voltage V_i and a resistor R_i as shown in Figure 23.26A, an inverting-voltage amplifier is obtained. The input voltage imposed across R_i causes $I_i = V_i/R_i$ to pass into point S and through the feedback resistor R_f. Since I_i is also equal to $-V_o/R_f$ as in the current follower, we see that

$$\frac{V_i}{R_i} = -\frac{V_o}{R_f} \quad \text{or, by rearrangement,} \quad \frac{V_o}{V_i} = -\frac{R_f}{R_i} \tag{23.32}$$

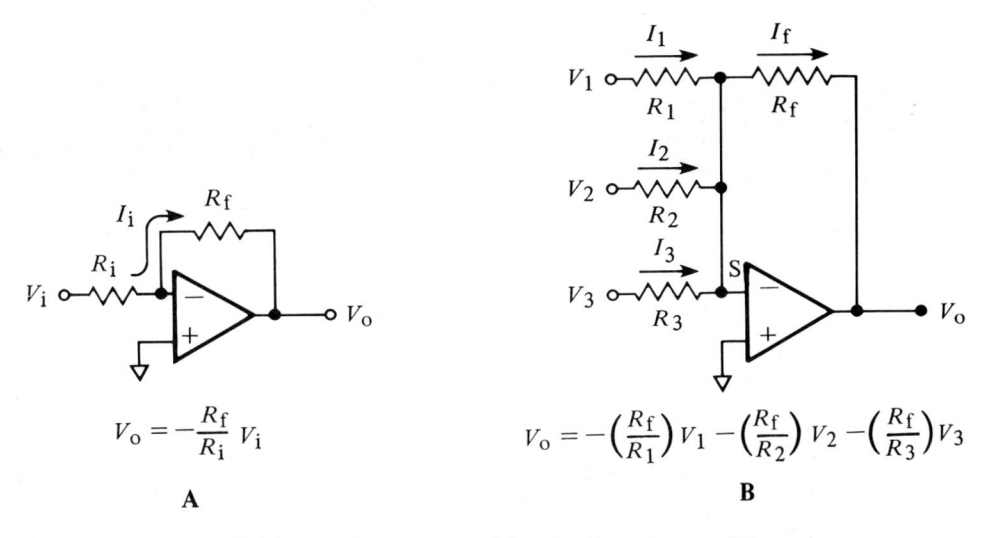

FIGURE 23.26. *A: Inverting amplifier. B: Summing amplifier.*

The transfer function of the inverting amplifier is, then, simply the negative ratio of the feedback resistor to the input resistor, and the *circuit gain*, or *signal gain*, is R_f/R_i. This circuit is applied in scaling circuits and in analog computation, particularly when it is advantageous to change the sign of a voltage.

An important variation on this theme is depicted in the weighted summing amplifier of Figure 23.26B. In this case, three voltage sources, V_1, V_2, and V_3, are connected across resistances R_1, R_2, and R_3 to generate currents I_1, I_2, and I_3 into point S. From Kirchhoff's current law, we conclude that $I_1 + I_2 + I_3 = I_f$; hence, point S is often called the *summing point*. If we substitute $I_1 = V_1/R_1$, $I_2 = V_2/R_2$, and $I_3 = V_3/R_3$ into this equation, and recognize that $I_f = -V_o/R_f$, we obtain

$$\frac{V_1}{R_1} + \frac{V_2}{R_2} + \frac{V_3}{R_3} = -\frac{V_o}{R_f} \tag{23.33}$$

On rearrangement, this yields

$$V_1\left(\frac{R_f}{R_1}\right) + V_2\left(\frac{R_f}{R_2}\right) + V_3\left(\frac{R_f}{R_3}\right) = -V_o \tag{23.34}$$

Inspection of Equation 23.34 reveals that each of the input voltages is weighted by the ratio of the feedback resistor to the corresponding input resistor.

Difference Amplifier

Another important application of op-amps is illustrated in the basic difference amplifier of Figure 23.27A. In this circuit, both the inverting input and the noninverting input are used to provide a voltage output proportional to the difference between two input voltages. The transfer function for this circuit is given by

$$V_o = K(V_2 - V_1) \tag{23.35}$$

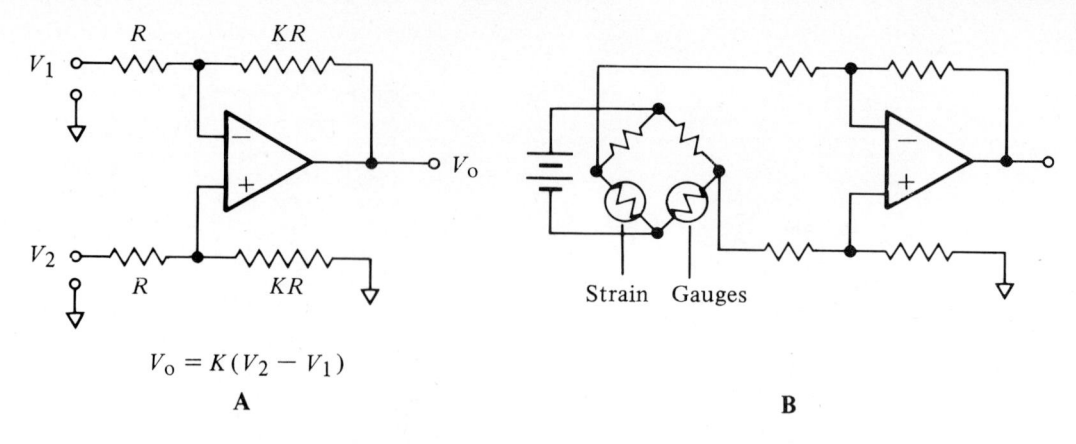

FIGURE 23.27. *A:* Op-amp difference amplifier. *B:* Wheatstone bridge application. Amplification of the off-balance bridge voltage.

$$V_o = K(V_2 - V_1)$$

where K = circuit gain
V_1 and V_2 = the two input voltages

The circuit provides high accuracy, which depends essentially on the accuracy of the resistors used in the input and feedback networks, and gain, which depends only on the value of K. It is useful when common-mode signals are to be rejected, such as in the Wheatstone bridge circuit of Figure 23.27B. The matched strain gauges in the bridge might be located in different parts of an instrument or transducer when a voltage output that is proportional to differential strain is desired. A common example of this type of measurement system is found in pressure transducers. Small differences in pressure across a thin stainless-steel diaphragm cause flexing of strain gauges attached on both sides of the diaphragm, thus producing changes in the resistances of the gauges. These resistance changes produce corresponding small changes in the voltage across the bridge, which must be amplified in order to drive some readout device. The difference amplifier serves to buffer the measurement circuit from the readout device.

By clever use of difference amplifiers, inverting amplifiers, and weighted summing amplifiers, any of the four basic mathematical operations can be carried out on analog signals so the signals can be scaled in virtually any desired manner.

Integrator

Up to this point, only resistive feedback and input elements have been considered. The addition of a capacitor to the feedback loop of an op-amp produces the charge-to-voltage converter shown in Figure 23.28A. When a charge source is connected to the summing point S of the op-amp, it is as though the source is being shorted to ground or, in this case, to circuit common. As you recall, the op-amp must then control V_o so that S is maintained at virtual ground. In so doing, the relationship $Q = CV$ for the capacitor must be satisfied, and thus the voltage output of the op-amp is given by $V_o = -Q_i/C$.

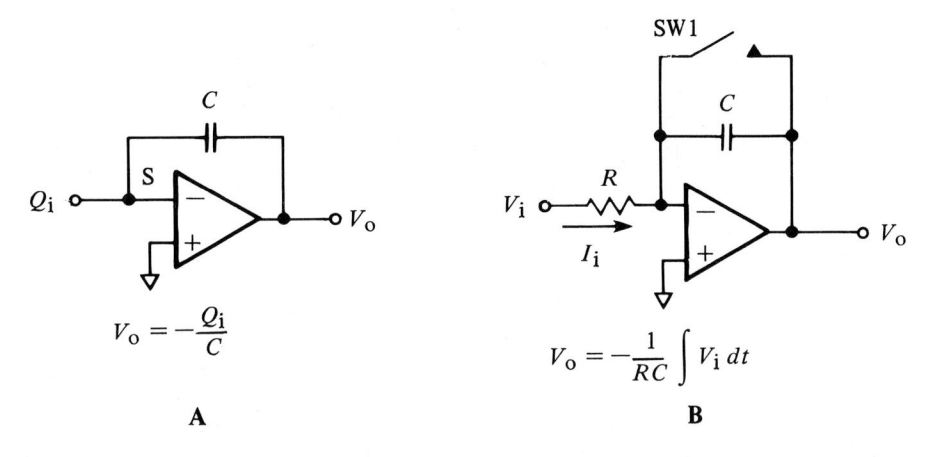

FIGURE 23.28. A: Charge-to-voltage converter. B: Basic integrator with reset switch.

When the charge source is replaced by a current source or a voltage source across an input resistance as illustrated in Figure 23.28B, the charge amplifier becomes an integrator. The total quantity of charge accumulated by the capacitor in time t is given by

$$Q = \int_0^t I_i \, dt = \int_0^t \frac{V_i}{R} \, dt = -CV_o \qquad (23.36)$$

On rearrangement of this equation, the transfer function of the op-amp integrator becomes

$$V_o = -\frac{1}{RC} \int_0^t V_i \, dt \qquad (23.37)$$

The switch SW1 is used to momentarily short the feedback capacitor to initialize the output to 0 V. The voltage output is then directly proportional to the integral of the input voltage since the most recent opening of SW1. Perhaps the most common application of analog integration is in NMR spectrometers. The integral of the proton NMR signal is plotted continuously above the NMR spectrum to provide a measure of the number of protons in a particular chemical environment.

Op-amp integrators are widely used in modern instrumentation because of their ability to decrease the noise level in signals. Random electronic noise is often of high frequency, relative to the "true" signal of interest, and has equal probability of being positive or negative. Thus, integration of the signal for a time much longer than the period(s) of the noise will tend to diminish the net effect of the noise. Therefore, many instruments are designed to continuously and repetitively integrate a DC or slowly varying signal for a fixed period, and the voltage output is sent to a digital readout device. The digits displayed may be arbitrary units, and must be related empirically to fundamentally meaningful units via a calibration curve or, more commonly, can be internally converted by adjustable zero and gain controls to read directly in some desirable unit—milligrams per liter, for example.

Differentiator

When the input and feedback elements of the op-amp integrator are exchanged, the op-amp differentiator, shown in Figure 23.29A, is obtained. Exchanging the resistor and capacitor produces the inverse mathematical function, and so

$$V_o = -RC\frac{dV_i}{dt} \tag{23.38}$$

for the op-amp differentiator. Differentiators are used when inflections in instrumental response are significant, that is, when abrupt changes in the slope of a signal are expected. Thus, the differentiator is sensitive to high-frequency components of signals. Electronic noise is often of very high frequency, however, which would result in large output. For this reason, the frequency response of the differentiator is deliberately limited by inserting a resistor R_i into the input network as depicted in Figure 23.29B. The upper cutoff frequency of the input network is then $f = 1/2\pi R_i C$. The time constant $R_i C$ should be arranged to be approximately equal to the period of the highest frequency signal to be differentiated.

Logarithmic Amplifier

Nonlinear functions such as logarithms can be implemented using op-amp circuitry. The necessary exponential feedback element is provided by a pn junction as suggested by Equation 23.24. This is depicted in Figure 23.30. The Shockley equation (Eqn. 23.24) can be rewritten as

$$I \cong c_1 e^{c_2 V} \tag{23.39}$$

where I = current through the forward-biased diode
   V = voltage across the diode
   c_1 and c_2 = temperature-dependent constants

In this circuit, $I = I_i = V_i/R$, and $V = -V_o$. Therefore,

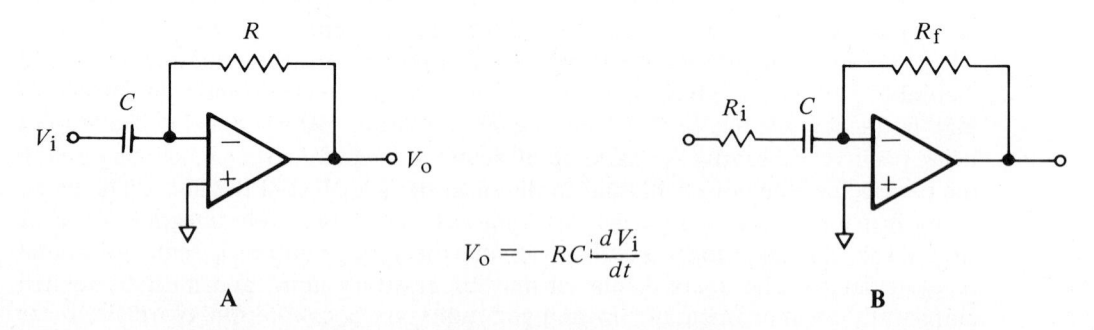

$$\frac{V_i}{R} = c_1 e^{-c_2 V_o} \quad \text{and} \quad \ln\frac{V_i}{R} = -c_2 V_o + \ln c_1 \tag{23.40}$$

FIGURE 23.29.   A: Basic differentiator. B: Differentiator with low-pass filter at the input.

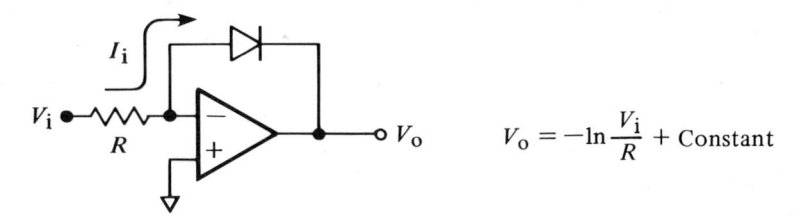

FIGURE 23.30. *Logarithmic amplifier.*

$$V_o = -\ln\frac{V_i}{R} + \text{Constant}$$

Equation 23.40 reveals a linear dependence of V_o upon ln V_i. Furthermore, the transfer function of the circuit can be scaled simply by changing R. Note also that there is a voltage offset due to the preexponential factor c_1.

Practical logarithmic circuits have sophisticated temperature- and frequency-compensation networks to ensure stability. Although the simple circuit of Figure 23.30 can be assembled for less than one dollar, at best it is accurate to only a few percent over one or two decades of input voltage. On the other hand, compensated logarithmic-function modules costing $20 to $100 are accurate to a few tenths of a percent, and are stable over four to six decades of input current and voltage.

Logarithmic-function modules based on op-amps find use in a variety of instruments in which signals are expected to have a wide dynamic range or in which the desired output is a logarithmic function of a measurable signal. One such application is in spectrophotometry, in which photocurrent from a photomultiplier or a photodiode can be processed by a current-to-voltage converter and then by a logarithmic amplifier to provide a voltage directly proportional to absorbance. Alternatively, photocurrent can be passed directly into the summing point of a logarithmic amplifier, thus circumventing the current-to-voltage conversion and noise sources inherent in the process. Recently, direct logarithmic conversion has become unfashionable because of the widespread use of integral microcomputers in chemical instrumentation. The devices make it possible to compute logarithms accurate to six to eight significant figures.

23.6 CHARACTERISTICS OF REAL OP-AMPS

We have deliberately avoided consideration of the nonidealities inherent in the operation of real op-amps. These characteristics are most important when the devices are operated at the threshold of their capabilities. In this last major section of the chapter, several of the most important of these characteristics are discussed, with special emphasis on the limitations that they place on the measurement process.

Gain and Frequency Response of Op-Amps

For most of the op-amp circuits discussed thus far, we have developed transfer functions that involve only DC quantities. Only in the development of the differentiator was the detection and measurement of AC quantities implied. However, one of the important properties of op-amps is that they are capable of processing a broad range of frequencies. This property is illustrated in the Bode plot of op-amp open-loop frequency response in Figure 23.31A.

The solid curve in Figure 23.31A represents the open-loop gain of a typical op-amp as a function of frequency. At low input-signal frequency, $A = 10^5$ or 100 dB, but as the frequency increases above 10 Hz, the gain rolls off at -20 dB/decade in the same fashion as in a low-pass filter. The frequency range from DC to the frequency at which $A = 1$ is called the *unity-gain bandwidth*, which for this amplifier is 1 MHz. This point also fixes the *gain bandwidth product* at 1 MHz, which is constant at all frequencies above 10 Hz. This fact enables the calculation of the bandwidth of a given amplifier configuration.

As an example, consider an inverting amplifier with a signal gain $A_{f_1} = -R_f/R_i = 1000$, as indicated by the upper dashed line in Figure 23.31A. The bandwidth of the amplifier is then 1 MHz/1000 = 1 kHz. An inverter with $A_{f_2} = 10$ then has a bandwidth of 100 kHz, and so on for other values of signal gain that might be chosen for a given op-amp. Hence, we see that negative feedback increases amplifier bandwidth, which can be calculated from the signal gain and the unity-gain bandwidth of the op-amp.

Two other parameters that relate to the speed or bandwidth of an amplifier are illustrated in Figure 23.31B. The output response of a voltage follower to a step input is characterized by the *rise time t_r*, which is the time required for the output to change from 10% to 90% of the total change. It can be shown that

$$t_r = \frac{0.35}{\text{Bandwidth}} \qquad (23.41)$$

For the amplifier with $A_{f_2} = 10$ discussed above, $t_r = 0.35/1$ MHz $= 0.35$ μsec. From the rate of change of voltage at the output during the transition, the *slew rate* can be

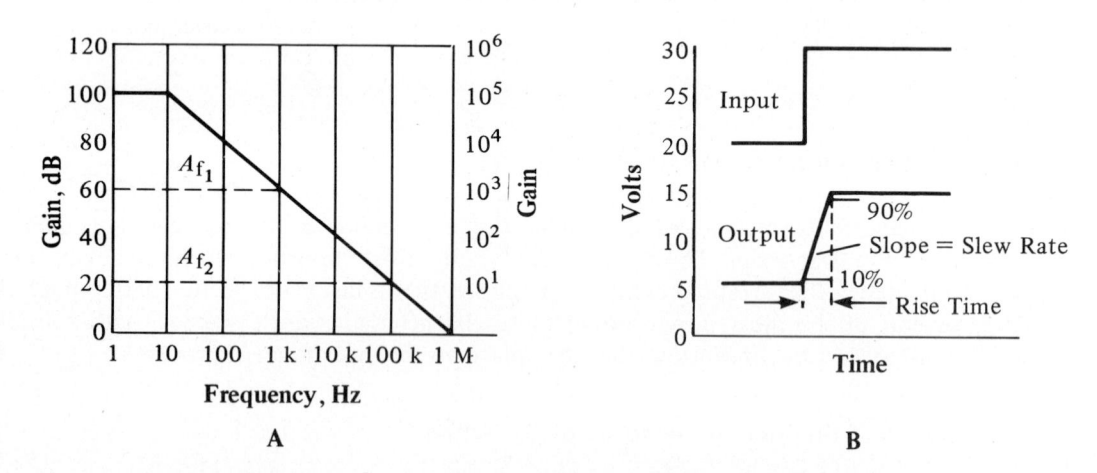

FIGURE 23.31. *Frequency characteristics of op-amps. A: Bode diagram. The dashed lines represent amplifiers with closed-loop gains of 1000 (A_{f_1}) and of 10 (A_{f_2}). B: Op-amp response to a step input in voltage.*

calculated as follows:

$$\text{Slew Rate} = \frac{\Delta V}{\Delta t} = \frac{5\text{ V}}{0.35\ \mu\text{sec}} = 14\text{ V}/\mu\text{sec} \qquad (23.42)$$

These parameters are of relatively little importance at DC or low frequency, but when op-amp circuits are required to amplify high-frequency signals or faithfully reproduce high-speed pulsed signals, such parameters must be considered.

Input Bias Current and Offset Voltage

We began our discussion of op-amps by assuming that the input resistance of an op-amp is infinite and therefore draws zero current. The input resistance of typical integrated-circuit op-amps ranges from 10^{12} to $10^{14}\ \Omega$. When a finite voltage appears at either input of an op-amp, a small *input bias current* results from the *IR* drop across the input resistance. These currents range from 10^{-8} to 10^{-15} A, but typically they are of the order of 50 pA for JFET input op-amps. Input bias currents generally are not significant except when very small currents are being measured. If there is a mismatch between the inputs of an op-amp, the difference between the two input bias currents is called the *input offset current*. Both of these nonideal characteristics can result in voltage errors at the output, particularly when low-level signals are being processed.

Mismatched components at the input of an op-amp can also generate an *offset voltage*. This offset causes the transfer function of Figure 23.22B to be shifted to the right or to the left depending on the sign of the offset. This, of course, results in an error at the output, which in most cases can be compensated for. Manufacturers of op-amps recommend simple offset-nulling circuits that can be used to adjust the output to zero when the input is connected to circuit common. Recently, the convenience of op-amp use has been increased considerably by the use of *laser trimming*. In this process, the manufacturer uses a laser to trim components on the integrated-circuit op-amp under dynamic conditions before it is encapsulated. In this way, the offset voltage is reduced to 0.5 to 5 mV depending on the model and cost.

In the next chapter, digital circuits, microcomputers, and interdomain converters will be explored. As will be seen, a special type of op-amp, the *comparator*, plays a central role in the implementation of these devices.

SELECTED BIBLIOGRAPHY

BERLIN, H. M. *Design of Op Amp Circuits with Experiments.* Indianapolis, Ind.: Howard W. Sams, 1977. *A useful introduction to op-amps containing well-written experiments.*

DIEFENDERFER, A. J. *Principles of Electronic Instrumentation.* Philadelphia: Saunders, 1979. *A textbook for a course in scientific instrumentation organized in the traditional format with references to manufacturers' literature on specific integrated circuits.*

HIGGINS, R. J. *Electronics with Digital and Analog Integrated Circuits.* Englewood Cliffs, N.J.: Prentice-Hall, 1983. *Similar to Diefenderfer but at a slightly more advanced level. Written in a light and lively style.*

HOLLER, F. J., AVERY, J. P., CROUCH, S. R., and ENKE, C. G. *Experiments in Electronics, Instrumentation, and Microcomputers.* Menlo Park, Calif.: Benjamin, 1982. *Written to accompany Malmstadt et al. Contains a large* number of experiments from simple circuits to microcomputer data acquisition.

JUNG, W. G. *IC Op-Amp Cookbook*. Indianapolis, Ind.: Howard W. Sams, 1974. Contains a wealth of practical information regarding practical op-amp circuits. A bit dated but still useful.

MALMSTADT, H. V., ENKE, C. G., and CROUCH, S. R. *Electronics and Instrumentation for Scientists*. Menlo Park, Calif.: Benjamin, 1981. Textbook for a course in scientific instrumentation. Organized according to the "top-down" approach.

SMITH, J. I. *Modern Operational Circuit Design*. New York: John Wiley, 1971. A sometimes whimsical approach to the basics of op-amps. Some parts are out of date, but the style is refreshing and the concepts are timeless.

REFERENCES

1. C. G. ENKE, *Anal. Chem.*, 43, 69A (1971).

PROBLEMS

1. A 25.0-kΩ resistor has a voltage of 5.00 V applied across it. (a) What is the current in the resistor? (b) How much power does the resistor consume?

2. An 8.00-V battery is connected across the series combination of a 0.100-kΩ, a 0.500-kΩ, and a 1.00-kΩ resistor. (a) What is the total resistance of the network? (b) What is the current in the loop? (c) What is the *IR* drop across each of the resistors?

3. Suppose that the three resistors of Problem 2 are connected in parallel to the same battery. Calculate (a) the total current in the circuit; (b) the current in each of the resistors.

4. The 8.00-V battery of Problems 2 and 3 is now connected directly to the 0.100-kΩ resistor, and the 0.500- and 1.00-kΩ resistors form a parallel network in series with the first resistor. Calculate (a) the total resistance of this network; (b) the current through each resistor; (c) the voltage drop across each resistor; (d) the power dissipation in each resistor.

5. Design a voltage divider that will give voltages of 0.5, 1.0, 2.0, and 5.0 V from a 10-V power source, and that has a total resistance of 20 kΩ.

6. Prove that the average voltage for a sinusoidal waveform is $V_{ave} = 0.637 V_{p-p}$.

7. Resistances in series are additive (Eqn. 23.6); resistances in parallel are inversely additive (Eqn. 23.11). Derive equations showing the total capacitance of (a) capacitors in series, and (b) capacitors in parallel.

8. Ordinary line voltage is 60 Hz. Calculate (a) the frequency in radians per second, and (b) the period in seconds.

9. A fairly typical value for the double-layer capacitance of an electrode in aqueous solution is about 0.20 F/m². Conductance bridges often come with a choice of two frequencies: 60 and 1000 Hz. Calculate the capacitive impedance at these two frequencies of a cell with two 1-cm² electrodes.

10. Determine the Thévenin equivalent circuit for a voltage divider made of a 3.00-kΩ and a 6.00-kΩ resistor across a 6.00-V battery. The output of the divider is across the 6-kΩ resistor.

11. Consider a high-pass filter made of a 15-kΩ resistor in series with a 0.01-µF capacitor. (a) Sketch the circuit. (b) What is the lower cutoff frequency of the filter? (c) What is the impedance of the circuit at 100 Hz? At 1 kHz? At 100 kHz? (d) Calculate the phase angle at 100 Hz and at 100 kHz.

12. Construct a Bode plot for the high-pass filter of Problem 11. Label the axes and any points or regions of interest.

13. Calculate the impedance at 60 Hz of (a) a series circuit consisting of a 10-kΩ resistor, a 1-µF capacitor, and a 1-mH inductance, and (b) the same components in parallel.

14. Consider the following data for a solid-state device.

I, A	V, V
1.88×10^{-5}	0.1
9.23×10^{-4}	0.2
4.43×10^{-2}	0.3
0.213	0.4
10.2	0.5
491	0.6

(a) Plot I versus V and $\ln(I)$ versus V. (b) Find the intrinsic current I_i for the diode. (c) Find I for $V = -2.00$ V, i.e., under reverse-bias conditions. (d) Is the diode made of silicon or germanium? (e) Is the value of I for $V = 0.6$ V reasonable? Explain.

15. (a) Calculate the ripple factor for an unregulated supply with $R_L = 50\,\Omega$ and $C = 2000\,\mu F$. (b) Calculate V_{ave} for the same power supply if $V_{rms} = 8.7$ V.

16. What is the maximum current that a 1/4-W, 5.0-V zener diode could be expected to carry?

17. Estimate the collector current for the emitter follower of Figure 23.19A if $V_C = 6$ V, $R_L = 500\,\Omega$, and $I_B = 40\,\mu A$.

18. The gain A of a typical op-amp is 320 V/mV. (a) Calculate the minimum voltage V_S required to cause V_o to change from $+V_1$ to $-V_1$ if $V_1 = 12$ V. (b) Given that the slew rate of the op-amp output is 12 V/μs, what is the minimum time required for the voltage change in part (a)?

19. (a) Design an op-amp circuit with a noninverting gain of 80. (b) Design a similar amplifier with three switch-selectable gains of 1, 10, and 100.

20. One of the disadvantages of the op-amp inverting amplifier as contrasted with the voltage follower is that the signal being ampli- fied must be capable of driving the input resistance of the inverter, which is simply R_i. Suggest a method for overcoming this problem.

21. Suppose a thermistor has a transfer function $R_T = k/T$, where k is a constant characteristic of the device, and T is the absolute temperature. Design an op-amp circuit whose output voltage depends linearly on absolute temperature. The circuit symbol for a thermistor is like that of a strain gauge.

22. The excitation waveform used in differential-pulse polarography is synthesized by summing waveforms A and B shown below. (a) Design an op-amp circuit to generate B, and perform this function. (b) How can the relative magnitudes of components A and B be changed in the resultant waveform?

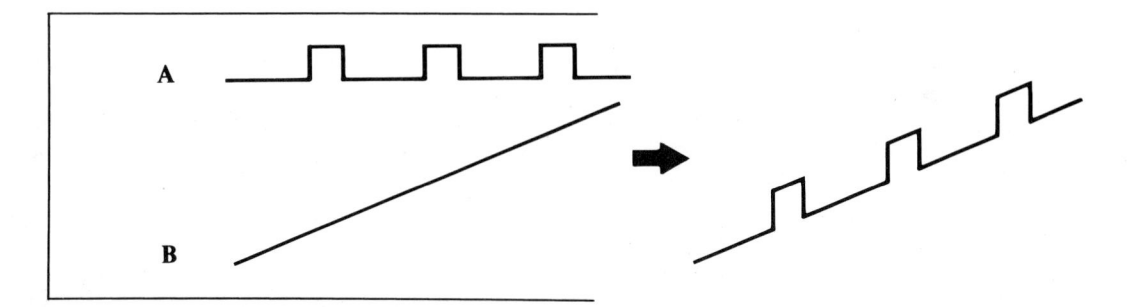

23. Derive the transfer function for the op-amp differentiator of Figure 23.29A.

24. (a) Design a circuit capable of processing the signals from two photodiodes to obtain a voltage proportional to the logarithm of the ratio of the light intensities incident on the photodiodes. (b) Derive the transfer function for the circuit.

25. Sketch a data-domains map similar to that of Figure 23.1. Assume that a DVM has been connected to the output of the circuit of Figure 23.25B to provide a voltage readout. Trace a path on the data-domains map to indicate each data-domain conversion that takes place during the measurement process. Specify the device that accomplishes each conversion.

26. An extremely fast op-amp has a rise time of 5 nsec at unity gain. (a) Calculate the unity-gain bandwidth. (b) Find the gain-bandwidth product. (c) Given that the open-loop gain of the

op-amp is 70 dB, sketch a Bode plot for the device. (d) Determine the bandwidth of the amplifier when the amplifier is connected as a follower with gain with $R_1 = 99\ \text{k}\Omega$ and $R_2 = 1\ \text{k}\Omega$.

27. Analog-to-digital converters often come as 8-, 10-, or 12-bit models. If analog data are converted to digital binary signals by these devices, what is the relative uncertainty or precision attainable with each one?

CHAPTER **24**

Digital Electronics, Data-Domain Conversions, and Microcomputers

F. James Holler

This chapter continues the journey from discrete circuits to high-level systems and devices. The discussion of data domains and analog circuits has laid a foundation for discussing digital logic, and basic gates and flip-flops. The operation of simple counters, analog switches, comparators, and timing devices will be explored as we approach the interface between the analog and the digital domains. This discussion will be followed by a description of large-scale integrated (LSI) circuit counters and a number of their functions including counting, period, frequency, and timing measurements.

    The last half of this chapter treats a variety of useful types of data-domain conversions, such as voltage-to-frequency conversions, and will close with the exploration of how microcomputers are interfaced with the real world and how data are acquired and processed by them.

24.1 DIGITAL LOGIC AND GATES

As was suggested in Chapter 23, digital information is represented by one of two allowed states, HI or LO. These states can represent the binary numbers 1 and 0 or the logic conditions TRUE and FALSE, respectively. The amplitudes of electrical signals used to represent binary states vary with the logic family in use and its particular electrical characteristics. Average voltages corresponding to HI and LO logic levels are illustrated in Figure 24.1 for three common logic families. TTL (transistor-transistor logic) and its variations constitute probably the most common logic family. However, as very large scale integrated-circuit (VLSI) technology has evolved, CMOS (complementary metal-oxide semiconductor) and its variations have become commonplace. ECL (emitter-coupled logic) devices exhibit very high speed and correspondingly high power consumption; these have been used particularly in high-speed computers and counting applications.

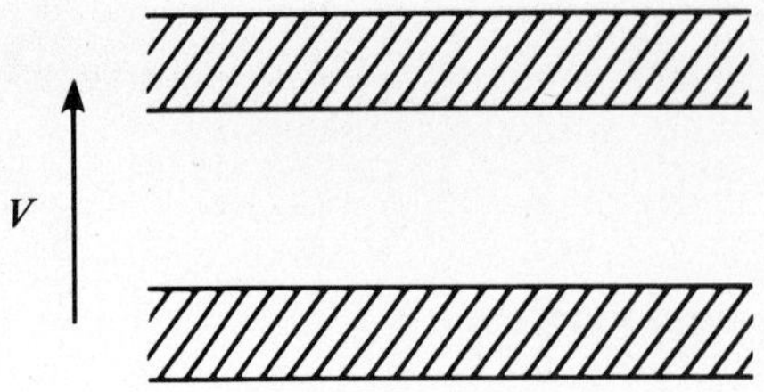

	Logic Family		
	TTL	CMOS	ECL
HI	3.0 V	V_{CC}	−0.9 V
LO	0.2 V	0 V	−1.75 V

FIGURE 24.1. *Electrical characteristics of three common logic families. Voltages corresponding to logic states HI and LO. V_{CC} is the supply voltage for CMOS, which varies from +3 V to +15 V.*

At the outset of our discussion, it is important that we agree to assign HI to TRUE (or 1) and LO to FALSE (or 0). Although it is often useful to make the opposite assignment, we will stick to the *positive-true logic* or *HI-true logic* convention.

All logic functions, including the simple examples described here, can be represented and manipulated according to the rules of Boolean algebra. Boolean algebra is useful both for representing complex logical functions in their simplest form and for the synthesis of logical functions with available functional building blocks.

The AND Gate

Consider the circuit of Figure 24.2A, in which two switches are connected in series with a battery and a lamp. The function of the circuit can be represented by the following statement: If switch *A and* switch *B* are closed, then the lamp will light. According to the rules of Boolean algebra, this statement can be written

$$A \cdot B = Y \quad \text{or} \quad AB = Y \tag{24.1}$$

The implication is that a switch being closed is a TRUE condition, as is the lighted lamp, and that a switch being open is a FALSE condition, as is the lamp being off. The logical AND operation is represented in Equation 24.1 by "·" or by no symbol at all, which is analogous to ordinary algebraic multiplication.

The logic states for the lamp circuit are given by the *truth table* of Table 24.1, which shows that the lamp is OFF if either or both switches are OFF and is ON if both switches are ON.

A similar function is performed by the diode circuit of Figure 24.2B. In this example, HI = TRUE = 1 = 5 V and LO = FALSE = 0 = 0 V, as shown in Table 24.1. If either input *A* or input *B* is connected to ground (0), then at least one of the diodes is forward biased and the output *Y* is near ground potential (0). But if *both* *A* and B are at 5 V, then *Y* will be near 5 V, or logic 1. The circuit symbol for the AND function is illustrated in Figure 24.2C, and this symbol is used regardless of how the circuit is implemented electronically.

One of the significant advantages of integrated circuits is that their use does not require an understanding of the details of their operation. For instance, the

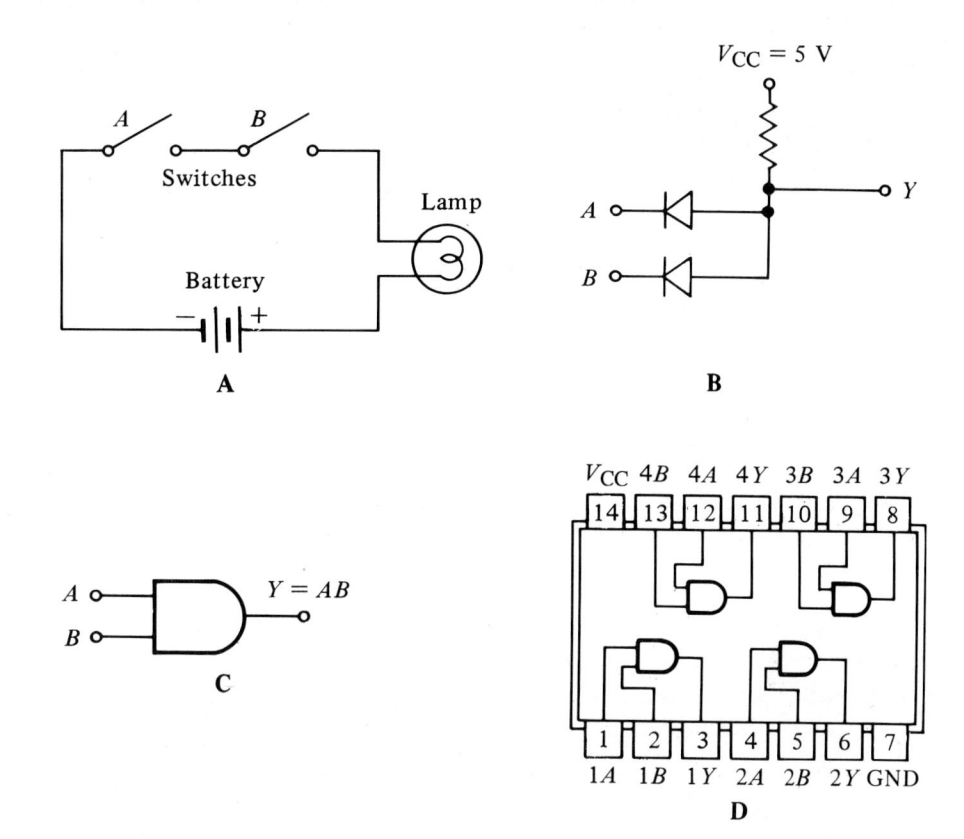

FIGURE 24.2. *Four implementations of the AND function. A: Simple lamp circuit. When switch* **A** *and switch* **B** *are closed, the lamp lights. B: Diode AND gate. C: AND gate symbol. D: 7408 integrated-circuit package containing four AND gates.* $V_{CC} = 5$ V.

TABLE 24.1. *Truth Tables for the AND Function*[a]

Switch A B	Lamp $Y = A \cdot B$	Inputs A B	Output $Y = A \cdot B$	Inputs A B	Output $Y = A \cdot B$
OFF OFF	OFF	0 0	0	FALSE FALSE	FALSE
ON OFF	OFF	1 0	0	TRUE FALSE	FALSE
OFF ON	OFF	0 1	0	FALSE TRUE	FALSE
ON ON	ON	1 1	1	TRUE TRUE	TRUE

a. $0 = $ LO $ = 0$ V; $1 = $ HI $ = 5$ V.

common 7408 quad AND gate, whose pin diagram is shown in Figure 24.2D, can be used quite easily after a 5-V power supply is connected to pin 14 of the integrated circuit, and ground to pin 7. Various combinations of logic signals applied to the input of any of the four gates produce output signals according to the truth tables of Table 24.1.

SEC. 24.1 Digital Logic and Gates

The OR Gate

The circuit of Figure 24.3A illustrates the OR function. When *either* switch A or switch B is closed, the lamp is lighted. This function is written as

$$A + B = Y \tag{24.2}$$

The symbolic representation of the OR gate is presented in Figure 24.3B. Table 24.2 shows the truth table for the OR gate. Whenever *either* input A or input B is brought to logic 1, the output is also logic 1.

The Inverter, NAND, and NOR Gates

It is often useful to be able to change the sense of a logic signal, that is, to change a HI to a LO or a LO to a HI. This function is called the NOT function and it is accomplished by the use of the *inverter* shown in Figure 24.4. The NOT function is written as NOT A, or $\bar{A}$.

When the NOT function is combined with the AND function, we obtain the NOT AND, or NAND, function illustrated in Figure 24.5. Symbols for the AND

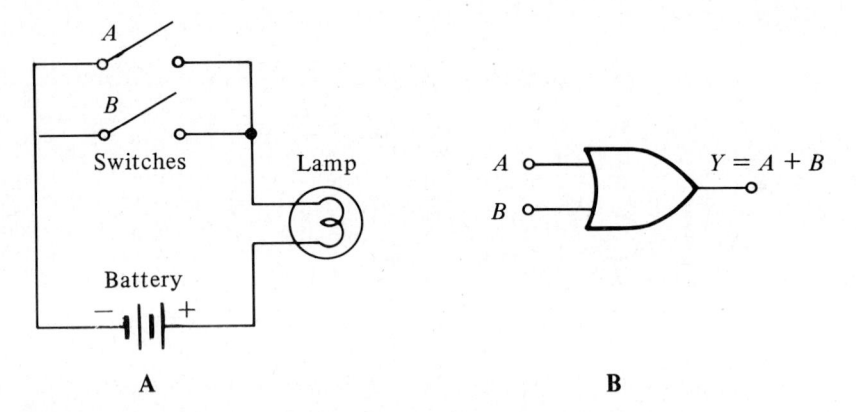

FIGURE 24.3. *OR function. A: Lamp circuit. When switch A or switch B is closed, the lamp lights. B: OR gate symbol.*

TABLE 24.2. *Truth Table for the OR Function*

Inputs		Output
A	B	$Y = A + B$
0	0	0
1	0	1
0	1	1
1	1	1

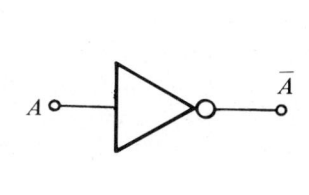

Input	Output
A	$\bar{A}$
0	1
1	0

FIGURE 24.4.   NOT function. An inverter and its truth table.

Inputs		Output
A	B	$Y = \overline{AB}$
0	0	1
1	0	1
0	1	1
1	1	0

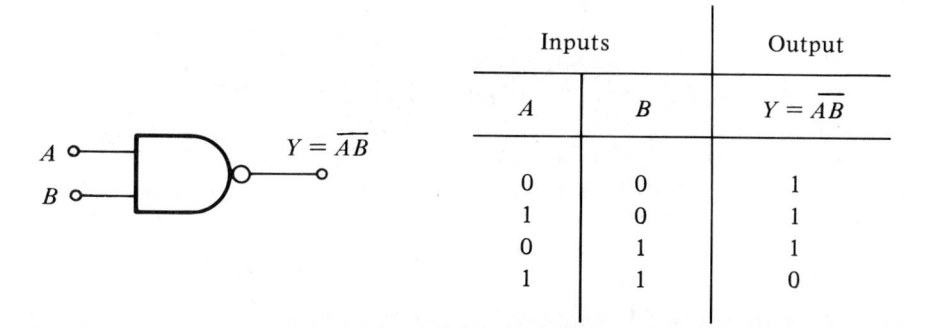

FIGURE 24.5.   NAND function. A NAND gate and its truth table.

gate and the NAND gate differ by the small circle, or "bubble," at the output of the NAND gate. The bubble indicates an inversion of the normal AND function such that the output is $Y = \overline{AB} = \overline{A \cdot B}$.

Similarly, the NOR gate of Figure 24.6 is formed by inverting the output of an OR gate, giving the output $Y = \overline{A + B}$.

Boolean Algebra

Careful scrutiny of the truth tables for the AND, OR, NAND, and NOR functions suggests that there are relatively simple relationships between AND and NOR and

Inputs		Output
A	B	$Y = \overline{A + B}$
0	0	1
1	0	0
0	1	0
1	1	0

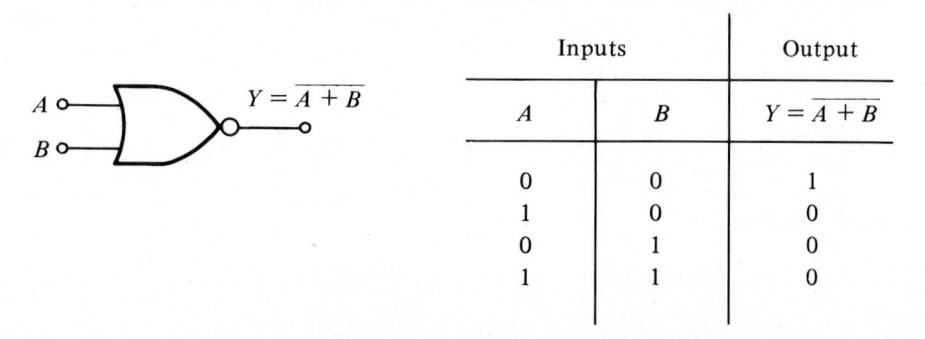

FIGURE 24.6.   NOR function. A NOR gate and its truth table.

SEC. 24.1   Digital Logic and Gates   813

between NAND and OR. To illustrate, consider the truth table of Table 24.3, in which the inputs to an OR gate are inverted before processing by the gate. The resulting outputs are identical to the outputs of a NAND gate for the same inputs. Thus, we have proven one of De Morgan's theorems: $\overline{A} + \overline{B} = \overline{AB}$. Gate implementations of De Morgan's theorems are shown in Figure 24.7; these theorems and a number of other important theorems of Boolean algebra are presented without proof in Table 24.4.

From a practical perspective, these theorems and other mathematical techniques permit the design of complex logic circuits using the minimum number of gates and other functional circuits. In fact, much of today's logic design is done by computer programs. The designer submits circuit requirements to the programs, and

TABLE 24.3. *Truth Table Illustrating De Morgan's Theorem and Equivalent Gates*

Inputs		Outputs			
A	B	$\overline{A}$	$\overline{B}$	$\overline{A} + \overline{B}$	$\overline{AB}$
0	0	1	1	1	1
1	0	0	1	1	1
0	1	1	0	1	1
1	1	0	0	0	0

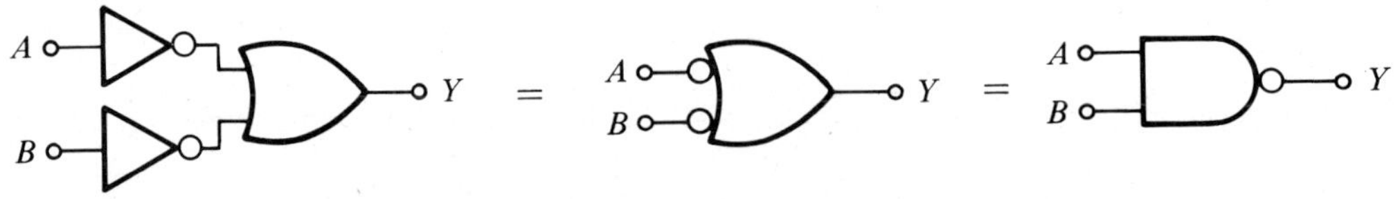

$$Y = \overline{A} + \overline{B} = \overline{AB}$$

A

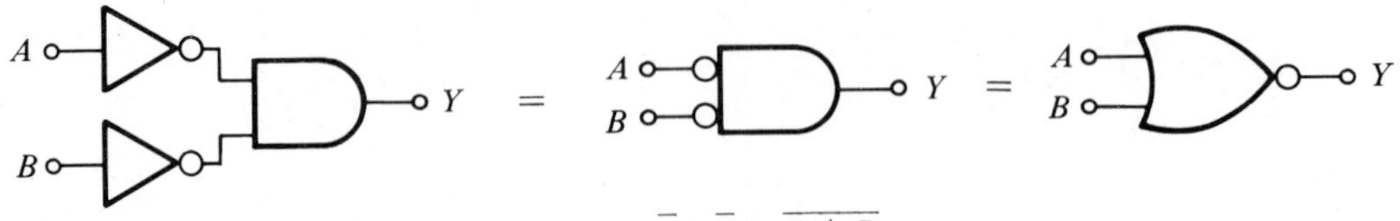

$$Y = \overline{A} \cdot \overline{B} = \overline{A + B}$$

B

FIGURE 24.7. *Illustration of De Morgan's theorems with gates.*

TABLE 24.4. *Theorems of Boolean Algebra*

AND theorems	$0 \cdot 0 = 0$	$A \cdot 1 = A$
	$1 \cdot 1 = 1$	$A \cdot A = A$
	$1 \cdot 0 = 0$	$A \cdot \overline{A} = 0$
	$A \cdot 0 = 0$	
OR theorems	$1 + 1 = 1$	$A + 0 = A$
	$0 + 0 = 0$	$A + A = A$
	$1 + 0 = 1$	$A + \overline{A} = 1$
	$A + 1 = 1$	
NOT	$\overline{\overline{A}} = A$	
Commutation	$A + B = B + A$	$AB = BA$
Absorption	$A + AB = A$	$A(A + B) = A$
Association	$A + (B + C) = (A + B) + C$	$A(BC) = (AB)C$
Distribution	$A + BC = (A + B)(A + C)$	$A(B + C) = AB + AC$
De Morgan's theorems	$\overline{A + B} = \overline{A} \cdot \overline{B}$	$\overline{AB} = \overline{A} + \overline{B}$

the program outputs the most efficient design and even specifies which integrated-circuit packages to use. Thus, by the iterative application of these techniques, it is now possible to create circuits and computers of higher and higher complexity without direct knowledge of the millions of details that may be incorporated in a given device.

Exclusive OR (XOR) and Equality Gates

The last of the simple gates that we will consider are the exclusive OR (XOR) gate and the equality gate, shown in Figure 24.8. Note that the XOR gate produces a 1 at its output if one and only one of the inputs is 1. The equality gate is formed when the output of the XOR gate is inverted. Its output is 1 when both inputs are 1 or when both inputs are 0. In effect, it detects the equality of the two inputs. The equality gate

$Y = A \oplus B$

XOR

$Y = \overline{A \oplus B}$

Equality

A

Inputs		Outputs	
A	B	$A \oplus B$	$\overline{A \oplus B}$
0	0	0	1
1	0	1	0
0	1	1	0
1	1	0	1

B

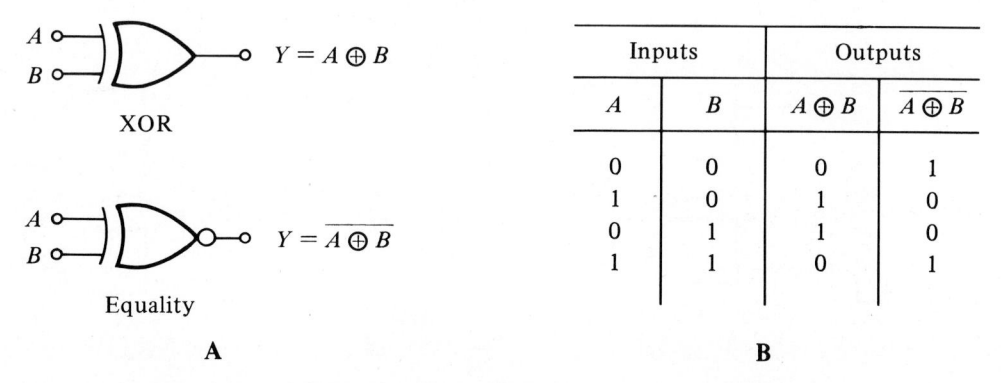

FIGURE 24.8. *Exclusive OR (XOR) and equality functions. A: Gate implementation. B: Truth table.*

is also called a *coincidence gate*, or a *comparator gate*. As might be expected, several equality gates can be used to compare two groups of bits for equality.

Although the gates presented thus far have had no more than two inputs, gates can have any number of inputs. Therefore, a four-input NAND gate must have all four inputs at 1 to produce a 0 at the output.

Latches and Registers

It is often necessary to store a bit of data or a group of bits for later retrieval. The basic *bistable latch* of Figure 24.9A, formed from two cross-coupled NAND gates, is capable of carrying out this task. To examine the operation of this circuit, let us assume that both the S (set) and the R (reset) inputs are at logic 1 and, furthermore, that output Q is at logic 1. Because the second input to NAND gate 2 is coupled to Q, it follows that output $\bar{Q}$ must then be 0. This signal is coupled to the input of gate 1, ensuring that Q remains at 1, and the data are said to be *latched*. Had we assumed that Q was at 0, we would have found that $\bar{Q}$ was at logic 1. Thus, if both the R and S inputs are at 1, Q and $\bar{Q}$ are in opposite states and are said to be complementary.

If, as before, Q is at 1 and the R input is momentarily brought to 0 (as shown in Fig. 24.9A), then $\bar{Q}$ must go to 1. Because $\bar{Q}$ is coupled to gate 1, Q must then go to 0 so that when R is brought back to 1, $\bar{Q}$ remains at 1. To change the state or "set" the latch, S is momentarily brought to 0, Q is set to 1, and $\bar{Q}$ becomes 0.

Though simple, the bistable latch finds wide application, particularly in eliminating mechanical switch "bounce," as illustrated in Figure 24.9B. The latch inputs are "pulled up" to logic 1 by the resistors connected to the power supply and to opposite sides of a single-pole, double-throw (SPDT) switch. The movable contact of the switch is connected to ground. When the switch is in the position shown, $\bar{Q}$ must be 1, so that if the switch is moved to position A, $\bar{Q}$ immediately goes to 0 and Q goes to 1. No matter how many times the contact bounces when it hits A, Q will remain at 1 because $\bar{Q}$ is coupled to gate 1 and remains at 0 until the next time the switch is thrown back to position B. The action of the latch is to provide absolutely

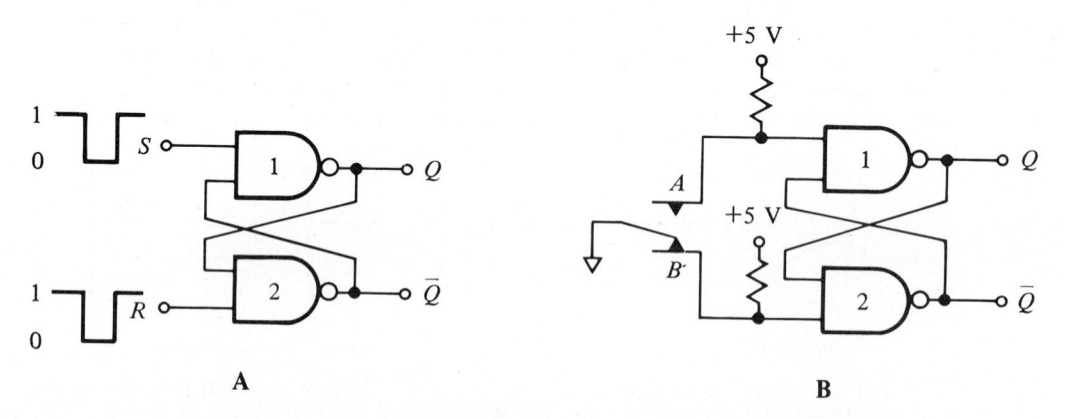

FIGURE 24.9. *A: Basic latch formed from cross-coupled NAND gates. B: Debouncing a single-pole double-throw switch using the latch.*

CHAP. 24 Digital Electronics, Data-Domain Conversions, and Microcomputers clean logic-level transitions each time the switch is "toggled," or thrown back and forth.

The next logical extension of the basic latch is the *gated-latch* circuit of Figure 24.10A. This circuit adds input gates 3 and 4 so that the data provided by the S and R inputs can be gated by the common *Clock* input. As before, the S, R, $\overline{Pr}$, and $\overline{Clr}$ inputs are at logic 1 when open. When both inputs (S and R) are at 0, no change occurs at the outputs on the application of a Clock pulse, as indicated in the truth table of Figure 24.10B. When the data inputs are in opposite states, one HI and one LO, these data are transferred to their respective outputs on application of the Clock pulse. An additional feature of this circuit is that the application of momentary 0 at the $\overline{Pr}$ input immediately produces a 1 at Q. Likewise, a momentary 0 at $\overline{Clr}$ produces a 0 at Q. The Clock input is also referred to as the *enable* input because the appearance of data at the output is "enabled" by the application of the Clock pulse.

A common form of the gated latch, called the *data latch*, is illustrated in Figure 24.11. A bit of information appearing at the Data input is transferred to Q on application of a Clock pulse. The information present at the Data input is acquired

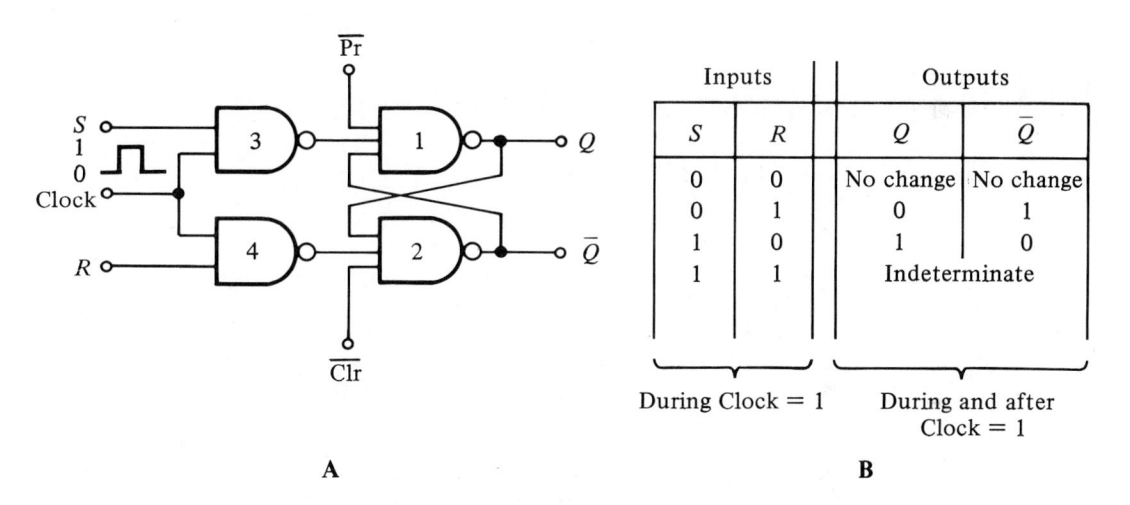

Inputs		Outputs	
S	R	Q	$\overline{Q}$
0	0	No change	No change
0	1	0	1
1	0	1	0
1	1	Indeterminate	

During Clock = 1   During and after Clock = 1

A   **B**

FIGURE 24.10. *A: Gated latch formed from cross-coupled NAND gates with two additional input gates.* $\overline{Pr}$ *and* $\overline{Clr}$ *are the* not preset *and* not clear *inputs, respectively. B: State table for the gated latch.*

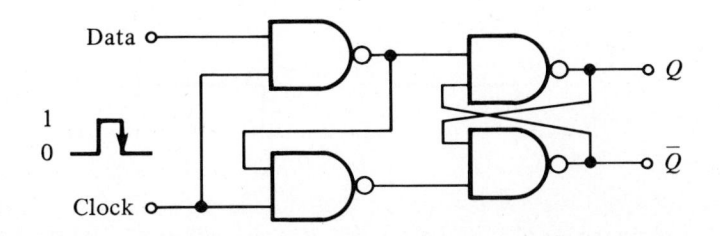

FIGURE 24.11. *Common data latch. Information at the Data input is acquired when the Clock input is at* 1.

SEC. 24.1   **Digital Logic and Gates**   **817**

while the Clock input is at 1, and is "latched" on the 1 → 0 transition. The information at Q remains stable until the next application of a Clock pulse. This circuit is useful for temporary storage of information, particularly in interfaces between computers and their peripheral devices.

A similar but somewhat more sophisticated circuit is the *J-K flip-flop*, illustrated in Figure 24.12A. As can be seen in the state table in Figure 24.12B, the action of the *J-K* flip-flop is not unlike that of the gated latch described above. The difference is that when both the J and the K inputs are at 1, the outputs Q and $\bar{Q}$ change state when the device is triggered by a Clock pulse. The bubble at the input of the *J-K* flip-flop indicates that the action of the device is triggered by a negative-going signal. The ">" at the Clock (Ck) input of the flip-flop indicates that the action is edge triggered, in this case by a 1 → 0 transition. If there were no bubble at the Clock input, the device would be triggered by a 0 → 1 transition.

Thus, if the Clock signal is alternately changed from 0 to 1 and 1 to 0, or "toggled", Q will alternately change state from 0 to 1 and then 1 to 0, as shown in the waveform diagram of Figure 24.12C. The result is that the output produces exactly half as many cycles as are applied to the Clock input. This is a very useful characteristic of J-K flip-flops that will be applied in the construction of counters and other

Inputs		Outputs
J	K	Q
0	0	No change
0	1	0
1	0	1
1	1	Change state

During Clock = 1   After Clock = 0

A

B

C

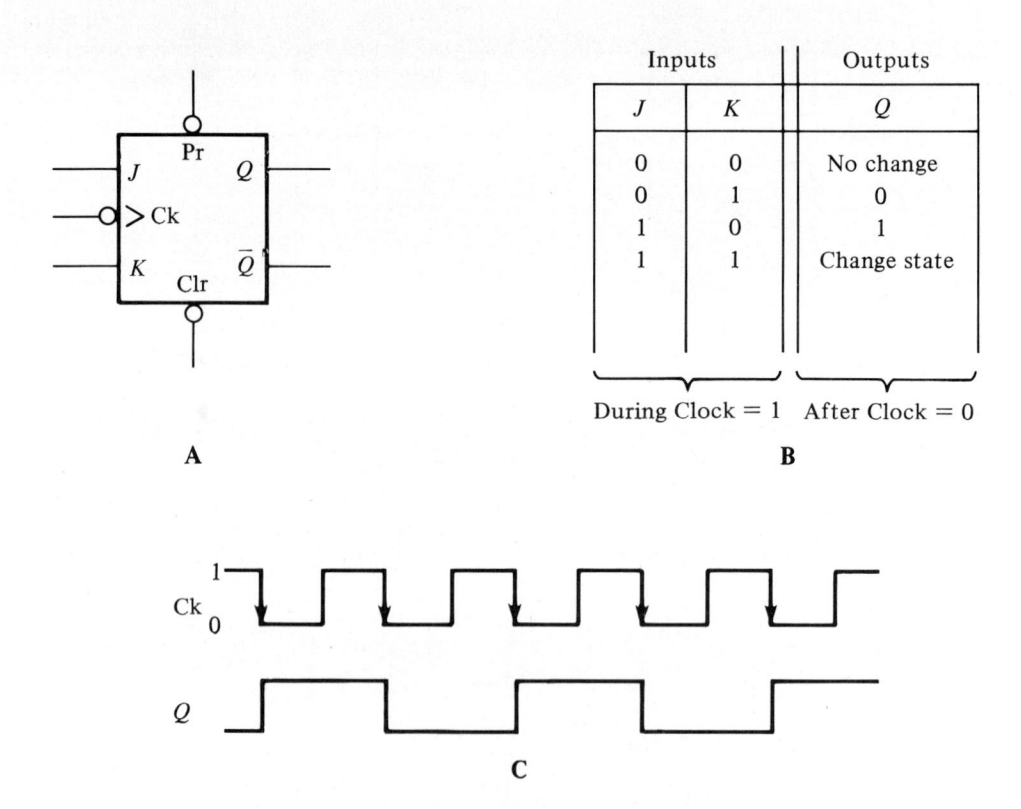

FIGURE 24.12. *A: J-K flip-flop. B: State table for the J-K flip-flop. C: Waveform diagram showing the transitions of the Q output with successive 1 → 0 transitions of the Clock input.*

useful devices. Of course, the application of a 0 to either the *Pr* input or the *Clr* input has the immediate effect of setting Q to 1 or to 0.

Simple Counters

When J-K flip-flops are combined in the manner illustrated in Figure 24.13A, a simple 4-bit asynchronous binary counter is formed. The operation of this counter is most easily understood by considering the waveforms of Figure 24.13B. The input waveform is a simple unipolar square wave of constant frequency. Flip-flop A (FF$_A$)

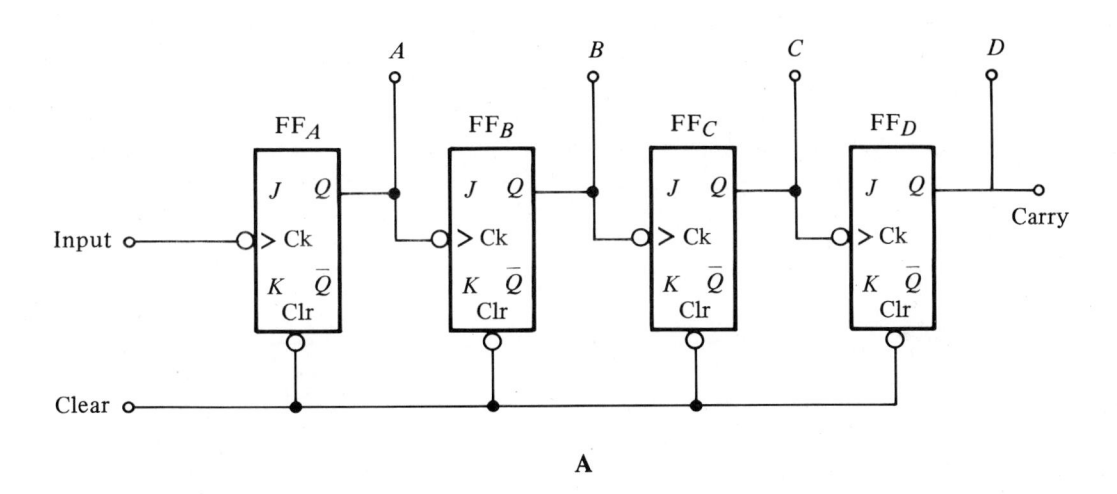

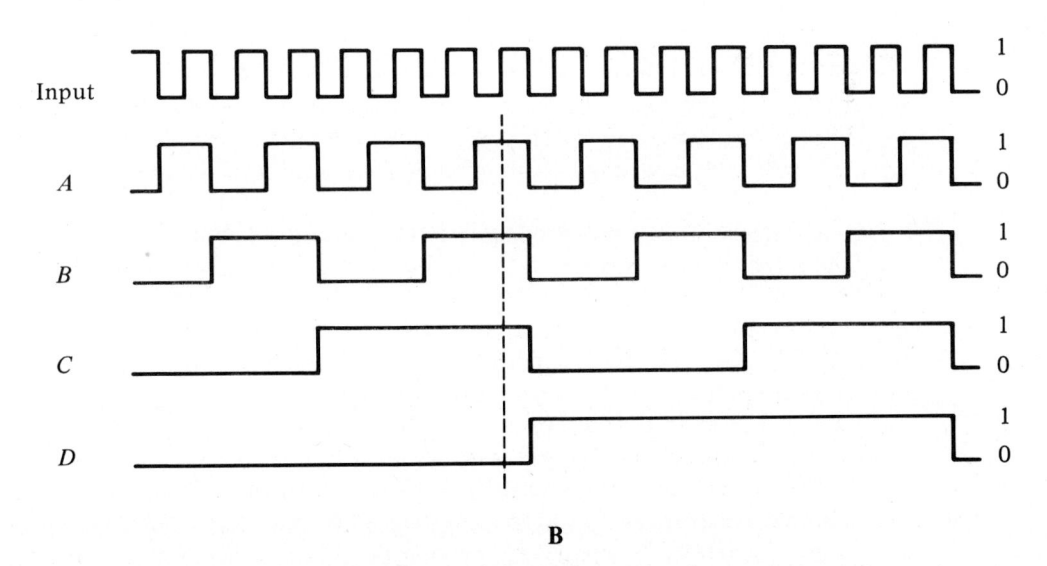

FIGURE 24.13. *A: Four-bit binary counter made from J-K flip-flops. B: Waveform diagrams for the input and outputs.*

SEC. 24.1 Digital Logic and Gates

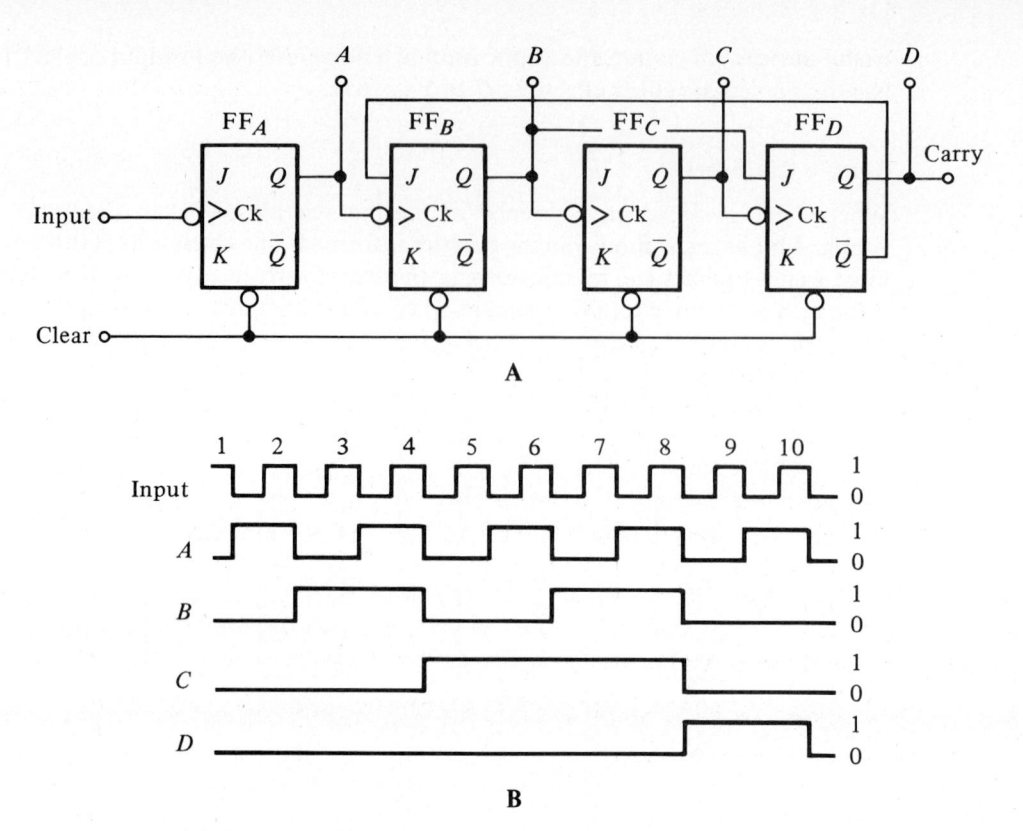

FIGURE 24.14. *A:* Decade-counting unit made from J-K flip-flops. *B:* Waveform diagrams.

divides the input frequency by 2, to produce waveform A at Q_A. This waveform is sent on to the Clock input of FF_B, which produces waveform B, with exactly half the frequency of the signal at A. This process continues for succeeding stages of the counter, with each new stage "scaling" down the output of the previous stage by a factor of 2.

At any point in time, the states of the outputs ($A, B, C, \ldots$) give the binary representation of the number of cycles of the input signal. If, for example, the counter is read at a time corresponding to the vertical dashed line in Figure 24.13B, $Q_D = 0$, $Q_C = 1$, $Q_B = 1$, and $Q_A = 1$. Therefore, the counter reads 0111 in base 2, or 7 in base 10 ($0111_2 = 7_{10}$). The number of distinct states that can be represented by this four-stage counter is $2^4 = 16$ states, corresponding to 0_{10} thru 15_{10}. When the counter contains all 1s, the next input cycle causes the counter to "roll over" to all 0s, whereupon the carry output Q_D changes state.

The capacity of such a counter is limited only by the number of flip-flops that are available. Because each flip-flop represents a power of 2, the capacity of a counter of n stages is given by $2^n - 1$.

This type of counter has the advantage of simplicity, but it suffers from the fact that each stage must wait for the previous stage to toggle before it can respond.

Thus, this type of counter is sometimes called a *ripple counter*. Because the various stages of the counter are not clocked simultaneously, it is also termed *asynchronous*. For more elaborate discussions of many types of counters, including synchronous binary counters, the reader is directed to the titles in the Selected Bibliography.

Decade Counter

Except for those few individuals who have become accustomed to thinking in binary, most people continue to think in decimal. Thus, most instruments and computers must communicate information to their users in decimal. For this reason, the BCD coding scheme introduced in Chapter 23 is used in counting as well. A 4-bit *decade-counting unit* (DCU) is created from the simple binary counter of Figure 24.13 by making two small modifications, which are shown in Figure 24.14A.

By connecting $\bar{Q}$ of FF_D to J of FF_B, Q_B is prevented from going to 1 on the 10th count. In addition, Q_B is connected to J of FF_D so that Q_D must go to 0 on the 10th count. These conditions result in the waveform diagram of Figure 24.14B. DCUs similar to this simple asynchronous counter, and counters with far more sophisticated functions such as preset capability and up/down counting, are available in medium-scale integrated (MSI) circuits. Complete counting systems that include many DCUs and additional control functions are also available in large-scale integrated (LSI) circuits (see Sec. 24.3).

24.2 SWITCHING

To this point in our discussion, switches have been represented by familiar mechanical toggle switches. Although these mechanical devices have many advantages, they exhibit a number of disadvantages, including low speed, inability to be easily controlled by electronic circuits, and bouncing when contact is made or broken. A partial solution to this problem was alluded to when we discussed NAND gates, but that solution applied only to digital logic signals.

Integrated-Circuit Analog Switches

The advent of FET technology (see Chap. 23) has permitted the fabrication of *analog switches*, which now possess nearly ideal switching characteristics. Switching times are quite rapid, of the order of tens to hundreds of nanoseconds, OFF resistances are of the order of 10^9 to $10^{12}\,\Omega$, and ON resistances are as low as 10 to 100 Ω. Analog switches are often fabricated to include four switches on a single integrated-circuit package, so the cost per switch can be quite low.

A function diagram of a typical analog switch is shown in Figure 24.15A. As usual, the input, the output, and the control signals are measured with respect to circuit common. By and large, these circuits can be used nicely without knowledge of the detailed circuitry within the integrated circuit. These devices usually require analog supply voltages (typically $+15$ V and -15 V), and perhaps a logic-level power supply (usually $+5$ V), although such requirements are seldom indicated in the basic

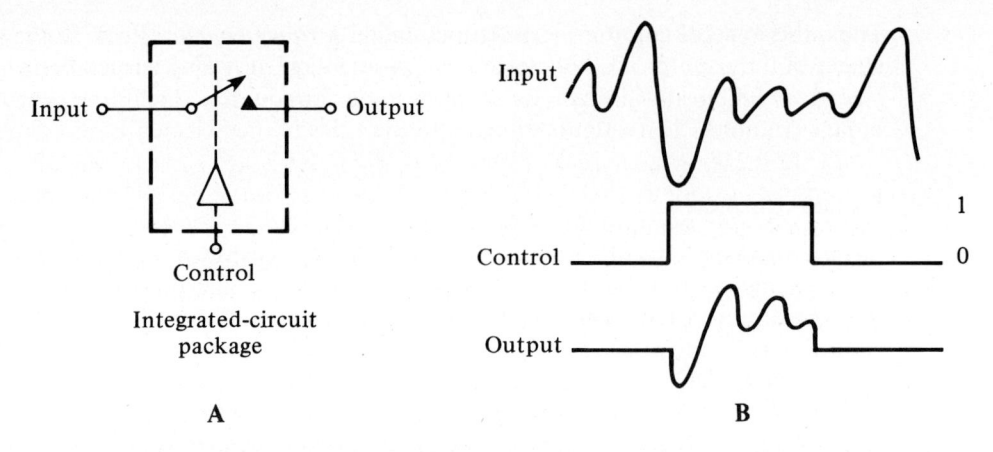

FIGURE 24.15. *A: Diagram of a simple analog switch. B: The input signal is passed along to the output only when the switch is closed, that is, at logic level 1.*

schematic. The control input is usually compatible with TTL circuits. For the example shown, this means a logic 0 opens the switch and a logic 1 closes the switch. The effect of the control input is clearly shown in Figure 24.15B. The analog input signal is passed to the output only when the control input is at logic 1.

Sample-and-Hold Amplifier

A very useful application of analog switches is in the sample-and-hold (or track-and-hold) amplifier depicted in Figure 24.16A. The charge on the capacitor at the input of the follower amplifier remains constant so long as the analog switch remains open. Thus, the voltage on the capacitor and v_o are identical. If the control input of the

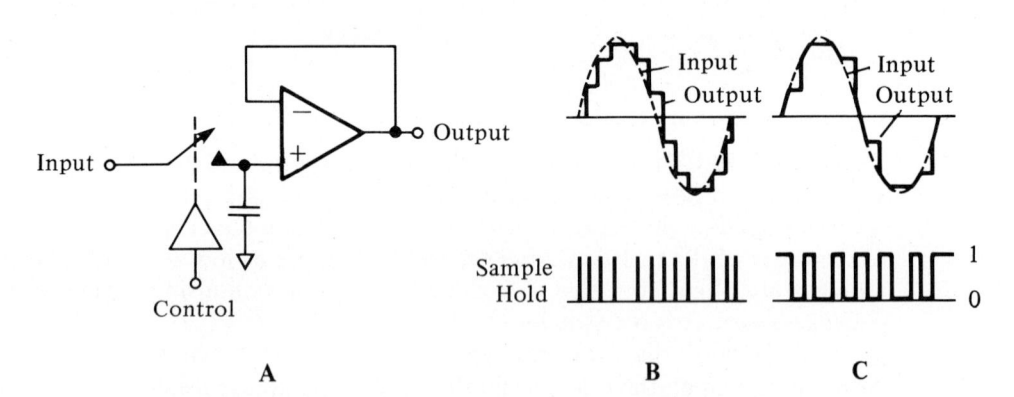

FIGURE 24.16. *A: Circuit for a sample-and-hold amplifier. B: Sample mode: The signal is sampled with a narrow pulse at the control input. C: Track Mode: The voltage output tracks the input signal while the switch is closed (logic level 1) and holds the final value when the switch is opened.*

822   CHAP. 24   Digital Electronics, Data-Domain Conversions, and Microcomputers

analog switch goes to logic 1, as shown in Figure 24.16B, the switch closes and the capacitor charges to whatever value v_i may take on during the period the switch is closed. When the switch is opened by bringing the control to logic 0, v_o holds the value of v_i at the time of the switch opening.

Small errors can appear in v_o due to the finite switch-closure time and to offset errors in the amplifier. These errors are generally around 0.02% for commercial integrated sample-and-hold amplifiers; they depend on the rate at which v_i changes as well as on the speed at which the analog switch can be opened. Sample-and-hold amplifiers are widely used in conjunction with successive-approximation analog-to-digital converters (see Sec. 24.4).

Comparators and Schmitt Triggers

A special type of op-amp, the *comparator*, is the key circuit element in carrying out analog-to-digital data-domain conversions. The comparator, depicted in Figure 24.17A, differs from the op-amp only in that it is designed to have a very large open-loop gain and thus a relatively narrow linear operating range. The device is used in the open-loop configuration so that it responds extremely rapidly to small differences in the two input signals. The response time, that is, the time between the application of a step voltage change at the input and when the output crosses a logic threshold, is about 200 nsec for a typical comparator. In addition, the output transistors of comparators are often manufactured with their collectors left open at the output so that an external resistor can be used to "pull up" the output to the logic supply voltage. This makes the output compatible with whatever logic family is in use.

The operation of the comparator is quite simple. When v_i crosses a defined reference voltage V_{ref}, the output changes state, as illustrated in Figure 24.17B. Thus, in effect, the output of the comparator provides one bit of digital information, to indicate whether v_i is greater than or less than V_{ref}. It is this property that allows the construction of several useful analog-to-digital converters.

However, a problem arises when v_i is a noisy signal, as illustrated in Figure 24.18. Each time the noise crosses V_{ref}, a transition is produced at the output, as

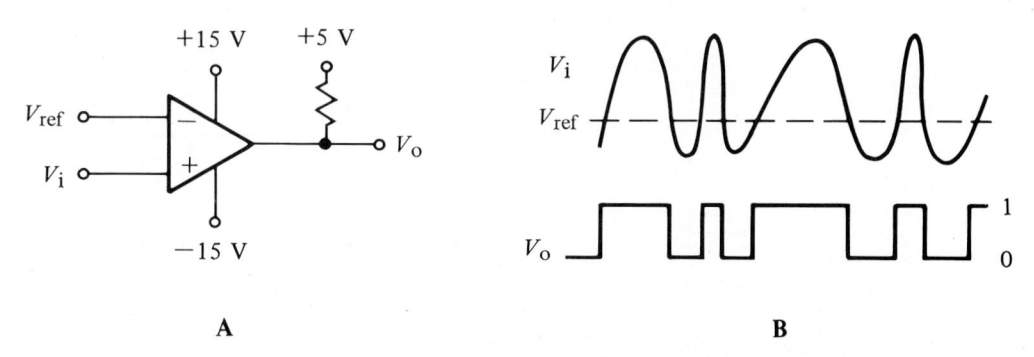

FIGURE 24.17. *A: Symbol for an op-amp comparator. B: Waveform diagram illustrating operation. The output voltage goes to a HI state (logic level 1) when the input voltage exceeds a fixed reference voltage V_{ref}, and to a LO state at inputs less than V_{ref}.*

SEC. 24.2 Switching

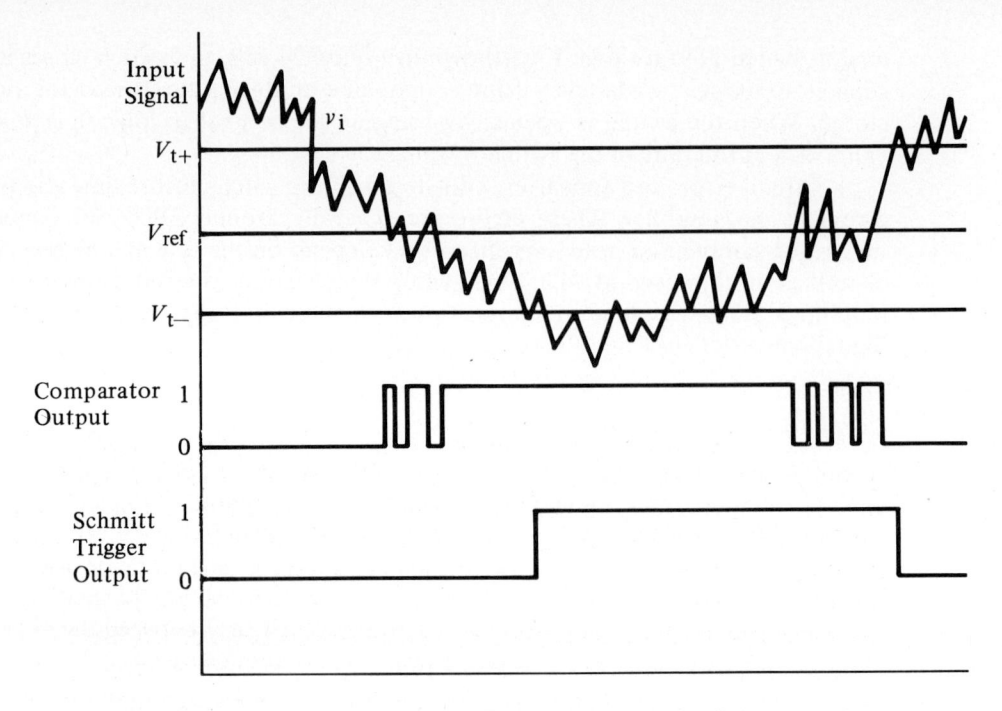

FIGURE 24.18. *Comparator and Schmitt-trigger outputs for a noisy input signal.*

shown by the comparator waveform. Such spurious transitions are often undesirable because the exact point of crossing the threshold is not clearly defined.

A possible solution to this problem is shown in the *Schmitt-trigger* circuit of Figure 24.19A. In this circuit, a fraction of the difference between v_o and V_{ref} is applied to the noninverting input of the comparator to provide *positive feedback*. The effect of positive feedback is shown in the plot of v_o versus v_i of Figure 24.19B, and also in Figure 24.18 for the noisy signal discussed above. The signal decreases past V_{t+} (the upper threshold voltage) and V_{ref}, but the $0 \rightarrow 1$ transition at the output occurs only

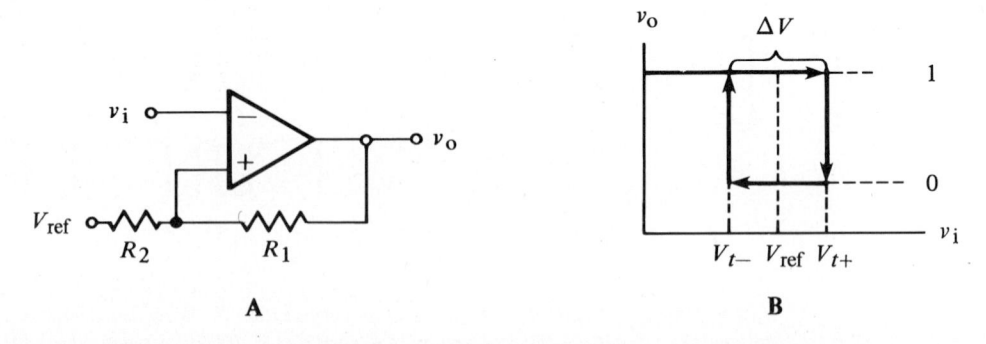

FIGURE 24.19. *A: Schmitt-trigger comparator circuit. B: Voltage hysteresis plot for a Schmitt trigger.*

824 CHAP. 24 Digital Electronics, Data-Domain Conversions, and Microcomputers when the signal crosses V_{t-}, the lower threshold voltage. As soon as this occurs, the voltage at the noninverting input increases to V_{t+}, which is by then well above the noisy signal so that multiple transitions do not occur. The signal must then increase to V_{t+} before the reverse transition can occur. The quantity $\Delta V = V_{t+} - V_{t-}$ is called the *hysteresis*. For a given comparator, the hysteresis depends on the relative values of R_1 and R_2. Schmitt triggers are used widely in situations involving noisy analog signals from which well-defined logic transitions must be produced.

24.3 RC CIRCUITS AND MULTIVIBRATORS

In Chapter 23, some of the properties of *RC* circuits were explored. We examined the reactive behavior of such circuits, characterized their use as filters, and showed how they can be used with op-amps to carry out analog differentiation and integration. In this section, we will investigate further the nature of *RC* circuits and show how they can be utilized to carry out accurate and precise timing functions in both the analog and digital domains.

Consider the simple series *RC* circuit of Figure 24.20A. The SPDT switch is arranged so that the *RC* can be connected either to the voltage source V_S (position 1) or to circuit common (position 2). If we begin with the switch in position 2, the capacitor is completely discharged. When the switch is thrown to position 1, the entire applied voltage instantly appears across the *RC* network so that $V_S = v_C + v_R$. At the instant of switch closure, no charge q has had time to accumulate on the capacitor, so $v_C = q/C = 0$, and thus $v_R = V_S$. The manner in which v_R and v_C change with time is clearly shown in the plot of Figure 24.20B.

As the capacitor is charged, v_C increases exponentially, and v_R decreases

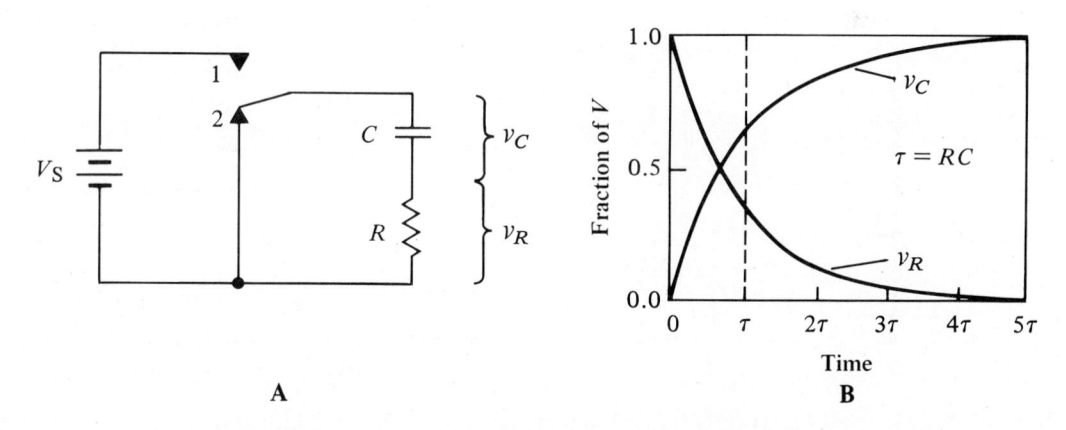

FIGURE 24.20. *A: Simple series RC circuit. B: Voltages across the capacitor, v_C, and the resistor, v_R, as a function of time after applying a step change in voltage V_S.*

exponentially, according to the following relationships:

$$v_C = V_S(1 - e^{-t/\tau}) \quad \text{and} \quad v_R = V_S e^{-t/\tau} \tag{24.3}$$

where $\tau = RC$ in seconds (ohms × farads)

Note that at $t = \tau = RC$, $v_R = V_S(1/e) = 0.368 V_S$, and likewise $v_C = V_S(1 - 1/e) = 0.632 V_S$. Of course, the sum of the two voltages at all points in time is V_S. With each time increment of τ seconds, v_R decreases and v_C increases to within $1/e$ of the difference between their values at the beginning of the increment and their absolute final values, which in this example are 0 and V_S, respectively. This is a property of all RC circuits and, in fact, of all devices possessing a purely exponential response. The quantity $RC = \tau$ is often referred to as the *time constant* of the circuit.

Monostable Multivibrator

RC circuits, comparators, analog switches, and flip-flops can be cleverly combined to produce some very sophisticated timing circuits such as the basic *monostable multivibrator*, or *one shot*, depicted in Figure 24.21A. In this circuit, a voltage divider is arranged to apply $V_+ = 0.667 V_S$ to the noninverting input of the comparator. Timing components R_t and C_t are connected so that the voltage across the capacitor v_- is applied to the inverting input, and so that C_t is discharged when the analog switch (SW) is closed. The output of the comparator is connected to the Clock input of the negative edge-triggered flip-flop, and the analog switch is controlled by $\bar{Q}$, which can be preset or cleared by the use of momentary switches.

Assume that the circuit has been initialized by depressing the appropriate switch. At this point, the flip-flop has been cleared, switch SW is closed so that v_- is near ground potential, and the output of the comparator is therefore at logic 1. The timer is started by momentarily depressing the start switch, which immediately

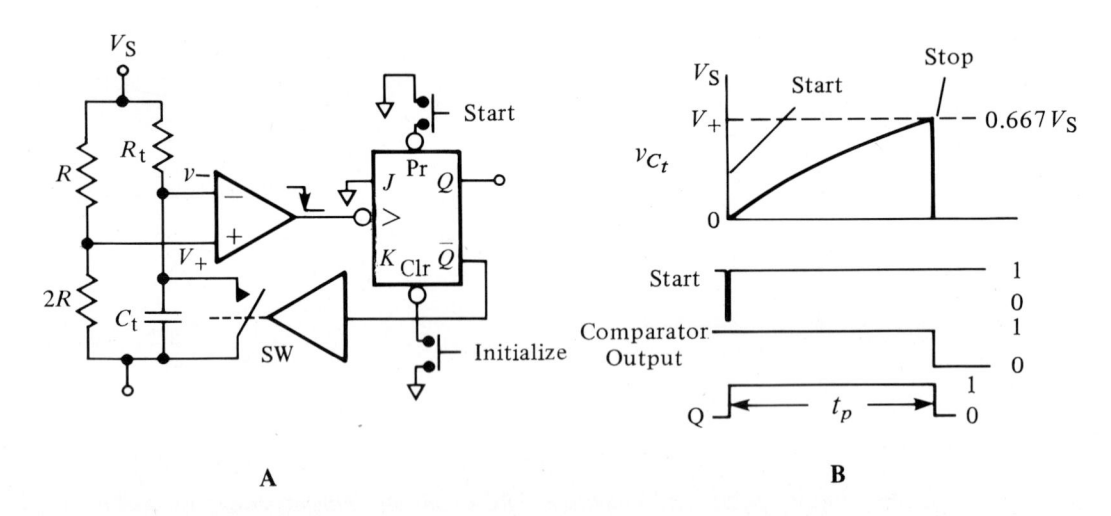

FIGURE 24.21. *A:* Monostable multivibrator. *B:* Waveform diagrams illustrating operation.

826   CHAP. 24   Digital Electronics, Data-Domain Conversions, and Microcomputers

sets Q of the flip-flop to 1 so that switch SW opens. The timing capacitor C_t begins to charge until its voltage reaches $0.667V_S$, whereupon the output of the comparator goes to logic 0. This sequence is illustrated in the waveform diagrams of Figure 24.21B. This action triggers the flip-flop so that the logic 0 at the J input is transmitted to Q, which reinitializes the timer and discharges C_t. Note that the comparator threshold is very nearly the $0.632V_S$ that would appear across C_t in time $t = R_tC_t$. In fact, the time duration of the pulse (t_p) that appears at Q is equal to $1.1R_tC_t$. Although relatively simple, this circuit allows the production of precise pulses merely by the appropriate selection of R_t and C_t.

Integrated circuits such as the versatile 555 timer are similar in principle to the circuit we have just discussed. They are widely used in timing and sequencing applications. For example, several monostables with different selected pulse widths can be connected in series to produce a sequence of pulses.

Astable Multivibrator

The 555 timer can be used as an *astable multivibrator*, as illustrated in Figure 24.22A. In this configuration, two reference voltages are used with two comparators to alternately set and reset a reset-set (*RS*) flip-flop to provide astable operation. The voltage divider consists of three equal resistors R, so a reference voltage of $2V_S/3$ is provided for comparator 1, and $V_S/3$ is connected to comparator 2. The timing portion consists of two resistors, R_1 and R_2, in series with capacitor C. When the analog switch SW is open, C charges with a time constant of $(R_1 + R_2)C$, but when SW is closed, C discharges only through R_2, making the discharge time constant R_2C.

An astable sequence is begun by resetting the flip-flop so that the voltage on the capacitor, v_C, begins to decay with time constant R_2C, as shown in Figure 24.22B. When v_C reaches $V_S/3$, the output of comparator 2 goes to logic 0, which sets Q and opens SW. The capacitor then begins to charge immediately with time constant

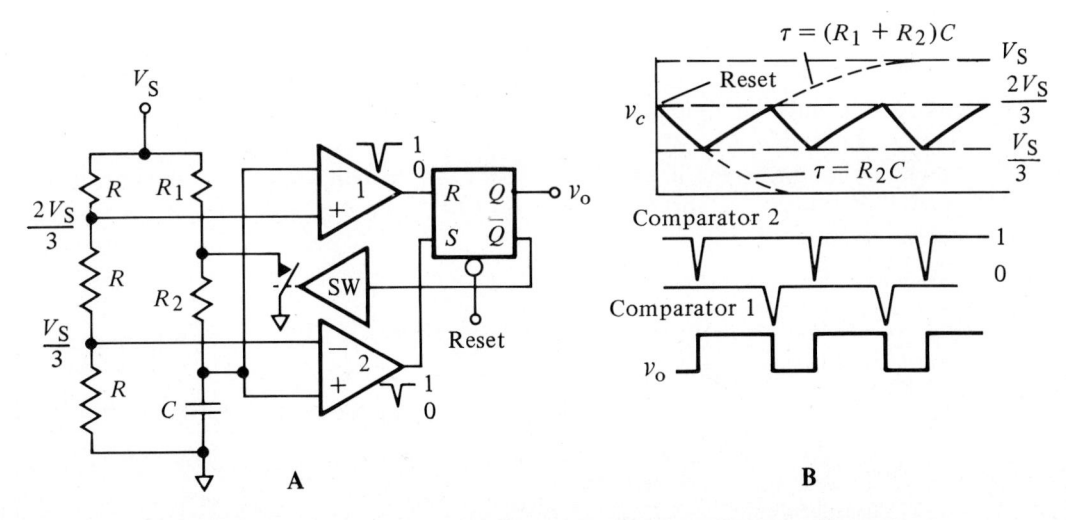

FIGURE 24.22. *A: Astable multivibrator. B: Waveform diagrams illustrating operation.*

$(R_1 + R_2)C$, and the output of comparator 2 returns to 1. When v_C reaches $2V_S/3$, the output of comparator 1 goes to logic 0, Q is reset, SW closes as before, and thus the cycle repeats itself to produce the waveform v_o, as shown.

The 555 timer is an extremely useful device. Because R_1, R_2, and C are supplied by the user, the period of oscillation can range from microseconds to hours. The *duty cycle*, or fraction of a cycle that v_o is at logic 1, can be varied from nearly 0.5 when $R_2 \gg R_1$ to a sharp pulse when $R_1 \gg R_2$. The frequency instability resulting from voltage and temperature drift in most cases amounts to only a few tenths of a percent, which is adequate for most applications.

24.4 COUNTING AND TIME-DOMAIN MEASUREMENTS

The high quality of modern analytical measurements is due in large measure to the availability of low-cost, accurate, and precise counting instrumentation. In the sections that follow, we will analyze modern electronic counting systems and show how a variety of time-domain measurements can be accomplished.

Fundamental Counting Measurements

The elements of an electronic counting system are shown in the block diagram of Figure 24.23. The object of a counting measurement is some characteristic of an electronic signal, usually the pulses derived from physical events by input transducers. The signal is first passed into a signal conditioner whose purpose is twofold: It discriminates against undesirable signals or noise, and it converts the voltage levels of the input signal to be compatible with those of the remaining circuits of the counter.

The definition of what constitutes a "countable" event depends on the nature of the chemical or physical system under investigation and the transducer being employed. At times, the definition may seem somewhat arbitrary. Basically, the input signal must exceed some *threshold amplitude* to be considered as a countable event.

Once the signal of interest has been shaped to produce clean signals that are compatible with the logic family in use (e.g., TTL), a *counting gate* is used to define the *boundary conditions* of the counting process. Perhaps the most frequently used boundary condition is a fixed period of time. As the name suggests, when the gate is open, logic pulses are allowed to pass on to the counter, and when it is closed, no pulses are counted. The opening and closing of the gate are controlled by signals that

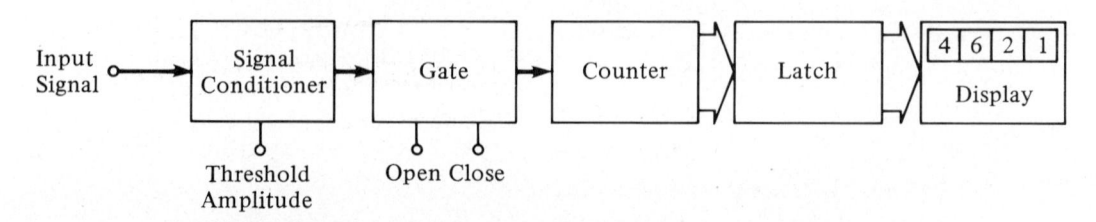

FIGURE 24.23. *Block diagram of a basic counting system.*

in some way define the boundary conditions of interest. These signals are usually logic-level transitions or very short pulses whose leading or trailing transitions define "open" and "close."

A simple gate can be implemented with the two-input AND gate of Figure 24.2C. The conditioned signal of interest is connected to one input, and the gate-control signal is connected to the second input. As long as the gate-control signal is at logic 1, the gate is open; when it is at logic 0, the gate is closed. This simple device does not provide separate "open" and "close" inputs, and so slightly more complicated devices are generally used.

Once the signal has been conditioned and gated according to a specific set of boundary conditions, the selected events must be counted. The nature of the counter depends heavily on its anticipated use. If the counter is to be used by human experimenters, it will likely be a BCD counter consisting of several DCUs such as the one shown in Figure 24.14A. The actual number of DCUs depends on the range of counts anticipated, but for many instruments it is either four or eight, with which counts of up to 99999999 can be accumulated. If the count information is to be used only by a computer or other data-processing device, the counter will probably be a binary counter consisting of some number of 4-bit stages. Because many computers handle information in 8-bit or 16-bit chunks, 4-stage counters with a maximum capacity of 65535 are typical.

As Figure 24.23 suggests, the outputs of the BCD or binary counter are connected to a latch consisting of as many 1-bit latches (e.g., that shown in Fig. 24.11) as are necessary to accommodate the capacity of the counter. At the end of the gate period, the data are stored temporarily in the latch so that the experimenter (or computer) has time to read them before the succeeding gate period ends.

Finally, the latched output of the counter is connected to a display or a decoder-display to provide a numerical output for the user. In the case of the binary counter, each bit can be used to drive a single light-emitting diode (LED) (as shown in Ch. 23, Fig. 23.5D), thus giving a direct binary readout. However, in most cases in which a user requires a readout, BCD encoding is employed. Thus, the four bits corresponding to each digit are normally decoded into seven logic-level signals to directly drive a seven-segment display, as illustrated in Figure 24.24. This process is shown for BCD 5 to correspond to the situation in which segments a, c, d, f, and g are lighted.

When all the blocks of the generalized counter of Figure 24.23 have completed their respective functions, the result is N counted events per boundary condition B. For example, nuclear spectroscopists often measure the number of gamma-ray emissions per unit time, and biochemists determine the number of platelets per unit volume of blood. The uncertainty in such measurements depends on the uncertainties in both N and B. The availability of high-precision time-base generators can make the error in N or in B negligibly small.

Precision Time Base

The only component remaining to be specified in order to construct a time-domain measurement system is a *high-precision time base*. Such a time base is shown in block-diagram form in Figure 24.25. The heart of the time base is a 10-MHz quartz-crystal oscillator, which is widely available and quite inexpensive. The frequency of a crystal

FIGURE 24.24. *BCD output of a decade-counting unit (DCU) is decoded to provide seven logic-level signals to directly drive a single seven-segment numeric display.*

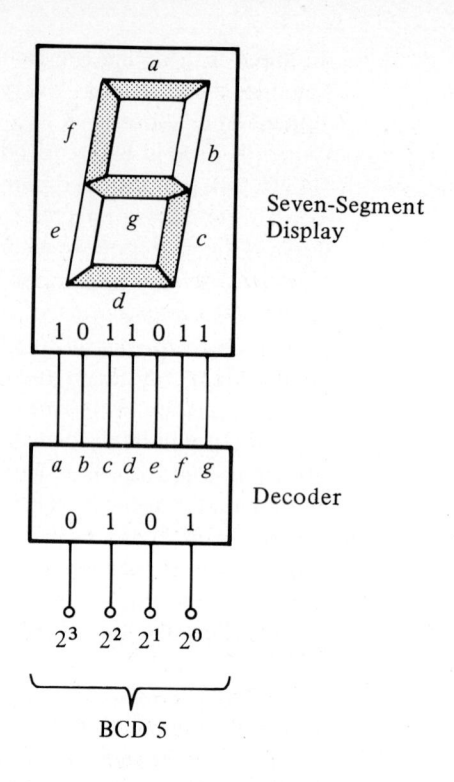

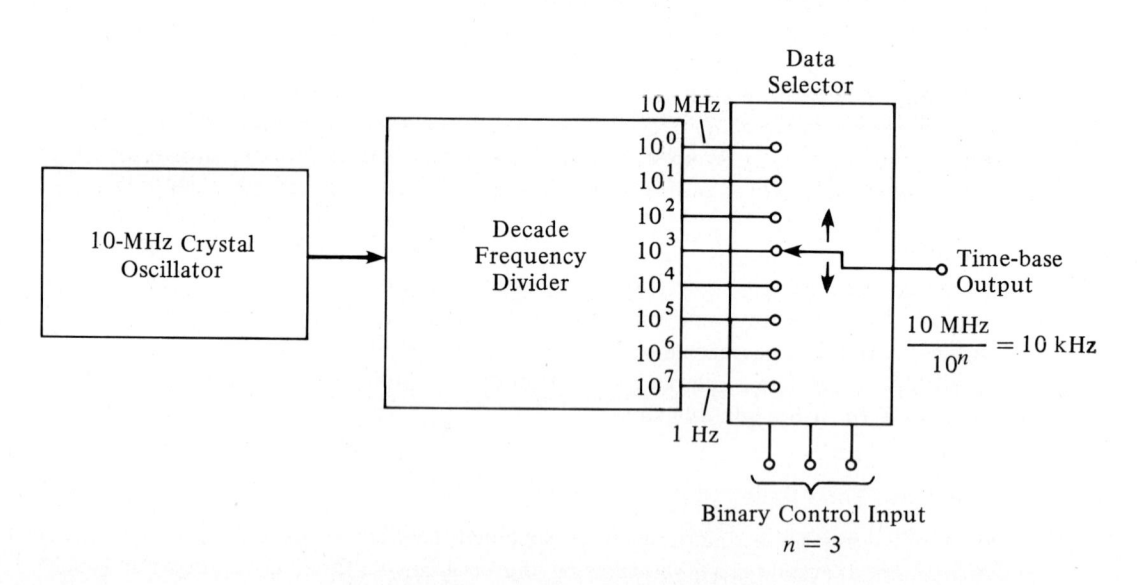

FIGURE 24.25. *High-precision time base.*

oscillator is usually accurate to a few parts in 10^4 and precise to a few parts in 10^6. The uncertainty in the time base is determined by the uncertainty in the crystal oscillator, so the precision and accuracy of time-domain measurements are seldom limited by errors in the time base.

The oscillator is connected to a *decade frequency divider*, which can consist of seven DCUs connected in series. Each successive DCU divides its input frequency by 10 to provide a frequency at its output that is a decade submultiple of 10 MHz. In the example shown in Figure 24.25, the eight frequencies from 10 MHz to 1 Hz are then connected to a *data selector*. This device allows only one of its eight inputs to pass through to the output of the time base. The frequency is selected by applying the proper binary number to the three control inputs. In this example, $n = 3_{10} = 011_2$. The output frequency is then given by

$$f_o = \frac{10 \text{ MHz}}{10^n} = \frac{10 \text{ MHz}}{10^3} = 10 \text{ kHz} \qquad (24.4)$$

Alternatively, the data selector can be a simple single-pole, eight-throw switch, an arrangement that is often found in old instruments.

Frequency Measurement

The components of the counting systems described above can be assembled in a variety of ways to carry out time-domain measurements. Frequency measurements can be made using the arrangement illustrated in Figure 24.26. Here, a gate is used in conjunction with the precision time base to generate an accurately known counting period. If the time-base control is adjusted to give a 1-Hz output, the counting gate is open for exactly 1 sec and the counter will display the number of cycles of the shaped input signal per second, which is, of course, the frequency. For counting periods shorter than 1 sec, the user (or the instrument) need only move the decimal point in the readout to obtain the frequency.

In general, the input signal will be asynchronous with the time-base output, and this results in an inherent error of ± 1 count in the counting process. To minimize the effect of this error, the counting period is made as long as possible without causing the counter to overflow. For very low frequency signals, the length of the counting

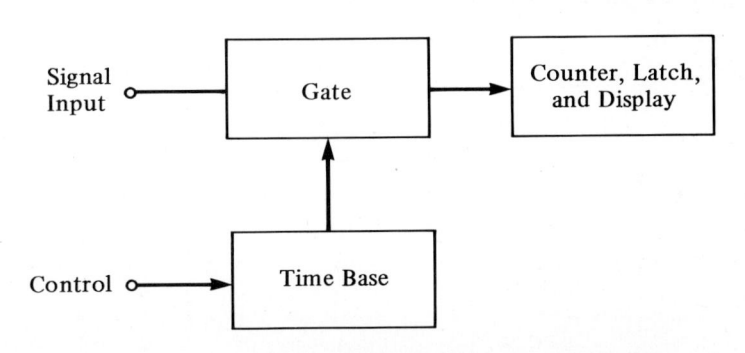

FIGURE 24.26. *Block diagram of a frequency-measuring device.*

period required to yield a sufficiently small error places a practical limitation on the frequency mode.

Period Measurement

In the period-measurement mode, the roles of the signal input and the time base are reversed, as depicted in Figure 24.27. The input signal is of relatively low frequency, and the time base is operated at the highest frequency possible without counter overflow. The counter then displays the number of standard clock cycles per period of the input signal. To further enhance this mode of operation, a decade divider can be placed between the shaped signal input and the gate. The effect of this is to open the gate for 10, 100, 1000, etc. cycles of the input signal, making it possible to accumulate a larger number of time-base cycles. The actual period is easily found by moving the decimal point the proper number of decimal places. This method of decreasing the effect of the inherent counter error is called *multiple-period averaging*.

The time-interval mode, or time *A-B* mode, is accomplished with the same array of counting modules as the period mode, but the open/close gate has two inputs, and is used so that event *A* opens the gate and event *B* closes the gate. For example, in constant-current coulometry, the signal used to start the electrolysis can be used to open the gate, and the stop signal used to close the gate. At the end of the electrolysis period, the number of moles of electrons is equal to the product of the counter readout, the magnitude of the constant current, and the faraday.

Universal Counting System

The tremendous increase in the sophistication and complexity of integrated circuits has made it possible to fabricate a complex instrumental system on a single silicon chip, the Intersil ICM 7216B, for example. With such an integrated circuit and an amazingly small number of external components—a few switches, a crystal oscillator, some seven-segment LEDs, and a few other passive components—a universal counter can be assembled in the laboratory for less than $50. Commercially assembled counters with similar capabilities are available for two or three times that amount.

Modern counting instrumentation is inherently useful for making basic time-domain measurements. But it is their use at the interface between the analog domain and the digital domain that clearly demonstrates that counters are ubiquitous.

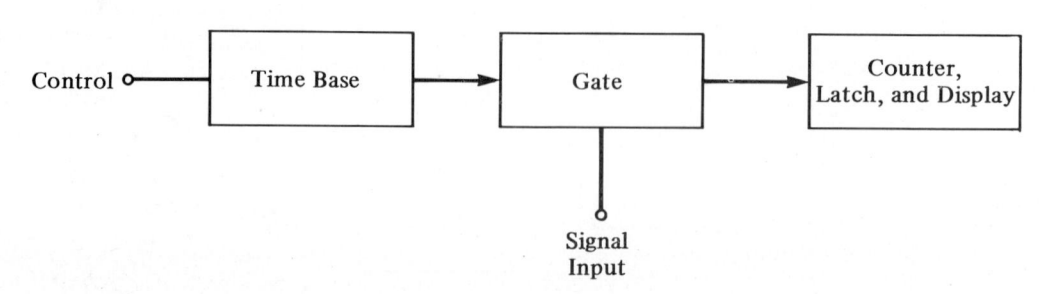

FIGURE 24.27. *Block diagram of a period-measuring device.*

24.5 DATA-DOMAIN CONVERSIONS

There are a variety of ways to convert data from one domain to another. At the heart of most data-domain conversions lies the fundamental concept of measurement science, illustrated by the block diagram of Figure 24.28. An unknown quantity Q_u is compared to a known standard Q_s using a difference detector, which provides a readout of $Q_u - Q_s$. In a *null-comparison measurement*, the standard is made variable so that it can be adjusted to give a difference output of zero so that $Q_u = Q_s$. A familiar example of this is the double-pan analytical balance. Standard masses are piled on one pan until the balance reads zero, at which point the unknown mass is taken to be the sum of the standard masses.

At the other extreme, Q_s is set to zero so that the difference detector output is taken to be Q_u directly. D'Arsonval current meters are examples of this approach, which for obvious reasons is called a *direct measurement*.

Finally, an intermediate approach allows the calculation of $Q_u = (Q_u - Q_s) + Q_s$ from the difference readout $(Q_u - Q_s)$ and the variable standard quantity Q_s. The single-pan automatic analytical balance, for example, has coarse adjustments that provide Q_s, normally to tenths of a gram, and the optical or electronic scale is the difference detector that provides the last three decimal places, $Q_u - Q_s$.

Analog-to-digital data-domain converters are generally based on the null-comparison concept. They can be broadly classified into two types: *charge-comparison* converters and *voltage-comparison* converters. First, we will discuss charge-comparison converters, including current-to-frequency, voltage-to-frequency, and dual-slope analog-to-digital converters. After the digital-to-analog converter (DAC) is described as a digitally controlled variable reference, voltage-comparison converters will be discussed.

Current-to-Frequency Converter

A basic *current-to-frequency* converter (IFC) is illustrated in Figure 24.29. In this circuit, an unknown current i_u is connected to the summing point of op-amp OA. In addition, a constant current of magnitude i_p is connected to the summing point through an analog switch as shown. Whereas i_u tends to charge the feedback capacitor C, i_p tends to discharge it. Assume that the output of the op-amp is positive, so that i_u causes the voltage across the capacitor, V_C, to decrease toward 0 V. When V_C

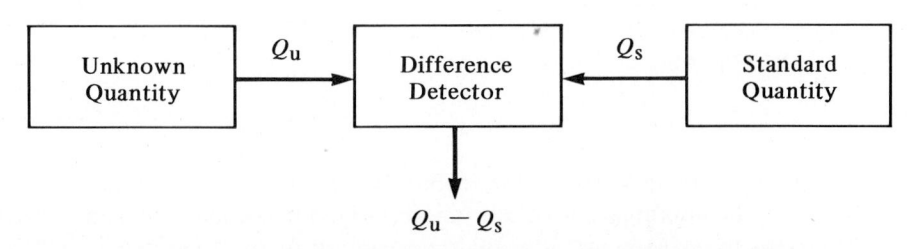

FIGURE 24.28. *The fundamental concept of measurement science: A difference detector compares an unknown quantity Q_u with a standard quantity Q_s.*

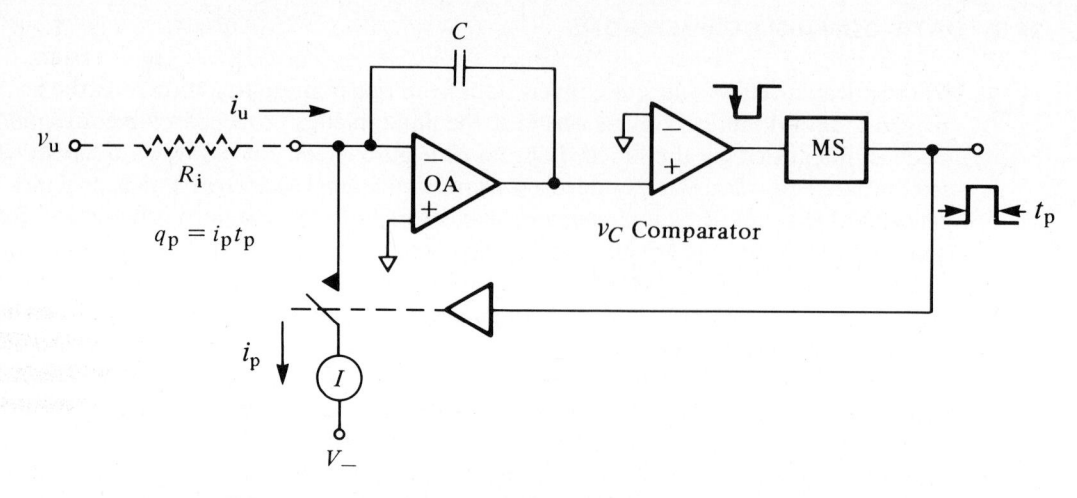

FIGURE 24.29. *Current-to-frequency converter. An unknown current i_u is converted to a frequency from the monostable MS. By addition of a known input resistor R_i (dashed line), an unknown voltage v_u can be converted.*

crosses 0 V, the output of the comparator goes LO, thus firing the monostable multivibrator MS, which generates a pulse of pulse width t_p. The analog switch closes for a period of time t_p and removes an amount of charge $q_p = i_p t_p$ from the summing point.

Assuming that t_p has been made sufficiently long, the removal of q_p from the capacitor causes V_C to become positive again, which allows the comparator output to return HI. This should rearm the circuit so that when i_u has supplied sufficient charge, the cycle will repeat itself. The result is that a constant frequency f_o is generated at the output of the MS for a constant i_u. Note, then, that the *average* current out of the summing point is $i_a = q_p f_o$, but when the circuit has stabilized, $i_a = i_u$. Thus,

$$i_u = i_a = q_p f_o = t_p i_p f_o \quad \text{or} \quad f_o = \frac{i_u}{t_p i_p} \qquad (24.5)$$

The frequency is directly proportional to the unknown current, and the proportionality constant is $1/t_p i_p$. This value obviously depends on circuit parameters such as the RC time constant of the MS, so the full-scale range of this device can be made any reasonable value by proper selection of circuit components. Full-scale frequencies for typical input currents range from 10 to 100 kHz.

This device is easily converted to a *voltage-to-frequency converter* (VFC) by adding an input resistor R_i so that $i_u = V_u/R_i$. The equation describing the behavior of a VFC is then

$$f_o = \frac{V_u}{R_i t_p i_p} \qquad (24.6)$$

Thus, R_i becomes a part of the scaling factor for the converter.

The combination of the op-amp and the comparator is a charge-difference detector. The MS and the analog switch provide a reference standard charge-increment generator that removes standard packets of charge q_p in order to balance the charge resulting from the unknown input current i_u or input voltage V_u.

CHAP. 24 Digital Electronics, Data-Domain Conversions, and Microcomputers

Connecting the output of a VFC to a frequency meter results in a *voltage-to-frequency-to-digital* converter. The circuit components can be arranged, for example, so that for $V_i = 10$ V, $f_o = 10$ kHz. Thus, if the counting period is 1 sec, 10^4 counts will be accumulated in a single period for a full-scale input. It is important to realize that this arrangement constitutes an *integrating converter*. The output of the meter will therefore display the average voltage over the counting period.

Dual-Slope Analog-to-Digital Converter

A second type of integrating charge-comparison converter is the dual-slope analog-to-digital converter (ADC), depicted in Figure 24.30A. In this device, an analog integrator is used to compare the charge resulting from the input voltage v_i to the charge from a negative reference voltage V_r. The control logic sequences the three-stage operation of the converter. In the first or idle stage, switch SW is connected to ground as shown, and $v_C = 0$. When the conversion is initiated, SW switches to v_i, and the integrator integrates in the negative direction until the counter overflows (Figure 24.30B). In this example, the counter has a capacity of 2000 counts (0 → 1999), and the period of integration is $2000/f_o$.

At this point in the sequence, SW switches to the reference position (V_r) so that v_C then begins to increase toward zero, and the counter continues to count up from zero at the same rate as before. Now, however, the integrator output has the same slope regardless of the value of v_i. When v_C crosses zero, the control logic causes SW to switch to the idle state. The counter output is then latched to provide time for readout while the next cycle is carried out, and then reset in preparation for the next cycle.

The amount of charge q_u accumulated on the capacitor during the first integration period is given by the product of the integration time $2000/f_o$ and the current v_i/R. Similarly, during the second integration period of time duration N/f_o, $q_r = (V_r/R)(N/f_o)$. Because $q_u = q_r$, it can be shown that $N = v_i(2000/V_r)$, or

$$v_i = \frac{N}{2000} V_r \tag{24.7}$$

If V_r is arranged to be 2.000 V, the output of the converter will read directly in volts, with full scale being 0 to 1.999 V.

This type of converter is widely used and is found in most DVMs. Such converters are simple, inexpensive, and accurate ($\pm 0.1\%$ or better), and provide rejection of noise from power lines. As shown in Chapter 23, Figure 23.10, integrated-circuit DVM chips such as the Intersil 7107 are easily implemented with a few external components.

Digital-to-Analog Converter

A variation on the summing amplifier of Chapter 23, Figure 23.26B is used in the design of the one-digit BCD digital-to-analog converter (DAC) shown in Figure 24.31. When logic 1 is applied to the control input of switch A only,

$$V_o = -\frac{10 \text{ k}\Omega}{100 \text{ k}\Omega}(-10.00 \text{ V}) = 1.000 \text{ V} \tag{24.8}$$

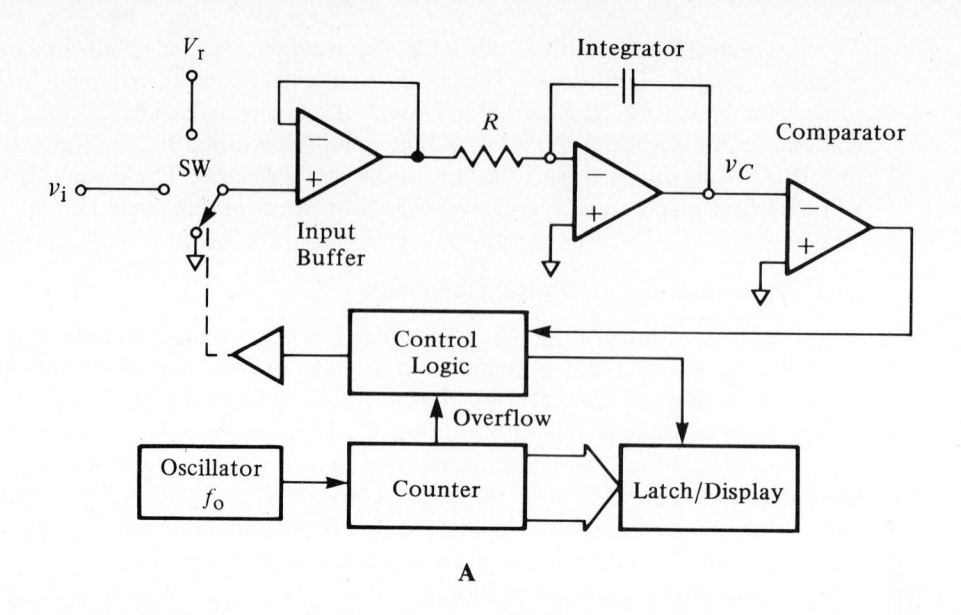

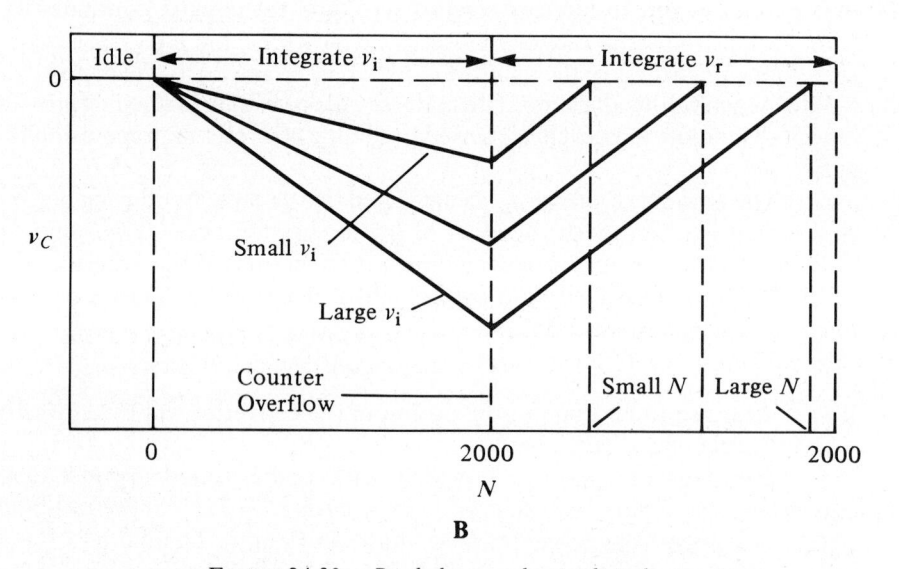

FIGURE 24.30. *Dual-slope analog-to-digital converter.*

Each successive input resistor is arranged to provide a factor of 2 more current into the summing point than the previous resistor. This yields values for V_o of 1 V, 2 V, 4 V, and 8 V for logic 1 applied to inputs *A*, *B*, *C*, and *D*, respectively. Clearly, BCD numbers 0 to 9 applied to the control inputs result in integer voltages from 0 to 9 V at the output.

The addition of four more resistors of 125 kΩ, 250 kΩ, 500 kΩ, and 1 MΩ and corresponding analog switches provides a second digit for the tenths place in voltage

836   CHAP. 24   Digital Electronics, Data-Domain Conversions, and Microcomputers

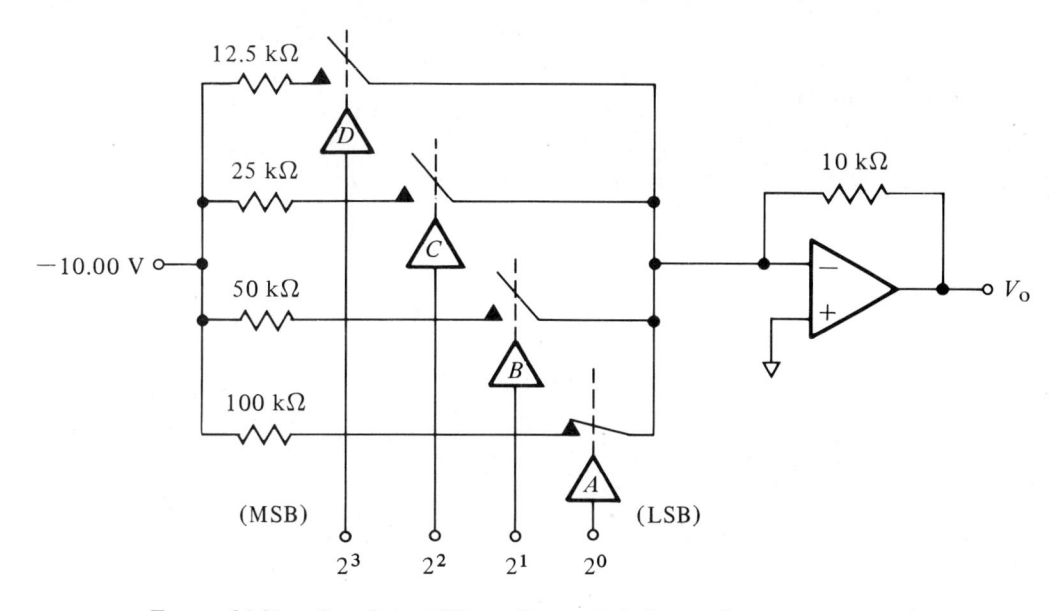

FIGURE 24.31. *One-digit BCD or binary digital-to-analog converter. Each switch closure yields a factor of 2 more current into the summing point. LSB and MSB least- and most-significant bit, respectively.*

output. This idea can be extended to produce BCD or binary DACs of any reasonable precision. Either the feedback resistor or the reference voltage can be varied in order to scale the output range to be any value within the limits of the op-amp.

Staircase Analog-to-Digital Converter

The availability of an n-bit DAC makes it possible to construct an n-bit voltage-comparison ADC. The simplest example of this type of converter is the staircase ADC depicted in Figure 24.32A. The input voltage v_i is compared to the output of the DAC, v_s, by the comparator. As long as $v_i > v_s$, the gate is open to allow oscillator pulses to clock the n-bit counter.

A conversion sequence begins when the counter is reset to zero, which causes the DAC output to go to zero, thus opening the gate. The counter output increases in magnitude by one least-significant bit (LSB) for each period t_p of the oscillator. As the plot of v_s versus time in Figure 24.32B shows, the DAC output then increases by a voltage corresponding to 1 LSB for each oscillator clock cycle, and forms the staircase-shaped waveform. When the staircase just crosses v_i, the gate is closed, and the digital output displays a number corresponding to $v_i \pm \frac{1}{2}$LSB.

Obviously, the higher the resolution of the DAC, the more precisely the number will represent v_i. This type of converter clearly illustrates the measurement process. The DAC serves as the reference standard that is compared to v_i by the difference detector (the comparator). The time of conversion is $t_C = nt_p$, where n is the counter output, which of course varies with v_i. This conversion time is advantageous if v_i is known to be relatively small most of the time. If v_i is large most of the

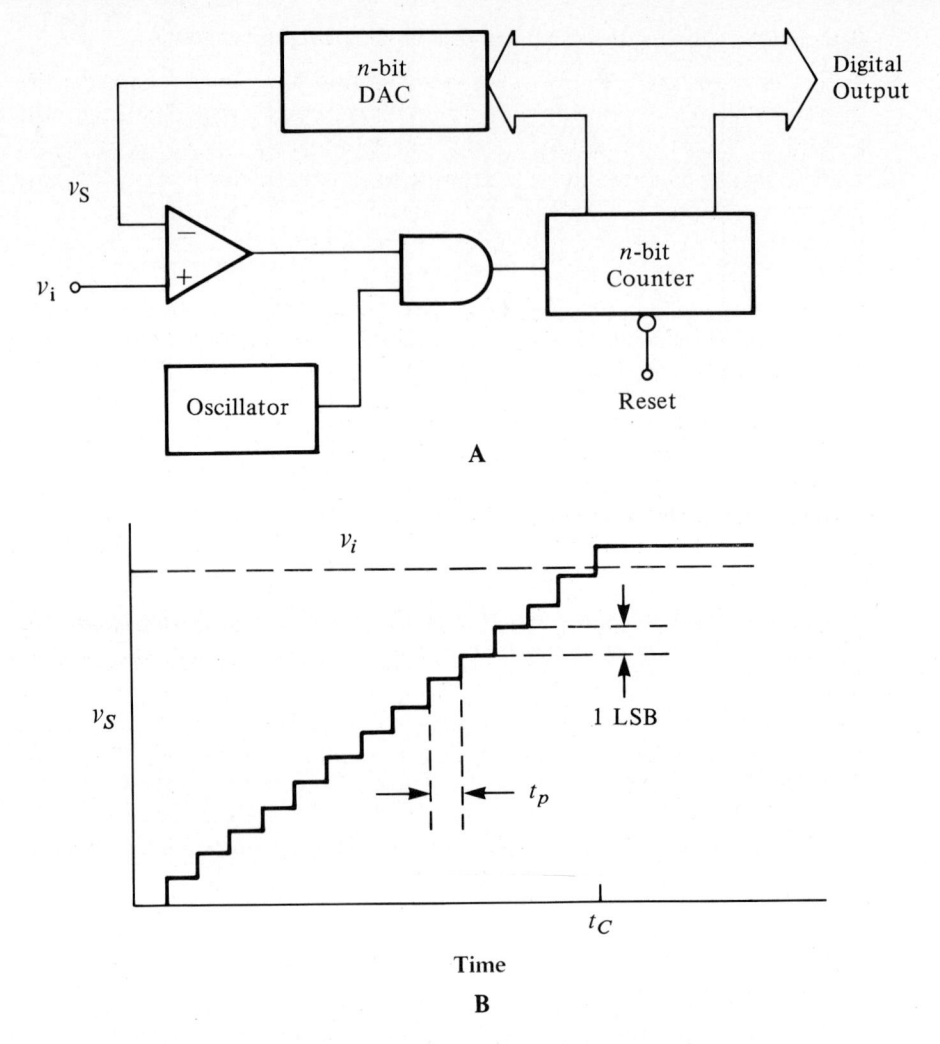

FIGURE 24.32. *Staircase ADC. The oscillator clocks the counter until $v_s > v_i$ by 1 LSB or less.*

time, then t_C will be proportionally longer. This type of ADC is often used in nuclear spectroscopy and related fields in which background signals of low intensity are encountered. The oscillator frequency can be as large as 100 MHz when high-speed counters, DACs, and comparators are used.

The operation of the staircase ADC can be made continuous by replacing the simple counter by an up/down counter controlled by the comparator. If v_i increases, the comparator output goes to logic 1 and the counter counts up, and if v_i decreases, the counter counts down. When the DAC output crosses v_i, the counter alternates between n and $n - 1$, a range which is within $\pm\frac{1}{2}$LSB of v_i. This type of ADC works well when v_i varies only slowly or when a continuous readout is important.

Successive-Approximation Analog-to-Digital Converter

To understand how a successive-approximation ADC works, consider the following question: What is the minimum number of trials necessary to determine, *with certainty*, a number N that lies between 0 and 15? Assume that following each guess you are told whether your number was high or low. The answer is that no more than four guesses are required. To illustrate, suppose that 10 is the target number. Let us first divide the range in half and guess that the unknown number is 7. The number 7 is less than 10, so the upper half of the range is halved again and added to 7, to obtain a second trial number, $N = 7 + 4 = 11$. This number is too large, so 4 is dropped, and half of 4 is added to 7, to yield $N = 9$, which is low. Finally, half of 2 is added to 9, to obtain the target value, $N = 10$. The stepwise approach to the final value is shown in the plot of Figure 24.33A. The rules for the successive approximation are as follows:

1. Begin with a guess of one-half of full range.
2. If too large, drop guess.
3. If too small, retain guess.
4. Add half of previous increment.
5. Repeat steps 2 through 4 until finished.

Note that n guesses are required to determine a number in the range 0 to $2^n - 1$. For example, 12 guesses are required to determine with certainty, a number between 0 and 4095.

The successive-approximation ADC uses exactly the same logic to arrive at a binary (or BCD) number to represent an unknown voltage v_i, as depicted in Figure

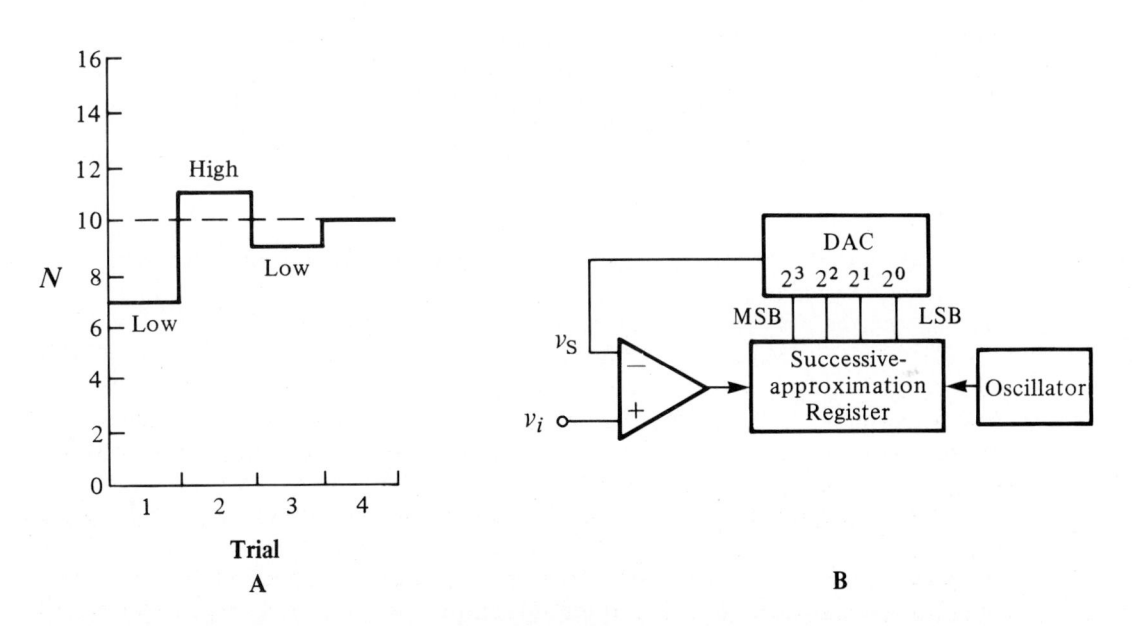

FIGURE 24.33. *A: Successive-approximation procedure applied to guessing a number. B: Successive-approximation ADC.*

SEC. 24.5 Data-Domain Conversions

24.33B. Here, the 4-bit DAC of Figure 24.31 is used to illustrate how the successive-approximation process can be carried out. Assume that $v_i = 5.1$ V and that all bits are initially set to 0. The first cycle of the oscillator sets the MSB ($= 2^3$) to 1, which causes the DAC voltage v_S to change to 8 V. Because $v_S > v_i$, the successive-approximation register (SAR) clears the 2^3 bit. The next cycle of the oscillator causes the 2^2 bit to be set to give $v_S = 4$ V $< v_i$, which causes the comparator output to go to logic 1. The SAR then sets bit $2^2 = 1$ before proceeding to the next cycle, which sets bit $2^1 = 1$, resulting in $v_S = 6$ V $> v_i$. The comparator output goes to 0, and thus the SAR clears the 2^1 bit. Finally, the SAR sets $2^0 = 1$, to give $v_S = 5$ V $< v_i$ which keeps bit $2^0 = 1$. The resulting binary number, 0101, represents the input voltage 5 V $\pm$ 0.5 V. Note the resolution is $\pm \frac{1}{2}$LSB, which, in this case, is $\pm \frac{1}{2}$V.

To increase the resolution of the ADC, a DAC of the required resolution must be provided and the SAR must have a correspondingly larger number of bits. Twelve-bit ADCs with input ranges of ± 5 V, ± 10 V, or 0 to 10 V are typical. Such converters have a fixed conversion time, usually 2 to 8 μsec for 12 bits. Successive-approximation converters of this type are widely used for computerized timed data acquisition. Because it is important that the voltage to be measured does not vary during the conversion process, a sample-and-hold amplifier is almost always used to sample the signal of interest before it is passed to the successive-approximation ADC.

24.6 MICROCOMPUTERS

We have already described most of the devices, components, and circuits that are necessary to explore some of the fundamentals of computerized data acquisition, data analysis, and experimental control. We will begin the present discussion with a brief description of a typical laboratory computer, and its internal structure and operation will be defined. A glossary of acronyms, buzzwords, and terms is then presented, along with a definition of each. Finally, the *interface* and the interactions between experiments and the computer will be explored, and data acquisition described.

The advent of the personal computer has profoundly influenced our lives and revolutionized the way in which science is done. The laws of supply and demand have forced the cost of highly sophisticated microcomputers such as the one illustrated in Figure 24.34 to less than $2000. Thus, any instrument of moderate complexity that could conceivably utilize the capabilities of such a computer for storage and retrieval of data, for data manipulation and computation, or for experimental control, is likely to be attached to a microcomputer. In fact, one or more microcomputers are often included as an integral part of some instruments.

The typical computer is built around a *system unit*, which contains a power supply, a main printed circuit board containing the *central processing unit* (CPU), memory, and various circuits and connectors for attaching the computer to peripheral devices. Space may be available in the system unit for disk drives for temporary or archival storage of computer programs, data, text, and so on.

The computer is equipped with a keyboard for manual entry of alphanumeric information and a display monitor for viewing the information. The monitor is often

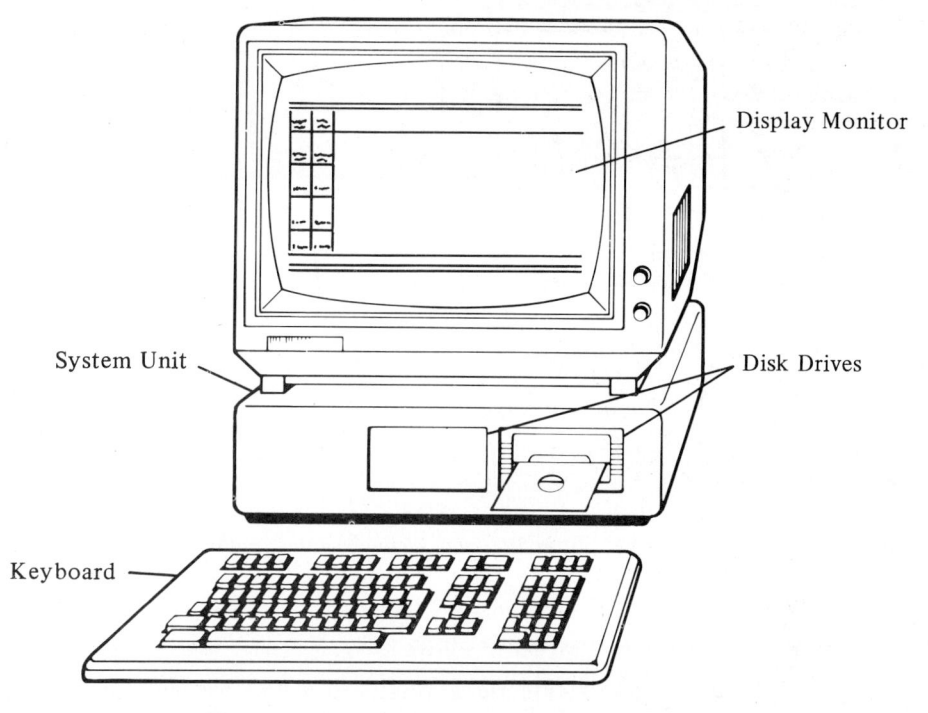

FIGURE 24.34. *Basic laboratory microcomputer.*

capable of displaying data plots and other useful graphics. To understand how these components are interconnected to form the working computer, consider the block diagram of the internal structure of the computer shown in Figure 24.35.

The CPU consists of a control-logic section to sequence the operations of the computer, and to control the flow of data within the CPU and between the CPU and external devices. Information is transferred to or from the CPU in parallel on a set of conductors or leads called the *data bus*. The data bus often has eight leads and can therefore carry eight bits of information simultaneously. The data bus is bi-directional, as indicated by the arrowheads in the figure.

The location of the source of data or the destination of data is indicated by signals on the *address bus*, which is often 16 bits wide. This permits $2^{16} = 65536$ distinct addresses to be accessed by the CPU. The synchronization and direction of the flow of data are determined by the signals of the *control bus*. For example, if data are transmitted by the CPU, a *write* signal is issued. If data are received by the CPU, a *read* signal is issued.

Both the programs to be executed and the data to be manipulated are contained in the memory. The location in memory where the processor is operating is indicated by the *program counter*. Most computers have a set of specialized memory locations within the CPU called *registers*. The registers can be used for temporary storage of data, or they can indicate (or "point to") addresses of data or instructions stored in memory. The CPU also contains an *arithmetic/logic unit*, which performs arithmetic or logic operations on data stored in the registers or in memory.

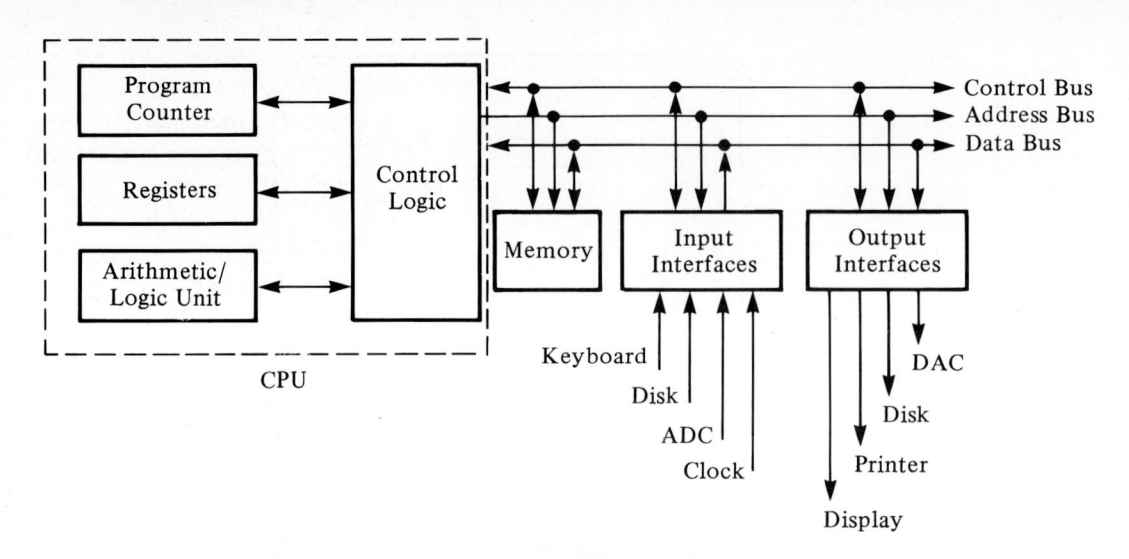

FIGURE 24.35. *Structure and communication paths of a microcomputer.*

Glossary of Computer Terminology

ASCII. Acronym for *A*merican *S*tandard *C*ode for *I*nformation *I*nterchange. Provides a binary representation of characters, numbers, and symbols of the English language.

Assembly Language. The set of mnemonic symbols for the fundamental binary instruction set of a computer. For example, for the 8080 family of microprocessors, JMP means jump to a specified location in memory. The CPU does not recognize JMP, but does recognize its binary equivalent, 11000011.

BASIC. Acronym for *B*eginners *A*ll-purpose *S*ymbolic *I*nstruction *C*ode. A high-level language provided for almost all small computers. A simple instruction in BASIC, such as PRINT, may require many assembly-language instructions to implement.

Baud Rate. The rate of flow of data on a serial transmission line in bits per second. Typical values are 110, 300, 600, 1200, 2400, 4800, 9600, and 19200 baud.

Bit. The smallest piece of digital information (short for *B*inary dig*it*)—logic level HI or LO, 1 or 0.

Boot/Bootstrap. A program stored in ROM for initializing a computer and loading the operating system into memory.

Buffer. A temporary storage area in memory for information being transferred to or from a peripheral device.

Byte. The fundamental unit of character storage, which is eight bits wide. The memory capacity, or disk capacity, of a computer is specified as some number of thousands of bytes—actually 2^{10} or 1024 bytes. Called kilobyte, K-byte, KB, or simply K. A computer with 4 K of memory has $2^{12} = 4096 = 4$ K-bytes of memory. A computer with a 16-byte address bus can address $2^{16} = 65536$ bytes $= 64$ K of memory.

Compiler. A computer program that translates programs written in high-level languages such as BASIC, COBOL, FORTRAN, and PASCAL into machine code, which can be loaded directly into the memory of the computer and executed.

Driver or Handler. A subprogram or subroutine that controls the operation of and the data flow to and from a peripheral device.

Floppy Disk. Low-cost, flexible, iron-oxide-coated plastic disk on which digital information can be stored. The storage capacity of these devices is increasing rapidly as the technology improves, and can be as high as 1 M-byte under special circumstances. Areas on the disk are randomly accessible by the CPU for reading and writing data.

Hard Disk. Metallic version of the floppy disk, capable of higher speed and greater capacity—up to several hundred M-bytes. These devices are considerably more expensive than floppies, and are generally housed in a sealed cabinet for protection from the environment.

Hardware. The physical devices that make up the computer and its peripherals.

Interface. The set of circuits or wires necessary to connect a computer to a peripheral device or an experiment.

Keyboard Monitor. A special device driver that manages communication between the operator and the CPU via the keyboard and display.

Machine Code/Machine Language/Machine Instructions. Binary numbers that the CPU interprets in order to perform its functions.

Modem. *M*odulator/*dem*odulator. Encodes binary data as audio frequencies that can be transmitted over telephone lines. Also decodes received audio frequencies into binary information usable by the computer.

Operating System. A program that is loaded into memory by the bootstrap. At least a part of it resides in memory at all times following the bootstrap procedure. This program controls all interactions with the computer via the keyboard monitor (for human interaction) and device drivers (for peripheral devices).

Peripheral/Peripheral Device. Device attached to the computer via an interface for communication, information storage, and so forth. Examples are printers, disk drives, and modems.

RAM. Acronym for *R*andom *A*ccess *M*emory. The randomly addressable main memory of the computer. In older computers, magnetic cores were used to store individual bits of information so that when the power to the computer was shut off, the information stored in the sense of the magnetic cores remained there indefinitely. Modern solid-state MOS memories are approximately a factor of 10^3 less expensive than core memory, so we are willing to accept the volatile nature of solid-state RAM or provide batteries to preserve the information when line voltage is off.

ROM. Acronym for *R*ead *O*nly *M*emory. Memory that cannot be written into by the CPU. The information stored in ROM is not lost when power is removed. Information is written into ROM during manufacture or by a special device known as a ROM programmer. Programs or data that do not change, such as the bootstrap or the operating system, can be stored in ROM. When the computer is turned on, the CPU automatically executes the bootstrap program stored in ROM. This program usually reads the operating system from ROM into RAM so that storage areas within the operating system can be read or written.

RS232-C. A serial-communications standard defined by the Electronics Industry Association (EIA). Probably the most popular such standard at present. A set of signals is specified for the receiver and transmitter of binary information. Nominal signal levels are logic $1 = -12$ V and logic $0 = +12$ V.

Software. Programs, written in assembly language or some higher-level language such as BASIC or FORTRAN, that are stored in memory to direct the operations of the computer and its peripherals.

Microcomputer Input and Output

The general availability of commercial serial and parallel input/output (I/O) integrated circuits has greatly simplified the task of building I/O interfaces. Because the technology is changing rapidly, we will discuss interfaces in a general way and avoid specific devices. A simple analog I/O system for a typical microcomputer is depicted in Figure 24.36.

The analog input (v_i) on the left of the figure is connected to an 8-bit ADC and an 8-bit latch/buffer to temporarily store the digital output until it is transmitted to the CPU via the bidirectional data bus. The outputs of the latch/buffer are *tri-state logic.* This means that when the outputs are not enabled via the EN signal, they are in the so-called third logic state, or the high-resistance state; that is, they are essentially disconnected from the bus so that they do not load it.

A data-acquisition sequence is initiated when a program running in the CPU

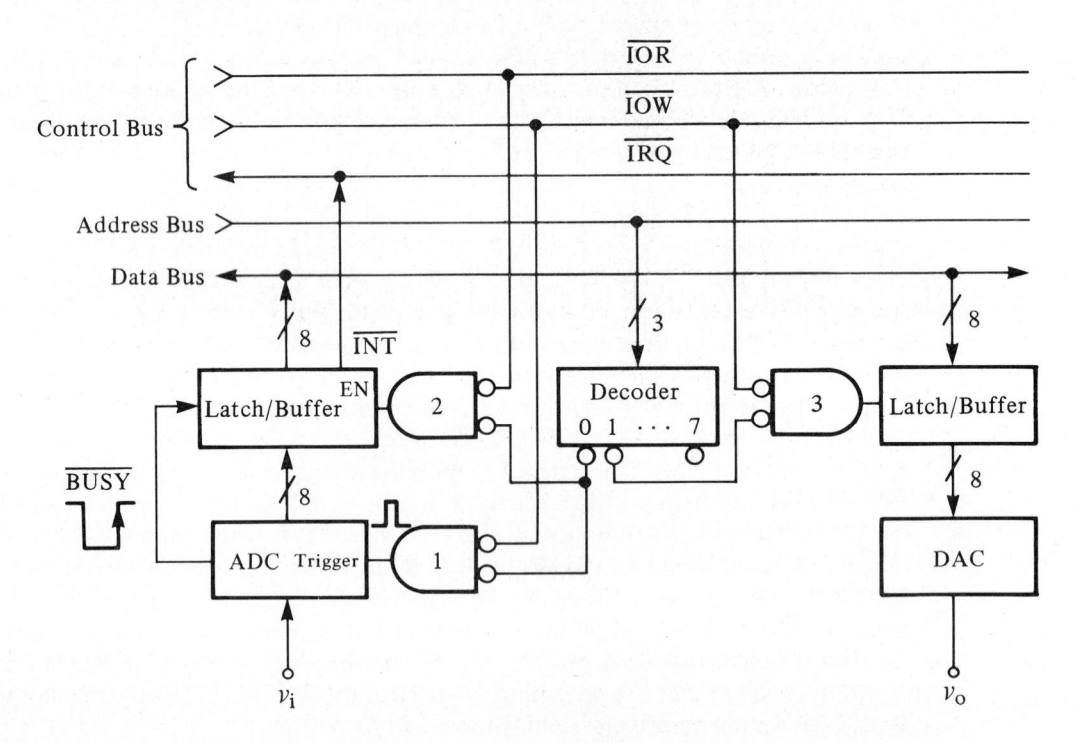

FIGURE 24.36. *Simplified 8-bit ADC and DAC interfaces for a microcomputer.*

844 CHAP. 24 Digital Electronics, Data-Domain Conversions, and Microcomputers issues an output instruction with an address corresponding to the ADC. This causes the address of the ADC (000 in this example) to appear on the address bus. At nearly the same time, the I/O write ($\overline{\text{IOW}}$) signal on the control bus of the computer goes to logic 0. The address decoder produces a logic 0 at its output that corresponds to the binary address present at its input. Because these signals appear simultaneously for only a few microseconds, they produce a sharp pulse at the output of gate 1. This pulse triggers the analog-to-digital conversion and causes $\overline{\text{BUSY}}$ to go to logic 0, indicating that a conversion is in progress. At the end of the conversion, the rising edge of $\overline{\text{BUSY}}$ latches the data into the latch/buffer and causes $\overline{\text{INT}}$ to go to logic 0, thus indicating to the CPU that data are available in the latch.

Two possible modes of data acquisition are available to the CPU. The data can be acquired by the *interrupt-driven mode*, as shown in Figure 24.36, or by the *programmed mode*. In the interrupt mode, the $\overline{\text{INT}}$ signal is connected to the interrupt request input line of the CPU, $\overline{\text{IRQ}}$. If the CPU is given the instruction to enable interrupts, and if $\overline{\text{IRQ}}$ goes to logic 0, the program will immediately jump to a location in memory reserved for a short program called the interrupt service routine. This program then instructs the CPU to place the ADC address on the bus and issue and I/O read ($\overline{\text{IOR}}$) signal to indicate that the CPU is prepared to accept data. The output of the address decoder and $\overline{\text{IOR}}$ are gated to enable the buffer outputs. The outputs are released from the third state, and the data are then available on the data bus. The data are then retrieved by the CPU and stored in memory. The interrupt approach is most advantageous when the CPU is performing more than one task and when the data from the ADC appear asynchronously.

In the programmed mode of data acquisition, the program in the CPU continuously monitors the $\overline{\text{INT}}$ line using its logical instruction set until it goes to logic 0. The CPU then enables the buffer and reads the data as before. This mode is appropriate when only data acquisition is being performed or when the desired rate of acquisition approaches the rate at which the CPU can retrieve data from the bus and store it in memory.

In many scientific experiments, data acquisition must be synchronized with an external event such as the flash of a flashlamp or pulsed laser. In such instances, data acquisition is initiated by a signal from the experiment itself. The CPU simply waits until the external signal is detected and then triggers the ADC.

It is often desirable to acquire data at an accurately and precisely controlled rate. This is easily accomplished using a precision time base (Fig. 24.25). The time-base output is connected to the trigger input of the ADC, and data are acquired at the desired decade frequency. The binary control inputs of the time-base generator can also be interfaced to the data bus so that the computer program controls the rate of data acquisition.

A widely available and far more sophisticated solution to the timing problem is the integrated-circuit *real-time clock*, or precision timer. Such circuits are designed to interface directly to the computer bus to provide program control over one or more crystal-controlled time bases.

The analog output (v_o) section on the right-hand side of Figure 24.36 requires considerably less interaction with the CPU for proper operation. When the computer program determines that a given voltage should appear at v_o, the binary number corresponding to the voltage is placed on the data bus, the address of the DAC (001_2)

is placed on the address bus, and the $\overline{\text{IOW}}$ signal is issued by the CPU. The logic 0 at the decoder output is gated with $\overline{\text{IOW}}$ to latch the data on the bus into the DAC. The voltage v_o changes to the new value and remains at that value until another value is transmitted by the CPU.

For the purposes of illustration (and for many applications), the 1-part-in-256 (0.4% of full scale) resolution of 8-bit ADCs and DACs is perfectly adequate. However, most scientific applications require 12-bit resolution (0.024% of full scale). Such converters are widely available at low cost and are relatively easily implemented.

In practice, the experimenter need not generally be involved in building direct interfaces to the computer bus. A number of companies specialize in fabricating printed-circuit cards that plug directly into the system unit and that carry complete analog I/O systems. These cards cost only a few hundred dollars, and are available for a number of popular microcomputers. The principal components of these cards are often contained in a single hybrid integrated circuit.

24.7 DATA PROCESSING AND ENHANCEMENT OF SIGNAL-TO-NOISE RATIO

Once data are stored in the memory of a laboratory computer, they can be treated in a variety of ways depending on the requirements of the experiment and the ingenuity of the experimenter. The computer is sometimes used to temporarily store the data prior to printing or plotting. In this application, the computer is simply a *data logger*. More often, however, the computer is used to manipulate the data in some fashion so as to extract the desired information from undesirable signals or noise. Moreover, derived information is often sent back to the experiment in order to control experimental variables and/or enhance the quality of the data.

The Nature of Noise

All electronic measurements contain noise that results from a broad range of sources. Noise can be classified into four categories: Johnson or thermal noise, shot noise, environmental noise, and $1/f$ or flicker noise. *Johnson noise* results from the thermal motion of electrons in resistive circuit elements. Because its magnitude is independent of frequency, it is said to be *white noise*. *Shot noise* results when discrete charges randomly move across junctions, such as the pn junctions in semiconductors or between the anode and cathode in a vacuum photodiode. Shot noise is also independent of frequency and is therefore white noise. Johnson noise and shot noise are termed *fundamental noise* because they depend on the nature of the devices in which they appear.

Environmental noise and $1/f$ *noise* are referred to as *excess noise* because they can be minimized by careful instrument design. Figure 24.37 represents a typical power-density spectrum for environmental noise. Note particularly that nearly all regions of the spectrum contain noise. The 60-Hz line-frequency component and its harmonics are quite troublesome. At high frequencies, radio and TV signals can interfere with measurements. Only the region between 10^3 and 10^6 Hz is relatively free of the noise. The $1/f$ component of the noise spectrum is surprisingly large.

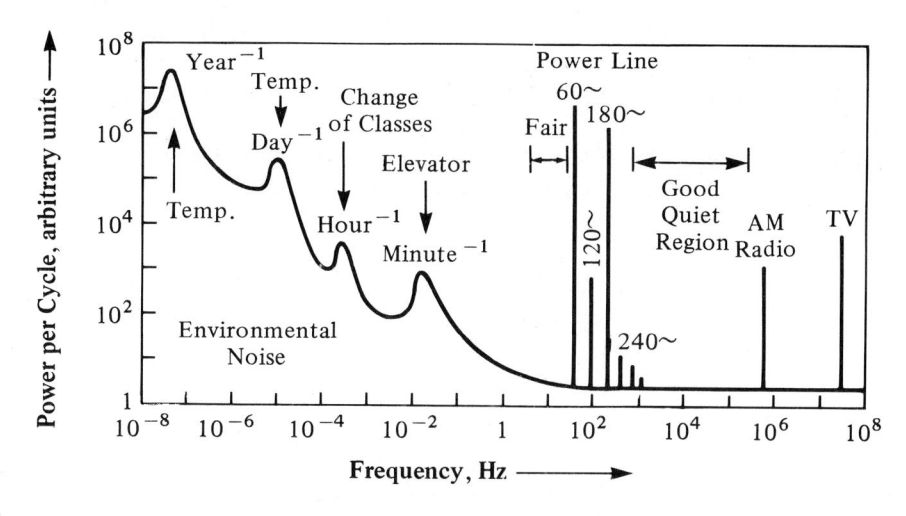

FIGURE 24.37. *Power-density spectrum of environmental noise. From T. Coor, J. Chem. Educ., 45, 533 (1968). Used with permission.*

Although environmental and $1/f$ noise are not well understood, one thing is obvious from Figure 24.37. Low-level DC measurements can be extremely difficult to make, and are impossible in many cases. There are a variety of artful means for extracting such signals, including lock-in amplification and boxcar integration, and these are discussed elsewhere [1].

Signal Averaging

If the noise on a signal is essentially white and is asynchronous with the signal, the noise is proportional to the square root of the time over which the signal is integrated, whereas the signal itself is directly proportional to the time. Thus, the *signal-to-noise ratio* (SNR) increases (improves) as the square root of the time. Similarly, for sampled measurements such as those carried out by a computerized data-acquisition system, the signal magnitude is proportional to N, the number of samples of the *same* repetitive signal, and the noise is proportional to $N^{1/2}$; so the SNR increases as $N^{1/2}$. This effect is shown by the kinetics data of Figure 24.38. Data were sampled by an ADC at equal time intervals beginning with the mixing of the reagents. The data for multiple experiments were summed point-by-point and normalized to fit on the same scale. The SNR of the averaged curve improves with increasing N.

Digital Filtering

Undesirable noise can sometimes be eliminated by the application of appropriate high-pass, low-pass, or bandpass filters. One of the advantages of computerized data acquisition is that digital filtering can be applied directly to the data as soon as they are acquired. Because filtering the data can be considered as a process of weighting or convolution, virtually any weighting function can be applied in software without changing or manipulating circuits in any way.

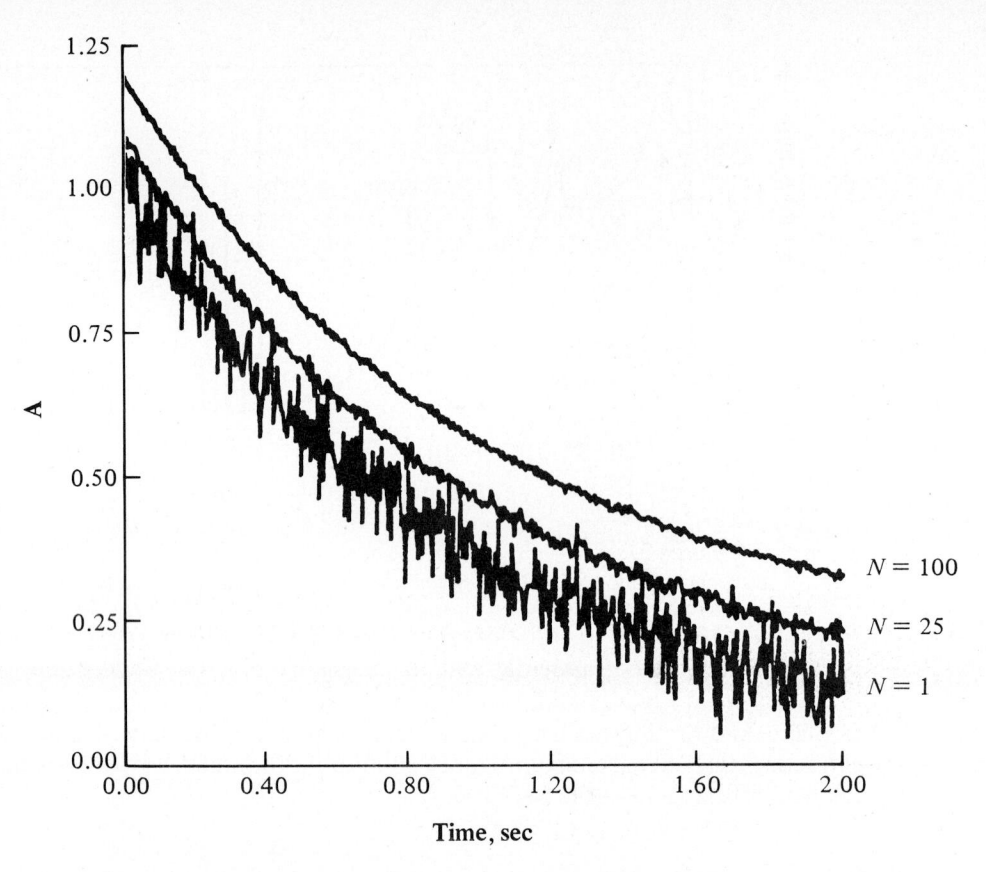

FIGURE 24.38. *Effect of signal averaging on noisy data. For raw data ($N = 1$), $SNR = 20$. For $N = 100$, $SNR = 200$.*

When digital filtering is carried out in the time domain, it is normally referred to as *smoothing*. The simplest approach is the *moving-average smooth*. In this approach, a datum and *n* data on either side of it are averaged, and the average is substituted for the central point. This smoothing window, $2n + 1$ points wide, is then moved to center on the next point in the array, and the process is repeated for all but the last *n* points. If there are *N* points in the entire data array, $N - 2n$ smoothed data result. The data of Figure 24.39 illustrate the obvious SNR improvement imparted by simple moving-average smoothing. The raw data constitute a gaussian-shaped peak with a full-width at half-maximum of 8 points. The raw data were smoothed with a 5-point and an 11-point width.

A somewhat more sophisticated approach is *least-squares polynomial smoothing* [2, 3]. The procedure involves weighting adjacent data with coefficients corresponding to a polynomial of any desired degree that provides the best fit of the points encompassed by the specified width of the smooth. This procedure is computationally simple; and with the proper choice of coefficients, it can also provide the first or second derivative of the data.

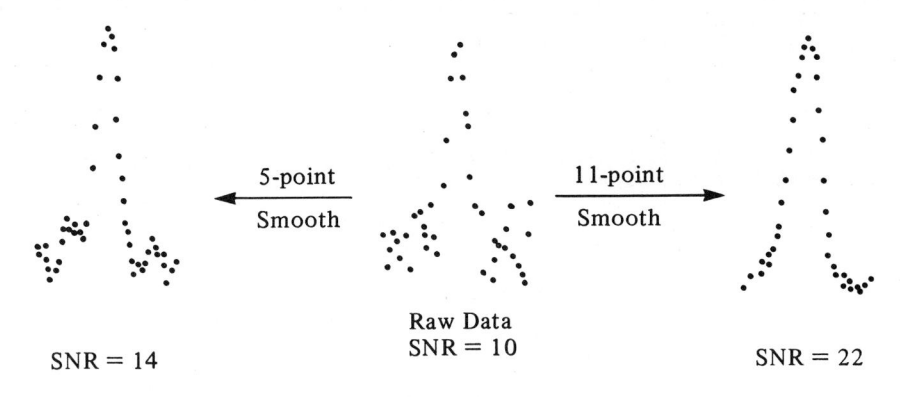

FIGURE 24.39. *Effect of smoothing on noisy data.*

The Fourier Transform

All amplitude-versus-time waveforms, $f(t)$, possess a frequency spectrum, $G(\omega)$. The relationship between the time-domain signal and the frequency spectrum is given by

$$f(t) = \frac{1}{2\pi} \int_{-\infty}^{\infty} G(\omega) e^{j\omega t} \, d\omega \tag{24.9}$$

and the inverse relationship by

$$G(\omega) = \int_{-\infty}^{\infty} f(t) e^{-j\omega t} \, dt \tag{24.10}$$

where $j = \sqrt{-1}$

These two integrals are said to be *Fourier transform pairs*. Quantity $G(\omega)$ is best viewed as the power density function $P(\omega) = G(\omega) \cdot G^*(\omega)$, where $G(\omega)$ and $G^*(\omega)$ are complex conjugates. The power-density spectrum of environmental noise (Fig. 24.37) is a function of this type.

An example of the use of the Fourier transform is illustrated in Figure 24.40. The exponentially damped sinusoidal waveform in Figure 24.40A is of a single frequency ω_0, so that when the integral of Equation 24.10 is evaluated, the power-density spectrum in Figure 24.40B shows a single peak at ω_0. Likewise, the time-domain signal in Figure 24.40C is the exponentially damped sum, $\sin(\omega_0) + \sin(10\omega_0)$. The frequency spectrum in Figure 24.40D shows a peak at ω_0, and a second peak at $10\omega_0$. Clearly, the frequency composition of virtually any time-domain waveform can be obtained by this means.

The power of the Fourier transform is in its ability to limit the bandwidth of noisy signals. This process is shown in Figure 24.41. The time-domain signal in Figure 24.41A undergoes Fourier transformation to yield the frequency spectrum of Figure 24.41B. The low-pass filter function of Figure 24.41C, with cutoff frequency ω_0, is multiplied times the frequency spectrum to give the spectrum in Figure 24.41D, which is free of high-frequency components. The inverse transformation finally gives a relatively noise-free time-domain signal. It is important to recognize that this type

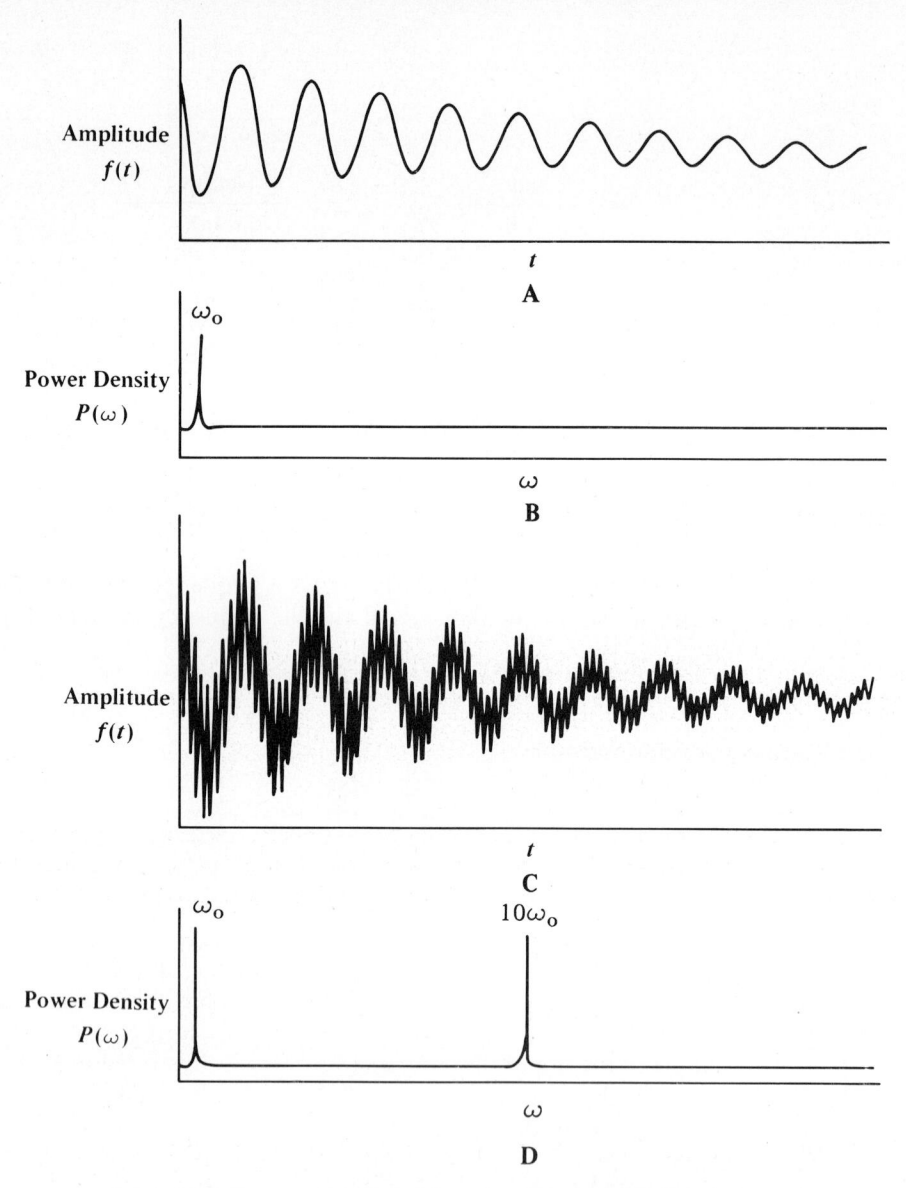

FIGURE 24.40. *A: Simple exponentially damped sinusoid of frequency ω_0 in the time domain. B: Fourier transform showing a single frequency. C: Time-domain plot of the sum of two damped sinusoids, ω_0 and $10\,\omega_0$. D: Frequency spectrum with two peaks, at ω_0 and $10\,\omega_0$.*

of operation in the frequency domain is exactly analogous to smoothing in the time domain.

Efficient high-speed programs have been written for computing the Fourier transforms, so that this type of numerical analysis can be carried out on an 8-bit microcomputer with a BASIC interpreter and only 4 K-bytes of memory. These

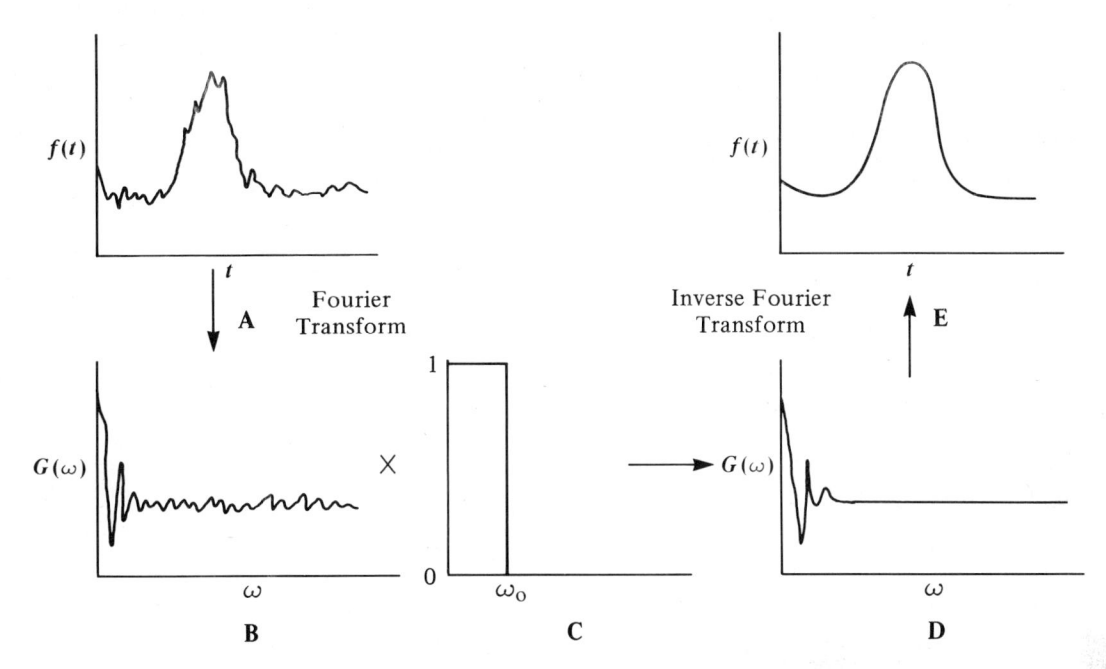

FIGURE 24.41. *A: Noisy spectral peak. B: Real portion of the frequency-domain spectrum of A. C: Digital low-pass filter function. D: Product of B times C. E: Inverse transformed data with (high-frequency) noise mostly removed.*

techniques are used in infrared spectroscopy, nuclear magnetic resonance spectrometry, and more recently in mass spectrometry. The outputs of these spectrometers are time-domain signals, which are transformed into the frequency domain to yield spectra, often with astoundingly large SNRs.

SELECTED BIBLIOGRAPHY

BERLIN, H. M. *The 555 Timer Applications Sourcebook, with Experiments*. Derby, Conn.: E & L Instruments, 1976.

DIEFENDERFER, A. J. *Principles of Electronic Instrumentation*. Philadelphia: Saunders, 1979.

GRIFFITHS, P. R., ed. *Transform Techniques in Chemistry*. New York: Plenum Press, 1978.

HIGGINS, R. J. *Electronics with Digital and Analog Integrated Circuits*. Englewood Cliffs, Calif.: Prentice-Hall, 1983.

HOLLER, F. J., AVERY, J. P., CROUCH, S. R., and ENKE, C. G. *Experiments in Electronics, Instrumentation, and Microcomputers*. Menlo Park, Calif.: Benjamin, 1982.

KORN, G. A. *Microprocessors and Small Digital Computer Systems for Engineers and Scientists*. New York: McGraw-Hill, 1977.

LYNN, P. A. *An Introduction to the Analysis and Processing of Signals*. New York: John Wiley, 1973.

MALMSTADT, H. V., Enke, C. G., and CROUCH, S. R. *Electronics and Instrumentation for Scientists*. Menlo Park, Calif.: Benjamin, 1981.

PERONE, S. P., and JONES, D. O. *Digital Computers in Scientific Instrumentation*. New York: McGraw-Hill, 1973.

ZUCH, E. L., ed. *Data Acquisition and Conversion Handbook: A Technical Guide to A/D and D/A Converters and Their Applications*. Mansfield, Mass.: Datel-Intersil, 1979.

REFERENCES

1. G. M. HIEFTJE, *Anal. Chem.*, **44**(6), 81A; **44**(7), 69A (1972).
2. A. ŠAVITSKY and M. J. E. GOLAY, *Anal. Chem.*, **36**, 1627 (1964).
3. C. G. ENKE and T. A. NIEMAN, *Anal. Chem.*, **48**(8), 705A (1976).

PROBLEMS

1. Design simple switch circuits that are analogs of the following statements. (a) The sports car will start when all doors are closed and all seat belts are fastened. (b) A spectrophotometer will not operate if cooling water is not flowing or if the sample compartment is not closed. Assume a flow switch is available that is closed when water flows.
2. Construct the four basic logic functions—AND ($Y = A \cdot B$), OR ($Y = A + B$), NAND ($Y = \overline{A \cdot B}$), and NOR ($Y = \overline{A + B}$)—using only NAND gates.
3. Construct the four basic logic functions in Problem 2 using only NOR gates.
4. By means of a truth table, prove that $\overline{A \cdot B} = \overline{A} + \overline{B}$.
5. Construct a truth table for $Y = A \cdot B + C$.
6. Construct a truth table for $Y = A \cdot \overline{B} \cdot C$.
7. Design a gate circuit to compare four digital signals—A_1, A_2, A_3, and A_4—with four other signals—B_1, B_2, B_3, and B_4. The circuit should produce 1 at the output with $A_1 = B_1, A_2 = B_2$, and so forth, and 0 at the output when any one of the four conditions is not TRUE.
8. Construct a table showing the level (0 or 1) of the four outputs (Q_D, Q_C, Q_B, and Q_A) of the 4-stage binary counter of Figure 24.13 for the first 16 input cycles. (Note that this will form a binary counting table for the numbers 0 to 15.)
9. Construct a table showing the level (0 or 1) of the four outputs (Q_D, Q_C, Q_B, and Q_A) for the first 10 input cycles of the 4-bit decade-counting unit of Figure 24.14. (Note that this will form a binary counting table for the numbers 0 to 9.)
10. For the Schmitt trigger of Figure 24.19, calculate V_{t+}, V_{t-}, and the hysteresis if the limiting voltages for $v_o = \pm 13$ V, $V_{ref} = 5$ V, $R_2 = 10 \Omega$, and $R_1 = 1$ kΩ.
11. An RC circuit such as that shown in Figure 24.20 is charged to 5 V, and then allowed to discharge. If $R = 1$ kΩ and $C = 1$ μF, what fraction of 5 V remains after three time constants?
12. Find the relative uncertainty resulting from the ± 1 count error in making the following frequency measurements. A 10-sec counting period derived from a crystal oscillator (1 ppm precision) is used for frequencies of (a) 1.5 Hz, (b) 1.5 kHz, and (c) 15 kHz.
13. Determine the clock frequency necessary to measure the period of a 500-Hz signal with 0.2% precision.
14. (a) If a VFC has a 10-V full-scale frequency of 100 kHz, what frequency will be measured for a 6.74-V input? (b) How many digits will be displayed if the counter has a 0.1-sec gate? (c) What is the minimum detectable voltage for this gate time?
15. For a dual-slope ADC, what capacity must the counter have to ensure 0.02% resolution?
16. Find the values of the resistors necessary to add a third digit to the BCD DAC of Figure 24.31.
17. What is the minimum conversion time for a 100-MHz 14-bit staircase ADC? What is the maximum conversion time?
18. Refer to the successive-approximation approach discussed in Section 24.5. (a) What is the minimum number of trials needed to determine, with certainty, a number N that lies between 0 and 31? (b) Using this range, what is the succession of guesses if the target number is 20? (c) If the target number is 6? (d) What is the minimum number of trials needed to determine, with certainty, a number N that lies between 0 and 63? What is the succession of guesses for this range if the target numbers are (e) 40, (f) 20, and (g) 6?
19. What is the conversion time for a 10-bit successive-approximation ADC if the oscillator frequency is 150 kHz?

20. The SNR of a signal that is the sum of 100 measurements is 10. Assuming the noise is not excess noise, how many measurements must be made to increase the SNR to (a) 50, (b) 100, and (c) 1000?

21. The output of an ADC gave the following list of numbers as measurements of a fluctuating signal: 3190, 3205, 3218, 3234, 3159, 3258, 3163, 3234, 3272, 3180, 3190, 3183, 3182, 3179, 3168, 3243, 3175, 3171, 3278, 3209. (a) Calculate the standard deviation, the relative standard deviation, and the SNR. (b) Estimate the standard deviation from one-fifth of the peak-to-peak fluctuation. (c) Compare the results from parts (a) and (b).

22. How many bits are required to produce an ADC of (a) 0.2% resolution, (b) 0.02% resolution, and (c) 0.002% resolution?

CHAPTER 25

Automation in Analytical Chemistry

Kenneth S. Fletcher III
Nelson L. Alpert

Automated instruments are classed as *continuous* or *discrete* (*batch*), depending on the nature of their operation. A continuous instrument senses some physical or chemical property by directly observing the sample, yielding an output that is a smooth (continuous) function of time. A discrete instrument works upon a batch-loaded sample and supplies information only after each batch. Each derives its operating principles from conventional analytical procedures, and must include provision for continuous unattended operation: receiving samples, performing selective chemical analyses under uncontrolled environmental conditions, and communicating with monitoring or control equipment.

A clear distinction should be made between *automatic* and *automated* devices [1]. Automatic devices cause required acts to be performed at given points in an operation without human intervention. For instance, an *automatic titrator* records a titration curve or simply stops a titration at an endpoint by mechanical or electrical means (e.g., a relay) instead of manually. Automated devices, on the other hand, replace human manipulative effort by mechanical and instrumental devices regulated by *feedback of information*; so, the apparatus is self-monitoring or self-balancing. An *automated titrator* may be intended to maintain a sample at some preselected (set point) state—for example, at pH = 8. To do this, the pH of the solution is sensed and compared to a set point of pH = 8, and acid or base is added continuously to keep the sample pH at the set point. This type of automated titrator is called a *pH stat* [2].

In the past, automated instruments were not well accepted because of their limited capability and reliability. However, due to the increased complexity and number of clinical, industrial, and other types of samples requiring analysis, classical (nonautomated) techniques, as well as automated techniques, have been improved in capability and reliability. Well-established instruments such as infrared analyzers, gas chromatographs, ion-selective electrode (ISE) systems, and automatic wet-chemical analyzers can now measure quite complex species and mixtures. Reliability has also increased, because the maturity of solid-state electronics has brought easier data handling and equipment maintenance along with it.

This chapter presents some basic considerations encountered in automating analytical instruments and illustrates some of the important interactions between the instrument and the system of which it is a part. The first portion of the chapter will deal with general concepts and some automated industrial applications in which continuous monitoring and feedback control is important. The second portion will deal with the approaches used in the clinical laboratory, where literally billions of tests are performed annually using automatic instruments.

25.1 INSTRUMENTAL PARAMETERS FOR AUTOMATED INSTRUMENTS

Several instrumental parameters need to be evaluated when unattended operation is proposed for a particular instrument. The definitions that follow have been accepted by the Scientific Apparatus Makers Association (SAMA) [3], and endorsed by the Instrument Society of America (ISA).

Sensitivity

Sensitivity is specified by the relationship between concentration and instrument output and hence by the slope of the instrument-response curve. Sensitivity also specifies the minimum detectable change in concentration, governed by the signal-to-noise ratio of the instrument. Sensitivity is generally defined as the concentration required to give a signal equal to twice the root mean square of the baseline noise. The ISA has recommended that this second definition be denoted by the term *dead band*, which is the range over which an input to an instrument can be varied without detectable response. A change in the slope of the instrument-response curve will generally result in a change in the level of detectable response—that is, the signal-to-noise ratio.

Inconspicuous instrumental, environmental, or chemical effects often cause a loss of instrument response. In atomic emission spectroscopy, for example, sensitivity is affected by such instrumental factors as flame temperature, aspiration rate, and slit width. In amperometric measurements, diffusion currents vary with temperature, and a significant loss in sensitivity may occur with a drop in sample temperature. In ISE measurements, sensitivity may be affected by chemical effects, such as changes in ionic strength or pH.

Accuracy

Accuracy indicates how close a measured value is to an accepted standard or true value. Statements of accuracy should be a percentage of the upper-range value of the reading, or (preferably) an absolute number of measured units. In each case, accuracy is measured in terms of the largest *error* occurring when the device is used under described operating conditions.

Reproducibility

Reproducibility, or precision, differs from accuracy. A poorly calibrated instrument may be inaccurate, but these inaccurate results may nonetheless be reproduced well. Although an automated instrument used primarily as a monitor may require high accuracy (and hence precision), an instrument used for control purposes may only need high reproducibility. The first kind of instrument is used, for instance, in cost control or in clinical testing, where accuracy is paramount; the second kind is used in process control, where a particular factor is to be held at a stable set point. There are various measures of precision. One of the most common is the *standard deviation s*, given by

$$s = \sqrt{\frac{\Sigma (\bar{x} - x_i)^2}{n - 1}} \qquad (25.1)$$

where $\bar{x}$ = mean of the individual measurements x_i
n = number of measurements

(See also Sec. 19.6 for further discussion of statistical considerations.)
Defining objectives in terms of accuracy, reproducibility, and sensitivity is only part of the task. The effect of ambient environmental factors, such as temperature, pressure, humidity, and supply-voltage stability, must also be evaluated. A common problem leading to loss of accuracy and sensitivity in real-world applications is the buildup of deposits on measurement transducers, such as electrodes, which then have to be periodically cleaned, either manually or automatically.

Selectivity

The selectivity of the chemical transducer is its ability to discriminate between the species of interest and possible interferences. Because commercially available instruments are designed for the largest possible number of applications, the instrument may not operate optimally in a particular situation. Therefore, both the measuring method and the chemical system being measured should be thoroughly understood.

Range and Span

Range is defined as the interval over which a quantity is measured. The *span* is simply the width of the range, the difference between the upper- and lower-range values. Consider the following example. For a typical pH meter, 0 mV corresponds to pH 7, and (at 30°C) each unit change in pH produces a change in potential of approximately 60 mV. In Figure 25.1A, pH is measured over the range pH 7 to 9 (0 to +120 mV); the span is 2 pH units (120 mV), and the zero is normal. If one wishes to measure pH over the range pH 8 to 10 (Fig. 25.1B), the 2-pH span is maintained, but the zero is suppressed and the millivolt range becomes +60 to +180 mV. The official glossary of the Instrument Society of America [3] defines a suppressed-zero range as a range in which the zero value of the measured variable is smaller than the lower-range value. Likewise, an elevated-zero range is a range in which the zero value of the measured variable is greater than the lower-range value; in Figure 25.1C, the pH range is pH 4 to 6 (−60 to −180 mV), and the span is again 2 pH units. Note in Figure 25.1D that,

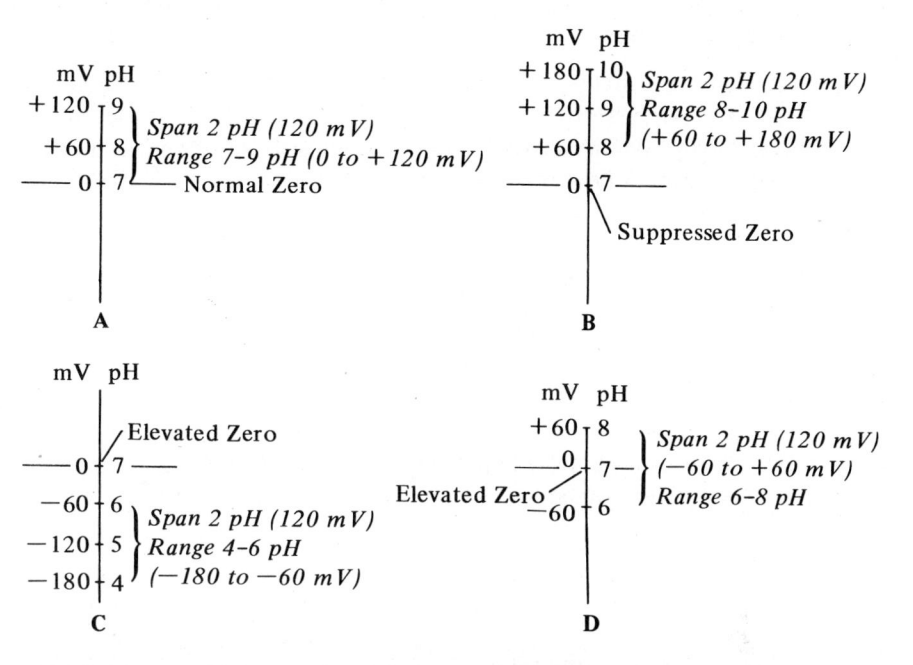

FIGURE 25.1. *Illustrative relationship between range, span, elevated-zero and suppressed-zero. Courtesy of the Foxboro Company.*

if the zero value falls between the upper- and lower-range values, there is again an elevated zero.

Because a change in span requires a change in the per-unit response of the instrument, it requires a change in gain; likewise, since a change in range requires a change in the zero point of the instrument, it requires a change in bias.

The dynamic (working) range of an automatic or automated instrumental technique must obviously fit the working range of concentrations to which it is being applied. In a continuous or automatic titration, for example, the dynamic range and span are governed by the sample size (the volume of the sample and the concentration of desired species in it), the size of the buret, and the concentration of the titrant. The presence of a second titratable species (interference) in the system reduces the usable span by the amount of the second species, since the titration will measure both species.

For cases in which the amount of the interference is known to be constant, a blank correction (zero elevation) may be possible if a small enough portion of the span is consumed. Unfortunately, since the concentrations of most interferences are subject to the same variations as those of the species of interest, one is usually left with three alternatives.

1. Find a method more selective for the species of interest.
2. Find a selective measurement for the interfering species and use it to correct the primary measurement signal.
3. Use chemical conditioning to remove the interference from the sample.

Speed of Response

In electronic instruments, there is always some lag between the physical change being measured and the recorded signal. Speed of response is usually defined as the time required for the instrument to reach a specified percentage of the total change observed. It may be stated in terms of *rise time*, the time required for an instrument output to change from, say, 10% of the ultimate value to 90%. Another measure is the *time constant*, defined as the time required for the response to build from 0% to 63% ($100 - 100/e$) of the ultimate value when a step-function signal is received at the detector. *Response time* can be defined as four or five time constants, representing 98 or 99% of the final value. Increasing the amplification of a system increases its noise in proportion to the square root of the amplification. This will require a larger response time to damp out the fluctuations, and so the response time should be increased in proportion to the square root of the amplification. If the amplification is increased 10-fold, for example, the response time should be increased about 3-fold. In a scanning instrument, the scan rate should be correspondingly reduced.

Dead-Time

Dead-time is often of great importance in automated systems. This is defined as the time interval, after alteration of the parameter being measured, that elapses before a change is observed at the detector. It can be minimized by placing transducers properly, keeping the sample lines short, and using high flow rates. It is never entirely absent. Batch-sampling analyzers introduce additional dead-time, since delay inherently exists between the time at which a sample is taken and that at which a signal is generated. The response of a gas chromatograph in an automated process-control analyzer is a good illustration of this type of delay, since the elution time of the species in the column is dead-time. In automatic batch samplers, times between sample injections can be much shorter than the dead-time so that after the first sample is detected, data output for subsequent samples is obtained in rapid succession.

25.2 SAMPLE CONDITIONING

Sample conditioning is the physical or chemical change needed to render samples suitable for measurement. Ignoring this basic requirement has caused many serious problems for automated instruments, particularly those designed for unattended operation.

From the physical point of view, the most important considerations are temperature, pressure, and sample cleanness. Suspended solids in liquid samples and dust in gas samples often interfere with transducers in continuous-sampling instruments and with the volumetric sampling techniques used in batch-sampling instruments. Separation of sample from debris-laden material can be accomplished simply using filters, but this generally does not meet the requirement of long-term unattended operation. Successful techniques either avoid filters or use designs that permit automatic backflushing. Some methods are based on density differences; separations using gravitational forces or the centrifugal forces created by cyclones are quite successful. Alternately, for gases and liquids, head-space techniques and diffusion through membranes are useful.

If the physical state of the sample (gas, liquid, or solid) is different from that required by the measurement, a phase conversion may be required. Some instruments (e.g., gas chromatographs and infrared analyzers) utilize either gas or liquid samples; the appropriate phase may be chosen to optimize any of the instrumental parameters discussed earlier. If the measurement is temperature sensitive and the sample temperature may vary, temperature compensation or temperature control must be provided. Temperature compensation is usually supplied as part of the measuring system; the sample temperature is continuously monitored and the measurement signal electrically corrected to offset the temperature effects. When temperature control is required, a batch sample or a sample side stream is used, and the sample temperature is adjusted prior to measurement. This procedure adds dead-time to the measurement because the sample has to be in the temperature converter long enough to come to thermal equilibrium.

The sample pressure (in a gas, for instance) may vary widely; it is often restricted by the mechanical constraints of the transducer. If a pressure drop is required in the measurement, phase changes or degassing (in liquids) may also occur and interfere severely with the measurement.

Chemical sample conditioning or reagent addition is often more difficult to apply on a practical scale than is physical conditioning, but can be invaluable in optimizing instrumental parameters. In many cases, it makes otherwise-impossible measurements feasible. Three types of reagent addition are utilized. In the first, the sample is diluted to lower the concentration of the species to be analyzed into the dynamic range of the measuring instrument. In the second, reagent addition converts a species for which no useful measurement technique exists into one amenable to measurement. In the third, reagent addition is used to suppress the effect of interfering species and thus render the measurement more selective.

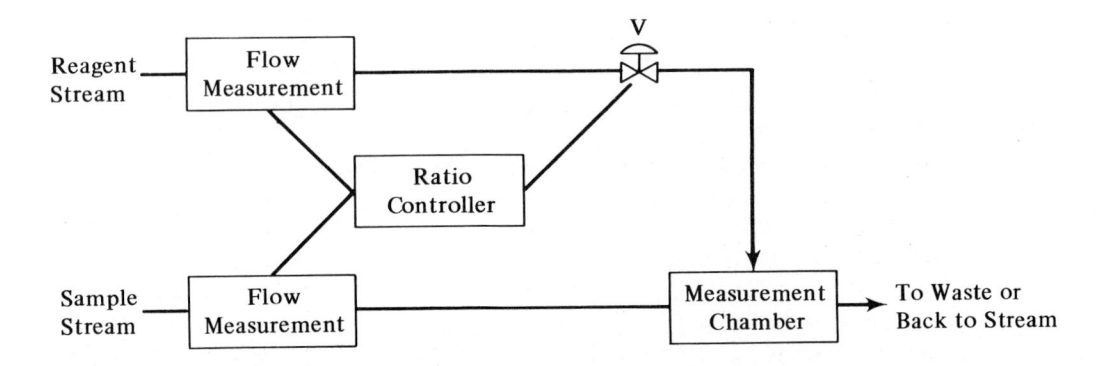

FIGURE 25.2. *Reagent addition system. The ratio controller compares reagent flow to sample flow and maintains a preset ratio by operating control valve V on the reagent stream. The streams are then mixed in (or before) the measurement chamber where the desired analytical measurement is performed. Courtesy of the Foxboro Company.*

The quantity of reagent added may not be critical. Frequently, only a moderate excess of reagent is required and, therefore, a 10 to 50% accuracy in this quantity suffices. The practical requirement in flowing systems is to maintain a constant ratio between the reagent flow and the sample flow. A simple system for accomplishing this is shown in Figure 25.2. In this system, the ratio controller maintains a flow of reagent to the sample stream at a level proportional to sample flow. This proportion is important in maintaining chemical stoichiometry as well as a known, constant dilution at the measurement point. Addition of reagent by diffusion through a membrane is a very attractive alternative in view of its simplicity. Alteration of the pH of a sample by diffusion of ammonia through Teflon membranes has been successfully used in ISE analysis. Widespread application of this technique awaits development of membranes suitable for use with other reagents.

25.3 AUTOMATED PROCESS CONTROL

The successful implementation of automated instruments in routine monitoring functions has led to their increased use in automated process control. Process control is accomplished by means of the *control loop*, which contains at least three parts:

1. An instrument that senses the value of the variable being regulated.
2. A controller that compares the measured variable to a reference value (set point) and produces an output proportional to the difference.
3. A final operator (controlled by the output of the controller) that actuates some mechanism to reduce the difference.

The two basic types of control loops applied to automated systems are termed *feedback* and *feedforward*, and are shown in Figure 25.3. The fundamental difference between these two types lies in the position of the measuring instrument. In feedback control, the measurement is performed either within or at the output of the process, and deviation from the set point causes an operation at the process input. Thus, an error must occur before corrective action can be initiated. In feedforward control, measurement is made at the input to the process, and any deviation from the set point is fed forward to initiate corrective action prior to occurrence of the error. Thus, feedforward systems are theoretically capable of perfect control.

Since a control loop is a dynamic system, its efficiency is governed by the response time of the entire system [4], which includes that of the measuring instrument, the controller, the final operator, and the process itself. Two time-response characteristics of the process are important in selecting the proper control strategy: the *dead-time* and the *resistance-capacitance (RC) time constant*. Dead-time was discussed above; an example is the time required to initiate many polymerization reactions. The *RC* time constant results from the ability of the process to absorb a change in input without an immediate proportionate change in output.

The dynamic characteristics of processes, whether electrical, thermal, or material (e.g., liquid or pressure systems), share common response characteristics. Thermal and material systems resist change in much the same way that electrical systems respond to resistance and capacitance. Table 25.1 presents the analogous

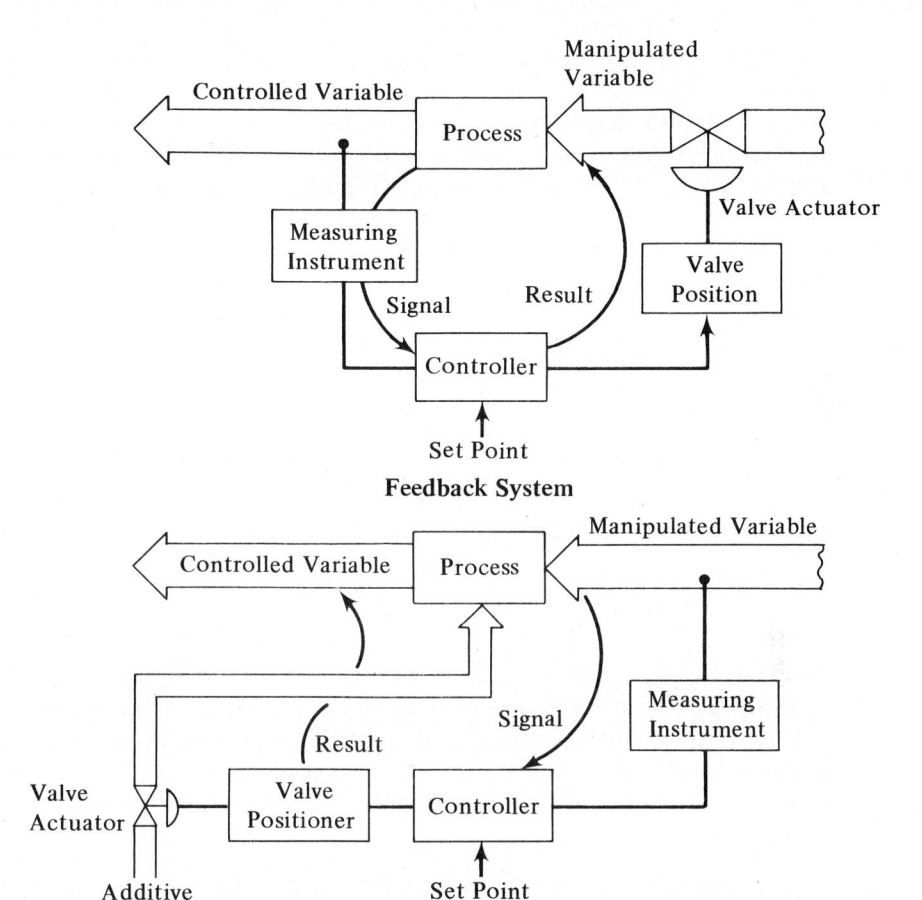

Controlled Variable ← Process ← **Manipulated Variable** ← Valve Actuator — Valve Position

Measuring Instrument → Signal → Controller → Result → Valve Position

Set Point → Controller

Feedback System

Controlled Variable ← Process ← **Manipulated Variable**

Measuring Instrument → Signal → Controller → Result → Valve Positioner → Valve Actuator — Additive

Set Point → Controller

Feedforward System

FIGURE 25.3. *Feedback and feedforward control systems. In feedback control, a measuring instrument obtains information at the output of a process, the signal obtained is compared to a set point, and the difference (or result) is applied to a final actuator. The result is ultimately detected by the measuring instrument, and closed-loop control results. In feedforward control, a measuring instrument obtains information at the input of a process, the signal obtained is again compared to a set point, but now the result is applied to an actuator that controls another input to the process. The result is not detected by the measuring instrument, and open-loop control results. Courtesy of the Foxboro Company.*

TABLE 25.1. *Dimensional Relationship of Systems*

	Electrical	Thermal	Material
Quantity	coulomb	joule	gram
Potential	volt	degree	pascal
Flow	coulomb/sec (ampere)	J/sec	g/sec
Resistance	volt-sec/coulomb (ohm)	deg-sec/J	Pa-sec/g
Capacitance	coulomb/volt (farad)	J/deg	g/Pa

relationships for thermal and material systems, as compared to the more familiar electrical system.

Process capacitance represents the ability of systems to store energy or material, and *resistance* represents the opposition to flow of energy or material into (or out of) a system. The chemical composition of a process can also introduce RC time constants. The presence of undissolved solid reactants, for example, represents a chemical capacitance; and the rate of dissolution of sparingly soluble reactants represents a resistance.

Modes of Process Control

Automated process control can use any of several modes of control, the choice being dictated by the dynamic characteristics of the process. Several of these methods are discussed below.

Two-Position Control. The simplest case is two-position (on/off) control. Here, any deviation of the measured value from a set point drives the final control operator to either a full-on or full-off position. This forces the measured value back and forth across the set point, and the measurement signal cycles about this point. The amplitude and frequency of this cycle depend on the response characteristics of the process. As the process dead-time becomes small, the frequency of the cycle becomes high; likewise, as the process capacitance becomes high, the amplitude of the cycle becomes small. This mode of control is used only for processes in which this cycling effect can be tolerated; it is most successful with those having large capacitance.

Proportional Control. As the capacitance of the process decreases, on-off control leads to increasing amplitude of oscillation. In proportional control, a continuous linear relationship between the value of the measured quantity and the position of the final control operator is established. This control scheme is shown schematically in Figure 25.4. The *proportional band* is determined by the amount of feedback around the amplifier (the value of R_f in Fig. 25.4). The proportional band is inversely proportional to the gain and is expressed as a percentage of the measurement span; it is defined as the change in the property measured that will cause the control operator to move between the fully open and fully closed positions. A narrow proportional band gives a full swing of control-operator position for a small change in the measured value, whereas a wide proportional band requires a large deviation from the set point for a full swing.

The proportional band (PB) is related to the gain by

$$\text{Gain} = \frac{100}{\%\text{PB}} \qquad (25.2)$$

Controller gain is given by the ratio of change in output to change in input:

$$\text{Gain} = \frac{\Delta\text{Output}}{\Delta\text{Input}} = \frac{\text{Output}}{e} \qquad (25.3)$$

where e = error between the set point and the measurement

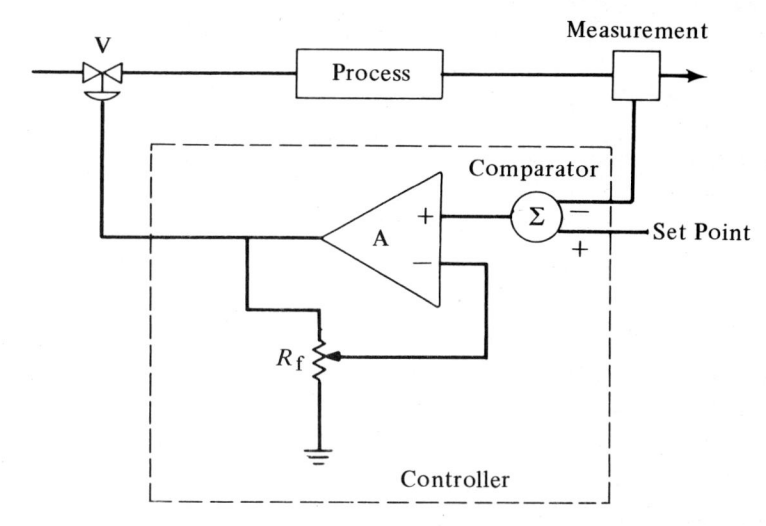

FIGURE 25.4. *Proportional controller. The feedback resistor R_f determines the gain or proportional band of the proportional controller. The symbol V means a valve, and A is an amplifier. Courtesy of the Foxboro Company.*

A *bias adjustment* is usually included to allow the controller output to be set at 50% of span when the measurement equals the set point. Thus,

$$\text{Output} = \frac{100}{\%\text{PB}} e + b \qquad (25.4)$$

The bias b is equal to the output when the error is zero. From Equation 25.4, it is clear that proportional action is not capable of perfect control since, after a load upset, the controller output cannot track the error as e approaches zero. The difference between the resulting measured value and the set point is called *offset*, Δe. This is shown schematically in Figure 25.5.

Under this new operating condition,

$$\text{Gain} = \frac{\text{Output}}{\Delta e} \qquad (25.5)$$

and from Equation 25.2,

$$\Delta e = \frac{\%\text{PB}}{100} (\text{Output}) \qquad (25.6)$$

Equation 25.6 shows that the offset is directly related to the proportional band. In the limit as %PB approaches zero (gain approaches infinity), the offset approaches zero. Pure proportional action is therefore adequate for processes that require proportional bands no wider than a few percent—that is, to easily controlled processes in which load changes are moderate.

Proportional-plus-Integral Control. Addition of integral action to proportional action is necessary for processes requiring wide proportional bands. In integral

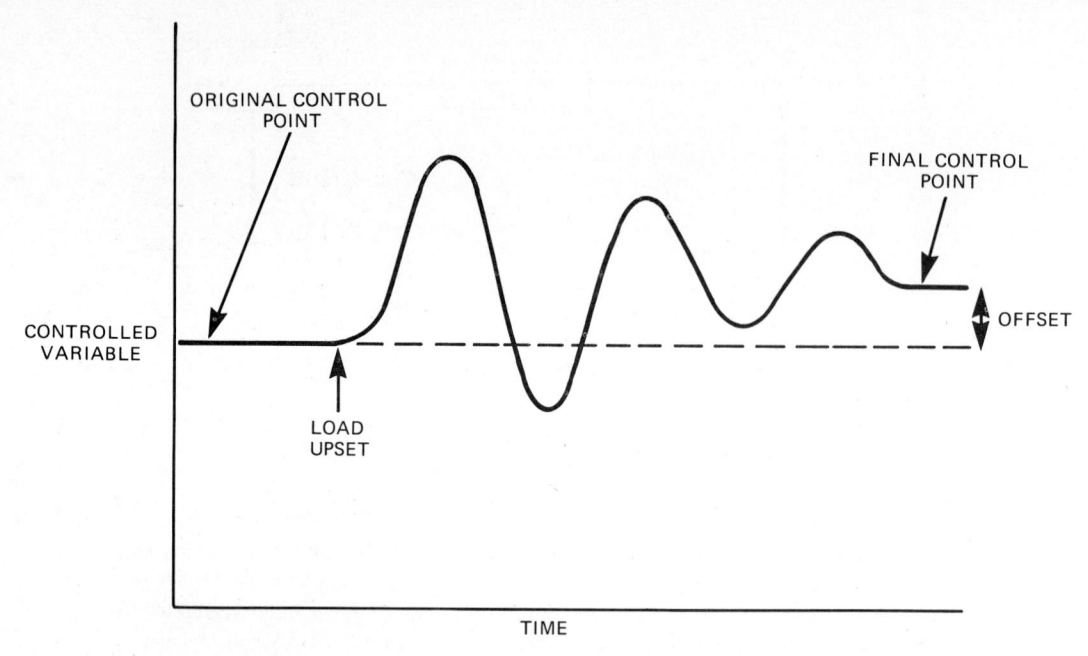

FIGURE 25.5. *Proportional control action. After a load upset, the controlled variable deviates from the set point. The new control point may then differ from the set point by the offset. Courtesy of the Foxboro Company.*

control, the time integral of the offset is fed back, thereby forcing the deviation to zero. This control scheme is shown in Figure 25.6. At balance, the error signal is zero and point *a* is maintained at ground potential. If an offset is present, a voltage develops across the capacitor, which charges at a rate proportional to R_rC. This signal is fed back to the control operator to eliminate the offset and thus return point *a* to ground potential.

The response equation for the proportional-plus-integral controller may be written

$$\text{Output} = \frac{100}{\%\text{PB}} \left(e + \frac{1}{R_rC} \int \Delta e \, dt \right) + b \qquad (25.7)$$

The time constant of the controller R_rC is called the *reset time* and is the time interval in which the controller output changes by an amount equal to the input change or deviation. Note that, when the offset returns to zero, Equation 25.7 reduces to that describing pure proportional control, Equation 25.4.

Proportional-plus-integral control is the most generally useful control mode and therefore the one usually applied to automated process control. Its major limitation is in processes with large dead-time and capacitance; if reset time is faster than process dead-time, the controller-response changes are faster than the process, and cycling results. In these cases, derivative control is beneficial.

Proportional-plus-Derivative Control. Here, derivative action is added to proportional controllers for processes with large capacitance and appreciable dead-time.

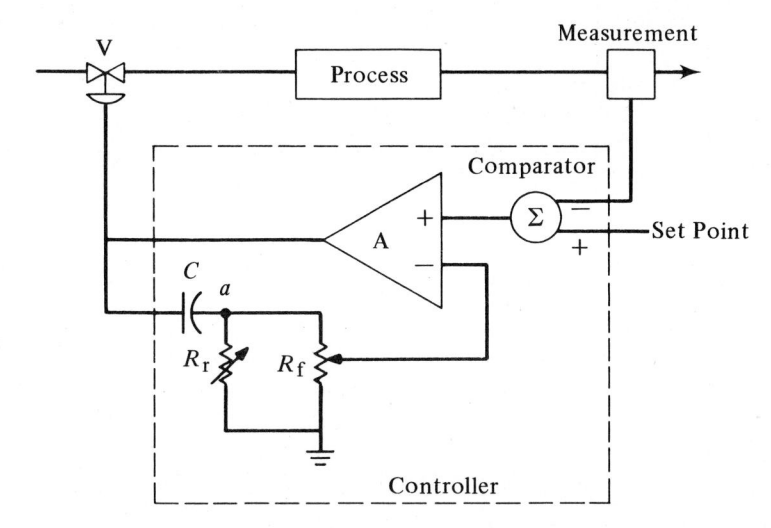

FIGURE 25.6. *Proportional-plus-integral controller. Integral action is accomplished by the series capacitor C in the feedback loop. The presence of offset causes this capacitor to charge at a rate that depends on R_rC, where R_r is the reset resistor. This forces the output to change in such a manner as to drive the offset to zero at a rate determined by the adjustment of the reset resistor. Courtesy of the Foxboro Company.*

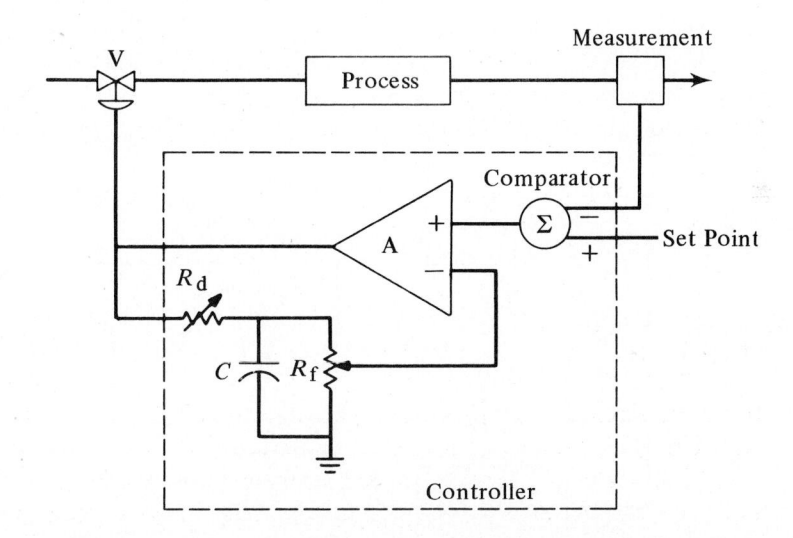

FIGURE 25.7. *Proportional-plus-derivative controller. Derivative action is accomplished by a shunt capacitor C across R_f. When deviation from the set point is rapid, the low reactance of the capacitor causes less negative feedback—hence, greater amplifier gain. The derivative time resistor R_d allows adjustment of the magnitude of derivative control action to a given rate of change of the error signal. Courtesy of the Foxboro Company.*

Control action is now proportional to the rate of change (the time derivative) of the error signal. The response equation is written as

$$\text{Output} = \frac{100}{\%\text{PB}}\left(e + t_D \frac{de}{dt}\right) + b \qquad (25.8)$$

where t_D = *derivative-action time*

Derivative-action time is defined as the amount of lead, in seconds, that the derivative action advances the effect of pure proportional action. Figure 25.7 illustrates a proportional-plus-derivative controller.

Derivative action is accomplished by placing a capacitor across the gain resistor. This capacitor has low reactance and reduces the feedback when the error signal is changing. Thus, a rapidly changing input signal increases the controller gain, producing a larger corrective output. When the time rate of change of the error signal becomes zero, derivative action ceases and Equation 25.8 reduces to that for pure proportional control. Note that derivative action is anticipatory since, if the rate of change of a load upset on the process is rapid, the controller can take large corrective action even though the magnitude of the load change is small, thus overcoming the inertia of the process.

Cascade Loops. A cascade loop is a useful variation on the simple feedback loop discussed earlier. As is shown in Figure 25.8, the cascade loop involves two controllers and two measurement instruments. The secondary (inner) loop directly controls an intermediate variable, such as stream flow or reagent flow. The primary (outer) loop manipulates the set point of the inner loop based on the variable of interest, perhaps temperature or composition. The primary controller is said to be *cascaded* into the secondary.

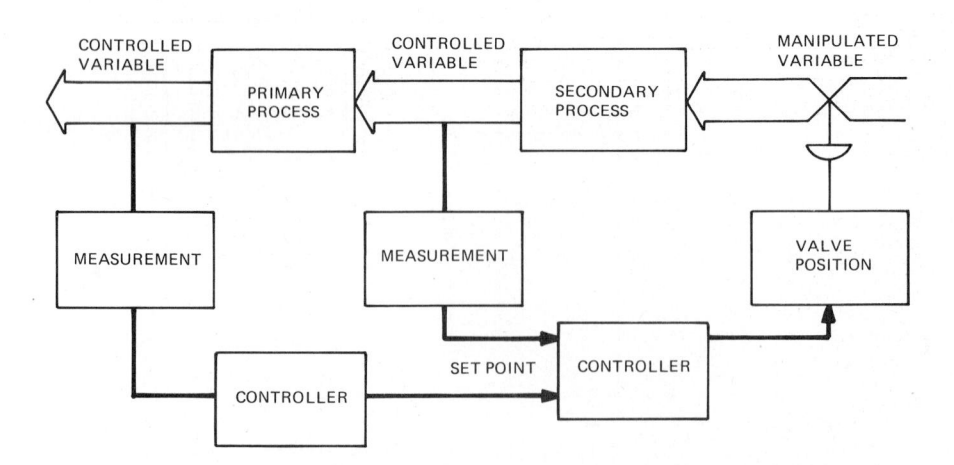

FIGURE 25.8. *Cascade control loop. The primary loop is said to be cascaded into a fast-acting secondary loop—the inner loop to the right in the diagram. The fast-acting loop, based on an independent measurement, corrects for load upsets before they affect the primary process.*

Such loops are most effective where gain variations in the process are troublesome, or where mass of energy flow needs to be directly manipulated. Cascade loops are also used to ensure process security. In this case, the outer loop might be an analytical instrument and the inner loop a traditional flow loop. Should the analytical instrument fail, the process will continue, although no claims can be made about the quality of control.

The Controller

Analog control. An example of a typical commercial controller is shown in Figure 25.9. The Foxboro SPEC 200 PID Controller and Display provides proportional-plus-integral (PI) or proportional-plus-integral-plus-derivative (PID) control. The set-point dial on the display unit places the set point at any level between 0 and 100% of the measurement range. A second pointer indicates the actual measurement value on the same scale. Additionally, an output meter indicates the percentage of output

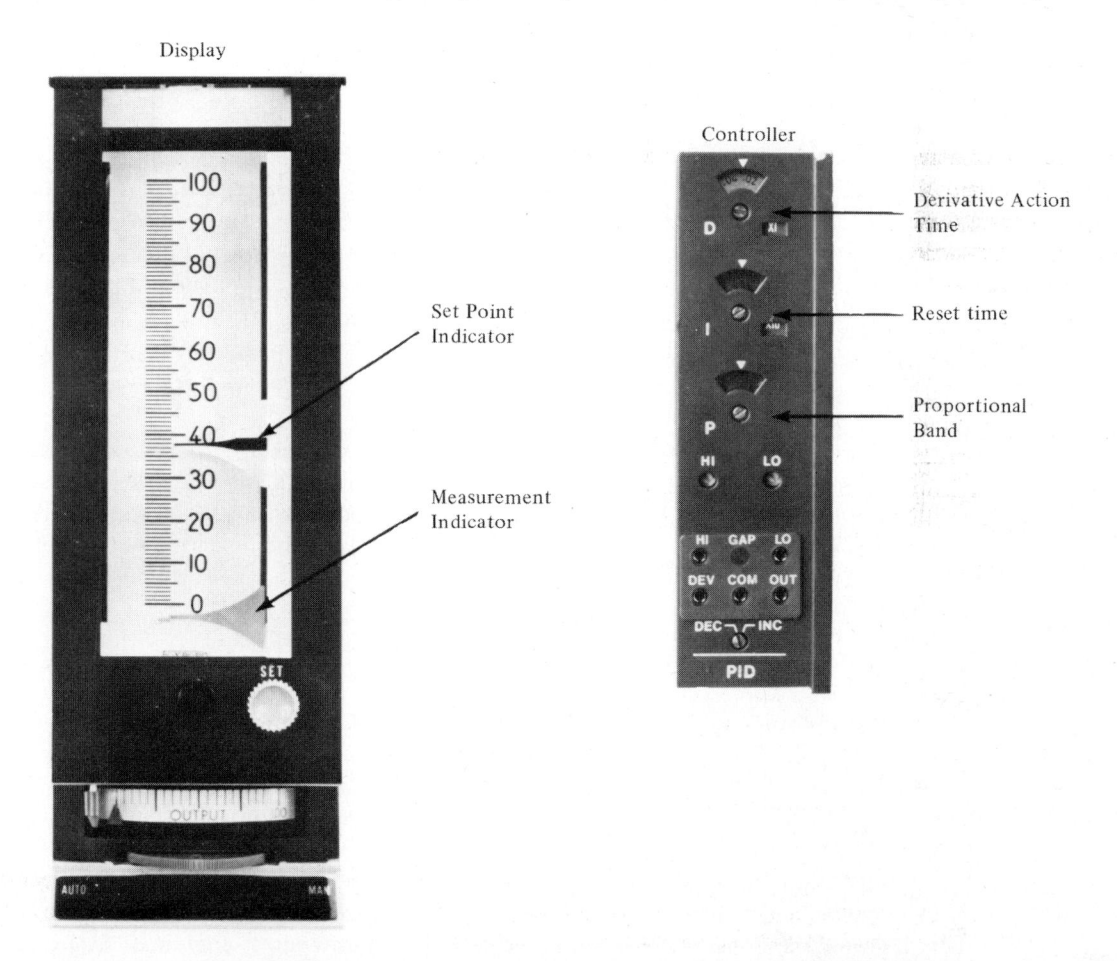

FIGURE 25.9. *Foxboro SPEC 200 Controller and Display. Courtesy of the Foxboro Company.*

being applied to the final control operator. The control unit holds the adjustments for setting proportional band (R_f in Fig. 25.4), reset time (R_r in Fig. 25.6), and derivative-action time (R_d in Fig. 25.7). Process controllers are also manufactured by Minneapolis Honeywell, General Electric, Leeds & Northrup, and others.

Digital Control. Digital computers were first used for process control in the early 1960s. Since that time, the costs of computing equipment have fallen so drastically that the computer has become an essential cost-cutting tool for control in both the modern plant and the chemistry lab.

The principal difference between digital control and analog control is in the nature of the digital processor. Analog solutions to the control modes discussed earlier are continuous and parallel. Digital solutions are discontinuous and sequential, and some care must be used in selecting implementation algorithms.

The control of a process by computer is termed *direct digital control* (DDC) if the results of the control calculation are applied directly to the process, and *set-point control* (SPC) if the calculation causes the set point of an analog controller to change. Under DDC, the control calculation is being executed in a digital processor. This is a sampled-data system, and the user must choose a sampling interval that is small with respect to the process time constant.

The SPC loop, with its analog controller being externally set, is really a cascade loop. Although expensive (two controllers), it does provide analog backup control for process security.

The advent of the microprocessor began a revolution in digital technology that led to inexpensive, but computationally significant, systems finding their way into the laboratory. For a minimal outlay, today's chemist can assemble a system comprising a processor, floppy-disk drives, a cathode ray tube (CRT), a printer, and an input/output module. Most systems use the BASIC programming language, and tend to be more "user friendly" than earlier systems.

The real value of digital control in the laboratory is the ability to do time- and event-based control of an experiment, collect and distill the data, and automatically report results. In reality, what is available should be viewed as a reconfigurable instrument synthesizer—a highly flexible tool to enhance the development of automated monitors and analyzers to be used in process control.

Discrete Instruments

Automatic instruments with discrete (batch) sample handling and analysis have special problems when used for automated process control. These instruments consist of a sampling system, an analyzer, and a memory device that maintains the output at a fixed level until the next signal appears (a trend output). Their output-response behavior is shown in Figure 25.10.

Two sources of dead-time are obvious. *Sampling dead-time* t_s is the time elapsed between the instant the sample is taken and the instant it enters the analyzer. This is shortened by using short sample lines and high flow rates. *Analytical dead-time* t_a is the time elapsed between the instant the sample enters the analyzer and the instant a new output value is displayed.

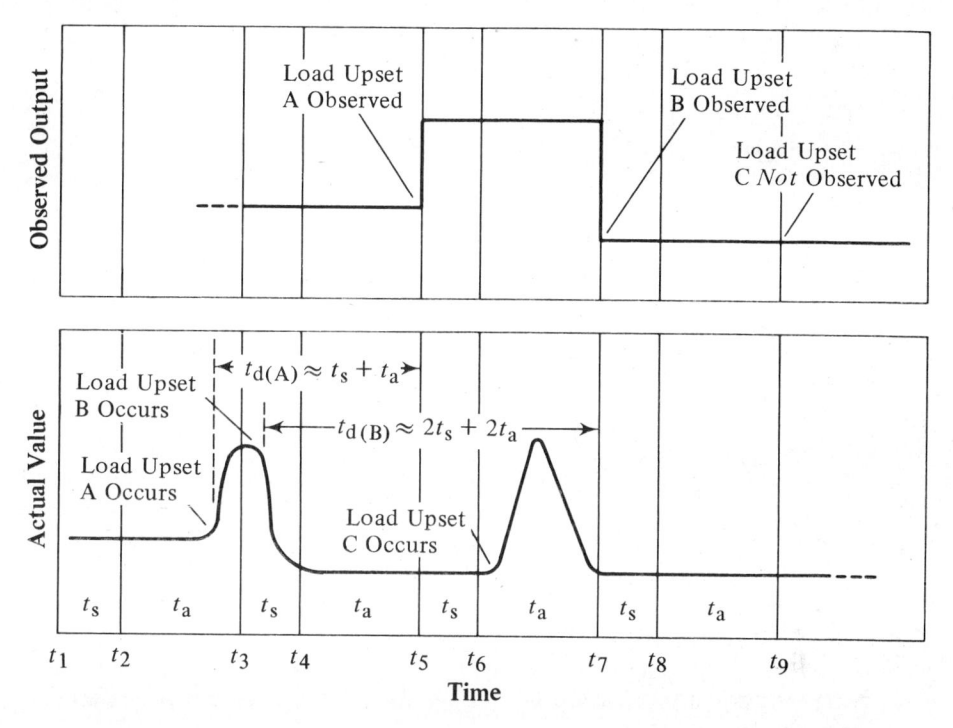

FIGURE 25.10. *Output-response behavior of discrete instruments.* t_s = *sampling dead-time* = $t_2 - t_1 = t_4 - t_3 = \ldots$; t_a = *analytical dead-time* = $t_3 - t_2 = t_5 - t_4 = \ldots$; $t_s + t_a < t_d < 2t_s + 2t_a$, *where t_d is the total measurement dead-time. Courtesy of the Foxboro Company.*

The total measurement dead-time is not constant. In the best case, a load upset that occurs just prior to sampling (load upset A in Fig. 25.10) may be detected with the least possible dead-time—that is, $t_{d(A)} \approx t_s + t_a$. In the worst case (load upset B in Fig. 25.10), a load upset occurs just after a sampling instant, and its detection is delayed for about twice the interval observed above—that is, $t_{d(B)} \approx 2t_s + 2t_a$. Finally, a load upset that occurs entirely within an analyzer interval t_a may be completely missed (load upset C in Fig. 25.10).

Whereas derivative action is of great value in the control of continuous processes with dead-time, it is useless for batch instruments because the instrument output changes in steps. This local high rate of change produces pulsing of the manipulated variable, the effect of which cannot be seen because of the measurement dead-time.

A *sampling controller* using PI action is most generally useful with discrete-sampling instruments. This is illustrated for a process using a gas chromatograph in Figure 25.11. A logic signal starts a timer that allows the controller to operate for only a fraction of the analytical dead-time. After this control interval, the error signal is removed, the controller output is held, and further corrective action is prevented until the next control interval. The time the controller waits before applying further corrective action is the total dead-time t_d. The major limitation of this system is that

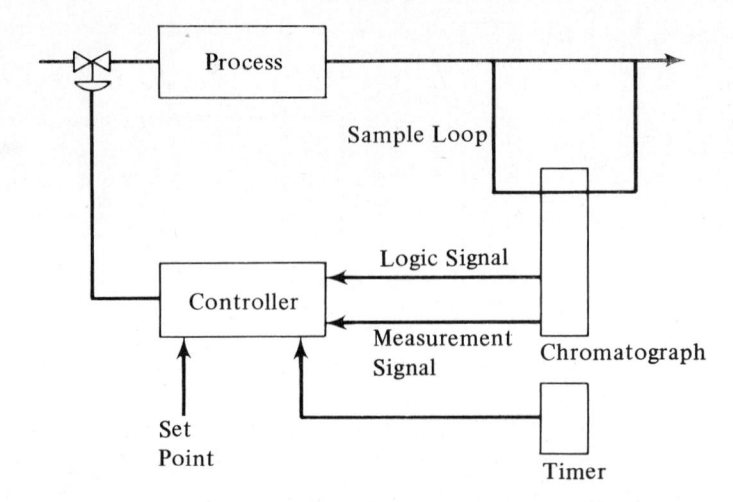

FIGURE 25.11. *Sampling controller. The sampling controller accepts a logic signal from a discrete analyzer (the chromatograph) and applies corrective action for a control interval established by a timer. Courtesy of the Foxboro Company.*

it is impossible to synchronize sampling instants with changes in the variable being controlled.

25.4 AUTOMATED INSTRUMENTS IN PROCESS-CONTROL SYSTEMS

The increasing use of automated instruments in process control is primarily driven by economic factors that demand increased efficiency in usage of raw materials, time, and energy. Desire for improvement in product quality, and social pressures for reduction of waste and pollution, are also very important. Two factors are required for long-term unattended operation of instruments: reliability and operational simplicity. Both favor single-component analysis or dedicated instruments that use probe-type sensors inserted directly into the process. Electrochemical techniques such as ISE potentiometry and conductivity are good examples of this. At the other extreme, difficult analytical measurements on complex samples may require sophisticated instruments and sampling systems. Examples are gas chromatography and mass spectrometry, for which reliability and operational simplicity have been greatly improved in recent years by developments in microprocessor and microelectronic technology. While any of the instruments discussed in earlier chapters of this text can be adapted for use in process control and monitoring, the following discussion will be limited to a few important instruments and some special considerations necessary for their successful implementation.

Spectroscopic Instruments

Molecules and atoms absorb or emit light at characteristic wavelengths. This both identifies the species and can be used for quantitative analysis. Molecular vibrations and rotations are excited by infrared absorption, and the outer electrons of molecules or atoms are excited by ultraviolet or visible absorption. Elemental analysis can also be performed using the ultraviolet and visible light emitted from a plasma of the sample.

The most accessible spectral ranges are 185 to 400 nm for the ultraviolet and 2.5 to 25 μm for the infrared, although an important class of near-infrared analyses exists in the molecular-vibrational overtone region (between 1.5 and 2.5 μm). The infrared absorption spectra of liquids and gases generally yield greater specificity than the ultraviolet because of less spectral overlap between species. Ultraviolet absorption is generally stronger, however, leading to high sensitivity for trace analysis or permitting use of shorter path-length cells. Ultraviolet and near-infrared detectors and sources are also of higher performance than their infrared counterparts.

There are many designs of spectrometers in use. Absorption measurements are the most common; the concentration of a component is computed using the Beer-Lambert law (see Chap. 7). The spectroscopic analyzers to be considered here perform analyses continuously and automatically. The data may simply be reported for monitoring purposes or may be used as input to a process-control system. Examples are given below.

Infrared Analyzers

Nondispersive infrared analyzers (NDIR) are used widely in industry for measuring simple gas streams. Their output is the concentration of a single component, typically CO, CO_2, or CH_4. The instruments have both sample and reference beams. The detector is sensitized by filling a heat-absorbing chamber with the gas of interest. Imbalance between the two beams due to absorption of light at the specific wavelengths characteristic of the sample gas results in an output indication. These instruments are low cost, simple, often available in an explosion-proof enclosure, and ubiquitous. A typical control application is the atmosphere in heat-treating furnaces. Manufacturers of this type of instrument are Beckman, Bendix, and Horiba.

More versatile single-component analyzers are made by alternately switching two narrow-band filters into the beam that passes through the cell to the detector. The wavelength of one filter coincides with an absorption band of the sample, and that of the other filter is used as a reference in a nonabsorbing region. The latter serves as the measurement of the baseline. These instruments work well for a wide range of process gases and liquids because more flexibility is possible in designing the sample-containing cell when only a single beam is involved. A typical application is analysis of CO_2 at wavelengths near 4.2 μm in beverages such as soft drinks, beer, and wine. The analyzer output can be used to control the addition of CO_2 to reach the desired level. Single-component analyzers are made by Anacon, Feedback Engineering (United Kingdom), and Foxboro.

Multicomponent analyses can be made if additional filters at different analytical wavelengths are moved sequentially into the beam. A series of absorbance measurements at different wavelengths can then be used to compute, in real time, the concentrations of several components in a mixture. Usually, a microcomputer is used to sort out the effects of spectral overlap. In one particularly versatile instrument, the discrete filters are replaced by a circular variable filter in which the transmitted wavelength is continuously selectable and depends on the angular position of the filter.

Several applications of multicomponent analyses are noteworthy. Milk analyses, in the mid-infrared, give butter-fat, lactose, and protein content. Cereal and grain analyses give fat, carbohydrate, and protein. In this case, near-infrared light is reflected from a finely ground specimen of the sample, but all other operations of the multiwavelength analysis and data computation are handled automatically. Finally, automatic monitoring for toxic vapors in workroom air, to comply with OSHA requirements, is an important application. Usually, samples are drawn on a programmed basis from different points in the factory. Manufacturers of multicomponent analyzers are Foxboro, Neoteck, and Technicon.

Analysis for moisture content, using the near-infrared, is an important application. Measurement is made using discrete-filter analyzers, by reflection from solids (e.g., cereals, grain, wood chips in paper manufacturing), or by transmission to measure traces of moisture in organic solvents such as dichloroethylene, which tend to be highly transparent in this spectral region.

Analysis of stack gas for control of combustion processes or for compliance with EPA emissions standards is another major area of applications. It is standard practice to control the fuel-to-air ratio in large power-generating boilers by measuring the oxygen remaining after the combustion process. Excess air represents waste heat vented to the atmosphere. Even more precise combustion control can be attained if the CO is measured as well as the O_2. A technique called *gas filter correlation* (GFC) in the mid-infrared is useful for CO; it is related to NDIR and has good specificity. GFC analyzers or ultraviolet analyzers, operating across stacks, can also monitor SO_2 and NO emissions. Emissions of particulates from stacks are monitored using the attenuation of visible light. Manufacturers of stack-gas analyzers are Measurex, Princeton Sensors, EDC, and others.

Gas Chromatographs

The chromatograph is a discrete instrument, because it operates as a batch-sampling device requiring that one take samples, inject them into the chromatograph, and record the chromatograms. A *process gas chromatograph* (PGC), therefore, requires an automatic sampling valve. These valves must reproducibly add samples to the column for millions of cycles. They are electrically or pneumatically actuated, using linear or rotating sliding seals. Despite efforts to improve the reliability of these components, they continue to be the most vulnerable part of the system. Currently, valve faces are fabricated from ceramic, glass-filled Teflon, Rulon, and other such materials in an effort to improve their long-term performance.

Sampling valves for both liquids and gases are available, as well as several detectors, such as thermal-conductivity and flame-ionization detectors, by which PGC can be applied over a range of concentrations from trace ppm to percent levels.

An added restraint in PGC is that the analysis must be completed within a specified time interval. This usually requires column switching, backflushing, and multiple injections of sample—timesharing the same detector. *Column switching* allows chromatogram development using different columns during each analysis, thereby enhancing the separation of the desired components and allowing for rejection (often at an early stage of analysis) of unwanted components. *Backflushing* is useful for eliminating components with long elution times. In one case, the flow of carrier gas through the column is reversed after the last component to be measured has been eluted. The slow-eluting components emerge from the input end of the column in a single band when the flow is reversed because they leave the column at the same relative rates at which they entered. Thus, they may be used to form a composite peak on the chromatogram for "total heavies," or vented, as desired.

An alternative solution is to dedicate the PGC to the analysis of only one or two components of the process stream. This is usually all that is required for control purposes.

The operation of the chromatograph makes data utilization difficult, for two reasons. First, much effort is needed to extract information from the complex chromatographic display. Automated procedures for doing this usually employ a time-base generator that selects a predetermined portion of the chromatogram for integration. When this window includes only the peak of interest, the answer can be automatically presented at periodic intervals. However, the position of this window relies on the fixed time interval between sample-injection and component-retention time. Because retention time is affected by column loading, temperature, and flow rate, these factors must be rigorously controlled. In one PGC system, this problem was eliminated (or at least minimized) by using programming techniques tied to a chromatogram time base rather than an absolute time base [5]. With the availability of inexpensive microprocessors, this technique will probably become more popular in future PGC.

Second, the application of chromatography to dynamic systems is limited by the time interval between analyses. This is particularly serious when chromatography is used in automated control systems. The solution that seems to be emerging is improvement of speed of analysis through miniaturization and dedication of instruments to single-component analysis.

The Daniel chromatograph developed by Honeywell in the early 1970s was inspired by the Micro Chromatograph developed for the Mars probe [6]. This instrument uses microcolumns packed with 50-μm-diameter supports and is reportedly capable of analysis times as short as 10 to 20 sec.

Application of short, small-bore capillary columns to PGC should also achieve the same goal. A significant recent development is the fabrication of a complete chromatographic system consisting of a 1.5-m capillary column, a gas-control valve, and a detector element integrated on a 5-cm^2 silicon wafer using photolithography and standard silicon-processing techniques [7].

Chromatographs are perhaps one of the most widely applied automated instruments in process analysis, particularly in the nonaqueous-chemical and petrochemical industries [8]. In petroleum refining, for example, the crude petroleum, containing hundreds of chemicals from methane to asphalt, is converted to salable cuts by distillation. Further processing by catalytic reforming, distillation, and chemical reaction yields materials used for fuels, lubricants, petrochemical feedstock, and other applications.

An illustration of the use of chromatography in this industry is in the control of distillation towers. Distillation uses the difference in composition between a liquid and the vapor formed from that liquid as the basis for separation. The efficiency of the process is affected by temperature, pressure, feed composition, and feed flow rate. Chromatography is used to monitor the composition of the feedstock and to apply feedforward control of the heat input (temperature) to the tower, or to monitor and control the composition of the product. In the latter case, the chromatograph output is simply compared with a set point, and the controller (using feedback) manipulates the temperature, pressure, or feed flow rate by activating the appropriate final operator. Both types of distillation control are widely used in petroleum refining.

In the petrochemical industries, hundreds of materials are produced using catalytic reforming, isomerization, and polymerization. Tower-distillation monitoring and control using the PGC is of great importance here also. Other specific applications include monitoring the purity of monomers used in manufacturing such polymers as those from vinyl chloride, vinyl acetate, ethylene, and styrene; and monitoring chlorinated hydrocarbons produced by chlorination and oxychlorination reactions—chlorinated solvents, weedkillers, pesticides, and many intermediates that are polymerized directly into plastics. Accurate determination of the BTU content of fuel gas, both for custody transfer and for furnace control, is accomplished using PGC to measure the concentration of all components, followed by calculation of the result. Additional applications of PGC are found in the pharmaceutical and food industries and, to a more limited extent, in analyzing furnace flue gas for combustion control.

Mass Spectrometers

Process mass spectrometry (PMS) has emerged as a reliable, rugged, multicomponent analysis technique, with application in many of the areas traditionally served by PGC. The driving force for this change is the increased speed of analysis, 10 to 40 sec per analysis compared to 30 sec to several minutes for PGC. Advances in vacuum systems, increased ease of computation and data manipulation, use of long-life filament materials with automatic filament replacement, and selection of alternative ionization sources have greatly aided acceptance of this relatively sophisticated analyzer as a process instrument. Long-term maintenance of high vacuum, essential to both operation and a low and stable background signal, is perhaps the critical requirement of PMS. This need has been met with the turbomolecular pump, a mechanical pump using no oil, that is capable of long-term operation with only occasional maintenace of bearings.

For process work, electron-impact sources with filaments of tungsten, rhenium, or thoriated iridium are usually chosen, depending on the application. The selection is influenced by the operating temperature, presence or absence of oxygen, and catalytic activity of the filament material to the sample components. Redundancy of filaments results in only minor interruption if a filament fails, although the mean time between failures has greatly increased owing to filaments having lifetimes in excess of 20,000 hr. Photoionization sources, though presently yielding lower sensitivity than other choices, are attractive and should find increased use because of their long-term reliability.

Time-of-flight, magnetic-sector, and quadrupole instruments offer competitive advantages, and the choice is dictated by the application. The quadrupole technique, which uses only electric fields to determine which m/z ratios will be analyzed, seems most attractive operationally.

Difficulty exists at the sample interface where the process gas or liquid must be transferred to the high vacuum of the ion source. Very small samples of the process stream, which is usually at or near atmospheric pressure, must be delivered continuously to the metering orifice of the instrument. Usually, a small sample stream is constructed: The sample is drawn through a chamber having input and output flow restrictors sized to maintain an internal pressure of about 1 torr. Controlled addition of sample across the metering orifice to the instrument is controlled by the pressure drop maintained between this chamber and the vacuum in the instrument. Fouling and dirt buildups on these orifices can affect results, and adequate maintenance is essential.

The most successful applications reported for PMS to date are for determination of low-molecular-weight gases: in the ethylene oxide process for C_2H_4, O_2, HCl, and Cl_2; in the ammonia process for H_2, N_2, O_2, Ar, CO, and CO_2; in fermentation for N_2, O_2, and CO_2; and in heat treating for CO, CO_2, H_2, N_2, and water vapor. Process instruments are currently available from Perkin-Elmer, Balzers, Anacon, Extranuclear, and others.

Electrochemical Instruments

Several types of electrochemical techniques have been used in automated systems—potentiometry, voltammetry, coulometry, conductivity, amperometry. At first glance, their use in instrument systems appears straightforward, because each transducer converts chemical information directly into an electrical signal. Unfortunately, few applications are found for those methods involving net current flow (e.g., amperometry) because the rate of mass transfer (and hence the current) depends on the sample flow rate, which may vary, and on how clean the electrode surface is. The present discussion will therefore be restricted to potentiometry, a zero-current technique, and conductivity. Both are important in process monitoring and control.

Conductance Analyzers

Conductivity is a nonselective technique; its ease of use and low cost often make it the method of choice for inferential measurement of ionic composition. For this technique, 2-electrode, 4-electrode, and noncontacting or electrodeless cells are available, the choice being dictated by the magnitude of conductivity, the accuracy required, and the degree of fouling expected from the process solution.

In 2-electrode measurements, solution conductivity is sensed by measuring the voltage drop between two electrodes across which is imposed a small alternating current (sinusoidal or other waveform) in order to reduce iR effects due to polarization. Because this cell senses the total iR drop between the electrodes, the measurement is quite susceptible to inaccuracy because of buildup of insulating films from substances in the process solution or their electrolysis products. This problem is largely avoided by using 4-electrode and electrodeless cells.

In 4-electrode cells, one pair of electrodes is used to drive a controlled current through the cell, and the second pair is used to sense the resulting iR drop. This second pair of electrodes is connected to a high-impedance voltage-measuring amplifier so that current drawn through them is minimal. The measurement is insensitive to fouling so long as the resistance of the films does not increase beyond the ability of the current source to provide the controlled current.

An interesting example of the use of contacting-electrode conductivity in process monitoring uses a microprocessor to increase versatility. Extremely high water purity, approaching the ultimate resistivity value of 18.18 MΩ-cm at 25.0°C, is required in semiconductor processing. Because required process monitoring is at elevated temperature, a problem develops because of the difference in temperature coefficients of resistivities of pure water and of water containing trace ionic impurities. One solution to this problem is to incorporate a temperature sensor in the conductivity cell and couple both measurements with a microprocessor that has data for the temperature coefficients stored in memory. This *intelligent instrument* can measure the resistivity, say at 55°C, compare this value to the resistivity of pure water at this temperature, and calculate the contribution due to impurities (assumed to be NaCl). Finally, the resistivity of the process water, referred to 25.0°C, or the concentration of impurity, as ppb NaCl, is calculated from the stored temperature-coefficient data and displayed. Other applications include monitoring of boiler water for steam turbines and nuclear-reactor water. Process-instrument suppliers of contacting cells include Beckman, Foxboro, and Rosemount.

Electrodeless conductivity uses a pair of coils, often toroidally wound, encapsulated in a nonconductive material (usually an inert plastic like Teflon) that isolates them chemically from the solution that couples the two coils. When an alternating current is passed through the *primary* coil, it induces a current flow in the solution loop proportional to the solution conductivity. A current is thus induced in the *secondary* coil, which serves as the detector. The technique is restricted to solutions of high electrical conductivity (1 to 100 mmho/cm), but these cells have a remarkable insensitivity to fouling. An important application is in monitoring of "effective alkali" in black liquor, the hot caustic that is used to convert wood chips into pulp. The process fluid is hot (180°C), corrosive (1 to 2 M NaOH), under high pressure (250 psig), and loaded with entrained fibers. Process instruments are made by Beckman, Foxboro, and Uniloc.

Potentiometric Analyzers

The glass electrode used for measuring pH is one of the most successful examples of potentiometry in automated instruments. Modern glass electrodes are highly reliable; they give selective, sensitive, and stable response to acidity over a very wide range of pH, and have been widely applied in industrial monitoring and control.

Using ISEs in automated systems has several advantages. In general, many wide-range electrodes are available for several types of ions, and in many cases they provide the only practical method for determining ionic activities in solution [9]. They generally exhibit fast response and can be used continuously with small samples, and with many types of samples with no pretreatment (colored solutions, slurries, etc.).

The general acceptance of ISEs in automated instruments, however, has been somewhat limited; this can be attributed to the fact that accuracy is strongly affected by chemical and environmental effects. Since these electrodes measure activity rather than concentration, factors such as ionic strength, complex formation, and pH need to be carefully controlled. In addition, few of these electrodes are perfectly selective for the ion of interest, and the presence of interfering ions must be considered before every application. The accuracy attainable may further suffer from temperature variations. For a monovalent ion, the Nernst equation shows that a change in temperature of $1°C$ causes a change in potential of 0.2 mV for a 0.1 M solution of the ion of interest, and a 1-mV change for a 10^{-5} M solution of the same ion. Electrode accuracy may suffer from drift because of pressure changes, flow changes, or electrode poisoning; the relative concentration error is 3.9% per millivolt uncertainty in measurement for a monovalent ion. When high accuracy is required, all of the factors must be carefully controlled. A side stream and flow-through cell allowing temperature control, flow control, adjustment of pH, and reagent addition to eliminate interfering ions may be required. An example of a flow-through cell that contains pH and reference electrodes, temperature compensator, and ultrasonic cleaner is shown in Figure 25.12.

Ultrasonic electrode cleaning has contributed significantly to the use of potentiometry in automated instruments. For example, during sugar refining, the raw washed sugar liquor is treated with phosphoric acid. Subsequently, lime is added to neutralize the acid; this quickly coats the pH electrodes, increasing the response time and finally snuffing the response entirely. Ultrasonic cleaning eliminates these problems and results in a substantially improved automated instrument in terms of reduced maintenance, increased reliability, and more efficient process control.

Some important automated instruments based on ISEs measure sodium in boiler feedwaters, fluoride in public water supplies, and water hardness (Ca^{2+} and Mg^{2+}) in water-conditioning systems. The sulfide electrode is extensively used in the paper industry for monitoring and control of sulfidity in paper pulping liquors (e.g., the Kraft process), for waste-treatment control of excess sulfide using a process that oxides sulfide with air, and for monitoring the level of sulfide being discharged from the plants. Other applications include the use of the cyanide electrode for measuring free and total cyanide in metal-plating baths, the use of the silver electrode for measuring silver ion in photographic emulsions and spent fixing solutions, and many others.

In some processes, potentiometry is used for controlling ionic species for which electrodes are not available. For example, the production of chlorine and caustic soda using mercury electrolysis cells results in wastes containing toxic levels of mercury. This waste is treated by precipitating the mercury as insoluble mercurous sulfide using sodium bisulfide, and then filtering the precipitate. Both pH and S^{2-} concentration should be controlled for efficient operation. At high pH, HgS forms soluble polysulfides; at low pH, sulfide is tied up as HS^- and H_2S. The waste liquor is therefore controlled at a pH near 7, and sulfide is added in a controlled manner to

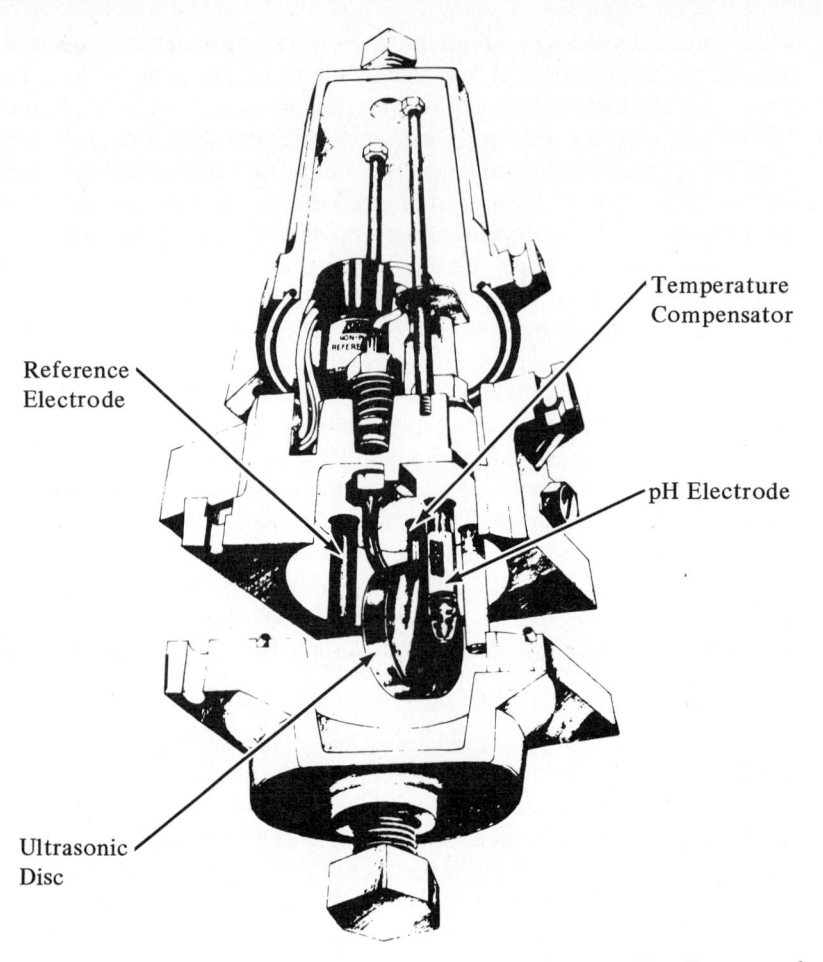

FIGURE 25.12. *Foxboro Model 85A pIon Electrode Assembly. Courtesy of the Foxboro Company.*

maintain a small residual level of about 1 ppm as H_2S. In this way, the very small residual concentration of mercury ion remaining in solution (governed by the common-ion effect and the solubility product of HgS) is effectively removed by means controlled by the sulfide electrode.

When the chemical composition of a stream is to be controlled using potentiometry, a significant problem is encountered because of the logarithmic relationship between the measured potential and the concentration of the species of interest. Consider a pH-control loop (a pH measurement, reagent valve, and controller); for a neutralization reaction, the gain of the pH measurement is the incremental change in pH caused by a particular quantity of added reagent. This is the slope of the titration curve; since it is inversely related to the buffer capacity of the system, it may vary over three or four orders of magnitude. It is apparent that the gain will be affected by the pH set point selected and by the type of acids or bases (i.e., strong or weak) in the system. Since efficient control requires that the total gain of the loop be less than 1, the extreme range of gain resulting from the potentiometric measurement must be accommodated by the other two elements in the loop.

Three mode controllers having proportional, reset, and derivative action and adaptive controllers having nonlinear or logarithmic control action have been applied successfully. The limited rangeability (50:1 to 100:1) of commercially available proportioning valves restricts their ability in this application, and a sequencing arrangement of suitably sized valves can be employed that selects the appropriate valve based on the measured pH. Alternatively, metering pumps can be used but are more costly and tend to have limited reliability. A clear commercial need exists for valves having the 1000:1 or greater rangeability required for this application.

25.5 AUTOMATION IN CLINICAL CHEMISTRY

Billions of tests are run annually in clinical chemistry laboratories; automation has therefore played a large role there. In the preceding sections, automated process-control systems were described. The first part of the present section considers the needs of the clinical chemistry laboratory as they relate to automation. The remainder will be devoted primarily to how clinical instruments are automated and which instrumental methods are most commonly used. Selected instruments will be described.

Automation Needs in Clinical Chemistry

The most significant factor that distinguishes the needs of the clinical laboratory from those of others is that the test results can directly affect the welfare and even the survival of a human being. Some of the factors influencing the design of clinical laboratory instruments are discussed below.

The Clinical Laboratory Environment. The task of the clinical chemist is to perform chemical analyses for diagnostic purposes. The concepts described in the first part of this chapter can be applied because of the common theme of automation for chemical analysis; however, the automation requirements in the clinical laboratory differ significantly from those in either process control or industrial analytical chemistry.

1. The sample is a natural biological material. It is not synthesized, and it cannot be controlled. Automation may control the testing process to some degree, but the sample itself cannot be controlled—except perhaps for automatically rejecting a sample that is too small or has been damaged—for instance, hemolyzed (cells ruptured).

2. The most common clinical sample, blood, is probably one of the most complex substances a chemist is called upon to analyze. He or she is expected to determine the concentrations of a few specific components out of thousands present without interference from any of the other components. This selectivity requirement is usually called *specificity* by clinical chemists.

3. The analytical problem is made still more difficult by the usually limited volume of the available sample, with particular limitations for pediatric and geriatric patients. The situation is further exacerbated when the physician requests a large number of tests on the sample.

4. In interpreting the results, the clinical chemist must be aware that the components to be measured may be affected by the recent history of the patient—ingestion of food or drugs (prescribed or not), physical exertion, and the degree of physical trauma or psychological reaction to the circumstances under which the sample was obtained. This exacerbates the problem of specificity and makes the interpretation of results critically dependent on skilled judgment, which must be applied both by responsible personnel in the laboratory and by the physician requesting the analysis.

5. Partly because of the complexity of the sample and problems relating to specificity, few standards are available. An expanding number of pertinent Standard Reference Materials is becoming available from the National Bureau of Standards. Even when available, cost dictates that these be used only to calibrate other reference materials. The latter include so-called standards purchased from reagent suppliers and control samples that are frequently derived from a carefully stored pool of blood samples. The precision of results can be assured by good instruments, properly selected methods, and meticulous protocol; it can be assessed by measuring replicate samples. Accuracy, however, is much more difficult to attain.

There are several strong motives for high accuracy in the clinical laboratory:

1. To ensure the diagnostic value of the test results
2. To monitor the progress of a patient under therapy
3. To follow the state of health of an individual, even though the tests may have been performed at different times and perhaps in different laboratories.

It is significant that, although "normal ranges" are developed statistically, these vary widely with sex, age, geographical and ethnic differences, and other factors. The situation is further complicated by the fact that the "normal range" for some components may have wide limits in the statistical population, but narrow limits in each individual biological system. Small deviations from the normal value will then indicate illness. A notable example of this is the concentration of calcium in serum, which is maintained constant to within about 1% by a healthy body, although the "normal" range for the adult population spans $\pm 13\%$.

Clinical chemists must meet all of these challenges in an environment that places increasing emphasis on controlling and minimizing the cost of tests. Economy is one of the important motivations for adopting automation in the clinical laboratory.

A second factor that has spurred automation in clinical chemistry is the steady rise both in the use of clinical chemistry tests by physicians and in the variety of diagnostic and monitoring tests. The annual growth rate was in the range 10 to 15% throughout the 1960s and early 1970s, but this has settled back to 5 to 8% more recently. The growth continues, but one factor in the decreased rate of growth is emphasis on diagnostically related testing as a means of cost containment, discouraging wide use of screening or "shotgun" tests.

The growing scarcity of skilled technologists is also a driving force, because a prime objective of automation is to eliminate the need for human intervention in a process. Although this may be applied literally in, for example, process control, the clinical laboratory environment described above requires more constraints. Skilled human judgment is essential for monitoring the viability of the sample and the validity and significance of the results. Therefore, automation is aimed at aiding the clinical analyst in the exercise of these skills.

Clinical Chemistry Tests. The diversity of tests that the clinical chemistry laboratory may be called upon to perform is continually expanding. Very few older tests are displaced by the newer ones developed. A reasonably sized laboratory will be prepared to perform over 60 different tests routinely, and a regional reference laboratory will offer between 200 and 300. However, many of the latter tests are performed infrequently and do not justify automation. Table 25.2 lists the tests that have been commonly automated. Recent additions of automated tests include those for therapeutic drugs and hormones, such as thyroid factors.

One indication of the general level of performance achieved in clinical chemistry laboratories is the fact that control samples show a relative standard deviation of between 1 and 3%. Extenuating circumstances, such as a required solvent-extraction procedure, may lead to relative deviations greater than 10%. Although automation may result in improved precision, the degree attained depends on the skills and motivation of the operator and on the general quality control prevalent in the laboratory.

Automation in the Clinical Laboratory

The first automation in clinical chemistry laboratories was applied primarily to sample handling and processing in the late 1950s. This emphasis can be attributed both to the quantity of specimens and to the state of technology.

TABLE 25.2. *Blood Tests Commonly Automated*

Acid phosphatase	Glucose
Albumin	Iodine—protein bound
Alcohol	Iron
Alanine transaminase	Lactic acid
Alkaline phosphatase	Lactic dehydrogenase
Aspartate transaminase	Lithium
Bilirubin—direct	Oxygen
Bilirubin—total	pH
Calcium	Phosphate—inorganic
Carbon dioxide	Potassium
Chloride	Protein—total
Cholesterol	Sodium
Creatine phosphokinase	Triglycerides
Creatinine	Urea nitrogen
Free fatty acids	Uric acid

Sampling Automation. In the terminology given in the first part of the chapter, instruments that emulate manual sample handling and processing without the use of control loops are *automatic* (mechanized), but not automated. Among the automatic functions are:

1. Sample pickup (from a container such as a small cup)
2. Sample dispensing
3. Dilution
4. Deproteinization
5. Reagent addition
6. Incubation
7. Insertion of the reacted sample into the detection system

It is interesting to note that, when blood is the sample, almost all of the automatic instruments require the use of serum or plasma; none automate the separation of the serum or plasma from the whole blood.*

It is customary to refer to instruments lacking the automatic functions on the above list as *manual* instruments. If these incorporate extensive electronic data-processing, they are called *semiautomatic*.

Discrete and Continuous-Flow Sampling. In the clinical laboratory, the terms *discrete* and *continuous flow* are applied somewhat differently than in process control. In discrete sampling, each sample undergoes a reaction measured in a cuvette not shared by other reactants. In continuous-flow sampling, successive samples pass into the same length of tubing, reagents are added and reactions occur, and finally the samples flow continuously into a cuvette for detection. To isolate successive samples, one or more air bubbles are pumped into the flow line between samples.

Since successive samples must be kept isolated from each other in order to avoid cross-contamination, discrete sampling is the natural method for automatic clinical sample processing. Surprisingly, the first successful automation of clinical sampling was a flow-sampling system, the AutoAnalyzer®, marketed in 1957 by Technicon Instruments Corporation. A single-channel Auto Analyzer is illustrated in Figure 25.13. Consistent with the growth patterns in clinical chemistry tests, this evolved in the late 1960s into a multiple-channel system, the SMA® (Sequential Multiple Analyzer), in which the sample stream was split into multiple streams, each for a different test, phased so that the results were generated sequentially. Note that such a multichannel system performs all tests incorporated in the instrument on each sample and precludes test selectivity except by exclusion of unrequested results from a patient report.

SMA systems were widely embraced for more than a decade without significant competition from a multiple-channel discrete-analysis system. As a result, Technicon dominated the field of clinical chemistry automation. Only since the late 1960s have

---

*Serum is the clear portion of blood remaining after the blood is allowed to clot and the clot containing the red cells and fibrin is separated out by centrifuging. Plasma is identical to serum except that it still contains fibrinogen, which is normally converted to the insoluble protein fibringogen to form the clot. Plasma is obtained by adding an anticoagulating agent to prevent the clotting reaction, and then centrifuging out the red cells.

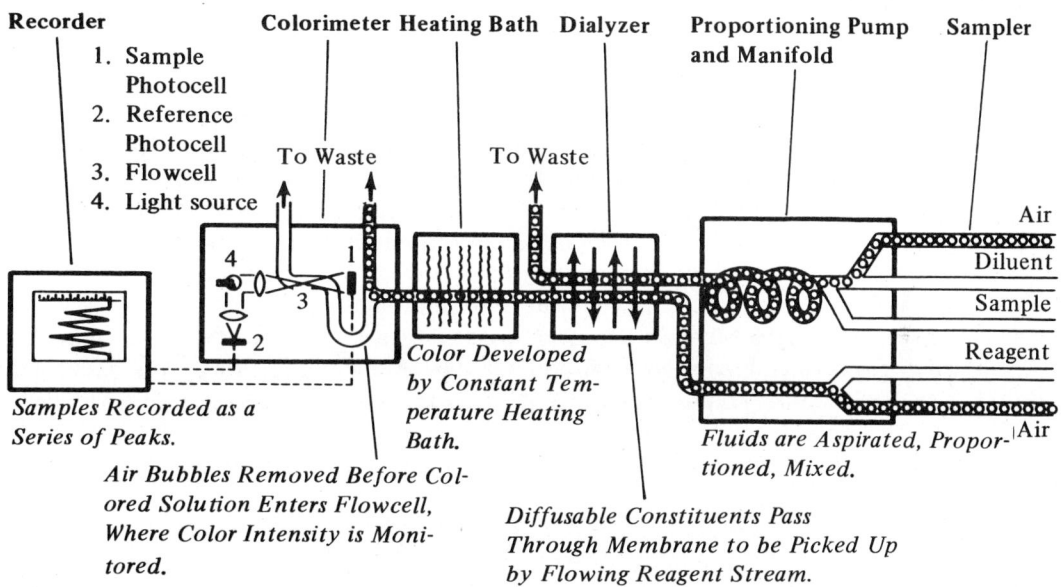

FIGURE 25.13. Single-channel AutoAnalyzer®. Courtesy of Technicon Instruments Corporation.

discrete-sampling instruments begun to play a significant role in automated clinical chemistry.

Instrument Categories. Automatic chemical analyzers are classified by function as follows:

1. *Multichannel.* Multichannel instruments analyze each sample for many different components—in parallel for a discrete analyzer, and sequentially for a continuous-flow analyzer.

2. *Random Access.* Random-access analyzers are a burgeoning variation of discrete multichannel analyzers. They contain no dedicated channels or test cuvettes for an incorporated test. Multiple tests may be selectively performed on each sample on any of a battery of, for example, 16 tests out of an available menu of many more than 16. The currently available battery can be readily redefined from the menu.

Although some attention was paid to test selection in older multichannel discrete analyzers, microprocessors and computer control have made random-access analyzers practical.

3. *Batch.* Batch instruments analyze each sample for a single component at a time, but can be readily changed to analyze other components one at a time. These are also called *single-channel analyzers.*

4. *Parallel Fast.* Parallel-fast analyzers are a special variation of batch analyzers, based on the use of a centrifuge. They are sometimes called *centrifugal fast analyzers.* The principle is illustrated in Figure 25.14. A central, removable disk has radial slots, with two or more wells molded into each slot. An automatic pipet dispenses microsamples into the outer well and reagents into the inner wells of each slot. The disk is placed in a centrifuge rotor with the slots oriented in line with cuvettes around the outer circumference of the rotor. When the rotor spins, the reagents simultaneously wash all of the samples into the cuvettes, which rotate sequentially under a fixed photometer. The transmittances of all the reaction mixtures are recorded and sorted out by a dedicated computer, which also reduces these signals to final test results. Because data may be collected over an extended time period, as the multiple cuvettes spin through the fixed photometer beam, parallel-fast analyzers are well suited to kinetic or reaction-rate tests as well as to endpoint tests. Thus, enzyme-activity determinations are popular tests on such analyzers. Centrifuge speeds around 600 rpm are customary during the data-collection period. Some

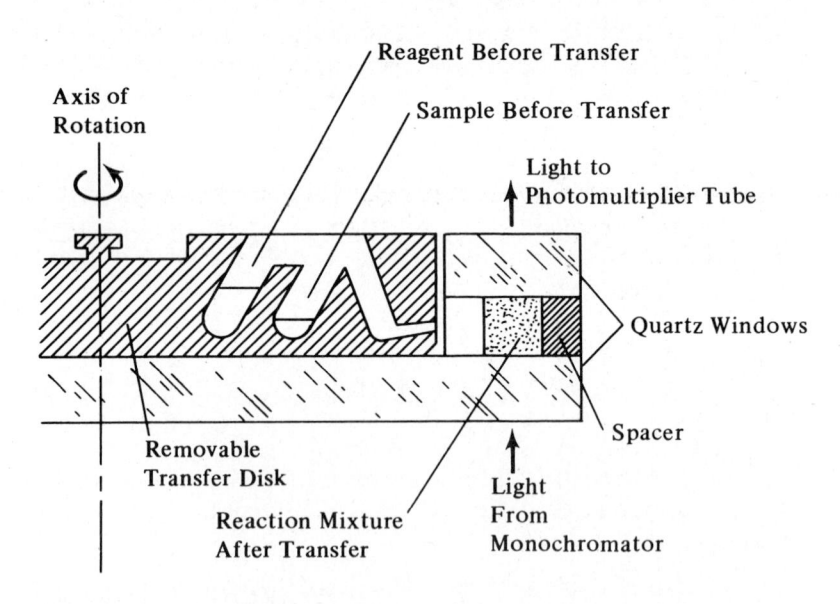

FIGURE 25.14. *Cross section of the sample disk and centrifugal-analyzer section of a parallel-fast analyzer. From R. C. Coleman, W. D. Schultz, M. T. Kelly, and J. A. Dean,* Amer. Lab., 3(7), 26 (1971), *by permission of International Scientific Communications, Inc.*

parallel-fast analyzers use plastic disks that can be discarded after a run. Others have a built-in wash system so that the disks can be reused.

5. *Dedicated.* A dedicated instrument analyzes for only a specified component or a limited number of diagnostically related components; generally, it is not adaptable to other applications.

More examples of the above classes of analyzers will be given below.

The Impact of Modern Electronics

Since the 1970s, modern electronics has led clinical automation in a completely new direction. The impact has been both on the instruments with automatic sample handling and on the manual instruments. Some of the newer instruments are truly automated, as well as automatic. Most of the advantages of modern electronics derive from applications of microelectronic digital circuitry crammed into a remarkably small space—a boon to overcrowded laboratories. In particular, microprocessors have replaced mechanical and electromechanical control of automation, enhancing reliability. Furthermore, miniaturization and cost reduction of electronics have enabled incorporation of impressive computer power in clinical instruments.

Data Readout. Prior to this technological change, manual instruments, such as spectrophotometers/colorimeters, had meter readouts with a linear transmittance (energy) scale and, in some cases, a nonlinear (logarithmic) absorbance scale, which in principle saved one computational step in deriving results in concentration units. The readout of automatic instruments was generally a strip-chart recorder, which traced a series of peaks for the successive samples. In some cases, the chart had a nonlinear concentration scale, which was often shaded or otherwise marked to indicate "normal" physiological ranges. To use these features, the deflection had to be calibrated with a sample of known concentration.

The first evident characteristic of modern automatic clinical instruments is the digital display of data. Most instruments have illuminated numerical displays; an instrument that performs more than one determination simultaneously may have several. For example, a digital flame photometer commonly has two displays for the simultaneous readout of Na and K, and a blood gas analyzer may have three, for pH and for the partial pressures of CO_2 and O_2.

Other types of data readout include printer-listers, high-speed line printers that enable formatting, and (in some cases) data storage for later retrieval—on floppy disks, for instance.

Data Processing. The above are superficial benefits of modern electronics. Of even greater value is the electronic processing of the raw data. A readout directly proportional to concentration can easily be produced, although this requires a logarithmic conversion for colorimetric or spectrophotometric systems. If the instrument is calibrated with a standard of known concentration, the readout can also be made directly in reportable units. This is a major time saver and, in addition, reduces errors in numerical manipulation and transcription.

As a further refinement, the electronics system can determine how long to wait before the readout signal is acceptably stable, and then lock the displayed result until the operator has recorded it and is ready to initiate the next reading. This type of automation is common to otherwise "manual" instruments. Other niceties are also available, such as automatic integration or averaging of the signal, automatic correction for nonlinear working curves, and automatic blank subtraction, all of which save time and minimize the chance for human error.

One of the most valuable applications of electronic data processing is in enzyme assays. Kinetic measurements of enzyme activity in which the rate of reaction is monitored (usually using ultraviolet measurement) are more specific than endpoint colorimetric methods, in which the development of color in a coupled reaction is measured after a fixed time. Modern systems continuously or intermittently monitor the growth in concentration of the reaction product or the decrease in concentration of one of the reactants (the substrate). From the rate of change in concentration or the average change in concentration over several fixed time intervals, the circuitry calculates the activity in reportable units. Other ramifications of these systems will be discussed below.

Some modern clinical instruments are dedicated either to endpoint colorimetric or to kinetic determinations; others allow selection of either mode. Some program the electronic data processing for the desired mode by means of a punched-card or other coded system that comes with the prepackaged reagents specifically made for this type of instrument. A design objective of this last type of system is to minimize the training needed by the operator.

Data Evaluation. In addition to data processing, electronic systems are being applied to evaluate the data as a further aid to the operator. For example, in an enzyme assay the system will evaluate the linearity of the reaction, which bears on the expected validity of the end result. Also, for samples of very high activity, the substrate may be prematurely exhausted. The electronics are often designed to warn of this dangerous possibility in which a low-level extraneous reaction may persist, yielding an erroneous reading of low activity for a serious case of high activity. Such samples may then yield useful answers if a second aliquot is run at a much higher dilution.

In addition, the instrument may be programmed with the range of normal values so that the display will automatically *flag* abnormals. The flag may be an asterisk, or an *H* for high and an *L* for low. Before the data are released to the requesting physician, flagged results in particular must be evaluated by the laboratory director to determine if the abnormality results from the patient, the instrument, or the reagent. This type of flagging is also helpful to the physician. Some laboratories have extended the principle of flagging by programming their own in-house computer systems to display a statistically likely diagnosis based on the out-of-normal-range data. This is an aid to the physicians, who base diagnoses on their examinations and on the patients' histories, as well as on the test results.

Another type of data evaluation performed on some automatic instruments is statistical analysis of a series of data, computing such functions as standard deviation and coefficient of variation (the relative standard deviation).

Instrument Monitoring. In addition to monitoring and evaluating data, circuitry may be arranged to monitor itself and other functions of the instrument. Among the variables that may be monitored are amplifier range, temperature, source operation, speed, reagent supplies, and waste level. The built-in electronics may also diagnose malperformance or misfunction of the instrument. Many recent clinical instruments have been designed so that clinical laboratory personnel who have no special instrument skills or little special training can apply self-diagnostic instrument features to determine which subsystem is at fault. These instruments have been designed in such a way that printed circuit boards and other subsystems can be replaced by laboratory personnel without use of special tools. This minimizes downtime, which is a highly desirable benefit, and reduces the need for high-cost repair specialists.

A further useful feature is the automatic control and sequencing of multiple functions of the instrument by the built-in electronics, which are effectively a dedicated microcomputer. In some cases, a closed loop is involved, so that certain functions are truly automated. For example, an instrument involved in enzyme determinations will monitor the temperature of the reaction cuvette and correct the assayed activity for deviations from the nominal temperature. Also, after initial calibration, an instrument may compare subsequent standards or control samples with the initial value and automatically correct for calibration drift, as well as alert the operator to excessive drift.

Overview. The above description of automation as applied to clinical instruments leads to an interesting conclusion. Some of the current instruments that do not automate sample handling or processing, and which are referred to as manual or semiautomatic, may be more effective in saving skilled labor and time and reducing human error than some of the automatic instruments of the 1950s and 1960s.

For reference, a spectrophotometer/colorimeter with direct concentration readout costs on the order of $2000; a digital Na/K flame photometer with an automatic sample diluter, about $6000; a flexible, computer-controlled, single-channel analyzer under $25,000; a computer-based parallel-fast analyzer, about $60,000, and a multichannel analyzer from about $80,000 to well over $200,000.

Automatic Instrumental Methods

Because the majority of tests in the clinical chemistry laboratory are colorimetrically based, the greatest effort toward automation has been with colorimetric methods. As previously discussed, most of the automation classifies as automatic rather than automated.

A natural extension, because of similar sample processing, is the use of a fluorimeter in place of the colorimeter. On the data-handling side, the fluorimeter usually provides a signal inherently linear with concentration, which avoids logarithmic conversion.

Enzyme Assays. More recently, the automatic colorimetric systems have been extended to provide kinetic enzyme determinations (or determinations of enzyme substrates). This required several key changes in the colorimetric systems:

1. Although sample processing is similar, temperature is a far more critical factor, since enzyme activity changes at a rate of about 7% per °C. Therefore, temperature equilibrium and constancy in the reaction cuvette are critical. Temperature control to within ±0.1°C is commonly specified.

2. The detection system must be sensitive at 340 nm, where many enzyme-activity assays are performed. This generally requires sources, filters, gratings, and detectors different from those used in colorimeters.

3. There are no recognized enzyme standards or reference materials. The accepted basis for measurement is the rate of reduction of the substrate, commonly nicotinamide adenine dinucleotide (NAD) for many reactions. The reduced form, NADH, absorbs at 340 nm, and the rate of change of this absorbance is measured in an enzyme-activity assay. The absorptivity (a in Beer's law) is known for NADH; from this and the rate of change of absorbance per unit time, the activity of the enzyme can be calculated in micromoles of substrate converted per minute. This is referred to as an *international unit* (IU), expressing the activity as IUs per liter. Therefore, an accurate, absolute absorbance scale must be established in each case in order to make a valid assay.

4. Many enzyme reactions require the measurement of high absorbance values (in the range 1.0 to 1.6), in contrast to colorimetry, for which most measurements are made at absorbances no higher than 0.7. This distinction requires a more stable photometer with an increased linearity range.

5. Many reactions, particularly for samples of lower activity, cause very small changes in absorbance during a reasonable observation period. Therefore, the photometer must have high sensitivity and low noise. Some instruments have sensitivity in the range of 10^{-4} absorbance units.

Atomic Emission Spectroscopy. Another commonly automated spectroscopic method is atomic emission spectroscopy (flame photometry). Because sample processing is less elaborate—usually only a dilution—atomic emission spectroscopy was the first technique to be adapted to the modern methods of data readout and processing. Instruments were available in 1964 that gave simultaneous numerical readout of the concentrations of Na and K in directly reportable units.

Electrochemical Methods. The prime candidates for electrochemical automated methods are blood analyzers which measure pH, P_{CO_2}, and P_{O_2}. These are, by the nature of the data, "stat" instruments frequently used in emergencies. (The term *stat* is applied to a test demanding immediate measurement and expeditious reporting of the result to the attending physician.) The pH is measured potentiometrically by means of the conventional glass electrode, P_{CO_2} is measured with a pH electrode covered with a plastic membrane that is permeable to CO_2, and P_{O_2} is measured amperometrically with a polarographic oxygen electrode (Pt wire covered with an O_2-permeable membrane). Blood gas determinations have depended heavily on the skill of the analyst, which held back the growth in demand. However, as instrumentation improved in the early 1970s, the test volume grew accordingly. Since 1974, the degree of automation of blood gas analyzers has rivaled that of other automatic clinical analyzers.

ISEs, perfected in the late 1970s, have been combined with modern electronics in routine clinical instruments for the measurement of Na, K, Cl, CO_2, and ionized Ca. These instruments frequently replace flame-emission spectrometers (flame photometers) for Na and K. They are uniquely used as stat instruments in operating-room environments where the flame in a flame photometer would be unacceptable.

Electrochemical analyzers based on the amperometric measurement of oxygen are used to measure the rate of oxidase enzyme reactions. For example, the substrate glucose is determined by measuring the rate of oxygen consumption in the presence of glucose oxidase. Results are obtained in less than a minute. Similarly, urea is determined enzymatically by measuring the rate of conductance change during urea hydrolysis in the presence of urease.

Immunoassay Methods. A pronounced trend in clinical chemistry has been a push toward the measurement of smaller concentrations. One motivation has been the interest in basic components, such as hormones, that directly relate to the causes of disease. Concentration variations of nanomoles or picomoles per liter are clinically significant. In a different direction, and somewhat less demanding, is the measurement of therapeutic drugs in serum, requiring measurements often in the range of micromoles but sometimes down to picomoles per liter. Drug measurements are important for two reasons:

1. For a given dosage of a drug, the concentration in the patient's blood depends on the patient's weight, other ingested drugs (prescribed or not), physical activity, diet, and many additional factors, all studied in the field of pharmacokinetics.
2. There is a well-documented therapeutic range of concentration. For some drugs, this range is fairly narrow, and exceeding the range by small amounts can lead to toxic side-effects.

The smaller concentrations of natural compounds and drugs generally cannot be measured by the more customary colorimetric methods. Immunoassays have enabled the extension of sensitivity by three to six orders of magnitude and have become an important analytical method in clinical laboratories.

The basic premise of an immunoassay is that the antibody-antigen reaction is specific. Many immunoassay methods use a limited quantity of antibodies mixed with both a pure, tagged antigen and a patient sample (serum or urine). Antigen in the latter competes with the tagged antigen for binding sites on the antibodies. After reaction, the bound and unbound fractions are separated by centrifugation, filtering, or other means. Measurement of the tagged antigen in either the bound or unbound fraction is monotonically related to the concentration of antigen (e.g., hormone or drug) in the specimen, but the relationship is not linear, nor is it a simple function. Microcomputers have been a boon to immunoassay data reduction.

The most prevalent immunoassay method is radioimmunoassay (RIA), in which a radioactive tag is used. Iodine-125 is the most widely used tag. RIA is highly sensitive for almost all sizes of antigens and antibodies.

A great deal of effort has been applied to finding and refining alternative methods of immunoassay, to eliminate both the use of radioactive reagents and the need for specialized, usually bulky counting equipment. Many of these methods are limited to larger antigens—for example, drugs—and are sometimes susceptible to chemical interferences. Although newer nonisotopic methods compete with RIA, the universality and sensitivity of the latter, and the frequently lower cost of reagents for RIA, leave this the dominant method of immunoassay. The average growth of immunoassay testing is between three and five times as great as clinical laboratory testing in general.

Table 25.3 summarizes the most prevalent nonisotopic methods of immunoassay as of this writing.

Selected Automatic and Automated Clinical Chemistry Analyzers

The great number of samples analyzed in the typical clinical chemistry laboratory justifies specialized instrumentation. A typical clinical instrument for common and routine tests may be capable of 200 to 1000 tests per hour, and some instruments can perform nearly 10,000 tests per hour. Virtually all new clinical chemistry instruments are microprocessor controlled, permitting great versatility, increased speed, and improved ease of use. Many instruments are "selective" in that only requested tests are performed, thus consuming less sample and reagent. Emergency or stat tests can be indicated, and performed rapidly without seriously interrupting routine or batch processing.

Because of the nearly weekly advances in modern clinical chemistry instrumentation, any description of "current" instruments is outdated as soon as it is written. Nevertheless, selected examples of several types of instruments will be discussed to provide some idea of the general types available, and some of their capabilities.

TABLE 25.3. *Nonisotopic Immunoassay Methods*

Tag	Name	Company	Principle	Applications
Enzyme	EMIT®	Syva	Antibody-antigen reaction suppresses enzyme activity	Mostly drugs
Enzyme	ELISA	Numerous	Antibody-antigen reaction causes enzyme-linked color development	Hormones, drugs, others
Fluorescence	FPIA	Abbott	Antibody-antigen reaction changes degree of fluorescence polarization	Mostly drugs
Fluorescence	TDA®	Ames	Change in fluorescence due to competitive binding	Mostly drugs
None	Nephelometry	Beckman, Hyland, Behring	Antibody-antigen complexing causes increase in light scatter	Immunoproteins, some drugs

Multichannel Analyzers. The first widely adopted multichannel system was the SMA 12/60® (sequential multiple analyzer) introduced by Technicon in 1967. Built on the continuous-flow principle, it is similar in operation to the single-channel analyzer shown in Figure 25.13, but with the sample split into 12 channels. This generates results sequentially on each sample at the rate of 60 samples, calibrators, and controls per hour. Although the 12/60 has largely been replaced, it set the standard in clinical laboratories for nearly a decade, and helped to establish the role of the screening battery, or "profile," in diagnostic medicine.

The next generation of this product was the computer-controlled Technicon SMAC®, which first became available in 1974, and has had a number of more refined models since. This instrument has 20 channels and generates results on about 150 samples, calibrators, and controls per hour. The effective rate on actual samples averages 90 to 100 per hour (about 2000 diagnostically significant tests per hour), after certain conditions are taken into account, including calibrators, controls, and samples that must be rerun because of out-of-range or questionable results. The latest model essentially consists of two analyzers in tandem. One analyzer can be processing routine samples in the "foreground," while the second is available for stat or special analyses.

SMAC offers significant advances over its predecessors besides productivity. Less sample is consumed (about 700 μL), and reagent volumes are smaller. The system utilizes more modern chemical methods, including kinetic assays and ISEs for Na and K.

An entirely different discrete analyzer, designed for a different application, is the DuPont Automatic Clinical Analyzer (ACA). This instrument's goal is to generate good answers whether or not the operator is highly skilled. The ACA is well suited for automating the off-hours stat testing in larger hospitals and for providing a wide selection of assays for smaller laboratories. Without such a system, the small laboratories would have to maintain personnel skilled in such a wide selection of assays, a difficult if not impossible task.

The operator of the ACA loads the sample into a well in a rigid header that fits on a track. Hanging from this header is a form on which the operator may enter sample and patient ID information, which is reproduced automatically with the test results. Then the operator loads a reagent pack for each test requested on that sample, placing each on the track just in front of the sample header. After separate aliquots of the sample are automatically dispensed into these packs, each pack enters the main body of the instrument, where the pack also serves as a purification column, as a mixing and reaction chamber, and finally as the test cuvette in the filter photometer. The operator need only prepare and dispense the serum, record ID data, and select the appropriate reagent packs. The expense of skilled labor is traded off for slow speed and the cost of reagent packs. The testing rate averages about one result every 55 sec.

Random-Access Analyzers. Random-access analyzers are discrete analyzers, with each test reaction occurring in a separate cuvette. One widely accepted line of random-access analyzers has been developed by Eastman Kodak Company around a new technology for clinical chemistry: dry-reagent pads. These pads are incorporated in a frame similar to a mount for a photographic transparency. Tests available with this technology have been based on standard testing methods, and the automated systems that run these slides are called Ektachem analyzers.

The system dispenses a 10-μL drop of serum on a spreading layer on the top of the pad. In colorimetric tests, the sample contacts the lower layers, which contain dry reagents for the selected test. When color has developed after a programmed time period, the slide shuttles into a filter reflectometer, which detects the degree of color development. The integral computer converts this measurement to concentration based on a stored calibration. Figure 25.15 is a functional diagram of the Ektachem 400, a system designed to run a combination of colorimetric (CM) and potentiometric (PM) slides, the latter available for Na, K, Cl, and CO_2.

The Ektachem system enables random selection of tests for each of a group of samples. Between 1 and 16 tests may be selected for each sample. Tests are completed at a rate of between 300 and 600 results per hour, depending on the mix of tests selected.

Once the operator loads slide cartridges and samples, and keys in desired tests for each sample, sample and data processing are completely automatic. The slide cartridges store compactly in a refrigerator, and have a shelf life of about a year. The system requires calibration with a reference material no more frequently than once a week, and a control sample need be run only once during an 8-hr operating period.

The key to all random-access analyzers is the combination of the random selection of tests for each of a number of samples, and the report of results in an ordered, collated form for each sample.

Parallel-Fast Analyzers. There are a number of commercial versions of centrifugal analyzers, which differ in such details as: (a) the number of samples accommodated on the rotor (15 to 40); (b) the means of setting variables for a run (manual or preprogrammed on paper tape or tape cassette); (c) the automating of such steps as the wash at the end of a run; and (d) the degree of sophistication in data generation, result listing, automatic evaluation of reaction linearity, flagging, collation of results from several runs, and so on.

In these analyzers, only one determination can be made for each sample during a given run. However, determinations are made in a matter of seconds after samples and reagents are loaded, and only microliter quantities of sample and reagents are required for each determination. Procedures can be changed simply by changing the reagents in the rotor and changing the wavelength setting. The latest versions can perform tests based on fluorescence or degree of light scattering, in addition to the conventional colorimetric procedures.

Dedicated Analyzers. A wide choice of such systems, with varying degrees of automation, are available for clinical applications. One of the most prevalent is an atomic emission spectrometer that determines both Na and K concentrations simultaneously on separate readouts a matter of seconds after a sample is aspirated. Automatic diluters are commonly built in. Auto samplers and printers are generally available as optional attachments. Some of these systems are readily converted to Li assays when needed.

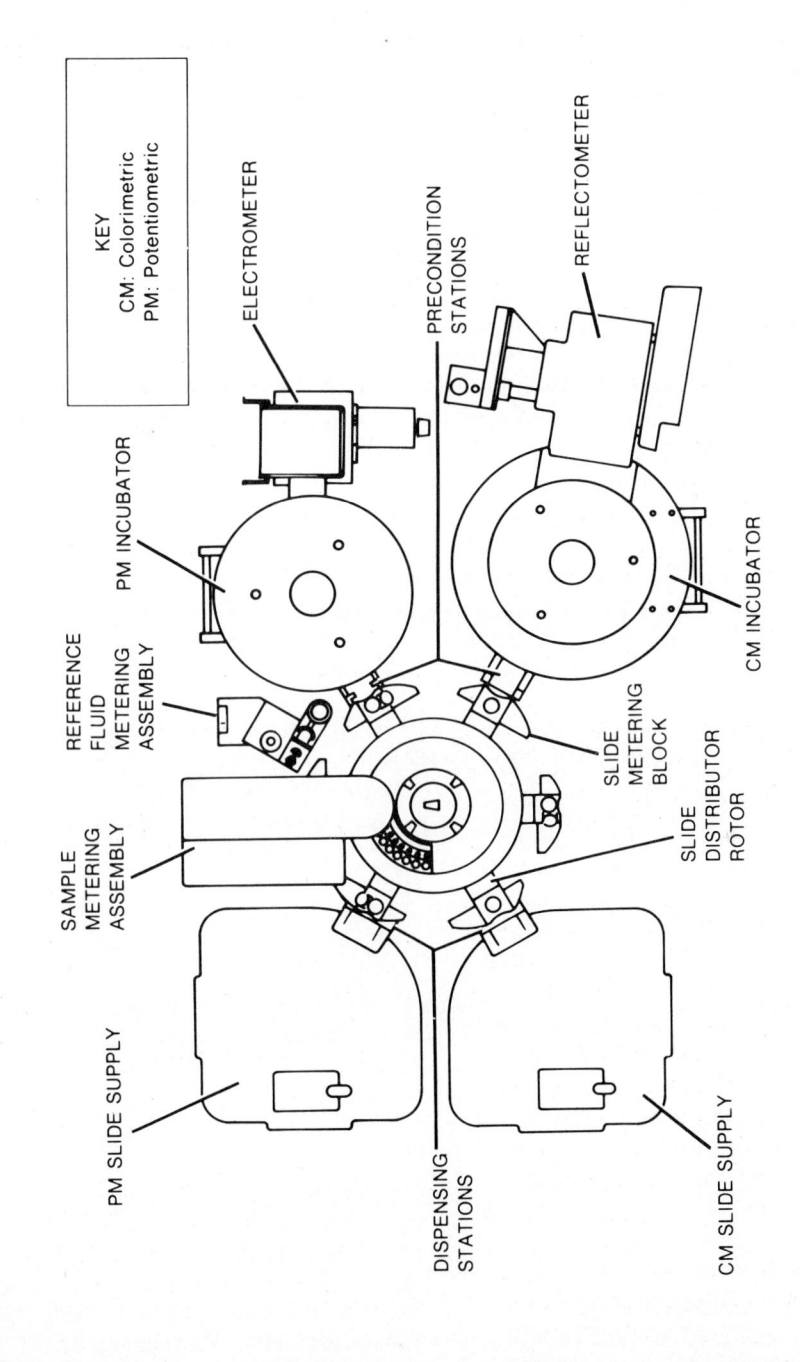

FIGURE 25.15. Schematic diagram of the Kodak Ektachem 400 Analyzer. Courtesy of Eastman Kodak Company.

Electrochemically based semiautomatic glucose and blood urea nitrogen (BUN) analyzers are quite common. They "walk the operator" through the manually initiated steps required.

Another type of widely used dedicated system is the blood gas analyzer. These are electrochemical instruments that measure pH, P_{CO_2}, and P_{O_2}, and sometimes K^+ in whole blood, either simultaneously or sequentially. The recent, more automatic systems operate completely "hands off" after the sample is aspirated. Because the measurements are generally made under stat circumstances, these newer systems periodically recalibrate themselves while on "standby," so that a sample can be run as soon as it arrives in the laboratory.

The growing impact of ISEs is evident in the proliferation of dedicated analyzers dependent on them. Such analyzers commonly measure Na, K, Cl, and optionally CO_2. Orion offers an ionized-calcium analyzer and a sodium/potassium analyzer, both using ISEs. Recently, some instruments have been introduced that measure ammonium ion, but their technology is not as well developed as that for other ions. Electrochemical instruments for Na and K, for example, offer no real measurement advantages over atomic emission; but they eliminate the need for using combustible gases, an obvious advantage in, say, an operating room, and they are substantially quieter in operation.

SELECTED BIBLIOGRAPHY

Books

CLEVETT, K. J. *Handbook of Process Stream Analysis.* New York: Halsted Press, 1973. *A handbook that describes many types of process analyzers.*

HOUSER, E. A. *Principles of Sample Handling and Sample Systems Design for Process Analysis.* Pittsburgh: Instrument Society of America, 1972. *Contains good information on the design of sample-handling systems.*

SHINSKEY, F. G. *pH and pIon Control in Process and Waste Streams.* New York: John Wiley, 1973. *Provides an excellent description of the application of potentiometry in process control.*

SHINSKEY, F. G. *Process Control Systems.* New York: McGraw-Hill, 1967. *All aspects of process control are covered.*

SMITH, D. E., and ZIMMERLI, F. H. *Electrochemical Methods of Process Analysis.* Pittsburgh: Instrument Society of America, 1972. *A good compilation of the electrochemical instrumentation available for use in industrial applications.*

Clinical Chemistry

ALPERT, N. L. *Clinical Instrument Reports.* Philadelphia: North American, 1975.

HICKS, R., SCHENKIN, J. R., and STEINRAUF, M. *Laboratory Instrumentation.* New York: Harper and Row, 1974.

LEE, L. W. *Elementary Principles of Laboratory Instruments*, 3rd ed. St. Louis, Mo.: C. V. Mosby, 1974.

WHITE, W. L., ERICKSON, M. M., and STEVENS, S. C. *Practical Automation for the Clinical Laboratory*, 2nd ed. St. Louis, Mo.: C. V. Mosby, 1972.

Articles

BOWERS, G. N., JR. "Analytical Problems in Biomedical Research and Clinical Chemistry." In W. W. Meinke and J. K. Taylor, eds., *Analytical Chemistry: Key to Progress in National Problems*, Chap. 3. NBS Special Publication 351. Washington, D.C.: U.S. Government Printing Office, 1972.

HOLLOWELL, C. D., and MCLAUGHLIN, R. D. "Instrumentation for Air Pollution Monitoring." *Environ. Sci. Tech.*, 7, 1011 (1973).

LIGHT, T. S. "Industrial Analysis and Control with Ion Selective Electrodes." In R. A. Durst, ed., *Ion Selective Electrodes*, Chap. 10. NBS Special Publication 314. Washington, D.C.: U.S. Government Printing Office, 1969.

Process Measurement and Control Terminology.

SAMA Standard PMC20-2-1970. Scientific Apparatus Makers Association, 370 Lexington Ave., New York, N.Y. Pub. No. 219.

SOULE, L. M. Basic Concepts of Industrial Process Control. *Chem. Eng.*, Sept. 22, 1969.

SOULE, L. M. Basic Control Modes. *Chem. Eng.*, Oct. 20, 1969.

REFERENCES

1. IUPAC Information Bulletin No. 26, International Union of Pure and Applied Chemistry, Oxford, England.
2. R. G. BATES, *Determination of pH: Theory and Practice*, New York: John Wiley, 1964, pp. 382–83.
3. SAMA Standard PMC-20-1-1973, Scientific Apparatus Makers Association, 370 Lexington Ave. New York, N.Y.
4. F. G. SHINSKEY, *Process Control Systems*, New York: McGraw-Hill, 1967, Chap. 1.
5. R. ANNINO, *J. Chromatog. Sci.*, 8, 288 (1970).
6. W. F. WILHITE, *J. Gas Chromatogr.*, 4, 47 (1966).
7. S. C. TERRY, J. H. JERMAN, and J. B. ANGELL, *IEEE Trans. Electron. Devices* ED-26, 1880 (1979).
8. R. VILLALOBOS, *Anal. Chem.*, 47(11), 983A (1975).
9. R. A. DURST, *Amer. Sci.*, 59, 353 (1971).

PROBLEMS

1. Calculate: (a) the thermal capacitance of 1 L of water; (b) the thermal resistance of a brick wall 1 m^2 in area and 5.0 cm thick; (c) the gas capacitance of a liter of air at room temperature.
2. Construct pH-mV diagrams showing the span, range, and zero elevation (or suppression) of a pH meter used over the full-scale ranges from pH 6 to 9, pH 4 to 7, and pH 8 to 10.
3. Describe (a) positive and (b) negative feedback. (c) What would happen if a temperature controller with positive feedback were used to control the flow of heat to a room?
4. Give several examples of automatic control devices used at home.
5. Calculate the pH to give 1 N acetate at neutrality for a solution of acetic acid titrated by a strong base. (a) What is the process gain at that point? (b) What are the pH and gain of the system when the acetic acid is half-neutralized? What is the relationship between buffer index and gain of this system?
6. What are the sensitivity, gain, and proportional band of a pneumatic temperature controller whose full-scale output changes linearly from 3 to 15 psi over the full-scale temperature interval 25 to 75°C? If the proportional band were decreased to 50%, what would be the sensitivity and gain for this controller?
7. What would be the effect on the gain of halving the proportional band and (a) doubling the input span; (b) leaving the input span unchanged?
8. A magnetic amplifier requires a 10–20-V input signal from a 4–20-mA control signal loaded with a 100-Ω resistor. Write an equation relating controller output to required output voltage. (Hint: Note that both gain and bias are required.)
9. A proportional controller is used to control the fluid level in a tank. Fluid is admitted to the tank through a valve with a flow factor of 20 m^3/hr/% controller output. A load change

occurs, and demand increases from 1000 to 1200 m³/hr. Assume the controller proportional band is 10. Calculate the new controller output and offset error. What would be the effect on offset if the proportional band were increased to 20?

10. Assume for the controller in Figure 25.4 that 0 to 100% output corresponds to 0 to 10 V error signal. (a) If the proportional band is 20%, what is the output voltage if, at zero error, the controller output is at 50% of scale? (b) If the load changes so that a new controller output of 40% is required, what is the offset error?

11. For the PI controller in Figure 25.6, select values for R_r and C and calculate the fractional value of R_f that needs to be fed back to give a composite controller having 25% proportional band and a reset time of 10 sec.

12. An instrument has a specified accuracy of ±2.0% full scale for a measurement range of 50 to 200°C. What is the uncertainty of an indicated temperature reading of 100°C?

13. (a) Name the five categories of automatic chemical analyzers used in the clinical laboratory. (b) Describe each category.

14. What are the key benefits derived from modern electronics in clinical instruments?

15. What are the key instrumental requirements for enzyme assays?

16. Which clinically significant components can be measured using ISEs?

17. What motivates the use of immunoassay methods?

18. Name the prevalent nonisotopic methods of immunoassay.

APPENDIX A

Units, Symbols, and Prefixes

Units in the text correspond to those in common usage. Many are gradually being replaced by the *Système International (SI)* or *International System of Units*. These recommended units, their symbols, and prefixes indicating multiples and fractions of units, are listed here.

SI Units

Quantity	Name	Symbol
Length	meter	m
Mass	kilogram	kg
Time	second	s
Electric current	ampere	A
Thermodynamic temperature	kelvin	K
Luminous intensity	candela	cd
Amount of substance	mole	mol
Plane angle	radian	rad
Solid angle	steradian	sr

Other Units in Use with SI

Quantity	Name	Symbol	Value in SI Unit
Time	minute	min	1 min = 60 s
	hour	h	1 h = 3600 s
	day	d	1 d = 86,400 s
Volume	liter	L	1 L = 1 dm^3 = 10^{-3} m^3

SI Derived Units

Quantity	Name	Symbol	Units	Special Multiples
Frequency	hertz	Hz	s^{-1}	—
Force	newton	N	$kg \cdot m \cdot s^{-2}$	10^{-5} N = 1 dyne (dyn)
Pressure[a]	pascal	Pa	$kg \cdot m^{-1} \cdot s^{-2} = N \cdot m^{-2}$	10^5 Pa = 1 bar
Power, radiant flux	watt	W	$kg \cdot m^2 \cdot s^{-3} = J \cdot s^{-1}$	—
Electric charge, quantity of electricity	coulomb	C	$A \cdot s$	—
Electric potential, potential difference, electromotive force	volt	V	$kg \cdot m^2 \cdot s^{-3} \cdot A^{-1} = J \cdot C^{-1}$	—
Electric resistance	ohm	Ω	$kg \cdot m^2 \cdot s^{-3} \cdot A^{-2} = V \cdot A^{-1}$	—
Electrical capacitance	farad	F	$A^2 \cdot s^4 \cdot kg^{-1} \cdot m^{-2}$	—
Conductance	siemen	S	$kg^{-1} \cdot m^{-2} \cdot s^3 \cdot A^2 = \Omega^{-1}$	—
Energy, work, quantity of heat[b]	joule	J	$kg \cdot m^2 \cdot s^{-2} = V \cdot C$	10^{-7} J = 1 erg
Magnetic flux	weber	Wb	$kg \cdot m^2 \cdot s^{-2} \cdot A^{-1} = V \cdot s$	10^{-8} Wb = 1 maxwell (Mx)
Inductance	henry	H	$kg \cdot m^2 \cdot s^{-2} \cdot A^{-2} = Wb \cdot A^{-1}$	—
Magnetic-flux density	tesla	T	$kg \cdot s^{-2} \cdot A^{-1} = Wb \cdot m^{-2}$	10^{-4} T = 1 gauss (G)
Luminous flux	lumen	lm	$cd \cdot sr$	—
Illumination	lux	lx	$cd \cdot sr \cdot m^{-2}$	—

a. 101,325 Pa = 1 atmosphere (atm) = 760 millimeters of mercury (mm Hg)
   133.322 Pa = 1 torr = 1 millimeter of mercury (mm Hg)
b. 3.6×10^6 J = 1 kilowatt-hour (kWh)
   1055.056 J = 1 British thermal unit (BTU)
   4.184 J = 1 thermochemical calorie (cal_{th})

Prefixes Indicating Multiples and Fractions of Units

Multiple	Prefix	Symbol	Fraction	Prefix	Symbol
10^{18}	exa	E	10^{-1}	deci	d
10^{15}	peta	P	10^{-2}	centi	c
10^{12}	tera	T	10^{-3}	milli	m
10^9	giga	G	10^{-6}	micro	μ
10^6	mega	M	10^{-9}	nano	n
10^3	kilo	k	10^{-12}	pico	p
10^2	hecto	h	10^{-15}	femto	f
10	deka	da	10^{-18}	atto	a
			10^{-21}	flato	ϕ

Common Exceptions to the SI System

Quantity	Name	Symbol	Units
Concentration	molal	m	mol/kg
	molar	M	mol/L
Conductance	mho	mho	Ω^{-1}
Density	gram per centimeter cubed	g/cm^3	g/cm^3 (also g/mL)
Energy	electron-volt	eV	eV (also keV, MeV)
Length	angstrom	Å	10^{-10} m
Plane angle	degree	°	$2\pi/360$ rad
	minute	′	1/60 degree
	second	″	1/60 minute
Pressure	atmosphere	atm	101,325 Pa
	bar	bar	10^5 Pa
	torr	torr	133.322 Pa
	millimeter of mercury	mm Hg	1 torr, 1/760 atm
Radioactivity	disintegrations per second	dps	dps
Temperature	degree Celsius	°C	
Volume	liter	L	
	milliliter	mL	
	microliter	μL	
Wavenumber	1 per centimeter	cm^{-1}	

APPENDIX B
Selected Fundamental Physical Constants

Numbers in parentheses refer to standard-deviation uncertainties in the last digit, computed on the basis of internal consistency.

Quantity	Symbol	Value	Error, ppm	SI	cgs
Velocity of light	c	2.9979250(10)	0.33	10^8 m·s^{-1}	10^{10} cm·s^{-1}
Electron charge	e	1.6021917(70)	4.4	10^{-19} C	10^{-20} emu
		4.803250(21)	4.4	—	10^{-10} esu
Planck's constant	h	6.626196(50)	7.6	10^{-34} J·s	10^{-27} erg·s
	$\hbar = \dfrac{h}{2\pi}$	1.0545919(80)	7.6	10^{-34} J·s	10^{-27} erg·s
Electron-volt	eV	1.60210	—	10^{-19} J	10^{-12} erg
		3.827	—	—	10^{-20} cal
Avogadro's number	N	6.022169(40)	6.6	10^{26} kmol^{-1}	10^{23} mol^{-1}
Atomic mass unit	amu	1.660531(11)	6.6	10^{-27} kg	10^{-24} g
Proton mass	M_p	1.672614(11)	6.6	10^{-27} kg	10^{-24} g
	M_p^*	1.00727661(8)	0.08	amu	amu
Electron mass	m_e	9.109558(54)	6.0	10^{-31} kg	10^{-28} g
	m_e^*	5.485930(34)	6.2	10^{-4} amu	10^{-4} amu
Neutron mass	M_n	1.674920(11)	6.6	10^{-27} kg	10^{-24} kg
	M_n^*	1.00866520(10)	0.10	amu	amu
Faraday constant	F	9.648670(54)	5.5	10^7 C·mol^{-1}	10^3 esu·mol^{-1}
		2.892599(16)	5.5	—	10^{14} esu·mol^{-1}
Gas constant	R	1.9872	—		cal·K^{-1}·mol^{-1}
		8.3143	—	J·K^{-1}·mol^{-1}	10^7 erg·K^{-1}·mol^{-1}
		8.2054	—	—	10^{-2} l·atm·K^{-1}·mol^{-1}
Rydberg constant	R_∞	1.09737312(11)	0.10	10^7 m^{-1}	10^5 cm^{-1}
Bohr magneton	μ_B	9.274096(65)	7.0	10^{-24} J·T^{-1}	10^{-21} erg·G^{-1}
Boltzmann constant	k	1.380622(59)	43	10^{-23} J·K^{-1}	10^{-16} erg·K^{-1}
Stefan-Boltzmann constant	σ	5.66961(96)	170	10^{-8} W·m^{-2}·K^{-4}	10^{-5} erg·s^{-1}·cm^{-2}·K^{-4}

Source: Adapted in part from "Reference Guide to Optical Energy Measurements," Princeton Applied Research Corp., 1974, by permission of the publisher.

APPENDIX C
Some Standard Electrode Potentials

Half-Reaction	Standard Potential, volts vs. NHE
$Ag^+ + e = Ag$	$+0.7994$
$AgBr + e = Ag + Br^-$	$+0.071$
$AgCl + e = Ag + Cl^-$	$+0.2224$
$AgI + e = Ag + I^-$	-0.152
$Ag_2O + H_2O + 2e = 2Ag + 2OH^-$	$+0.342$
$2AgO + H_2O + 2e = Ag_2O + 2OH^-$	$+0.57$
$Ag_2O_3 + H_2O + 2e = 2AgO + 2OH^-$	$+0.74$
$Ag_2S + 2e = 2Ag + S^{2-}$	-0.71
$Al^{3+} + 3e = Al$	(-1.66)
$As + 3H^+ + 3e = AsH_3$	(-0.60)
$As_2O_3 + 6H^+ + 6e = 2As + 3H_2O$	$+0.234$
$H_3AsO_4 + 2H^+ + 2e = HAsO_2 + 2H_2O$	$+0.559$
$Ba^{2+} + 2e = Ba$	(-2.90)
$BiO^+ + 2H^+ + 3e = Bi + H_2O$	$+0.32$
$BiOCl + 2H^+ + 3e = Bi + H_2O + Cl^-$	$+0.16$
$Bi_2O_3 + 3H_2O + 6e = 2Bi + 6OH^-$	-0.46
$Br_2 + 2e = 2Br^-$	$+1.087$
$2HOBr + 2H^+ + 2e = Br_2 + 2H_2O$	$(+1.6)$
$2BrO_3^- + 12H^+ + 10e = Br_2 + 6H_2O$	$+1.52$
$C_2N_2 + 2H^+ + 2e = 2HCN$	$(+0.37)$
$2CO_2 + 2H^+ + 2e = H_2C_2O_4$	(-0.49)
$Ca^{2+} + 2e = Ca$	(-2.87)
$Cd^{2+} + 2e = Cd$	-0.402
$Cl_2 + 2e = 2Cl^-$	$+1.359$
$2HOCl + 2H^+ + 2e = Cl_2 + 2H_2O$	$(+1.63)$
$ClO_3^- + 2H^+ + e = ClO_2 + H_2O$	$(+1.15)$
$ClO_4^- + 2H^+ + 2e = ClO_3^- + H_2O$	$(+1.19)$
$Co^{2+} + 2e = Co$	-0.28
$Co(NH_3)_6^{3+} + e = Co(NH_3)_6^{2+}$	$(+0.1)$
$Co(OH)_3 + e = Co(OH)_2 + OH^-$	$(+0.17)$
$Cr^{2+} + 2e = Cr$	-0.56
$Cr(III) + e = Cr(II)$	-0.41

Half-Reaction	Standard Potential, volts vs. NHE
$Cr_2O_7^{2-} + 14H^+ + 6e = 2Cr^{3+} + 7H_2O$	+1.33
$Cu^{2+} + 2e = Cu$	+0.337
$Cu(II) + e = Cu(I)$	(+0.153)
$2Cu^{2+} + 2I^- + 2e = Cu_2I_2$	+0.86
$Fe^{2+} + 2e = Fe$	−0.440
$Fe(III) + e = Fe(II)$	+0.771
$Fe(CN)_6^{3-} + e = Fe(CN)_6^{4-}$	+0.356
$Ga^{3+} + 3e = Ga$	−0.56
$Ge^{2+} + 2e = Ge$	(0.0)
$H_2GeO_3 + 4H^+ + 4e = Ge + 3H_2O$	−0.13
$2H^+ + 2e = H_2$	0.0000
$Hg_2^{2+} + 2e = 2Hg$	+0.792
$2Hg^{2+} + 2e = Hg_2^{2+}$	+0.907
$Hg_2Br_2 + 2e = 2Hg + 2Br^-$	+0.1392
$Hg_2Cl_2 + 2e = 2Hg + 2Cl^-$	+0.2680
$Hg_2I_2 + 2e = 2Hg + 2I^-$	−0.040
$I_2 + 2e = 2I^-$	+0.536
$HOI + H^+ + 2e = I^- + H_2O$	(+0.99)
$2IBr_2^- + 2e = I_2 + 4Br^-$	(+0.87)
$2ICN + 2H^+ + 2e = I_2 + 2HCN$	+0.63
$2IO_3^- + 12H^+ + 10e = I_2 + 6H_2O$	+1.19
$H_5IO_6 + H^+ + 2e = IO_3^- + 3H_2O$	(+1.6)
$In^{3+} + 3e = In$	−0.33
$IrBr_6^{2-} + e = IrBr_6^{3-}$	+0.99
$K^+ + e = K$	−2.925
$Mg^{2+} + 2e = Mg$	(−2.37)
$Mn^{2+} + 2e = Mn$	(−1.19)
$MnO_2 + 4H^+ + 2e = Mn^{2+} + 2H_2O$	+1.23
$MnO_4^- + 8H^+ + 5e = Mn^{2+} + 4H_2O$	+1.51
$MnO_4^- + 4H^+ + 3e = MnO_2 + 2H_2O$	+1.69
$MnO_4^- + e = MnO_4^{2-}$	+0.56
$Mo(VI) + e = Mo(V)$	+0.48
$N_2H_5^+ + 3H^+ + 2e = 2NH_4^+$	(+1.27)
$2NH_3OH^+ + H^+ + 2e = N_2H_5^+ + 2H_2O$	(+1.42)
$N_2 + 2H_2O + 4H^+ + 2e = 2NH_3OH^+$	(−1.87)
$3N_2 + 2H^+ + 2e = 2HN_3$	(−3.1)
$HNO_2 + H^+ + e = NO + H_2O$	(+0.99)
$2NO_3^- + 4H^+ + 2e = N_2O_4 + 2H_2O$	(+0.80)
$Na^+ + e = Na$	−2.713
$Ni^{2+} + 2e = Ni$	−0.23
$Ni(OH)_2 + 2e = Ni + 2OH^-$	−0.72
$NiO_2 + 4H^+ + 2e = Ni^{2+} + 2H_2O$	+1.68
$H_2O_2 + 2H^+ + 2e = 2H_2O$	(+1.77)
$O_2 + 4H^+ + 4e = 2H_2O$	(+1.229)
$OsO_4 + 4H^+ + 4e = OsO_2 + 2H_2O$	+0.96
$H_3PO_3 + 2H^+ + 2e = H_3PO_2 + H_2O$	(−0.50)

Half-Reaction	Standard Potential, volts vs. NHE
$H_3PO_4 + 2H^+ + 2e = H_3PO_3 + H_2O$	(-0.276)
$Pb^{2+} + 2e = Pb$	-0.126
$PbO_2 + H_2O + 2e = PbO + 2OH^-$	$+0.28$
$PbSO_4 + 2e = Pb + SO_4^{2-}$	-0.356
$PdCl_4^{2-} + 2e = Pd + 4Cl^-$	$+0.623$
$PdCl_6^{2-} + 2e = PdCl_4^{2-} + 2Cl^-$	$+1.29$
$Pt^{2+} + 2e = Pt$	$(+1.2)$
$PtCl_4^{2-} + 2e = Pt + 4Cl^-$	$+0.73$
$ReO_4^- + 4H^+ + 3e = ReO_2 + 2H_2O$	$+0.51$
$RhCl_6^{3-} + 3e = Rh + 6Cl^-$	$(+0.44)$
$RuO_4^- + e = RuO_4^{2-}$	$+0.59$
$RuO_4 + e = RuO_4^-$	$+1.00$
$S + 2H^+ + 2e = H_2S$	$(+0.14)$
$2SO_3^{2-} + 2H_2O + 2e = S_2O_4^{2-} + 4OH^-$	(-1.12)
$S_4O_6^{2-} + 2e = 2S_2O_3^{2-}$	$(+0.09)$
$SO_4^{2-} + 4H^+ + 2e = SO_2 + 2H_2O$	$(+0.17)$
$S_2O_8^{2-} + 2e = 2SO_4^{2-}$	$(+2.0)$
$(SCN)_2 + 2e = 2SCN^-$	$(+0.77)$
$Sb + 3H^+ + 3e = SbH_3$	(-0.51)
$SbO^+ + 2H^+ + 3e = Sb + H_2O$	$+0.212$
$Sb_2O_3 + 6H^+ + 6e = 2Sb + 3H_2O$	$+0.152$
$Sb_2O_5 + 6H^+ + 4e = 2SbO^+ + 3H_2O$	$(+0.58)$
$Se + 2H^+ + 2e = H_2Se$	(-0.40)
$H_2SeO_3 + 4H^+ + 4e = Se + 3H_2O$	$(+0.74)$
$SeO_4^{2-} + 4H^+ + 2e = H_2SeO_3 + H_2O$	$(+1.15)$
$Sm^{3+} + e = Sm^{2+}$	-0.8
$Sn^{2+} + 2e = Sn$	-0.140
$TcO_4^- + 4H^+ + 3e = TcO_2 + 2H_2O$	$+0.74$
$Te + 2H^+ + 2e = H_2Te$	(-0.72)
$TeOOH^+ + 3H^+ + 4e = Te + 2H_2O$	$(+0.56)$
$Ti^{3+} + e = Ti^{2+}$	-0.37
$Tl^+ + e = Tl$	-0.336
$Tl^{3+} + 2e = Tl^+$	$+1.28$
$Tl_2O_3 + 3H_2O + 4e = 2Tl^+ + 6OH^-$	$+0.02$
$V^{3+} + e = V^{2+}$	-0.255
$VO^{2+} + 2H^+ + e = V^{3+} + H_2O$	$+0.337$
$VO_2^+ + 2H^+ + e = VO^{2+} + H_2O$	$+0.9994$
$W(CN)_8^{3-} + e = W(CN)_8^{4-}$	$+0.457$
$W_2O_5 + 2H^+ + 2e = 2WO_2 + H_2O$	(-0.04)
$2WO_3 + 2H^+ + 2e = W_2O_5 + H_2O$	(-0.03)
$Zn^{2+} + 2e = Zn$	-0.7628

Note: Values in parentheses are calculated from calorimetric data.
Source: From L. L. Meites, ed., *Handbook of Analytical Chemistry*, Tables 5-1 and 5-12. Reprinted with permission of McGraw-Hill Book Company.

APPENDIX D
Answers to Selected Problems

Chapter 1

1. mM
2. $0.05420/n$ V (0°C); $0.06154/n$ V (98.6°F); $0.07404/n$ V (100°C)
3. $q = 2nFAc(Dt)^{1/2}/\pi^{1/2}$
4. $i = Cv$
5. $q = 10\,\mu C$ at -1.0 V; approx. $0.19 e^-$/atom
6. (Red)/(Ox) = 1 if $E = E^0$; surface ratio = 1; (Red)/(Ox) = 1.5, 49, 2400, 2.8×10^8, 8.0×10^{16}
7. 5, 50, 500 μsec; 100, 10, 1 mA; 92, 806 6900 μsec

Chapter 2

2. (a) 0.03; (b) 0.05; (c) 0.3; (d) 0.006
3. (a) $f_{Mg^{2+}} = 0.70$, $f_{Cl^-} = 0.91$; (b) $f_{Mg^{2+}} = 0.56$, $f_{K^+} = 0.86$, $f_{Cl^-} = 0.86$; (c) $f_{Mg^{2+}} = 0.22$, $f_{K^+} = 0.69$, $f_{Cl^-} = 0.69$
4. (a) pH = 4.41; (b) pH = 10.7
5. 0.544 V vs. SCE
6. -0.949 V
7. (a) $1.2 \times 10^{-4} M$; (b) 0.320 V; (c) $4 \times 10^{-6} M$; (d) $4 \times 10^{-10} M$
8. 7.5%
9. For a 1% interference, Zn^{2+}, $3.1 \times 10^{-7} M$; Fe^{2+}, $1.25 \times 10^{-6} M$; Pb^{2+}, $1.6 \times 10^{-6} M$; Mg^{2+}, $7.1 \times 10^{-5} M$; Na^+, $1.8 \times 10^{-2} M$. For a 10% interference, all answers are 10 times larger except for Na^+, $5.8 \times 10^{-2} M$.
10. 0.0795%
11. 0.57×10^{-6} g I per mL
12. (a) $3 \times 10^{-5} M$; (b) $4 \times 10^{-5} M$
15. 5×10^{-4}
16. 47.93 mL
17. Zn^{2+}, 50.7; Pb^{2+}, 20.2; Mg^{2+}, 0.01; H^+, 1000; Na^+, 2.9×10^{-3}; K^+, 9.7×10^{-4}

Chapter 3

1. 7.2 μA
3. $n = 2$, $E_{1/2} = -0.500$ V, reversible
4. (a) calibration curve; (b) standard addition; (c) standard addition; (d) calibration curve
6. (a) $I_d = 3.58$; (b) $D = 8.7 \times 10^{-6}$ cm^2/sec
7. 0.032 cm^2
8. 0.156 mM, or 5.0 ppm
9. 2
10. 1.3×10^7
11. (a) effective D changes; uncertainty of i_c; (b) no; (c) yes; (d) no affect on DC polarography; decreases sensitivity of differ-

ential pulse polarography; (e) not in all cases since background uncertainty may change
12. 0.0015, or 0.15%
13. (a) linear decrease to end point and constant thereafter; (b) V-shaped;
    (c) 100 mV: $Ag^+ + e \rightarrow Ag$, and the reverse, before endpoint; no defined reaction after endpoint
    600 mV: same reactions before endpoint, but after endpoint, $Ag + Cl^- \rightarrow AgCl + e^-$ and discharge of the supporting electrolytes
14. $A = 4\pi(3mt/4\pi d_{Hg})^{2/3}$ mm$^2 = 0.852(mt)^{2/3}$
15. $A = 3.96 \times 10^{-2}$ cm^2, $c = 0.156$ mM; $i_p = 6.84, 21.6$ μA; 3.2 and 10.2 times that of the maximum i_l
16. (a) $i_c = 0.079, 0.79$ μA; (b) $i_p/i_c = 86$ and 27

Chapter 4

1. 1.797×10^{-4} μg Cr
2. 82.8 g Br$_2$/100 g sample
3. (a) 2.49×10^{-3} N; (b) 9.34 mg; (c) 0.558 mL H$_2$ gas
4. 96.485 mA
5. $i = 96.485\phi^2$ A, where $\phi =$ flow rate in mL/sec
6. 1.036×10^{-2} (i/v) eq/L
7. 3.25 μm
8. $t/t_{1/2} = 1, 3.3, 6.7, 10.0, 13.3$
9. (a) 405 sec; (b) 7.0 sec
10. 0.1116% Fe
11. 4.65 ppm SO$_2$
12. 7.70 mg N

Chapter 5

1. $1.0 \times 10^{-5}, 3.0 \times 10^{-5}, 1.5 \times 10^{-4}$ N
2. 890 ppm
3. (a) 25 cm^{-1}, solution resistances will vary between 73 and 86 Ω; (b) 100,000 μmho (midpoint equivalent to 100 Ω)
4. 110 to 1130 Ω, or 9100 to 885 μmho

5. (a) $\Lambda = 6.73$; (b) $\alpha = 0.0176$; (c) $K_a = 6.3 \times 10^{-6}$
6. 2033 Ω
7. 0.0175 N
9. $\kappa_{AgCl} = 1.81 \times 10^{-6} \Omega^{-1}$ cm^{-1}; $K_{sp,AgCl} = 1.72 \times 10^{-10}$ M^2
10. 2.4% hydrocarbon in glycol; 1.5% glycol in hydrocarbon
13. 12.4 volume percent

Chapter 6

1. (a) 5.000 μm; (b) 0.2480 eV/molecule; (c) 5718 cal/mole; (d) 3.972×10^{-13} ergs/molecule
2. (a) 2.22×10^3 cm^{-1}; (b) 6.66×10^{13} Hz; (c) 4.50×10^4 Å; (d) 4.41×10^{-13} ergs/molecule; (e) 0.275 eV/molecule
3. (a) 6.00×10^{11} Hz; (b) 20.0 cm^{-1}; (c) 2.48×10^{-3} eV/molecule
4. (a) 5.827×10^{14} Hz; (b) 1.944×10^4 cm^{-1}; (c) 0.5145 μm; (d) 2.410 eV/molecule
5. (a) 0.2 cm^{-1}; (b) 10 cm
6. (a) 23.6°; (b) infrared; (c) 20 μm
7. (a) 32.8°; (b) 34.1°

Chapter 7

1. 588.997 nm; 5.08960×10^{14} Hz
2. -4.68 eV; 265 nm
3. -4.78 eV; 259 nm
4. 4.72×10^{-10} M; 0.502 ng
5. 2.42×10^4 M^{-1} cm^{-1}
6. ortho—23.7%; para—56.4%
7. 6.25
8. (a) 210 L/g-cm; (b) 3.15×10^4 M^{-1} cm^{-1}; (c) 0.0298 mg; (d) 1.19 ppm
9. 0.260% Cu
10. 0.533% Mn
11. 2.0×10^6
12. 120 g/mole
13. (a) 1:1; (b) 1.8×10^6
14. blank absorbance
15. (a) 8.87 mM; (b) 6.44 mM; 7.7, approx. 8 waters

Chapter 8

1. (a) 0.7 to 2.5 μm; 2.5 to 50 μm; 50 to 1000 μm; (b) 14285 to 4000 cm^{-1}; 4000 to 200 cm^{-1}; 200 to 10 cm^{-1}
2. (a) 14,285 cm^{-1}; (b) 0.7 μm; (c) 1.77 eV
3. (a) 9; (b) 30; (c) 7
4. 1200 cm^{-1}; 1680 cm^{-1}; 2050 cm^{-1}
5. 2130 cm^{-1}
6. 564.8 nm; 543.2 nm; 533.0 nm; 497.3 nm
7. 0.384; 0.0566; 0.00836
9. 0.753 (depolarized); 0.074 (polarized); 0.016 (polarized); 0.783 (depolarized)
10. 50 MW; 20 MW
11. AgCl in sealed or demountable cell; use of D$_2$O

Chapter 9

8. about 0.1 M
12. about 10:1
14. 70 ppm
15. (a) about 14:1

Chapter 10

9. 0.34 ppm
10. 7.1 ppm
11. (a) 37.0 ± 1.0% Ba by weight (about 2.7% relative standard deviation). (b) Because the pure compound should contain 32.12% Ba, and the analytical result is about five standard deviations removed, there is a high probability that the compound is not 100% pure. Most probably, there is an excess of a barium salt in the crystalline compound. (c) 0.2 ppm
19. (a) Ca: $g_u/g_0 = 3/1$; Na: $g_u/g_0 = 4/2$ and 2/2 for the shorter- and longer-wavelength transitions, respectively; K: same as Na; (b) Ca: 3.06 × 10^{-8}, 4.14 × 10^{-6}, 7.89 × 10^{-5}; Na: 3.68 × 10^{-6}, 1.25 × 10^{-4}, 1.03 × 10^{-3} for 588.9963 nm, and 1.87 × 10^{-6}, 6.30 × 10^{-5}, 5.21 × 10^{-4} for 589.5930 nm; K: 7.63 × 10^{-5}, 1.15 × 10^{-3}, 5.87 × 10^{-3} for 764.494 nm, and 4.10 × 10^{-5}, 6.07 × 10^{-4}, 3.06 × 10^{-3} for 769.901 nm; (c) Na doublet: 1.974, 1.980, 1.985; K doublet: 1.862, 1.898, 1.920
20. (a) S/N approx. 7, 100; (b) 0.3 and 0.02 of the concentrations giving rise to the two signals, respectively

Chapter 11

6. Mg, Zn, Cu, Th, Zr, Mn, Ca, Si, Al (a Mg alloy)
7. 0.04 μg/mL
8. σ = ±0.0023 μg/mL; relative standard deviation = ±6.6%
10. fraction ionized at 2500 K is 1.24 × 10^{-9}; at 5000 K is 1.11 × 10^{-3}
11. 0.13% Si, 1.4% Na

Chapter 12

1. (a) +0.86 ppm (0.86 ppm downfield); (b) 1174 Hz
2. A, e; B, f; C, c; D, d; E, b; F, a
3. Acetone: singlet, 2.11 ppm. Methylethyl ketone: singlet, 2.05 ppm (area 3); quartet, 2.40 ppm (area 2); triplet, 0.99 ppm (area 3)
4. Propane: triplet, 0.99 ppm (area 6); multiplet, 1.29 ppm (area 2). 1-Nitropropane: triplet, 1.01 ppm (area 3); multiplet, 2.00 ppm (area 2); triplet, 4.31 ppm (area 2)
5. (a) 4258 Hz/gauss; (b) 100 MHz; (c) 20.1 MHz
7. (a) 3.55 ppm; (b) 3.66 ppm; (c) 2920 Hz
8. Hindered rotation; therefore, essentially two different environments for the methyls
11. benzene:toluene = 2:1
12. C$_3$H$_7$Cl = 2-chloropropane; C$_7$H$_{16}$O$_3$ = (CH$_3$CH$_2$O)$_3$CH; C$_7$H$_7$ClO = 1-chloro-4-methoxybenzene; C$_9$H$_9$ClO = para-isomer of CH$_3$CH$_2$C(=O)—C$_6$H$_4$—Cl
13. CH$_3$(CO)OCH$_2$CH(CH$_3$)$_2$
14. A = HCFCl—CFCl$_2$
    B = HCCl$_2$—CF$_2$Cl

Chapter 13

1. $H = 2399, 5869, 8931$ gauss for $g = 2.800$; $H = 3420, 8367, 12{,}733$ gauss for $g = 1.964$
2. $\Delta H = 174.9, 427.8, 651.1$ gauss for $g_1 = 2.045$, $g_2 = 2.160$; $\Delta H = 243.6, 596.0, 907.0$ gauss for $g_1 = 1.960$, $g_2 = 2.110$
3. $0.314, 0.767, 1.168$ cm^{-1}/molecule
4. ratio $m_s = +\tfrac{1}{2}/m_s = -\tfrac{1}{2} = 0.9985$, $0.9942, 0.9777, 0.8933$ for 9.4 GHz; $0.9963, 0.9858, 0.9463, 0.7588$ for 23 GHz; $0.9944, 0.9784, 0.9194, 0.6571$ for 35 GHz
5. 64 lines for 1,2- and 1,3- compounds (no Hs or Fs are equivalent); 27 or 64 lines for the 1,4- compound depending on whether the Fs are *trans-trans* or *trans-cis* substituted
6. 28 lines
7. 60 lines for the Cu compound; 120 lines for the V compound
8. 5 lines; 1:4:6:4:1
9. 8 lines intensity 1, 12 lines intensity 2, 6 lines intensity 4, 1 line intensity 8; 27 lines total
10. $3.2, 7.6, 11.6 \times 10^{-17}$ erg; $0.45, 1.1, 1.6$ cal/mole
11. (a) 1.5×10^{16} spins/cm^3; (b) 4.8×10^{-5} M
12. (a) 1.1×10^{11} spins; (b) 36 pg; (c) 0.2 ng/mL

Chapter 14

11. $2\theta = 34.34°$
12. $\lambda = 2.287$ Å. This corresponds to the Kα line of chromium.
13. $\lambda = 0.248$ Å
14. $T = 0.633$
15. PET with $2d$ spacing of 8.742
16. 1.2 ppb Ni; 2.2 ppb Pb; 5.0 ppb Zn
18. $W_{Ni} = 0.7608$, $W_{Fe} = 0.2267$ after 7 iterations

Chapter 15

1. O 2s, 25 eV; Si 2p, 103 eV; Si 2s, 155 eV; C 1s, 285 eV; Ag 3d, 368 and 375 eV; O 1s, 530 eV.
2. P 2p, 135 eV; S 2p, 165 eV; C 1s, 285 eV; N 1s, 400 eV; O 1s, 530 eV; F 1s, 690 eV
3. Binding energy for Fe $2p_{3/2}$ photoelectron = 714 eV in Fe$_2$O$_3$; 717 eV in FeF$_2$; 710 eV in Fe. From Table 15.1, the "nominal" $2p_{3/2}$ binding energies for Mn, Fe, and Co are 641, 710, and 779 eV, respectively.
4. See C. A. Evans, Jr., *Anal. Chem.*, 47(9), 819A, 855A (1975).
5. (a) In a close-packed arrangement, one silver atom would occupy *about* 6.5×10^{-16} cm^2 of the surface area of a (flat) silicon substrate, approximately the cross-sectional area of a silver atom. For 0.001 monolayer, this would correspond to about 1.2×10^{10} atoms. (b) 2×10^{-12} g. (c) 1.4×10^{-3} g/cm^3; 600 ppm.
6. (a) 1153 eV; (b) 920 eV
7. C 1s in CH$_4$, CO, CO$_2$; O 1s in CO, CO$_2$.
9. 83%, 68%, 30% MoO$_2$
10. Top spectrum: $E_b = 1086, 954, 933, 117, 64$ eV. Bottom spectrum; $E_b = 1102, 955, 930, 112, 68$ eV, within measuring error. The two sets are the same, within error, as would be expected. $L_{II} = 2p_{1/2}$; $L_{III} = 2p_{3/2}$

Chapter 16

1. $H = 1052$ gauss, $r = 29.05$ cm
2. $H = 2975$ gauss, $r = 29.88$ cm
3. voltage should be reduced to 1765 V to just observe $m/z = 850$ at highest magnetic field
4. chamber: +1765 V; exit slit: ground; repellers: 1775 V; filament: 1715 V; target: 1845 V
5. exact mass = 220.1679; formula = C$_{11}$H$_{22}$O$_4$
6. CO$^+$/C$_2$H$_4^+$: $R = 770$; C$_{20}$H$_{40}^+$/C$_{19}$H$_{36}^+$: $R = 7700$
7. C$_2$H$_2^+$: 6.70×10^{-6} sec; C$_6$H$_5^+$: 1.14×10^{-5} sec; C$_6$H$_6^+$: 1.15×10^{-5} sec
8. 153 kHz
9. C$_6$H$_6$: 6.6, 0.22%; C$_2$H$_4$O$_2$: 2.2, 0.41%, C$_2$H$_8$N$_2$: 2.9, 0.02%; C$_3$H$_7$Cl: 3.3, 33,

1.09%; C_4H_4S: 5.2, 4.5%; $C_{16}H_{34}$: 17.6, 1.5%
11. $\%d_3 = 90; \%d_2 = 6; \%d_1 = 4$
12. chlorobenzene
13. *sec*-butylamine
14. perbromic acid ($HBrO_4$)
15. ketone 1: 3-methyl-2-pentanone; ketone 2: 2-methyl-3-pentanone
16. propylphenyl ether
17. $R = \dfrac{M}{\Delta M} = \dfrac{32}{2(15.994914) - 31.972074} = 1800$
18. mother ion = CH_3O^+ ($m = 31$); daughter ion = CHO^+ ($m = 29$)
19. (A) (a) $C_2H_5O^+ \to H_3O^+ + C_2H_2$, $m^* = 8.02$
    (b) $C_2H_5O^+ \to CHO^+ + CH_4$, $m^* = 18.69$
    (B) H_3COCH_2Y, since a rearrangement is necessary to facilitate elimination of C_2H_2
20. $m/z = 114$: molecular (parent) ion
    $m/z = 99$: 114 − 15 (loss of CH_3)
    $m/z = 71$: 114 − 43 (loss of propyl)
    $m/z = 57$: $C_4H_9^+$ (*t*-butyl ion)
    $m/z = 43$: $C_3H_7^+$ (*i*-propyl ion)
    2,2,4-trimethylpentane spectrum is in column (a)
21. *t*-butyl amine; $C_4H_{11}N$
22. Since m/e for P is an even number, there must be 0 or an even number of nitrogens (eliminating C_4H_8NO and $C_5H_{12}N$). Among the remaining, $C_5H_{10}O$ gives the best correspondence.
23. order of appearance = CO, N_2, C_2H_4; the order of increasing exact nuclidic masses

Chapter 17

23. 2 urea → biuret + ammonia; or 3 urea → cyanuric acid + 3 ammonia
24. $CaC_2O_4 \cdot H_2O \to CaC_2O_4 + H_2O \to CaCO_3 + CO \to CaO + CO_2$; and equivalent reactions for $BaC_2O_4 \cdot H_2O$—all at their appropriate temperatures. Associated weight losses are 6.16, 9.59, 15.07 mg for the Ca salt; and 3.70, 5.76, 9.05 mg for the Ba salt.
27. (c) 0.113°C and 0.0186°C for the two reactions assuming a heat capacity due only to the solutions mixed
28. 2
29. $Cu(NH_3)_6Cl_2$
32. $C_p = 1.78$ cal/g-°C; overall $C_p = 4.51$ cal/g-°C
33. 11.2 cm^2
35. $\Delta H_{ionization} = +3.15$ kcal/mole
36. (a) 151 cal/°C; (b) 15.20 Ω; too high
37. $K = 4980$ at 25 sec; $\Delta G = -5.05$ kcal; $\Delta H = -10.00$ kcal; $\Delta S = -16.6$ kcal/mole-K
38. $K = 9.43 \times 10^5$ at 25 sec; calculated $\Delta G, \Delta H,$ and ΔS vary with time, indicating incorrect stoichiometry assumed

Chapter 18

1. $[C] = 1.3 \times 10^{-5}$ M
2. $t = 4.7 \times 10^3$ sec
9. $[\text{8-quinolinol}]_0 = 5.7_1 \times 10^{-7}$ M; $[\text{5,7-dibromo-8-quinolinol}]_0 = 1.1_3 \times 10^{-5}$ M
10. $k_{Zn}^{NiL} = 1.5 \times 10^{-4}$ M^{-1} min^{-1}; $k_{Cu}^{NiL} = 0.93$ M^{-1} min^{-1}

Chapter 19

1. 286 ppm Al
2. (a) The 5.62-MeV peak is due to the pair-production interaction of the primary 6.13-MeV gamma ray with the crystal and the loss of *one* 0.511 position annihilation photon from the crystal without interaction. This is the "1st escape peak." (b) The 5.11-MeV peak is due to loss of *both* 0.511-MeV annihilation photons resulting from a pair-production event in the crystal. This is the "2nd escape peak." (c) The 0.511-MeV peak is due to pair-production interaction of the 6.13 MeV primary gamma ray in the *surroundings and shielding* of the detector. The 0.511-MeV position annihilation photons generated in the surrounding

materials then intersect the crystal, yielding the observed 0.511-MeV peak.
3. 0.926 g Mn/g Fe
4. 0.8 ppm I$^-$
5. $N_0 = 2.60 \times 10^5$
7. Possible reactions:
   (a) $^{64}_{28}$Ni (n, γ) $^{65}_{28}$Ni

   $^{65}_{28}$Ni $\xrightarrow{2.55\,hr}$ β^- and γ (1.48 MeV, etc.)

   (b) $^{58}_{28}$Ni (n,γ) $^{59}_{28}$Ni

   $^{59}_{28}$Ni $\xrightarrow[8 \times 10^4\,yr]{EC}$ $^{59}_{27}$Co

   (c) $^{62}_{28}$Ni (n,γ) $^{63}_{28}$Ni

   $^{63}_{28}$Ni $\xrightarrow{92\,yr}$ β^- (no γ)

   (d) $^{58}_{28}$Ni (n,p) $^{58}_{27}$Co

   $^{58}_{27}$Co $\xrightarrow{71.4\,days}$ β^+ and several γs

8. 31.4% Co
9. net rate = 1400 ± 50 counts/min
10. ratio A/B = 0.41 ± 0.03
11. 1.9×10^4 yr
12. (a) 3.70×10^{10} atoms; (b) 1600 yr
13. (a) 1.00 microcurie; (b) 2.10×10^{-6} sec^{-1}; (c) 1.76×10^{10} atoms; (d) 6.60×10^{-12} g
14. 725 curies
15. 1.77 disintegrations per second
16. 1 half-life
17. average lifetime of an atom = Σ (individual atomic lifetimes)/(total number of atoms) = $1.443 t_{1/2}$
18. 102 and 755 μCi/μg; ^{131}I

Chapter 21

1. (a) $R_f = 1/(1 + k')$; (b) $k' = 2.3$; (c) yes, $\Delta R_f = 0.03$
2. (a) k' for o-, m-, and p-nitroaniline = 1, 2, and 3.5, respectively; (b) α for m- and p-nitroaniline = 2.0 and 3.5, respectively
3. (a) R_f for 3'- and 2'-GMP = 0.50 and 0.58, respectively; (b) α = 1.4
4. (a) 64,000; (b) yes, $R = 2.7$
5. (b) molecular weight = 3.2×10^3; (c) about 1000
6. (a) 27.6 mg; (b) 3.8 mg; (c) 360 cm; 11
7. (a) $t_r = N H (1 + k')/v$; (b) $t_r = 50$ min
8. (a) $L_2 = 26.7, L_1 = 26.7$ m, pressure drop = 267 atm; (b) $t_{r_2} = t_{r_1}/0.004 = 7500$ min; (c) $d_{p_2} = 8\ \mu$m, pressure drop = 250 atm. Option (c) is best.
9. (a) log $k' = 1.55, k' = 35.5$
10. (a) 1.2×10^{-3} cm/sec
11. $V_m = 1.26$ cm^3 $V_r = 13.6$ cm^3
12. (a) 5.4 μg; (b) 0.56
13. (a) $R = 5.3, 1.8$; column 1 shows better resolution; (b) $R = 7.4, 19.7$; column 2 now shows better resolution
14. 1.65
15. (a) 60 μL; (b) 0.285, 1.29 mL; (c) 15.1, 17.6, 19.5
16. 0.35 L ACN and 0.84 L H$_2$O gases; 4× and 20× of those volumes for columns B and C, respectively

Chapter 22

3. (a) $I = 776$. (b) Only that n-butylacetate behaves in this system as if it were a hydrocarbon of 7.75 carbon number. Structural information is better derived from the index increment ΔI ($I_{polar} - I_{nonpolar}$).
7. toluene, 724; cyclohexane, 690
8. (a) 1770, 2810, 3840, 4880 mm Hg; (b) $j = 0.558, 0.374, 0.280, 0.223$
9. 29.6, 50.0, 64.6 mL/min
10. (a) $V'_r = 59.6, 226.8, 606.9$ mL; (b) $V_n = 16.7, 63.5, 170.0$ mL; (c) $V_g = 10.7, 40.9, 109.5$ mL; (d) 8.7, 9.2; (e) 9.2
12. (a) 4500; (b) 6300; (c) 9800; (d) 39000; (e) 4400; (f) 0.9, 1.3, 98, 390 m
13. (a) $t'_r = 14.5, 15.2, 16.6, 17.2, 18.4, 19.4$ min for the six peaks; (b) $\alpha = 1.05, 1.04, 1.06$ for the three pairs of enantiomers; (c) $\alpha_{1,3} = 1.14, \alpha_{3,5} = 1.11, \alpha_{2,4} = 1.13, \alpha_{4,6} = 1.13$; (d) $R_{1,2} = 2.3, R_{1,3} = 6.8$; (e) $N \sim 57,000$; $H \sim 0.7$ mm; $k' = 4.9$; (f) $R = 2.3$

Chapter 23

1. (a) 0.200 mA; (b) 1.00 mW

2. (a) $R_S = 1.60$ kΩ; (b) 5.00 mA; (c) 0.500, 2.50, 5.00 V
3. (a) $I_{total} = 104$ mA; (b) $I = 80, 16, 8$ mA
4. (a) $R_{total} = 433$ Ω; (b) $I = 18.5, 12.3$, and 6.2 mA through the 0.1-, 0.5-, and 1.0-kΩ resistors, respectively; (c) 1.85, 6.15, 6.15 V, respectively; (d) 38, 76, 38 mW, respectively
7. (a) $1/C_{series} = \Sigma 1/C_i$; (b) $C_{parallel} = \Sigma C_i$
8. (a) 377 radians/sec; (b) 16.7 msec
9. 265 and 16 Ω at 60 and 1000 Hz
10. $R_{th} = 2.0$ kΩ, $V_{th} = 4.00$ V
11. (b) 1060 Hz; (c) 1.60×10^5, 2.19×10^4, 1.50×10^4 Ω; (d) 84.6°, 0.61°
13. (a) 12.65 kΩ; (b) 0.377 Ω
14. (b) 40 nA; (c) ≈ 40 nA; (d) silicon; (e) no
15. (a) 2.40%; (b) 11.7 V
16. 50 mA
17. approx. 3.4 mA
18. (a) 0.0075 mV; (b) 2.0 μsec
23. $V_o = -RC \dfrac{dV_i}{dt}$
26. (a) 70 MHz; (b) 70 MHz; (d) 700 kHz
27. 0.39, 0.10, 0.024%

Chapter 24

2. The AND function will require 2 NAND gates; the OR function, 3; the NAND function, 1; the NOR function, 4—in appropriate configurations, and with appropriate numbers of gate inputs.
3. The same four logic functions will require, respectively, 3, 2, 4, and 1 NOR gates, appropriately configured.
7. $Y = \overline{(A_1 + B_1)} \cdot \overline{(A_2 + B_2)} \cdot \overline{(A_3 + B_3)} \cdot \overline{(A_4 + B_4)}$
8. For $N = 0$ to 15, the outputs Q_D, Q_C, Q_B, and Q_A will produce the binary counting table 0000 to 1111.
9. For $N = 0$ to 9, the outputs Q_D, Q_C, Q_B, and Q_A will produce the binary counting table 0000 to 1001.
10. $\Delta V = v_{t+} - v_{t-} = 5.079$ V $- 4.822$ V $= 0.257$ V
11. 0.050
12. (a) 0.067; (b) 6.7×10^{-5}; (c) 6.7×10^{-6}
13. 250 kHz
14. (a) 67.4 kHz; (b) 4 digits; (c) 0.001 V
15. 5000 counts
16. 10, 5, 2.5, and 1.25 MΩ for R_0 through R_3
17. 10 nsec, 163.84 μsec
18. (a) 5 trials; (b) 16, 24, 20; (c) 16, 8, 4, 6; (d) 6 trials; (e) 32, 48, 40; (f) 32, 16, 24, 20; (g) 32, 16, 8, 4, 6
19. 66.7 μsec
20. (a) 2500; (b) 10,000; (c) 10^6
21. (a) standard deviation = 45.2, relative standard deviation = 1.41%, SNR = 71.1; (b) 23.8
22. (a) 9 bits; (b) 13 bits; (c) 16 bits

Chapter 25

1. (a) 4184 J/K; (b) 0.07°C sec/J [beginning with 5 BTU/hr ft^2 (°F/in.)]; (c) 1.19 g/Pa
5. (a) pH = 9.4; buffer index = $\beta = 1.2 \times 10^{-4}$, gain = 8300; (b) pH = 4.76; $\beta = 0.6$, gain = 1.7
6. sensitivity = 0.24 psi/°C, gain = 1, % PB = 100; at 50% PB, gain = 2, sensitivity = 0.48 psi/°C
8. $V_o = 625\,I + 7.5$
9. 60, 1%; 2%
10. (a) 5 V; (b) -2%
11. for $C = 100$ μF, $R = 100$ kΩ; 1/4 fed back
12. ± 3°C

Index

0-0 transition, 250
$1/f$ noise, 846

A

AB quartet, 377
Absolute activity, radiochemical, 597
Absorption, x-ray, 445
Absorption coefficient
  linear, 441
  mass, 442
Absorption edge, 418
Absorption spectrophotometers, in process control, 871
Absorptivity, 175. *See also* Molar absorptivity
AC signals, 780
Accelerator, nuclear, 620
Accuracy, 855
Acid error, 27
Activation analysis, 614
Activation energy, of reactions, 564
Activation polarization, 7
Activity, 15
  enzymatic, 574
  nuclear, 597
Activity coefficient, 15
ADC, 835
Address bus, 841
Adjusted flow rate, 729
Adjusted retention time, 729

Adjusted retention volume, 729
Adsorbent, 661
Adsorption chromatography, 661, 697
Adsorption current, 59
  tests for, 60
Aerosol, 285
Affinity chromatography, 724
Air-acetylene flame, 280
Air peak, in gas chromatography, 739
Alkaline error, 27
Allylic coupling, 381
Alpha decay, 603
Alpha particle, 596
Amperometric titrations, 88
  applications of, 92
Amperometry, 88
Amphiprotic solvents, 46
Amplifiers. *See also* Operational amplifier
  difference, 793
  sample-and-hold, 822
  track-and-hold, 822
Amplitude-time waveform, 849
Analog control, 867
Analog domain, 768
Analog switch, 821
Analog-to-digital converter, 835
  staircase, 837
  successive-approximation, 837
Analytical concentration, 641
AND gate, 810
Angular momentum vector, 396

Annihilation, of positrons, 603
Anode, 3
Anodic stripping voltammetry. *See* Stripping voltammetry
Anti-Stokes fluorescence, 316
Anti-Stokes lines, 217
Antibody, 630, 724
Antigen, 630, 724
Antineutrino, 601
Aprotic solvents, 47
Arithmetic/logic unit, 841
Arrhenius equation, 563
ASCII, 842
Assembly language, 842
Astable multivibrator, 827
Asynchronous counter, 821
Atom formation in flames, 283
Atomic absorption spectrometry, 280, 295
    applications of, 311
    background correction in, 301
    electrothermal atomization in, 304
    secondary absorption lines, 299
    sensitivity of, 304
    use of organic solvents in, 313
    Zeeman background correction in, 310
Atomic emission spectrometry, 280, 291
    in gas chromatography, 751
Atomic fluorescence spectrometry, 280, 315
    instrumentation for, 317
Atomic mass, 595
Atomic number, 595, 600
Atomization, electrothermal, 304
ATR. *See* Attenuated total reflectance
Attenuated total reflectance, 237
Auger electrons, 602
Auger spectra, 457, 469
Auger spectroscopy, 451
    scanning microprobe, 469
AutoAnalyzer®, 882
Automated instruments, 854
    parameters for, 854
    in process control, 870
Automated process control, 860
Automated titrator, 854
Automatic coulometric titrations, 105
Automatic devices, 854
Automatic titrator, 854
Automation
    clinical chemistry requirements for, 879
    in clinical laboratories, 881
    of kinetic methods, 583
    of sampling, 882
Auxochrome, 169
Avalanche breakdown, of diodes, 787
Average optical power, 226
Average polarographic current, 57
Averaging, multiple period, 832

B

Babington nebulizer, 340
Backflushing, gas chromatographic, 873
Background correction, atomic absorption spectroscopic, 301
Band broadening, chromatographic, 667, 735
Bandpass, 178, 804
Bandwidth, 178, 804
Barn, 618
Barrier layer cell, 193
Base peak, in mass spectrometry, 496
Base, in transistors, 790
Baseline method, 241
Baseline resolution, 671
BASIC, 842
Batch extraction, 644
Batch instruments, 854, 883
Bathochromic shift, 170, 180
Baud rate, 842
BCD, 772, 821, 829
Beer's law, 114, 174, 250, 300, 302, 445
    assumptions of, 176
    deviations from, 176
Beta decay, 601
Beta emission, 601, 749
Beta particle, 595
Beta ray, 595
Biamperometric titrations, 90
Bias, 863
Bias current, op-amp, 805
Binary-coded decimal, 772, 821, 829
Binding energy
    nuclidic, 616
    photoelectron, 452
    table of photoelectron, 457
Bipolar square-wave pulse conductance, 131
Bipolar transistor, 790
Bistable latch, 816
Bit, 771, 842
    least significant, 837
Bjerrum method, 187
Blackbody radiation, 146
Blackbody source, 192
Blue shift, 180
Bode plot, 784
    for op-amps, 804
Bohr magneton, 396
Bolometer, 158, 229
Boltzmann equation, 363, 401
Boltzmann excess, in NMR, 363
Bonded-phase chromatography, 663, 707
Boolean algebra, 810, 813
Boot, 842
Bootstrap, 842
Born-Oppenheimer approximation, 166
BPC. *See* Bonded-phase chromatography

Bragg diffraction, 414
Bragg's law, 426
Branching ratio, 597
Bremsstrahlung radiation, 416, 603
Buffer
  computer, 842
  ionization, 289
  potential, 118
  spectrochemical, 334
Byte, 842

C

Cadmium arc lamp, 691
Cadmium-109, 425
Calibration curve
  atomic absorption, 303
  polarographic, 68
Californium-252, 513, 619
Calomel electrode, 21
Capacitance, 772, 782
  double-layer, 77
  electrode, 6
  process, 862
Capacitive current, 55, 77
Capacitive reactance, 782
Capacity
  ion exchange, 713
  sample, 674
Capacity factor, 665
Capillary columns, 740
  tandem, 762
Capillary gas chromatography, 740
Capture reaction, of neutrons, 616
Carbon dioxide laser, 513
Carbon-13, 495
Carbon-14, 601, 609, 637
Carrier distillation method, 334
Cascade loop, 866
Catalytic current, 58
  tests for, 60
Catalytic determinations, 585
  table of selected, 562
Catalytic methods, 570
Catalytic reactions
  activators of, 587
  inhibitors of, 587
Catalytic titrations, 587
Catalyzed reaction, 570
Cathode, 3
Cathodic stripping voltammetry, 85
Cell
  conductance, 128
  electrochemical, 14
  electrolytic, 14
  galvanic, 14
  schematic representation of electrochemical, 18
  voltaic, 14
Cell constant, 123
Cell factor, 124
Central processing unit, 840
Centrifugal fast analyzers, 884
Charge, electrical, 6, 96, 782
Charge coupled detector, 198
Charged particle activation analysis, 615
Charging current, 5, 55, 77
Chelate extraction systems, 648
Chelation, in solvent extraction, 643
Chemical ionization mass spectrometry, 509
Chemically equivalent nuclei, 371
Chemical potential, 8
Chemical shift
  in NMR, 365
  physical causes of, 367
  table of carbon-13 NMR, 371
  table of proton NMR, 370
Chemiluminescence, 247
Chromatogram, 660
Chromatography, 658
  adsorption, 661, 697
  affinity, 724
  bonded-phase, 663, 707
  elution, 660
  gel-filtration, 663
  gel-permeation, 663
  high-performance liquid, 659, 681
  ion, 715
  ion-exchange, 663, 712
  ion-pair, 711
  liquid-liquid, 658, 706
  liquid-solid, 661, 698
  and mass spectrometry, 479, 506
  normal-phase, 708
  paper, 659, 675
  partition, 658, 663, 706
  principles of liquid, 659
  reverse-phase, 663
  size-exclusion, 663, 718
  thermodynamics of gas, 729
  thin-layer, 659, 678
  two-dimensional paper, 678
Chromophore, 169
  in NMR, 365
Circuit
  equivalent, 774
  gain of, 799
  series RC, 783, 825
  Thévinin equivalent, 779
Clock input, 817
CMOS, 809
Cobalt-60, 595, 611
Cockroft-Walton neutron generator, 620

Coherent radiation, 222
Coincidence detection, 609
Coincidence gate, 816
Collector, in transistors, 790
Colorimeter, 192, 194
    Duboscq, 181
Colorimetry, 150, 161
Colors of light, 163
Column
    capillary gas chromatographic, 740
    gas chromatographic, 740
    liquid chromatographic, 685
    microbore, 686
    open-tubular gas chromatographic, 738
    packed gas chromatographic, 743
    surface-coated open tubular, 741
    tandem capillary, 762
    wall-coated open tubular, 741
Column back pressure, equation for, 682
Column efficiency, 666
Column packing, gas chromatographic, 743
Column switching, gas chromatographic, 873
Comparator gate, 816
Comparators, 823
Compiler, 843
Complementary metal-oxide semiconductor, 809
Compound nucleus, 616
Compressibility factor, gas chromatographic, 729
Compton continuum, 610
Compton edge, 610
Compton radiation, 425
Compton scattering, 604
    in x-ray spectrometry, 436
Computer terminology, 841
Computers, 766
Concentration gradient, 8
Concentration polarization, 7
Concentration-response detectors, 762
Conductance
    bipolar square-wave pulse, 131
    for chromatography detectors, 139
    detector, 717
    electrical, 777
    electrodeless, 876
    equivalent, 123
    limiting ionic equivalent, 126
    process control, 875
    solubility determination by, 138
    specific, 123
Conductance cells, 128
Conductance detector, 129
Conductance titrations
    acid-base, 132
    complexation, 137
    nonaqueous, 138
    precipitation, 137
Conductivity, thermal, 748
Conductivity detector, 129, 717
Constant, nuclear decay, 596
Constant-current coulometric titrations, 98
    automatic, 105
    primary, 99
    secondary, 100
Constants
    Rohrschneider, 743
    table of physical, 900
Constructive interference, 232
Contact shift, 384
Continuous extraction, 644
Continuous flow instruments, 882
Continuous flow methods, 565
Continuous instruments, 854
Continuous variation method, 186
Continuous wave laser, 225
Continuous wave NMR spectroscopy, 362
Continuum x-rays, 416
Control bus, 841
Control loop, 860
Controlled-potential coulometry, 105
    determination of n values, 107
    in flowing stream, 109
    predictive, 108
    spectro-, 113
    in stirred solutions, 106
    thin-layer cell, 112
Controllers, 867
Convection, 8
Conversion factors, 898
    spectroscopic, 146
Converter
    analog-to-digital, 835
    current-to-frequency, 833
    digital-to-analog, 835
    integrating, 835
    voltage-to-frequency, 834
    voltage-to-frequency-to-digital, 835
Coolidge x-ray tube, 416
Corrected fluorescence spectra, 261
Correlation chart, 214, 238
    infrared spectroscopic, 238
Cottrell equation, 9, 78
Coulombic barrier, 618
Coulometric analysis, 97
Coulometric titrations. *See* Constant-current coulometric titrations
Coulometry, controlled potential. *See* Controlled-potential coulometry
Counter electrode, 3
Counter-current distributions, 644
Counter-current movement, 640

Counters, 819
Counting gate, 828
Coupling constant, 373
   correlation table, 379
   sign of in NMR, 379
CPAA. *See* Charged particle activation analysis
CPU, 840
Cross section, nuclear reaction, 618
Crown compounds, 37
Crystal oscillator, 829
Curie, 637
Curium-242, 425
Current, 772
   adsorption, 59
   average polarographic, 57
   capacitive, 55, 77
   catalytic, 58
   charging, 5, 55, 57
   diffusion, 57
   eddy, 340
   faradaic, 9, 56
   intrinsic diode, 786
   kinetic, 58
   limiting, 56, 100
   maximum polarographic, 57
   NMR ring, 368
   op-amp bias, 805
   op-amp offset, 805
   residual, 55
Current efficiency, 97
Current follower, 797
Current gain, for transistors, 790
Current-to-frequency converter, 833
Czerny-Turner mounting, 326

D

DAC. *See* Digital-to-analog converter
Daniel chromatograph, 873
Data acquisition, in kinetic methods, 583
Data bus, 841
Data-domain conversions, 833
Data domains, 767
   electrical, 767
   nonelectrical, 767
Data latch, 817
Data processing, 846
   in clinical chemistry, 885
Dating
   by radionuclide activity, 597
   rocks by mass spectrometry, 508
Daughter nuclide, 595
DC arc, 332
DCP. *See* Direct current plasma jet

DC polarography. *See* Polarography
DCU, 821, 829
Dead band, 854
Dead-stop titration, 92
Dead time, 429, 858, 860, 868
   of detectors, 607, 625
Debye-Hückel theory, 16
Decade counter, 821, 829
Decade frequency divider, 831
Decade voltage divider, 775
Decay curve, for radioactive isotopes, 599
Dedicated analyzers, 892
Dedicated instruments, 885
Degenerate transitions, in NMR, 376
Degrees of freedom, 213
Deionization, 724
Delta scale, in NMR, 366
DeMorgen's theorems, 814
   table of, 815
Densitometer, 344
Depolarization ratio, 218
Depolarized electrode, 6
Depolarizer, 6, 118
Derivative-action time, 866
Derivative approach, in kinetic methods, 569
Derivative spectroscopy, 196
Derivatization, in gas chromatography, 731, 751, 756
Desalting, 724
Destructive interference, 232
Detection limit
   in atomic absorption spectroscopy, 304
   in atomic fluorescence spectroscopy, 315
   in electron-probe microanalysis, 444
   in fluorescence and phosphorescence spectroscopy, 263
   in instrumental neutron activation analysis, 625
   table of for atomic absorption, 282, 306
   table of for atomic fluorescence, 318
   table of for emission spectroscopy, 342
   table of for flame emission, 294
   x-ray fluorescence spectrometric, 432
Detection, in paper chromatography, 678
Detector dead-time, 607, 625
Detector photopeak efficiency, 621
Detector resolution, 622
Detector. *See also* Detectors
   absorbance, 691
   atomic absorption, 751
   atomic emission, 751
   barrier-layer, 193
   bolometer, 158, 229
   charge-coupled, 198
   coincidence, 609
   concentration-response, 762

Index   915

Detector (cont'd.)
  conductance, 129, 717
  diode array, 312, 691
  efficiency of gamma ray, 614
  electron-capture, 749
  electrooptic, 198
  flame photometric, 751
  flame-ionization, 748
  gas-ionization, 605
  Ge(Li), 611, 614
  Geiger counter, 428, 608
  Golay, 158, 229
  Hall conductivity, 129
  ion-chamber, 606
  liquid-scintillation, 609
  mass-response, 762
  microphone, 203
  multichannel, 199, 609
  NaI(Tl), 609, 611, 614, 622
  photoconductive, 158
  photoelectric, 158
  photographic plate, 158
  photomultiplier, 193, 581
  photon, 228
  phototube, 193, 581
  photovoltaic, 193, 582
  pneumatic, 158, 229
  proportional counter, 429, 607
  pyroelectric, 229
  resolution of gamma-ray, 614
  scintillation counter, 429, 608
  semiconductor, 429, 612
  Si(Li), 430
  solid-state, 612
  thermal, 228
  thermal conductivity, 747
  thermistor, 229
  thermocouple, 158, 229
  thermopile, 229
  Vidicon tube, 198, 312
Detectors. *See also* Detector
  design of gas chromatographic, 762
  electrochemical, 693
  emission spectroscopic in chromatography, 352
  ESCA, 463
  Fourier transform infrared spectroscopic, 230
  gas chromatographic, 747
  infrared spectroscopic, 228
  liquid chromatographic, 690
  mass spectrometric, 493
  nuclear spectroscopic, 604
  optical, 157
  table of liquid chromatographic, 690
  for ultraviolet-visible spectroscopy, 193
  x-ray spectrometric, 428

Deuterium, in NMR, 383
Deuterium arc lamp, 192, 301
Diamagnetism, 402
Diatomite, 740
Dichroism, 181
DIDA, 627
DIE. *See* Direct-injection enthalpimetry
Dielectric constant, 141
Diethyldithiocarbamate, 653
Difference amplifier, 793, 799
Differential-pulse polarography. *See* Pulse polarography
Differential reaction-rate methods, 574, 587
Differential scanning calorimetry, 523, 542, 545
Differential spectrophotometry, 206
Differential thermal analysis, 523, 532, 539, 541
Differentiator, op-amp, 802
Diffraction grating, 153
  order of, 154
  resolving power of, 154
  suppression of orders from, 155
Diffraction, 145
  x-ray, 426
Diffuse reflectance, 201
Diffusion, 8
  eddy, 668, 735
  longitudinal, 669, 735
  molecular, 669, 735
  solute in gas chromatography, 735
Diffusion coefficient, 8, 57
Diffusion control, 63
Diffusion-controlled electrolysis, 105
Diffusion current, 57
  temperature dependence of, 65
  tests for, 60
Diffusion-current constant, 94
Diffusion layer, 9, 113
Digital control, 868
Digital domain, 771
Digital filtering, 847
Digital logic, 809
Digital multimeter, 779
Digital-to-analog converter, 835
Digital voltmeter, 777
Diluents, in DTA, 539
Diode, 785, 802
  fluid, 785
  germanium, 786
  light-emitting, 829
  silicon, 786
  zener, 787, 789
Diode array detector, 312
Dipole moment, 215
Direct current plasma jet, 336
Direct digital control, 868

Direct-injection enthalpimetry, 524, 546
   instrumentation, 547
Direct isotope dilution analysis, 627
Direct-line fluorescence, 316
Direct-reading spectrometer, 330
Discrete instruments, 854, 868, 882
Dispersion, 153
Displacement chromatography, 655
Distribution coefficient, 641
   chromatographic, 659
   gas chromatographic, 730
   ion-exchange chromatographic, 714
Distribution ratio, 641
Distribution
   gaussian, 632
   normal, 632
   Poisson, 633
Dithizone, 653
DME. See Dropping mercury electrode
DMM. See Digital multimeter
Domain conversion, 833
Dopant, 785
Double-layer capacitance, 77
Double-resonance spectra, 383
Drain, in transistors, 792
Driver, 843
Dropping mercury electrode, 52
   construction of, 65
DSC, 523, 542. See also Differential scanning calorimetry
   applications of, 546
   calculations in, 545
   instrumentation for, 543
DTA, 523, 532. See also Differential thermal analysis
   applications of, 539
   calculations in, 539
   instrumentation for, 534
Duboscq colorimeter, 181
Duty cycle, 828
DVM, 777, 835
Dye laser, 199, 226

E

$E_{1/2}$, 56
ECD, 749
Echelle grating, 326
ECL. See Emitter-coupled logic
Eddy current, 340
Eddy diffusion, 668, 735
Effective plates, 738
Efficiency, chromatographic, 671, 683
Einstein coefficient, 222
Einstein-Planck law, 146
Einstein's equation, 603

Elastic neutron scattering, 616
Elastic scattering, 215
Electric quadrupole, 359
Electrical discharge lamp, 192
Electrical double layer, 5
Electrical gradient, 8
Electroactive species, 70
Electrochemical cell, 14
Electrochemical detectors, 129, 717
   in chromatography, 693
Electrochemical methods
   advantages of, 2
   classification of, 3
Electrochemical reversibility, 63
Electrochemical techniques, table of, 4
Electrodeless conductance, 139, 876
Electrodeless discharge lamp, 299
Electrode potentials
   effect of complexation on, 17
   formal, 16, 116
   standard, 15
   table of standard, 901
Electrodes
   antimony, 29
   capacitance of, 6
   counter, 3
   depolarized, 6
   dropping mercury, 52
   emission spectroscopic, 342
   enzyme, 38
   fluoride, 32
   gas-sensing, 38
   glass, 31
   glassy carbon, 695
   graphite, 305, 332
   hydrogen gas, 21
   indicator, 3
   liquid membrane, 35
   mercury pool, 106, 117
   micro, 39
   minigrid, 116
   nonpolarizable, 7
   optically transparent, 116
   oxygen, 76
   pH glass, 27
   platinum gauze, 106
   polarized, 6
   pressed pellet, 32
   quinhydrone, 28
   redox, 20
   reference, 3
   reticulated vitreous carbon, 110
   silver/silver chloride, 22
   silver sulfide, 34
   solid state, 32
   Thalamid, 24
   thallium amalgam/thallous chloride, 24

Electrodes (cont'd.)
  tin dioxide, 116
  working, 3
Electrode sign convention, 7
Electrogravimetry
  controlled-current, 117
  controlled-potential, 116
Electrolysis, diffusion-controlled, 105
Electrolytic cell, 14
Electromagnetic radiation, 145
Electromagnetic spectrum, 147
Electron capture, in radioactivity, 595, 602
Electron-capture detector, 749
Electron gun, 480
Electronic spectra, 161
Electronic transitions, 163, 167
Electronics, 766
Electron impact source, 480, 874
Electron multiplier, 493
Electron-probe microanalysis, 443
Electron spectrometer, 462
Electron spectroscopy for chemical analysis. See ESCA
Electron spin resonance spectroscopy. See ESR
Electrooptic detector, 198
Electrophoresis, 722
  paper, 659
  thin-layer, 659
Electrophoretic effect, 125
Electrostatic analyzer, mass spectrometric, 485
Electrothermal atomization, 304
Elementary particles, 595
Eluotropic solvent series, 699
Elution, 655
  gradient, 701
  isocratic, 699
Elution chromatography, 655, 660
Elution time, 729
Elution volume, 729
Emission intensity, general equation for, 324
Emission spectrometer, 326
Emission spectroscopy, 322
  accuracy and precision of, 348
  applications of, 349
  in chromatography, 352
  detection limits in, 342
  instrumentation for, 325
  for qualitative analysis, 343
Emission spectrum, 254
Emitter, in transistors, 790
Emitter-coupled logic, 809
EMPA. See Electron-probe microanalysis
Emulsions, 647
ENAA. See Epithermal neutron activation analysis

Enable input, 817
Enantiomers, 731, 758
Endoergic reactions, 600, 618
Endothermic reactions, 532, 545
Energy dispersion, x-ray, 426
Enthalpogram, 551
Environmental noise, 846
Enzymatic methods, 572
Enzyme activity, 574
Enzyme assays, 887
Enzyme electrodes, 38
Enzymes, 724
Enzyme unit, 574
Epithermal neutron activation analysis, 615
EPMA. See Electron-probe microanalysis
Equality gate, 815
Equation. See also Law
  Beer's, 300, 302
  Boltzmann, 363, 401
  Cottrell, 9, 78
  Debye-Hückel, 16
  ESR hyperfine splitting, 404
  Heisenberg uncertainty, 401
  Ilkovic, 57
  Karplus, 380
  Larmor, 360
  Maxwell-Boltzmann, 292
  McConnell, 406
  Nernst, 7, 15, 98
  Onsager, 126
  Randles-Sevcik, 83
  Saha, 323
  Shockley, 786, 802
  Stern-Volmer, 273
  total angular momentum, 396
  Van Deemter, 670, 735
Equilibrium constants, determination by thermometric titration, 553
Equivalent circuit, 774
Equivalent conductance, 123
  limiting ionic, 126
Error, 855
  propagation of, 635
ESCA, 451
  determination of valence states by, 466
  instrumentation for, 460
  principles of, 451
  resolution of, 456, 463
  sampling systems for, 464
Escape depth, ESCA and Auger spectroscopic, 453
Escape peak, x-ray, 420
ESR, 159, 395
  instrumentation for, 399
  sensitivity of, 400
  uses of, 395
Exact mass measurements, 487

Excess noise, 846
Exchange capacity, 713
Excitation sources, 332
Excitation spectrum, 254
Exclusion limit, in size-exclusion chromatography, 719
Exclusive OR gate, 815
Exoergic reactions, 600
Exothermic reactions, 532, 545
Extraction. *See* Solvent extraction

F

FAB MS, 513
Faradaic current, 9, 56
Faradaic processes, 4, 6
Faraday's law, 96
   for flowing streams, 109
Fast-atom bombardment mass spectrometry, 513
Fastie-Ebert mounting, 326
Fast neutron activation analysis, 615
Fast neutrons, 624
Fast reactions, 563, 565
Feedback, in operational amplifiers, 796
Feedback control, 860
Feedforward control, 860
Fellgett advantage, 234
   in FTMS, 493
FEP. *See* Full-energy peak
FET, 792
FIA. *See* Flow-injection analysis
FID, 748
   in NMR, 362
Field desorption mass spectrometry, 513
Field-effect transistor, 792
Field ionization mass spectrometry, 509
Filter. *See also* Optical filter
   digital, 847
   electrical, 788
   high-pass, 784
   interference, 241, 258
   low-pass, 784
   narrow-bandpass, 258
   sharp-cut, 258
   transmission x-ray, 420
First-order decay, 596
First-order reactions, 569
First-order spectra
   rules for interpreting, 375
   in NMR, 374
Fission, 595, 619
Fixed-time procedures, in kinetic methods, 569
Flame emission spectrometry. *See* Atomic emission spectrometry

Flame photometry. *See* Atomic emission spectrometry
Flame-ionization detector, 748
Flames, types of for atomic spectroscopy, 280
Flicker noise, 846
Floppy disk, 843
Flow-injection analysis, coulometric, 112
Flowing streams, coulometry in, 109
Fluid diode, 785
Fluorescence, 250
   anti-Stokes, 316
   in chromatography, 268
   direct-line, 316
   of inorganic compounds, 265
   instrumentation for, 256
   kinetics of, 251
   linearity of, 253
   nonresonance, 316
   of organic compounds, 263
   sensitized, 317
   sources for, 256
   stepwise-line, 316
   thermally assisted, 316
   quenching of, 251, 272
Fluorescence detectors, in chromatography, 692
Fluorescence quenching, 251
   dynamic, 273
   static, 272
Fluorescence spectroscopy, 250
   multicomponent, 268
Fluorescent tracers, 271
Fluoride electrode, 32
Fluorimeter, 256
Fluors, 609
FNAA. *See* Fast neutron activation analysis
Forbidden transitions, 165
Fore-prism, 155
Formal electrode potential, 16, 116
Formation constant, 648
   polarographic determination of, 64
Forward bias, 785
Fourier transform mass spectrometry, 491
Fourier transform spectrometry, 155, 849
   advantages of, 234
   detector for infrared, 230
   in NMR, 362
Fragment ions, 480
   in mass spectrometry, 496
Franck-Condon principle, 166
Free-induction decay, 362
Fremy's radical, 395
Frequency, 145, 780
   decade divider, 831
Frequency-domain spectrum, 156, 231
Frequency factor, 564
Frequency measurement, 831

Frequency spectrum, 849
Frontal development, 655
FTMS. *See* Fourier transform mass spectrometry
Full-energy peak, 610
Full-wave rectifier, 789
Full width at half maximum, 611
Functional group analysis, 243
Fundamental noise, 846
Fundamental physical constants, table of, 900

G

g-value, 396
Gain, 862
  circuit, 799
  open-loop, 795, 804
  power, 784
  signal, 799
  transistor current, 790
Gain-bandwidth product, 804
Galvanic cell, 14
Galvanic-cell reactions, 18
Gamma-ray detectors
  efficiency of, 614
  resolution of, 614
Gamma rays, 595, 600, 601, 603
Gamma-ray spectrometry, 604
Gas chromatographic stationary phases, 743
Gas chromatographic supports, 741
Gas chromatography
  applications of, 754
  capillary, 740
  columns for, 740
  detectors for, 747
  dynamics of, 734
  instrumentation for, 739
  in process control, 871
  qualitative, 751
  quantitative, 753
  thermodynamics of, 729
Gas chromatography-mass spectrometry. *See* GCMS
Gas-ionization detector, 605
Gas-sensing electrodes, 38
Gate
  AND, 810
  coincidence, 816
  comparator, 816
  equality, 815
  exclusive OR, 815
  inverter, 812
  NAND, 812
  NOR, 812
  OR, 812
  in transistors, 792
  XOR, 815
Gated latch, 817
Gauss, 359
Gaussian distribution, 632
Gaussian profile, 666
GCMS, 479, 491, 506
Geiger counter, 428, 608
Gel electrophoresis, 723
Gel-filtration chromatography, 663
Ge(Li) detector, 611, 614
Gel-permeation chromatography, 663
General differential rate expressions, 568
General elution problem, 701
Germanium diode, 787
Gibbs-Helmholtz law, 524
Gibbs phase rule, 639
Glass pH electrodes. *See* pH electrodes
Glass transitions, in polymers, 546
Glassy carbon electrode, 695
Globar, 228
Golay detector, 229
Goniometer, 414
Gradient elution, 655, 701
Gran's method, 43
Graphite electrode, 304, 332
Grating. *See also* Diffraction grating
  echelle, 326
Grotrian diagram, 291
Ground state, 248
Group frequencies, 213
Guard columns, 684
Gyromagnetic ratio, 359

H

H-value, 667, 734
Half-life, 578
  of radionuclides, 597
Half-wave potential, 56
Hall electrolytic conductivity detector, 129
Hard disk, 843
Hardware, computer, 843
Heat capacity, 538, 539, 544, 546, 551
  measurement of, 541
Heat-leak modulus, 548
Heat of reaction, 529, 554
HECD. *See* Hall electrolytic conductivity detector
Height equivalent to a theoretical plate, 667, 734
Heisenberg uncertainty principle, 401
Helmholtz coils, 399
HETP, 667, 734
High-pass filter, 784

High-performance liquid chromatography, 659, 681
High-speed liquid chromatography. *See* High-performance liquid chromatography
Hildebrand solubility parameter, 647
  in chromatography, 707
Hollow cathode lamp, 297, 751
HPLC. *See* High-performance liquid chromatography
HSLC. *See* High-performance liquid chromatography
Hydrogen lamp, 192
Hyperchromic effect, 170
Hyperfine coupling, 402
Hyperfine coupling constant, 406
Hypochromic effect, 170
Hypsochromic shift, 170

I

IC. *See* Ion chromatography
ICP. *See* Inductively coupled plasma discharge
ICR. *See* Ion cyclotron resonance
IFC. *See* Current-to-frequency converter
IIDA. *See* Inverse isotope dilution analysis
Ilkovic equation, 57
Immunoassay, 630, 889
Impedance, 782
INAA. *See* Instrumental neutron activation analysis
Incandescent lamp, 192
Incoherent radiation, 222
Indicator electrode, 3
Indicator reaction, 561
Inductively coupled plasma discharge, 339
Inductive reactance, 783
Inelastic scattering, 217
  of neutrons, 616
Influence coefficient, in x-ray fluorescence spectrometry, 437
Infrared analysis, in process control, 871
Infrared region, 215
Infrared spectrometer
  dispersive, 230
  Fourier transform, 231
Infrared spectroscopy
  cells for, 234
  correlation chart for, 238
  detectors for, 228
  instrumentation for, 227
  sources for, 228
Initial-rate methods, 568
Injector. *See also* Inlet
  liquid chromatographic sample, 688
  sample, 688, 745

Inlet
  flash evaporation, 746
  gas chromatographic, 745
  splitter, 746
Inner-filter effect, 272
INSECT. *See* Yates
Instrumental neutron activation analysis, 615
  theory of, 620
Instrumentation
  for DSC, 543
  for DTA, 534
  for fluorescence, 256
  for phosphorescence, 256
  for TT and DIE, 547
Integral approach, in kinetic methods, 569
Integrated-circuit analog switch, 821
Integrated circuits, 810
Integrating converter, 835
Integrator, op-amp, 800
Interdomain conversions, 767
Interface, 843
Interference filters, 241, 258
Interferences, in neutron activation analysis, 622
Interferogram, 156, 233
Interferometer, 231
Internal conversion, 250, 601
Internal reference method. *See* Internal standard method
Internal standard method. *See also* Pilot ion method
  in emission spectroscopy, 347
  in flame emission spectrometry, 294
  in NMR analysis, 366
  in x-ray fluorescence spectrometry, 435
International system of units, tables of, 897
International unit, 574, 888
Interstitial volume, 665
Intersystem crossing, 250
Intrinsic current, of diodes, 786
Inverse isotope dilution analysis, 629
Inverter gate, 812
Inverting amplifier, 798
Iodine-125, 631, 889
Iodine-131, 631
Ion association, in solvent extraction, 643
Ion-association extraction systems, 652
Ion chambers, 606
Ion chromatography, 715
Ion cyclotron resonance spectrometry, 491
Ion gun, 482
Ion-exchange chromatography, 663, 712
  uses of, 714
Ion exchangers, 713
Ionic strength, 16
Ionization buffer, 289

Ionization interference, 288
Ionization potential, by mass spectrometry, 509
Ion microprobe, 516
Ion-pair chromatography, 710
Ion-pair formation, 606
Ion-probe analysis, 515
Ion scattering spectrometry, 516
Ion selective electrodes
    advantages and disadvantages of, 44
    errors of, 30
    glass, 31
    interferences in, 30
    liquid membrane, 35
    micro, 39
    pressed pellet, 32
    in process control, 876
    properties of, 30
    quantification of, 42
    solid state, 32
Ion sputtering, 515
    of surfaces, 469
IR drop, 774
Irradiation time, 624
Irreversible electrode behavior, 7
Isoabsorptive wavelength. See Isosbestic point
Isocratic elution, 699
Isoelectric point, 723
Isoionic point, 723
Isomation, 206
Isosbestic point, 180, 185
Isotope dilution analysis, 627
Isotope-ratio mass spectrometry, 508
Isotopes, 495, 595
    radioactive, 595
    stable, 595
Isotopic abundance, 495, 595
    table of for selected elements, 495
Isotopic labeling, 508
ISS. See Ion scattering spectrometry
IU, 574, 888

J

Jablonski energy-level diagram, 249
JFET, 792
J-K flip-flop, 818
Job's method, 186
    for equilibrium constants, 190
    general equations for, 190
Johnson noise, 581, 846
Junction field-effect transistor, 792

K

Karplus equation, 380
KBr-pellet technique, 235
K electron capture, 425
Kinetic current, 58
    tests for, 60
Kinetic energy
    of Auger electrons, 460
    of ions, 483
Kinetic methods. See also Catalytic methods; Enzymatic methods; Reaction rates
    automated, 583
    differential reaction-rate, 574
    instrumentation requirements of, 580
    mathematical basis of, 567
    table of classification of, 562
    temperature control in, 582
    types of, 561
Kinetics, of extraction, 652
Kirchhoff's current law, 776
Kirchhoff's voltage law, 774, 791
Klystron, 399
Kovats index, 752

L

Lambert's law, 174
Laminar flow, 565
Lanthanide-induced shift, 384
Larmor equation, 360
Laser action, principle of, 221
Laser microprobe, 345
Laser pumping, 223
Lasers, 220
    in atomic fluorescence spectroscopy, 317
    carbon dioxide, 513
    dye, 226
    gas, 224
    ionic-crystal, 224
    molecular, 226
    Nd-YAG, 224, 513
    pumping schemes for, 223
    ruby, 224
    tunable, 199
    YAG, 224, 513
Latches, 816
Law. See also Equation
    Arrhenius, 563
    Beer's, 114, 174, 250, 445
    Bragg's, 426
    Einstein's, 603
    Einstein-Planck, 146

Faraday's, 96, 109
Gibbs phase, 639
Gibbs-Helmholtz, 524
Kirchhoff's current, 776
Kirchhoff's voltage, 774, 791
Lambert, 174
Mosley's, 416
Nernst distribution, 641
Newton's cooling, 536, 548
Ohm's, 123, 131, 773
Snell's, 152
Stokes, 255
LC. See Liquid chromatography
LCD, 777
LCEC, 695
LCMS, 695
Least significant bit, 837
Least-squares polynomial smoothing, 848
LED, 829
LEED, 463, 468
Left-hand rule, 358, 367
Light
  colors of, 163
  polarized, 253
  speed of, 145
  stray, 179
Light-emitting diode, 829
Light-source stabilization, 580
Limiting current, 56, 100
Limiting ionic equivalent conductance, 126
Linear absorption coefficient, in x-ray fluorescence spectrometry, 434, 440
Linear scan voltammetry, 81
  detection limit in, 83
Liquid chromatography
  columns for, 685
  detectors for, 690
  principles of, 659
  pumps for, 686
  and solvent extraction, 655
Liquid chromatography-mass spectrometry, 695
Liquid-crystal display, 777
Liquid crystals, 385
Liquid-junction potential, 18
Liquid-liquid chromatography, 658, 706. See also Partition chromatography
Liquid scintillation detector, 609
Liquid-solid chromatography, 661, 698
LIS. See Lanthanide-induced shift
LLC. See Liquid-liquid chromatography
Logarithmic amplifier, op-amp, 802
Logarithmic extrapolation method, 576
Longitudinal diffusion, 669, 735
Longitudinal relaxation, 364

Loop injector, liquid chromatographic, 688
Low-energy electron diffraction, 463, 468
Low-pass filter, 784
LSB, 837
LSC. See Liquid-solid chromatography
Luminescence, 247
L'vov platform, 308

M

Machine code, 843
Machine language, 843
Magnetically anisotropic group, 368
Magnetically equivalent nuclei, 371
Magnetically isotropic group, 369
Magnetic moment, 358, 396
Magnetic resonance, 360
Magnetic sector mass spectrometer, 483
Magnetogyric ratio, 359
Manual instruments, 882
Masking, 183
Masking agents, 650
Mass, measurement of exact, 487
Mass absorption coefficient, 145
  in XRF, 442
Mass filter, 490
Mass number, 595, 600
Mass-response detectors, 762
Mass spectrometers
  detectors for, 493
  double-focusing, 485
  high-resolution, 485
  inlets for, 477
  instrumentation for, 477
  Mattauch-Herzog, 482, 487
  Nier-Johnson, 487
  quadrupole, 490
  single focusing, 485
  spark source, 482
  time-of-flight, 489
Mass spectrometry
  chemical ionization, 509
  compound identification by, 505
  detection in gas chromatography, 751
  electron-impact, 505
  fast-atom bombardment, 513
  field desorption, 513
  field ionization, 509
  Fourier transform, 491
  isotope-ratio, 508
  in liquid chromatography, 695
  of metastable ions, 504
  mixture analysis by, 506
  negative ion, 504

Mass spectrometry (cont'd.)
  plasma desorption, 513
  in process control, 874
  secondary-ion, 515
  spark-source, 514
  tandem, 517
  thermal desorption, 513
  thermospray ionization source, 695
Mass spectrometry-mass spectrometry, 517
Mass spectrum, interpretation of, 494
Mass transfer, 8, 105
  in gas chromatography, 736
  of solutes, 661, 668, 736
Mass transport, 8
Matrix effect, in x-ray fluorescence spectrometry, 433
Matrix modifier, 308
Mattauch-Herzog mass spectrometer, 482, 487
Maxima, polarographic, 59
Maximum polarographic current, 57
Maxwell-Boltzmann equation, 292
McConnell relation, 406
McLafferty rearrangement, 502
McReynolds constants, 743
Medium-scale integrated circuit, 821
Mercury arc lamp, 192, 257, 691
Mercury pool electrode, 106, 117
Mercury-vapor lamp. See Mercury arc lamp
Metastable ions, in mass spectrometry, 504
Methane time, 739
Method of continuous variations, 186
Method of proportional equations, 578, 587
Michaelis constant, 572
Michaelis-Menten mechanism, 572
Michelson interferometer, 231
Microbore columns, 686
Microcomputers, 840
  input and output, 844
Microphone detector, 203
Microphotometer, 332
Microwave spectroscopy, 159
Microwave-induced plasma, 352
Migration, electrochemical, 8, 722
Minigrid electrode, 116
Mixing jet, in kinetic methods, 566
Mixing methods, 565
Modem, 843
Moderators, nuclear, 619
Molar absorptivity, 168, 175. See also Absorptivity
Molar-ratio method, 186
Molecular absorption, 301
Molecular diffusion, 669, 735
Molecular ion, 480
  determination of composition of, 494
Molecular laser, 226
Molecular vibrations, 213

Molecular weight, determination by spectrophotometry, 191
Monitor, computer, 843
Monochromatic radiation, 146
Monochromators, 152, 194, 326
Monostable multivibrator, 826
Mosley's law, 416
Most-significant bit, 837
Moving-average smooth, 848
MS/MS, 517
MSB. See Most-significant bit
MSI. See Medium-scale integrated circuit
Multichannel analyzers, 891
  in radiochemistry, 609
Multichannel detectors, 199
Multichannel instruments, 883
Multichannel spectrometers, 348
Multiple ion monitoring, 506
Multiple-period averaging, 832
Multiplicity, 248
Multiply charged ions, in mass spectrometry, 504

N

NAA. See Neutron activation analysis
NaI(Tl) detector, 609, 611, 614, 622
NAND gate, 812
Narrow-bandpass filter, 258
NDIR. See Nondispersive infrared analyzers
Near-ultraviolet region, 192
Nebulization, of sample in atomic spectroscopy, 285
Nebulizer, Babington, 340
Negative feedback, 796
Negatrons, 596. See also Beta particle; Beta ray
  emission of, 601
Nematic phases, NMR spectra in, 385
Nernst distribution law, 641
Nernst equation, 7, 15, 98
Nernst glower, 228
Nessler tubes, 181
Net retention volume, 729
Neutrino, 602
Neutron activation analysis, 615
  interferences in, 622
  principles of, 616
Neutron capture reaction, 616
Neutron flux, 618, 622
Neutrons
  epithermal, 615
  fast, 624
  thermal, 614, 619, 624
Newton's law of cooling, 536, 548

Nickel-63, 749
Nier-Johnson mass spectrometer, 487
Nitrogen rule, in mass spectrometry, 494
Nitrogen ultraviolet region, 192
Nitrous oxide-acetylene flame, 280
NMR, 158, 357
  chemically equivalent nuclei in, 371
  conformational changes in, 373
  magnetically equivalent nuclei in, 371
  molecular weight determination by, 387
  proton exchange in, 372
  quantitative, 386
  sample preparation for, 388
  sample size for, 366
  sensitivity of, 366
  solvents for, 388
  spin system nomenclature in, 374
  theory of, 358
  time scale of, 371
NMR spectra
  carbon-13, 382
  deuterium, 383
  errors in analysis of, 378
  fluorine-19, 382
  high resolution, 364
  in nematic phases, 385
  phosphorus-31, 382
  of solids, 364
  wide-line, 364
NMR spectrometers
  commercial, 385
  field sweep, 360
  Fourier transform, 362
  frequency sweep, 360, 362
  high-field, 374
NOE. *See* Nuclear Overhauser effect
Noise
  $1/f$, 846
  electronic, 801, 846
  environmental, 846
  excess, 846
  flicker, 846
  fundamental, 846
  Johnson, 581, 846
  shot, 846
  thermal, 846
  white, 846
Nomenclature, spectroscopic, 145, 166
Nonaqueous titrations, conductance, 138
Nondispersive infrared analyzers, 871
Nonfaradaic processes, 5
Nonpolarizable electrode, 7
Nonpolarized light, 253
Nonresonance fluorescence, 316
NOR gate, 812
Normal distribution, 632
Normal-phase chromatography, 708

n-type semiconductor, 612
Nuclear accelerator, 620
Nuclear decay constant, 596
Nuclear fission, 595, 619
Nuclear isotopic sources, 619
Nuclear magnetic resonance spectroscopy. *See* NMR
Nuclear Overhauser effect, 385
Nuclear reactions, 600
Nuclear reactors, 619
Nucleons, 595
Nuclide, 595, 600
Nuclidic mass, 595
Nujol, 235

O

OA. *See* Operational amplifiers
Odd-electron molecules, 396
Offset, 863
Offset current, op-amp, 805
Offset voltage, op-amp, 805
Ohm's law, 123, 131, 773
Onium, 643
Onsager equation, 126
Op-amps. *See* Operational amplifiers
Open-loop gain, 795, 804
Open tubular columns, 738
Operating system, 843
Operational amplifiers. *See also* Amplifier
  characteristics of, 803
  comparator, 823
  current follower, 797
  difference, 799
  differentiator, 802
  integrator, 800
  inverting, 798
  logarithmic, 802
  principles of, 794
  summing, 798
  voltage follower, 796
Optical detectors, 157
Optical emission spectroscopy, 322
Optical filters, 155, 258. *See also* Filters
  restrahlen, 155
Optical isomers, 731, 758
Optically transparent electrode, 116
Optically transparent thin-layer electrode, 116
Optical resolution, 156
Optical resonator, 224
Optical spectrometer, 150
Optoacoustic effect, 202
Orders of grating, 154
  suppression of, 155
OR gate, 812

Oscillator, 130
 quartz-crystal, 829
Oscillometry, 139, 876
OTE. *See* Optically transparent electrode
OTTLE. *See* Optically transparent thin-layer electrode
Overall extraction constant, 649
Overpotential, 7
Overvoltage, 7
Oxine, 653
Oxygen, removal from solution, 56
Oxygen electrode, 76

P

PAA. *See* Photon activation analysis
Packed columns, 743
Packings
 gas chromatographic column, 740
 liquid chromatographic column, 682, 698, 713, 722
 pellicular, 684
Pair production, 604
Paper chromatography, 268, 659, 675
 quantitative, 681
Paper electrophoresis, 659
Parallel data transmission, 772
Parallel-fast analyzers, 884, 892
Parallel resistors, 776
Paramagnetic ions, 267
Parent ion, 480
Parent nuclide, 595
Partition chromatography, 658, 663, 706. *See also* Liquid-liquid chromatography
Partition factor, 736
PC. *See* Paper chromatography
Peak optical power, 226
Peak-to-peak voltage, 780
Pellicular packings, 684
Period, 780
Period measurement, 832
Peripheral device, 843
PGAA. *See* Prompt gamma activation analysis
PGC. *See* Process gas chromatograph
pH
 definition of, 24
 effect on color formation, 183
 operational definition of, 25
$pH_{1/2}$, 188
Phase angle, 782
Phase rule, 639
pH buffer solutions
 primary standard, 25
 secondary standard, 26
 uncertainty of, 26

pH electrodes
 acid error of glass, 27
 alkaline error of glass, 27
 antimony, 29
 glass, 27
 quinhydrone, 28
pH measurement
 accuracy of, 26
 in process control, 875
pH meters, 40
Phosphorescence, 251
 in chromatography, 268
 instrumentation for, 256
 kinetics of, 251
Phosphorimeter, 260
Phosphors, 270, 609
Photoacoustic spectroscopy, 202
Photoconductive cell, 158
Photodetector, 193
Photodiode array detector, 312
 in chromatography, 691
Photoelectric detector, 158
Photoelectric effect, in radiochemistry, 604
Photoelectron binding energy, 452
 table of, 457
Photoelectron spectroscopy, 451
Photographic emulsion, 330, 347
Photographic plate detector, 158
Photoluminescence, 247
 principles of, 248
Photometer, 192
Photometric errors, 184
Photomultiplier tubes, 193, 581
 spectral response of, 260
Photon activation analysis, 616
Photon counting, 206
Photon detectors, 228
Photons, 145
Phototube, 193, 581
Photovoltaic cell, 193, 582
pH stat, 45, 854
Physical constants, table of, 900
Pilot-ion method, in polarography, 69
PIXE. *See* Proton induced x-ray emission analysis
Plasma, 882
Plasma desorption mass spectrometry, 513
Plasmon, 453
Plate, theoretical, 667, 734
Plate height. *See* Height equivalent to a theoretical plate
Platinization, 128
Platinum gauze electrode, 106
Platinum resistance thermometer, 528
Pleochroism, 181
Pneumatic detector, 158, 229

pn junction, 785
Point-to-plane spark technique, 349
Poisson distribution, 633
Polarizability, 217
Polarization, 253
Polarized electrode, 6
Polarized light, 253, 311
Polarogram, 53
Polarograph, 65
Polarographic maxima, 59
Polarographic wave, 56
   shapes of, 62
Polarography, 53. *See also* Amperometry; Pulse polarography
   accuracy of, 72
   applications of, 70, 75
   detection limit for, 72
   interferences in, 73
   precision of, 72
   selectivity of, 74
Polonium-210, 425
Polychromatic radiation, effect of on Beer's law, 177
Population inversion, 222
Positive feedback, 824
Positive-ion excitation, in XRF, 442
Positrons, 595, 601
Postcolumn reaction
   in chromatography, 692
   in ion-exchange chromatography, 715
Potassium-40, 595, 614
Potential buffer, 118
Potentials, table of standard electrode, 901
Potentiometer, 775
Potentiometric titrations
   acid-base, 45
   complexometric, 48
   oxidation-reduction, 47
   precipitation, 48
Potentiometry, 13
   instrumentation for, 40
   in process control, 876
Potentiostat, 65
Power, 772, 773
   average optical, 226
   electrical, 544
   peak optical, 226
   radiant, 173
Power-density spectrum, 849
Power gain, 784
Power supplies, 787
Precision, 856
Precision spectrophotometry, 206
Precision time base, 829
Preconcentration, x-ray spectrometric, 432
Pretitration, 102

Primary ion-pairs, 606
Prisms, 152
   fore-, 155
Process capacitance, 862
Process control
   automated, 860
   automated instruments for, 870
   and conductance analyzers, 875
   and gas chromatographs, 871
   and infrared analyzers, 871
   and mass spectrometers, 874
   and potentiometric analyzers, 876
Process gas chromatograph, 872
Process resistance, 862
Prompt gamma activation analysis, 616
Propagation of error, 635
Proportional band, 862
Proportional control, 862
Proportional counter, 429, 607
Proportional-plus-derivative control, 864
Proportional-plus-integral control, 863
Protogenic solvents, 46
Proton induced x-ray emission analysis, 616
Proton magnetic resonance, 365. *See also* NMR
Protophilic solvents, 47
Pseudo-contact shift, 384
Pseudo-first-order reactions, 569, 576
Pseudo-zero-order reactions, 568, 576
p-type semiconductor, 612
Pulse height, in nuclear detectors, 605
Pulse NMR, 156
Pulse polarography, 77
   applications of, 86
   detection limit in, 72, 81
   differential, 80
   instrumentation for, 86
Pump, liquid chromatographic, 686
Pumping schemes, for lasers, 223
Pyroelectric detector, 230

Q

Quadrupole mass spectrometer, 490
   resolution of, 491
Quantum counter, in fluorescence spectroscopy, 262
Quantum theory, for spectroscopy, 149
Quartz-halogen lamp, 192
Quenching
   dynamic, 273
   of fluorescence, 251
   static, 272

R

Radiant power, 173
Radioactive dating, 508, 597, 637
Radioactive decay, types of, 600
Radioactive isotopes, 595
Radioactive sources, for x-ray fluorescence, 425
Radioactivity, 594
Radiochemical analysis, statistics of, 632
Radiochemical neutron activation analysis, 615
Radioimmunoassay, 630, 889
Radioisotopic sources, for x-ray fluorescence, 425
Radioluminescence, 247
Radiorelease methods, 630
RAM, 843
Raman effect, 215
Raman spectroscopy, 215
   instrumentation for, 220
   sampling devices for, 226
   sources for, 220
Randles-Sevcik equation, 83
Random-access analyzers, 883, 891
Random access memory, 843
Range, 856
Rapid-scanning spectroscopy
   dispersive, 197
   multiplex, 198
Rate constant, 563
Rate expressions, general differential, 568
Rayleigh scattering, 215, 217
RC time constant, 12, 77, 826, 860
Reactance
   capacitive, 782
   inductive, 783
Reaction
   endoergic, 600
   exoergic, 600
Reaction cross section, 618
Reaction rates, *See also* Kinetic methods
   adjustment of, 563
   concentration effects, 564
   effect of temperature on, 563
   fast, 563
   ionic strength effects, 564
   measurement of, 563
   slow, 563
   solvent effects, 564
Reactors, nuclear, 619
Read only memory, 843
Rectifier, 787
Red shift, 180
Redox electrodes, 20
Reduced mass, 213

Reference electrodes, 3, 21
   calomel, 21
   hydrogen gas, 21
   silver/silver chloride, 22
   thallium amalgam/thallous chloride, 24
Reflectance
   diffuse, 201
   spectrometer, 200
   specular, 200
Refraction, 153
Refractive index, 153, 177
Registers, 816, 841
Regulator, voltage, 788
Relative retention, 671
Relaxation time, in NMR, 363
Reproducibility, 856
Reset time, 864
Residual current, 55
   correction for, 67
Resistance
   parallel, 776
   process, 862
   series, 774
   specific, 123
Resistivity, 772
Resolution, 671
   chromatographic, 670, 673, 736
   in ESCA and SAM, 472
   ESCA spectral, 456, 463
   in FTMS, 493
   in mass spectrometry, 485
   of nuclear detectors, 611, 622
   optical, 156
   of quadrupole mass spectrometers, 491
   in size-exclusion chromatography, 720
   of TOF mass spectrometers, 489
Resolution factor, 736
Resolving power, 154
Resonance, NMR, 360
Resonance fluorescence, 316
Resonance frequency, NMR, 360
Resonance lines, 291
Resonance Raman spectroscopy, 219
Resonance transitions, 291
Response time, 858
Restrahlen filter, 155
Retention index, 753
Retention time, 665
Retention volume, 664
Reticulated vitreous carbon, 110
Reverse bias, 785
Reverse-phase chromatography, 663, 708
Reversibility, electrochemical, 63
Reversible electrode behavior, 7
R_f value, 666, 678
Right-hand rule, 358, 367

Ring current, 368
Ripple counter, 821
Ripple factor, 789
Rise time, 804, 858
 instrumental, 582
rms, 780
RNAA. See Radiochemical neutron activation analysis
Rohrschneider constants, 743
ROM, 843
Root mean square voltage, 780
Rotating can phosphorimeter, 260
Rotational transitions, 163, 213
Rowland circle, 326
RS232-C, 844
Russell-Saunders coupling, 292
RVC. See Reticulated vitreous carbon
R_x value, 678

S

Saha equation, 323
Salting-out agents, 652
SAM. See Scanning Auger microprobe
Sample-and-hold amplifier, 822
Sample capacity, 674
Saturation,
 in ESR, 401
 in NMR, 364
Scanning Auger microprobe, 469
Scanning electron microscopy, 443
Scattering
 Compton, 604
 elastic, 214
 elastic neutron, 616
 inelastic, 217
 inelastic neutron, 616
 Rayleigh, 215, 217
 Tyndall, 215
 x-ray, 445
Schmitt trigger, 824
Scintillation detector, 429, 608
SCOT columns, 741
Secondary absorption lines, atomic spectroscopic, 299
Secondary electrons, 602
Secondary fluorescence effect, 439
Secondary-ion mass spectrometry, 515
Secondary ion-pairs, 606
Secondary x-ray emission, 414
Secondary x-ray target, 419
Second-order reactions, 576
Second-order spectra, NMR, 374
Selection rules, 165
 for infrared absorption, 214
 for Raman activity, 217
Selectivity, 856
 electron-capture detector, 750
 ion-exchange chromatographic, 714
 polarographic, 74
Selectivity coefficient, 31
Selectivity factor, 671, 731
Semiautomatic instruments, 882
Semiconductor detector, 612
 x-ray spectrometric, 429
Semiconductors
 n-type, 612, 785
 p-type, 612, 785
Sensitivity. See also Detection limit
 atomic absorption, 304
 instrumental, 854
 polarographic, 72, 81
Sensitized fluorescence, 317
Separation, baseline in gas chromatography, 737
Separation factor, 731
Separations. See also Chromatography
 in x-ray spectrometry, 432
Serial data transmission, 771
Series RC circuit, 783, 825
Series resistors, 774
Serum, 882
Set point, 860
Set-point control, 868
Sharp-cut filter, 258
Shielding constant, NMR, 367
Shift reagents, 384
Shockley equation, 786, 802
Shot noise, 846
Signal averaging, 847
Signal gain, 799
Signal-to-noise ratio, 846, 847, 851, 855
Sign conventions, in potentiometry, 7
Silanization
 of gas chromatographic supports, 742
 of liquid-chromatographic supports, 708
Silicon diode, 786
Silver/silver chloride electrode, 22
Silver sulfide electrode, 34
SIMPLEX factor analysis, 673
SIMS, 515
Single-channel analyzers, 884
Singlet state, 248
SI units, tables of, 897
Size-exclusion chromatography, 663, 718
 uses of, 720
Slew rate, op-amp, 805
Slit width. See also Bandpass; Bandwidth
 spectral, 178
Slow reactions, measurement of, 563
SMA, 882, 891

SMAC, 891
Snell's law, 152
SNR. *See* Signal-to-noise ratio
Software, 844
Solid state detector, 612
  x-ray spectrometric, 429
Solvent effects
  in NMR spectroscopy, 370, 388
  spectrometric, 172
Solvent extraction
  backwashing, 647
  chelate systems, 648
  choice of solvent for, 646
  fraction extracted in, 642
  ion-association systems, 652
  kinetics of, 652
  and liquid chromatography, 655
  principles of, 640
  stripping, 647
  systems for, 642
Solvent programming, 701
Solvent-strength parameter, 700
Source. *See also* Sources
  blackbody, 192
  cadmium arc, 691
  DC arc, 332
  deuterium arc, 192, 301, 691
  direct current plasma, 336
  electrical discharge, 192
  electrodeless discharge, 299
  electron impact, 480, 874
  for ESCA, 461
  globar, 228
  helium resonance, 462
  Hg arc, 192, 257, 691
  hollow cathode, 297, 751
  hydrogen arc, 192
  incandescent, 192
  inductively coupled plasma, 339
  klystron, 399
  mercury-arc, 192, 257, 691
  microwave-induced plasma, 352
  Nernst glower, 228
  nuclear isotopic, 619
  quartz-halogen, 192
  radioactive for x-ray fluorescence, 423
  radioisotopic for x-ray fluorescence, 423
  spark discharge, 336, 482
  thermospray ionization, 695
  in transistors, 792
  tungsten filament, 192
  for UPS, 462
  xenon-arc, 256
  x-ray tube, 416
  zinc arc, 691
Sources. *See also* Source
  for atomic fluorescence spectroscopy, 317
  of electromagnetic radiation, 150
  excitation, 332
  for fluorescence spectroscopy, 256
  for infrared spectroscopy, 228
  optical, 256
  for ultraviolet-visible spectroscopy, 192
Span, 856
Spark discharge, 336
Spark source, 482
Spark-source mass spectrometry, 482, 514
Specific activity
  of enzymes, 574
  in radiochemistry, 627
Specific conductance, 123
Specific heat, 546
Specific resistance, 123
Specific retention volume, 730
Specificity, 879
Spectra. *See* Spectrum
Spectral bandwidth, 178
Spectral slit width, 157, 178
Spectrochemical buffer, 334
Spectrocoulometric titrations, 113
Spectroelectrochemistry, 116
Spectrograph, 150, 326
  recording of spectrum with, 330
Spectrometer, 326. *See also* Spectrograph; Spectrophotometer
  direct-reading, 330
  double beam, 194
  electron, 462
  multichannel, 348
  optical, 150
  single beam, 194
Spectrometry, 144. *See also* Spectroscopy; Spectrum
  nomenclature of, 166
  solvent effects in, 172
  steric effects in, 173
  x-ray absorption, 445
Spectrophotometer, 150, 192
  components of, 192
  dual wavelength, 195
Spectrophotometry
  differential, 206
  microprocessor-controlled, 204
  of mixtures, 184
  photon counting in, 206
  precision, 206
Spectroscopic conversion factors, table of, 146
Spectroscopic splitting factor, 396
Spectroscopic states, 248
Spectroscopy, 144
  derivative, 196
  dual-wavelength, 195
  electron spin resonance, 159
  Fourier transform, 155

microwave, 159
nomenclature for, 145, 166
photoacoustic, 202
and quantum theory, 149, 161
rapid-scanning, 197
x-ray, 159
Spectrotitrimeter, 188
Spectrum
of air-acetylene flame, 284
Auger, 457, 469
benzene absorption, 179
beta ray energy, 602
corrected fluorescence, 261
diffuse reflectance, 210
electronic, 161
emission, 331
flame emission, 295
fluorescence emission, 254, 269
fluorescence excitation, 254, 269
frequency, 849
frequency-domain, 156, 231
gamma ray, 610, 613
mass, 497, 511
of mercury arc lamp, 257
of nitrous oxide-acetylene flame, 284
power-density, 849
resonance Raman, 219
second derivative absorption, 197
time-domain, 156, 231
total luminescence, 268
of xenon arc lamp, 257
Spectrum stripping, in x-ray spectrometry, 430
Specular reflectance, 200
Speed of light, 145
Spin decoupling, 383
Spin degeneracy, 398
Spin flip, 360
Spin labels, 409
Spin-lattice relaxation, 364
Spin quantum number, 248, 358
Spin-spin coupling, 373
structure effects in, 379
Spin-spin multiplets, 374
Spin-spin relaxation, 363
Spontaneous emission, 222
SSD. *See* Solid state detector
SSMS. *See* Spark-source mass spectrometry
Staircase ADC, 837
Stallwood jet, 335
Standard addition method
with ion-selective electrodes, 42
in polarography, 68
in x-ray fluorescence spectrometry, 436
Standard deviation, 632, 856
Standard deviation of the mean, 632
Standard electrode potentials, 15
table of, 901

Standards, for emission spectroscopy, 346
Stationary phase
gas chromatographic, 743
liquid chromatographic, 660
table of commercial gas chromatographic, 732, 744
Statistical weight, 292
Statistics, in radiochemistry, 632
Stepwise-line fluorescence, 316
Steric effects, spectrometric, 173
Stern-Volmer equation, 273
Stimulated emission, 222
Stoichiometry, determination of by spectrophotometry, 185
Stokes law, 255
Stokes lines, 217
Stopped-flow methods, 565
Stray light, 179
Stripping technique, in radiochemistry, 599
Stripping voltammetry, 84
applications of, 86
detection limit in, 72
differential pulse, 85
Successive-approximation ADC, 837
Summing amplifier, 798
Summing point, 799
Sum peak, x-ray, 420
Superhyperfine splitting, 406
Supporting electrolyte, 53
Supports, gas chromatographic, 741
Suppressor column, 715
Surface analysis, 517
Surface charging, in ESCA, 464
Surface-coated open tubular columns, 741
Système International (SI) units, tables of, 897

T

TA. *See* Thermal analysis
Tandem capillary columns, 762
Tandem mass spectrometry, 517
Tau scale, NMR, 366
Tau-value, 159
TCD. *See* Thermal conductivity detector
Temperature control, in kinetic methods, 582
Temperature programming, in gas chromatography, 752, 762
Terminology
computer, 842
spectroscopic, 145, 166
table of computer, 841
Tetramethylsilane, 366
TG. *See* Thermogravimetry
Thalamid electrode, 24
Theoretical plate, 666, 734, 738
Thermal analysis, 532

Thermal conductivity, 537
Thermal conductivity detector, 747
Thermal desorption mass spectrometry, 513
Thermal detectors, 228
Thermalization, of neutrons, 616
Thermal lag, 529
Thermally assisted fluorescence, 316
Thermal neutrons, 614, 619, 624
Thermal noise, 846
Thermistor, 229, 549
Thermocouple, 158, 229, 528, 534
Thermogram, 525
Thermogravimetry, 523, 525
    applications of, 531
    calculations for, 531
    instrumentation for, 526
    theory of, 528
Thermometer, Pt resistance, 528
Thermometric titrations, 524, 546, 552, 554
    instrumentation for, 547
Thermopile, 229
Thermospray ionization source, 695
Thévinin equivalent circuit, 779
Thickness determination, by XRF, 440
Thin films, in x-ray fluorescence spectrometry, 434
Thin-layer chromatography, 268, 659, 678
    quantitative, 681
Thin-layer electrophoresis, 659, 724
Three-spin systems, in NMR, 377
Threshold energy, nuclear reaction, 617
Threshold reactions, 618
Time base, precision, 829
Time constant, 826, 858, 860
Time domain, 769
Time-domain signal, 849
Time-domain spectrum, 156, 231
Time-of-flight mass spectrometer, 489
Time of relaxation, 125
Tin dioxide electrode, 116
TLC. *See* Thin-layer chromatography
TMS, 366
TOF mass spectrometer, 489
Total ionic strength adjustment buffer, 41
Total luminescence spectrum, 268
Track-and-hold amplifier, 822
Transducers, 768
Transfer function, 572, 783
    of operational amplifiers, 795, 801
Transformer, 130, 787
Transistor, 790
    bipolar, 790
    field-effect, 792
    junction field-effect, 792
Transistor-transistor logic, 771, 792, 809
Translational motion, 213
Transmission x-ray filter, 420

Transmittance, 174
    internal, 174
Transverse relaxation, 363
Tri-state logic, 844
Triplet state, 248
Triplet-state molecules, 396
Tritium, 601, 609, 749
Truth table, 810
TT. *See* Thermometric titrations
TTL, 771, 792, 809
Tunable laser, 199
Tungsten filament lamp, 192
Turbulent flow, 565
Two-dimensional paper chromatography, 678
Two-position control, 862
Two-spin systems, NMR, 376
Tyndall scattering, 215

U

Ultraviolet photoelectron spectroscopy, 451
Ultraviolet region, 192
    detectors for, 158, 193, 198, 312, 581, 691
    near, 192
    nitrogen, 192
    sources for, 192, 256, 297, 301, 691
    vacuum, 168, 192
Units, tables of, 897
Unity-gain bandwidth, 804

V

Vacuum ultraviolet region, 168, 192
Valence-band energies, determination of by ESCA, 472
Valence states, determination of by ESCA, 466
Van Deemter equation, 670, 735
Vaporization interference, 286
Variable-time procedures, in kinetic methods, 569
Variance, chromatographic zone, 734
Very large scale integrated circuit, 809
VFC. *See* Voltage-to-frequency converter
Vibrational coupling, 214
Vibrational relaxation, 250
Vibrational transitions, 163, 213
Vicinal coupling, 380
Vidicon tube detector, 198, 312
Virtual ground, in operational amplifiers, 798
Virtual state, 215
Visible region, 192
    detectors for, 158, 193, 198, 312, 581, 691
    sources for, 192, 256, 297, 691
VLSI. *See* Very large scale integrated circuit

Void volume, 665, 718
Voltage, 772
   op-amp offset, 805
   peak-to-peak, 780
   root mean square, 780
Voltage divider, 775, 779
   decade, 775
Voltage follower, 796
Voltage regulator, 788
Voltage-to-frequency converter, 834
Voltage-to-frequency-to-digital converter, 835
Voltaic cell, 14
Voltammetry, 52. *See also* Linear scan voltammetry; Stripping voltammetry
   applications of, 86
   upper concentration limit, 55
Voltammogram, 52
Voltmeter, digital, 777

W

Wall-coated open tubular columns, 740
Waveguides, 399
Wavelength, 145
Wavelength dispersion, x-ray, 426
Wavenumber, 145
WCOT. *See* Wall-coated open tubular columns
Wheatstone bridge, 130, 549, 748, 800
White noise, 846
Work function, of ESCA spectrometers, 465
Working electrode, 3

X

X-band spectrometer, 399
Xenon-arc lamp, 256

XOR gate, 815
XPS, 451. *See also* ESCA
x-ray absorption, 414
x-ray absorption spectrometry, 445
x-ray diffraction, 414
x-ray fluorescence spectrometry, 412
   excitation in, 442
   film thickness determination by, 440
   fundamental-parameter method, 437
   positive-ion excitation, 442
   quantitative, 433
   sample preparation for, 430
   scatter correction in, 436
x-ray photoelectron spectroscopy, 451. *See also* ESCA
x-ray scattering, 445
x-ray spectrometer
   energy dispersive, 414
   wavelength dispersive, 414
x-ray spectroscopy, 159, 412
x-ray tube, 416
   for ESCA, 462
x-rays, 595
   continuum, 416
XRF. *See* x-ray fluorescence spectrometry

Y

YAG laser, 513

Z

Zeeman effect, 310
Zeeman splitting, in ESR, 398
Zener diode, 787, 789
Zinc arc lamp, 691

Element	Symbol	Atomic Number	Atomic Mass
Actinium	Ac	89	227.0278[c]
Aluminum	Al	13	26.98154
Americium	Am	95	(243)[a]
Antimony	Sb	51	121.75
Argon	Ar	18	39.948
Arsenic	As	33	74.9216
Astatine	At	85	(210)[a]
Barium	Ba	56	137.33
Berkelium	Bk	97	(247)[a]
Beryllium	Be	4	9.01218
Bismuth	Bi	83	208.9804
Boron	B	5	10.81
Bromine	Br	35	79.904
Cadmium	Cd	48	112.41
Calcium	Ca	20	40.08
Californium	Cf	98	(251)[a]
Carbon	C	6	12.011
Cerium	Ce	58	140.12
Cesium	Cs	55	132.9054
Chlorine	Cl	17	35.453
Chromium	Cr	24	51.996
Cobalt	Co	27	58.9332
Copper	Cu	29	63.546
Curium	Cm	96	(247)[a]
Dysprosium	Dy	66	162.50
Einsteinium	Es	99	(252)[a]
Erbium	Er	68	167.26
Europium	Eu	63	151.96
Fermium	Fm	100	(257)[a]
Fluorine	F	9	18.998403
Francium	Fr	87	(223)[a]
Gadolinium	Gd	64	157.25
Gallium	Ga	31	69.72
Germanium	Ge	32	72.59
Gold	Au	79	196.9665
Hafnium	Hf	72	178.49
Helium	He	2	4.00260
Holmium	Ho	67	164.9304
Hydrogen	H	1	1.0079
Indium	In	49	114.82
Iodine	I	53	126.9045
Iridium	Ir	77	192.22
Iron	Fe	26	55.847
Krypton	Kr	36	83.80
Lanthanum	La	57	138.9055
Lawrencium	Lr	103	(260)[a]
Lead	Pb	82	207.2
Lithium	Li	3	6.941
Lutetium	Lu	71	174.967
Magnesium	Mg	12	24.305
Manganese	Mn	25	54.9380
Mendelevium	Md	101	(258)[a]
Mercury	Hg	80	200.59

Atomic masses ($^{12}_{6}C = 12.00000$)